MW00782982

NINTH EDITION

DRAFTING & DESIGN
FOR ARCHITECTURE & CONSTRUCTION

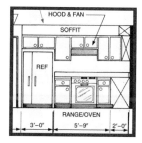

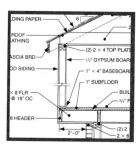

DANA J. HEPLER

Principal, Hepler Associates,
Architects and Land Planners,
New York City and Massapequa,
New York, and Adjunct Professor
of Architecture, New York
Institute of Technology,
Westbury, New York

PAUL ROSS WALLACH

Architecture Instructor, Cañada
College, Redwood City, California,
and Technical Writing Consultant,
Burlingame, California

DONALD E. HEPLER

President, Technical Writing
and Design Service Inc.,
Somers, Connecticut

DELMAR
CENGAGE Learning

Australia • Brazil • Japan • Korea • Mexico • Singapore • Spain • United Kingdom • United States

DELMAR
CENGAGE Learning

DRAFTING & DESIGN FOR ARCHITECTURE
& CONSTRUCTION NINTH EDITION
Dana J. Hepler, Paul Ross Wallach, and Donald E. Hepler

Vice President, Editorial: Dave Garza

Director of Learning Solutions: Sandy Clark

Associate Acquisitions Editor: Kathryn Hall

Managing Editor: Larry Main

Senior Product Manager: Sharon Chambliss

Editorial Assistant: Cristopher Savino

Vice President, Marketing: Jennifer Baker

Marketing Director: Deborah Yarnell

Associate Marketing Manager: Jillian Borden

Production Director: Wendy Troeger

Production Manager: Mark Bernard

Content Project Manager: Sara Dovre Wudali

Art Director: Casey Kirchmayer

Technology Project Manager: Christopher Catalina

Production Technology Analyst: Joe Pliss

For product information and technology assistance, contact us at
Cengage Learning Customer & Sales Support, 1-800-354-9706
For permission to use material from this text or product,
submit all requests online at **www.cengage.com/permissions.**
Further permissions questions can be e-mailed to
permissionrequest@cengage.com

Library of Congress Control Number: 2011939833
ISBN-13: 978-1-111-12813-5
ISBN-10: 1-111-12813-8
Delmar
5 Maxwell Drive
Clifton Park, NY 12065-2919
USA

Cengage Learning is a leading provider of customized learning solutions with office locations around the globe, including Singapore, the United Kingdom, Australia, Mexico, Brazil, and Japan. Locate your local office at: **international.cengage.com/region**
Cengage Learning products are represented in Canada by Nelson Education, Ltd.

To learn more about Delmar, visit **www.cengage.com/delmar**

Purchase any of our products at your local college store or at our preferred online store **www.cengagebrain.com**

Printed in the United States of America
1 2 3 4 5 6 7 15 14 13 12

CONTENTS

Preface vii

PART 1
INTRODUCTION TO ARCHITECTURE   1

1—Architectural History and Styles   2

Development of Architectural Forms 2
Development of Architectural Styles...........7
Influences on Early American Architecture......7
Early American Styles 10
Later American Styles 12
Architectural Design—The Future 16

2—Fundamentals of Design   20

Architecture and Design 20
Elements of Design 22
Principles of Design 26

PART 2
ARCHITECTURAL DRAFTING FUNDAMENTALS   33

3—Scales and Measurements   34

Architect's Scale.........................34
Engineer's Scale.........................38
Metric Scales...........................38

4—Conventions and Procedures   43

Architectural Drawings 43
Architectural Conventions 51

Architectural Drawing Techniques 54
Drafting Pencils and Pens 55
Guides for Straight Lines..................57
Instruments for Curved Lines 58
Correction Equipment 59
Aids and Devices for Drafting..............59

5—Introduction to Computer-Aided Drafting and Design   63

CAD Characteristics......................63
CAD Functioning........................65
CAD Components 65
Using CAD Systems 68
Types of CAD Drawings...................82
CAD Applications........................91

PART 3
BASIC AREA DESIGN   97

6—Environmental Design Factors   98

Orientation.............................98
Ergonomic Planning.....................109
Ecology and the Environment..............110

7—Indoor Living Areas   118

Living Area Plans 118
Living Rooms 121
Dining Rooms and Areas 127
Family Rooms 130
Great Rooms...........................133

Special-Purpose Rooms 133
Fireplace Design 139

8—Outdoor Living Areas   143

Porches 143
Patios. 147
Lanais 151
Swimming Pools 153
Outdoor Recreation Facilities 158

9—Traffic Areas and Patterns   160

Traffic Patterns 160
Entrances 160
Halls 169
Stairs 170
Elevators 171

10—Kitchens   174

Kitchen Design Considerations. 174
Kitchen Planning Guidelines 187

11—Service Areas   194

Utility Rooms 194
Garages and Carports 197
Mud Rooms 200
Driveways 201
Workshops 201
Storage Areas. 203
Specialized Areas 206

12—Sleeping Areas   209

Bedrooms 209
Baths. 219

PART 4
BASIC ARCHITECTURAL DRAWINGS 227

13—Site Development Plans   228

Site and Environmental Analysis. 228
Zoning Ordinances. .229
Topographic Drawings 236
Survey Plans 238
Plot Plans 251

Landscape Plans 257
Rendering Site Drawings. 264
Site Details and Schedules. 271

14—Designing Floor Plans   275

Floor Plan Development. 275
The Design Process. .275
Functional Space Planning 286
Plans for Special Needs 295

15—Drawing Floor Plans   303

Types of Floor Plans .304
Floor Plan Symbols. .306
Steps in Drawing Floor Plans 315
Drawing Floor Plans on CAD. 315
Steps in Drawing Cad Floor Plans 315
Size and Scale in Floor Plans. 319
Three-Dimensional Floor Plans 319
Multiple-Level Floor Plans 319
Reversed Plans. 324
Reflected Ceiling Plans 324
Floor Plan Dimensioning 324
Building Information Modeling 332

16—Designing Elevations   334

Floor Plan Relationship. 334
Elements of Design in Elevations 335
Elevation Design Sequence 337

17—Drawing Elevations   353

Elevation Projection .353
Projecting Elevations. 355
Elevation Symbols. 369
Interior Elevations. 375
Elevation Dimensioning 380
Building Information Modeling 382
Presentation Elevation Drawings. 382

18—Sectional, Detail, and Cabinetry Drawings   388

Sectional Drawings 388
Full Sections 388
Detail Sections 396
Cabinetry and Built-In Component Drawings . .404

PART 5
PRESENTATION METHODS   415

19—Pictorial Drawings   416

Types of Pictorial Projection Drawings 416
Perspective Drawings. .418
Projection Methods. 426

20—Architectural Renderings   431

Rendering Media. 431
Effects of Light 437
Texture 438
Entourage 440
Landscape 441

21—Architectural Models   443

Design Study Models .443
Presentation Models 447
Steps in Constructing a Model 448

PART 6
FOUNDATIONS AND CONSTRUCTION SYSTEMS   455

22—Principles of Construction   456

Structural Design 456
Modular Construction .463

23—Foundations and Fireplace Structures   472

Foundation Materials and Components 472
Types of Foundations .477
Fireplace Construction 490
Foundation Drawings. .497

24—Wood-Frame Systems   506

Skeleton-Frame Construction 506
Post-and-Beam Construction 513

25—Masonry and Concrete Systems   527

Masonry Construction Systems 527
Concrete Construction Systems 535

26—Steel and Reinforced-Concrete Systems   545

Steel Building Construction 545
Steel Structural Members 547
Steel Fasteners and Intersections 552
Structural Steel Drawing Conventions 557

27—Disaster Prevention Design   561

Preventing Wind Damage. 561
Preventing Earthquake Damage 565
Gas Containment and Venting 565
Fire Prevention and Control 567
Air Purification 567
Water Control. 567
Controlling Toxic Pollutants 567
Pest Control 569
Electrical Hazards 569
Preventing Personal Injury 569
Security. 569

PART 7
FRAMING SYSTEMS   571

28—Floor Framing Drawings   572

Types of Platform Floor Systems. 572
Floor Framing Members 573
Floor Framing Plans 586
Floor Framing Plans for Steel 594

29—Wall Framing Drawings   598

Exterior Walls 598
Interior Walls 619
Stud Layouts 628

30—Roof Framing Drawings   633

Roof Function. 633
Roof Framing Members 633
Roof Framing Components 636
Roof Framing Drawings 648
Roof Slope and Pitch. .652
Roof Framing Methods with Wood 654
Roof Framing Types 655

Roof Framing Methods with Steel. 659
Roof Covering Materials. 661
Roof Appendages 666

PART 8
ELECTRICAL AND MECHANICAL DESIGN DRAWINGS 671

31—Electrical Design and Drawing 672

Energy Conservation, Efficiency, and Sustainability. 672
Electrical Principles. .674
Lighting Design 681
Developing and Drawing Electrical Plans. 687
Electrical Working Drawings. 693
Electronic Building Systems 700

32—Comfort Control Systems (HVAC) 704

HVAC Plans and Conventions 704
Principles of Heat Transfer 705
Insulation 708
Heating Systems 713
Cooling Systems 720
Environmental Comfort Factors. 721
Passive Solar Systems. 723
Active Solar Systems 725

33—Plumbing Drawings 730

Plumbing Conventions and Symbols 730
Plumbing Systems. 732
Plumbing Drawings 738

PART 9
DRAWING MANAGEMENT AND SUPPORT SERVICES 747

34—Drawing Management 748

Drawing Sequence 748

Management Methods 748
Drawing Coordination 749
Checking Drawings. 749
Corrections and Changes 754

35—Schedules and Specifications 761

Schedules 761
Material Lists. 769
Specifications. 770

36—Building Costs and Financial Planning 775

Building Costs. 775
Financial Planning 778

37—Codes and Legal Documents 782

Building Codes 782
Legal Documents 785

APPENDIXES

A—Careers in Architecture and Related Fields 788

B—Mathematical Calculations 793

Arithmetic Calculations 793
Structural Calculations 796
Geometric Calculations. 813

C—Architectural Abbreviations and Professional Organizations 817

D—Architectural Synonyms 821

E—Glossary 823

Index 836

PREFACE

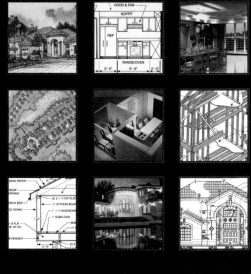

Drafting & Design for Architecture & Construction is a comprehensive textbook designed for use in a first course in architectural or construction drafting. Its purpose is to help students learn the fundamental skills and concepts necessary for architectural planning, designing, and drawing.

The ninth edition covers all introductory areas of both drafing and design in architecture and construction. Emphasis is placed on the appropriate balance between these two elements. The two major subject matter components of this text, drafting and design, are intricately connected. The design component needs the drafting function as a method of communication. Drafting procedures alone, without clear design input, can result in drawings without function and/or aesthetic appeal. Both are needed to create full functional architectural plans. Likewise the two related career fields represented in this text, architecture and construction, are linked through a common body of knowledge of construction types, details and materials.

ORGANIZATION AND CONTENT

Material in this ninth edition has been organized into nine major parts consisting of thirty-seven chapters plus appendixes.

Parts

> **Part One,** "Introduction to Architecture," provides background information on the history and development of major architectural styles, with excellent examples shown of both past and present designs. It also covers the basic principles and elements of architectural design.

> **Part Two,** "Architectural Drafting Fundamentals," provides basic information on the use of scales, drafting instruments, and CAD systems, and explains the various architectural drafting conventions used in creating working drawings. Information in this part is needed to apply the information covered in subsequent specific drafting and design chapters. CAD information has been rewritten, updated, and expanded for this edition.

> The ninth edition emphasizes the understanding of CAD fundamental concepts and applications to architecture and construction. CAD is introduced in Chapter Five through the coverage of common and generic CAD tools and operations. This is followed by specific applications and examples throughout the entire text. Because of the wide and diverse range of CAD systems and software versions, the duplication of user's manual explicit procedures, which are software specific, have been minimized.

> **Part Three,** "Basic Area Design," covers the environmental and functional design factors needed to plan specific areas of a structure. This includes the design considerations necessary for effective solar orientation, efficient energy use, and ergonomic and ecological planning. Major considerations include the function, location, decor, size, and shape of all residential areas.

Coverage includes all aspects of environmental safety and protection that can be controlled or advanced through effective architectural design. The negative effects on the architectural environment created by human behavior(s) i.e., social issues, are not covered. This includes such post construction voluntary practices as smoking and the use of portable polluting and energy draining appliances and devices.

> **Part Four,** "Basic Architectural Drawings," presents the design process and drafting methods used to combine areas into composite, functional, and effective architectural plans. Procedures for designing and drawing floor plans, elevations, detail, cabinetry, and site development drawings are explained. Guidelines for designing structures for persons with physical impairments are also included in this part. The information on site development and design factors and procedures has been completely revised and updated. Because site design should precede floor plan design, site plan coverage is now loacated in Chapter 13, in order to precede Floor Plan Design in Chapter 14. New information on cabinetry drawings has also been added.

> **Part Five,** "Presentation Methods," shows the different methods used to present architectural designs to nontechnical personnel such as marketing staffs, financial supporters, and prospective buyers. Step-by-step instructions for preparing one- and two-dimensional drawings, renderings, and three-dimensional models are provided.

> **Part Six,** "Foundations and Construction Systems," begins with an overview of the basic scientific and modular principles on which construction systems are based. Each major construction system is then explained as students are introduced to the specialized drawings needed to complete detailed descriptions of the structural design. Types of drawings included are those used to describe foundations and fireplaces and wood-frame, masonry, concrete, steel, and reinforced-concrete systems. This part includes a chapter that covers disaster prevention design features needed to reduce structural failure due to earthquakes, tornadoes, floods, and hurricanes.

> **Part Seven,** "Framing Systems," explains and shows in detail how to design and draw the framing systems for the major construction components of a building: floors, walls, and roofs.

> **Part Eight,** "Electrical and Mechanical Design Drawings," includes the principles and procedures for preparing working drawings to describe the electrical, comfort control (HVAC), and plumbing systems of a structure. Passive and active solar heating and cooling systems are also explained.

> **Part Nine,** "Drawing Management and Support Services," describes how architectural plans are checked and combined into sets and how drawings are interrelated to other drawings, details, and documents such as schedules, specifications, cost estimates, financial plans, codes, and contracts. A complete set of working drawings is presented.

Appendices

The appendices include material that is applicable and frequently used in the study and practice of architectural drafting, design, and construction but does not completely or sequentially fall exclusively into one of the thirty-seven basic content chapters. Coverage here includes architectural and construction career information, types of mathematical calculations, standard abbreviations used on architectural drawings and by professional organizations, architectural synonyms employed in different related fields and or geographical areas, and a glossary of architectural and construction terms.

> **Appendix A Careers in Architecture and Related Fields** describes the many careers that require knowledge of some aspect of architectural drafting, design, and construction. This information includes educational requirements, essential skills and knowledge needed, job descriptions, and activities. Careers covered here include not only basic architectural drafting and design areas, but also related construction careers for which a knowledge of architectural drafting and design is important or helpful.

> **Appendix B Mathematical Calculations** contains a summary of the mathematical calculations most frequently needed and used in architectural drafting and design and in related construction fields. This includes calculations in basic arithmetic and geometry and also the mathematical operations necessary to calculate structural forces and materials as applied to light construction.

> **Appendix C Architectural Abbreviations and Professional Organizations** lists abbreviations that are used on all types of architectural drawings and documents to conserve space while providing consistency in interpretation and communication. It also lists the acronyms used by professional organizations that serve the architecture and construction fields.

> **Appendix D Architectural Synonyms** lists synonyms that are used by drafting and design professionals. Because the use of some terms varies between different fields and geographical areas, an understanding of the various use of terms is essential.

> **Appendix E Glossary** of architectural and construction terms that allows students to cross-reference terms found in various parts of this text and which may be found on all types of architectural drawings and documents.

Chapter Organization

The sequence of chapters is generally organized in the order usually practiced in the architectural design and development process. However, this order can be rearranged to accommodate different course goals and priorities.

Chapter concepts are organized such that the student progresses from the familiar to the unfamiliar and from the simple to the complex. Where possible, related drawings are shown together or cross-referenced to provide a broader understanding of the relationship between drawings and documents. This is done on a single illustration, within the same chapter, and among cross-reference drawings in different chapters. To further reinforce the relationship of drawings, numerous symbol charts are provided. These show the plan and elevation symbol, abbreviation, and a pictorial drawing of each architectural feature.

Each chapter is introduced with listings of the major objectives and important terms defined and explained in the chapter. Each chapter ends with a set of exercises.

Because communication in the field of architectural drafting and design depends largely on understanding the vocabulary of architecture, new terms, abbreviations, and symbols are defined or explained where they first appear. This learning is reinforced throughout the remainder of the text.

Exercises

The exercises that appear at the end of each chapter are organized to provide the maximum amount of reinforcement of the concepts covered. Exercises are flexible, ranging from the very simplest, which can be completed in a few minutes, to the more complex, which require considerable time and detailed application of the principles of architectural drafting and design. Completion of all these exercises by the student will result in the creation of a complete set of related architectural plans and documents.

Drafting and Design for Architecture and Construction is illustrated with more than *1,500* drawings, photographs, and charts. Every illustration has been specifically selected and/or prepared to reinforce and amplify the principles and procedures described in the text.

EXERCISES

OBJECTIVES

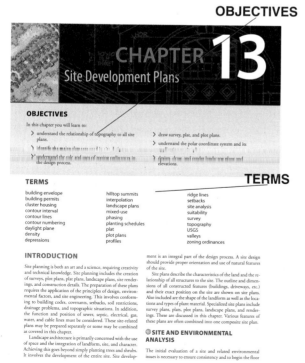

TERMS

Special Features

Chapter Six is devoted to a comprehensive introduction of the environmental design factors that relate to architecture and construction. Environmental design details and applications are also integrated into the subject matter through the entire ninth edition. To identify the locations of this coverage, a green symbol, as shown here, 🌐 is positioned next to the appropriate section head.

ANCILLARY MATERIALS

Ancillary materials related to this edition include:

The Student Workbook

The contents of the Student Workbook are related to the text by chapter and include drawing exercises to be used as tests, assignments, or teaching aids. An interpretation test involving an eleven plan set of residential drawings is included in the workbook to aid learning provide testing of the agreement among drawing in a set. An additional 83 pages of Chapter Review Tests has also been added to the ninth edition. (ISBN: 9781111128159)

Instructor's Guide

An Instructor's Guide is available which includes: teaching strategies, career information and suggestions for using the text, Instructor Resources CD, plan sets, tests and workbook. It also includes an interpreting architectural drawing test and answer key, plus the answers and solutions to text, workbook, and instructor's guide exercises and tests. (ISBN: 9781111128166)

The Instructor Resources CD

This educational resource creates a truly electronic classroom. It contains tools and instructional resources that enrich the classroom and make the instructor's preparation time shorter. The elements of the Instructor Resources link directly to the text to provide a unified instructional system. With Instructor Resources you can spend your time teaching, not preparing to teach. (ISBN 9781111128142)

Features contained in the Instructor Resources CD include the following:

> **Course Outlines.** The skills and material to be learned are outlined by text chapters for a 9-, 18-, and 36-week course. Recommended electives for vocational and professional career preparation are also provided.

> **Program Development.** This section includes instructional suggestions, information on classroom organization, and developing student skills. Work-school relationships, methods of conducting a course, and out-of-class activities are also covered.

> **Handout Masters.** The handout masters are designed to be printed, photocopied, and distributed to students. They extend and enrich the text content, sometimes with a drawing, sometimes with written information. Each handout is keyed to a specific text chapter. These may also be used with PowerPoint and Image Library illustrations for class presentations.

> **PowerPoint Presentations.** These images provide the basis for a lecture outline to present concepts and material. Key points and concepts can be graphically highlighted for student retention.

> **Image Library.** This database of key images taken from the text can be used in lecture presentations, tests and quizzes, and PowerPoint presentations. Additional Image Masters and Color Image Masters, which tie directly to lesson plans provided in the "Teaching Strategies" section, are also provided.

> **ExamView Test Bank.** Questions of varying levels of difficulty are provided in true/false, multiple choice, fill-in-the-blank, and short answer formats so you can assess student comprehension. This versatile tool enables the instructor to manipulate the data to create original tests.

> **Metric Applications in Architecture.** An introduction to the metric system as applied to architectural drafting plus metric drafting conventions is presented here. A set of metric working drawings and related questions and answers is also provided.

> **Professional References.** These references include professional organizations related to all phases of architecture, construction, and engineering. A list of accredited schools of architecture in the United States is provided.

Architectural Plan Set

A full set of residential plans is available to provide classrooms with examples of actual architectural drawings prepared using the same standards as presented in the text. The set is comprised of ten C size (24 × 36) sheets prepared at a scale of ¼"=1'-0". Plan, elevation, and detail drawings in the set are identified with many illustrations in the text.

ABOUT THE AUTHORS

Dana J. Hepler received his bachelor's degree from Ohio State University and master's degree in architecture from New York Institute of Technology. He has been associated with several of the largest architectural firms in the world as designer, director of planning, and construction manager. He has received national and international awards for his designs. Presently he is principal of Hepler Associates, Architects and Land Planners, New York City and Massapequa, New York, and adjunct professor of architecture at the New York Institute of Technology, Westbury, New York.

Paul Ross Wallach received his undergraduate education at the University of California at Santa Barbara and did his graduate work at California State University, Los Angeles. He has acquired extensive experience in the drafting, designing, and construction phases of architecture and has taught architecture and engineering drawing for many years in Europe and California at the secondary and postsecondary levels. He currently teaches architecture at Cañada College in Redwood City, California, and also does technical writing and consulting.

Donald E. Hepler completed his undergraduate work at California State College, California, and his graduate work at the University of Pittsburgh, Pennsylvania. He has been an architectural designer and drafter for several architectural firms, has served as an officer with the United States Army Corps of Engineers, and has taught architecture, design, and drafting at both the secondary and college levels. He is the former publisher of McGraw-Hill's technical education program and is currently devoting full time to technical authorship and consulting.

ACKNOWLEDGMENTS

The publisher and authors gratefully acknowledge the cooperation and assistance received from many individuals and companies during the development of *Drafting & Design for Architecture & Construction*. Thanks are due to the drafting teachers who reviewed this edition and previous editions of the work and helped us develop excellent educational materials into a more comprehensive, up-to-date, and effective architectural drafting and design program: Susan Campbell, Glendale Community College, Glendale, Arizona; Robert J. Duering, Central Community College, Grand Island, Nebraska; Elizabeth H. Dull, High Point University, High Point, North Carolina; Donald W. Hain, Orleans/Niagara BOCES, Sanborn, New York; Catherine L. Kendall, Pellissippi State Technical Community College, Knoxville, Tennessee; Robert Potts, Schuykill Institute of Business and Technology, Pottsville, Pennsylvania; Phillip A. Reed, Old Dominion University, Norfolk, Virginia; and Joe Swantek, Schuykill Institute of Business and Technology, Pottsville, Pennsylvania; Nicholas Kosloski, Roy Slater, Somers High School, Somers, Connecticut; Don Russell, Westwood College, Denver, Colorado; Kent Hadnot, Westwood College, Houston, Texas; Moema Shortridge, Westwood College, Denver, Colorado; John Musolino Westwood College—South Bay Campus, Torrance, California. Somers High School students Megan Davis, Maddi Dawson, Justin Liquori, Catherine Machnicki, and Sam Smith. Credits for contributing architects, illustrators, and photographers are shown directly under each illustration.

A special thanks is given to contributors of multiple illustrations. These include John B. Schols, David Karram, Alfred Karram, Home Planners Inc., John Henry, Jenkins and Shue, The Western Pennsylvania Conservancy, Lindal Cedar Homes, Scholz Designs, Marc Michaels, Interior Design, Trus-Joist MacMillan, Eagle Windows & Doors, La Strada Interior Design, Artesanos Models, Chief Architect Software, Roman Pulaski Cabinetry, Pedini Designs, Feather Line Coachs, Marol-Radziner Modular Homes, Timberpeg Homes, Hon Hofmann Architect, Hanles Wood LLC, Auto Desk Software, Oakleaf Conservatories, D.K. Designs, Robert Ruschak, and Hepler Associates, PC

Special thanks are also extended to financial consultant John Kingston, CPA, CVA who prepared the contents of the financial related topics.

The authors extend particular appreciation to Donna Hepler for her efforts in coordinating the processing of manuscripts and illustrations; and for the managing of the essential communication links among authors, designers, drafters, photographers, manufacturers, and the publisher.

PART 1

INTRODUCTION TO ARCHITECTURE

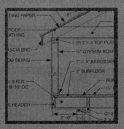

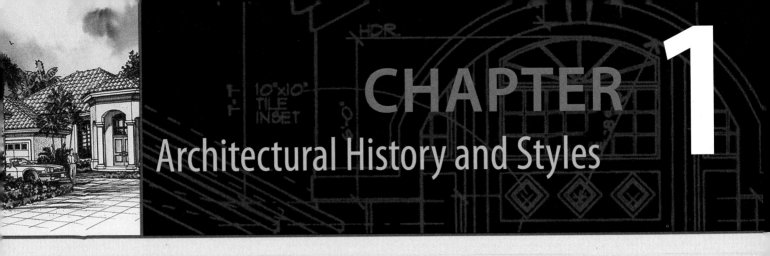

CHAPTER 1

Architectural History and Styles

OBJECTIVES

In this chapter you will learn to:

> recognize historical architectural styles and identify several distinct characteristics of each style.

> relate how the development of materials and construction methods influenced the future of architectural styles.

TERMS

arch
bearing walls
buttress
Cape Cod style
dome
domed huts
Dutch Colonial style
Early American style
English style

French style
Gothic arch
infrastructure
Italian style
keystone
Mediterranean style
Mid-Atlantic style
New England Colonial
Oriental architecture

pit houses
post-and-lintel construction
ranch style
smart growth
Southern Colonial
Southwestern ranch
sprawl
vault
Victorian era style

INTRODUCTION

The study of architectural history is more vast and complex than can be covered in one textbook—much less in one chapter. However, an overview of architectural forms is an excellent way to begin. Architecture is dynamic. As societies change and develop, so does architecture. How buildings have changed over the centuries—from covered pits, to domed huts, to solid walls, to open framed structures as shown in Figure 1.1—is the story of architectural design and construction. This chapter provides a background for evaluating and studying the broad range of architectural styles and forms.

DEVELOPMENT OF ARCHITECTURAL FORMS

When humans were nomadic, shelter consisted of natural caves or portable tents made of animal skins. Depressions, either natural or excavated, were later spanned with logs and covered with vegetation or skins to create the first roofed **pit houses.** As people began to settle in fixed locations, the need to draw or plan the construction of dwellings arose. People began to construct permanent tents and adobe huts and to modify caves or shelters with existing natural materials. This all began at least 10,000 years ago and by 7800 B.C. the first known town of Jericho had developed on the north end of the Dead Sea. By 4500 B.C. towns of **domed huts** were built as shown in Figure 1.2.

The addition of more permanent dwellings near fertile areas gave rise to villages. Village life created a need for still more planning, such as for public areas. The art and science of architecture began with the planning and construction of the first dwellings and public areas. The field of architectural drafting began when people first drew the outline of a shelter or a village in the sand or dirt. They planned how to build structures with existing materials. From these early beginnings to the present, the development of architecture has spanned more than 12,000 years.

As centuries passed and civilizations developed, human needs expanded. Lifestyles and cultures began to develop and change. More complete, accurate, and detailed drawings

FIGURE 1.1 > This contemporary structure represents the culmination of 12,000 years of architectural development.
Photo courtesy Lindal Cedar Homes. Lindal.com

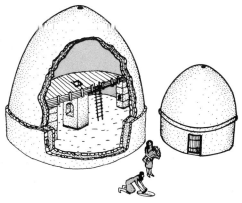

FIGURE 1.2 > Early domed huts. *Hepler/Wallach/Hepler*
© Cengage 2013

became necessary and basic principles of architecture began to be developed. The first known architect was Imhoptep, who designed the step pyramid in Egypt in 2780 B.C. This pyramid was an unoccupied solid structure. Most occupied buildings at that time were bearing wall types. **Bearing walls** are solid walls that provide support for each other and for the roof. Figure 1.3 shows an early Egyptian application of bearing wall construction. This is the temple built at Ereda in 3500 B.C. Note the comparison of this design with many contemporary structures. One of the first major problems in architectural design and construction was how to provide door and window openings in these supporting walls without sacrificing the needed support.

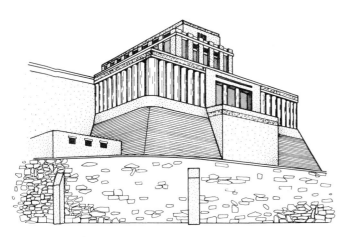

FIGURE 1.3 > Bearing wall construction in 3500 B.C. *Hepler/ Wallach/Hepler © Cengage 2013*

Post and Lintel

One solution to the problem was simply to place a horizontal beam, called a *lintel,* across two vertical posts. This early type of construction became known as **post-and-lintel construction.** This method (now called post and beam) was used by the Egyptians and later by the ancient Greeks. See Figure 1.4.

Because most ancient people used stone as their primary building material, architectural designs were limited. Because of the great weight of the stone, stone post-and-lintel construction could not support wide openings. Therefore, many posts (or columns) were placed close together to provide the needed support. The Greeks and, later, the Romans expanded their architectural designs by creating several styles of these columns. See Figure 1.5.

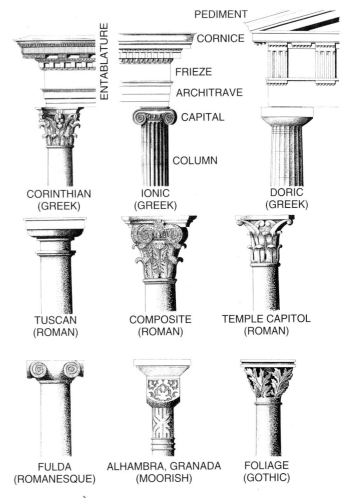

CORINTHIAN (GREEK)   IONIC (GREEK)   DORIC (GREEK)

TUSCAN (ROMAN)   COMPOSITE (ROMAN)   TEMPLE CAPITOL (ROMAN)

FULDA (ROMANESQUE)   ALHAMBRA, GRANADA (MOORISH)   FOLIAGE (GOTHIC)

FIGURE 1.5 > Greek and Roman column orders of architecture. *Hepler/Wallach/Hepler © Cengage 2013*

FIGURE 1.4 > The Parthenon (447 B.C.) is an early post-and-beam structure. *North Wind Picture Archives*

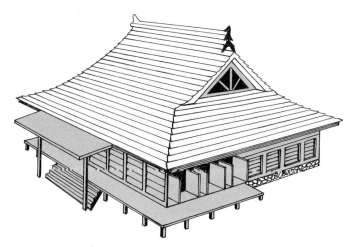

FIGURE 1.6A > Features of Oriental architecture. *Hepler/ Wallach/Hepler © Cengage 2013*

FIGURE 1.6B > Example of Oriental residential architecture. *Yo Shin So, Kenneth Mishimoto, AIA, Architect*

Oriental architects also made effective use of the post-and-lintel method. They were able to construct buildings with greater space between the posts under the lintel because they used lighter materials, such as wood. The use of these lighter materials allowed them to develop a style of architecture that was very open and graceful. Classical oriental homes are built of wood, paper, plaster, and stone. Roofs are covered with tile, thatch, or wood shingles. Sliding wall panels (*shoji*), which are used for doors, and ornamental gardens are also features of **Oriental architecture** as shown in Figure 1.6A and 1.6B.

The Arch

The Romans, who used stone, began a new trend in the design of wall openings when they developed the **arch.** Arch construction overcame several limitations of the post and lintel. Arches were easier to erect because they are constructed from many smaller, lighter blocks of stone. Each stone is supported by leaning on the keystone in the center. The **keystone** is a wedge-shaped stone that locks the other stones in place. See Figure 1.7. This construction has the advantage of spanning greater areas instead of being limited by the size of the one stone used for the lintel. The Romans combined arches and columns extensively in their architecture.

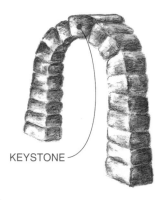

KEYSTONE

FIGURE 1.7 > A keystone supports both sides of an arch. *Hepler/Wallach/Hepler © Cengage 2013*

The Vault

The success of the arch led to the development of the vault. A **vault** (Figure 1.8) can be viewed as a series of arches that forms a continuous arched covering. The term may also refer to an arched underground passageway or a space, such as

FIGURE 1.8 > A barrel vault is a series of connected arches. *Hepler/Wallach/Hepler © Cengage 2013*

a room, that is covered by arches. When two barrel vaults intersect as shown in Figure 1.9, a cross vault is created.

The Dome

A **dome** is a further refinement of the arch. A dome is made of many arches arranged so that their bases form a circle and the tops meet in the center. See Figure 1.10. The Romans viewed the dome as a symbol of power. The Pantheon is an example of a dome used in Roman architecture about 27 B.C. Throughout the world, domes have often been used in religious and governmental structures, as in the U.S. Capitol building (Figure 1.11).

The Gothic Arch

A variation of the arch was a defining characteristic of the Gothic style of architecture that spread throughout Europe during the Middle Ages. The pointed arch was called the **Gothic arch** and became a very popular feature in cathedrals. The emphasis on vertical lines created a sense of height and aspiration. However, the pointed arch posed the same problem as did other arches: spreading at the bottom because of the weight above.

To add support to an arch or bearing wall, a protruding structure called a **buttress,** or pilaster, was added at the base, as shown in Figure 1.12. As the style evolved, buttresses were connected to higher areas of the walls and came to be called flying buttresses. A flying buttress helps support the sides of

FIGURE 1.11 〉 The U.S. Capitol building is a contemporary domed structure. *David R. Frazier Photolibrary, Inc./Mark Burnett*

FIGURE 1.9 〉 A cross vault is the intersection of two barrel vaults. *Hepler/Wallach/Hepler © Cengage 2013*

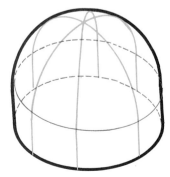

FIGURE 1.10 〉 Arches spaced in a circle form a dome. *Hepler/Wallach/Hepler © Cengage 2013*

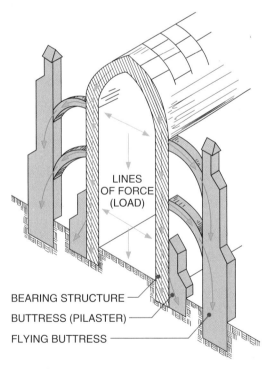

LINES OF FORCE (LOAD)

BEARING STRUCTURE
BUTTRESS (PILASTER)
FLYING BUTTRESS

FIGURE 1.12 〉 Buttresses (pilasters) support bearing walls. Arrows represent lines of force. *Hepler/Wallach/Hepler © Cengage 2013*

FIGURE 1.13 > Flying buttresses support the thin walls of Notre Dame Cathedral. *David R. Frazier Photolibrary, Inc.*

a wall without adding additional weight, allowing thinner walls and more windows to be used. Large structures such as the high-arched cathedral in Figure 1.13 incorporate flying buttresses in their design.

DEVELOPMENT OF ARCHITECTURAL STYLES

The Gothic (pointed) arch is one of many features that were developed over time. The development of one architectural solution and a resulting style in one culture often causes changes in the architecture of another culture. Transitions occur from one time period to another as well as from one part of the country and world to another.

Few structures, past or present, are pure examples of one specific style. In fact, to identify an architectural style as one that originated in only one country or in only one time period is difficult, if not inaccurate. Nonetheless, architectural styles *are* categorized and labeled by their most common and outstanding features. The label *Early American style,* for example, is something of a misnomer, because all styles that found their way to America during colonial times can be labeled Early American. The overlapping of characteristics of architectural design is also typical among European styles.

Nonetheless, labels are applied and used as a frame of reference with which to study and compare architectural styles.

The term *Colonial* can be applied to most early American homes. The progression of design developed in the following order:

1. *Early Colonial* is the first true Colonial style of homes in America.
2. *Classical Colonial* is the adoption of the ancient and classical architectural forms.
3. *Traditional Colonial* is the adoption of the older style of Colonial architecture with modernizing features.
4. *Contemporary Colonial* is the adoption of clean contemporary lines with much glass and an open (informal) floor plan.

INFLUENCES ON EARLY AMERICAN ARCHITECTURE

To understand the development of American architecture, an overview of the following European styles provides an important background. The English, French, Spanish, and Italians have provided the most significant influences on Early American architecture.

FIGURE 1.14 › Elizabethan style. © *used under license from Shutterstock, 2011/Charlie Edward*

English Architecture

English style architecture includes several variations of some common architectural features. For example, some features that many English structures share are high-pitched roofs, massive chimneys, half-timber siding, small windows, and exterior stone walls. Within this frame of reference, variations in architectural style range from the very simple to the very lavish. Wood may replace the stone on the exterior walls. The most commonly adapted English styles include the Elizabethan and the Tudor.

An example of Elizabethan style with its characteristic half-timber walls is shown in Figure 1.14. Figure 1.15 shows a Tudor-style residence with its multiple gables, small leaded windows, and large free-standing chimneys.

French Architecture

Because the **French styles** were brought to this country much later than the English styles, their impact on colonial residential architecture was far less pronounced. However, some French styles, such as Regency, Mansard, Provincial, and Chateau, were accepted and used in many areas. French Mansard architecture can be identified by the Mansard roof. See Figure 1.16. This roof design was developed by the French architect François Mansard. On a French Provincial house, the roof is high pitched with steep slopes and rounded dormer windows projecting from the sides. The features of a French Provincial design are shown in Figure 1.17.

FIGURE 1.15 › Tudor style. © *used under license from Shutterstock, 2011/Denise Kappa*

FIGURE 1.16 › Contemporary application of a Mansard roof. *Used by permission Hanley Wood, LLC*

FIGURE 1.17 › French Provincial style. *Hepler/Wallach/Hepler © Cengage 2013*

Spanish and Italian Architecture

Spanish and Italian architecture share several similarities—arches, low-pitched roofs of ceramic tile, and stucco exterior walls, as shown in Figure 1.18. A distinguishing feature of a Spanish home is an open courtyard patio. Two-story Spanish homes contain open balconies often with grillwork trim.

Although **Italian style** architecture is very similar to Spanish architecture, a few features are particular to Italian styles.

Columns and arches are generally part of an entrance, as in Figure 1.19, and windows or balconies open onto a loggia. A loggia is an open passage covered by a roof.

Classical moldings around first-floor windows also help to distinguish the Italian style from the Spanish. Despite these distinguishing features, both of these styles are generally classified as **Mediterranean style** or Southern European style architecture. Figure 1.20 shows a contemporary adaptation of the Mediterranean style.

FIGURE 1.18 › Southwest style derived from Spanish architecture. © *J. Scott Smith/Beateworks/Corbis*

FIGURE 1.20 › Contemporary adaptation of the Mediterranean style. *Tucson Realty & Trust Co.*

FIGURE 1.19 › Contemporary adaptation of Italian architecture. *John Henry, Architect*

EARLY AMERICAN STYLES

The early colonists came to this country from many different cultures and were familiar with many different styles of architecture. **Early American style** architecture refers to all styles that developed in various regions of the colonies.

The European styles that primarily dominated Early American residential architecture were brought from England and France. French styles, however, were brought to this country in the eighteenth century, much later than English styles. The English styles had greater impact on colonial residential architecture.

FIGURE 1.21 ❯ Cape Cod style. *Used by permission Hanley Wood, LLC*

New England Colonial

The colonists who settled the New England coastal areas brought the strong influence of English architectural styles with them. Because they lacked time and equipment and depended on the locally available building materials, the colonists had to greatly simplify the English styles. This adapted style came to be known as **New England Colonial** architecture.

One of the most popular of the New England Colonial styles was called the **Cape Cod style.** See Figure 1.21. This one-and-one-half-story gabled-roof house has a central front entrance, a large central chimney, exterior walls of clapboard or beveled siding, and may include dormers. The floor plan is generally symmetrical. Cold New England winters also influenced the development of varied design features that added warmth, such as window shutters, small window areas, and enclosed breezeways.

Mid-Atlantic Colonial

The availability of brick, a seasonal climate, and the influence of the architecture of Thomas Jefferson led to the development of the **Mid-Atlantic style** of architecture. In colonial days, Mid-Atlantic style buildings, located from Virginia to New Jersey, were formal, massive, and ornate. This style is also known as *classical revival* because it was influenced by early Greek and Roman architecture. It also included adaptations of many urban English designs, such as the symmetrical, hip-roofed Mid-Atlantic style shown in Figure 1.22. The *Georgian* style is a simplified version of this design that includes many elements of the New England Colonial style.

Dutch Colonial

Many colonists from the Netherlands and Germany settled in New York and Pennsylvania. A gambrel roof was a typical part of their colonial style of architecture. This style was originally described as "Deutsch." However, the German term was given the English form by the colonists who called the masonry farmhouse style **Dutch Colonial.**

Southern Colonial

When the early settlers migrated south, warmer climates and outdoor living activities led them to develop the **Southern Colonial** style of architecture. The house became the center of plantation living. Southern Colonial homes were usually larger than most English houses. A second story was added, often with a veranda or porch. See Figure 1.23.

FIGURE 1.22 > Mid-Atlantic Colonial style. © *used under license from Shutterstock, 2011/Robynrg*

FIGURE 1.23 > Southern Colonial style. *Used by permission Hanley Wood, LLC*

LATER AMERICAN STYLES

After the colonial period, other architectural styles continued to evolve. Styles were influenced by climate, availability of land, and industrial developments—as well as by other architectural styles.

Victorian Era

During the early reign of Queen Victoria (1839 to 1901), architecture reflected the past with Greek and Gothic revival and Renaissance styles. Later more ornate designs such as Queen Anne became popular. The new machinery

FIGURE 1.24 > This design contains styles characteristic of the Queen Anne–Victorian era. *Used by permission Hanley Wood, LLC*

developed during the Industrial Revolution in the late eighteenth century led to the addition of intricate house decorations (gingerbread). Ornate finials, lintels, parapets, and balconies were added to existing designs, as shown in Figure 1.24. During the late 1800s **Victorian era styles** became more ornate with elaborate trim, steep gables, and tall pinnacled towers.

Ranch Style

As settlers moved west of the colonies, they adapted architectural styles to meet their needs. The availability of land eliminated the need for second floors. Because the needed space was spread horizontally rather than vertically, a rambling plan called **ranch style** resulted.

The Spanish and Mexican influence in the Southwest led to the popularization of the **Southwestern ranch,** with a U-shaped plan and a patio in the center. One-story Spanish/Mexican homes were the forerunners of the present ranch-style homes that were developed in the Southwest. See Figure 1.25A. Figure 1.25B shows a contemporary adaptation of the rambling ranch style.

Influences on Contemporary Styles

Advances in architecture throughout history have depended on the use of available building materials. In American colonial times, builders had only wood, stone, and some ceramic materials, such as glass, with which to work. With today's improved technological developments, lighter and safer buildings can be designed in forms, sizes, and shapes never before possible. See Figure 1.26. With so many choices, designers can create many new combinations of styles and materials. These must be carefully combined using the basic principles and elements of design presented in Chapter 2.

Historical styles continue to influence contemporary architecture. Advancements in technology, however, have freed designers from many design restrictions of the past. Present-day designers must often decide how many contemporary features can or should be incorporated into the design of a particular architectural style. The house design shown in Figure 1.27 contains many historic architectural elements yet incorporates numerous contemporary features such as large glass areas.

Contemporary Materials

Advances in architecture throughout history have depended on using locally available building materials. Early American architecture reflects the use of these materials. These materials and bearing wall construction continue to be used in contemporary buildings; however, great changes are being seen in buildings with the development of steel, aluminum, structural glass, reinforced and prestressed concrete, wood laminates, plastics, and other new synthetics.

Today's buildings are stronger, safer, and more maintenance free than they were just a few years ago. New developments

FIGURE 1.25A > Southwestern ranch style. *Used by permission Hanley Wood, LLC*

FIGURE 1.25B > Contemporary ranch style. *Used by permission Hanley Wood, LLC*

FIGURE 1.26 › Contemporary style incorporating new materials and shapes. © *used under license from Shutterstock, 2011/Krsmanovic*

FIGURE 1.27 › Traditional design incorporating contemporary components. *Isleworth—Tom Price, Architect; Phil Eschbach, Photographer*

FIGURE 1.28A > Contemporary modular ranch home. *Marmol Radziner Prefab, Joe Fletcher Photographer*

FIGURE 1.28B > Module placement with a crane. *Marmol Radziner Prefab*

in central heating and cooling and energy efficiency have enabled buildings to meet the heavy demands of new and more complex appliances and electronic devices. Due to the development of new building materials and the elimination of dangerous materials such as asbestos and lead, contemporary buildings provide a more healthy environment for the occupants. New materials and construction methods also make roofs and floors stronger, safer, and quieter than ever before under high-wind or earthquake conditions. Even devices for sensing problems such as detectors for smoke, radon, and break-ins have become much more efficient and effective.

Contemporary Construction Methods

The development of new materials is usually not possible without the development of new construction methods. For example, large glass panels could not have been used in the eighteenth century even if they had been available, because no large-span lintel-support system had been developed. Only when both new materials and new methods exist can the design be completely flexible.

Present-day structures are usually a combination of old and new. In a modern building, examples of the old post-and-lintel method can be used together with skeleton-frame, curtain-wall, or cantilevered construction.

Today, large premanufactured components can be used. The modular home shown in Figure 1.28A, although prefabricated, is custom designed to the owners' wants and needs. Modules fit together (see Figure 1.28B) and combine to produce a finished house in half the time. Also, contemporary structures tend to have fewer structural restrictions. Lines may be simpler, bolder, and less cluttered. Other contem-

porary buildings can be constructed with more diversified structural shapes. No longer are architectural shapes simply squares or cubes. Shapes such as triangles, octagons, pentagons, circles, and spheres are now used extensively. Without these advancements, the major structures of today, like those shown in Figure 1.29, could not be built—and the planned buildings of tomorrow extend the boundaries even further.

ARCHITECTURAL DESIGN— THE FUTURE

Today's architecture stems from a rich historical background. Architectural styles continually change and develop. The future of architecture will certainly continue to be influenced by the development of new materials and new construction methods, which support the way people live in society.

With technological advancements, designers have greater freedom of choice to create diverse and exciting architectural designs. Greater freedom and more choices, however, also mean greater challenges. One of the primary challenges of architectural design is to blend art and technology. The role of the designer is to create a relationship between art and technology that enables all types of buildings to be both technically appropriate and aesthetically acceptable. This is accomplished through the creative use of new materials and new technologies while including provisions for environmental protection and the advancement of new concepts such as smart growth.

🌐 Smart Growth

During the past 50 years, community growth and development have often been uncontrolled. This has resulted in unmanageable growth (**sprawl**) in many communities.

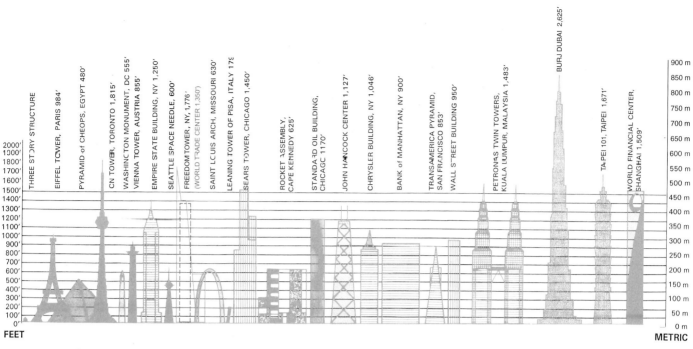

FIGURE 1.29 ⟩ Comparison of the world's largest structures. *Hepler/Wallach/Hepler © Cengage 2013*

Smart growth master plans provide a set of development and construction standards that can be used to preplan and manage desirable growth goals. Although the specific contents of smart growth plans may vary slightly throughout the country, the principles of smart growth include the following:

1. *Control development to strengthen existing communities.* This principle requires the creation of a land use master plan and smart growth building codes to which all developers and builders must adhere.

2. *Encourage mixed land use and mixed-use buildings.* The goal of this principle is to create zoning codes that encourage the development of a mixture of residential, retail, commercial, and recreational areas. This zoning is designed to overcome the effects of previous developments that were unregulated or based on single-use separate zoning districts, which resulted in undesirable urban sprawl, repetitive strip malls, traffic congestion, pollution, overextended **infrastructure,** automotive dependence, social and economic segregation, and the elimination of open green spaces.

3. *Encourage consultation among communities.* This goal is established to keep architectural styles consistent among adjoining communities. This also involves creating and maintaining high levels of excellence in architectural design.

4. *Take advantage of compact building sizes and create a range of housing opportunities.* Creating a wide range of building opportunities should result in the develop-

ment of a mixture of homes with workforce housing within walking distances from shopping and schools.

5. *Provide a variety of transportation choices.* This goal sets standards for the creation of efficient pedestrian and automotive linkages between adjacent communities. This involves the development of walkable safe streets, parking areas, safe bicycle and pedestrian pathways, sidewalks, and public transportation. Landscaping, lighting, seating, and consistent unobtrusive signage are also important components of this plan.

6. *Create pleasant environments and attractive communities.* Providing multipurpose civic facilities for such functions as entertainment and recreation are the main goal of this principle. This also encourages the creation of such architectural features as plazas, niches, and tree-lined streets.

7. *Preserve open space and natural resources.* Avoiding excessive sprawl and the related environmental deterioration is accomplished through maintaining or expanding green open spaces as a preserve. This goal directs architects to avoid filling every available space with a structure.

8. *Make development decisions predictable, fair, and cost effective.* To make smart growth possible and practical, architects, developers, and municipalities must keep their goals and designs realistic. This means keeping within financial constraints, construction capabilities, scheduling limits, and zoning restrictions.

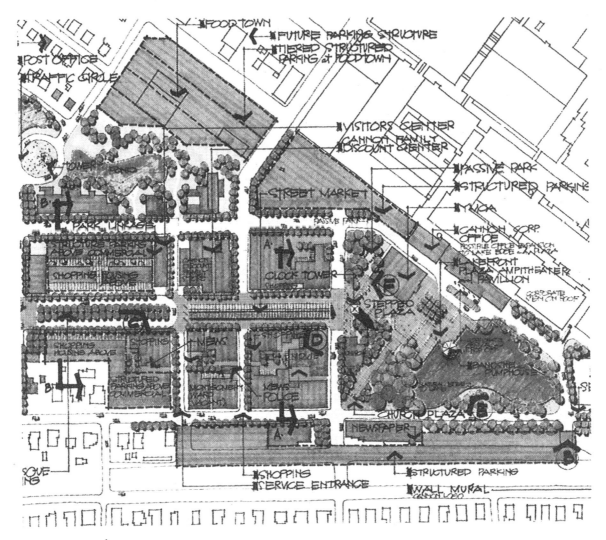

FIGURE 1.30A > Example of a smart growth community plan. *Hepler Associates, LA PC*

By adhering to these principles, smart growth projects meet the needs of local and regional planning authorities. Under this new operating system, architects, city planners, and citizens properly plan and manage the decisions needed to provide smart growth for their communities. Figure 1.30A shows a community plan that illustrates the application of many smart growth principles. Figure 1.30B shows a streetscape rendering that illustrates the eye-level interpretation of a portion of a smart growth plan. More details and procedures for developing smart growth plans are found in Chapter 13, *Site Development Plans.*

FIGURE 1.30B > Streetscape rendering of a portion of a smart growth area. *Sears Architects, Grand Rapids, Michigan*

Architectural History and Styles Exercises

1. Explain what is meant by the term *architectural styles*.

2. Describe post-and-lintel construction. Give examples of it in three different cultures. Compare it to the development of the arch.

3. Find examples of several different styles of architecture in your area. Photograph or sketch the structures and list their distinguishing characteristics.

4. Which European countries and styles had the greatest impact on Early American architecture? List two or three characteristics of each architectural style.

5. Describe a contemporary structure—imagined or real. Tell why you consider it to be a contemporary style.

6. Considering the styles in this chapter, which do you prefer for your home? Explain a few advantages and disadvantages of building this style.

7. Do research to identify and prepare a report on an architectural style not mentioned in this text. Share what you learn with your class.

8. Identify houses in your town or city that are examples of various architectural styles.

9. Compare the tallest structure in your area to the structures in Figure 1.29.

10. List the major construction styles of ancient and modern structures.

11. Describe how smart growth concepts can be applied to your community.

OBJECTIVES

In this chapter you will learn to:

> relate design concepts to architecture.

> understand why form follows function.

> identify six elements of design.

> apply design principles to a work of architecture.

TERMS

aesthetic value	intensity	rhythm
balance	interior decoration	space
creativity	interior design	subordination
eclectic design	line	transition
elements of design	opposition	unity
emphasis	organic design	variety
form	principles of design	value
functionalism	proportion	
hue	repetition	

INTRODUCTION

The reason one building is considered attractive and another unattractive is related to how well the fundamental principles and elements of design are applied. These fundamentals are used in developing architectural designs that are both attractive and functional. This applies to all types and styles of architecture although the applications may vary depending on individuality and creativity.

ARCHITECTURE AND DESIGN

Design activities are either formal or informal. Informal design occurs when a product is made by the designer without the use of a plan. Formal design involves the complete preparation of a set of working drawings. The working drawings are then used in constructing the product. Architectural design is nearly always of the formal type.

Ideas in the creative state are recorded by sketching basic images. These sketches are then revised until the ideas have crystallized and are given final form. First sketches rarely produce a finished design. Usually, many revisions are necessary.

A basic idea, regardless of how creative and imaginative, is useless unless the design can be constructed successfully. Designing involves the transfer of basic sketches into architectural working drawings. Every useful building must perform a specific function. Every part of the structure should also be designed to perform a specific function. Today's buildings are designed to be functional and aesthetic and for both purposes the elements and principles of design are applied.

Louis Sullivan, an important American architect in the late 1800s and early 1900s, wrote, "Our architecture reflects us as truly as a mirror." Architecture reflects the people, society, and culture of a given time. For example, modern architecture reflects our freedom and our technological advances.

Form Follows Function

"Form follows function" is a design concept conceived by Louis Sullivan but largely identified with Frank Lloyd Wright, probably the most famous of all twentieth-century architects. One of the earliest residential applications of this concept is the Robie House (Figure 2.1), which was designed by Wright more than a century ago and is considered one of

FIGURE 2.1 ⟩ Frank Lloyd Wright's Robie House, a cornerstone of modern architecture.
© Jon Arnold Images Ltd./Alamy.

the "cornerstones of modern architecture." Today's effective designers continue to follow this concept.

"Form follows function" means that any architectural form (shape, object) should have an intended practical purpose and should perform a function. This concept distinguishes architecture from other art forms, such as sculpture. A sculp-

ture's primary purpose is its **aesthetic value.** Its value is in the appreciation of its form, its beauty, or its uniqueness. **Functionalism** is the quality of being useful, of serving a purpose other than adding beauty or aesthetic value. Functionalism in architecture led to the development of the organic concept. In the **organic design** approach, all materials, functions, forms, and surroundings are completely coordinated and in harmony with nature.

Interior Design

The overall architectural style of a structure is an important consideration in developing the design of the interior. This means that the design elements and principles are matched to achieve a finished product that is consistent in style both inside and outside. Notice how the repeated use of the pointed arch in Figures 2.2A and 2.2B creates a design that efficiently relates the interior to the exterior.

Interior designers create all aspects of a building's interior. **Interior decoration** is not the same as **interior design.** Interior decorators design and select wall, floor, and ceiling coverings and finishes. They also may select window treatments,

FIGURE 2.2A ⟩ Exterior lines designed for consistency with interior line shapes. *John Henry, Architect*

FIGURE 2.2B ⟩ Interior lines designed for consistency with the exterior. *John Henry, Architect*

furniture, and accessories. However, they do not design the structural components of the interior. Interior designers may work these areas, but they may also create the plans and details necessary for the construction of interior walls, floors, ceilings, cabinetry, electrical, plumbing, and HVAC fixtures.

Eclectic design can be very interesting and attractive although it may appear inconsistent if architectural periods or themes are mixed in one area. Whenever this mixing is poorly done, the results are often more eccentric than a pleasing combination of architectural periods. When the elements and principles of design are followed, however, the results can be very effective. Eclectic design is not eccentric design.

Designers need to consider that styles and individual tastes change. An effective, creative designer recognizes the difference between *trends* (general developments) and *fads* (temporary popular fashions).

Creativity in Architectural Design

Creativity in architecture involves the ability to imagine forms before they exist. Creative imagination often involves arranging familiar objects and patterns in new ways. Creativity and imagination are both needed to bring many isolated and unrelated factors together into arrangements of cohesive unity and beauty. Every part of a building should be designed in relation to its function. Architects and interior designers apply the elements and principles of design to a building's function to make it aesthetically pleasing as well.

ELEMENTS OF DESIGN

Like a mixture composed of many ingredients, a design is composed of many elements. The basic **elements of design** are the tools of design. These include *line, form, space, color, light* (value), *texture,* and *materials.*

Line

Lines enclose space and provide the outline or contour of forms. Straight lines are either horizontal, vertical, or diagonal. Curved lines have an infinite number of variations. The design element of **line** can produce a sense of movement or produce a greater sense of length or height.

Horizontal lines emphasize width as the eye moves horizontally expanding the perception of space. Vertical lines create the impression of height because they lead the eye upward. These lines create a feeling of strength and alertness. Diagonal lines create a feeling of restlessness or transition. Vertical and horizontal lines tend to dominate architectural designs, giving a sense of stability. Curved lines indicate soft, graceful, and flowing movements. The curved and repeated arch lines of the ceiling in Figure 2.3 blend well with the

FIGURE 2.3 > Dramatic use of repeated ceiling arch design.
Marc-Michaels Interior Design, Inc.

floor-level horizontal lines to create a dramatic effect. As in any art form, it is often the combination of straight and curved lines in patterns that creates the most pleasing design.

Form

Lines joined together can produce a **form** and create the *shape* of an area. More than two straight lines joined together can produce triangles, rectangles, squares, and other geometric shapes. Closed curved lines can form circles, ovals, and ellipses, as well as free-form closed curves. The relationships of these forms or shapes is an important factor in design.

Circles and ovals convey a feeling of completeness. Squares and rectangles produce a feeling of mathematical precision. Repeated rectangles, created by the contrast of light and dark in Figure 2.4A, create a strong and precise impression of this building. Figure 2.4B reveals the interior view of this exterior, which displays a blend of the interior and exterior design elements. Whether closed, open, solid, or hollow, the form of a structure should always be determined by its function.

Space

Space surrounds form and is contained within it. A design can create a feeling of space. Architectural design includes the art of defining space and space relationships. Space is as important a consideration as the actual objects and materials.

Color

Choices of *color* have a strong influence on the final appearance of any design. In architecture, color can strengthen or diminish

FIGURE 2.4A 〉 Rectangular shapes define form. *Eagle Window & Door Inc.*

FIGURE 2.4B 〉 Matching interior design elements. *Eagle Window & Door Inc.*

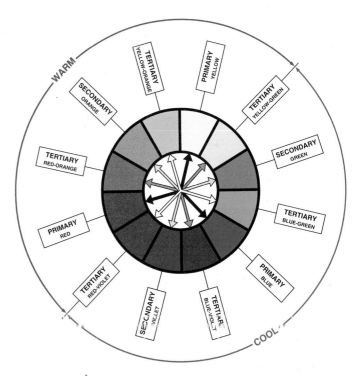

FIGURE 2.5 〉 Color wheel based on the triadic color system of the three primary colors: red, yellow, and blue. *Hepler/Wallach/Hepler © Cengage 2013*

interest. It can also distinguish one part from another. Color may be an integral part of an architectural material such as natural wood. Manufactured products, such as synthetic wall panels, may have color added to create a desired effect. To create effective designs, designers need to understand the nature and relationships of colors.

The Color Spectrum

Colors in the spectrum are divided into primary, secondary, and tertiary colors. *Primary* colors are red, yellow, and blue. These cannot be made from any other color. The primary colors and combinations of colors are illustrated on the color wheel in Figure 2.5.

A *secondary* color can be made from equal mixtures of two primaries. Green is a combination of yellow and blue. Violet is a combination of blue and red. Orange is a combination of red and yellow.

A *tertiary* color is the combination of a primary color and a neighboring secondary color. The tertiary colors are red-orange, yellow-orange, yellow-green, blue-green, blue-violet, and red-violet.

A *neutral* does not show color in the ordinary sense of the word. The neutrals are white, gray, and black. The three primary colors, if mixed in equal strengths, will produce black. When colors cancel each other out in this manner, they are neutralized.

Color Quality

For greater accuracy in describing a color's exact appearance, colorists distinguish three qualities: hue, value, and intensity.

The **hue** of a color is its basic consistent identity. A color hue may be identified as being yellow, yellow-green, blue, blue-green, and so forth. Even when a color is made lighter or darker, the hue remains the same.

The **value** of a color refers to the lightness or darkness of a hue. See Figure 2.6. A great many degrees of value can be obtained. Varying the value of colors can dramatically change the mood of a room.

A *tint* is lighter (or higher) in value than the normal value of a color. It is produced by adding white to a color. A lighter tint of a hue will make a room look larger.

A *shade* is darker (or lower) in value than the normal value of the color. A shade is produced by adding black to the normal color. A dark shade will often make a room look smaller.

A *tone* is usually produced by adding gray to the normal color. Each color on a color wheel can have a value that is equivalent to another color if both colors have the same amount of gray in them.

The **intensity** (strength) of a color is its degree of purity (or brightness), that is, its freedom from neutralizing factors. This quality is also referred to in color terminology as *chroma*. A color entirely free of neutral elements is called a saturated color. The intensity of a color can be changed, without changing the color's value, by mixing that color with a gray of the same value.

Color *harmonies* are groups of colors that relate to each other in a predictable manner. The basic color harmonies are *complementary, monochromatic,* and *triadic.* Complementary colors are opposite each other on the color wheel. Monochromatic colors are side by side. Triadic colors are three colors that are an equal distance apart on the color wheel.

Uses of Color

The use of color has a very strong effect on the atmosphere of a building in several ways. The perceived level of formality, temperature, and mood are all influenced by the color design. The combination of soft pastel yellows and oranges creates a relaxed and quiet atmosphere in the room shown in Figure 2.7. Colors such as red, yellow, and orange create a feeling of warmth, informality, cheer, and exuberance. Colors such as blue and green create a feeling of quiet, formality, restfulness, and coolness.

Color is also used to change the apparent visual dimensions of a building. It is used to make rooms appear higher or longer, lower or shorter. Warm bold colors, such as red, create the illusion of advancement. Cool and pale colors (including pastels) tend to recede. See Figure 2.8. The predominant use of white on the exterior of Figure 2.9 creates a quiet impression. The repeated use of glass block also contributes to this theme. Interior designers often use neutral or solid surface colors and finishes to create the impression of a larger space. Carefully located built-in mirrors are also used for this purpose.

Light and Shadow

Light reflects from the surfaces of forms. Shadows appear in areas that light cannot reach. Light and shadow both give a sense of depth to any structure. The effective designer plans the relationship of light and dark areas accordingly.

FIGURE 2.7 > Pastels and soft materials and lighting create a quiet atmosphere. *Marc-Michaels Interior Design, Inc.*

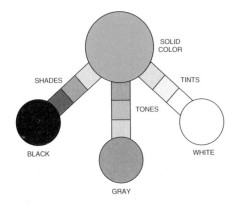

FIGURE 2.6 > Color combined with black or white creates tints, tones, or shades. *Hepler/Wallach/Hepler © Cengage 2013*

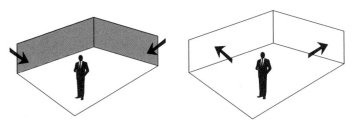

FIGURE 2.8 > Bold colors advance—pale colors recede. *Hepler/Wallach/Hepler © Cengage 2013*

FIGURE 2.9 > Predominance of white and glass on an exterior. *Pittsburgh Corning Corp.*

To achieve a dramatic effect, the designer must consider which surfaces reflect (instead of absorb) light and which surfaces refract (bend) light as it strikes the surface of materials. The designer must also remember that with continued exposure to light, visual sensitivity decreases. People become adapted to degrees of darkness or lightness after extended exposure. Thus, a designer should plan for a variety of levels of light in a room or building. Light and shadow patterns are used with high-contrast materials to create the dramatic visual effect shown in Figure 2.10. Light and shadow patterns at different times of the day or night must also be considered. Computer programs allow designers to study the relationship of design elements to the environment with high levels of accuracy and flexibility.

Texture and Materials

Materials are the raw substances with which designers create. Materials possess their own unique properties, such as color, form, dimension, degree of hardness, and texture. *Texture* is a significant factor in the selection of appropriate materials. *Texture* refers to the surface finish of an object—its roughness, smoothness, coarseness, or fineness. Surfaces of materials

such as concrete, stone, and brick are rough and dull and suggest strength and informality. Smoother surfaces, such as those of glass, aluminum, and plastics, create a feeling of luxury and formality. The designer must be careful not to

FIGURE 2.10 > Light and shadow patterns create this dramatic effect. *Hamid Rafiei Dreamarch3D.com*

FIGURE 2.11 > Masonry textures dominate this design.
Robert P. Ruschak Photography

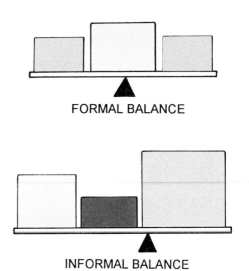

FIGURE 2.12 > Two types of balance. *Hepler/Wallach/Hepler*
© *Cengage 2013*

include too many textures of a similar nature, such as stone and brick. When positioned close together, they tend to compete with each other. Textures, such as wood and stone, are more pleasing when contrasted with other surfaces. The rough texture of stone is the dominant feature of the view shown in Figure 2.11.

Rough surfaces reduce the apparent height of a ceiling and walls may appear darker. Smooth surfaces increase the apparent height of a ceiling or wall and reflect more light, thus making colors appear brighter.

PRINCIPLES OF DESIGN

The basic **principles of design** are the guidelines for determining how to combine the elements of design. For buildings to be aesthetically pleasing, as well as functional, the basic principles of design should be applied. These are *balance, rhythm, repetition, emphasis, subordination, proportion, unity, variety, opposition,* and *transition*.

Balance

Equilibrium (feeling of stability) in design is known as **balance.** Buildings are *informally balanced* if they are asymmetrical and *formally balanced* if they are symmetrical. The balance scale in Figure 2.12 illustrates this difference. The exterior of the building shown in Figure 2.13 is formally balanced as is the interior of the bath in Figure 2.14. The exterior of the building in Figure 2.15 is informally balanced. The room in Figure 2.16 is formally balanced. Buildings or rooms can be balanced in different ways. Likewise architectural components such as the symmetrical window assembly in Figure 2.17 can add reinforcement to a building or a room's balance.

FIGURE 2.13 > Formally balanced design. *John Henry, Architect*

Whether a design is formal or informal, balance requires a harmonious relationship in the distribution of line, form, space, color, light, texture, and materials.

Rhythm and Repetition

When lines, planes, or surface treatments are repeated in a regular sequence, the order or arrangement creates a sense of rhythm. **Rhythm** creates motion and carries the viewer's eyes to various parts of the space. This may be accomplished by the repetition of lines, colors, and patterns. **Repetition** is designed into the structure of the building shown in

FIGURE 2.14 > Interior symmetrical design. *LaStrada Furniture and Interiors, Inc.*

FIGURE 2.15 > Informally balanced design. *Hepler/Wallach/ Hepler © Cengage 2013*

FIGURE 2.16 > Formally balanced living room. *Marc-Michaels Interior Design, Inc.*

FIGURE 2.17 > Symmetrically balanced window assembly. *Eagle Window & Door Inc.*

Figure 2.18 by repeating structural window and roof line shapes. The design in Figure 2.19 illustrates a very creative application of the principle of rhythm and repetition. This design is not only aesthetically attractive, it is also extremely ergonomically functional.

Emphasis and Subordination

The principle of **emphasis,** or giving something importance, means drawing a viewer's attention to an area or subject. In architectural design, some emphasis or *focal point* (center of attention) should be designed into each exterior and interior space. Directing attention to a point of emphasis, the focal point, can be accomplished by arrangement of features, contrast of colors, line direction, variations in light, space relationships, and changes in materials or texture. The point of emphasis in Figure 2.20 is the entertainment center; all

other features are subordinate. In Figure 2.21 the point of emphasis is created beyond the room by the angular position of the window wall. **Subordination** occurs when emphasis is achieved through design. Other features become subordinate. They have less emphasis or importance.

FIGURE 2.18 > Structural repetition patterns create rhythm. *Photo courtesy Lindal Cedar Homes. Lindal.com*

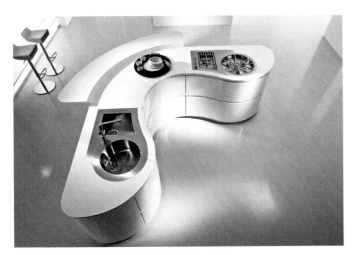

FIGURE 2.19 > Application of rhythm and repetition. *Pedini*

FIGURE 2.20 > Screen position creates a focal point for this media-oriented room. *Audio Tec Design*

Proportion

The term **proportion** refers to the relationship of one part to another, or ratio. The early Greeks found that the proportions of a rectangle in the ratio of 2 to 3, 3 to 5, 5 to 8, and 6 to 10 were more pleasing than other ratios. For example, a room or a rug with dimensions of 9′ × 15′ or 10′ × 16′ will have the proportions of 3 to 5 and 5 to 8.

Figure 2.22 shows several classical systems used to create desirable proportions in a design. In the sixteenth century scientists observed that many natural features related to each other at a ratio of 1 to 1.618, (near 3:5) which they called the "golden rule" or Phi. Closely related proportions have been used by artists and architects for centuries to create attractive shapes.

The proportion (ratio) of interior space, furniture, and accessories should be harmonious. Large bulky components in small rooms should be avoided, just as small components in large rooms should not be used. Areas can appear completely different depending on how the proportional division of space within the area is allocated. The total volume of space within the two rectangles shown in Figure 2.23 is the same; however, the proportions appear different due to the division of space.

Unity

Unity is the expression of the sense of wholeness in a design. Every structure should appear complete. No parts should appear as appendages or afterthoughts. Designers achieve unity through the use of consistent line and color, even though the building is composed of many different parts. Unity, or harmony, as the name implies, is the joining together of the basic elements of good design to form one

FIGURE 2.21 > Point of emphasis created beyond the room. *Interiors by Steven G*

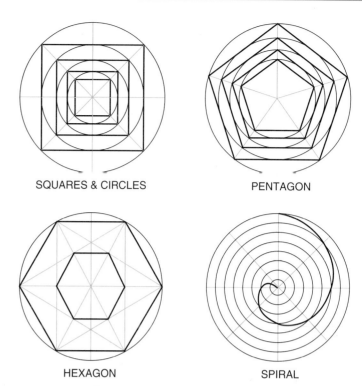

SQUARES & CIRCLES   PENTAGON

HEXAGON   SPIRAL

FIGURE 2.22 > Proportional systems used in two-dimensional design. *Hepler/Wallach/Hepler © Cengage 2013*

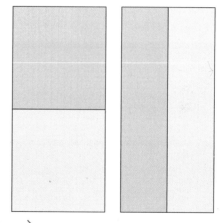

FIGURE 2.23 > The two rectangles are the same although proportionally they appear different. *Hepler/Wallach/Hepler © Cengage 2013*

harmonious, unified whole. Unity can be achieved by using any of the elements of design consistently throughout the entire design.

Unity is often achieved through the use of basic geometric shapes. For centuries, the architectural development of livable structures has been restricted by the use of the square and the cube. Other shapes, such as the triangle, octagon, parabola, pyramid, pentagon, circle, and sphere are now used extensively. This is possible due to the development of materials that are stronger and lighter and have a variety of uses, such as those used in the geodesic dome shown in Figure 2.24.

Variety and Opposition

Too much unity, too much rhythm, or too much repetition can ruin a sense of **variety** or contrast. Likewise, too little of any of the elements of design will also result in a lack of variety. Without variety, any area can become dull and tiresome to the eye of the observer.

Variety can often be achieved with **opposition.** Opposition involves contrasting elements such as short and long, thick and thin, straight and curved, light and dark. Changes in color are used to achieve variety. The building shown in Figure 2.25 illustrates opposition created by a lighting effect. The building lines in Figure 2.26 provide line opposition.

Transition

The change from one color to another or from a curved to a straight line, if done while maintaining the unity of

FIGURE 2.24 > A geodesic dome structure possesses mass and unity. *James Eismont, Photographer*

FIGURE 2.26 > Opposition created with building lines. *Diane Kingston, Photographer*

FIGURE 2.25 > Opposition created with light. *Pittsburgh Corning Corp.*

Transition in architecture is also considered in relation to changes in the surroundings. For example, where extreme climate changes occur, the transition between seasons needs to be considered so that a design functions well under all conditions. For example, Frank Lloyd Wright's design for the "Fallingwater" residence is just as dramatic in winter (Figure 2.27) as in summer (Figure 2.28).

Design Applications

Many illustrations throughout the remainder of this text contain features that show the application of the elements and principles of design. One typical example for each is shown in the following illustrations:

Elements of Design

Line	Figure 8.16
Form	Figure 7.10
Space	Figure 7.1
Color	Figure 8.27
Light and shadow	Figure 8.24
Texture and materials	Figure 7.1

the design, is known as **transition.** Transition may be the change from a curved molding on the floor to the flat wall or a change from one floor covering to another in adjoining rooms.

FIGURE 2.27 ❯ "Fallingwater" in winter. *Thomas A. Heinz, Courtesy of Western Pennsylvania Conservancy*

Principles of Design

Balance	Figure 4.6
Rhythm and repetition	Figure 7.8
Emphasis and subordination	Figure 7.11
Unity	Figure 7.28
Variety and opposition	Figure 7.13
Transition	Figure 9.9

FIGURE 2.28 ❯ "Fallingwater" in summer. *Thomas A. Heinz, Courtesy of Western Pennsylvania Conservancy*

Fundamentals of Design Exercises   CHAPTER 2

1. Choose a figure from Chapter 1 or 2 and discuss how the concept of "form follows function" relates to it.

2. Choose an architectural design from Chapter 1 or 2. Describe the six elements of design in the figure you selected.

3. How would you make a small room look larger and a large room look smaller? Explain the reasons for your decisions.

4. Describe a building in your neighborhood or in a magazine in terms of design. Tell what elements and principles of design are its most outstanding features. (You might include a photograph or magazine illustration.)

5. List each element of design and describe your preference for applying each to a residence of your own design.

6. Describe the difference between a *shade,* a *tone,* and a *tint.*

7. Describe primary, secondary, and tertiary colors.

8. Describe complementary, monochromatic, and triadic colors.

9. List the basic elements of design.

10. List the basic principles of design.

PART 2

ARCHITECTURAL DRAFTING FUNDAMENTALS

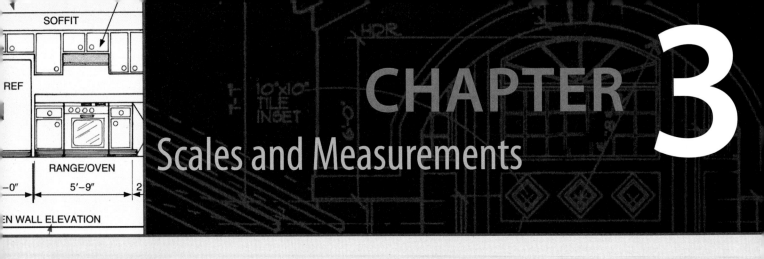

CHAPTER 3
Scales and Measurements

OBJECTIVES

In this chapter you will learn:

> how to use an open-divided scale.

> how and where to use different architect's scales.

> how to read scales from opposite sides.

> how to use an engineer's scale.

> how to read a metric scale.

TERMS

architect's scale
engineer's scale
enlarged scale

full scale
fully-divided scale
metric scale

open-divided scales
reduced scale

INTRODUCTION

In architectural drawing the term *scale* may refer to the proportional (ratio) size of a drawing or to an architectural measuring instrument. When making a scaled (proportional) drawing, one measurement is used to represent another. Scaled drawings allow objects of all sizes to be proportionally reduced or enlarged to show the correct relationship of all parts. Different ratios are required depending on the size of the object and the drawing format size.

Without the use of **reduced scales,** no object larger than a sheet of paper could be accurately drawn. For example, if the earth were drawn to the same scale used on a typical architectural drawing, 1/4″ = 1'-0″ (or 1″ = 4'-0″, a ratio of 1 to 48), the drawing sheet would need to be 165 miles wide! See Figure 3.1.

Conversely, small objects, even those that may be invisible or barely visible to the human eye, can be proportionally enlarged to show details. In architectural drawing, a small item such as a hinging mechanism may need to be drawn at an **enlarged scale.**

In the preparation of scaled drawings, instruments called scales are used. The three main types of scales are the architect's scale, civil engineer's scale, and metric scale.

ARCHITECT'S SCALE

The ability to use **architect's scales** accurately is required not only when preparing drawings but also when checking existing architectural plans and details. The architect's scale is also needed in a variety of related architectural jobs such as bidding, estimating, and model building.

Whether to reduce a structure's size so that it can be drawn to fit on paper or to enlarge a small detail for clarity and accurate dimensions, a drafter needs to use the appropriate scale divisions.

Architect's scales are either open divided or fully divided. In **fully-divided scales,** each main unit on the scale is fully subdivided into smaller units along the full length scale. On **open-divided scales,** only the main units of the scale are graduated (marked off) all along the scale. There is a fully subdivided/one foot (12″) unit at the start of each scale. See Figure 3.2.

The main function of an architect's scale is to enable the architect, designer, or drafter to plan accurately and make drawings in proportion to the actual size of the structure. For example, when a drawing is prepared to a reduced scale of 1/4″ = 1'-0″, a line that is drawn 1/4″ long is thought of by the drafter as 1'-0″, not as 1/4″. See Figure 3.3.

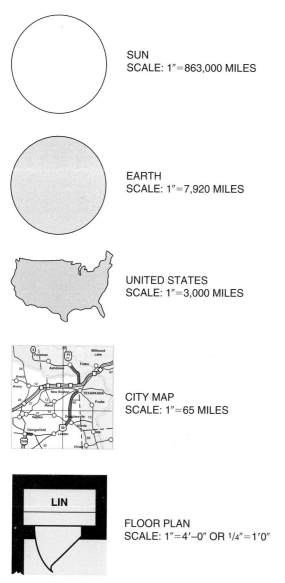

FIGURE 3.1 > Scaled drawings are needed to show the size and shape of large objects. *Hepler/Wallach/Hepler © Cengage 2013*

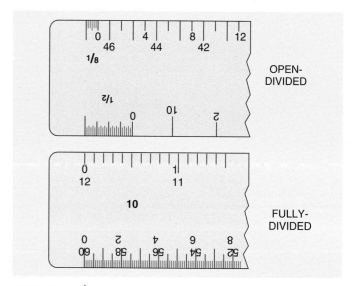

FIGURE 3.2 > Open-divided and fully-divided scales. *Hepler/Wallach/Hepler © Cengage 2013*

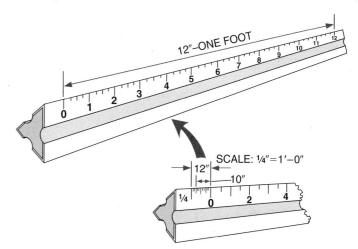

FIGURE 3.3 > In scaled drawings, one measurement represents another. *Hepler/Wallach/Hepler © Cengage 2013*

Using a Scale

Architect's scales may be either bevel or triangular style. See Figure 3.4. Note that the triangular style has three sides and six edges. It accommodates 11 different scales (marked units of measure). One edge is a full-size scale of 12″, divided into 16 parts per inch. The other five edges contain open-divided scales paired to include 3/16 with 3/32, 1/4 with 1/8, 3/4 with 3/8, 1 with 1/2, and 3 with 1 1/2. Locating two scales on each edge maximizes the use of space. One scale reads from left to right. The opposite scale, which is twice as large, reads from right to left. For example, the 1/4″ scale and the 1/8″ scale are placed on the same edge but are read

from opposite directions. Be sure you are reading the scale numbers in the correct direction when using an open-divided scale. Otherwise, your measurement could be wrong. See Figure 3.5

The architect's scale can be used to make the divisions of the scale equal 1″ or 1′-0″. For example, 1/2″ can equal 1″ or 1′-0″ or any unit of measurement such as yards or miles. See Figure 3.6.

Because buildings are large compared to the size of a person or appliance, most major architectural drawings use a scale that relates the parts of an inch to 1′-0″. Architectural details, such as cabinet construction and joints, often use the parts of an inch to represent 1″. On open-divided scales, the divided section at the end of the scale is not a part of the numerical

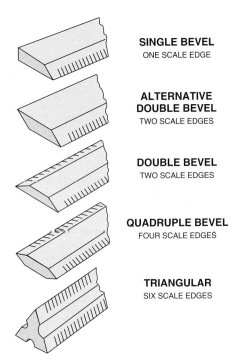

FIGURE 3.4 > Architectural scale shapes. *Hepler/Wallach/ Hepler © Cengage 2013*

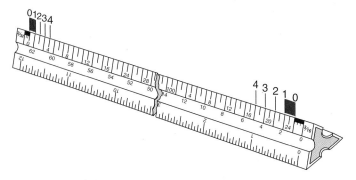

FIGURE 3.5 > Scales that read from right to left are twice the size as those that read from left to right. *Hepler/Wallach/Hepler © Cengage 2013*

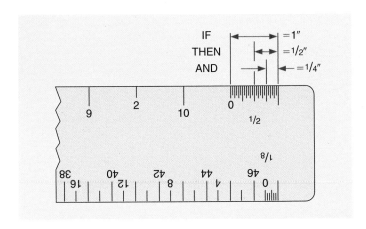

FIGURE 3.6 > On a half-inch scale, 1/2″ may equal 1″ or 1′-0″. *Hepler/Wallach/Hepler © Cengage 2013*

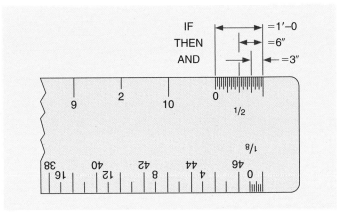

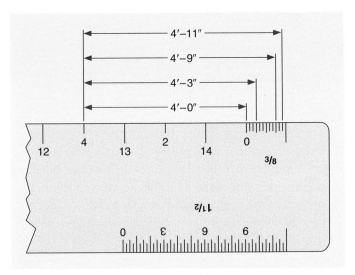

FIGURE 3.7 > Subdivisions are used for inches or inch fractions. *Hepler/Wallach/Hepler © Cengage 2013*

scale. This divided section is an additional length to show smaller subdivisions (inches) of the larger unit (feet). When measuring with the scale, start at zero, not at the end of the fully divided section. First measure the number of larger units (for example, feet) and then measure the additional smaller units (inches) in the subdivided area. Look at Figure 3.7. The distance of 4′-11″ is established by measuring from the division line 4 to 0 for feet. Then, measure on the subdivided area 11″ past 0. On this scale, each line in the subdivided part equals 1″. On smaller scales, these lines may equal only 2″. On larger scales, they may equal a fractional part of an inch. Figure 3.8 shows a further use of the architect's scale.

To further understand the architect's scale, compare one specific distance shown on different scales. Compare the actual length of the 5′-6″ dimension on the four different scales shown in Figure 3.9. Figure 3.10 shows the reduction on six different scales.

The architect's scale is only as accurate as its user. In using the scale, always lay out the overall dimensions of the drawing first. If the width and length are correct, only minor errors in subdimensions may occur. Moreover, if overall dimensions are correct, it is easier to check subdimensions. If one is inaccurate, another will be also.

Remember, an architect's scale is a measuring device, not a drawing instrument. Never use a scale as a straightedge for drawing. The fine increment lines on a scale will be worn down or removed if misused, making accurate measuring difficult.

Selecting a Scale (Proportion)

If the structure to be drawn is extremely large, a small scale must be used. Small structures can be drawn to a larger scale, because they will not take up as much space on the drawing sheet. Most plans that show major parts of residences (floor plans, elevations, and foundation plans) are drawn to 1/4″ scale. Construction details pertaining to these drawings are often drawn to 1/2″, 3/4″, or 1″ = 1′-0″.

Remember that as the scale changes, not only does the length of each line increase or decrease, but the width of each wall also increases or decreases, as shown in Figure 3.11. A wall drawn to the scale of 1/16″ = 1′-0″ is small and little detail can be shown. The 1/2″ = 1′-0″ wall would probably cover too large an area on the drawing if the building were very large. Therefore, the 1/4″ and 1/8″ scales are used most often for drawing floor plans and elevations.

The **full scale** in fractional inches may be used to draw objects full size 1″ = 1″ or 1′ = 1′ or to a scale of 1″ = 1′-0″. An open-divided full scale is 12″, and each inch is divided into 16 units throughout.

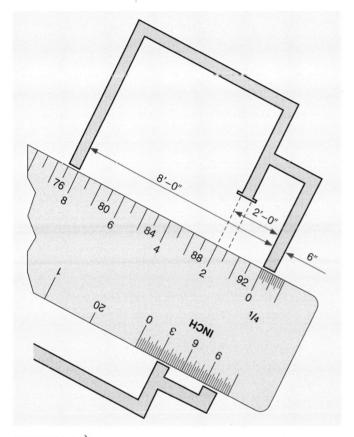

FIGURE 3.8 > Scale used to measure feet and inches. *Hepler/Wallach/Hepler © Cengage 2013*

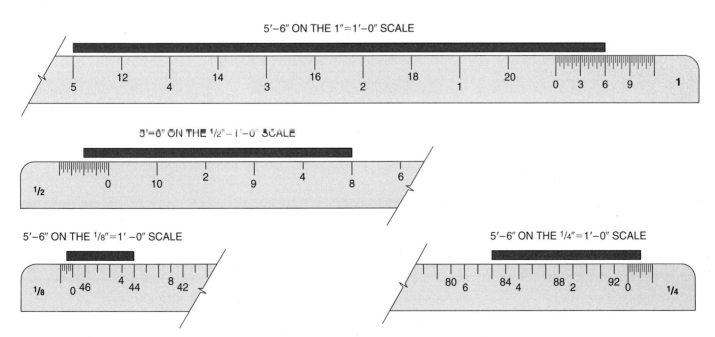

FIGURE 3.9 > The distance 5′-6″ measured on different scales. *Hepler/Wallach/Hepler © Cengage 2013*

1″ = 1′-0″ will **decrease** drawing size by 12 times							
1½″ = 1′-0″	″	″	″	″	″	″ 8	″
3″ = 1′-0″	″	″	″	″	″	″ 4	″
½″ = 1′-0″	″	″	″	″	″	″ 24	″
¼″ = 1′-0″	″	″	″	″	″	″ 48	″
⅛″ = 1′-0″	″	″	″	″	″	″ 96	″

FIGURE 3.10 > Resulting reduction of the different scales. *Hepler/Wallach/Hepler © Cengage 2013*

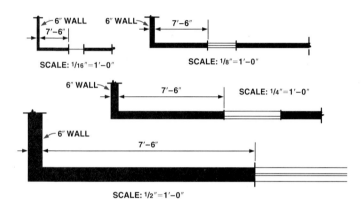

FIGURE 3.11 > Floor plan corner wall drawn at different scales. *Hepler/Wallach/Hepler © Cengage 2013*

ENGINEER'S SCALE

Although the architect's scale is used for most architectural drawings, the **engineer's scale** is often used for plans that show the size and features of the land surrounding a building (plot plans, site plans, landscape plans). An engineer's scale divides the inch into decimal parts. These parts are 10, 20, 30, 40, 50, and 60 parts per inch. See Figure 3.12. Each one of these units can represent any distance, such as an inch, a foot, a yard, or a mile, depending on the final drawing size.

The engineer's scale can also be used to draw floor plans. With its 40 divisions per inch, the engineer's scale can be used at a scale of 1″ = 4′ (1:48), which is the same ratio as 1/4″ = 1′-0″. An engineer's scale does not use feet and inches; it instead uses feet and the decimal parts of a foot. Thus, 2′-6″ reads 2.5′ and 7′-3″ reads 7.25′ on an engineer's scale. See Figure 3.13.

An engineer's scale of 1″ = 10′ is normally used for small sites. If a land site is very large, a scale of 1″ = 20′ or 1″ = 30′ may be needed to allow the plan to fit the sheet size. See Figure 3.14.

Selecting a Scale (Instrument)

Different scales are designed for a broad range of applications. Before a drawing is started, determine the actual size of the area to be covered. Then select the scale—whether an architect's or engineer's scale—that will provide the greatest detail and yet fit completely on the drawing sheet. See Figure 3.15.

METRIC SCALES

Metric scales such as those shown in Figure 3.16 are used in the same manner as the architect's scale to prepare

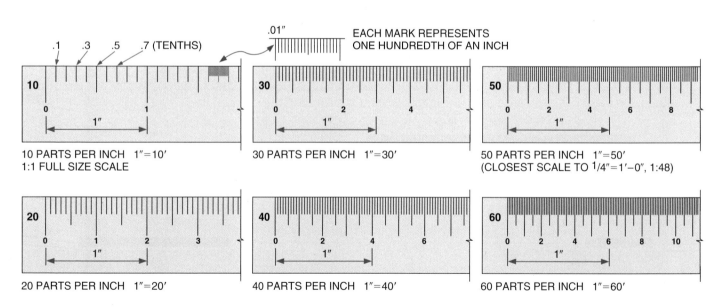

FIGURE 3.12 > Engineer's (decimals) scales. *Hepler/Wallach/Hepler © Cengage 2013*

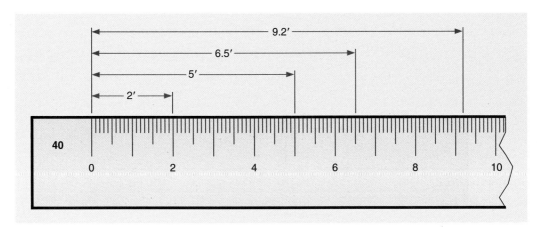

FIGURE 3.13 ＞ The number 40 on an engineer's scale represents the same ratio as 1/4″ = 1′-0″ on an architect's scale. *Hepler/Wallach/Hepler © Cengage 2013*

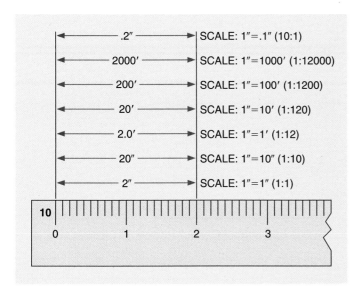

FIGURE 3.14 ＞ Distances represented by 2″ on different engineer's scales. *Hepler/Wallach/Hepler © Cengage 2013*

reduced-size drawings. Metric scales, however, use ratios in increments of 10 rather than the fractional ratios of 12 used in architect's scales. The metric system of measure is a decimal system. Units are related by tens. Most measurements used on architectural drawings are linear distances. The basic unit of measure in the metric system for distance is the *meter* (m). Prefixes are used to change the base (meter) to larger or smaller amounts by *units of 10.* Prefixes that represent subdivisions of less than 1 meter are *deci-, centi-,* and *milli-.* A decimeter equals one-tenth (0.1) of a meter. A centimeter equals one one-hundredth (0.01) of a meter. A millimeter equals one one-thousandth (0.001) of a meter. See Figure 3.17. The most commonly used subdivisions of a meter are the centimeter and the millimeter. The most commonly used multiple of the meter is the *kilo*meter. A kilometer equals 1,000 meters.

The numbers on a meter scale mark every tenth line to represent centimeters. Each single line represents millimeters. Note that there are 10 millimeters within each centimeter.

DRAWING TYPE	U.S. CUSTOMARY ARCHITECT'S SCALES (FEET/INCHES)	ISO METRIC SCALES (MILLIMETERS)	CIVIL ENGINEER'S SCALES (FEET/DECIMAL)
Site plans	1/8″ = 1′-0″ THRU 1/32″ = 1′-0″	1:100 THRU 1:500	1″ = 10′ THRU 1″ = 200′-0″
Floor plans	1/4″ = 1′-0″ or 1/8″ = 1′-0″	1:50 or 1:00	1″ = 10′ THRU 1″ = 30′-0″
Foundation plans	1/4″ = 1′-0″ or 1/8″ = 1′-0″	1:50	1″ = 10′ THRU 1″ = 30′-0″
Exterior elevations	1/4″ = 1′-0″ or 1/8″ = 1′-0″	1:50 or 1:00	1″ = 4′-0″ or 2′-0″
Interior elevations	1/2″ = 1′-0″	1:20	1″ = 4′-0″ or 2′-0″
Construction details	1 1/2″ = 1′-0″ THRU 3/4″ = 1′-0″	1:5 THRU 1:10	1″ = 1′-0″ or 2′-0″
Cabinet details	1/2″ = 1′-0″	1:20	1″ = 2′-0″

FIGURE 3.15 ＞ Range of scales used on architectural drawings. *Hepler/Wallach/Hepler © Cengage 2013*

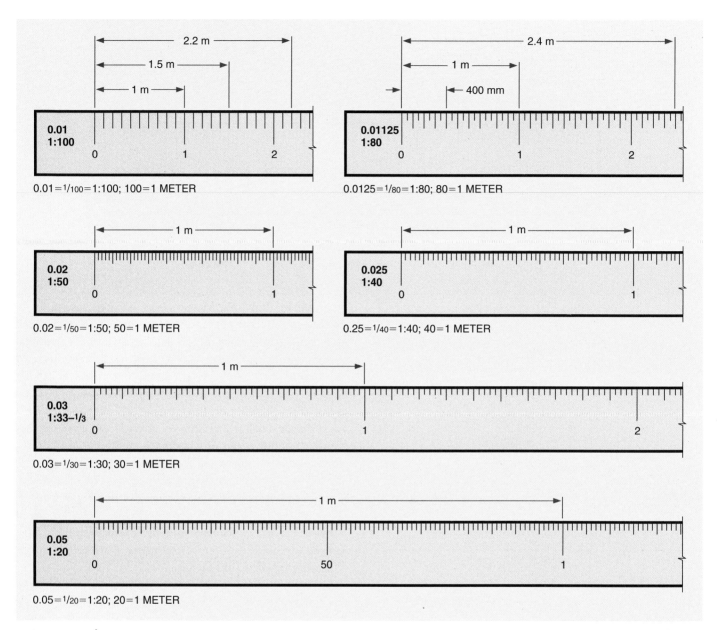

FIGURE 3.16 > Metric scales. *Hepler/Wallach/Hepler © Cengage 2013*

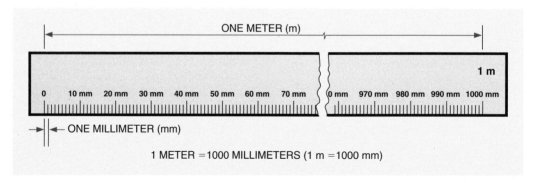

FIGURE 3.17 > A millimeter is one one-thousandth of a meter. *Hepler/Wallach/Hepler © Cengage 2013*

USE	RATIO	COMPARISON TO 1 METER
City map	1:2500 1:1250	(0.4 mm equals 1 m) (0.8 mm equals 1 m)
Plat plans	1:500 1:200	(2 mm equals 1 m) (5 mm equals 1 m)
Plot plans	1:100 1:80	(10 mm equals 1 m) (12.5 mm equals 1 m)
Floor plans	1:75 1:50 1:40	(13.3 mm equals 1 m) (20 mm equals 1 m) (25 mm equals 1 m)
Details	1:20 1:10 1:5	(50 mm equals 1 m) (100 mm equals 1 m) (200 mm equals 1 m)

FIGURE 3.18 > Use of metric ratios. *Hepler/Wallach/Hepler*
© Cengage 2013

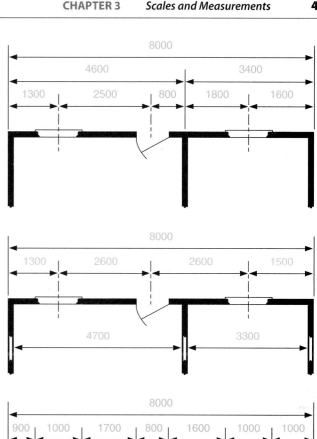

Just as with any other scale, the ratio chosen depends on the size of the drawing compared with the full size of the object. Figure 3.18 shows some common metric ratios and the various types of architectural drawings for which they are used.

In architectural drawing, ISO standards recommend that only millimeters be used for all architectural and engineering drawings. See Figure 3.19. This eliminates the use of decimal points as shown in Figure 3.20.

Some drawings prepared with fractional dimensions need to be converted to metric dimensions. Figure 3.21 shows the conversion of inches to millimeters. To convert inch dimensions to millimeter dimensions, use the following formula:

$$\text{Formula: in.} \times 25.4 = \text{mm}$$
$$(\text{inches} \times 25.4 = \text{millimeters})$$

$$\text{Example: } 6'\text{-}6'' = 78''$$
$$78'' \times 25.4 = 1981.2 \text{ mm}$$

Prepare all drawings in a set using either metric ratios or the customary fractional system. *Do not mix metric and customary units.* If approximate conversion from one system to the other is necessary, refer to the appendix. When very accurate conversion from customary to metric units is necessary, consult a handbook or use the *Metric Practice Guide* from the American Society of Mechanical Engineers.

CAD Scales

Knowledge of architect's and engineer's scales is necessary for both manual and CAD drafting. In CAD drawing the

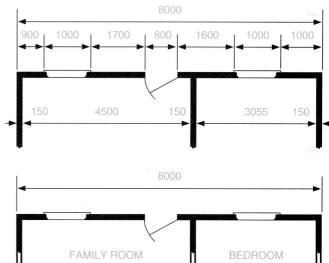

FIGURE 3.19 > Dimensions in millimeters. *Hepler/Wallach/Hepler* © Cengage 2013

Scale command is used to set a drawing object at a specific architect's, engineer's, or metric scale size. First the drawing sheet size is determined by selecting *Layout* and right mouse clicking to *Page Setup Manager.* A viewport is then set up to scale the image based on magnification (*Zoom*), which allows the drawing to fit on the paper size available. Smaller scales or larger sheet sizes are options that can ensure a proper fit using a new viewport (drawing area).

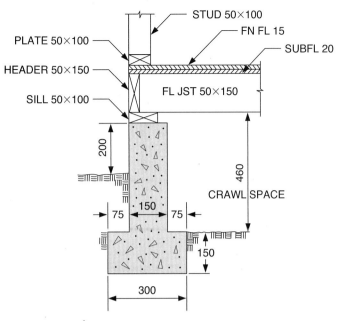

STUD 50×100
FN FL 15
SUBFL 20
PLATE 50×100
HEADER 50×150
SILL 50×100
FL JST 50×150
CRAWL SPACE
200
460
150
75   75
150
300

ALL DIMENSIONS FOR ARCHITECTURAL DRAWINGS ARE IN MILLIMETERS (mm)		
CONVERSIONS		
Convert from:	To:	Multiply by:
Feet	Millimeters	304.8 (305)
Millimeters	Feet	25.4 (25)

The physical scale of a drawing object can be changed by using the *Scale* command found under the *Modify* pull-down menu. During drawing, the *Measure* command can be used to locate a specific size on a segment. This command can also be used to draw a specific line length by inputting the line length at the scale of the drawing.

Scales and Measurements Exercises

CHAPTER **3**

1. Draw the following four lines using a scale of 1/4″ = 1′-0″:
   a. 5′-0″
   b. 7′-6″
   c. 9′-10″
   d. 11′-3″

2. Measure the distances you drew in Exercise 1, using the 1/8″ = 1″ scale.

3. Using a 2D CAD system, learn the commands for producing different scaled drawings.

4. Measure a book, desk, car, and room using a metric scale. Record your results. Compare your measurements with customary measurements.

5. Convert the following dimensions to millimeters: 5′-6″, 6′-8″, 10′-4″, 11′-7″, 15′-3″.

6. List the scales (proportional measures) you will use in drawing plans of a residence you are designing. Explain why you chose a particular scale in relation to the size of the paper you are using.

SOFFIT

EF

RANGE/OVEN
5'-9"

WALL ELEVATION

CHAPTER 4
Conventions and Procedures

OBJECTIVES

In this chapter you will learn to:

> understand the architectural drawing system.

> differentiate between the types and purposes of architectural drawings.

> produce the line conventions used on architectural drawings.

> develop good lettering techniques.

> sketch lines, patterns, and a floor plan.

> choose appropriate pens, pencils, and drafting media.

TERMS

coding system	line conventions	T square
compass	models	technical pens
construction documents	overlay	template
detail drawings	parallel slide	title block
dividers	plans	trace
drafting machine	protractor	triangles
elevations	renderings	underlays
flexible curve	sections	vellum
layering	stampat	working drawings

INTRODUCTION

Builders follow a set of working drawings in order to make a designer's idea become a reality. To clearly communicate information about the project to be built, standards and conventions in the preparation of these drawings have been established. Drafters and designers must understand and apply these conventions to make the plans consistently readable and understandable. The ability to understand and use a wide range of instruments and equipment is vital for this function.

Other design and construction professionals also need to understand the language of architectural drawing to effectively use sets of plans in their work.

ARCHITECTURAL DRAWINGS

The design of a structure is interpreted through the use of several types of architectural drawings, including floor plans, elevations, construction details, and pictorial drawings. Drawings may vary from simple to complex. The number of drawings needed to construct a building depends on the complexity of the structure and on the degree to which the designer needs or wants to control the methods and details of construction. A *minimum* set of plans provides the builder with great latitude in selection of materials and processes. A *maximum* set of plans will ensure, to the greatest degree possible, agreement between the wishes of the designer and the final constructed building. See Figure 4.1.

Even though a building may be relatively simple, as many drawings as necessary should be prepared. Any detailed working drawing that is omitted forces the builder into the role of the designer. For some buildings and some builders, this may be acceptable. For others, this is highly unacceptable. The more plans, details, and specifications that are accurately developed for a structure, the closer the finished building will be to what was conceived by the designer.

DRAWINGS AND DOCUMENTS	SIZE OF SET OF PLANS		
	Min.	Aver.	Max.
Floor plans	X	X	X
Front elevation	X	X	X
Rear elevation		X	X
Right elevation	X	X	X
Left elevation		X	X
Auxiliary elevations			X
Interior elevations		X	X
Exterior pictorial renderings		X	X
Interior renderings			X
Plot plan (site)	X	X	X
Landscape plan			X
Survey plan	X	X	X
Full section	X	X	X
Detail sections		X	X
Floor-framing plans			X
Exterior-wall framing plans			X
Interior-wall framing plans			X
Stud layouts			X
Roof-framing plan			X
Electrical plan		X	X
Air-conditioning plan			X
Plumbing diagram			X
Schedules			X
Specifications			X
Cost analysis			X
Scale model			X
Reflected ceiling plan			X
Code compliance document		X	X

FIGURE 4.1 ⟩ Types of drawings in a plan set. *Hepler/Wallach/Hepler © Cengage 2013*

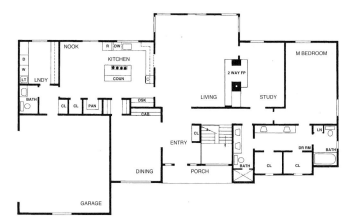

FIGURE 4.2 ⟩ Floor plan. *Hepler/Wallach/Hepler © Cengage 2013*

Types of Drawings

Architectural drawings are often called "the plans." However, specific architectural drawings show certain views of a structure and only some of those details are called *plans.* The following list includes brief descriptions of the various types of architectural drawings:

- **Plans** (or plan views) are views from the top down, a "bird's-eye" view. An example is the floor plan shown in Figure 4.2.
- **Elevations** are flat two-dimensional views of all vertical walls. See Figure 4.3.
- **Sections** show a view of one "slice" of a planned structure. It's as if an imaginary line were cut vertically at a particular place, showing the internal parts of the structure and components along the plane of that cut. See Figure 4.4.
- **Detail drawings** are prepared at a larger scale than other types of drawings. They are drawn to reveal precise infor-

mation about construction methods and materials. Details may be prepared in plan view, in pictorial form, or as sections. See Figure 4.5.

- **Renderings** are usually one-, two-, or three-point perspective drawings. Often called *pictorials,* these show how the finished product is expected to look. Renderings are made of both the building and its site. See Figures 4.6 and 4.7.
- **Models** are constructed as three-dimensional reduced-scale replicas of a structure or structures as shown in Figure 4.8 on page 47, and detailed in Chapter 21. CAD-generated 3D drawings, as shown in Figure 4.9 on page 48, are also classified as models although prepared on a two-dimensional surface.

Drawings and Documents

Information about an architectural design and its construction is provided basically in three ways: general-purpose drawings, working drawings, and construction documents.

Architectural drawings used for sales promotion or preliminary planning purposes are known as general-purpose drawings. Drawings of this type usually consist of only approximate room sizes and dimensions on a single-line floor plan. See Figure 4.10 on page 48. Pictorial drawings of exterior or front elevation views are also used for these purposes.

Drawings used during the building process are known as **working drawings.** Working drawings should contain all of the information needed to completely construct a building: the dimensions, materials, and drawings of the building's shape. Complete floor plans and elevations are required. A full set of working drawings also includes specialized drawings, such as framing, electrical, plumbing, and landscape plans.

Even with a set of drawings, the building information is still incomplete. So much information is necessary for building a structure that not all of it can be put on a set of working drawings. For this reason **construction documents** are

FRONT ELEVATION

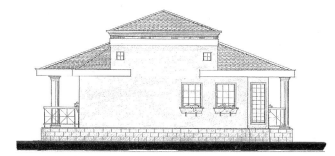

REAR ELEVATION

LEFT ELEVATION

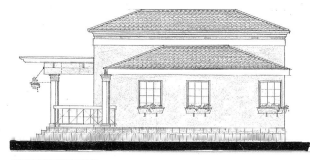

RIGHT ELEVATION

FIGURE 4.3 > Elevation drawings. *Hepler/Wallach/Hepler*
© Cengage 2013

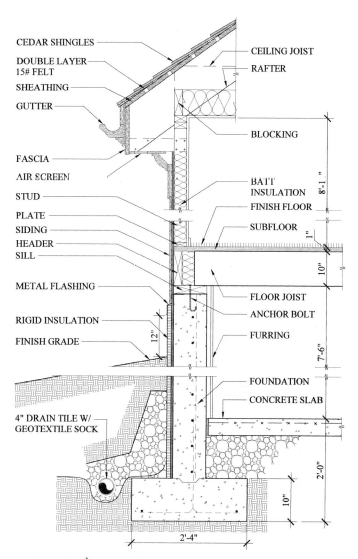

CEDAR SHINGLES
DOUBLE LAYER 15# FELT
SHEATHING
GUTTER
FASCIA
AIR SCREEN
STUD
PLATE
SIDING
HEADER
SILL
METAL FLASHING
RIGID INSULATION
FINISH GRADE
4" DRAIN TILE W/ GEOTEXTILE SOCK

CEILING JOIST
RAFTER
BLOCKING
BATT INSULATION
FINISH FLOOR
SUBFLOOR
FLOOR JOIST
ANCHOR BOLT
FURRING
FOUNDATION
CONCRETE SLAB

FIGURE 4.4 > Sectional drawing. *Hepler/Wallach/Hepler*
© Cengage 2013

prepared that contain hundreds of facts and figures, plus legal and financial information related to the building process. Documents such as building specifications and schedules eliminate guesswork and specify exactly which processes, materials, and building components are to be used.

Reading Architectural Drawings

A small number of working drawings and documents may be sufficient and easy to use for one residence. However, for a very large project, several sets of complicated plans may be needed that require many different views, as well as documents containing very specific details and information.

Coding System

To make a large number of drawings manageable and easy to use, a **coding system** is often necessary. The coding system identifies every specific drawing and detail. It is also a method of keeping similar drawings together and organized in a working drawing set.

Most architects follow the American Institute of Architects' (AIA) alphanumeric coding system. In the AIA's coding system, drawings are identified by letters and numbers for

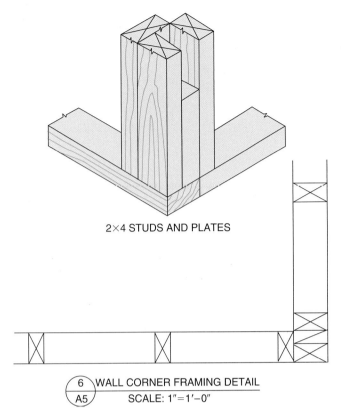

2×4 STUDS AND PLATES

6 / A5 WALL CORNER FRAMING DETAIL
SCALE: 1"=1'-0"

FIGURE 4.5 ＞ Detail drawing. *Hepler/Wallach/Hepler © Cengage 2013*

FIGURE 4.6 ＞ Exterior rendering. *Used by permission Hanley Wood, LLC*

ease of referencing. Figure 4.11 illustrates the breakdown of these letters into codes, which identify the group and drawing number. For example, in the code shown in Figure 4.11, "A" indicates that the drawing belongs to a set of architectural working drawings. The number "2" identifies the group to which the drawing belongs within the set, and the "1" after the period shows that the drawing is the first one in the group.

The group number always remains the same, no matter how many drawings the group contains. More drawings may be added within groups without interrupting the alphanumerical order in the set.

Coding CAD Drawings

Use of the AIA alphanumeric coding system on CAD drawings enables these categories to be used to create layers assigned to each discipline, drawing, or group. Numbering systems for mechanical drawings are coordinated primarily through the individual companies in accordance with standards set by the American National Standards Institute.

Title Blocks

Similar to other kinds of written information, architectural drawings are identified by titles. In any drawing system **title blocks** identify drawings in a consistent and convenient format. Figure 4.12 lists the information that appears on most title blocks, and Figure 4.13 shows a typical completed title

block. Title blocks are usually located on the bottom and/or right side of each drawing sheet.

Because the number of revisions varies, revision entries are made in sequence from bottom to top. Revisions are shown with a number inside a triangle on each drawing change.

Cross-Referencing

Drawing all views, sections, or details of features on one floor plan or one elevation is usually impossible. For example, floor plans do not show height details and dimensions. Elevation drawings do not show all horizontal dimensions. Therefore, cross-references are often necessary to guide the reader from one drawing to another. Numbered symbols are generally used for this purpose.

As shown in Figure 4.14 on page 50, a circle with a directional arrow is drawn on a plan view. The arrow points in the direction of the area to be referenced elsewhere and shows the drawing sheet number and the detail number, which helps the user locate the referenced elevation or detail.

Layering is a method of aligning related plan drawings to ensure accuracy and eliminate much duplication of effort. First, a base drawing is prepared. In architectural work the base drawing is usually a floor plan. Then related drawings are prepared directly over the base and aligned. Layering is sometimes called *overlay drafting* or *pin drafting*.

Aligning specialized plans in a set of drawings, as shown in Figure 4.15 on page 51, ensures that all structural features such as walls and columns align. It also eliminates the potential problems of overlapping mechanical, electrical, piping, and other facilities on the same drawing base. When this aligning is done on a CAD system, the layering task is used. Most complete CAD systems allow for the use of 255 layers.

Layering also simplifies the interpretation of drawings by subcontractors. For example, a plumbing contractor can be given only the floor plan level and the plumbing level. This eliminates the clutter of HVAC, electrical, etc., plans, although the contractor may be given a complete set of plans for cross-referencing.

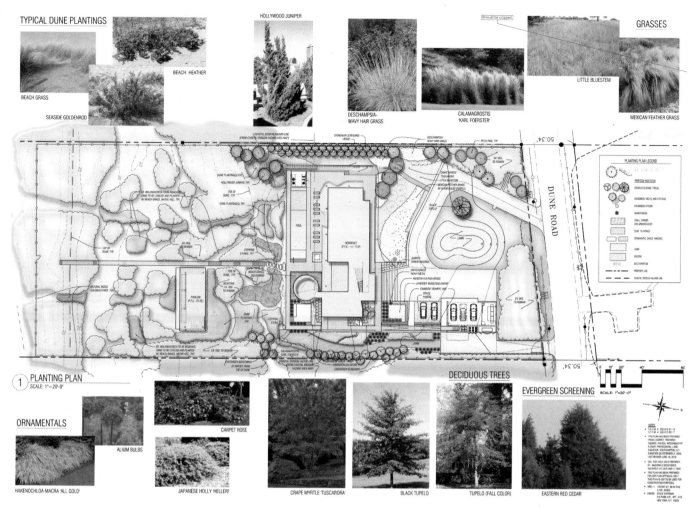

FIGURE 4.7 ❯ Rendering of an estate master plan. *Timothy A. Rumph, RLA*

CAD Layering

Layering is a method of organizing a drawing in which specific line types, colors, or entities are drawn separately so they can be removed from or added to a drawing. These include dimensions, construction lines, details, symbols, grids, furnishings, equipment, materials, and fixtures. Preparing many layers provides ease and flexibility in managing and manipulating drawings and details later in the design and drawing process.

Each color can also be assigned a line weight or type. This results in a colored line being shown on the monitor, but the line assigned to each color or line weight will print or plot. Each layer can be printed separately or in any combination.

Layers within layers are also possible. For example, a floor plan may contain many layers representing a series of floor plan levels—second, third, and so on—that is used to ensure the alignment of features. First activate the *Layer* (*LA*) command from the *Format* pull-down menu. Then you can create

FIGURE 4.8 ❯ Presentation models. *Hepler Associates, LA PC*

FIGURE 4.9 > CAD-created and -rendered model. *Robert McNeel & Associates*

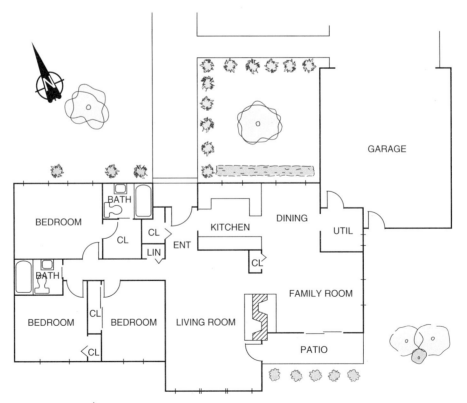

FIGURE 4.10 > Single-line floor plan. *Hepler/Wallach/Hepler © Cengage 2013*

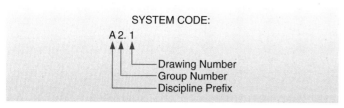

SYSTEM CODE:

A 2 . 1

— Drawing Number
— Group Number
— Discipline Prefix

FIGURE 4.11 > Coding system used to identify discipline, drawing number, and group. *Hepler/Wallach/Hepler © Cengage 2013*

- Project title and number
- Drawing sheet title
- Name and address of client
- Name and address of architect or firm
- Name and address of contractor (if known)
- Initials of designer
- Initials of drafter
- Initials of checker
- Revision block including
  - Title
  - Number
  - Preparer
- Professional seal space
- Scale
- Date
- Sheet number (using AIA code & showing number of sheets in the set)
- Key plan (if needed to identify location)

FIGURE 4.12 > Title block information. *Hepler/Wallach/Hepler © Cengage 2013*

a new layer or edit an existing layer using the *Layer Properties Manager.* The changeable items include the line weights, line types, color, plotting visibility, and screen visibility.

Pages with sections and detail drawings that have been referenced *from* other plans also need to be cross-referenced *back* to the original drawing. In other words, cross-referencing needs to work two ways, so that a person reading the drawings knows where each drawing belongs in relation to the entire structure.

Callouts

A set of architectural drawings also needs to contain the information that identifies many of a building's components, such as doors, windows, rooms, and equipment. A different geometric form designates each component. For example, a door may be shown as a square, a window as a small circle, a room as a rectangle, and equipment as an octagon. These shapes become labels known as *callouts.* See Figure 4.16. To show visually separate components, numbers or letters are usually shown inside the geometric shape.

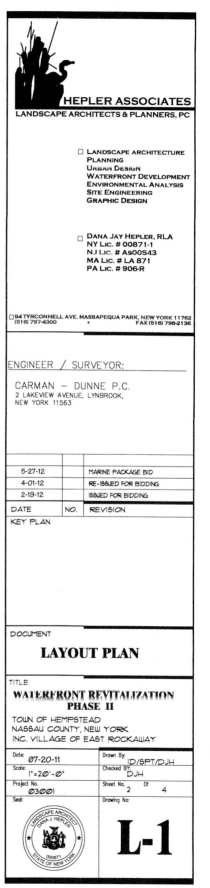

HEPLER ASSOCIATES
LANDSCAPE ARCHITECTS & PLANNERS, PC

☐ LANDSCAPE ARCHITECTURE
PLANNING
URBAN DESIGN
WATERFRONT DEVELOPMENT
ENVIRONMENTAL ANALYSIS
SITE ENGINEERING
GRAPHIC DESIGN

☐ DANA JAY HEPLER, RLA
NY LIC. # 00871-1
NJ LIC. # AS00543
MA LIC. # LA 871
PA LIC. # 906-R

☐ 94 TYRCONNELL AVE. MASSAPEQUA PARK, NEW YORK 11762
(516) 797-4300   •   FAX (516) 798-2136

ENGINEER / SURVEYOR:

CARMAN — DUNNE P.C.
2 LAKEVIEW AVENUE, LYNBROOK,
NEW YORK 11563

DATE	NO.	REVISION
5-27-12		MARINE PACKAGE BID
4-01-12		RE-ISSUED FOR BIDDING
2-19-12		ISSUED FOR BIDDING

KEY PLAN

DOCUMENT

LAYOUT PLAN

TITLE
WATERFRONT REVITALIZATION PHASE II
TOWN OF HEMPSTEAD
NASSAU COUNTY, NEW YORK
INC. VILLAGE OF EAST ROCKAWAY

Date: 07-20-11	Drawn By: ID/SPT/DJH
Scale: 1"=20'-0"	Checked By: DJH
Project No. 03001	Sheet No. 2 Of 4
Seal:	Drawing No:

L-1

FIGURE 4.13 > Typical title block. *Hepler/Wallach/Hepler © Cengage 2013*

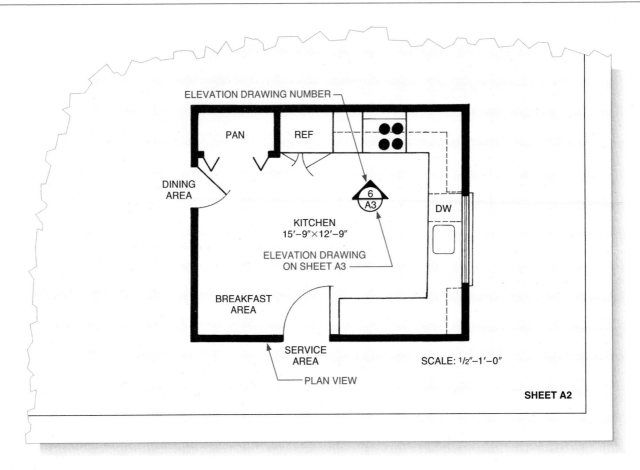

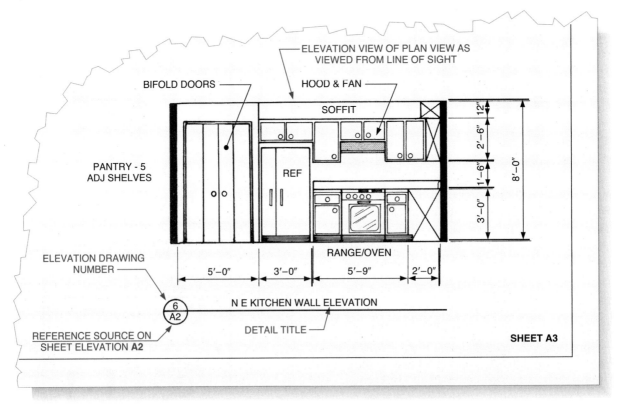

FIGURE 4.14 > Use of cross-referencing symbols. *Hepler/Wallach/Hepler © Cengage 2013*

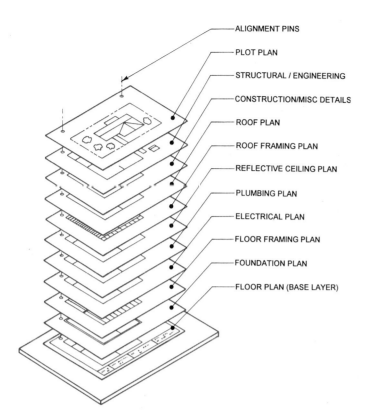

FIGURE 4.15 ❯ Layering of plan drawings in a set. *Hepler/ Wallach/Hepler © Cengage 2013*

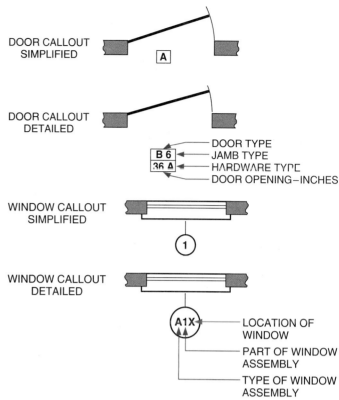

FIGURE 4.16 ❯ Use of callout symbols on doors and windows. *Hepler/Wallach/Hepler © Cengage 2013*

ARCHITECTURAL CONVENTIONS

Architectural Line Conventions

Most drafting and CAD equipment is aimed toward helping an architect, drafter, or designer produce the highest quality line work. Architectural drawings are mainly communicated through a language of lines referred to as **line conventions.** The lines have meaning and can be read like the letters of an alphabet. In fact, the term *alphabet of lines* is sometimes used to denote line conventions used on architectural drawings. These are shown in Figure 4.17 with the pencil grades. CAD adjustments are needed to produce these lines. Many types of lines are found on a single drawing. See Figure 4.18.

Just as different line patterns are used to represent certain features of a drawing, various *line weights* are used to emphasize or de-emphasize areas of a drawing. Architectural line weights are standardized to provide for consistent interpretation of architectural drawings. All architectural line weights must be very dark (opaque) so that they will reproduce well. The only lines that should remain very light are layout construction lines and guidelines so they will *not* be seen on the reproductions. A list follows of the types of lines in the alphabet of lines:

1. *Object lines,* also called *visible lines,* are used to show the main outline of the building, including exterior walls, interior partitions, porches, patios, driveways, and walls. These lines should be drawn wide (thick) to stand out on the drawing.

2. *Dimension lines* are thin unbroken lines on which building dimensions are placed.

3. *Extension lines* extend from the object lines to the dimension lines. They are drawn thin to eliminate confusion with the object outlines.

4. *Hidden lines* are used to show areas that are not visible on the surface, but that exist behind the plane of projection. Hidden lines are also used in floor plans to show objects above the floor section, such as wall cabinets, arches, and beams. Hidden lines are drawn thin.

5. *Center lines* denote the centers of symmetrical objects such as exterior doors and windows. These lines are usually necessary for dimensioning purposes. Centerlines are drawn thin.

6. *Cutting-plane lines* are very wide (thick) lines used to denote an area to be sectioned. In this case, the only part of the cutting-plane line drawn is the extreme ends of the line. This is because the cutting-plane line would interfere with other lines on the drawing.

NAME OF LINES	LINE SYMBOLS	LINE WIDTH	PENCIL	PEN SIZES	
Object lines	—————————	Thick	H,F	2	0.50 mm
Hidden lines	– – – – – – –	Medium	2H,H	0	0.35 mm
Center lines	—— – —— – —— – —	Thin	2H,3H,4H	0	0.35 mm
Long break lines	———⋀——⋀———	Thin	2H,3H,4H	0	0.35 mm
Short break lines	∼∼∼∼∼	Thick	H,F	2	0.50 mm
Phantom lines	—— – – —— – – —	Thin	2H,3H,4H	0	0.35 mm
Stitch lines	— – – — – – —	Thin	2H,3H,4H	0	0.35 mm
Border lines	▬▬▬▬▬▬▬	Very thick	F,HB	3	0.00 mm
Extension lines Dimension lines	|←——————→|	Thin	2H,3H,4H	00	0.25 mm
Leader lines		Thin	2H,3H,4H	00	0.25 mm
Cutting-plane lines		Very thick	F,HB	3	0.80 mm
Section lines	////////	Thin	2H,3H,4H	00	0.25 mm
Layout lines Guidelines		Very thin light	4H		
Lettering	ARCHITECTURAL	Thick	H,F	1	0.40 mm

FIGURE 4.17 ❯ Architectural line conventions. *Hepler/Wallach/Hepler © Cengage 2013*

7. *Break lines* are used when an area cannot or should not be drawn entirely. A ruled line with freehand breaks is used for long, straight breaks. The long break line is thin. A wavy and thick, uneven freehand line is used for smaller, irregular breaks.

8. *Phantom lines* are used to indicate alternate positions of moving parts, adjacent positions of related parts, and repeated detail. The phantom line is thin with a long dash and two short dashes.

9. *Fixture lines* outline the shape of kitchen, laundry, and bathroom fixtures, or built-in furniture. These lines are thin to eliminate confusion with object lines.

10. *Leaders* are used to connect a note or dimension to a symbol or to part of the building. They are drawn thin and sometimes are curved to eliminate confusion with other lines.

11. *Section lines* are used to indicate the cut surface in sectional drawings. A different symbol pattern is used for each building material. The section lining patterns are drawn thin.

12. *Border lines* are the heaviest lines used on a drawing and are often preprinted with the title block. Border lines define the active area of a drawing sheet.

13. *Guidelines* are drawn to provide a horizontal guide for lettering to keep letters and numbers aligned. These are very light lines so they do not reproduce on a finished blueprint.

14. *Construction (layout) lines* are very light preliminary layout lines that do not become part of the finished drawing when reproduced. These lines are the lightest on any drawing.

CAD Line Weights and Types

CAD software programs allow the user to change the individual line weight (the thickness of the line in either inches or millimeters) and the style of the line type (center, hidden, dashed, phantom, and many others). In addition, the user can choose from a wide range of colors. The changes should be coordinated through the *Layer* command tools for improved manageability of drawings.

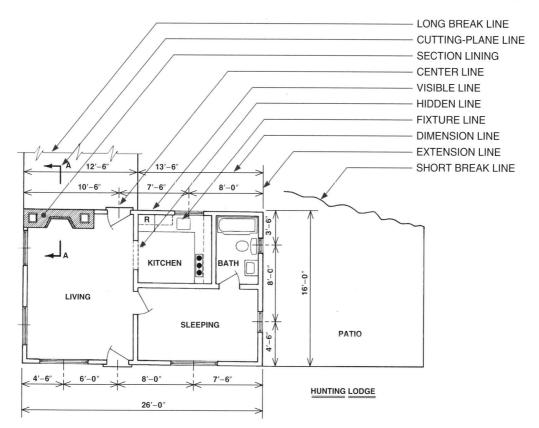

FIGURE 4.18 > Application of line conventions. *Hepler/Wallach/Hepler © Cengage 2013*

Architectural Lettering

Styles

Without lettering, a plan does not communicate a complete description of the materials, type, size, and location of the various components. All labels, notes, dimensions, and descriptions must be legible on architectural drawings if they are to be an effective means of graphic communication.

Architectural designs are often personalized. Likewise, lettering styles may reflect the individuality of various architects and drafters. Architectural drafters often develop their own style of lettering to work quickly, yet maintain accurate and attractive drawings. Nevertheless, personalized styles are all based on the American National Standard Alphabet shown in Figure 4.19.

No personalized style should be used that is difficult to read or easily misinterpreted. Errors of this type can be very costly, especially if numbers used for dimensioning are misread.

Developing Lettering Skills

Practice is necessary to develop the skills needed to letter effectively. Although architectural lettering styles may be very different, all professional drafters follow certain basic techniques for lettering. Although finished drawings may be

Straight

ABCDEFGHIJKLMNOPQRSTUVWXYZ   1234567890

Inclined

ABCDEFGHIJKLMNOPQRSTUVWXYZ   *1234567890*

FIGURE 4.19 > The American National Standard Alphabet.
Hepler/Wallach/Hepler © Cengage 2013

prepared with a CAD program, readable and understandable labels and dimensions are nevertheless essential for design and field sketches.

1. Always use guidelines when lettering. See Figure 4.20.
2. Choose one style of lettering, and practice the formation of the letters of that style until you master it.
3. Make letters bold and distinctive. Avoid a delicate, fine touch.
4. Make each line quickly from the beginning to the end of the stroke. Do not try to develop speed at first. Make each stroke quickly, but take your time between letters and between strokes until you have mastered each letter.
5. Practice with larger letters (about 1/4″, or 6 mm), and gradually reduce the size until you can letter effectively at 1/8″ (3 mm).

USE GUIDELINES FOR GREATER

ACCURACY IN LETTERING.

LETTERING WITHOUT GUIDELINES

LOOKS LIKE THIS.

FIGURE 4.20 ❯ Use of lettering guidelines. *Hepler/Wallach/ Hepler © Cengage 2013*

6. Aim for uniform and even spacing of areas between letters by practicing words and writing sentences, not alphabets.

7. Practice lettering whenever possible as you take notes, address envelopes, or write your name.

8. Use only the CAPITAL alphabet. Lowercase letters are rarely used in architectural work.

9. If your lettering has a tendency to slant in one direction or the other, practice making a series of vertical and horizontal guidelines.

10. If slant lettering *is* desired, practice slanting the horizontal strokes at approximately 68°.

11. Letter the drawing last to avoid smudges and overlapping with other areas of the drawing. This procedure will enable you to space out your lettering and to avoid lettering through important drawing details.

12. Use a medium-soft pencil, preferably an HB or F. A medium-soft lead pencil will glide and is more easily controlled than a hard lead pencil.

13. Numerals used in architectural drawing should be adapted to the same style as the letters. Fractions also should be made consistent with the style. Fractions are 1 2/3 times (slightly larger than) the height of the whole number. The numerator and the denominator of a fraction are each 2/3 (slightly smaller than) the height of the whole number as shown in Figure 4.21. Notice also that in the expanded style, the fraction is slashed to conserve vertical space.

14. The size of the lettering should be related to the importance of the labeling. See Figure 4.22.

15. Specialized lettering templates can also be used.

CAD Lettering

In CAD, lettering is produced using the keyboard and the CAD program's "text" features. Menu items include the option of selecting type font (style), slope angle, line weight (pen size), width, height, and spacing of characters.

Two types of lettering are available in most CAD programs: multiline text and single-line text. Multiline text looks and acts like text produced by a word processor. The popup interface allows all aspects of the text to be modified or en-

Proportions for fractions

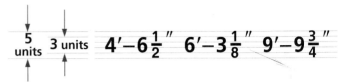

$4'-3\frac{1}{2}''$ $6'-6\frac{3}{4}''$ $8'-9\frac{1}{4}''$

Alternate fraction style used to conserve space

5 units   3 units $4'-6\frac{1}{2}''$ $6'-3\frac{1}{8}''$ $9'-9\frac{3}{4}''$

FIGURE 4.21 ❯ Lettering fractions. *Hepler/Wallach/Hepler © Cengage 2013*

$\frac{1}{4}''$ **TITLES–LABELS**

$\frac{3}{16}''$ TITLES–LABELS

$\frac{1}{8}''$ GENERAL LETTERING

$\frac{1}{16}''$ NOTES IN SMALL AREAS

FIGURE 4.22 ❯ Letter height related to label importance. *Hepler/Wallach/Hepler © Cengage 2013*

hanced including underlining, bolding, italicizing, and font style and letter height changes. The single-line text tool is the older style of text with fewer graphics but the functionality is still available. Both text options are found under the *Draw* pull-down menu under *Text*.

ARCHITECTURAL DRAWING TECHNIQUES

Drawing (as a verb) is an overall term for the creation of all types of graphic forms. *Drafting* is drawing with the use of mechanical devices.

Rendering and Sketching Techniques

In addition to the precise technical line work on floor plans and elevations, other line techniques are used for rendering and sketching. Some drawings are *rendered* to provide the prospective customer with a better idea of the final appearance of the building. These drawings show no dimensions but may include items such as plantings, floor surfaces, shade and shadows, and material textures.

Some of the line techniques used for renderings are simply variations—in the distance between lines, the width of lines, or the blending of lines. Dots, gray tones, or solid black areas are other ways of showing materials, texture, contrast between areas, or light and shadow patterns. See Figure 4.23. Pictorial renderings are covered in Chapter 20.

Sketching is a means of communicating that is used constantly by designers. In fact, most designers begin with a sketch. Sketches, or rough drafts drawn freehand, are used to record dimensions and the placement of existing objects and features prior to beginning a final drawing. Many times alternatives to a design problem are shown with sketches. Sketches also help record ideas on the job site and help the designer remember unique features about a structure or site. Then the actual design activity can continue in a different location.

When sketching, use a soft pencil. Hold the pencil comfortably. Draw with the pencil; do not push it. Position the paper so your hand can move freely. Sketch in short, rapid strokes. Long, continuous lines tend to arc when drawn freehand. Sketching on graph paper helps increase speed and accuracy, as shown in Figure 4.24.

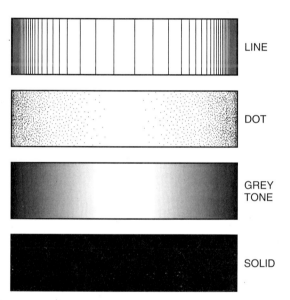

FIGURE 4.23 > Shading methods. *Hepler/Wallach/Hepler © Cengage 2013*

DRAFTING PENCILS AND PENS

Drafting pencils and pens are used with other drafting instruments to produce accurate, readable, and consistent architectural lines and symbols.

Drafting Pencils

Pencils used for drafting are either wood encased or mechanical, as shown in Figure 4.25. The width and density of the line produced depends on the degree of hardness and the point of the pencil's lead. Although referred to as "lead," the core of a drawing pencil is composed mainly of graphite. Hard pencils (3H, 4H) are often used to begin architectural layout work. Medium pencils (2H, H) are used for most of the lines in a completed drawing. Soft pencils (HB, B, 2F) are used for lettering and thick cutting-plane lines, as well as for shading in pictorial drawings. Figure 4.26 shows the range of lead hardness and the resulting line weights.

Drafting pencils can be sharpened to several types of points depending on the type of line desired. Regardless of the type of pencil point used, care must be taken to produce an even point. When uneven points are used, such as a chisel point, uneven lines will result. See Figure 4.27. Pencil lines used on architectural drawings vary in width, but should not vary in density. Thin lines should be just as black and dense as thick lines.

Floor plans and elevation drawings are prepared primarily for the builder. These drawings must be accurately scaled and dimensioned. The accuracy, effectiveness, and appearance of a finished drawing depend largely on the selection of the correct pencil and the point of that pencil.

FIGURE 4.24 > Graph paper sketching. *Hepler/Wallach/Hepler © Cengage 2013*

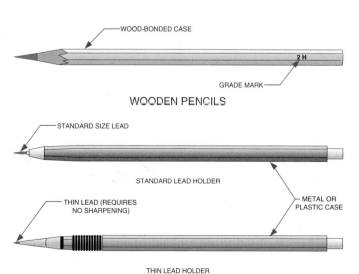

FIGURE 4.25 > Drafting pencil types. *Hepler/Wallach/Hepler © Cengage 2013*

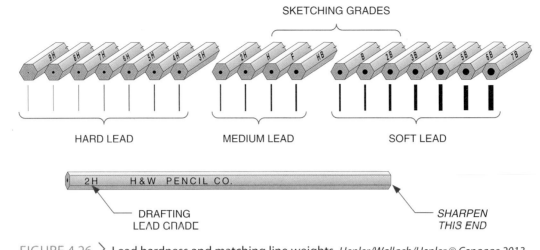

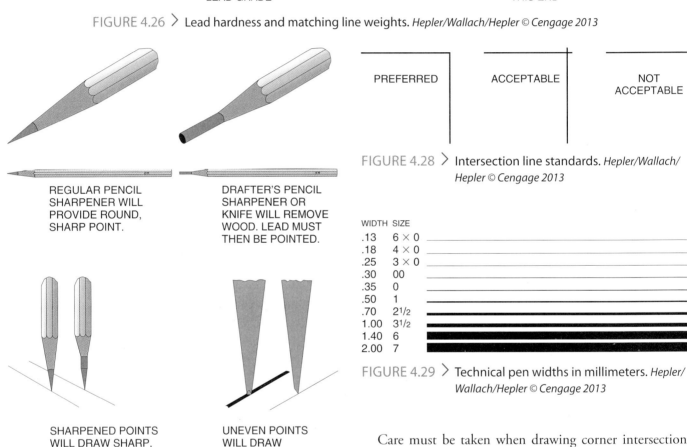

REGULAR PENCIL SHARPENER WILL PROVIDE ROUND, SHARP POINT.

DRAFTER'S PENCIL SHARPENER OR KNIFE WILL REMOVE WOOD. LEAD MUST THEN BE POINTED.

SHARPENED POINTS WILL DRAW SHARP, EQUAL LINES.

UNEVEN POINTS WILL DRAW UNEVEN LINES.

PREFERRED	ACCEPTABLE	NOT ACCEPTABLE

WIDTH	SIZE	
.13	6 × 0	
.18	4 × 0	
.25	3 × 0	
.30	00	
.35	0	
.50	1	
.70	2½	
1.00	3½	
1.40	6	
2.00	7	

Care must be taken when drawing corner intersections. Overlapping corner lines may intersect another material part or dimension and create confusion. When corner lines do not meet, no corner exists for interpretation or measurement. See Figure 4.28 for the preferred method.

Technical Pens

Ink pens used for drafting are called **technical pens.** Their points range in thickness from 0.13 mm to 2.00 mm. See Figure 4.29. One reason for using pens rather than pencils is to create very dense and consistent lines. Working with pens manually, however, tends to slow down drawing speed, and ink lines are difficult to erase.

The degrees of hardness of drawing pencils range from 9H, extremely hard, to 7B, extremely soft. Pencils in the hard range are used for layout work. Basic architectural drawings are usually drawn with pencils in a medium range. If the pencil is too soft, it will produce a line that smudges. Mechanical pencils with leads of 0.3, 0.5, 0.7, and 0.9 mm do not need to be sharpened because the lead width matches the correct line width. The 0.5 mm pencil is usually used for thin lines and the 0.9 mm pencil used for thick lines.

GUIDES FOR STRAIGHT LINES

T Square

The **T square** serves several purposes. It is used primarily as a guide for drawing horizontal lines. It also serves as a base for a triangle that is used to draw vertical and inclined lines. Common T-square lengths for architectural drafting are 18, 24, 36, and 42 inches.

T squares must be held tightly against the edge of the drawing board, and triangles must be held firmly against the T square to ensure accurate horizontal and vertical lines. See Figure 4.30. Because only one end of the T square is held against the drawing board, some sag may occur when long T squares are not held securely. Figure 4.30 also shows the difference between right-handed and left-handed use of a T square. Horizontal lines are drawn from left to right by right-handed people and right to left by left-handed people. Vertical lines are made by pulling the pencil or pen upward.

Parallel Slide

The **parallel slide** (or parallel rule) performs the same function as the T square. Extremely long lines are common in many architectural drawings such as floor plans and elevations. Because most of these lines should be drawn continuously, the parallel slide is used extensively by many architectural drafters.

A parallel slide is anchored at both sides of a drawing board. This attachment eliminates the possibility of sag at one end, which is a common objection to the use of the T square. Another advantage of using the parallel slide is that the drawing board can be tilted to a very steep angle without causing the slide to slip to the bottom of the board. If the parallel slide is adjusted correctly, it will stay in the exact position in which it is placed.

Triangles

Triangles are used to draw vertical and diagonal or inclined lines with either a T square or other parallel slide. A variety of combinations produces numerous angles, as shown in Figure 4.31.

The 8-inch 45° triangle and the 10-inch 30°–60° triangle are preferred for architectural work. Adjustable triangles are used to draw angles that cannot be laid out by combining the 45° and 30°–60° triangles.

Drafting Machine

A **drafting machine** is a mechanical tool that can serve as an architect's scale, triangle, protractor, T square, or parallel slide

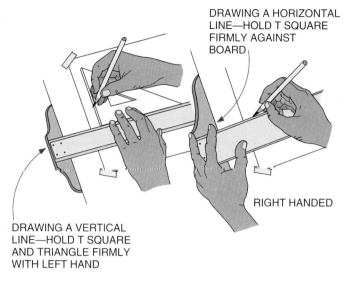

DRAWING A HORIZONTAL LINE—HOLD T SQUARE FIRMLY AGAINST BOARD

RIGHT HANDED

DRAWING A VERTICAL LINE—HOLD T SQUARE AND TRIANGLE FIRMLY WITH LEFT HAND

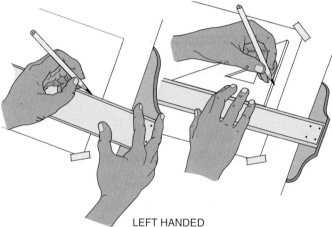

LEFT HANDED

FIGURE 4.30 > Methods of T-square drawing. *Hepler/Wallach/Hepler © Cengage 2013*

all in one. A drafting machine consists of a "head" to which two graduated scales are attached perpendicular to each other. The scales (arms) of the drafting machine are usually made of aluminum or plastic. The horizontal scale performs the function of a T square or parallel slide in drawing horizontal lines. The vertical scale performs the function of a triangle in drawing vertical lines.

Large drawings can be made using track drafting machines in which a protractor head is mounted on a movable track. These machines are smoother and faster than elbow-type machines. However, the operation of the head is identical. The protractor head of a track machine is mounted on a vertical track that is attached to a horizontal track.

CAD Line Tasks

Preparing architectural drawings requires the use of the *Line* command more than any other command. Click on the *Draw* pull-down menu, then select *Line*. Lines can have a wide variety of characteristics applied to them including colors,

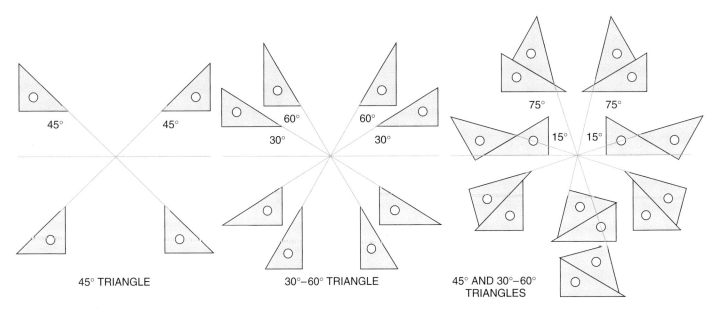

FIGURE 4.31 > Angles possible using 45°, 60°, and 30° triangle combinations. *Hepler/Wallach/Hepler © Cengage 2013*

weights, style, or type. The creation of the line is done by picking a starting point, pointing in a specific direction for the line to follow, and typing in a distance. This is known as a *distance and direction* style of input. A line can also be created by picking points to begin and end the line segment. The type size and color are determined when the *Line* command is used.

Protractor

A **protractor** provides a graduated scale applicable to all 360° of a circle. To draw an angle, align the center of the protractor on the vertex of the angle and mark the degree located on the circumference of the protractor. Then draw a line between the center and the degree mark.

INSTRUMENTS FOR CURVED LINES

Compasses

A **compass** is used in architectural work to draw circles, arcs, radii, and parts of many symbols. Small circles are drawn with a *bow compass.*

Large circles on architectural drawings, such as those used to show the radius of driveways, walks, patios, and stage outlines, are drawn with a large *beam compass.* Very small circles on architectural drawings are drawn with either a *drop-bow compass* or a *circle template.*

Irregular Curve Instruments

Many architectural drawings contain irregular lines. A **flexible curve** is used to repeat irregular curves that have no true radius or series of radii and cannot be drawn with a compass.

Curved lines that are not part of an arc can also be drawn with a French (irregular) curve.

Dividers

Dividing an area into an equal number of parts is a common task performed by architectural drafters. In addition to the architect's scale, **dividers** are used for this purpose.

To divide an area equally by the trial-and-error method, first adjust the dividers until they appear to represent the desired division of the area. Then place one point at the end of the area and "step off" the distance with the dividers. If the divisions turn out to be too short, increase the opening on the dividers. If the divisions are too long, decrease the setting. Repeat the process until the line is equally divided.

Dividers are also used frequently to transfer dimensions and to enlarge or reduce the size of a drawing.

CAD Arcs, Circles, and Curves

Arcs, circles, and curves are used frequently on architectural drawings for windows, doors, pool shapes, electrical connections, and so forth. Arcs can be created by choosing the *Draw* pull-down menu, selecting *Arc,* and then choosing the specific style of arc to create. These styles include the options for angle, length of chord, and radius. Locating the start point, center point, and end point will also yield an arc.

The specific type of arc style that is selected will determine how the arc is created. Arcs require three points to create and are created in a counterclockwise rotation. Irregular curves are drawn using the *Spline* command and locating several points on a curve with a cursor. A spline is a smooth curve that passes through a series of connected points. Increasing the number of points will improve the accuracy and smoothness of the curve. Ellipses are drawn by locating the ends of the major and minor axis points. The *Ellipse* command is then used to select the center of the ellipse, point in the major axis direction, and type the major axis radius, then the minor radius.

CORRECTION EQUIPMENT

Mistakes and corrections are part of every drawing process. Designers employ a variety of erasers and ways of keeping drawings and sketches clean. *Basic erasers* are used for general purposes. *Gum erasers* are used for light lines. *Electric erasers* are very fast and do not damage the surface of the drawing paper. A very light touch is used to eradicate lines. *Kneaded erasers* pick up loose graphite by dabbing.

To keep drawings and sketches clean, *dry cleaner bags* are used to remove smudges. *Powder* sprinkled on the drawing reduces smudging and keeps instruments clean. It also enables drafting instruments to move freely.

Erasing shields are thin pieces of metal or plastic with a variety of small, different-shaped openings. The appropriate opening is positioned over a line to be erased. The shield covers the surrounding area, allowing lines to be erased without disturbing nearby lines that are to remain on the drawing.

A *drafting brush* is used periodically to remove eraser and graphite particles and to keep them from being redistributed on the drawing. Do not blow on a drawing or use your hand to remove debris.

Making Changes on CAD Drawings

Lines, blocks, or layers can be deleted by using the *Undo* command. This command can be used to erase the last entry or previous entries by clicking the *Undo* command's backward arrow until the desired entries are eliminated. This command can also be used to eliminate an entire group or block. To reverse deletions and restore the entries as first drawn, the *Redo* command—or the forward arrow—can be used. After deletions are made, drawing can proceed as before. The erase command can be used to delete single lines; or may delete an area by placing a "box" around the material to be deleted.

AIDS AND DEVICES FOR DRAFTING

Construction often begins immediately upon completion of the working drawings. Under these conditions, speed in the preparation of drawings is of utmost importance. To work quickly, many time-saving devices are employed by architectural drafters. These devices eliminate unnecessary time on the drawing board without sacrificing the quality of the drawing.

Architectural Templates

Templates are usually made of sheet plastic. Openings in the template are shaped to represent various symbols and fixtures. A symbol or fixture is traced on the drawing by following the outline with a pencil or pen. This procedure eliminates the repetitious task of measuring and laying out the symbol each time it is to be used on the drawing. Note that the template scale must always be the same as the scale of the drawing.

A wide assortment of templates is available. Many are used to draw only one type of symbol. Some are designed specifically for furniture, doors, windows, or landscape features. Others provide electrical or plumbing symbols. Some serve as lettering, circle, or ellipse guides.

Overlays

An **overlay** is any sheet that is placed over an original drawing. The information placed on an overlay becomes a visual part of the original drawing. Some overlays remain separate sheets and some are permanently affixed to the base drawing.

Stampat

A solid overlay cannot be used on a drawing unless the drawing is to be photocopied only. A solid overlay on vellum will appear as a solid blue surface in some reproduction processes. A **stampat** is a transparent sheet onto which a drawing has been photocopied on one surface. The other surface is adhesive. The stampat sheet is adhered to the original drawing surface before the drawing is reproduced. The stampat drawing will appear on the reproduced print looking as if it had always been part of the original drawing.

Sheet Overlays

Most separate *sheet overlays* (**trace**) are made by drawing on transparent or translucent material such as acetate or drafting film or vellum. Overlays are used in the design process to add to or change features of the original drawing without marking the original drawing.

Overlays are also used to add features to a drawing that would normally complicate the original drawing. Lines that are hidden and many other details can be made clear by drawing this information on an overlay. See Figure 4.32.

Pressure-Sensitive Overlays

Pressure-sensitive overlays adhere directly to the surface of the drawing. Details that are often repeated on other drawings or projects are frequently reproduced on pressure-sensitive stampat "appliqué paper." The transparent appliqué is then attached to any drawing without repeated redrawing.

Tapes

Many types of *printed pressure-sensitive tapes* can be substituted for drawn lines and symbols on architectural drawings.

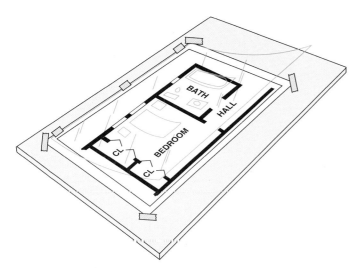

FIGURE 4.32 ⟩ Use of transparent or translucent trace overlay. *Hepler/Wallach/Hepler © Cengage 2013*

FIGURE 4.33 ⟩ Typical press on or stamp patterns used on architectural drawings. *Hepler/Wallach/Hepler © Cengage 2013*

These are used to produce lines and symbols that otherwise would be difficult and time consuming to construct. When using a CAD system, these symbols are located in the line symbol library. *Drafting tape* is used to attach drawings to a drawing board.

Stamps

For architectural features that are often repeated, stamps are effective time-savers. Stamps can be used with any color ink. Stamps are used most often for symbols that do not require precise positioning on the drawing, such as landscape features, people, and cars, as shown in Figure 4.33. Stamps may also be used for furniture outlines and labels. Stamps that use nonreproducible blue ink are used to stamp outlines on drawings, which are then traced over, altered, or rendered.

Underlays

Underlays are drawings or parts of drawings that are placed under the original drawing and traced onto the original. Architects often use them as master drawings. To be effective, a master drawing must be prepared to the correct scale and aligned carefully each time it is used.

Many symbols and features of buildings are drawn more than once. Many drafters prepare a series of underlays of the features repeated most often on their drawings. These are traced exactly or altered on the original drawing. When using a CAD system, these symbols or images can be called up from the graphics library and positioned anywhere on a drawing at any size or angle. Underlays are commonly prepared for doors, windows, fireplaces, trees, walls, and stairs.

Guidelines for lettering are frequently prepared on underlays. When placed under the drawing, the drafter can trace the lines instead of measuring each one.

Squared Paper

Graph or squared paper is often used to provide underlay guidelines for architectural drawings. These grid sheets are printed in gradations of 4, 8, and 16 squares per inch. Squared paper is also available in decimal-divided increments of 10, 20, and 30 or more squares per inch.

Grid underlays are optional when drawing on a CAD system. Grids can be added and removed at any time by assigning a layer to a grid that is different than the drawing layers.

Burnishing Plates

Burnishing plates are embossed sheets with raised areas that represent an outline of a symbol or texture. The plates are placed under a drawing. Then a soft pencil is rubbed over the raised portions of the plate, transferring the symbol or texture onto the surface of the drawing. The use of burnishing plates allows the drafter to quickly create consistent symbols or texture lines throughout a series of drawings.

Drafting Media

Architectural drawings are prepared on paper, vellum, or polyester film depending on whether pencil or ink is used. **Vellum** is a good-quality tracing paper. Preliminary design work and progressive sketches are usually done on extremely thin tracing paper ("bum wad," "flimsy," "trash"). These preliminary drawings are eventually discarded.

Drawing sheets are available in two American Standard size series, as shown in Figure 4.34. Because architectural drawings

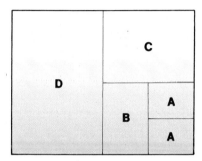

LETTER SIZES	STANDARD SIZES	
A	9x12	8.5x11
B	12x18	11x17
C	18x24	17x22
D	24x36	22x34
E	36x48	34x44
F	28x40	

FIGURE 4.34 > Drafting media sizes. *Hepler/Wallach/Hepler © Cengage 2013*

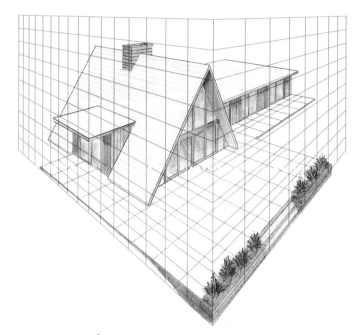

FIGURE 4.35 > Use of preprinted perspective grid. *Hepler/Wallach/Hepler © Cengage 2013*

are prepared on such a large reduction scale, large sheets, usually D or E sizes, are required. For example a 120′-0″ building drawn on a 1/4″ = 1′-0″ scale requires 30″ of actual drawing space.

A wide variety of drafting papers is available. Some vellum papers have nonreproducible grid lines that do not show when the original drawing is duplicated. Grid papers are also printed with nonreproducible angles and lines for perspective drawings (pictorial drawings). See Figure 4.35.

The size of the drawing surface is determined at the beginning of a project. The drawing format selected should be larger than the largest drawing in the set. Figure 4.36 shows the standard sizes of paper or vellum used for architectural drawings.

The type of paper, as well as the drawing pencil or pen, greatly affects the line quality. Different pencil grades of hardness or softness are needed for different papers. Weather conditions also affect line quality. During periods of high humidity, harder pencils must be used. Drawing paper with a hard surface helps to produce distinct, clean lines, especially when using technical pens. Soft surfaces absorb too much ink and result in feathered lines which do not erase well.

CUSTOMARY (INCHES)	METRIC (mm)
8″ × 10″	
8″ × 11″	
*8.5″ × 11″ (A size)	210 × 297 mm (A4)
*9″ × 12″	
11″ × 14″	297 × 420 mm (A3)
*11″ × 17″ (B size)	
*12″ × 18″	
14″ × 17″	
15″ × 20″	
*17″ × 22″ (C size)	420 × 594 mm (A2)
*18″ × 24″	
19″ × 24″	
21″ × 27″	
*22″ × 34″ (D size)	594 × 841 mm (A1)
*24″ × 36″	
*34″ × 44″ (E size)	841 × 1189 mm (A0)
*36″ × 48″	

* Most commonly used.

FIGURE 4.36 > Standard drawing paper sizes. *Hepler/Wallach/Hepler © Cengage 2013*

Conventions and Procedures Exercises

CHAPTER **4**

1. Describe six types of architectural drawings in terms of the type of information that is communicated in each type. List the ones you would use for your own set of drawings.

2. How are drawings used during the planning and construction of a building?

3. Explain the purpose of a coding system and cross-referencing.

4. Using a CAD system, draw a reference symbol and a callout. Label them.

5. Find and obtain a sample title block that is used by a local design or architectural office. Design your own version of a title block.

6. Select a lettering style from a CAD text menu and create all of the data you would use in a title block.

7. Practice drawing each of the lines shown in Figure 4.17.

8. Use three different grades of pencil to draw five of the lines in Figure 4.17. Compare your results.

9. Copy the rules for lettering using any lettering style you choose.

10. Select a lettering style to be used on your own set of plans. Complete three practice sheets. Critique and improve each.

11. Draw the line and shade forms shown in Figure 4.23.

12. Practice drawing the architectural line conventions shown in Figure 4.17 freehand using a soft lead pencil.

13. On quarter-inch grid paper, design and sketch a small floor plan.

14. Sketch the floor plan symbols shown in Figure 4.18.

15. Sketch the corner framing detail in Figure 4.5.

16. Sketch the four elevations in Figure 4.3.

17. Sketch the sectional detail drawing in Figure 4.4.

SOFFIT

REF

RANGE/OVEN
5'-9"

WALL ELEVATION

Introduction to Computer-Aided Drafting and Design

OBJECTIVES

In this chapter you will learn:

> the different kinds of CAD hardware and software and their functions.

> the basic CAD drawing commands.

> how a CAD system is used to create architectural drawings.

TERMS

AEC
block
building information modeling (BIM)
CADD
Cartesian coordinate system
central processing unit (CPU)
commands
compact disks (CDs)
computer-aided drafting (CAD)

display
entities
graphics card
hardware
ink-jet printers
laser printers
layers
modem
monitor

networking
objects
plotters
rendering
software
solid models
surface models
symbol library
wireframe drawing

INTRODUCTION

The preparation of architectural drawings has progressed from using a soil grooving stick to stone etchings to a soft surface stylus and pens and pencils and, finally, to computer-aided drawings.

Computer-aided drafting (CAD) and design is a process through which architectural and engineering drawings and documents are prepared on a computer. Basically, a CAD system is a combination of computer software (programming) and hardware (equipment) that allows designers and drafters to create drawings and store them electronically. All phases of the design and drawing process, from preliminary concept drawings through the completion of final working drawings and documentation, can be completed using a CAD system. Note that the acronym **CADD** is often used to identify programs that can be used not only for drawing but also for design; thus CADD, instead of CAD.

The fields of architecture engineering and construction are classified as **AEC.** CAD provides an opportunity to quickly and accurately show all areas of a design as shown in Figure 5.1.

Engineering CAD programs generate drawings through the use of separate geometric forms such as lines, circles, and arcs. This is similar to manual drafting, except that it is faster, more accurate, and consistent. Architectural CAD systems use blocks of commands to create walls, roofs, elevations, foundations, and so on, which further accelerates the drawing process.

This chapter introduces the basic principles and practices used in computer-aided drafting for architecture. More specific reference information is provided throughout the text where applications are needed. Nevertheless, this coverage is not intended to replace a user's manual, which should be the student's guide of choice to the step-by-step sequences and procedures necessary for successful CAD operations.

CAD CHARACTERISTICS

A drawing prepared using a CAD system should appear identical to a good-quality drawing prepared manually. A computer-aided drafting system is an electronic drafting tool that uses the speed and accuracy of a computer to produce clear, accurate, and consistent drawings. This is done by

FIGURE 5.1 ❯ Different CAD drawings of the same area. *Hepler/Wallach/Hepler © Cengage 2013*

automatically performing repetitious tasks such as symbol insertion, line drawing, automatic dimensioning, and lettering. By automating these tasks, drawing productivity and accuracy are greatly increased. This results in the preparation of more drawings with fewer errors, leaving more time available for the creative process.

CAD systems have many advantages. Making changes and alterations to existing CAD drawings saves much time and money. Revisions that once took hours can now be completed within minutes, once any design change decisions have been made. Hard copies (prints) of drawings can be produced as needed, again saving much time.

CAD drawings are consistent. The lines, symbols, and lettering are always uniform. This greatly reduces the probability of misinterpretation of the drawing's content. The use of symbol libraries not only adds consistency but also eliminates the need to draw often-used details over and over. The creation of symbol libraries by the designer and the availabil-

ity of manufacturers' detail libraries and embedded software libraries make this possible.

Through electronic transmission of files, such as PDFs, JPEGs, or BITMAP and TIFF files, CAD drawings can be sent and received in minutes by clients or other professionals. This speeds up the design process and ensures consistency in communication. Because CAD drawings are stored electronically, there is no deterioration of the original drawing as a result of duplication processes. Figure 5.2 shows a comparison between a CAD drawing and a manual drawing of the same detail.

The ease and speed with which drawings can be electronically stored and retrieved result in fast, clear, and accurate changes and revisions. This is accomplished through the use of editing functions that allow drawings to be quickly deleted, redrawn, rotated, mirrored, stretched, or otherwise manipulated electronically. The speed with which two-dimensional drawings can be used to create three-dimensional drawings

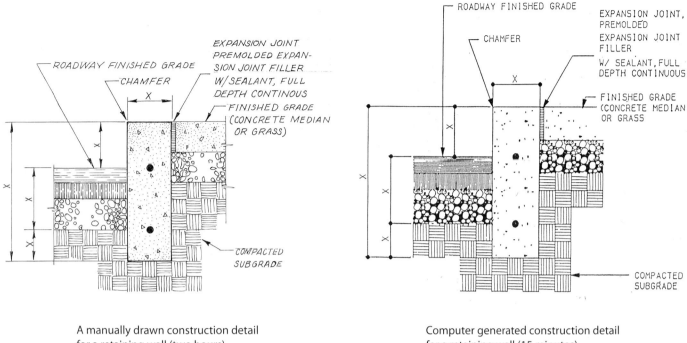

A manually drawn construction detail
for a retaining wall (two hours)

Computer generated construction detail
for a retaining wall (15 minutes)

FIGURE 5.2 > Comparison of a manual and a CAD drawing of the same object. *Hepler/Wallach/Hepler © Cengage 2013*

is one of the most convenient and time-saving capabilities of CAD. One of the greatest features of CAD is the capacity to quickly, accurately, and consistently alter, combine, and add to drawings.

In addition to producing drawings, CAD systems can electronically produce related construction documents such as schedules, specifications, budgets, structural calculations, and graphics analyses. These can all be transmitted worldwide via the Internet.

CAD systems and humans are compatible. Each excels in what the other cannot do well. Humans can think, create, visualize, design, reason, and make decisions—CAD systems cannot. Humans are slow, inaccurate, inconsistent, and error prone compared to CAD systems, which are extremely fast, accurate, and consistent. However, CAD systems are only as accurate as the information supplied. Consequently, to use a CAD system effectively, a thorough understanding of the principles and practices of architectural drafting and design is essential. This includes a working knowledge of the design process, drafting standards and procedures, projection methods, drawing types, documentation, and construction systems.

Always remember that CAD systems cannot design and draw any more than keyboards can write stories. People design and draw *using* drawing tools such as CAD. Therefore, the development of architectural design skills is vital to the effective use of CAD. Producing an attractive drawing of a nonfunctional design is a total waste of time.

CAD FUNCTIONING

CAD systems function via interactions among CAD hardware, software, and the operator. Information entered into a computer is known as *input.* A CAD drafter communicates with the computer through the use of input devices, usually a keyboard, mouse, and cursor. When a drafter requests a specific function such as a line, arc, or numeral, the operator then indicates specifically—through a keyboard, grid system, mouse, or stylus pick—where the line, arc, or numeral is to be placed on the monitor. Once the function is completed on the monitor's screen, the next task is performed in the same manner. Each task is performed in sequence until the drawing is complete. More advanced programs can also generate construction documents by interfacing with a coded CAD drawing.

CAD COMPONENTS

CAD systems are divided into two categories; stand-alone or networked. Stand-alone systems consist of dedicated hardware and software serving a single unit. Networked systems are comprised of several individual computer stations that share a printer, plotter, and **central processing unit (CPU).** A CPU for a large networked system is known as a mainframe. In either system the main components are comprised of both hardware and software categories. Figure 5.3 shows students using a CAD workstation.

FIGURE 5.3 > Student working at a CAD station. *Somers High School students Catherene Machnicki and Justin Liquori*

CAD Hardware

The **hardware** for a CAD system includes the CPU, input devices such as a keyboard and a mouse, storage devices, and output devices such as monitors and printers. Figure 5.4 shows the typical hardware components of a CAD stand-alone work-

station. Figure 5.5 illustrates the relationship and interaction among the CPU, storage, and input and output functions.

CPU

The central processing unit is the "brain" or engine in all CAD systems. The main component of the CPU is a microprocessor chip. This chip controls the speed and power of the system, both of which are important for adequately operating the software. The CPU also houses the computer processor, memory, and interfaces to the input and output terminals.

The Monitor

A **monitor** is a hardware device resembling a television screen. It allows the operator to see the results of commands given to the computer. CAD systems should be equipped with the largest monitors possible to avoid eyestrain while working with complex drawings. A 17-inch monitor is the smallest monitor recommended for long-term CAD work. Larger monitors (over 21 inches) provide more viewing area because less of the viewing area is covered by menus and toolbars. A **graphics card** designed specifically for CAD use and a

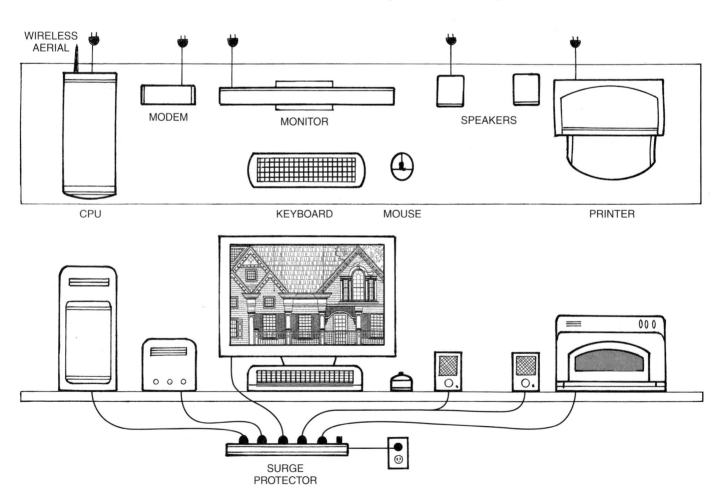

FIGURE 5.4 > Typical hardware components of a stand-alone CAD system. *Hepler/Wallach/Hepler © Cengage 2013*

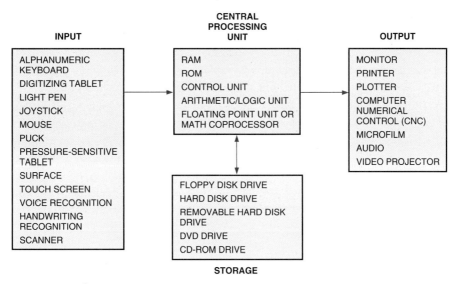

FIGURE 5.5 > Relationship among CAD system components. *Hepler/Wallach/ Hepler © Cengage 2013*

high-resolution liquid-crystal **display** (LCD) screen are the essential ingredients of a robust CAD system.

The resolution of a monitor depends on the number of pixels (dots of light) it uses. More pixels are needed to produce a sharper line image. A 19-inch or larger screen should have 1600×1200 or more pixels for CAD work. Color monitors are needed to effectively use many CAD functions. The colors selected to show on a monitor can be assigned to specific **layers.** This allows drawings to be plotted with various line weights, colors, and types assigned to each layer.

Memory

Large amounts of graphics, numerical data, and text can be stored in a computer's memory. Computers contain two types of memory: ROM (read-only memory) and RAM (random-access memory). ROM contains the fixed data that the computer uses while it is operating. It contains instructions that keep the "operating system," which coordinates instructions between the software and hardware, operating smoothly. RAM is the computer's temporary memory. The amount of RAM determines the amount of software data the CPU can process at one time.

Data Storage Devices

Storage capacity is not the same as memory. Memory enables a computer to function, whereas storage devices are used to electronically store software programs and CAD drawing files. Information stored in memory is intended for future use, or as a safety backup. In either case it is important to specifically identify stored data by project, date, and subtopic. For stored drawings the AIA classification, drawing, and detail numbers should be used. *Flash*, *zip*, and *thumb drives*

are portable storage devices with memory chips on which data are stored. The most common storage devices are hard disk drives and CD drives. These are standard equipment on almost all computer models.

Hard disk drives are generally the main storage device on a computer. They store the main operating system, the software, and any electronic files, such as drawing files, that the operator creates and saves. CAD programs generally take up a large amount of hard disk space, and architectural CAD files can take up a large percentage also. Therefore, most CAD systems include a minimum of an 18-gigabyte (GB) disk drive. Hard disk capacities of up to 250 GB are common on large CAD systems. Some may have a capacity of 1,000,000 GB (or terabyte). (One GIGAbyte $= 10^9$) (One Terabyte $= 10^{12}$).

Compact disks (CDs) are used for added storage and/or data backup for hard disk drives. The use of laser disk technology provides greater speed and increased capacity compared to earlier types of computer disks. DVD technology is similar to that of CDs but through compression, DVDs can store 8 to 16 times more data than a CD depending on whether single-layer or dual-layer DVDs are used.

Safety-conscious companies keep a complete, current backup of the files on their hard disks. Then, if a hard disk becomes corrupted or fails, the company can retrieve its important files from the backup. To solve the need for comprehensive backups, many companies use an additional hard drive or CD (CDR, CDRW, DVD), which can hold from 700 MB to 4.7 GB of memory.

Input Devices

Data or information (*input*) can be entered into the computer using various input devices. The alphanumeric *keyboard* is a standard input device that resembles a typewriter. In addition to using the keyboard to type words and numerals on drawings or documents, the keyboard can also be used to enter drawing commands. Most computer systems also include a "pointing device" (*mouse*) that can be used to move an arrow (*cursor*) around the screen. The mouse is also used to control the crosshair pointer's location on the monitor and to select drawing commands and locate points on a monitor.

Another development in computer technology provides a different means of entering unique data—by *voice*. Voice recognition devices enable a computer operator to enter commands by speaking into a microphone connected to the computer. Voice recognition also provides an alternative for people with physical challenges who cannot easily manipulate a mouse or keyboard.

Output Devices

Monitors are considered temporary output devices because information viewed on a monitor is intended for use while at the computer workstation and in architectural work. Drawings are saved and reproduced primarily for use on construction sites. This requires multiple sets of hard copy (prints) architectural drawings. Hard copies are reproduced on paper or vellum through the use of printers or plotters. Drawing sheet size is the determining factor in selecting the output device.

Laser printers offer a fast method of producing high-quality graphics and typeset reproductions. Their small format, however, is a drawback for most architectural work that requires large sheets. Most laser printers can print an area no larger than 11 × 14 inches. Large laser printers are available but extremely expensive. Large photocopiers are the most popular devices used to convert digital information into drawings.

Ink-jet printers offer good graphic quality at low cost. These printers spray tiny jets of black and/or colored ink to produce a good-quality drawing. Desktop ink-jet printers, however, are restricted to small sizes and cannot handle the large D or E size sheets needed to reproduce most architectural drawings. This restricts their architectural use to small-scale drawings, details, and check prints. Large ink-jet **plotters** use both sheet-fed and roll media, which allows for the duplication of large sizes.

Modems

A **modem** is a telecommunications device that allows computer operators to send and receive information over standard telephone lines, optic cable lines, or a wireless system. In this sense, modems are both input devices and output devices. Modems are often installed in CAD systems, particularly in architectural offices that may have more than one location. They allow large CAD drawings to be sent directly from one branch to another or from the designer to a major client for approval. This reduces the amount of time involved in creating a hard copy and transporting it to its destination. The **networking** of computers can be accomplished using either wired or wireless technology.

Multimedia Computers

Multimedia is a term used to indicate the combining of more than one medium.

To expand a conventional CAD system into a multimedia system, several components must be added. These include a DVD read/write drive and also a digital camera and digital video drives.

CAD Software

CAD **software** contains programs that instruct the hardware devices to perform specific tasks as directed by the drafter. Therefore, drafters must give directions to the software that will produce a functional design on a technically correct drawing. This not only requires proficiency in CAD operation but also a working knowledge of the principles and practices of architectural and construction design and drafting.

CAD software is judged by the number of functions (tasks) the program can perform, ease of operation, accuracy, line resolution, and flexibility. These functions vary widely. Low-end programs often do not contain the necessary commands to effectively create a full set of drawings. They also require more steps to complete a drawing and the image quality is imperfect. High-end (robust) programs offer combinations of tasks that produce a high-quality image much faster. To produce technically valid architectural drawings, software programs must produce drawings that conform to acceptable architectural conventions, as covered in Chapter 4. Selecting an appropriate CAD software program depends largely on the type and complexity of the drawings and documents to be produced. Other factors to be considered include cost, manufacturer's support, hardware compatibility, and warranties.

USING CAD SYSTEMS

Commands provide the link between drafters and software. Commands are used to activate specific CAD functions, such as drawing lines or circles, in precise locations. Commands can be entered using a keyboard or a pointing device such as a mouse. A mouse controls a cursor or crosshair pointer on a monitor. Most CAD software contains on-screen menus, which are used to select commands, as shown in Figure 5.6. Regardless of the method used to enter commands, the drawing outcome is the same. Command methods and labels

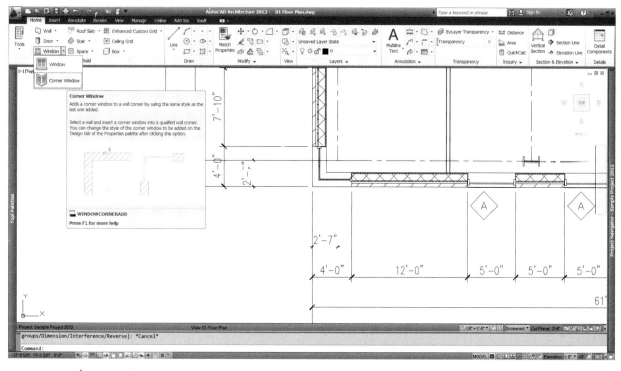

FIGURE 5.6 > Use of an AutoCAD® ribbon menu in drawing a window. *Image courtesy of Autodesk, Inc.*

differ slightly among software systems, but the functions are very similar.

Different software programs use different terms to describe the same function or command. To be consistent, AutoCAD® terms are used throughout this text. Three general categories of commands contained in a CAD system are drawing commands, editing commands, and utility commands.

Drawing Commands

Drawing commands (also called *tasks*) are the commands used to create geometry. Individual geometric forms, such as lines, circles, and polygons, are known as **entities** or **objects.** Most CAD programs have hundreds of commands. The commands presented here are the necessary ones that are used most often. Many seldom used commands are not essential, for actual drawing but increase the speed and convenience of operation when used.

Many procedures can be used to complete a given task. The most basic are included in the following descriptions of drawing commands:

- The *Line* command is probably the most used CAD command. Using this command, the drafter can draw lines by connecting points with a cursor (controlled by a mouse) or by entering the line's Cartesian coordinates at the keyboard, as shown in Figure 5.7. Two general categories of lines are available in most CAD systems: *single lines* and *polylines.* A wide variety of line types, widths,

and styles can be selected from the *Object-Properties* toolbar. Figure 5.8 shows several different types of lines.

Lines have a wide variety of characteristics. Types of line options include object, layout, dashed, phantom, center, dimension, and cutting plan lines. Line weights (thickness)

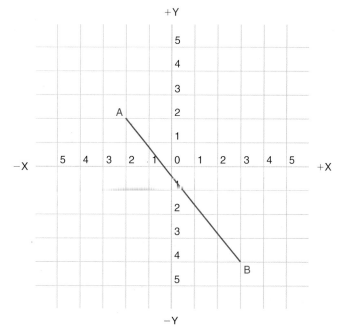

POINT A COORDINATES ARE: −2X, +2Y
POINT B COORDINATES ARE: +3X, −4Y

FIGURE 5.7 > Line completion using X-Y coordinates. *Hepler/Wallach/Hepler © Cengage 2013*

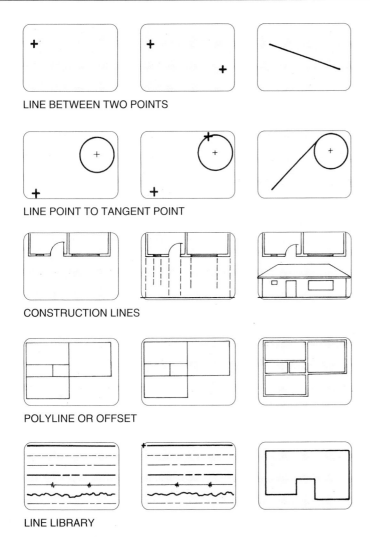

LINE BETWEEN TWO POINTS

LINE POINT TO TANGENT POINT

CONSTRUCTION LINES

POLYLINE OR OFFSET

LINE LIBRARY

FIGURE 5.8 > Different line types. *Hepler/Wallach/Hepler*
© Cengage 2013

Circles, arcs, and curves are used frequently on architectural drawings for windows, doors, pool shapes, and electrical connections.

- The *Arc* command creates arcs of any radius and any length. To create an arc, drafters can specify the center and the angle of two radii, or they can specify the start, middle, and end points of the arc. Most CAD software provides several other options in addition to these basic ones. See Figure 5.10. Arcs can be created by selecting the *Draw* menu, then *Arc*, and then the specific style of arc to be created. The specific type of arc style selected determines how the arc is created. At least three points are required to create an arc. Arcs are created in a counterclockwise rotation.

- The *Text* command is used for adding notes and dimensions to a drawing by positioning the cursor on the selected location and inputting letters and/or numerals with the keyboard, as shown in Figure 5.11. Two types of lettering are available in most CAD programs; *multiline* text and *single-line* text. Multiline text is text that can be modified or enhanced, including underline, bold, letter height, and italic and font style changes. This text can also be mirrored, compressed, or stretched. Text can also be justified right, left, or centered. Figure 5.12 shows a comparison of multiline and single-line lettering. Uppercase letters are used on architectural drawings except where bodies of notes are inserted. Bold type is used on major labels and architectural fonts are preferred.

- The *Spline* command is used to produce irregular curves by locating a series of cursor points and connecting them with lines as shown in Figure 5.13. Increasing the number of points will improve the accuracy and smoothness of a curve.

- The *Ellipse* command is used to complete an ellipse inside a rectangle by inputting the major and minor axes of the ellipse as shown in Figure 5.14. This is done by selecting the center of the ellipse, and typing the major axis radius, then the minor axis radius.

- Rectangles are drawn with the *Rectangle* command by locating two opposite corners (Figure 5.15), by typing the width and length, or by dragging the cursor to opposite

can be varied in inch or metric increments. Color options are virtually limitless. Line commands can also create a double line function. This is most helpful when drawing floor plans, walls, and other details that contain close parallel lines.

- The *Circle* command creates circles of any size using various methods. For example, drafters can specify a circle using the center and a radius or diameter, or by specifying two or three points on the edge of the circle. See Figure 5.9.

PICK CENTER
POINT AND
ENTER RADIUS
OF 0.5

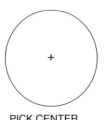

PICK CENTER
POINT AND
ENTER
DIAMETER OF 1.0

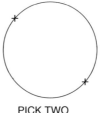

PICK TWO
POINTS THAT
DEFINE THE
DIAMETER

PICK TWO OR
THREE POINTS
THAT DESCRIBE
THE CIRCUMFERENCE

FIGURE 5.9 > CAD commands for drawing circles. *Hepler/Wallach/Hepler* © Cengage 2013

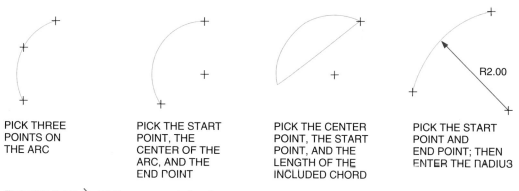

PICK THREE
POINTS ON
THE ARC

PICK THE START
POINT, THE
CENTER OF THE
ARC, AND THE
END POINT

PICK THE CENTER
POINT, THE START
POINT, AND THE
LENGTH OF THE
INCLUDED CHORD

PICK THE START
POINT AND
END POINT; THEN
ENTER THE RADIUS

R2.00

FIGURE 5.10 ⟩ CAD commands for drawing arcs. *Hepler/Wallach/Hepler © Cengage 2013*

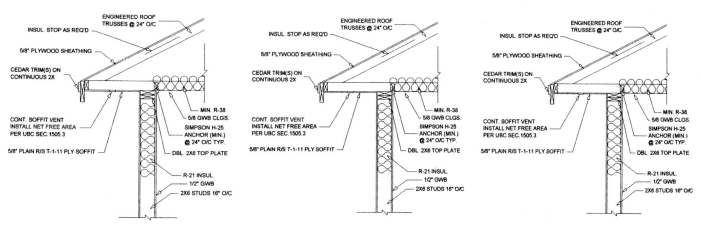

STEP 1: LOCATION MARKED

BOXED EAVE DETAIL
STEP 2: TEXT INSERTED

FIGURE 5.11 ⟩ Methods of text insertion. *Hepler/Wallach/Hepler © Cengage 2013*

FINISHED FLOORING IS INSTALLED OVER THE SUBFLOOR.

CONVENTIONAL RESIDENTIAL FRAMING NORMALLY USES
SOLID LUMBER JOISTS MADE FROM DOUGLAS FIR, PINE,
SPRUCE, OR HEMLOCK.

FIGURE 5.12 ⟩ Comparison of multiline and single-line
lettering. *Hepler/Wallach/Hepler © Cengage 2013*

corners. Squares or rectangles can also be drawn using the
Line command and drawing outlines at 90°.

- Polygons are completed with the *Polygon* command. Regular polygons are objects with equal sides and included
angles. Polygons can be created by inscribing their shape
within a circle or circumscribing the shape outside a circle.
Polygons can also be created by locating the center and
radius and inputting the number of sides as shown in
Figure 5.16. Another method of drawing a polygon is to

 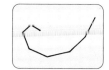 

FIGURE 5.13 ⟩ Method for drawing curves with the spline
command. *Hepler/Wallach/Hepler © Cengage
2013*

  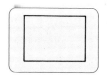

FIGURE 5.15 ⟩ One method for drawing a rectangle. *Hepler/
Wallach/Hepler © Cengage 2013*

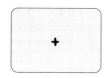  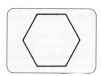

FIGURE 5.14 ⟩ Method for drawing an ellipse. *Hepler/
Wallach/Hepler © Cengage 2013*

FIGURE 5.16 ⟩ One method for drawing a polygon. *Hepler/
Wallach/Hepler © Cengage 2013*

enter the center of a circle on the monitor and keyboard the number of sides as shown in Figure 5.17.

- A *fillet* is a small arc located on the inside of two intersecting lines after the line intersections are trimmed. Rounded room corners are drawn using a *Fillet* command.

- A *round* is a small arc placed on the outside of two intersecting lines, or corners. As with fillets, the intersecting lines (corners) are trimmed.

FIGURE 5.17 > Polygon construction using a keyboard.
Hepler/Wallach/Hepler © Cengage 2013

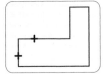 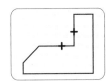 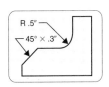

FIGURE 5.18 > Fillet and chamfer construction. *Hepler/ Wallach/Hepler © Cengage 2013*

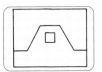 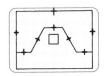 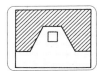

FIGURE 5.19 > Method for identifying hatched areas.
Hepler/Wallach/Hepler © Cengage 2013

- The *Chamfer* command is used to connect two lines to create a cut-off corner. This connects the two lines in a manner similar to that for a fillet. This command eliminates the need to delete two lines and draw two new lines. In review the major methods of connecting corners includes the use of the arc, rectangle, fillet, round, or chamfer. The method for drawing fillets and chamfers is shown in Figure 5.18.

- The *Hatch* command is used to fill in areas of a drawing with surface patterns such as crosshatching or solid colors. Hatching is done by using a pointing device on all border lines of the area to be hatched, as shown in Figure 5.19. Figure 5.20 shows an interior perspective drawing with hatched shading, and Figure 5.21 shows examples of hatching to represent different surface materials. Figure 5.22 shows a typical variety of the color patterns available for hatched areas.

Hatching is used to show materials on siding, sectional views, and ground or floor surfaces. Hatch patterns are provided in CAD software programs but can be created by the designer and stored in a hatch library. More intricate patterns for specialized areas can also be purchased. Many manufacturers allow access to their hatch pattern software which displays their product materials. Very intricate hatch patterns would be nearly impossible to accurately draw in a reasonable amount of time without the use of CAD hatching as shown in Figure 5.23. The creation of realistic materials in pictorial drawings is also made practical through the use of hatching as shown in Figure 5.24.

- Dimensioning commands control the dimension style (architectural or engineering), positioning (vertical, angular,

FIGURE 5.20 > Hatched interior surfaces and shadowing. *Hamid Rafiei Dreamarch3D.com*

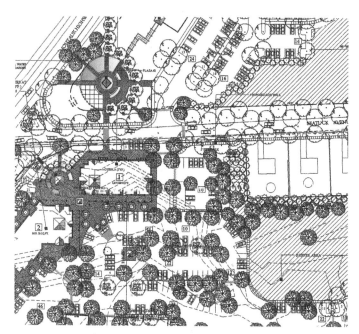

FIGURE 5.21 > Site plan with multiple surface hatching.
Hepler/Wallach/Hepler © Cengage 2013

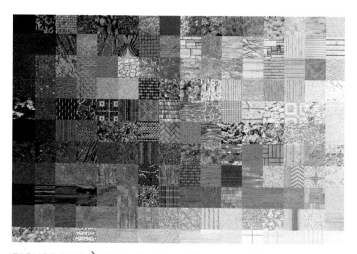

FIGURE 5.22 > Color hatching library sample. *Hepler/
Wallach/Hepler © Cengage 2013*

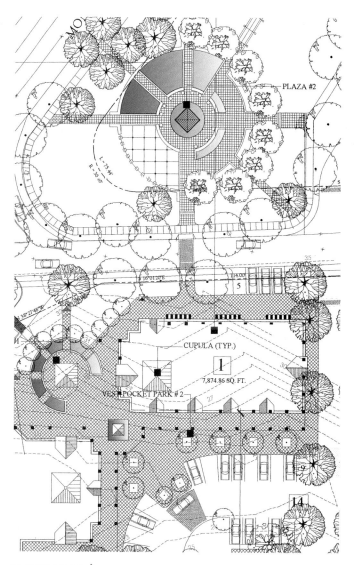

FIGURE 5.23 > Hatch patterns on a site plan. *Hepler/Wallach/
Hepler © Cengage 2013*

FIGURE 5.24 > Material patterns on an exterior perspective
drawing. *Christofer Polaske Designs*

or horizontal), and mode (customary, metric, decimal, or fractional). Dimensions are entered by placing the cursor on the end points of the distance to be dimensioned as shown in Figure 5.25.

The manual dimensioning of an architectural drawing may require as much time as the actual drawing. Dimensioning practices are also a major source of errors and omissions. Automatic CAD dimensioning, using dimensioning commands, dramatically reduces the amount of

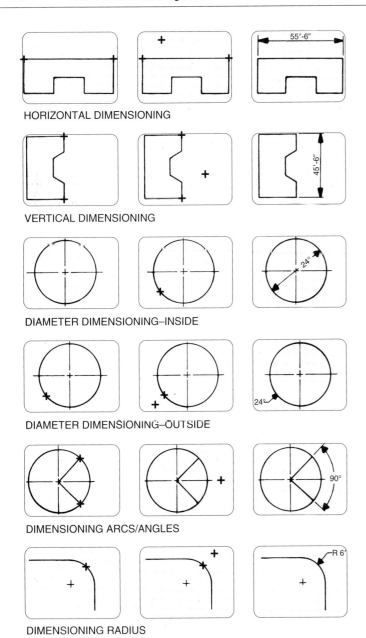

HORIZONTAL DIMENSIONING

VERTICAL DIMENSIONING

DIAMETER DIMENSIONING–INSIDE

DIAMETER DIMENSIONING–OUTSIDE

DIMENSIONING ARCS/ANGLES

DIMENSIONING RADIUS

FIGURE 5.25 > Methods of entering dimensions. *Hepler/Wallach/Hepler © Cengage 2013*

time devoted to dimensioning and decreases the magnitude and frequency of dimensioning errors. Architectural CAD drawings are dimensioned using a combination of tools found in the *Dimension* menu. The dimension style must be set up using standards from the AIA—not the setups used for mechanical, structural, or electrical styles of drawings. The settings found in the dimensional style software are extensive, including arrowhead styles, text types, and the space from the end of the dimensional line to the text.

Dimensions that define length may be horizontal, vertical, or angular. Regardless, dimensions must always be placed to be read from the bottom or the right; never from the left or top of a drawing. Architectural dimensions are always placed above and parallel to the dimension line and

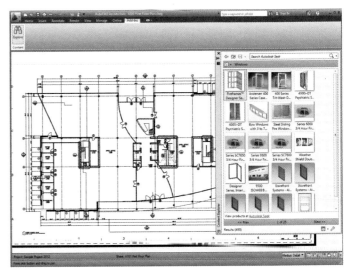

FIGURE 5.26 > Use of Autodesk® Seek library to search design files, content, and objects related to drawing details. *Image courtesy of Autodesk Inc.*

never perpendicular to a dimension line. Linear dimensions are the most commonly used dimension in all types of drawings. These are applied by selecting a line for the dimension, picking two extension lines, and then locating the dimensional placement. The numerical value of the dimension is automatically placed, because dimensions always represent the actual size of a building or object. Guidelines for dimensioning each type of architectural drawing are found in Chapters 13 through 18.

CAD Libraries

Symbol libraries provide designers with a most productive tool that is efficient and saves time. Before CAD many commonly used symbols were redrawn, traced, or photo inserted onto drawings over and over. This practice was very time consuming and error prone. CAD symbols or details are now included in CAD software, provided by manufacturers, or created by designers and stored in a symbol library for future use.

Symbols are either in plan, elevation, or pictorial form. Symbols libraries relate to specific drawing types and become part of a finished drawing. These include electrical, plumbing, HVAC, welding, structural, appliance, doors and windows, and plants and trees libraries. Figure 5.26 shows examples of these options. See details in chapters covering these specialized topics.

Electrical plans are prepared using a CAD layered floor plan. On simple plans, electrical symbols are included as a separate layer on the floor plan. For larger or more complex structures, a separate plan is prepared. In either case electrical plans should be prepared on a separate CAD layer. Fixtures, switches, and devices are moved from the electrical symbol library to their position on a drawing then connected with a curved line using the *Spline* command.

Plumbing drawings contain a series of lines representing pipes. Special line types for gas, water, oil, etc., can be selected using the *Line* command. The *Break* command is also used to interrupt a solid line at regular intervals by inputting the gap size, spacing, and number. Once lines have been drawn and broken, symbols for valves, unions, elbows, tees, and joints are added. This may be done by first breaking the line using the *Trim* command. Then symbols from the plumbing symbol library are inserted in the gap with the cursor. Plumbing libraries also include symbols for bath, kitchen, and utility room fixtures and often appliances.

Because most HVAC symbols are located on or near duct work, all duct lines should be drawn first. This is done using the *Line* task with the *Offset* command because parallel duct lines are usually too far apart to use the *Polyline* command. Once the ducts have been drawn, the symbols can be moved from the HVAC library and placed on the drawing using the *Insert* command.

Section symbols are used to represent a slice through a material. Symbols that represent a wide variety of materials are stored and accessed using the *Hatch* command. After the hatch pattern and boundary have been selected, the material is added to a drawing by identifying the hatch border with a cursor. Hatch patterns should be applied to enable the pattern to be stretched or scaled without changing the pattern makeup.

Welding symbol libraries contain symbols for each type of weld. They also include the symbol line on which the welding symbols can be located with the cursor. These are stored as base symbols *blocks*. The numerals related to each weld are then added using the *Text* command.

Structural symbol libraries consist of plan and elevation details of wood and steel members and components. Some also include details of common structural components. Sections of structural steel members are provided in most steel member libraries and include drawings of the fasteners used in steel construction. These are stored in both plan and sectional views. The *Copy-Multiple* command is used on structural steel drawings to locate and draw evenly spaced rows of bolts and rivets. This command allows an evenly spaced member to be repeated by inputting the spacing, location, and number of copies needed.

Door and window symbol libraries include plan elevation and detail symbols for all types of windows and doors. Manufacturers' libraries enable very detailed symbols to be inserted into a plan using no more time than that needed to draw abbreviated symbol forms.

The most used site plan symbols are plan views of plants and trees. Site plans also include symbols that represent land surface material, plant material, and site-related fixtures and devices. These are included in many CAD software programs, but most are created, stored, and retrieved as blocks. Care must be taken to adjust the scale of these stored symbols to match the scale of each specific drawing. Surface materials are available in blocks and are inserted into the drawing using the *Hatch* command. Site plan symbol libraries also contain instructional details on planting and erosion control.

Elevation symbols are of two types: individual and surface symbols. Individual symbols, such as doors and windows, are stored in blocks and inserted using the *Insert* command onto the elevation drawing in the same manner as floor plan symbols are applied. Surface symbols such as siding and roofing materials are applied using the *Hatch* command. Hatch boundaries are identified and must totally enclose a space with no gaps. The specific hatch pattern is then selected from a hatch library. Hatch blocks, which make up the hatch library, can be drawn and stored by the user or selected from the system's library. Many component and material manufacturers supply software containing blocks of their products, which can be stored and used in combination with other blocked symbols.

Entourage symbol libraries contain objects that are not part of the drawing itself, such as people, vehicles, and furniture. These are movable items and not included on working drawings. Identification symbols for compass direction (north arrow), scales, titles, and revision boxes are also stored in libraries because they are part of every drawing sheet.

Editing Commands

CAD software provides many commands that allow the drafter to change the size, shape, and position of entities. In addition to being critical for making revisions, many of these commands are used by drafters to help create the original drawing. Editing commands are used not only to correct errors, but also to make changes and adjustments during the process of design and drawing. Therefore, the editing functions are one of the factors that separate low-end from high-end systems.

Editing commands include the following:

- The *Undo, Delete,* and *Erase* commands are used to eliminate a single item, group, area, or the last entry drawn by the cursor placement (Figure 5.27). Figure 5.28 shows one method used to delete an entire segment of a drawing. This is done by clicking the backward arrow until the desired entries are eliminated. This removes each entry progressively backward, but cannot eliminate an entry out of the original sequence of drawing. To reverse deletions and restore the entries as first drawn, the *Redo* command or the forward arrow can be used. After deletions are made, drawing can proceed as before. These commands can also

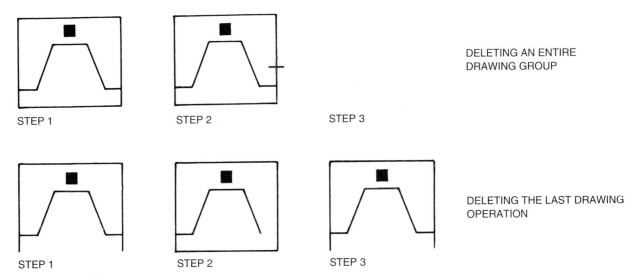

FIGURE 5.27 > Methods of deleting a drawing, group, or area. *Hepler/Wallach/Hepler © Cengage 2013*

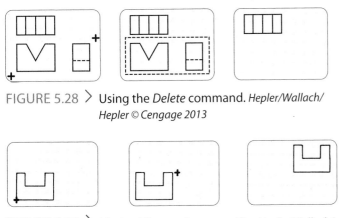

FIGURE 5.28 > Using the *Delete* command. *Hepler/Wallach/Hepler © Cengage 2013*

FIGURE 5.29 > Method for moving an entity. *Hepler/Wallach/Hepler © Cengage 2013*

be used to eliminate an entire group or block. The *Erase* command removes either the last entry or a selected entry permanently.

- The *Move* command allows the drafter to select one or more entities and move them to another location in the drawing, as shown in Figure 5.29. This is done by identifying the area to be moved, then picking a starting point and a final destination point.

- The *Copy, multiply,* or *repeat* command produces a copy of the selected entities and places them elsewhere in the drawing. The original entities remain where they were originally placed. This is done by inputting the spacing and number of copies needed, as shown in Figure 5.30.

- The *Rotate* command allows one or more selected entities to be rotated around a point defined by the drafter. This is accomplished by identifying the rotation point with a cursor and specifying the amount of rotation. Figure 5.31 shows the rotation of a window symbol, which allows a window to be placed on another wall. Figure 5.32 shows how an object is rotated about its point of origin. Rotated plans are not the same as reflected plans. Reflected plans, such as a reflected ceiling plan, are prepared by drawing ceiling features on a layer aligned with the floor plan. The layer is then printed separately to produce a separate plan.

- The *Mirror* command produces an exact mirror image of selected entities. A mirror image can be drawn and reflected left-right, up-down, or on any X-Y axis. Figure 5.33 shows a plan application, and Figure 5.34 shows a roof truss that has been mirrored. If a mirror is placed on its edge and perpendicular to a drawing, the reverse image of the drawing will show in the mirror. CAD drawings can be reversed on the X or Y axis by clicking the *Modify* menu, then the *Mirror* command. Plans are often mirrored to create variety and options in a design or to adjust a plan to better fit a site or to reduce the amount of drafting time. To ensure that labels are not reversed, the *Set Variable* command and the *Mirror Text* setting should be used. This will return the text portion of the drawing to its original readable position; otherwise, the mirror command must be executed before labeling or dimensioning. When floor plans are drawn with

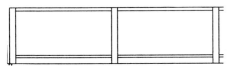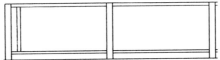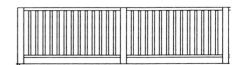

FIGURE 5.30 > Method for multiplying entities. *Hepler/Wallach/Hepler © Cengage 2013*

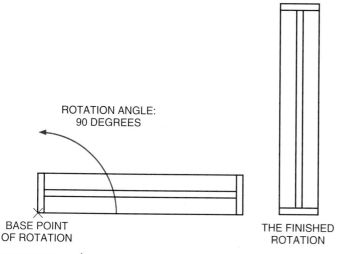

FIGURE 5.31 > Method for rotating a plan view symbol.
Hepler/Wallach/Hepler © Cengage 2013

 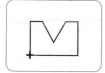 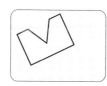

FIGURE 5.32 > Object rotated around a point of origin.
Hepler/Wallach/Hepler © Cengage 2013

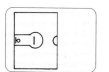 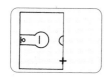 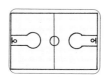

FIGURE 5.33 > Plan view mirroring. *Hepler/Wallach/Hepler © Cengage 2013*

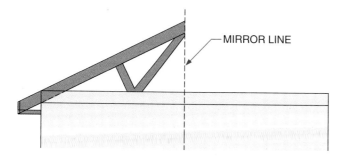

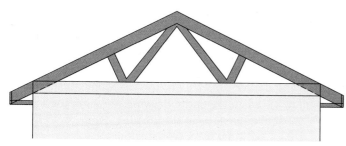

FIGURE 5.34 > Roof truss mirroring. *Hepler/Wallach/Hepler © Cengage 2013*

a mirrored version, the two versions are often labeled left hand and right hand.

- The *Offset* command produces an exact parallel copy of a line at a specified distance from a given line at all points on the line. Figure 5.35 shows a curved offset line that is used to create the ledge thickness of a swimming pool.
- The *Align* command enables objects to be aligned in a straight row. This is done by identifying each object to be aligned using the cursor (Figure 5.36).
- The *Array Polar* command allows objects to be repeated and evenly distributed in a circular or rectangular pattern by entering the center location and the number of entries as shown in Figure 5.37.
- The *Stretch* command is used to elongate a feature. The cursor is used to select the portions to be lengthened (Figure 5.38).
- The *Join* command creates a corner where two lines are aligned to intersect by placing the cursor on both lines (Figure 5.39).

  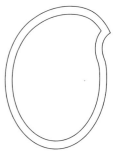

FIGURE 5.35 > Use of the *Offset* command to create parallel curved lines. *Hepler/Wallach/Hepler © Cengage 2013*

 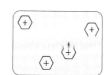 

FIGURE 5.36 > Method for aligning objects. *Hepler/Wallach/Hepler © Cengage 2013*

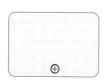 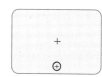 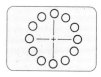

FIGURE 5.37 > Method for repeating objects in a circular pattern with the *Array Polar* command.
Hepler/Wallach/Hepler © Cengage 2013

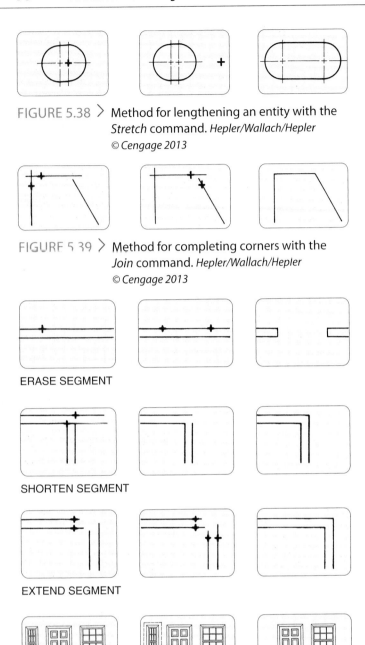

FIGURE 5.38 > Method for lengthening an entity with the *Stretch* command. *Hepler/Wallach/Hepler © Cengage 2013*

FIGURE 5.39 > Method for completing corners with the *Join* command. *Hepler/Wallach/Hepler © Cengage 2013*

ERASE SEGMENT

SHORTEN SEGMENT

EXTEND SEGMENT

DELETE SEGMENT

FIGURE 5.40 > Method for using the *Segment* functions of the *Lengthen* command. *Hepler/Wallach/Hepler © Cengage 2013*

- The *Lengthen* command can either shorten or lengthen a segment of a line by placing the cursor on the part of the line to be changed. Figure 5.40 shows the use of the *Delete (erase) Segment, Shorten Segment,* and *Extend Segment* functions. The *Lengthen* or *Extend* command adjusts an object's length to end at a predescribed point. This is an end point on a single object or one object among multiple objects.
- The *Gap* or *Trim* command is used to interrupt a solid line. Multiple gaps at regular intervals are produced by inputting the gap size, spacing, and number (Figure 5.41).

FIGURE 5.41 > Method for creating gaps in a line. *Hepler/Wallach/Hepler © Cengage 2013*

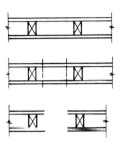

FIGURE 5.42 > Method for creating a break in a drawing. *Hepler/Wallach/Hepler © Cengage 2013*

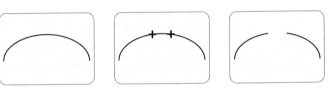

FIGURE 5.43 > Use of the *Break* command. *Hepler/Wallach/Hepler © Cengage 2013*

- The *Break* command interrupts a portion of a drawing by identifying the ends of the area to be interrupted, as shown in Figure 5.42. Figure 5.43 shows the application of the *Break* command on a curved line.
- The *Redraw* command removes extraneous lines and smudges and redraws the image. The *Regenerate* command goes further by recalibrating all entries and then redraws the image.

Utility Commands

Some of the commands available in a CAD program do not manipulate the drawing directly. Instead, they assist the drafter in various ways. Without utility commands, the drawing and editing commands would not function effectively. These commands clarify, simplify, and greatly speed up the process of drawing while ensuring the accuracy of finished drawings. Examples of these utility commands are listed next:

- The *Zoom* command magnifies or reduces the size of a drawing on the screen. See Figures 5.44A and 5.44B. It does not affect the actual dimensions of the drawing, however. The drafter can *zoom in* to view details that are too small to see when the entire drawing shows on the screen, or *zoom out* to see more of the drawing. Figure 5.45 shows an extreme example of zooming.

The *Zoom* command is used constantly in preparing all types of architectural drawings. For example, site plans

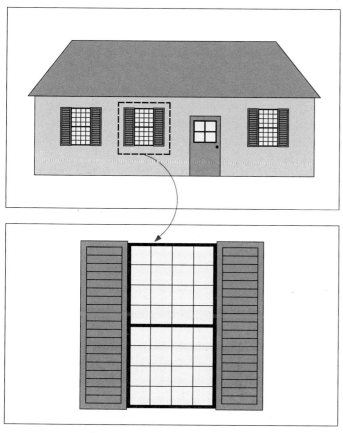

FIGURE 5.44A › Use of the *Zoom* command to magnify (zoom in on) a window detail. *Hepler/Wallach/Hepler © Cengage 2013*

  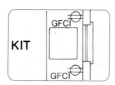

FIGURE 5.44B › Steps in zooming in. *Hepler/Wallach/Hepler © Cengage 2013*

often cover a D-size sheet (24″ ×36″). This is more than 10 times the size of a CAD monitor. The *Zoom* command is similar to the zoom function of a camera. It allows the entire drawing to be viewed and also enables small details or parts to be magnified or demagnified while drawing. This means that small details can fill a computer monitor or an entire housing subdivision can be made visible. A window is created around the area to be magnified by picking opposite corners of the window. A specific magnification scale is required, which can be typed in as part of the *Zoom* command. There is also a zoom capability that allows a drawing to be magnified only through the mouse control. If repeated zooming to a certain size is necessary. The *View* command is used to hold a view at a specific magnification factor. This view can then be held while other screens are in use and returned to the original size on command.

- The *Pan* command moves the drawing horizontally and vertically on the screen without changing the current zoom percentage. Figures 5.46A and 5.46B show examples of how the *Pan* command can be used on an elevation and floor plan. This moves the viewport, which is the rectangular window through which a drawing is viewed. The viewport is controlled by using the *Zoom* and *Pan* commands together. The *Pan* command allows a drafter to move to any location on a large drawing at a consistent selected magnification.

- The *Plot-Print* command allows the drawing to be plotted with a pen plotter or printer. Entire drawings, blocks, or layers can be plotted together or separately.

- The *Layer* command allows different segments of a drawing to be drawn separately and then combined into a single drawing (Figure 5.47A). An individual layer can be used for separate parts of a drawing such as dimensions, grids,

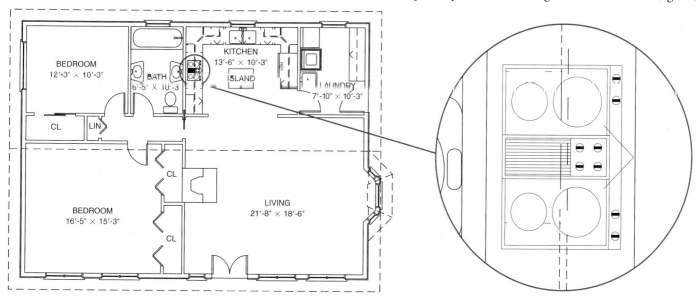

FIGURE 5.45 › Example of extreme zoom. *Hepler/Wallach/Hepler © Cengage 2013*

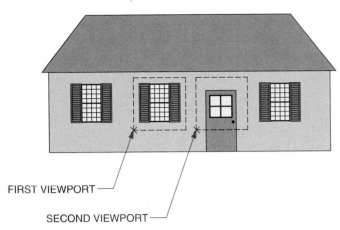

FIRST VIEWPORT
SECOND VIEWPORT

FIGURE 5.46A > Use of the *Pan* command to move a drawing segment. *Hepler/Wallach/Hepler © Cengage 2013*

 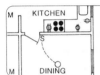 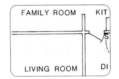

FIGURE 5.46B > Panning a floor plan. *Hepler/Wallach/Hepler © Cengage 2013*

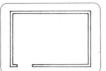 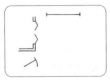 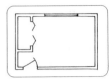

FIGURE 5.47A > Application of the *Layer* command to separate floor plan segments. *Hepler/Wallach/Hepler © Cengage 2013*

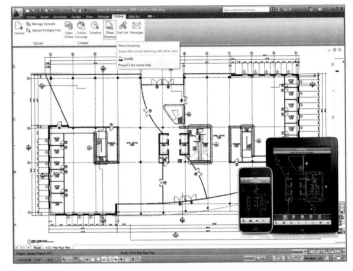

FIGURE 5.47B > Use of a color layer to separate and add details to a drawing. *Image courtesy of Autodesk Inc.*

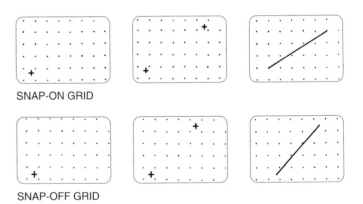

SNAP-ON GRID

SNAP-OFF GRID

FIGURE 5.48 > Use of the *Grid* command to snap on and snap off grids. *Hepler/Wallach/Hepler © Cengage 2013*

furnishings, or labels. Multiple building levels and specialized plans can also be layered using the same base drawing. Figure 5.47B shows the use of the color layer to view, edit, and share drawings detail through a web browser.

In using the *Layer* command, drawings are organized into specific entities. These include dimensions, construction lines, details, symbols, grids, furnishings, equipment, materials, and fixtures. Preparing many layered segments provides ease and flexibility in managing and manipulating drawings and details later in the design and drawing process. Different line weights, types, and colors can be assigned to each layer later. Each color can also be assigned a line weight or type. This results in a colored line layer on the monitor. Each layer can be printed separately or in any combination.

- The *Grid* command creates Cartesian coordinate X and Y dots to be superimposed on a drawing. The *Snap-On* function moves the cursor to the nearest grid intersection or dot. The *Snap-Off* function allows the cursor to be moved to any location regardless of grids, as shown in Figure 5.48.

Using the *Grid* command for a CAD drawing is like hand drawing on graph paper. There are two types of CAD grid systems; *Snap Grid* and *Display Grid*. The *Snap Grid* command gives the designer two choices: The *Snap-On* command allows the crosshairs to move from one grid point to the nearest grid intersection. The *Snap-Off* command allows a point to remain exactly where placed by the

cursor without regard to the grid lines. The *Display Grid* command allows the grids to remain on the monitor while other drawing operations are performed. The display grid can be turned on or off repeatedly during the drawing process. Figure 5.49 shows various grid patterns.

Grid spacing and points can be selected to align with any modular scale and spacing. Grids should be selected to represent the smallest grid unit to be used on the drawing. Typical grid sizes for floor plans and elevations are 12″, and 4″ or 6″ for details. Most modular grids are available

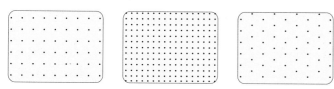

FIGURE 5.49 › Grid patterns. *Hepler/Wallach/Hepler © Cengage 2013*

FIGURE 5.50 › Use of the *Ortho* command to align (orthogonalize) on the X and Y axes at 90° angles. *Hepler/Wallach/Hepler © Cengage 2013*

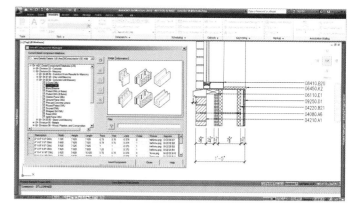

FIGURE 5.51A › AutoCAD® symbol library options with related details. *Image courtesy of Autodesk Inc.*

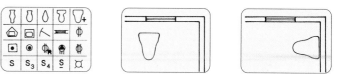

FIGURE 5.51B › Identifying and positioning library items on a plan. *Hepler/Wallach/Hepler © Cengage 2013*

ORIGINAL SCALE   SCALE DOWN   SCALE UP

FIGURE 5.52 › Use of the *Scale* command to increase and decrease the original drawing size. *Hepler/Wallach/Hepler © Cengage 2013*

with major and minor module lines in different weights and colors. Square or rectangular floor plans can be drawn using two-dimensional grid systems, which align the plan on X and Y Cartesian coordinate axis lines. Grid lines or dots that represent grid intersections can be superimposed on the drawing using different color lines or line weights. Assigning grid lines to a different layer helps separate the grid lines from other elements of the plan. Lines can also be drawn on absolute X and Y coordinates when the exact X and Y coordinates are known. Relative coordinates can also be used by referencing the last used coordinates as a base.

- The *Ortho* command (orthogonalize) adjusts all lines to a vertical or horizontal position at 90° angles, as shown in Figure 5.50. This can be done with or without the display of grids on the monitor. Because all ortho lines are either parallel or located 90° apart, the *Ortho* command must be turned off to draw any other angle.

- The *Block* command allows access to combination symbols or detail blocks electronically stored in many categories as plan, section, and elevation views. Some libraries are embedded in the original CAD software (Figure 5.51A); others can be created and stored by the drafter. The size and position of library items can be changed to match the scale and view of the drawing. Figure 5.51B shows how library items are positioned. Symbol libraries were discussed in more detail earlier in this chapter.

A *block* is a group of entities that are stored as a single object, typically as a separate drawing file. Drawings selected to be stored within a current drawing block that cannot be used outside the drawing are created using the *Block* command. Drawings stored as a separate drawing file can be used in any other drawing by using the *Wblock* command. Typical items that are stored in blocks include construction details, windows, doors, map symbols, fix-

tures, and entire drawings. A **block,** which is the generic term used for both *Block* and *Wblock* creations, can be placed into drawings using the *Insert* command.

- The *Scale* command is used to establish the scale of each drawing or increase or decrease the scale of an existing drawing (Figure 5.52). Full-size dimensions are always used. In CAD drawing the *Scale* command is used to establish a drawing at a specific architect's metric or engineer's scale. The physical scale of a drawing object can be changed by using the *Scale* command. While drawing, the *Measure* command can be used to locate a specific size on a segment. This command can also be used to draw a specific line length by inputting the line length at the scale of the drawing.

- The *Drag* command allows a line or entity, such as a symbol, to be moved (dragged) to a different location by the cursor.

- The *Save* command allows drawings to be filed on the hard disk or CD.

TYPES OF CAD DRAWINGS

The type of CAD drawing a drafter creates depends on the purpose of the drawing. The two major types of drawings are two-dimensional (2D) drawings and three-dimensional (3D) drawings.

Two-Dimensional Drawings

Traditional architectural working drawings are usually two dimensional. Only two dimensions (width and length, width and height, or length and height) are shown on one drawing. Two-dimensional drawings created on CAD systems include floor plans, elevation drawings, detail drawings, and site drawings.

Cartesian Coordinate Systems

Two-dimensional CAD drawings are based on a **Cartesian coordinate system,** which is based on an *X axis* and *Y axis.* These two axes are at right angles to each other. The point at which they meet is called the *origin.* See Figure 5.53. Note also that each axis has a positive (+) side and a negative (−) side.

In Cartesian geometry, the axes divide a drawing into four imaginary quadrants. To locate a point in any of these quadrants, you need only specify two numbers. The two numbers are called a *coordinate pair.* Point 1 in Figure 5.54 is located in quadrant 1 two units to the right of the origin on the X axis and one unit above the origin on the Y axis (+2, +1). Point 2 is located in quadrant 3, two units to the left of the origin on the X axis and one unit below the origin on the

Y axis (−2, −1). Note that the X coordinate is always the first number in a coordinate pair, and the Y coordinate is always the second number. Polygons, as shown in Figure 5.55, are constructed by repeating this process in the sequence shown

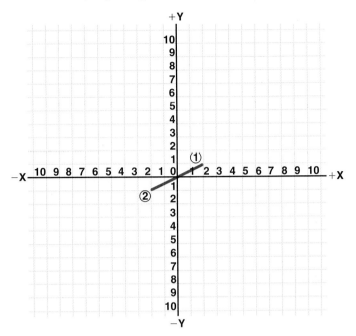

FIGURE 5.54 > Locating points on a Cartesian coordinate grid. *Hepler/Wallach/Hepler © Cengage 2013*

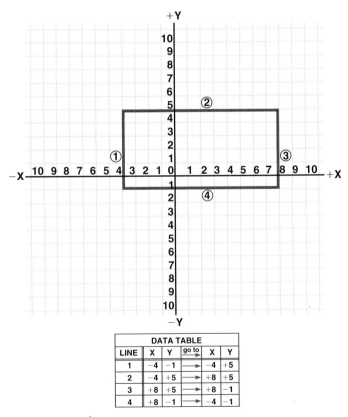

DATA TABLE					
LINE	X	Y	go to	X	Y
1	−4	−1	→	−4	+5
2	−4	+5	→	+8	+5
3	+8	+5	→	+8	−1
4	+8	−1	→	−4	−1

FIGURE 5.55 > Using the Cartesian coordinate system to plot a rectangle. *Hepler/Wallach/Hepler © Cengage 2013*

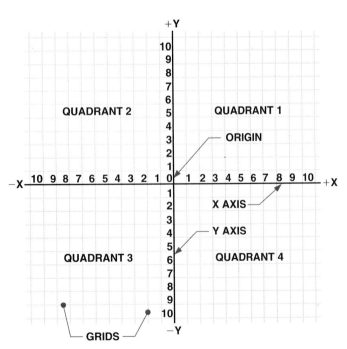

FIGURE 5.53 > Parts of a Cartesian coordinate system. *Hepler/Wallach/Hepler © Cengage 2013*

in the figure's data table. The data table shows the coordinates of the four lines on the drawing.

Unlike polygons, the development of curves requires the use of more closely spaced coordinate points. With that exception, the drawing process is the same as in the *Spline* command. Figure 5.56 shows a curve and a related data table containing coordinates for the development of a 20-point curve. The use of more points produces a smoother curve. The use of less frequent points produces a flatter, segmented curve.

An alphanumeric keyboard can also be used to create lines by using the keys corresponding to the X-Y points, once the original point has been established. For example, in Figure 5.57, point B is established by striking left bracket,

−3 (X axis), comma, −5 (Y axis), and right bracket as shown in Figure 5.57.

CAD Floor Plans

CAD floor plans range from fully dimensioned working drawings, as shown in Figure 15.41 in Chapter 15, to multicolored, layered plans (Figure 5.58) or surface textured and rotated plans, as shown in Figure 15.2A. Site, HVAC, plumbing, floor framing, and electrical plans are specialized floor

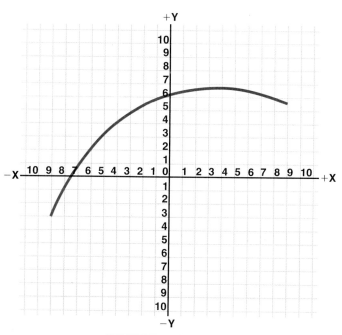

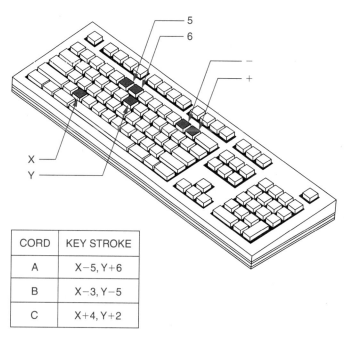

CORD	KEY STROKE
A	X−5, Y+6
B	X−3, Y−5
C	X+4, Y+2

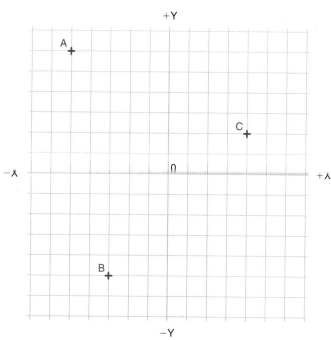

DATA TABLE					
SEGM	X	Y	go to	X	Y
1	−9	−3	→	−8.7	−2
2	−8.7	−2	→	−8.2	−1
3	−8.2	−1	→	−7.7	0
4	−7.7	0	→	−7	1
5	−7	1	→	−6	2.1
6	−6	2.1	→	−5	3.1
7	−5	3.1	→	4	4
8	−4	4	→	−3	4.8
9	−3	4.8	→	−2	5.5
10	−2	5.5	→	−1	5.8
11	−1	5.8	→	0	6.2
12	0	6.2	→	1	6.4
13	1	6.4	→	2	6.6
14	2	6.6	→	3	6.8
15	3	6.8	→	4	6.9
16	4	6.9	→	5	6.8
17	5	6.8	→	6	6.6
18	6	6.6	→	7	6.4
19	7	6.4	→	8	6
20	8	6	→	9	5.5

FIGURE 5.56 〉 Using the Cartesian coordinate system to plot a curve. *Hepler/Wallach/Hepler © Cengage 2013*

FIGURE 5.57 〉 Keyboarding the X-Y coordinates. *Hepler/Wallach/Hepler © Cengage 2013*

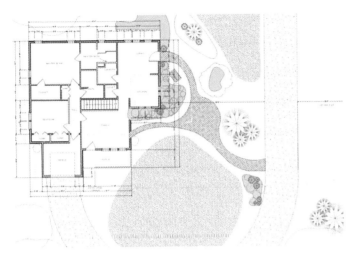

FIGURE 5.58 ❯ Use of color to show different floor plan conventions. *Chief Architect Software*

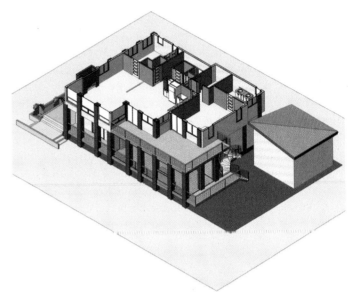

FIGURE 5.59 ❯ Example of a pictorial floor plan. *Swain CAD Solutions*

plans and are prepared as outlined in Chapters 13 through 18. These plans are drawn using the *Line* command for walls; the *Block* command for symbols such as doors, windows, and fixtures; the *Dimension* command to describe sizes; and the *Text* command for notes and dimension numerals. Colored lines as seen on a monitor may be converted to different line weights and types when printed.

CAD floor plans can also be drawn in pictorial form as shown in Figure 5.59. Floor plan walls are drawn using the *Line* or *Multiline* command. The *Multiline* command task can only be used to draw a linear wall style. There is little flexibility in creating floor plans using the *Multiline* command. The preferred method is to use the *Line* command along with the *Offset* command to draw exterior and interior walls. This is done after setting the wall thickness and then copying the line at a preset distance to complete a two-line system. The *Join* command is also used to create a corner by placing the cursor on two perpendicular lines. After walls have been drawn, symbols for doors, windows, and fixtures are added from symbol libraries or blocks. Dimensions are then added using the *Dimension* command, and the *Text* command is used to add notes.

Symbols are added by selecting each symbol from a symbol library block using the *Block* command and locating the position of each on the drawing with the cursor. A block can be part of the CAD software or created by drawing and filing it in a group block. Figure 5.60 shows the application of a furniture symbol block on a floor plan. Blocks contain different levels of detail. Compare the CAD simplified symbol with the detailed symbol plan shown in Figure 5.61.

Using an exterior or interior wall elevation layer as a guide, repetitive vertical wall framing members, such as studs, can be drawn on a superimposed layer using the *Copy-Multiple*

command. This is done using the *Single-Line* or *Polyline* command as with floor framing plans. The *Trim* command can be used to eliminate framing members from openings for doors, windows, or chimneys. Lintels, jambs, and plates can then be added separately.

Because foundation drawings align with the floor above, floor plans are used to establish the basic lines of the foundation plan. This is done by drawing the foundation plan on a separate layer over the floor plan layer, which can be shown with lighter lines or colors. The floor plan layer can be periodically removed to check progress. The use of layers in this way enables drafters to accurately draw different layers of a total design and coordinate their work.

Floor framing plans are prepared on a separate layer using the corresponding floor plan layer as the base drawing. The joist framing can be represented as either a single-line or a double-line drawing. The *Line* command is used to create both the *Offset* command and to create the second line of the joist. The *Copy-Multiple* command can be used to space the joists evenly. Interruptions for such features as stairwell, windows, doors, and chimney openings can be deleted by identifying the segment to be removed and using the *Trim* command.

The top-level floor plan layer is used as a guide in preparing roof framing plans. The *Line* command or the *Polyline* command is then used to draw rafters. The *Trim* or *Break* command is then used to remove portions of rafters for openings such as chimneys, skylights, and dormers. Double rafters and special framing for intersections are then added using

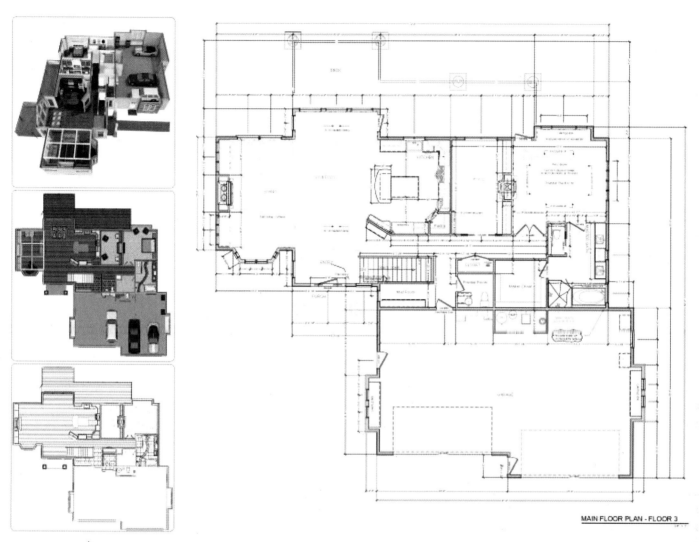

FIGURE 5.60 > Floor plan with detailed entities. *Chief Architect Software*

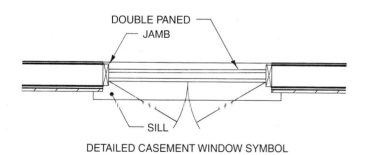

DETAILED CASEMENT WINDOW SYMBOL

DOUBLE PANED
JAMB
SILL

SIMPLIFIED WINDOW SYMBOL

FIGURE 5.61 > Simplified and detailed symbol. *Hepler/ Wallach/Hepler © Cengage 2013*

the *Offset* command. Rafters may be drawings with a single or double line.

CAD Layering

All elements of a floor plan should be layered so that layers can be turned on or off, moved, or become a separate drawing at any time. The entire floor plan can also be a layer that contains sublayers such as doors, windows, dimensions, fixtures, furniture, and even movable objects such as people and automobiles. With the *Layer* command, designers can group entities together to make identification and editing easier and more effective. This also controls what is sent to the plotter, including the coloring and line types used on entities. Floor plans that align vertically with other plans can also be layered (first and second floor, etc.) by assigning different colors to each layer. This may also involve the layered separation of such entities as outside walls, stairwells, and chimney locations. Preparing layered plans eliminates errors of alignment between plan drawings, including the placement of load-bearing walls and supports. Specialized floor plans such as plumbing, electrical, and HVAC plans are also layered to ensure agreement with the basic plan and to prevent lines and fixtures from being located in the same space.

Floor Plan Calculations

Without entering a dimension, the *Distance* command can be used to calculate the distance or angle between two points. The *Area* command can be used to calculate areas of any defined and closed space. This is done by identifying and adding corner points. Most CAD programs will also calculate perimeters using the same data.

CAD Elevation Drawings

Elevation drawings are two dimensional, and they represent length and height compared to floor plans, which show the width and length of a structure. CAD elevations are created from a floor plan in the manner described in Chapter 16.

First, the horizontal ends of doors, windows, chimneys, offsets, and building ends are projected from the floor plan using the *Construction Line* command on a separate layer. This is done in a distinct color and on a separate layer. Next, different color horizontal construction lines are drawn, representing the heights of these features plus the heights of the ground line, floor lines, and ridge lines. A heavier or different color line task is then used to connect the major lines of the elevation over the construction line layer. Elevation features that should align can be automatically aligned using the *Align* command. Symbols representing doors and windows are brought from the symbol library and placed in the construction line intersections. Object lines are then added and the construction lines removed using the *Erase* command. Then surface patterns as shown in Figure 5.62, can be added using the *Hatch* command. Last, dimensions and notes are added using the *Dimension* and *Text* commands.

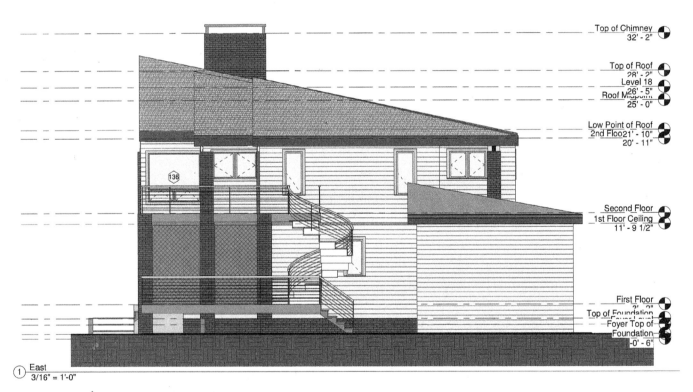

FIGURE 5.62 > Elevation drawing with hatch layers added for siding, roof, cornice, and chimney materials. *Swain CAD Solutions*

Interior wall elevation drawings are prepared with CAD by treating a specific floor plan wall the same as an outside wall and by following the same drawing sequence. This involves assigning a separate layer to the floor plan and each elevation. Here the height of all features—doors, windows (top and bottom), ledges, railings, cabinets, counters, etc.—must be established and drawn before floor plan key intersections are projected as shown in Figure 5.63.

CAD Detail Drawings

Multicolored layers are used extensively in detail drawings to show different materials, notes, and dimensions. The *Hatch* command is often used to show material sections (see Figure 5.64). Other than hatch areas, construction details contain few entities that can be stored in a library. Most stored details are those created by the designer and used frequently as an entire block.

Three-Dimensional Drawings

Most current architectural software programs can create three-dimensional drawings directly from two-dimensional drawings. This is done by adding a third axis (Z) to the Cartesian coordinate system, as shown in Figure 5.65. The Z coordinate allows the addition of depth to the drawing. Once a floor plan has been completed on a CAD system, a 3D drawing can be created of the exterior or interior by using 3D commands. This also applies to creating 3D plumbing, HVAC, electrical, and site profiles from plan views. The 3D construction and shading capabilities provide a depth to imagery. The main advantage of working in 3D is the accuracy and special relationships that can be created between objects. These models can be used to check the structural stability, orientation, and pictorial appearance of the design. This provides opportunities for some elements of the design to be changed after the model is studied. Figure 5.66 shows a one-point perspective drawing of this type.

FIGURE 5.63 > Interior elevation hatch rendering. *Swain CAD Solutions*

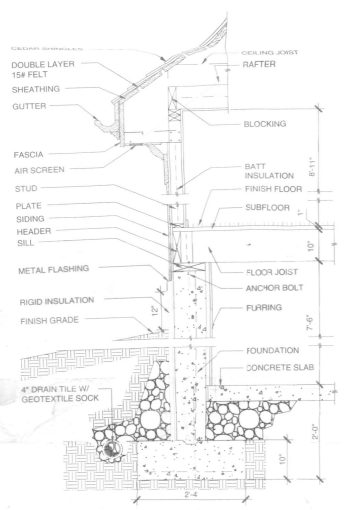

CEDAR SHINGLES
DOUBLE LAYER 15# FELT
SHEATHING
GUTTER
FASCIA
AIR SCREEN
STUD
PLATE
SIDING
HEADER
SILL
METAL FLASHING
RIGID INSULATION
FINISH GRADE
4" DRAIN TILE W/ GEOTEXTILE SOCK

CEILING JOIST
RAFTER
BLOCKING
BATT INSULATION
FINISH FLOOR
SUBFLOOR
FLOOR JOIST
ANCHOR BOLT
FURRING
FOUNDATION
CONCRETE SLAB

8'-11"   1"   10"   12"   7'-6"   2'-0"   10"   2'-4

FIGURE 5.64 ⟩ Detail drawing with hatched areas and object lines, dimensions, and notes in different color layers. *Hepler/Wallach/Hepler © Cengage 2013*

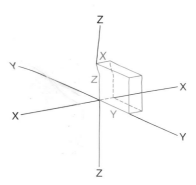

FIGURE 5.65 ⟩ X, Y, and Z axes are needed to produce three-dimensional drawings. *Hepler/Wallach/ Hepler © Cengage 2013*

FIGURE 5.66 ⟩ Pictorial drawings created from a plan view. *Chief Architect Software*

Three different types of three-dimensional drawings can be produced on a CAD system. These include wireframe drawings, surface models, and solid models.

Wireframe Drawings

Wireframe drawings are basically see-through stick drawings in which some or all hidden areas are exposed. They show an object's width, length, and height (X, Y, and Z) dimensions. These can be difficult to understand unless some or all of the hidden lines are "removed" (temporarily not displayed).

The drawing shown in Figure 5.67 shows the width, length, and height with dominate visual lines and with selected hidden lines removed. In Figure 5.68 the hidden lines are totally exposed.

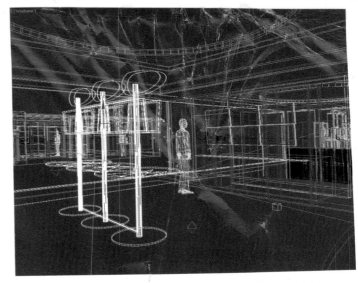

FIGURE 5.67 ⟩ Wireframe drawing with selected hidden lines removed. *Jack Hale Drawing*

FIGURE 5.68 > Wireframe drawing with exposed hidden lines. *Jack Hale Drawing*

FIGURE 5.70 > Surface model drawing. *Chief Architect Software*

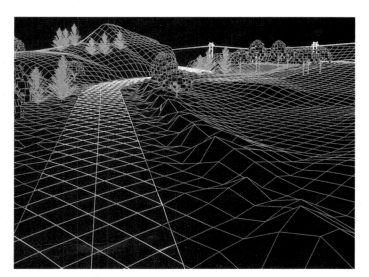

FIGURE 5.69 > Wireframe drawing showing site contours. *Autodesk, Inc.*

Surface Models

Surface models, as shown in Figure 5.70, are drawings that consist of solid plane surfaces instead of the connected lines used in wireframe drawings. These models can be visualized as a wireframe model with the outside surface covered with a siding.

Surface models are often rotated to reveal all surfaces in 3D, as shown in Figure 5.71. The buildings represented by the surface model shown in Figure 5.72 is the building shown in the elevation drawing in Figure 5.62. The surface model shown in Figure 5.73 includes multiple surface textures and colors. A full size detail of this corner is shown in Figure 5.6.

Solid Models

3D **solid models** are the same as surface models except the interior of the structure can be shown as in Figure 5.74. Although a solid model may look similar to a surface model or a wireframe with hidden lines removed, it is actually quite different. A solid model has mass properties, which other forms of 3D drawings do not have. For example, you can assign a material, such as copper, aluminum, or steel, to a solid model. Then the CAD software can measure the mass, density, and other properties of the model. Architects often use solid models to perform structural analyses on buildings and other structures. Unlike surface models, solid models can be used to create sectional drawings in 3D.

In addition to their mass properties, solid models can be rendered so that they look almost like photographs, as shown in Figure 5.75. In CAD, **rendering** is the process of adding shading and lights to a drawing so that it looks more realistic.

Virtual Reality Systems

Virtual reality is a natural extension of three-dimensional drawing technology. A *virtual reality* system creates a computer-generated "world" or "reality" in which the user seems to be immersed in the computer images.

A virtual reality (VR) program is created by completing a 3D model drawing of a building, then rendering walls and adding details using specialized software. The user, wearing a helmet with a viewing screen and a data glove, can move freely (as if in the building) to inspect walls, heights, and the location of features such as doors and windows. VR is rarely used due to its great expense and complexity.

Some architectural applications allow the architect to make real-time changes to the building. For example, if a

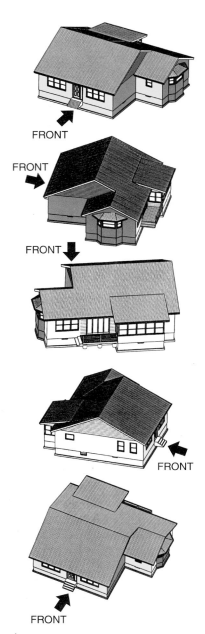

FIGURE 5.71 › Examples of rotation positions. *Hepler/ Wallach/Hepler © Cengage 2013*

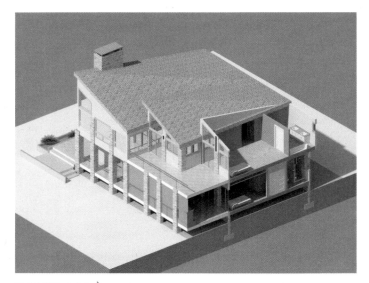

FIGURE 5.72 › Surface model of the building shown in Figure 5.62. *Swain CAD Solutions*

FIGURE 5.73 › Solid model developed from a wireframe drawing. *Chief Architect Software*

FIGURE 5.74 › Solid floor plan model rendering. *Hamid Rafiei Dreamarch3D.com*

client decides a window is too high, the architect can "reach out" within the virtual environment and move it down. The change is reflected in the CAD drawing file as well as in the virtual environment.

CAD Rendering

Surface materials can be added to pictorial drawings using the Photoshop color library and the *Hatch* command. Trees and other landscape features are also available in 3D form, but must be adjusted to fit the scale of the drawing. The *Material* or *Hatch* commands are used to add surface textures, and the *Render* command is used to add specific colors to a drawing surface. Shadows can be automatically generated by establishing the angle of the light source, or sun. Figure 5.76 illustrates the use of CAD to show the effects of sunlight, shadow, and color patterns.

Nothing adds more realism than the inclusion of movable entities in a drawing. CAD entourage libraries or blocks contain people, animals, and vehicles in a wide variety of sizes, shapes, and positions. The size of entourage symbols can be manipulated using the *Stretch* or *Scale* command and can be located using the *Move* command. The *Stretch* command can be used to either elongate or condense a feature. The *Move*

command allows entities to be moved to a new location by placing the cursor on the object and dragging it to the new position. If more than one entity is needed in a pattern, the *Copy-Multiple* command can be used by inputting the location, spacing, and number of entries. Note the application of these features throughout the remainder of this text.

CAD APPLICATIONS

In addition to preparing architectural drawings, CAD and other computer systems are used to produce many documents related to architecture. These include the processing of information to produce schedules, specifications, budgets, contracts, codes, engineering analyses, and construction management forms. See Chapters 34 through 37 for the application of these systems.

Computer programs are used to create, store, and retrieve specification data. Programs can be developed by designers or master specifications are available from the Construction Specifications Institute. These masters can be altered based on the content of the design and the standard of construction.

Manufacturers' websites contain valuable design information that can shorten and improve the design and drawing process. Manufacturers' websites represent a reliable source for locating detailed information about products, including sizes, specifications, and detail drawings. This information

FIGURE 5.75 〉 Pictorial model rendering. *Hamid Rafiei Dreamarch3D.com*

FIGURE 5.76 〉 CAD rendering using multiple commands and media. *Hamid Rafiei Dreamarch3D.com*

takes the form of drawings and data. DWG file drawings, when downloaded, can be altered to fit specific design needs. Details of this type are often altered to be consistent with the base drawing in line weight, dimensioning practice, or application. Imported PDF or JPEG files usually cannot be altered and, therefore, must be used as supplied.

Manufacturers of building materials and components often provide designers with compact disks that contain drawings of their product. Some details can be stored and inserted into any drawing and altered for consistency with the conventions of the base drawings. Some imported files cannot be altered and must be used as received.

Computer spreadsheets are used extensively to prepare building materials and components schedules. Information such as cost, square footage, volume, and model numbers, colors, and materials are input. The totals for each item and/ or location can be calculated. The *Bill of Materials* command is used to prepare spreadsheet outputs. This enables items to be entered and their cost, square footage, volume, model numbers, or quantity to be tracked by *attribute* within the inserted blocks on a drawing. Once loaded into a spreadsheet format, the values can be mathematically tabulated. Many CAD programs can also alter items in the specifications when any drawing detail is changed.

Drawings can be coded to create a database that can automatically generate project cost estimates. These databases contain flexible prices, labor costs, and detailed descriptions of thousands of building materials and components. These figures can be combined with wage rates for each trade within every zip code. Other factors such as building size, shape, and type can then be used to compute an estimate for an entire project. If component symbols are coded when entered into a drawing, the final set of drawings can be scanned and entered into a computer database to automatically produce a precise take-off estimate.

Many municipalities store their code information including zoning, planning, and land-use regulations on a website. Master municipal contracts and bidding documents can be downloaded by approved licensed professionals.

Engineering Studies

More sophisticated programs are used to produce engineering-related data and simulations such as structural analyses, environmental effects, and energy requirements. For example, computer programs are used to create pictorial contour drawings as shown earlier in Figure 5.69. This is done by combining X-Y coordinate input data and the corresponding datum level (Z) for each coordinate point. These points are then connected using surface modeling techniques to create the pictorial contour lines. The individual data points are connected using the *Polyline* or *Spline* command to create the individual lines. The lines are than meshed together to create a surface model of the area being deigned.

Once the lines have been created or a surface mesh generated, the designer's viewpoint (using the *VPOINT* command) can be changed to get a true 3D perspective. Use the *3D Orbit* command for the easiest control. Digital graphic files are available from the U.S. Geological Survey (USGS) and contain information from USGS maps that can be used for large site plans. Local municipal planning departments can often supply digital map and/or aerial photographs with property lines superimposed.

Building Information Modeling

All of these CAD functions are interfaced with working drawings through the use of **building information modeling (BIM).** BIM systems use parametrics that connect drawings with a wide variety of construction documents. BIM drawings are created in transparent 3D form, which produces and/or changes 2D working drawings as the design process progresses. All BIM written documentation and calculations are stored in a database and simultaneously linked (shared) with all other entities in the system. This includes the architect, designer, engineer, owner, contractor, and developer.

BIM is changing the way architecture and construction is practiced. The American Institute of Architects describes BIM as "a model based technology linked with a database of project information." By this definition all BIM drawings and documents will be capable of searching and linking regional, national, and international standards. This use of digital information technology covers all areas of the design, construction, and operation of all building facilities. BIM systems are of three types; *3D BIM* includes only 3D shape information, *4D BIM* uses 3D BIM and adds a cost dimension, *5D BIM* adds scheduling, estimating, analysis, and time constraint data.

Through the simultaneous sharing of database information all areas of a design are updated each time any addition or change is made in any area. This includes all categories of drawings, schedules, specifications, code data, legal documents, cost, and time estimates. This is accomplished by adding very detailed information to the database as each document or drawing is completed. For example; when a window symbol is added to a drawing, all information relating to that window is added to the database. This includes the maximum amount of information usually found on a window schedule, window detail drawing, and manufacturers detail. This is well beyond the normal dimensions and notes included on a basic floor plan or elevation drawing. As a result the drawings, and documents used by many different managers are always current and consistent. This facilitates the coordination of designers, contractors, and support services personal. These procedures not only apply to the design and construction phase but also apply to the total life cycle of a building, including future renovation projects.

BIM drawings are created from information transmitted from floor plans, elevations, sections, and details. The development of a BIM set of drawings starts with the development of a basic floor plan as shown in Figure 5.77. From this plan elevation drawings are developed which contain related alpha-numeric coding. Figure 5.78A shows the right elevation of this plan which also includes detailed datum information. A color hatched version of this elevation is shown in Figure 5.78B. From the data transmitted from the floor plan and all elevations, three dimensional drawings are created as illustrated in Figure 5.79. Pictorial drawings can then be created as viewed from any angle. Figure 5.80 shows a pictorial section of the same design.

Abbreviated wire frame pictorial BIM models are usually created to include transparent interior and exterior views of the building. When these drawings are layered, vertically and horizontally; structural, electrical and mechanical systems can be shown on the same 3D drawing. This exposure reveals and specifically identifies conflicts for space, thus discrepancies among drawings become immediately apparent. For example; rather than separately studying many different drawings, 3D transparent drawings allow the viewing of several entities at the same time. This can reveal such problems as plumbing lines occupying the same space as an electrical conduit or HVAC ducts crossing through cable channels.

Other versions of a drawing and document control system are imbedded in to some CAD software. For example the AutoCAD® Architecture allows the designer or drafter to export drawings in an Industry Foundation Class (IFC) file format. This software helps the designer to create, manage and share drawing and document data with team members during the

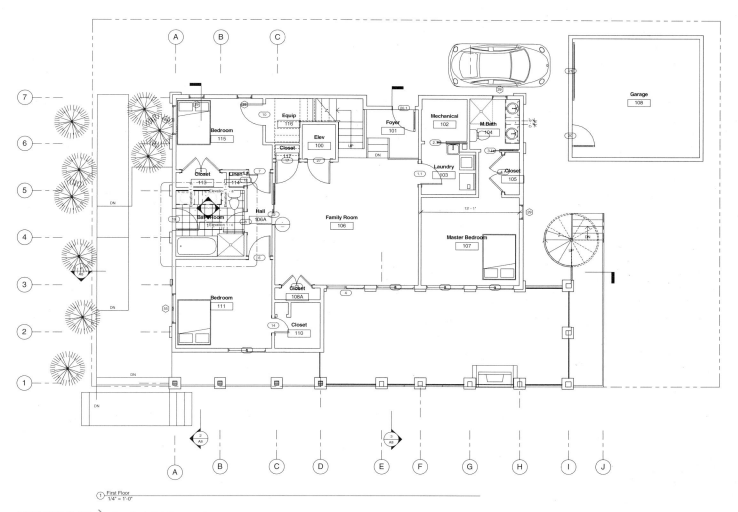

FIGURE 5.77 > Basic BIM floor plan. *Swain CAD Solutions*

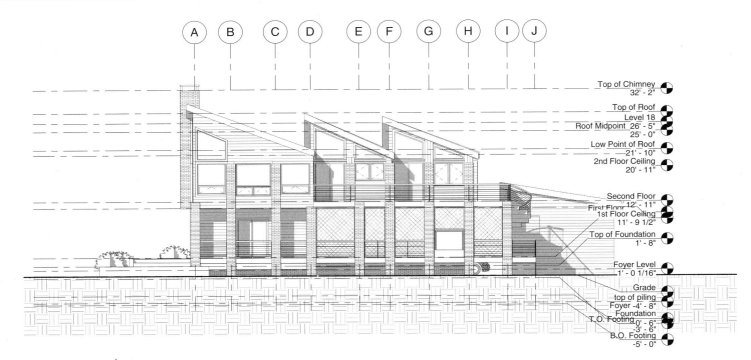

FIGURE 5.78A > BIM elevation derived from floor plan. *Swain CAD Solutions*

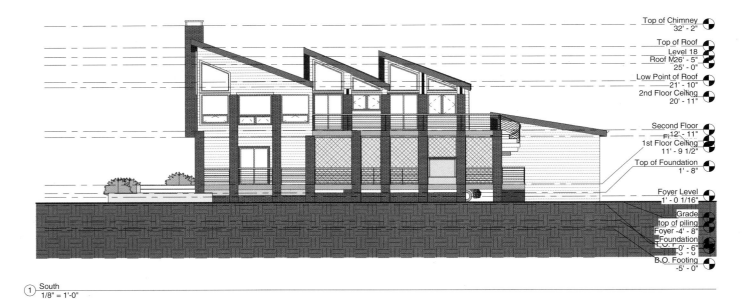

FIGURE 5.78B > Hatched color version of the elevation shown in Figure 5.78A. *Swain CAD Solutions*

FIGURE 5.79 > BIM perspective images. *Swain CAD Solutions*

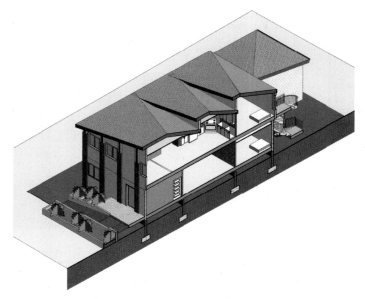

FIGURE 5.80 > BIM pictorial section. *Swain CAD Solutions*

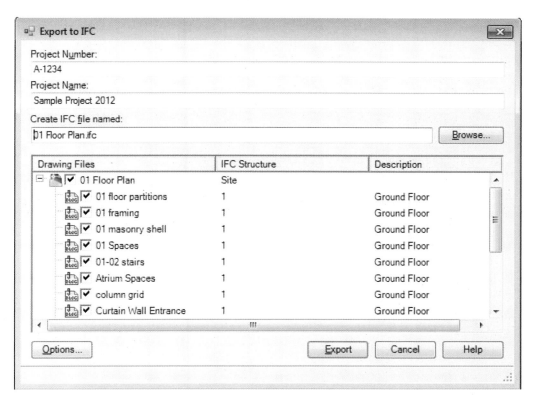

FIGURE 5.81 > Drawing management support document. *Image courtesy of Autodesk Inc.*

entire design process. Figure 5.81 shows an example of a control document used to organize drawing files with this software.

Continuing Education

CAD systems continue to change and improve. To adjust to these improvements a basic knowledge of architectural drafting, design and construction is necessary. Consequently this chapter and this text provide only an introduction to CAD and CADD. Many CAD software developers offer continuing education opportunities such as the Autodesk® Exchange. This program, described in Figure 5.82, allows users to access an integrated web-based site for assistance in content and procedures which support continued learning.

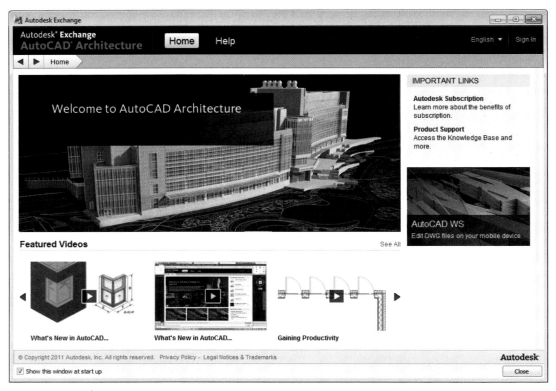

FIGURE 5.82 > Autodesk® Exchange. *Image courtesy of Autodesk Inc.*

Introduction to Computer-Aided Drafting and Design Exercises

CHAPTER **5**

1. List the hardware components in a typical CAD system.

2. Explain why the basic principles and practices of architectural drafting must be learned in addition to CAD operations.

3. List the major commands used for drawing and editing.

4. Describe three utility commands.

5. Describe the difference between memory and storage.

6. Describe the relationship between the BIM floor plan in Figure 5.77 and the elevation drawing shown in Figure 5.78A.

7. Define the following terms: software, hardware, input, output, computer, flash drive, hard disk, menu, cursor, plotter, enter, command, mode, monitor, hard copy, RAM, mouse, library, block, Cartesian coordinates.

BASIC AREA DESIGN

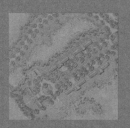

CHAPTER 6

Environmental Design Factors

OBJECTIVES

In this chapter you will learn:

> the principles of LEED.

> the basics of environmental planning.

> methods of sun protection.

> to orient a house on a lot to take best advantage of solar energy and features of the lot.

> to design structures ergonomically.

> ways to prevent pollution (ecology).

TERMS

active solar design systems
baffle
deciduous tree
earth-sheltered homes
ecological planning

ergonomics
land integrity
LEED
lot

noise abatement
orientation
overhang
passive solar design systems

⊕ INTRODUCTION

A wide range of factors must be considered to develop a fully functional architectural design—from a building's geographical area to the dimensions of an average adult. The designer must carefully study such environmental factors as the climate, sun orientation, site characteristics, potential hazards, and energy sources. A building's design is also influenced by the needs of the people who will occupy the building. Considering these factors, the designer should also attempt to protect and improve the environment. This is accomplished through the creation of designs that control or mitigate the negative effects of the pollutants and hazards that can be present in buildings and building sites. The design methods and materials used in this process are described in numerous chapters in this text. This chapter introduces these concepts; later chapters describe the specific design practices used.

ORIENTATION

A building must be positioned to maximize desirable features and minimize the negative aspects of the environment. This is accomplished through effective **orientation.** A building's

orientation is its relationship to its environment. These include the relationship to the sun, view, noise sources, wind and breeze directions, and adjacent structures and facilities.

⊕ Energy Orientation

Local resources and climatic conditions have always affected uses of energy—heating, cooling, and lighting. For example, early Native Americans built adobe houses under overhanging cliffs. The cliffs provided protection from the hot sun during summer and the cool wind during winter. See Figure 6.1. The adobe material in these houses absorbed heat during the day and released it at night to warm the area. People soon found they could use other natural resources for heat. By burning renewable fuel such as wood, twisted grass, or blubber, people became less dependent on architectural designs and materials for protection from harsh environmental conditions. Later, when inexpensive fossil fuels (e.g., coal) appeared to be an endless energy source, people began to rely almost exclusively on those. Designers controlled inside environments artificially.

Today we know that the supply of fossil fuels is finite, so we need to apply energy-efficient principles in building design to

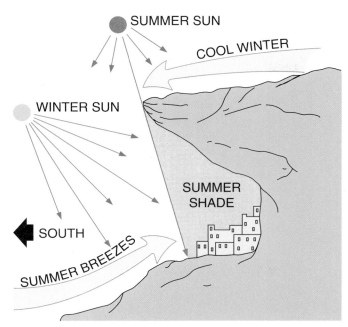

FIGURE 6.1 > Early methods of solar heating and cooling. *Hepler/Wallach/Hepler © Cengage 2013*

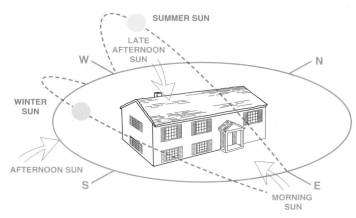

FIGURE 6.2 > Summer and winter sun paths. *Hepler/Wallach/ Hepler © Cengage 2013*

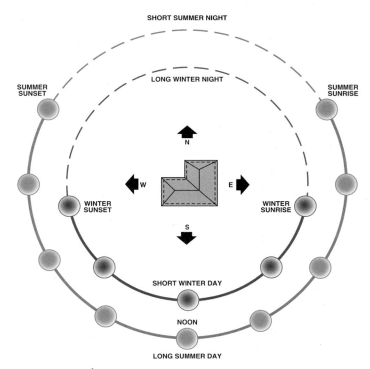

FIGURE 6.3 > Basis of seasonal solar planning. *Hepler/ Wallach/Hepler © Cengage 2013*

control indoor environments. Energy must be obtained from sources and methods besides the burning of fossil fuels. Solar, wind, hydroelectric, and nuclear energy are available for use today. The combustion of natural materials such as reclaimed waste, wood, and other organic materials can also be used. Except for energy from the sun, all of these sources require special equipment to effectively heat and cool buildings. Mechanical or electrical devices can be added to control the sun's energy. Such systems are called **active solar design systems.** However, carefully designed buildings can use the power of the sun without mechanical devices and still provide much environmental control. These systems are called **passive solar design systems.** They use only the design features and orientation of a building to gain and control the sun's energy. (Detailed discussions of both types of solar systems are presented in Chapter 32.)

🌐 Solar Orientation

The first step in designing the orientation of a building is to consider its relationship to the sun. Because the earth's axis is tilted, the angle of sunlight changes from summer to winter as the earth revolves around the sun. See Figure 6.2. In the northern hemisphere, the south and west sides of a structure are warmer than the east and north sides. The south side of a building is the warmest side, because of the nearly constant exposure to the sun during its periods of intense radiation. Ideally, a building should be oriented and windows positioned to absorb this southern-exposure heat in winter and to repel the excess heat in summer. See Figure 6.3.

A structure needs to be located and oriented to ensure that the areas requiring the most solar exposure will be correctly positioned in reference to the sun. Figure 6.4 shows a model that can be used to check sun and shade patterns produced at different hours or during different seasons. Many zoning codes limit the height of new structures that block sunlight from reaching existing buildings. Keep in mind that effective solar orientation should not only provide the greatest heating or cooling effect, but should also be planned to provide the greatest amount of natural sunlight where needed. For example, designers should consider the placement of windows and skylights as a source of both heat and light. See Figure 6.5.

Walls, floors, and furniture can absorb and store heat from the sun. This heat is naturally released when the temperature is cooler

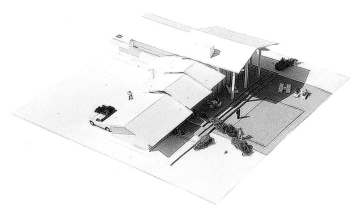

FIGURE 6.4 > Model used to study shade patterns. *Hepler/ Wallach/Hepler © Cengage 2013*

FIGURE 6.5 > Extensive use of skylights. *Klumb residence— Preston Bolton Architects; Balthazar Karab, Photographer*

or at night. Open interiors and high ceilings encourage ventilation and cooler temperatures. Low ceilings and closed floor plans tend to increase temperatures. Chapter 32 contains more detailed information on features that take advantage of passive solar principles in the design of structures. Figure 6.6 illustrates the major

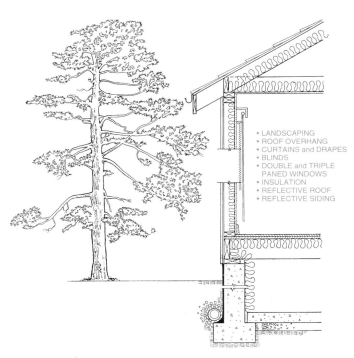

- LANDSCAPING
- ROOF OVERHANG
- CURTAINS and DRAPES
- BLINDS
- DOUBLE and TRIPLE PANED WINDOWS
- INSULATION
- REFLECTIVE ROOF
- REFLECTIVE SIDING

FIGURE 6.6 > Common deterrents to heat transfer through walls. *Hepler/Wallach/Hepler © Cengage 2013*

methods of reducing the effects of solar radiation passing through windows on an interior space.

Room and Outdoor Area Locations

Rooms should be located to absorb the heat of the sun or to be baffled (shielded) from the heat of the sun. Consider not only the function of the room, but the seasons and the time of day the room will be used most. The location of each room should also make maximum use of the light from the sun. See Figure 6.7.

Generally, sunshine should be available in the kitchen and dining areas during the early morning and should reach the living room by afternoon. To accomplish this, kitchen and dining areas should be placed on the south or east side of the house. Living room areas are placed on the south or west side to receive the late-day rays of the sun, when the room is most likely to be used. The north side is the most appropriate side for placing sleeping areas. It provides the greatest darkness in the morning and evening and is also the coolest side. Northern light is also consistent and diffused and has little glare.

The same principles apply to planning outdoor living areas. Those areas that require sun in the morning should be located on the east side. Those requiring the sun in the evening should be placed on the west side. For both indoor and outdoor areas, remember that in the northern hemisphere, the north sides of buildings receive no direct sunlight.

Overhang and Baffle Protection

Roof **overhangs** and **baffles** (shields) should be designed to allow the maximum amount of sunlight and heat to penetrate

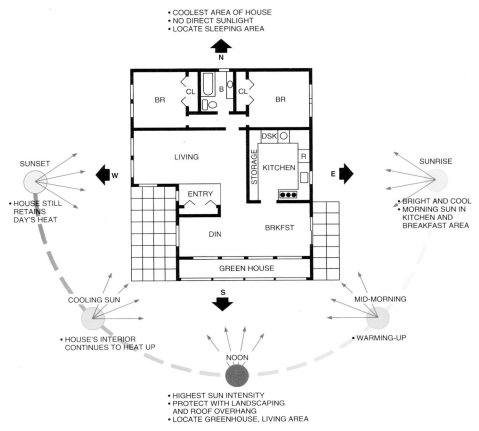

FIGURE 6.7 > Guidelines for room orientation. *Hepler/Wallach/Hepler © Cengage 2013*

the inside of a building in winter. Conversely, the maximum amount of sun and heat should be shielded from entering the interior during a summer midday.

Because the angle of the sun differs in summer and in winter, roof overhangs can be designed with a length, height, and angle that will shade windows in summer, yet allow the sun to enter during the winter. See Figure 6.8. The edge of the overhang also needs to be related to the height of the window. As shown in Figure 6.9, more summer sun rays will reach

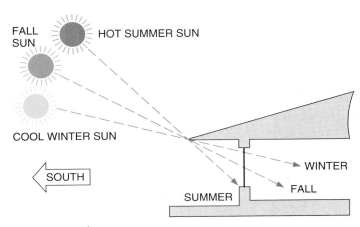

FIGURE 6.8 > Overhang designed to control seasonal sun exposure. *Hepler/Wallach/Hepler © Cengage 2013*

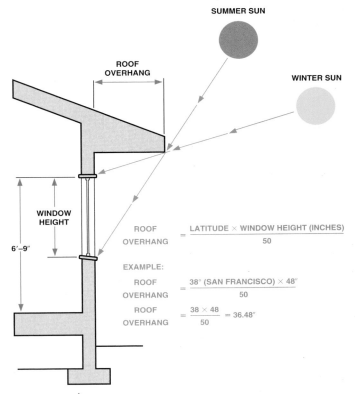

FIGURE 6.9 > Calculating southside roof overhang. *Hepler/ Wallach/Hepler © Cengage 2013*

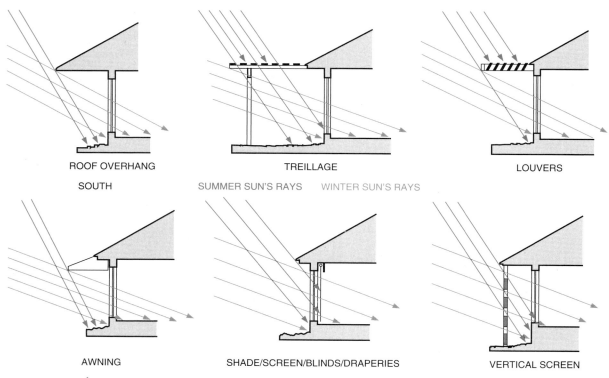

FIGURE 6.10 > Types of sun baffling. *Hepler/Wallach/Hepler © Cengage 2013*

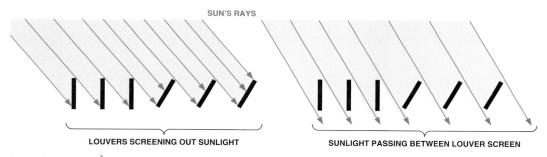

FIGURE 6.11A > Use of louvers to screen sunlight. *Hepler/Wallach/Hepler © Cengage 2013*

the window area if the overhang is smaller. Larger overhangs result in fewer sun rays reaching the window area. Figure 6.10 shows several effective overhang baffling systems. Baffling should be accomplished without blocking out natural light.

Figure 6.11A shows how fixed louvers can be used to block sunlight. Figure 6.11B shows a trellis system that allows the slats to be moved 90° to block summer sun or admit winter rays. Not all radiation and light from the sun can be reduced with overhangs or baffles. Figure 6.11C shows the interior effect of both direct and indirect radiation through windows.

🌐 Land and a Structure

A particular plan may be compatible with one site and yet appear totally out of place in another location. The environmental success of a design depends on how well the structure is integrated with its surroundings. In addition to orientation, other factors that affect site design include topography, size, vegetation, precipitation, temperature, wind, vistas, and noise.

Characteristics of a Site

Every structure should be designed as an integral part of the site, regardless of the shape or size of the terrain (land surface). Buildings should not appear as appendages ("add-ons") to the land but as a functional part of the landscape. For the indoor and outdoor areas to function effectively as parts of the same plan, the building and the site must be designed together.

Before orienting and designing a structure, consider the specific physical characteristics of the site, such as hills, valleys, fences, other buildings, and trees. Physical features such as these may affect wind patterns and the amount and direction of available sunlight in different seasons.

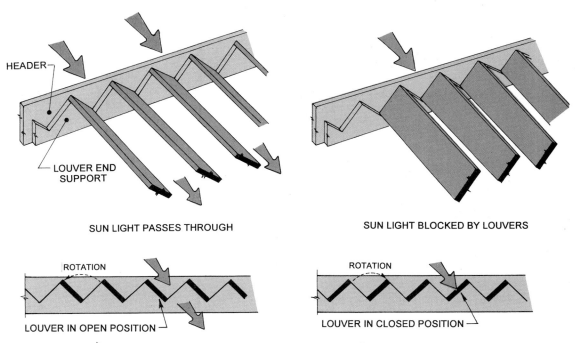

HEADER

LOUVER END
SUPPORT

SUN LIGHT PASSES THROUGH

SUN LIGHT BLOCKED BY LOUVERS

ROTATION

ROTATION

LOUVER IN OPEN POSITION

LOUVER IN CLOSED POSITION

FIGURE 6.11B > Convertible sun baffle louvers. *Hepler/Wallach/Hepler © Cengage 2013*

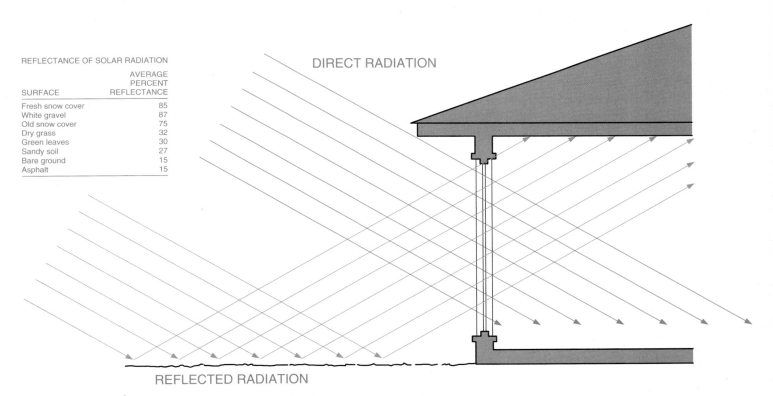

REFLECTANCE OF SOLAR RADIATION	
SURFACE	AVERAGE PERCENT REFLECTANCE
Fresh snow cover	85
White gravel	87
Old snow cover	75
Dry grass	32
Green leaves	30
Sandy soil	27
Bare ground	15
Asphalt	15

DIRECT RADIATION

REFLECTED RADIATION

FIGURE 6.11C > Effect of direct and reflected radiation. *Hepler/Wallach/Hepler © Cengage 2013*

Large bodies of water may also affect air temperature and air movements. Also, surrounding pavement areas and buildings can raise or lower temperatures, because concrete and asphalt collect and store the sun's heat.

Consider the view options when orienting a house. Orientation of specific areas toward the best view, or away from an objectionable view, usually means careful planning of the position of the various areas of the building. The house shown

FIGURE 6.12 > Orientation designed to capture sunset vistas. *Hepler/Wallach/Hepler © Cengage 2013*

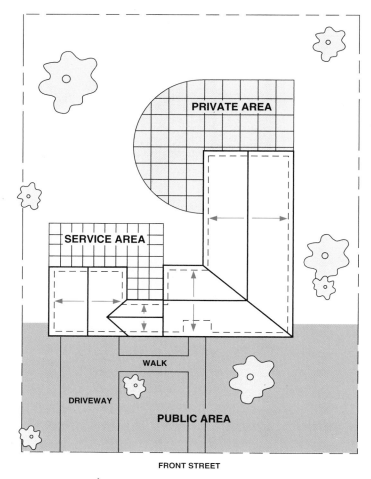

FIGURE 6.13 > Residential lot areas. *Hepler/Wallach/Hepler © Cengage 2013*

in Figure 6.12 is oriented to take advantage of sunset vistas. The plan of this design is shown in Chapter 13.

Lot Areas

Building sites are sold and registered as **lots.** The size and shape of a lot affects the flexibility of choice in locating structures. For planning purposes, lots are divided into three areas according to their function: public, private, and service areas. See Figure 6.13. The placement of all structures on a lot determines the relative size and relationship of the three areas. Zoning ordinances also affects the size and placements of structures on a site. Remember, the features of the lot should be an integral part of the total design. The site design is as important as the basic floor plan design of a structure.

🌐 Earth-Sheltered Homes

Characteristics of the land (site) and soil conditions are of utmost importance in the design of **earth-sheltered homes.** These types of homes are designed to be partially covered with earth. See Figure 6.14.

The thought of living partly underground may at first seem oppressive. However, with careful planning and proper orientation, adequate light can be achieved.

Earth-sheltered homes have several advantages. Regardless of how high or low the outside temperature is, the soil just a short distance below the surface remains at a comfortable and constant temperature. Heating and cooling units can be smaller and are used less often than those in conventional structures. The underground location also avoids the problems of wind exposure and winter storm winds.

Construction costs for earth-sheltered homes can be less than those for conventional types of construction if experienced builders are employed. The major concern in building

earth-sheltered homes is waterproofing, but this can be accomplished with waterproof paint, sealants, membrane blankets, and proper drainage. The structure of an earth-sheltered home also needs to be heavier than that of conventional buildings, to support the heavy soil loads.

Earth-sheltered structures generally require little maintenance. Groundwater accounts for most maintenance problems. However, with careful study of drainage patterns during the planning phase and effective waterproofing during construction, such problems can be avoided.

Soil is an important consideration for designers of earth-sheltered buildings. Soils used to cover roofs and walls must be carefully selected to avoid frost heave, excessive swelling, and runoff tendencies. Soil that supports vegetation (organic soil) must be used for surface areas.

The best site location for an earth-sheltered home is on a gentle downward slope. See Figure 6.15. The exposed walls should make maximum use of glass to capture as much light and heat as possible. This requires a southern exposure. Windowed areas should also be oriented away from prevailing winds but should face the best possible view.

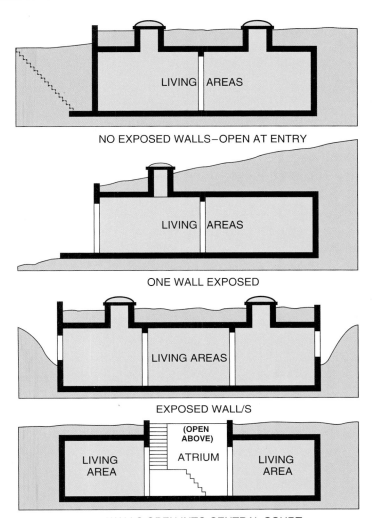

NO EXPOSED WALLS–OPEN AT ENTRY

ONE WALL EXPOSED

EXPOSED WALL/S

WALLS OPEN INTO CENTRAL COURT

FIGURE 6.14 > Types of earth-sheltered homes. *Hepler/Wallach/Hepler © Cengage 2013*

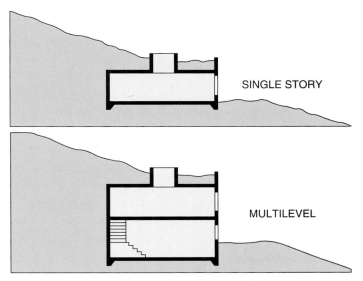

SINGLE STORY

MULTILEVEL

FIGURE 6.15 > Single-level and multilevel earth-sheltered homes on sloping lots. *Hepler/Wallach/Hepler © Cengage 2013*

Rooms requiring the most light, such as the living room, should be located in windowed areas. Seldom-used rooms or rooms not requiring windows can be located in ground-locked areas as shown in Figure 6.16. Skylights can be used in otherwise dark areas if they are effectively sealed and drained.

Although sloping lots are best, earth-sheltered designs can be placed on relatively level lots. Mounds of earth called *berms* can be created to provide protection.

Figure 6.17 shows a detail wall section of an earth-sheltered home wall. Figure 6.18 is a sectional detail of a sod-covered roof.

Other nonconventional building types include rammed earth, adobe block, and straw-bale construction, which are detailed in Chapter 22. *Rammed earth* construction uses a mixture of damp soil that is rammed into a compacted form to a wall of sedimentary rock. These walls have high insulating qualities. They also contribute to protecting the environment through reducing the use of wood, steel, and cement in home construction. *Adobe construction* uses a mixture of sand, clay, and straw to create blocks that are stacked in layers to create solid masonry walls. Because of the density and tubular structure of straw, straw bales can be compressed into strong structural shapes to create very stable and rigid walls. *Straw-bale* walls are covered with stucco on the exterior and plaster for the interior.

Vegetation

Vegetation of all types greatly aids in heat, light, wind, humidity, and noise control. Trees, shrubs, and ground-cover foliage, when effectively used, can also baffle undesirable views and enhance attractive scenes.

Deciduous trees maximize summer cooling and winter heating. They provide shade in the summer and then lose their leaves in winter, which allows the sun's warmth to penetrate the building. See Figure 6.19. Dense, coniferous (evergreen) trees and shrubs are most effective for blocking or redirecting north or northwest storm winds to help protect a building during all seasons.

Vegetation is an important design consideration. However, it should not be used as a substitute for appropriate orientation design to control a building's environment. Carefully planned and selectively preserved vegetation also contributes to the absorption of solar heat through evaporation. Vegetation also helps control excessive runoff and humidity. Chapter 13, *Site Development Plans,* covers the positive effects of maintaining the natural environment of a site while avoiding the removal of vegetation and topsoil, which leads to site destruction.

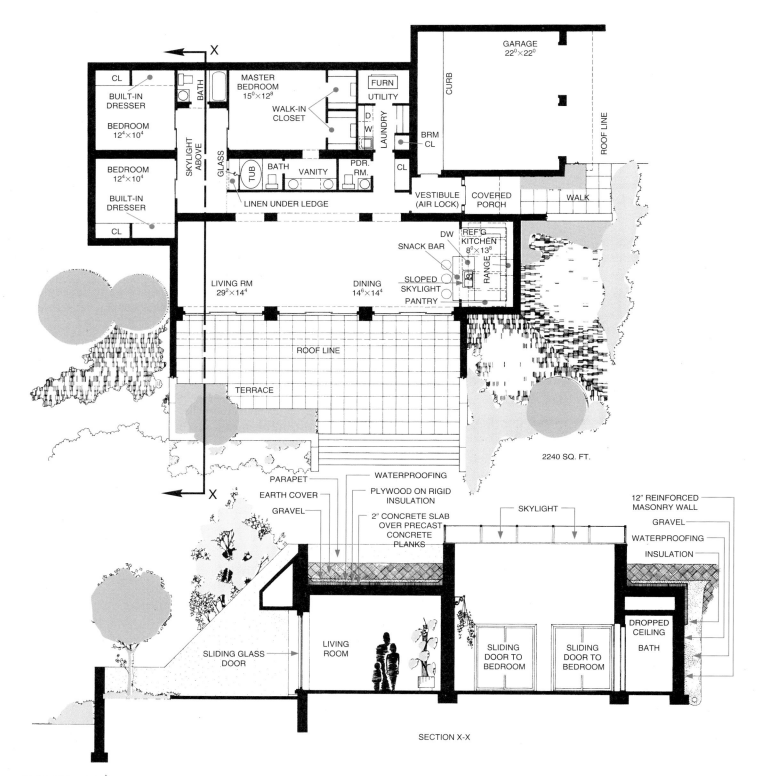

FIGURE 6.16 > Floor plan and section of an earth-sheltered home. *Hepler/Wallach/Hepler © Cengage 2013*

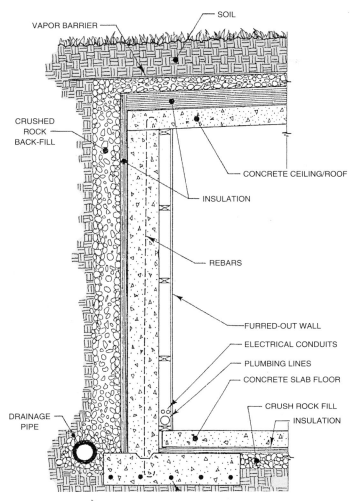

FIGURE 6.17 > Earth sheltered wall section. *Hepler/Wallach/ Hepler © Cengage 2013*

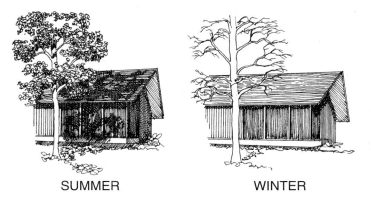

SUMMER   WINTER

FIGURE 6.19 > Deciduous trees help maximize summer cooling and winter solar heating. *Hepler/ Wallach/Hepler © Cengage 2013*

Wind Control

One of the functions of effective orientation is wind control. Although the sun can provide natural energy to a structure, wind can easily diminish the sun's effect. Outdoor living can also be seriously curtailed by excessive wind.

Existing indoor heat can be lost very rapidly when cold air is forced into buildings through minute crevices, usually around doors and windows. Heat also escapes by *windchill* loss through walls, windows, roofs, and foundations. The windchill effect is the loss of internal stored heat. Building orientation, vegetation, and the features of the land can be used to control or minimize the effects of prevailing winds. The direction and velocity of prevailing winds also affect the temperature, ventilation, live loads, and uplift on a structure.

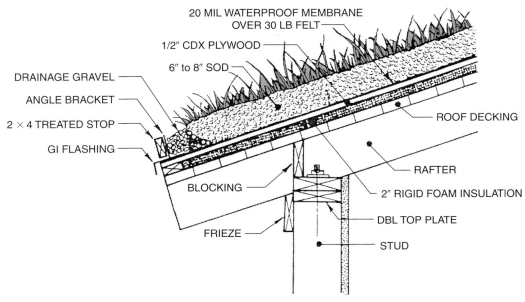

FIGURE 6.18 > Sod roof detail. *Hepler/Wallach/Hepler © Cengage 2013*

Wind Patterns

Air movements at a site should be studied during different parts of the day and during different seasons. Once wind patterns are known, buildings should be oriented to take full advantage of (or offer full protection from) the cooling effect of wind directions.

Some wind patterns are relatively common. Desirable summer breezes usually flow from one direction and winter winds from the opposite direction. Also, cool air will always move to replace rising warm air. For this reason, air above a body of water usually moves toward land during the day and from land toward water at night. See Figure 6.20. The wind pattern on southern sloping sites is generally a movement up the slope during the day and down the slope at night. See Figure 6.21. Strong prevailing winds can change these movements, however.

🌐 Protective Measures

Gentle breezes are usually desirable, but harsh winds are not. Protection from wind can be provided by locating buildings in sheltered valleys or opposite the windward side of hills. Wind velocity differences are often created by the positioning of trees and lower vegetation. Existing wooded areas can have a baffling effect on wind. See Figure 6.22. If no wooded areas exist, or if vegetation is young or not available, construction baffles such as fences or walls may be necessary. Figure 6.23 illustrates several methods of baffling and the effect on wind patterns.

A building can be oriented at an angle to present a narrow side or angle to the wind and avoid direct, right-angle wind impact. This effect is shown in Figure 6.24. Low roof angles can also help deflect wind over a structure.

Wind Effects

In planning urban buildings or isolated clusters of buildings, care must be taken to avoid the creation of turbulent wind eddies. This effect is caused by high-velocity winds striking the upper floors of high-rise buildings and being forced downward and back against lower buildings, creating turbulence on the surface. See

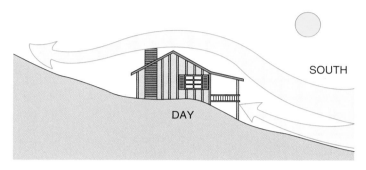

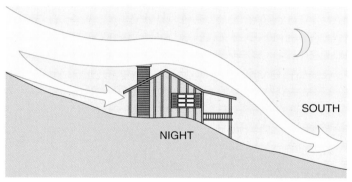

FIGURE 6.21 > Air movement on a southern sloping site. *Hepler/Wallach/Hepler © Cengage 2013*

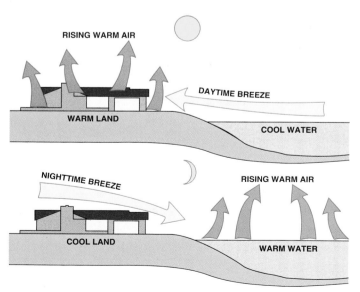

FIGURE 6.20 > Effect of large bodies of water on air movement. *Hepler/Wallach/Hepler © Cengage 2013*

FIGURE 6.22 > Wooded areas help baffle wind. *Energy Focus, Inc.*

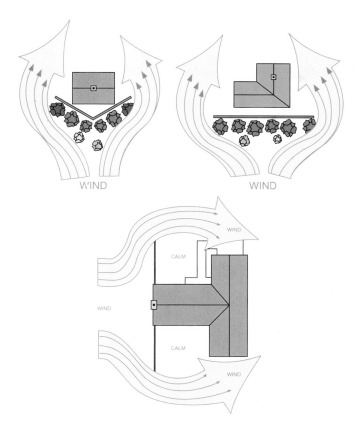

FIGURE 6.23 > Walls and fences are used to deflect winds.
Hepler/Wallach/Hepler © Cengage 2013

NARROW SIDE OF HOUSE EXPOSED TO WIND
—LIMIT WINDOWS ON WINDWARD SIDE

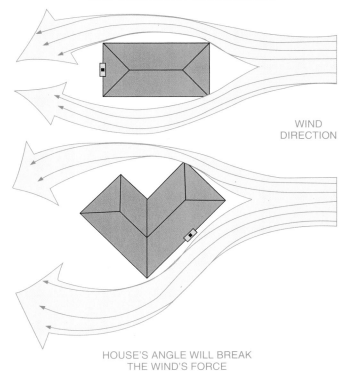

WIND
DIRECTION

HOUSE'S ANGLE WILL BREAK
THE WIND'S FORCE

FIGURE 6.24 > Use of building position to reduce wind effect.
Hepler/Wallach/Hepler © Cengage 2013

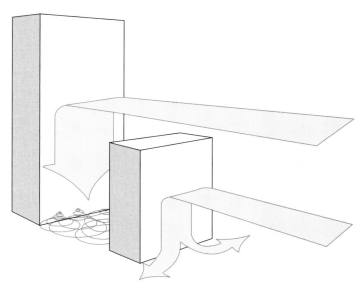

FIGURE 6.25 > Structure size and position affect wind
patterns. *Hepler/Wallach/Hepler © Cengage 2013*

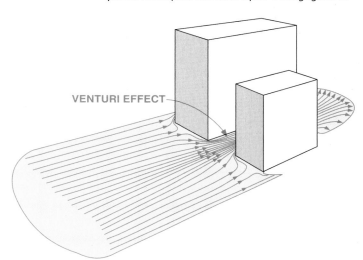

FIGURE 6.26 > Venturi effect created by building positions.
Hepler/Wallach/Hepler © Cengage 2013

Figure 6.25. This effect could also be caused on a smaller scale by winds trapped in courtyards and patios.

In urban situations, the *venturi effect* also complicates wind control. The venturi (wind-tunnel) effect is created as large amounts of moving air are forced into narrow openings. The reduced area through which the wind must pass creates a partial vacuum, and the air picks up speed as it is pulled through the opening. See Figure 6.26. The venturi effect can be partially controlled by avoiding the alignment of streets or buildings with the direction of prevailing winds.

ERGONOMIC PLANNING

Buildings are for people. Therefore, buildings must be ergonomically (biotechnically) planned. **Ergonomics** is a science that deals with designing and arranging things that people use. In architectural design, this means the design must match the size, shape, reach, and mobility of all residents.

Human Dimensions

Human dimensions are especially critical in planning the size and position of cabinets, shelves, and work counters. See Figure 6.27. Traffic areas, door openings, and windows are all based on human dimensions. When buildings (such as schools) are designed primarily for children, obviously the scale must be adjusted.

Ergonomic planning also applies to persons with physical disabilities, such as those with hearing, speech, visual, or mobility impairments. For example, door openings need to be wider and countertops lower for convenient use by people in wheelchairs. (Detailed design requirements for persons with disabilities are provided in Chapter 14.)

Safety Factors

Safety factors in design cover a vast array of concerns for any person who occupies a building, whether it be a home or a public or commercial structure. The designer must ensure that no design feature creates a health or safety risk, such as hidden steps or low headroom clearances. Air for an environmentally safe building should be electronically filtered to provide for the elimination of harmful pollutants. Safety precautions include specifying appropriate mechanical equipment, such as gas furnaces, electrical wiring, devices, and machinery. Hazardous materials (such as asbestos) and accident-causing materials, such as extra smooth floors, thin glass, or unstable ceiling coverings, must be avoided.

Architectural design also involves the safe arrangement of outdoor traffic areas. Adequate vehicular turning angles, fire lanes, and exit signs are a few examples. Of course, safety in design also implies that a building will be structurally sound and will adequately support all anticipated weight.

The increasing use of technology is creating another concern. Before the design process is begun, an assessment must be made to determine what special technology needs are anticipated. Automated buildings (called "smart buildings") now contain such built-in electronic features and accommodations as TV cables, high- and low-voltage circuits, and computer equipment. Magnetic or radio-wave interferences may need to be blocked to enable sensitive electronic equipment to operate effectively. Distance between persons and electromagnetic forces should be considered for safety.

⊕ ECOLOGY AND THE ENVIRONMENT

In the fields of architecture, engineering, and construction, the term *environment* is used to describe the physical conditions that influence the life of organisms. Ecology is the

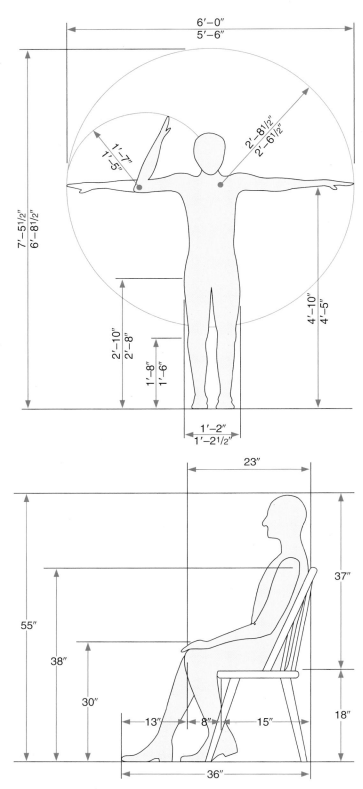

FIGURE 6.27 > Average human dimensions. *Hepler/Wallach/ Hepler © Cengage 2013*

science that describes the relationship between all living organisms and their environment. Organisms include all plants and animals, including people. In the past 100 years, the number of people on the earth has more than tripled. This population increase, combined with the shift from an agrarian (farming) society to an industrial one, has led to the creation of previously unknown environmental problems. The ever-increasing material needs of our technological economy have created enormous pollution problems that must be solved if humanity is to survive. Designers must plan in ways that eliminate or reduce pollutants.

A prime requirement in the creation of every design is to preserve our supply of clean air, pure water, and fertile land. The contemporary architect or designer must be sure that structures do not interfere with natural ecological balances. Good **ecological planning** means protecting or improving the environment without sacrificing the qualities of good design. It requires knowledge of the problems of pollution and possible solutions.

Natural forces continually work to maintain a balance between all elements of the environment. This minimizes the dominance of one element over another. Likewise, responsible architectural designers and planners attempt to protect the environment from damage caused by irresponsible building and site desecration. Because society's need for buildings and site development for homes, stores, offices, recreation, and transportation facilities is constant, construction activities cannot be stopped. Responsible planners therefore strive to achieve a workable balance between new building needs and environmental preservation. *Smart growth* practices, described in Chapter 1, are used to apply this effort to community planning. Building and site design activities, as covered in the remainder of this text, show how architectural design drawings and documents are used to protect the environment.

🌐 Land Integrity

Land is polluted by the discharge of solid and liquid wastes on land surfaces. Pollutants originate from industrial, agricultural, and residential waste. When they exist in excessive quantities, pollutants create health hazards, contribute to soil erosion, cause unpleasant odors, and overwork sewage-treatment plants. Land is also degraded by excessive removal of vegetation and by destroying the natural contours of a site.

Achieving solutions to these problems starts during the design process and continues through all construction phases with attention paid to **land integrity.** Design and construction professionals can prevent, correct, or avoid these problems in many ways, such as these:

- Reserve open, green spaces in site development projects.
- Minimize the removal of trees and major vegetation.
- Recycle topsoil removed during the building process.
- Minimize recontouring of natural land slopes, which results in tree root destruction.
- Use engineered landfills for disposing of excess land material.

FIGURE 6.28 > Residence integrated into a wooded site.
Hepler/Wallach/Hepler © Cengage 2013

- Preserve natural wildlife habitats wherever possible.
- Minimize the use of insecticides and pesticides and use natural biodegradable products.
- Erect barriers to prevent soil erosion during the land contouring process.
- Integrate buildings into the site with minimum contour changes as shown in Figure 6.28.
- Maximize the addition of trees and other vegetation in landscaping.
- Create a site drainage plan to reduce excessive runoff or accumulation of water.
- Test to ensure that soil compaction is adequate to support foundation and building weights.

Sanitary land filling is a practice that involves compacting solid waste material into layers (about 10′ thick), and covering it with clean soil. Thorough landfill projects also include the planting of ground cover—trees and shrubs—on the filled area. Such practice restores the land to its original condition (or better), both ecologically and aesthetically. As discussed in Chapter 13, site development plans should be designed to take advantage of the natural features of the land while minimizing the negative environmental characteristics.

🌐 Air Quality

The two types of air pollutants are those that originate outside a building or site and those that originate inside a building or site. Outside pollutants include industrial and vehicular emissions and naturally occurring irritants such as vegetation pollen, wind-borne dust, and mold spores and mildews from ponds, lakes, and streams. Materials that release gases or fine particles into the air are the primary cause of indoor pollution. Inadequate ventilation can increase pollutants to unhealthy levels by not allowing outside air to dilute toxic

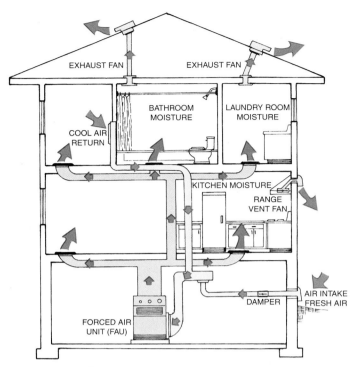

FIGURE 6.29 > Heating and cool-air return system. *Hepler/ Wallach/Hepler © Cengage 2013*

emissions. Because high temperatures and humidity levels increase the concentration of some pollutants, controlling these through design is important. *Sick building syndrome is the term used when several pollutants are present in a building and affect many building occupants at the same time.* The use of a continually operating exhaust system, as shown in Figure 6.29, is the most effective way to ensure that contaminants do not accumulate in a building.

Pollutants found inside a structure include toxic gases such as carbon monoxide and nitrous dioxide derived from unvented gas heaters, leaking chimneys and furnaces, chimney downdrafts, and gas stoves. Another gas, radon, may come from the earth beneath buildings and from unfiltered well water.

Another source of dangerous interior pollutants includes the airborne particles that off-gas from some plywoods, particle boards, foam insulation, textiles, and glues. These sources include formaldehyde, asbestos, wood preservatives, and paints. Biological pollutants—mostly mold and dust mites—can also be created in wet or moist areas and breathing them can lead to serious respiratory aliments.

Building occupants often create their own pollutants by adding tobacco smoke, aerosol spray, and cleaning agent vapors to the interior environment.

Designers and construction managers can reduce or eliminate many air pollution problems through the following actions:

- Specify products that do not include excessive amounts of vapor-producing formaldehyde.

- Thoroughly insulate integral garages from the remainder of the building.

- Specify non–lead-based paints.

- In renovation work, ensure that insulation does not contain asbestos before removal.

- Provide maximum ventilation, air conditioning, and dehumidifiers in areas where moisture and mold may accumulate such as laundries, indoor pools, or spas.

- Specify and indicate the location of all smoke, carbon monoxide, and radon alarms on plan and elevation drawings.

- If radon is present in the area, provide an internal exhaust from beneath the foundation to above the roof.

- If possible, include a mud room and half bath at a service entrance to eliminate the transfer of outside debris and pollutants to the inside.

- Design or specify fireplaces with maximum draw and air supply to eliminate downdrafts.

- Orient the structure to minimize the exposure to any excessive outside pollutants.

- Provide maximum air filtration and air cleaners to collect outside pollutants, especially airborne pollens.

- Design maximum exhaust to areas that may contain secondhand smoke.

- Design adequate hood exhausts over cooktops.

- Adhere to U.S. Environmental Protection Agency (EPA) standards for testing, interoperating, and maintaining emission and material use.

- Vent all furnaces, wood stoves, and gas appliances to the outside and ensure sufficient air supply.

- Thoroughly insulate and ventilate shop areas that may create metal, plastic, or wood dust or fumes.

- Design adequate attic or crawl space ventilation through cornice and roof framing design.

Water Quality

Natural water (H_2O) usually contains minerals, salts, trace metals, bacteria, organic matter, and nutrients. At low levels these elements are harmless, but at high levels they may become toxic pollutants. Once pollutants enter the water supply they are difficult to remove or mitigate. See Figure 6.30.

The quality of water entering and surrounding a building is often compromised by many factors. Water pollution is often caused by the dumping or leakage of sewage, industrial chemicals, and agricultural wastes into oceans, lakes, rivers, and streams. These wastes include pathogens (such as disease-causing bacteria), unstable organic solids, mineral and petroleum compounds, plant nutrients, and insecticides. Water pollution results in the destruction of marine life and

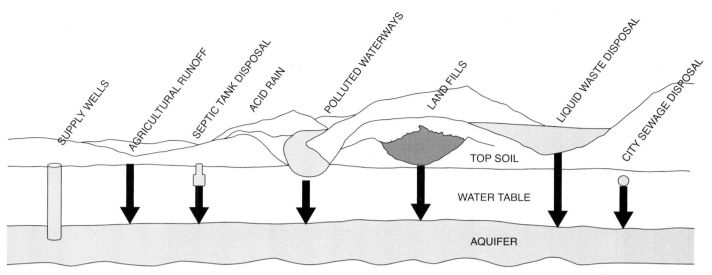

FIGURE 6.30 > Effects of water pollutants on wells. *Hepler/Wallach/Hepler © Cengage 2013*

presents very serious potential health hazards to animal and human life.

The excessive use of insecticides, fertilizers, roadway salts, and pesticides, as well as existing forms of natural bacteria and heavy metals in surrounding aquifers, also contributes to water pollution. Even safe water can become hazardous if surface water accumulation results in flooding or provides a stagnant pool ideal for the growth of mosquito larvae, harmful bacteria, fungi, and molds.

Designers can help maintain good water quality and safety through the following actions:

- Design the site topography by directing water flow away from all structures by general contouring or, if necessary, through the use of swales, trenches, or berms.

- Design gutters and downspouts to exhaust water away from structures.

- Specify drainage tile around the perimeter of foundation walls.

- For large areas install area drains to interrupt massive water runoffs.

- Install the largest possible septic tank and leach field based on the lot size.

- Design gray-water receptacles to diminish the amount of dishwasher and bath water exhaust into septic or municipal sewer systems.

- Locate the septic system at least 50′ or as far away as possible from bodies of water.

- In designing the plumbing system, include water filter devices such as charcoal and ultraviolet filters and water softeners.

- Include a water purification system, such as a reverse osmosis or ultraviolet system, specifically for drinking water.

- In locating structures on a site, avoid wetlands that offer a habitat for wildlife.

- Avoid removal of vegetation from very steep slopes to prevent excessive runoff.

- Minimize the effect of acid rain by specifying covers for all outside water containers such as pools and spas.

- Design the site irrigation system to utilize gray water.

- Specify 1.6-gallon tanks for toilets.

- Locate the water meter in a place that is convenient for reading, but out of obvious sight.

- Locate all structures and facilities above the 100′ flood plane line. See Chapter 13.

- Locate private wells on the site plan according to EPA and local codes.

Visual Perception

Many air, water, and land pollutants, such as unsanitary garbage dumps and smog-producing agents, are not only unhealthy but also visually undesirable. Other sources of visual pollution include junkyards, exposed utility lines, public litter, barren land, and large billboards. These are aesthetically objectionable and should be avoided in the architectural design process. Designers can ensure maximum visual attraction for structures and sites through adhering to the following:

- Follow the basic elements and principles of design as outlined in Chapter 2.

- Design building elevations that are consistent with the architectural character of the community as illustrated in Chapters 1 and 2.

- Develop the site plan according to the guidelines shown in Chapter 13.
- Use as much of the natural terrain and vegetation as possible.
- Specify underground lines for power, phone, and cable delivery. A housing development with underground utility lines is safer and much more appealing than a development with overhead lines.
- Design trash storage facilities to eliminate visible exterior trash.
- Use foliage and structures to visually screen undesirable sights.
- Include a total landscape and planting plan with the original plan set.
- Orient the site and buildings to maximize exposure to attractive vistas.
- Design mailboxes, newspaper receptacles, and similar facilities to visually recede and not distract from the property and the major features of the building design.

Noise Abatement

Unwanted sound is called noise. When sound is extremely offensive it is called noise pollution. The effects of sound depend on the time, location, duration, source, and listener control. Undesirable sounds should be controlled at their source. When this is ineffective the use of surface shapes, absorption materials, or sound insulation construction can be used. Sound requires a path. Sound energy travels in all directions from a source. It weakens through distance or by the interruption of its path with a physical screen. As a general rule, hard surfaces reflect sound; porous (soft) materials absorb sound.

Sound levels are measured in *decibels*. Exposure to excessive decibel levels, over 80, or to even moderate levels, over 60, for long periods creates stress and can result in neurosis, irritability, and hearing loss. See Figure 6.31. Excessive noise can also create hazardous environments by eliminating people's ability to identify and discriminate among sounds, especially those that warn of danger.

Many preventive measures can be achieved through effective architectural design. Some of the **noise abatement** methods designers can use to keep noise at acceptable levels include the following:

- Orient floor plans to locate quiet areas away from outside noise sources.
- Use foliage to provide sound barriers and buffers.
- Add insulation and thickness to interior walls that have noise sources.
- Maximize the use of sound-absorbing materials such as carpets, drapes, upholstery, and tapestries.
- Specify double or triple glazed windows.

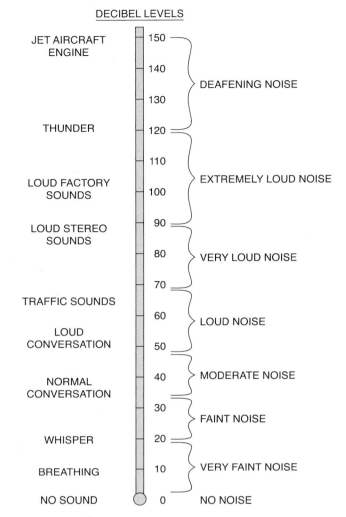

FIGURE 6.31 > Decibel levels. *Hepler/Wallach/Hepler © Cengage 2013*

- Specify building materials that absorb sound, not reflect sound.
- Specify shock-absorbing footings under vibrating stationary equipment and appliances.
- Specify appliances that contain maximum insulation, sound barriers, and quiet operating motors.
- Avoid designing parallel hard surfaces in small confined areas, which can cause sound waves to reflect and result in echoes or flutters.
- Use concave (outward curved) surfaces to direct sound waves to specific locations.
- Use convex (inward curved) surfaces to diffuse sound waves.
- Specify fibrous sound-absorbing construction materials, such as acoustical ceiling tiles, or build sound-absorbing wall coverings to deaden sound.
- Design sound insulation construction to maximize the sound loss and heat transfer through floors, walls, and ceilings. See Figure 6.32.

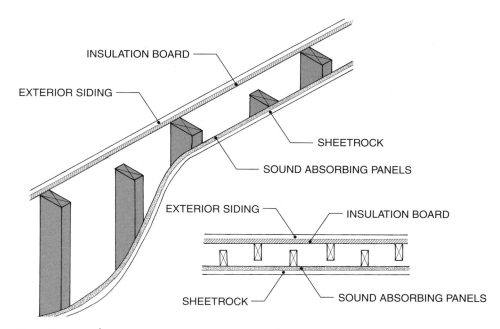

FIGURE 6.32 > Sound isolation wall construction. *Hepler/Wallach/Hepler © Cengage 2013*

- Design insulation around sound-producing mechanical equipment.
- Specify glass-fiber insulation in HVAC ducts.
- Specify expansion valves and piping loops in the plumbing system to reduce pipe vibrations.

Fire Protection

All materials (solids, liquids, or gases) will burn, explode, or become seriously damaged if exposed to a sufficiently high heat source, open flame, or spark. Elements with the greatest combustibility include solids such as wood and wood products, liquids such as gasoline and kerosene, and gases such as propane and natural gas. High heat, flame, or spark sources include heating appliances, furnaces, fireplaces, stoves, electrical overloads, tools and machinery, gas-operated vehicles and unvented areas, which can cause spontaneous combustion.

Architectural designers must create building designs that separate heat, flame, and spark sources from flammable materials. In addition building features must be planned to control the spread of a fire, increase the time an area can withstand combustion, and plan egress (exit) patterns for all areas. These protections can be accomplished through the following design measures:

- Prepare all electrical plans to conform to Underwriters Laboratories (UL) standards.
- Plan for hard-wired installation of smoke and gas detectors.
- Where codes permit, design for underground service drops and site wiring.
- Adhere to all local electrical and gas line codes.
- Include vent plans for all furnace and heater locations.

- Include gas shut-off valve and switch location on floor plan.
- Specify only UL-approved appliances, furnaces, and built-in electrical devices.
- Ensure that all fireplace designs have an ample source of oxygen and chimney draw (vacuum).
- Design walls with fire stop members to reduce air flow through walls. Insulating air should not move.
- Select fire-resistant material for walls behind cook tops and wood stoves or space heaters.
- Specify fire-resistant material for working surfaces in the kitchen, laundry, and shop areas.
- Where possible, building designs should include fire-resistant material such as concrete and chemically treated wood.
- Plan for vents in confined areas that could cause spontaneous combustion.
- Ensure that gas pipe joint locations are not confined, which could cause leaks when surrounding materials expand.
- Define specific fixed locations for fire extinguishers to be accessible but not obtrusive.
- For high roof pinnacles (peaks) add a lightning rod to divert lightning strikes to the ground.
- Design a fire-resistant and vented container area for the storage of flammable materials such as paints and cleaning supplies.
- Design fire-rated walls, ceilings, and doors for fire-prone areas such as shops and garages. A fire code resistance rating is based on the time a surface can withstand high temperatures without igniting or disintegrating. The rating time for walls is from 1/2 to 4 hours; for floors and ceiling, from 1 to 4 hours; and for doors, 3/4 to 3 hours.

Electrical and Electronic Hazards

In addition to the danger of electrical shock from faulty electrical products or connections, continuous exposure to excessive amounts of electromagnetic force (EMFs) is considered unsafe. Living in proximity to high-power voltage lines, faulty microwave appliances, and concentrations of television cables and wiring may pose a threat to occupants' health. Designers must therefore practice the following safeguards:

- Notes in the electrical plan should prohibit the use of wiring harness concentrations that could create electronic reception interference.
- Specify lighting switch locations that provide convenient access in a dark room. See Chapter 31.
- Provide adequate electrical circuits and outlets to prevent overloading and to eliminate the future need for long and potentially hazardous extension cords. See list in Chapter 31.
- Locate GFCI (ground fault circuit interrupter) receptacles at every location potentially exposed to water.
- Avoid siting a residence under or near high-voltage power lines.
- Design shields between any potential source of radiation and people.
- If an aboveground service drop is to be used, dimension the minimum height, according to code, on an elevation drawing.
- Design the electrical service distribution to provide adequate wattage for current and future loads.

🌐 LEED: Leadership in Energy and Environmental Design

The energy and environmental design concepts presented in this chapter and in other locations in this text support the goals of **LEED.** The U.S. Green Building Council created the LEED standards to encourage architects, engineers, and construction firms to produce more cost-effective, energy-efficient, environmentally friendly, and sustainable designs. LEED standards cover all aspects of architecture: building materials, components, and design. The "green" symbol throughout this text indicates the application of LEED coverage.

LEED accredited design professionals have demonstrated their knowledge and application of LEED environmental standards.

LEED provides third-party verification that a building was designed and built using strategies aimed at improving performance in the areas of energy savings, water efficiency, CO_2 emissions reduction, improved indoor environmental quality, and stewardship of resources and sensitivity to their impacts.

Figure 6.33 shows an elevation drawing of a design which maximizes the application of LEEDs concepts and practices. The leaders on this drawing indicate specific LEEDs features. These include:

- The use of shade trees over paved areas, light colored walkways, and roof systems with highly reflective qualities, assist in reducing the heat island effect that can disturb microclimates and wildlife habitats.
- The building was designed to improve energy efficiency, and through the use of computer modeling, the facility is expected to achieve greater energy and cost savings per year when compared to current energy code requirements.
- The heating and cooling equipment installed in the building minimizes or eliminates the emission of compounds that contribute to ozone depletion.
- The roof system uses a combination of spray foam insulation and lightweight insulating concrete that provides a consistent high insulation value that reduce the energy needed for cooling and heating the interior.
- The exterior walls are constructed of insulated concrete forms (ICF) filled with concrete and have a higher insulation value and greater thermal mass, as compared to conventional exterior wall assemblies.

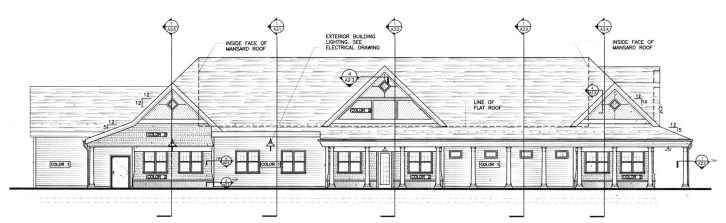

FIGURE 6.33 > LEED features identified in this elevation drawing; United States Post Office at Southampton, NY.
WFC Architects, Paul K. Gartelmann, AIA LEEDS AP

- Many of the materials used in the building are made from recycled content and are environmentally compatible, such as the steel super structure and exterior wall assembly, interior materials and finishes and the HVAC system that minimizes the direct impact on ozone depletion.

- The plumbing fixtures are low-flush or reduced-flow, contributing to substantial water savings per year.

- The light fixtures are high-efficiency and the intensity of light is adjustable to reduce overall electrical consumption; timers and motion sensors throughout the building automatically turn off lights when specific areas become unoccupied.

- During construction, procedures were used to significantly reduce contaminant sources and air-born pollutions inside the building, through the use of walk-off mats at entrances, carbon dioxide monitoring, improved indoor air filtering, and low emitting building materials and finishes.

- The site lighting design reduces sky glow light pollution, light spillover onto adjacent properties, wattage per square foot and overall energy consumption.

- The site is designed to be bicycle and pedestrian accessible and has preferred parking for low-emitting and fuel efficient vehicles.

- A large percentage of the site has been preserved as vegetated space which promotes biodiversity and habitat for natural vegetation and wildlife through the planning of native and adaptive materials which minimize the need for an irrigation system.

- All of the rainwater from the building and parking areas is recharged into the groundwater supply by using vegetated swales and surface retention areas, in lieu of using traditional underground storm water structures which minimizes the disturbance of the site with excessive excavation.

- During building construction, a waste management plan diverted over 50% of the debris from entering landfills.

Environmental Design Factors Exercises   CHAPTER 6

1. Explain why certain side(s) of a house receive the most light and heat. Compare winter and summer changes.
2. Draw a rectangle to represent a floor plan of a house. Label each side N, E, S, and W. Then indicate where you would orient and place each room.
3. Draw the outline of an earth-sheltered home.
4. Sketch a 75′ × 110′ property. Label one side as a hill with a 10° slope upward. Sketch a house on the property in the most desirable location.
5. Tell how you would use solar planning, wind control, and vegetation for a residence of your own design.
6. Find magazine and newspaper articles or advertisements about products for buildings that are designed to control pollution.
7. Identify buildings that emit pollutants into the air, water, or land. List ways of correcting these conditions.
8. List the ecological factors to be considered in planning a residence of your own design.
9. Make a bar graph showing the decibel levels of deafening and very loud noises, pleasant sounds, and quiet.
10. List the major architectural design related goals of LEED.
11. Identify the LEEDs features shown in Figure 6.33 and 6.34.

FIGURE 6.34 > United States post Office at Southampton, NY. *WFC Architects, Paul K. Gartelmann, AIA LEEDS AP*

Indoor Living Areas

OBJECTIVES

In this chapter you will learn:

> to identify the functions of indoor living areas.

> to design the location, decor, size, and shape of indoor living areas.

> how a room's orientation, walls, floors, windows, ceilings, lighting, and furniture can contribute to room function and appearance.

> to design indoor living areas and work them into a convenient floor plan.

TERMS

closed plan
decor
egress
entertainment room
family room
furniture templates

great room/gathering room
home office
home theater
living area
media room
open plan

parlor
partition
recreation room
serving walls
studio
study

INTRODUCTION

Your first impression of a home is probably the image you retain of the *living area*. In fact, this is the only indoor area of the home that most guests observe. The **living area** is where the family entertains, relaxes, dines, listens to music, watches television, enjoys hobbies, and participates in other recreational activities.

LIVING AREA PLANS

In most two-story dwellings, the living area is normally located on the first floor and is adjacent to the foyer or entrance. However, in split-level homes or one-story homes with basements, part of the living area may be located on the lower level.

The total living area is divided into rooms or smaller areas that serve specific purposes. These subdivisions may include the living room, dining room, family room, great room, recreation room, and special-purpose rooms such as a den, home office, media room, entertainment room, or studio. In some homes, particularly smaller ones, rooms may serve two or more functions. For example, the living room and dining room are often combined. In other homes, the entire living area may be one room.

The subdivisions of a living area are called rooms even though they are not always separated by a **partition** or a wall. The subdivision areas perform the function of a room, whether there is a complete separation, a partial separation, or no separation.

When partitions do not totally divide the rooms of an area, the arrangement is called an **open plan** or informal plan. When rooms are completely separated by partitions and doors, the plan is known as a **closed plan** or formal plan.

Open Plan

In an open plan living area, the living room, dining room, and entrance may be part of one open area. Instead of walls, separation of the living room from other rooms is accomplished in different ways. Figure 7.1 shows an example of the open plan living area in Frank Lloyd Wright's "Fallingwater."

Placement of area rugs or furniture will not completely separate areas of a room, but will create a functional and visual separation. If occasional privacy is desired in an open plan, folding accordion doors can be installed to close off part of the area.

An open plan effect can be designed with glass walls to separate functions. See Figure 7.2. The glass walls separate

FIGURE 7.1 > Open plan living area. *Christopher Little, Courtesy of Western Pennsylvania Conservancy*

FIGURE 7.2 > Glass walls create an open plan effect. *Chief Architect Software*

FIGURE 7.3 > Two-sided fireplace separating a living room and dining room. *James Eismont, Photographer*

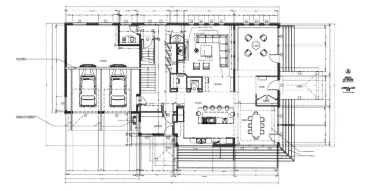

FIGURE 7.4 > Use of openings and levels to separate room functions. *Chief Architect Software*

FIGURE 7.5 > A closed plan living room. *John Berenson Interior Design Inc.*

the living room from the outdoor patio and yet maintain the open view and allow light to come into the room.

In the open plan shown in Figure 7.3, a two-sided fireplace with a two-story chimney is used to separate the living room and dining room. Figure 7.4 shows an open floor plan with openings and levels used to separate the living room, dining room, family room, and foyer.

Closed Plan

In a closed plan, rooms are completely separated from the other rooms by means of walls. See Figure 7.5. Access is through doors, arches, or relatively small openings in partitions. Closed plans are found most frequently in traditional or period-type homes.

Combined Plans

Some large contemporary living area plans include both closed and open rooms. In combined plans there is a closed,

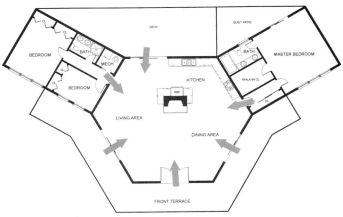

FIGURE 7.6 > Completely open great room. *Used by permission Hanley Wood, LLC*

FIGURE 7.7 > Centrally located living room. *© used under license from Shutterstock, 2011/Jorge Salcedo*

formal living room and an open plan living area that functions as an all-purpose *great room*. (Great rooms are discussed later in this chapter.) The living area shown in Figure 7.6 has no separations between the living, dining, and kitchen areas although the kitchen is not directly visible from the living room.

LIVING ROOMS

The living room is the formal center of the living area in most homes. Thus its functions, location, decor, size, and shape are extremely important and affect the design and appearance of the entire residence. The popularity of formal living rooms in a closed plan has been diminishing in recent years. This was partially caused by the increased inclusion of great rooms in open plans.

Function

A designer begins a living area design by first determining the functions of rooms required in terms of the residents' needs. A living room is the key room in the design because it can serve many purposes. For example, it can be an entertainment center, a library, a social room, and perhaps a dining center. Its particular functions depend on the living habits of the residents. A living room needs to be designed in relation to its functions and activities.

Sometimes a living room can be designed by the process of elimination. For example, if a separate recreation room is planned, then planning for a TV in the living room might not be necessary. If a separate den or study is provided for reading and for storing large numbers of books, then this function might also be eliminated from the living room. On the other hand, if many living area activities are to be combined in one room, then a great room may be designed. Regardless of the exact, specific activities anticipated, the living room should be planned as an integral part of the home for family and guests.

Location

Ideally, the living room should be centrally located and adjacent to the outside entrance. See Figure 7.7. In smaller residences, the entrance may open directly into the living room. Whenever possible, however, this arrangement should be avoided. The living room should not be the only "traffic lane" to the sleeping and service areas of the house. Guests could be disturbed.

Orientation

Careful consideration should be given to the placement of the living room in relation to its surroundings, including other rooms. The living room should be oriented to take full advantage of the position of the sun and the most attractive view. Because the living room is used primarily in the afternoon and evening, it should be located to receive the afternoon sun in the southwest sky. See Figure 7.8. The window patterns in Figure 7.9 are designed to admit the maximum amount of light and also provide the most desirable view. Figure 7.10 reveals the exterior view of this window assembly.

Decor

The general **decor** (pattern of decoration) of the living room depends primarily on the tastes, habits, and personalities of the residents. If their tastes are contemporary, then the wall, ceiling, window, and floor treatments should be consistent with the clean, smooth lines often found in contemporary architecture and contemporary furniture. If the residents prefer colonial or another style of architecture, then this style should be the theme of the decor.

FIGURE 7.8 > A sun- and view-oriented living area. *Eagle Window & Door Inc.*

FIGURE 7.9 > Window assembly designed to maximize natural light. *Eagle Window & Door Inc.*

For example, the decor of the living room in Figure 7.11 is consistently traditional; the living area in Figure 7.12 is distinctly contemporary with hard reflective surfaces. Both are totally different examples of the effective use of the principles and elements of good design presented in Chapter 2.

FIGURE 7.10 > Exterior view of Figure 7.9 window assembly. *Eagle Window & Door Inc.*

Appropriate color choices, effective lighting techniques, and the tasteful selection of materials for walls, floor covering, and ceiling can make a room appear inviting. The selection and placement of well-designed furniture can contribute to the appearance of comfort in a living room. The use of mirrors, floor-to-ceiling drapes, and arrangements of furniture can create a spacious effect in a relatively small room. The interior design of a room, like selecting clothing, should minimize faults and emphasize good points.

Walls

The appearance of walls depends on more than wall coverings. The design and placement of doors, windows, fireplaces, chimneys, or built-in furniture along the walls of the living room will influence the entire room's appearance and should be designed as integral parts of the room. A designer considers these features in conjunction with the kind of wall-covering materials selected. Many materials are available, including plaster, gypsum, wallboard, wood paneling, brick, stone, tile, plastics, paper, and glass.

Windows

Just as the placement of a window in a living room wall should become an integral part of a wall, the view from the

FIGURE 7.11 > An example of excellence in traditional design and decor. *Marc-Michaels Interior Design, Inc.*

FIGURE 7.12 > An example of excellence in contemporary design and decor. *Chief Architect Software*

window should become part of the living room decor. Notice how the window shapes in Figure 7.13 conform to the ceiling line and also provide an uninterrupted transition to the outside through the sliding glass doors. Also notice how the window mullion columns align with the domed roof beams in Figure 7.14. The exterior of this window design (Figure 7.15) shows how this pattern is extended to include the aligned overhang supports. When planning windows, also consider the various seasonal changes in landscape features.

FIGURE 7.13 > Effective relationship between ceiling lines and window shapes. *Photo courtesy Lindal Cedar Homes. Lindal.com*

FIGURE 7.14 > Integration of window mullions with roof beams. *Eagle Window & Door Inc.*

Although windows themselves can be decorative items, the primary function of a window is to admit light ventilation and to provide a pleasant view of the landscape. Under some conditions, however, only the admission of light is desirable. If the view from the window is unpleasant or is restricted by other buildings, translucent glass can be used to allow the natural light to enter without showing the unwanted view.

Fireplaces

The primary function of a fireplace is to provide heat, but it can also be used as a room partition or as a major decorative feature. A fireplace and the chimney masonry may cover an entire wall and can become the focal point of the living area, as shown in Figure 7.16. A massive freestanding fireplace, however, also can function as a partition between rooms in an open floor plan. Like all elements of a room's decor, a fireplace should correlate with the architectural style of the room.

Floors

The living room floor should reinforce and blend with the color scheme, textures, and overall style of the living room.

FIGURE 7.15 > Exterior view of window patterns shown in Figure 7.14. *Eagle Window & Door Inc.*

Exposed hardwood flooring, room-size carpeting, wall-to-wall carpeting, throw rugs, and sometimes polished flagstone and tile are appropriate for living room use.

FIGURE 7.16 > Fireplace and chimney used as a major focal point in living area design. *Photo courtesy Lindal Cedar Homes. Lindal.com*

Ceilings

Most conventional ceilings are flat surfaces covered with plaster or gypsum board. However, new building materials, such as laminated beams and arches, and new construction methods have resulted in greater varieties and improvements in ceiling design. Higher ceilings allow for better air circulation and warm air exhaustion. They also create a feeling of spaciousness that a low ceiling over the same amount of floor space does not. Crown molding also adds a decorative design element to a ceiling. See Chapter 16 for ceiling design details.

Lighting

Appropriate lighting is essential to a room's atmosphere and comfort. Living room lighting generally comprises three types of lighting arrangements: general lighting, local lighting, and decorative lighting. See Chapter 31 for lighting details.

General lighting refers to illuminating the entire room and is often accomplished through the use of ceiling fixtures, wall spotlights, and cove lighting. *Local lighting* is light for a specific purpose, such as reading, drawing, or sewing. Local lighting may be provided by table lamps, wall lamps, pole lamps, or floor lamps. *Decorative lighting* is used to improve the appearance of a room, create a mood, or to enhance a particularly attractive feature in the room.

Furniture

Furniture for the living room should reflect the motif (theme) and architectural style of the home. Whether freestanding or built-in, furniture should maintain lines consistent with

FIGURE 7.17 > Contemporary furniture located adjacent to a *growing green glass wall. Chris Polanski Designs in conjunction with K. Curran Designs*

the entire wall treatment and blend functionally into the total decor of the room. For example, Colonial-style furniture is chosen to be consistent with Colonial architecture. Likewise, contemporary interiors are complemented with contemporary furniture to maintain a consistent design theme. Figure 7.17 illustrates this concept with the inclusion of Polanski chairs in a minimalistic contemporary setting. Figure 7.18 shows how furniture locations are planned to allow for different functions.

A living room designed primarily for conversation is often called a *formal* living room (formerly called a **parlor**). This type of living room is usually closed and small and would, of course, require furniture different from an *informal* living room designed for television viewing, dining, and other activities. Houses that have formal living rooms usually also have family rooms with informal furniture.

Size and Shape

Ideally, when designing a living room, the type and amount of living room furniture should be determined *before* the size of the living room is established. One of the most difficult aspects of planning the size and shape of a living room, or any other room, is to provide sufficient wall space for the effective placement of furniture. Continuous wall space is needed for the placement of many kinds of furniture, especially musical equipment, bookcases, chairs, and sofas. The placement of fireplaces, doors, or openings to other rooms should be planned to conserve as much wall space as possible for furniture placement.

Figure 7.19 shows typical ranges of size for living room furniture. In addition to standard sizes, furniture may be custom designed (specially made) for specific locations and shapes. See Figure 7.20.

FIGURE 7.18 > Multifunctional living area. *LaStrada Furniture and Interiors, Inc.*

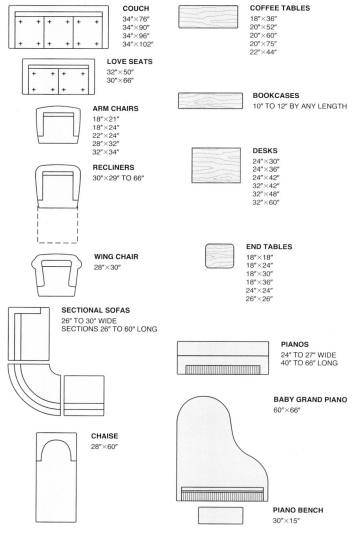

COUCH
34″×76″
34″×90″
34″×96″
34″×102″

LOVE SEATS
32″×50″
30″×66″

ARM CHAIRS
18″×21″
18″×24″
22″×24″
28″×32″
32″×34″

RECLINERS
30″×29″ TO 66″

WING CHAIR
28″×30″

SECTIONAL SOFAS
26″ TO 30″ WIDE
SECTIONS 26″ TO 60″ LONG

CHAISE
28″×60″

COFFEE TABLES
18″×36″
20″×52″
20″×60″
20″×75″
22″×44″

BOOKCASES
10″ TO 12″ BY ANY LENGTH

DESKS
24″×30″
24″×36″
24″×42″
32″×42″
32″×48″
32″×60″

END TABLES
18″×18″
18″×24″
18″×30″
18″×36″
24″×24″
26″×26″

PIANOS
24″ TO 27″ WIDE
40″ TO 66″ LONG

BABY GRAND PIANO
60″×66″

PIANO BENCH
30″×15″

FIGURE 7.19 > Typical sizes of living room furniture. *Hepler/ Wallach/Hepler © Cengage 2013*

FIGURE 7.20 > Furniture designed to match wall and window contours. *Eagle Window & Door Inc.*

To ensure that the room will accommodate the necessary furniture, **furniture templates,** as shown in Chapter 14 and a room plan are made to the same scale. The templates represent the width and length of each piece of furniture. The room plan represents the size and shape of the room or area. By seeing the amount of space occupied by the furniture, a designer can then more effectively evaluate the design of the entire room.

Living rooms vary greatly in size. A room $12' \times 18'$ would be considered a small or minimum-sized living room. A living room of average size is approximately $16' \times 20'$, and a large living room would be $20' \times 26'$ or more.

DINING ROOMS AND AREAS

The design of dining facilities for a residence depends greatly on the dining habits of the family. A separate dining room may be large and formal. An informal dining area may consist of a dining alcove in a living area or even a breakfast nook in the kitchen. Large homes may contain several dining facilities in different areas.

Function and Location

The function of a dining area is to provide a place for the family and guests to eat breakfast, lunch, or dinner, whether in casual or formal situations. When possible, a separate dining area capable of seating from 6 to 12 persons for dinner should be provided in addition to breakfast and lunch facilities.

Dining facilities may be located in many different areas, depending on the residents' needs and preferences. Regardless of the exact position of the dining area, it should be adjacent to the kitchen. The ideal dining location is one that requires few steps from the kitchen to the dining table.

In a closed plan, a separate dining room is usually located between the living room and kitchen, as shown in Figure 7.21. In an open plan, several different dining locations are possible. Open-area dining facilities can be in the kitchen (Figure 7.22) or the living room. However, the preparation of food and other kitchen activities should not be in direct view from the dining area. Figure 7.23 shows a dining area separated from the kitchen with a work-service counter. Figure 7.24 illustrates how locating a dining room adjacent to a glass-walled atrium creates the impression of more space. Figure 7.25 shows the effect of including a dining area integrated into an open plan which includes a kitchen and living area.

To baffle (separate or hide) an area, many design options are possible. For example, folding or sliding doors can separate

FIGURE 7.21 > Effective location of a dining room and family room. *Used by permission Hanley Wood, LLC*

FIGURE 7.22 > Dining area located in a kitchen. *Marvin Windows & Doors*

FIGURE 7.23 > Serving counter separating kitchen and dining area. *Bisult Kitchens, Garden City, NY*

FIGURE 7.25 > Dining area separated by level and distance. *Hamid Rafiei Dreamarch3D.com*

FIGURE 7.24 > Dining room located adjacent to a space-expanding atrium. © *Amit Geron/Arcaid/Corbis*

FIGURE 7.26 > Sliding doors used as optional separators between the dining room and the kitchen. *Hepler/Wallach/Hepler © Cengage 2013*

cabinets are also effective in partially separating the dining and living areas as shown in Figure 7.27.

Some families enjoy dining outdoors on a porch or a patio. If so, the porch, deck, or patio should be near the kitchen and directly accessible to it. Locating the patio or dining porch directly outside the dining room or kitchen provides maximum use of the facilities. This location minimizes the distance from the kitchen and the possible inconveniences of outside dining facilities.

and hide the kitchen from the dining area. See Figure 7.26. A two-sided fireplace offers another option for separating the dining room from other parts of the living area.

When dining facilities are not located in the living room, the dining area should be located adjacent to it. Family and guests normally enter the dining room from the living room and use both rooms jointly.

A partial separation of the dining room and the living room can be accomplished by different floor levels or by dividing the rooms with common half walls. **Serving walls** also provide a functional separation between the dining room and kitchen. Fireplaces, entertainment enclosures, and storage

Decor

The decor of the dining room should blend with the remainder of the house. The floor, walls, and ceiling treatment of the dining area should work well with the decor of the living area.

If a dining porch or a dining patio is used, its decor should also be considered part of the dining room decor.

To create a partially closed dining area, the decor of any kind of partial divider wall should be considered as well as its purpose. A divider may be a planter wall; glass wall; half wall of brick, stone, or wood panels; fireplace; or grillwork.

Controlled lighting is another means to greatly enhance the decor of the dining room. General illumination that can be subdued or intensified can provide the appropriate atmosphere for any occasion. This type of lighting is controlled by a *rheostat,* commonly known as a dimmer switch. In addition to general illumination, local lighting should be provided directly over the dining table. See Figure 7.28.

Size and Shape

The size and shape of the dining area are determined by the size of the family, the size and amount of furniture, and the clearances and traffic areas needed between pieces of furniture. The dining area should be planned for the largest group that will dine in it. There is little advantage in having a dining room table that expands if the room is not large enough to accommodate the expansion. One advantage of the open plan is that the dining facilities can be expanded into the living area.

Dining room furniture may include an expandable table, side chairs, armchairs, buffet, server or serving cart, china closet, and serving bar. In most situations, a rectangular dining room will accommodate the furniture better than a square dining room.

Regardless of the furniture arrangement, a *minimum* space of 2′ (610 mm) should be allowed between a chair and the wall or other furniture when the chair is pulled out. This allowance will permit serving traffic behind chairs, and will allow persons to approach or leave the table without difficulty.

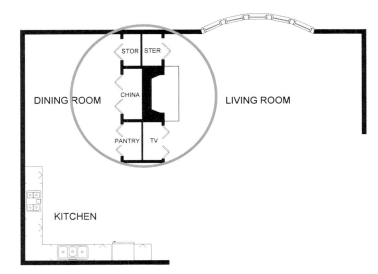

FIGURE 7.27 > Use of fireplace and cabinetry to separate areas. *Used by permission Hanley Wood, LLC*

FIGURE 7.28 > The effective use of local and general lighting in a dining area. *Chief Architect Software*

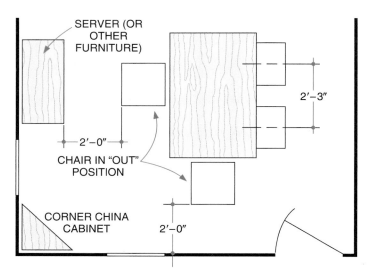

FIGURE 7.29 > Dining area space allowances. *Hepler/ Wallach/Hepler © Cengage 2013*

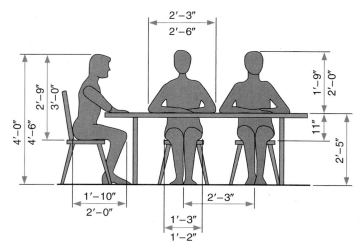

FIGURE 7.30 > Space requirements for dining room seating. *Hepler/Wallach/Hepler © Cengage 2013*

Another space consideration is the distance *between* people when seated at the table. This spacing is accomplished by allowing 27″ (686 mm) from the centerline of one chair to the centerline of another. See Figure 7.29. Figure 7.30 illustrates the average space required for adults when seated for dining. The shapes and sizes of typical dining room furniture are shown in Figure 7.31.

A dining room that would accommodate the minimum amount of furniture—a table, four chairs, and a buffet—would be approximately 10′ × 12′ (3,048 mm × 3,658 mm). A minimum-sized dining room that would accommodate a dining table, six or eight chairs, a buffet, a china closet, and a server would be approximately 12′ × 15′ (3,658 mm × 4,572 mm). A large dining room would be

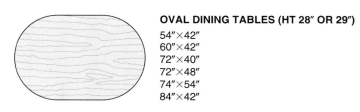

OVAL DINING TABLES (HT 28″ OR 29″)
54″×42″
60″×42″
72″×40″
72″×48″
74″×54″
84″×42″

ROUND TABLES (HT 28″ OR 29″)
DIAM 30″, 32″, 36″, 42″, 48″

RECTANGULAR DINING TABLES

34″×30″	60″×40″
42″×30″	60″×42″
48″×30″	72″×36″
48″×42″	

SERVING CARTS (HT 30″)

24″×16″	52″×18″
34″×20″	64″×16″
36″×16″	

CORNER CABINETS (HT 80″)
34″×34″
36″×36″
38″×38″

BUFFETS (HT 31″)

36″×16″	52″×18″
48″×16″	60″×20″

CHINA CABINETS (HTS 60″ TO 72″)

48″×16″	62″×16″
50″×20″	36″×18″

DINING ROOM CHAIRS (SEAT HT 16″)
(BACK HTS 29″ TO 36″)
17″×19″
20″×17″
22″×19″
24″×21″

FIGURE 7.31 > Typical sizes of dining room furniture. *Hepler/ Wallach/Hepler © Cengage 2013*

14′ × 18′ (4,267 mm × 5,486 mm) or larger. Figure 7.32 shows a dining room layout that was created using furniture templates.

FAMILY ROOMS

Because of the trend toward more informal living, the majority of homes today are designed to include a family room.

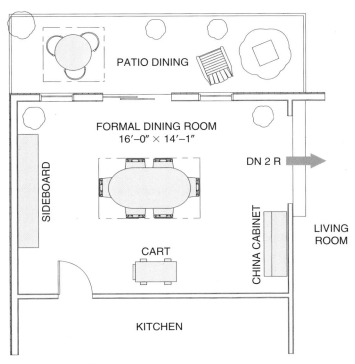

FIGURE 7.32 > Dining room furniture templates. *Hepler/Wallach/Hepler © Cengage 2013*

FIGURE 7.33 > Kitchen extension into a family room.
© Jessie Walker/Lived In Images/Corbis

Function

The purpose of the **family room,** as the name implies, is to provide facilities for family-centered activities. In extremely large residences, special-purpose rooms may be provided for specific types of activities, such as playing music, sewing, or painting. Typically however, additional rooms are not part of the plan, and facilities and equipment must be provided for a wide variety of activities in the family room. For this reason, the family room is also known as the *activities* or *multiactivity room.*

Location

Activities in the family room often result in the accumulation of hobby materials and clutter. Thus, the family room should be easily accessible, but not visible, from the rest of the living area.

Commonly, the family room is located adjacent to the kitchen. This location revives and expands the idea of the old country kitchen as the room in which most family activities were centered. See Figure 7.33.

When the family room is located adjacent to the living room or dining room, it becomes an extension of those rooms for social affairs. In this location, the family room is often separated from the other rooms by folding doors, screens, or sliding doors. Figure 7.34 illustrates a family room with di-

rect access to the kitchen, garage, and lavatory and obliquely to the dining room.

Another popular location for the family room is between the service area and the living area.

Decor

The decor of the family room should provide a vibrant atmosphere. Ease of maintenance should be another chief consideration. Furniture materials such as plastic, leather, and wood are easy to care for and provide great flexibility in color and style. Family room furniture should be informal and suited to all members of the family.

The floor should be resilient—able to keep its original shape or condition despite hard use. Linoleum or tile made of asphalt, rubber, or vinyl will best resist the abuse normally given a family room floor. If rugs are used, they should be the kind that will stand up under rough treatment. They should also be washable.

For walls, soft, easily damaged materials such as wallpaper and gypsum board should be avoided. Materials such as tile and paneling are most functional. Chalkboards, bulletin boards, cupboards, and toy-storage cabinets should be used when appropriate. Work areas that fold into the wall when not in use conserves space. Such areas perform a dual function if the wall cover can also be used as a chalkboard or a bulletin board.

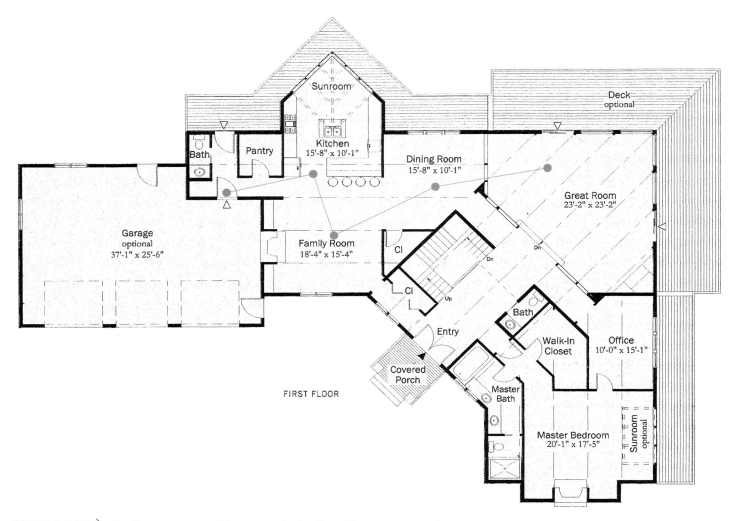

FIGURE 7.34 > Family room located between the kitchen, dining room, and garage. *Photo courtesy Lindal Cedar Homes. Lindal.com*

Because a variety of hobby and game materials will be used in the family room, sufficient storage space must be provided. This includes the use of built-in facilities such as cabinets, closets, and drawer storage.

Acoustical ceilings are recommended. These help keep the noise of the various activities from spreading to other parts of the house. This feature is especially important if the family room is located on a lower level.

Size and Shape

The size and shape of the family room depend on the equipment needed for the planned activities. The room may vary from a minimum-sized room of approximately 150 sq. ft. to a very spacious family room of 300 sq. ft. or more. Most family rooms require a size that ranges between these two extremes. See Figure 7.35.

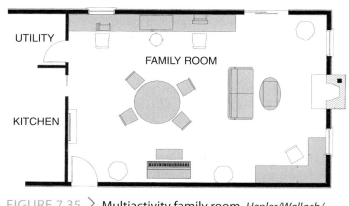

FIGURE 7.35 > Multiactivity family room. *Hepler/Wallach/Hepler © Cengage 2013*

GREAT ROOMS

Rooms that combine the functions of a living room, dining room, family room, and sometimes the kitchen are called **great rooms** or sometimes **gathering rooms.** Great rooms have gained in popularity because of the multiple functions that can be enjoyed at the same time. Conversations among people in different areas can occur without interrupting separate activities. Great rooms are not just large rooms. Great rooms are part of an open plan that integrates elements of the living, dining, and food preparation areas. Notice how the great room in Figure 7.36 includes these activities and how the functions are functional without walls. Figure 7.37 shows a completely open plan great room in which spaces above separate functions and help traffic flow.

SPECIAL-PURPOSE ROOMS

In addition to the traditional living room, dining room, family room, and great room, the living area may also include rooms devoted to specific activities. These rooms are designed for recreation, work, or entertainment, as discussed next.

Recreation Rooms

The **recreation room** may also be called a game room or playroom. As the name implies, it is a room designed specifically for active play, exercise, and recreation.

Recreation rooms may also include facilities for crafts and hobbies that do not require large power tools or equipment. Exercise equipment may also be included if a separate exercise room is not planned.

Function

The function of the recreation room often overlaps that of the family room. Overlapping occurs when a multipurpose room is designed to provide for recreational activities such as table tennis and billiards, but also includes facilities for quieter activities such as knitting, model building, and other hobbies.

The design of the recreation room depends on the number and arrangement of the facilities needed for the various pursuits. Activities for which many recreation rooms are designed include billiards (Figure 7.38), chess, checkers, table tennis, darts, watching television, eating, and dancing. See Figure 7.39.

FIGURE 7.36 > Great room including kitchen exposure. *Used by permission Hanley Wood, LLC*

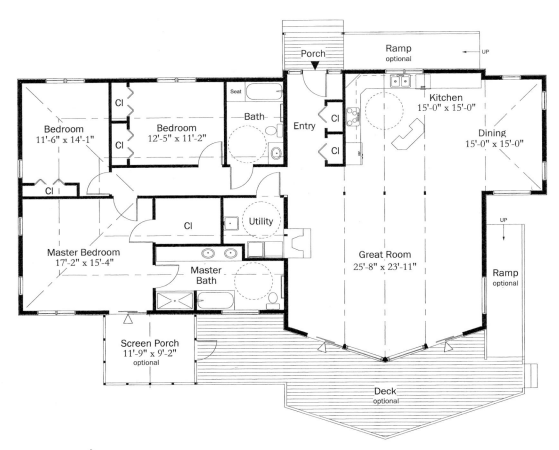

FIGURE 7.37 > **Open plan great room.** *Photo courtesy Lindal Cedar Homes. Lindal.com*

FIGURE 7.38 > **Recreation room featuring billiards and home theater facilities.**
Taylor Architectural Photography

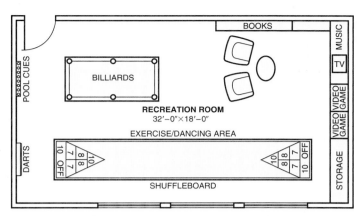

FIGURE 7.39 > Multiactivity recreation room. *Hepler/Wallach/Hepler © Cengage 2013*

Location

The recreation room should be located away from the quiet areas of the house. Most often, it is located in the basement or on the ground level.

A basement location uses space that might otherwise be wasted. Also, basement recreation rooms can often provide more space for activities that require large areas and equipment, such as table tennis, billiards, and shuffleboard. A good ground-level location would allow activities to be expanded onto a patio or terrace.

Decor

Designers take more liberties in decorating the recreation room than with any other room. They do so primarily because of the active, informal atmosphere that characterizes the recreation room. This atmosphere lends itself readily to unconventional furniture, fixtures, and color schemes.

Bright, warm colors can reflect a party mood. Furnishings and accessories can be used to accent a variety of central themes. Regardless of the theme, recreation room furniture should be comfortable and easy to maintain. The same decorating guidelines that apply to the family room also apply to recreation room walls, floors, and ceilings.

Size and Shape

The size and shape of a recreation room may depend on whether the room occupies basement space or an area on the main level. If basement space is used, the only restrictions on the size are the other facilities there, such as the laundry, utility, or workshop areas.

Home Offices

The most common work-related room in a residence is the **home office** (Figure 7.40). This can be a professional office or an office used for personal business. A home office may func-

FIGURE 7.40 > Home office with built-in storage facilities. *© Andrea Rugg Photography/Beateworks/Corbis*

tion as a **study** or **studio** depending on use. Large residences may include both; smaller homes may combine a home office with a family room, great room, or guest bedroom.

Major considerations in planning a home office include furniture size, type, and placement, and lighting, storage, and electronic facilities. Figure 7.41 shows the outline of typical office furniture and equipment. The number and size of each item depends on the type of activity planned. It is also important to consider the location of the office in relation to other rooms including the entry, **egress**, and traffic flow. If the home office is to be used by clients or guests, an outside or foyer access should be planned, as shown in Figure 7.42. If no direct outside access is required, the home office should be located near the entrance to provide client access without passing through other areas, as shown in Figure 7.43.

Electronic facilities for a home office range widely depending on the office function. For example, computer configurations are determined by the requirement of size, speed, drives, speakers, keyboards, mouse, modem, printer, scanner, and wire management needs. Wiring amperage must be adequate to serve all office functions simultaneously. Electrical surge protection is also included because office machines, especially computers and printers, have sensitive circuits that can be damaged by power fluctuations. (See Chapter 31.) Although the office of the future may be wireless, today wires and cables should be channeled in accessible walls or under floors or baseboard units to avoid the dangers of exposed wiring. Fixtures may include plug-ins for phones, fax machines, modem, and satellite, cable TV, Internet, and computer networks.

Although the maximum amount of natural lighting is desirable, general artificial, well-diffused lighting should be designed for nighttime use. Glare-proof local task lighting is also needed at each task station.

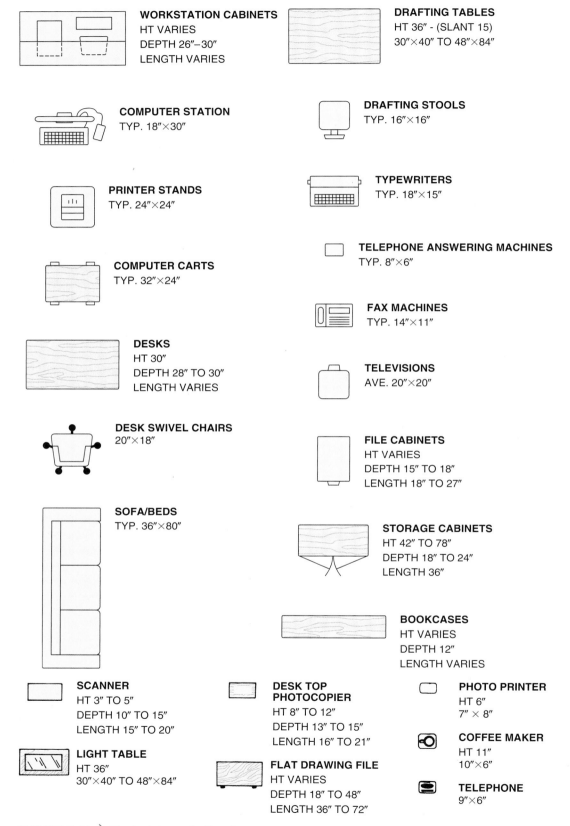

WORKSTATION CABINETS
HT VARIES
DEPTH 26"–30"
LENGTH VARIES

DRAFTING TABLES
HT 36" - (SLANT 15)
30"×40" TO 48"×84"

COMPUTER STATION
TYP. 18"×30"

DRAFTING STOOLS
TYP. 16"×16"

PRINTER STANDS
TYP. 24"×24"

TYPEWRITERS
TYP. 18"×15"

COMPUTER CARTS
TYP. 32"×24"

TELEPHONE ANSWERING MACHINES
TYP. 8"×6"

FAX MACHINES
TYP. 14"×11"

DESKS
HT 30"
DEPTH 28" TO 30"
LENGTH VARIES

TELEVISIONS
AVE. 20"×20"

DESK SWIVEL CHAIRS
20"×18"

FILE CABINETS
HT VARIES
DEPTH 15" TO 18"
LENGTH 18" TO 27"

SOFA/BEDS
TYP. 36"×80"

STORAGE CABINETS
HT 42" TO 78"
DEPTH 18" TO 24"
LENGTH 36"

BOOKCASES
HT VARIES
DEPTH 12"
LENGTH VARIES

SCANNER
HT 3" TO 5"
DEPTH 10" TO 15"
LENGTH 15" TO 20"

DESK TOP PHOTOCOPIER
HT 8" TO 12"
DEPTH 13" TO 15"
LENGTH 16" TO 21"

PHOTO PRINTER
HT 6"
7" × 8"

LIGHT TABLE
HT 36"
30"×40" TO 48"×84"

FLAT DRAWING FILE
HT VARIES
DEPTH 18" TO 48"
LENGTH 36" TO 72"

COFFEE MAKER
HT 11"
10"×6"

TELEPHONE
9"×6"

FIGURE 7.41 > Typical sizes of office furniture and equipment. *Hepler/Wallach/Hepler © Cengage 2013*

Entertainment Rooms

Entertainment rooms are also known as **media rooms** or **home theaters** and include big-screen TVs, VCRs, DVDs, stereo systems, and sometimes a piano or organ. Small residences may include these facilities in a family room or great

room; large residences may have a separate home theater or music room. The music center shown in Figure 7.44 is located in a separate area of a living room. The entertainment unit shown in Figure 7.45 is located on a great room wall with electronic components exposed. The entertainment center in Figure 7.46 is designed as a wall unit in a living room with all components concealed in a bookshelf cabinet design. In designing plans for entertainment rooms or centers, the location of both visual and auditory components is important. This includes the placement of screens, projectors, speaker systems, viewer and/or listener seating, wiring channels, and wiring access panels. Figure 7.47 illustrates the optimum angles and relative sizes and distances to be planned between screen and viewer.

The placement of television and film projection equipment depends on three factors; eye level, viewing distance, and screen size. Screens or monitors should be limited to a 15° viewing cone, vertically from the viewer's eye level. A viewing distance equal to twice the width of the screen, at a 30° side angle and a 15° vertical angle provides the most comfortable viewing conditions. However the viewing distance should be no more than five times the width of the screen at no more than a 30° side angle.

Entertainment or media rooms should be located with direct access to the living area and away from bedrooms or other private areas. Preplanning of all wiring, lighting, acoustics, soundproofing, seating, and built-in cabinetry is extremely important in designing an entertainment room or center. In developing floor plans for homes with few space restrictions, separate home theater rooms are often

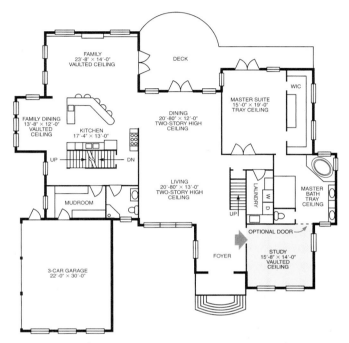

FIGURE 7.42 > Home office study located for convenient outside access. *Scholz Design Inc.*

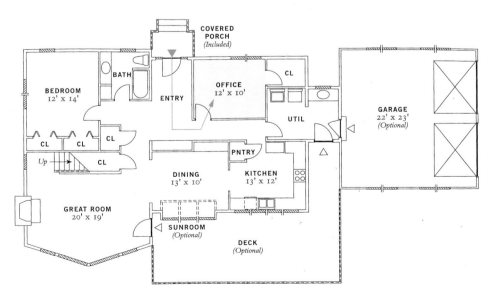

FIGURE 7.43 > Convenient inside access to a home office. *Photo courtesy Lindal Cedar Homes. Lindal.com*

FIGURE 7.44 > Music center located in a living area.
Marc-Michaels, Interior Design, Inc.

FIGURE 7.45 > Entertainment center with exposed components. © *Karen Melvin/Corbis*

included. The advantage of these rooms over comprehensive media rooms is the ability to accommodate an extremely large screen. This enables the viewing of films and videos made for theater and sporting events for a larger number of viewers. Figure 7.48 shows a home theater room with these capacities, and Figure 7.49 is a design concept sketch of this room.

FIGURE 7.46 > Entertainment center with concealed components. *Hamid Rafiei Dreamarch3D.com*

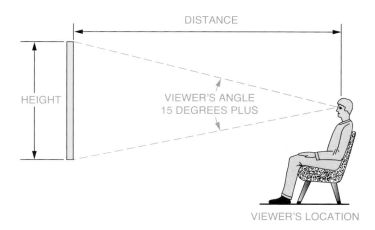

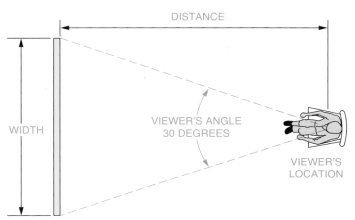

FIGURE 7.47 〉 Screen viewing distances and angles. *Hepler/Wallach/Hepler © Cengage 2013*

FIGURE 7.48 〉 Home theater facilities. *Cinemations Home Theaters*

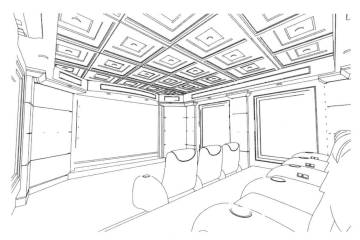

FIGURE 7.49 〉 Home theater concept sketch. *Cinemations Home Theaters*

FIREPLACE DESIGN

Fireplaces can be designed as an aesthetic and functional part of any living or sleeping area.

Because masonry fireplaces and chimneys are exceptionally heavy, they cannot be supported by the normal building footings. Special provisions must be made. The design of the fireplace influences the type of foundation that is needed. In designing fireplaces, the style, type, support, framing, size, materials, components, ratio of the opening flue and firebox, and the height of the chimney must all be considered to ensure adequate draw. Construction details relating to fireplace and foundation design are covered in Chapter 23. Fireplaces are classified by fuel type, type of opening, construction, and architectural style (contemporary, Spanish, colonial, etc.).

Fuel Types

Fireplaces burn either wood, natural gas, or synthetic materials. Each fuel type has advantages and disadvantages. Selec-

tion of the appropriate fuel type depends on the availability of fuel, venting restrictions, and local fire codes.

Wood

Oxygen is the vital ingredient needed for effective wood burning and proper functioning of a fireplace. Because warm air rises, air in the room is drawn into the fireplace, supplying the fire with needed oxygen. Because cold air continually replaces rising warm air, much of the heat produced by many fireplaces goes up the chimney. To reduce this heat loss and redirect some of the heat back inside, warm-air outlets that balance the inlet of cold air are effective. Use of outlets of this type allows heat to reenter the room, while smoke, debris, and toxic fumes are directed outside through the chimney.

Natural Gas

A gas fireplace offers maximum operating convenience and warm-air circulation. Concealed circulating fans keep air moving around the firebox and expel heated air into the room. Gas fireplaces have automatic pilot lights that are quickly turned off or on. Units can also be thermostatically controlled. "Logs" used in gas fireplaces resemble wood in appearance but are made of noncombustible ceramic or masonry materials.

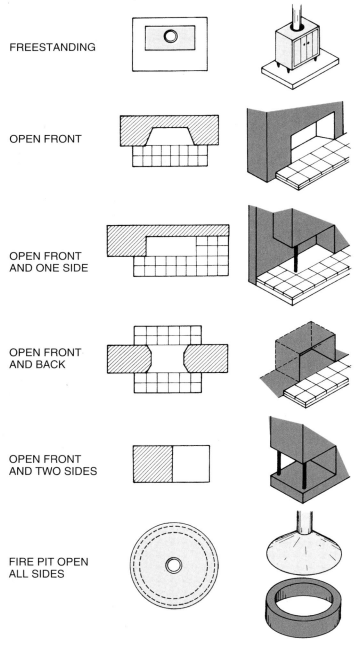

FREESTANDING

OPEN FRONT

OPEN FRONT AND ONE SIDE

OPEN FRONT AND BACK

OPEN FRONT AND TWO SIDES

FIRE PIT OPEN ALL SIDES

FIGURE 7.50 > Types of fireplace openings. *Hepler/Wallach/ Hepler © Cengage 2013*

Requiring neither heavy masonry and foundations nor a front hearth, prefabricated gas fireplaces can rest on any type of flooring with no limitations on enclosure size or trim. Gas fireplaces require no direct vertical flue and may be vented through a wall. Therefore, the area above the firebox need not be a chimney flue.

Synthetic Materials

Some fireplaces are designed to burn synthetic materials, such as gelled alcohol. These fireplaces require no venting to the outside, no hearth, and no special flooring.

Types of Fireplace Openings

Fireplaces are divided into six basic types of openings as shown in Figure 7.50. A flush opening is also called a single face and is shown in Figure 7.16. A two-sided or corner fireplace is used in L-shaped areas as shown in Figure 7.51. A three-sided or peninsula fireplace is used to partially separate two rooms as shown in Figure 7.52. The see-through fireplace allows the fireplace to be viewed from opposite sides. The see-through fireplace in Figure 7.53 is a gas fireplace without a vertically aligned chimney. Four-sided or open-pit fireplaces as sketched in Figure 7.54 require large hoods for venting smoke and fumes.

Freestanding metal or ceramic fireplaces are available in a variety of shapes as shown in Figure 7.55. They are relatively light wood-burning stoves and therefore need no concrete foundation for support. A stovepipe leading into the chimney or through the roof provides the exhaust flue. Because metal units reflect more heat than masonry, metal fireplaces are much more heat efficient, especially if centrally located. For safety, fire-resistant materials such as concrete, brick, stone, or tile must be used beneath and around these fireplaces.

See Chapter 23 for fireplace construction details.

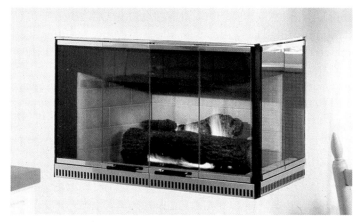

FIGURE 7.51 > Two-sided corner fireplace. *Heatilator, Inc.*

FIGURE 7.52 > Three-sided, peninsula fireplace. *Heatlor, Inc.*

FIGURE 7.53 > See-through gas fireplace. *Heatilator, Inc.*

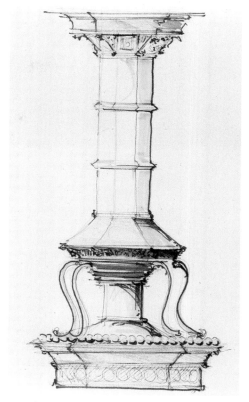

FIGURE 7.54 > Open-pit fireplace. *John Henry, Architect*

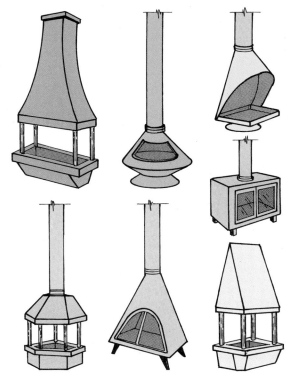

FIGURE 7.55 > Freestanding fireplaces and stoves. *Hepler/ Wallach/Hepler © Cengage 2013*

Indoor Living Areas Exercises   CHAPTER **7**

1. List the functions you want in a living room for yourself or an imaginary client.

2. Draw a simple sketch of an open plan living room for the functions you listed in Exercise 1. Indicate the locations of doors, windows, a fireplace, foyer, entrance, and dining room and label them accordingly.

3. Sketch a closed plan living room for the same functions as in Exercise 1. Show the position of adjacent rooms. Explain the reasons for your choices.

4. Using *Line* commands on a CAD system, draw the living room outline you prefer from those that you designed in Exercises 2 and 3. Indicate doors and windows.

5. Begin a picture file of your own for different rooms. Cut out pictures showing the decor you like in living rooms—including furniture, lights, etc.—from catalogs, newspapers, or magazines. Then list the furniture you would like to have for your design, including color and materials.

6. Sketch the dining area of your own home. Then make another sketch changing the design (open or closed plan) without changing any *outside* walls. (Make any needed templates to help create a different arrangement.)

7. Sketch a dining room to scale showing the position of all furniture you would like to include in the dining room of a house of your own design.

8. Collect pictures of dining areas and furniture and accessories.

9. Draw a floor plan for a dining room shown in this chapter.

10. Calculate the minimum-sized dining room you would need to seat six people at a 60″ × 42″ table.

11. Refer back to Figure 7.35. This plan is drawn to 3/16″ = 1′-0″ scale. Tell what the dimensions of the room are. Evaluate the size and the arrangement in relation to its functions.

12. Sketch a family room, recreation room, special-purpose room, or office you would like to include in a home of your own design. List the activities and furniture needed. Show the location of all furniture and facilities (scale 1/2″ = 1′-0″).

13. Design a family room primarily for children's activities. Include the furniture needed. Describe the colors and materials you would select for this room.

14. Redesign the room in Exercise 13 to accommodate teen activities. Describe the changes you made and explain why.

15. Draw an outline of a family room adjacent to a kitchen.

16. Draw a floor plan for the recreation room shown in Figure 7.38.

17. Use the typical dimensions of the office equipment listed in Figure 7.41 to sketch a plan to accommodate the following facilities: desk, chair, bookcases, drafting table, and lounge.

18. Collect pictures of rooms and furnishings that you would like to use in a house of your own design.

Outdoor Living Areas

OBJECTIVES

In this chapter you will learn to:

> design and sketch a porch, patio, and lanai.

> design and sketch a swimming pool.

> calculate the area and volume of swimming pools.

> plan outdoor recreation facilities.

TERMS

balcony
court
deck

lanai
patio
porch

stoop
swim-out
veranda

INTRODUCTION

A home's living areas may be extended to the outdoors. Porches, patios, decks, pools, and other features provide space for dining, entertaining, playing, exercising, or relaxing. When planning an outdoor living area, consider the area's function, location, decor, size, and shape. A well-designed area will look and function like a natural extension of the interior.

PORCHES

A **porch** is a covered platform leading into an entrance of a building. Porches may be enclosed by glass, screen, or posts and railings. Balconies and decks are actually elevated porches. (A **deck,** however, usually refers to an open, elevated platform.) Similar to a porch, a **stoop** is a projection from a building. However, a stoop does not provide sufficient space for activities. It provides only shelter and an access to or landing surface for the entrance of the building.

Porches are often confused with patios. Although a patio may also be adjacent to a house and seemingly attached to it, a patio is directly on the ground, even if it has a finished surface. The main difference between a porch and a patio is that a porch is attached *structurally* to a house. Figure 8.1 shows an example of a porch attached to a house that pro-

vides both traffic access and functional use. Figure 8.2 shows the floor plan for an extensive wrap-around porch and deck connecting the living with the service area. Figure 8.3 shows a porch that functions as a patio because of the large open walls on all sides.

FIGURE 8.1 > Porch with entry access to living area functions. *Hepler/Wallach/Hepler © Cengage 2013*

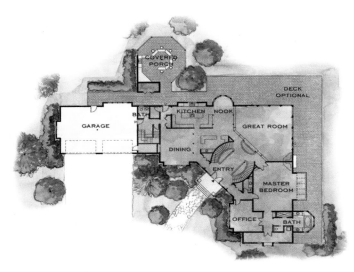

FIGURE 8.2 > Wrap-around porch and deck. *Photo courtesy Lindal Cedar Homes. Lindal.com*

FIGURE 8.4 > Example of a home with a two-level veranda. *Dixie Pacific*

FIGURE 8.3 > Porch with many patio features. *Marc-Michaels Interior Design, Inc.*

Function and Types

The classic front porch and back porch that characterized most homes built in this country during the 1920s and 1930s were built merely as places in which to sit. Little effort was made to use the porch for any other activities. However, Southern Colonial homes were designed with **verandas,** large porches that extended around several sides of the home. See Figure 8.4. Outdoor life often centered on the veranda. The multiple decks shown in Figure 8.5 serve a variety of purposes including dining, relaxing, and regulating traffic flow. These decks are connected at different levels to the living–dining area, entrance, bedrooms, and a wading pool.

A **balcony** is a porch suspended from an upper level of a structure. There is usually no access from the outside. Balconies often extend a living area. See Figure 8.6. Others serve as a private extension of a bedroom. The balcony protects the lower level from the sun and precipitation.

Hillside lots lend themselves to vertical plans and allow maximum flexibility for such outdoor living areas. See Figure 8.7. Spanish- and Italian-style architecture is typically characterized by numerous balconies that integrate indoor and outdoor living areas. New developments in building materials have increased the recent popularity of balconies in many styles of architecture.

Location

The location of the porch depends on its purpose or function. The family's preferences for use of the porch should also be considered when designing its orientation. For example, if daytime use is anticipated for the porch and direct sunlight is desirable, then a southern exposure should be planned. If little sun is wanted during the day, a northern exposure is preferable. If morning sun is desirable, an eastern exposure is best, and for the afternoon sun and sunset, a western exposure.

A continuous porch is often designed to function with the living area or with the sleeping area. The porch shown in Figure 8.8 continues on three sides of the house to include all three areas: living, sleeping, and service areas.

FIGURE 8.5 > Multilevel decks in Fallingwater are used for different functions. *Thomas A. Heinz, Courtesy of Western Pennsylvania Conservancy*

FIGURE 8.6 > Multiple balconies and deck extend living space. *Photo courtesy Lindal Cedar Homes. Lindal.com*

Decor

The porch should be designed as an integral and functional part of the total structure. A blending of roof styles and major lines of the porch roof and house roof is especially important.

A porch can be made consistent with the rest of the house by extending the lines of the roof to provide sufficient roof overhang, or projection, over the porch area. See Figure 8.9 and 8.10. A similar consistency should characterize the vertical columns or support members of the porch and the railings.

Porch railings can provide adequate ventilation and also offer semiprivacy and safety. Various materials and styles can be used, depending on the degree of privacy or sun and wind protection needed. Railings on elevated porches, such as balconies, should be higher than 3′ (914 mm) for general safety, as well as to discourage the use of the top rail as a place to sit. By code, most *balusters* (vertical posts) must be spaced closely enough (usually 4″) to prevent a child's head from going through.

Porch furniture should withstand any kind of weather. The covering material should be waterproof, stain resistant, and washable. Nonetheless, protection from wind and rain should be planned.

FIGURE 8.7 > Balconies serving as living and sleeping areas. *Hepler/Wallach/Hepler © Cengage 2013*

FIGURE 8.8 > Continuous porch with patio below. *Photo courtesy Lindal Cedar Homes. Lindal.com*

FIGURE 8.9 > Continuous deck with overhanging sun baffles. *Hepler/Wallach/Hepler © Cengage 2013*

Size and Shape

Porches range in size from the very large verandas to rather modest-sized stoops. A porch approximately 6′ × 8′ (1,829 mm × 2,438 mm) is considered minimum sized. An 8′ × 12′ (2,438 mm × 3,658 mm) porch is about average. Porches larger than 12′ × 18′ (3,658 mm × 5,486 mm) are considered large. The shape of the porch depends greatly on how the porch can be integrated into the overall design of the house.

PATIOS

The word **patio** is Spanish for courtyard, an open space enclosed wholly or partly by buildings. Courtyard living was an important part of Spanish culture. Therefore, courtyard design was an important component of early Spanish architecture.

Function and Types

The patio may perform all of the functions outdoors that the living room, dining room, recreation room, kitchen, and family room perform indoors. The patio may be referred to by other names, such as *loggia, breezeway, court,* or *terrace.*

Patios are divided into three main types according to function: living patios (including dining), play patios, and quiet patios. See Figure 8.11. Regardless of the type of patio, it should be secluded from the street and from neighboring residences if possible.

FIGURE 8.10 > Porch with railing and supported overhang. *Photo courtesy Lindal Cedar Homes. Lindal.com*

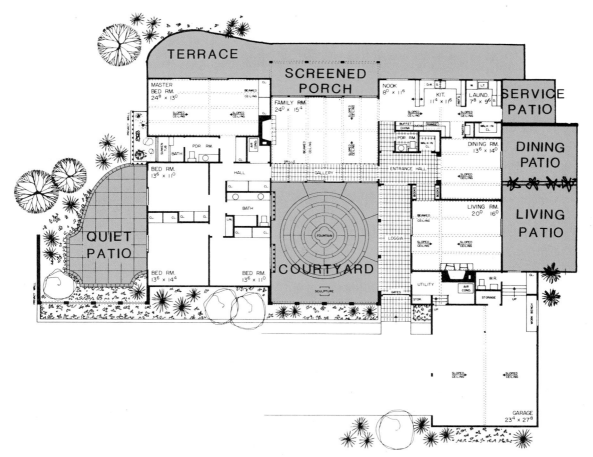

FIGURE 8.11 > Floor plan with different types of patios. *Used by permission Hanley Wood, LLC*

FIGURE 8.12 > Dining patio combined with lanais. *Wood residence—Taylor and Taylor, Architects*

A patio that is enclosed on three sides, as also shown in Figure 8.11, is called a **court.** Courts (or courtyards) are a characteristic of Spanish architecture. When all four sides are enclosed, the patio is similar to an interior atrium as described in Chapter 7. The patio shown in Figure 8.12 combines the facilities of a quiet patio and a dining patio. Dining patios may be designed for dining only and may include outdoor cooking facilities. These modular units include grills, side burners, sinks, refrigerators, serving bars, and storage units. Figure 8.13 illustrates the typical sizes of these components.

Location

The type of patio affects its location in relation to other rooms in the home. For instance, living patios should be located close to the living room or dining room. When dining is anticipated on the patio, access should be provided from the kitchen or dining room.

A children's play patio, or play terrace, for physical activities is not necessarily associated with the living area. Sometimes a play patio is located next to the service area so that it can double as a service terrace. Children's play patios should be located to allow for easy adult observation.

A quiet patio can become an extension of the bedroom for relaxation or sleeping. A quiet terrace should be secluded from the normal traffic of the home.

Often the design of the house will allow these separately functioning patios to be combined into one large, continuous patio. See Figure 8.14. Similar to a continuous porch, this type of patio may be accessible from the playroom, living room, bedrooms, or kitchen. Other designs divide large patios and porches into different areas by different levels.

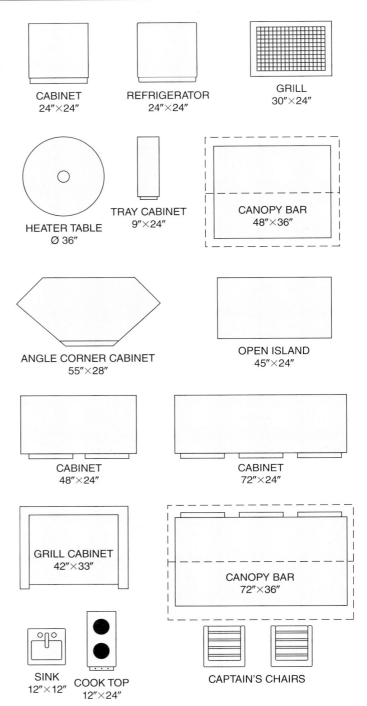

FIGURE 8.13 > Patio cooking equipment dimensions. *Vermont Islands*

Patios can be placed at the end of a building, between corners, or along the exterior form of the structure. They may also be placed in the center, such as in the center of a U-shaped house. A courtyard patio offers complete privacy from all sides.

A patio can be located completely apart from the house. A shady, wooded area, a beautiful view, or a unique feature

FIGURE 8.14 > Continuous patio designed for multiple functions. *John B. Scholz, Architect*

FIGURE 8.15 > Patio adjacent to a seawall. *Hepler/Wallach/ Hepler © Cengage 2013*

of terrain may determine an ideal location for a patio. The designer should take full advantage of the most pleasing view and should restrict the view of undesirable sights. If the patio is located a short distance from the house, it should still be easily accessible to the house with paths or walkways.

When the patio is placed on the north side of the house, the house itself can be used to shade the patio. If sunlight is desired, the patio should be located on the south side of the house. By locating patios on different sides of a building, sun exposure and sun protection can be available during some part of the day. Protection from the sun may also be controlled with planned landscaping and fences.

Patios located adjacent to a large body of water, as shown in Figure 8.15, must be designed to withstand storm surges and sea spray. They may also require boat docking facilities to be included in the design.

Decor

The materials used in the decor of the patio should be consistent with the lines and materials used in the home. Patios should not appear to be designed as an afterthought. They should appear and function as an integral part of the total design.

Patio Surface

The patio surface, or deck, should be constructed from materials that are permanent and maintenance free. Flagstone, concrete, composite, and brick are among the best materials for use on patio surfaces. Redwood, cedar, or engineered wood are the best surfaces for elevated wood decks. On wood decks, slats may be spaced to provide drainage.

Brick-surface patios are very popular because bricks can be placed in a variety of arrangements. The area between the bricks may be filled with concrete, gravel, sand, or grass. A concrete patio is effective when a smooth, unbroken surface is desired. Concrete works well for patios where bouncing-ball games are played or where a poolside cover (roof) is desired. However, where patio surfaces also function as pool decks, a non–heat-absorbing, non-slip material is preferable.

Patio Cover

The manner in which a patio is covered, or not covered, is closely related not only to the decor but also to the sunlight. Patios need not be covered if the house naturally shades the patio. Because a patio is designed to provide outdoor living, too much cover can defeat the purpose of the patio.

Coverings can be graded, or tilted, to allow light to enter the patio when the sun is high and block the sun's rays when the sun is lower. Straight or slanted louvers can be placed to admit the high sun and block the low sun or vice versa. See Chapter 7.

Plastic, fiberglass, and other translucent materials used to cover patios admit sunlight and yet provide protection from the direct rays of the sun and from rain. When such translucent coverings are used, it is often desirable to cover only part of the patio. This arrangement provides sun for part of the patio and shade for other parts and also allows rising heat to escape. Balconies can also be used effectively to provide shade and control light on a patio.

Conservatories

A conservatory is a glass-enclosed room originally intended for growing plants. Conservatories are also used as indoor patios as shown in Figure 8.16. They are hot in the summer and cold in the winter without maximum temperature control. For this reason conservatories are often called three-season rooms. Figure 8.17 shows the interior of a three-season room. A conservatory that is designed to provide maximum sun exposure is sometimes called a *solarium*. Figure 8.18 shows a large solarium with glass walls and ceiling.

Patio Walls and Baffles

Patios are designed for outdoor living, but outdoor living does not mean living in public. Some privacy is usually desirable. Natural landforms can sometimes provide privacy.

Walls can often be used effectively to baffle, or shield, the patio from a street view, from wind, and from direct and reflected sun rays. Baffling devices include solid or slatted fences, concrete blocks, post-and-rail, brick, or stone walls, and hedges or other landscaping. Figure 8.19 shows a patio with maximum protection both from the sides and overhead.

A solid baffle wall is often undesirable because it restricts the view, eliminates the circulation of air, and makes the patio appear smaller. When possible, natural vegetation is preferred.

In mild climates, a patio may be enclosed with solid walls to make the patio function as another room. In such an en-

FIGURE 8.16 > **Attached conservatory room.** *Barron and Jacobs, Design, Build, Remodelers*

FIGURE 8.18 > Solarium with maximum sun exposure. *Oak Leaf Conservatories, www.oakleafconservatories .com*

FIGURE 8.17 > **Three-season room interior.** *Barron and Jacobs, Design, Build, Remodelers*

FIGURE 8.19 > Protected patio. *Hepler/Wallach/Hepler*
© *Cengage 2013*

FIGURE 8.20 > Patio designed for night use. *Western Wood Products Association*

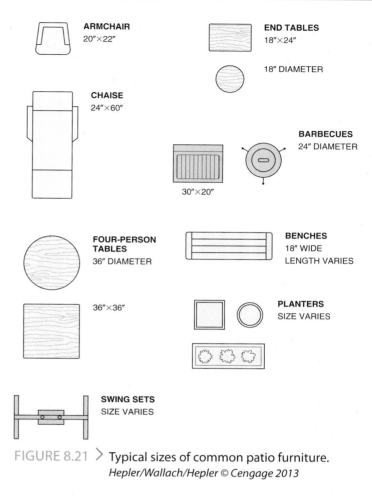

FIGURE 8.21 > Typical sizes of common patio furniture. *Hepler/Wallach/Hepler © Cengage 2013*

closed patio, some opening should nonetheless be provided to allow light and air to enter. Grillwork openings are an effective and aesthetically pleasing solution to this problem. Where wind and blowing sand or dust may be a problem, glass windscreens can be used to protect the patio.

Lighting

The patio should be designed so that it can be used both day and night. This means using general and local electrical lighting as well as natural lighting. If the windows and doors between the house and patio are designed correctly, light from inside the house can be utilized on the patio at night. See Figure 8.20. The combination of internal and external lighting design can extend the number of hours the patio can be used.

Size and Shape

As with other rooms, the function influences the size of a patio. Patios vary greatly in size. An oriental garden terrace, for example, can be small because it often has no furniture and is designed primarily to provide a baffle and a beautiful view. A patio used for recreation needs to be large enough for equipment and furnishings, such as picnic tables and benches, lounge chairs, serving carts, game apparatus, and barbecue pits. See Figure 8.21. Adequate space for the storage of games, apparatus, and fixtures also needs to be considered.

Patios tend to vary more in length than in width. Some patios may extend along the entire length of the house. A patio 12′ × 12′ (3,658 mm × 3,658 mm) is considered a minimum-sized patio. Patios with dimensions of 20′ × 30′ (6,096 mm × 9,144 mm) or more are considered large.

LANAIS

Lanai is the Hawaiian word for porch, but it also refers to a covered exterior passageway. Large lanais often double as patios.

Function

Lanais actually function as exterior hallways. They provide shelter for the exterior passageways of a building.

Lanais that are parallel to exterior walls are usually created by extending the roof overhang to cover a traffic area where people walk. Figure 8.22 shows the simplest type of lanais created by a large roof overhang. A typical lanai eliminates the need for more costly interior halls. Lanais are used extensively in warm climates. Figure 8.23 shows a lanai plan, and Figure 8.24 shows a photograph of the finished area. Note that the lanais shown in Figures 8.25 and 8.12 double as lounge areas.

Location

In residence planning, a lanai can be used most effectively to connect opposite areas of a home. Lanais are commonly located between the garage and the kitchen, the patio and the kitchen or living area, and the living area and service area. U-shaped buildings are especially suitable for lanais because it is natural to connect the ends of the U.

When lanais are carefully located, they can also function as sheltered access from inside areas to outside facilities such as patios, pools, outdoor cooking areas, or courtyards. A covered or partially covered patio is also considered a lanai when it doubles as a major access path from one area of a structure to another. A lanai can also be semienclosed and provide not only traffic access but privacy, as well as sun and wind shielding.

Decor

The lanai should be a consistent, integral part of the structure's design. The lanai cover may be an extension of the roof overhang or supported by columns. Columns also provide a visual boundary without blocking the view. If glass is placed between the columns, the lanai becomes an interior hallway rather than an exterior one. This feature is sometimes the only difference between a lanai and an interior hall.

It is often desirable to design and locate the lanai to provide access from one end of an extremely long building to the other end. The lines of this kind of lanai strengthen and reinforce the basic horizontal and vertical lines of the building.

If a lanai is to be utilized extensively at night, effective lighting must be provided. Light from within the house can be used when drapes are open, but additional lighting fixtures are used for the times when drapes are closed.

Size and Shape

Lanais may extend the full length of a building and may be designed for maximum traffic loads. They may be as small as the area under a roof overhang. However, a lanai at least 4' (1,219 mm) wide is desirable. The length and type of cover is limited only by the location of areas to be covered.

FIGURE 8.22 > Lanais created by a large roof overhang. *Hepler/Wallach/Hepler © Cengage 2013*

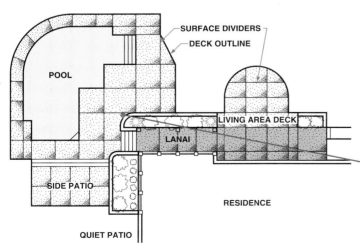

FIGURE 8.23 > Lanais connecting living area to pool deck. *Hepler/Wallach/Hepler © Cengage 2013*

FIGURE 8.24 > Photograph of the lanais shown in Figure 8.23. *Hepler/Wallach/Hepler © Cengage 2013*

FIGURE 8.25 > Lanais designed for multiple use. *John B. Scholz, Architect*

FIGURE 8.26 > Lighting enables pool area to be used at night. *Hepler/Wallach/Hepler © Cengage 2013*

SWIMMING POOLS

Swimming pools are an integral part of residential design. Designed for exercise and relaxation, pools can also enhance the design of a house. Pools add much to the initial cost of a residence. Pools also require expensive and continual maintenance. The addition of a pool and/or spa to a site design plan is therefore more common in warmer climates where year-round use is possible.

Function

The ideal pool should provide for all functions: exercise, relaxation, and enhancement of the site decor. Although pools are primarily used during daytime hours, a lighted pool, such as the very dramatic one shown in Figure 8.26, expands the living area by making the pool area inviting and usable at night. Different colors of lights can be used to create specific effects, as shown in Figure 8.27.

Location and Orientation

Several factors affect the location of residential pools: the relationship with the house, sun exposure, and privacy. The pool should be located as close to the living area as possible, allowing pool deck and patio space between the house and the pool.

Most building codes require controlled access to the pool. This means you must enter the pool area from the house or through a fence gate.

The orientation of a pool should be considered in relation to the sun. A pool should be positioned to allow the option of full sun exposure or partial shade and an adjacent shade-escape area. Shade for the pool deck or connecting patio may come from the north side of a house. On wooded sites, the orientation can be designed so trees can supply the needed shade. See Figure 8.28. However, the trees should not block the sun from the pool during most of the day.

Enclosure is a major consideration in the design of pools for privacy and safety. The pool in Figure 8.28 is built into an area surrounded by the natural vegetation of the dense woods, providing privacy as well as shade.

Screened walls and overhead enclosures have the advantage of blocking bugs and debris, but they also reduce the amount of direct solar heat on the water. Enclosures should be planned during the floor plan and elevation design phases (Chapters 14 and 16) to ensure consistency with the lines of the house.

A pool located near a large natural body of water must be designed to avoid interfering with the natural water table. Pools in some areas, as shown in Figure 8.29, must be seawall secured to prevent collapse and to keep out wildlife, such as alligators.

FIGURE 8.27 ❯ Color lighting options used to create different effects. *Fiberstars*

FIGURE 8.28 ❯ Pool located in wooded area, which provides shade and privacy. *Hepler/Wallach/Hepler © Cengage 2013*

Pool Construction

Materials

The frames of pools are constructed with concrete, wood, or steel. Pool surfaces that you see beneath the water may be covered with any of a variety of materials: vinyl sheets, marble composite, pebble aggregate, or paint.

Pool decks are constructed using non–heat-absorbing concrete mixtures, acrylic composites, or wood slats on elevated surfaces. The use of spray-on deck surfaces that are heat resistant, slip resistant, waterproof, and "mildew proof" are the most popular. Flagstone and pure concrete are not recommended for pool deck surfaces because these materials retain heat and become slippery when wet.

FIGURE 8.29 ❯ Pool located on a lakefront. *Hepler/Wallach/Hepler © Cengage 2013*

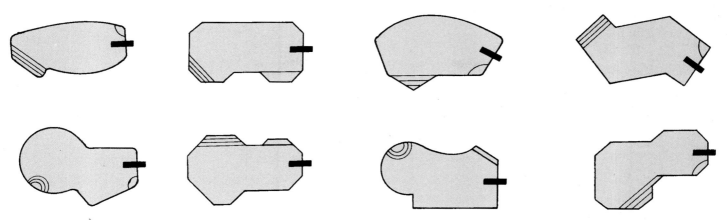

FIGURE 8.30 > Various pool shapes showing access steps, diving boards, and swim-outs. *Hepler/Wallach/Hepler © Cengage 2013*

Pool Shapes

With the development of dry-mix concrete, pool walls of almost any shape, including free-form shapes, can be created. See Figure 8.30. More than one shape can fit a pool site, depending on the contour of the site. Patio shapes around or near the pool also need to be considered.

Pools can also be designed and built to create the illusion of a waterfall as viewed from the top. This is done through the use of a recirculating spillway at one end, as shown in Figure 8.31.

In addition to the surface shape of a pool, the depth must be considered. It must first be determined if the pool is to be all shallow (3′ to 4′ deep), all deep (6′ to 10′), or a combination. Combinations of 3′ to 4′ on one end dropping to depths of 6′ to 10′ on the opposite end are most popular. If a diving board is to be included, then the deep end must include a *diving well* at least 8′ deep extending out 10′ horizontally from the end of the board as shown in Figure 8.32.

Calculating Pool Sizes

Residential pools range in size from about 200 to 800 sq. ft. or more. For example, a small 12′ × 18′ pool is 216 sq. ft. and a large 20′ × 40′ pool is 800 sq. ft. To calculate the area of a rectangular pool, use the following formula:

$$W \times L = A$$
(width × length = area)
Example: 14′ × 28′ = 392 sq. ft.

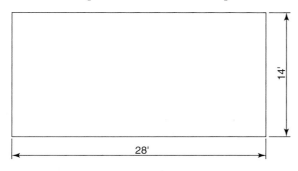

FIGURE 8.31 > Pool designed with recirculating waterfall. *Endless Pools*

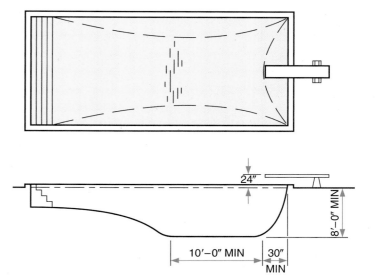

FIGURE 8.32 > Pool dimensions needed for diving board use. *Hepler/Wallach/Hepler © Cengage 2013*

To calculate the area of a round pool, use the formula for determining the area of a circle:

$$\pi r^2 = A$$
(pi × radius squared = area)
Example: 3.14 × (15′ × 15′) = 706.5 sq. ft.

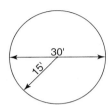

To calculate the area of a pool, such as the one in Figure 8.33, that has a combination of circular and rectangular areas, divide the entire pool area into smaller round and rectangular segments. First calculate the area of each part, round to even square feet, and add them together. For example, to find the area of the pool shown in Figure 8.33:

Area A (rectangle)	16 × 30	= 480
Area B (rectangle)	6 × 10	= 60
Area C (half-circle)	3.14 × 64 × 1/2	= 100
TOTAL SQ. FT.		640

For estimating cost and water capacity, cubic area is used to define the size of a pool. For example, an 18′ × 38′ pool that is 6′ deep contains 18′ × 38′ × 6′, or 4,104 cu. ft., of space. The same size pool 8′ deep contains 5,472 cu. ft. To find the volume of a container, use the following formula:

$$V = W \times L \times D$$
volume (cu. ft.) = width × length × depth
Example: 14′ × 28′ × 8′ = 3,136 cu. ft.

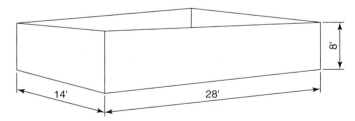

Because pools usually slope from shallow to deep, most cubic foot calculations involve dividing the pool into segments that have the same depth, or using average depth, as shown in Figure 8.34. The average depth of this pool is 6′-0″.

To determine the volume of a pool with a combination of cylindrical and cubic shapes, divide the pool area into separate cylindrical and cubic segments according to identifiable shapes. Then calculate the volume of each and add them together. For example, to find the volume of the pool shown in Figure 8.33, use the appropriate formula for each segment area. Remember, you can multiply in any sequence. Assume the bottom of each area is flat. The depth of area A is 8′, B is 3′, and C is 6′.

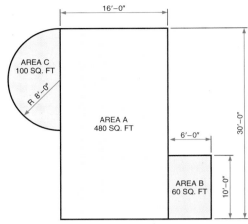

Area A (rectangle)	8′ × 16′ × 30′	= 3,840
Area B (rectangle)	3′ × 6′ × 10′	= 180
Area C (half-circle)	6′ × 3.14 × 32′	= 603
TOTAL CU. FT.		4,623

FIGURE 8.33 > Calculation of square footage for an irregularly shaped pool. *Hepler/Wallach/Hepler © Cengage 2013*

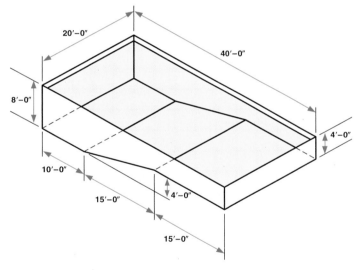

FIGURE 8.34 > Measurements needed to calculate pool volume. *Hepler/Wallach/Hepler © Cengage 2013*

Safety Devices

More than 600 people drown in residential pools each year. Visibility from house to pool is important, but certain pool design features can help reduce this number dramatically. These include minimum 4′ fencing, self-latching gates, latched house doors and windows, strong pool covers, alarms, ladders, steps, and/or swim-outs. A **swim-out** is an elevated platform below the water level that allows the swimmer to get out of the pool without using a ladder. In combination with fencing, a house can function as part of the barrier between the pool and site intruders as shown in Figure 8.35. If chil-

FIGURE 8.35 > A house can provide part of a pool safety barrier. *© Carl & Ann Purcell/Corbis*

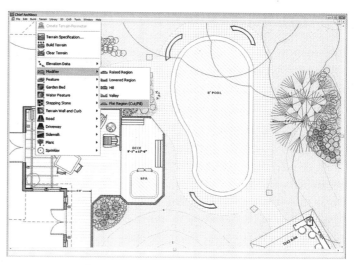

FIGURE 8.36 > Spa integrated with pool design. *Chief Architect Software*

dren live in the house that has the pool, other safety measures must be taken.

Other safety devices and equipment beyond the basic design features include clip-on child alarms, rope and float line, filter basket cover, posted emergency information, outside telephone, and portable infant fences.

Pool Equipment

Pools are simply cavities in the ground filled with water. To make a pool function properly, water must be circulated, filtered, purified, and sometimes heated. All of these functions require operating equipment.

Pool water is circulated through a series of filters and purifiers to keep the water sufficiently pure and clean for swimming. A water pump pulls water from the pool through a series of pipes connected to a skimmer device and drain. The pump moves the water through the filter, purifier, and sometimes a heater. After the water passes through these devices, the pure, clean, and heated water returns to the pool through pipes. These pipes are connected to *outlets* in the pool walls *under* the waterline. The number of outlets spaced throughout the pool determines the amount and balance of water circulation. Small pools may need only one outlet, while larger pools may need four or more.

Timing devices are recommended to control the amount of time the pump operates each day. Normally the pump is set to operate during the daylight hours because the pool equipment produces some noise. For this reason, it is better to locate the equipment away from lounging areas. The functioning of the plumbing system required to circulate, filter, and purify pool water is covered in Chapter 33.

Additional features to consider when designing a pool include a diving board, whirlpool spas, screened enclosures, and decorative fountains. Diving boards require additional foundation thickness.

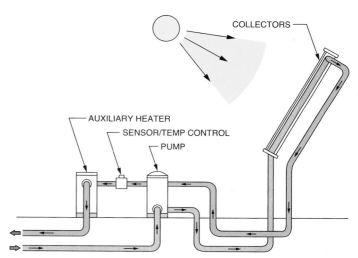

FIGURE 8.37 > Water temperature controlled with solar panels. *Hepler/Wallach/Hepler © Cengage 2013*

Spas can be designed into the total pool layout. They can be included in the same pool circulation system. However, if extremely hot water is anticipated (hot tub), a bypass system needs to be used to avoid overheating the pool water. Figure 8.36 shows a pool with an integrated spa. Spa water can be isolated by use of a spillway bypass. Figure 8.37 shows how solar collectors are used to control pool and spa water temperature.

Pool Water Purification

Chlorine-dispersing systems are the most widely used method of purifying swimming pool water. However, chlorine concentration and vapors negatively affect personal health and contribute to environmental pollution. Alternative methods of water purification that reduce the sole dependence on chlorine include ionizers, bromine generators, ozone generators, reverse osmosis, and ultraviolet systems.

OUTDOOR RECREATION FACILITIES

Information describing the facilities needed for outdoor recreation activities is included on architectural plans. These activities often include tennis, badminton, basketball, croquet, shuffleboard, table tennis, play gym, bocce ball, and horseshoes. The selection of specific activities depends on personal interests, the building budget, local codes, and the size, shape, slope, and location of available space.

Some activities require the design and construction of permanent facilities. Others may be planned to temporarily share common spaces. For example, tennis, handball, basketball, and play gyms require firm foundations and usu-ally paved surfaces, which can be shared. Activities such as badminton, croquet, and bocce ball can temporarily use the same flat lawn areas.

Architectural plans and details that describe the construction of outdoor permanent facilities are drawn on a site plan. Figure 8.38 includes the layout and dimensions of the most popular sport courts. This information is included in the set of plans submitted for municipal building permits. The position of temporary facilities is either drawn on the site plan using dashed lines or the location is only labeled on the plan. Chapter 13, *Site Development Plans*, covers the sizes, materials, and details needed for the construction of outdoor recreation facilities.

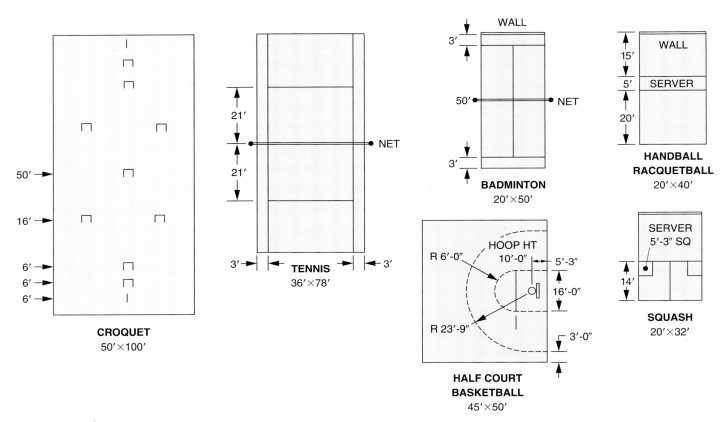

FIGURE 8.38 > Layout and dimensions of popular sports courts. *Hepler/Wallach/Hepler © Cengage 2013*

Outdoor Living Areas Exercises

1. From catalogs, newspapers, and magazines, cut out pictures of porch furniture that you particularly like (that is, that you would choose for your own porch).

2. Plan a porch or patio for a house of your own design. Sketch the basic outline and the facilities.

3. Draw the outline of the patio shown in Figure 8.11.

4. Draw the outline of a lanai you would plan for a U-shaped home of your own design.

5. Draw the outline of the lanai you designed in Exercise 4.

6. Sketch a floor plan of your own home. Add a lanai to connect two of the areas, such as the sleeping and living areas.

7. Explain the purposes of a lanai and describe two different plans where lanais would function well.

8. Name the required operating equipment needed for residential pools.

9. List the factors to consider in designing a pool.

10. Design a pool deck and patio area for a home you are designing. Locate the position of all operating equipment.

11. Draw the outline of a free-form pool shape with deck and patio areas.

12. Add recreation facilities to the plan shown in Figure 8.11.

CHAPTER 9

Traffic Areas and Patterns

OBJECTIVES

In this chapter you will learn:

> to determine the effectiveness of a traffic pattern in a house.

> the kinds and functions of entrances.

> guidelines for entrance design.

> to design a foyer and entry.

> to plan hallways that function efficiently.

> guidelines for designing stairs.

> to calculate the correct space needed for stairways and stairwells.

TERMS

apron
dividers
foyer
headroom

landings
L-winder
marquee
riser

spiral stairs
stairs
traffic areas
tread

INTRODUCTION

The **traffic areas** of any building provide passage from one room or area to another and within a room or area. Planning the traffic areas of a residence is not extremely complex because relatively few people are involved. Nevertheless, efficient allocation of space is important. The main traffic areas of a residence include the entrances or foyers, halls, stairs, and areas of rooms that are part of a traffic pattern.

TRAFFIC PATTERNS

Traffic patterns of a residence should be carefully considered when designing room layout. A minimum amount of space should be devoted to traffic areas. Extremely long halls and corridors should be avoided. These are difficult to light and provide no living space. Traffic patterns that require passage through one room to get to another should also be avoided, especially in the sleeping area.

The traffic pattern shown in the plan in Figure 9.1 is efficient and functional. It contains a minimum amount of hall space without creating a boxed-in appearance. It also provides

access between areas and from the entrance without passing through other areas.

One method of determining the effectiveness of the traffic pattern of a house is to imagine yourself moving through the house by placing your pencil on the floor plan and tracing your route through the house as you perform a whole day's activities. Do the same for other members of the household. You will be able to see graphically where the heaviest traffic occurs and whether the traffic areas have been planned effectively.

ENTRANCES

Entrances are divided into several different types: the site entrance, the main building entrance, the service entrance, and special-purpose entrances. Main entrances should have an outside waiting area (porch, **marquee**, lanai), a separation (door), and an inside waiting area (foyer, entrance hall). See Figure 9.2.

Function and Types

Entrances provide for and control the flow of traffic into and out of a building. Different types of entrances have different functions depending on the design of the structure.

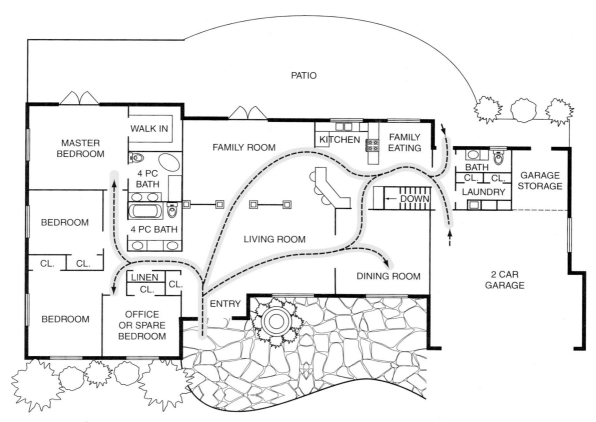

FIGURE 9.1 〉 Effective traffic pattern. *Hepler/Wallach/Hepler © Cengage 2013*

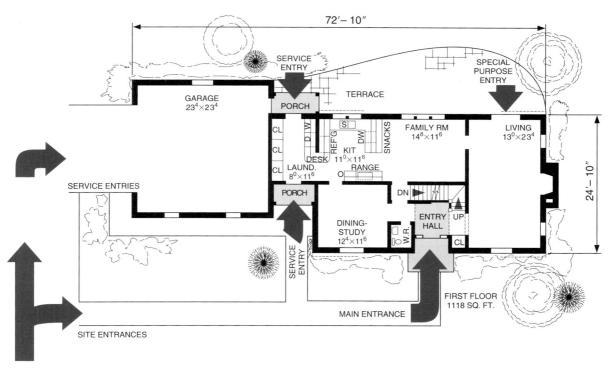

FIGURE 9.2 〉 Types of entrances. *Used by permission Hanley Wood, LLC*

Site Entrance

To design a site entrance, attention must first be given to the space from the street or road to the house. A site entrance includes the driveway, walkway, and adjacent parking or turnaround space for vehicles.

Driveways connect the street or road to a walkway and to a garage or carport and should be easily identified from the street. Driveways may be designed to lead directly to a garage, as in Figure 9.3. Some driveways are designed to intersect with the front entrance and connect with an attached, side, or rear garage, as in Figure 9.4. Driveways may be straight, curved, or circular. Figure 9.5 shows a circular driveway. Circular or semicircular driveways allow a car to return to the street without driving in reverse or turning around. In Figure 9.6 the circular drive allows for direct access to the main entrance and connects to a parking area.

FIGURE 9.3 > Straight driveway used as a residential entrance. *Hepler/Wallach/Hepler © Cengage 2013*

FIGURE 9.4 > Curved driveway leading to an attached garage and apron. *Hepler/Wallach/Hepler © Cengage 2013*

A turning and parking **apron** (area leading to garage) provides a means to exit a driveway without backing up onto the street. See Figure 9.7. To avoid double backing, the minimum turning radii shown must be strictly followed.

Some driveway entrances need to be gated for security reasons. For example, if a pool is not separately fenced and a perimeter fence along the property borders is used instead, the perimeter fence will need a drive entrance gate. See Figure 9.8.

Walkways leading to a front entrance may either connect the house entrance directly with the street or sidewalk, or lead to the driveway, or both. See Figure 9.9.

Main Entrance

The main entrance provides access to the house. It is the entrance through which guests are welcomed and from which all major traffic patterns radiate. The main entrance should be readily identifiable. It should provide shelter for anyone awaiting entrance. Figure 9.10 fulfills all of these requirements.

Some provision should be made in the main entrance wall to see callers from the inside before admitting them. Side panels, lights (panes) in the door, or windows that face the side of the entrance are used to allow a view of someone outside.

The main entrance should be planned to create a desirable first impression. A direct view of other areas of the house

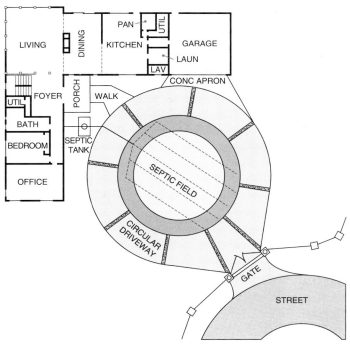

FIGURE 9.5 > Circular drive connected to entry and garage. *Hepler/Wallach/Hepler © Cengage 2013*

FIGURE 9.6 > Circular drive with direct front entrance. *UNI GROUP USA*

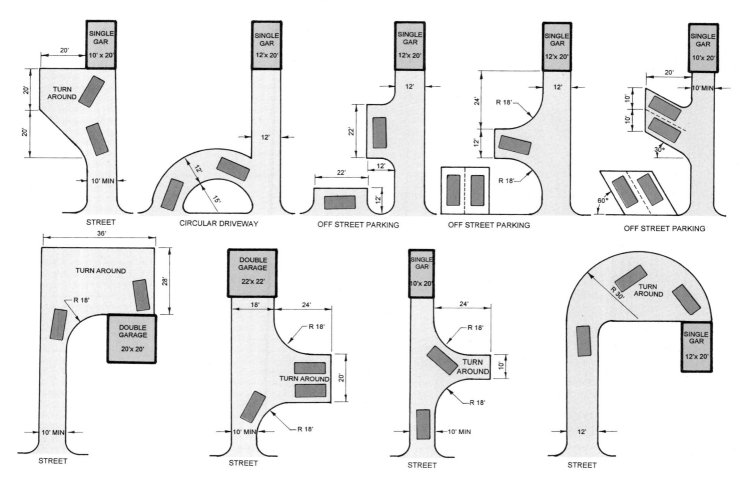

FIGURE 9.7 > Driveway dimensions for turning and parking. *Hepler/Wallach/Hepler © Cengage 2013*

FIGURE 9.8 > Gated drive. *Hepler/Wallach/Hepler © Cengage 2013*

FIGURE 9.9 > Overhang covered walkway connecting driveway and entrance. *Hepler/ Wallach/Hepler © Cengage 2013*

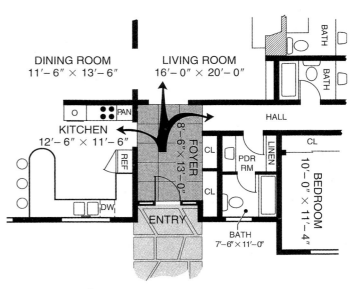

FIGURE 9.11 > Entrance foyer with efficient traffic flow. *Hepler/Wallach/Hepler © Cengage 2013*

- Consistent decor
- Outside weather protection
- Effective lighting day and night
- Avoidance of traffic through the living room center.

Service Entrance

The service entrance is to be used for any entry or exit that would be inappropriate and inconvenient at the main entrance. A person should be able to pass through the service entrance and enter parts of the service area, such as the garage, mud room, lavatory, laundry, or workshop. Supplies can also be delivered to the service areas without going through other parts of the house.

Special-Purpose Entrances

Special-purpose entrances and exits do not provide for outside traffic. Instead they are intended for movement from the inside living areas of the house to the outside living areas. A sliding door from the living area to the patio is a special-purpose entrance. It is not recommended as an entrance for street or sidewalk traffic. Figure 9.13 shows the location of service entrances and special-purpose entrances. Figure 9.14 shows a porch lounge adjacent to a bedroom special entrance.

Location

The main entrance should be centrally located to provide easy access to each area. It should be conveniently accessible from driveways, sidewalks, or street.

FIGURE 9.10 > Efficient and architecturally inviting entrance design. *Photo courtesy Lindal Cedar Homes. Lindal.com*

from the foyer should be baffled but not sealed off. The link between the main entrance and other interior areas is the foyer. The function of a **foyer** is to provide a connection between the main entrance and the living, sleeping, and service areas. This enables traffic to move to any area without passing through another room. Figure 9.11 shows a foyer entry plan that allows for a free flow of interior movement. Figure 9.12 shows several foyer plans designed for this purpose.

The entrance foyer should include a closet for the storage of outdoor clothing. This foyer closet should be large enough for both family and guests to use. A foyer arrangement must allow for the swing of the entrance door or doors. If the foyer is too shallow, passage will be blocked when the door is open, and only one person can enter at a time.

Use this checklist for the design of a main entrance:

- Adequate space to handle traffic flow
- Access to all living, sleeping, and service areas of a home
- A guest closet
- Bathroom access for guests

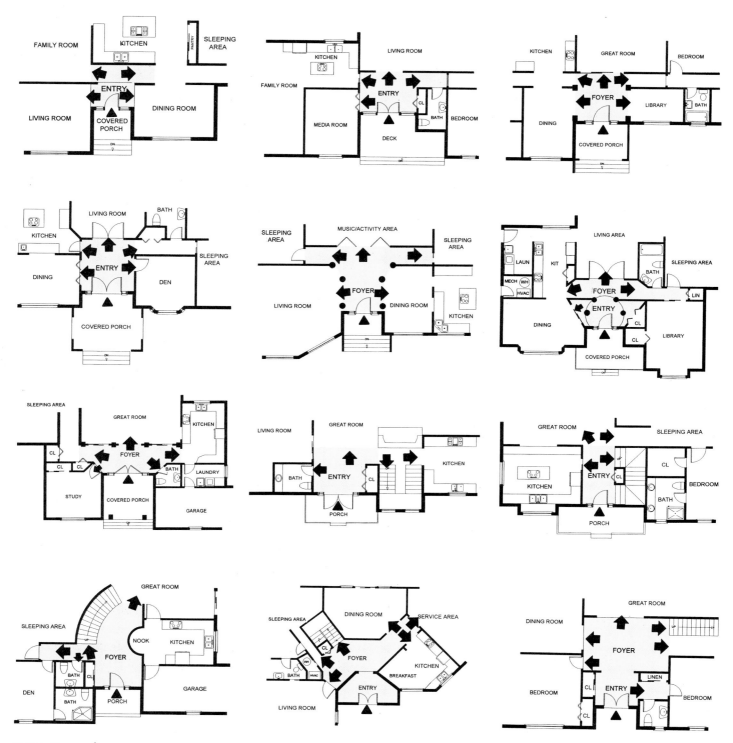

FIGURE 9.12 > Foyer plans designed to direct traffic. *Hepler/Wallach/Hepler © Cengage 2013*

The service entrance should be located close to the driveway and garage, and near the kitchen and food-storage areas.

Special-purpose entrances and exits are often located between the bedroom and the quiet patio, between the living room and the living patio, and between the dining room or kitchen and the dining patio.

Decor

To create a desirable first impression, a main entrance should be easily identifiable and yet be an integral part of the architectural style.

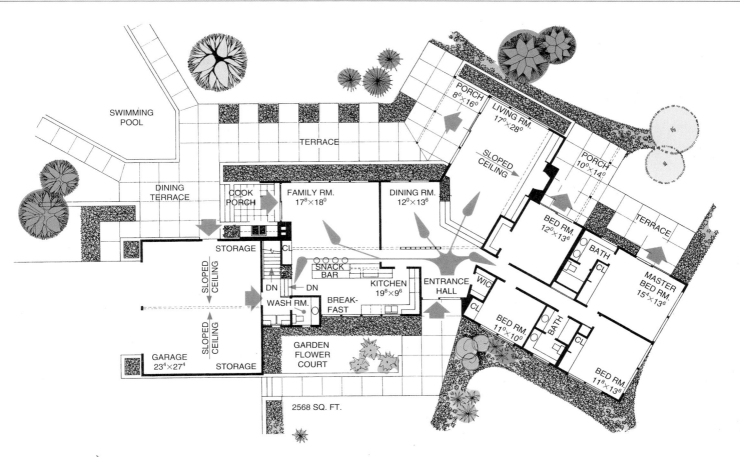

FIGURE 9.13 > Special-purpose and service entrances. *Used by permission Hanley & Wood, LLC*

FIGURE 9.14 > Porch adjacent to bedroom entrance.
California Redwood Association

The total design of the entrance should be consistent with the overall design of the house. That means the design of the door, the side panel, and the deck and cover should be directly related to the lines of the house.

Open and Closed Planning

Open planning is desirable for entrances. This means the view from the main entrance to the living area should be baffled without creating a boxed-in appearance. The foyer should not appear to be a dead end. The extensive use of glass, effective lighting, and carefully placed baffle walls can create an open and inviting impression. See Figure 9.15.

Open planning between the entrance foyer and the living areas can also be accomplished by the use of louvered walls, half walls (pony walls), or planter walls. These provide a relief or change in the line of sight but not a complete separation. Lowering or elevating the foyer or entrance can also produce the desired effect of separation without enclosing the area. Foyers are not enclosed in open planning. In formal or closed plans, the foyer is partially or fully closed off.

Surface Materials

The outside portion of the entrance should be weather-resistant wood, stone, brick, or concrete. The foyer deck should be easily

FIGURE 9.15 > Open plan foyer. *John Henry, Architect*

maintained and be resistant to mud, water, and dirt brought in from the outside. Asphalt, vinyl or rubber tile, stone, flagstone, marble, and terrazzo are most frequently used for the foyer deck. The use of a different material in the foyer area helps to define the area when no other separation exists.

Paneling, masonry, murals, and glass are used extensively for entrance foyer walls. The walls of the exterior portion of the entrance should be consistent with the other materials used on the exterior of the house.

Lighting

An entrance must be designed to function day and night. Natural lighting should be planned for lighting entrance areas during daylight hours. General lighting, spot lighting, and all-night lighting are effective after dark.

Lighting can be used to accent distinguishing features or to illuminate the pattern of a wall. This type of lighting actually provides more light by reflection and helps to identify and accentuate the entrance at night. See Figure 9.16.

Size and Shape

The size and shape of the areas inside and outside the entrance depend on the budget and the type of plan. The outside covered portion of the entrance should be large enough

FIGURE 9.16 > Illuminated entrance and façade. *John Henry, Architect*

to shelter several people and also provide the amount of space needed to open a storm door. Outside shelter areas are the same range in size and shape as porches and patios. (Refer back to Chapter 8.)

The inside of the entrance foyer should be sufficiently large to allow several people to enter at the same time, remove their coats, and put their things in the closet. A 6′ × 6′ (1,829 mm × 1,829 mm) foyer is considered minimum for this function. A foyer 8′ × 10′ (2,438 mm × 3,048 mm) is average, but a more desirable size is 8′ × 15′ (2,438 mm × 4,572 mm).

HALLS

Halls are the highways and streets inside the home. They provide a controlled path that connects the various areas of the house. Halls should be planned to eliminate or minimize the passage of traffic through rooms. Long, dark, tunnel-like halls should be avoided. Halls should be well lighted, light in color and texture, and planned with the decor of the whole house in mind.

The hall shown in Figure 9.17 doubles as a library and gallery. Light is provided by glass block windows on the top left and angular soffit lighting on the top right. See Figure 18.37A for details. The hallway shown in Figure 9.18 receives diffused light from both walls and cornice lighting on the curved ceiling. The surround lighting turns a pedestrian hall/walkway into an attractive gallery for display. Even in very large hallways, as seen in Figure 9.19, an abundance of well-diffused light is necessary.

Minimum hall widths (usually 2′-6″ to 3′-0″) are determined by building codes. Halls must also be wide enough for furniture movement and for wheelchair access. One method of channeling hall traffic without using solid walls is to use **dividers.** Planters, half walls, louvered walls, and even furniture can be used as dividers. The plans in Figure 9.20 illustrate some of the basic principles of efficient hall design.

FIGURE 9.17 ❯ Hallway doubling as a library and gallery.
Hepler/Wallach/Hepler © Cengage 2013

FIGURE 9.18 ❯ Hall decor blended with effective lighting design. *Marc-Michaels Interior Design, Inc.*

FIGURE 9.19 > Large dramatically lighted hallway and foyer. *Marc-Michaels, Interior Design, Inc.*

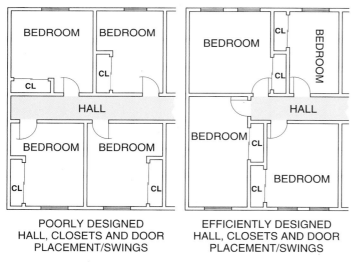

POORLY DESIGNED
HALL, CLOSETS AND DOOR
PLACEMENT/SWINGS

EFFICIENTLY DESIGNED
HALL, CLOSETS AND DOOR
PLACEMENT/SWINGS

FIGURE 9.20 > Methods of minimizing hall length. *Hepler/Wallach/Hepler © Cengage 2013*

STAIRS

Stairs are inclined hallways that provide access from one level to another. Stairs may lead without a change of direction, or they may turn 90° or 180° by means of **landings.** Figure 9.21 illustrates the most common types of stairs used in light construction. For accident prevention some **L-winder, spiral,** and open-riser stairs are not approved by some codes, depending on their size and location.

Figure 9.22 shows an example of a classical curved stair system with an intermediate landing and ornamental iron railing. For details relating to floor planning see Chapter 14 for stairwell openings and stair runs. Also refer to Chapter 28 for floor framing and Chapter 29 for wall framing of stairwell construction methods.

Materials and Lighting

With the use of newer, stronger building materials and new techniques, stairs can now be supported by many different devices. Stairs no longer need to be enclosed in areas that restrict light and ventilation.

Stairwells (areas for stairs) should be lighted at all times when in use. Natural light is the most energy efficient. Thus windows should be utilized to provide natural light for stairs wherever possible. Three-way switches should be provided at the top and bottom of the stairwell to control the stair lighting. (For details, see Chapter 31.)

Size and Shape

Many variables should be considered when designing stairs. The **tread** is the horizontal part of the stair, the "step," or the part on which you walk. Treads must be made of or covered with nonslip surfaces. The average depth, or distance from front to back, of the tread is 10″ (254 mm). The **riser** is the

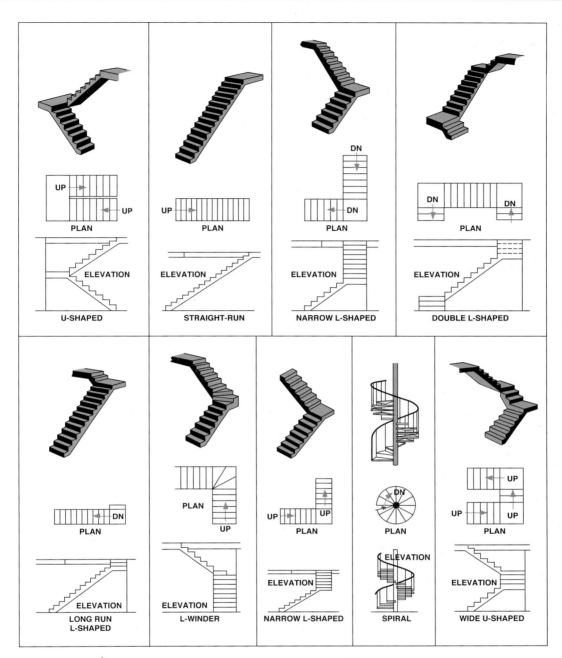

FIGURE 9.21 › Common types of stair systems. *Hepler/Wallach/Hepler © Cengage 2013*

vertical part of the stair. The average riser height is 7 1/4″ (184 mm). See Figure 9.23.

The overall width of the stairs is the distance between the stair railings. A minimum of 3′ (914 mm) should be allowed for the total stair width. However, a width of 3′-6″ (1,067 mm) or even 4′ (1,219 mm) is preferred to accommodate the movement of furniture. See Figure 9.24.

Headroom is the vertical distance between the top of each tread and the top of the stairwell ceiling. A minimum headroom distance of 6′-6″ (1,981 mm) should be allowed. However, a distance of 7′ (2,134 mm) is more desirable.

The tread width, the riser height, and the headroom all help to determine the total length of the stairwell. Landing dimensions are generally determined by the size of the stairs and the space for the stairwell. More clearance must be allowed where a door opens onto a landing. See Figure 9.25. A landing should be planned for stair systems that have more than 16 risers. It should be located at the center between levels to eliminate long runs. A minimal landing is 2′-6″ × 3′-0″.

ELEVATORS

The use of elevators in homes and light construction buildings is increasing significantly. A minimum of 3′ × 4′ of space on a floor plan is required. Elevators are designed into a floor plan primarily for use by people with physical

FIGURE 9.22 > Classical curved stair system. *Marc-Michaels, Interior Design, Inc.*

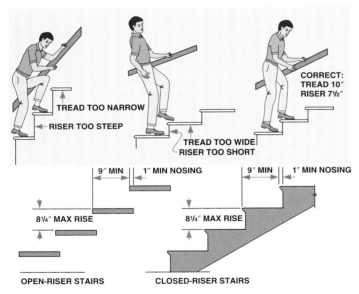

FIGURE 9.23 > The importance of correct tread and riser dimensions. *Hepler/Wallach/Hepler © Cengage 2013*

disabilities.* Smaller units (dumbwaiters) can be included for the vertical movement of household goods, food, fireplace wood, etc. The floor plan in Figure 9.26 includes elevators that can be used to move people or firewood or for food service.

*Stair side lifts are an alternative for this purpose.

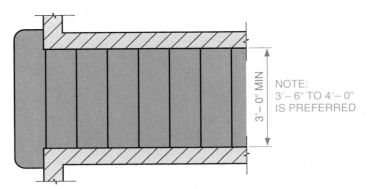

FIGURE 9.24 > Minimum stair width. *Hepler/Wallach/Hepler © Cengage 2013*

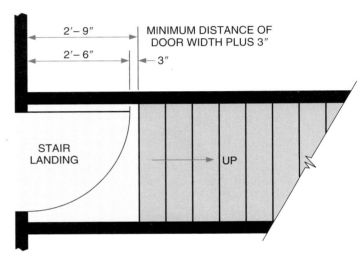

FIGURE 9.25 > Minimum landing dimensions. *Hepler/ Wallach/Hepler © Cengage 2013*

FIGURE 9.26 > Locations of personal, firewood, and food service elevators. *Hepler/Wallach/Hepler © Cengage 2013*

Traffic Areas and Patterns Exercises   CHAPTER **9**

1. Sketch the floor plan of a home of your design. Plan the most efficient traffic pattern by tracing the route of your daily routine.

2. Draw the plan view of one of the stair systems shown in this chapter.

3. Draw or sketch a plan view of a stair system in your home or school.

4. Name the types of stairs that turn 90°, 180°, 360°, and 0°.

5. List the types of entrances and tell the function of each type.

6. Redesign an entrance shown in this chapter. Add space that is consistent with the main lines of the house.

7. Redesign and enlarge the foyer for the living area shown in Figure 9.2. Label the materials you select for the outside deck, overhang, access walk, foyer floor, and walls.

8. Add a foyer to the plan of a house you are designing.

9. Draw a plan for the foyer you redesigned in Exercise 7.

10. Redesign the floor plan shown in Figure 9.27 to include a foyer with access to all areas.

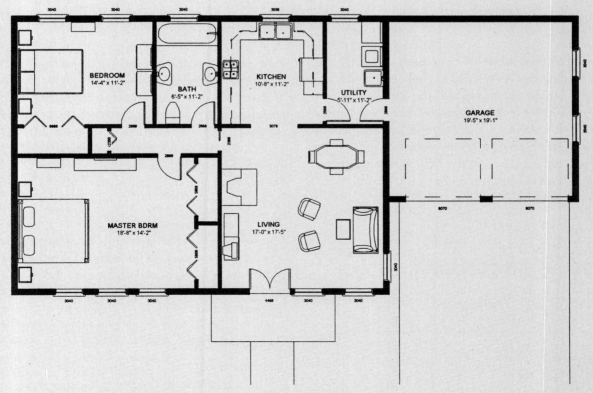

FIGURE 9.27 > Floor plan to be redesigned. *Hepler/Wallach/Hepler © Cengage 2013*

Kitchens

OBJECTIVES

In this chapter you will learn to:

> apply guidelines to efficient kitchen design.

> determine the best shape, size, and location for the kitchen.

> plan a work triangle for a kitchen.

> design an aesthetically consistent decor for a kitchen.

> sketch small and large kitchens using the basic kitchen shapes.

TERMS

corridor kitchen
downdraft exhaust
Energy Star
family kitchen
GFCI outlet

gas pilot
island kitchen
L-shaped kitchen
one-wall kitchen
peninsula kitchen

pilotless gas appliance
toxic by-product
U-shaped kitchen
work triangle
workstation

INTRODUCTION

A well-planned kitchen is one that functions efficiently, and yet is attractive and easy to maintain. To design an efficient kitchen, the designer must consider the room's function, location, decor, size, and shape. However, because a kitchen requires so much equipment, the design of a kitchen entails additional considerations and decisions. In addition to the kitchen design factors covered in this chapter, the integration of plumbing, electrical, heating, and ventilation requirements must also be considered.

KITCHEN DESIGN CONSIDERATIONS

Kitchen design involves planning space configurations, work surfaces, storage requirements, and the number, type, size, and location of all components. In addition, the relation of the kitchen to other rooms or areas, light sources, vistas, and traffic flow must also be considered.

Understanding the functions of a kitchen is the first step in planning a kitchen's design.

Functions

Food preparation is, of course, the primary function of the kitchen. However, the kitchen may also be used as a dining area. The proper placement of appliances is important in a well-planned kitchen. Locating appliances in an efficient pattern eliminates wasted motion. An efficient kitchen has three basic areas or centers: the storage center, preparation center, the cooking center. A fourth area, the cleanup center is combined into one or more of the others, usually preparation. See Figure 10.1.

Storage Center

The refrigerator is the major appliance in the storage center. The refrigerator may be freestanding, built-in, or even suspended from a wall. Cabinets for the storage of utensils and food ingredients, as well as a countertop work area for preparing food, are also included at this center.

Cooking Center

The major appliances in the cooking center are the range (cooktop), microwave oven, and oven. The range and oven may be combined into one appliance or be separated into two appli-

FIGURE 10.1 ＞ Kitchen functional areas: cooking, preparation, cleanup, and storage. © *Mark Hunt/Huntstock/Corbis*

FIGURE 10.2 ＞ The kitchen work triangle. *Norcraft Companies, Inc.*

ances, with the burners installed in the countertop (cooktop) as one appliance and an oven built into a cabinet. The cooking center should have countertop work space, as well as storage space for minor appliances and cooking utensils. An adequate supply of electrical or gas outlets for using appliances is necessary.

Cleanup Center

The sink is the major fixture in this center. Sinks are available in one-, two-, or three-bowl models with a variety of cabinet arrangements, countertops, and drainboard areas. The cleanup center may also include a waste disposal unit, an automatic dishwasher, a waste compactor, and cabinets for storing cleaning supplies.

The Work Triangle

If you draw a line connecting the three centers of the kitchen, a triangle is formed. See Figure 10.2. This is called the **work triangle.** The perimeter of an efficient kitchen work triangle should be no more than 22′ (6,706 mm) or less than 12′ (3,658 mm). Although the size of the work triangle is an indication of kitchen efficiency, the triangle is primarily useful as a starting point in kitchen design.

The arrangements of the three areas of the work triangle may vary greatly. However, efficient arrangements can be designed in each of the seven basic types of kitchens discussed next.

Types of Kitchens

U-Shaped Kitchen

The **U-shaped kitchen** is very efficient and popular. The sink is located at the bottom of the U, and the range and the refrigerator are at the opposite ends. In this arrangement, traffic

FIGURE 10.3 ＞ U-shaped kitchen. *Hepler/Wallach/Hepler*

passing through the kitchen is completely separated from the work triangle. The open space in the U between the sides should be 4′ (1,219 mm) or 5′ (1,524 mm). This arrangement produces a very efficient small kitchen. See Figure 10.3. Figure 10.4 shows various U-shaped kitchen designs and the planned work triangles.

When designing U-shaped kitchens, special attention must be given to door hinges and drawer positions. Design cabinet doors and drawers to open without interfering with each other, especially at cabinet corners.

Peninsula Kitchen

The **peninsula kitchen** is similar to the U-shaped kitchen, but one end of the U is not adjacent to a wall. It projects into

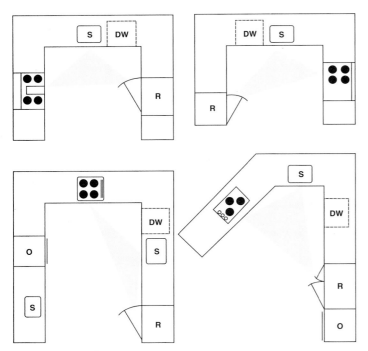

FIGURE 10.4 > U-shaped kitchen arrangements. *Hepler/ Wallach/Hepler © Cengage 2013*

FIGURE 10.5 > Peninsula kitchen. *© Treve Johnson/Lived In Images/Corbis*

the room like a piece of land (peninsula) into a body of water. This peninsula is often used for the cooking center. However, it may serve several other functions as well. The peninsula is often used for an eating area as well as for food preparation. See Figure 10.5. It may join the kitchen to the dining room or family room. Figure 10.6 shows various arrangements of peninsula kitchens and the resulting work triangles.

Most peninsula kitchens contain large countertops for work space. Peninsulas may contain only lower or base cabinets, but some may include upper cabinets suspended from ceilings. The peninsula kitchen shown in Figure 10.7 includes an adjacent breakfast area and utility and laundry facilities. To the right of this plan is the dining room and to the left, the garage.

L-Shaped Kitchen

The **L-shaped kitchen** has continuous counters, appliances, and equipment located on two adjoining, perpendicular walls. Two work centers are usually located on one wall and the third center is on the other wall. See Figure 10.8. The work triangle is not in the traffic pattern. If the walls of an L-shaped kitchen are too long, the compact efficiency of the kitchen is destroyed.

An L-shaped kitchen requires less space than the U-shaped kitchen. The remaining open space often created by an L-shaped arrangement can serve as an eating area adjacent to a family room, without taking space from the work areas. If the center area is used for eating, a minimum of 36″ (914 mm) must be allowed as an aisle between cabinets and chairs.

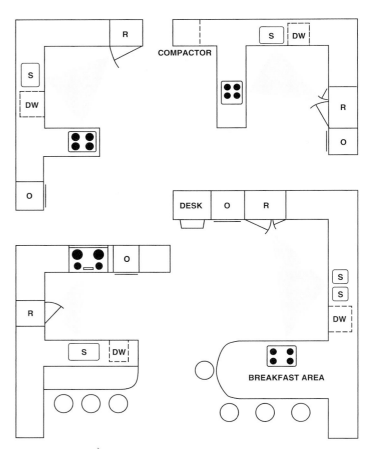

FIGURE 10.6 > Peninsula kitchen arrangements. *Hepler/ Wallach/Hepler © Cengage 2013*

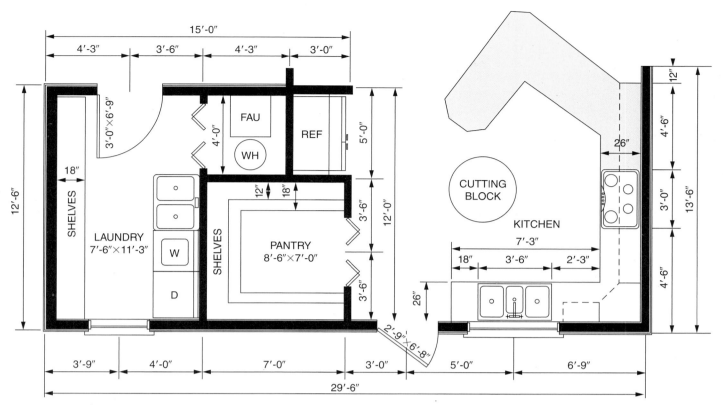

FIGURE 10.7 > Peninsula kitchen with adjacent utility and laundry rooms. *Hepler/Wallach/Hepler © Cengage 2013*

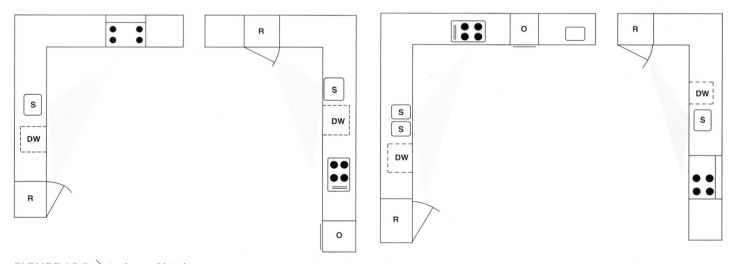

FIGURE 10.8 > L-shaped kitchen arrangements. *Hepler/Wallach/Hepler © Cengage 2013*

Corridor Kitchen

Two-wall **corridor kitchens** are very efficient arrangements for long, narrow rooms. See Figure 10.9. They are very popular for small apartments, but are used extensively anywhere space is limited. A corridor kitchen produces a very efficient work triangle, as long as traffic does not need to pass through the work triangle. The corridor space between cabinets (not walls) should be no smaller than 4' (1,219 mm). One of the best work arrangements locates the refrigerator and sink on one wall and the range on the opposite wall.

One-Wall Kitchen

A **one-wall kitchen** is an excellent plan for small apartments, cabins, or houses in which little space is available. The work

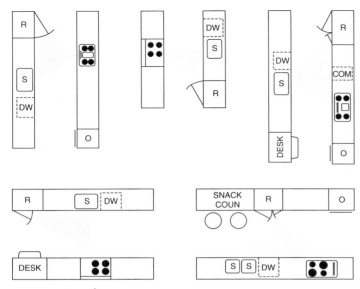

FIGURE 10.9 > Corridor kitchen arrangements. *Hepler/ Wallach/Hepler © Cengage 2013*

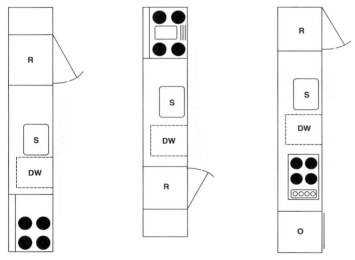

FIGURE 10.10 > One-wall kitchen arrangements. *Hepler/ Wallach/Hepler © Cengage 2013*

centers are located along one line rather than in a triangular shape, but this design still produces an efficient arrangement. See Figure 10.10.

When planning a one-wall kitchen, the designer must be careful to avoid creating walls that are too long. Adequate storage facilities need to be well planned also, because space is often limited in a one-wall kitchen.

Island Kitchen

The **island kitchen,** another geographically named arrangement, has a separate, freestanding structure that is usually located in the central part of the kitchen. An island in the

FIGURE 10.11 > Island including sink, work counter, and eating area. *Allmilmo USA Corporation*

FIGURE 10.12 > Island with range, sink, and eating area. *New England Cabinet Co.*

kitchen should be accessible on all sides but not intersect the work triangle. It usually has a range top or sink, or both. Other facilities are sometimes located in the island, such as a mixing center, work table, serving counter, extra sink, or snack center.

The kitchen shown in Figure 10.11 contains an island sink, work counter, and a lower level eating area. The island in Figure 10.12 includes a range, "prep" sink, and eating area.

FIGURE 10.13A ＞ Kitchen with two islands. *Russell MacMasters, Photographer*

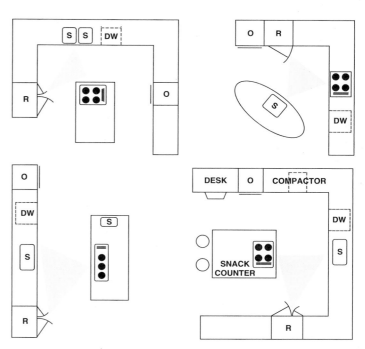

FIGURE 10.14 ＞ Island kitchen arrangements. *Hepler/ Wallach/Hepler © Cengage 2013*

FIGURE 10.13B ＞ Ninety-degree view of kitchen in Figure 10.13A. *Russell MacMasters, Photographer*

FIGURE 10.15 ＞ Contemporary style island kitchen. *DAL-Tile*

The kitchen shown in Figure 10.13A includes two islands. One island contains a sink and large work counter; the other contains extra burners and a large work counter that doubles as a serving counter. You can visualize this kitchen better by observing the 90° view in Figure 10.13B.

Figure 10.14 shows examples of other island facilities. The island design is especially convenient when two or more persons work in the kitchen at the same time.

When an island contains a range or grill, allow at least 16″ (406 mm) on the sides for utensil space. Also consider the use of a downdraft exhaust system that pulls vapors down and out rather than up to eliminate the need for overhead hooded vents. Allow at least 42″ (1,067 mm) on all sides of an island. If used for eating, also add the depth of the chair or stool.

Figure 10.15 illustrates a contemporary-style kitchen with a multifunction island used for preparation. Figure 10.16 shows a plan and elevation drawing of an island with a combination snack counter and cooktop. Note how the counter

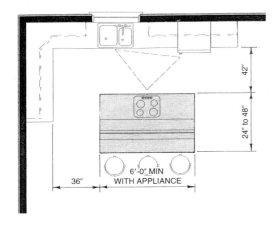

FIGURE 10.17 > Work island with cold storage and secondary oven. *Bisult Kitchens, Garden City, NY*

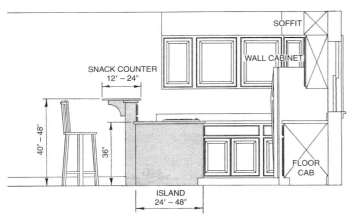

FIGURE 10.16 > Island kitchen plan and elevation. *Hepler/Wallach/Hepler © Cengage 2013*

FIGURE 10.18 > Food service island with cooktop. *Bisult Kitchens, Garden City, NY*

height is designed to keep the cooktop separate from the eating surface. The island in Figure 10.17 also includes a cold storage compartment and a secondary oven. In summary, islands can be any style consistent with kitchen decor and be designed for any kitchen function, including dining facilities as shown in Figure 10.18. The island shown in Figure 10.19 demonstrates an application of "form follows function." The curved edge not only allows for easy movement around the island but also provides a central cook station and an elevated eating ledge.

Family Kitchen

The **family kitchen** is an open kitchen using any kitchen shape. The function of an open kitchen is to provide a meeting place for the entire family—in addition to the usual kitchen services. A family kitchen often appears to have two parts in one room. The three food preparation work centers comprise one section. The dining area and family

room facilities comprise another section. See Figure 10.20. Figure 10.21 shows several typical arrangements for family kitchens. Open plan family room kitchens can also be adapted to serve as part of an open plan great room design. In this type of design, a complete visual opening between the living and dining facilities is planned. Kitchens of this type are designed in many configurations, as shown in Figure 10.22. Figure 10.23 on page 184 shows the dining part of a family kitchen which is usually open and separated only by a kitchen counter.

FIGURE 10.19 > Functional free-form island. *Pedini*

FIGURE 10.20 > Family kitchen. *Merrillat Industries Inc.*

ISLAND FAMILY KITCHEN

ONE-WALL FAMILY KITCHEN

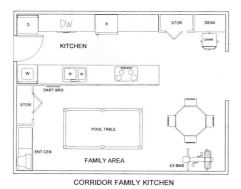

CORRIDOR FAMILY KITCHEN

PENINSULA FAMILY KITCHEN

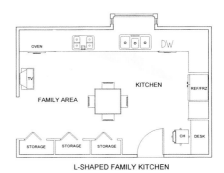

L-SHAPED FAMILY KITCHEN

U-SHAPED FAMILY KITCHEN

FIGURE 10.21 > Family kitchen arrangements. *Hepler/Wallach/Hepler © Cengage 2013*

Regardless of its shape, the kitchen is the core of the service area and should be located near the service entrance as well as near the utility area. The kitchen must be adjacent to eating areas, both indoors and outdoors. The children's play area should also be visible or easily accessible from the kitchen.

Family kitchens must be rather large to accommodate the necessary facilities. An average size for a family kitchen is 225 sq. ft. (20 sq. m). Eating areas can be designed with either tables and chairs or with chairs or stools at a counter. When counters are used for eating, allow at least 12″ (305 mm) for knee space between the end of the counter and the face of the base cabinet.

Multiple **workstation** kitchens are an adaptation of the family and island kitchen concept, as shown in Figure 10.24.

This plan type is appropriate when two or more persons routinely prepare meals at the same time. Islands in large multiple workstation kitchens can be designed to be movable when unlocked, and if no plumbing or electrical connections are required. This can be accomplished through the use of multiple wide ball-bearing rollers.

Decor

Kitchens cost more per square foot than any other room. Most of this cost relates to the selection of appliances, cabinetry, and fixtures. By selecting the least expensive models of appliances, hardware, and cabinetry, the same kitchen design can often be built for one-fourth the cost of a kitchen that contains the most expensive features. High end electrical and

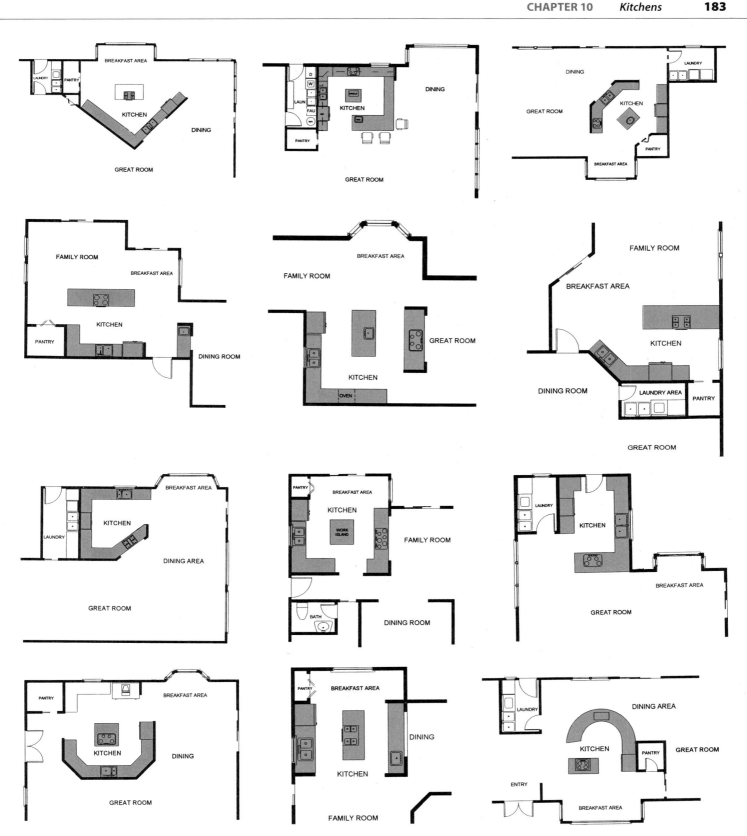

FIGURE 10.22 > Open plans for a great room kitchen. *Hepler/Wallach/Hepler © Cengage 2013*

FIGURE 10.23 > Dining area open to kitchen. *Marc-Michaels Interior Design, Inc.*

FIGURE 10.24 > Multiple workstation kitchen. *Hepler/ Wallach/Hepler © Cengage 2013*

plumbing fixtures and lines also adds much to the cost of a kitchen.

Even though most kitchen appliances are produced in contemporary designs, some clients and designers prefer to decorate kitchens with a traditional style as a motif or theme. The cabinets, floors, walls, and accessory furniture would then be selected according to that chosen theme. Designing a totally harmonious kitchen is made easier by the wide variety of appliance sizes, colors, and styles. Compare the bold and simple lines of the contemporary-style kitchen shown in Figure 10.25 with the elements of country style found in the kitchen in Figure 10.26. All aspects of each design are consistent with the overall decor.

Regardless of the style, the kitchen walls, floors, countertops, and cabinets should require a minimum amount of maintenance. Materials that are relatively maintenance free include stainless steel, stain-resistant plastic, ceramic tile, washable wall coverings, washable paint, vinyl, molded and laminated plastic countertops, doors, drawers, and cabinet bases.

Options in kitchen design have broadened because of new synthetic and composite materials and new construction methods for cabinets and countertops.

FIGURE 10.25 > Contemporary kitchen arrangement with island serving area. *Hepler/Wallach/Hepler © Cengage 2013*

FIGURE 10.26 > Colonial-style family room and kitchen. *Armstrong World Industries*

Size and Shape

Average human dimensions as described in Chapter 6 are a key factor in selecting the size and shape of kitchen components. Reaching distances are most important. Figure 10.27 shows typical reaching considerations from a standing position. Many component heights must be altered to accommodate wheelchair occupants as shown in Figure 10.28. Fortunately, components are manufactured in a wide range of sizes to accommodate almost any design configuration. Figure 10.29 includes standard horizontal (width) dimensions used in kitchen design. Figure 10.30 shows the dimensions of standard wall and base cabinets. These are the most commonly used sizes. Figure 10.31 shows the total range of cabinet sizes, and Figure 10.32 lists the range of sizes of major kitchen appliances. Manufactured, modular base cabinets are 24″ deep and modular wall cabinets are 12″ deep. Custom floor cabinets can be any size but often 26″ deep to match appliance depth.

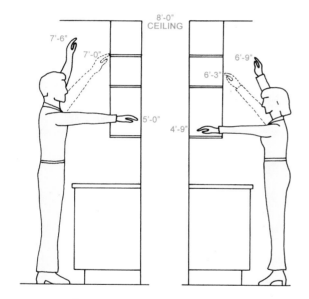

FIGURE 10.27 > Typical kitchen reaching heights. *Hepler/ Wallach/Hepler © Cengage 2013*

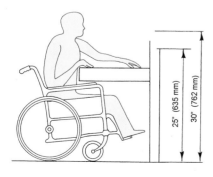

Counter top heights and thigh clearance.

FIGURE 10.28 > Wheelchair reaching heights for kitchens. *Hepler/Wallach/Hepler © Cengage 2013*

Kitchen Drawings

Drawings used to describe kitchen designs include floor plans elevations, pictorials, and construction details. A floor plan of the kitchen pictured in Figure 10.33A is shown in Figure 10.33B. This plan includes all width and length dimensions. Kitchen floor plans also show the horizontal distances between critical elements. Elevation drawings show the vertical alignment of cabinets, countertops, and appli-

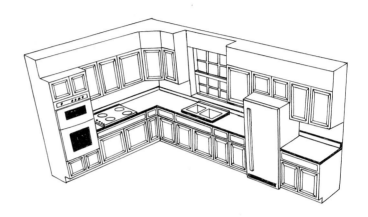

LEFT ELEVATION

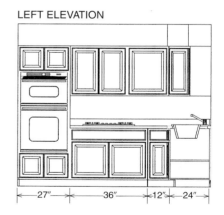

FRONT ELEVATION

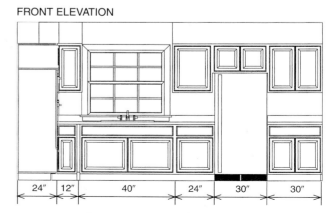

FIGURE 10.29 > Standard kitchen horizontal dimensions. *Hepler/Wallach/Hepler © Cengage 2013*

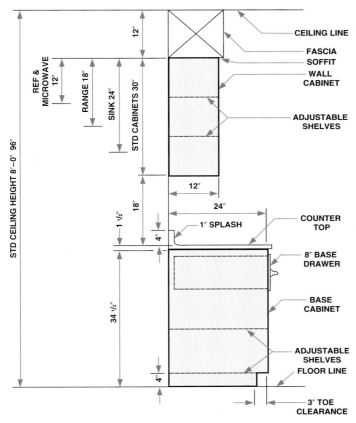

FIGURE 10.30 > Standard wall and base cabinet dimensions. *Hepler/Wallach/Hepler © Cengage 2013*

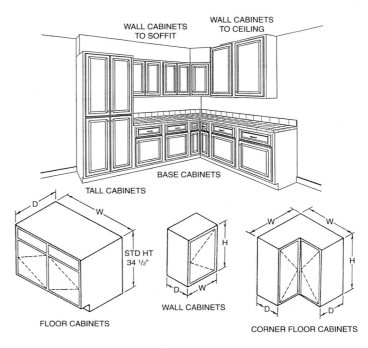

FIGURE 10.31 > Cabinet sizes and types. *Hepler/Wallach/Hepler © Cengage 2013*

STANDARD CORNER FLOOR CABINETS
• HEIGHT — 34 1/2″
• DEPTH — 24″
• WIDTHS — 24″ to 36″

TALL CABINETS
• HEIGHT — 84″
• DEPTHS — 12″ to 24″
• WIDTHS — 18″ to 24″

STANDARD WALL CABINETS
• HEIGHT 34 1/2″ (plus counter top = 36″)
• DEPTH — 24″
• WIDTHS — 9″ to 48″

STANDARD WALL CABINETS
• HEIGHTS — 12″ to 30″
• DEPTH — 12″
• WIDTHS — 12″ to 42″

ances. Figure 10.33C on page 190, is an elevation drawing of the wall that includes the refrigerator, cooktop, sink, and window. Although floor plans and elevations totally describe a kitchen design, pictorial drawings, such as that shown in Figure 10.33D on page 190, are often prepared to visually interpret the relationship of these drawings in one view.

Kitchen elevation drawings should be prepared for each wall containing cabinets and appliances. These are projected from a floor plan as described in Chapter 17 and as shown in Figure 10.34 on page 191. When this is done the side view of an intersecting wall elevation becomes a section, as shown in Figure 10.35 on page 192.

Drawing kitchen plans is covered in Chapter 15, *Drawing Floor Plans.* Drawing kitchen wall elevations is covered in Chapter 17, *Drawing Elevations.*

KITCHEN PLANNING GUIDELINES

The following guidelines are divided into categories; function, location, size and space, utilities, and appliances. Some of the guidelines overlap and not all apply to every design situation. In these cases, design compromises must be made.

Function

1. Plan specific centers for storage, preparation, cooking, and cleanup.

2. Ensure that all work areas include all of the necessary appliances.

3. Provide adequate storage facilities throughout the kitchen, including sealed containers for trash and garbage.

4. Plan sufficient counter and storage space for small portable appliances and devices.

5. Specify heat-resistant countertops such as stone, tile, cultured marble, Formica, or Corian.

6. Consider using an island for one or more work centers.

7. Plan multiple workstations if several persons routinely prepare food at the same time.

8. Specify all surfaces to be grease and water resistant. Floor surfaces should be nonslip and easy to maintain.

9. Plan all wall treatments and construction, including low partitions and pass-through counters.

TOP/BOTTOM REFRIGERATORS
HT 56″ TO 66″
DEPTH 24″ TO 27″
WIDTH 30″ TO 42″
(WITH BOTTOM DRAWER)

REF/FREEZERS
HT 56″ TO 66″
DEPTH 24″ TO 28″
WIDTH 30″ TO 42″

SINGLE SINKS
DEPTH 20″ TO 22″
WIDTH 24″ TO 30″

DOUBLE SINKS
DEPTH 20″ TO 22″
WIDTH 32″ TO 42″

TRIPLE SINKS
DEPTH 20″ TO 22″
WIDTH 42″ TO 55″

DBL SINKS/DRAIN BRDS
DEPTH 20″ TO 22″
WIDTH 50″ TO 60″

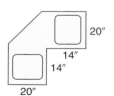

TYPICAL CORNER SINKS

CORNER SINKS
DEPTH 20″× TO 22″
WIDTH —varies

DISHWASHERS
DEPTH 24″
WIDTH 24″ TO 29″
(WITH DRAWERS)

TRASH COMPACTORS
DEPTH 21″
WIDTH 12″

SINK DISPOSAL UNITS
TYP. HT 14″, 8″ DIAM

FREESTANDING RANGES
HT 36″
DEPTH 24″ TO 27″
WIDTH 20″ TO 40″
(WITH SINGLE DRAWER)

DROP-IN RANGES
TYP. 22″ × 30″

COOKTOPS/GRILLS
TYP. 21″ × 18″

4-BURNER COOKTOPS
GAS/ELEC
TYP. 21″ × 26″

**FREE STANDING RANGE
WITH DRAWERS OR OVEN
4-GRIDDLE**
HT 36″
DEPTH 24″ TO 27″
WIDTH 20″ TO 40″

6-BURNER COOKTOPS
TYP. 21″ × 36″

SINGLE OVENS
HT 22″ TO 40″
DEPTH 16″ TO 25″
WIDTH 23″ TO 30″
(WITH DRAWER)

**DOUBLE OVENS
WITH MICROWAVE**
HT 40″ TO 58″
DEPTH 23″ TO 27″
WIDTH 22″ TO 30″

MICROWAVE OVENS
HT 18″ TO 20″
DEPTH 14″ TO 20″
WIDTH 20″ TO 30″

RANGE HOODS
HT VARIES
DEPTH 17″ TO 24″
WIDTH 30″ TO 72″

WASHERS/DRYERS
DEPTH 24″ TO 27″
WIDTH 24″ TO 29″

FIGURE 10.32 › Common sizes of kitchen appliances. *Hepler/Wallach/Hepler © Cengage 2013*

Location

1. Keep work triangle traffic lanes unobstructed.

2. Plan for direct access from the kitchen to the dining area and preferably also to the family room, utility room, and garage.

3. Design great room kitchens to provide support for great room functions.

4. Allow a minimum 3″ space between microwave ovens and food storage.

5. Locate refrigerators away from direct sunlight.

6. Locate trash and garbage disposal containers close to the preparation.

7. Plan cabinet and appliance locations according to manufacturers' specifications. Figures 10.30 and 10.32 show standard dimensions for each.

8. Arrange cabinets so that they create a consistent line without gaps, depressions, or awkward protrusions.

FIGURE 10.33A >
Picture of kitchen shown in Figures 10.33B, C, and D. *Roman Polaski, www.romanpolaski.com*

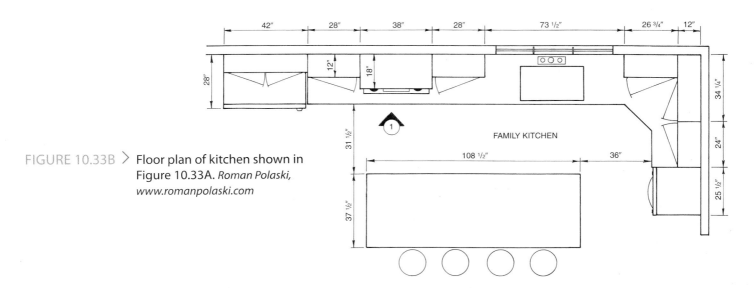

FIGURE 10.33B > Floor plan of kitchen shown in Figure 10.33A. *Roman Polaski, www.romanpolaski.com*

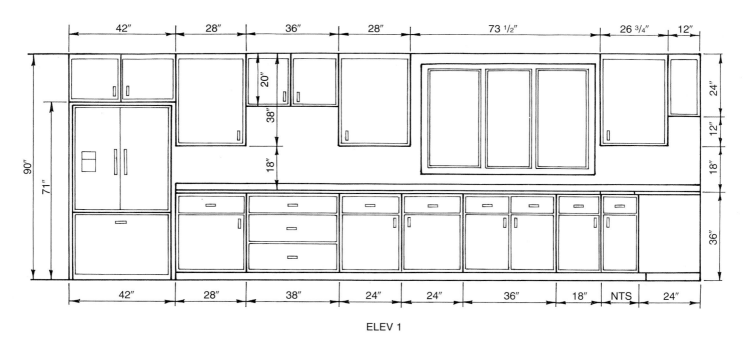

ELEV 1

FIGURE 10.33C > Elevation details of kitchen shown in Figure 10.33A. *Roman Polaski, www.romanpolaski.com*

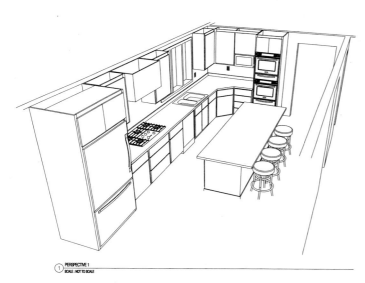

FIGURE 10.33D > Pictorial drawing of kitchen shown in Figure 10.33A. *Roman Polaski, www.romanpolaski.com*

Size and Space

1. Ensure that the work triangle measures no more than 22′ or less than 12′.

2. For kitchen eat-in areas, keep countertop heights between 28″ and 30″ for chairs and between 40″ and 48″ for stools.

3. Keep working heights for counters at 36″.

4. Keep detail dimensions in inches.

5. Allow at least 12″ for knee space if counters are used for eating.

6. If space allows, include a pantry to store staple foods.

7. Allow at least 4′ of aisle space between cabinets or appliances.

8. Allow at least 15″ on each side of an island cooktop for utensil storage.

9. Keep shelves within a reachable height (maximum height 72″).

10. Provide counter space between each appliance.

11. Allow a minimum of 18″ for work space on both side of a cooktop.

12. Keep the minimum distance between islands and base cabinets to between 32″ and 42″.

13. Avoid sharp corners on peninsula and island counters.

14. Allow traffic space around open dishwasher, refrigerator, and oven doors.

15. If eating areas are next to work center counters, use a raised counter to protect the eating counter.

16. Locate the dishwasher near storage of dishes and flatware.

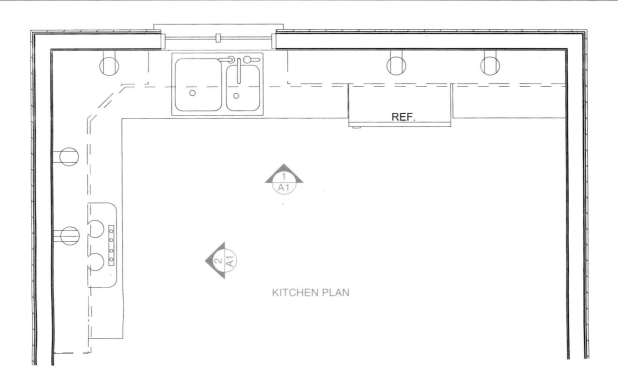

KITCHEN PLAN

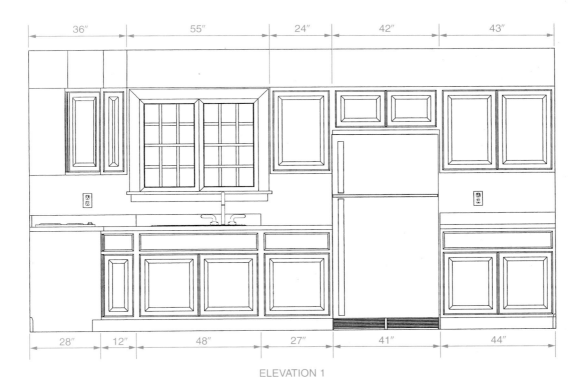

ELEVATION 1

FIGURE 10.34 > Relationship of kitchen plan and elevation. *Hepler/Wallach/Hepler © Cengage 2013*

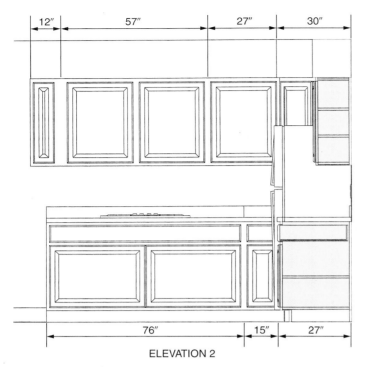

FIGURE 10.35 > Kitchen elevation showing a corner cabinet section. *Hepler/Wallach/Hepler © Cengage 2013*

🌐 Utilities

1. Provide an adequate number (minimum 4′ intervals) of **GFCI** (ground fault circuit interrupters) **outlets** for each work center's counter space.

2. Concentrate shadowless and glareless task lighting on each work area, including islands, sink, and cooktop.

3. Plan plumbing lines (supply and waste) for sinks, dishwasher, ice maker, and pure water systems.

4. Include adequate ventilation of cooking fumes by planning for hoods, ceiling fans and vents, HVAC ducting, and **downdraft exhausts.**

5. Provide access to mechanical and electrical control panels.

6. Plan to admit as much natural light as possible.

7. Design all electrical installations to be UL standards.

8. Provide heating and cooling controls for the kitchen area that do not rely on large-zone controls.

9. Provide electrical and or gas power supply for refrigeration, cooktop exhaust fans, waste disposal, waste compactor, dishwasher, and laundry appliances if located in the kitchen area.

10. Avoid the use of equipment or appliances that produce **toxic by-products.** These include carbon monoxide, nitric oxide, nitrogen dioxide, hydrogen cyanide, formaldehyde from gas appliances, ozone and non-shielded electromagnetic radiation from electrical appliances.

Appliances

1. Because major appliances are shown on floor plans and specifications lists, their selection is the responsibility of the designer. In addition to the basic dimension, other factors must be considered. These include energy efficiency and personal and environmental safety.

2. To avoid excessive heat exchange, separate heat-producing appliances such as ovens and cooktops from the refrigerator by a minimum of 6″.

3. Hang cabinet doors to open toward the work triangle (Figure 10.36A).

4. Place appliances so that open appliance doors do not block traffic as shown in Figure 10.36B.

5. Specify **pilotless gas appliances** to avoid toxic fumes. **Gas pilots** use gas resources even when appliances are not in use.

6. Specify energy-efficient appliances. The most energy-efficient appliances can be chosen by observing the **Energy Star** rating displayed on the appliance.

7. Specify dishwashers with heat boosters or no heat drying option for maximum energy saving and convenience.

8. Provide adequate ventilation around motorized appliances.

Kitchen design consideration for people with physical impairments is covered in Chapter 14, *Designing Floor Plans.*

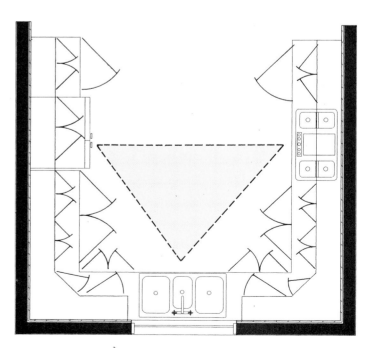

FIGURE 10.36A > Cabinet doors should open away from work locations. *Hepler/Wallach/Hepler © Cengage 2013*

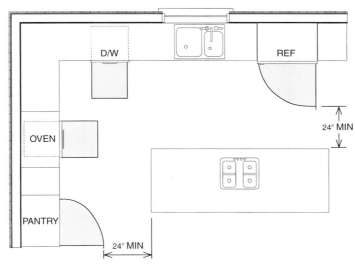

FIGURE 10.36B ❭ Open appliance doors should not block traffic. *Hepler/Wallach/Hepler © Cengage 2013*

Kitchens Exercises

CHAPTER 10

1. List the six types of kitchen shapes and give at least one advantage and one disadvantage of each.

2. Sketch a floor plan of one of the U-shaped kitchens shown in Figure 10.4. Show the position of the dining area in relation to this kitchen, using a scale of 1/2″ = 1′-0″.

3. Sketch a family kitchen using any of the seven kitchen types in your design.

4. Sketch a floor plan of the kitchen in your own home. Prepare a revised sketch to show how you would redesign this kitchen. Try to make the work triangle more efficient.

5. Sketch a floor plan of a kitchen you would include in a house of your own design, using a scale of 1/2″ = 1′-20″.

6. Draw one plan of any of the kitchen shapes shown in Figure 10.22.

7. Calculate the space needed and plan a kitchen with a work triangle in a 14′ × 16′ peninsula kitchen and in an 8′ × 14′ L-shaped kitchen.

8. Collect pictures of kitchens shown in magazines. Identify the kitchen type and find the work triangle in each. List the good points and bad points of each kitchen design.

Service Areas

OBJECTIVES

In this chapter you will learn to:

> determine what kinds of equipment are included in a utility room.

> evaluate locations for a utility room.

> sketch a garage and a carport.

> design storage facilities.

> calculate the area needed for garages and driveways.

> design and sketch an efficient and safe workshop area.

> design areas for pets, fitness, and trash handling.

TERMS

carport
detached garage
dropleaf workbench
hand tools
integral garage

mud room
peninsula workbench
power tools
subterranean garage
utility room

ventilated shelving
walk-in closet
wall closet
wardrobe closet

INTRODUCTION

Service areas include utility rooms, garages and carports, mudrooms, workshops, storage areas, and specialized areas such as fitness, pet, and trash and garbage facilities. Because a great number of different activities are related to these areas, they should be designed for great efficiency. Service areas should include facilities for the maintenance and servicing of the other areas of the home.

UTILITY ROOMS

The **utility room** may include facilities for washing, drying, ironing, sewing, and storing household cleaning equipment. It may contain heating and air-conditioning equipment and even pantry shelves for storing groceries. Other names for this room are the *service, mechanical,* or *all-purpose room.*

Function

The major function of most utility rooms is to serve as a laundry area. They may also accommodate water heating, water purification, and heating and air-conditioning equipment. Figure 11.1 shows utility rooms designed to function as a laundry, storage, or mechanical equipment room.

Laundry

To make laundry work as easy as possible, the appliances and working spaces in a laundry area should be located in the order in which they will be used. Such an arrangement will save time and effort during the steps in the laundering process: receiving and preparing, washing, drying, ironing, storage, and, often, sewing.

The first step in laundering—receiving and preparing the items—requires hampers or bins, as well as counters on which to collect and sort the articles. Storage facilities should be located nearby for laundry products such as detergents, bleaches, and stain removers.

The next step is the actual washing. Washing takes place in the area containing the washing machine (washer) and laundry tubs, trays, or sinks.

The equipment needed for the third step includes a dryer and indoor drying lines. Dryers require either a 220-volt

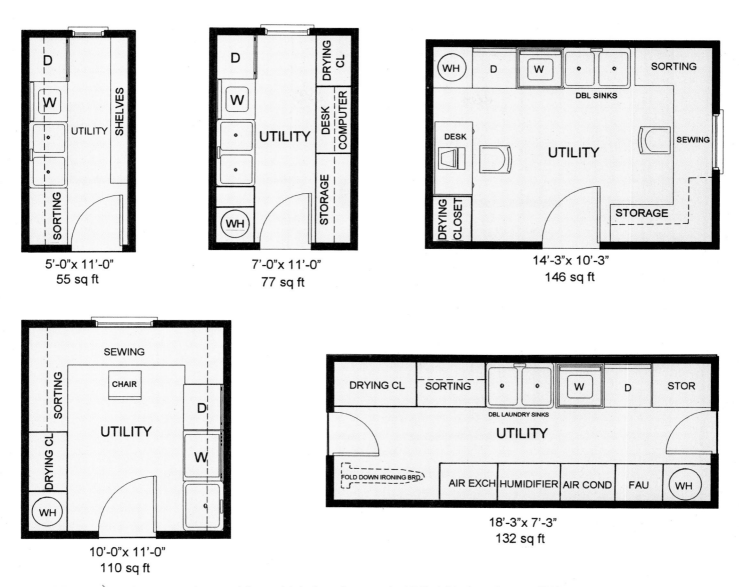

5'-0"x 11'-0"
55 sq ft

7'-0"x 11'-0"
77 sq ft

14'-3"x 10'-3"
146 sq ft

10'-0"x 11'-0"
110 sq ft

18'-3"x 7'-3"
132 sq ft

FIGURE 11.1 > Utility rooms designed for multiple functions. *Hepler/Wallach/Hepler © Cengage 2013*

(220V) outlet or access to gas. An exhaust duct must connect a dryer to the outside.

For the last step of the process, the required equipment consists of a counter for folding, an iron and ironing board, and a rack on which to hang finished ironing. Facilities for sewing and mending are also often included. A sewing machine may be portable or it may fold into a counter or wall.

Heating and Air Conditioning

If the utility room is used to house heating and air conditioning equipment, additional space must be planned for the furnace, heating and air-conditioning ducts, water heater, and any related equipment such as humidifiers or air purifiers.

Location

A separate utility room is desirable because all laundry functions and maintenance equipment can be centered in one place. Space is not always available for a separate utility room, however. Laundry facilities may need to be located in some other area. Plans for the location of laundry facilities in the garage, closet, service, or sleeping areas are shown in Figure 11.2.

Placing the laundry appliances in or near the kitchen puts them in a central location and near a service entrance. Plumbing facilities are nearby, and some kitchen counters may be used for folding. However, these advantages may be offset by

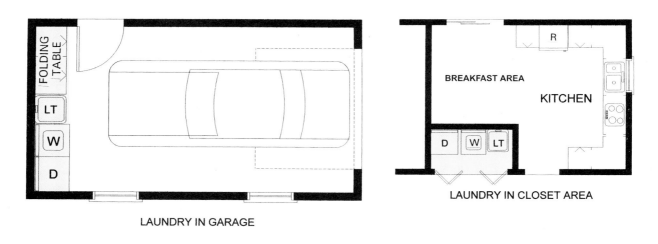

LAUNDRY IN GARAGE

LAUNDRY IN CLOSET AREA

LAUNDRY IN SERVICE AREA

LAUNDRY IN SLEEPING AREA

FIGURE 11.2 › Laundry room locations. *Hepler/Wallach/Hepler © Cengage 2013*

noise from the machines and odors from detergents, bleaches, and softeners near the kitchen. It is also desirable to keep the laundering process away from areas where food is prepared. Because clothing is changed in the sleeping area, laundry facilities in that area may be convenient.

Style and Decor

Style and decor in a utility room depend on the function of the appliances, which are themselves an important factor in the appearance of the room. Simplicity, straight lines, and

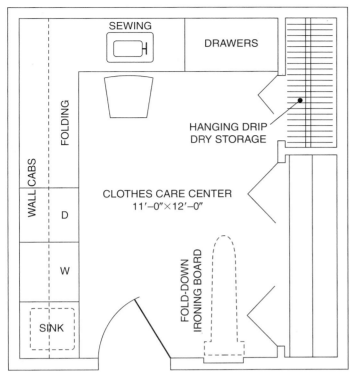

FIGURE 11.3 > Laundry including clothing maintenance functions. *Hepler/Wallach/Hepler © Cengage 2013*

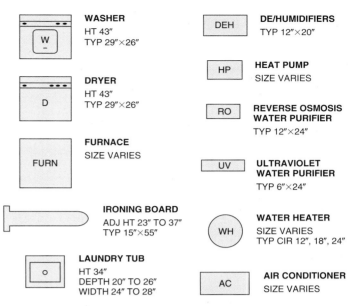

FIGURE 11.4 > Utility room equipment sizes. *Hepler/Wallach/ Hepler © Cengage 2013*

continuous counter spaces produce an orderly effect and permit work to progress easily. Such features also make the room convenient to maintain.

An important part of the decor is the color of the paint used for walls and cabinet finishes. Colors should harmonize with the colors used on the appliances. All finishes should be washable. The walls may be lined with sound-absorbing material.

The lighting in a utility room should be 48″ (1,219 mm) above any equipment used for washing, ironing, and sewing. However, lighting fixtures above work areas and laundry sinks can be farther from the worktop area. Lighting should be located to avoid shadows on work surfaces caused by the body of the person using the area.

Size and Shape

When space is available, all phases of laundering and clothing maintenance can be located in the laundry area. See Figure 11.3. When space is limited, moveable work center units can be used. For storage, the same sizes of cabinets used in kitchens and bathrooms can also work well in laundry areas.

Depending on what equipment is included, the shapes and sizes of utility rooms differ. See Figure 11.4. The average floor space required for appliances, counters, and storage areas is 100 sq. ft. (9.3 m²). However, this size may also vary according to the budget or needs of the household.

Appliances and Equipment

Utility room appliances include clothes washer, clothes dryer, hot water heater, furnace, and possibly air-conditioning equipment. The same standard of environmental safety and energy efficiency described for kitchen appliances applies to utility room appliances.

Specifically, front-loading washers are 50% more energy efficient than top loaders. The most efficient dryers contain a moisture sensor that stops when clothes are dry. Gas dryers are more expensive than electric dryers but less costly to operate. Water heater capacities range from 20 to 80 gallons. It is important to specify the appropriate size for the number of occupants. Oversize heaters cost more to operate but undersize heaters quickly run out of water. Tankless water heaters do not store water and instead heat water on demand. A domestic oil burner furnace produces hot water on demand in the furnace boiler.

GARAGES AND CARPORTS

Areas for parking and storing vehicles often make up a large percentage of the space available on a property. Therefore, the maximum utilization of space is important to consider when designing garages, carports, and driveways.

Function and Location

A garage is an enclosed structure designed primarily to shelter an automobile. It may be used for many secondary purposes—as a workshop, as a laundry room, or for storage space. A garage may be connected with the house, which is called an **integral garage** (Figure 11.5), or it may be a separate building, which is a **detached garage.** In any case, there should be easy access from the garage to the service area of the house. Figure 11.6 shows several possible garage locations in relation to the house.

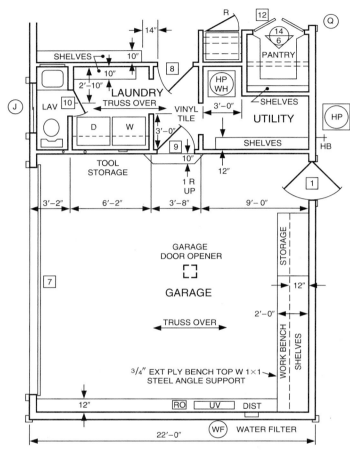

FIGURE 11.5 > Integral garage and adjacent utility room.
Hepler/Wallach/Hepler © Cengage 2013

Many building codes require an elevation change (step-up) of 16″ (406 mm) to 18″ (457 mm) from an integral garage floor to a residence floor. This is to prevent gasoline fumes or leaks from entering a house. Fireproof doors and fire-shielded sheathing or drywall are also required in common house–garage walls and ceilings.

A covered walkway or breezeway from the garage or carport to the house should be provided if the garage is detached. Often a patio or porch is planned for this area to help integrate the detached garage with the house.

A **carport** looks like a garage with one or more of the exterior walls removed. It may be completely separate from the house, or it may be built against the existing walls of the house or garage. Carports are most acceptable in mild climates where complete protection from cold weather is not needed. They offer protection primarily from sun and moisture.

Both the garage and the carport have distinct advantages. The garage is more secure and provides more shelter. However, carports lend themselves to open planning techniques and are less expensive to build than garages.

Decor

The lines of the garage or carport should be consistent with the major building lines and the architectural style of the house. See Figure 11.7. The garage or carport must never appear to be an afterthought.

Floor

The garage floor must be solid and easily maintained. A concrete slab 4″ (102 mm) thick and reinforced with welded wire mesh provides the best deck surface for a garage or carport.

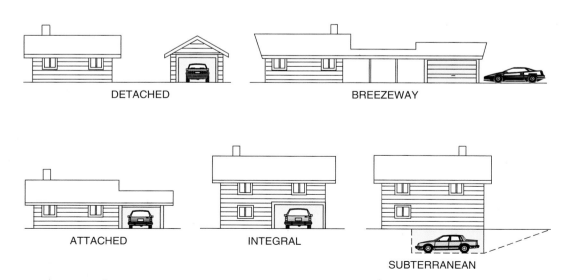

FIGURE 11.6 > Garage locations. *Hepler/Wallach/Hepler © Cengage 2013*

FIGURE 11.7 ❭ Garage style that matches home style.
© *Image Source/Corbis*

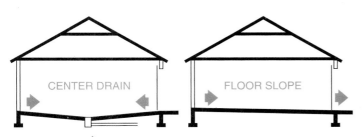

FIGURE 11.8 ❭ Garage floor drainage options. *Hepler/
Wallach/Hepler © Cengage 2013*

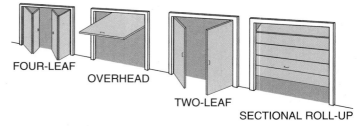

FIGURE 11.9 ❭ Common garage door types. *Hepler/Wallach/
Hepler © Cengage 2013*

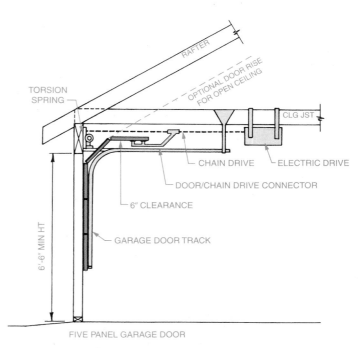

FIGURE 11.10 ❭ Space requirements for overhead garage
door openers. *Hepler/Wallach/Hepler
© Cengage 2013*

A vapor barrier of waterproof materials should be provided under the slab. The garage floor must have adequate drainage either to the outside or through drains located inside the garage. See Figure 11.8.

Doors

The design of the garage door greatly affects the appearance of the house. Several types of garage doors are available: four-leaf swinging, overhead, two-leaf swinging, and sectional roll-up. See Figure 11.9. Electronic devices are available for opening the door of the garage from the car.

For overhead doors, ceiling clearance must be planned to avoid interference between the opened door and light fixtures or other projections. Sectional roll-up doors require a 16″ (406 mm) clearance between the top of the door and the ceiling. Solid overhead doors require more clearance depending on the height of the door. Horizontal ceiling space must be at least 6″ (152 mm) above the height of the door. Figure 11.10 shows typical clearances required for overhead doors.

Materials for garage doors include steel, aluminum, fiberglass, or wood—usually redwood or cedar. Metal doors are insulated or hollow. Solid overhead or roll-up doors are either manually operated or electronically controlled by a radio transmitter that activates a motor and lights. Garage doors of all types are available in a variety of patterns, styles, and sizes. They can be purchased with or without windows.

Storage Design

Most garages are also used for storage space. Cabinets should be elevated above the floor several inches to avoid exposure to moisture and to facilitate cleaning the garage floor. Garden-tool cabinets can be designed to open from the outside of the garage.

Size

The size of the garage depends on the number of vehicles to be parked, plus space for any workshop facilities and storage.

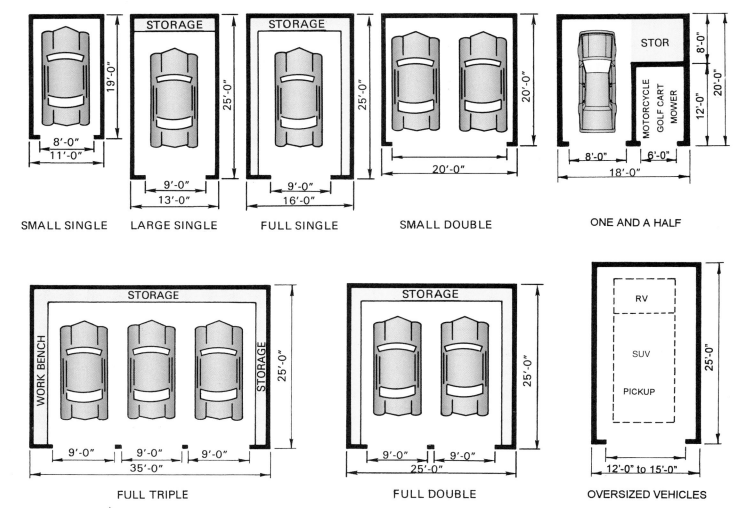

FIGURE 11.11 > Typical garage sizes. *Hepler/Wallach/Hepler © Cengage 2013*

For example, storage space for bicycles, lawnmowers, and other lawn and landscape maintenance equipment must be considered as well as space for vehicles. Typical garage sizes are shown in Figure 11.11. To allow for side mirror clearance, garage doors designed to house SUVs, pickup trucks, or large sedans must be at least 9′ (2,743 mm) wide and 8′ (2,438 mm) high. Garage door heights and widths up to 18′ (5,486 mm) are available for RV storage.

Although some variations exist among manufacturers, standard garage door heights are 6′-6″, 6′-8″, 6′-9″, 7′-0″, 7′-6″, and 8′-0″. Standard garage door widths are 7′, 8′, 9′, 10′, 12′, 14′, 15′, 16′, 17′, 18′, and 20′.

MUD ROOMS

A **mud room** in the service area allows access from the outside without passing through other rooms. The mud room often connects the garage with the utility room and should provide space for changing and storing outer garments. Figure 11.12 shows a typical mud room configuration.

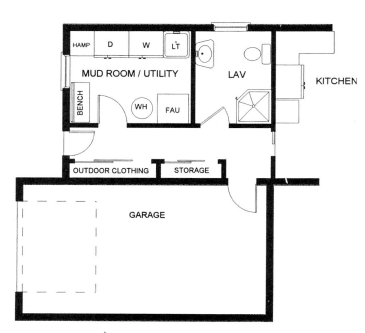

FIGURE 11.12 > Typical mud room configuration. *Hepler/Wallach/Hepler © Cengage 2013*

DRIVEWAYS

The main functions of a driveway are to provide access to all entrances and to the garage and to provide temporary parking space. See Figure 11.13. However, a driveway can serve other purposes, too. A wide apron at the door of the garage can become a useful area, whether for car washing or for children's games. It can also enable cars to turn around without backing out onto a main street.

The driveway should be of brick, stone, asphalt, or concrete construction. Concrete should be reinforced with welded-wire fabric to maintain rigidity and prevent cracking. Masonry pavers (Figure 11.14) are often used over a concrete base or compacted base material. The driveway should be designed at least several feet wider than the track of a car, which is approximately 5'-0" (1,524 mm). Slightly wider driveways of approximately 7' to 9' (2,134 to 2,743 mm) width are desirable for access and pedestrian traffic. To comfortably accommodate wheelchairs, a width of 10' (3,048 mm) may be needed. Sufficient space in the driveway should be provided for parking guests' cars. (Review Chapter 9, *Traffic Areas and Patterns*, for more details concerning driveway configurations and sizes.)

WORKSHOPS

The workshop is an area planned for working with equipment, tools, and materials.

Function and Location

A home workshop is designed for activities ranging from hobbies to home maintenance. See Figure 11.15. As part of the service area, a workshop may be located in the garage, in the basement, in a separate room, or even in an adjacent building.

Workbench space, power tools, hand tools, and the storage areas should be systematically planned to allow for the

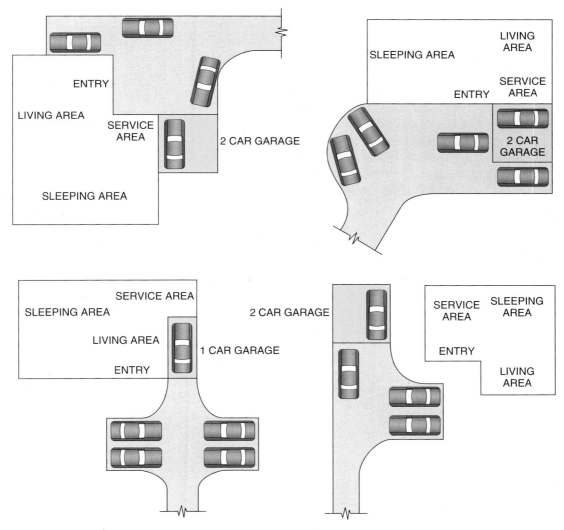

FIGURE 11.13 > Parking and turning aprons in the service area. *Hepler/Wallach/Hepler*
© *Cengage 2013*

FIGURE 11.14 > Driveway surface of masonry pavers. *Cornerstone Developers, Inc.*

FIGURE 11.15 > Home workshop for a variety of activities. *James Eismont*

maximum amount of work space. Tools and equipment for working with large materials should be placed where the material can be handled easily. Any flammable finishing material, such as turpentine or oil-based paint, should be stored in metal cabinets.

Workbench

A workbench, usually with a vise for holding materials in place, is a major component of a home workshop area. The average workbench is 36″ (914 mm) high.

Workbenches are available to suit different needs. A movable workbench is appropriate for large projects. A **peninsula workbench** has three working surfaces with storage compartments on each of the three sides. A **dropleaf workbench** is excellent for work areas where a minimum amount of space is available. The side portions, or "drop leaves," can be extended for increased work space or folded down for storage in a small space.

Hand Tools

Certain **hand tools** are necessary for any type of hobby or home maintenance work. These basic tools include a claw hammer, carpenter's square, files, drills, screwdrivers, planes, pliers, chisels, scales, wrenches, saws, mallets, and clamps.

Hand tools may be safely stored in cabinets that keep them dust free or hung on appropriate hooks on *perforated hardboard.* Tools too small to be hung should be kept in drawers.

Power Tools

Electrically operated tools are called **power tools.** Those commonly used in home workshops include electric drills, saber saws, routers, band saws, circular saws, radial-arm saws, jointers, belt sanders, lathes, and drill presses.

To conserve motors, separate-drive motors can be used to drive more than one piece of power equipment. Separate 110V and 220V electrical circuits for lights as well as power tools should be planned for the home workshop area. Figure 11.16 shows work space clearances and arrangements necessary for safe and efficient machine operation.

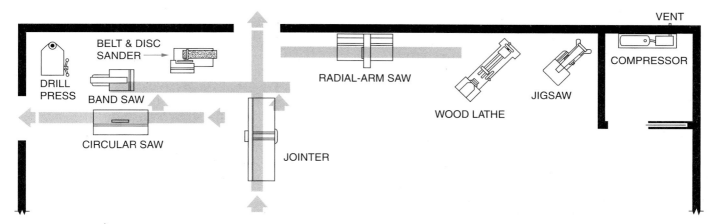

Multipurpose machines that can perform a variety of operations are convenient in a home workshop. Less equipment is necessary, and less space is required.

Decor

The work area should be as maintenance free as possible. Glossy paint, paneling, or tiles over drywall can retard the accumulation of shop dust on the walls. Exhaust fans help eliminate much of the dust and the gases produced in the shop. The shop floor should be concrete or linoleum. For safety, abrasive strips on floors around machines will eliminate the possibility of slipping.

Noise is a concern. Do not locate noisy equipment near the sleeping areas. Interior walls and ceilings should be soundproofed by offsetting studs and adding adequate and continuous insulation to produce a sound barrier. Figure 11.17 shows several types of wall configurations designed to abate sound passing through walls.

Light and color are very important factors in designing the work area. Pastel colors, which reduce eyestrain, should be used for the general color scheme of the shop. Extremely light colors that produce glare and extremely dark colors that reduce effective illumination should be avoided. Choose colors that not only create a pleasant atmosphere in the shop but also help to provide the most efficient and safe working conditions.

Workshops should be well lighted. General lighting should be provided in the shop to a high intensity level on machines and worktable tops. (Refer to Chapter 31 for electrical design considerations.)

Size

The size of the work area depends on the size and number of tools, equipment, the workbench, and the storage facilities provided or anticipated. Plan the size of the work area for maximum expansion. At first, only a workbench and a few tools may occupy the area. If space is planned for the maximum amount of facilities, new equipment, when added, will fit appropriately into the basic plan. The designer must also anticipate the types and amounts of materials that will require storage space in the future and design the space accordingly.

STORAGE AREAS

Areas should be provided for general storage as well as for specific storage within each room.

Function and Types

Storage facilities—whether closets, cabinets, furniture, or room dividers—should be designed for convenient retrieval of stored articles. Those articles that are used daily or weekly should be stored in or near the room where they are needed. Those used only seasonally should be placed in more permanent storage areas. Areas that would otherwise be considered wasted space can become general storage areas. Parts of the basement, attic, or garage often fall into this category.

Closets

There are three basic types of storage closets: wardrobe, walk-in, and wall. A **wardrobe closet** is a shallow clothes closet built into the wall. The minimum depth for this closet is 24" (610 mm) and can be any length.

Guest closets, normally located in the foyer area, are wardrobe closets designed to hold outdoor apparel, as shown in Figure 11.18. Wardrobe closets and walk-in closets, which also include space for stacked clothing, shoes, and drawer items, are usually located in bedrooms.

Depths of more than 24" (610 mm) makes reaching the back of the closet difficult. Swinging or sliding doors should

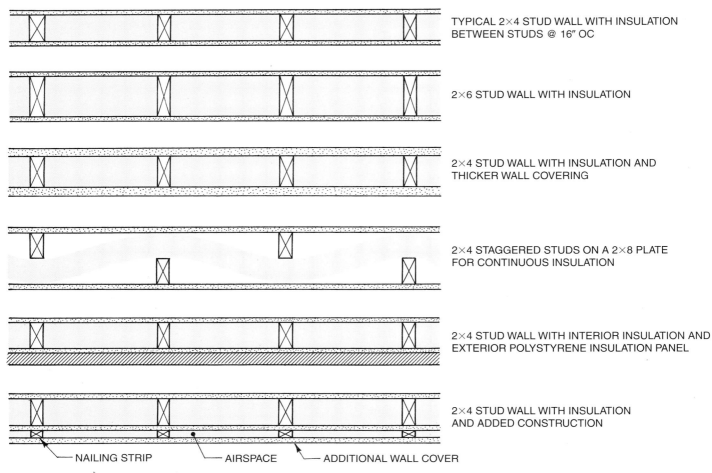

TYPICAL 2×4 STUD WALL WITH INSULATION BETWEEN STUDS @ 16″ OC

2×6 STUD WALL WITH INSULATION

2×4 STUD WALL WITH INSULATION AND THICKER WALL COVERING

2×4 STAGGERED STUDS ON A 2×8 PLATE FOR CONTINUOUS INSULATION

2×4 STUD WALL WITH INTERIOR INSULATION AND EXTERIOR POLYSTYRENE INSULATION PANEL

2×4 STUD WALL WITH INSULATION AND ADDED CONSTRUCTION

NAILING STRIP — AIRSPACE — ADDITIONAL WALL COVER

FIGURE 11.17 › Noise abatement walls. *Hepler/Wallach/Hepler © Cengage 2013*

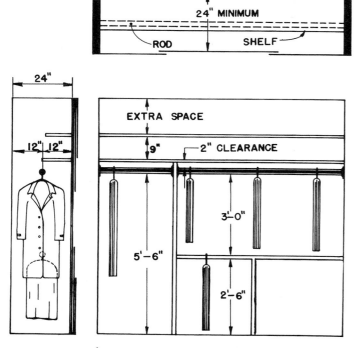

FIGURE 11.18 › Wardrobe closet dimensions. *Hepler/ Wallach/Hepler © Cengage 2013*

expose all parts of the closet that need to be within reach. A disadvantage of the wardrobe closet is the amount of wall space required for the doors.

A **walk-in closet,** as the name implies, is a closet large enough to enter and turn around. The area needed for this type of closet is equal to the amount of space needed to hang clothes plus space for a walkway. (See Chapter 12 for more details about these types of closets.) Although some space is wasted because of the walkway, the closet front area takes up less wall space in the room. Only one closet door is needed. The minimum width is 6′-0″ but can be any length.

A **wall closet** is a shallow closet in the wall for cupboards, shelves, and drawers. Wall closets are normally 18″ (457 mm) deep. This size provides access to all stored items without using an excessive amount of floor area.

Protruding closets that create an offset in a room should be avoided. By filling the entire wall of a room with closet space, a square or rectangular room can be designed without the use of offsets.

Doors on closets should be sufficiently wide to allow easy accessibility. Swing-out doors have the advantage of providing extra storage space on the back of the door. However, space must be allowed for the door swing. For this reason, sliding doors are often preferred. All closets, except very shallow wall closets, should be provided with lighting.

Furniture and Built-In Features

Chests and dressers are freestanding pieces of furniture used for storage, generally in the bedroom. They are available in a variety of sizes and usually contain shelves and drawers.

Window seats are hollow, chest-like structures that are built in below windows for persons to sit on. The hinged tops are often padded and can be raised to allow storage of items inside.

A *room divider* often doubles as a storage area. Room dividers often extend from the floor to the ceiling but may also be only several feet high. Many room dividers include shelves and drawers on both sides.

Shelves

Shelves for storage areas are available in a variety of sizes and materials including solid lumber, plywood, and hollow-core plywood. **Ventilated shelving** is made by welding steel rods together at 1/2″ or 1″ intervals. The rods are then coated with vinyl. Ventilated shelving is available in 9″, 12″, or 16″ depths and up to 12′ lengths.

Location

Different types and configurations of storage facilities are located throughout a dwelling. Figure 11.19 shows a storage configuration designed for many different areas. Besides furniture, the most appropriate types of storage facilities for each room in the house are as follows:

- *Living room:* room divider, wall cabinets, bookcases, window seats, entertainment center (Figure 11.20).
- *Dining area:* room divider, closet.
- *Family room:* built-in wall storage, window seats.
- *Recreation room:* built-in wall storage.
- *Porches:* storage under porch stairs.
- *Patios:* sides of barbecue, storage shed.
- *Outside:* storage areas built into the side of the house.
- *Halls:* wall closets, ends of blind halls, bookshelves.
- *Entrance:* room divider, closet.
- *Den:* wall closet, bookcases.
- *Kitchen:* wall and floor cabinets, room divider, wall closets.
- *Utility room:* floor and wall cabinets.
- *Garage:* cabinets built above the hood of a car, wall closets along sides, added construction on the outside of the ga-

FIGURE 11.19 > Comprehensive storage configuration. *LaStrada Furniture and Interiors, Inc.*

FIGURE 11.20 > Entertainment center storage. © *David Papazian/Beateworks/Corbis*

rage. Figure 11.21 shows an over-hood storage plan, and Figure 11.22 shows a plan for the maximum use of an upper garage level for living quarters.

- *Workshop:* tool board, closets, cabinets.
- *Bedroom:* closets; storage under, at foot, and at head of bed; cabinets; shelves.
- *Bathroom:* cabinets, room dividers.

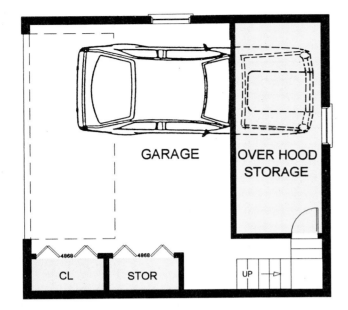

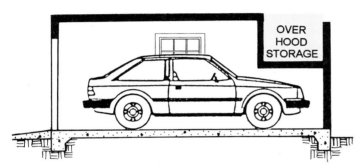

FIGURE 11.21 > Garage storage options. *Hepler/Wallach/ Hepler © Cengage 2013*

SPECIALIZED AREAS

The service area may also provide space and facilities for trash and garbage storage and disposal, fitness training, and pet support.

Trash and Garbage Facilities

Providing space and facilities for the temporary storage and disposal of trash and garbage is often neglected in the development of architectural plans. Trash is waste material such as paper, plastic, glass, and metal. Garbage contains food remains that decompose. If carefully separated and selected, food remains can be used for compost building.

Trash and garbage storage facilities should be hidden from view, secured from children's and pets' access as well as possible spillage, and should be easy to use and transport. Because most trash and garbage is generated in the kitchen, more collection locations are needed there. These may include a food compactor, waste disposal, or sealed containers in a cabinet or on a cabinet door. Other rooms require different numbers and types of storage facilities.

In addition to local storage, a holding area needs to be designated and reserved for the accumulated trash and garbage prior to either truck pickup or transport to a recycle center. This area should have individual bins or areas that separate material according to local environmental codes. Hazardous wastes must be stored separately for individual disposal. This

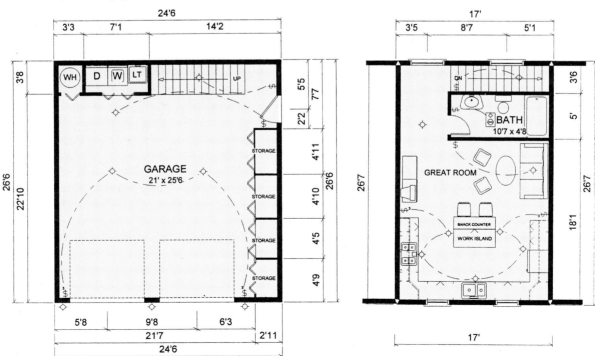

FIGURE 11.22 > Living quarters designed into a garage upper level. *Hepler/Wallach/Hepler © Cengage 2013*

MOVABLE EQUIPMENT	STATIONARY EQUIPMENT	INSTALLED FACILITIES
Punching Bag	Tread Mill	Full Length Mirror
Rings	Stationary Bike	Refrigerator
Free Weights	Stair Stepper	TV - Radio - Audio
Pull Up Bar	Rowing Machine	Shower
Picas	Massage Table	Lavatory
Climbing Rope	Universal Machine	Dressing Room
Weight Bench	Body Bag	Weight Rack
Dip Rack	Slant Board	Storage Space

FIGURE 11.23 › Fitness facilities options. *Hepler/Wallach/Hepler © Cengage 2013*

area should be located in the proximity of the garage or near convenient street access.

If architectural plans do not provide space and facilities for the storage and disposal of trash and garbage, the occupants may store these unsightly, insect-attracting and odorous materials in full view, indoors and outdoors.

Fitness Facilities

Space and facilities for fitness training range from the storage of a few items of equipment in an existing space to a totally equipped fitness room. A list of available fitness equipment and supporting facilities is shown in Figure 11.23. The selection of these facilities depends on personal preference, budget, and space available.

Pet Support Facilities

Pets like people—people like pets. Nevertheless, pets need a home base to be separated from people when they want—and vice versa. Accommodations can range from a designated area in a service area room to a separate room reserved solely for pets. The most popular and practical location is a planned area in a utility room. Regardless of the location the area should be accessible 24 hours a day. Specific plans and details should be developed during the floor plan design stage.

Facilities should include a bed or sleeping surface, food and water containers, storage space for pet toys and possessions, and a litter box for cats. If pets are not trained, a cage may also be needed. If the property is securely fenced or has an invisible fence, an outside access door can be added. To provide the maximum amount of freedom with minimum supervision, a property perimeter fence should be added to the site plan. The pet area should be easily cleanable with washable surfaces and adequate ventilation. The size of the area should be adjusted to the size and number of pets anticipated.

Service Areas Exercises

1. Design a utility room including a complete laundry facility within an area of 12′ × 12′ (144 sq. ft.). Show the location in relation to other areas of a house.

2. Design a utility room for the house you are planning.

3. Design a full double garage and driveway for the house of your design. Include storage, laundry facilities, and a workbench. Identify the type of door you would use.

4. Draw a plan for a garage and/or a workshop.

5. Design a work area for the house you are planning.

6. Add storage facilities to the house of your design.

7. Draw a walk-in closet plan.

8. Sketch plans to provide facilities for pets, exercise, and trash and garbage handling.

CHAPTER 12
Sleeping Areas

OBJECTIVES

In this chapter you will learn to:

> plan and draw bedrooms for a sleeping area.

> plan and draw baths appropriate to the size and arrangement of the floor plan.

> design an efficient bath.

> design a master bedroom with bath.

TERMS

bidet
central bath
compartment plan
cross ventilation
fixtures

half-bath
lavatory
master bath
master bedroom
sauna

walk-in closet
wall storage cabinet
wardrobe closet
water closet
whirlpool tub

INTRODUCTION

Approximately one-third of our time is spent sleeping. Therefore, the sleeping area should be planned to provide facilities for maximum comfort and relaxation. The sleeping area should be located in a quiet part of the house and include bedrooms and baths.

BEDROOMS

Houses are often categorized by the number of bedrooms. For example, a house may be described as a three-bedroom home or a four-bedroom home. A single person or couple with no children may require only a one-bedroom home. Three-bedroom homes are most common, since they accommodate most families. See Figure 12.1.

Function

The primary function of a bedroom is to provide facilities for sleeping. However, some bedrooms may also provide facilities for writing, reading, watching TV, listening to music, or relaxing. As with other rooms, the size and shape of each bedroom depends on the occupants, activities, and furniture designated for that room.

For babies and very young children, a bedroom may serve as a nursery, with a crib and related furniture and equipment. For older children, a bedroom may be a double room with twin beds and other furniture such as desks and entertainment equipment. A **master bedroom** for adults not only has a large bed or beds and other furniture, it usually also has an adjacent bath and, perhaps, a separate dressing room. See Figure 12.2.

Location

For sleeping comfort and privacy, bedrooms should be grouped in a quiet part of the house, as far from the living area as possible. If a further separation is wanted between the master bedroom and children's or guest bedrooms, locating the master bedroom on a different level is recommended for multilevel dwellings. Regardless of the location, all bedrooms must have access to a hall from which a bath is also accessible.

If space permits, some area, preferably a separate room, can be planned for solitary use. This can be for reading, meditation, or quiet study. The sleeping area is ideal for this type of room or area since it is located away from the active areas of the house.

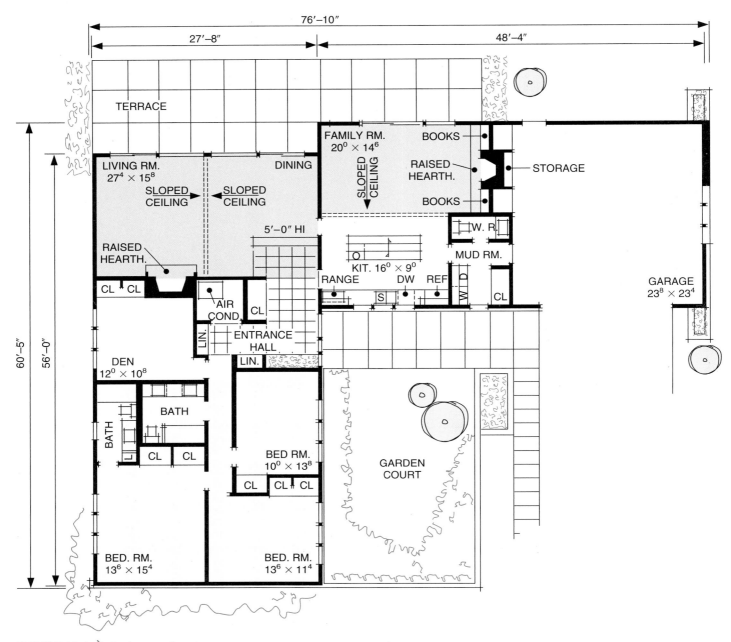

FIGURE 12.1 ❯ Bedroom sleeping areas. *Used by permission Hanley & Wood, LLC*

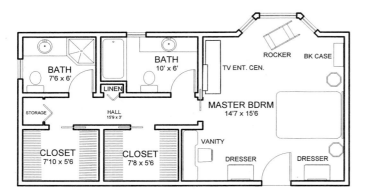

FIGURE 12.2 ❯ Master bedroom suite. *Hepler/Wallach/Hepler*
© Cengage 2013

Guidelines for Noise Control

Because noise contributes to fatigue, location is particularly important for a restful bedroom area. To eliminate as much noise as possible, planned locations and well-selected materials can help accomplish this. See Figure 12.3. The following guidelines are valuable for designing bedrooms that are quiet and restful:

1. The bedroom should be in the quiet part of the house, away from major street noises.

2. Air is a good insulator. Therefore, closets can be located to provide sound buffers. Clothing and other items stored in closets can also help muffle sound.

3. Carpeting or porous wall and ceiling panels absorb noises. Rooms above bedrooms should be carpeted.

4. Floor-to-ceiling draperies help to reduce noise.

5. Acoustical tile in the ceiling is effective in reducing noise.

6. Trees and shrubbery outside the bedroom help absorb sounds.

7. The use of double-glazed insulating glass for windows and sliding doors helps to reduce outside noise.

8. The windows of an air-conditioned room should be kept closed during hot weather. This eliminates noise and aids in keeping the bedroom free from dust and pollen.

9. In extreme cases when complete soundproofing is desired, the wall structure and materials may be designed to provide continuous sound insulation.

10. Placing rubber pads under appliances such as refrigerators, dishwashers, washers, and dryers often eliminates vibration and noise throughout the house.

Wall Space

Bedroom entrance doors, closet doors, and windows should be grouped to conserve wall space whenever possible. By minimizing the distance between doors and windows, the amount of usable wall space is expanded. Long stretches of wall space are best for efficient furniture placement.

Doors

Several types of doors may be used in bedrooms. Pocket, bypass (sliding), and bifold doors are used for closets. Swinging and pocket doors can be used for the entrance door. A swinging door should always swing into the bedroom against an adjacent wall, and not outward into the hall. See Figure 12.4. Figure 12.5 shows an efficient arrangement of closets, door locations, and door swings. The door connecting the bedroom with an outside deck, balcony, or patio should be French, swinging, or sliding glass to maximize light and ventilation. Bedroom closet doors can be specified as *mirrored,* which provides a dual function—as a door and as a full-length mirror—as shown in Figure 12.6.

Standard door heights range from 6′-8″ to 8′-0″. Standard door widths for bedroom doors range from 2′-6″ to 3′-0″. A width of at least 2′-8′ is needed to allow passage of some furniture, and a minimum width of 2′-9″ is needed for wheelchair entry.

Windows and Ventilation

Bedroom windows should be placed to provide air circulation, light, and solar heat from the south. Similar to door placement, designers must also consider the efficient use of wall space for window placement. One method of conserving wall space for bedroom furniture is to use high windows. High, narrow-strip windows, called ribbon windows, provide

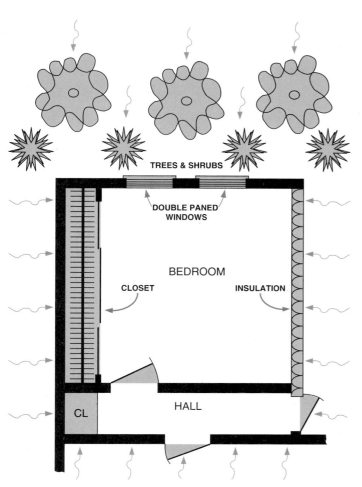

FIGURE 12.3 > Methods of minimizing noise. *Hepler/Wallach/ Hepler © Cengage 2013*

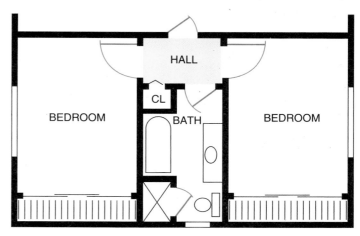

FIGURE 12.4 > Bedroom doors should swing into the bedroom and toward a wall. *Hepler/Wallach/ Hepler © Cengage 2013*

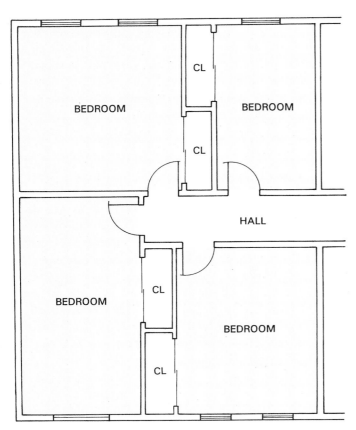

FIGURE 12.5 > Closet and door locations. *Hepler/Wallach/ Hepler © Cengage 2013*

FIGURE 12.6 > Mirrored closet door. *© Marc Gerritsen/Lived In Images/Corbis*

space for furniture to be placed underneath. They also ensure some privacy for the bedroom. Building and fire codes require an escape-size window in each room if no outside door exists. Bedroom windows should be large enough to allow firefighters with backpacks to enter.

Proper ventilation is necessary in bedrooms and is conducive to sound rest and sleep. When air conditioning is available, the windows and doors may remain closed. Central air conditioning and humidity control provide constant levels of temperature and humidity and are efficient methods of providing ventilation and air circulation.

Without air conditioning, windows and doors must provide the ventilation. Bedrooms should have **cross ventilation.** However, the draft must not pass over the bed. See Figure 12.7. High ribbon windows provide cross ventilation without causing a draft on the bed. Jalousie windows are also effective, because they direct air upward.

Storage Space

Storage space placed in or adjacent to bedrooms is needed primarily for clothing and personal accessories. Furniture, closets, cabinets, and dressing rooms serve as storage facilities. These should be located within easy reach and should be easy to maintain.

Freestanding furniture storage space in the bedroom area is usually found in dressers, chests, vanities, and dressing tables. However, most storage space should be provided in the closets. Closet types were introduced in Chapter 11. Figure 12.8 shows a bedroom **wardrobe closet** organized to store hanging and stacked clothing, shoes, and drawer items. Average dimensions are also included. Figure 12.9 is an example of a bedroom **wall storage cabinet,** which can be used to replace or reduce the amount of bedroom furniture. These closets should be located to avoid the creation of awkward wall offsets. See Figure 12.10.

The minimum amount of space required for bedroom **walk-in closets** includes the storage space plus the amount

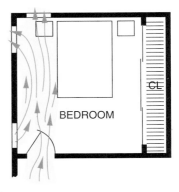

FIGURE 12.7 > Cross ventilation to avoid draft over bed. *Hepler/Wallach/Hepler © Cengage 2013*

of space required to walk in and turn around. Note the dimensions in Figure 12.11. Figure 12.12 shows several walk-in closet configurations. Figure 12.13 shows a portion of a walk-in closet devoted to shelf, cabinet, and drawer storage. The walk-in closet in Figure 12.14 features a center island and a mirrored dressing alcove. Compare the wardrobe closet and walk-in closet spaces shown in Figure 12.15.

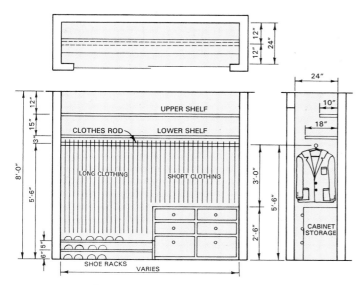

FIGURE 12.8 > Wardrobe closet. *Hepler/Wallach/Hepler © Cengage 2013*

FIGURE 12.9 > Enclosed storage within a walk-in closet. *LaStrada Furniture and Interiors, Inc.*

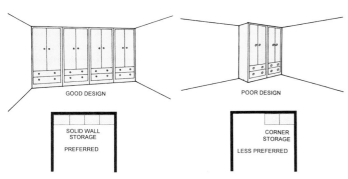

FIGURE 12.10 > Closets should not create room offsets. *Hepler/Wallach/Hepler © Cengage 2013*

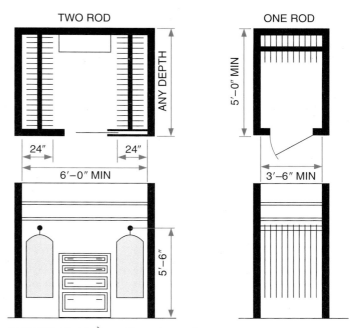

FIGURE 12.11 > Walk-in closet dimensions. *Hepler/Wallach/ Hepler © Cengage 2013*

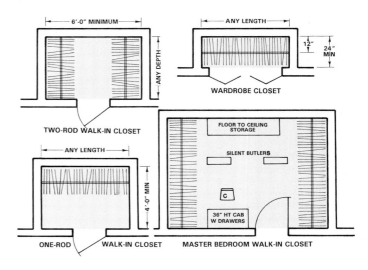

FIGURE 12.12 > Walk-in closet and wardrobe configurations. *Hepler/Wallach/Hepler © Cengage 2013*

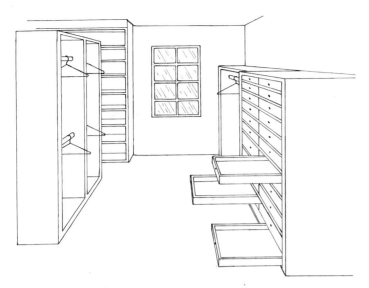

FIGURE 12.13 ⟩ Walk-in closet drawer and cabinet storage.
Hepler/Wallach/Hepler © Cengage 2013

Dressing Areas

A dressing area is usually located adjacent to the master bedroom. It may be a separate room, an alcove, or part of the bedroom separated by a divider that also provides storage space. Figures 12.16A and 12.16B show a dressing room, bath, balcony, and bedroom design, which comprises a master bedroom suite. The total floor plan, which includes the bath shown in Figure 12.16A and 12.16B, is found in Figure 15.28A and 15.28B.

Children's Rooms

Children's bedrooms and nurseries must be planned to be comfortable, quiet, and sufficiently flexible to allow change as the child grows and matures. See Figure 12.17. For example, storage shelves and rods in closets should be adjustable so that they can be raised as the child grows taller. Light switches should be placed low for small children and have a delay switch that allows the light to stay on for some time after the switch has been turned off.

Adequate facilities for study and hobby activities should be provided, such as a desk and worktable. Storage space for books, models, and athletic equipment is also desirable. Chalkboards and bulletin boards on the walls help make the child's room usable.

Grouping or dividing children's rooms into convertible double rooms is one method of designing for future change as children grow. See Figure 12.18. Eventually the divided room can become a single large guest bedroom or study.

The use of pull-down beds is effective for youth or guest bedrooms. This design, as shown in Figure 12.19, provides

FIGURE 12.14 ⟩ Walk-in closet with island and mirrored alcove. *LaStrada Furniture and Interiors, Inc.*

FIGURE 12.15 ⟩ Wardrobe and walk-in closet comparison.
Hepler/Wallach/Hepler © Cengage 2013

space for other activities when the bed is in the upright wall position. The use of attics (Figure 12.20) as a location for youth or guest bedrooms also makes maximum use of otherwise less used space.

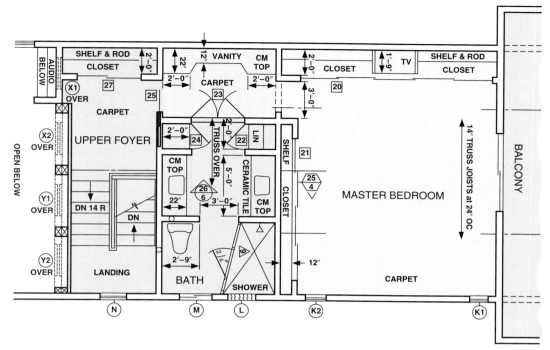

FIGURE 12.16A 〉 Master suite with dressing areas and adjoining bath. *Hepler/Wallach/Hepler*
© *Cengage 2013*

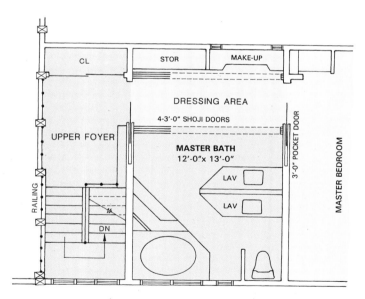

FIGURE 12.16B 〉 Alternative design for plan shown in Figure 12.16A. *Hepler/Wallach/Hepler*
© *Cengage 2013*

FIGURE 12.17 〉 Child-oriented bedroom. *Marc-Michaels Interior Design Inc.*

Decor

In general, bedrooms should be decorated in quiet, restful tones. Matching or contrasting bedspreads, draperies, and carpets help accent the color scheme. Uncluttered furniture with simple lines also helps to develop a restful atmosphere in

the bedroom. The bedroom shown in Figure 12.21 illustrates the effective blending of fabrics, colors, lighting, and architectural details to create a warm and relaxing environment.

Size and Shape

The type, size, and style of furniture to be included in the bedroom should be chosen before the size of the bedroom

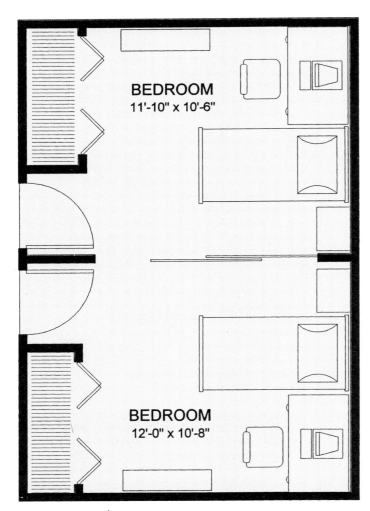

FIGURE 12.18 > Convertible double bedroom. *Hepler/Wallach/Hepler © Cengage 2013*

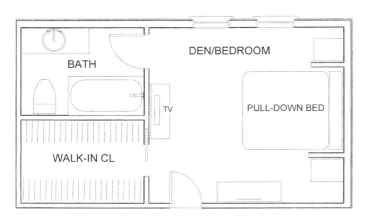

FIGURE 12.19 > Use of pull-down bed to maximize the use of space. *Hepler/Wallach/Hepler © Cengage 2013*

FIGURE 12.20 > Use of attic as youth or guest bedroom. *Western Wood Products Association*

FIGURE 12.21 > Effective use of fabrics, colors, lighting, and architectural details in a bedroom. *Fran Murphy Interiors*

is established. In a preliminary design, the sizes and amount of furniture determine the size of the room and not the reverse. Because the bed or beds require the most space, the room size must provide adequate space for and around the bed. Furniture size and the spacing between furniture has a significant effect on room size. Figure 12.22 shows the minimum spacing requirements between bedroom furniture. A minimum-sized bedroom should accommodate at least a

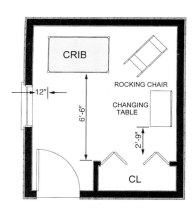

SMALL NURSERY

CRIB 28"x 52" (2'-4"x 4"x4")

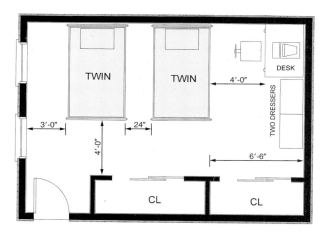

BEDROOM with TWIN BEDS

39"x 75" (3'-3"x 6'-3")

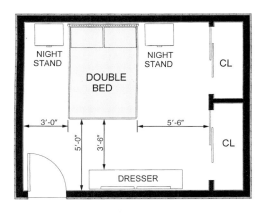

BEDROOM with DOUBLE BED

54"x 75" (4'-6"x 6'-3")

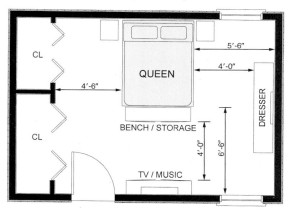

BEDROOM with QUEEN BED

60"x 80" (5'-0"x 6'-8")

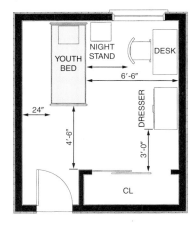

YOUTH BEDROOM

YOUTH BED 31"x 74" (2'-7"x 6'-2")

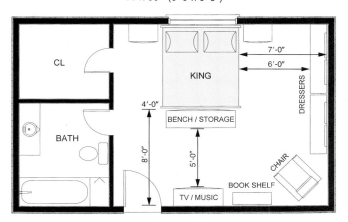

BEDROOM with KING BED

STANDARD KING 72"x 84" (6'-0"x 7'-0")
CALIFORNIA KING 76"x 80" (6'-4"x 6'-8")

FIGURE 12.22 > Minimum spacing between bedroom furniture. *Hepler/Wallach/Hepler © Cengage 2013*

FIGURE 12.23 > Built-in components that blend with architectural style. *LaStrada Furniture and Interiors, Inc.*

FIGURE 12.24 > Contemporary bedroom decor. *LaStrada Furniture and Interiors, Inc.*

single bed, bedside table, and dresser. In contrast, a larger, master bedroom suite may include a separate dressing area, vanity area, master bath, TV, DVD, radio, king or queen bed, dressers, armoire, chaise, chairs, walk-in closet, and built-in storage cabinets. Bedroom furniture and components may also be built in. Note how the built-in components in the bedroom in Figure 12.23 blend with the architectural style. Figure 12.24 also shows a contemporary bedroom with minimal lines, embellishments, and smooth contours.

The size of the furniture and the space between the furniture needs to be considered to determine the dimensions of the room. See Figure 12.25. Use of furniture templates, as with other rooms, is an effective method for designing bedroom sizes and shapes. With templates, the traffic patterns and the way the room functions can be foreseen.

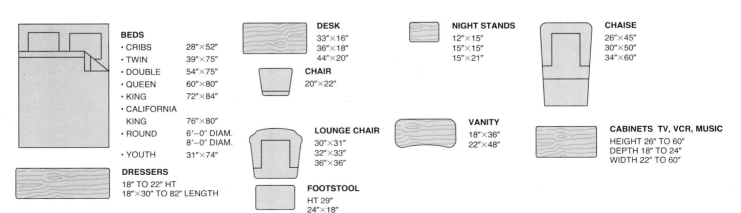

BEDS
- CRIBS — 28″×52″
- TWIN — 39″×75″
- DOUBLE — 54″×75″
- QUEEN — 60″×80″
- KING — 72″×84″
- CALIFORNIA KING — 76″×80″
- ROUND — 6′–0″ DIAM. 8′–0″ DIAM.
- YOUTH — 31″×74″

DRESSERS
18″ TO 22″ HT
18″×30″ TO 82″ LENGTH

DESK
33″×16″
36″×18″
44″×20″

CHAIR
20″×22″

LOUNGE CHAIR
30″×31″
32″×33″
36″×36″

FOOTSTOOL
HT 29″
24″×18″

NIGHT STANDS
12″×15″
15″×15″
15″×21″

VANITY
18″×36″
22″×48″

CHAISE
26″×45″
30″×50″
34″×60″

CABINETS TV, VCR, MUSIC
HEIGHT 26″ TO 60″
DEPTH 18″ TO 24″
WIDTH 22″ TO 60″

FIGURE 12.25 > Typical bedroom furniture sizes. *Hepler/Wallach/Hepler © Cengage 2013*

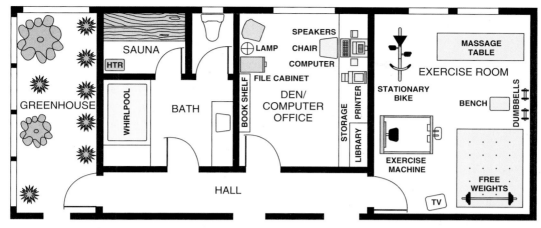

FIGURE 12.26 > Bath with sauna and whirlpool. *Hepler/Wallach/Hepler © Cengage 2013*

Bedrooms and all livable rooms need to be a minimum of 100 sq. ft. An average size bedroom is between 144 sq. ft. and 200 sq. ft. Bedrooms over 200 sq. ft. are considered large.

BATHS

Baths (or bathrooms) must be planned to be functional, attractive, and easily maintained. They can vary widely in size depending on the space available and additional areas included.

Function

Designing a bath involves the appropriate placement of fixtures, cabinets, accessories, and plumbing lines. Adequate ventilation, heating, and lighting also need to be planned. In addition to the normal functions, baths may also provide facilities for dressing, exercising, or laundering. Some may include a **sauna** and **whirlpool tub.** See Figure 12.26. Where space exists, more elaborate baths can be designed to function as relaxing lounges, as shown in Figure 12.27.

Fixtures

The basic bath **fixtures** (items connected to plumbing lines) included in most bathrooms are a **lavatory** (sink), a **water closet** (toilet fixture), and a bathtub or shower. The convenience of the bathroom depends on the arrangement of these fixtures.

Lavatories, or sinks, are available in a wide variety of colors, materials, sizes, and shapes. They are manufactured with porcelain-covered steel or cast iron, stainless steel, brass, copper, or acrylic or other composite materials such as cultured marble. Sinks are either set into an opening, molded with the countertop into one piece, or made into a freestanding fixture on a pedestal. Sink areas should be well lighted and free of traffic, and include a mirror on the wall over the sink. Side mirrors provide additional angles for viewing. A comfort-

FIGURE 12.27 > Lounge type bath in a contemporary style. *LaStrada Furniture and Interiors, Inc.*

able sink height for most people is between 34″ and 36″. For wheelchair access, a height of 36″ with knee access height of 27″ to 30″ is usually required.

Water closets are available in either one-piece or two-piece models. One-piece models are either mounted on the floor or on a wall. Water closets need a minimum of 18″ distance from the center to a side wall or to other fixtures. For wheelchair access, a clear 30″ radius opening is required around water closets. If space allows, the water closet should not be visible when the door to the bath is open. Some states have enacted codes that require all new residential construction to include low-flush, 1.6-gallon toilets. To share plumbing lines, **bidets** should be located close to toilets, with a minimum of 30″ between the two fixtures.

The variety of tubs and showers available in squares, rectangles, or irregular shapes allows a great amount of flexibility in fixture arrangements. Many small or average-size baths include a showerhead on a tub wall. This tub/shower combination then requires an enclosure or a shower curtain rod. Figure 12.28A shows a picture of the bath plan in Figure 12.28B. This plan includes a large oval whirlpool tub with shower and matching curved shower curtain enclosure. Grab bars and other fixtures installed for people with impairments must be secured to studs.

Tubs without a showerhead can be located in open areas. See Figure 12.29. Large tubs can include whirlpool outlets, jets, and controls for hydrotherapy. Special tubs with side openings that seal when closed are available for persons with physical impairments. Grab bars installed beside the tub are simple, but effective safety features, particularly for older people and those with physical impairments.

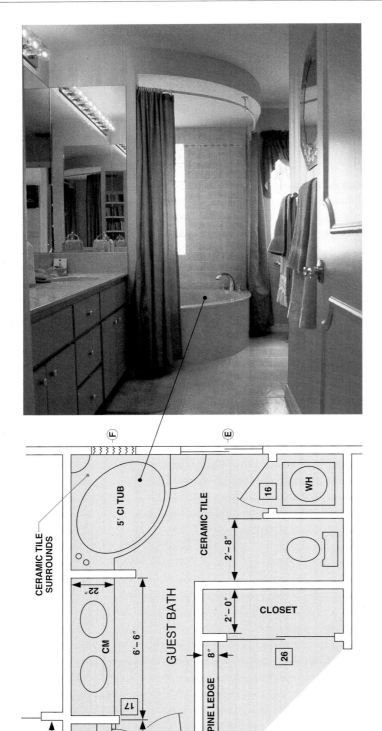

FIGURE 12.28A > Entrance view of the Figure 12.28B plan. *Hepler/Wallach/Hepler © Cengage 2013*

FIGURE 12.28B > Plan view of Figure 12.28A. *Hepler/Wallach/Hepler © Cengage 2013*

FIGURE 12.29 > Tub located for optimum outside exposure. *American Standard*

FIGURE 12.30 > Shower door and configuration. *LaStrada Furniture and Interiors, Inc.*

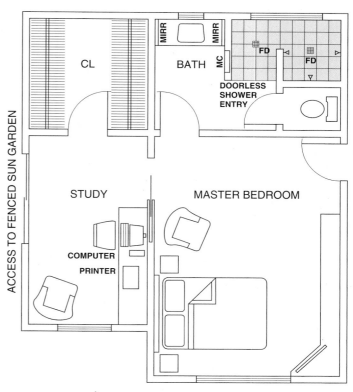

FIGURE 12.31 > Shower design without doors. *Hepler/ Wallach/Hepler © Cengage 2013*

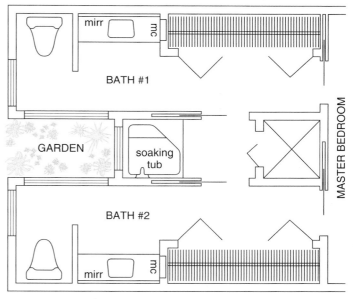

FIGURE 12.32 > Three-compartment bath. *Hepler/Wallach/ Hepler © Cengage 2013*

A separate shower stall is often preferred or substituted for a tub. One-piece prefabricated units are available that are made of materials such as fiberglass, acrylic, coated steel, or aluminum. Many shower stalls are constructed using ceramic tile, marble, or synthetic materials. Glass shower walls, used for panels and doors, are made from shatterproof glass mounted between metal frames as shown in Figure 12.30. Where space is available, showers can be designed without a door. See Figure 12.31.

Some showerheads can now equalize hot and cold water pressure. For example, if cold water pressure is reduced during shower use, the hot water pressure is also automatically reduced to avoid scalding.

Accessories

In addition to the three basic fixtures, many accessories help improve a bath's functions. Accessories range from various furnishings and plumbing devices to lights and heating devices. *Bath furnishings* include such items as a medicine cabinet, extra mirrors, a magnifying mirror, extra counter space, a dressing table, shelves for linen storage, and a clothes hamper. A bath designed for, or used in part by, children should include a low or tilt-down mirror, benches for reaching the lavatory, low towel racks, and shelves for bath toys.

Plumbing accessories such as a whirlpool bath or bidet might be installed. Figure 12.39 on page 224, shows several bidet arrangements. Foot-pedal controls for water and single-control faucets are other types of accessories.

Layout

There are two basic types of bath layouts: the compartment plan and the open plan. In the **compartment plan,** partitions (sliding doors, glass dividers, louvers, or even plants) are used to divide the bath into compartments: the water closet area, the lavatory area, and the bathing area. See Figure 12.32. In the open plan, all bath fixtures are completely or partially

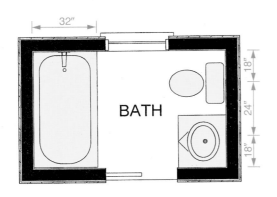

SMALL BATHROOM 5'-0"x 8'-0"

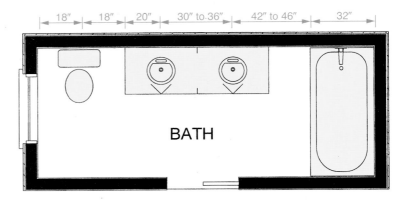

AVERAGE BATHROOM 5'-6"x 14'-0" (78.4 sq ft)

FIGURE 12.33 ⟩ Minimum spacing between bathroom fixtures. *Hepler/Wallach/Hepler © Cengage 2013*

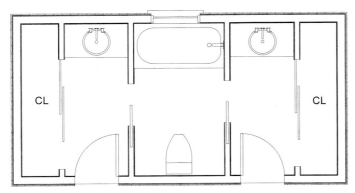

LARGE PARTITIONED BATH

FIGURE 12.34 ⟩ Compartment bath with shared and individual fixtures. *Hepler/Wallach/Hepler © Cengage 2013*

FIGURE 12.35 ⟩ Expansion of light and space with glass block and mirrors. *Pittsburgh Corning Corp.*

visible. The spacing between bath fixtures is vital for the convenient functioning of the entire bath. Figure 12.33 shows the minimum spacing allowed between fixtures and from fixtures to walls. See Chapter 14 for the spacing needed for fixtures for people with special needs. Figure 12.34 shows a compartment bath with separate and shared fixtures. Figure 12.35 illustrates the use of mirrors and glass block to add light and the feeling of expanded space.

Ventilation

Baths should have either natural ventilation from a window or forced ventilation from an exhaust fan. Care should be taken to place windows in a position where they will not cause a draft on the tub or interfere with privacy. A bath can be designed without windows. However, substitute sources of light and ventilation are then needed. This can be achieved by installing a light and ventilating fan combination that is controlled by a single switch.

Lighting

Lighting should be relatively shadowless in the area used for grooming. Shadowless general lighting can be achieved with fluorescent tubes installed on the ceiling and covered with glass or plastic panels. Skylights can also help provide general illumination.

Heating

Heating in the bath is especially important. In addition to the conventional heating outlets, an electric heater or heat lamp is often used to provide instant heat. A wall or floor heat source should be placed under the window to eliminate drafts. All gas and oil heaters should be properly ventilated. Heat lamps should be controlled by a timer to avoid overheating.

Location

Two factors in locating baths are positioning of plumbing lines and accessibility from other areas of the house.

Plumbing Lines

The plumbing lines that carry water to and from the bath's fixtures should be concealed and minimized as much as possible. When two baths are located side by side, placing the fixtures back to back on opposite sides of the plumbing wall reduces the length of plumbing lines. See Figure 12.36. In multiple-story dwellings, efficient use of plumbing lines can be accomplished if the baths are placed one directly above another. When a bath is placed on a second floor, a plumbing wall must be provided through the first floor for the soil and water pipes.

Accessibility

Ideally, a private bath should be located adjacent to each bedroom, as shown in Figure 12.37. Often, this is not possible, and a **central bath** is designed to meet the needs of the entire family. See Figure 12.38. The central or general bath should be in the sleeping area, accessible from all of the bedrooms. A bath for general use plus a bath adjacent to the master bedroom is a desirable compromise. A **master bath** is accessible only from the master bedroom.

Bathing or showering facilities are usually not needed in the living or service areas. **Half-baths** containing only a lavatory and water closet are designed for these areas, unless a full bath is conveniently located nearby.

Decor

Today's baths need not be strictly functional and sterile in decor. They can be planned and furnished in a variety of styles. Baths should be decorated and designed to provide the maximum amount of light and color.

Materials

Materials used in the bathroom should be water resistant, easily maintained, and easily sanitized. Tiles, linoleum, marble, plastic laminate, and glass are excellent materials for bath use. If wallpaper or wood paneling is used, it should be waterproof. If plaster or green drywall construction is exposed, a gloss or semigloss paint should be used on the surface.

Fixtures are now available in a variety of colors, so that they can be coordinated or even matched with accessories. Matching countertops and cabinets are also available.

New materials and components enable the designer to plan baths with modular units that range from one-piece molded showers and tubs to entire bath modules. In these units, plumbing and electrical wiring are connected after the unit is installed. The environmental issues described in

FIGURE 12.37 > Sleeping areas with baths designed for each bedroom. *Hepler/Wallach/Hepler © Cengage 2013*

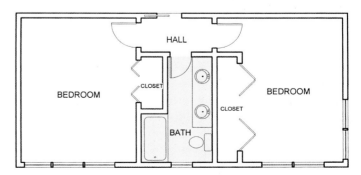

FIGURE 12.38 > Central bath serving two bedrooms. *Hepler/ Wallach/Hepler © Cengage 2013*

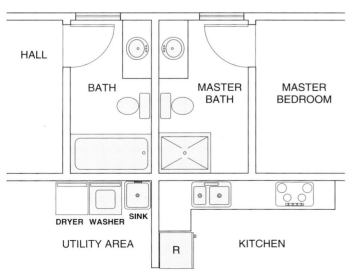

FIGURE 12.36 > Fixtures located to minimize plumbing lines. *Hepler/Wallach/Hepler © Cengage 2013*

HALF BATH 25 SQ FT +

HALF BATH 35 SQ FT +

HALF BATH 30 SQ FT +

SMALL FULL BATH 50 SQ FT +

FULL BATH 100 SQ FT +

PARTITIONED BATH 80 SQ FT +

BACK TO BACK BATHS 120 SQ FT +

LUXURY BATH 100 SQ FT +

LARGE PARTITIONED BATH 100 SQ FT +

LARGE LUXURY BATH 220 SQ FT +

LARGE BATH WITH PEDESTAL BATH 140 SQ FT +

LARGE BATH 150 SQ FT +

LARGE LUXURY BATH 160 SQ FT +

HERS & HIS BATH WITH GARDEN 200 SQ FT +

LARGE LUXURY BATH 350 SQ FT +

FIGURE 12.39 › A variety of bath shapes and configurations. *Hepler/Wallach/Hepler © Cengage 2013*

Chapter 6 also apply to the sleeping area. When selecting surface materials, it is important to avoid materials that produce toxic emissions. Organic or toxic–emission-free materials should be expressly specified as alternatives.

Size and Shape

Bath sizes and shapes are influenced by the size and spacing of basic fixtures and accessories. The type of plan—whether compartmentalized or open—and the relationship to other rooms in the house also influence size and shape. Additional space may be required to accommodate wheelchairs. Figure 12.39 shows a variety of bath shapes and arrangements.

The minimum size bath that can include all three basic fixtures is 5′ × 8′. Sizes range from this minimum to luxury compartmentalized baths. Figure 12.40 shows the standard sizes of bath cabinets. Although cabinets may be custom made to any size, choosing standard cabinet dimensions saves considerable time and expense. Figure 12.41 shows the various standard sizes of bath fixtures, and Figure 12.42 illustrates several configurations designed for residential saunas.

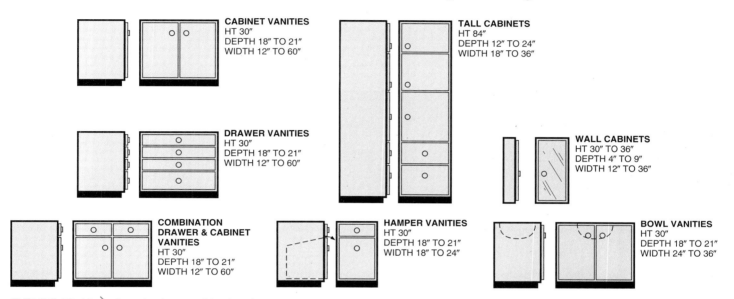

FIGURE 12.40 > Standard sizes of bath cabinets. *Hepler/Wallach/Hepler © Cengage 2013*

FIGURE 12.41 > Standard sizes of bath fixtures. *Hepler/Wallach/Hepler © Cengage 2013*

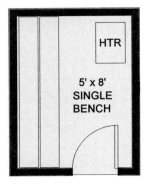
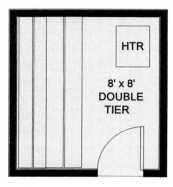
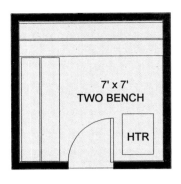

FIGURE 12.42 ⟩ Typical sauna shapes. *Hepler/Wallach/Hepler © Cengage 2013*

Sleeping Areas Exercises

CHAPTER **12**

1. Design a bedroom, 100 sq. ft. in size, for a 6-year-old child.
2. Design a bedroom, 150 sq. ft. in size, for a teenager.
3. Design a master bedroom with an adjoining bath that is 200 sq. ft. in size.
4. Draw a bedroom shown in this chapter.
5. Plan the bedroom areas for the home you are designing.
6. Using dimensions provided in this chapter, calculate the minimum size of a bedroom that could accommodate a king size bed, built-in TV, a dresser, and a lounge chair.
7. Collect pictures of bedrooms you like. Identify good and bad design features.
8. Draw a plan with a master bath and a central bath.
9. Draw a plan of a bath you think is poorly designed. Then draw a plan for remodeling the bath to make it more functional.
10. Draw the plans for the bath areas in a home of your own design.
11. Design and draw a bath that is $12' \times 8'$.
12. Calculate the dimensions needed for a bath you design with one cabinet vanity, two drawer vanities, one hamper vanity, one wall cabinet, a countertop lavatory, a bathtub, and a one-piece water closet. (Use the standard sizes given in Figures 12.40 and 12.41 for reference.)

PART 4

BASIC ARCHITECTURAL DRAWINGS

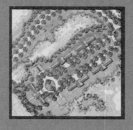

Site Development Plans

OBJECTIVES

In this chapter you will learn to:

> understand the relationship of topography to all site plans.

> identify the major elements used in site design.

> understand the role and uses of zoning ordinances in the design process.

> draw survey, plat, and plot plans.

> understand the polar coordinate system and its application to site plans.

> design, draw, and render landscape plans and elevations.

TERMS

building envelope
building permits
cluster housing
contour interval
contour lines
contour numbering
daylight plane
density
depressions

hilltop summits
interpolation
landscape plans
mixed-use development
phasing
planting schedules
plat
plot plans
profiles

ridge lines
setbacks
site analysis
suitability
survey
topography
USGS
valleys
zoning ordinances

INTRODUCTION

Site planning is both an art and a science, requiring creativity and technical knowledge. Site planning includes the creation of surveys, plot plans, plat plans, landscape plans, site renderings, and construction details. The preparation of these plans requires the application of the principles of design, environmental factors, and site engineering. This involves conforming to building codes, covenants, setbacks, soil restrictions, drainage problems, and topographic situations. In addition, the function and position of sewer, septic, electrical, gas, water, and cable lines must be considered. These site-related plans may be prepared separately or some may be combined as covered in this chapter.

Landscape architecture is primarily concerned with the use of space and the integration of landform, site, and character. Achieving this goes beyond simply planting trees and shrubs. It involves the development of the entire site. Site develop-

ment is an integral part of the design process. A site design should provide proper orientation and use of natural features of the site.

Site plans describe the characteristics of the land and the relationship of all structures to the site. The outline and dimensions of all constructed features (buildings, driveways, etc.) and their exact position on the site are shown on site plans. Also included are the shape of the landform as well as the locations and types of plant material. Specialized site plans include survey plans, plats, plot plans, landscape plans, and renderings. These are discussed in this chapter. Various features of these plans are often combined into one composite site plan.

SITE AND ENVIRONMENTAL ANALYSIS

The initial evaluation of a site and related environmental issues is necessary to ensure consistency and to begin the floor

- Slope
- Soils
- Vegetation
- Wildlife and factors related to habitat, especially if rare and endangered species inhabit the site
- Hydrology:
  – Surface water
  – Flood hazard
  – Groundwater
  – Wetlands (FWW/Tidal)
- Climate (regional) and microclimate (specific to site)
- Geology
- Visual character

FIGURE 13.1 > Environmental factors affecting site design. *Hepler/Wallach/Hepler © Cengage 2013*

- Existing site, street layout, and topography
- Existing land use and zoning
- Historical significance and preservation
- Available utilities
- External factors
  – Noise
  – Site accessibility
- Demographics (population characteristics)
- Socioeconomic forecasts

FIGURE 13.2 > Human factors affecting site design. *Hepler/Wallach/Hepler © Cengage 2013*

plan design process that follows. Completing a **site analysis** is the first step in producing an acceptable site design. The design should meet the needs of the user, as well as protect and enhance the environment. Future inhabitants of the site must also be considered. Both environmental (Figure 13.1) and human-related elements (Figure 13.2) that influence development and design are analyzed in the site analysis. The surrounding area of a site should also be considered when determining future plans. For example, commercial zoning changes may affect plans for constructing a residence.

Suitability Levels

Environmental and human-related elements must be analyzed to determine the level of a site's **suitability** for development and building.

- *High suitability:* Many favorable conditions exist to make this area relatively inexpensive to develop with a minimum amount of environmental impact.
- *Moderate suitability:* Some special design and construction measures will be needed to modify this land and preserve the environment.

- *Low suitability:* Conditions exist that place serious restrictions on building in this area. For example, either environmental damage will result if the site is disturbed, or high construction costs will be needed to avoid damage to ecosystems or to protect the public.
- *Not suitable:* Disturbance or impact to this area will cause significant environmental damage or adversely impact the public safety and welfare. Costs to develop the area are excessive.

Suitability levels may apply to the entire site or to selected zones of a site.

ZONING ORDINANCES

Before beginning the design concept, all local zoning ordinances must first be thoroughly checked. **Zoning ordinances** are laws or regulations designed to provide safety and convenience for the public and to preserve or improve the environment.

Local building codes must also be checked prior to beginning a design. Redesigning or redrawing plans that are in conflict with laws is very time consuming and costly. Working drawings should not be started until the basic site plan has been approved by the local zoning authorities.

Specific zoning ordinances may differ among communities. However, zoning categories are very similar. For example, zoning laws specify the type of occupancy, population density (number of persons in an area), land use, and the building type allowed in each zone of a community. *Tax maps,* which identify each property in a community are found in each community's building department.

Most codes divide municipalities (cities) into residential, commercial, and industrial zones. Residential zones are divided into single-family dwellings, multiple-family dwellings (duplex, triplex, and quadruplex), and apartments that include units for five or more families. Commercial zones include schools, offices, retail stores, and medical facilities. Industrial zones include factories, warehouses, or any facility requiring the movement or storage of large vehicles or manufacturing equipment. Where appropriate, separate zones are established for hazardous areas, such as those where the danger of flooding and earthquakes exists.

Structural Types

The types of structures allowed in each community are related to the designated use of each zone. A community committee often regulates these ordinances. Zoning ordinances may be intended to maintain a degree of architectural consistency within a given area. Ordinances of this type may restrict or allow only specific styles, periods, materials, landscaping, heights, colors, or sizes.

Maximum and sometimes minimum building sizes are often specified to control the use of space, traffic, and the impact on the environment. For example, a single-family residential zone in one community may require a minimum size home of 2,000 sq. ft. and a maximum of 5,000 sq. ft. Another community may require a minimum of 2,500 sq. ft. with an unlimited maximum area. In addition to the amount of square footage allowed, some codes specify maximum building size with reference to street frontage and property lines.

All codes now include maximum building heights to allow neighbors maximum access to views, air circulation, and sunlight. A maximum height of 35 ft. is often used for residential zones. Many newer codes now include limits on the amount of space used in upper floors. See Figure 13.3. In this illustration the X and Y percentages represent the variables among daylight plane laws. This pyramid principle, called the **daylight plane,** is required to allow more light to reach adjacent properties located on the north side. See Figure 13.4. The impact of shadowing is becoming more critical as building areas become more densely built. Using constructed models or computer-drawn models, shadow patterns are studied and checked during different seasons and times.

Land Coverage and Setbacks

Regardless of the square footage of a building, laws may also restrict the percentage of property allowed for building space. This is done to preserve as much *green space* as possible. Land coverage (footprint) percentages vary from one community to another, but are usually from 25% to 40%. For example, a small site, 50′ × 120′, with a coverage of 40% can contain a maximum size building of 2,400 sq. ft. See Figure 13.5.

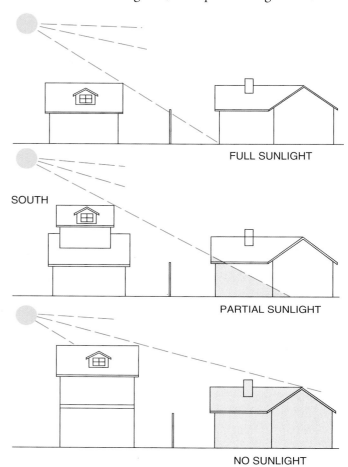

FIGURE 13.4 > Some codes restrict the blockage of sunlight on adjacent properties. *Hepler/ Wallach/Hepler © Cengage 2013*

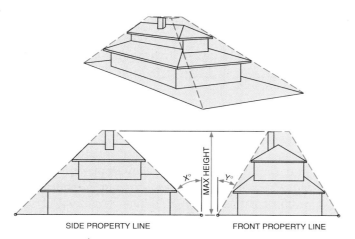

FIGURE 13.3 > The pyramid shape allows maximum sun penetration. *Hepler/Wallach/Hepler © Cengage 2013*

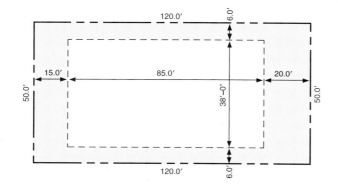

FIGURE 13.5 > Typical setback ordinance requirements. *Hepler/Wallach/Hepler © Cengage 2013*

- MINIMUM LOT SIZE IS 6,000 SQ FT
- FRONT SETBACK IS 15′
- REAR SETBACK IS 20′
- SIDE SETBACK IS 6′
- MAXIMUM LAND COVERAGE IS 40%
- MAXIMUM HEIGHT IS 30′

  6,000 SQ FT
  −2,400 SQ FT
  3,600 SQ FT OF THE BUILDING SITE THAT MAY NOT HAVE STRUCTURE COVERAGE

BUILDABLE AREA IS: 85′ × 38′ = 3,230 SQ FT
40% OF THE LOT IS: .4 × 6,000 = 2,400 SQ FT

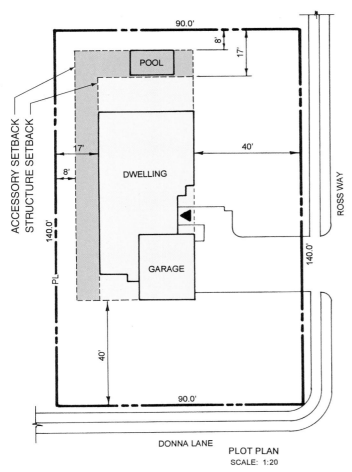

FIGURE 13.6 > Setback requirements applied to a corner lot. *Hepler/Wallach/Hepler © Cengage 2013*

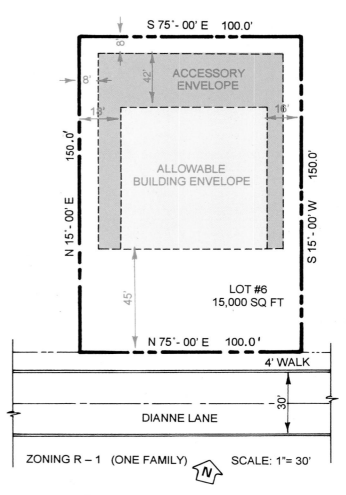

FIGURE 13.7 > Setback requirements applied to an interior lot. *Hepler/Wallach/Hepler © Cengage 2013*

Zoning laws include codes that dictate the distance required from any building to the property lines. These distances are known as **setbacks.** Refer again to Figure 13.5. Some codes require that a structure be placed no closer than 5′ from a side property line. Other codes may require 10′ or more. Setback distances may be different between front, side, and rear property lines.

Setback lines are drawn within and parallel to property lines to indicate the acceptable location of structures. Setbacks and other zoning dimensions also vary for pools, garages, corner lots, hillside sites, easements, parking areas, and walks, as shown in Figure 13.6. Figure 13.7 includes typical setback requirements for an interior lot, including adjacent roadway dimensions. Where lot shapes are irregular, setbacks must be held to the nearest building corner or tangent point to a property line, as dimensioned in Figure 13.8.

Typical single-family-dwelling zoning ordinances (often called R-1) limit the front, rear, and side setbacks. They also limit the lot size and maximum land coverage. Figure 13.5 shows an example of how to calculate the **building envelope** (buildable area) for a 50′ × 120′ lot.

Figure 13.9 shows a lot with a building and accessory envelope that is determined as follows:

1. Draw setback lines for front yard, rear yard, and side yard setbacks. Also add accessory (other buildings and features) setbacks.

2. The remaining area (63′ × 66′) shows the allowable building envelope (4,158.0 sq. ft.).

3. A typical residential (R) code states the maximum gross floor area can be no more than 25% of the lot area. Thus, $0.25 \times 15,000$ sq. ft. = 3,750 sq. ft. This is the maximum building envelope (or footprint) of the structure.

4. Note that the entire building envelope cannot be used. To calculate the amount that can be used:

Actual gross floor area = 3,750 sq. ft.

Allowable building envelope = 4,158 sq. ft.

$4,158 - 3,750 = 408$ must remain open in the building envelope.

5. Thus, 90% (3,750 ÷ 4,858 = 0.90) of the allowable building envelope can be covered by the house. This does not include decks, patios, garages, or other buildings.

Setback, usage, land coverage, structural type, and height restrictions must all be combined to determine the location, size, and height of structure allowed for a particular lot. See Figure 13.10. It is important to check with the local zoning department to determine the exact requirements for setbacks on all floor levels. Setback dimensions may be measured from the property line to the wall line, eave line, projecting fireplace, or cantilevered second story of the structure, depending on local setback codes.

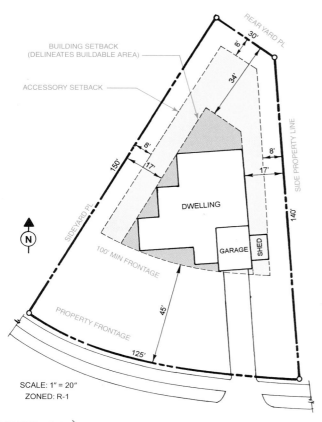

FIGURE 13.8 > Setback requirements applied to an irregularly shaped lot. *Hepler/Wallach/Hepler © Cengage 2013*

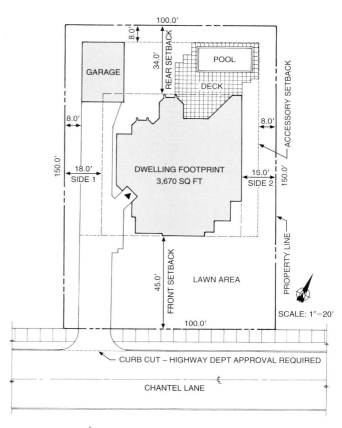

FIGURE 13.9 > Setback requirements applied to a lot with accessory structures. *Hepler/Wallach/Hepler © Cengage 2013*

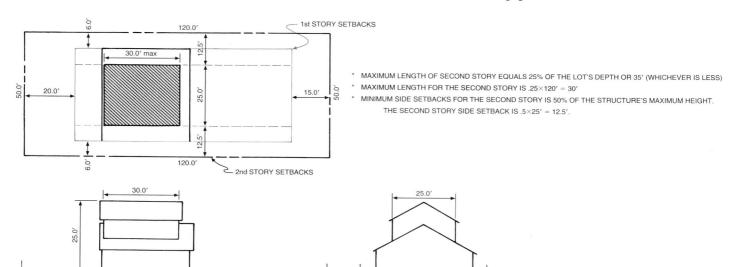

FIGURE 13.10 > Typical zoning ordinances, height restrictions, and setbacks for all levels. *Hepler/Wallach/Hepler © Cengage 2013*

⊕ Density Zoning

To create a good living environment and prevent overbuilding in an area, architects and builders who plan and build multiple-home developments must conform to density zoning laws. **Density,** in architectural terms, is the relationship of the number of residential structures and people to a given amount of space. The density of an area is the number of people or families per acre or square mile. For example, a town may have a density of 3 families per acre or 1,920 families per square mile.

The *average density* of an area is the ratio of all inhabitants to a specific geographic area. Density patterns may vary greatly within different parts of one area. Some parts of that area may be crowded, while other parts may be less populated. In other patterns, the population may be evenly distributed. Density planning for an area must be based on the maximum number of people who will occupy the area, regardless of the patterns.

Figures 13.11A through 13.11D show the same area with different housing patterns designed to achieve different densities. Figure 13.11A shows a conventional single-family development pattern that yields greater numbers of houses but is visually monotonous and socially undesirable. Automobile use here is essential. Figure 13.11B shows a **cluster housing** development that contains single, quad, and duplex units

to yield more open space. Curved streets add architectural interest by breaking up roadway monotony. Automobiles are also essential here. The cluster development in Figure 13.11C includes single and multiple residential types with a commercial, retail, and school component. This plan yields better opportunity for open spaces and a walkable, livable, workable community. The need for automotive traffic is reduced. Athletic fields, lawn strips, and an organic road configuration, as shown in the **mixed-use development** of Figure 13.11D, create interest and a feeling of more open space. Increased density is achieved here with the use of zero-lot-line (ZLL) duplexes and townhouses (see next paragraph). Auto traffic in this developed community is greatly reduced. Figure 13.11E

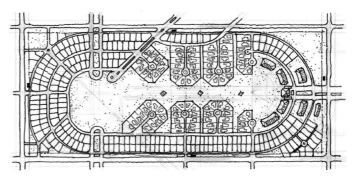

FIGURE 13.11C ⟩ Cluster housing development with expanded open space. *Hepler/Wallach/ Hepler © Cengage 2013*

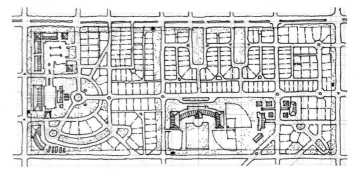

FIGURE 13.11D ⟩ Mixed-use community development. *Hepler/Wallach/Hepler © Cengage 2013*

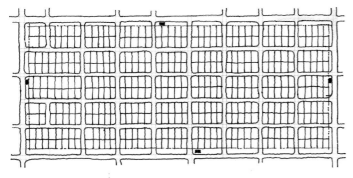

FIGURE 13.11A ⟩ Conventional housing development pattern. *Hepler/Wallach/Hepler © Cengage 2013*

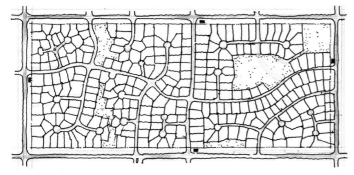

FIGURE 13.11B ⟩ Cluster housing pattern with limited open space. *Hepler/Wallach/Hepler © Cengage 2013*

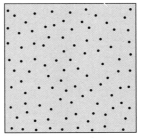 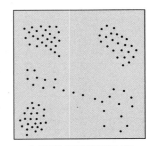

A. MOST PROPERTY PRIVATELY OWNED – 100 LIVING UNITS

B. OPEN SPACES WITH 100 CLUSTERED LIVING UNITS

FIGURE 13.11E ⟩ Comparison of areas with the same average density. *Hepler/Wallach/Hepler © Cengage 2013*

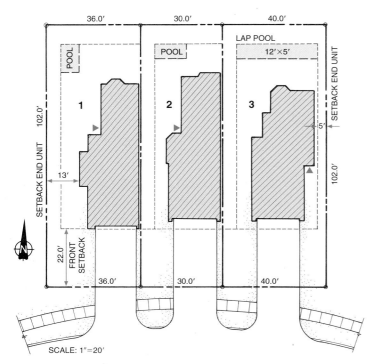

FIGURE 13.12 > Zero-lot-line properties. *Hepler/Wallach/ Hepler © Cengage 2013*

shows an example of an area with the same average density. Only the density pattern has changed.

To prevent overcrowding, local zoning ordinances may restrict the size of each building. This automatically restricts the number of families allowed to occupy a specific area. This approach also spreads the density patterns equally. As another approach, zoning laws may encourage clustering of many residents into fewer structures, such as high-rise apartments or townhouses (attached houses) with larger open public areas. In higher density developments, smaller size lots can be used more efficiently by eliminating one side yard and reducing the front, rear, and other side yard. This is termed a zero-lot-line (ZLL) property. See Figure 13.12.

A combination of plans involves zoning different parts of an area for single-family residences, townhouses, and high-rise apartments. The amount of space planned for each type of structure depends on the average density desired. Figure 13.13 shows a land area model used to demonstrate, study, or clarify density patterns. Zoning analysis drawings, or maps, are prepared to demonstrate and plan the design and configuration of "smart growth" plans. The zoning map shown in Figure 13.14 demonstrates the application of planned development district (PDD) concepts. PDD involves planning for a core in a development that includes, for example, a post office, housing, shops, churches, and a train station to create a workable community.

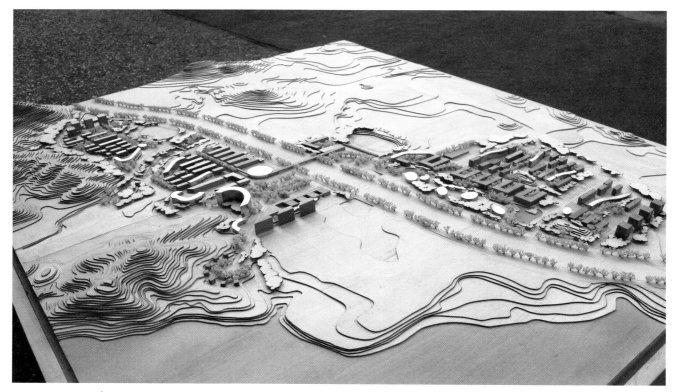

FIGURE 13.13 > Density study model. *Model Technology Co., Ltd. Beijing Jian Yi Xuan*

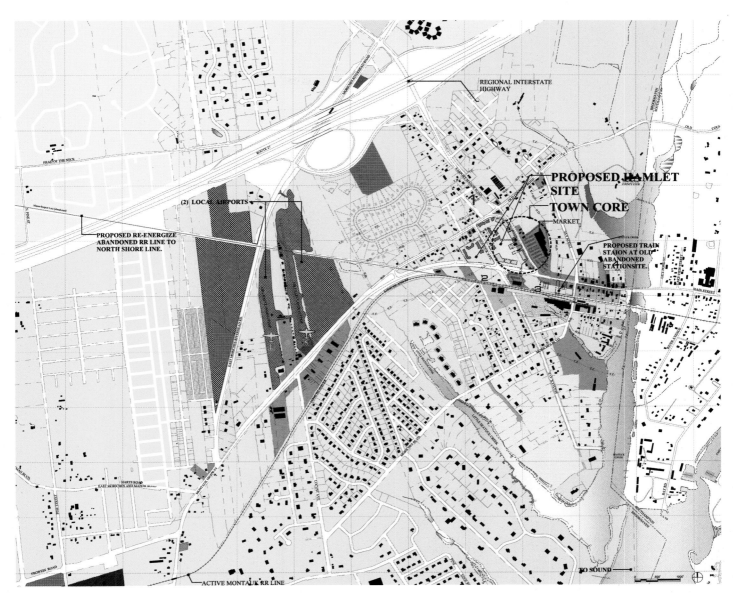

FIGURE 13.14 > Density analysis map. *Hepler Associates, PC*

🌐 Building Permits

Building permits are required to ensure that sites and buildings conform to community standards for structural types, land coverage, setbacks, and density zoning. A well-planned community, such as the one shown in Figure 13.15, has a master plan for growth that includes building codes and zoning regulations. See Chapter 37. In addition to residential needs, community plans include provisions for schools, parks, traffic, shopping, police, fire protection, and often architectural style consistency.

Before a public structure or dwelling can be built, a building permit must be obtained from the local building department. Once the working drawings for a project are complete, a municipal building inspector checks each area of the design. The design must be in compliance with all existing codes

and ordinances. If the drawings and specifications meet the code requirements, a building permit will be issued. See Figure 13.16. By carefully checking all local and regional code requirements before finalizing the design and preparing working drawings, revisions can be avoided.

In addition to local building departments, the administration of local codes may be co-regulated or controlled by other local or governmental agencies such as city planning commissions, air pollution control districts, fire departments, public health departments, water pollution control boards, and perhaps even art and design commissions or historical preservation societies. If a federal building is involved, the Department of Housing and Urban Development (HUD), Department of Health, Education, and Welfare (HEW), Federal Housing Authority (FHA), or other agencies may also be involved in the approval process.

FIGURE 13.15 > Community development plan. *Hepler/Wallach/Hepler © Cengage 2013*

Zoning ordinances and some building codes do contain allowances for exceptions to the law. If a building cannot be designed or sited to conform to all local laws, builders can request a *variance* from the building department. In making this request, the builder must show that the exemption will not harm or inconvenience neighbors, the community, or the environment in any way. Variances are often requested for setbacks, styles, and building sizes and types. Permitting is also based on the balance of housing types and other community facilities. Figure 13.17 is an example of a combination site plan that shows the location and relationship among mixed-use residential townhouses and work, play, and transportation areas.

TOPOGRAPHIC DRAWINGS

Topography refers to the general configuration of a surface, including its relief, that is, the horizontal width, length, and height at different locations on a landform. *Grading* is the act of altering the relief of the land to create a new topography to assist drainage, conservation, or road construction. Topographic drawings such as plan views, elevations, and profiles are used to specifically describe the existing topography of a site and the planned surface changes to be achieved by grading.

Site designers visualize the 3D appearance of a future landform and convert this vision into understandable working

PERMIT

Page 1 of 1

Permit #: 970012   Type: Residence   Issued: 1-3-11   by: JP

Job Location: 603   Issue Loc:

Lot: 8   Subdiv: Lake Estates

Parcel: 8A

Owner: Kingston   Elev: 6'   Fl Map: A4

Project: 95 K   Seawall Datum

Job Description: Residence & Pool

Applicant Name: J R Smith   Type: Skeleton Frame

Applied Date: 1-2---   Appl Oper: 4

Contract Phone: 364 8752   Inspector Area: 6   Work w/o Permit Fee: 50

Contractor Name: T. Jones   Cert Nbr: 336

Business Name: Capitol Builders   Septic Tank: Dwg A1

Setbacks Front: 50   Left: 35   Right: 35   Rear: 35

FCC Code: 329

Square Footage: 3560   Rate: TBD   Job Value: 195,000

Number of Units: 1   Floors: 2   Buildings: 1 + Dock

ROW: 24' C RD   Zoning: PUD   Map No: 00617

Minimum Floor Elevation: 6' Seawall Datum   Residential/~~Commercial~~

H. Mitchell _____   _1/3/11_ _____

Building Official or Authorized Signature   Date

FIGURE 13.16 > Sample building permit. *Lake County, Florida, Building Department*

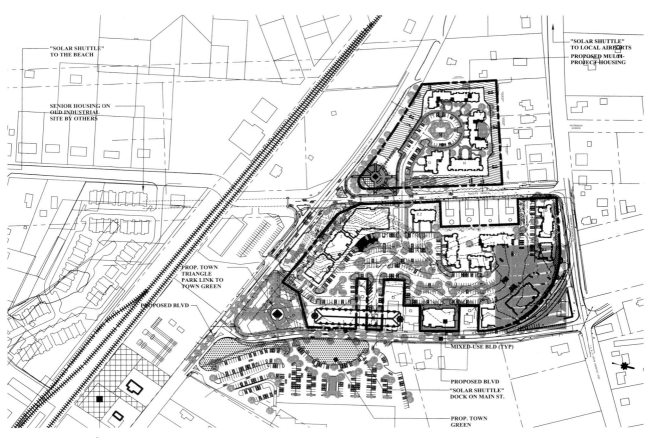

FIGURE 13.17 > Site plan with permitting information. *Hepler Associates, PC*

drawings. These include surveys, plat plans, plot plans, landscape plans, and details such as drainage plans, grading calculations, and instructional drawings. Symbols on these drawings represent structures, contours, vegetation, and water features.

Zoom and Pan on Site Plans

CAD *Zoom* and *Pan* commands are used constantly in preparing all types of architectural drawings. For example, site plans often cover a D-size sheet (24″ × 36″) and are therefore more than 10 times the size of a CAD monitor. The *Zoom* command is similar to the zoom function on a camera. It allows the entire drawing to be viewed and also enables small details or parts to be magnified or demagnified while drawing.

The area that is going to be magnified is selected by choosing the *View* pull-down menu, selecting *Zoom,* and then selecting the *Window* option. A window is created around the area that will be magnified by picking opposite corners of the window. A specific magnification scale is required, which can be typed in as part of the *Zoom* command. Real-time zoom capabilities allow the drawing to be magnified only through wheel mouse controls.

The *Pan* command shifts the placement of the image on the screen. It does not change the magnification, just the position of the "viewing window."

SURVEY PLANS

A **survey** is a drawing showing the exact size, shape, and levels of a property. When prepared by a licensed surveyor, a survey plan is used as a legal document to establish property rights. It is filed with the deed to the property. The lot survey includes the length of each boundary, tree locations, utility lines, corner elevations, contour of the land, and position of streams, rivers, roads, or streets. It also lists the owner's name and the owners or titles of adjacent lots. A survey drawing must be accurate and must also include a complete written description of the lot.

Establishing Dimensions

Surveying a site involves locating points (coordinates) on the earth's surface. To indicate and connect these points on a drawing, the *polar coordinate system* is used. In this system, a fixed, true north-south reference line, called a *meridian,* is established. The direction of a line on the survey drawing is given in relation to the meridian and expressed as the line's *bearing.*

The exact plan and shape of a lot is shown by property lines. Property line dimensions are shown directly on the line by length and angle. Angles are dimensioned using either the *American system* or the *azimuth system.* See Figure 13.18. In the *American system* a compass is divided into four quadrants:

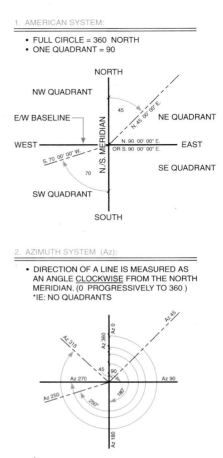

FIGURE 13.18 > American and azimuth property dimensioning systems. *Hepler/Wallach/ Hepler © Cengage 2013*

NE, SE, SW, and NW. Angular dimensions are shown by noting the degrees, minutes, and seconds from either N, E, S, or W and in a clockwise direction. There are 360 degrees (°) in a circle, 60 minutes (′) in a degree, and 60 seconds (″) in a minute. Thus a 45° line in the northeast quadrant is dimensioned N 45° 00′ 00″ E. This means the line is 45° from north, heading toward east.

Figure 13.19 illustrates how the azimuth system is used to calculate the bearings of four property lines enclosing a site. In the *azimuth system,* each line is dimensioned as an angle, reading clockwise from the north meridian, from 0° to 360°. These angular lines are drawn by aligning the 0° or 360° line on a protractor with the north meridian line. Place the center of the protractor at the intersection of the meridian and the east-west line. The degree is located on the circumference of the protractor and connected with a line to the protractor center.

These property dimensions are used on legal documents. Once established and certified by a licensed surveyor, a written description of the dimensions of a property can become the legal basis for real estate ownership. Figure 13.20 is an example of a legal property description, in this case, a deed description.

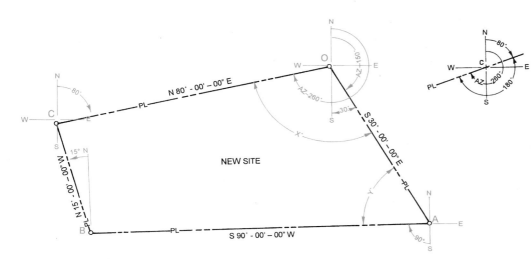

1. Bearing OA:
   AZ = 150 (REF to Pt "o")
   thus: AZ 180 - AZ 150 = 30
   OA: S 30 E or S 30 -00' - 00"E.

2. Bearing Ab:
   Using a protractor on the
   vertex, graphically determined
   due west or S. 90 - 00' - 00" W.
                  N. 90 - 00' - 00" W.
                  = AZ 270

3. Bearing bc:
   Put protractor on vertex,
   graphically determined
   N. 15 - 00' - 00" W.

4. Bearing CO:
   If AZ is 260 (Ref to point "0".
   Then AZ 260 - 180 = AZ 80 =
   N. 80 - 00' - 00" E.

FIGURE 13.19 > Steps in determining bearings using the azimuth system. *Hepler/Wallach/Hepler © Cengage 2013*

Beginning at a merestone in the northerly line of Old County Road, as shown on said map, said point being the southeasterly corner of land now or formerly of John Doe and the southwesterly corner of the premises herein described; thence running N 86°52'58" E, a distance of 192.57 feet to an iron pin thence turning and running S 81°37'12" E, a distance of 79.63 feet to an iron pin; thence turning and running N 70°31'32" E, a distance of 224.50 feet to an iron pin; the last three (3) courses being along Old County Road, as shown on said map; thence turning and running N 43°54'41" E, a distance of 158.31 feet to an iron pin; thence running N 38°39'13" E, a distance of 195.44 feet to a point; thence turning and running N 22°59'50" W, a distance of 148.71 feet to an iron pin; the last three (3) courses being along "Lot 1" as shown on said map; thence turning and running N 63°43'59" W along land now or formerly of Jane Doe, as shown on said map, a distance of 78.12 feet to a point; thence running N 64°35'53" W along land now or formerly of Jane Doe and land now or formerly of Mary Doe, as shown on said map, partly by each, in all a distance of 192.02 feet to a point; thence turning and running N 39°01'47" W along land now or formerly of Mary Doe, as shown on said map, a distance of 107.13 feet to a merestone; thence turning and running S 27°01'24" W along land now or formerly of John Doe, as shown on said map, a distance of 760.47 feet to a merestone, which point marks the point or place of beginning.

FIGURE 13.20 > Legal property description. *Hepler/Wallach/Hepler © Cengage 2013*

Elevations

The height of any point on a site is dimensioned from a fixed elevation point called a *datum*. Datum is always zero and all elevation measurements are measured up or down from the datum.

Several types of drawings and lines are used to show elevation distances and shapes on survey drawings. These include profile drawings (land sections), elevation point notations, and contour lines. Plan views contain a large amount of detail including all width and length dimensions. However, with the exception of key locations, the height (elevation) of the property is not shown throughout the entire drawing. This approach is satisfactory for a relatively flat site or building surface, but a more descriptive method of showing the relief of a landform is necessary for parcels with larger elevation differences. Contour lines and elevation notations are added to plan views to show the exact vertical dimensions of a site at all surface points. **Profiles** are elevation drawings that show a removed section of a landmass. Profiles are projected vertically from intersections with a cutting plane line and contour points.

Contours

In architectural terms contours represent the form of a terrain as identified by lines that represent distances above and below a given level.

Contour Lines

Contour lines are evenly spaced continuous lines that connect points on a terrain that lie on the same height above a fixed elevation surface, that is, a *datum*. As mentioned earlier, a datum is a fixed and numbered reference position from which distances, angles, and heights are measured. On site, landscape, and topographic plans and surveys, dimensions from the datum are noted as vertical distances. The standard and commonly used datum point is mean sea level. Any fixed point on a sidewalk, seawall, court, or patio surface can be used as the reference datum for small sites.

Because all points on a contour line represent the same identical vertical distance from the datum, these lines are always continuous, forming a circumscribed closed loop as shown in Figure 13.21. For example, the surface of a body of

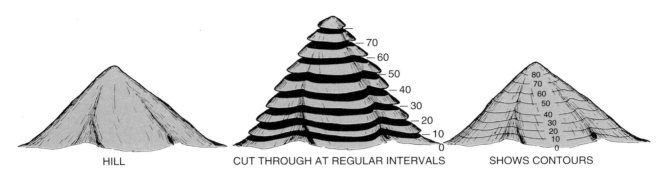

FIGURE 13.21 ❯ Contour lines are imaginary cuts through the terrain. *Hepler/Wallach/Hepler © Cengage 2013*

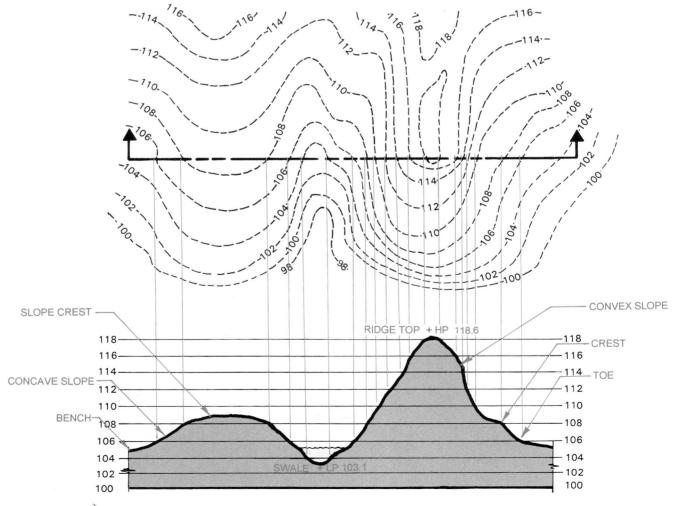

FIGURE 13.22 ❯ Elevation profile section related to contour lines. *Hepler/Wallach/Hepler © Cengage 2013*

water always remains level and at a constant elevation. Therefore, the perimeter, or water's edge, represents a consistent height above the datum. This creates a contour line that ends at the same level at which it began.

Some contour lines require a large amount of space to reconnect and may even exit and reenter a drawing sheet. Nevertheless these lines will eventually meet if sufficient space is available, although some may meet outside the drawing. The contour lines of a steep slope are close together as shown in Figure 13.22. The lines of a more level surface are further apart. Contour lines, because they represent exact elevations, can never merge or split, except where the contour lines are superimposed on each other. This occurs over vertical surfaces such as walls or cliffs.

Figure 13.23 is a landform model in which contour lines are represented by built-up horizontal layers of plywood or foamboard. The thickness of each layer is scaled to the depth of the contour interval. Smaller contour intervals obviously produce a more accurate description of a landform than larger intervals.

Figure 13.24 shows contour lines that are similar to those in Figure 13.25, except the contour numbers are in reverse order. The center elevation in Figure 13.25 represents a hill; the center elevation in Figure 13.24, however, represents a depression. Often the crest of a hill or the low point of a depression falls between contour intervals. In this case a plus sign (+) and the exact elevation level is noted if precision is required. The high point in Figure 13.25 would be referred to as hill 800.

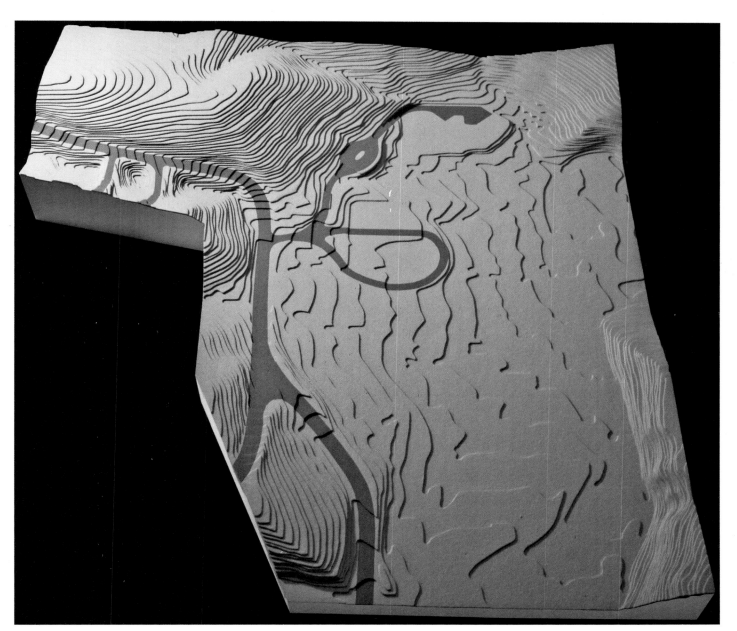

FIGURE 13.23 > Landform model with contours, structures, and foliage. *Jon Hoffman—LEED*

FIGURE 13.24 〉 Contour numbers reveal a depression.
Hepler/Wallach/Hepler © Cengage 2013

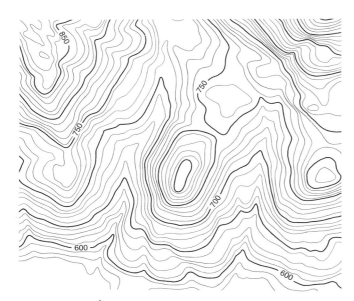

FIGURE 13.25 〉 Contour lines identified every 10'. *Hepler/ Wallach/Hepler © Cengage 2013*

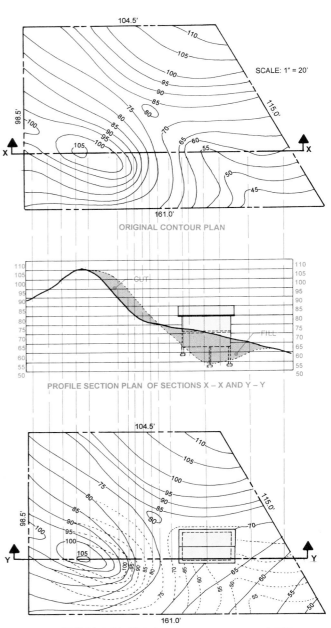

ORIGINAL CONTOUR PLAN

PROFILE SECTION PLAN OF SECTIONS X – X AND Y – Y

PROPOSED CONTOUR PLAN (OLD CONTOURS ARE DOTTED)

FIGURE 13.27 〉 Dashed contour lines show original contour grades. Solid contour lines show new contours.
Hepler/Wallach/Hepler © Cengage 2013

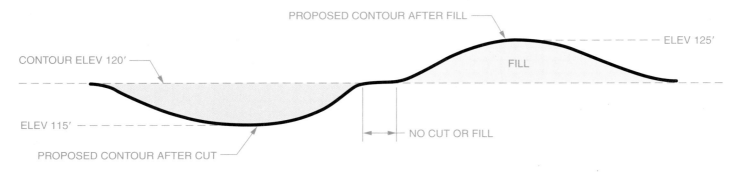

SECTIONAL VIEW OF CONTOUR REFORMING

FIGURE 13.26 〉 Profiles showing cuts and fills on a flat surface. *Hepler/Wallach/Hepler © Cengage 2013*

FIGURE 13.28 > Site model showing structures and related topography. *Model Technology Co., Ltd. Beijing Jian Yi Xuan*

Not every contour line includes an elevation number. The interval between numbers depends on the scale of the drawing. The interval noted in Figure 13.25 represents 10′ of elevation with heavy unnoted lines used every 50′.

Where land surfaces are to be altered during the construction process, proposed contour lines are added to the drawing. Land is either removed (*cut*) or added to (*filled*). Figure 13.26 shows the addition of contour lines to a flat surface to denote the amounts of cut or fill. Figure 13.27 shows how cuts and fills are shown on a contour drawing. The original contour lines are drawn as dashed lines, whereas the proposed contour is shown with solid lines, although some surveyors reverse the application of dashed and solid lines. This illustration also shows the projection of a profile elevation from these contour lines.

Contour altering lines are used when the original contour lines are shown on the same drawing with the contour lines of a surface to be altered. Surveyors usually supply survey drawings with solid contour lines. Site designers must therefore convert these to dashed lines if new solid contour lines are to be used to describe grading changes. Figure 13.28 illustrates the use of a contour landform model to study the location of structure as related to the surrounding topography.

Terrain texture can be indicated by simply drawing wavy contour lines to denote roughness. Figure 13.29 shows a contour drawing with both rough and smooth surfaces identified by this method.

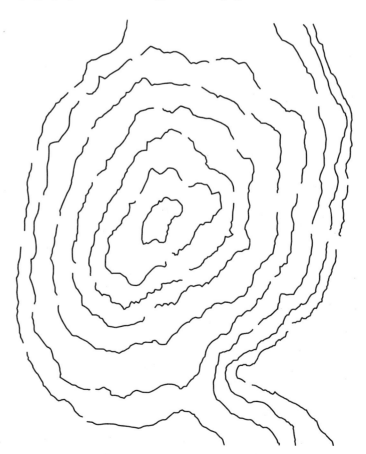

FIGURE 13.29 > Method of showing rough surface textures.
Hepler/Wallach/Hepler © Cengage 2013

Contour Intervals

The vertical distance between contour lines is known as the **contour interval.** To ensure the accuracy of any topographic plan or profile, contour intervals must always be evenly spaced; otherwise projections will be inaccurate. The size of contour intervals is largely dependent on the scale and complexity of the drawing. A contour interval can be any convenient distance, but is usually an increment of 5'. Contour intervals of 5', 10', 15', and 20' are common on large surveys. The use of smaller intervals (1' or 2') gives a more accurate description of the slope and shape of the terrain than does the use of larger intervals. Note the different levels of detail between the 10' and 2' intervals of the same terrain as shown in Figures 13.30 and 13.31, respectively. The size

of the interval depends on the scale of the drawing and on the size of the area to be shown. Large contour intervals are normally used on large regional maps.

Contour Anatomy

Different segments and patterns created by contour lines, such as hills, ridges, valleys, and depressions, can be easily confused; therefore, some feature identifications are necessary. These include the following.

Hilltop summits (or *crests*) are identified by closed contour lines in a small area at the top of a surrounding slope. Military engineers identify each hill with the highest contour interval number. For example, if the highest point of a hill is at an elevation of 112, between contour intervals 110 and 120, the hill would be labeled *hill 112*. The exact point of the highest elevation is marked accordingly with a plus (+) sign.

Depressions appear identical to hills because of the closed loop contours. Therefore, differences must be determined by referring to the interval numbers, which ascend to show hills and descend to show depressions. Graphically, depressions are often marked with hash lines on the inside of the internal loop. Note the contour descending numbers and the hash lines, which clearly label the loop as a depression, in Figure 13.24. Because they always represent a lower elevation than the surrounding terrain, small bodies of water will always appear as a depression.

Ridge lines connect the highest elevation points in a series of contour lines that align where a change of direction occurs as shown. The contour line points (V- or U-shaped areas) where the contour lines change direction always point downhill. Thus, ridges are the opposite of valleys as seen in Figure 13.25.

Valleys (or *swales*) on a contour drawing appear similar to ridges; however the points (directional change areas), which represent a valley, point uphill. Ridges and valleys are related since the descending sides of a ridge eventually form a valley, and then ascend to create another ridge. This can be observed in Figure 13.22 by following the contour line numbering sequences on the valley and ridge lines. This illustrates how runoff from the sides of ridges creates streams or rivers in valleys, specially where slope angles are severe.

Interpolation refers to the practice of creating contours from evenly spaced spot elevations and grids. This process involves the preparation of an elevation grid as shown in Figure 13.32. The elevation of each equally spaced grid square is noted at the corner of each square. Using a contour interval, which must be at the same scale as the grid, contour lines can be created by plotting each contour interval number as a point on the grid. This assumes the elevation change between intervals is uniform.

Contour numbering is essential for the correct interpretation of contours. Numbering contour lines identifies differences between hills and depressions and between valleys and

FIGURE 13.30 ❯ Large intervals show only general slope direction. *Hepler/Wallach/Hepler © Cengage 2013*

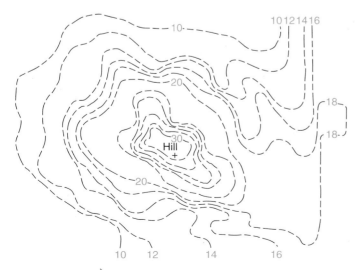

FIGURE 13.31 ❯ Small intervals reveal steep and gradual slope details. *Hepler/Wallach/Hepler © Cengage 2013*

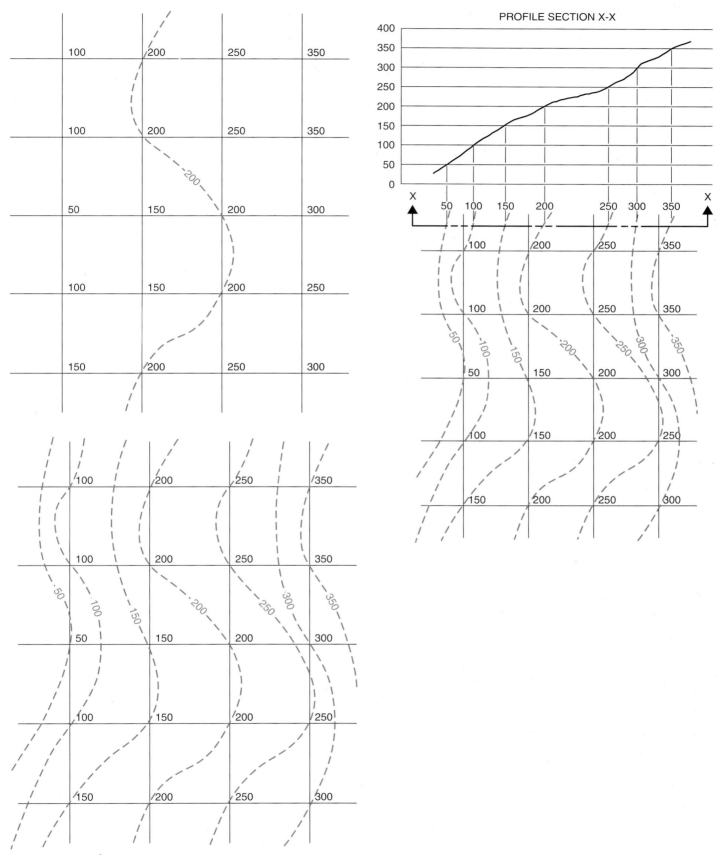

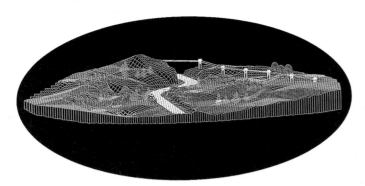

ridges. Numbering is also a key to the exact amount of vertical rise, or drop. Numbering each line eliminates the need to trace and follow lines to determine the height above datum at any given point.

Pictorial Contours on CAD

Computer programs are used to create pictorial contour drawings such as that shown in Figure 13.33. This is done by combining X-Y coordinate input data and the corresponding datum level (Z) for each coordinate point. These points are then connected using surface modeling techniques to create the pictorial contour lines.

The individual data points are connected using the *Polylines* or *Splines* command to create the individual lines. The lines are then meshed together to create a surface model of the area being designed. Once the lines are created or a surface mesh is generated, the designer's viewpoint can be changed with the *Vpoint* command to get a true three-dimensional perspective. Use the *3D Orbit* command for the easiest control.

Digital graphics files are available from the U.S. Geological Survey (USGS) that contain information from USGS maps, which can be used for large site plans. Local municipal planning departments can often supply digital map and aerial photographs with property lines superimposed.

For large, flat projects such as parking lots and airport tarmacs (runway areas, etc.), contour lines may not be used because the slope differential is less than a foot. Only the elevations of selected points are placed on the drawing. An arrow is drawn beside each point to show slope direction, as shown in Figure 13.34.

Symbols on Survey Drawings

Symbols are used extensively to describe the features of the terrain. Some symbols, such as tree symbols, resemble the appearance of a feature. Most survey symbols are graphic representations. The charts shown in Figure 13.35 include symbols and abbreviations used on survey drawings. The

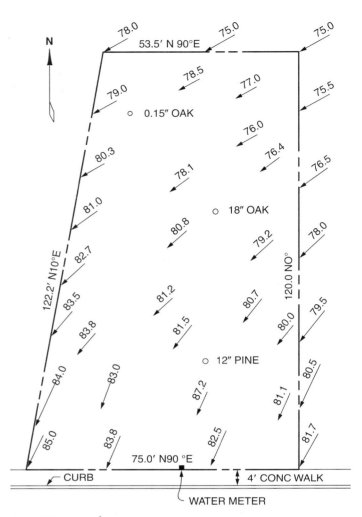

FIGURE 13.34 > Elevation notations showing slope direction. *Hepler/Wallach/Hepler © Cengage 2013*

survey plan in Figure 13.36 shows the application of the most common survey plan symbols.

CAD Site Symbols

Site plan symbols are included in many CAD software programs but most are created, stored, and retrieved as blocks using the *Block* command. Care must be taken to adjust the scale of these symbols to match the scale of the drawing. Land surface materials are available in blocks and are inserted into the drawing using the *Hatch* command. These symbols are also available for elevation drawings.

Many site details are "stand-alone" details that are not an integral part of a site plan. Site details include removed sections, such as curb details and lighting fixtures, and planting instruction drawings, which may be included with landscape plans. It is very efficient to include these drawings in a site detail library, which eliminates having to redraw them on every drawing each time they are needed.

NAME	ABBREV	SYMBOL	NAME	ABBREV	SYMBOL	NAME	ABBREV	SYMBOL	NAME	ABBREV	SYMBOL
TREES	TR		CULTIVATED AREA	CULT		BUILDINGS	BLDGS		LARGE RAPIDS	LRG RP	
GROUND COVER	GRD CV		WATER	WT		SCHOOL	SCH		WASH	WSH	
BUSHES SHRUBS	BSH SH		WELL	W		CHURCH	CH		LARGE WATERFALL	LRG WT FL	
OPEN WOODLAND	OP WDL		PROPERTY LINE	PR LN		CEMETARY	CEM	†CEM	BOUNDARY, U.S. LAND SURVEY TOWNSHIP	BND US LD SUR TWN	
MARSH	MRS		SURVEYED CONTOUR LINE	SURV CON LN	75	POWER TRANSMISSION LINE	PW TR LN		BOUNDARY, TOWNSHIP APPROXIMATED	BND TWN	
DENSE FOREST	DN FR		ESTIMATED CONTOUR	EST CON		GENERAL LINE LABEL TYPE	GN LN	oil line	BOUNDARY, SECTION LINE U.S. LAND SURVEY	BND SEC LN US LD SUR	
SPACED TREES	SP TR		FENCE	FN		BOUNDARY, STATE	BND ST		BOUNDARY, SECTION LINE APPROXIMATED	BND SEC LN	
TALL GRASS	TL GRS		RAILROAD TRACKS	RR TRK		BOUNDARY, COUNTY	BND CNTY		BOUNDARY, TOWNSHIP NOT U.S. LAND SURVEY	BND TWN	
LARGE STONES	LRG ST		PAVED ROAD	PV RD		BOUNDARY, TOWN	BND TWN		INDICATION CORNER SECTION	COR SEC	
SAND	SND		UNPAVED ROAD	UNPV RD		BOUNDARY, CITY INCORPORATED	BND CTY		U.S. MINERAL OR LOCATION MONUMENT	U.S. MIN MON	▲
GRAVEL	GRV		POWER LINE	POW LN		BOUNDARY, NATIONAL OR STATE RESERVATION	BND NAT OR ST RES		DEPRESSION CONTOURS	DEP CONT	
WATER LINE	WT LN		HARD-SURFACE HEAVY DUTY ROAD – FOUR OR MORE LANES	HRD SUR HY DTY RD		BOUNDARY, SMALL AREAS: PARKS, AIRPORTS, ETC	BND		FILL	FL	
GAS LINE	G LN		HARD-SURFACE HEAVY DUTY ROAD – 2 OR 3 LANES	HRD SUR HY DTY RD		LEVEE	LEV		CUT	CT	
SANITARY SEWER	SAN SW		IMPROVED LIGHT DUTY ROAD	IMP LT DTY RD		RIVER	RV		LAKE, INTERMITTENT	LK INT	
SEWER TILE	SW TL		TRAIL UNIMPROVED DIRT ROAD	TRL UNIM DRT RD		STREAM PERENNIAL	ST PER		LAKE, DRY	LK DRY	
PROPERTY CORNER WITH ELEVATION	PROP CR EL	EL 70.5	ROAD UNDER CONSTRUCTION	RD CONST		STREAM INTERMITTENT	ST INT		SPRING	SP	
SPOT ELEVATION	SP EL	+78.8	BRIDGE OVER ROAD	BRG OV RD		STREAM DISAPPEARING	ST DIS		PILINGS	PLG	
WATER ELEVATION	WT EL	80	ROAD OVERPASS	RD OVP		SMALL RAPIDS	SM RP		SWAMP	SWP	
BENCH MARKS WITH ELEVATIONS	BM/EL	BM ✕ 84.2 BM △ 84.2	ROAD UNDERPASS	RD UNP		SMALL WATERFALL	SM WT FL		SHORELINE	SH LN	

FIGURE 13.35 > Common survey symbols. *Hepler/Wallach/Hepler © Cengage 2013*

Guidelines for Drawing Surveys

The numbered arrows in Figure 13.37 correspond to the following guidelines for preparing survey drawings:

1. Record the elevation above the datum of the lot at each corner.

2. Represent the size and location of streams and rivers by wavy lines (blue lines on geographical surveys).

3. Use a + symbol to show the position of existing trees. The elevation at the base of the trunk is shown on some drawings.

4. Indicate the compass direction of each property line by degrees, minutes, and seconds.

5. Use a due north arrow to show compass direction.

6. Break contour lines to insert the height of each contour above the datum.

7. Show lot corners by small circles or overlapping property lines.

8. Draw the property line symbol by using a heavy line with two dashes repeated throughout.

9. Show elevations above the datum by contour lines.

10. Show any proposed change in grade line by contour lines. Use solid contour lines to show original grade and dashed contour lines to show the new proposed grading levels.

11. Show lot dimensions directly on the property line. The dimension on each line indicates the distance between property corners.

12. Give the names of owners of adjacent lots outside the property line. The name of the owner of the site is shown inside the property line.

13. Dimension the distance from the property line to all utility lines.

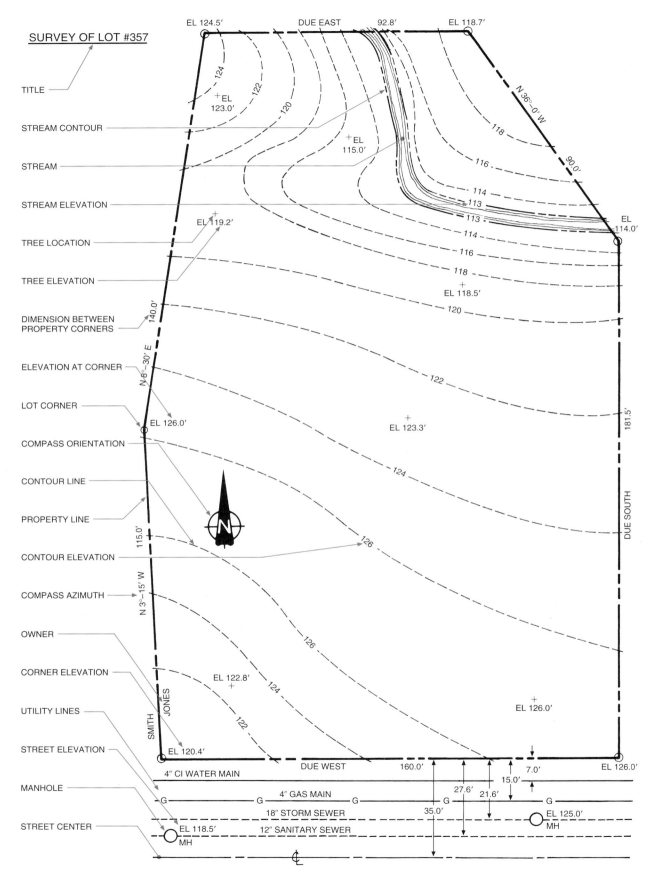

SURVEY OF LOT #357

TITLE

STREAM CONTOUR

STREAM

STREAM ELEVATION

TREE LOCATION

TREE ELEVATION

DIMENSION BETWEEN
PROPERTY CORNERS

ELEVATION AT CORNER

LOT CORNER

COMPASS ORIENTATION

CONTOUR LINE

PROPERTY LINE

CONTOUR ELEVATION

COMPASS AZIMUTH

OWNER

CORNER ELEVATION

UTILITY LINES

STREET ELEVATION

MANHOLE

STREET CENTER

EL 124.5'   DUE EAST   92.8'   EL 118.7'

+EL 123.0'

+EL 115.0'

N 36°-0' W   90.0'

EL 114.0'

EL 119.2'

124   122   120   118   116   114   113   113   114   116   118   120   122   124   126

+EL 118.5'

140.0'

N 8°-30' E

EL 126.0'

+EL 123.3'

181.5'

DUE SOUTH

115.0'

N 3°-15' W

126

SMITH   JONES

EL 122.8' +

+EL 126.0'

124   122

EL 120.4'

DUE WEST   160.0'   7.0'

15.0'

4" CI WATER MAIN

27.6'   21.6'

4" GAS MAIN

G   G   G   G   G   G

18" STORM SEWER   35.0'   EL 125.0' MH

12" SANITARY SEWER

EL 118.5' MH

FIGURE 13.36 > Application of survey plan symbols. *Hepler/Wallach/Hepler © Cengage 2013*

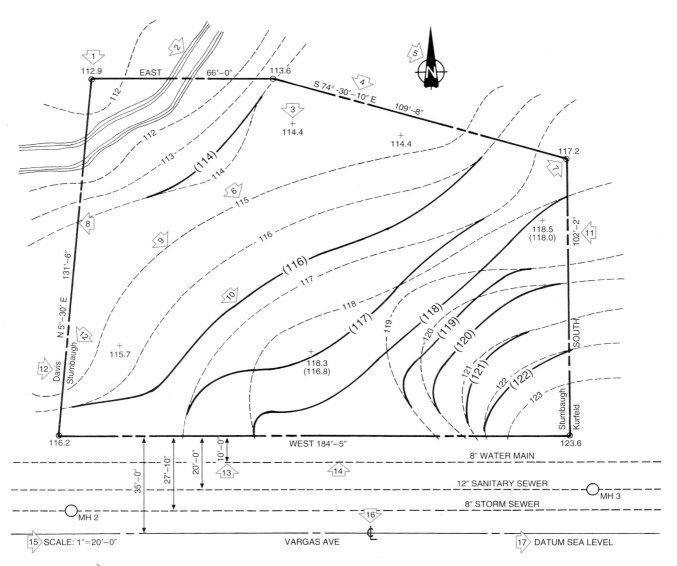

FIGURE 13.37 > Survey drawing keyed to survey guidelines. *Hepler/Wallach/Hepler © Cengage 2013*

14. Show the position of utility lines by dotted lines. Utility lines are labeled according to their function.

15. Draw surveys with an engineer's scale. Dimensions are shown as feet and decimal parts of a foot (for example, 6.5′). Common scales for surveys are 1″ = 10′ and 1″ = 20′. Scales of 1″ = 30′, to 1″ = 60′ may be used for very large sites.

16. Show existing streets and roads either by center lines or by curb or surface outlines. The center line symbol is cL and is used for reference when an actual centerline is not drawn.

17. Indicate the datum level used as reference for the survey.

In addition, show the outline of the permissible septic distribution field area, if appropriate (see Figure 13.38).

Geographical Survey Maps

U.S. Geographical Survey (**USGS**) maps are similar to property surveys except they cover extremely large areas. The entire world is divided into geographical survey regions. However, not all regions have been surveyed and mapped. Geographical survey maps show the general contour of the area, natural features of the terrain, and structures.

When large areas are covered, a large reduction scale is used. When smaller areas are covered, as shown in Figure 13.39A, a smaller reduction scale such as 1:24000 is used. Figure 13.39B shows a portion of a geological survey map that can be compared to Figure 13.39C, which shows an aerial photograph of the same area.

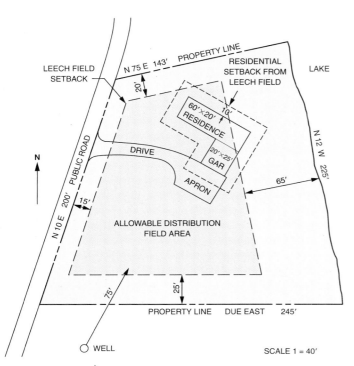

FIGURE 13.38 > Allowable distribution field area. *Hepler/ Wallach/Hepler © Cengage 2013*

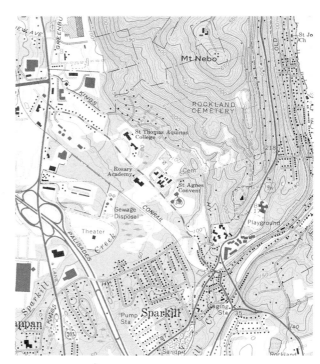

FIGURE 13.39A > Partial USGS map. *U.S. Geological Survey*

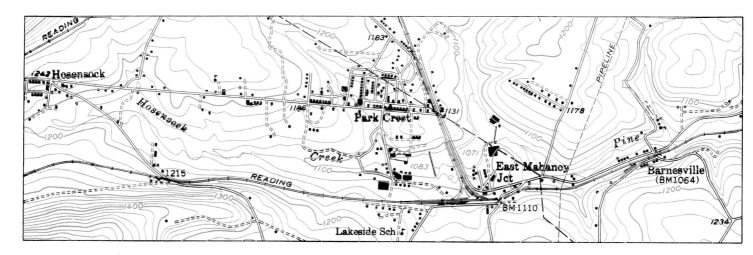

FIGURE 13.39B > Survey map related to aerial photograph in Figure 13.39C. *U.S. Geological Survey*

Plats

A **plat** is a survey (map, chart, or plan) of multiple connected properties. Plats are legal descriptions of a land site and are identified by plat name, section, township, county, and state. A plat is part of a geographical survey region, which is divided into areas that contain further subdivisions. See Figure 13.40. A plat plan with minimal information is shown in Figure 13.41. This includes only lot numbers and overall dimensions.

Plats may include the compass bearing (direction) of the plat area, dimensions of each property line, and the position of all roads, utility lines, and easements, as shown in Figure 13.42. Some show only lot shapes, as in Figure 13.43. Others identify lots or buildings by numbers that refer to a more detailed survey. See Figure 13.44 on page 254. Plats are prepared for residential developments, industrial parks, urban developments, and shopping complexes. Rendered plat plans, as shown in Figure 13.45 on page 254, are often

FIGURE 13.39C > Aerial photograph of area shown in Figure 13.39B. *U.S. Geological Survey*

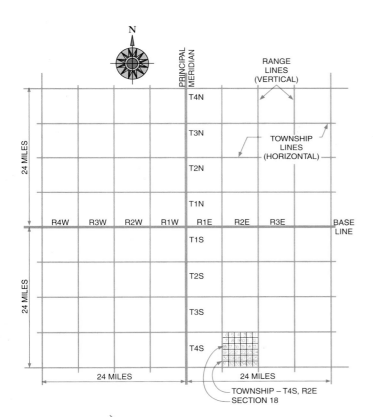

FIGURE 13.40 > Divisions of a 48-square-mile region.
Hepler/Wallach/Hepler © Cengage 2013

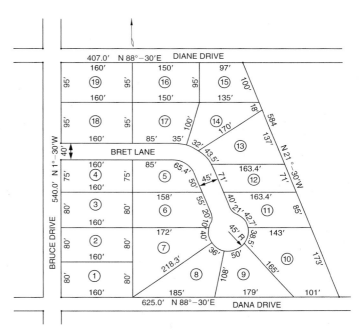

FIGURE 13.41 > Plat with minimal information. *Hepler/
Wallach/Hepler © Cengage 2013*

used for housing development marketing. For this use only general lot sizes and shapes are shown. A plat plan becomes a legal survey if fully dimensioned and stamped by a licensed surveyor. Plats of the type shown in Figure 13.45 must be prepared and submitted to municipal authorities as part of the permitting process. Aerial photographs are also used to prepare plats. This is done by superimposing property lines as shown in Figure 13.46 on page 255.

PLOT PLANS

Plot plans are used to show the size and shape of a building site and the location and size of all buildings on that site. The position and size of walks, drives, pools, streams, patios, and courts are also shown. Compass orientation of the lot is given, and contour lines are sometimes shown. Plot plans may also include details showing site construction features. See Figure 13.47 on page 255. If a separate survey or landscape plan is prepared, contour lines, utility lines, and planting details are usually omitted from a plot plan. Building foundations are flat; this means the elevation is the same at all locations.

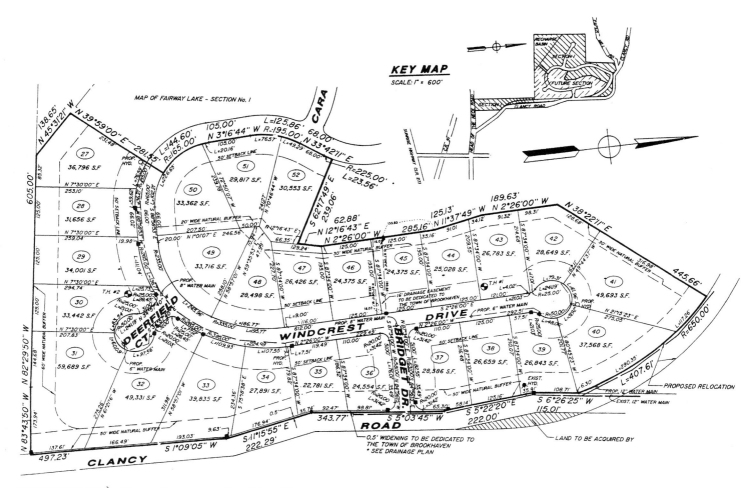

FIGURE 13.42 > Plat with property lines identified. *Hepler/Wallach/Hepler © Cengage 2013*

Therefore, contour lines stop where they intersect building lines. Contour lines reappear outside at some point on the building perimeter. This keeps the contour planes continuous. All design possibilities should be considered before a final design is chosen and drawn. Figure 13.48 on page 256, shows a variety of plot plan designs that balance open space, vegetation, and structural forms.

Guidelines for Drawing Plot Plans

When plot plans are prepared, the features indicated by the numbered arrows in Figure 13.49 on page 257, should be drawn according to these guidelines:

1. Draw the property lines and the outline of the main structure on the lot. Crosshatching is optional.
2. Draw the outlines of other buildings on the lot.
3. Show overall building dimensions.
4. Locate each building by dimensioning perpendicularly from the property line to the closest point on

a building. On curved or slanted property lines, dimension to points of tangency (touching but not intersecting).

5. Show the position and size of driveways.
6. Show the location and size of walks.
7. Indicate elevations of key surfaces such as floors, ground line, patios, driveways, and courts.
8. Outline and show the symbol for surface material used on patios and terraces.
9. Label streets adjacent to the site.
10. Place overall lot dimensions on dimension lines outside the property line.
11. Show the size and location of constructed recreation areas, such as tennis courts. (None on this plan.)
12. Show the size and location of pools, ponds, or other bodies of water.
13. Indicate the compass orientation of the lot with a north arrow.

FIGURE 13.43 〉 Rendered plat plan showing lot shapes and positions. *Hepler/Wallach/Hepler © Cengage 2013*

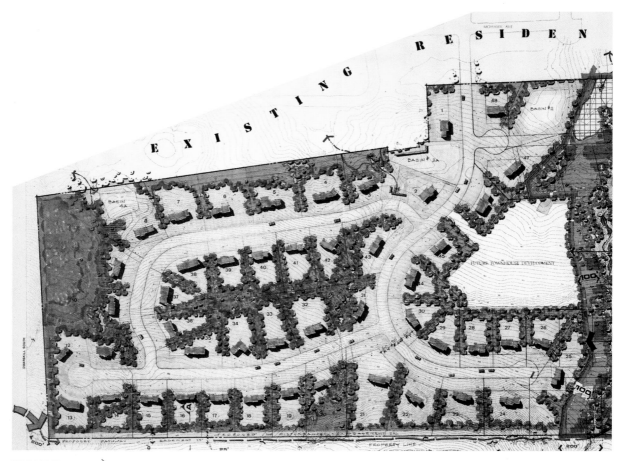

FIGURE 13.44 > Plat plan showing numbered lots and vegetation. *Hepler/Wallach/Hepler © Cengage 2013*

FIGURE 13.45 > Development marketing plat. *Hepler/Wallach/Hepler © Cengage 2013*

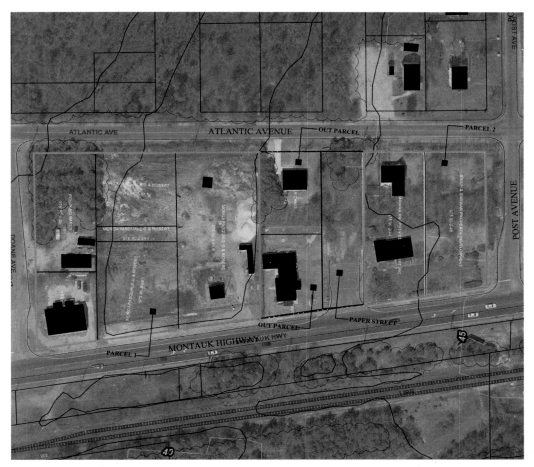

FIGURE 13.46 > Property lines superimposed on an aerial photograph. *Hepler/Wallach/Hepler © Cengage 2013*

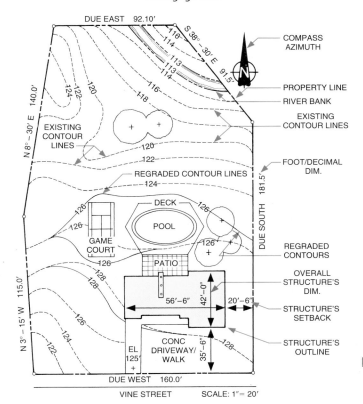

FIGURE 13.47 > Application of plot plan symbols. *Hepler/Wallach/ Hepler © Cengage 2013*

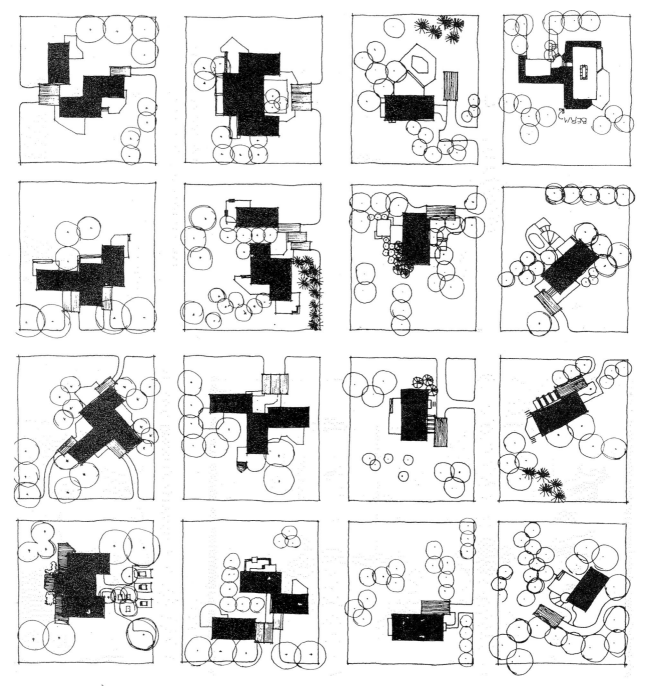

FIGURE 13.48 > Optional lot plan designs. *Hepler/Wallach/Hepler © Cengage 2013*

14. Use a decimal (civil engineer's) scale, such as $1'' = 10'$, or $1'' = 20'$, for preparing plot plans.

15. Show the position of utility lines.

16. Include the compass direction with the perimeter dimensions for each property line.

17. Show trunk base location and coverage of all major trees. Trunk diameter may also be noted.

18. Label and dimension all landscape construction features and auxiliary structures.

19. Identify the location of entrances with symbols. Arrows are most often used. See Figure 13.50.

In addition, if a septic system is used, draw and dimension the location and minimum distance allowed from system components to the nearest building.

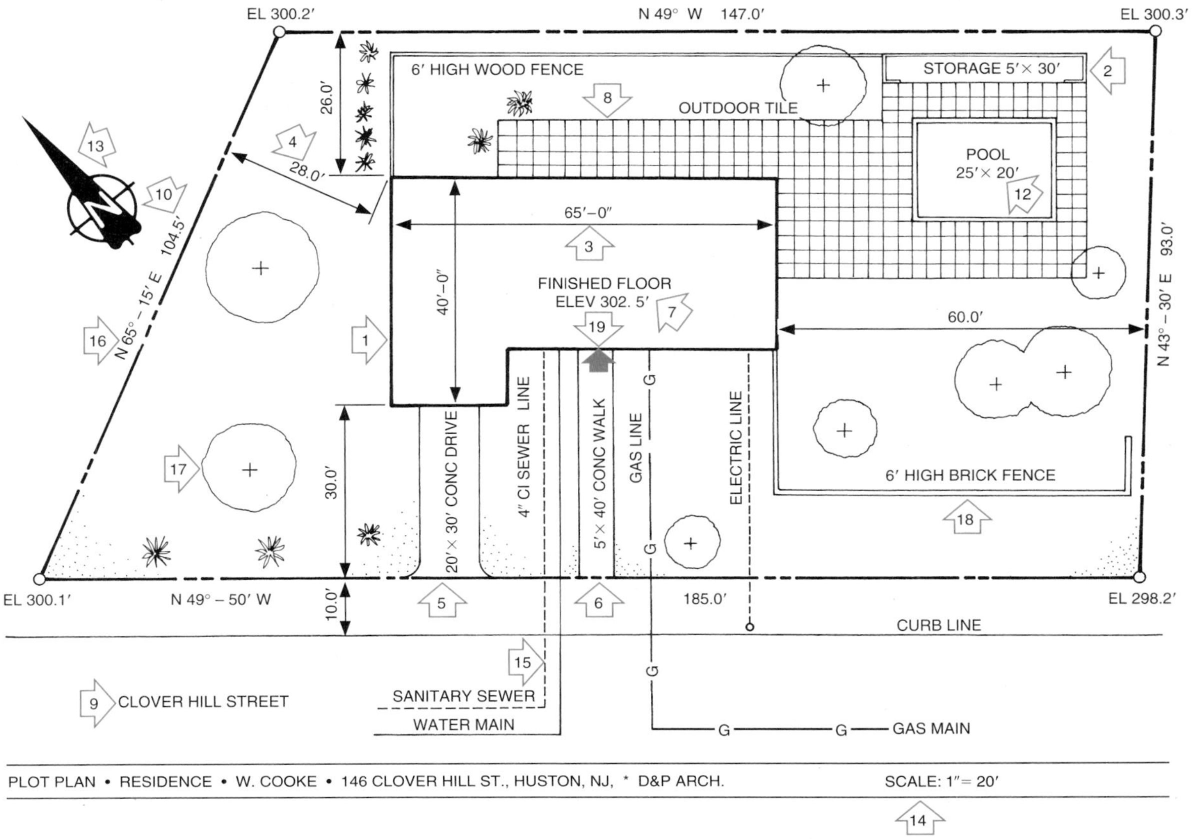

EL 300.2′   N 49° W   147.0′   EL 300.3′

6′ HIGH WOOD FENCE

26.0′

STORAGE 5′ × 30′   2

8   OUTDOOR TILE

4

28.0′

POOL
25′ × 20′   12

13

10

65′–0″

16   N 65° – 15′ E   104.5′

3

FINISHED FLOOR
ELEV 302. 5′   7

40′–0″

1

19

60.0′

N 43° – 30′ E   93.0′

17

4″ CI SEWER LINE

20′ × 30′ CONC DRIVE

5′ × 40′ CONC WALK

GAS LINE

ELECTRIC LINE

6′ HIGH BRICK FENCE

30.0′

18

5   6   185.0′

10.0′

EL 300.1′   N 49° – 50′ W   CURB LINE   EL 298.2′

15

9   CLOVER HILL STREET   SANITARY SEWER
WATER MAIN   G   G   GAS MAIN

PLOT PLAN • RESIDENCE • W. COOKE • 146 CLOVER HILL ST., HUSTON, NJ, * D&P ARCH.   SCALE: 1″ = 20′

14

FIGURE 13.49 > Plan keyed to plot plan guidelines. *Hepler/Wallach/Hepler © Cengage 2013*

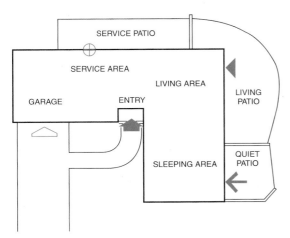

SERVICE PATIO

SERVICE AREA

LIVING AREA

LIVING
PATIO

GARAGE   ENTRY

QUIET
PATIO

SLEEPING AREA

FIGURE 13.50 > Types of entrance symbols. *Hepler/Wallach/ Hepler © Cengage 2013*

Variations

Although plot plans should be prepared according to the standards shown in Figure 13.49, many optional features may also be included in plot plans. For example, a plot plan may show only the outline of the building (Figure 13.50), or it may include shading, outlines of the roof intersections, or crosshatching. Sometimes the interior partitions of a building are drawn to reveal connections between outside living areas and the inside rooms. Figures 13.51A through 13.51E shows a comparison between an undeveloped parcel of land and the completely developed property. Also shown is the site plan, one of 40 drawings, used to complete this project. A red arrow identifies the same location as each drawing.

LANDSCAPE PLANS

Landscape plans are drawings that show the types and locations of vegetation. They may also show contour changes

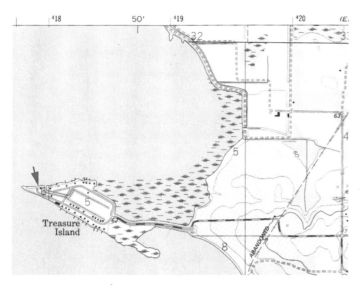

FIGURE 13.51A > USGS map of the site in Figure 13.51B. *Hepler/Wallach/Hepler © Cengage 2013*

FIGURE 13.51B > Site before development. *Hepler/Wallach/Hepler © Cengage 2013*

FIGURE 13.51C > Site after development. *Hepler/Wallach/Hepler © Cengage 2013*

and the position of buildings. Such features are necessary to make the placement of the vegetation meaningful. Symbols are used on landscape plans to show the position of trees, shrubbery, flowers, vegetable gardens, hedges, and ground cover. See Figure 13.52. Estate plans as shown in Figure 4.7 incorporate many landscape plan features but omit contour and detail dimensions.

Landscape plans are also used to show the landscaping possibilities of large parcels such as the 20-acre plan in Figure 13.53. This plan shows incorporation of community athletic facilities, retail shops, restaurants, lakes, boating, bioretention, rain, gardens, and storm water management. These features are added while preserving over a third of the area in its natural condition.

FIGURE 13.51D > Dimensioned site plan. *Hepler/Wallach/ Hepler © Cengage 2013*

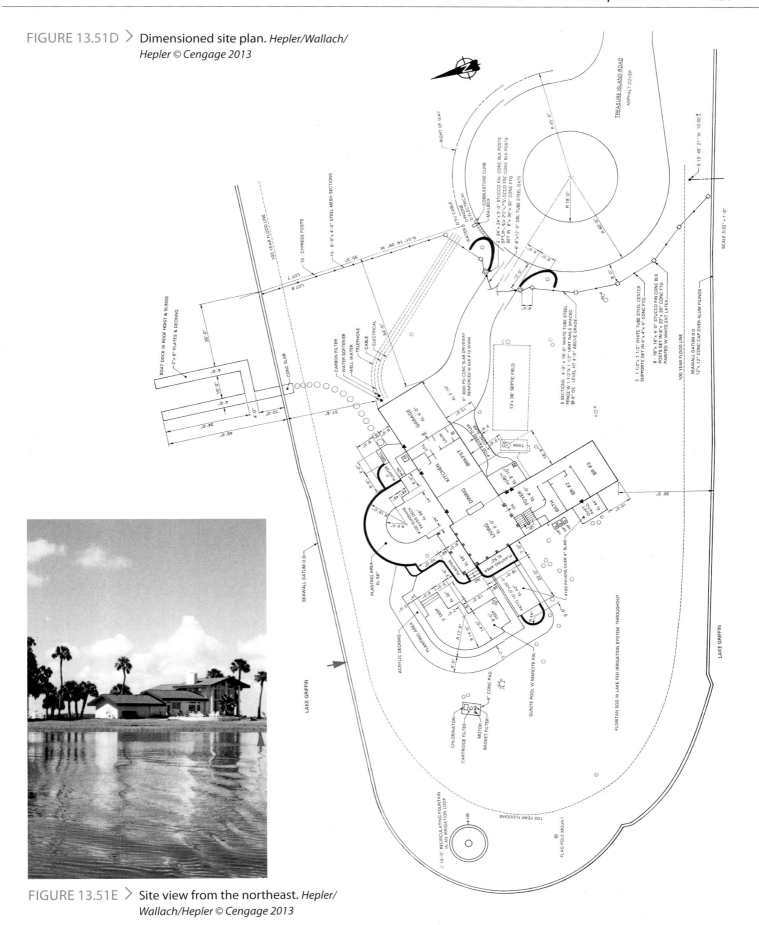

FIGURE 13.51E > Site view from the northeast. *Hepler/ Wallach/Hepler © Cengage 2013*

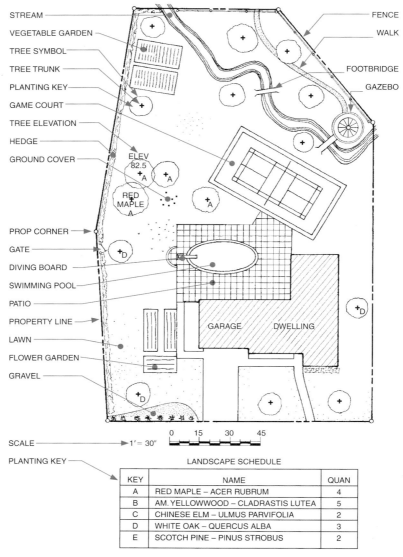

STREAM
VEGETABLE GARDEN
TREE SYMBOL
TREE TRUNK
PLANTING KEY
GAME COURT
TREE ELEVATION
HEDGE
GROUND COVER
PROP CORNER
GATE
DIVING BOARD
SWIMMING POOL
PATIO
PROPERTY LINE
LAWN
FLOWER GARDEN
GRAVEL
SCALE
PLANTING KEY

FENCE
WALK
FOOTBRIDGE
GAZEBO

GARAGE   DWELLING

ELEV 82.5
RED MAPLE A

SCALE 1' = 30"
0 15 30 45

LANDSCAPE SCHEDULE

KEY	NAME	QUAN
A	RED MAPLE – ACER RUBRUM	4
B	AM. YELLOWWOOD – CLADRASTIS LUTEA	5
C	CHINESE ELM – ULMUS PARVIFOLIA	2
D	WHITE OAK – QUERCUS ALBA	3
E	SCOTCH PINE – PINUS STROBUS	2

FIGURE 13.52 > Landscape plan keyed to landscape plan guidelines.
Hepler/Wallach/Hepler © Cengage 2013

Guidelines for Drawing Landscape Plans

The following guidelines for drawing landscape plans are illustrated by the numbered arrows in Figure 13.54.

1. Existing tree data including the existing topographic elevation, caliper (trunk diameter approximate; measured 1' above the finished grade), type, and the tree outline canopy are shown with a number keyed to an existing tree schedule. This data is usually provided as part of the survey, which is used as a "base" for various site plans.

2. Existing topographic lines (dashed) are added, with numbers shown on or above the line.

3. Proposed topographic lines (solid) are also known as contour lines. Numbers are in brackets to differentiate existing and proposed data.

4. Property lines define site limits.

5. Landscape plans and site plans are usually prepared using an engineer's scale although an architect scale of 3/32" = 1'-0" are sometimes used.

6. Buildings are outlined, crosshatched, or shaded depending on scale and use of the site plan document. A landscape plan at 1" = 10' may show more detail (such as windows and door openings) than a 1" = 100' scale plan.

7. Proposed trees and shrubs are shown by drawing an outline of the canopy (extent of branches).

8. Proposed trees and shrubs are identified by quantity and type, such as "(1 AMC)," which is labeled on or beside the tree symbol. The *1* refers to quantity, and *AMC* is a key or an abbreviation that is indexed on the plant list.

FIGURE 13.53 > Multiuse master plan. *Hepler Associates, PC*

9. Plant lists indicate quantity, key, botanical names, common name, size, type of rootball, and other pertinent data the landscape architect deems necessary for the proper identification.

10. Shrubs are sometimes shown as masses. The quantity of plants is determined by the mature size of the specified plant. In some cases some plants are drawn less than mature size to provide a denser planting pattern. This is done to solve a multitude of design problems such as accentuating a pleasant view or screening an unpleasant view.

11. Outlines of constructed recreation areas are shown.

12. Mass plantings of shrubs, ground cover perennials, and annuals are drawn as outlines. The complexity of the outline shapes is dependent on scale. In complex plantings, areas are crosshatched or "textured" to differentiate adjacent areas. The number and type are keyed in the same way trees are keyed. Mass perennial plantings may list numerous plants to be "placed in the field" by the landscape architect. Enlarged plans may be prepared at a larger scale for perennial beds. This is done to show individual plant locations.

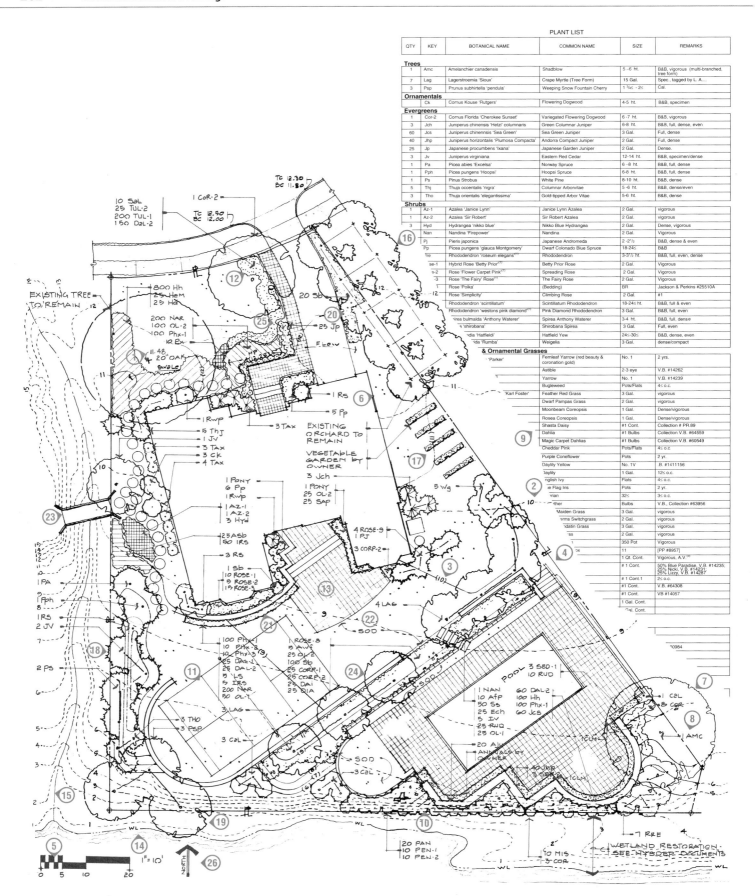

PLANT LIST

QTY	KEY	BOTANICAL NAME	COMMON NAME	SIZE	REMARKS
Trees					
1	Amc	Amelanchier canadensis	Shadblow	5 –6 ht.	B&B, vigorous (multi-branched, tree form)
7	Lag	Lagerstroemia 'Sioux'	Crape Myrtle (Tree Form)	15 Gal.	Spec., tagged by L. A...
3	Psp	Prunus subhirtella 'pendula'	Weeping Snow Fountain Cherry	1 ¾≤ - 2≤	Cal.
Ornamentals					
	Ck	Cornus Kouse 'Rutgers'	Flowering Dogwood	4-5 ht.	B&B, specimen
Evergreens					
1	Cor-2	Cornus Florida 'Cherokee Sunset'	Variegated Flowering Dogwood	6 -7 ht.	B&B, vigorous
3	Jch	Juniperus chinensis 'Hetzi' columnaris	Green Columnar Juniper	6-8 ht.	B&B, full, dense, even
60	Jcs	Juniperus chinennsis 'Sea Green'	Sea Green Juniper	3 Gal.	Full, dense
40	Jhp	Juniperus horizontalis 'Plumosa Compacta'	Andorra Compact Juniper	2 Gal.	Full, dense
25	Jp	Japanese procumbens 'Ixana'	Japanese Garden Juniper	2 Gal.	Dense.
3	Jv	Juniperus virginiana	Eastern Red Cedar	12-14 ht.	B&B, specimen/dense
1	Pa	Picea abies 'Excelsa'	Norway Spruce	6 –8 ht.	B&B, full, dense
1	Pph	Picea pungens 'Hoopsi'	Hoopsi Spruce	6-8 ht.	B&B, full, dense
1	Ps	Pinus Strobus	White Pine	8-10 ht.	B&B, dense
5	Thj	Thuja occentalis 'nigra'	Columnar Arborvitae	5-6 ht.	B&B, dense/even
3	Tho	Thuja orientalis 'elegantissima'	Gold-tipped Arbor Vitae	5-6 ht.	B&B, dense
Shrubs					
1	Az-1	Azalea 'Janice Lynn'	Janice Lynn Azalea	2 Gal.	vigorous
1	Az-2	Azalea 'Sir Robert'	Sir Robert Azalea	2 Gal.	vigorous
3	Hyd	Hydrangea 'nikko blue'	Nikko Blue Hydrangea	2 Gal.	Dense, vigorous
	Nan	Nandina 'Firepower'	Nandina	2 Gal.	Vigorous
	Pj	Pieris japonica	Japanese Andromeda	2 -2½	B&B, dense & even
	Pp	Picea pungens 'glauca Montgomery'	Dwarf Colonado Blue Spruce	18-24≤	B&B
	re	Rhododendron 'roseum elegans'[1]	Rhododendron	3-1½ ht.	B&B, full, even, dense
	se-1	Hybrid Rose 'Betty Prior'[1]	Betty Prior Rose	2 Gal.	Vigorous
	e-2	Rose 'Flower Carpet Pink'[1]	Spreading Rose	2 Gal.	Vigorous
	-3	Rose 'The Fairy' Rose[1]	The Fairy Rose	2 Gal.	Vigorous
		Rose 'Polka'	(Bedding)	BR	Jackson & Perkins #25510A
	-12	Rose 'Simplicity'	Climbing Rose	2 Gal.	#1
		Rhododendron 'scintillatum'	Scintillatum Rhododendron	18-24≤ ht.	B&B, full & even
		Rhododendron 'westons pink diamond'	Pink Diamond Rhododendron	3 Gal.	B&B, full, even
		pirea bulmalda 'Anthony Waterer'	Spirea Anthony Waterer	3-4 ht.	B&B, full, dense
		a 'shirobana'	Shirobana Spirea	3 Gal.	Full, even
		dia 'Hatfield'	Hatfield Yew	24≤-30≤	B&B, dense, even
		ida 'Rumba'	Weigelia	3 Gal.	dense/compact
& Ornamental Grasses					
	'Parker'		Fernleaf Yarrow (red beauty & coronation gold)	No. 1	2 yrs.
	ti		Astible	2-3 eye	V.B. #14262
			Yarrow	No. 1	V.B. #14239
			Bugleweed	Pots/Flats	4≤ o.c.
	'Karl Foster'		Feather Red Grass	3 Gal.	vigorous
			Dwarf Pampas Grass	2 Gal.	vigorous
			Moonbeam Coreopsis	1 Gal.	Dense/vigorous
			Rosea Coreopsis	1 Gal.	Dense/vigorous
			Shasta Daisy	#1 Cont.	Collection # PR.89
			Dahlia	#1 Bulbs	Collection V.B. #64559
			Magic Carpet Dahlias	#1 Bulbs	Collection V.B. #60549
			Cheddar Pink	Pots/Flats	4≤ o.c.
			Purple Coneflower	Pots	2 yr.
			Daylily Yellow	No. 1V	B. #1411156
			Daylily	1 Gal.	12≤ o.c.
			nglish Ivy	Flats	4≤ o.c.
			e Flag Iris	Pots	2 yr.
			rian	32≤	3≤ o.c.
		ther		Bulbs	V.B., Collection #63956
		Maiden Grass		3 Gal.	vigorous
		erms Switchgrass		2 Gal.	vigorous
		ndatin Grass		3 Gal.	vigorous
		ss		2 Gal.	vigorous
				350 Pot	Vigorous
		x		11	[PP #8957]
				1 Qt. Cont.	Vigorous, A.V.[1]
				# 1 Cont.	50% Blue Paradise, V.B. #14235; 25% Nicki, V.B. #14231; 25% Lizzy, V.B. #14287
				# 1 Cont.1	2≤ o.c.
				#1 Cont.	V.B. #64308
				#1 Cont.	VB #14057
				1 Gal. Cont.	
				Gal. Cont.	

FIGURE 13.54 > Application of landscape symbols. *Hepler/Wallach/Hepler © Cengage 2013*

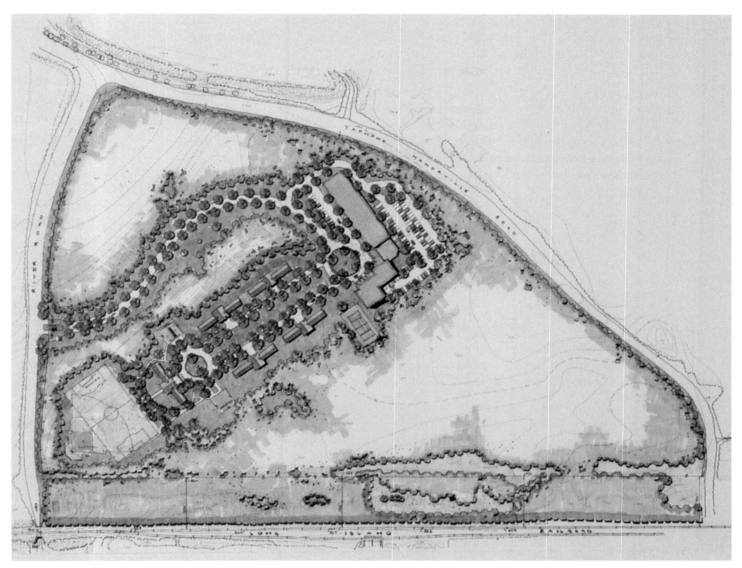

FIGURE 13.55 〉 Typical landscape plan. *Hepler/Wallach/Hepler © Cengage 2013*

13. Paved areas (hardscapes) are outlined and sometimes hatched or textured to indicate scale and enhance the "readability" of the drawing. See Figure 13.55.

14. Water edge is indicated by the label "WL" (waterline). The body of water is also labeled.

15. Streams can be depicted by a double line or a labeled center line in the case of intermittent (seasonal) streams.

16. Orchards or allées are shown by outlining the tree canopy as one mass with a center line connecting the trunks.

17. Vegetable gardens are shown by outlining the plant furrows.

18. Evergreen mass plantings, such as trees or hedges, are used to provide privacy and winter windbreaks.

19. Trees provide shade and scale to the site. At times landscape architects specify very large trees with calipers (trunks) over 30″ to solve various design issues.

20. Shrubs are used to define spaces, outline walks, and conceal foundation walls. Shrubs can also accentuate or de-emphasize the landform/building relationship.

21. Flower gardens are shown by outlining their shapes.

22. Lawns are shown as outlines with or without "stippling" (sparsely placed dots).

23. Bridges and other structures are shown with a standard bridge symbol.

24. Landscape plantings should enhance the function and aesthetics of the site. Ideal landscape plans work in conjunction with the grading of the site to enhance the site opportunities and minimize the constraints.

25. Plantings in front of the house should address the major architectural "lines" of the structure. Generally, landscape design provides for a mixture of plant types, forms, and colors to allow for the experience of *progressive realization* as one approaches the house. In other words, the entire house or site is not seen all at once.

26. A north arrow is always shown because building orientation can have a major effect on landscape design.

Phasing

The process of completely landscaping a site may be prolonged over several years because of a lack of time or money. **Phasing** spreads the project over a large period of time. Parts of the plan are completed at different times.

When a landscape plan is phased, the total plan is drawn. Then different shades or colors are used to identify the items that will be planted in the first month or year, and in successive months or years.

RENDERING SITE DRAWINGS

Landscape renderings are used by landscape architects, architects, planners, and environmental professionals to realistically communicate the scale and visual character of a project. They visually convey the integration of the architecture and the site. Proper landscape design is accomplished by using a variety of plants, pavings, water, topography, scale, form, color, and texture.

Rendering Media

Landscape drawings require the use of many different types of inks, markers, pens, pencils, paints, and papers. You will learn more about these and techniques for using them in Chapter 20.

If rendering techniques are to be added directly to a print, a heavy blackline or brownline print, not a blueline print, should be used. Blueline prints make a poor rendering medium if defined shapes and lines are to be maintained. Blue lines do not stand out when other colors are added.

Plan Rendering

Plan views are the most common form of site illustration because plans best define the overall scope of a project or development. A rendered landscape plan can be used to describe the landscaping of large parcels or small residential sites. A landscape rendering can instantly tell the viewer whether the site is a hilly, wooded area or a grassy knoll. A rendered landscape plan typically shows trees, shrubs, ground cover, grass, walks, driveways, curbs, steps, pools, patios, rock outcroppings, walls, and bodies of water.

The amount of rendering and intensity of detail should be consistent with the scale of the drawing. To describe a large site development, a minimum amount of detail rendering and detail is often adequate because of the small scale used. See Figure 13.56.

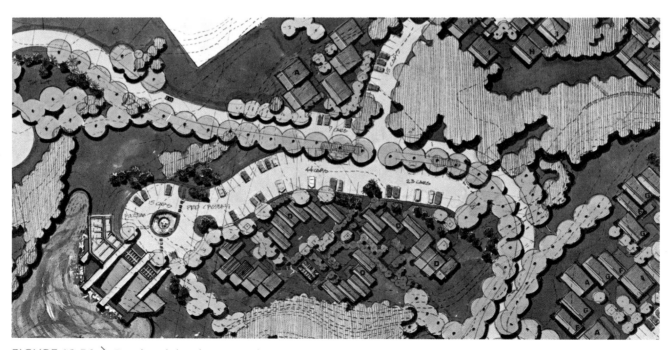

FIGURE 13.56 > Rendered development plan. *Hepler/Wallach/Hepler © Cengage 2013*

A large site at the scale of 1″ = 200′ can show trees and existing vegetation in simple groups or masses. A small residential site plan at a scale of 1″ = 10′ can show a considerable amount of detail. Individual trees and even branches can be shown, as well as textured paving and ground cover. Elements such as pools, arbors, shrubs, and decks can also show some amount of texture. Figure 13.57 shows a site plan with foliage labeled to make this plan more readable.

Before rendering a full landscape plan, the individual elements should be mastered. The illustrations discussed next show a progression of rendering techniques designed to build specific skills before rendering a complete plan. In these illustrations note how the image changes when color, value, texture, and shadow all come together.

Paving

Figure 13.58 shows three steps in plan rendering of different paving types. Practice these steps with successive sheets of tracing paper until you are satisfied.

Trees and Shrubs

In a plan view, tree branches may hide plant material, ground cover, and some structural areas located below. Because of their shape, the angle of the sun, and resulting shadows, treetops in a plan view contain a mixture of light, shade, and shadow. These are rendered by adding value, color, and texture. See Figure 13.59. Also note that larger trees cast larger shadows, and conifers (pine trees) cast shadows of a different shape. Deciduous trees cast different shadows in summer than in winter. Figure 13.60 illustrates the sequence of rendering several types of trees on a landscape plan.

Water

The rendering of water is challenging because water tends to be monochromatic (one color) and, hence, appears flat. Many aspects of the appearance of water result from its reflective qualities, which are difficult to render. Several techniques can be used to illustrate the reflections and movement of pool and stream water. See Figure 13.61.

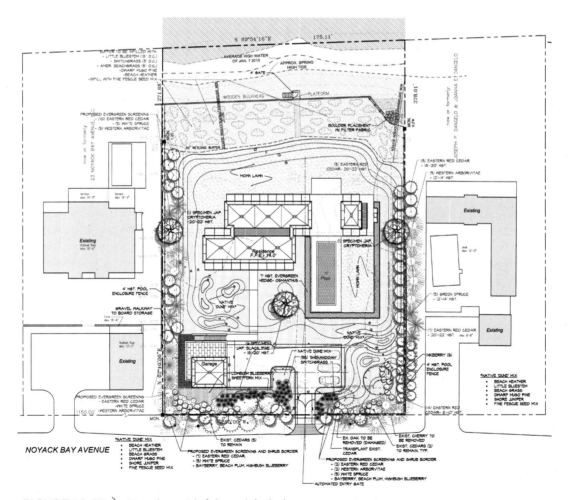

FIGURE 13.57 〉 Site plan with foliage labeled. *Timothy A. Rumph, RLA*

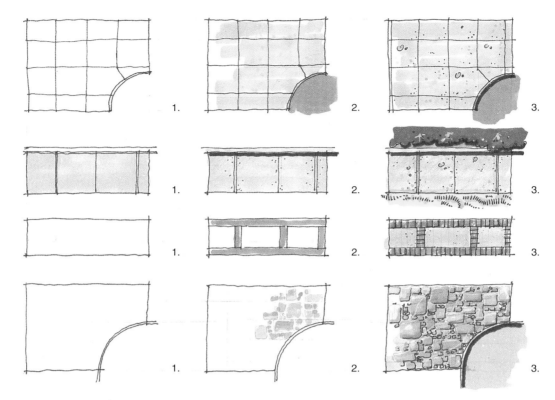

FIGURE 13.58 ⟩ Steps in rendering pavement surfaces in plan view. *Hepler/Wallach/Hepler © Cengage 2013*

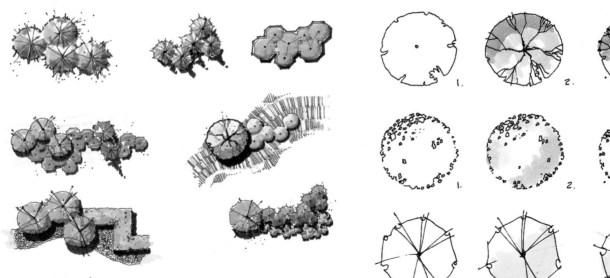

FIGURE 13.59 ⟩ Plan view of rendered trees and shrubs.
Hepler/Wallach/Hepler © Cengage 2013

FIGURE 13.60 ⟩ Steps in rendering trees. *Hepler/Wallach/ Hepler © Cengage 2013*

Entourage

Including people, vehicles (especially automobiles), and other familiar objects in a rendering helps establish the size of the project and its various components. The key to effectively rendering people and objects is the proper use of scale and shadow. Everything in the rendering must be drawn at the same scale. In plan views, shadows are also very important. See Figure 13.62.

Rendering a Complete Landscape Plan

The techniques just described were applied in rendering the complete landscape and seascape plan shown in Figure 13.63. To prepare a complete landscape rendering, begin by determining the color scheme on a separate but similar print or

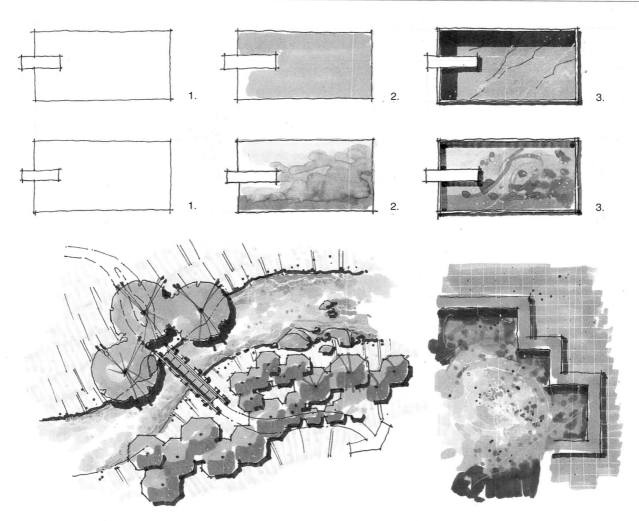

FIGURE 13.61 ⟩ Plan view techniques for rendering water surfaces. *Hepler/Wallach/Hepler © Cengage 2013*

FIGURE 13.62 ⟩ People's shadows add realism to this plan. *Hepler/Wallach/Hepler © Cengage 2013*

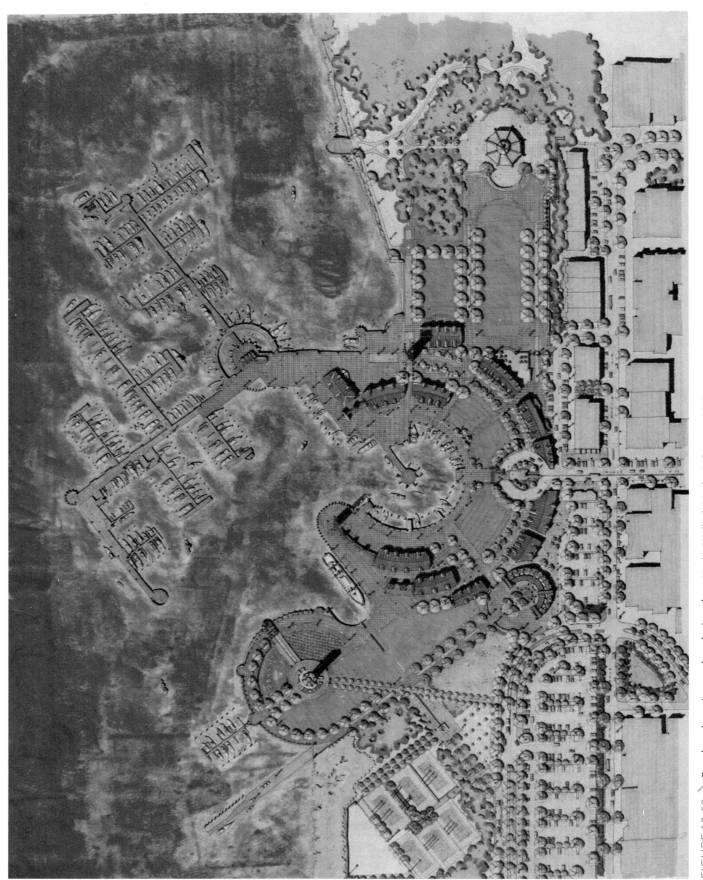

FIGURE 13.63 ⟩ Rendered marina and park site plan. *Hepler/Wallach/Hepler © Cengage 2013*

tracing. Then render higher elements, such as trees, arbors, and trellises for plants. Next, render lower elements, such as shrubs, decks, and ground cover. Render people, vehicles, and other objects last. Finally, add texture and shadows.

Elevation Rendering

Plan views may be best for showing all the features of a site, but elevation drawings are also needed to show landscape features as realistically viewed from eye level. Landscape elevation renderings combine drawings of trees, shrubs, walls, topography, vehicles, and people with the structure to produce a realistic view of the site. These drawings are used by design professionals to reveal form and scale of the project. The addition of color, value, texture, shades, and shadows provides depth to elevation drawings. Rendered elevation drawings are effective tools for sales and client-approval presentations because they help viewers perceive spatial depth and the final appearance of the project. See Figure 13.64.

FIGURE 13.64 ⟩ Rendered elevation of a natural waterfall.
Hepler/Wallach/Hepler © Cengage 2013

Trees and Shrubs

Vegetation sizes, shapes, colors, and textures vary greatly. These differences must be considered in rendering. See Figure 13.65A. Trees and shrubs are located and rendered to show them at full maturity even though they will be planted at much smaller sizes. Trees especially should help define space without totally blocking desirable natural views outside and eye-level views from inside structures. See Figure 13.65B. Where dense vegetation is used as a baffle, outlining or rendering the winter form of deciduous trees and shrubs is recommended.

Water

Water in elevations is found in the form of natural waterfalls, constructed waterfalls, and fountains. Water as a focal point in an elevation should be vibrant and alive. This is accomplished through the use of color and texture, with a proper mix of current flow and froth (bubbling). Figure 13.66 shows a waterfall as part of an elevation profile drawing. Figure 13.67 shows techniques for adding the illusion of water action to the elevation drawing.

Entourage

As with complete plan renderings, adding people and familiar objects to an elevation rendering adds both realism and scale. The addition of both people and the automobile to Figures 13.62 and 13.66 provides much needed size comparisons in these illustrations.

Rendering Walls

Retaining walls are a common element in most multilevel landscape designs. These are constructed from wood timbers, brick, stone, and cast-concrete forms. Effective renderings duplicate the wall size, color, joints, texture, shades, and shadows. See Figure 13.68.

FIGURE 13.65A ⟩ Rendering techniques for different trees in elevation. *Hepler/Wallach/Hepler © Cengage 2013*

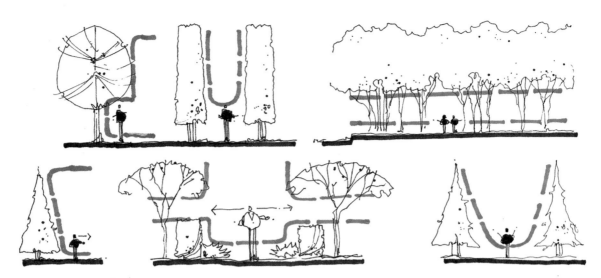

FIGURE 13.65B ⟩ Open space and views are considered when selecting and locating trees.
Hepler/Wallach/Hepler © Cengage 2013

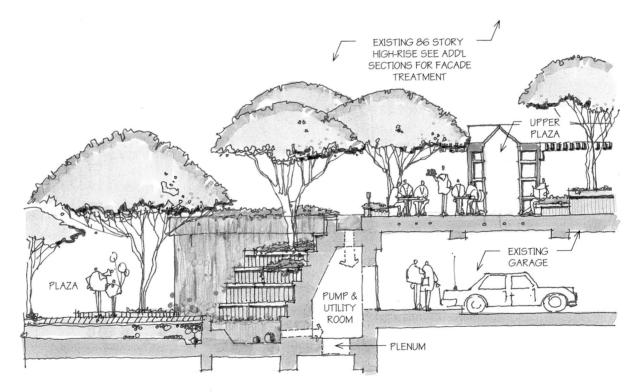

EXISTING 86 STORY HIGH-RISE SEE ADD'L SECTIONS FOR FACADE TREATMENT

UPPER PLAZA

EXISTING GARAGE

PLAZA

PUMP & UTILITY ROOM

PLENUM

FIGURE 13.66 ⟩ Rendering of a waterfall in a profile drawing. *Hepler/Wallach/Hepler © Cengage 2013*

Rendering Elevation Sections

When a design contains multiple levels, a rendered section is often the best way to describe the relationship of the levels. Adding vegetation, people, and automobiles also provides a sense of scale and proportion. Refer to Figure 13.66. Techniques for rendering water action is shown in Figure 13.67 and rendering masonry surfaces is shown in Figure 13.68.

Rendering Site Structure Elevations

When adding landscape features to a structure elevation, care must be taken to not hide the major lines of the building.

First add color, texture, and shadows to the building. Then begin rendering the foreground and work toward the background. Notice how this was accomplished in Figure 13.69. Here the vegetation and boulder placement are rendered without hiding the basic elements of the design.

SITE DETAILS AND SCHEDULES

Site development involves the design of landscape features, such as plants, with structural elements. Details, usually sections or profiles, are needed to ensure that construction is completed as designed. Unlike structural details, many site details are prepared to illustrate construction and planting

methods. These include tree and foliage planting, preservation, drainage, and irrigation, as well as structure and surface construction methods. Figure 13.70 shows typical pavement and wall construction sectional details. Instructional details are keyed to a related plan and or planting schedule. This is shown in the tree preservation plan in Figure 13.71. Some landscape details are combined in one plan as shown in the irrigation plan of Figure 13.72, which also includes planting details and site construction details.

Planting schedules are prepared and indexed with numbers corresponding to a landscape plan. These schedules function as a guide for the purchase and placement of each size and species of plant material. Refer back to Figure 13.54 for a partial planting list that is keyed to the related planting drawing. See Chapter 35 for a description of all types of schedules.

FIGURE 13.67 > Rendering technique to show water action. *Hepler/Wallach/Hepler © Cengage 2013*

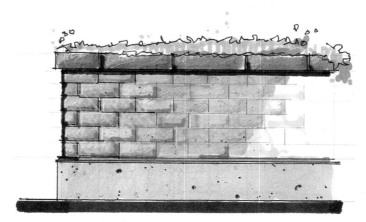

FIGURE 13.68 > Techniques for rendering a masonry wall elevation. *Hepler/Wallach/Hepler © Cengage 2013*

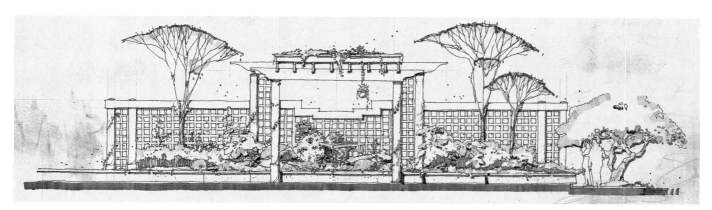

FIGURE 13.69 > Landscape features added to a fence and trellis. *Hepler/Wallach/Hepler © Cengage 2013*

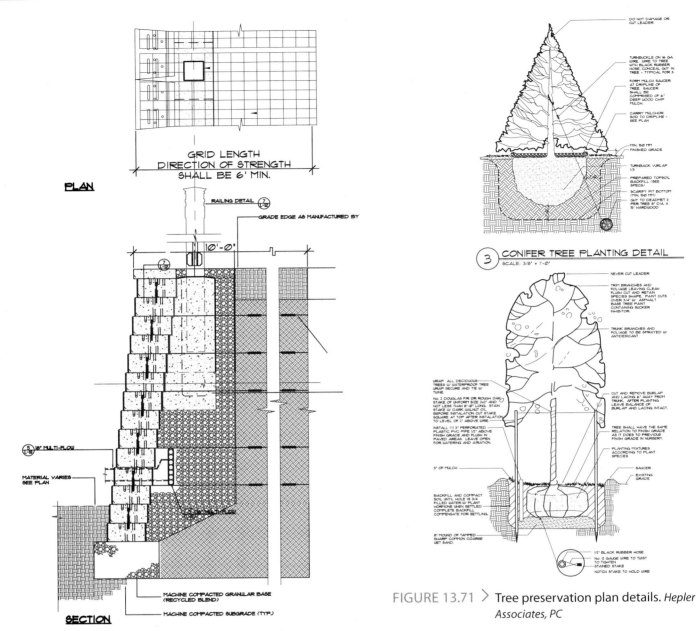

FIGURE 13.70 > Pavement and wall details. *Hepler Associates, PC*

FIGURE 13.71 > Tree preservation plan details. *Hepler Associates, PC*

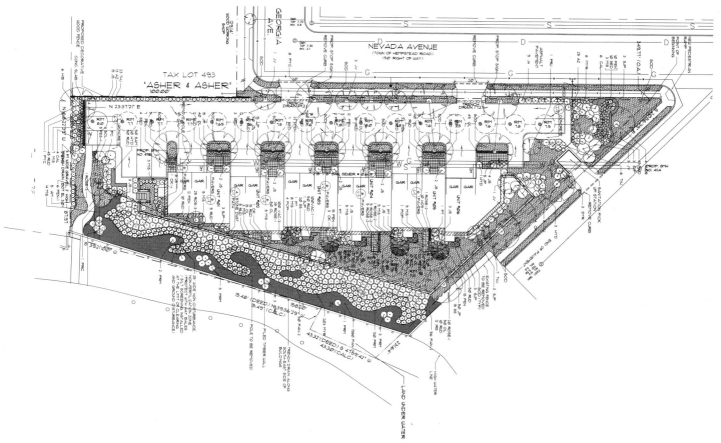

FIGURE 13.72 › Irrigation plan details. *Hepler Associates, PC*

Site Development Plans Exercises

CHAPTER **13**

1. Identify and discuss the environmental and human-related influences that affect site design.
2. Describe the zoning daylight plane ordinances for second-story setbacks. List why these are important.
3. Draw the setback and building area for a lot 130′ × 65′ according to the zoning requirements in your community. Determine the maximum size building possible for a site with 35% land coverage. Determine the maximum size of a house for that site.
4. Describe the zoning laws and the density pattern you would prefer for an area in which you wish to locate a house of your design.
5. Locate a house on a 60′ × 120′ lot. Setbacks are 5′ on the sides and 20′ front and rear.
6. Prepare a survey plan for a home you are designing.
7. Determine the bearing of property lines for a lot in your area using both the azimuth and American systems. Estimate the contour lines and include them in a sketch of this property. Complete a profile view.
8. Study a plot plan in this chapter and identify the highest and lowest levels above datum. What changes and/or structures would you recommend for this site?

9. Sketch a plat of your neighborhood using roads or streets as the outer boundaries.

10. Draw a plot plan of a house you are designing.

11. Draw a plot plan of a property in your area.

12. Draw and render a landscape plan for a house and property of your own design.

13. Redesign, draw, and render a landscape plan for a property in your area.

14. Find 10 drawings in this text that are part of the design shown in Figure 13.51.

15. Add landscape features to the sketches shown in Figure 13.73. Add details showing the construction of prefabricated components as shown in Figure 13.74.

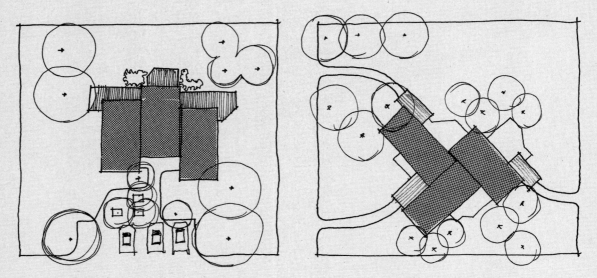

FIGURE 13.73 > Add symbols representing additional landscape features to these plans. *Hepler Associates, PC*

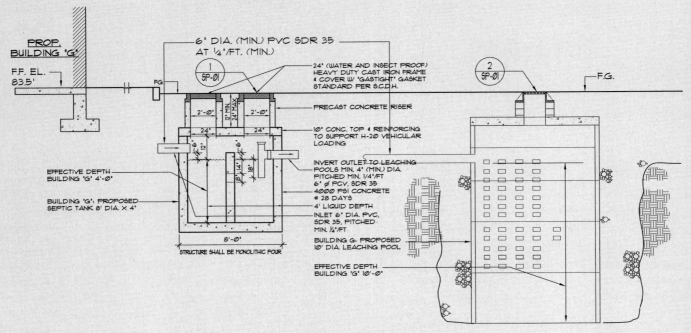

① COMPOSITE SANITARY DETAIL
SCALE: N.T.S.

FIGURE 13.74 > Typical sanitary detail. *Hepler Associates, PC*

Designing Floor Plans

OBJECTIVES

In this chapter you will learn to:

> gather information from a client that is needed to design an architectural project.

> analyze a building site.

> use the design process to prepare for drawing accurate and functional floor plans.

> create floor plan sketches.

> design floor plans to accommodate the needs of people with physical impairments.

TERMS

base map
building codes
composite analysis
conceptual design
easements
feng shui
floor plans

idealized drawings
room template
setbacks
single-line drawing
site analysis
site-related drawing

situation statement
slide-out
slope
templates
user analysis
zones

INTRODUCTION

The most commonly used architectural drawings are floor plans. Designing floor plans is not the same as drawing floor plans. Designing involves performing the procedures through which an architectural plan is systematically developed to satisfy specific wants and needs. This is accomplished by creating plans that provide for the efficient functioning of a building. These plans should also enhance the positive features of an existing site and minimize its negative characteristics.

In developing floor plan designs, refer to the specific room and area guidelines found in Part 3. Figure 14.1 shows a basic floor plan created to satisfy many wants and needs including the major categories of entertainment, family activities, sleeping accommodations for three family members, plus master bedroom, and quarters for two separate guests.

This chapter outlines procedures used in developing floor plan designs, beginning with a presentation of the steps in the design process. Guidelines for developing architectural designs to accommodate special needs are also included.

FLOOR PLAN DEVELOPMENT

Final **floor plans** contain more specific information about an architectural design than any other type of drawing. They include descriptions of locations, sizes, materials, and components contained in the design. Floor plans serve as a point of reference for other drawings in a set. For this reason the first phase of the architectural design process leads to the development of *basic floor plans.*

THE DESIGN PROCESS

The architectural design process involves many personal, social, economic, and technical variables that work together to create detailed working drawings. To effectively apply the principles and elements of design to an architectural project, established design sequences and procedures must be followed. This process is a logical sequence of thought and activities that begins with an inventory and analysis of the project. This process continues through the design of basic site and floor plans and proceeds through the completion of all working drawings. See Figure 14.2.

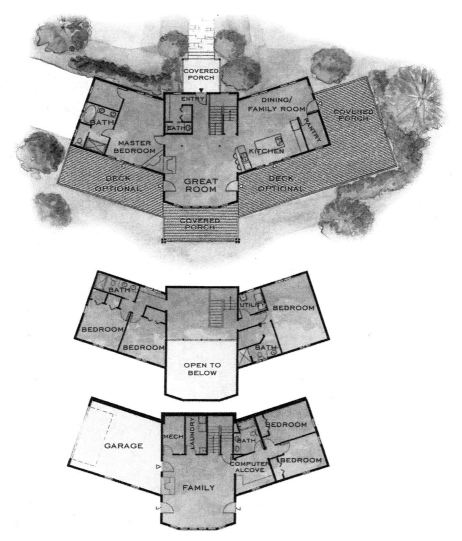

FIGURE 14.1 > Basic floor plan without dimensions and details. *Photo courtesy Lindal Cedar Homes. Lindal.com*

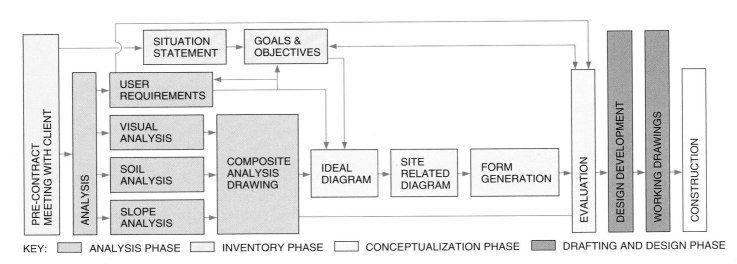

KEY: ▢ ANALYSIS PHASE   ▢ INVENTORY PHASE   ▢ CONCEPTUALIZATION PHASE   ▢ DRAFTING AND DESIGN PHASE

FIGURE 14.2 > The architectural design process. *Hepler/Wallach/Hepler © Cengage 2013*

Defining the Project

The first step in the design process is to define the project. An agreement established between a client and a designer involves the purpose, theme, scope of the project, budget, and schedules. This agreement is then translated into a **situation statement** as shown in Figure 14.3. This statement identifies and records the client's major requirements and any special design requirements and problems. Any subsequent design drawings relate to this situation statement.

Needs and Wants

The success of any design depends on how well the finished product meets the needs of the residents. During the entire process, the needs of the residents—and their "wants" as well—must be kept clearly in mind. This includes physical and lifestyle considerations.

A *need* is an absolute requirement and must be implemented in the design. A *want* is a desirable feature, but not absolutely required. Wants can be compromised because of budget constraints, space, or code restrictions. An effective design must meet all of the client's needs and as many wants as possible.

The designer and the client prepare needs and wants lists. See Figure 14.4. If the "wants" are listed in priority order, items on the list can be cut beginning at the bottom of the list and proceeding to the top, if necessary. In this way, what is wanted most is more likely to remain in the final design.

Goals and Objectives

Once the client and designer agree on a situation statement and create a list of the client's needs and wants, major goals and specific objectives can be developed. This is an important step. Constant reference is made to these goals and objectives during all phases of the design process. They provide the focus of the project for the designer and become a basis for evaluating all aspects of the design. See Figure 14.5. Once this is completed the analysis phase can begin.

Needs
- Living area open plan
- Contemporary design
- Home office or study which can be shared
- Living area fireplace
- Courtyard patio
- Pool with large living deck
- Boat dock
- Drive apron for parking
- Private MBR deck
- Privacy from road
- Three bedrooms including master suite
- Maximize view of mountains
- Dining facilities for 10 guests maximum

Wants
1. Jogging track
2. Shop area near garage
3. Badminton court
4. Minimum lawn area
5. Bridge connecting deck and dock
6. Whirlpool
7. Basketball court
8. Minimize tree removal
9. Maximize solar use
10. Separate home office for Mr. & Mrs. Smith
11. MBR fireplace
12. Greenhouse
13. Billiards room
14. Cabana near pool

FIGURE 14.4 > Design needs and wants list. *Hepler/Wallach/Hepler © Cengage 2013*

Major goals
Design a contemporary residence for the existing site with good visual profiles, aesthetic appeal and emphasis on functional, non-destructive use of all site features, including maximum use of solar energy. Plan working, living, and recreation areas to conform to space and priority needs. Position all facilities so that all are not visible from one vantage point.

Objectives
1. Provide stimulating, casual, and open atmosphere.
2. Locate private and public areas to avoid user conflicts.
3. Position all facilities for minimum environmental impact and minimum maintenance.
4. Orient structures for maximum solar use.
5. Building areas to blend with existing site landform.
6. Relate interior living areas to exterior space.
7. Residence not to be completely visible from access road.
8. Plan circulation patterns for both vehicular and pedestrian traffic with parking area.
9. Plan facilities for badminton, swimming, basketball, jogging, and whirlpool.
10. Provide courtyard for seasonal use.
11. Interior and exterior dining facilities for maximum 10 guests.
12. Provide Mr. Smith with an office for evening and weekend use.
13. Provide Mrs. Smith with an accessible office area to meet clients daily.
14. Use natural contemporary lines and materials consistent with site.
15. Design a focal point fireplace for living area.
16. Design gradual realization for vehicular approaching traffic.
17. Keep total cost within limits established by clients.
18. Boat dock to have access from deck area.
19. Provide three bedrooms, including master suite.

Situation
Mr. and Mrs. John Smith have acquired a five acre parcel of wooded land. They want the site developed as a residence and also for home office use. The Smith family of four have strong feelings for environmental preservation and are very fitness oriented. They want a high degree of privacy and need the capacity to entertain weekend guests. The total cost cannot exceed $60,000 more than the sale price of their existing home. Completion of construction (closing) must be no later than one year from the signing date of the design contract.

FIGURE 14.3 > Design situation statement. *Hepler/Wallach/Hepler © Cengage 2013*

FIGURE 14.5 > Design goals and objectives. *Hepler/Wallach/Hepler © Cengage 2013*

Analyzing the Project

After the project is defined, an organized and sequenced analysis must be made of user needs and wants, site features, soil conditions, the slope of the land, and the views. These separate analyses lead to the development of a comprehensive analysis.

User Analysis

In a **user analysis,** each goal is further refined into descriptions of space elements, usage, size, and the relationships between areas. With a user analysis, a designer can break down each design element into manageable parts. To make evaluation, verification, and discussion easier, a chart is usually prepared. See Figure 14.6.

The user analysis has great influence on the development of a design. No area should be omitted. If the user analysis is inadequate or contains erroneous information, the final design will not reflect the major goals and objectives of the project.

The user analysis is particularly important when designing a project that will be used by a person or persons with physical impairments. Design considerations and guidelines for developing designs to accommodate special needs are presented at the end of this chapter.

🌐 Site Analysis

An architectural project should be developed to take advantage of a site's positive features and minimize its negative features. Completing a **site analysis** not only helps the de-

Space elements	Primary users	Min. size	Notes and relationships
Living room	8–16 adults	16' X 22'	Pool view–fireplace. Access to foyer & dr.
Dining room	6–12 adults	14' X 20'	Access to kit & lr.
Study #1	Mr. Smith	14' X 20'	Private–quiet.
Study #2	Mrs. Smith	12' X 14'	Private–client. Accessible–joint office w/Mr. S?
Entry	Family–guests	8' X 10'	Visible from drive–baffle from street.
Parking	2 family cars / 6 guest cars	9' X 10' stalls	Access to main entry.
Decks or terrace	Family–guests	6' X 20'	Overlook pool–next to living area.
Courtyard	Family–guests	200 sq. ft.	For casual entertainment next to kit.
Kitchen	Family	12' X 16'	Access to deck, lr, & laundry.
Garage	Family	20' X 24'	Access to kit–convert to shop.
Service pickup	Service pers.	40 sq. ft.	Screen from living areas.
Master bedroom	2 adults	16' X 24'	Morning sun–king bed–access to pool–suite w/ bath–quiet area.
Bedrooms	2 children	16' X 18'	Plan for teen growth–away from living & master br.
Baths	Children & guests	2' 8' X 10'	Access from children's rooms & guests.
Guest bedroom	Guests	12' X 16'	Bath access–or convertible study?
Site considerations	All	Entire site	Solve sitting water problem. Use rock formations & add foliage for visual appeal.
Solar considerations	All	Bldgs. & site	Use passive techniques–care in orientation of facilities.
Recreation facilities	All	Courts, pool	Orient w/sun & screen from residence.

FIGURE 14.6 ❯ Building user analysis. *Hepler/Wallach/Hepler © Cengage 2013*

signer make proper design decisions, but also helps ensure appropriate land use.

Three types of site analyses are used to develop a final site analysis drawing: soil analysis, slope analysis, and visual analysis. Each of these three distinct factors affects the potential use of the different areas of the site. There are five phases in the development of a site analysis:

- *Phase 1—Development of a Base Map:* A **base map** shows all fixed factors related to the site that must be accommodated in the site plan. It includes topographical features; the outline and location of property lines, adjacent streets, existing structures, walkways, paths, terraces, and utility lines; easements; setback limits; and the north compass direction. **Easements** are rights-of-way across the land, such as for utility lines. **Setbacks** are minimum distances structures must be located from property lines as set by the local government. (See Chapter 13 for setback details.)

    Base maps are usually prepared to a scale of 1″ = 10′, 1″ = 20′, or 1″ = 30′ on an engineers scale; or 1/8″ = 1′-0″ up to 1/32″ = 1′-0″ on an architect's scale. Scale

selection depends on the site size and drawing format. Many copies of this map will be used during the design process for analysis and development. See Figure 14.7.

- *Phase 2—Soil Analysis:* Soil is composed of rocks, organic materials, water, and gases. Variations in the percentage of these ingredients determine the physical characteristics of the soil and its capacity to support the weight of structures. In general, coarse-grained soils, because of their drainage and bearing capacity, are preferred for buildings but not for plants. Conversely, fine-grained soils with high organic content are preferred for plants but not for buildings. The U.S. Department of Agriculture (USDA) classifies four types of soil according to their quality for building a structure:

Type 1. *Excellent:* Coarse-grained soils—no clays, no organic matter

Type 2. *Good to fair:* Fine, sandy soils (minimum organic and clay content)

Type 3. *Poor:* Fine-grained silts and clays (moderate organic content)

Type 4. *Unsuitable:* Organic soils (high clay and peat content).

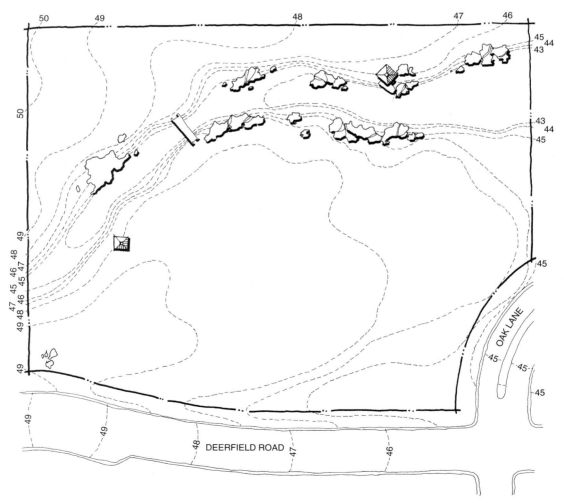

FIGURE 14.7 > Property base map. *Hepler/Wallach/Hepler © Cengage 2013*

To prepare a soil analysis drawing, follow these steps:

1. Obtain a soil classification for the site from a county soil survey or from private borings.

2. Draw areas on the base map representing the different soil types, as shown in Figure 14.8.

3. Note the bearing capacity and depth to bedrock for each soil category. This information is given in kilopounds (kips, or K) in the USDA survey book (1 K = 1,000 lb. for 1 sq. ft. of soil area).

4. Provide a legend showing the categories of soil types and describe the soil characteristics of each type. On the drawing, note where soil conditions can or cannot be used for building.

5. Color-code each soil capacity type in the legend to match the drawing.

- *Phase 3—Slope Analysis:* The **slope** of a particular site greatly affects the type of building that can or should be designed for it. The slope percentage also determines what locations are acceptable, preferred, difficult, or impossible

for building. See Figure 14.9. The cost of building may be greatly affected by excessive slope angles.

To complete a slope analysis drawing, refer to Figure 14.10 and complete the following steps:

1. To the base map, add *contour lines* (lines connecting points that have the same elevation) derived from a U.S. Geological Survey (USGS) map of the area. If the site is very hilly, a surveyor may add more closely spaced contours to provide a more detailed description of the slope of the site. Existing contour lines are dashed, since the finished contour grade lines will later be drawn solid. (See Chapter 13 for more information on contour lines.)

2. Identify the classification of slopes on the drawing:
   - 0% to 5%—excellent
   - 5% to 10%—good to fair
   - 10% to 25%—poor
   - Over 25%—unsuitable.

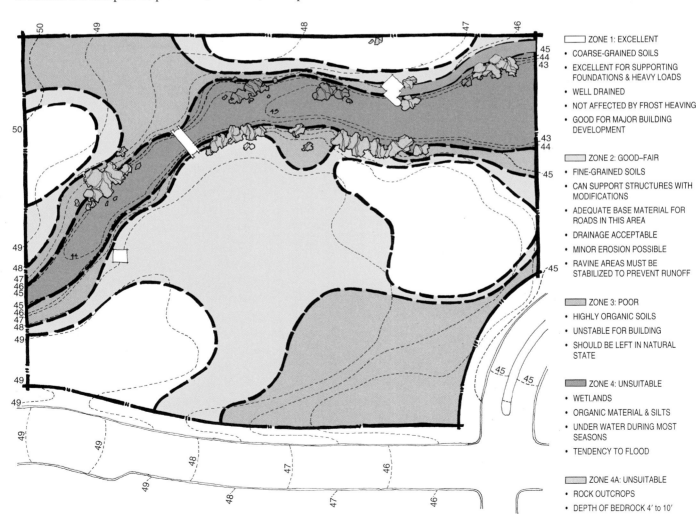

FIGURE 14.8 > Soil analysis drawing. *Hepler/Wallach/Hepler © Cengage 2013*

0–2% FLAT

2–3%
SLIGHT
SLOPE

3–7%
MODERATE
SLOPE

7–10%
MEDIUM
SLOPE

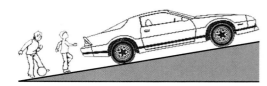

10–15%
STEEP
SLOPE

15–30%
VERY
STEEP SLOPE

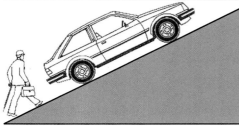

30–50%
EXTREMELY
STEEP SLOPE

FIGURE 14.9 > Slope percentages. *Hepler/Wallach/Hepler*
© Cengage 2013

3. Identify each slope category using colors or tones to show the degree of development potential of each section. Generally, light colors are used for areas suitable for development and dark colors are used for less suitable areas.

4. Provide a color-keyed legend of slope categories and note both potential and constraints for development of each slope category. Note the assets and limitations for each slope category. Note erosion or drainage problems, if any, for each category.

- *Phase 4—Visual Analysis:* Analyze the aesthetic and environmental potential of a site visually. Because visual observations and aesthetic qualities are often subjective and elusive, an organized method of recording and analyzing is important. Refer to Figure 14.11 and follow these steps to prepare a visual analysis drawing that can be used to provide input for future design phases:

  1. On the base map, locate the direction of the best views from each important viewer position. Label the nature of each view, and rate it as good, fair, or poor. Make recommendations for the treatment of each view such as "enhance" or "screen."

  2. Identify existing structures on the base map and describe their condition as good, fair, poor, unsound, or hazardous. Note suggestions to enhance, remove, or rehabilitate.

  3. Draw the outline and location of all existing and significant plant material, such as large shrubs and trees. Label the type, and indicate the condition of each as good, fair, or poor. Also locate, draw, and indicate large stands of ground vegetation to be saved.

  4. Identify any wildlife population and habitat areas to be saved. Indicate animal food and water sources.

  5. With directional arrows, show the direction of prevailing winter winds. Also show the direction of prevailing summer breezes.

  6. Find and label the source of any desirable fragrances or undesirable odors. For the latter, indicate possible solutions, such as minimizing with aromatic vegetation, screening, or removal of the source.

  7. Locate and label exposed open space, semienclosed public space, and private space.

  8. Note and record the path of the sun.

- *Phase 5—Composite Analysis:* Once the soil, slope, and visual analysis drawings have been completed, this information is combined into a single composite analysis drawing. A **composite analysis** drawing is prepared to determine the best location **zones** (areas) for the placement of structures on the site. Location zones are judged and numbered for development potential:

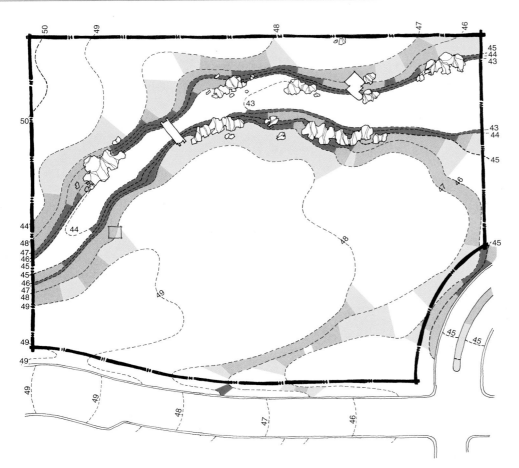

FIGURE 14.10 > Slope analysis drawing. *Hepler/Wallach/Hepler © Cengage 2013*

☐ 0–5%: EXCELLENT
- FLAT TO MODERATELY FLAT
- EASY TO DRAIN
- RESISTS EROSION
- REQ MIN GRADING
- IDEAL FOR RECREATION
- ALL TYPES ROADS FEASIBLE

☐ 5–10%: GOOD–FAIR
- SLOPING
- DRAINS EASILY
- MIN EROSION
- SOME GRADING REQ
- SLOPE RANGE EXCELLENT FOR BUILDING
- DRIVEWAYS SHOULD BE PLACED PARALLEL TO SLOPE

☐ 10–25%: POOR
- HILLY
- RAPID RUNOFF CREATES EROSION PROBLEMS
- STABILIZATION NECESSARY ON UNDEVELOPED SLOPES
- EXCESSIVE CUT & FILL NEEDED FOR STRUCTURES
- SINGLE-STORY STRUCTURES IMPRACTICAL
- ROADS REQUIRE REINFORCEMENT & BASE STRUCTURES

☐ 25%: UNSUITABLE
- SEVERE
- SERIOUS RUNOFF & EROSION PROBLEMS
- WILL NOT SUPPORT STRUCTURES WITHOUT MAJOR ALTERATIONS WHICH WOULD DESTROY THE ENVIRONMENT
- EXCELLENT FOR HIKING

1. Excellent
2. Good or fair
3. Poor
4. No development potential.

To prepare a composite analysis drawing, refer to Figure 14.12 and follow these steps:

1. Place the soil analysis drawing directly over the slope analysis. Align the property lines with the base map and tape the base map to the drawing board.

2. Attach tracing paper over the slope and soil drawings and trace a line around each distinct area.

3. Determine which development zone each outlined area represents. For example, if a 0% to 5% slope area overlaps with a coarse-grained soil area, the zone is labeled "1" (excellent potential). Another example would be a poor, clay soil area that overlaps with a 20% slope area; such a zone is labeled "3" (poor). Label each zone on an overlay drawing.

4. Place the overlay drawing over the visual analysis drawing and repeat the same outlining of areas covered in

step 3 to complete the composite analysis drawing as shown in Figure 14.12. Apply judgment concerning priorities when there is an overlapping area conflict.

Developing a Conceptual Design

A **conceptual design** represents the best response to the information on the site analysis and in the user analysis chart. Two types of sketches are created to develop a conceptual design: idealized and site-related drawings.

Idealized Drawings

Idealized drawings or diagrams are a series of study sketches (usually drawn on inexpensive tracing paper, nicknamed "trash" or "bum wad"). These designate the ideal spatial relationships of the user elements from the user analysis. Ideal diagrams are freehand, bubble-like sketches that show how the separate user elements fit together. The bubbles are not used to show the *sizes* of the areas, only their *spatial* relationships. Designers usually complete a number of studies or sketches. They do as many sketches as necessary until they achieve one

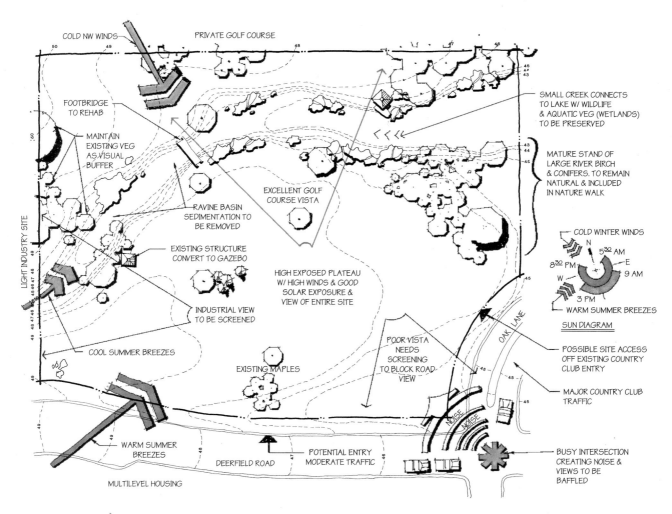

FIGURE 14.11 > Visual analysis notes. *Hepler/Wallach/Hepler © Cengage 2013*

study sketch that provides the best possible spatial relationship between the different elements. See Figure 14.13.

Site-Related Drawings

A **site-related drawing,** such as that shown in Figure 14.14, is one that matches the idealized drawing to the site and introduces size requirements. The main effort at this stage of design is concentrated on "fitting" the various elements of the user analysis onto the site, while maintaining the most ideal spatial relationships.

The scale (size) of each element is first introduced at this phase of the design process. The approximate position, size, and shape of each room, area, or feature are sketched on tracing paper that is placed over the composite site analysis drawing. Now the floor plan design and position begin to take physical form in relation to the land and the surroundings. Several site-related studies are usually completed to integrate the design with the site. At this stage, a designer needs to focus on the specific characteristics of the site—its constraints and opportunities.

From all of the site-related diagrams and sketches, one is chosen that becomes the floor plan conceptual design. The designer now begins to generate the form of the design. Drawings are refined into a loose graphic drawing for evaluation. See Figure 14.15. This drawing may not be exactly to scale, but becomes a preliminary floor plan.

Evaluating Preliminary Designs

Evaluation is always needed to determine the degree of excellence of a design. Self-evaluation of a design is critical and necessary. This requires checking the quality to see if it measures up to the predetermined goals and objectives and to the user analysis requirements.

The conceptual design must be evaluated and necessary revisions made before beginning the final design development phase. To redesign some elements at the conceptual design stage is easier, cheaper, and more time efficient than later in the process.

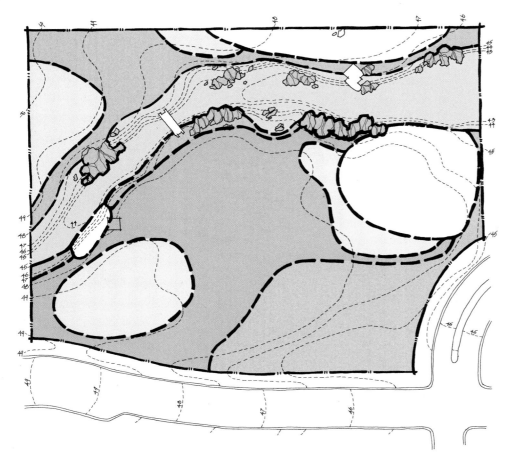

ZONE 1: EXCELLENT
- PRIME DEVELOPMENT LOCATION
- COMPACTABLE SOILS
- EXCELLENT DRAINAGE
- SLOPE 0–5%
- MINIMUM EROSION PROBLEMS
- ROAD & SERVICES EXCELLENT
- SOLAR ORIENTATION OF STRUCTURES POSSIBLE

ZONE 2: GOOD–FAIR
- GENTLE SLOPES 0–10%
- COMPACTABLE SOIL
- GOOD DRAINAGE
- EROSION PROBLEMS MINIMAL EXCEPT NEAR RAVINE
- PRIME AREA FOR SITING STRUCTURES

ZONE 3: GOOD TO FAIR WITH MODIFICATIONS
- VARIABLE SLOPE 0 TO 10%
- SOIL SUITABLE FOR STRUCTURES WITH MODIFICATIONS
- HIGH RUNOFF RATES NEAR BANKS
- EROSION RISKS AT BANKS
- VEG REMOVAL AT CREEK BANKS COULD DAMAGE ECOTONE
- EXCELLENT FOR NATURAL USE FOR PATHS

ZONE 4: RESTRICTED DEVELOPMENT
- ORGANIC SOIL
- POOR DRAINAGE
- GENERALLY UNDER WATER
- FOG POCKET DANGERS
- LEAVE UNDERDEVELOPED AND AS NATURAL AS POSSIBLE

ZONE 4A: RESTRICTED DEVELOPMENT
- ESTABLISHED WOODED STANDS
- WILDLIFE HABITAT & FEEDING GROUNDS AT WATERS EDGE
- PROVIDES BUFFER FROM ROAD

FIGURE 14.12 ⟩ Composite analysis of site. *Hepler/Wallach/Hepler © Cengage 2013*

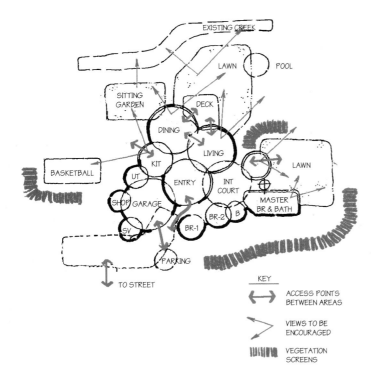

KEY
⟷ ACCESS POINTS BETWEEN AREAS
↗ VIEWS TO BE ENCOURAGED
⫶⫶⫶ VEGETATION SCREENS

FIGURE 14.13 ⟩ Idealized "bubble" diagram. *Hepler/Wallach/ Hepler © Cengage 2013*

Many details and sizes will not be determined at this point. However, the position of the structures and the relationship between the design elements should not change significantly after this evaluation is completed.

To evaluate a design, the contents of conceptual design must be compared with *each* specific goal and objective in the user analysis. If a goal has not been accomplished, then that part of the design must be altered to achieve the desired result. A well-developed design will contain very few discrepancies between the goals of the user analysis and the conceptual design.

Design Development

After the necessary changes have been made in the conceptual design as a result of evaluation and client feedback, the final design development phase begins. During this phase, details are added to the site-related diagram in progressive sketches. Sketches are redone until the outlines of the design parts fit together without overlapping and without awkward offsets.

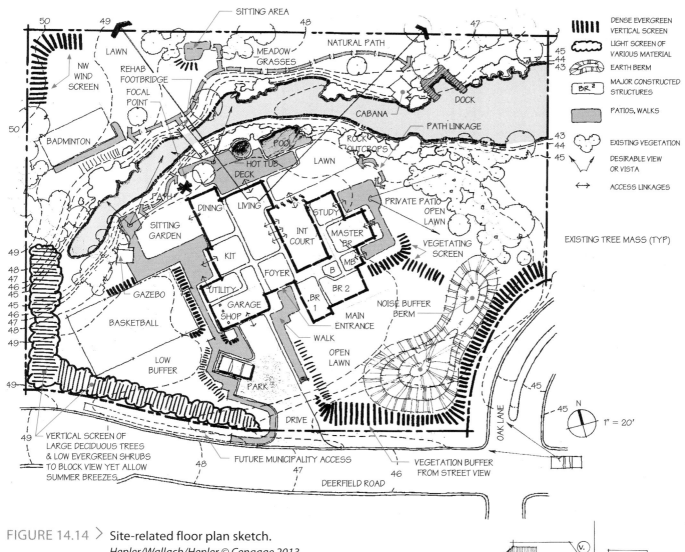

FIGURE 14.14 > Site-related floor plan sketch.
Hepler/Wallach/Hepler © Cengage 2013

Once the design is "smoothed out," a scaled, **single-line drawing** is prepared, as shown in Figure 14.16. This drawing includes both floor plan and site features. After this drawing is completed, the designer can concentrate on refining interior building floor plans. See Figure 14.17.

Designing with CAD Symbols

Drawing symbols are entities or groups that are used repeatedly and are stored in libraries as blocks. Typical items that are blocks include windows, doors, map symbols, and fixtures. A *block* is a group of entities that is stored as a single object, typically as a separate drawing file. Drawings stored only within the current drawing that cannot be used outside the current drawing are created by the *Block* command. Drawings can also be stored as a separate drawing file that can be used in any drawing; this is accomplished by using the *Wblock* command.

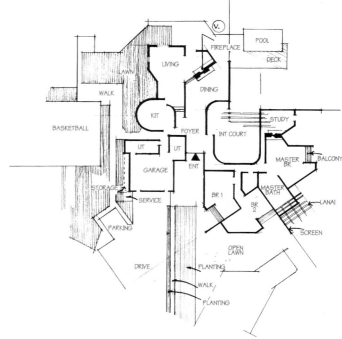

FIGURE 14.15 > Floor plan conceptual study drawing.
Hepler/Wallach/Hepler © Cengage 2013

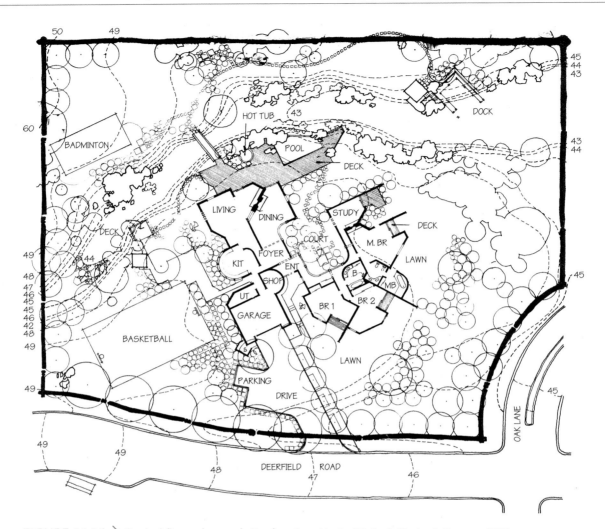

FIGURE 14.16 > Scaled floor plan and site drawing. *Hepler/Wallach/Hepler © Cengage 2013*

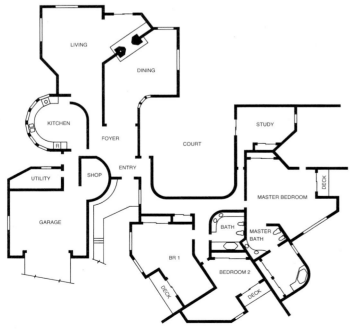

FIGURE 14.17 > Revised and refined floor plan. *Hepler/ Wallach/Hepler © Cengage 2013*

All drawing files (.dwg) can be used like a *Wblock*. Blocks (the generic term used for both *Block* and *Wblock* creations) are placed back into drawings by using the *Insert* command. Choose the file to be inserted, enter the scale of the drawing), and then enter the insertion point on the drawing. The insertion point on the inserted object was created as the *Block* insertion point in the *Block/Wblock* command.

Blocks can also be used to replicate whole drawings like an apartment building. CAD programs include libraries of both architectural and engineering drawing symbols.

FUNCTIONAL SPACE PLANNING

Once a conceptual plan has been developed, the required space for each area needs to be finalized. Experienced designers can determine the relationships of areas and record design ideas through the use of progressive sketches. Students and inexperienced designers should proceed through a process of determining final space requirements through the use of **templates.** The work done with templates, as presented in this chapter, can be performed manually or on a CAD system.

Floor Plan Sketches

A floor plan design sketch must satisfy all original goals and objectives. Once this has been accomplished, many more sketch versions of the floor plan may be necessary. Think of the first sketch as only the beginning. Many sketches are usually necessary before a designer achieves an acceptable floor plan. Through successive sketches, costly and poor design features can be eliminated. By planning to use standard building materials and furnishings, many sizes are established. The exact positions and sizes of doors, windows, closets, and halls should be determined at this point. You may wish to consider both open and closed types of floor plans. Refer back to Chapter 14.

Further refinement of the design is done by resketching until a satisfactory design is reached. Except for very minor changes, making a series of sketches is always better than erasing and changing the original sketch. Many designers use tracing paper to trace the acceptable parts of the design and then add design improvements on the new sheet. This procedure provides the designer with a record of the total design process. Early sketches sometimes contain ideas and solutions to problems that might develop later in the final design process.

A final scaled sketch should be prepared on grid paper to provide a more accurate and detailed sketch. It should also include the locations of shrubbery, trees, patios, walks, driveways, pools, and gardens. Once a final sketch is complete, three-dimensional conceptual CAD models may also be developed to aid in interpreting the conceptual design.

At this point in the design sequence, a consultation between the general contractor and the designer is recommended. This may also involve a review of the drawings set by subcontractors in the areas of framing, plumbing, electrical, and HVAC. This review is intended to discover any potential obstacles to the completion of work in each area. If conflicts arise, then adjustments must be made to the plans or to the contractors' methods before proceeding. Preliminary approval of the plans by municipal zoning and building departments is also advisable before the preparation of a final working drawing set.

After plans at this stage are approved by the client, contractor, zoning and building departments, and possibly a civic review committee, the preparation of working drawings can begin. Working drawings and documents are prepared to further refine the basic design concepts into very exact plan sets that can be used for bidding, budgeting, and construction purposes. You will learn more about these processes later in this text.

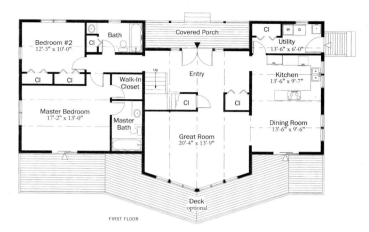

FIGURE 14.18A > Basic concept floor plan. *Photo courtesy Lindal Cedar Homes. Lindal.com*

FIGURE 14.18B > Embellished presentation floor plan. *Photo courtesy Lindal Cedar Homes. Lindal.com*

At this point the final position of walls, partitions, stairs, fixtures, cabinets, porches, and decks is nearly finalized. Figure 14.18A shows a final concept floor plan used as a base for the finished floor plan, which will be used for construction reference. Floor plans of this type, without dimensions and fine details, are often rendered and embellished for presentation purposes, as shown in Figure 14.18B.

Variations in Developing Floor Plans

The design process and sequence of preparing floor plans have been considered from the "inside-outside" point of view. The needs of the inside areas determine the size and shape of the outside. However, some design situations require a plan to be developed within a given, predetermined outside area. Apartment, modular unit, mobile, and manufactured

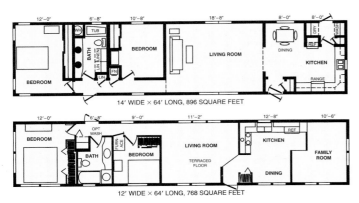

FIGURE 14.19 > Mobile homes are designed from the "outside in". *Hepler/Wallach/Hepler © Cengage 2013*

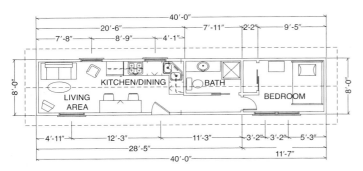

FIGURE 14.20 > Shipping container house plan. *Hepler/Wallach/Hepler © Cengage 2013*

home designs fall into this category. Figure 14.19 shows how "outside-in" applies to mobile home design. This concept can also be applied to the conversion of shipping containers into houses as shown in Figure 14.20.

The size and shape of mobile (motor) homes is very restricted due to roadway clearances. One method of expanding their original dimension is through the use of **slide-out** rooms. These rooms can be expanded outward once the vehicle is parked off-road. The interior of a motor home with three slide-out areas is shown in Figure 14.21. Figure 14.22 illustrates the use of a motor home floor plan to show walls, slide-out positions, furniture, and appliance locations. Manufactured homes that are delivered to a permanent building site in pieces can be designed with more flexibility than totally preassembled homes. When delivered in parts, much larger dimensions can be used, as shown in Figure 14.23.

Although limited in size and shape, modular prefabricated homes can be customized through innovative design. Figure 14.24 illustrates how a modular home can be designed to integrate with its natural environment. Figure 14.25 shows one module being set onto an existing foundation.

FIGURE 14.21 > Mobile home slide-out model. *Featherlite Coaches*

Apartment floor plans are restricted to the size and shape of the apartment building. Therefore, the size is determined by the number of apartments planned for each floor. In such a plan the arrangement of features is determined by the location of the steel and concrete columns. These structural bearing members cannot be removed or repositioned; only nonbearing walls can be moved or relocated. Figure 14.26 shows an apartment floor plan that represents one of six units on a floor. The location and shape of this apartment, and others, is shown on the key plan of Figure 14.27. The floor plan shown in Figure 14.28 shows the development of a plan which conforms to a pre-established shape and dimensions.

Because of limitations of time or money, it may be desirable to construct a house over a period of time. A house can be built in several steps. The basic part of the house can be constructed first. Then additional rooms (usually bedrooms) can be added in future years as the need develops.

When future expansion of the plan is anticipated, the complete floor plan should be drawn before the initial construction begins, even though the entire plan set may not be completed at that time. If only part of the building is planned and built and a later addition is made, the addition will invariably look "tacked on." This appearance can be avoided by designing the original floor plan for expansion. See Figure 14.29.

Planning Space for Rooms and Areas

Furniture and Equipment

Selecting the style, size, and amount of furniture needed for a room is the first step in determining a room's space require-

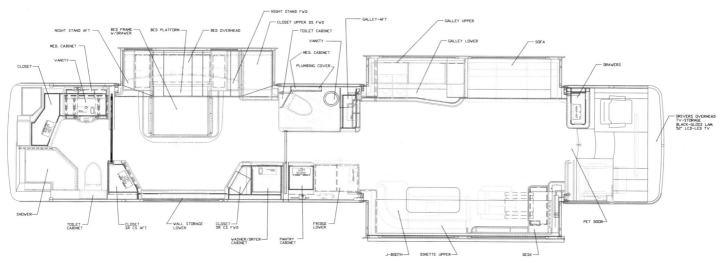

FIGURE 14.22 > Mobile home detailed floor plan. *Featherlite Coaches*

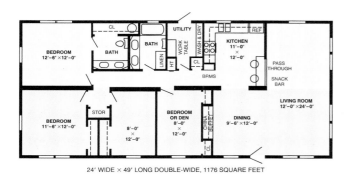

24' WIDE × 49' LONG DOUBLE-WIDE, 1176 SQUARE FEET

FIGURE 14.23 > Manufactured home split for shipping.
Hepler/Wallach/Hepler © Cengage 2013

ments. Furniture should be chosen according to the needs of the residents—whether that means including a piano for someone interested in music or a large amount of bookcase space for an avid reader. The artist, drafter, or engineer may require drawing furniture or computer hardware. These individual pieces of furniture affect each room's specific size and shape. Figure 14.30 shows the common sizes of residential furniture.

After furniture dimensions have been established, furniture templates can be made and arranged in functional patterns. As you learned in previous chapters, furniture templates are thin pieces of paper, cardboard, plastic, or metal

FIGURE 14.24 > Modular home integrated with a natural site. *Marmol Radziner, Prefab; Joe Fletcher, Photograph*

FIGURE 14.25 > Module placement on a foundation.
Marmol Radziner, Prefab

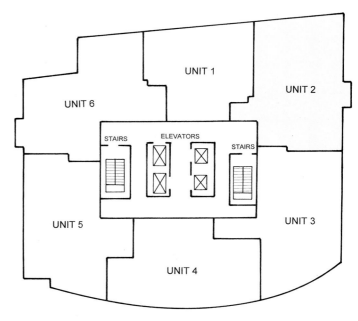

KEY PLAN CONDOMINIUMS
One Rincon Hill, San Francisco

FIGURE 14.27 > Apartment key plan. *One Ricon Hill*

FIGURE 14.26 > Apartment Unit 2 floor plan. *One Ricon Hill*

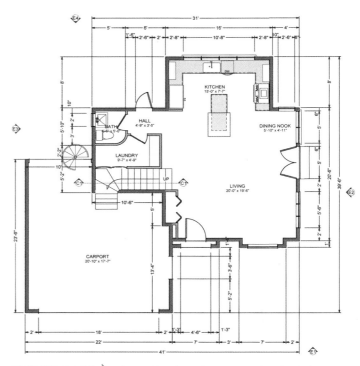

FIGURE 14.28 > Floor plan fitted to an established shape.
Chief Architect Software

that are used to determine exactly how much floor space each piece of furniture will occupy. See Figure 14.31.

Templates are always selected to the scale that will be used in the final drawing of the floor plan. The scale most frequently used on floor plans is 1/4″ = 1′-0″. Scales of 3/16″ = 1′-0″ and 1/8″ = 1′-0″ are usually used for larger buildings.

Wall-hung furniture, or any projection from furniture, even though it does not touch the floor, should be included as a template. This is necessary because the floor space under this furniture is not usable for any other purpose.

FIGURE 14.29 > Three-stage expansion floor plans. *Used by permission Hanley & Wood, LLC*

Figure 14.32 shows the use of furniture templates on a floor plan. Furniture templates are placed in the arrangement that will best fit the living pattern anticipated for the room. Space must be allowed for the free flow of traffic, as well as for opening and closing doors, drawers, and windows.

Room Sizes and Shapes

After suitable furniture arrangements have been established, room dimensions can be determined by drawing an outline around the scaled furniture templates. Then a **room template** can be made by cutting around the outline of the room.

Because the cost of a home is largely determined by the size and number of rooms, room sizes must also be adjusted to conform to an acceptable price range. Figure 14.33 shows area sizes for each room in small, medium, and large dwellings. These areas represent only average sizes. Even where no financial restrictions exist, room sizes should be limited by the functional requirements of the room. Just as a room can be too small, it can also be too large to function well for its intended purpose.

Visualizing the exact amount of real space that will be occupied by furniture or that should be allowed for traffic through a room is sometimes difficult. One device used to give a point of reference is a template of a human figure. See Figure 14.34. This template will help you see how a person would move throughout the room. With a human figure template, you can check the appropriateness of furniture size, number, placement, and the adequacy of traffic allowances.

ITEM	LENGTH, IN (mm)	WIDTH, IN (mm)	HEIGHT, IN (mm)
COUCH	72(1829)	30(762)	30(762)
	84(2134)	30(762)	30(762)
	96(2438)	30(762)	30(762)
LOUNGE	28(711)	32(813)	29(737)
	34(864)	36(914)	37(940)
COFFEE TABLE	36(914)	20(508)	17(432)
	48(1219)	20(508)	17(432)
	54(1372)	20(508)	17(432)
DESK	50(1270)	21(533)	29(737)
	60(1524)	30(762)	29(737)
	72(1829)	36(914)	29(737)
STEREO CONSOLE	36(914)	16(406)	26(660)
	48(1219)	17(432)	26(660)
	62(1575)	17(432)	26(660)
END TABLE	22(559)	28(711)	21(533)
	26(660)	20(508)	21(533)
	28(711)	28(711)	20(508)
TV CONSOLE	38(965)	17(432)	29(737)
	40(1016)	18(457)	30(762)
	48(1219)	19(483)	30(762)
SHELF MODULES	18(457)	10(254)	60(1524)
	24(610)	10(254)	60(1524)
	36(914)	10(254)	60(1524)
	48(1219)	10(254)	60(1524)
DINING TABLE	48(1219)	30(762)	29(737)
	60(1524)	39(914)	29(737)
	72(1829)	42(1067)	28(711)
BUFFET	36(914)	16(406)	31(787)
	48(1219)	16(406)	31(787)
	52(1321)	18(457)	31(787)
DINING CHAIRS	20(508)	17(432)	36(914)
	22(559)	19(483)	29(737)
	24(610)	21(533)	21(787)

ITEM	DIAMETER, IN (mm)	HEIGHT, IN (mm)
DINING TABLE (ROUND)	36(914)	28(711)
	42(1067)	28(711)
	48(1219)	28(711)

FIGURE 14.30 > Common furniture sizes. *Hepler/Wallach/ Hepler © Cengage 2013*

FIGURE 14.31 ⟩ Templates representing furniture width and length. *Hepler/Wallach/Hepler © Cengage 2013*

Combining Areas into a Floor Plan

Students and inexperienced designers often prefer to create floor plans through the use of templates rather than use the idealized (bubble) diagram method shown in Figure 14.13. Figure 14.35 shows the sequence of creating or evaluating space, starting with furniture needs through the development of a scaled sketch. As areas are combined, adjustments are made to allow space for such features as fireplaces, traffic flow, and storage space. Unlimited design variations may be possible within one defined area. The next step is to sketch these areas and rooms into a floor plan in the same manner as "fitting" progressive sketches into an idealized diagram.

🌐 Final Design Evaluation Guides

Corrections and changes made after a floor plan is complete, and especially after the electrical, plumbing, and HVAC plans have been added, require much added time and many extra steps and procedures. To avoid excessive changes, the following design guidelines should be addressed and cor-

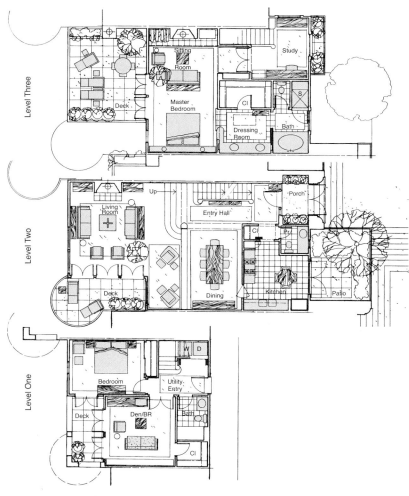

FIGURE 14.32 ⟩ Furniture templates used to check adequate room sizes. *Hepler/Wallach/Hepler © Cengage 2013*

ROOM	SMALL	MEDIUM	LARGE
Formal Living	160	200	400 +
Dining	120	200	300 +
Kitchen	100	160	250 +
Utility-Laundry	40	60	120 +
Master Bedroom	150	250	320 +
Bath	40	80	120 +
Den/office	80	120	200 +
Family-great room	190	250	470+
Foyer	60	100	160 +
Porch	50	100	200 +
Closet	10	20	30 +
Walk-in Closet	30	50	100 +
Halls	3' wide	3'-6"	4'-0" +
2-Car Garage	440	540	680 +
3-Car Garage	640	800	940 +
Bedrooms	120	170	250 +

FIGURE 14.33 > Common room sizes in square feet. *Hepler/ Wallach/Hepler © Cengage 2013*

rected before a final working drawing floor plan is begun. Coverage of these items is contained in Chapter 34.

1. Are traffic patterns between rooms, and within rooms, adequate, convenient, and efficient? Are halls too long?
2. Are rooms and area locations positioned to maximize their intended use?
3. Does the plan contain spaces that are wasted that could instead be utilized or rearranged for better use?
4. Does the main entrance lead to an area, such as a foyer, which allows traffic to flow directly to the living, service, and sleeping areas without passing through another area?
5. Does the kitchen layout provide for the most efficient functioning of tasks that are connected to the three points of a work triangle?
6. Are bath features and fixtures appropriately arranged and integrated to provide the most efficient functioning and use of space?
7. Are adequate storage spaces provided for both local and general storage, including garbage and trash disposal?
8. Are room sizes, shapes, and proportions adequate and appropriate for the convenient use of furniture? Is there enough space around beds and behind table chairs in their usable position? Are there awkward room offsets without a purpose? Are room ratios approximately 3:5?
9. Do door swings obstruct the use of furniture when in the open position?
10. Do all entrances provide an effective and convenient connection between interior and exterior space?
11. Are all laundry fixtures, heating and cooling equipment, and electrical panels located on the plan?
12. Are required special needs facilities fully noted on the plan?
13. Depending on climate, are rooms oriented to maximize or minimize the effect of sun exposure? Are passive solar features included where possible? Does the plan include a north arrow?

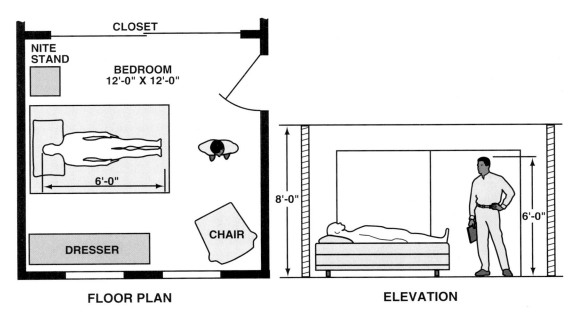

FLOOR PLAN   **ELEVATION**

FIGURE 14.34 > Human templates used to check space needs. *Hepler/Wallach/Hepler © Cengage 2013*

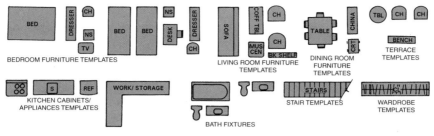

STEP 1. CUTOUT SCALED FURNITURE TEMPLATES

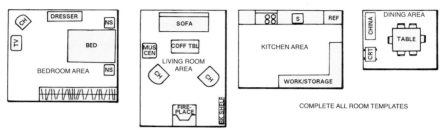

STEP 2. POSITION TEMPLATES

COMPLETE ALL ROOM TEMPLATES

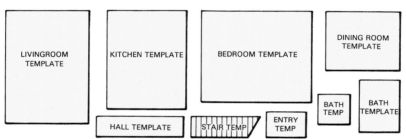

STEP 3. "WRAP-AROUND" WALLS FOR ROOM TEMPLATES

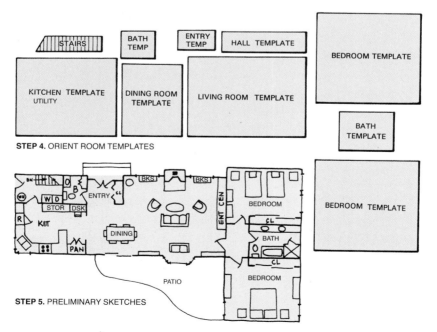

STEP 4. ORIENT ROOM TEMPLATES

STEP 5. PRELIMINARY SKETCHES

FIGURE 14.35 > Sequence of designing with floor plan templates. *Hepler/ Wallach/Hepler © Cengage 2013*

14. Is the plan oriented or baffled to provide protection from wind effects, air pollution, and objectionable noise levels? Does the plan block undesirable visual sights and reveal attractive vistas?

These areas represent the most common errors found in floor plan designs. Correcting these errors and omissions in the early design stage saves considerable time, frustration, and expense during the final working drawing stages.

🌐 PLANS FOR SPECIAL NEEDS

Special design provisions must often be made to enable people with physical impairments that restrict mobility, hearing, or vision to use areas and facilities within and around buildings. Design requirements for full accessibility in public buildings are found in **building codes.** Information for special residential designs can be obtained from federal, state, and local governmental agencies and some private organizations and companies. Information can also be accessed online and from libraries. Following these guidelines also aids in designing residences that promote "aging in place." This concept allows older citizens to live in their homes for many added years.

Following are two lists of some of the many design guidelines for planning buildings that can be fully, safely, and conveniently used by persons with impairments. One list applies to public buildings and the other to residences. However, the same general principles and practices often apply to both. Always follow federal regulations and check with the state and local agencies in the area in which the structure will be built for regulations that apply to the development of plans for any specific project. Also, talk with people who have impairments. Find out what kinds of problems they encounter. Then find ways to eliminate the problems by utilizing safe and acceptable design alternatives.

Public Buildings

All building codes contain design requirements for public buildings that prohibit the use of structural barriers or infringements on the comfort and safety of persons with impairments.

Outdoor Considerations

1. A passenger loading zone for automobiles must be at least 4′ × 20′ and near an accessible entrance.

2. The minimum width of ramps is 3′-0″, with a maximum slope of 1:12 and a maximum rise of 2′-6″. Handrails must be provided for ramps longer than 6′-0″, and at least 6′-0″ of level area must be provided

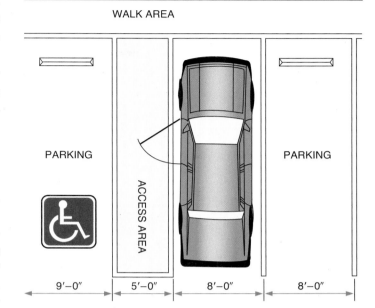

FIGURE 14.36 ⟩ Standard handicapped parking and access space. *Hepler/Wallach/Hepler © Cengage 2013*

at the top of each ramp. Nonskid surfaces must be used on all ramps.

3. Parking facilities must be provided in the parking area nearest the building and marked with the international access symbol. This area must be out of the main traffic flow and connected to the building by a ramp if the level changes. Parking spaces for persons with impairments must be at least 9′-0″ wide and have an adjacent access area of at least 4′-0″ or 5′-0″ wide depending on local code. See Figure 14.36. Many municipalities require even wider loading zones.

4. For people with visual impairments, walkways must be at least 4′-0″ wide, with no less than 6′-8″ headroom clearances. Walks must be level and ramps must be used when it is necessary to change levels. Walks must be free of obstructions and be surfaced with nonskid material. At least 32″ must be allowed for a cane sweep width, as shown in Figure 14.37.

Entrances and Doors

5. At least one entrance must be accessible to wheelchair traffic and provide access to the entire building.

6. Interior doors must open at least 90° and be 2′-8″ (min.) to 3′-0″ wide, with threshold heights of no more than 3/4″ on exterior doors and 1/2″ on interior doors. Walls or objects can be no closer than 4′-0″ from the door hinge, with 12″ push-side clearances. See Figure 14.38.

FIGURE 14.37 ﹥ Cane sweep distance. *Hepler/Wallach/Hepler*
© *Cengage 2013*

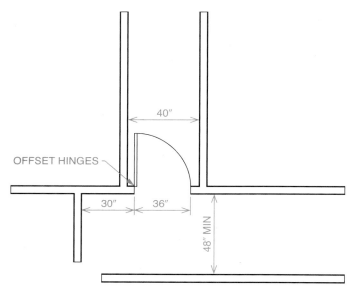

FIGURE 14.38 ﹥ Door and wall clearance for people with impairments. *Hepler/Wallach/Hepler*
© *Cengage 2013*

7. Doors should be provided with handles that can be opened with a closed fist.

8. Thresholds should not be higher than 1/2″ on swinging doors and 3/4″ on sliding doors.

9. Door kickplates should cover the bottom 10″ of a door accessed by wheelchair users.

Floors and Pathways

10. Texture changes using raised strips, grooves, rough, or cushioned surfaces should be used to warn people with visual impairments of an impending danger area, including ramp approaches.

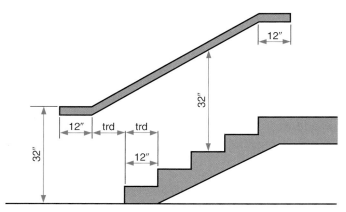

FIGURE 14.39 ﹥ Handrail dimensions. *Hepler/Wallach/Hepler*
© *Cengage 2013*

11. Floors should have nonslip surfaces even when wet, or be covered with carpeting with a pile thickness of no more than 1/2″. Floors should have contrasting color and texture borders to warn people with limited vision of a level change ahead.

Indoor Traffic Areas

12. Hall widths must be at least 3′-0″, with 5′-0″ provided in all turning areas. Halls must provide access to all areas of the building without the need to pass through other rooms.

13. There should be a minimum of three treads in a series of stairs. Treads and risers should be uniform and treads should have a minimum width of 11″ with round nosings. A landing should be planned for stair systems that contain more than 16 risers. Handrails must be at least 32″ above the floor and tread height. See Figure 14.39.

14. At least one ramp or stair lift must be provided as an alternative to stairs.

15. Wall projections, if located between 27″ and 80″ from the floor, cannot extend more than 4″ from a wall. Objects mounted below 27″ from the floor may project any amount. Freestanding objects between 27″ and 80″ may only project 12″ from their support, as shown in Figure 14.40.

Signs, Alarms, and Lighting

16. Public phone amplifiers, eye-level warning lights to augment audio alarms, and high-frequency alarms should be provided for people with hearing impairments.

17. Braille signs, level-change warning surfaces, and restrictions on wall protrusions over 4″ must be provided for people with visual impairments. Elevator buttons must also include braille. Elevator buttons and office signs must be accessible from wheelchair and contain large braille letters and numerals.

18. Protruding signs must be at least 7′-6″ from the floor.

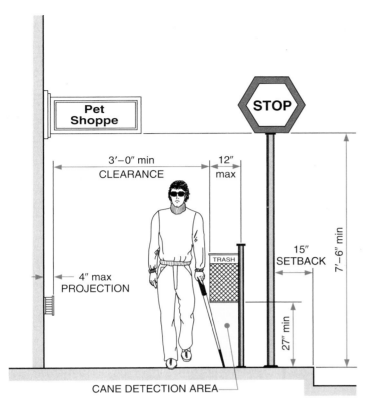

FIGURE 14.40 > Projection limits of objects from walls.
Hepler/Wallach/Hepler © Cengage 2013

19. Emergency warning alarms should be both visible and audible.

20. Lighting should be free of glare or deep shadows.

🌐 Lavatory Facilities

21. Lavatory facilities must include a 30″ wheelchair turning radius. Water closet seat tops must be 1′-6″ from the floor. Figure 14.41 illustrates the minimum floor plan dimensions for water closet enclosures. Figure 14.42 includes the minimum vertical dimensions. A 27″ minimum knee-room height must be provided under sinks and drinking fountains allow wheelchair clearance for the tub and lavatory. (See Figure 14.43.) Grab bars must be provided near water closets, sinks, and bath areas.

22. Sinks should be mounted no closer to the floor than 30″ and should extend a minimum of 17″ from the wall to provide adequate knee space, as shown in Figure 14.44. A minimum floor space area of 2′-6″ × 4′-0″ must be provided around sinks for wheelchair access. See Figure 14.45.

23. The controls for all fixtures and appliances should be within reach from a wheelchair.

24. Door handles and controls for fixtures should be the lever type, which allows people with little strength to operate them.

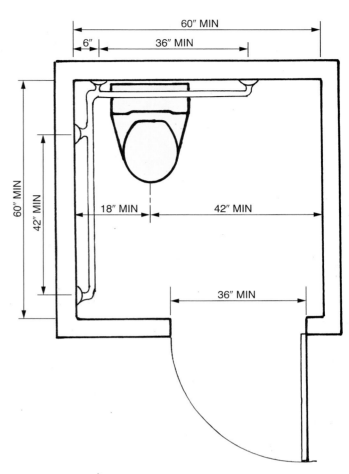

FIGURE 14.41 > Minimum horizontal dimensions for water closet enclosures. *Hepler/Wallach/Hepler © Cengage 2013*

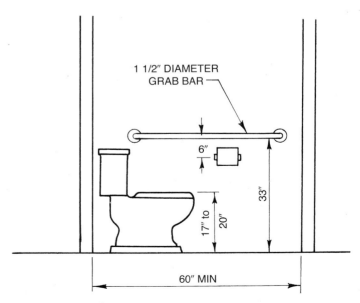

FIGURE 14.42 > Minimum vertical heights for water closet enclosures. *Hepler/Wallach/Hepler © Cengage 2013*

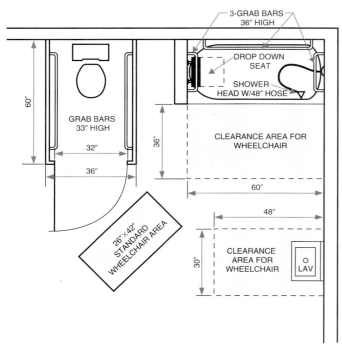

FIGURE 14.43 > Lavatory clearances for people with impairments. *Hepler/Wallach/Hepler © Cengage 2013*

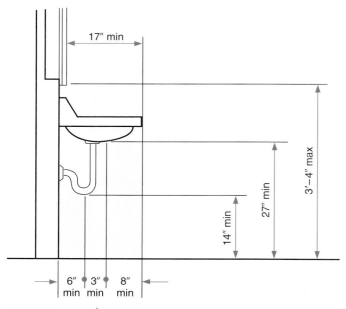

FIGURE 14.44 > Under sink knee-space requirements for wheelchair use. *Hepler/Wallach/Hepler © Cengage 2013*

🌐 Residences

Designers must carefully follow building codes when designing a residence for someone with an impairment. Beyond that, working closely with the client is important. Except in extreme cases, people with impairments have abilities as well. By identifying the specific needs and capabilities of the client,

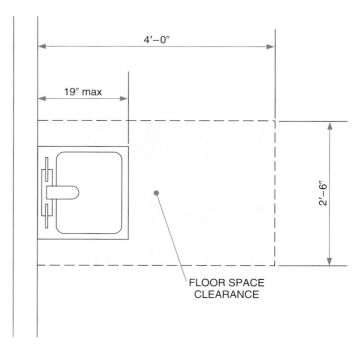

FIGURE 14.45 > Floor space clearances needed around sinks for wheelchair use. *Hepler/Wallach/ Hepler © Cengage 2013*

"overdesigning" can be avoided and a safe, convenient, and comfortable design can be achieved.

Design features that affect people with mobility limitations usually relate to providing for wheelchair use. Because wheelchairs require the greatest amount of space compared to other disability apparatus, plans that accommodate wheelchairs will easily function for other design conditions. Figure 14.46 shows the dimensions and turning radius of a *standard* wheelchair. Wheelchairs vary in size, however. When designing a floor plan for use by someone in a wheelchair, use the dimensions of the largest model. Figure 14.47 shows an application of a 30″ turning radius in a bath, and Figure 14.48 illustrates how this applies to wheelchair rotation under a sink.

Some general guidelines for designing residences are provided in the following list. Unless stated otherwise, design guidelines apply to wheelchair use.

Outdoor Entrances/Exits

1. To accommodate wheelchairs, the pathway or ramp to the entrance should be 36″ to 48″ wide and have a nonslip surface such as outdoor carpeting or sand paint.

2. If a ramp is used, a minimum landing platform of 5′ × 5′ should be located in front of the door. Plan the shape of the platform to accommodate the door swing and still allow easy access. For long ramps, more landings may be needed. A covering for protection from the weather is recommended. If room is unavail-

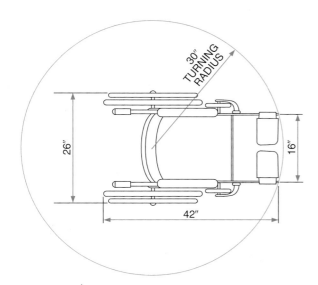

FIGURE 14.46 > Standard wheelchair turning radius. *Hepler/ Wallach/Hepler © Cengage 2013*

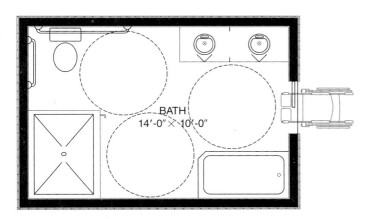

FIGURE 14.47 > Wheelchair turning radius for baths. *Hepler/ Wallach/Hepler © Cengage 2013*

able for a ramp, consider planning for the use of a mechanical lift.

3. The vertical rise of a ramp should be 1:12. Handrails should be 32″ to 36″ high and extend 1′-0″ past the end of the ramp.

4. The height of the doorknob from the floor should be 36″ or less. Threshold height should be 1/2″ or less.

Indoor Traffic Areas and Floors

5. Levers on doors can be operated more easily than doorknobs.

6. Doorways need to be 32″ to 36″ wide. If a turn is required for a wheelchair to pass through a doorway, be sure the doorway is wide enough or provide extra turn space in front of the door.

7. For ease of use by people in wheelchairs and people with limited vision, doorways should not have raised thresholds.

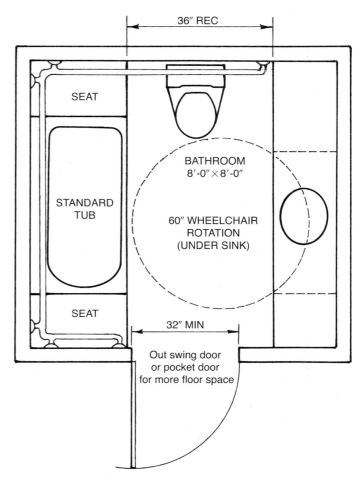

FIGURE 14.48 > Wheelchair rotation under sink. *Hepler/ Wallach/Hepler © Cengage 2013*

8. Hallways must be 36″ to 48″ wide.

9. Hardwood floors or tiled surfaces are best. If carpeting is preferred, use carpet that has short, dense pile.

Living Areas

10. Rooms should have 5′-0″ or more of clear area for turning a wheelchair.

11. Furniture planning and placement should allow adequate area for wheelchairs to move through the room. People with limited vision should have a clear passageway through rooms and not be required to walk around articles of furniture.

12. Height of tables and work areas, such as desks, should be approximately 30″.

Kitchens

13. Allow adequate work area space for turning a wheelchair or for using crutches or a walker, usually 16 to 25 sq. ft. Important factors are the shape of the kitchen and the arrangement of appliances that create the work triangle.

14. Appliance cooking controls should be placed in front of the burners. Ovens should have side-hinged doors. All controls must be operable with a closed fist.

15. Braille control panels for people who are blind or have extremely limited vision are available from some appliance manufacturers.

16. A clear 28″, 31″, or 36″ of floor space should be provided under selected base cabinets or next to appliances.

17. Countertops should be 30″ to 33″ from the floor, with 27″ to 29″ for knee clearance. Cabinet pulls should be recessed.

18. Dishwashers, washers, and dryers should be front loading and have front controls.

19. Side-by-side refrigerator/freezers are most convenient.

20. Sinks need to be 34″ or less from the floor.

Baths

21. Water closet seats should be 1′-6″ from the floor.

22. Lavatories should not be higher than 34″ from the floor and should have adjoining counter space. They need to be open underneath. Exposed pipes should be insulated. A single faucet with a lever control is preferred.

23. The bottom of the mirror should be 40″ from the floor. (Consider the eye level of a person in a wheelchair.)

24. The top of a medicine cabinet shelf should not be more than 50″ from the floor.

25. Space should be allowed near the bathtub, shower, and water closet to allow for transfer from a wheelchair. Reinforced grab bars should be installed. Space requirements for wheelchair transfers are illustrated in Figure 14.49.

26. A shower should be at least 5′ × 4′ in size, and the floor should have a nonslip surface. Reinforced grab bars should be installed. There should be no lip on the floor surface entrance to the shower.

Bedrooms and Storage Areas

27. Bedrooms should be designed to allow wheelchair maneuverability and access to the bed, as shown in Figure 14.50. The top surface of bedroom furniture should be a maximum of 34″ above the floor.

28. Sliding doors are preferred for closets. Hang rods should be 4′-6″ or less from the floor.

29. Storage facilities should be designed for easy reach from a wheelchair. See Figure 14.51.

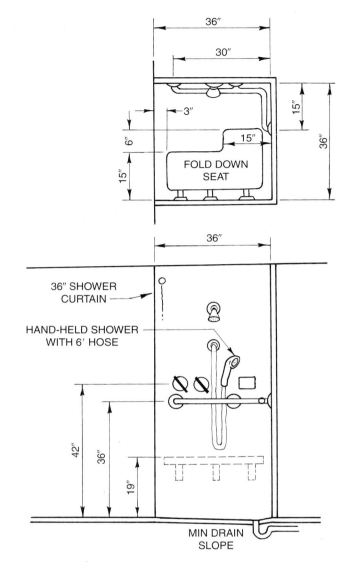

FIGURE 14.49 > Wheelchair transfer requirements for showers. *Hepler/Wallach/Hepler © Cengage 2013*

Electrical Considerations

30. Switches and outlets should be 40″ from the floor.

31. Lights or other visual cues (aids) should be connected to the telephone, doorbell, and other devices as needed for people with hearing impairments.

32. Locate switches on the latch side of doors.

33. Plan a switch to control at least one light in each room.

34. Outlets should average one for every 6 feet of wall space.

35. Kitchen appliance outlets should average one for every 4 feet of wall space.

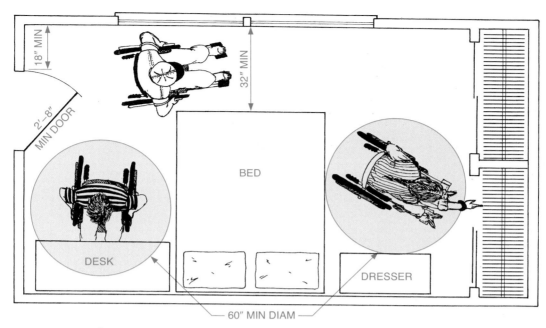

FIGURE 14.50 > Bedroom wheelchair clearances. *Hepler/Wallach/Hepler © Cengage 2013*

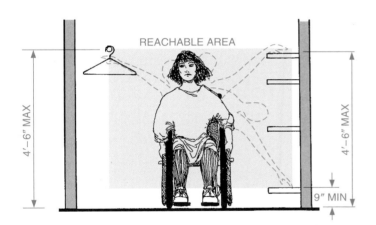

FIGURE 14.51 > Reaching distances from wheelchairs. *Hepler/Wallach/Hepler © Cengage 2013*

36. GFCI outlets should be installed near any water source.
37. All outside outlets must be weatherproof outlets.
38. Specify alarms to detect smoke, gas, sound, and movement.
39. Provide a braille-marked master distribution panel.
40. Timer and motion sensors should be programmed to turn off lights after a specified amount of time lapse without movement in a room.

Feng Shui

The design process presented in this chapter is based on analyzing personal wants and needs, site characteristics, and the development of spatial relationships in a floor plan. Other design practices such as feng shui are also used in developing floor plans. **Feng shui** is an ancient form of design based on the concepts of energy flow, balance, and harmony in the relationship of natural elements. This involves the systematic orientation and arrangement of many design components such as furniture placement, landforms, waterways, materials, color, and floor plan layout.

Designing Floor Plans Exercises   CHAPTER **14**

1. List the design steps necessary to design a residence through the development of a conceptual design.

2. Prepare a situation statement and set goals and objectives for a house of your own design.

3. Explain how a composite analysis is prepared and used to create a plan of a design.

4. Prepare room templates and use them to make a functional arrangement for the living area, service area, and sleeping area of a house.

5. Arrange templates for a sleeping area, service area, and living area of your own design in a total plan.

6. Make room templates of each room in your own home. Rearrange these templates according to a remodeling plan, and make a sketch.

7. Make a list of furniture you would need for a home you will design. The list should include the number of pieces and size (width and length) of each piece of furniture.

8. Make a furniture template $1/4'' = 1'-0''$ for each piece of furniture you will include in a home of your design.

9. Develop a floor plan for a family of four, including two small children and someone who uses a wheelchair.

10. Gather pictures of floor plans from real estate magazines and home-planning catalogs. Evaluate them in terms of their space planning arrangements.

11. Choose a floor plan in Chapter 14 or 15 and redesign it in two ways. One plan should accommodate a person in a wheelchair and the other one a person who has a visual impairment.

12. Change the first-floor master suite in Figure 14.52 into a recreation room. Design and sketch a second-floor master suite.

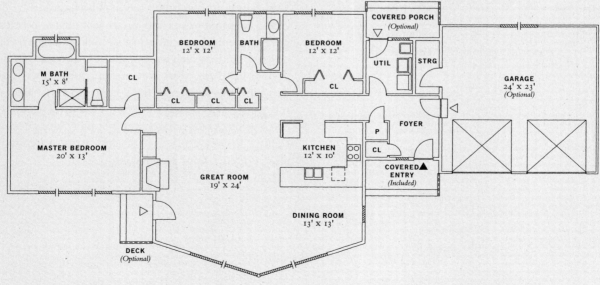

FIGURE 14.52 > Floor plan to be revised and second floor added. *Photo courtesy Lindal Cedar Homes.*
Lindal.com

OBJECTIVES

In this chapter you will learn to:

> use information on a scaled floor plan to draw a complete floor plan.

> name and explain the types of floor plans.

> use graphic symbols to communicate information on a floor plan.

> draw a floor plan according to a sequence of CAD and manual steps.

> draw dimensions that convey precise, accurate information for builders.

TERMS

abbreviated floor plans
break line
dimension lines
door swing
extension lines
floor plan sketches
metric
modular

mullion
multiple-level floor plans
muntins
object lines
overall dimensions
reflected ceiling plan
reversed plans

rise
run
schedules
subdimensions
symbols
working drawings

INTRODUCTION

A complete floor plan is a scaled drawing of the outline and partitions of a building as seen if the building were cut (sectioned) horizontally about 4′ (1,219 mm) above the floor line. See Figure 15.1. *Drawing* floor plans is not the same as *designing* floor plans. A CAD system will follow drawing instructions and will aid in the fast and accurate completion of a floor plan. A CAD system, however, cannot initiate or create a floor plan design.

The final design of a floor plan, as covered in Chapter 14, should be completed prior to actually drawing a final floor plan. During the design process changes can be made at any time without creating serious problems. However, design changes made while drawing a floor plan often create problems that can require the redesign of related areas. This may mean returning to the design process and redrawing all or some of the plan, resulting in much wasted time and mistakes. For example, the size and location of all components

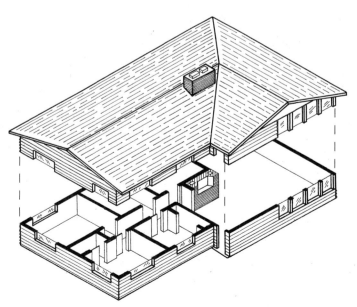

FIGURE 15.1 > A floor plan is a section through a building.
Hepler/Wallach/Hepler © Cengage 2013

(cabinets, appliances, fixtures) should be determined *before* drawing, not *when* drawing these features. If sizes are not fixed during the design phase, walls or other components may need to be moved because these types of components may not fit where placed. Small wall adjustments may be possible but large movements (over 4″) can cause a domino effect.

As discussed in Chapter 14, there are many types of floor plans, ranging from very simple sketches to completely dimensioned and detailed floor plan working drawings. In this chapter, how to draw floor plans is covered.

TYPES OF FLOOR PLANS

Floor plans are classified by the amount and type of information each conveys. **Floor plan sketches** are used in the design process or on a construction site and contain minimal details. Single-line floor plans are used for the same purpose but are more accurate in scale. **Abbreviated floor plans,** which include minimal dimensions or labels, are used primarily for marketing and display purposes. See Figure 15.33 on p. 322. Abbreviated floor plans are often drawn with furniture and surface treatments to add realism. The floor plan in Figure 15.2A is identical to the plan in Figure 15.2B, but the plan in Figure 15.2A is a pictorial that includes furniture and surface textures, whereas Figure 15.2B has no textures added. For maximum realism most presentation floor plans are produced in 3D full color as shown in Figure 15.3. Floor plans of this type may also be used as a base for construction models as displayed in Chapter 21.

Drawings that contain all of the information needed to construct a structure are called **working drawings.** Completely dimensioned and accurately scaled floor plans are working drawings necessary for construction. Basic information included shows the size and position of all exterior walls, interior partitions, fireplaces, doors, windows, stairs, built-in furniture, appliances, cabinets, connecting walks, patios, lanais, and decks. Wall and surface construction materials are also shown.

The prime function of floor plans is to communicate information to building contractors. Complete working-drawing floor plans prevent misunderstandings between designers and builders. Contractors should be able to correctly interpret working drawings without consultation.

Specialized floor plans are developed from basic working-drawing floor plans. For construction and installation of electrical, plumbing, and HVAC (heating, ventilating, air-conditioning) systems, separate specialized plans are drawn with specific symbols added. On very small projects, these symbols may all be included on the basic plan. However, this often makes the drawing too crowded and difficult to read. For most projects, electrical, HVAC, and plumbing plans are separate drawings. These are covered in detail in Chapters 31, 32, and 33, respectively.

The amount of detail used in drawing floor plans depends on the scale of the plan. Figure 15.4 shows the difference

FIGURE 15.2A > A presentation floor plan with furniture and floor materials shown. *Hepler/Wallach/Hepler © Cengage 2013*

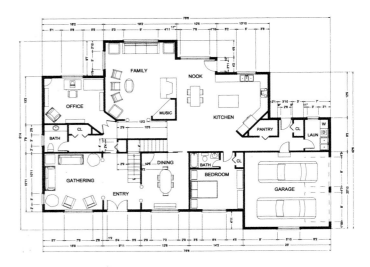

FIGURE 15.2B > Floor plan with dimensions and without textures. *Hepler/Wallach/Hepler © Cengage 2013*

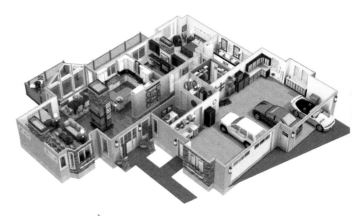

FIGURE 15.3 > Presentation floor plan in three dimensions. *Chief Architect Software*

between the amount of detail possible on three plans at different scales. If more information is needed than the scale of the plan allows, a removed detail drawing should be prepared. The ½″ = 1′-0″ plan shown in Figure 15.4 can include overall horizontal dimensions. If more detail dimensions are required a larger scaled plan, such as 1′ = 1′-0″, is needed. This design is pictured in Figure 15.5. Notice how the specific plan details compare with the features revealed in the picture. Figure 15.6 shows how room sizes appear at different scales.

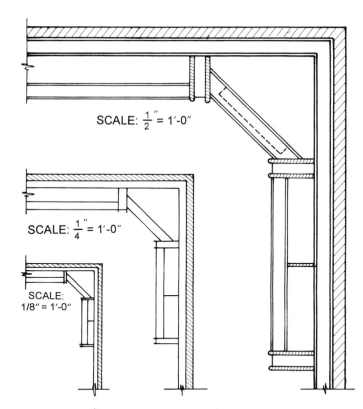

FIGURE 15.4 > Floor plan details at different scales. *Hepler/Wallach/Hepler © Cengage 2013*

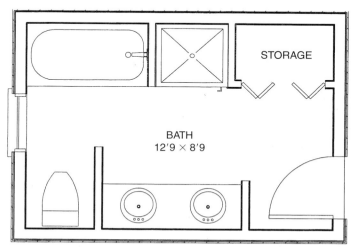

SCALE: 1/4″ = 1′-0″

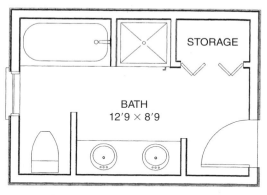

SCALE: 3/16″ = 1′-0″

FIGURE 15.5 > Finished cabinetry shown in Figure 15.4. *Hepler/Wallach/Hepler © Cengage 2013*

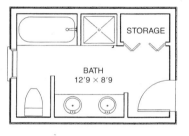

SCALE: 1/8″ = 1′-0″

SCALE: 1/16″ = 1′-0″

FIGURE 15.6 > Room sizes at different scales. *Hepler/Wallach/Hepler © Cengage 2013*

FLOOR PLAN SYMBOLS

On drawings, drafters use **symbols** to identify construction materials such as fixtures, doors, windows, stairs, and partitions. The use of symbols saves time and space. Imagine trying to repeat a description every time that a material or component is used!

Common symbols used on floor plans include symbols for walls, doors, windows, appliances, fixtures, sanitation facilities, and building materials. See Figure 15.7. Floor plan symbols for plumbing, heating, air-conditioning, and electrical components are covered in later chapters. Architectural symbols are standardized. However, some variations of symbols are used in different parts of the country.

Wall Symbols

Different types of wall construction are represented by different floor plan wall symbols. On simple plans, walls are represented by single lines. However, on working-drawing floor plans, the actual scaled width of each wall is drawn. Figures 15.8 and 15.9 show methods of representing different types and variations of wall construction on floor plans.

Door Symbols

Floor plan door symbols show the top view of a door and the width of each doorway. Door symbols usually show each door open 30° to 90° and connected to an arc that represents

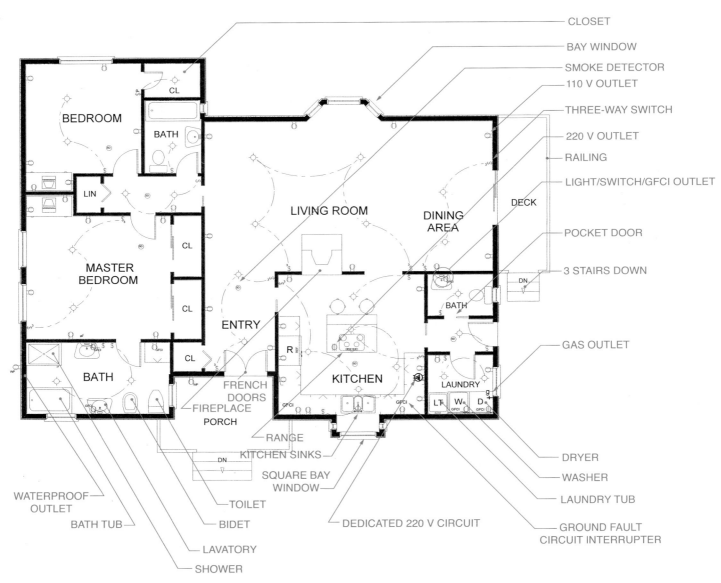

FIGURE 15.7 > Floor plan symbols in place. *Hepler/Wallach/Hepler © Cengage 2013*

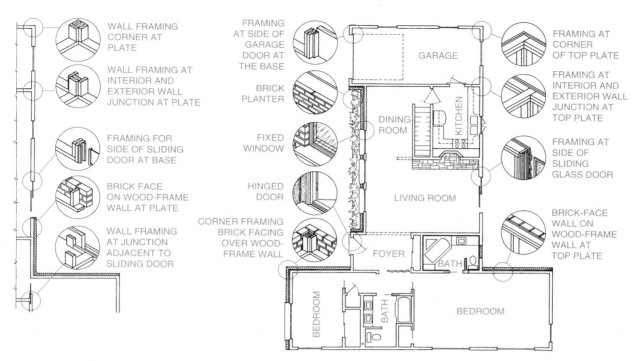

FIGURE 15.8 > Floor plan symbols visualized. *Hepler/Wallach/Hepler © Cengage 2013*

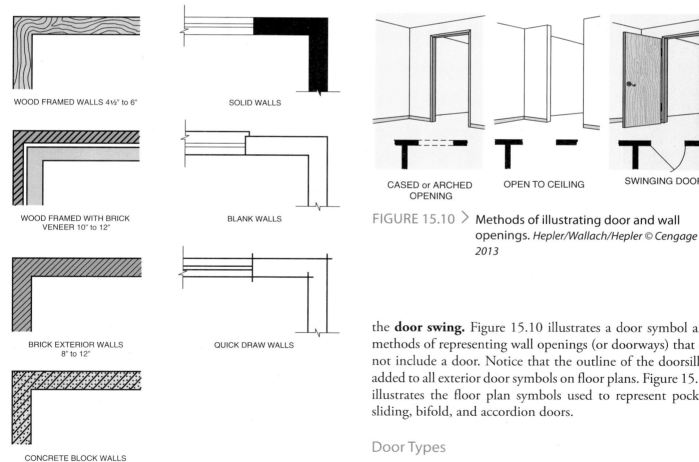

FIGURE 15.10 > Methods of illustrating door and wall openings. *Hepler/Wallach/Hepler © Cengage 2013*

WOOD FRAMED WALLS 4½" to 6"

WOOD FRAMED WITH BRICK VENEER 10" to 12"

BRICK EXTERIOR WALLS 8" to 12"

CONCRETE BLOCK WALLS 4" to 12"

SOLID WALLS

BLANK WALLS

QUICK DRAW WALLS

CASED or ARCHED OPENING

OPEN TO CEILING

SWINGING DOOR

FIGURE 15.9 > Methods of drawing different wall types. *Hepler/Wallach/Hepler © Cengage 2013*

the **door swing.** Figure 15.10 illustrates a door symbol and methods of representing wall openings (or doorways) that do not include a door. Notice that the outline of the doorsill is added to all exterior door symbols on floor plans. Figure 15.11 illustrates the floor plan symbols used to represent pocket, sliding, bifold, and accordion doors.

Door Types

Interior doors, those located within interior partitions, and exterior doors, those that lead outdoors, are generally flush, paneled, or louvered. Interior flush doors have a hollow core

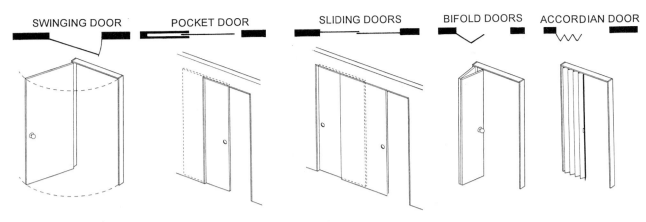

FIGURE 15.11 ＞ Illustrated floor plan door symbols. *Hepler/Wallach/Hepler © Cengage 2013*

covered with a thin wood, plastic, or metal veneer. The core of exterior flush doors is solid to provide strength and insulation, and to prevent warping from moisture exposure. Panel doors are constructed from vertical stiles and horizontal rails. Thin panels of wood, plastic, metal, or glass cover the stiles and rails.

Door Styles

Interior doors are manufactured in many different configurations to serve a variety of needs. Exterior doors are generally single- or double-swing doors for entrances and bypass sliding glass doors for patio or deck traffic. See Figure 15.12. Figure 15.13 shows swinging, sliding, and folding door patterns. Many added configurations are available for three-unit components, as illustrated in Figure 15.14.

Door styles are also indicated on door schedules and on elevation drawings that are cross-referenced with floor plans. **Schedules** are detailed lists that contain information such as size and type. Often on a floor plan, a number in a square near the door symbol identifies the door style shown on a door schedule or detail. See Figure 15.15.

Door Sizes

Different door types and styles are available in many size ranges for width, height, and thickness. The sizes shown in Figure 15.16 are included on all working-drawing floor plans. These sizes are used to draw the openings in floor plan walls. On larger scaled drawings, or plan details, additional information is needed. Because vertical sizes are shown on elevation drawings, and not on floor plans, the height dimension

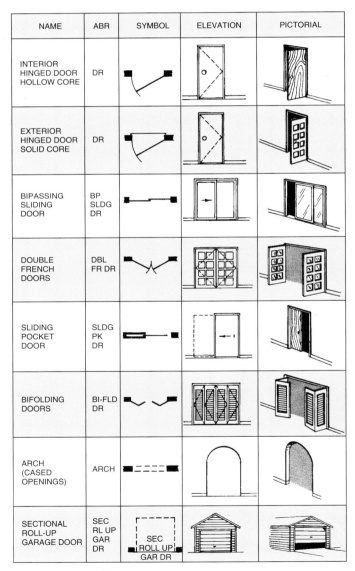

FIGURE 15.12 ＞ Common door symbols. *Hepler/Wallach/ Hepler © Cengage 2013*

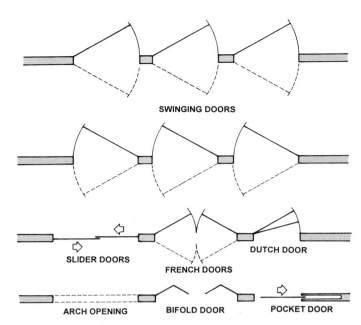

FIGURE 15.13 > Swinging, sliding, and folding door configurations. *Hepler/Wallach/Hepler © Cengage 2013*

must be included on a door schedule. If a door schedule is not used, door width and height dimensions are often shown directly on the door symbol. When this is done, the foot and inch dimensions are abbreviated, as shown in Figure 15.17. Exact door framing information should always be determined from the manufacturer's data.

Most building codes require that exterior doors be solid and at least 3′-0″ wide. Interior doors must be at least 2′-6″ wide but can be hollow core. Bathroom doors must be at least 2′-2″ but wider widths are recommended. The minimum width of wheelchair access doors is 2′-8″, with 3′-0″ being preferable.

Window Symbols

Floor plan window symbols show the outline of the sash, glass position, and any mullions and muntins. A **mullion** is a vertical member separating multiple windows. **Muntins** are vertical and horizontal framing strips that separate window panes (lights). See Figure 15.18. Windows are often distinguished by the manner in which they open. For example, on casement windows, the direction of swing is indicated much like it is in a door symbol. On awning windows, the outline of the open window position is shown with dashed lines. On small-scale drawings, often only the sash outline or glass position is shown. Figure 15.19 shows the plan and elevation symbols for common window types. The point of the dashed line represents the hinge size. Arrows show the slide direction.

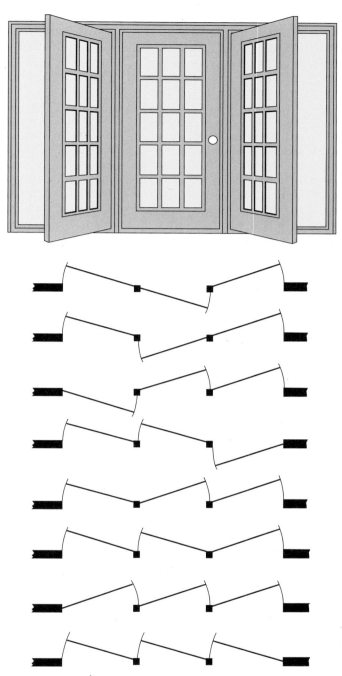

FIGURE 15.14 > Types of triple swinging door configurations. *Hepler/Wallach/Hepler © Cengage 2013*

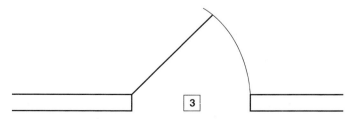

FIGURE 15.15 > Door schedule key on a floor plan. *Hepler/Wallach/Hepler © Cengage 2013*

TYPE	WIDTH	HEIGHT
Exterior Swing	2/6 2/8 3/0 3/6 4/0	6/8 7/0 8/0
Sliding Glass:		
Single Unit	3/6 3/0 3/4 4/0 4/4	6/8 6/10 7/0 8/0
Double	5/0 6/0 8/0	6/8 6/10 7/0 8/0
3 Panel	7/0 8/0 9/0 12/0	6/8 6/10 7/0 8/0
4 Panel	10/0 12/0 16/0	6/8 6/10 7/0 8/0
Sidelights	1/0 1/2 1/4 1/6	6/8 7/0 8/0
Transom	2/6 2/8 3/0 3/6 4/0 6/0 8/0 9/0 10/0 12/0	1/0 1/2 1/6
Double French	3/0 4/0 5/0 6/0	6/8 7/0 8/0
Garage Roll-up	8/0 9/0 10/0 12/0 14/0 15/0 16/0 17/0 18/0 20/0	6/6 6/8 6/9 7/0 7/6 8/0
Interior Swing	2/4 2/6 2/8 3/0 3/6 4/0	6/8 7/0 8/0
Closet Slide Panel	1/0 1/2 1/4 1/6 1/8 1/10 2/0 2/4 2/6 2/8 3/0 3/6 4/0	6/8 7/0
Bifold	2/0 2/4 2/6 2/8 3/0	6/8 7/0
Pocket	2/6 2/8 2/10 3/0	6/8
Folding	4/0 6/0 8/0 12/0 16/0 20/0 24/0	6/8 7/0

FIGURE 15.16 > Common light construction door sizes. *Hepler/Wallach/Hepler © Cengage 2013*

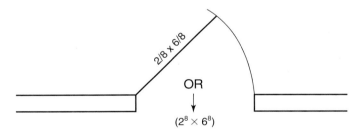

FIGURE 15.17 > Methods of labeling door sizes on floor plans. *Hepler/Wallach/Hepler © Cengage 2013*

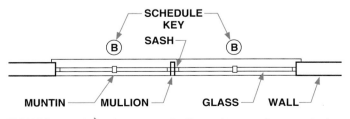

FIGURE 15.18 > Elements of a floor plan window symbol. *Hepler/Wallach/Hepler © Cengage 2013*

Only the width of windows is needed to draw window symbols on floor plans. Height dimensions are shown either on elevation drawings or stated in a window schedule that contains size, style, type, and manufacturer's information. The most common window sizes, as shown in Figure 15.20, are included on all working-drawing floor plans, as are the common door sizes. Listing only the width and height of windows and doors is sufficient for most floor plans. For detailed plans, more information is needed as shown in Chapter 17. Exact sizes for rough framing must always be secured from manufacturers' data. On a floor plan, a letter in a circle is provided as a key or cross-reference to the window schedule.

Appliance and Fixture Symbols

Figure 15.21 shows the plan and elevation symbols used for common appliances and fixtures. Overall width and length dimensions (as shown previously in Chapters 10 and 11) can be helpful for drawing floor plan symbols.

When preparing detail drawings, manufacturers' specifications must always be used. These show the exact dimensions

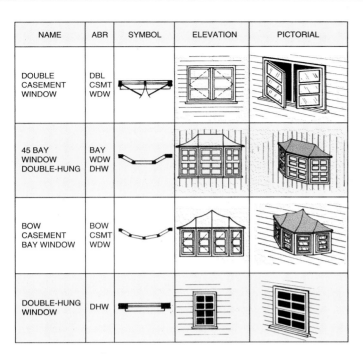

NAME	ABR	SYMBOL	ELEVATION	PICTORIAL
DOUBLE CASEMENT WINDOW	DBL CSMT WDW			
45 BAY WINDOW DOUBLE-HUNG	BAY WDW DHW			
BOW CASEMENT BAY WINDOW	BOW CSMT WDW			
DOUBLE-HUNG WINDOW	DHW	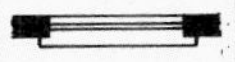		

NAME	ABR	SYMBOL	ELEVATION	PICTORIAL
HORIZONTAL SLIDING WINDOW	SLD WDW		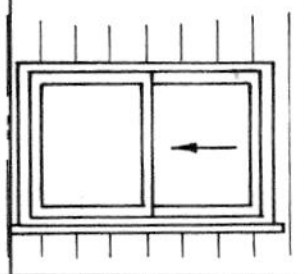	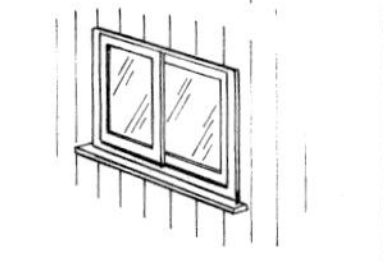
SWINGING CASEMENT WINDOW	CSMT WDW	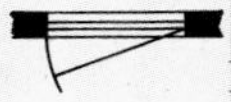		
HOPPER WINDOW	HOP WDW	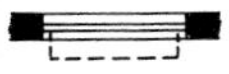	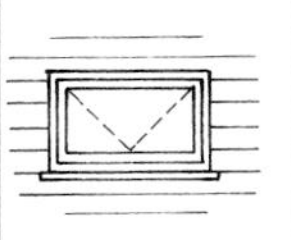	
DOUBLE DOUBLE-HUNG WINDOW	DBL DHW			

FIGURE 15.19 > Common window symbols. *Hepler/Wallach/Hepler © Cengage 2013*

TYPE	WIDTH	HEIGHT
Double-Hung	1/10 2/0 2/4 2/6 2/8 2/10 3/0 3/4 3/6 3/8 3/10 4/0 4/6	2/6 3/0 3/2 3/6 4/0 4/6 4/10 5/0 5/6 5/10 6/0
Horizontal Slider	3/8 4/8 5/8 6/0 6/6	2/10 3/6 4/2 4/10 5/6 6/2
Casement	1/6 2/0 3/0 3/4 4/0 6/0 8/0 10/0 12/0	1/0 2/0 3/0 3/4 3/6 4/0 4/6 5/0 5/4 5/6 6/0
Fixed	1/0 1/6 2/0 2/6 3/0 4/0 4/6 5/0 5/10 6/0	4/6 4/10 5/6 6/6 7/0 7/6 8/0
Hopper	2/0 2/8 3/6 4/0	1/4 1/6 1/8 2/0 4/0 6/0
Jalousie	1/8 2/0 2/6 3/0	3/0 4/0 5/0 6/0
Bay-Bow	4/0 6/0 8/0 10/0 12/0	3/0 3/4 4/0 5/0 6/0
Awning	2/0 2/6 2/8 3/0 3/4 4/0 5/4 6/0 6/8 8/0 10/0 12/0	1/6 2/0 3/0 3/6 4/0 5/0 6/0
Half-Elliptical	5/0 6/0 8/0	1/6 1/0 1/10
Half-Round Top	2/0 2/4 2/6 2/10 3/0 3/6 4/0 4/8 5/0 5/4 6/0 R	—
Quarter-Round	1/6 2/0 2/6 3/0 R	—

FIGURE 15.20 > Common light construction window sizes. *Hepler/Wallach/Hepler © Cengage 2013*

NAME	ABR	SYMBOL	ELEVATION	PICTORIAL
SINK	S	S		
FLOOR CABINETS	FL CAB			
WALL CABINETS	W CAB			
RANGE	R			
REFRIGERATOR	REF	R		
DISHWASHER	DW	DW		
OVEN BUILT-IN	O	O		

NAME	ABR	SYMBOL	ELEVATION	PICTORIAL
WASHER	W	W		
DRYER	D	D		
LAUNDRY TUB	LT	LT		
WATER HEATER	WH			
COOK TOP RANGE	CK TP			
RANGE WITH OVEN COVER	R			
FOLD-UP IRONING BOARD	I BRD			

FIGURE 15.21 > Common appliance and fixture symbols. *Hepler/Wallach/Hepler © Cengage 2013*

of each unit plus the clearance dimensions for needed installation. Appliance and fixture details are listed on schedules, as well as on floor plans. Schedules provide more information for purchasing, related cabinet design, and installation. Separate details are often prepared to illustrate the exact relationship of appliances and fixtures to adjacent cabinetry. These are covered in Chapters 10, 11, and 35. For marketing purposes, presentation floor plans often include appliances and fixtures in detail and color.

Bathroom Symbols

Figure 15.22 shows symbols for common bath fixtures. Symbols for freestanding units are usually drawn using a fixture template. Fixtures that align with cabinets must be carefully positioned. The type, style, and size specifications for each fixture must be taken from manufacturers' data to ensure a proper fit. This information is included in the fixture schedules specifications or detail drawing.

Furniture Symbols

Complete working-drawing floor plans do not usually include furniture symbols because they interfere with construction notes and dimensions. Furniture symbols are used mostly by interior designers on abbreviated floor plans to

represent the width and length of each furniture piece. These symbols are either drawn with drafting instruments, furniture templates, or obtained from a computer software library. Furniture is not included on working-drawing floor plans because it is not part of the permanent structure and interferes with necessary labeling and dimensioning. Furniture outlines are often included and highlighted on presentation plans to add clarity and show space relationships. Figures 15.34 and 15.35 on p. 323, show floor plans with only furniture added.

CAD Floor Plan Symbols Blocks

After wall and other line work is completed on a floor plan, symbols such as doors, windows, fixtures, and fireplaces can be inserted. These symbols are added by selecting each symbol from a symbol block using the *Block* command and locating the position of each on the drawing with the cursor. A block can be part of the CAD software or created by drawing it and filing it in a group block. Figure 15.23 shows a typical CAD floor plan symbol block.

Some blocks contain different levels of detail. In Figure 15.24, compare the CAD simplified fireplace symbol with the detailed fireplace symbol. Symbol libraries can be accessed by clicking on the *Tools* menu and clicking on *AutoCAD® Design Center.*

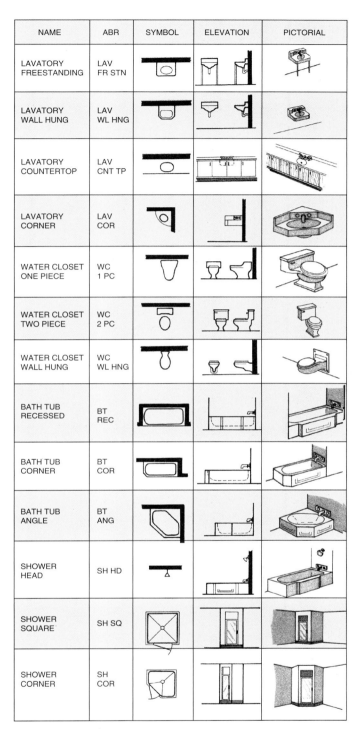

FIGURE 15.22 > Common bath fixture symbols. *Hepler/ Wallach/Hepler © Cengage 2013*

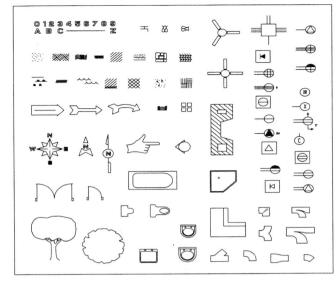

File Name: 4000-ART.DWG

ARCHITECTURAL SYMBOL LIBRARY
DOORS AND WINDOWS (PLAN VIEW)

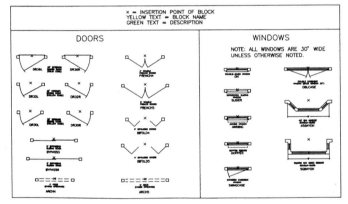

File Name: DRWNLIB.DWG

FIGURE 15.23 > Typical CAD library symbols. *Hepler/Wallach/ Hepler © Cengage 2013*

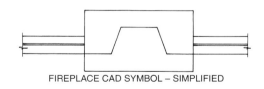

FIREPLACE CAD SYMBOL – SIMPLIFIED

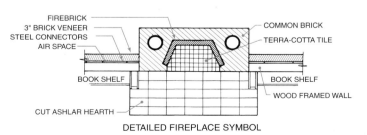

DETAILED FIREPLACE SYMBOL

FIGURE 15.24 > Standard fireplace symbol compared to a detailed symbol. *Hepler/Wallach/Hepler © Cengage 2013*

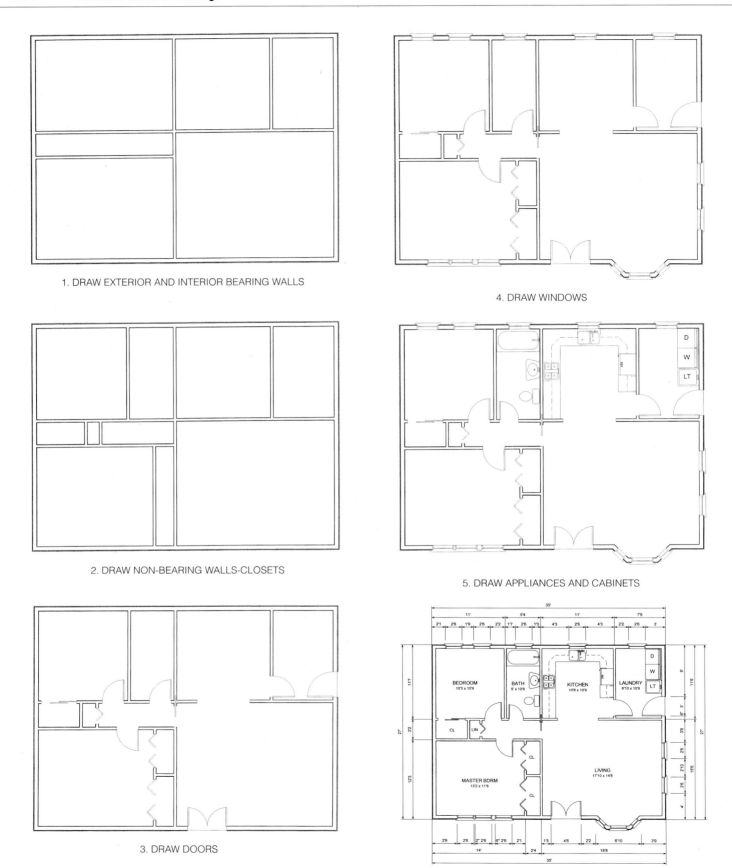

1. DRAW EXTERIOR AND INTERIOR BEARING WALLS

2. DRAW NON-BEARING WALLS-CLOSETS

3. DRAW DOORS

4. DRAW WINDOWS

5. DRAW APPLIANCES AND CABINETS

6. DRAW DIMENSIONS AND LABELS

FIGURE 15.25 > Sequence of drawing floor plans. *Hepler/Wallach/Hepler © Cengage 2013*

STEPS IN DRAWING FLOOR PLANS

For maximum speed, accuracy, and clarity, the following steps should be observed in laying out and drawing floor plans. See Figure 15.25.

1. Block in the overall dimensions of the house, and add the thickness of the outside walls.

2. Lay out the position of interior partitions.

3. Locate the position of doors and windows by center line and by their widths.

4. Darken the **object lines** (visible lines), such as the main exterior walls and interior partitions.

5. Add door and window symbols with a 2H pencil. Draw the door to swing open toward a perpendicular wall to provide the most convenient access. See Figure 15.26.

6. Add symbols for stairwells, if applicable.

7. Erase extraneous layout lines if they are too heavy.

8. Draw the outlines of kitchen and bath fixtures.

9. Add the symbols and sections for any masonry work, such as fireplaces and planters. The outlines of the chimney, firebox, and hearth with material symbols are shown on floor plans. Fireplace construction is shown on detail drawings.

10. Dimension the drawing as described in Figures 15.42 through 15.49.

DRAWING FLOOR PLANS ON CAD

Floor plan walls are drawn using the *Line* or *Multiline* command. The *Multiline* command task can only be used to draw a linear wall style. Creating a wall using the *Multiline* command allows the designer to lay out the specifics (wall thickness) of the wall at one time.

The preferred method is to use the *Line* command to draw the exterior or interior wall and then use the *Offset* command, after setting the wall thickness, to copy the line at a preset distance to complete the two-line wall system.

CORRECT ACCESS   DIFFICULT ACCESS

FIGURE 15.26 > Doors should swing open toward a perpendicular wall. *Hepler/Wallach/Hepler © Cengage 2013*

The *Join* command is also used to connect two perpendicular lines to create a corner by placing the cursor on both lines. After walls are drawn, symbols for doors, windows, and fixtures are added from symbol libraries or blocks. Dimensions are then added using the *Dim* command, and the *Text* command is used to add notes.

Floor Plan Layering

All elements of a floor plan can and should be layered so that layers can be turned on, off, moved, or become a separate drawing at any time. The entire floor plan can also be a layer that contains sublayers such as doors, windows, dimensions, fixtures, furniture, and even movable objects such as people and automobiles. Using the *Layer* command allows designers to group entities together in layers to make identification and editing easier, control what will be sent to the printer, and control the coloring and line types used on entities.

Floor plans that align vertically with other plans can also be layered (first and second floors of a house) by specific colors related to what is being drawn. This usually involves outside walls, stairwells, and chimney locations. Preparing layered plans eliminates errors of alignment between plan drawings, such as when aligning load-bearing walls and supports from the foundation to the first floor to the second floor through to the roof. Specialized floor plans such as plumbing, electrical, and HVAC plans are also layered to ensure agreement with the basic plan by preventing lines and fixtures from being located in the same space.

STEPS IN DRAWING CAD FLOOR PLANS

CAD systems are used to draw floor plans accurately and quickly. CAD systems do not create designs. As in conventional drafting the design process should proceed final drawing activity. The steps used to draw floor plans using CAD are similar to those used for conventional drafting sequences, with some exceptions. CAD programs differ in their terminology and use of commands but most generally follow the steps shown in Figure 15.27.

- Draw a block drawing, which is a scaled outline of each room, with closet and stair openings placed in their relative locations to each other.

- Draw the position of all walls representing the lines shared by blocks. This is done using a double line or equivalent command. Establish the wall thicknesses separately for different size walls. For example interior walls are usually 4 1/2″ wide. Outside walls include an additional width for insulation, drywall, studs, and siding so outside walls are drawn 5 1/4″ to 5 3/8″ thick. Walls should be drawn on level 2. At this point walls can be moved to align or intersect other

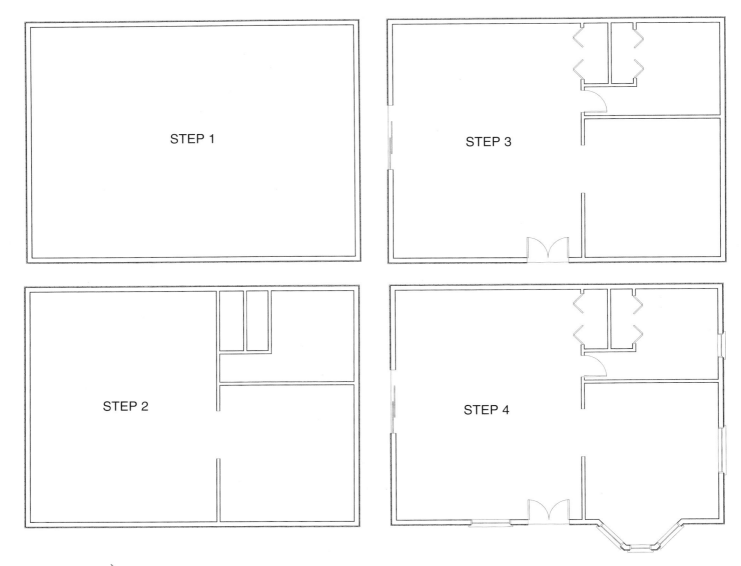

FIGURE 15.27 ⟩ Sequence of drawing a CAD floor plan. *Hepler/Wallach/Hepler © Cengage 2013*

walls to create the final room, area, and building width and length dimensions.

- Adjust wall intersections to eliminate crossover lines using the editing command. Eliminate the portion of each wall that represents the width of each door, window, arch, or other wall break.
- From the symbol library, insert the symbol for each door or window type into the open space. Assign the window and door symbols to a separate layer to enable plotting with lighter lines in a different color.
- Draw walled rectangles to represent stairwell openings. Put the stairwell opening on a separate layer to aid in calculating the stair details.

- Draw the outline of any element that is to be built outside the now established wall configuration. This include fireplaces, built-in furniture cabinets, fixtures, and appliances. If existing symbols from the symbol library are not satisfactory, create the desired design using a single-line command or a combination of lines and symbols. Do not be satisfied with a library symbol if it does not match the established design.
- Draw the outline of outside features such as driveways, walkways, patios, porches, decks, and pools. Assign a separate layer to these elements so they can be removed, revealing only the floor plan.

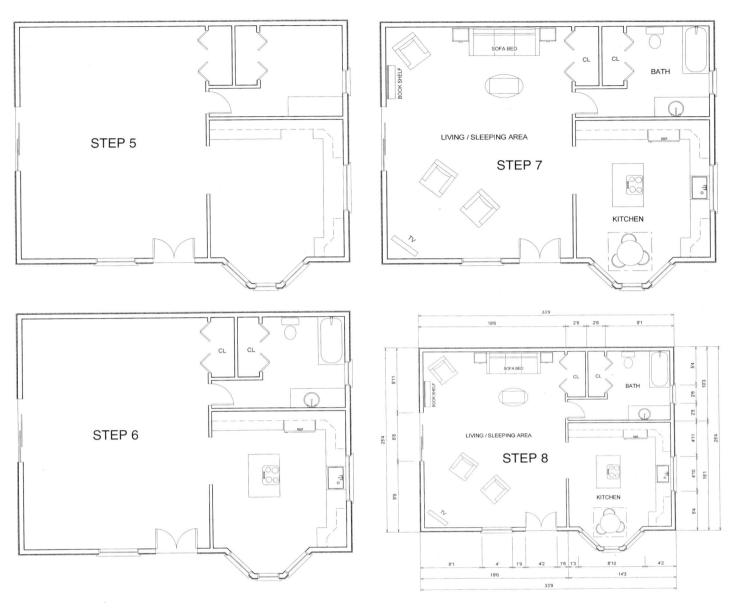

FIGURE 15.27 > Continued

- Dimension the plan to locate all horizontal distances needed for construction. Follow the dimensioning guidelines found in Figures 15.42 through Figure 15.49. These guidelines are based on American Standard conventions for architectural drawings. If an automatic CAD dimensioning program does not conform to these standards, adjustments should be made to produce acceptable dimensions. If this is not possible, dimensions should be added using line and text commands. Assign a separate layer so dimensions can be removed or plotted in a different color and line weight.

- Using the text command add all labels, titles, notes, and the scale. Assign a separate layer to all text input.

- Add material symbols from the library to represent patterns that cover surface areas. These include the texture color and patterns of carpeting, furniture, decking, paving, and landscape features. Because these patterns are usually added for presentation purposes, a separate layer is necessary to allow the basic plan to be plotted without them.

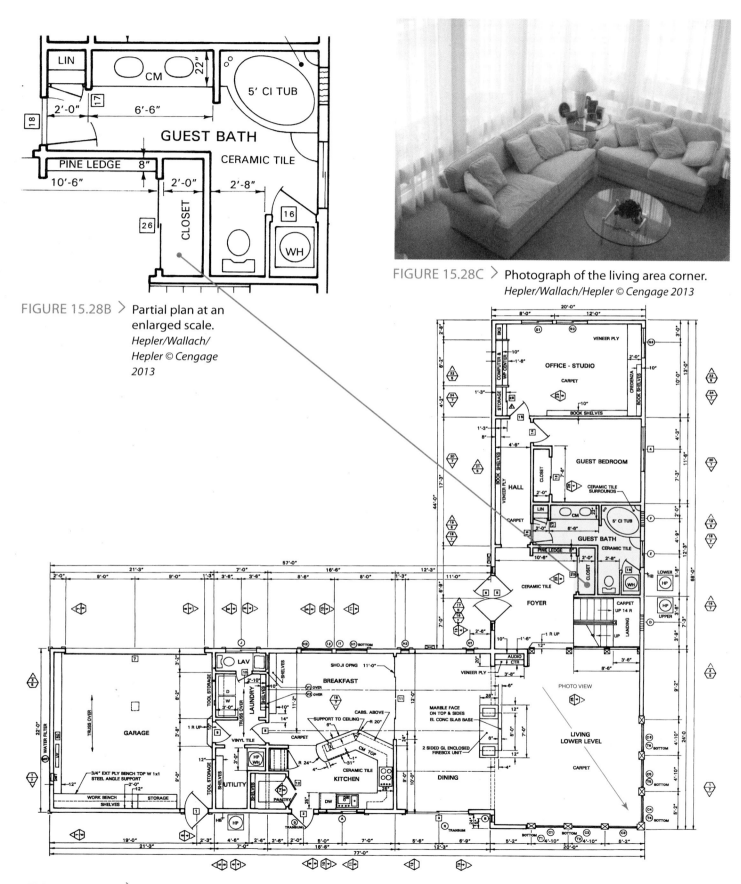

FIGURE 15.28B > Partial plan at an enlarged scale. *Hepler/Wallach/ Hepler © Cengage 2013*

FIGURE 15.28C > Photograph of the living area corner. *Hepler/Wallach/Hepler © Cengage 2013*

FIGURE 15.28A > Reduced scale floor plan. *Hepler/Wallach/Hepler © Cengage 2013*

SIZE AND SCALE IN FLOOR PLANS

Floor plans for large commercial or industrial buildings may be drawn at a scale of 1/8″ = 1′-0″ or less. Most residential floor plans are drawn to a scale of 1/4″ = 1′-0″.

Figure 15.28A shows a complete floor plan. This plan was prepared at a scale of 1/4″ = 1′-0″. The size was reduced to fit the book page. A portion of this drawing is shown at the original 1/4″ = 1′-0″ scale in Figure 15.28B. Figure 15.28C is a photograph with the plan area indicated by a red arrow.

THREE-DIMENSIONAL FLOOR PLANS

Two-dimensional floor plans function best when used for the exact description of scale, dimensions, and details. For presentation purposes 3D floor plans provide more realism with familiar viewing angles. Figure 15.29 shows a floor plan prepared in a pictorial format. The addition of walls or partial walls to a 3D floor plan provides more structure to a plan as shown in Figure 15.30.

MULTIPLE-LEVEL FLOOR PLANS

Drawing Separate Plans

Bilevel, two-story, one-and-one-half-story, and split-level homes require a separate floor plan for each level. The separate plans of **multiple-level floor plans** are prepared on tracing paper and drawn at the same scale as the first-floor plan. The tracing paper needs to be placed directly over the first-floor plan to ensure alignment of exterior walls, partitions, and vertical features. Once the major outline has been traced, the first-floor plan is removed.

Alignment of features, such as stairwell openings, outside walls, plumbing walls, vents, and chimneys, is critical in preparing second-floor plans. Where no second floor exists over part of a first floor, the outline of the first-level roof is shown. See Figure 15.31.

Figure 15.32 on page 322, shows a typical second-floor plan of a one-and-one-half-story house. This drawing, in addition to revealing the second-floor plan, shows the outline

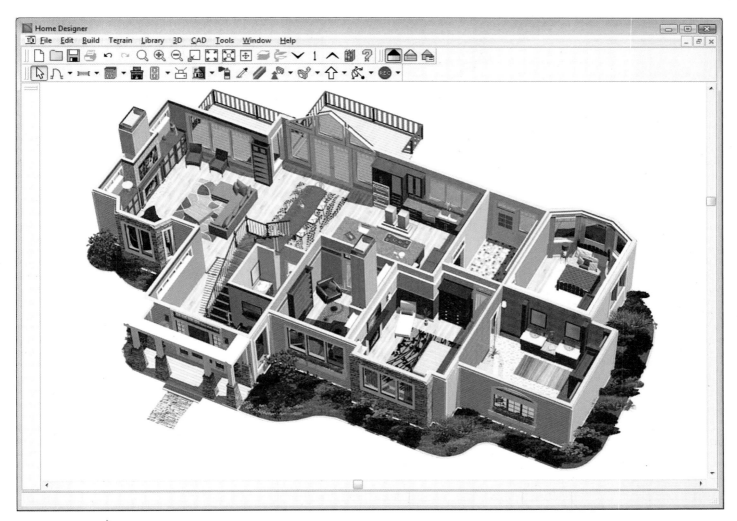

FIGURE 15.29 > Perspective floor plan in 3D. *Chief Architect Software*

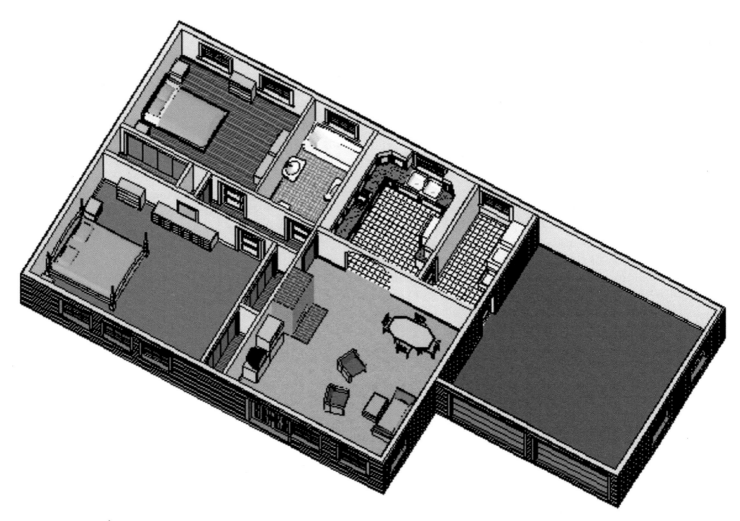

FIGURE 15.30 > Walled isometric floor plan. *Hepler/Wallach/Hepler © Cengage 2013*

of the roof as a single line, and the outline of the building as dotted lines under the roof.

Figure 15.33 shows a first-, second-, and third-floor plan of a three-level house. Visualize the position of each level by referring to the pictorial rendering of this house. Note that the three different levels of this plan do not stack evenly on top of each other. Upper floors are sometimes smaller in area. Compare the alignment of the first-level floor plan in Figure 15.28A with the upper level floor plan shown in Figure 31.39. In drawing multiple floor plans with a CAD computer program, the layering command places each plan on a separate layer, which can be viewed separately or combined with other layers.

Figures 15.34 and 15.35 show a first- and a second-floor presentation plan with only furniture added. In this design an area of the first floor extends to the roof. This area is therefore labeled "open to below" on the second floor plan and "open to above" on the first floor plan.

Figures 15.36 and 15.37 show the same plans with more details and with dimensions added. A comparison of these two plans illustrates the different uses of each type. In drawing higher or lower floor plans, structured features must be exactly aligned and consistently dimensioned. This includes such items as stairwells, chimneys, interior bearing walls, and outside walls. Layering practice and dimension checking are the keys to accurate alignments. To orient and aid in visualizing the vertical appearance of these plans, refer to the matching elevation drawings in Chapter 16.

Calculating Dimensions for Stair Systems

Floor plan stair symbols show the width and depth of each tread beginning at the plan level. A **break line** (see Figure 4.17) is used to eliminate the need to draw every stair to the next level, either up or down. (See Chapter 9 for information on different types of stairs.)

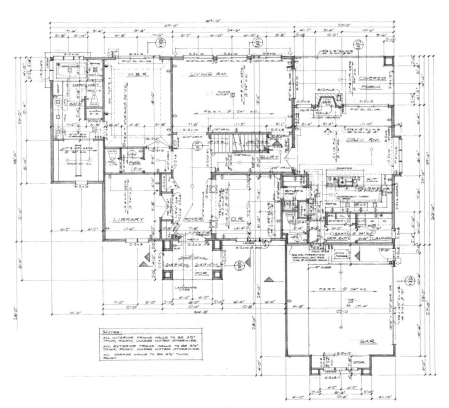

FIRST FLOOR PLAN

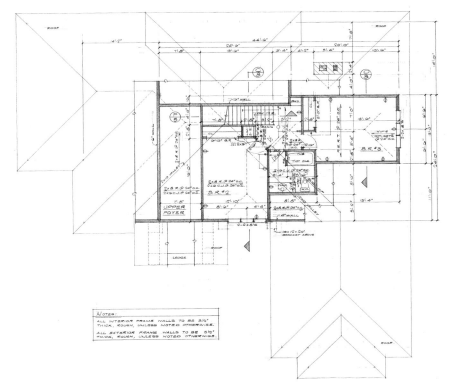

SECOND FLOOR PLAN

FIGURE 15.31 > Alignment of first- and second-floor plans. *Used by permission Hanley Wood, LLC*

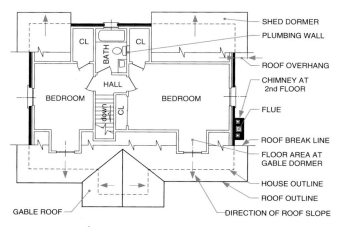

FIGURE 15.32 > Method of drawing one-and-one-half-story floor plan. *Hepler/Wallach/Hepler*
© *Cengage 2013*

When drawing multiple-level floor plans, stair systems must align on all levels. Also the number and width of treads (or steps) must be calculated and shown on all levels. To calculate stair tread width, use the following formula:

$$\frac{\text{stair run (inches)}}{\text{number of treads}} = \text{tread width (inches)}$$

In addition to indicating the stair system, plans of second floors or higher levels must also show the outline of the stairwell opening. To determine the size of the stairwell opening, the complete stair system should be designed before the final floor plan is prepared. Figure 15.38A illustrates the common stair terminology used in calculating stair dimensions.

FIGURE 15.33 > Alignment of a three-level plan. *Winter Park Design, Inc.*

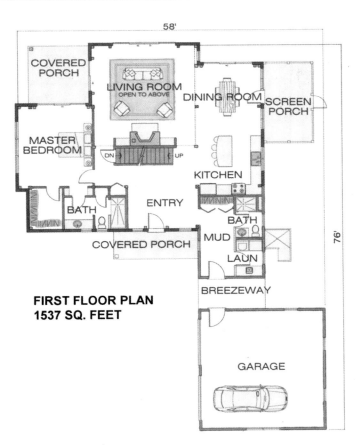

FIGURE 15.34 > First-floor plan with furniture only. *Timberpeg Inc.*

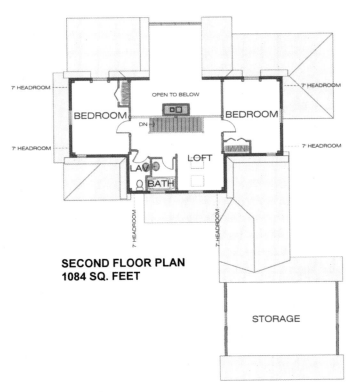

FIGURE 15.35 > Second-floor plan with furniture only. *Timberpeg Inc.*

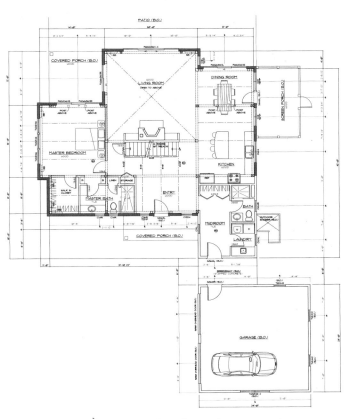

FIGURE 15.36 > Detailed first-floor plan to match Figure 15.34. *Timberpeg Inc.*

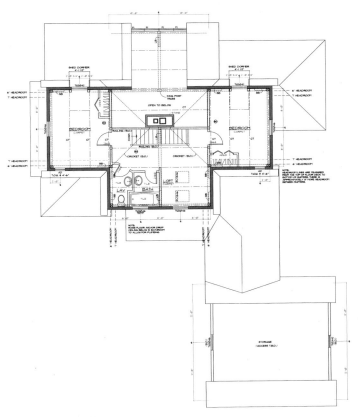

FIGURE 15.37 > Detailed second-floor plan to match Figure 15.35. *Timberpeg Inc.*

Figure 15.38B shows the sequence followed to arrive at a stair system **rise, run,** stairwell opening, and head clearance. Figure 15.38C is a stair run scale that illustrates the relationship of run, rise, and treads.

REVERSED PLANS

Floor plans offered as options in the development stage are often reversed to provide more plan choices. Reversing plans alters the appearance of a house and relocates rooms to avoid or take advantage of environmental and orientation factors and street locations. **Reversed plans** are accomplished by turning a plan over to provide a mirror image, as shown in Figure 15.39. The plan can be traced in this position or a print created by feeding the drawing into a print machine with the front side facing down. On a finished plan, this should be done before lettering to avoid printing reversed letters. On a CAD system, a mirror command is used to produce a reversed plan. The two versions are often labeled left-hand or right-hand plans.

Reversing Plans on CAD

If you place a mirror on its edge on a drawing you will see the reverse image of the drawing in the mirror. CAD drawings can be reversed on the X or Y axis by clicking the *Modify* menu, then the *Mirror* command. Plans are often mirrored to create variety and options in a development, to adjust a plan to better fit a site, or to reduce the amount of drafting time. To ensure that labels are not reversed the *Set Variable* command and the *Mirror Text* setting should be used. This will return the text portion of the drawing to its original readable position. Otherwise the mirror command must be done before labeling or dimensioning.

REFLECTED CEILING PLANS

Complex ceiling designs and multiple-lighting fixtures or levels often require the preparation of a **reflected ceiling plan.** See Figure 15.40. These plans are drawn using a floor plan as a base. Floor plan walls and partitions are traced and symbols of ceiling features are drawn as the ceiling would be viewed if the floor were a mirror.

Reflecting Plans on CAD

Reflected plans such as a reflected ceiling plan are prepared by drawing ceiling features on a layer aligned with the floor plan. The layer is then printed separately to produce a separate plan.

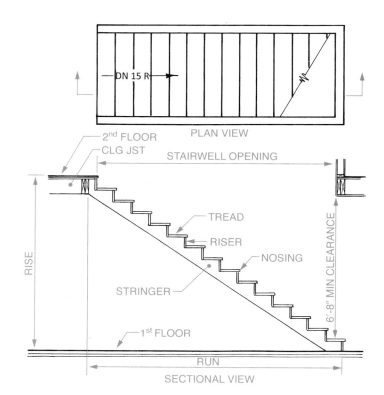

FIGURE 15.38A › Stair system terminology. *Hepler/Wallach/ Hepler © Cengage 2013*

FLOOR PLAN DIMENSIONING

A completely dimensioned drawing is necessary to complete any building exactly as designed. Dimensions on the floor plan show the builder the width and length of the building. They show the location of doors, windows, stairs, fireplaces, planters, and so forth. Just as symbols and notes show exactly what materials are to be used in the building, dimensions show the sizes of materials and exactly where they are located.

Because a large building must be drawn on a relatively small sheet, a small scale such as 1/4″ = 1′-0″ or 1/8″ = 1′-0″ must be used. The use of such a small scale means that many dimensions must be crowded into a very small area. Therefore, only major dimensions such as the overall width and length of the building and of separate rooms, closets, halls, and wall thicknesses are usually shown on the floor plan. Figure 15.28A shows a fully dimensioned first-floor plan, and Figure 15.41 shows a second-level floor plan with many details and dimensions. Dimensions too small to show directly on the floor plan are described either by a note on the floor plan or by separate, enlarged details.

Enlarged details, or detail drawings, are sometimes merely enlargements of a portion of the floor plan. They may also be

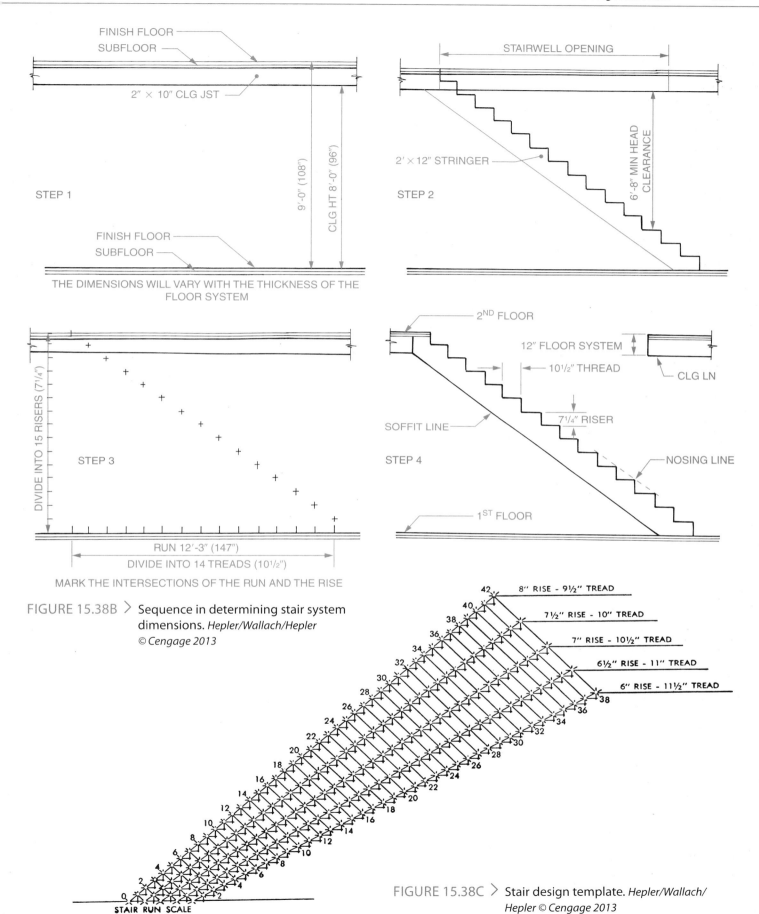

FINISH FLOOR
SUBFLOOR

2" × 10" CLG JST

STEP 1

9'-0" (108")

CLG HT 8'-0" (96")

FINISH FLOOR
SUBFLOOR

THE DIMENSIONS WILL VARY WITH THE THICKNESS OF THE
FLOOR SYSTEM

STAIRWELL OPENING

2' × 12" STRINGER

6'-8" MIN HEAD CLEARANCE

STEP 2

DIVIDE INTO 15 RISERS (7¼")

STEP 3

RUN 12'-3" (147")
DIVIDE INTO 14 TREADS (10½")

MARK THE INTERSECTIONS OF THE RUN AND THE RISE

2ND FLOOR

12" FLOOR SYSTEM

10½" THREAD

CLG LN

SOFFIT LINE

7¼" RISER

STEP 4

NOSING LINE

1ST FLOOR

FIGURE 15.38B ❯ Sequence in determining stair system
dimensions. *Hepler/Wallach/Hepler*
© Cengage 2013

8" RISE - 9½" TREAD
7½" RISE - 10" TREAD
7" RISE - 10½" TREAD
6½" RISE - 11" TREAD
6" RISE - 11½" TREAD

STAIR RUN SCALE

FIGURE 15.38C ❯ Stair design template. *Hepler/Wallach/
Hepler © Cengage 2013*

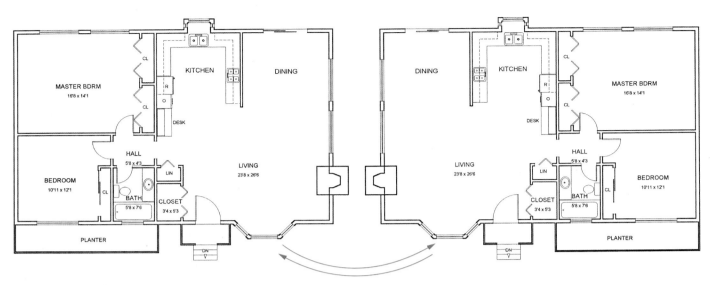

FIGURE 15.39 > Reversed floor plan. *Hepler/Wallach/Hepler © Cengage 2013*

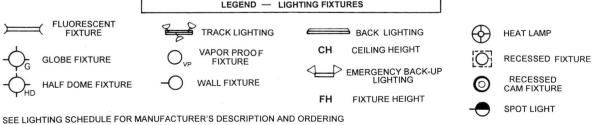

LEGEND — LIGHTING FIXTURES			
FLUORESCENT FIXTURE	TRACK LIGHTING	BACK LIGHTING	HEAT LAMP
GLOBE FIXTURE	VAPOR PROOF FIXTURE	CH CEILING HEIGHT	RECESSED FIXTURE
HALF DOME FIXTURE	WALL FIXTURE	EMERGENCY BACK-UP LIGHTING	RECESSED CAM FIXTURE
		FH FIXTURE HEIGHT	SPOT LIGHT

SEE LIGHTING SCHEDULE FOR MANUFACTURER'S DESCRIPTION AND ORDERING

FIGURE 15.40 > Reflected ceiling plan. *Hepler/Wallach/Hepler © Cengage 2013*

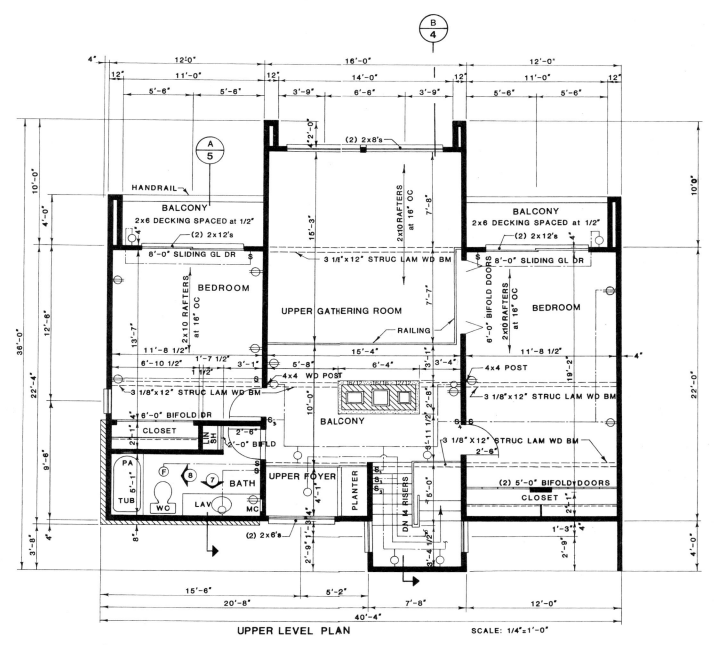

FIGURE 15.41 > Fully dimensioned floor plan. *Hepler/Wallach/Hepler © Cengage 2013*

section drawings cross-referenced on the floor plan. Separate drawings are usually necessary to communicate adequately dimensions for fireplaces, planters, built-in cabinets, door and window details, stair framing details, or any unusual construction methods.

CAD Floor Plan Grids

Square or rectangular floor plans can be drawn using a two-dimensional grid system that aligns the plan on X and Y Cartesian coordinate axis lines. Grid lines or dots that represent grid intersections can be superimposed on the drawing using different color lines or line weights. Assigning grid lines to a different layer helps separate the grid lines from other elements of the plan. Grid line spacing must be in increments that are compatible with the scale of the drawing. Grid size and spacing can be set to any size.

Grids are frequently used in preparing modular drawings. The most commonly used grid spacing is 12″. The use of grids allows the design process to proceed without the use of scales. Lines can also be drawn on absolute X and Y coordinates when exact X and Y coordinates are known. Relative coordinates can be used by referencing the last used coordinates as a base.

The two types of CAD grid systems are *Snap Grid* and *Display Grid.* The *Snap-On* command allows the crosshairs to move from one grid point to the nearest grid intersection. The *Snap-Off* command allows a point to remain exactly where placed by the cursor without regard to the grid lines. The *Display Grid* command allows the grids to remain on the monitor while other drawing operations are performed. The displayed grid can be turned on or off repeatedly during the drawing process. The use of the *Ortho* command will align all drawing lines either vertically or horizontally (90°) without the use of a grid. *Polar* increments can also be set to draw angular lines to any selected degrees of a circle.

Dimension Selection

A floor plan must be completely dimensioned to ensure that the house will be constructed precisely as designed. Complete dimensions convey the exact wishes of the architect and owner to the builder. If adequate dimensions are not provided, the builder is placed in the position of a designer. A good builder is not expected to be a good designer. Supplying complete dimensions will eliminate the need for the builder to guess or interpret the size and position of the various features of this plan. All needed information about each room, closet, partition, door, and window is included in working drawings, schedules, specifications, and documents.

A floor plan with only limited dimensions shows just the overall building dimensions and the width and length of each room. It is sufficient to summarize the relative sizes of the building and its rooms for the prospective owner, but insufficient for building purposes.

CAD Dimensioning

Floor plans are dimensioned by using a combination of tools that are found under the *Dimension* pull-down menu. The dimension style needs to be set up using standards from the American Institute of Architects. Other setups are used for mechanical, structural, and electrical styles of drawings. The settings found in the dimensional style area are extensive—everything from arrowhead styles and text types down to the space from the end of the dimensional line to the text can be set.

Linear dimensions are the most common for all types of drawings. These are applied by selecting a line and selecting the dimension placement or by picking two **extension lines** and then the dimensional placement. The value of the dimension is automatically placed so accuracy when creating the drawing is important. Circular objects are dimensioned using the radius or diameter tools; angles are dimensioned using the *Angular Dimension* command.

Guidelines for Dimensioning

Many construction mistakes result from errors made in architectural drawings. Most errors in architectural drawings are the result of mistakes in dimensioning. Dimensioning errors are therefore costly in time, efficiency, and money. Familiarization with the following guidelines for dimensioning floor plans will eliminate much confusion and error.

These guidelines are illustrated by the numbered arrows in Figure 15.42. The guidelines also reference Figures 15.42 through 15.49.

1. Architectural **dimension lines** are unbroken lines with dimensions placed above the line. Dimension lines should be located no closer than 1/4″ from object lines. Up to 1″ is preferred. Dimension lines should be spaced at least 1/4″ between rows.

2. Foot and inch marks are normally used on architectural dimensions. Sometimes these marks are omitted and a dash or slash is used. For example 8-4 or 8/4 means 8′-4″. If metric measures are used, the dimensions are always in millimeters. Therefore, size unit notations are not needed. However, a note should be placed on the plan stating that all dimensions are in millimeters.

3. Dimensions over 1′ are expressed in feet and inches. Detail drawings often contain only inch dimensions regardless of size.

4. Dimensions less than 1′ are shown in inches.

5. A slash is often used with fractional dimensions to conserve vertical space.

6. Vertical dimensions should be placed to read from the right side of the drawing. Horizontal dimensions read from the bottom of the drawing.

7. **Overall dimensions** for the length and width of a building are placed outside other dimensions.

8. Line and arrowhead weights for architectural dimensioning are thin and dark. Arrowhead styles are optional. See Figure 15.43.

9. Room sizes may be shown by stating width and length on abbreviated plans.

10. When the area to be dimensioned is small, numerals may be placed outside the extension lines.

11. Framed interior walls are dimensioned to the center of partitions. Figure 15.44 illustrates the methods of dimensioning wood-framed walls, openings, and partitions.

12. Window and door sizes may be shown directly on the door or window symbol, or may be indexed to a door or window schedule with a reference callout.

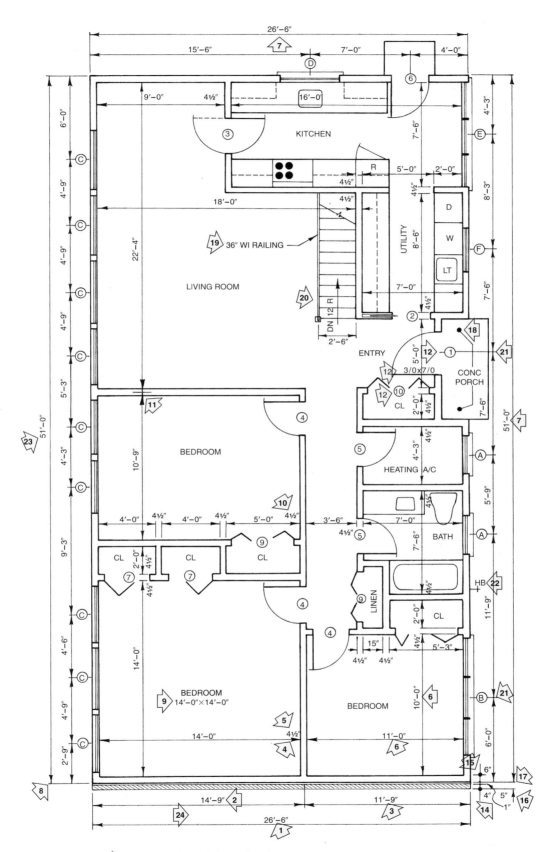

13. Solid concrete walls are dimensioned from wall to wall, exclusive of wall surfaces coverings. See Figure 15.45.

14. Curved leaders are sometimes used to eliminate confusion with other dimension lines.

15. When areas are too small for arrowheads, dots may be used to indicate dimension limits.

16. The dimensions of brick or stone veneer must be added to the framing dimension. See Figure 15.46.

17. When the space is small, arrowheads may be placed outside the extension lines.

18. A dot on the end of a leader refers to the entire area noted.

19. Dimensions that cannot be seen on the floor plan or those too small to place on the drawn object are noted with leaders for easier reading.

20. In dimensioning stairs, the number of risers is placed on a line with an arrow indicating the direction down (DN) or up (UP).

21. Windows, doors, pilasters, beams, construction members, and areaways are dimensioned to their center lines. (*Areaways* are the sunken areas in front of basement doors and windows that allow light and air to reach the basement or crawl space.)

22. Use notes or abbreviations when symbols do not show clearly what is intended.

23. Architectural dimensions always refer to the actual size of the building regardless of the scale of the drawing. The building in Figure 15.42 is 51′-0″ in length.

24. **Subdimensions** must add up to overall dimensions. For example: 14′-9″ + 11′-9″ = 26′-6″. Most rows of dimensions include both feet and inches and may include fractional inches. In adding rows of mixed numbers such as these, add the inches separately, convert the inch total to feet and inches, and then re-add the foot total as follows:

 • Total number of inches ÷ 12 = feet and inch fractions

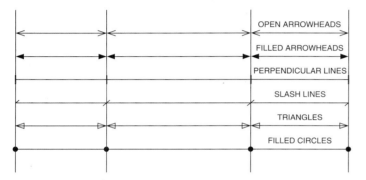

FIGURE 15.43 > Dimensional arrowhead styles. *Hepler/ Wallach/Hepler © Cengage 2013*

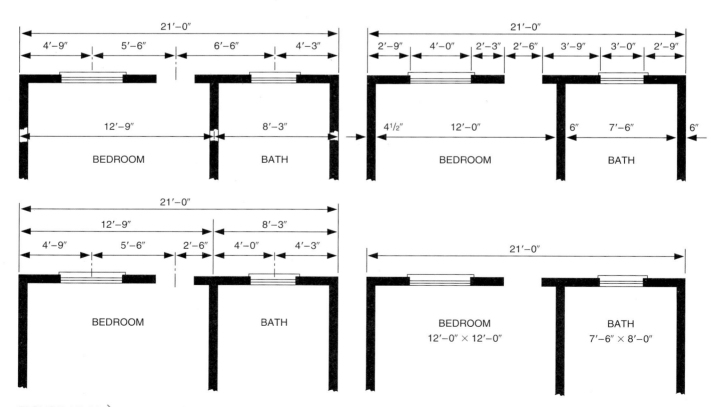

FIGURE 15.44 > Methods of dimensioning frame construction. *Hepler/Wallach/Hepler © Cengage 2013*

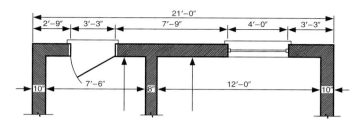

FIGURE 15.45 > Methods of dimensioning concrete walls. *Hepler/Wallach/Hepler © Cengage 2013*

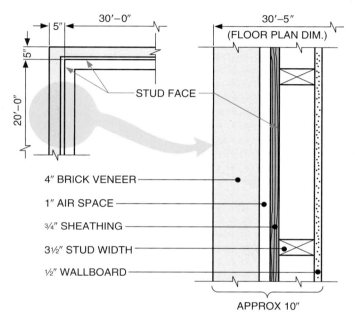

4" BRICK VENEER

1" AIR SPACE

¾" SHEATHING

3½" STUD WIDTH

½" WALLBOARD

STUD FACE

APPROX 10"

FIGURE 15.46 > Method of dimensioning a brick veneer wall. *Hepler/Wallach/Hepler © Cengage 2013*

- LCD = lowest common denominator in a series into which all denominators can be divided.

Example:	To add:	Follow these steps:
	1'-7⅞"	Step 1: 1'-7¹⁴/₁₆"
	2'-8⅛"	2'-8⁴/₁₆"
	6'-10⁹/₁₆"	6'-10⁹/₁₆"
	11'-2¹¹/₁₆"	11'-2¹¹/₁₆"
		Step 2: 20'-27¹⁸/₁₆"
		Step 3: 22'-5⅜"

The following guidelines are not illustrated in Figure 15.42.

25. When framing dimensions alone are desirable, rooms are dimensioned by distances to the outside face of the studs in the partitions. See Figure 15.47.

26. Because building materials vary somewhat in size, first establish the thickness of each component of the wall and partition, such as plaster, brick, or tile thicknesses. Add these thicknesses together to establish the total wall thickness. Common thicknesses of wall and partition materials are shown in Figures 15.46 and 15.47.

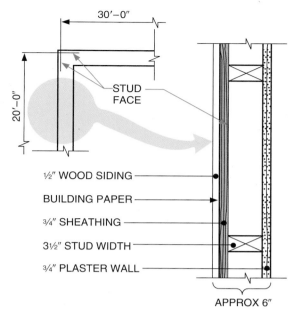

STUD FACE

½" WOOD SIDING

BUILDING PAPER

¾" SHEATHING

3½" STUD WIDTH

¾" PLASTER WALL

APPROX 6"

WOOD FRAME WALL

FIGURE 15.47 > Method of dimensioning to the stud face on a wood-framed wall. *Hepler/Wallach/Hepler © Cengage 2013*

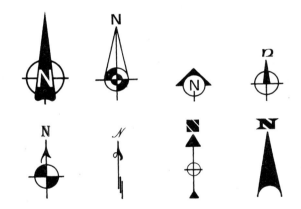

FIGURE 15.48 > Styles of north arrows used on plan views. *Hepler/Wallach/Hepler © Cengage 2013*

27. The scale of each drawing can be noted in the title block. If separate drawings on a single drawing sheet are drawn at different scales, each drawing must be labeled with the appropriate scale.

28. An arrow showing the direction of north is placed on each floor plan unless the plan is prepared without any site identification. Various types of north arrows are shown in Figure 15.48.

29. All outside buildings and definable features should also be dimensioned using the same basic guidelines where they apply. Figure 15.49 shows a dimensioned pool area that connects to the plan shown in Figure 15.28A.

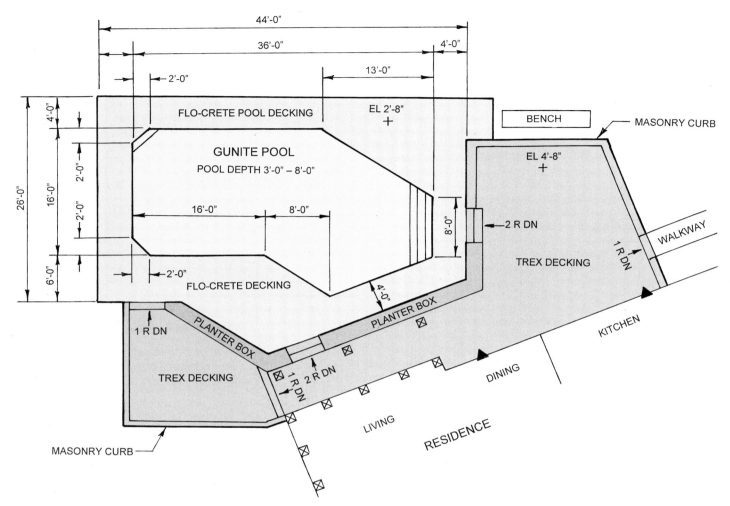

FIGURE 15.49 > Pool area dimensioning. *Hepler/Wallach/Hepler © Cengage 2013*

METRIC DIMENSIONING

If a building is designed using **metric** sizes, all dimensions are shown in millimeters (mm), as shown in Figure 15.50. Refer to Chapter 3 to review metric scales.

MODULAR DIMENSIONS

An attempt should always be made to establish major dimensions to conform to 16″, 24″, and 48″ increments. However, if a building is designed to totally conform to **modular** grids, all dimensioning *must* conform to these standards as covered in Chapter 22.

BUILDING INFORMATION MODELING

Building information modeling (BIM) drawings are 3D CAD views of the exterior and interior space of a building. These include 3D drawings of walls, doors, windows, and plumbing, electrical, and HVAC systems. Through the use of building modeling software, these drawings are layered to reveal the relationship between building systems and structural components. BIM drawings are created from information contained on floor plans, elevations, sections, and details. This results in the creation of multiple 3D drawings that show volume and provide for the simultaneous observation of potentially conflicting systems.

BIM drawings are also linked to the generation of supporting documents such as specifications, purchase orders, cost estimates, scheduling data, and financial analysis. Construction management and environmental code monitoring are also included. And it all starts with the basic floor plan.

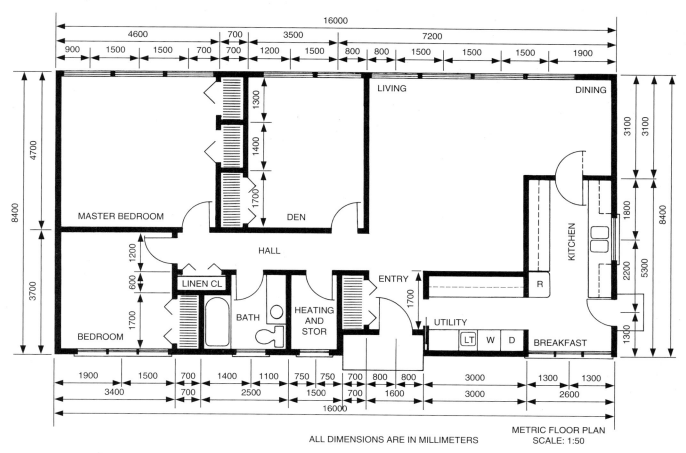

FIGURE 15.50 > Metric floor plan dimensioning. *Hepler/Wallach/Hepler © Cengage 2013*

Drawing Floor Plans Exercises

CHAPTER 15

1. Draw a complete floor plan, using a sketch of your own design as a guide, and using a scale of 1/4″ = 1′-0″.

2. Dimension an original scaled floor plan that you have completed for a previous assignment.

3. Draw and dimension the floor plan of your own home.

4. Design and draw a studio apartment under 800 sq. ft.

5. Design and draw a cabin plan under 800 sq. ft.

6. Design and draw a three-bedroom, two-bath home under 1500 sq. ft.

7. Design and draw a two-story, four-bedroom home under 2,500 sq. ft.

CHAPTER 16
Designing Elevations

OBJECTIVES

In this chapter you will learn to:

> apply the principles and elements of design (Chapter 2) to creating elevation drawings.

> recognize different roof styles as options for roof design.

> select and design window styles in relation to elements of design and window functions.

> locate doors on an elevation design, considering style, size, and types of doors.

TERMS

awnings
casements
clerestory windows
dormers
double hung windows
Dutch hip
eave line
elevation drawings
fascia
fenestration
flush

French door
gable
gambrel roofs
ground line
hip roofs
hopper windows
jalousie
mansard roofs
mullions
muntins
overhang

panel doors
pitch
ridge line
rise
run
shed roof
skylights
slope (roof)
valley
vaulted roofs

INTRODUCTION

Elevation drawings, or elevations, show the vertical surfaces of a structure. Exterior elevation drawings show the entire front, sides, and rear of a structure. Interior elevations show vertical surfaces of interior walls. In this chapter, you will learn how to apply the elements of design to creating the exterior form of a building, including the selection of roof, window, and door styles.

Designing the elevations of a structure is only one part of the total design process. However, the elevation design reflects the plane of the building that people see. The entire structure may be judged by the elevations.

FLOOR PLAN RELATIONSHIP

Because a structure is designed from the inside out, the design of the floor plan normally precedes the design of the elevation. The complete design process requires a continual relationship between the elevations and the floor plan.

Flexibility is possible in the design of elevations, even in those designed from the same floor plan. Once the location of doors, windows, and chimneys has been established on the floor plan, the development of an attractive and functional elevation for the structure depends on various factors. Roof style, overhang, grade-line position, and the relationship of windows, doors, and chimneys to the building line must be considered. Choosing a desirable elevation design is not an automatic process that follows the floor plan design, but a creative process that requires knowledge and imagination.

The designer should keep in mind that only horizontal distances can be established on the floor plan. However, on an elevation, vertical heights, such as heights of windows, doors, and roofs, must also be shown. As these vertical heights are established, the appearance of the outside and the way that the heights affect the internal functions of the building must be considered.

ELEMENTS OF DESIGN IN ELEVATIONS

Creating floor plans is a process of allocating interior space to meet functional needs. Designing elevations involves combining the elements of design to create functional and attractive building exteriors.

The principles and elements of design were defined in Chapter 2. In this chapter, the elements of design (line, form, space, color, light, materials, and texture) are applied to the creation of elevations. The total appearance of an elevation depends on the relationship among its component parts, such as surfaces, roofs, windows, doors, and chimneys. The balance of these parts, the emphasis placed on various components of the elevation, the texture of the surfaces, the light, the color, and the shadow patterns all greatly affect the general appearance of an elevation.

Lines

The lines of an elevation tend to create either a horizontal or vertical emphasis. The major horizontal lines of an elevation are the **ground line, eave line,** and **ridge line.** If these lines are accented, the emphasis will be placed on the horizontal as shown in Figure 16.1. The deck lines of this building add to the horizontal emphasis. In contrast the dominance of the vertical window mullions and the extreme high pitch of the roof give the building in Figure 16.2 a vertical emphasis. In general, a low building will usually appear longer and even lower if the design consists mostly of horizontal lines. The reverse is true for tall buildings with vertical lines.

Building lines should be consistent. The lines of an elevation should appear to flow together as one integrated line pattern. Continuing a line through an elevation for a long distance is usually better than breaking the line and starting it again. Rhythm can be developed with lines, and lines can be repeated in various patterns.

When additions are made to an existing design, care must be taken to ensure that the lines of the addition are consistent with the established lines of the structure. The lines of the component parts of an elevation should relate to each other, and the overall shape should reflect the basic shape of the building.

Because many factors affect the appearance of an elevation, the relationship of these factors is important. For example, observe how the elements of design are combined in Figure 16.3 to produce two different elevations, although both are projected from the same floor plan. The

FIGURE 16.2 > Vertical emphasis in design. *Photo courtesy Lindal Cedar Homes. Lindal.com*

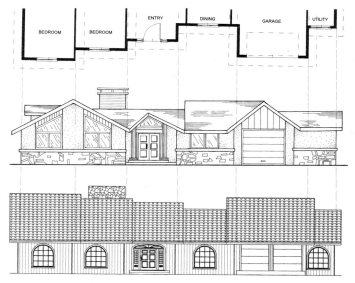

FIGURE 16.3 > Many factors affect elevation appearance. *Photo courtesy Lindal Cedar Homes. Lindal.com*

FIGURE 16.1 > Horizontal emphasis in design at "Fallingwater". *Western Pennsylvania Conservancy—Christofer Little, Photographer*

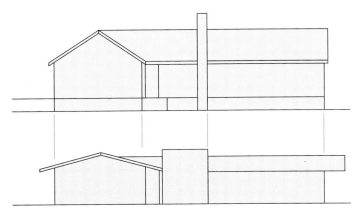

FIGURE 16.4 > Height factors affecting elevation appearance. *Hepler/Wallach/Hepler © Cengage 2013*

FIGURE 16.5 > Formally balanced elevations. *Oak Leaf Conservatories, www.oakleafconservatories.com*

difference between the two elevations in Figure 16.3 is primarily due to changes in the roof styles, siding materials, chimney shapes, and door and window styles. Changes in the height of the ground line, eave line, and ridge line alone can greatly alter the appearance of an elevation as shown in Figure 16.4.

Form and Space

Lines combine to produce form and create the geometric shape of an elevation. Elevation shapes should be balanced. The term *balance* refers to the symmetry of the elevation. Formal balance is used extensively in Colonial and period styles of architecture. Informal balance is more widely used in contemporary residential architecture and in ranch and split-level styles. (Refer back to Chapter 2.) Figure 16.5 shows a formally balanced elevation design. All elements of this design are symmetrical including the building components, the arch, and the formal gardens. Figures 16.1 and 16.2 are informally balanced.

Vertical or horizontal emphasis, or accent, can be achieved by several different devices. An area may be accented by mass, color, or material.

In addition to the elements of design, the basic architectural style of a building needs to be considered when designing elevations. A building's style is closely identified by the elevation design.

The type of building structure must also be compatible with the architectural style of the elevation. However, within basic styles of architecture, there is considerable flexibility in the type of structure. Figure 16.6 shows the basic types of residential structures.

Elevations should appear as one integral and functional façade, rather than as a surface in which holes have been cut for windows and doors and other structural components. Doors, windows, and chimney lines should be part of a continuous

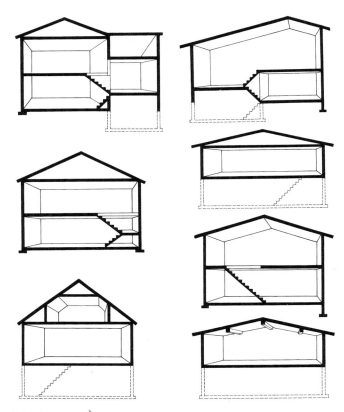

FIGURE 16.6 > Basic types of residential structures. *Hepler Associates, PC*

pattern of the elevation and should not appear to exist alone. See Figure 16.7.

Light and Color

An elevation that is composed of all light areas or all dark areas tends to be uninteresting and neutral. Some balancing of light, shade, and color is desirable in most elevations.

FIGURE 16.7 〉 Consistent pattern of related door and window lines in balance. *Chief Architect Gallery*

FIGURE 16.8 〉 Effect of light and contrast on elevation design. *Pittsburgh Corning Corp.*

Materials and Texture

An elevation may contain various types of materials, such as glass, wood, masonry, and ceramics. These must be carefully and tastefully balanced for the design to be effective. An elevation composed of too many similar materials is ineffective and neutral. Likewise, an elevation that uses too many different materials is equally objectionable. In choosing materials for elevations, designers should not mix horizontal and vertical siding or different types of masonry. If brick is the primary masonry, brick should be used throughout. It should not be mixed with stone. Mixing masonry siding materials and patterns is inconsistent with the way natural stone appears in nature. Note how the texture of the siding material affects the elevation appearance in Figure 16.10. Siding materials should also be resistant to mold, fungi, wood-boring insects, and bacteria. The use of off-gasing toxic materials and lead paints or sealers should also be avoided.

ELEVATION DESIGN SEQUENCE

The first step in elevation design is to choose an architectural style. (Refer to Chapter 1.) Then sketch the outline of an exterior wall showing the roof shape and the position of doors, windows, and other key features such as chimneys or dormers.

Next, create a series of progressive sketches to develop an elevation design. Experiment with different roof styles, door and window designs, siding materials for the exterior walls, overhangs, chimney shapes, roof materials, and trim variations. Sketches can also show various architectural styles

Shadow patterns can be created by depressing specific areas, using overhangs, texturing, and varying the colors. Door and window trim, columns, battens (strips covering joints), and overhangs can be used to create most shadows. Figure 16.8 demonstrates the effect light and color have on the appearance of an elevation. Glass block at night can emit light while providing interior privacy. Figure 16.9 shows the realism gained by adding texture, shadows, foliage, and finally color to an elevation drawing. This figure also shows the horizontal relationship of key features on the elevation drawing to the floor plan.

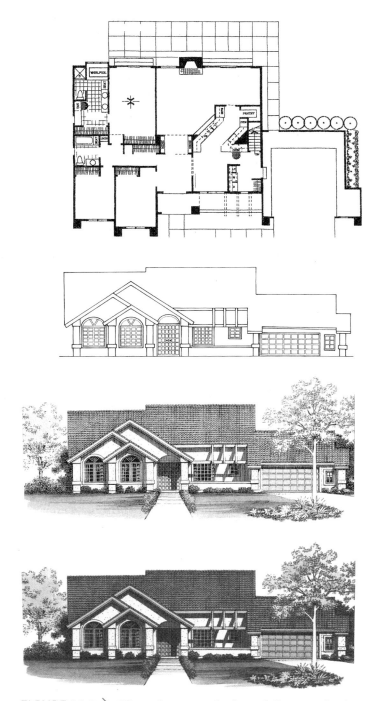

FIGURE 16.10 > Use of texture contrasts. © *Ocean/Corbis*

FIGURE 16.9 > Effect of texture, shadows, foliage, and color on elevation appearance. *Used by permission Hanley Wood, LLC*

derived from the same floor plan. The sketches shown in Figure 16.11 represent part of the elevation design process, which resulted in the design shown later in Figure 16.22 and Figure 13.51. All four sketches in Figure 16.11 are projected from the same floor plan as shown. These elevations are all different because of changes in roof type and pitch, window shape, chimney exposure, and surface textures.

Roof Types

To design elevations, a designer needs to know roof styles and which style best matches the building's overall style. There are many styles of roofs. The gable, hip, flat, and shed styles are the most popular. See Figure 16.12. Other features that affect the appearance of the roof must also be considered. These include the size and shape of dormers, skylights, vents, chimneys, and cupolas. In addition to style, the overhang size and the roof pitch (angle) must be determined during the design process.

Roof framing plans are subsequently developed from the basic roof design. A roof plan shows the outline of the top view of a roof with solid ridge, **valley,** and chimney lines. Dashed lines represent the outline of the floor plan under the roof. Small arrows show the downward slope. (Detailed information on roof framing drawings is presented in Chapter 30.)

Gable Roofs

A **gable** is the triangular end of a building. Roofs that fit over this area are gable roofs. Gable roofs are the most common roof style because of their adaptability to a wide variety of architectural styles, from Colonial to contemporary. They also drain and ventilate easily. Figure 16.13 illustrates a gable

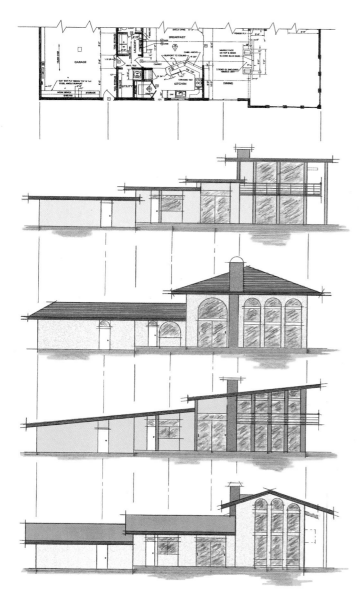

FIGURE 16.11 > Progressive elevation design sketches.
Hepler/Wallach/Hepler © Cengage 2013

roof in a Colonial structure, and Figure 16.14 shows a contemporary gable application. Elevation drawings, as introduced in chapter 4, are flat, two-dimensional orthographic views of vertical walls. Most elevations such as the views in Figure 16.12 are of this type. Many designs, have exterior surfaces that recede from the orthrographic plane; the lines are therefore foreshortened on an elevation drawing.

Variations of gable roofs include A-frames, winged, and pleated gables. A-frame roofs extend to the floor line, creating continuous ceilings and walls inside. Winged gable roofs are created by extending the ridge overhang further than the overhang at the corners. See Figure 16.14. Pleated (folded plate) roofs consist of a series of aligned and connected small gable roofs.

Hip Roofs

Hip roofs provide eave-line protection around the entire perimeter of a building. The hip roof overhangs shade windows that would not be shaded at a gabled end. For this reason, hip roofs are very popular in warm climates. See Figure 16.15. Another variation of the hip roof is the **Dutch** (Deutsch) **hip.** A Dutch hip is created by extending the ridge outward to make a partial gable end at the top of the hip.

Flat Roofs

When a low building silhouette is desirable, flat roofs are ideal. Flat roofs have a slight slope (1/8″ to 1/2″ per foot) for drainage, unless water is used as an insulator. They can also function as decks on multilevel structures. Flat roofs do not have the structural advantage normally gained by rafters leaning on a ridge board. Heavier rafters (ceiling joists) are needed. Because of snow-load problems in cold climates, flat roofs are more popular in warm climates. See Figure 16.1.

Shed Roofs

A **shed roof** is a flat roof that is slanted. If the down slope faces south, shed roofs are ideal for solar panels. When **clerestory windows** are added between offsetting shed roofs, light can be provided for the center of a building. Many industrial buildings use multiple shed roofs and clerestory windows in a sawtooth pattern to maximize center light.

Butterfly Roofs

Two shed roofs that slope to the center create a butterfly roof. This roof style allows for higher outside walls, which can provide more light access.

Gambrel Roofs

Gambrel roofs are double-pitched roofs. They are also known as barn roofs. Gambrel roofs are the distinguishing feature of Dutch Colonial houses. They are used to create more headroom in one-and-one-half-story homes.

Mansard Roofs

Mansard roofs are double-pitched hip roofs with the outside constructed at a very steep pitch. This type of roof is used on formally designed French Provincial homes.

Vaulted Roofs

Vaulted roofs are curved panel roofs. They are composed of a series of manufactured curved panels that are erected side by side between two bearing walls. This arrangement

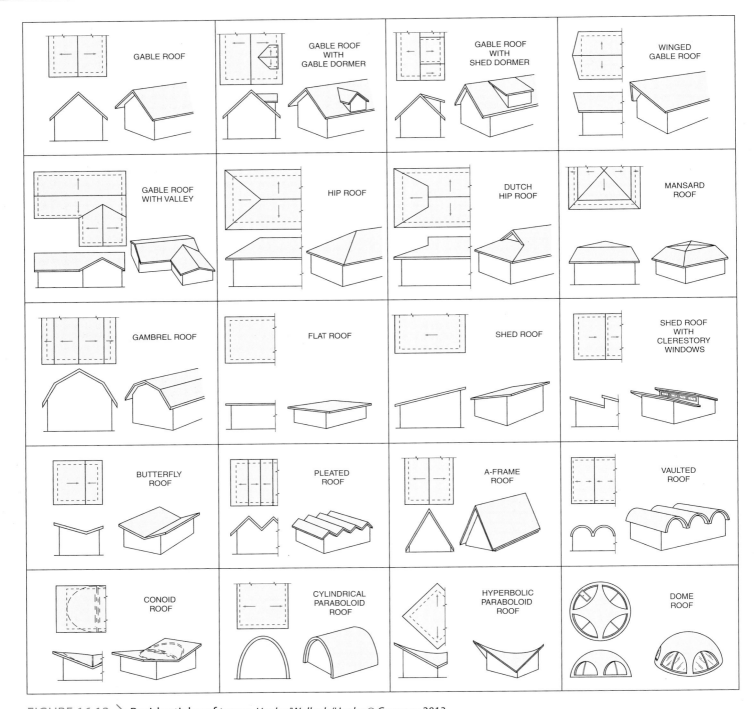

FIGURE 16.12 > Residential roof types. *Hepler/Wallach/Hepler © Cengage 2013*

allows for larger open areas since the curved construction is structurally stable.

Bermuda Roofs

Bermuda roofs originated on the island of Bermuda in the Caribbean Sea. There the large **fascia** areas characteristic of the design are used to collect rainwater. This design effect is often used as a feature in other areas of the world. See Figure 16.16.

Dome and Dome-Shaped Roofs

As you learned in Chapter 1, domes have been used in architecture for centuries. Dome roofs, like A-frames, provide both roof and walls in one structurally sound unit. Because there is no need for internal support walls or columns, completely open floor space and flexible room sizes are possible.

The geodesic dome, developed by R. Buckminster Fuller, can be inexpensively mass produced at relatively low cost.

FIGURE 16.13 > Colonial (Greek) style gable roof. *Hepler/ Wallach/Hepler © Cengage 2013*

Actually, geodesic domes are not true "domes." They are series of triangles that are combined to form hexagons and pentagons.

Several restrictions must be kept in mind when designing dome structures. The use of domes for residential construction restricts the design to a predetermined area. Working with walls that are not plumb (vertical) creates problems in fitting cabinets, fixtures, appliances, and furniture effectively into the design. True domes, as found in cathedrals, are rarely used for light construction buildings. In addition to geodesic domes, many variations of triangular framing are frequently used to form roofs. This type of framing provides wider unobstructed space than any other type. The most common use of triangular roof framing is used to create a base for glass panels, as shown in Figure 16.17.

New Technology and Roof Styles

The development of new building materials and methods in molded plywood, plastics, and reinforced concrete has

FIGURE 16.14 > Contemporary winged gable roof. *Hepler/Wallach/Hepler © Cengage 2013*

FIGURE 16.15 > Elevation drawing with hip and gable roofs. *John Henry, Architect; Jones Clayton, Construction*

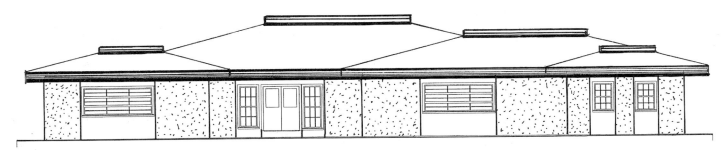

FIGURE 16.16 > Front view of a Bermuda roof. *Hepler/Wallach/Hepler © Cengage 2013*

FIGURE 16.17 > Conservatory glass roof. *Oak Leaf Conservatories, www.oakleafconservatories.com*

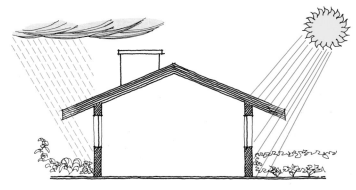

FIGURE 16.18 > Effects of long and short overhangs. *Hepler/Wallach/Hepler © Cengage 2013*

led to the development of many different roof shapes. The conoid, cylindrical parabolic, and hyperbolic parabolic roofs are among the most recent designs.

Roof Overhang

The **overhang** is the portion of the roof that projects past the outside walls. Sufficient roof overhang should be provided to afford protection from the sun, rain, and snow. See Figure 16.18. The length and angle of the overhang greatly affect a roof's appearance and ability to provide protection. Figure 16.19 shows that when the pitch is steep, a larger overhang is needed to provide protection. However, if the overhang of a low-pitch roof is extended to equal the protection of the high-pitch overhang, it may block the view

from the windows. To provide protection and, at the same time, allow sufficient light to enter the windows, slatted overhangs may be used. The roof design shown in Figure 16.20 includes openings that allow light to penetrate while providing extensive roof overhang protection for the structure.

The fascia edge of an overhang does not always need to parallel the sides of a house. See Figure 16.21. The amount of overhang is also determined by architectural style. Large

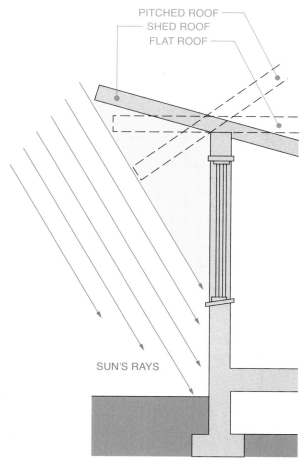

PITCHED ROOF
SHED ROOF
FLAT ROOF

SUN'S RAYS

FIGURE 16.19 > Overhang effect at different angles. *Hepler/ Wallach/Hepler © Cengage 2013*

FIGURE 16.20 > Overhang openings admit additional light. *Robert P. Ruscuak, Photography*

gable end overhangs, such as the 9′ overhang shown in Figure 16.22, must be supported by columns. Overhangs may be enclosed to form a soffit as shown earlier in Figure 16.13. Where overhangs cannot protect a south-facing façade, windows can be recessed through the use of structural elements as shown in Figure 16.23.

Roof Pitch

In designing roofs, the pitch or slope of the roof must be determined. The **slope** is the relation of the horizontal distance (run) to the vertical distance (rise). The **run** is the horizontal distance between the ridge and the outside wall. The **rise** is the vertical distance between the top of the wall and the ridge. The run is always shown in units of 12. Therefore, a slope of 6/12 means the roof rises 6″ for every 12″ of run. The **pitch** is the ratio of the rise over the span. Pitch is covered in more detail in Chapters 17 and 30.

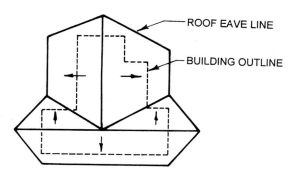

ROOF EAVE LINE

BUILDING OUTLINE

FIGURE 16.21 > Roof outline not aligned with building perimeter. *Hepler/Wallach/Hepler © Cengage 2013*

Dormers

The design of **dormers** greatly affects the silhouette of elevations. Dormer design is influenced by the floor plan outline and by the need for added light, space, and ventilation. Figure 16.24 shows the most common types of residential dormers. Dormer construction is covered in Chapter 30.

FIGURE 16.22 > Large overhang requiring column support.
Hepler/Wallach/Hepler © Cengage 2013

FIGURE 16.23 > Sun-blocking structural design.
BPA Architecture Planning Interiors

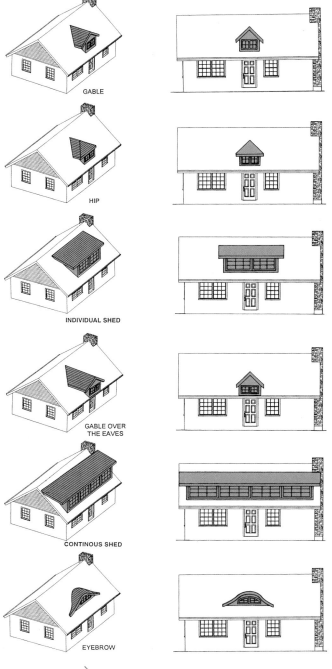

GABLE

HIP

INDIVIDUAL SHED

GABLE OVER
THE EAVES

CONTINOUS SHED

EYEBROW

FIGURE 16.24 > Common dormer types. *Hepler/Wallach/
Hepler © Cengage 2013*

Chimneys

The outline of any structure is influenced by chimney size and shape. Large chimneys create the feeling of power and mass. In Figure 16.25 note how chimney size and shape, combined with roof pitch, overhang, and grade-line differences, can create totally different elevations using the same basic floor plan. Chimney construction details related to furnace exhaust and fireplace heat and smoke controls are covered in Chapter 23. The actual appearance of fireplaces and chimneys cannot be shown on floor plans, only their plan view shape. Interior elevations reveal the details of a fireplace. Exterior elevations show the shape and style of chimneys that are located on an outside wall as shown in Figure 16.26. Common construction and surfacing materials for chimneys are illustrated in Figure 16.27.

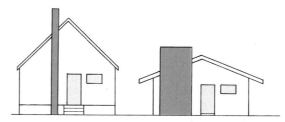

FIGURE 16.25 > Different effects of chimney profile, grade line, and roof pitch on an elevation's appearance. *Hepler/Wallach/Hepler © Cengage 2013*

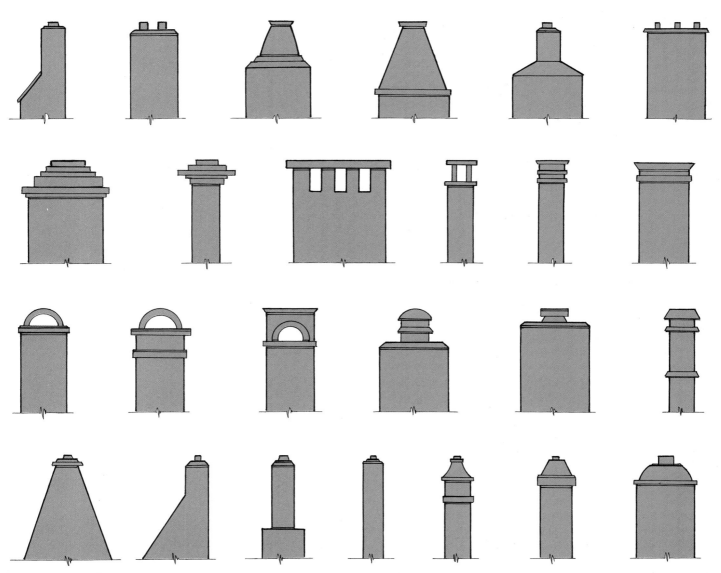

FIGURE 16.26 > Chimney elevation designs. *Hepler/Wallach/Hepler © Cengage 2013*

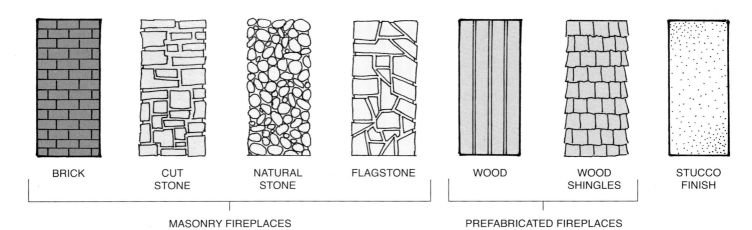

BRICK   CUT STONE   NATURAL STONE   FLAGSTONE   WOOD   WOOD SHINGLES   STUCCO FINISH

MASONRY FIREPLACES   PREFABRICATED FIREPLACES

FIGURE 16.27 > Chimney structure and surface materials. *Hepler/Wallach/Hepler © Cengage 2013*

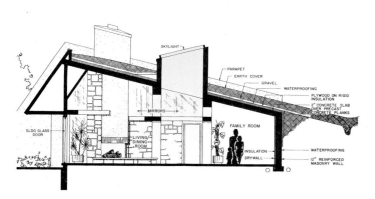

FIGURE 16.28 > Skylights provide added light where needed. *Hepler/Wallach/Hepler © Cengage 2013*

Skylights

Roof windows are commonly called **skylights.** See Figure 16.28. Skylights are either domed or flat. Skylights require a cathedral ceiling unless access walls are built to connect the roof opening with a flat ceiling. Light shafts can be used to transmit light from the roof through an attic or crawl space to a flat ceiling. Figure 16.29 shows several types of light shafts. Manufactured skylight units are available in a variety of shapes (Figure 16.30), but can also be constructed into the roof. Many skylight components allow remote opening capabilities and also provide control of shading devices.

Skylights are available in clear, tinted, and tempered or wire glass. Acrylic skylights are also available in sheet or sandwich panels. Most on-site constructed skylights are made for atrium or conservatory applications as shown in Figure 16.31. The design and location of skylights is directly related to the type of ceiling planned. Figure 16.32 shows the common types of ceilings used in a variety of architectural styles.

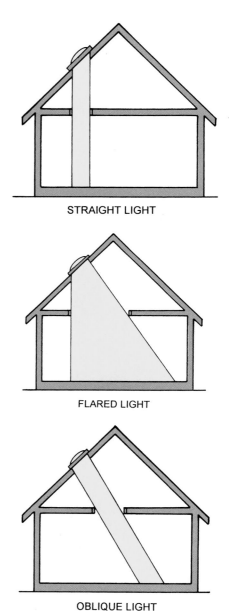

STRAIGHT LIGHT

FLARED LIGHT

OBLIQUE LIGHT

FIGURE 16.29 > Types of light shafts. *Anderson Windows Inc.*

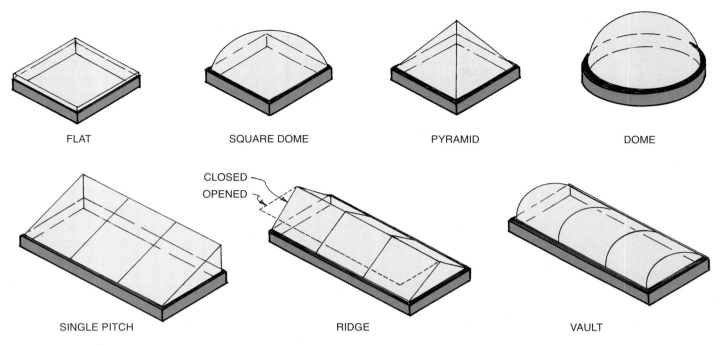

FLAT   SQUARE DOME   PYRAMID   DOME

SINGLE PITCH   CLOSED / OPENED   RIDGE   VAULT

FIGURE 16.30 > Common skylight types. *Hepler/Wallach/Hepler © Cengage 2013*

Decks, Porches, and Balconies

Decks, porches, and balconies are primarily designed with the floor plan. If covered or extended beyond the outside walls of a structure, they can totally change the silhouette of a building. The pictorial elevation view of the site in Figure 16.33 reveals the scope of the total design not apparent in the Figure 16.1 view.

Window Styles

Windows are designed and located to provide light, ventilation, a view, and—in some climates—heat. To accomplish these goals, the size, location, and shape of each window must be planned according to the following guidelines:

1. Relate window lines to the elevation shape, as shown in Figure 16.7, to avoid a tacked-on look.

2. Plan window height to allow for furniture and built-in components that are placed near windows. See Figure 16.34.

3. Plan window sizes to match available standard sizes.

4. Decide which windows need to open for ventilation and which should be fixed.

5. Be sure each window functions from the inside as required.

6. Position windows to access the best views. Avoid window placement that exposes undesirable views. Avoid **mullions** and **muntins** if they restrict views. See Figure 16.35.

FIGURE 16.31 > On-site constructed skylight. *Oak Leaf Conservatories, www.oakleafconservatories.com*

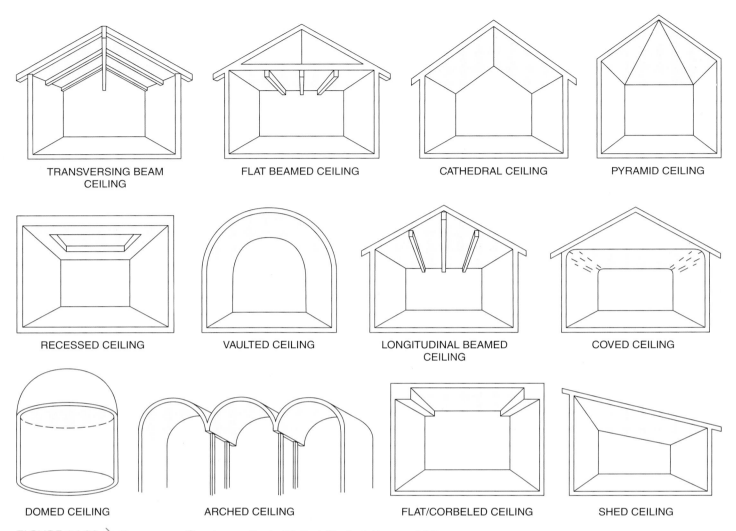

TRANSVERSING BEAM CEILING

FLAT BEAMED CEILING

CATHEDRAL CEILING

PYRAMID CEILING

RECESSED CEILING

VAULTED CEILING

LONGITUDINAL BEAMED CEILING

COVED CEILING

DOMED CEILING

ARCHED CEILING

FLAT/CORBELED CEILING

SHED CEILING

FIGURE 16.32 > Common ceiling types. *Hepler/Wallach/Hepler © Cengage 2013*

FIGURE 16.33 > Elevation view of "Fallingwater." *Thomas A. Heinz, Courtesy of Western Pennsylvania Conservancy*

7. In warm climates, minimize the amount of window space on the south and maximize north-facing windows. Do the reverse in cold climates.

8. Keep the window style consistent with the architectural style of the house.

9. Where possible, align the tops of all windows and doors in each elevation.

10. If the building has more than one level, vertically align the sides of windows for each level where possible.

11. Don't allow small areas between windows and other major features. Balance the wall spaces between windows, doors, and chimneys according to the principles of balance and proportion described in Chapter 2.

12. If windows are to provide the entire light source during daylight hours, 20% of the room's floor area

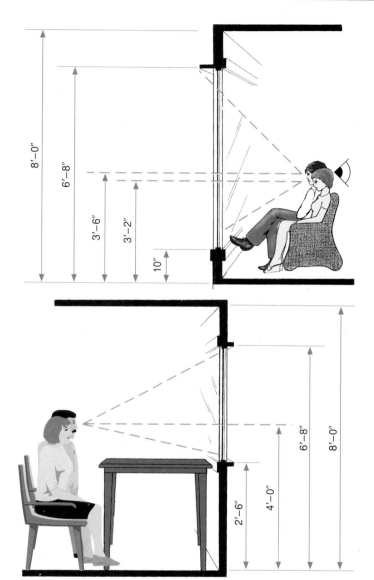

FIGURE 16.34 〉 Window height effect on view. *Small Homes Council*

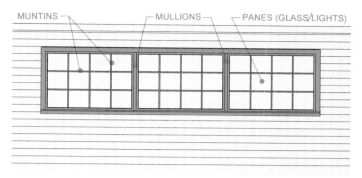

FIGURE 16.35 〉 Mullions, muntins, and panes illustrated.
Hepler/Wallach/Hepler © Cengage 2013

FIGURE 16.36 〉 Integrated door and window wall design.
Marvin Windows & Doors

should be windowed. Ten percent is considered the minimum.

13. Windows that provide ventilation should be located to capture prevailing breezes and provide the best air circulation.

14. If possible, locate windows on more than one wall in each room, to provide for the best distribution of light and ventilation.

15. **Fenestration** is the arrangement of windows or openings in a wall. Arrange fenestration patterns to conform to the elements of design. See Figure 16.36.

Window Types

Windows slide, swing, pivot, or remain fixed. Choosing the right window type for each need requires a knowledge of the function and operation of each type. See Figure 16.37. Figure 16.38 shows the elevation symbols for the most common window types. Fixed windows consist of a frame and a glazed fixed glass. **Casements** are outward-swinging, side-hinged windows, **awnings** are top hinged and called **hopper windows** if hinged at the bottom. **Double-hung windows** have two sashes that move vertically and hold their positions by friction. Louvered, or **jalousie,** windows are similar to awning with smaller segments. Pivoting windows are similar to casement but pivot (hinge) at the center rather than at the

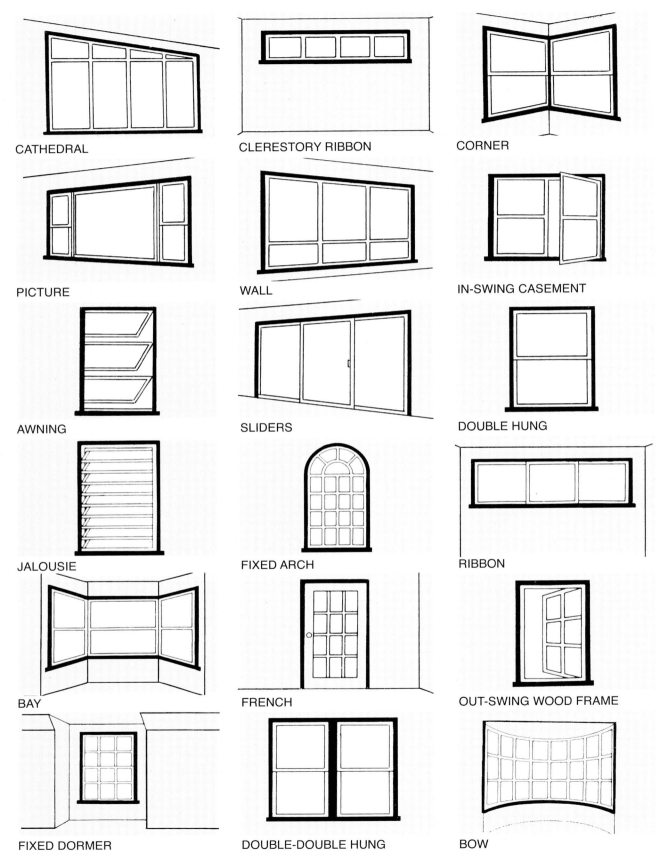

CATHEDRAL

CLERESTORY RIBBON

CORNER

PICTURE

WALL

IN-SWING CASEMENT

AWNING

SLIDERS

DOUBLE HUNG

JALOUSIE

FIXED ARCH

RIBBON

BAY

FRENCH

OUT-SWING WOOD FRAME

FIXED DORMER

DOUBLE-DOUBLE HUNG

BOW

FIGURE 16.37 > Interior view of common window types. *Hepler/Wallach/Hepler © Cengage 2013*

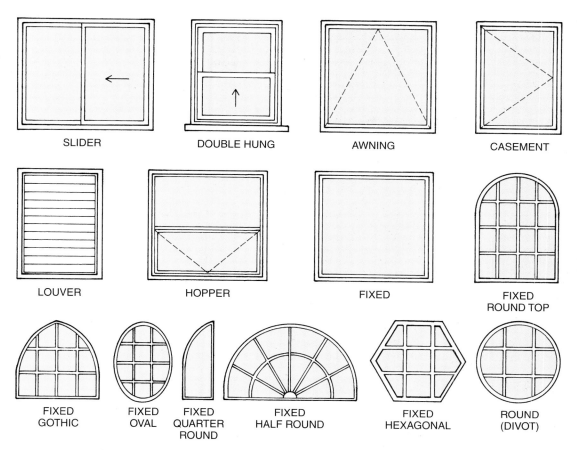

FIGURE 16.38 > Elevation symbols for the common window types. *Hepler/Wallach/Hepler © Cengage 2013*

sides. Sliders may consist of two sashes that slide horizontally, or one sash may be fixed and the other slides.

Door Styles

The style, size, and location of doors do not have as great an effect on elevation design as windows. This is because of the limited options among door sizes and for door locations. Usually an elevation will either have only one door or will contain no doors at all. Nevertheless, the principles of design relative to other elevation features are the same for doors as for windows.

Door types fall into three main categories: exterior, interior, and garage doors. Exterior doors provide security and visual privacy. Interior doors provide privacy and sound control between rooms. Figure 16.39 shows an elevation view of the most common types of interior and exterior doors. Light construction doors fall into four main categories; solid wood, wood **flush** (hollow core), wood rail and stile, and hollow metal doors (steel or aluminum). These are divided into types such as **panel doors,** solid surface, full glass **(French door)** or Multi-lite, double (Dutch), louvered, or screened. Common garage doors sizes range in height from 6′-6″ to 8′-0″ at 6″ intervals. Widths range from 8′-0″ to 18′-0″ at 1′ intervals.

Detailed information about doors is contained in a door schedule and cross-referenced to floor plans and elevations. Door framing information is presented in Chapter 35.

FIGURE 16.39 › Common exterior door styles. *Hepler/Wallach/Hepler © Cengage 2013*

Designing Elevations Exercises   CHAPTER 16

1. Sketch an elevation of your own design. Trace the elevation with in a flat roof, gable roof, shed roof, and butterfly roof. Choose the one you like best and the one that is most functional for your design. Explain why you made that choice.

2. Sketch the front elevation of your home or a home you like. Change the roof style, but keep it consistent with the major lines of the elevation. Move or change the doors and windows to improve the design.

3. Redesign the elevation from Exercise 2, moving the doors and windows and changing the materials. Be sure the door and window lines relate to the major lines of the building.

4. Collect pictures of roofs, windows, and doors that you particularly like. Try to identify house styles for which they are best suited.

CHAPTER 17

Drawing Elevations

OBJECTIVES

In this chapter you will learn:

> to follow steps to project elevations from a floor plan and complete an elevation drawing.

> to mathematically establish the pitch of a roof.

> symbols used on elevations.

> to draw accurately scaled and dimensioned elevations.

> shading and rendering techniques to use on elevations.

> the concept of Building Information Modeling

TERMS

auxiliary elevation
building information modeling (BIM)
ceiling lines
datum line
exterior elevation drawings
finished dimensions

floor lines
foreshortened
framing dimensions
ground lines
interior elevation drawings
orthographic projection

pitch
presentation drawings
profile drawings
slope diagram
span
true size

INTRODUCTION

The main features of the outside of a building are shown on elevation drawings. **Exterior elevation drawings** are two-dimensional orthographic representations of the exterior of a structure. These drawings are prepared to show the design, materials, dimensions, and final appearance of the structure's exterior components. In a building, these components include doors, windows, the surfaces of the sides, and the roof. Interior as well as exterior elevation drawings are projected from floor plans. Dimensions are used to show sizes, and elevation symbols are used to indicate various features on the drawings.

ELEVATION PROJECTION

In **orthographic** (multiview) **projection,** related views of an object are shown as if they were on a two-dimensional, flat plane. To visualize and understand orthographic projection, imagine a building surrounded by a transparent box, as shown in Figure 17.1. If you draw the outline of the structure

on the transparent planes that make up the box, you can create several orthographic views. For example, the front view is on the front plane, the side view on the side plane, and the top view on the top (horizontal) plane.

If the planes of the top, bottom, and sides were hinged and swung out away from the box, as shown in Figure 17.2, six views of the house would be created. Note how each view is positioned on an orthographic drawing. Study the position of each view as it relates to the front view. The right side is to the right of the front view. The left side is to the left, the top (roof) view is on the top, and the bottom view is on the bottom. The rear view is placed to the left of the left-side view, since, if this view were hinged around to the back, it would fall into this position.

Notice that the length of the front view, top (roof) view, and bottom view are exactly the same as the length of the rear view. Notice also that the heights and alignments of the front view, right side, left side, and rear view are the same.

As with all orthographic drawings, surfaces that recede at an angle other than 90° from the projection plane are **foreshortened.** That is, the distance between the two ends is not

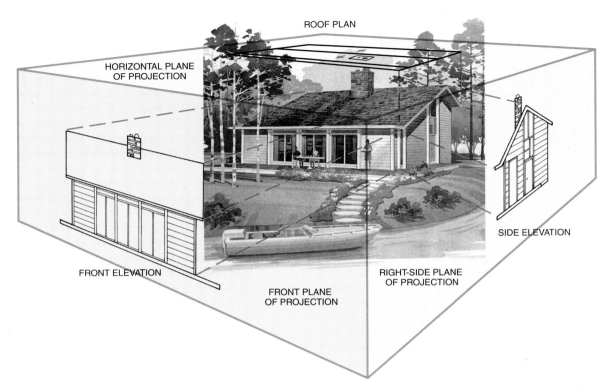

FIGURE 17.1 ⟩ Visualizing elevation planes through a projection box. *Used by permission Hanley Wood, LLC*

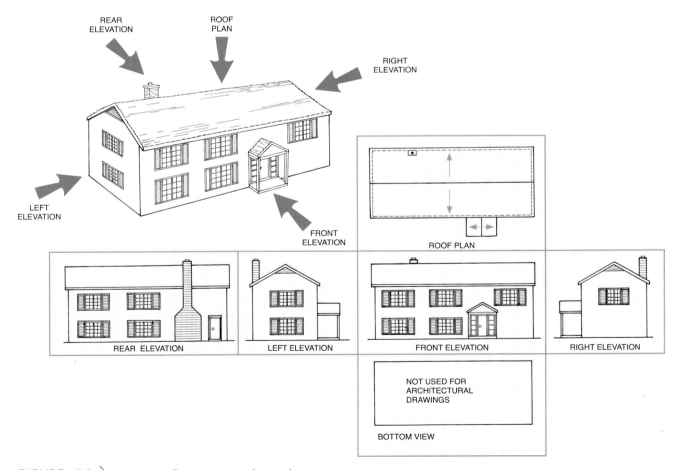

FIGURE 17.2 ⟩ Open and flat projection box sides. *Used by permission Hanley Wood, LLC*

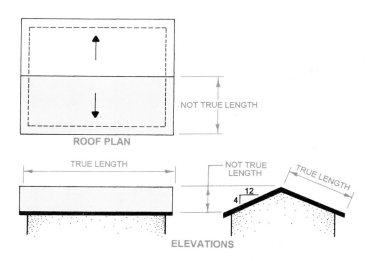

FIGURE 17.3 > True and foreshortened surface views.
Hepler/Wallach/Hepler © Cengage 2013

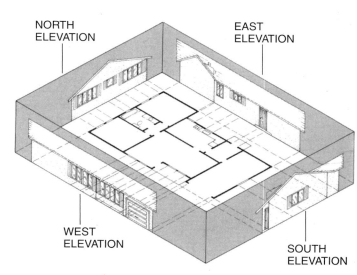

FIGURE 17.4A > Floor plan projected to elevation planes.
Hepler/Wallach/Hepler © Cengage 2013

the true size. Figure 17.3 shows how the vertical and horizontal roof surface distances from roof ridge to eave are not true on the top and front views, but are true on the side view.

All six views are rarely used to depict architectural structures. Instead, only four elevations (sides) are usually shown. The top roof view is used to create roof plans. The bottom view of a floor is not developed. Instead the foundation underneath the structure is described by foundation plan and construction sections.

Figure 17.4A shows how elevations are projected from the floor plan. The positions of the chimney, doors, windows, overhang, and building corners are projected directly from the floor plan to the elevation plane. The four elevation drawings shown in Figure 17.4B illustrates the relationship between elevation views and the floor plans as also shown in Figure 15.31. Observe how the common features such as the dormer, chimney, eave, and ridge lines appear when viewed from different sides of the structure.

Elevation drawings are used to show the design of the finished appearance of a structure. Elevations are drawn to an exact scale, usually the same as the floor plan. Elevations accurately represent all height dimensions that are not shown on floor plans. The styles of windows, doors, and siding are also indicated on elevation drawings. The vertical position of all horizontal planes, such as **ground lines, floor lines, ceiling lines,** deck or patio lines, and roof lines, is only revealed on elevation drawings. Lines below the ground line such as foundation and footing lines are drawn with dashed lines.

Only through the use of elevation drawings can the vertical relationship of buildings be visualized. For example, on the site profile in Figure 17.5, heights are shown by elevation notations. However, little height detail is apparent without elevation drawings. Elevation drawings of a site are known as **profile drawings.** These drawings show a section cut through the terrain. The site profile drawing in Figure 17.5 on page 358, is cross-referenced to a callout cutting-plane line on a site plan, which shows the identification and location of this drawing as noted.

PROJECTING ELEVATIONS

Think of an elevation as a drawing placed on a flat, vertical plane. Figure 17.6 on page 358, shows how a vertical plane is related to and projected from a floor plan. To relate the four elevations of a building to an orthographic form, observe the configuration of views shown in Figures 17.7A through 17.7D. Figure 17.7A on page 359, shows the front elevation and Figure 17.7B on page 359, shows the right-side elevation of the plan in Figure 15.34 and 15.35. Figures 17.7C on page 359, and 17.7D on page 360, show the left-side and rear elevations, respectively, of this plan. To better observe and understand the relationship of these elevations, observe how the picture in Figure 17.8A on page 360, shows the areas covered by the front and right-side elevations. Figure 17.8B on page 360, also provides an interior picture to help orient and visualize several exterior elevations as related to the chimney location and dimensions.

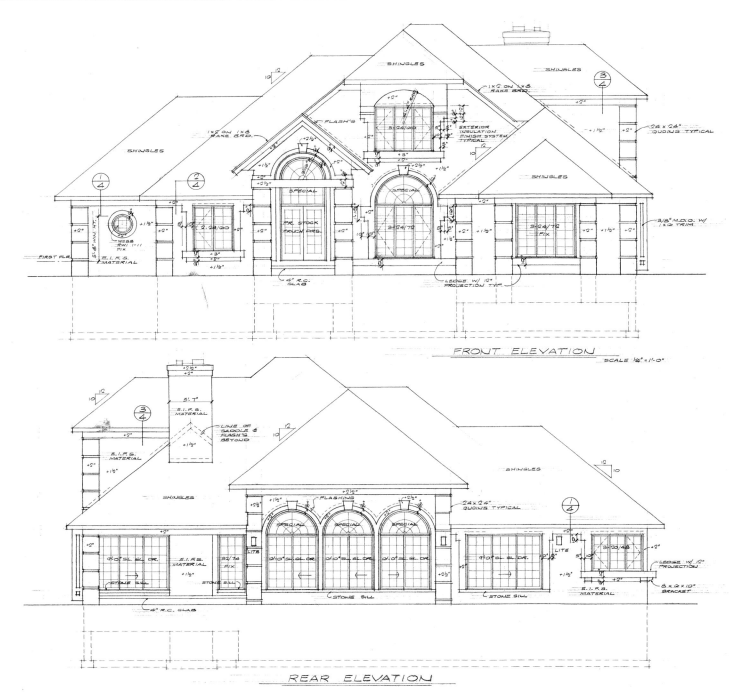

FIGURE 17.4B > Four elevations revealing common features. *Used by permission Hanley Wood, LLC*

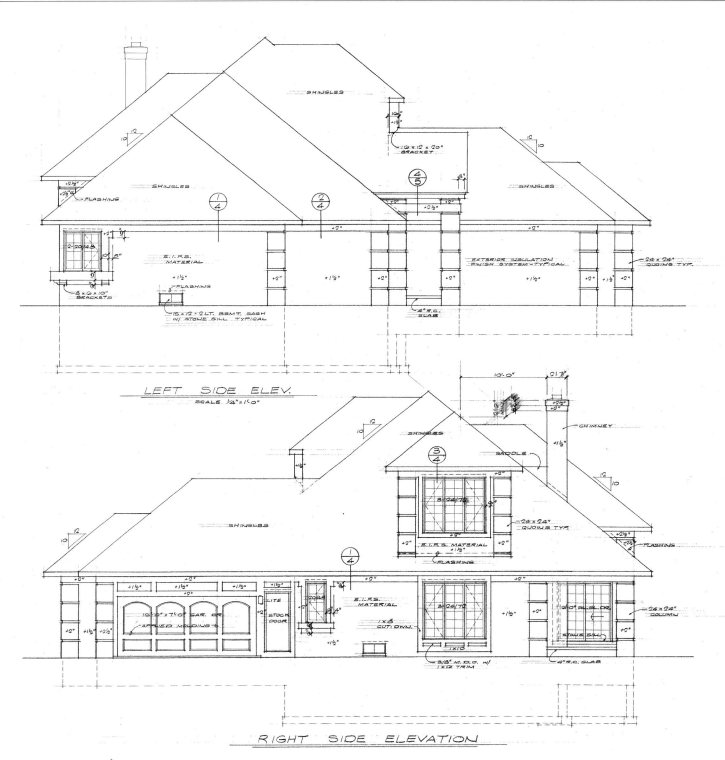

LEFT SIDE ELEV.
SCALE ¼"=1'-0"

RIGHT SIDE ELEVATION

FIGURE 17.4B > Continued

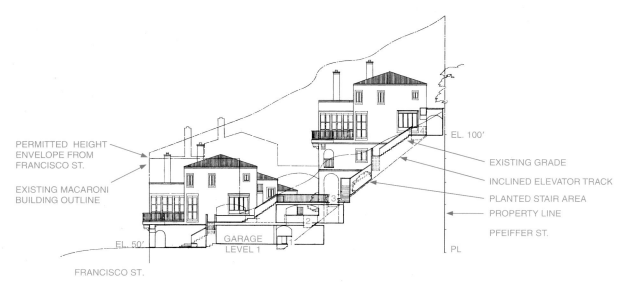

PERMITTED HEIGHT
ENVELOPE FROM
FRANCISCO ST.

EXISTING MACARONI
BUILDING OUTLINE

EL. 100'

EXISTING GRADE

INCLINED ELEVATOR TRACK

PLANTED STAIR AREA

PROPERTY LINE

PFEIFFER ST.

PL

EL. 50'

GARAGE
LEVEL 1

FRANCISCO ST.

FIGURE 17.5 > Site profile. *BAR Architects, Inc.*

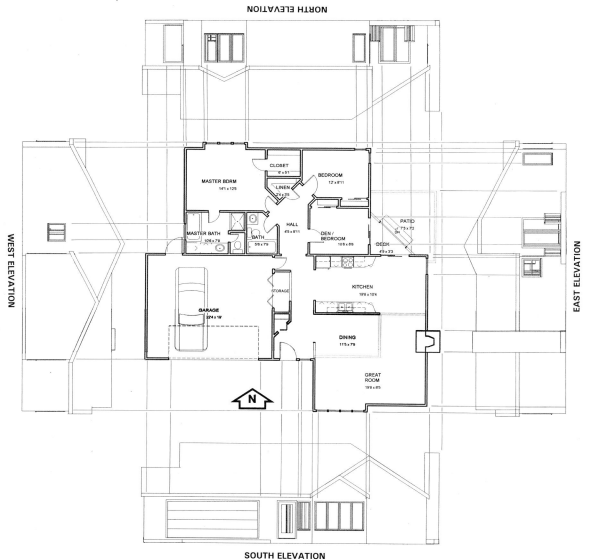

NORTH ELEVATION

WEST ELEVATION

EAST ELEVATION

SOUTH ELEVATION

MASTER BDRM
14'1 x 12'5

CLOSET
8' x 5'1

LINEN
2'4 x 3'5

BEDROOM
12' x 8'11

MASTER BATH
10'6 x 7'9

BATH
5'6 x 7'9

HALL
4'5 x 8'11

DEN/
BEDROOM
10'6 x 8'8

PATIO
7'3 x 7'2
DN

DECK
4'9 x 3'3

STORAGE

GARAGE
22'4 x 19'

KITCHEN
19'8 x 10'4

DINING
11'5 x 7'9

GREAT
ROOM
19'8 x 8'5

N

FIGURE 17.6 > Projection of floor plan features to elevation drawings. *Hepler/Wallach/Hepler © Cengage 2013*

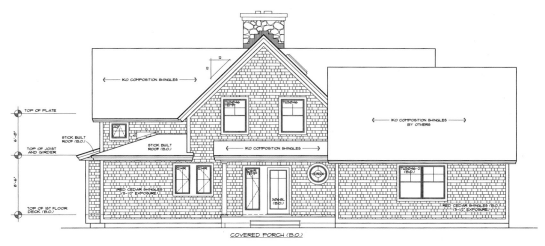

FRONT

FIGURE 17.7A 〉 Front elevation. *Timberpeg Inc.*

RIGHT SIDE

FIGURE 17.7B 〉 Right-side elevation. *Timberpeg Inc.*

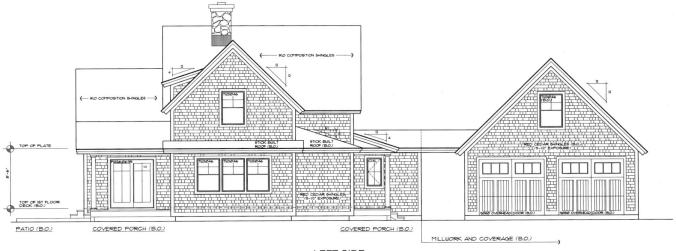

LEFT SIDE

FIGURE 17.7C 〉 Left-side elevation. *Timberpeg Inc.*

FIGURE 17.7D 〉 Rear elevation. *Timberpeg Inc.*

FIGURE 17.8A 〉 View of the front and left elevation planes. *Timberpeg Inc.*

Orientation

Four elevations are normally projected by extending lines outward from each wall of the floor plan. As discussed earlier, these elevations are classified according to their location and are called the front, rear (or back), right, and left elevation. When these elevations are projected on the same drawing sheet, the rear elevation appears to be upside down and the right and left elevations appear to rest on their sides. See Figure 17.9. Because of the large size of most combined floor plan and elevation drawings, and because of the need to show elevations as normally seen, the elevation drawing is rotated so each elevation can be drawn with the ground line on the bottom. See Figure 17.10.

FIGURE 17.8B 〉 Chimney view to orient elevations. *Timberpeg Inc.*

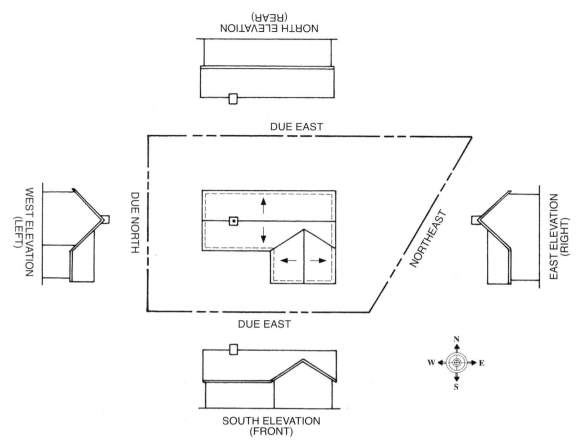

FIGURE 17.9 > Compass and building orientation of elevations. *Hepler/Wallach/Hepler © Cengage 2013*

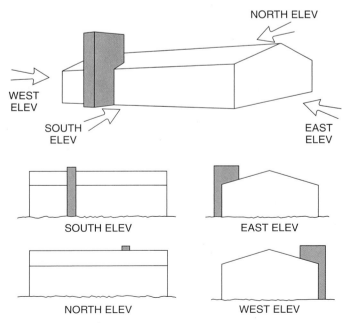

FIGURE 17.10 > Ground-line positioning of elevations.
Hepler/Wallach/Hepler © Cengage 2013

The north, east, south, and west compass points are often used to describe and label elevation drawings. This method is preferred because it reduces the chance of elevation callout error. When this method is used, the north arrow on the floor plan or site plan is the key. For example, in Figure 17.9, the rear elevation is facing north. Therefore, the rear elevation is also the north elevation. The front elevation is the south elevation, the left elevation is the west elevation, and the right elevation is the east elevation. Figure 17.10 shows the alignment of elevations by compass orientation.

When elevations do not align exactly with the four major compass points, a split compass reading may be used. Figure 17.11 shows a split labeled "Southwest Elevation." This illustration also shows the relationship of the elevation drawing to the construction process and to the finished building.

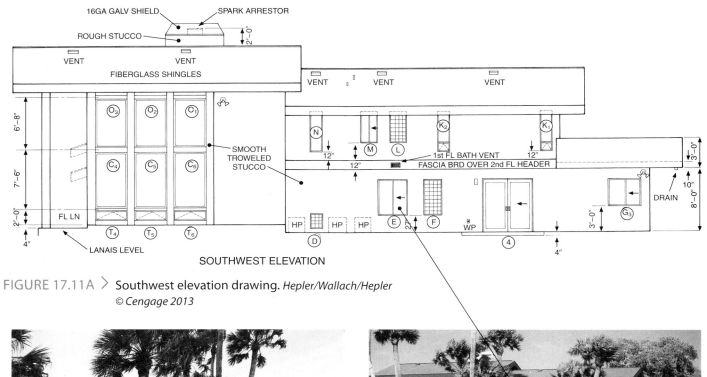

FIGURE 17.11A 〉 Southwest elevation drawing. *Hepler/Wallach/Hepler © Cengage 2013*

FIGURE 17.11B 〉 Southwest view of construction. *Diane Kingston, Photographer*

FIGURE 17.11C 〉 Southwest view of completed structure. *Hepler/Wallach/Hepler © Cengage 2013*

The elevation drawing in Figure 17.12 shows the opposite (northeast) side of the house in Figure 17.11. Located at a right angle to the southwest elevation is the southeast elevation. A partial southeast elevation is shown in Figure 17.13. *Part One* is the southeast elevation and *Part Two* shows the completed structure as viewed from the southeast. Notice a major design feature that has been changed between the drawing and the completed structure.

In the same way that the southwest compass direction is the opposite of the northeast direction, the southeast elevation is the opposite (180°) of the northwest elevation. A partial northwest elevation is shown in Figure 17.14. To better understand the relationship among elevation views, observe the common chimney, roof, and balcony lines. Study Figure 13.51 to see the site plan and views of the finished structure. Also observe the indexing of these elevations to the floor plans in Figure 15.28.

Unique buildings or buildings with atypical wall and roof construction, as shown in Figure 17.15, require very detailed elevation drawings. In addition to the four exterior elevations (Figure 17.16), a very detailed elevation is provided as shown in Figure 17.17. Note the related floor plan detail provided in Figure 17.18.

Auxiliary Elevations

In such cases some lines and surfaces on the elevations may appear shortened because of the receding angles. An **auxiliary elevation** view may then be necessary to clarify the **true size** of the elevation where walls are not perpendicular. To project an auxiliary elevation, follow the same projection procedures as for other elevation drawings. When an auxiliary elevation is drawn, it is prepared in addition to—and does not replace—other standard elevation drawings. Where inclined

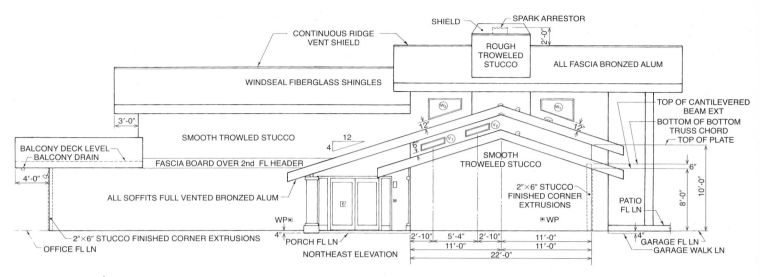

CONTINUOUS RIDGE VENT SHIELD

SHIELD

SPARK ARRESTOR

ROUGH TROWELED STUCCO

ALL FASCIA BRONZED ALUM

WINDSEAL FIBERGLASS SHINGLES

SMOOTH TROWLED STUCCO

TOP OF CANTILEVERED BEAM EXT

BOTTOM OF BOTTOM TRUSS CHORD

TOP OF PLATE

BALCONY DECK LEVEL
BALCONY DRAIN

FASCIA BOARD OVER 2nd FL HEADER

SMOOTH TROWELED STUCCO

2"×6" STUCCO FINISHED CORNER EXTRUSIONS

ALL SOFFITS FULL VENTED BRONZED ALUM

PATIO FL LN

WP

WP

2"×6" STUCCO FINISHED CORNER EXTRUSIONS

OFFICE FL LN

PORCH FL LN

NORTHEAST ELEVATION

GARAGE FL LN
GARAGE WALK LN

W_2 W_1 V_2 V_1

12
4

3'-0" 4'-0" 4" 2'-10" 5'-4" 2'-10" 11'-0" 11'-0" 11'-0" 11'-0" 22'-0" 6" 8'-0" 10'-0" 4" 12" 6"

FIGURE 17.12 > Northeast elevation. *Hepler/Wallach/Hepler © Cengage 2013*

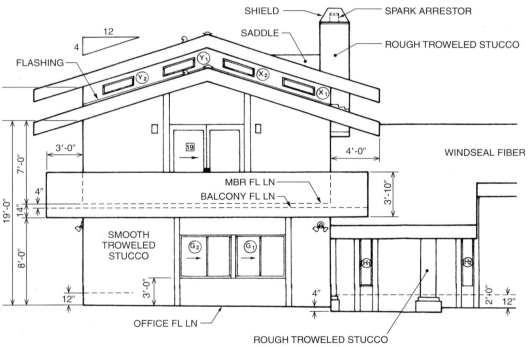

SHIELD

SPARK ARRESTOR

SADDLE

ROUGH TROWELED STUCCO

FLASHING

WINDSEAL FIBER

SMOOTH TROWELED STUCCO

MBR FL LN
BALCONY FL LN

OFFICE FL LN

ROUGH TROWELED STUCCO

12
4

Y_1 Y_2 X_2 X_1

19 G_2 G_1 H_1 H_2

19'-0" 7'-0" 4" 14" 8'-0" 12" 3'-0" 3'-0" 4'-0" 3'-10" 4" 2'-0" 12"

FIGURE 17.13 > Top: southeast elevation; Bottom: completed southeast side. *Hepler/Wallach/Hepler © Cengage 2013*

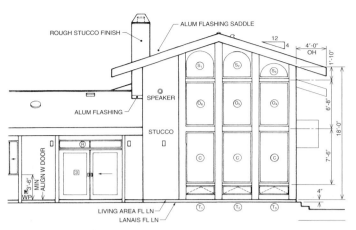

FIGURE 17.14 > Northwest elevation. *Hepler/Wallach/Hepler*
© *Cengage 2013*

FIGURE 17.15 > Conservatory requiring detailed
elevation. *Oak Leaf Conservatories,*
www.oakleafconservatories.com

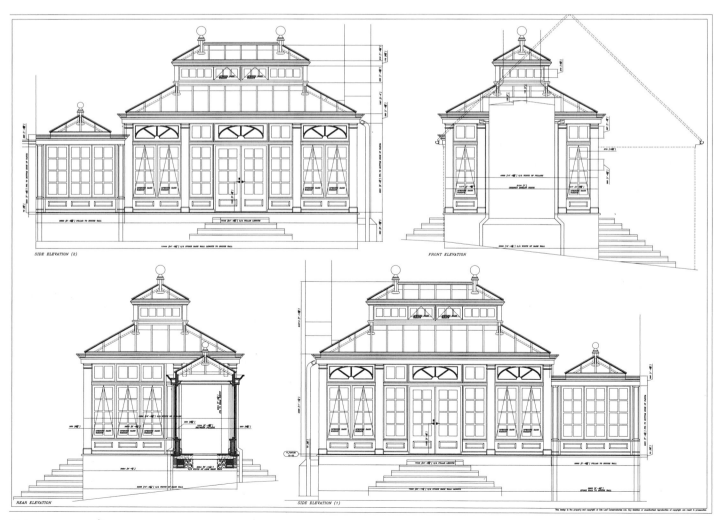

FIGURE 17.16 > Exterior elevations of Figure 17.15. *Oak Leaf Conservatories, www.oakleafconservatories.com*

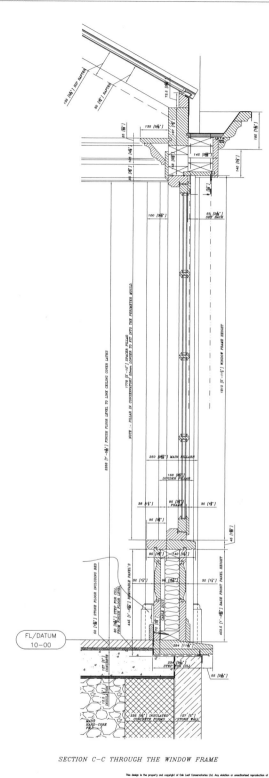

SECTION C-C THROUGH THE WINDOW FRAME

FIGURE 17.17 > Dimensioned elevation section of Figure 17.15. *Oak Leaf Conservatories, www.oakleafconservatories.com*

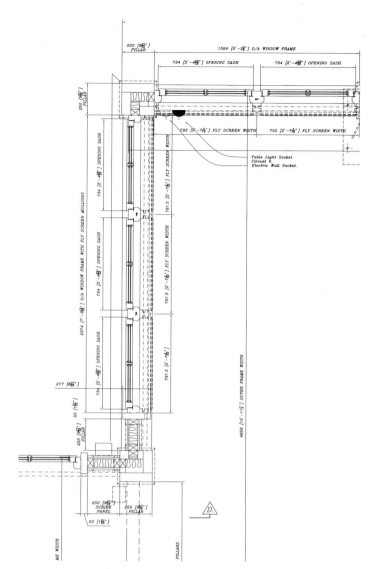

FIGURE 17.18 > Related floor plan details for Figure 17.15. *Oak Leaf Conservatories, www.oakleafconservatories.com*

surfaces intersect more than one right-angled elevation, as in Figure 17.19, several auxiliary elevations may be needed to show the true length of all surfaces. Auxiliary elevations are also used to describe vertical design features of separate structures that cannot be shown on plan views. Figure 17.20 shows a separate elevation drawing that details the gate pictured in Figure 13.51D.

Elevations and Construction

Framing elevations show the position, type, and size of members needed for constructing the framework of a structure. When these are not prepared, builders rely solely on exterior and interior elevations for the height of framing members. This means that precise dimensions on the elevation drawings are crucial for accurate construction.

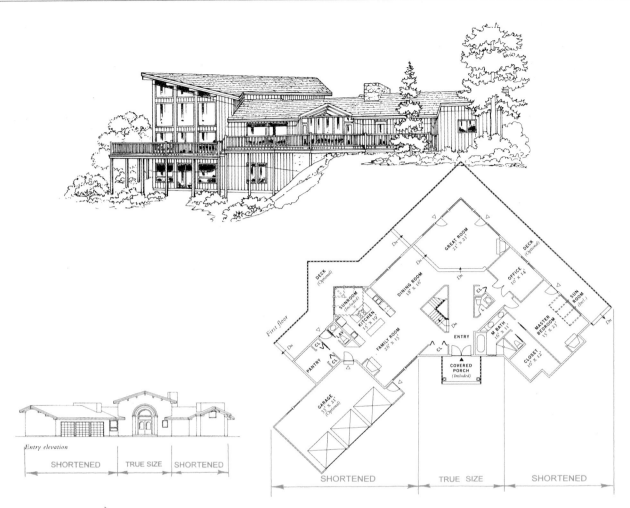

FIGURE 17.19 > Auxiliary view with true and foreshortened distances. *Photo courtesy Lindal Cedar Homes. Lindal.com*

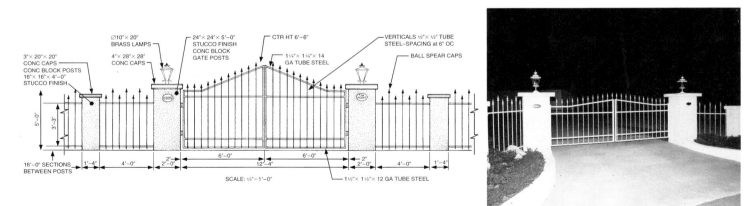

FIGURE 17.20A > Dimensioned elevation of gate and fence. *Hepler/ Wallach/Hepler © Cengage 2013*

FIGURE 17.20B > Completed gate from Figure 18.37D. *Hepler/ Wallach/Hepler © Cengage 2013*

Steps in Projecting Elevations

The major lines of an elevation drawing are derived by projecting vertical lines from the floor plan to the elevation drawing plane and measuring the position of horizontal lines from the ground line. To develop an elevation drawing that exactly reflects the features of a floor plan, refer to Figure 17.21 and follow these steps:

STEP 1   Using the floor plan, *project the vertical lines* that represent the main lines of the building. These lines show the overall length or width of the building. They also show the width of doors, windows, and the major parts or offsets of the building.

       When projecting an elevation on a CAD system, use the *Grid* function to project the major lines from the floor plan to the elevation plane. During the drawing process, floor plans can be rotated 90° to position each elevation with the ground line on the bottom during the drawing process. High-end architectural software can create elevations from floor plans automatically when height dimensions are properly entered.

STEP 2   *Measure and draw horizontal lines* that represent the height of the ground line, footing, doors, tops and bottoms of windows, chimney, siding, breaks, planters, and other key features. To eliminate the repetition of measuring each of these lines for each elevation, a sheet showing the scaled lines is often prepared. See Figure 17.22.

STEP 3   *Complete the basic elevation.* First, develop the roof elevation projection. To determine the height of the eave and ridge line, the roof slope (angle, roof pitch) must be established. A high-slope roof has a greater distance between the ridge line and the eave line than does a low-slope roof. See Figure 17.23. **Pitch** is the angle of the roof described in terms of the ratio of the rise over the span (rise/span). Span is the horizontal distance covered by a roof. Rise is the vertical distance. The run is always expressed in units of 12. The span is the run doubled. Therefore, the span is always expressed in units of 24.

       After the pitch has been established, a **slope diagram** must be drawn on the elevation, as shown in Figure 17.24. The slope diagram is developed on the working drawings by the drafter. The carpenter must work with the pitch fraction (ratio) to determine the angle of the rafters from a pitch angle table, so the ends of the rafters can be correctly cut. Double the run to find the span. The

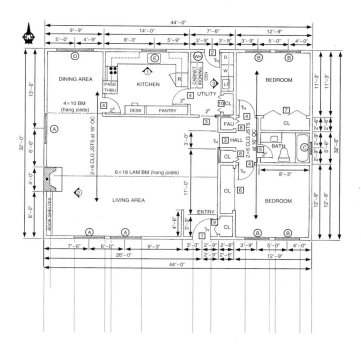

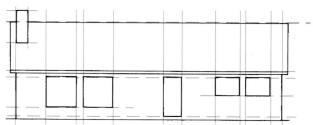

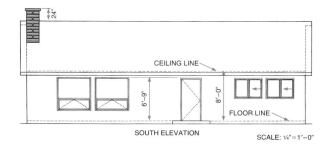

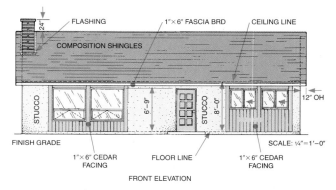

FIGURE 17.21 > Steps in projecting and drawing elevations. *Hepler/Wallach/Hepler © Cengage 2013*

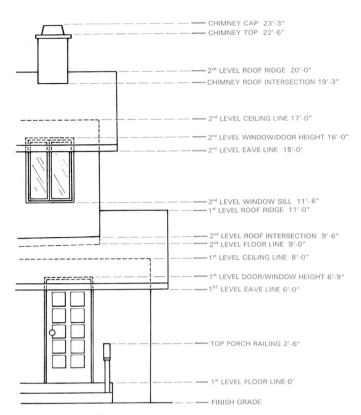

FIGURE 17.22 〉 Key elevation height lines. *Hepler/Wallach/ Hepler © Cengage 2013*

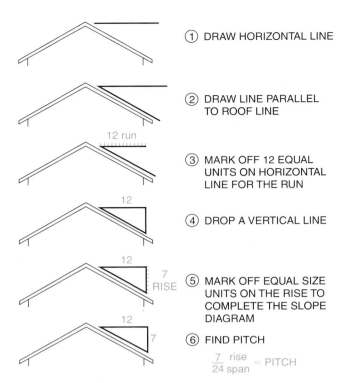

① DRAW HORIZONTAL LINE

② DRAW LINE PARALLEL TO ROOF LINE

12 run

③ MARK OFF 12 EQUAL UNITS ON HORIZONTAL LINE FOR THE RUN

12

④ DROP A VERTICAL LINE

12

7
RISE

⑤ MARK OFF EQUAL SIZE UNITS ON THE RISE TO COMPLETE THE SLOPE DIAGRAM

12

7

⑥ FIND PITCH

$\dfrac{7}{24}$ $\dfrac{\text{rise}}{\text{span}}$ = PITCH

FIGURE 17.24 〉 Steps in drawing a roof slope diagram. *Hepler/Wallach/Hepler © Cengage 2013*

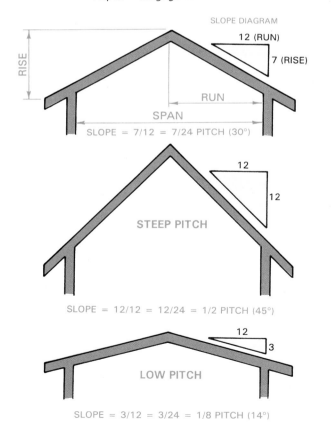

SLOPE DIAGRAM

12 (RUN)

7 (RISE)

RISE

RUN

SPAN

SLOPE = 7/12 = 7/24 PITCH (30°)

12

12

STEEP PITCH

SLOPE = 12/12 = 12/24 = 1/2 PITCH (45°)

12

3

LOW PITCH

SLOPE = 3/12 = 3/24 = 1/8 PITCH (14°)

FIGURE 17.23 〉 Roof pitch description. *Hepler/Wallach/Hepler © Cengage 2013*

span is the distance between the supports of the roof. It is a constant of 24. Place the rise over the span (24) and reduce if necessary. This fraction available in charts is used by carpenters to determine the rafter angle in degrees.

A roof elevation can be projected from a roof plan. See Figure 17.25. Note that the end of every eave and every valley and ridge intersection are projected at a right angle to the plan view outline. Figure 17.26 shows a comparison of X, Y, and Z roof dimensions between elevation views of common roof types.

STEP 4   *Establish the intersection of all vertical and horizontal lines,* including the eave and ridge line. These represent the outline of all features to be shown on the elevation. After they are established, darken the lines to identify the position of each.

STEP 5   *Add details and symbols,* such as indicating door and window trim, mullions, muntins, siding, and roofing materials. The amount of detail included on an elevation depends on the scale and on whether separate details are to be prepared. Figure 17.27 shows two extremes in drawing cornice molding.

STEP 6   *Add final dimensions, labels, and notes.*

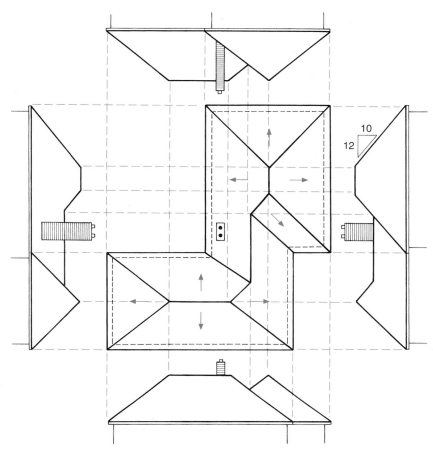

FIGURE 17.25 › Relation of roof plan and roof elevations. *Hepler/Wallach/ Hepler © Cengage 2013*

Many different elevation styles can be projected from one floor plan. The roof style, pitch, overhang, grade level, windows, chimney, and doors can all be manipulated to create different effects.

Exterior Elevation Projection on CAD

Projection (*Construction*) lines are used to project an elevation drawing from a floor plan by creating a projection line layer using the *Draw* and then the *Construction Line* commands. First the horizontal lines of the elevation must be drawn with construction lines. These represent the location of all elevation heights above grade. Then key intersections from the floor plan are located with the cursor and projected to each related horizontal line. These represent the sides of window, door, chimney, and other offsets on the plan. A heavier or different color line task is then used to connect the major lines of the elevation over the construction line layer. Then the construction lines are removed. Elevation features that should align can be automatically aligned using the *Align* command.

ELEVATION SYMBOLS

Symbols are needed to clarify and simplify elevation drawings. They help to describe the basic features of an elevation. Some symbols identify door and window styles and positions. Standardized patterns of dots, lines, and shapes show the types of building materials used on exterior walls. These kinds of symbols on an elevation make the drawing appear more realistic.

Material Symbols

Most standard architectural symbols resemble the material they represent. However, in many cases the symbol does not show the exact appearance of the material. For example, the symbol for brick does not include all of the lines shown in a pictorial drawing. Representing brick on an elevation drawing exactly as it appears is a long, laborious, and unnecessary process. Many elevation symbols appear as if the material were viewed from a distance.

When using a CAD system, elevation symbols such as doors and windows can be stored in symbol libraries.

PROJECTION AND RELATIONSHIP FOR ELEVATIONS AND ROOF PLANS

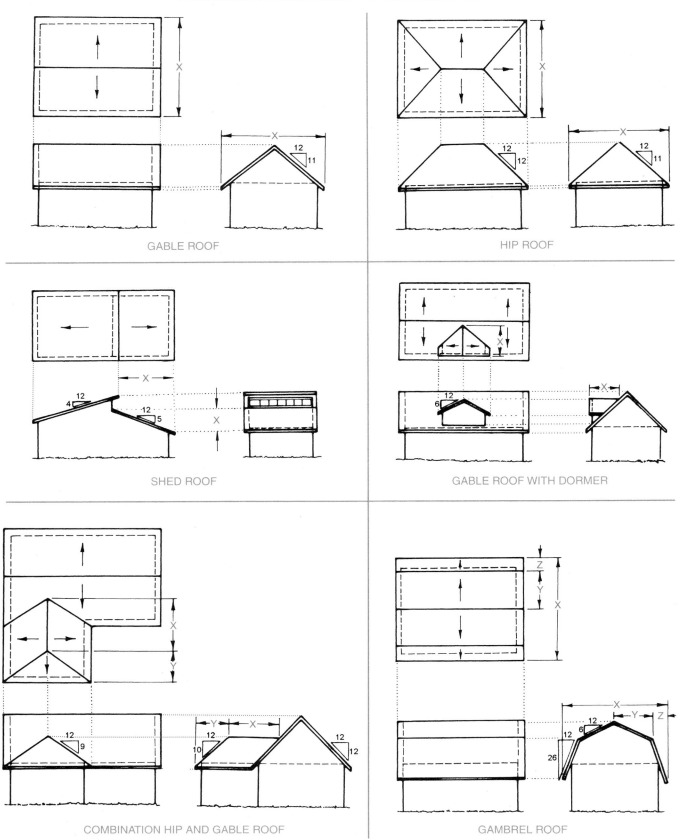

FIGURE 17.26 ⟩ Related distances on roof style elevations. *Hepler/Wallach/Hepler © Cengage 2013*

For example, the *Material Symbols* function can be used to add siding material symbols on elevation surfaces. See Figure 17.28.

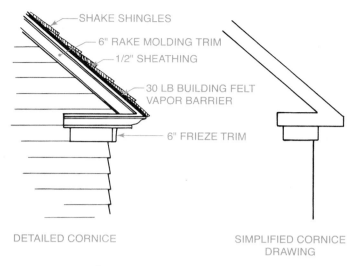

DETAILED CORNICE

SHAKE SHINGLES
6" RAKE MOLDING TRIM
1/2" SHEATHING
30 LB BUILDING FELT VAPOR BARRIER
6" FRIEZE TRIM

SIMPLIFIED CORNICE DRAWING

FIGURE 17.27 > Detailed and simplified cornice drawing. *Hepler/Wallach/Hepler © Cengage 2013*

Ecological Elevation Materials

Elevation drawings show exterior and interior wall covering material, door, window, roofing, and surface finishes. Materials selected for these components should be toxin free, biodegradable, natural, and recyclable if possible.

Window Symbols

The position and style of windows greatly affect the appearance of elevations. Windows are, therefore, drawn on the elevation with as much detail as the scale of the drawing permits. Parts of windows that should be shown on all elevation drawings include the sill, sash, mullions, and muntins, if any. See Figures 17.29 and 15.19.

In addition to showing the parts of a window, it is also necessary to show the direction of the hinge for casement and awning windows. Dotted lines are used on elevation drawings, as shown in Figure 17.29. The point of the dashed line shows the part of the window to which the hinge is attached.

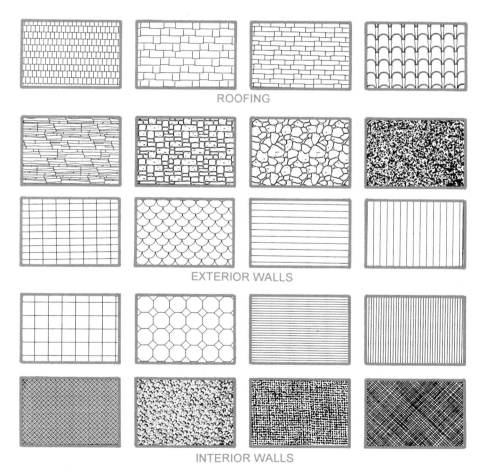

ROOFING

EXTERIOR WALLS

INTERIOR WALLS

FIGURE 17.28 > Common elevation material symbols. *Hepler/Wallach/Hepler © Cengage 2013*

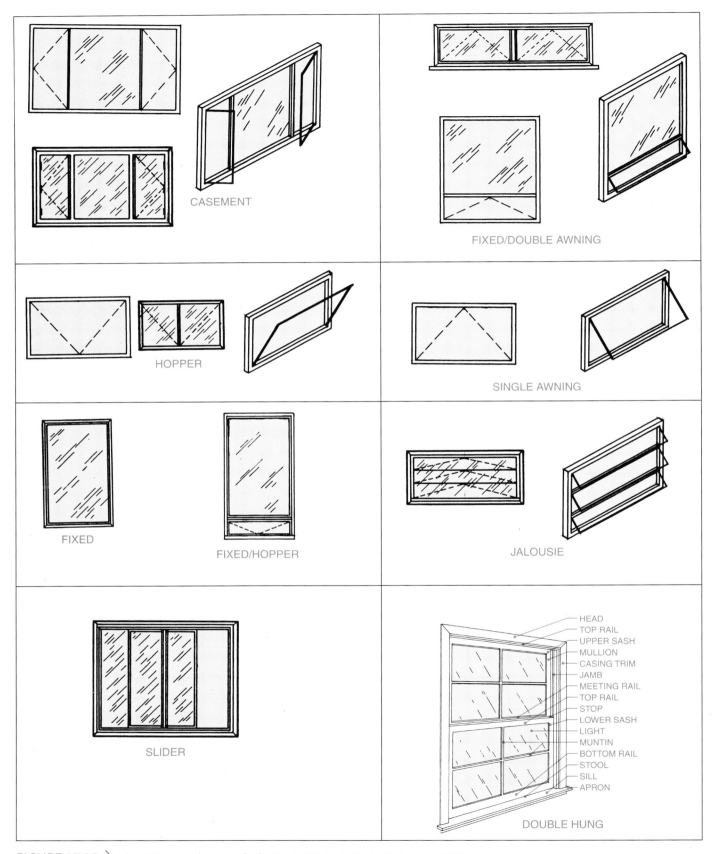

CASEMENT

FIXED/DOUBLE AWNING

HOPPER

SINGLE AWNING

FIXED

FIXED/HOPPER

JALOUSIE

SLIDER

HEAD
TOP RAIL
UPPER SASH
MULLION
CASING TRIM
JAMB
MEETING RAIL
TOP RAIL
STOP
LOWER SASH
LIGHT
MUNTIN
BOTTOM RAIL
STOOL
SILL
APRON

DOUBLE HUNG

FIGURE 17.29 > Elevation window symbols. *Hepler/Wallach/Hepler © Cengage 2013*

COMPLETED WINDOW DETAIL –
ONE DRAWING FOR EACH
TYPE OF WINDOW USED
ON THE STRUCTURE

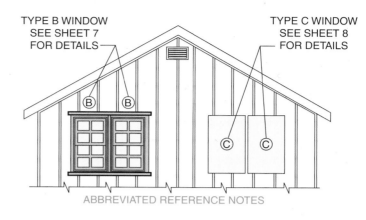

FIGURE 17.30 ❯ A single window detail may be prepared to show a window style. *Hepler/Wallach/Hepler © Cengage 2013*

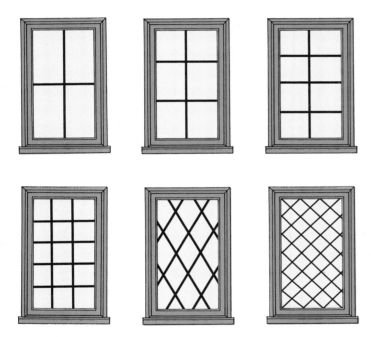

FIGURE 17.31A ❯ Common muntin grid designs. *Hepler/ Wallach/Hepler © Cengage 2013*

Many different styles of windows are available. Refer back to Chapter 16. These illustrations show the normal amount of detail used in drawing windows on elevations. An alternative method of showing window styles on elevation drawings involves including just one window detail for each style on the plan drawn to a larger scale. See Figure 17.30. When the elevation drawing is prepared, the size and outlined position of the window are shown with a letter or number to refer to a detail drawing. This detail drawing is also indexed to a window schedule that contains complete purchasing information, framing, and installation data for each window. Many window types contain optional muntins. Some muntins snap onto the surface of the glass and do not actually separate panes of glass. Figure 17.31A shows common muntin grid designs. Figure 17.31B shows the common types of muntin grid patterns. (See Chapter 35 for examples of schedules.)

Selecting the style, type, material, color, glazing, and dimensions for each window involves hundreds of combination options. For this task designers refer to manufacturers' catalogs or websites to find the appropriate combination of spec verifications for each window. Figure 17.32 shows a sample

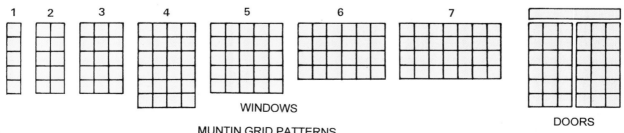

WINDOWS

MUNTIN GRID PATTERNS

DOORS

FIGURE 17.31B ❯ Common muntin grid patterns. *Hepler/Wallach/Hepler © Cengage 2013*

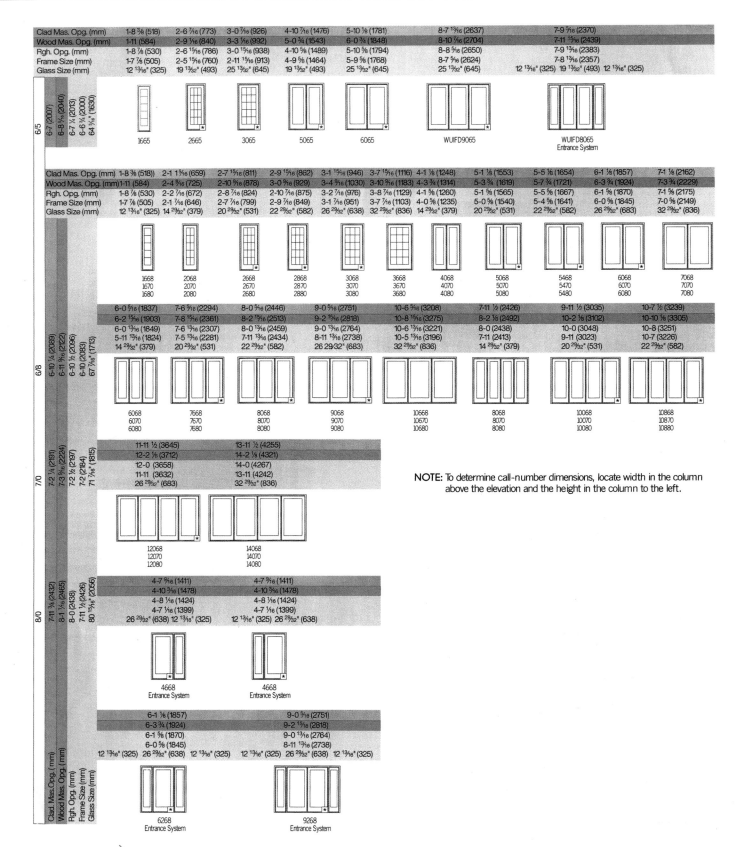

NOTE: To determine call-number dimensions, locate width in the column above the elevation and the height in the column to the left.

FIGURE 17.32 > Typical window manufacturer's catalog page. *Marvin Windows & Doors*

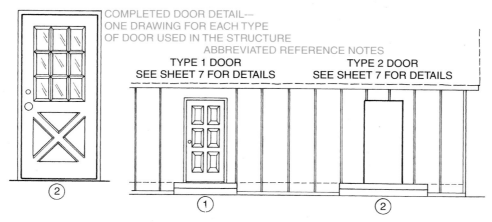

FIGURE 17.33 > Use of door codes to identify door styles. *Hepler/Wallach/Hepler*
© Cengage 2013

page from a large window manufacturer's catalog containing hundreds of combinations. Once selected this information is transferred to a window schedule as shown in Figure 35.2. The dimensional information is then used to place each window onto an elevation. Window information stored in a database can be retrieved and transferred to each selected elevation location.

Door Symbols

Doors are shown on elevation drawings by methods similar to those used for illustrating window styles and positions. They are either drawn completely, if the scale permits, or shown in abbreviated form. Sometimes door codes are indexed to a door schedule. See Figure 17.33. The complete drawing of the door, whether shown on the elevation drawing sheet or as a detail drawing on a separate sheet, should show the division of panels and lights, sill, jamb, and head-trim details.

Many exterior door styles are available. Refer back to Figure 16.39. The total relationship of the door and trim to the entire elevation cannot be seen unless the door trim is also shown. Exterior doors are normally wider than interior doors. Exterior doors must provide access for larger amounts of traffic and be large enough to permit the movement of furniture. They must also be thick enough to provide adequate safety, insulation, and sound barriers. Refer back to Chapter 16 for common door sizes.

Selecting the appropriate combination of door specifications is similar to the process for window selection except door swing, thickness, and threshold information must be added. Figure 17.34 is a sample page from a door manufacturer's catalog that shows hundreds of optional combinations. Just as with windows, stored door information can be ac-

cessed from a manufacturer's database and transferred directly to an elevation drawing.

CAD Elevation Symbol Libraries

Elevation symbols are of two types: individual and surface symbols. Individual symbols such as doors and windows are stored in *Blocks* and inserted (*Insert* command) onto an elevation drawing in the same manner as floor plan symbols are applied. Surface symbols such as siding and roofing materials are applied using the *Hatch* command. Hatch boundaries must totally enclose a space with no gaps. The hatch pattern is then selected from a hatch library. Hatch blocks can be drawn and stored by the user or selected from the system's library. Many component and material manufacturers supply software containing blocks of their products that can be stored and used in combination with other blocked symbols.

INTERIOR ELEVATIONS

Just as exterior elevations illustrate the outside walls, **interior elevation drawings** are necessary to show the design of interior walls (vertical planes). Because of the need to show cabinet height and counter arrangement details, interior wall elevations are most often prepared for kitchen and bathroom walls. See Figure 17.35. An interior wall elevation shows the appearance of the wall as viewed from the center of the room.

A coding system is used to identify the walls on the floor plans for which interior elevations have been prepared. The code symbol tells the direction of the view, the elevation detail number, and the page or sheet number. See Figure 17.36. If only a few interior elevations are prepared, then the title of

RUSTIC	DRA1P	DRA2A & DRA2A80	DRA2C & DRA2C80	DRA2P80	DRA1A80	DRA1C80	DRATP &DRATP70 DRATP80	DRAPC &DRAPC70 DRAPC80	DRAPR &DRAPR70 DRAPR80
SIZES	6/8 & 8/0	6/8 & 8/0	6/8 & 8/0	6/8 & 8/0	8/0	8/0	6/8, 7/0 & 8/0	6/8, 7/0 & 8/0	6/8, 7/0 & 8/0
DOORS	0.16 0.01	0.16 0.01	0.16 0.01	0.16 0.01	0.16 0.01	0.16 0.01	0.16 0.01	0.16 0.01	0.16 0.01

RUSTIC	DRA2B	DRA2D	DRA2E	DRA2R	DRA2F	DRA4R	DRA4C
SIZES	6/8	6/8	6/8	8/0	8/0	8/0	8/0
DOORS	0.16 0.01	0.16 0.01	0.16 0.01	0.16 0.01	0.16 0.01	0.16 0.01	0.16 0.01

DRF3C & DRF3C80

FIR GRAIN	SDL 3 Lite	SDL 6 Lite
SIZE	6/8 & 8/0	6/8 & 8/0
DOORS	0.27 0.11	0.27 0.11

MAHOGANY	DRM41-36	DRM60	DRM6080 3/6	DRM13	DRM13 4 Lite	DRM13 8/10 Lite	DRM13 9 Lite
SIZE	6/8	6/8	8/0	6/8 & 8/0	6/8 & 8/0	6/8 & 8/0	6/8 & 8/0
DOORS	0.16 0.01	0.16 0.01	0.16 0.01	0.26 0.11	0.26 0.11	0.26 0.11	0.26 0.11

TRIMMABLE	DRS00T	DRS41T	DRS61T	DRS2GT	DRS2BT	DRS2DT	DRS90T
SIZES	6/8	6/8	6/8	6/8	6/8	6/8	6/8
DOORS	0.16 0.01	0.16 0.01	0.16 0.01	0.16 0.01	0.16 0.01	0.16 0.01	0.16 0.01

FIGURE 17.34 > Typical door manufacturer's catalog page. *Plastpro, Inc.*

the room and the compass direction of the wall are the only identification needed.

Interior elevations can also be indexed to a cutting-plane callout on a full or room floor plan. For example, the southwest kitchen elevation (B/3) in Figure 17.37 is indexed to the floor plan in Figure 15.28 and is pictured in Figure 10.5. As a comparison, the northeast dining room elevation (C/3) in Figure 17.38 is indexed to the same plan and shows the reverse side of elevation B/3. Cutting-plane callouts can be used to intersect through several horizontal or vertical levels. The interior wall elevation shown in Figure 17.39 illustrates how an interior elevation is drawn through two levels with one section code (C/7) in Figure 17.39. Floor plan details of the elevation area are located in Figure 15.28.

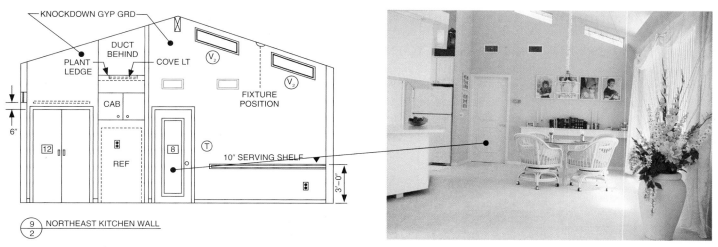

FIGURE 17.35A › Breakfast room wall elevation. *Hepler/Wallach/Hepler © Cengage 2013*

FIGURE 17.35B › Completed wall shown in Figure 17.35A. *Hepler/Wallach/Hepler © Cengage 2013*

FIGURE 17.36 › Numerical codes used to identify wall elevations. *Hepler/Wallach/Hepler © Cengage 2013*

Interior Elevation Projections on CAD

Interior wall elevation drawings are prepared on CAD by treating a specific floor plan wall the same as an outside wall and by following the same drawing sequence. This involves assigning a separate layer to the floor plan and each elevation.

Here the height of all features—doors, windows, ledges, railings, cabinets, counters, and so on—must be established and drawn before floor plan key intersections are projected.

Steps in Drawing an Interior Elevation

The following steps in drawing an interior elevation are outlined in Figure 17.40.

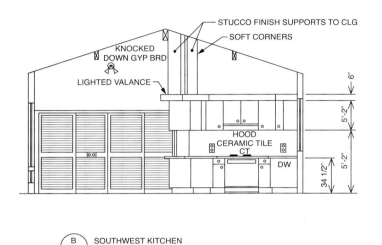

FIGURE 17.37 > Interior room elevation. *Hepler/Wallach/ Hepler © Cengage 2013*

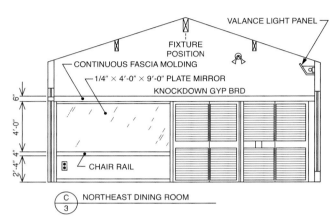

FIGURE 17.38 > Interior room elevation—other side of elevation shown in Figure 17.37. *Hepler/ Wallach/Hepler © Cengage 2013*

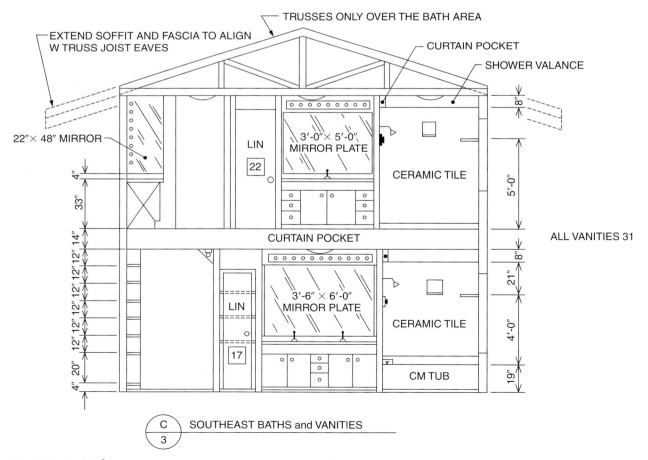

FIGURE 17.39 > Two-level interior elevation. *Hepler/Wallach/Hepler © Cengage 2013*

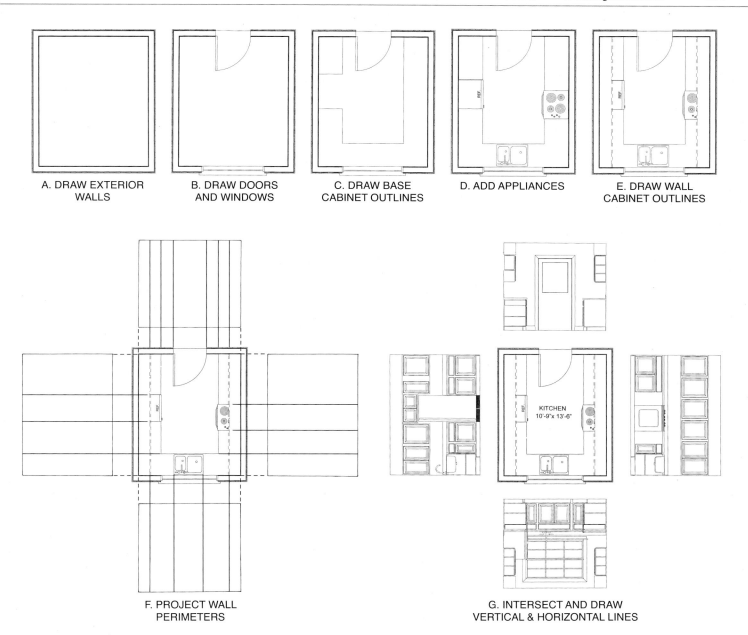

A. DRAW EXTERIOR WALLS

B. DRAW DOORS AND WINDOWS

C. DRAW BASE CABINET OUTLINES

D. ADD APPLIANCES

E. DRAW WALL CABINET OUTLINES

F. PROJECT WALL PERIMETERS

G. INTERSECT AND DRAW VERTICAL & HORIZONTAL LINES

KITCHEN 10'-9"x 13'-6"

FIGURE 17.40 > Steps in drawing interior elevations. *Hepler/Wallach/Hepler © Cengage 2013*

STEP 1   Draw the floor plan outline and project lines outward from each wall.

STEP 2   Connect projected lines to represent the floor line and ceiling line of each elevation. Refer to Chapters 9 and 15 to determine the size of any stair run and configuration.

STEP 3   Add details, cabinets, and fixtures to the floor plan and project lines to represent the height of each feature on each elevation.

Projecting the interior elevation in this manner is appropriate for accurate drawing, but results in an elevation drawn on its side or upside down. Therefore, interior elevation drawings, like exterior elevations, are not left in the position in which they were originally projected from the floor plans. Interior elevations are repositioned so that each floor line appears on the bottom as a room would normally be viewed.

Once the features of the wall are projected to the elevation from the floor plan, dimensions, instructional notes, and additional features can be added to the drawing. See Figure 17.41.

Interior elevations provide a great amount of detail: the height of all cabinets, shelving, ledges, railings, wall lamps, fixtures, valances, mirrors, chair rails, electrical outlets, switches, landings, and stair profiles. Elevation drawings also include wall surface treatment labeling. Using a common floor line and ceiling line for several elevations eliminates much layout work. In some situations, an interior elevation can span several levels. Check the floor plan in Figure 15.28 for the sources of

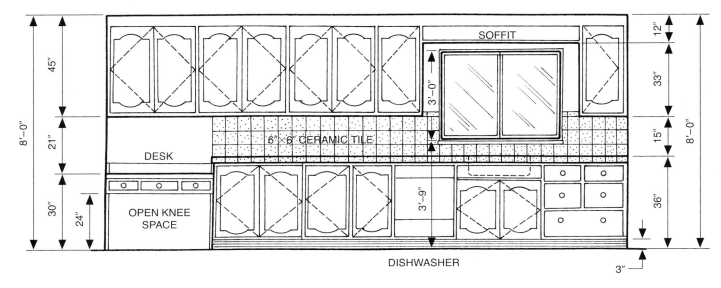

FIGURE 17.41 > Kitchen wall interior elevation. *Hepler/Wallach/Hepler © Cengage 2013*

elevations shown in Figures 17.42. Single room elevations are often drawn at a larger scale than ¼″ = 1′-0″.

ELEVATION DIMENSIONING

The vertical (height) dimensions are as important on elevation drawings as horizontal (width and length) dimensions are on floor plans. Many dimensions on elevation drawings show the vertical distance from a datum line. The **datum line** is a reference that remains constant. Sea level is commonly used as the datum or basic reference for many drawings. However, any given line can be conveniently used as a base or datum line for vertical reference.

Dimensions on elevation drawings show height above the ground line. They also show the vertical distance from the floor line to the ceiling and roof ridge and eave lines, and to the tops of chimneys, doors, and windows. Distances below the ground line are shown by dotted lines.

Standards and Guidelines for Dimensioning

Elevation dimensions must conform to basic standards to ensure consistency of interpretation. The numbered arrows on the elevation drawing in Figure 17.43 show the applications of the following guidelines for elevation dimensioning:

1. Vertical elevation dimensions should be read from the right side of the drawing.
2. Levels to be dimensioned should be labeled with a note, abbreviation, or term.
3. Room heights are shown by dimensioning from the floor line to the ceiling line.
4. The depth of footings is dimensioned from the ground line.
5. Heights of windows and doors are dimensioned from the floor line to the top of the windows or doors.
6. Elevation dimensions show only vertical distances (height). Horizontal distances (length and width) are shown on floor plans.
7. Windows and doors may be indexed by a code or symbol to a door or window schedule, if the style of the windows and doors are not shown on the elevation drawing. See also Figure 17.44.
8. The slope of the roof is shown by indicating a slope diagram.
9. Dimensions for small, complex, or obscure areas should be indexed to a separate detail.
10. Ground-line elevations are expressed as heights above a datum point (for example, sea level).
11. Heights of chimneys above the ridge line are dimensioned.
12. Floor and ceiling lines are shown with hidden lines.
13. Heights of planters, fences, and walls are dimensioned from the ground line.
14. Thicknesses of slabs are dimensioned.
15. Overall height dimensions are placed on the outside of subdimensions.
16. Thicknesses of footings are dimensioned.

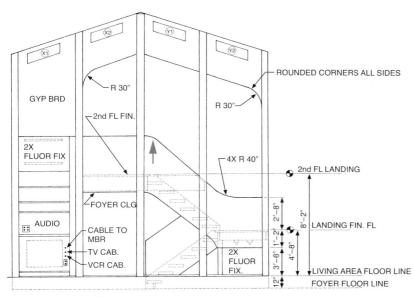

FIGURE 17.42A > Two-level interior elevation. *Hepler/Wallach/Hepler © Cengage 2013*

FIGURE 17.42C > Completed wall shown in elevation drawing of Figure 17.42B. *James Eismont, Photographer*

FIGURE 17.42B > Elevation view under construction. *James Eismont, Photographer*

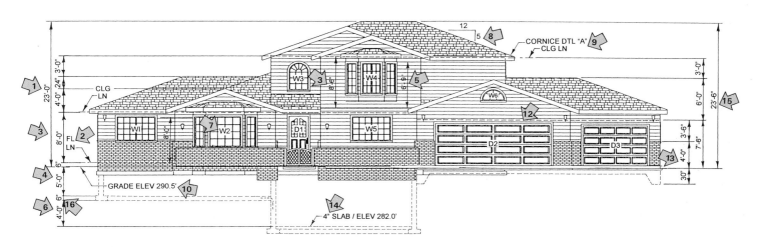

FIGURE 17.43 > Numbers indicate specific guidelines for dimensioning. *Hepler/Wallach/Hepler © Cengage 2013*

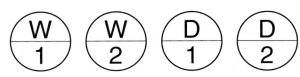

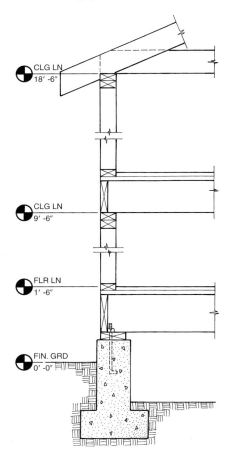

17. Refer to Figure 17.45. When the level to be dimensioned is obscure or extremely close to other dimensions, use an elevation line symbol and label the level line.

18. The datum must be identified with a note if not part of the elevation drawing.

Types of Dimensions

Two types of elevation dimensions are used: framing dimensions and finished dimensions. **Framing dimensions** show the actual distances between framing members. This is the most common method and is preferred by most builders. To avoid an accumulation of measuring errors and variations in

member thickness, framing member dimensions are often dimensioned to their centers.

Finished dimensions show the actual vertical distances between finished features, such as from the finished floor to finished ceiling levels. Interior dimensions may be added to a full section drawing as done in Figure 17.46. Placing dimensions on a separate interior elevation drawing as done in Figure 17.47 is preferred where possible. These two types of dimensions should not be alternately used on the same drawing, unless the exception is clearly noted. A note on each drawing or set of drawings should indicate which method was used.

Features that are shared by an interior and exterior wall, such as doors and windows, are usually shown on both interior and exterior elevations, but dimensioned only on the exterior elevation drawings. Many other features or components—such as cabinets, shelving, counters, ledges, railings, wall lamps, switches, and receptacles—can only be dimensioned on interior elevations.

BUILDING INFORMATION MODELING

Once floor plan and elevation designs are completed **building information modeling (BIM)** drawings can be prepared. BIM drawings are three-dimensional views that are prepared using 3D building modeling software, as shown in Figure 17.48. BIM is the process of creating and managing construction data of all types through linkage to a project information database. BIM drawings provide 3D semitransparent views that reveal the total volume of a building. These drawings also show simultaneously potential conflicts among elements such as plumbing, electrical, and HVAC systems. Revealing the 3D spatial relationships of defined spaces is a valuable component of BIM drawings.

One of the most significant aspects of a BIM system is the electronic generation of construction documents such as specifications, purchase orders, cost estimates, scheduling details, construction management data, financial analyses, and environmental monitoring. This is done through automatic electronic links from BIM drawings to various documents.

Although floor plan and elevation designs may precede the development of BIM drawings, experienced designers create BIM drawings and documents throughout the entire design process.

PRESENTATION ELEVATION DRAWINGS

Dimensioned elevation drawings, prepared to reveal every line and detail, lack realism. To add a more natural look to elevation drawings, landscape features are added to exterior

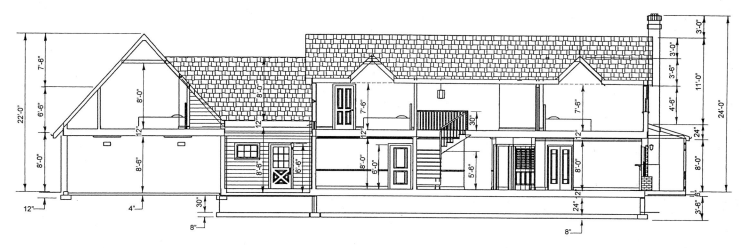

FIGURE 17.46 > Method of dimensioning both interior and exterior elevations. *Hepler/Wallach/Hepler © Cengage 2013*

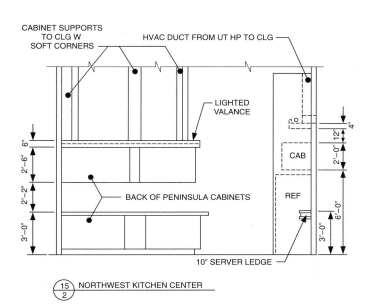

FIGURE 17.47 > Typical interior elevation dimensions. *Hepler/Wallach/Hepler © Cengage 2013*

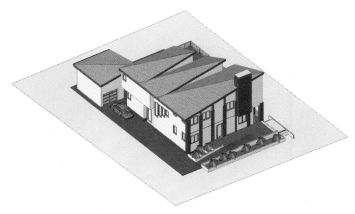

FIGURE 17.48 > 3D BIM drawing. *Rivit pictorial by Swain CAD Solutions*

elevations and wall textures and color are added to interior elevation drawings to create **presentation drawings.**

Exterior Presentation Elevations

Adding landscape features to an elevation usually hides some key lines and features. Therefore, trees, plants, and other landscape features should not be added to elevation working drawings. A separate elevation should be traced or plotted for this purpose. The realistic appearance of the elevation shown in Figure 17.49 is created by adding foliage, cars, people, and shadows. Adding color and texture to siding, doors, windows, and roofs also gives realism to the drawing. Heavy foliage around the elevation shown in Figure 17.50 adds depth and texture to the drawing without hiding the main features of the design. Figure 17.51 shows some of the methods of drawing landscape features on an elevation. Figure 17.52 shows the use of a combination of water colors and dry markers to add texture and color to siding and windows. These can be added in color or black depending on the duplicating method to be used. Hard line drawings are recommended if the presentation drawing is to be reproduced with the entire set of plans.

CAD elevations can also be rendered as shown in Figures 17.53 and 17.54. These elevation renderings were completed using a combination of CAD line tasks, photos shop methods, and hand-drawn lines. In both of these renderings, note how the people and tree outlines are made semitransparent to avoid hiding the building details. By adding only color, conventional elevations can be converted to a presentation elevation. For example, note the difference between the black-and-white elevations shown in Figure 17.7, compared to the same elevations as rendered in color in Figures 17.55 and 17.56 on page 386. Even more

FIGURE 17.49 › Exterior presentation elevation drawing. *Hepler/Wallach/Hepler © Cengage 2013*

FIGURE 17.50 › Rendered exterior elevation. *Hepler/Wallach/Hepler © Cengage 2013*

FIGURE 17.51 › Landscape features for elevation drawings. *Hepler/Wallach/Hepler © Cengage 2013*

reality is added to the rear-side elevation by including more intricate details, as in Figure 17.57.

Elevation working drawings can also be converted to presentation elevations by adding an oblique side to the drawing. For example, right-side walls were added to the drawing in Figure 17.58 to reveal two sides of the house for presentation purposes. If viewed carefully the original orthographic elevation outline can be seen.

Interior Presentation Elevations

Converting interior elevation working drawings into presentation elevations involves adding texture, color, and shadows to wall surfaces, doors, windows, appliances, built-ins, and sometimes furniture. Figure 17.59 shows a rendered interior elevation used for presentation purposes. The rendering methods for pictorial drawings presented in Chapter 20 also apply to presentation elevation drawings. Rendered elevations are more accurate and easier to prepare than pictorial drawings, but pictorial drawings appear more realistic.

FIGURE 17.52 > Siding and window rendering. *Hepler/ Wallach/Hepler © Cengage 2013*

FIGURE 17.54 > CAD elevation rendering. *Hepler/Wallach/ Hepler © Cengage 2013*

FIGURE 17.53 > Combination CAD and hand drawn rendering. *Hepler/Wallach/Hepler © Cengage 2013*

FRONT ELEVATION

© 2007 TPEG INC.

FIGURE 17.55 > Elevation with only color added.
Timberpeg Inc.

FIGURE 17.57 > Elevation with details added. *Timberpeg Inc.*

RIGHT SIDE ELEVATION

© 2008 TPEG INC.

FIGURE 17.56 > Color added to conventional elevation.
Timberpeg Inc.

FIGURE 17.58 > Oblique depth added to an orthographic
elevation. *Used by permission Hanley Wood, LLC*

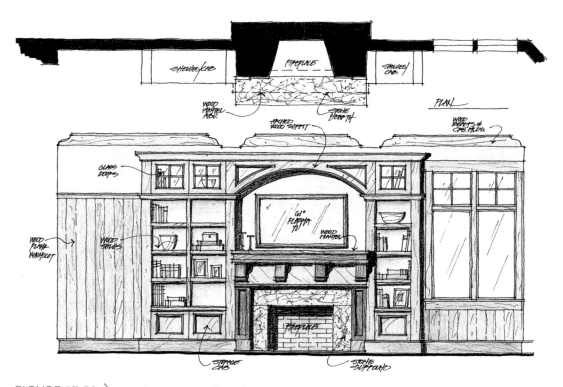

FIGURE 17.59 > Interior presentation elevation. *Hepler/Wallach/Hepler © Cengage 2013*

Drawing Elevations Exercises

1. Project the front, rear, right, and left elevations of a floor plan of your own design. Add elevation symbols.

2. Draw a kitchen wall elevation for a kitchen of your design.

3. Sketch and dimension the front elevation of your home or another home with which you are familiar.

4. Using a CAD system, develop a library of symbols for elevations to use in drawing your own design. Note or list which symbols are already included in the CAD program.

5. Project and sketch or draw the front elevation suggested in one of the pictorial drawings in Chapter 19.

6. Copy five of the trees, shrubs, or plants shown in Figure 17.51 as practice for creating a landscape rendering. Then create one with shadows for an elevation of your own design.

7. Create an exterior elevation drawing from the south elevation shown in Figure 17.6.

8. Draw a front elevation for the plan shown in Figure 15.33.

9. Draw an interior elevation of the living room fireplace wall shown in Figure 15.20.

10. Create an interior elevation presentation drawing from the elevation shown in Figure 17.6.

CHAPTER 18

Sectional, Detail, and Cabinetry Drawings

OBJECTIVES

In this chapter you will learn to:

> describe types of sectional drawings.

> communicate views of sections based on a cutting plane.

> draw sections, using correct codes and proper dimensioning.

> evaluate when a detail sectional drawing is needed.

> read and prepare detail drawings.

> design and prepare cabinet drawings.

TERMS

break lines	detail section	modular cabinetry
break-out sections	footing section	molding
built-in components	full section	removed sections
cabinet coding system	head sections	sectional drawings
concept study	horizontal wall sections	sill sections
cutting plane	jamb sections	transverse section
cutting-plane line	longitudinal section	vertical wall sections

INTRODUCTION

Architectural sections are drawings that are important in the design process. They show details not visible on floor plans or elevations. Most architectural sections contain symbols, reference codes, and dimensions to indicate construction information. Other sections may only be sketches to compare heights, while still others may be pictorial drawings. Sections are used to show the exact details of construction. The ability to prepare technical architectural drawings depends on a thorough understanding of sectional drawings.

SECTIONAL DRAWINGS

Sectional drawings reveal the internal construction of an object. An architectural sectional drawing (or an architectural section) that is prepared for the entire structure is called a **full section.** A sectional drawing that shows only specific parts of a building is a **detail section.** The size and complexity of the parts usually determine whether a full section and/or detail

sections are needed. Because they provide all information needed for construction, architectural sections are used as working drawings. Sections are also useful as design concept drawings and presentation drawings.

Architectural designs that involve multiple levels and variations in height often require the preparation of full-section sketches (in addition to floor plans) during the design process. Sections are often needed to clarify the elevation relationships of ground, footings, floors, and roof lines.

When a presentation drawing is needed to show the general appearance of both the interior and exterior of a building in one drawing, a *pictorial section* is often completed. This is usually done using one-point perspective methods. See Figure 18.1.

FULL SECTIONS

In full-section drawings, entire buildings are drawn as if they were cut in half. The purpose of full sections is to convey how a building is constructed from the foundation through the roof.

FIGURE 18.1 ❯ Presentation section. *Hepler/Wallach/Hepler © Cengage 2013*

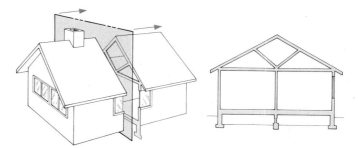

FIGURE 18.2 ❯ Transverse section with cutting plane. *Hepler/Wallach/Hepler © Cengage 2013*

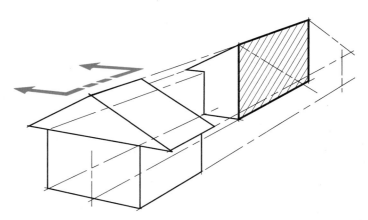

FIGURE 18.3 ❯ Longitudinal section with cutting plane. *Hepler/Wallach/Hepler © Cengage 2013*

The Cutting Plane

A **cutting plane** is an imaginary plane that passes through a building. The position of a cutting plane is shown by the **cutting-plane line** that is drawn as a long heavy line with two dashes. Sectional drawings may show either transverse or longitudinal sections. A **transverse section** shows a cutting plane across the shorter or minor axis of a building, as shown in Figure 18.2. A **longitudinal section** shows a cutting plane along the length or major axis of a building. See Figure 18.3. Sections convey information about the inner construction. Both transverse and longitudinal full sections share the same external outlines as elevation drawings.

The cutting-plane line is usually placed on a floor plan to tell which part is drawn as a section. The arrows at the ends of the line tell the direction of the view. See Figure 18.4.

Because the cutting-plane line can easily interfere with dimensions, notes, and details, only the extreme ends of the cutting-plane line are indicated on most architectural drawings. The part of the line that is omitted on the drawing is then assumed to be a straight line between the ends of the cutting-plane line. There is no limit to the number of cutting planes that can be drawn. Each section must be separately identified.

Symbols

When sections are referenced to another drawing, the symbol shown in Figure 18.5 should be used. This referencing method is the same method introduced in Chapter 4. The top part (sheet A3) of Figure 18.6 illustrates the use of a section identification symbol on a cutting-plane line. The referenced section detail is shown on sheet A8 at the bottom. Figure 18.7A shows a partial floor plan with coded cutting-plane lines. The cutting plane B4 is indexed to the longitudinal full section shown in Figure 18.7B. The trans-

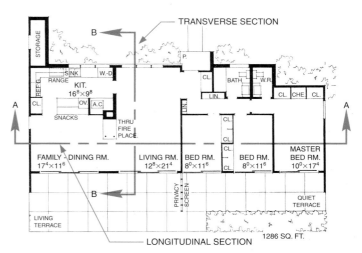

FIGURE 18.4 ❯ Transverse and longitudinal cutting-plane lines on a floor plan. *Hepler/Wallach/Hepler © Cengage 2013*

verse full section A5 shown in Figure 18.7C is also from the same floor plan.

A floor plan is a small-scale horizontal section. Therefore, many floor plan symbols are also used in plan detail drawings.

Many *section-lining symbols* appear realistic, as though the solid materials were cut through. Others are simplified in order to save time on the drawing board.

A building material is only sectioned if the cutting-plane line passes through it. The outline of all other materials visible behind the plane of projection must also be drawn in the proper position and scale. Symbols for building materials are shown in Figure 18.8.

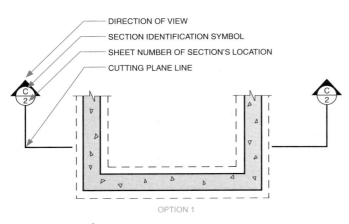

FIGURE 18.5 ❯ Cutting-plane reference symbol. *Hepler/Wallach/Hepler © Cengage 2013*

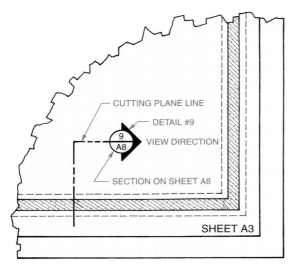

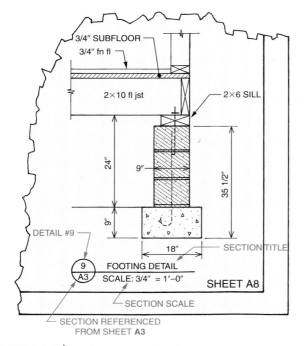

FIGURE 18.6 ❯ Relationship of reference symbol to detail. *Hepler/Wallach/Hepler © Cengage 2013*

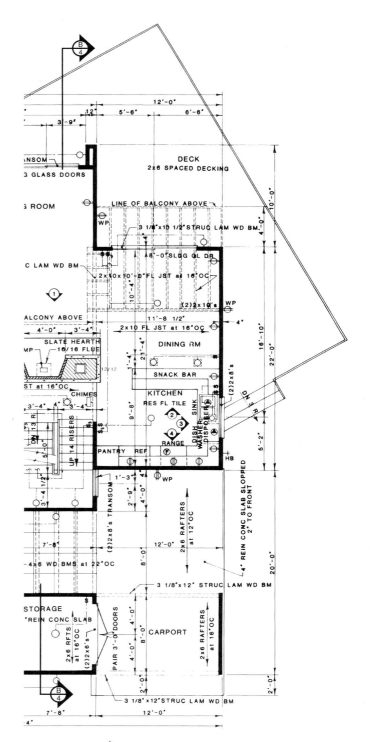

FIGURE 18.7A ❯ Floor plan cutting plane referenced to the full section in Figure 18.7B. *Home Planners, Inc.*

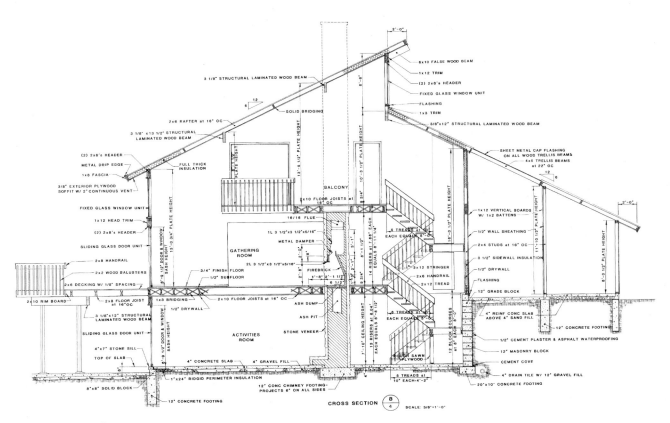

FIGURE 18.7B > Longitudinal section referenced from the floor plan in Figure 18.7A. *Home Planners, Inc.*

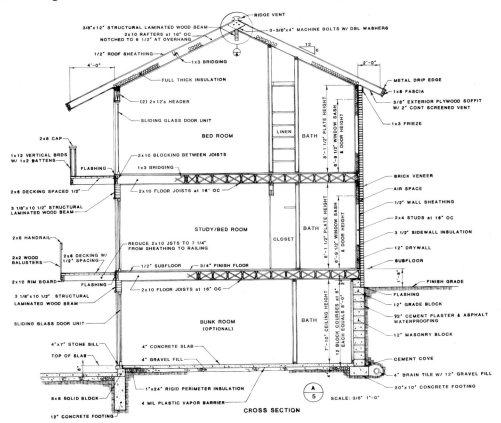

FIGURE 18.7C > Transverse section referenced from floor plan in Figure 18.7A. *Home Planners, Inc.*

NAME	ABBRV	SECTION SYMBOL	ELEVATION	NAME	ABBRV	SECTION SYMBOL	ELEVATION
COMMON BRICK	COM BRK	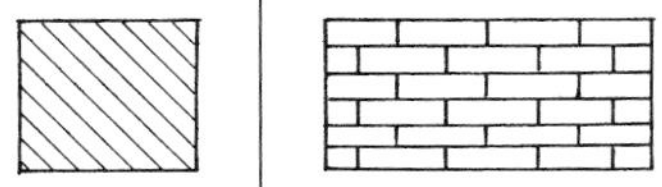		WELDED WIRE MESH	WWM	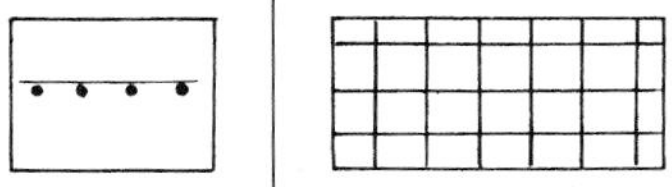	
FACE BRICK	FC BRK	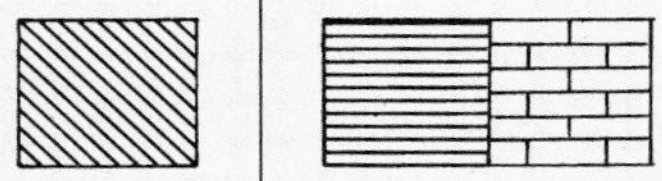		FABRIC	FAB	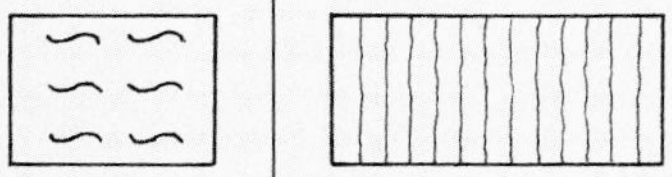	
FIREBRICK	FRB	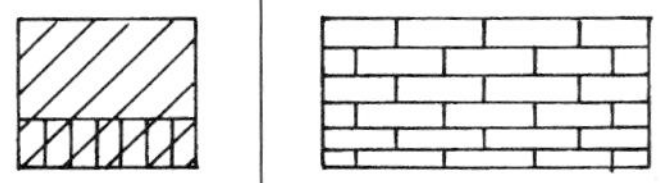		LIQUID	LQD	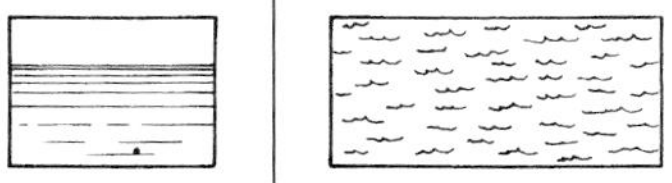	
GLASS	GL	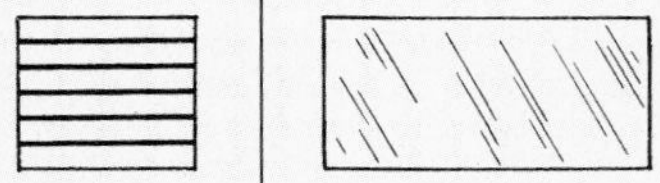		COMPOSITION SHINGLE	COMP SH	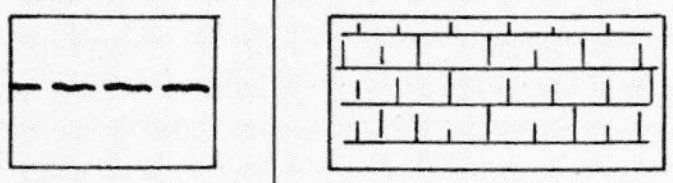	
GLASS BLOCK	GL BLK	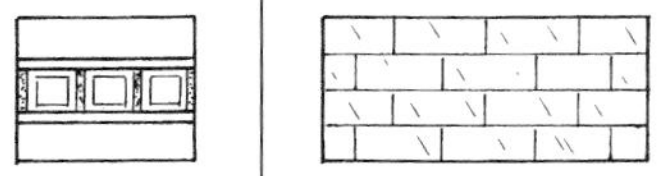		RIGID INSULATION SOLID	RDG INS	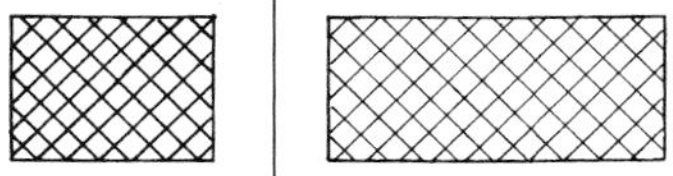	
STRUCTURAL GLASS	STRUC GL	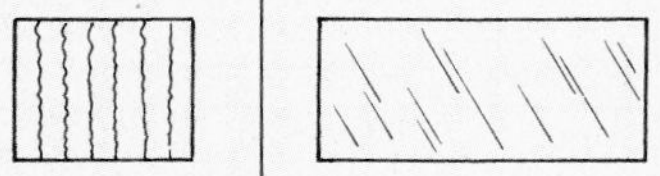		LOOSE-FILL INSULATION	LF INS	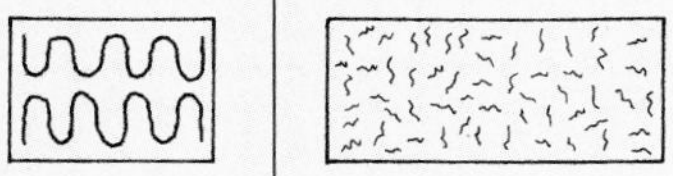	
FROSTED GLASS	FRST GL	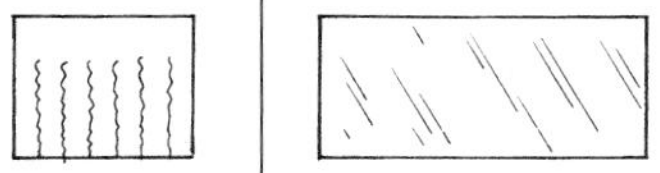		QUILT INSULATION	QLT INS	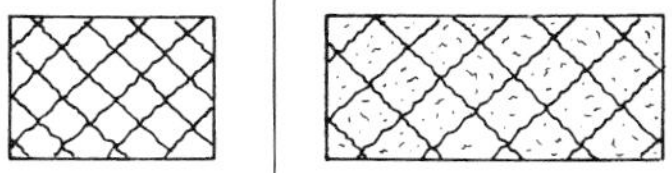	
STEEL	STL	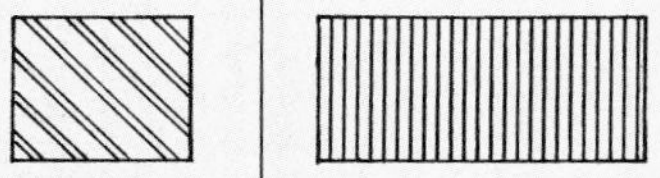		SOUND INSULATION	SND INS	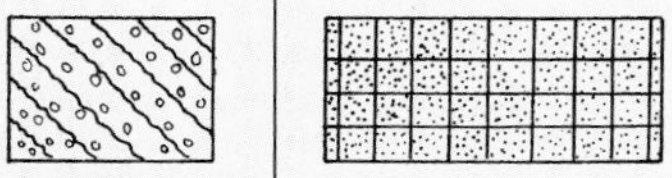	
CAST IRON	CST IR	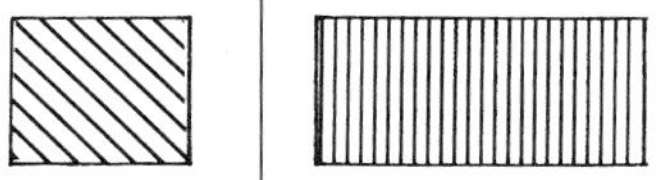		CORK INSULATION	CRK INS	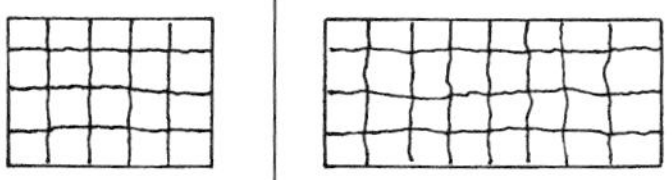	
BRASS & BRONZE	BRS BRZ	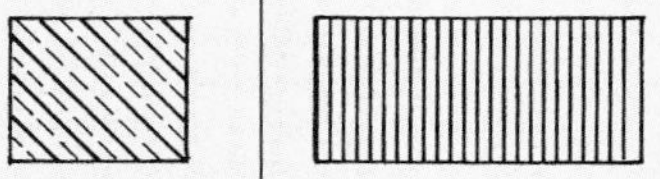		SHEET METAL (FLASHING)	SHT MTL FLASH		
ALUMINUM	AL			REINFORCING STEEL BARS	REBAR	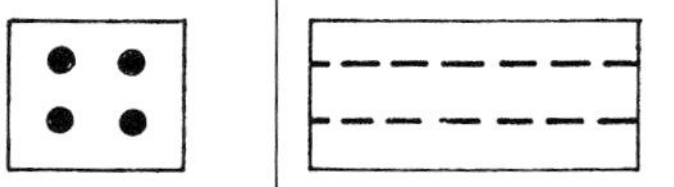	

FIGURE 18.8 ⟩ Section symbols for building materials. *Hepler/Wallach/Hepler © Cengage 2013*

CAD Section Symbol Library

Section symbols are used to represent a slice through a material. Symbols that represent a wide variety of materials are stored and accessed using the *Hatch* command. After the hatch pattern and boundary have been selected, the material is added to a drawing by identifying the hatch border with a cursor. Hatch patterns should be applied to enable the pattern to be stretched or scaled without changing the pattern makeup.

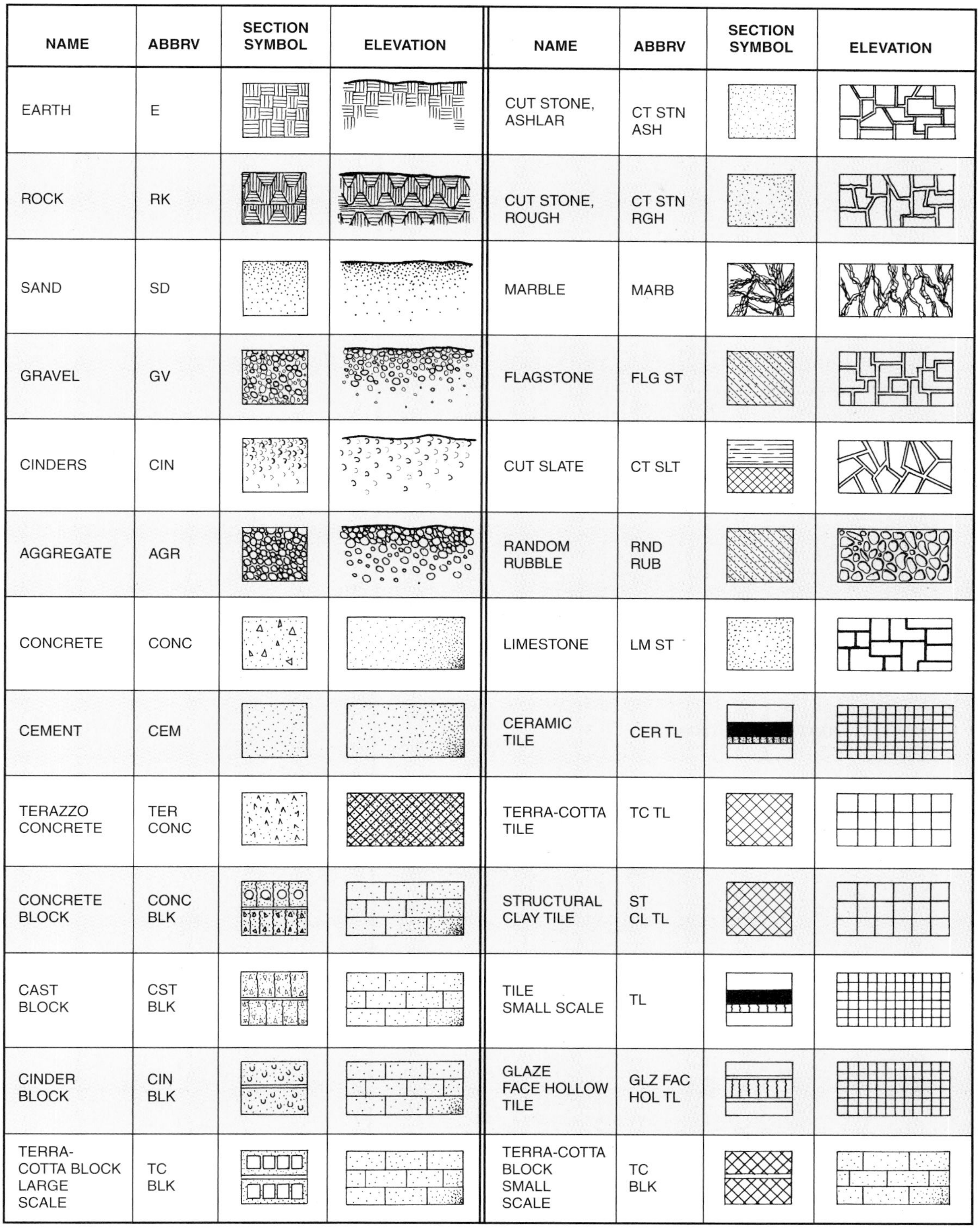

NAME	ABBRV	SECTION SYMBOL	ELEVATION	NAME	ABBRV	SECTION SYMBOL	ELEVATION
EARTH	E			CUT STONE, ASHLAR	CT STN ASH		
ROCK	RK			CUT STONE, ROUGH	CT STN RGH		
SAND	SD			MARBLE	MARB		
GRAVEL	GV			FLAGSTONE	FLG ST		
CINDERS	CIN			CUT SLATE	CT SLT		
AGGREGATE	AGR			RANDOM RUBBLE	RND RUB		
CONCRETE	CONC			LIMESTONE	LM ST		
CEMENT	CEM			CERAMIC TILE	CER TL		
TERAZZO CONCRETE	TER CONC			TERRA-COTTA TILE	TC TL		
CONCRETE BLOCK	CONC BLK			STRUCTURAL CLAY TILE	ST CL TL		
CAST BLOCK	CST BLK			TILE SMALL SCALE	TL		
CINDER BLOCK	CIN BLK			GLAZE FACE HOLLOW TILE	GLZ FAC HOL TL		
TERRA-COTTA BLOCK LARGE SCALE	TC BLK			TERRA-COTTA BLOCK SMALL SCALE	TC BLK		

FIGURE 18.8 > Continued

Scale

Because full sections show construction methods used in the entire building, they are drawn to a relatively small scale (1/4″ = 1′-0″). However, a scale that is too small of-ten makes the drawing and interpretation of smaller details extremely difficult. One method of maintaining a larger scale and drawing area for large sections is to insert break lines. See Figure 18.9. **Break lines** indicate that much of the repetitive portion of the building was removed from the

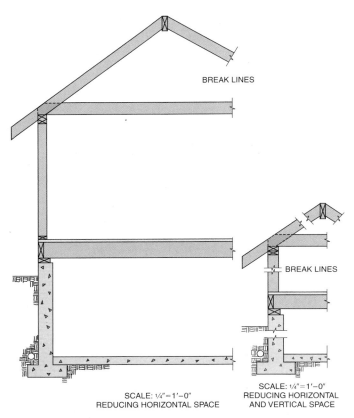

FIGURE 18.9 ❯ Use of break lines on small-scale sections.
Hepler/Wallach/Hepler © Cengage 2013

drawing. With break lines, a very large area can be drawn at a larger and more readable scale and still fit on one sheet.

Another method is to draw detail sections of the removed portions of the building separate from the full-section drawing. **Removed sections** drawn at a larger scale are used to clarify small details.

Abbreviated full sections are often prepared as part of the design process or for presentation purposes. The section shown in Figure 18.10A is prepared to show only the outline and relationship of horizontal and vertical components, without dimensions or details. Figure 18.10B shows the elevation design **concept study** relating to this section including circulation patterns, access concepts, and a partial roof plan.

Sectional Dimensions

Full sections expose the size and shape of building materials and components not revealed on floor plans and elevations. These sections are an excellent place on which to locate many detail dimensions. Full-section dimensions primarily show specific elevations, distances, and the exact size of building materials. See Figure 18.11. The guidelines for dimensioning elevation drawings apply also to full-elevation sections and to detail sections. (See Chapter 17.)

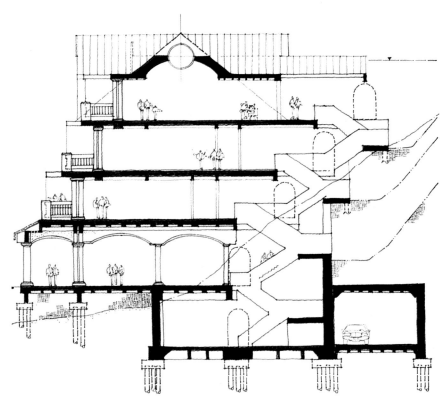

FIGURE 18.10A ❯ Presentation section showing only major outlines. *© Dana Hepler*

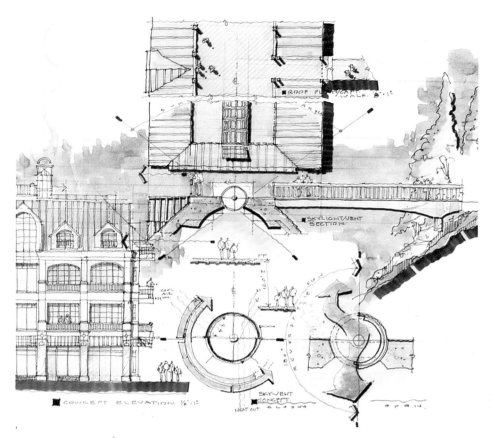

FIGURE 18.10B 〉 Elevation concept study relating to the section in Figure 18.10A. © *Dana Hepler*

Steps in Drawing Full Sections

In drawing full sections, the architect "constructs" the framework of a house on paper. Figure 18.12 shows the progressive steps in the layout and drawing of a section of the side edge of a house.

STEP 1   Lightly draw the finished floor line approximately at the middle of the drawing sheet.

STEP 2   Measure the thickness of the subfloor and of the joist and draw lines representing these members under the floor line.

STEP 3   From the floor line, measure up and draw the ceiling line.

STEP 4   Measure down from the floor line to establish the top of the basement slab and footing line, and draw in the thickness of the footing.

STEP 5   Draw two vertical lines representing the thickness of the foundation and the footing.

STEP 6   Construct the sill detail and show the alignment of the studs and top plate.

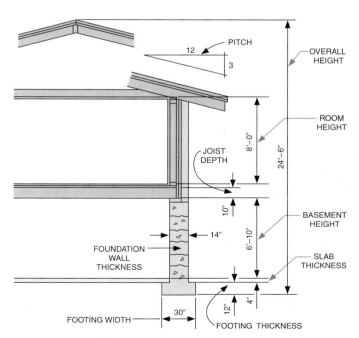

FIGURE 18.11 〉 Major dimensions used on a vertical section. *Hepler/Wallach/Hepler © Cengage 2013*

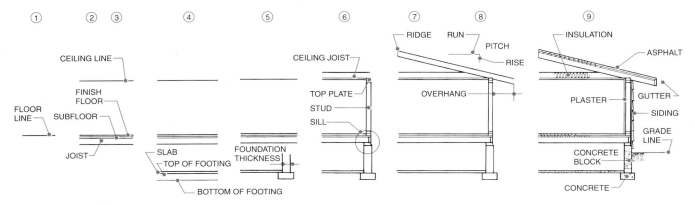

FIGURE 18.12 ⟩ Sequence for drawing an elevation section. *Hepler/Wallach/Hepler © Cengage 2013*

STEP 7 Measure the overhang from the stud line and draw the roof pitch by projecting from the top plate on the angle determined by the slope diagram.

STEP 8 Establish the ridge point by measuring the distance from the outside wall horizontally to the center of the ridge line. This is usually at the center of the structure.

STEP 9 Add details and symbols representing siding and interior finish.

DETAIL SECTIONS

Because full sections are usually drawn to a small scale, many small parts are difficult to interpret. To reveal the exact position and size of many of these small parts, enlarged detailed sections are prepared. Detail sections clarify any construction feature that could not be described on the basic floor plans, elevations, or full sections. Detail sections may be prepared on a vertical (elevation) plane or a horizontal (plan) plane. Like full sections, detail sections are keyed to a plan or elevation view. See Figure 18.13. Figure 18.14 shows a typical dimensioned detail section of a fireplace and chimney assembly.

Most detail sections are prepared in orthographic form. When multiple use is planned, pictorial details may be drawn as shown in Figure 18.15.

Vertical Wall Sections

Vertical wall sections show exposed construction members on a vertical plane. They are prepared for exterior and interior walls.

Exterior Walls

Elevation drawings do not reveal construction details because of their small scale and because many details are hidden by

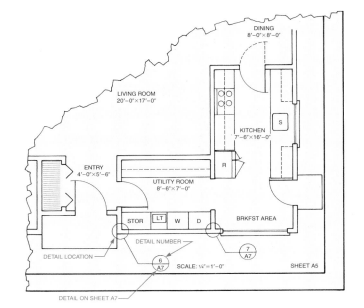

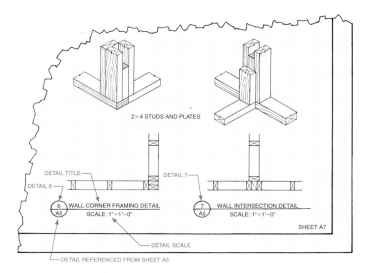

FIGURE 18.13 ⟩ Cross-referencing symbols used on detail drawings. *Hepler/Wallach/Hepler © Cengage 2013*

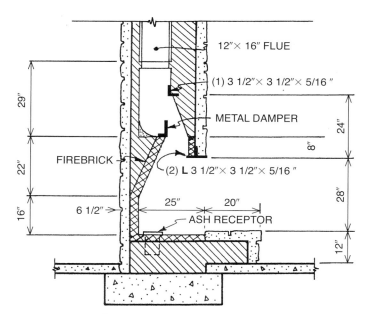

FIGURE 18.14 > Typical construction section detail. *Hepler/ Wallach/Hepler © Cengage 2013*

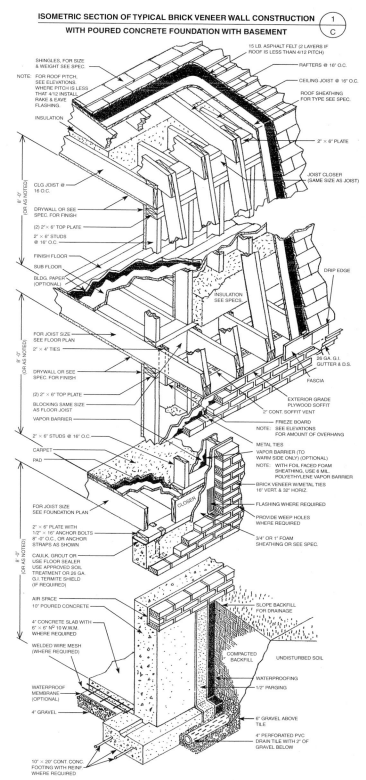

FIGURE 18.15 > Pictorial detail drawing. *Used by permission Hanley Wood, LLC*

siding materials. Several other methods are used to produce exterior wall section drawings large enough to show construction details and dimensions. These include the use of break lines and removed sections to reduce the length of the drawing.

Break Lines

Similar to full-section drawings, break lines are used on detail sections to reduce vertical distances. As discussed earlier, using break lines allows the area to be drawn larger than would be possible if the entire distance were included in the drawing. Break lines are used where the construction does not change over a long distance. Figure 18.16 shows the use of break lines to enlarge a frame-wall section by reducing the length of the drawing.

Removed Sections

Sometimes it is impossible to draw an entire wall section to a large enough scale to show needed information, even when using break lines. In these cases, a removed section is drawn at a larger scale, separate from the original location. Removed sections are frequently drawn for the ridge, cornice, sill, footing, and intersecting beam details.

Cornice sections are used to show the relationship between the outside wall, top plate, and rafter construction. See Figure 18.17. Some cornice sections show gutter details.

Sill sections, as shown in Figure 18.18, show how the foundation supports and intersects with the floor system and the outside wall.

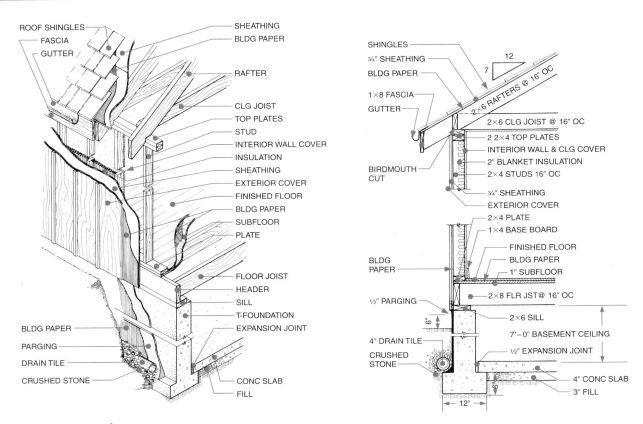

FIGURE 18.16 › Use of break lines to reduce vertical drawing height. *Hepler/Wallach/Hepler © Cengage 2013*

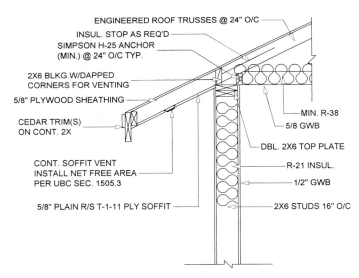

FIGURE 18.17 › Cornice section detail. *Hepler/Wallach/Hepler © Cengage 2013*

A **footing section** is needed to show the width and height of the footing, the type of material used, and the position of the foundation wall on the footing. Figure 18.19 shows several footing details and the pictorial interpretation of each type.

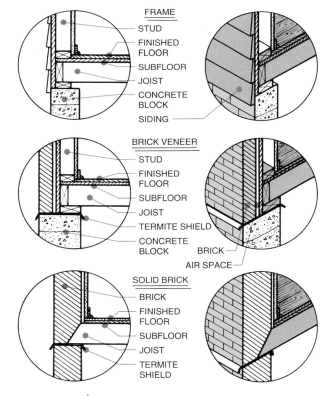

FIGURE 18.18 › Detailed sill sections. *Hepler/Wallach/Hepler © Cengage 2013*

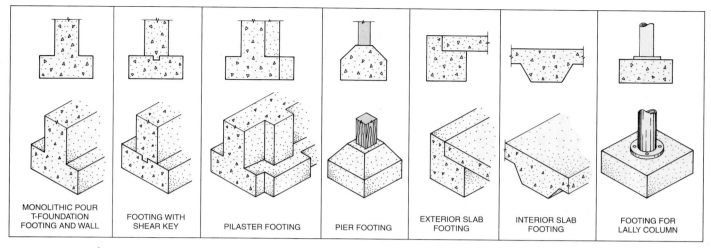

| MONOLITHIC POUR T-FOUNDATION FOOTING AND WALL | FOOTING WITH SHEAR KEY | PILASTER FOOTING | PIER FOOTING | EXTERIOR SLAB FOOTING | INTERIOR SLAB FOOTING | FOOTING FOR LALLY COLUMN |

FIGURE 18.19 > Detailed footing sections. *Hepler/Wallach/Hepler © Cengage 2013*

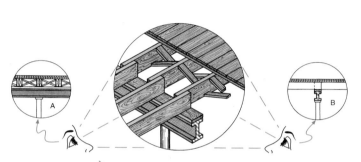

FIGURE 18.20 > Two views of a section detail. *Hepler/ Wallach/Hepler © Cengage 2013*

Beam details are necessary to show how the joists are supported by beams and how the columns or foundation walls support the beams. As for all sections, the position of the cutting-plane line is extremely important. Figure 18.20 shows two possible positions of the cutting plane. If the cutting-plane line is placed parallel to the beam, you see a cross section of the joist, as shown in drawing A. If the cutting-plane line is placed perpendicular to the beam, you see a cross section of the beam, as shown in drawing B.

Interior Walls

To illustrate the methods of constructing inside partitions, sections are often drawn of interior walls at the base and at the ceiling. A base section shows how the wall-finishing materials are attached to the studs and how the intersection between the floor and wall is constructed. A crown section shows the intersection between the ceiling and the wall and how the finished construction materials of the wall and ceiling are related. Sections may also be prepared for stair and fireplace details.

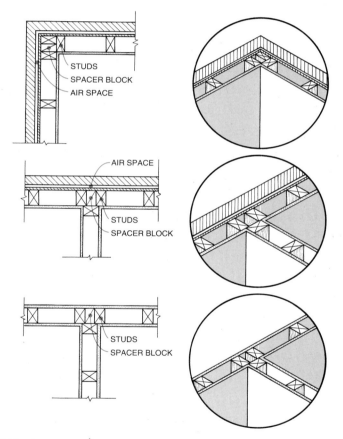

FIGURE 18.21 > Horizontal wall sections. *Hepler/Wallach/ Hepler © Cengage 2013*

Horizontal Wall Sections

Horizontal wall sections of exterior and interior walls are drawn to clarify how walls are constructed. Walls in these sections are similar to those on floor plans, but are drawn at a larger scale in order to show a horizontal sectional view of each construction member as shown in Figure 18.21.

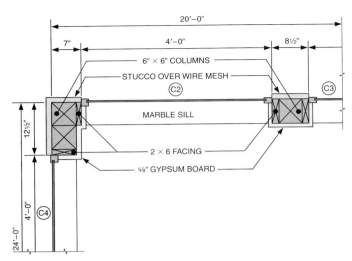

FIGURE 18.22A 〉 Corner wall section detail. *Hepler/Wallach/ Hepler © Cengage 2013*

FIGURE 18.22B 〉 Corner under construction. *Hepler/ Wallach/Hepler © Cengage 2013*

FIGURE 18.22C 〉 Completed corner of living room shown in Figure 15.28. *Hepler/Wallach/Hepler © Cengage 2013*

Exterior Walls

Although a floor plan is a horizontal section, many construction details are omitted because of the small scale used. Therefore, larger horizontal sections are drawn to show the more exact construction details of corners, intersections, and window and door framing, as shown in Figure 18.22. This section represents the corner of the living room in Figure 15.28.

Interior Walls

Typical horizontal sections of interior wall intersections indicate construction methods. For example, horizontal sections are needed to show the inside and outside corner construction of a paneled wall and how paneled joints and other building joints are constructed. See Figure 18.23. Figures 18.24A and 18.24B illustrate how crown **molding** detail sections and base molding detail sections are drawn. Figure 18.24C links these intersections to an exterior wall section. The cross section of standard molding shapes used on interior wall sectional drawings is shown in Figure 18.25. Molding and paneling are used to create paneled walls as shown in Figure 18.26. Note also the use of molding to provide a finished appearance to the beams in this illustration.

A typical wall panel sectional drawing is shown in Figure 18.27. Horizontal sections are also very effective for illustrating various methods of attaching building materials. Interior wall features that require off-site construction must be detailed to ensure aesthetic consistency with the room decor. Dimensional control is also necessary to accurately integrate a separate component with established sizes as shown in the fireplace mantel in Figure 18.28.

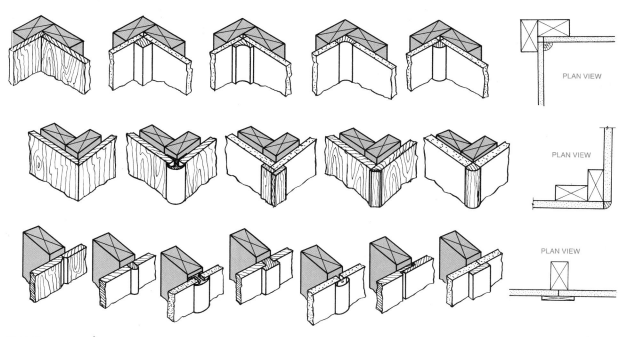

FIGURE 18.23 > Wall panel joints and corners. *Hepler/Wallach/Hepler © Cengage 2013*

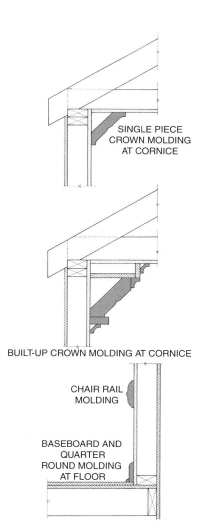

SINGLE PIECE CROWN MOLDING AT CORNICE

BUILT-UP CROWN MOLDING AT CORNICE

CHAIR RAIL MOLDING

BASEBOARD AND QUARTER ROUND MOLDING AT FLOOR

FIGURE 18.24A > Common molding and trim sections. *Hepler/Wallach/Hepler © Cengage 2013*

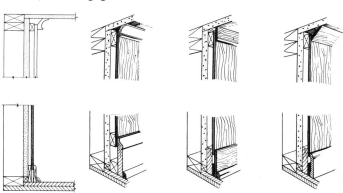

FIGURE 18.24B > Common crown and base molding sections. *Hepler/Wallach/Hepler © Cengage 2013*

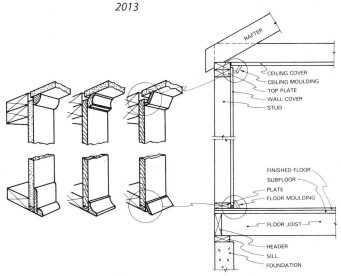

RAFTER

CEILING COVER
CEILING MOULDING
TOP PLATE
WALL COVER
STUD

FINISHED FLOOR
SUBFLOOR
PLATE
FLOOR MOULDING

FLOOR JOIST

HEADER
SILL
FOUNDATION

FIGURE 18.24C > Crown and base molding shown on a wall section. *Hepler/Wallach/Hepler © Cengage 2013*

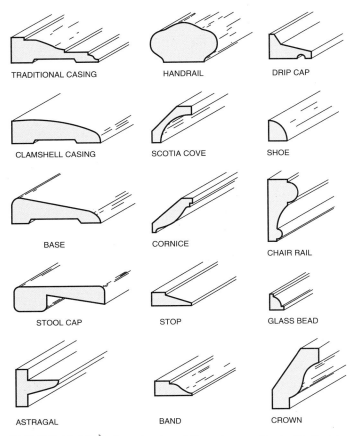

TRADITIONAL CASING   HANDRAIL   DRIP CAP

CLAMSHELL CASING   SCOTIA COVE   SHOE

BASE   CORNICE   CHAIR RAIL

STOOL CAP   STOP   GLASS BEAD

ASTRAGAL   BAND   CROWN

FIGURE 18.25 > Standard molding shapes. *Hepler/Wallach/Hepler © Cengage 2013*

Window Sections

Because window construction is hidden, sectional drawings are needed to show construction details. Figure 18.29 shows both horizontal and vertical window construction areas that are commonly sectioned. These include head, sill, and jamb construction. Window manufacturers generally use pictorial sections to illustrate key features and methods of installation. See Figure 18.30.

Head sections and **sill sections** are vertical sections. Preparing head and sill sections on the same drawing is possible only when a small scale is used. If a larger scale is needed, the head and sill must either be drawn independently or break lines must be used.

When a cutting-plane line is extended horizontally across the entire window, the resulting sections are known as **jamb sections.** Jamb details (the horizontal section) are projected from the window-elevation drawing. The construction of both jambs is usually the same, with the right jamb drawing being the reverse of the left. Only one jamb detail is normally drawn. The builder then interprets one jamb as the reverse of the other.

Door Sections

A horizontal section of all doors is indicated on a floor plan. However, a floor plan does not include sufficient detail for installation. Similar to window construction sections, sill, head, and jamb sections are necessary to show door construction.

FIGURE 18.26 > Use of molding on paneled walls, bookshelves, and beams. *© Robert Trachtenberg/Corbis Outline*

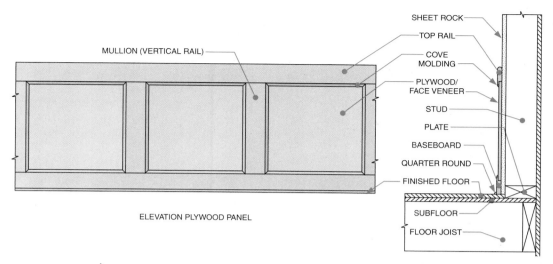

FIGURE 18.27 > Paneling elevation and section. *Hepler/Wallach/Hepler © Cengage 2013*

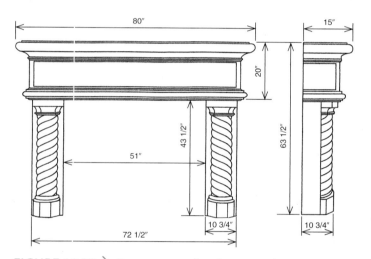

FIGURE 18.28 > Precast stone fireplace mantle. *Hepler/ Wallach/Hepler © Cengage 2013*

When a cutting-plane line is extended vertically through the sill and head, a vertical full section is created. These sections are often too small to show the desired degree of detail necessary for construction. A removed section can be drawn at a larger scale to show head or sill sections.

Because doors are normally not as wide as they are high, an adequate jamb detail can be projected without the use of break lines or removed sections.

Sectional drawings of the framing details of the door sill, head, and jamb, exclusive of the door and door frame assembly, are often prepared for framing purposes. In such drawings, the framing section is drawn separately, with the door frame and door removed. Door manufacturers often prepare pictorial **break-out sections** to show the internal construction of components. These drawings also show the framing, trim, and door in their proper locations. Drawings

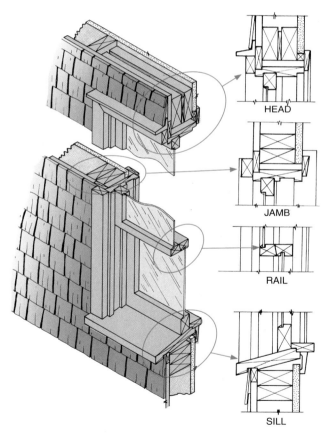

FIGURE 18.29 > Sectioned window construction details. *Hepler/Wallach/Hepler © Cengage 2013*

of garage doors and industrial-size doors are usually prepared with sections of the brackets and apparatus necessary to house the door assembly. See Chapter 29 for horizontal and vertical door sections that include the head, sill, and jamb. Door manufacturers also use pictorial sections to illustrate important features and methods of installation.

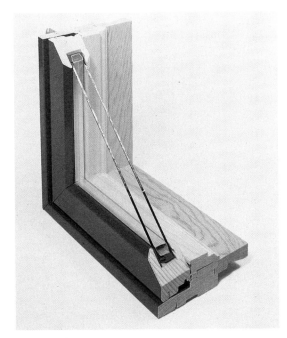

FIGURE 18.30 > Pictorial section showing window construction. *Kolbe and Kolbe*

CABINETRY AND BUILT-IN COMPONENT DRAWINGS

Basic architectural plans are prepared at a scale too small to show the exact size and construction details of many cabinets and built-in components. On simple designs, details are often explained with a note or reference to a manufacturer's product. On larger, more complex designs, separate details are provided in pictorial, plan, elevation, and/or sectional drawings.

Cabinet Construction and Types

Cabinets are either custom-built (made to order) or manufactured using modular sizes. In either case, the quality of the finished product depends on the materials, joints, hardware (hinges, pulls, latches), finish, and the accuracy of construction and installation. The quality of components ranges widely from economy units, which use the least expensive materials and methods, to premium components, which resemble fine furniture.

CAD Detail Blocks

Many architectural details are used repeatedly in otherwise original designs. These symbols can be created and stored as a block by using the *Wblock* command and selecting the perimeter of the detail with the pick box.

Cabinets are either wall hung or positioned on the floor (base cabinets). See Figures 18.31A and 18.31B. Numerous

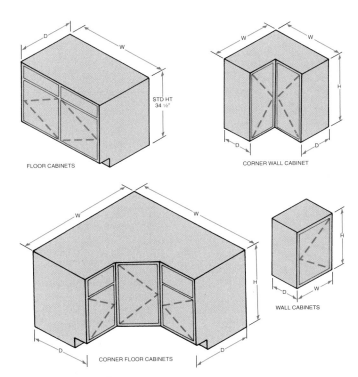

FIGURE 18.31A > Typical floor and wall cabinets. *Hepler/Wallach/Hepler © Cengage 2013*

BASE CABINETS				
SYM	AMT	DEPTH	HEIGHT	WIDTH
A	1	24"	34 1/2"	36"
B	2	24"	34 1/2"	36"
C	1	24"	34 1/2"	21"
D	1	24"	34 1/2"	48"
E	1	24"	34 1/2"	9"
WALL CABINETS				
1	1	12"	36"	36"
2	1	12"	18"	42"
3	2	12"	36"	12"
4	2	12"	36"	24"
5	1	12"	36"	30"
6	1	12"	36"	18"
7	1	12"	18"	30"
8	1	12"	30"	21"

FIGURE 18.31B > Common kitchen cabinet dimensions. *Hepler/Wallach/Hepler © Cengage 2013*

WALL CABINETS   Wall Cabinets are 12" deep (excluding doors). Most wall cabinets available in 3" width increments from 9" to 48".

Single Door

Available in 24", 30", 36", 42" heights.

Double Door

Available in 24", 30", 36", 42" heights.

Wall End

Available in 30", 42" heights.

45° Corner Glass Mullion Door

Available in 30" height.

18" High Double Door

18" High Double Door

Available in 30" width.

BASE CABINETS   Base Cabinets are 24" deep (excluding doors) and 34½" high except where noted.

Base Tray

Available in 9" width Left or right hinging.

Single Door

Available in 12", 15", 18", 21", 24" widths. Left or right hinging.

Double Door

Available in 27", 30", 33", 36", 39", 42", 45", 48" widths.

Single Drawer

Available in 30", 36" widths.

Base Blind Corner

Available in 36", 39", 42", 45", 48" widths.

Sink Base Double Door

Available in 24", 27", 30", 33", 36", 39", 42", 48" widths.

TALL CABINETS   Tall Cabinets are 24" deep (excluding doors) except where noted.

Single Oven

Available in 27", 30", 33" widths. Available in 84", 90", 96" heights.

96" High Utility Cabinet

Available in 18", 24" widths. Available in 12" or 24" depths.

90" High Pantry Cabinet

Available in 36" width.

VANITY CABINETS   Vanity Base Cabinets are 31½" high and 21" deep except where noted.

Vanity Bowl

Available in two door 24" to 42" widths in 3" increments. Three door available in 48" width. Four door available in 60" width.

Vanity Bowl-Two Drawer

Available in 24", 30", 36" widths. Available in 18" (space saver) depth.

84" Vanity Linen

Available 18" wide. Left or right hinging.

FIGURE 18.32 > Standard modular cabinet dimensions. *Merillat Industries Inc.*

styles and sizes are available, as shown in Figure 18.32. Most modular base cabinet sizes are standardized at 34 1/2″ in height and 24″ deep. Modular wall cabinets are usually 12″ deep. Custom-built wall cabinets are usually 15″ deep, and floor cabinets are 26″ deep. Cabinets are manufactured for baths, kitchens, and laundry rooms.

Materials used in the construction of cabinets and built-in components include hardwood, softwood, pressboard, stranded lumber, laminates, ceramic tile, marble, and synthetic materials. Through the creative use of recycled wood, curved shapes can be created for cabinets and countertops. Figure 18.33 illustrates a unique and functional application of curved lines for cabinetry and countertops that are totally formaldehyde free.

Though not always seen, the types of joints and fasteners used to attach parts have an important effect on the overall

FIGURE 18.33 > Curved cabinetry created with recycled wood laminates. *Pedini*

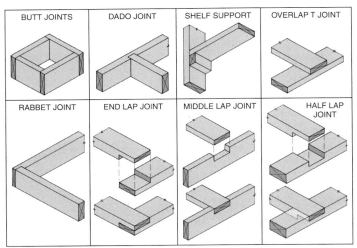

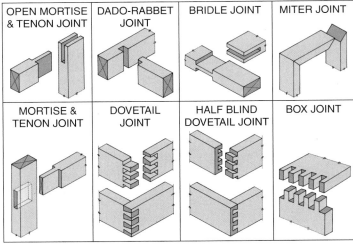

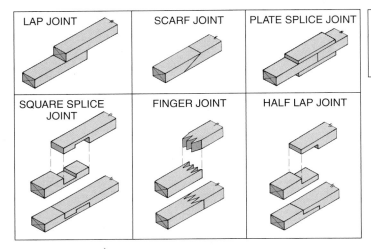

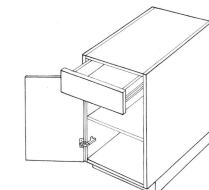

FIGURE 18.34 > Common cabinetry joints. *Hepler/Wallach/ Hepler © Cengage 2013*

appearance and durability of cabinets. For example, simple butt joints are typically used on economy cabinets, whereas joints such as dovetails and mortise and tenons are used in premium units. See Figure 18.34. High-quality cabinetry includes many natural materials with minimal use of synthetic materials that may contain toxic amounts of petroleum-based adhesives and plastic, as covered in Chapter 6. Quality is also based on the methods of construction. Figure 18.35 illustrates the accepted methods of constructing **modular cabinetry.**

Manufacturers' Product Libraries

Many manufacturers supply customers with compact disks (CDs) that contain detail drawings of their products shown in various architectural applications. These fall primarily into two categories: DWG files or PDF files. CDs supplied

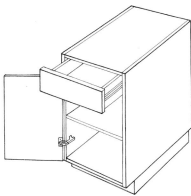

EUROPEAN STYLE CABINET

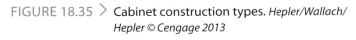

AMERICAN FACE FRAME
STYLE CABINET

FIGURE 18.35 > Cabinet construction types. *Hepler/Wallach/ Hepler © Cengage 2013* in the DWG format can be stored, inserted into any drawing, and then altered. Details of this type are often altered to be consistent with the base drawing in line weight, dimensioning practice, or application. Imported PDF files cannot be altered and must therefore be used as supplied.

Built-In Components

Because of the intricate details involved, only the outlines of most **built-in components** are drawn on floor plans and elevation drawings. Separate, large-scale dimensioned detail drawings are used to show construction and installation information.

Most built-in component designs include precise joinery, hidden hinges, roller sliding parts, and special hardware. Manufactured units for built-in products are usually prefabricated in modular units. The surrounding finish carpentry must be designed to blend with the unit. See Figure 18.36. This requires framing and detail drawings.

When prefabricated units are not used, special framing drawings of the walls, shelves, fascia, and soffits are needed. Built-ins of simpler design, such as shelves, mantels, and planters, are usually built on site. These may only require an interior elevation plus floor plan notes and dimensions. See Figure 18.37.

Dimensioning Cabinetry and Built-Ins

When cabinets or built-ins are custom made or built on site, normal dimensioning practices are used. When factory-produced components are used, they must be precisely positioned with other components, cabinets, and framing. Appliances and plumbing fixtures are designed to stand alone, slide into a space between cabinets, or "drop into" (fit within) a countertop space. In all of these cases, adequate space must be dimensioned on plans and elevations or detailed to ensure a proper fit. Openings that are dimensioned to center lines, such as doors and windows, need not be precise

FIGURE 18.36 > Modular built-in components. *Audio Tec Designs*

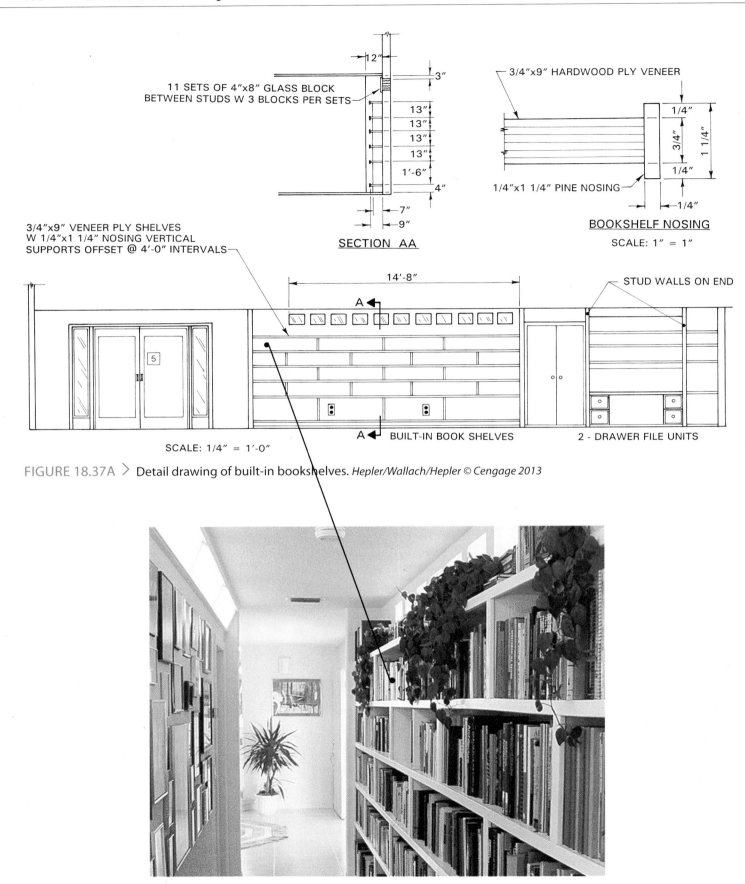

11 SETS OF 4"x8" GLASS BLOCK
BETWEEN STUDS W 3 BLOCKS PER SETS

12"

3"

13"
13"
13"
13"
1'-6"

4"

7"
9"

SECTION AA

3/4"x9" HARDWOOD PLY VENEER

1/4"
3/4"
1 1/4"
1/4"

1/4"

1/4"x1 1/4" PINE NOSING

BOOKSHELF NOSING

SCALE: 1" = 1"

3/4"x9" VENEER PLY SHELVES
W 1/4"x1 1/4" NOSING VERTICAL
SUPPORTS OFFSET @ 4'-0" INTERVALS

14'-8"

A

STUD WALLS ON END

5

A   BUILT-IN BOOK SHELVES

2 - DRAWER FILE UNITS

SCALE: 1/4" = 1'-0"

FIGURE 18.37A > Detail drawing of built-in bookshelves. *Hepler/Wallach/Hepler © Cengage 2013*

FIGURE 18.37B > Completed built-in bookshelves. *Hepler/Wallach/Hepler © Cengage 2013*

for construction purposes. Openings that are dimensioned between the sides denote a finished ("must hold") opening. This is necessary to ensure that appliances or prefabricated components will fit into the space. The word *finished* (or "FIN") may also be added to a "must hold" dimension as in Figure 18.38. To ensure proper fitting of components, all information also needs to be entered into the appropriate schedule. See Chapter 35.

Some cabinetry must be built to accommodate all component dimensions (width, height, and depth). A drawing that shows the dimensions of the component and housing is then prepared for accurate positioning. See Figure 18.39. Positioning drawings are often included in the manufacturer's specifications.

Usually many sizes of models of a manufactured product are available. An alphabetical dimensioning system is used, like the one charted in Figure 18.40 to show model size differences. In dimensioning a cabinet or appliance opening, wall-to-wall, not stud-to-stud, dimensions must be used or the item may not fit into the space. Adding the word "*HOLD*" under a dimension is often done to emphasize the importance of a precise dimension.

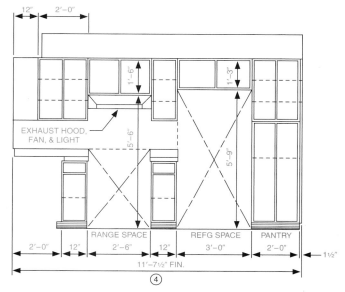

FIGURE 18.38 > Dimensioning practice for reserved appliance spaces. *Hepler/Wallach/Hepler* © Cengage 2013

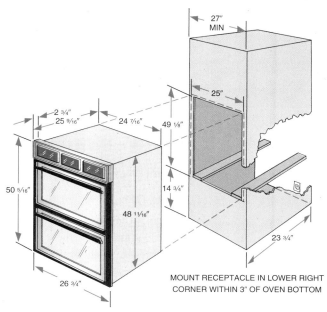

FIGURE 18.39 > Appliance positioning dimensions. *Hepler/Wallach/Hepler* © Cengage 2013

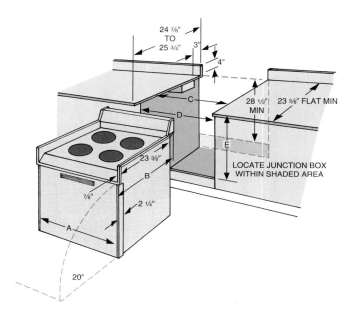

FIGURE 18.40A > Alphanumeric system of dimensioning cabinets. *Hepler/Wallach/Hepler* © Cengage 2013

MODEL	A	B	C	D	E
1	26 3/4"	24 1/4"	26"	27"	26 1/16"
2	26 3/4"	24 1/2"	26"	27"	28 7/16"
3	24 3/8"	24 3/4"	22 1/2" min.	27"	34 1/2"
4	44 1/2"	24 3/4"	22 1/2" min.	27"	14 3/4"
5	28 1/8"	22"	24" min.	24"	32 1/2"
6	28 1/8"	24 3/4"	24" min.	27"	32 1/2"
7	49 5/8"	24 3/4"	24" min.	27"	13 1/4"
8	47 1/4"	24 3/4"	24" min.	27"	15 5/8"

FIGURE 18.40B > Cabinet and cutout dimensions. *Hepler/Wallach/Hepler* © Cengage 2013

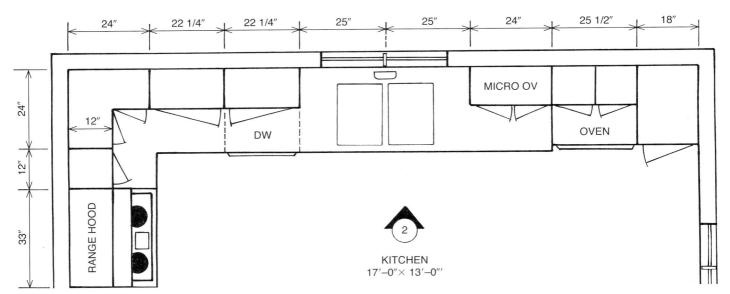

FIGURE 18.41A › View described in plan, elevation, and pictorial drawings. *Roman Polaski, www.romanpolaski.com*

FIGURE 18.41B › Cabinetry floor plan. *Roman Polaski, www.romanpolaski.com*

Drawings used to describe cabinet designs include floor plans, elevations, pictorials, and construction details. A floor plan of the kitchen pictured in Figure 18.41A is shown in Figure 18.41B. This plan includes all width and length dimensions. Elevation drawings show the vertical alignment of cabinets and countertops. Figure 18.41C is an elevation drawing of the wall that includes the sink and window. Although floor plans and elevations describe cabinet layout and design, pictorial drawings, such as that shown in Figure 18.41D, are often prepared to visually interpret the relationship of these drawings.

When unusual design features are used, drawings of the surrounding framing must be detailed. See Figure 18.42.

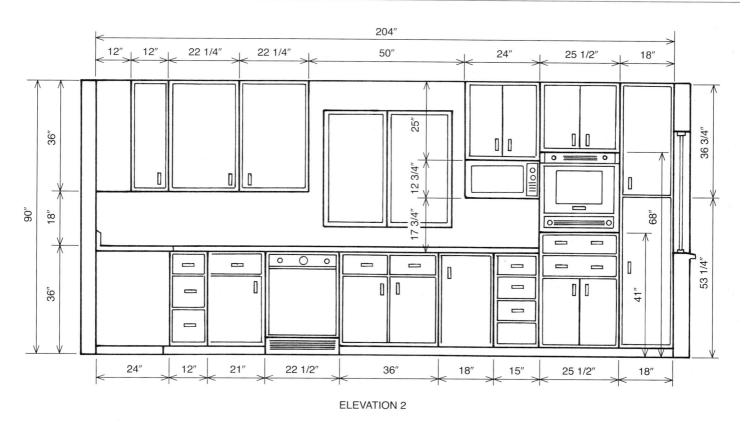

ELEVATION 2

FIGURE 18.41C > Cabinet elevation of view 2 in Figure 18.41A. *Roman Polaski, www.romanpolaski.com*

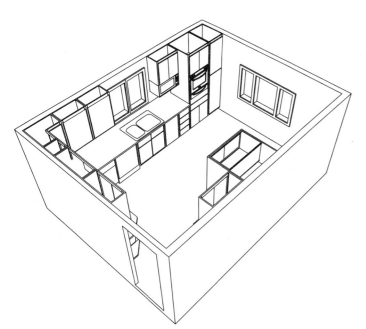

FIGURE 18.41D > Pictorial interpretation of plan and elevation drawings. *Roman Polaski, www.romanpolaski.com*

This is necessary whether the components are site built or factory built. Normally cabinets are factory built and assembled on site. Figure 18.43A is a dimensioned plan view of the completed cabinetry shown in Figure 18.43B. Many detailed drawings were prepared to complete this design.

A shortcut method of dimensioning cabinets is the use of a **cabinet coding system.** The manufacturer's code number for standard modular units is shown on the cabinet outlines drawn on a floor plan. See Figure 18.44. This system is also used for custom-made cabinets by referencing the code number to a detailed drawing of each cabinet.

FIGURE 18.42A > Completed kitchen area shown in Figure 18.42C. *Hepler/Wallach/Hepler © Cengage 2013*

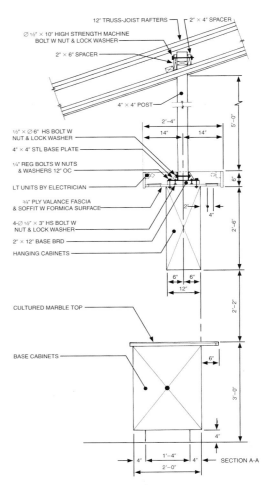

12" TRUSS-JOIST RAFTERS — 2" × 4" SPACER

Ø ½" × 10" HIGH STRENGTH MACHINE BOLT W NUT & LOCK WASHER

2" × 6" SPACER

4" × 4" POST

5'-0"

2'-4"

14" 14"

½" × Ø 6" HS BOLT W NUT & LOCK WASHER

4" × 4" STL BASE PLATE

¼" REG BOLTS W NUTS & WASHERS 12" OC

LT UNITS BY ELECTRICIAN

¾" PLY VALANCE FASCIA & SOFFIT W FORMICA SURFACE

4-Ø ½" × 3" HS BOLT W NUT & LOCK WASHER

2" × 12" BASE BRD

HANGING CABINETS

6"

2'

4"

2'-6"

6" 6"

12"

2'-2"

CULTURED MARBLE TOP

BASE CABINETS

6"

3'-0"

4"

4" 1'-4" 4"

2'-0"

SECTION A-A

VALANCE & HANGING CABINET HANGER DETAILS
SCALE 1"=1'-0"

FIGURE 18.42B > Sectional drawing of the kitchen cabinet design shown in Figure 18.42A. *Hepler/ Wallach/Hepler © Cengage 2013*

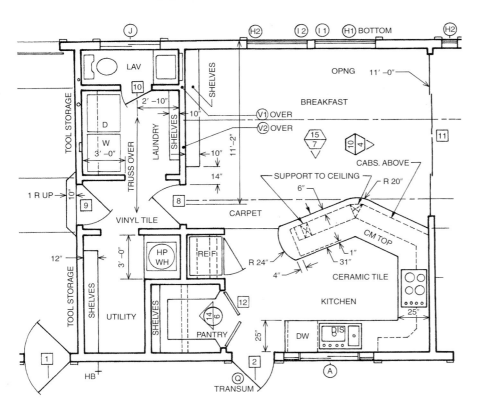

FIGURE 18.42C > Floor plan design showing cabinetry detailed in Figure 18.42B. *Hepler/ Wallach/Hepler © Cengage 2013*

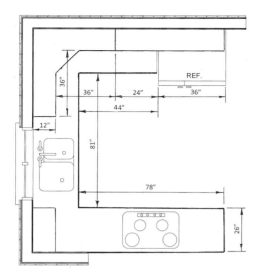

FIGURE 18.43A ⟩ Cabinet plan view. *Hepler/Wallach/Hepler*
© Cengage 2013

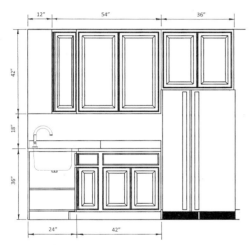

FIGURE 18.43B ⟩ East elevation. *Hepler/Wallach/Hepler*
© Cengage 2013

FIGURE 18.43C ⟩ View toward Figure 18.43B. *Roman Polaski,*
www.romanpolaski.com

FIGURE 18.43D ⟩ North elevation. *Hepler/Wallach/Hepler*
© Cengage 2013

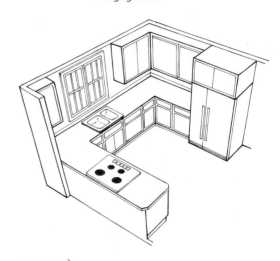

FIGURE 18.43E ⟩ Perspective view of the plan shown in
Figure 18.43A. *Hepler/Wallach/Hepler*
© Cengage 2013

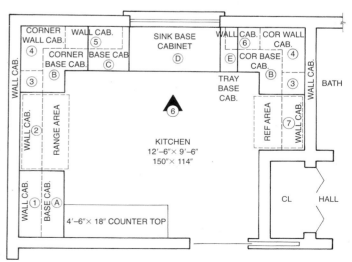

FIGURE 18.44 ⟩ Cabinet coding system example of
dimensional data base. *Hepler/Wallach/*
Hepler © Cengage 2013

Sectional, Detail, and Cabinetry Drawings Exercises CHAPTER 18

1. Draw a full section of a house you have designed.

2. Draw a section through the view shown in Figure 18.5, revolving the cutting-plane line 90°.

3. Draw a head, jamb, and sill section of a typical window and a typical door for the house you have designed.

4. Draw a sill, cornice, and footing section for the house you have designed.

5. Name two methods of cabinet construction and describe what affects their quality.

6. List materials used in cabinet construction.

7. Draw a coded cabinet plan for the kitchen and bath for the house you are designing.

8. Cutting-plane line A5 is missing from the floor plan in Figure 18.7A. Find the location of A5 on this floor plan.

9. Select the best location and draw a longitudinal and transverse section through the plan shown in Figure 18.45.

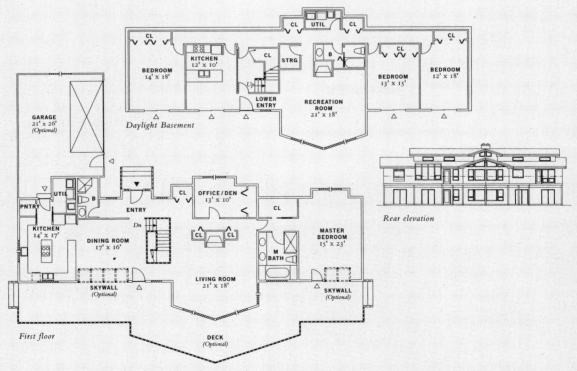

FIGURE 18.45 > Draw a longitudinal section A-A and a transverse section B-B through this plan.
Photo courtesy Lindal Cedar Homes. Lindal.com

PART 5

PRESENTATION METHODS

OBJECTIVES

In this chapter you will learn:

> to differentiate between oblique, isometric, and perspective drawings.

> geometric principles involved in projecting lines (from a given point or at a constant angle) to create 3D images.

> to apply principles of perspective drawing to create interior and exterior pictorial drawings.

> projection methods for drawing pictorials.

TERMS

bird's-eye view
horizon line
isometric drawings
line of sight
oblique drawings
one-point perspective

parallel angle projection
perspective drawings
picture plane
solid models
station point

surface models
three-point perspective
two-point perspective
vanishing point
wireframe

INTRODUCTION

Pictorial drawings are picture-like drawings. Unlike elevation drawings that reveal only one side of an object, pictorials show several sides of an object in one drawing.

TYPES OF PICTORIAL PROJECTION DRAWINGS

Three types of pictorial drawings are used in architecture—oblique, isometric, and perspective drawings, as shown in Figure 19.1. Oblique and isometric drawings are created by **parallel angle projection.** Perspective drawings are more complex to draw but appear more realistic.

Oblique Drawings

Oblique drawings are created by projecting parallel lines from an elevation drawing at any angle from 10° to 45° from the horizontal plane. Because the front view of oblique drawings are prepared to scale, dimensions can be added, as shown in Figure 19.2. They are often used as a substitute for or to clarify working drawings that may be difficult to understand.

Isometric Drawings

Isometric drawings are prepared with parallel receding lines drawn at 30° from the horizontal plane. Because of the visual distortion created by receding parallel lines, isometric drawings are usually used for details, as shown in Figure 19.3, or to describe small areas. Oblique drawings are sometimes dimensioned and used as working drawings because they include true dimensions as shown in Figure 19.4.

If used to draw an entire building, parallel angle drawings appear distorted. Therefore, pictorial drawings of total structures are usually prepared in perspective form, which results in a more realistic view.

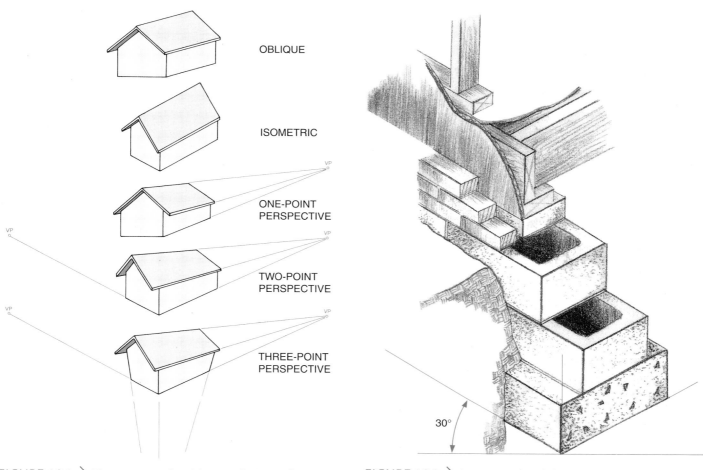

OBLIQUE

ISOMETRIC

ONE-POINT
PERSPECTIVE

VP

TWO-POINT
PERSPECTIVE

VP

THREE-POINT
PERSPECTIVE

VP

30°

FIGURE 19.1 > Three types of architectural pictorial drawings. *Hepler/Wallach/Hepler © Cengage 2013*

FIGURE 19.3 > Isometric detail drawing. *Hepler/Wallach/Hepler © Cengage 2013*

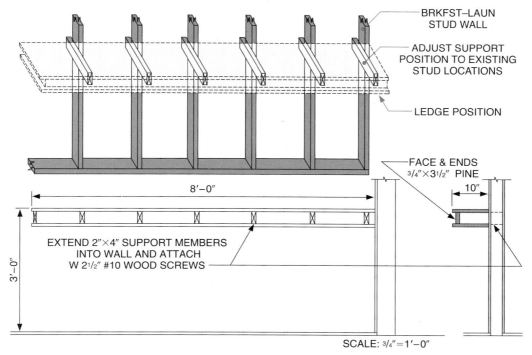

BRKFST–LAUN
STUD WALL

ADJUST SUPPORT
POSITION TO EXISTING
STUD LOCATIONS

LEDGE POSITION

FACE & ENDS
3/4″×3 1/2″ PINE

10″

8′–0″

3′–0″

EXTEND 2″×4″ SUPPORT MEMBERS
INTO WALL AND ATTACH
W 2 1/2″ #10 WOOD SCREWS

SCALE: 3/4″=1′–0″

FIGURE 19.2 > Oblique construction detail drawing as seen in Figure 17.35. *Hepler/Wallach/Hepler © Cengage 2013*

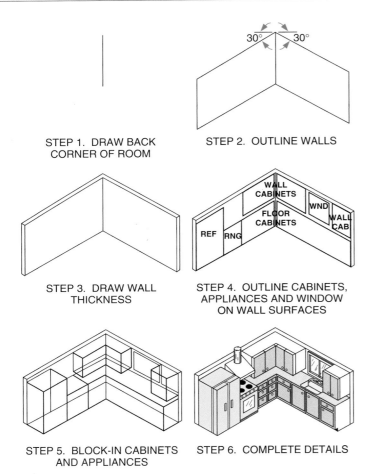

STEP 1. DRAW BACK CORNER OF ROOM

STEP 2. OUTLINE WALLS

STEP 3. DRAW WALL THICKNESS

STEP 4. OUTLINE CABINETS, APPLIANCES AND WINDOW ON WALL SURFACES

STEP 5. BLOCK-IN CABINETS AND APPLIANCES

STEP 6. COMPLETE DETAILS

FIGURE 19.4 > Interior isometric drawing. *Hepler/Wallach/ Hepler © Cengage 2013*

PERSPECTIVE DRAWINGS

In **perspective drawings,** receding lines of a building appear to converge. They are not drawn parallel. A perspective drawing, more than any other kind of drawing, most closely resembles the way people actually see an image. If you look down railroad tracks, the parallel tracks appear to come together and vanish at a point on the distant horizon. Similarly, horizontal lines on a perspective drawing appear to meet at a distant point. The point at which these lines seem to meet and disappear is known as the **vanishing point.**

On two-point exterior perspective drawings, a **horizon line** is established. This line is the observer's eye level. The location of the observer is the **station point,** as shown in Figure 19.5. A **picture plane** is an imaginary plane between the station point and the object on which a perspective view is observed. The vanishing points in a perspective drawing are always placed on the horizon line. If the horizon line is placed through the center of a building, the building will appear to be at eye level (Figure 19.6). For this reason the horizontal line on a two-point interior perspective wing is usually placed at or near eye level as shown in Figure 19.7. If the horizon line is placed low or below a building, the building will appear as

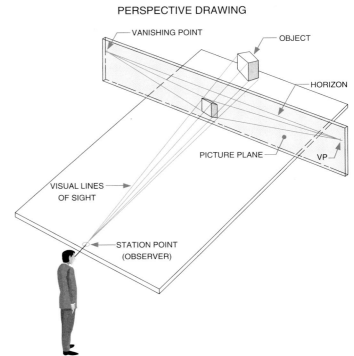

PERSPECTIVE DRAWING

FIGURE 19.5 > Visualization of a two-point perspective shape on a picture plane. *Hepler/Wallach/ Hepler © Cengage 2013*

FIGURE 19.6 > Effect of horizon line placed through the center of a building. *Jenkins & Chin Shue, Inc.*

if you were looking up at it (Figure 19.8). If the horizon line is placed above a building, the building will appear to be below your **line of sight** (Figure 19.9). Objects placed close to or on the horizon line are less distorted than objects placed a greater distance from the horizon.

We are accustomed to seeing areas decrease in depth as they recede from our point of vision. That is why the sides of an isometric drawing (prepared with the true dimensions of a building) appear distorted. To make perspective drawings appear more realistic, the actual lengths of the receding side

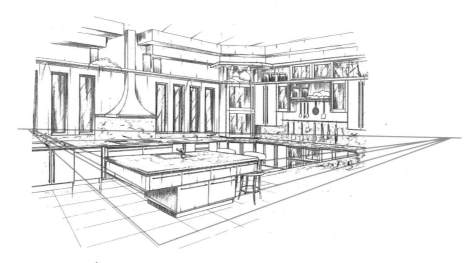

FIGURE 19.7 > Interior horizon line at eye level. *LaStrada Furniture and Interiors, Inc.*

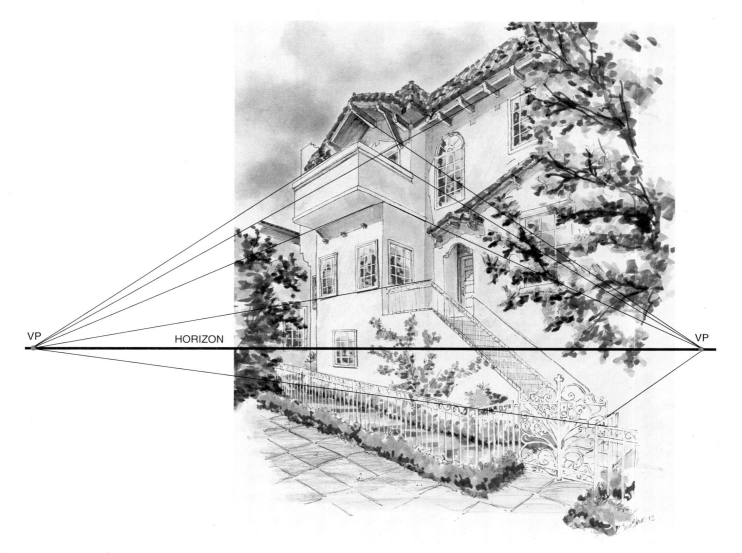

VP   HORIZON   VP

FIGURE 19.8 > Effect of horizon line placed low on a building. *Jenkins & Chin Shue, Inc.*

HORIZON

FIGURE 19.9 > Effect of horizon line placed above a building. *Turner, Lechmer & Romero, Architects*

lines of the building are shortened. Because perspective draw-ings do not reveal the true size of the building, they are not used as working drawings.

Exterior Perspective Drawings

Perspective drawings of building exteriors are either one-, two-, or three-point perspective drawings depending on the size and proportion of the structure.

One-Point Perspective

A **one-point perspective** is a drawing in which the front view is drawn to its true elevation scale and all receding sides are projected to a single vanishing point. If the vanishing point is placed directly to the right side of an interior or enclosed space as shown in Figure 19.10, the left side will show. If the vanishing point is placed on the left side of an interior or enclosed space, as shown in Figure 19.11, the right side will show. If the object is placed above the horizon line and vanishing point, more of the bottom of the object will show. If the object is placed below the horizon line and vanishing point, more of the top of the object will show. Vanishing points need not always fall outside the building outline. When vanishing points are located within the building out-line, only the frontal plane will show. Figure 19.12A shows an elevation drawing that has been converted to a one-point perspective by projecting elevation lines to a center vanish-ing point. Compare the photograph of the house shown in Figure 19.12B with the drawing in Figure 19.12A.

Follow these steps when drawing or sketching a one-point perspective. See Figure 19.13.

STEP 1   Draw the horizon line and mark the position of the vanishing point. If the vanishing point is to the left and outside the building, the left side of the building will show. If the vanishing point is to the right and outside the building, the right side of the building will show.

STEP 2   Draw the front view of the building to a conve-nient scale.

STEP 3   Project all visible corners of the front view to the vanishing point.

STEP 4   Estimate the length of the house. Draw lines par-allel with the vertical lines of the front view to indicate the back of the building.

STEP 5   Make all object lines heavy, such as roof overhang. Erase the projection lines leading to the vanishing point.

Two-Point Perspective

A **two-point perspective** drawing is one in which the re-ceding sides are projected to two vanishing points, one on opposite ends of the horizon line. See Figure 19.14. In a two-point perspective, no sides are drawn exactly to scale. All sides recede to vanishing points. Therefore, the only true-length line, the one that is to scale, on a two-point perspective may be the vertical line in the corner of the building from which both sides are projected.

When the vanishing points are placed very close together on the horizon line, considerable distortion results because of the acute receding angles, as shown in Figure 19.15. When the vanishing points are placed farther apart, the drawing looks more realistic. Placing the drawn object closer to the horizon also helps create a more realistic appearance.

One vanishing point is often placed farther from the sta-tion point than the other vanishing point. This placement allows one side of the building to recede at a sharper angle

FIGURE 19.10 > Effect of the vanishing point placed on the right. *Jenkins & Chin Shue, Inc.*

FIGURE 19.11 > Effect of the vanishing point placed on the left. *Jenkins & Chin Shue, Inc.*

FIGURE 19.12A 〉 Elevation converted to a one-point perspective. *Hepler/Wallach/Hepler © Cengage 2013*

FIGURE 19.12B 〉 Photograph of the drawing in Figure 19.12A. *Photo courtesy Lindal Cedar Homes. Lindal.com*

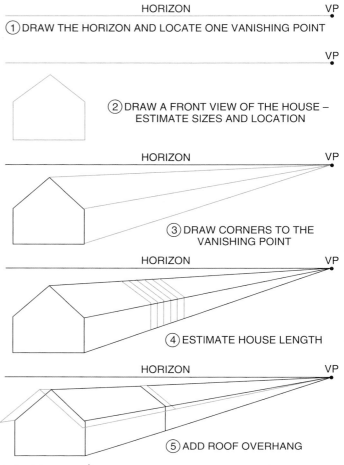

HORIZON   VP

① DRAW THE HORIZON AND LOCATE ONE VANISHING POINT

VP

② DRAW A FRONT VIEW OF THE HOUSE – ESTIMATE SIZES AND LOCATION

HORIZON   VP

③ DRAW CORNERS TO THE VANISHING POINT

HORIZON   VP

④ ESTIMATE HOUSE LENGTH

HORIZON   VP

⑤ ADD ROOF OVERHANG

FIGURE 19.13 〉 Sequence for projecting a one-point perspective drawing. *Hepler/Wallach/Hepler © Cengage 2013*

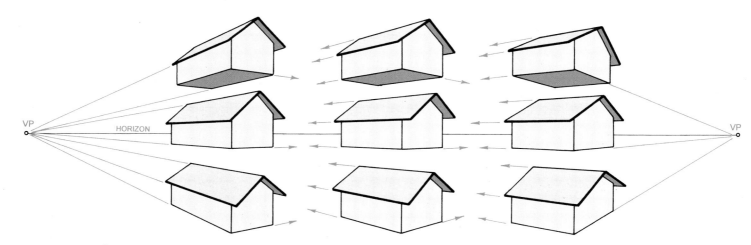

FIGURE 19.14 〉 Vertical and horizontal position options on a two-point perspective. *Hepler/Wallach/Hepler © Cengage 2013*

than the other. The left vanishing point in Figure 19.16 is placed closer to the left of the building than the right vanishing point. This exposes more of the right side of the structure. The reverse is true in Figure 19.17 where the right vanishing point is closer to the building. In drawing or sketching a simple two-point perspective, the steps outlined in Figure 19.18 can be followed.

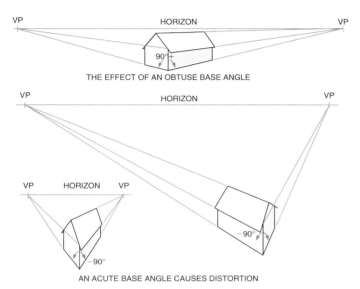

FIGURE 19.15 > Effect of vanishing point and horizon line positioning. *Hepler/Wallach/Hepler © Cengage 2013*

Three-Point Perspective

Three-point perspective drawings are used to overcome the height distortion of tall buildings. In a one- or two-story building, the vertical lines recede so slightly that, for practical purposes, they are drawn parallel. However, the top or bottom of extremely tall buildings appears smaller than the area nearest the viewer. A third vanishing point may be added to provide the desired recession. See Figure 19.19. The farther away the third vanishing point is placed from the object, the less the distortion. If the lower vanishing point is placed so far below or above the horizon that the angles are hardly distinguishable, then the advantage of a three-point perspective may be lost because the vertical lines are almost parallel.

Interior Perspective Drawings

A pictorial drawing of the interior of a building may be an isometric drawing, a one-point perspective, or a two-point perspective drawing. Pictorial drawings may be prepared for an entire floor plan. More commonly, however, a pictorial drawing is prepared for a partial view of a single room or to show a particular interior detail.

One-Point Perspective

A one-point perspective of the interior of a room is a drawing in which all of the intersections between walls, floors, ceilings, and furniture are projected to one vanishing point. Drawing

FIGURE 19.16 > Effects of placing vanishing point closer to the left side of a building. *Jenkins & Chin Shue, Inc.*

HORIZON

FIGURE 19.17 > Effects of placing vanishing point closer to the right side of a building.
Jenkins & Chin Shue, Inc.

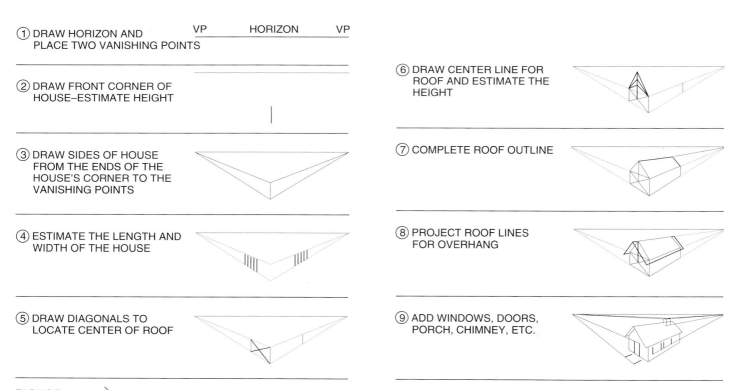

① DRAW HORIZON AND PLACE TWO VANISHING POINTS

VP   HORIZON   VP

② DRAW FRONT CORNER OF HOUSE–ESTIMATE HEIGHT

③ DRAW SIDES OF HOUSE FROM THE ENDS OF THE HOUSE'S CORNER TO THE VANISHING POINTS

④ ESTIMATE THE LENGTH AND WIDTH OF THE HOUSE

⑤ DRAW DIAGONALS TO LOCATE CENTER OF ROOF

⑥ DRAW CENTER LINE FOR ROOF AND ESTIMATE THE HEIGHT

⑦ COMPLETE ROOF OUTLINE

⑧ PROJECT ROOF LINES FOR OVERHANG

⑨ ADD WINDOWS, DOORS, PORCH, CHIMNEY, ETC.

FIGURE 19.18 > Steps in drawing a two-point perspective. *Hepler/Wallach/Hepler © Cengage 2013*

ONE-POINT
PERSPECTIVE

TWO-POINT
PERSPECTIVE

THREE-POINT PERSPECTIVE

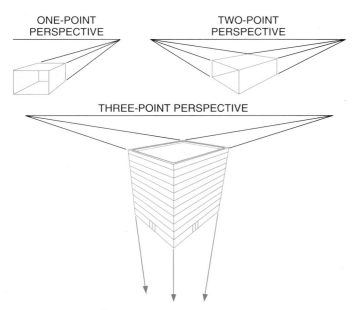

FIGURE 19.19 > Comparison of one-, two-, and three-point perspective drawings. *Hepler/Wallach/Hepler © Cengage 2013*

a one-point perspective of the interior of a room is similar to drawing the inside of a box with the front of the box removed. In a one-point interior perspective, walls perpendicular to the plane of projection, such as the back wall, are drawn to scale. The vanishing point on the horizon line is then placed somewhere on this wall. In the sketch shown in Figure 19.20, the vanishing point is located vertically just above eye level and horizontally centered in the room. The points where this wall intersects the ceiling and floor are then projected from the vanishing point to form the intersection between the side walls and the ceiling and the side walls and the floor.

Vertical Placement

Vertical placement of the vanishing point in a one point perspective greatly affects the appearance of a drawing. If the vanishing point is placed high, as in Figure 19.21A, very little of the ceiling will show in the projection, but much of the floor area will be revealed. If the vanishing point is placed near the center of the back wall, as in Figure 19.21B, an equal

FIGURE 19.20 > Vanishing point lines projected to an interior single point. *John Henry, Architect*

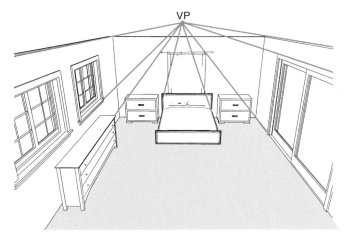

FIGURE 19.21A ＞ Effects of high vanishing point height.
Hepler/Wallach/Hepler © Cengage 2013

FIGURE 19.21B ＞ Effect of a middle vanishing point height.
Hepler/Wallach/Hepler © Cengage 2013

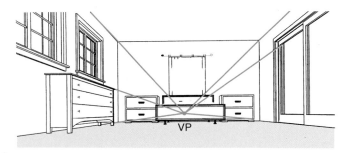

FIGURE 19.21C ＞ Effect of a low vanishing point height.
Hepler/Wallach/Hepler © Cengage 2013

amount of ceiling and floor will show. If the vanishing point is placed low on the wall, as in Figure 19.21C, much of the ceiling but very little of the floor will be shown. Because the horizon line and the vanishing point are at your eye level, you can see that the position of the vanishing point affects the angle from which you view the object.

Moving the vanishing point horizontally from right to left on the back wall has an effect on the view of the side walls.

If the vanishing point is placed toward the left, more of the right wall will be revealed. See Figure 19.22. Conversely, if the vanishing point is placed near the right side, more of the left wall will be revealed in the projection. This was done in Figure 19.23 to expose the left wall. The arrow in the floor plan shows the viewer's direction and location.

When a vanishing point is placed in a floor plan drawing, a perspective view from above is created. These are often called a **bird's-eye view.** Unlike vertical perspective views, these horizontal views show all four walls of a room. To show all four interior walls of a room on the same drawing, the vanishing point must be placed near the center of the room as shown in Figure 19.24.

When drawing wall offsets and furniture, always block in the overall size of the item to form a perspective view. The steps are shown in Figure 19.25. The details of furniture or closets or even of persons can then be completed within this blocked-in cube or series of cubes.

Two-Point Perspective

Two-point perspectives are normally prepared to show the final design and decor of two walls of a room. The vertical base line on an interior two-point perspective is one rear corner of the room. On an interior drawing, this line may be a corner of a room, an article of furniture, or other vertical line. Not only are the walls projected to the vanishing points in a two-point perspective, but each object in the room is also projected to the vanishing points. The sequence of steps in drawing two-point interior perspectives is shown in Figures 19.26A through 19.26C.

PROJECTION METHODS

Several methods can be used to project lines to vanishing points on perspective drawings. These include connecting lines with a straight edge, drawing over underlay perspective grid sheets (Figures 19.27A and 19.27B).

CAD Pictorials

As covered in Chapter 5, CAD pictorial drawings are created from two-dimensional drawings. This is accomplished through the use of computer programs that add the third Cartesian axis (Z). The three main types of CAD pictorial drawings are **wireframe, surface models,** and **solid models.** Wireframe models are see-through stick drawings in which some or all hidden lines are visible. Figure 19.28 shows a wireframe drawing with selected hidden lines removed. Surface model drawings are created with solid plane

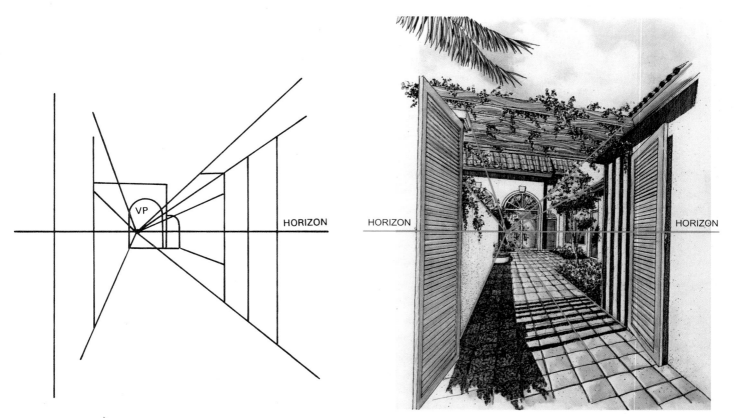

FIGURE 19.22 > Effect of left-side vanishing point placement. *Jenkins & Chin Shue, Inc.*

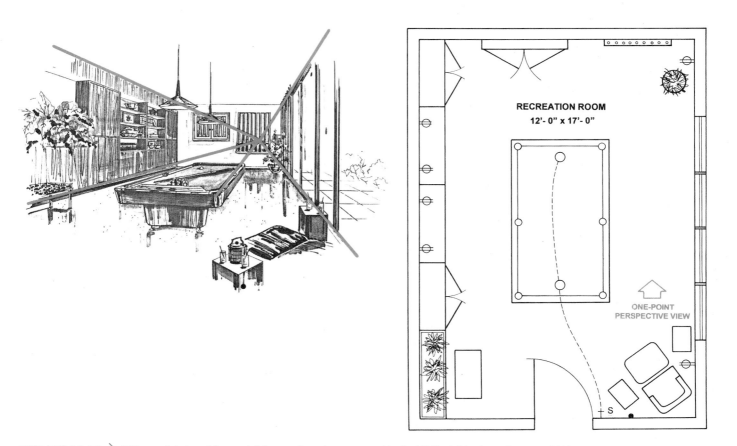

FIGURE 19.23 > Effect of right-side vanishing point placement. *Hepler/Wallach/Hepler © Cengage 2013*

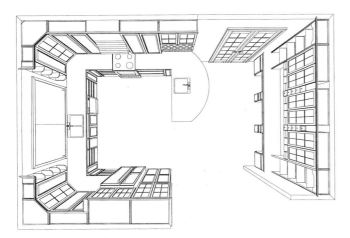

FIGURE 19.24 › One-point plan perspective; also called a bird's-eye view. *Hepler/Wallach/Hepler © Cengage 2013*

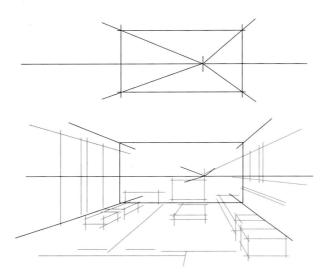

FIGURE 19.25 › Steps in developing an interior one-point perspective. *Used by permission Hanley Wood, LLC*

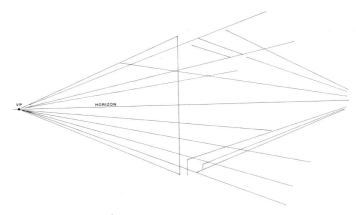

FIGURE 19.26A › Layout step for developing a two-point interior perspective. *Jenkins & Chin Shue, Inc.*

FIGURE 19.26B › Major outlines added. *Jenkins & Chin Shue, Inc.*

FIGURE 19.26C › Completed interior two-point perspective. *Jenkins & Chin Shue, Inc.*

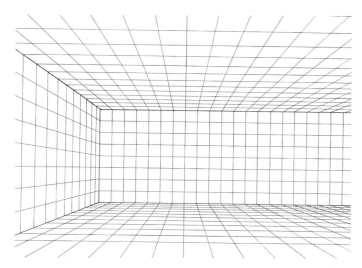

FIGURE 19.27A > One-point perspective grid. *Hepler/ Wallach/Hepler © Cengage 2013*

FIGURE 19.28 > Wireframe drawing with exposed hidden lines. *Jack Hale Drawing*

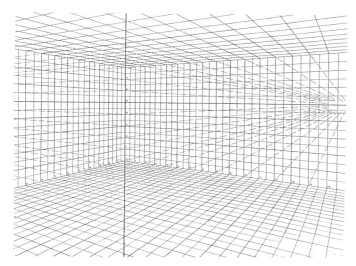

FIGURE 19.27B > Two-point perspective grid. *Hepler/ Wallach/Hepler © Cengage 2013*

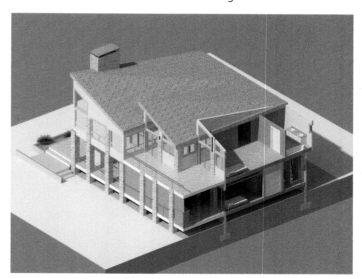

FIGURE 19.29 > Surface model developed from a wireframe drawing. *Swain CAD Solutions*

surfaces. The drawing shown in Figure 19-28 is prepared with most hidden lines removed. Note the difference in Figure 19-29 where all hidden lines are exposed. The use of different color layers can be used to separate and better define the visible lines compared to hidden lines.

Like conventional pictorial drawings, CAD perspective drawings can be created with the vanishing point located in a wide range of locations. Figure 19.30 shows a variation of a bird's-eye view with three vanishing points located below and outside the room plan.

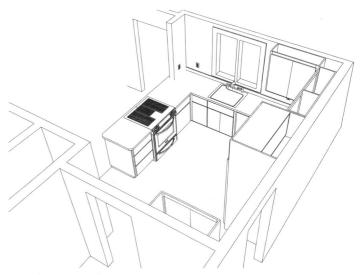

FIGURE 19.30 > Bird's-eye view with vanishing point outside room. *Christopher Polanski Design*

Pictorial Drawings Exercises

1. Draw a two-point perspective of a building of your own design.

2. Draw a one-point and a two-point perspective of your own home.

3. Sketch a three-point perspective of the tallest building in your community.

4. Prepare a one-point interior perspective of your own room.

5. Prepare a one-point perspective of a room in a house of your own design.

6. Draw a one-point interior perspective of a classroom. Prepare one drawing to show more of the ceiling and left wall. Prepare another drawing to show more of the floor and right wall.

CHAPTER 20

Architectural Renderings

OBJECTIVES

In this chapter you will learn to:

> recognize the wide selection of media available for renderings.

> evaluate when to use which media to achieve an artistic effect.

> add realism to drawings by the use of shading, shadows, texture, entourage, and landscapes.

> follow the correct sequence for preparing a rendering.

TERMS

acrylics
entourage
felt markers
line drawings

overlay
pastels
render

texture
wash drawing
watercolors

INTRODUCTION

Because pictorial drawings are three dimensional, their shape resembles a realistic view of a building. Our eyes see more than shape though. We see color, texture, shades, shadows, people, and landscape features. In a rendering, these features are added to a pictorial drawing. To **render** a drawing is to make the drawing appear more realistic—whether it is a plan, an elevation drawing, or a perspective drawing. Drawings are rendered by adding realistic textures and by establishing shade and shadow patterns. This may be done using a variety of media.

CAD Rendering

Architectural rendering using CAD involves a wide variety of tasks and commands to manipulate library resources. These include adding lines, color, and features such as foliage, doors, windows, and sky scenes.

Combining these into a finished CAD rendering involves many steps. As covered in Chapter 5, the preparation of CAD renderings begins with the completion of a solid model drawing. Surface textures and materials are then added from the textures and materials library to each appropriate building plane. This also includes the selection of exterior siding, roofing, and interior surface treatments for each plane. Light and shadow elements are then added through the use of photoshop and *Hatch* commands. Finally, exterior landscape features, including sky, vegetation, and interior furniture and details are added for realism.

Some renderings are a combination of a CAD drawing with hand-rendered final additions and blending of features. A skillfully created CAD rendering may appear identical to a hand-prepared rendering. A knowledge of media, light, and textures is necessary for both types of renderings. Figure 20.1 shows a CAD rendering with landscape and street features added.

RENDERING MEDIA

A rendering may utilize only one medium, or several media may be combined to create various images. Media used to render drawings include pencils, charcoal, pastels (light-colored, water-based chalk), ink, watercolors, oil paint and **acrylics** (water-based permanent paints), felt markers, and overlays.

FIGURE 20.1 > CAD rendering with features added. *Sears Architects, Grand Rapids, Michigan*

Pencil Renderings

Soft (B) pencils are effective media for rendering architectural pictorials. Changes in the weight and density of lines create many tones. See Figure 20.2. Variations in the spacing of pencil lines and in the pressure of the pencil can create different values and contrasting effects. Smudge blending to add tone is accomplished by rubbing a finger over soft penciled areas. To add surface realism, extremely soft charcoal pencils can be used over a textured surface, as done in Figure 20.3.

Pencil renderings are popular because shading and texture can easily be added to penciled pictorial outlines. Colored or **pastels** can be used to make the various colors and values of building surfaces appear very realistic.

FIGURE 20.2 > Pencil rendering techniques. *Hepler/Wallach/Hepler © Cengage 2013*

FIGURE 20.3 > Charcoal rendering. *Avery Architectural and Fine Arts Library; Hugh Ferris, Illustrator*

Ink Renderings

Because ink lines cannot be blended, the distance between lines is controlled to create the appearance of texture, light, shade, and density. Figure 20.4 shows ink line patterns as rendering techniques. Ink lines and strokes placed close together produce dark effects. Farther apart, they create lighter effects. See Figure 20.5.

Watercolor and Wash Drawings

The use of wash drawings or watercolors is a fast and effective method of adding realism to pictorial drawings. When only black and gray tones are used, this type of rendering is called a **wash drawing.** When color is added, these drawings are known as **watercolors.** Watercolor paints blend to create a variety of attractive color and gradation effects. Therefore, watercolors are used extensively for presentations and for advertising. Perhaps the most effective use of watercolor techniques is for the pictorial combination of landscape settings with structures. Watercolors are also used to embellish floor plans as shown in Figure 20.6.

Oil and Acrylic Renderings

Architectural renderings in oil paints or **acrylics** are more time consuming and expensive than any other medium. For this reason, they are rarely prepared, except as works of fine art.

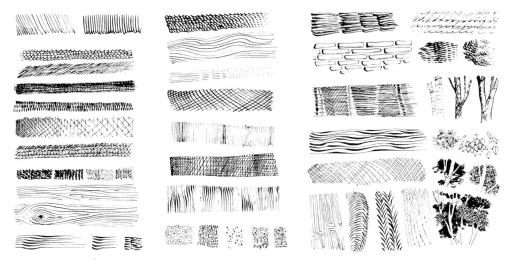

FIGURE 20.4 > Common ink strokes. *Hepler/Wallach/Hepler © Cengage 2013*

FIGURE 20.5 > Ink rendering. *Used by permission Hanley Wood, LLC*

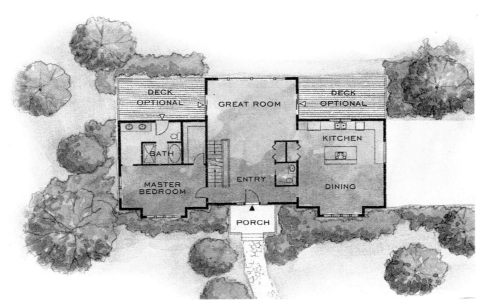

FIGURE 20.6 ＞ Watercolor rendering. *Photo courtesy Lindal Cedar Homes. Lindal.com*

Felt-Marker Renderings

The use of **felt** (or felt-tip) **markers** is a fast way of adding color to pictorial drawings. Stroke lines do not blend easily, so this method is usually restricted to adding patches of color to existing drawings.

Pressure-Sensitive Overlays

A technique that is used to convert a perspective drawing into a rendering is to apply a pressure-sensitive **overlay** to the drawing. Preprinted pressure-sensitive screens are used to add tones, texture, and shadow. These effects are created by variations in the distance between lines or dots, in the width of lines, and in the blending of lines. Pressure-sensitive overlays can be used to create gray tones or solid black areas for contrast or for light and shadow patterns.

Media Combinations

Most architectural renderings include a variety of media, depending on the cost and the emphasis desired. Architectural illustrator Mark Englund (Figure 20.7) is shown preparing a magazine rendering using layers of ink and watercolors.

FIGURE 20.7 ＞ Architectural illustrator at work. *Photo by Mark Englund*

FIGURE 20.8 › Rendering with pastels and watercolors over a line drawing. *Jenkins & Chin Shue, Inc.*

Interior design features are frequently rendered using watercolor over line work to reveal the color and texture of materials. See Figure 20.8. Other media normally used in combination with pencil or ink **line drawings** include airbrush, pastels, and felt markers. Figure 20.9 illustrates the relationship of a preliminary line drawing to a finished rendering. As the line drawing "morphs" into a rendering it could be a CAD rendering or prepared in any other media. Figure 20.10 also shows a rendering and the pictorial line drawing that was used as its base.

Another method of preparing pictorial renderings is to use colored or gray illustration board. When this method is used, all lines, shades, or tints are added with a variety of white watercolor or acrylic paint. Notice how the black, white, and gray tones in Figure 20.11 emphasize texture and depth.

The effectiveness of any medium is related to how skillfully it is used to create realistic textures, shades, shadows, and landscape features. Figure 20.12 illustrates the artful blending of these elements to produce a realistic and dramatic rendering.

FIGURE 20.9 > Relationship of a line drawing with a rendering. *LaStrada Furniture and Interiors, Inc.*

FIGURE 20.10 > Comparison of a line drawing with a rendering. *LaStrada Furniture and Interiors, Inc.*

FIGURE 20.11 > Rendering over gray stock. *John Henry, Architect*

FIGURE 20.12 > Dramatic multimedia rendering. *John Henry, Architect*

EFFECTS OF LIGHT

Light Source and Shade

When shading a building, consider the location of the sun or other light sources. Areas exposed to the light source should appear lighter. Areas not exposed to a light source should be shaded or darkened. See Figure 20.13. When an object with sharp corners is exposed to a light source, one side may be extremely light and the other side of the object extremely dark. However, objects and buildings often have areas that are round (cylindrical). These areas change gradually from dark to light. A gradual shading from extremely dark to extremely light must be made. Figure 20.14 shows several methods of freehand shading of sharp corners and rounded corners and how these methods affect the way light is rendered.

Light and shadow patterns at different times of the day or night must also be considered. Computer programs allow designers to study the relationship of design elements to the environment with high levels of accuracy and flexibility. For example, Figure 20.15 simulates the effect of sunlight and artificial light on an architectural design.

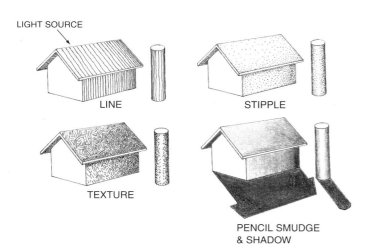

FIGURE 20.13 > Shading methods. *Hepler/Wallach/Hepler © Cengage 2013*

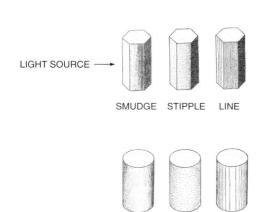

FIGURE 20.14 > Effect of light on different surface shapes. *Hepler/Wallach/Hepler © Cengage 2013*

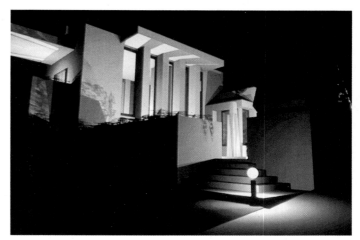

FIGURE 20.15 > Computer manipulation used to produce different effects by means of high, medium, and low lighting. *Hepler/Wallach/Hepler © Cengage 2013*

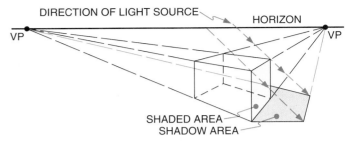

FIGURE 20.17A > Shadow pattern related to vanishing point VP and light source. *Hepler/Wallach/Hepler © Cengage 2013*

FIGURE 20.16A > Shadow effect from a low light source. *Hepler/Wallach/Hepler © Cengage 2013*

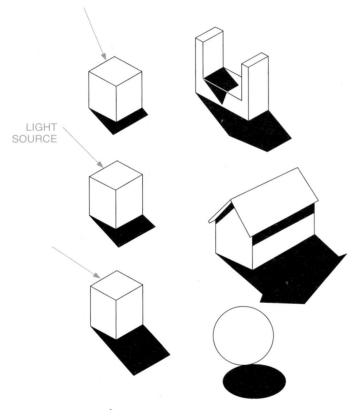

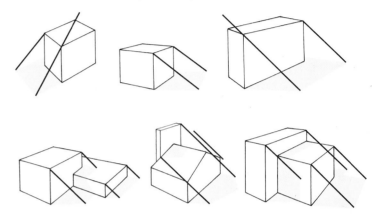

FIGURE 20.17B > Shadows reflect building shapes. *Hepler/Wallach/Hepler © Cengage 2013*

hang depth, building depressions, offsets, and extensions. Figure 20.17A illustrates how shadow patterns are related to vanishing points. Figure 20.17B reveals how various building type drawings are affected by shadow patterns.

Various techniques can be used to indicate shadows. Figure 20.18 shows how depth and shadows are rendered on windows. Some windows show reflected light, and others are drawn to reveal the room behind, as though the window were open. Keep in mind that most windows look dark during the day because the inside of the building is darker than the outside. To produce realistic-looking windows, dark colors or even black surfaces are often used, as shown earlier in Figure 20.5. Foliage and blue sky reflections on windows are often used to depict a very sunny site, as done in Figure 20.19.

FIGURE 20.16B > Shadow effect from a high light source. *Hepler/Wallach/Hepler © Cengage 2013*

Shadow

To determine which areas of a building should be drawn darker to indicate shadowing, the angle of the light source must first be established. Once the angle of the light source has been established, then all shading should be consistent with the direction and angle of the shadows, as shown in Figures 20.16A and 20.16B. In addition to the light source, consider the building outline and site contours when drawing shadows on buildings. Note the connection between the light source angle, building outlines, and site contours. Also, shadowing is often used to reveal hidden features such as over-

TEXTURE

Giving **texture** to an architectural drawing means making building materials appear as smooth or as rough as they actually are. Smooth surfaces are very reflective and hence are very light. Only a few reflection lines are usually necessary to illustrate the smoothness of surfaces such as aluminum, glass, and painted surfaces. For rough surfaces, the material can often be shown by shading and texturing. A variety of texturing methods and rendering techniques for the exterior

FIGURE 20.18 > Shadows used to render windows. *Hepler/Wallach/Hepler © Cengage 2013*

FIGURE 20.19 > Window reflections used to denote brightness. *Jenkins & Chin Shue, Inc.*

FIGURE 20.20A > Finely detailed pictorial line drawing. *Sears Architects, Grand Rapids, Michigan*

FIGURE 20.20B > Color rendering of the line drawing in Figure 20.20A. *Sears Architects, Grand Rapids, Michigan*

FIGURE 20.21 > Pencil rendered textures. *Hepler/Wallach/ Hepler © Cengage 2013*

materials used on siding and roofs can be used. The effect of a color rendering compared to a line drawing is shown in Figures 20.20A and 20.20B. Figure 20.20A shows a line drawing with very fine detailing of features without color. Observe the different appearance when only color is added to this same streetscape in Figure 20.20B. This mixed-use design is part of a *smart growth* community development as covered in Chapter 1.

Figure 20.21 is a pencil rendering that emphasizes the different textures of the siding and roof. Notice how the shading of the texture depends on the light source. Also note how the shadows from trees and overhangs are incorporated into the texture and values to add more realism to the drawing.

Siding materials may be added directly to drawings through various media or through the use of CAD material libraries.

Surface materials can also be added to pictorial drawings using the Photoshop library and the *Hatch* command. Trees and shrubs are also available in 3D form but must be adjusted to fit the scale of the drawing as shown in Figure 20.22.

Rendering "stand-alone" features such as fences, walls, and chimneys requires close attention to be paid to independent shadow and shade effects on texture. See Figure 20.23.

ENTOURAGE

The term **entourage** refers to the people or objects that are part of a building's surroundings and are used to enhance the size, distance, and reality of renderings. Sketches of people—sitting, standing, and walking—are often necessary to show the relative size of a building and to put the total drawing in proper perspective. Because people should not interfere with the view of the building, architects frequently draw people in outline or in extremely simple form. In a drawing, people may be used to indicate pedestrian traffic patterns or to provide a feeling of perspective and depth. Automobiles are also added to architectural renderings to indicate relative size and to give a greater feeling for external traffic patterns.

People, boat, and automobile outlines in a variety of settings, angles, and scales are available on pressure-sensitive sheets or in traceable entourage publications. See Figure 20.24. Entourage figures are often reduced or enlarged to fit within a drawing. Figures closer to the vanishing point are progressively smaller than figures near the station point.

FIGURE 20.22 > CAD rendering with foliage, sky, and shadow details. *Hamid Rafiei Dreamarch3D.com*

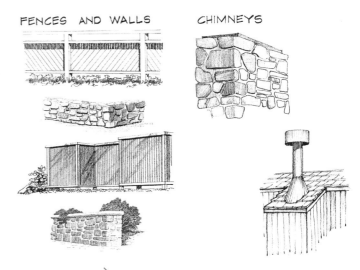

FIGURE 20.23 > Rendering of stand-alone features. *Hepler/ Wallach/Hepler © Cengage 2013*

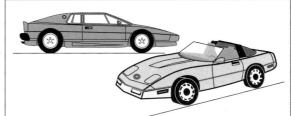

FIGURE 20.24 > Entourage sample images. *Hepler/Wallach/Hepler © Cengage 2013*

CAD Entourage Libraries

CAD entourage libraries or blocks contain people, animals, and vehicles in a wide variety of sizes and shapes, including their walking, sitting, and standing positions. In using entourage symbols, their size can be manipulated using the *Stretch*

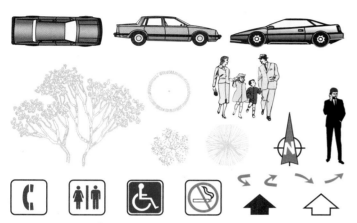

FIGURE 20.25 > Entourage images at different scales and angles. *Hepler/Wallach/Hepler © Cengage 2013*

command and they can be located using the *Move* command. The *Stretch* command can be used to either elongate or condense a feature. The *Move* command allows entities to be moved to a new location by pointing the cursor to the object, then dragging it to the new position.

If more than one entity is needed in a pattern, the *Copy-Multiple* command can be used by inputting the location, spacing, and number of entries. Figure 20.25 shows how a variation in scale and viewing angles can be adjusted to fit into any scaled drawing with the vanishing point and horizon in any position.

LANDSCAPE

In rendering, adding landscape features—whether to pictorial or elevation drawings—involves drawing trees, shrubs, ground cover, driveways, and walkway surfaces. Trees should be placed so as not to block the view of the buildings. See Figure 20.26.

Rendering landscape pictorial drawings also involves using the same techniques to render accessory structures, water features, retaining walls, trellises for vines or other plants, and many other features. See Figure 20.27. Also, refer back to Chapter 13.

FIGURE 20.26 > Pictorial rendering techniques used for landscape features. *Hepler/Wallach/Hepler © Cengage 2013*

FIGURE 20.27 > CAD rendering with multiple buildings and dense landscape patterns. *Mardian Development Company*

Architectural Renderings Exercises

CHAPTER 20

1. Render a perspective drawing of your own house.
2. Render a perspective drawing of a house of your own design.
3. Render a perspective sketch of your school. Choose your own medium: pencil, pen and ink, watercolors, pastels, or airbrush.
4. Collect illustrations that could be adapted for use on renderings: drawings of people, cars in different sizes and positions, landscapes, plants, and so forth.

CHAPTER 21

Architectural Models

OBJECTIVES

In this chapter you will learn to:

> describe architectural models made for design study purposes.

> explain the differences between presentation and design study models.

> tell what input is needed to create a computer model.

> construct an architectural model.

TERMS

basic layout models
CAD models
design study models

detailed model
interior design model
landform model

presentation model
solid form model
structural model

INTRODUCTION

The two general types of architectural models are constructed models, which are covered here, and 3D **CAD models,** which were covered in Chapter 5. Constructed models are three-dimensional replicas of a building or site made to scale. These models reveal real and accurate relationships between elements as actually viewed with human eyes. Single buildings or large developments can be viewed from any location and from any distance by many persons at the same time or at different times. This fosters construction or design-related discussions without the need for a formal presentation. Constructed models may also include movable features (cars, people, furniture) that can be repositioned for design study at any time.

Conversely CAD models can be developed quickly from floor plans and elevations and can include very small details. CAD models can also show interior views at eye level or from any position without viewers having to change their locations. Computer presentations can be controlled to emphasize selected features and to eliminate distractions.

The two basic types of models studied in this chapter are *design study models* and *presentation models*. To check basic design ideas or construction methods during the design process, a design study model is made. A **presentation model** is used

for sales purposes because people can understand a design more easily by looking at a model than they can by looking at architectural working drawings.

DESIGN STUDY MODELS

Architectural models have been used for centuries and up to the present to visualize the final appearance, function, and construction of a design (Figure 21.1).

Design study models are helpful during the design process. They are used to check the form of a structure, verify the basic layout, clarify construction methods, or show interior design options. They can also be used to finalize the orientation of a structure on a site. Models of this type are used to study sun angles and show shadow patterns at different times and on different days. See Figure 21.2.

Solid Form Models

Before final dimensions are applied to a design, a **solid form model** like the one shown in Figure 21.3 is often made. It is used to check the overall proportions of a building. Solid form models that contain no details can help designers to study the size and relationship of building clusters. Solid form models are made from Styrofoam™, balsa, clay, or soap.

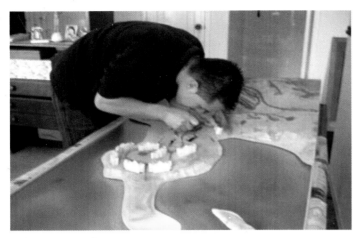

FIGURE 21.1 > Building architectural models. *Artesanos Architectural Models*

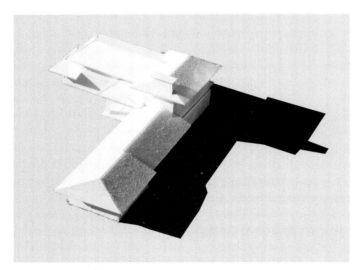

FIGURE 21.2 > Design study model. *Hepler/Wallach/Hepler*
© Cengage 2013

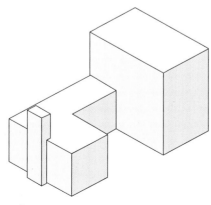

FIGURE 21.3 > Solid form model. *Hepler/Wallach/Hepler*
© Cengage 2013

For buildings designed for high-density areas, solid form models of adjacent buildings are often needed to show the relative size and position of all buildings. Solid form models are also used in community and city planning. Figure 21.4A shows a model of this type that was used to study a renewal proposal for a portion of Manhattan. Figure 21.4B shows the related plan of this area.

FIGURE 21.4A > City planning model. *Hepler/Wallach/Hepler*
© Cengage 2013

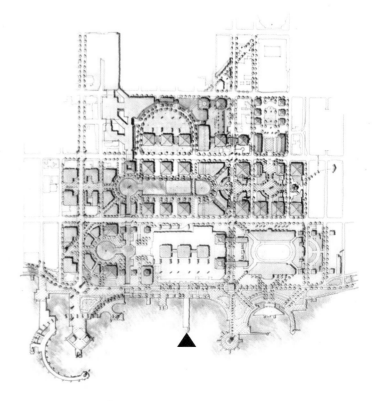

FIGURE 21.4B > Concept drawing of the area shown in Figure 21.4A. *Hepler/Wallach/Hepler*
© Cengage 2013

Basic Layout Models

Design study models also provide a means to check the overall layout and function of a design. First the preliminary floor plan and elevations must be available. Then **basic layout models** are made. These models are constructed to the same scale as the floor plans and elevations, usually 1/8″ = 1′-0″ or 1/4″ = 1′-0″. They are not finely detailed because they will probably be revised and altered many times. The layout model shown in Figure 21.5A is shown in front of the completed house it represents. This model was used by contractors to help explain design details to subcontractors during construction (Figure 21.5B).

Basic layout models are especially useful for illustrating the relationship of units in a townhouse master planned community, as shown in Figure 21.6. Models can also be placed di-

rectly on a site and photographed from different angles. This reveals more realistically how the completed structure will appear when built on the intended site. Figure 21.7 contains a preliminary design model situated on the actual location of the future building.

Structural Models

Only the structural members of a building are shown on a **structural model.** Builders use these models to check unique structural methods or to study framing options. These checks are especially important if many houses with the same structural design are to be built. Building a model of this type is also a good way to learn framing methods.

Structural models are usually built to a scale of 1″ = 1′-0″, 3/4″ = 1′-0″, or 1/2″ = 1′-0″. Smaller scales of 1/8″ = 1′-0″ or 1/4″ = 1′-0″ are difficult to construct.

Structural members are cut to scale or purchased from model stock. Some parts are assembled into panels by placing

FIGURE 21.5A > Basic layout model with completed structure. *James Eismont, Photographer*

FIGURE 21.6 > Townhouse master plan models. *Artesanos Architectural Models*

FIGURE 21.5B > Model used to check progress on site. *Hepler/Wallach/Hepler © Cengage 2013*

FIGURE 21.7 > Model positioned to simulate reality. *Hepler/ Wallach/Hepler © Cengage 2013*

FIGURE 21.8 > Structural model used for client orientation. *Photo courtesy Lindal Cedar Homes. Lindal.com*

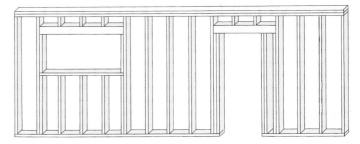

FIGURE 21.9 > Scaled model wood wall framing. *Hepler/Wallach/Hepler © Cengage 2013*

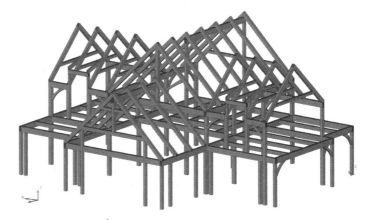

FIGURE 21.10 > Structural model framing. *Timberpeg Inc.*

them directly over a wall-framing drawing or over elevation drawings. The panels and other components are then assembled with glue and pins as shown in Figure 21.8. A structural model is built in the same sequence as a full-size house.

The steps in constructing a structural model follow these sequences:

STEP 1   *Foundation plan base.* Attach a print of the foundation plan to a sheet of rigid foamboard or Styrofoam sheet with rubber cement or use a preglued sheet. Allow some space on all sides for future site features.

STEP 2   *Foundation walls.* Cut foundation walls to scale using wood, foamboard, or Styrofoam. Glue these to the foundation plan base. Glue beams and columns on the plan using dowels or precut model components. If a slab foundation is planned, construct the footing similar to the walls and use plywood to represent the slab.

STEP 3   *Prepare materials.* Cut wood strips or procure model kit materials to represent studs, rafters, sills, and joists.

STEP 4   *Wall framing.* Using an elevation drawing or elevation framing drawing as a pattern, glue strips together to make a wall-framing panel as shown in Figure 21.9. Repeat for each interior partition and exterior wall. Fasten wall-framing plans to a wood base and cover with wax paper. Strips of scaled wood members are then aligned with members shown on the plan and glued together.

STEP 5   *Floor framing.* Place scaled floor joists on a floor framing plan, or floor plan, using predesigned spacing. Frame around stairwell, chimney/fireplace, or other openings as described in Chapter 28.

STEP 6   *Roof framing.* Refer to the roof framing plan for the size and spacing of rafters as shown in Chapter 30. Place scaled roof rafters on an exterior elevation plan and glue each to a ridge board and to the top exterior wall plate to create a gable roof. Using a 90° elevation, cut and glue descending-length rafters to hip rafters and top plates to create a hip roof. Trusses can be used as an alternative.

STEP 7   *Floor and siding.* Subflooring and finished flooring, plus exterior siding and insulation, are partially added in areas such that they do not hide the basic framing methods. See Figure 21.10.

Structural models with partially completed and removable walls and roofs are often used to verify the relationship of finished areas with structural elements as shown in Figure 21.11.

Interior Design Models

An **interior design model** is used to show individual room designs. These can be shown effectively with one-room models. Individual room models usually include the floor, ceiling, and three walls. One wall and ceiling remains open for

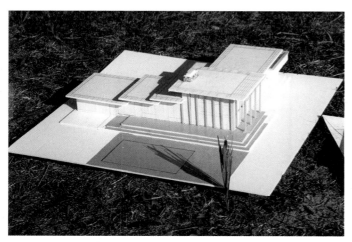

FIGURE 21.11 > Combination layout and structural model.
Hepler/Wallach/Hepler © Cengage 2013

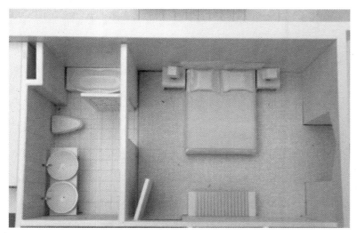

FIGURE 21.12 > Four wall room model. *Artesanos
Architectural Models*

FIGURE 21.13 > Three wall room model. *Artesanos
Architectural Models*

FIGURE 21.14 > Presentation model. *Model Technology Co.,
Ltd. Beijing Jian Yi Xuan*

viewing, like a dollhouse. One-room models are usually built to a scale of 1″ = 1′-0″ or 1 1/2″ = 1′-0″. Scaled human figures, decor, and furniture add to the realism of interior design models. Figure 21.12 shows a four wall room model and Figure 21.13 is a three wall model which allows top and also right side viewing.

PRESENTATION MODELS

Most presentation models are used to promote the sale of a building, land parcels, or community development projects. Presentation models replicate (copy) the actual appearance of the real project in as much detail as the scale allows. Presentation models, such as the one shown in Figure 21.14, are often used for design comparisons and competitions.

Landform Models

Presentation models are frequently used to show the landform around a building. A **landform model** represents the shape

and slope of a site. In this model, the thickness of each layer is equal (in scale) to the different levels of the land. The shape of each layer is the same shape as the contour lines on a survey drawing. Where contour intervals are large, landform layers are added to create a smoother contour. See Figure 21.15. Landform models are needed to show relationships between different ground line elevations. This assists in clarifying the differences in elevations at all sides of a building. Landform models are also extensively used for community development as shown in Figures 21.16 and 21.17.

To develop smooth contours, small posts are used to represent the contour height. These posts are spaced throughout the model site. Flexible screens are then laid on the posts and covered with papier-mâché, which can then be smoothed. See Figure 21.18.

Detailed Models

City and housing tract developers are the largest users of presentation models. To show building relationships, developers

FIGURE 21.15 > Landform model using contour interval construction. © *Dana Hepler*

FIGURE 21.16 > Community development models. *Artesanos Architectural Models*

FIGURE 21.17 > Landform model for community development. *Model Technology Co., Ltd. Beijing Jian Yi Xuan*

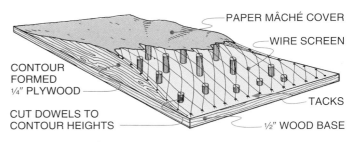

PAPER MÂCHÉ COVER

WIRE SCREEN

CONTOUR FORMED ¼" PLYWOOD

TACKS

CUT DOWELS TO CONTOUR HEIGHTS

½" WOOD BASE

FIGURE 21.18 > Smooth landform construction. *Hepler/Wallach/Hepler © Cengage 2013*

often surround a **detailed model** with solid form models of adjacent and nearby buildings, as shown in Figure 21.19. Constructed models of this type are often used to create a subject for photographs. A photograph can be retouched later to add or eliminate details and features.

Housing developers use landform models with small-scale solid form models to show specific lot locations and their relationship to other features. The placement of people, cars, and trees on these models adds realism and better defines the scale of the project. See Figure 21.20.

STEPS IN CONSTRUCTING A MODEL

An accurate and realistic model is constructed to a precise scale with careful attention to detail. Many materials and items used to build models, such as those shown in Figure 21.21A and 21.21B, may not be available in the scale needed. These must then be constructed by the model maker. Model makers may use manufactured parts or may fabricate materials.

Methods of model construction vary depending on the material and the amount of detail required. The following procedures represent a typical sequence for constructing solid wall models:

FIGURE 21.19 〉 Detailed model with surrounding buildings in solid form. © Dana Hepler

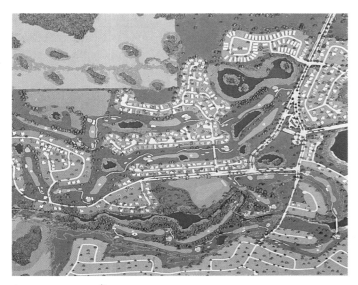

FIGURE 21.20 〉 Housing development model. *The Plantation at Leesburg*

STEP 1   *Floor plan base.* Attach a print of all floor plans to a sheet of rigid foamboard or Styrofoam sheet with rubber cement or use a preglued sheet. Allow some space on all sides.

STEP 2   *Wall construction.* Glue a print of all exterior elevations to a sheet of foamboard or softwood. This foamboard should be the same scaled thickness as the outside walls indicated on the floor plan. Cut

PART	MODEL MATERIALS	METHODS OF CONSTRUCTION
■ STRUCTURE		
base	plywood; particleboard hollow core panel	Cut to maximum, but convenient, size for structure and site.
walls	softwood; cardboard; acrylic; matboard; foamboard; Styrofoam™; wallpaper; fabric; plywood	Cut walls to exact dimensions of elevations. Allow for overlapping of joints at corners. Wall thickness must be to scale.
floors	flocking for carpet; printed paper of floor type; thin wood veneer; vinyl scraps	Paint floor area with slow-drying colored enamel and apply flock. Remove excess when dry. With paper, glue in place. With wood veneer, rule black lines for strip effect and glue in place.
windows and doors	purchased, premade strips of wood or plastic; thin, clear acetate; acrylic	Glue premade windows and doors in place. Cut strips for casing, sill, and window frames, and glue in place. Draw the windows and doors directly on the walls.
roofs	thin, stiff cardboard or wood; paint; colored sand; wood pieces for shingles; premade/printed roof coverings; sandpaper	Cut out roof patterns and assemble. For roof coverings, glue on sand, wood pieces for shingles, or preprinted roof coverings.
■ BUILDING MATERIALS		
siding materials	scored sheets of balsa wood; foamboard; preprinted paper patterns; wood strips for board and batten and horizontal siding	Glue or paint siding materials to model walls.
stucco	spackle; plaster of paris; sandpaper; sand and thick paint	Mix and dab on with a brush leaving a rough texture.
brick and stone	printed paper; embossed plastic sheets; thin softwood	Glue paper in place. Cut grooves in wood, and paint color of bricks or stone.
wood paneling	printed paper; thin veneer wood; molded plastic sheets	Glue paper or plastic sheet in place. With veneer wood, rule lines for strip effect.

FIGURE 21.21A 〉 Model materials for structure and building surfaces. *Hepler/Wallach/Hepler* © Cengage 2013

PART	MODEL MATERIALS	METHODS OF CONSTRUCTION
■ FURNISHINGS furniture, appliances, fixtures	commercial models; doll furniture (to scale); cardboard; softwood; clay; Styrofoam™; soap; fabric; paint	Purchase commercial model furniture or carve/sculpt to shape. Paint or flock for a finish.
fireplaces	(Refer to brick and stone materials.)	Carve fireplace, and simulate finish.
■ BUILDING SITE topography	wood base; wire screen; papier-mâché; fine gravel	Build up sloped areas with sticks and wire screen. Place papier-mâché over wire, and glue gravel for soil effect.
geologic features, terrain, water	stones; sand; colored gravel and sand	Glue small rocks for boulders. Paint high-gloss blue paint for water.
swimming pools	wood strips; blue paint or paper; sheet glass; acrylic	Outline pool with glued wood strips. Paint or glue paper for water and attach clear glass. Ripple acrylic surface while drying for ripple effect.
■ LANDSCAPE grass	green paint and flock	Paint grass area and apply flock. Remove excess when dry.
trees and bushes	sponges; lichen; small twigs	Grind up sponges and paint shades of green. Glue pieces to twigs for trees and bushes. Lichen may be purchased in model stores in bulk or as model trees.
wood fences	wood strips	Glue together to form a fence.
masonry fences	(Use same materials as for masonry walls.)	Form the same as walls to fence size.
gazebo	wood strips	Assemble from a working drawing.
■ MISCELLANEOUS automobiles	commercial models; toys; clay; soap; Styrofoam™	Purchase or shape from soft materials and paint.
people	commercial models; toys; clay; soap; Styrofoam™	Purchase or shape from soft materials and paint.
■ DRIVEWAYS	sandpaper, sand over glue, paint	

FIGURE 21.21B > Model materials for furnishings, site, landscape, and movable items. *Hepler/ Wallach/Hepler © Cengage 2013*

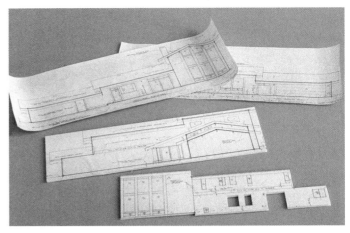

FIGURE 21.22 > Elevation drawings glued to foamboard. *Hepler/Wallach/Hepler © Cengage 2013*

the foamboard to create a wall for each elevation as shown in Figure 21.22. Be sure to add sufficient material to allow exterior corners to overlap unless corners are mitered.

STEP 3 Repeat this procedure for all interior partitions. If interior elevations are not available, the outline of each partition must be drawn onto and cut from foamboard. Usually all windows and doors are cut out of interior and exterior walls, as shown in Figure 21.23. An alternative, particularly for design study models, is to simulate them using paint or a suitable material.

STEP 4 The next step is to attach window and door trim and acetate. Figure 21.24 shows one method of constructing model window trim. Figure 21.25 shows the application of acetate to an inside wall with transparent tape. Doors may be cut from thin, stiff cardboard and hinged with transparent tape. If windows and doors are not cut out, they could just be shaded black or gray for effect, as in Figure 21.26, or the selected material should be glued in place.

STEP 5 *Wall attachment.* Glue the exterior walls to the floor plan base and to each other at the corners. Fit the corners carefully so that the overall outside wall lengths align properly with the floor plan corners. Use a 90° triangle to check the plumb and squareness of the tops of corner intersections. Glue interior partitions to the appropriate partition lines indicated on the floor plans. See Figure 21.27. Glue intersecting corners to outside walls and to other partitions. Figure 21.28 illustrates howexterior walls are positioned on a floor plan in preparation for adding interior partitions. Next glue the interior partitions onto the base and to other interior partitions or exterior

FIGURE 21.23 > Cutting windows and doors in elevation panel. *Hepler/Wallach/Hepler © Cengage 2013*

FIGURE 21.25 > Attaching acetate windows. *Hepler/Wallach/ Hepler © Cengage 2013*

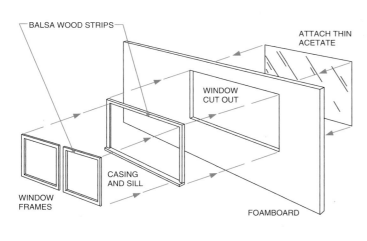

BALSA WOOD STRIPS

ATTACH THIN ACETATE

WINDOW CUT OUT

WINDOW FRAMES

CASING AND SILL

FOAMBOARD

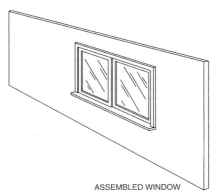

ASSEMBLED WINDOW

FIGURE 21.24 > Window trim assembly. *Hepler/Wallach/ Hepler © Cengage 2013*

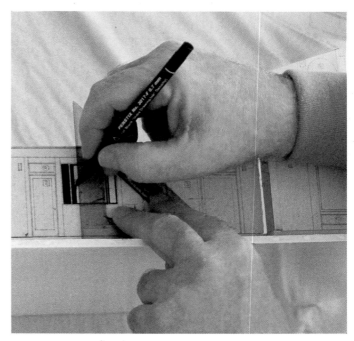

FIGURE 21.26 > Darkening windows on a solid panel. *Hepler/Wallach/Hepler © Cengage 2013*

FIGURE 21.27 ❯ Attaching walls to base and adjoining walls. *Hepler/Wallach/Hepler © Cengage 2013*

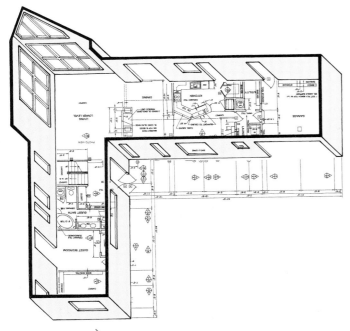

FIGURE 21.28 ❯ Position of external model walls on a floor plan. *Hepler/Wallach/Hepler © Cengage 2013*

walls. Figure 21.5 shows the completed model described in the preceding steps.

STEP 6 *Wall finishing.* Paint interior walls and floors with white or lightly tinted tempera paint. Apply texture or simulated (imitation) coverings to represent floor tiles, masonry surfaces, fireplaces, and chimneys. Add siding materials or textures to exterior surfaces. See Figure 21.29. Spraying the finished surfaces with fixative will help surface treatments adhere better.

FIGURE 21.29 ❯ Adding surface texture to walls. *Hepler/Wallach/Hepler © Cengage 2013*

STEP 7 *Cabinetry and fixtures.* Make solid 3D forms to represent built-ins. These may include kitchen and bath cabinet fixtures, fireplaces, chimneys, and bookshelves. If desired, add color to cabinets, countertops, or fixtures. Figure 21.30 shows a model with all fixtures and furniture added. Figure 21.31 shows a close-up view of the kitchen and dining area of this model.

STEP 8 *Roof construction.* Roofs should be constructed separately so that they can easily be removed to reveal the interior. Flat roofs can be constructed directly from a roof plan. The pitched roof in Figure 21.32 must be cut to align with the pitch indicated on an elevation drawing. Using true length lines construct the roof in panels to represent continuous flat surfaces. Then glue the panels together. Other roof components and simulated roof coverings can then be added. Construct the chimney and glue it to the top of the roof to align with the interior chimney. This is done to allow the part of the chimney above the roof to be removed along with the roof. Figure 21.33 shows a townhouse model with a removable roof and upper level.

STEP 9 *Outdoor areas.* After the structural part of the model is complete, remove any extra base material from the floor plan. Proceeding outward from the house, add outdoor areas, such as patios, pools, ponds, decks, or lanais. Include all of the features that are within the property's perimeter.

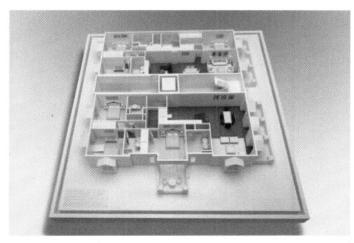

FIGURE 21.30 > Interior model with all fixtures and furniture. *Artesanos Architectural Models*

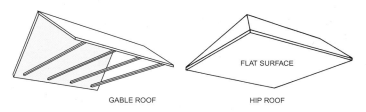

FIGURE 21.32 > Model roof construction. *Hepler/Wallach/ Hepler © Cengage 2013*

FIGURE 21.31 > Model room close-up. *Artesanos Architectural Models*

STEP 10 *Landscape features.* A complete architectural model includes landscaping and details of the site. Construct walkways, driveways, and steps to connect different levels of the property. Retaining walls might need to be added to separate parts of the site. Add landscape features such as trees, fountains, shrubbery, and ground cover. Cars, people, and furniture can be included to help define the scale and the areas.

FIGURE 21.33 > Removable roof and upper level model. *Archi-Vision Scale Models*

Architectural Models Exercises

CHAPTER **21**

1. What are the basic purposes of architectural models?
2. Describe the types of design study models and explain their functions.
3. What features does a presentation model usually include?
4. List the steps for constructing a model.
5. Construct a basic layout model for a house shown in Chapter 1.
6. Construct a model of a house you designed.
7. Complete a computer model of a house you designed.

PART 6

FOUNDATIONS AND CONSTRUCTION SYSTEMS

CROSS E
HEADER

CHAPTER 22
Principles of Construction

OBJECTIVES

In this chapter you will learn to:

> name and define physical forces that act on a building.

> describe the factors that determine the strength of structural components.

> draw a modular floor plan, elevation, and detail drawing.

TERMS

bearing-wall structures
building load
cantilever
compression forces
dead loads
deflection
lateral (horizontal) loads

live loads
modular components
module
prebuilt homes
prebuilt modules
prefabricated homes

reinforcement
shear forces
skeleton-frame structures
stability
tension forces
torsion forces

INTRODUCTION

New construction materials and new methods of using conventional materials provide designers with great flexibility in construction design. Stronger buildings can now be erected with lighter and fewer materials.

Although the basic principles for preparing construction drawings are the same for all types of construction, the use of symbols, conventions, and terms changes from system to system. Construction systems are broadly divided into four material groups: wood, steel, masonry, and concrete. Specific drawings for these groups will be discussed in detail in subsequent chapters. Most contemporary designs include a combination of these systems and materials. Sometimes parts are preassembled and brought to the building site. Information about construction using prefabricated components is included in this chapter.

Structural design concepts are based primarily on Newton's third law of motion, which states that "for every force there is an equal and opposite force." These forces produce stresses and deformation of all building materials. Such forces include live loads, dead loads, impact loads, seismic loads, and wind loads. These loads are affected by gravity, temperature, and humidity.

Other than fixed dead loads, all forces defined in this chapter are variable in location, magnitude, and time. These factors must therefore be integrated into the structural design of all buildings.

Materials and practices used in construction must be structurally sound and also environmentally sustainable and safe. Where appropriate and possible, materials should also be biodegradable, recyclable, and energy efficient and adhere to LEEDS recommended practices and standards.

STRUCTURAL DESIGN

Structurally, buildings are divided into two types: bearing-wall structures and skeleton-frame structures. **Bearing-wall structures** have solid walls that support the weight of the walls, floors, and roof.

As covered in Chapter 1, the early stone buildings of antiquity and, later, log cabins were bearing-wall structures as shown in Figure 22.1. With the development of post-and-lintel construction and the arch, larger openings could be de-

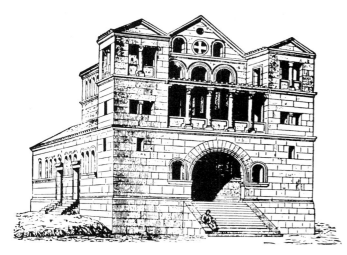

FIGURE 22.1 > Bearing-wall construction. *GAF Corp.*

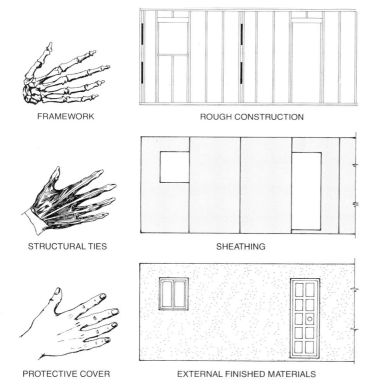

FRAMEWORK

ROUGH CONSTRUCTION

STRUCTURAL TIES

SHEATHING

PROTECTIVE COVER

EXTERNAL FINISHED MATERIALS

FIGURE 22.2 > Skeleton frame compared to vertebrate structure. *Hepler/Wallach/Hepler © Cengage 2013*

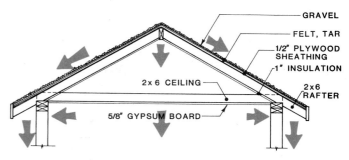

GRAVEL

FELT, TAR

1/2" PLYWOOD SHEATHING

1" INSULATION

2 x 6 CEILING

2 x 6 RAFTER

5/8" GYPSUM BOARD

FIGURE 22.3 > Major lines of force. *Hepler/Wallach/Hepler © Cengage 2013*

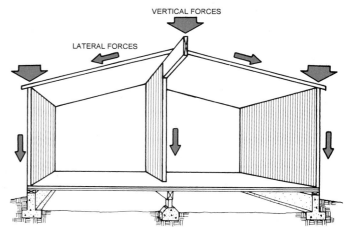

VERTICAL FORCES

LATERAL FORCES

FIGURE 22.4 > Interior partitions help support roof loads. *Hepler/Wallach/Hepler © Cengage 2013*

In both types of construction, structural **stability** is based on the strength and placement of the structural members.

The design of structures involves the laws of physics. Not only must a foundation be designed to support the loads (weight, or mass) of a structure, but the structure itself must also withstand the forces acting on it. Figure 22.3 shows a building's major lines of force supported by a foundation through the exterior walls. Figure 22.4 illustrates how an interior bearing partition helps support roof loads.

The walls of a structure are given stability by their attachment to the ground and to the roof. The roof is supported by wall framework, interior partitions, or columns. Each exterior wall and bearing partition is supported by the foundation, which in turn is supported by footings. Footings distribute building loads over a wide area of load-bearing soil and thus tie the entire structural system to the ground. In A-frame or continuous-arch construction, the roof is supported by direct connection with the foundation, as shown in Figure 22.5.

Every architectural designer needs to understand the relationships among loads, forces, and strength of materials. The proper selection and use of materials depend on understanding the loads and forces acting on these materials. Regardless of the materials and methods used, the physical principles of structural design remain constant.

signed into solid walls. Later, with the development of lighter and stronger materials such as steel and structural lumber, **skeleton-frame structures** could be constructed. Skeleton-frame structures have an open, self-supporting framework covered by an outer, nonbearing surface. Most contemporary homes are of the skeleton-frame type. Figure 22.2 illustrates the concept of skeleton-frame construction compared to the human anatomy. Skeleton-frame structures using steel are known as *steel-cage* or *curtain-wall construction.*

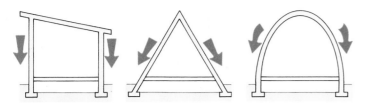

FIGURE 22.5 > Roof load forces transferred directly to the foundation. *Hepler/Wallach/Hepler © Cengage 2013*

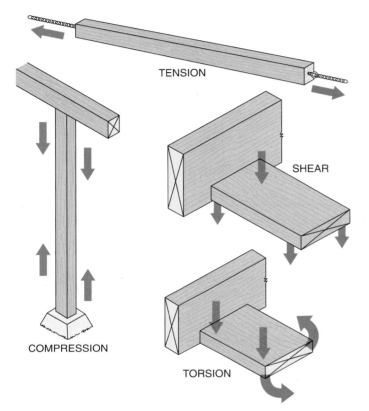

FIGURE 22.6 > Types of forces. *Hepler/Wallach/Hepler © Cengage 2013*

Structural Forces

Four types of force that exert stress on building materials are compression, tension, shear, and torsion. See Figure 22.6. **Compression forces** push on objects. They tend to flatten materials. **Tension forces** pull on objects. They stretch materials. A supporting chain on a hanging light fixture is in tension. **Shear forces** tend to make one part of an object slide past another part. Excessive shear loads may cause material fractures by abrupt action, as scissors cut a piece of paper. **Torsion forces** twist an object. They can twist a member out of shape or fracture it completely by overloading an end or by movement of a connecting member. In wood construction the direction of grain is important because compression and tension forces are minimized when these forces are parallel

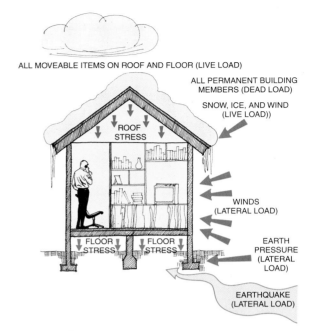

FIGURE 22.7 > Types of loads and forces. *Hepler/Wallach/ Hepler © Cengage 2013*

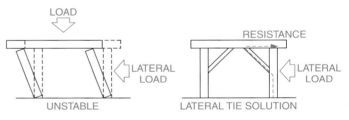

FIGURE 22.8 > Diagonal ties provide rigidity. *Hepler/Wallach/ Hepler © Cengage 2013*

to the grain. The shear strength of wood is greatest when perpendicular to the grain.

Loads

Loads supported by buildings include live loads and dead loads. **Live loads** are the weight of all movable objects, such as people and furniture. Live loads also include the weight of snow and the force of wind. **Dead loads** are the weight of building materials and permanently installed components. Every piece of lumber, brick, glass, and nail adds to the dead load of a structure. The total weight or mass of all live and dead loads is known as the **building load.** See Figure 22.7.

Most loads follow lines of gravity. However, wind, earth (next to the foundation), and earthquakes also act on a building. These can exert **lateral (horizontal) loads.** There are several ways to counteract the force of these lateral loads. The early Egyptians recognized that the triangle provided the most rigidity with the fewest number of members. Likewise, creating triangular support for right-angle intersections stabilizes a structure, as shown in Figure 22.8. This principle also

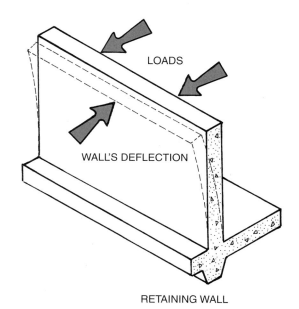

FIGURE 22.9 > Effect of lateral wall support. *Hepler/Wallach/ Hepler © Cengage 2013*

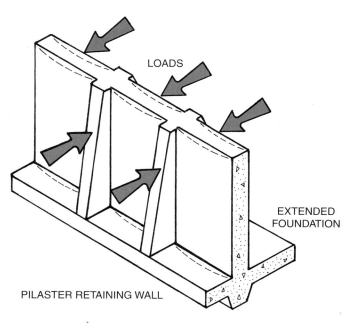

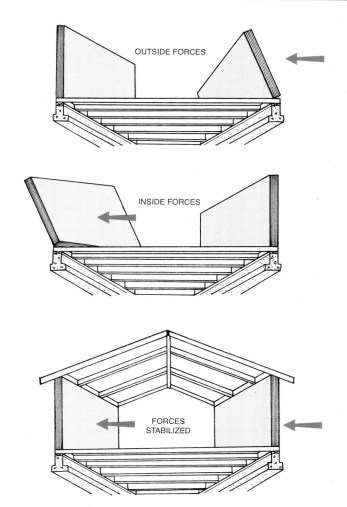

FIGURE 22.10 > A roof adds stability to walls. *Hepler/Wallach/ Hepler © Cengage 2013*

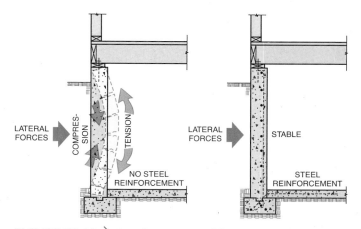

FIGURE 22.11 > Reinforcement adds to wall stability. *Hepler/ Wallach/Hepler © Cengage 2013*

applies to the stability of freestanding walls as illustrated in Figure 22.9.

Figure 22.10 shows how a roof acts to stabilize exterior walls, and Figure 22.11 describes how **reinforcement** provides protection from compression and tension forces.

Roof loads are measured in pounds per square foot (PSF). See Figure 22.12. For all practical purposes, snow and wind loads are combined and considered as one total live load. For example, the combined wind and snow loads in the South Pacific are 20 pounds per square foot (PSF) compared to 30 PSF in the central/western parts of the United States and 40 PSF in the northern parts of the United States.

The design of a structure affects its ability to withstand loads. For example, the pitch of a roof helps determine its ability to withstand snow and wind loads. Snow loads are exerted in a vertical direction. Wind loads are exerted in a horizontal direction. Therefore, a high-pitched roof will withstand snow

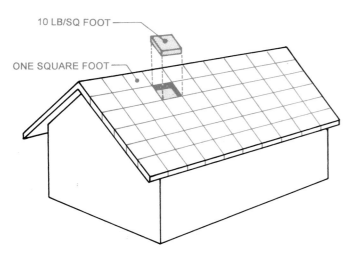

FIGURE 22.12 > This roof load is 10 pounds per square foot. *Hepler/Wallach/Hepler © Cengage 2013*

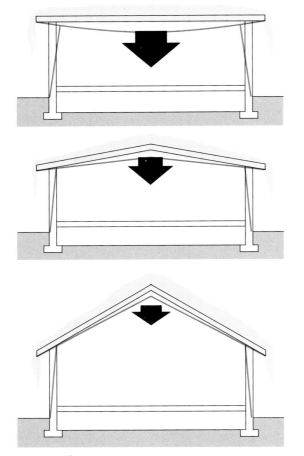

FIGURE 22.13 > Snow loads related to roof pitch. *Hepler/Wallach/Hepler © Cengage 2013*

loads better than a low-pitched roof (see Figure 22.13), but the reverse is true of wind loads (see Figure 22.14).

Buildings must support their own weight. Dead loads can be calculated for each building or specified in precalculated building codes. Material types, their size, and the

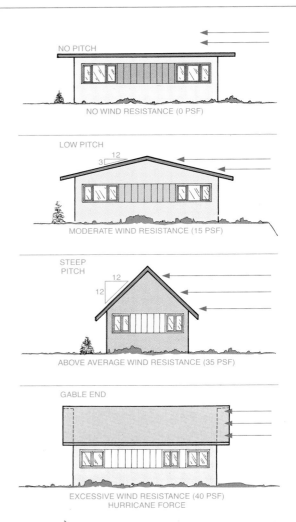

FIGURE 22.14 > High-pitched roofs create high wind resistance. *Hepler/Wallach/Hepler © Cengage 2013*

spacing of members are also calculated or specified by code to prevent the building of structures that are unstable due to excess loading.

Roof loads include live loads (snow and wind) as well as the dead loads of shingles, sheathing, and rafters. All loads are computed on the basis of pounds per square foot, or kilograms per square meter if metric measurements are used. The typical asphalt- or composition-shingle roof weighs approximately 10 to 12 PSF. Thus a 40′ × 20′ (800 sq.ft.) asphalt-shingle roof should be designed to carry a dead load of 8,000 to 9,600 pounds (800 sq.ft. × 12 PSF = 9600 lbs.). A Spanish tile roof weighs 17 PSF.

Resistance

Compression, tension, shear, and torsion forces create enormous stress on building materials. If a material is sufficiently strong, this stress will create little or no damage. However, if the material is weaker than the forces applied, the resulting

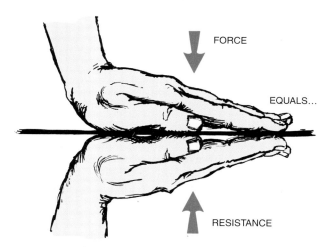

FIGURE 22.15 > Resistance must be equal to or greater than the force to provide stability. *Hepler/ Wallach/Hepler © Cengage 2013*

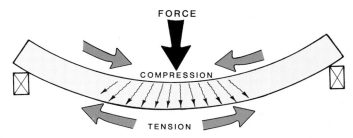

FIGURE 22.16 > The effect of force on compression and tension. *Hepler/Wallach/Hepler © Cengage 2013*

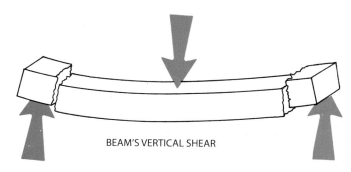

FIGURE 22.17 > Vertical shear on a beam. *Hepler/Wallach/ Hepler © Cengage 2013*

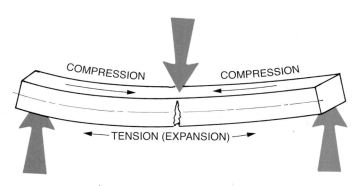

FIGURE 22.18 > Excessive load results. *Hepler/Wallach/Hepler © Cengage 2013*

stress can compress, stretch, slice, twist, or completely fracture a member. Therefore, the resistance of each structural member must always be equal to or greater than the force applied. See Figure 22.15.

To ensure that every structural member can resist the forces and stresses created by building loads, the material, size, shape, placement, and spacing must be carefully planned. If any of these factors change during the building process, another design element may need to be changed to maintain the structural integrity of the building. For example, if a load is increased, stronger materials or a shorter distance between supporting members would be needed.

Strength of Materials

The *strength* of a construction material is the material's capacity to support loads by resisting compression, tension, shear, and torsion forces or stresses. The structural strength of a member, or a construction component, depends on the type, size, and shape of the material. Figure 22.16 illustrates how compression and tension forces act on a member to produce deflection. A vertical shear action occurs when the load and the support react against each other. If the member is not strong enough to support the load, the obvious result is a fractured member as shown in Figure 22.17. A member will also fail if the load creates too great an expansion and compression, as is visible in Figure 22.18. Different structural materials have varying capacities to resist stress and to support building loads. For example, a steel member can support more weight than a wood member of the same size. The load-bearing capacity will also vary among species of wood because different fibers have different stress levels.

Deflection, or bending, stress results from both compression and tension forces acting on a member at the same time.

See Figure 22.19. Reinforcing the bottom half of a member balances this stress. Deflection is sometime called *deformation* and is the measure of how much a member bends under a specific load. The measured amount of deflection is used to determine the required size for a member that must support a given load. Figure 22.20 illustrates the deflection of a beam where the loaded side is compressed and the bottom side is tensed.

Larger members can obviously support greater loads than smaller members of the same material. Material sizes greatly affect load resistance. Inadequate size selection can result in material failure. However, choosing oversized materials can result in extreme material waste as shown in Figure 22.21. Builders must avoid excess notching or drilling of structural members, such as for inserting pipes, because notching and drilling reduce the structural strength of the member.

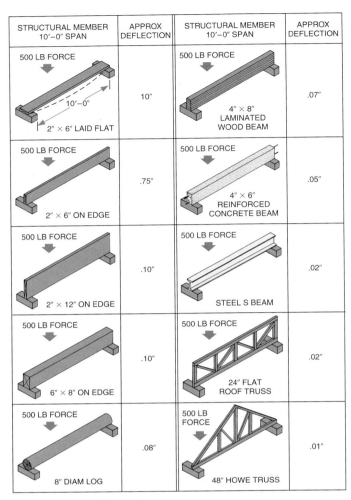

STRUCTURAL MEMBER 10'-0" SPAN	APPROX DEFLECTION	STRUCTURAL MEMBER 10'-0" SPAN	APPROX DEFLECTION
500 LB FORCE — 10'-0" — 2" × 6" LAID FLAT	10"	500 LB FORCE — 4" × 8" LAMINATED WOOD BEAM	.07"
500 LB FORCE — 2" × 6" ON EDGE	.75"	500 LB FORCE — 4" × 6" REINFORCED CONCRETE BEAM	.05"
500 LB FORCE — 2" × 12" ON EDGE	.10"	500 LB FORCE — STEEL S BEAM	.02"
500 LB FORCE — 6" × 8" ON EDGE	.10"	500 LB FORCE — 24" FLAT ROOF TRUSS	.02"
500 LB FORCE — 8" DIAM LOG	.08"	500 LB FORCE — 48" HOWE TRUSS	.01"

FIGURE 22.19 > Deflection differences among structural members. *Hepler/Wallach/Hepler © Cengage 2013*

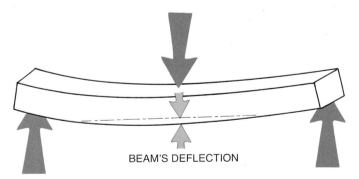

BEAM'S DEFLECTION

FIGURE 22.20 > Beam deflection. *Hepler/Wallach/Hepler © Cengage 2013*

Shape

The different shape of a material influences its ability to support loads. For example, a sheet of thin paper will not stand up by itself. When the paper is folded in half, however, it can support itself plus a light object. See Figure 22.22. Folded several times, this same piece of paper will support a heavier object. Figure 22.23 shows how this principle relates to de-

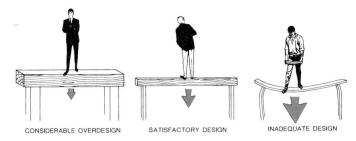

CONSIDERABLE OVERDESIGN   SATISFACTORY DESIGN   INADEQUATE DESIGN

FIGURE 22.21 > Relationship of member size to resistance. *Hepler/Wallach/Hepler © Cengage 2013*

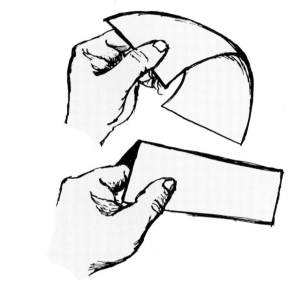

FIGURE 22.22 > Member stability related to its shape. *Hepler/Wallach/Hepler © Cengage 2013*

flection, rigidity, and the load-bearing capacity of a structural member.

Placement

The strength of a building material is significant only when it becomes an integral part of a structure. Most materials are somewhat flexible until tied in to a structure. Orientation, or position, is also an important factor. See Figure 22.24. A structural member placed on its narrower edge will increase the horizontal deflection but the vertical deflection will be unaffected. Members with their widest dimension positioned parallel to the load direction will resist greater loads than members placed with their smallest dimension in the load direction. Combining the horizontal and vertical components of the member in the form of a channel or I-beam reduces both the vertical and the horizontal deflection.

Spans and Spacing

Spacing is the distance between parallel structural members. *Span* is the distance a member extends between vertical supports. The maximum allowable span of a member is directly related to the loads applied to it and the strength of the

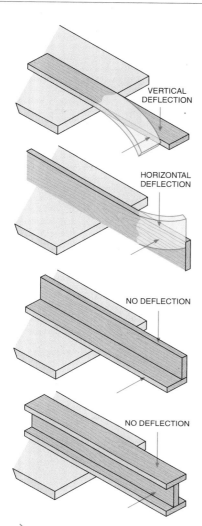

FIGURE 22.23 > Member shape related to deflection. *Hepler/Wallach/Hepler © Cengage 2013*

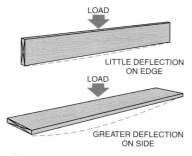

FIGURE 22.24 > Member position related to deflection. *Hepler/Wallach/Hepler © Cengage 2013*

structural member. Obviously, stronger materials can support greater loads at greater distances with less deflection. Decreasing the span while using the same structural members can increase the load-bearing capacity of each member. Figure 22.25 shows the variable span and spacing factors affecting the structural stability of a building.

Cantilever is the term used when only one end of a horizontal structural member is supported. Cantilevered mem-

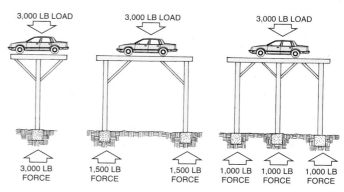

FIGURE 22.25 > Span and spacing factors. *Hepler/Wallach/Hepler © Cengage 2013*

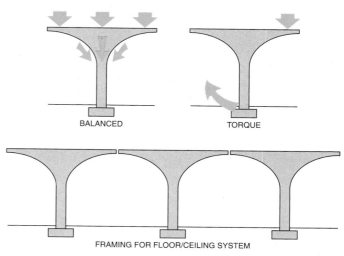

FIGURE 22.26 > Center-supported cantilevered members. *Hepler/Wallach/Hepler © Cengage 2013*

bers can be center supported (Figure 22.26) or eccentric (off center). When the center supports equal dead loads on all sides, the member has equilibrium. Supporting on one side only would cause a torque (torsion stress). Eccentric cantilevered members are supported on one side opposite an unsupported end. Eccentric cantilevering requires stronger materials and stronger anchorage on the supported side. Deflection increases as the distance from the support or the amount of load increases as shown in Figure 22.27. Roof overhangs, balconies, and decks are structural features that are often cantilevered (Figure 22.28).

MODULAR CONSTRUCTION

One of the most significant advances in structural design and construction is the manufacture of preconstructed parts, or components, of a building. The construction business is one of the last segments of industry to fully use standardized, interchangeable parts. Most basic materials—wood, steel, and masonry—are now available in standardized sizes. The sizes may vary, but modular materials and components are designed to fit together with precision.

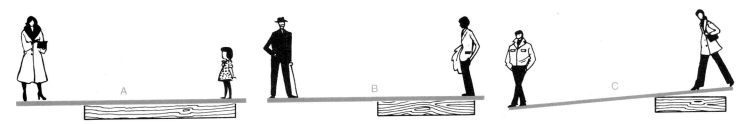

FIGURE 22.27 ❯ Effect of loads and overhang length on cantilevered member. *Hepler/Wallach/Hepler © Cengage 2013*

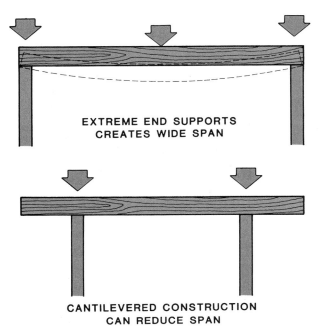

EXTREME END SUPPORTS CREATES WIDE SPAN

CANTILEVERED CONSTRUCTION CAN REDUCE SPAN

FIGURE 22.28 ❯ Advantage of cantilevering. *Hepler/Wallach/ Hepler © Cengage 2013*

Modular Components

Modular components are designed as parts or sections to be constructed away from the building site. This eliminates much on-the-job construction work. Typical components may range from preassembled wall sections and windows to molded bathrooms.

As more components are used, on-site construction work changes from piece-by-piece building to the assembling of factory built components.

Size Standardization

Designing with modular components means the designer must adhere strictly to standard sizes in creating an architectural plan. Sizes of the components are uniform, with many different interchangeable parts.

Modular building design is based on measurements that are divisible by the same base unit. The base unit is known as a **module.** For example, a system based on a module of 4″, 16″, 24″, and 48″ is used for U.S. customary measurements.

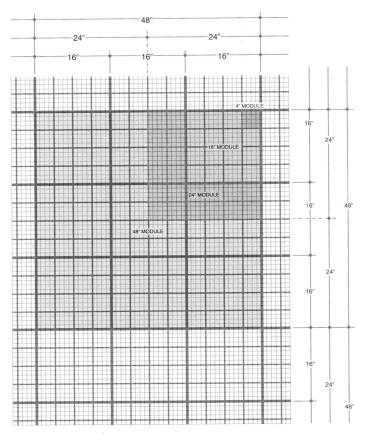

FIGURE 22.29 ❯ Two-dimensional modular grid. *Hepler/ Wallach/Hepler © Cengage 2013*

The metric modular system is based on 100, 300, 600, and 1200 millimeters (mm).

Modular design and construction involves all three dimensions: length, width, and height. The overall width and length dimensions are the most critical in the planning process. The modular planning grid shown in Figure 22.29 is a horizontal plane divided into equal spaces in length and width. It provides the basic control for the architectural modular coordination system. The entire grid is divided into equal spaces of 4″, 16″, 24″, and 48″. See Figure 22.30. All module sizes divide equally into 4′ × 8′ panels. The 16″ unit is used in multiples for wall, window, and door panels to provide an increment small enough for flexible planning. Increments of 24″ and 48″ are used for overall dimensions.

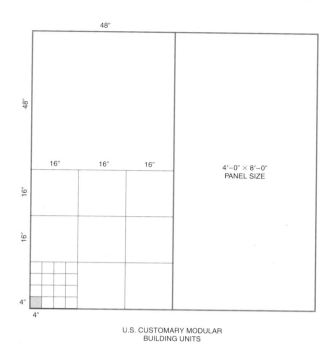

U.S. CUSTOMARY MODULAR
BUILDING UNITS

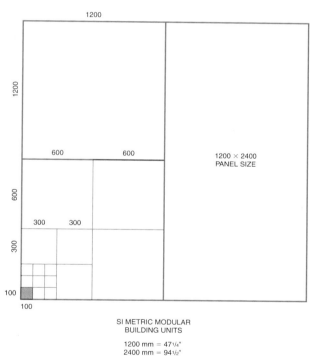

SI METRIC MODULAR
BUILDING UNITS

1200 mm = 47¼"
2400 mm = 94½"

FIGURE 22.30 > U.S. customary and S.I. metric modular
building units. *Hepler/Wallach/Hepler*
© *Cengage 2013*

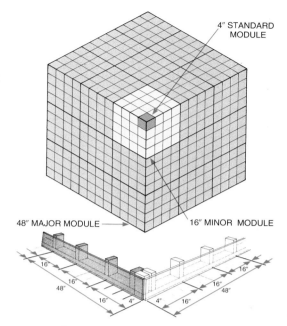

FIGURE 22.31 > Three-dimensional view of the standard,
major, and minor modules. *Hepler/Wallach/
Hepler* © *Cengage 2013*

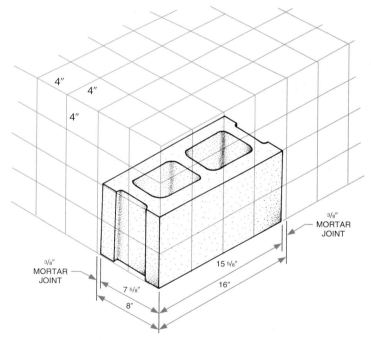

FIGURE 22.32 > Concrete block dimensions based on a 4"
modular unit. *Hepler/Wallach/Hepler*
© *Cengage 2013*

The 24" module is called the *minor module*. The 48" module is called the *major module*. See Figure 22.31.

Many building materials, such as lumber, plywood, brick, tile, and concrete block, are available in modular units. See Figure 22.32. The use of small modular materials, usually based on a 4" module, saves time and expense. The use of large modular units, such as wall panels, door assemblies, and window assemblies, can save much more time and resources. See Figure 22.33. To ensure that all components fit as planned, components must be designed to align with modular grid lines. This requires that architectural drawings be prepared such that building dimensions align with established modular grids.

Modular Drawings

Modular drawings are developed just like other architectural drawings, except both horizontal and vertical members are designed to align with modular grid lines. When drawing a

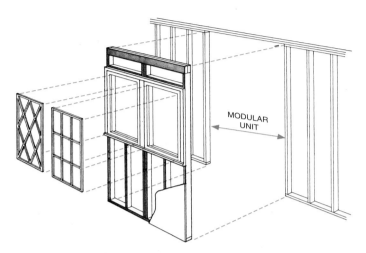

FIGURE 22.33 > Components of a manufactured window assembly align with the same modular standard. *Hepler/Wallach/Hepler © Cengage 2013*

modular design, space must still be provided for such items as doors and windows, plumbing runs, medicine cabinets, closets, and fireplaces.

Framing for the modular components is then extremely critical. Any variation in the framing opening will result in a misfit when the modular units are positioned into the structure.

Modular Floor Plans

To build a modular structure, all drawings must conform to the same modular standards. For example, basic floor plans should be drawn on a modular grid. All nonmodular dimensions need to be aligned or converted to the nearest modular grid line. Figure 22.34 shows a floor plan aligned with a 16″ modular grid. Likewise, stud layouts are also drawn on a modular grid, as are the studs in Figure 22.35. If a window or door assembly is not available in a modular width, align one side on a 16″ grid to reduce the number of studs required.

Dimensions required by existing building laws or from built-in equipment must still be incorporated and coordinated with any prebuilt modular system. This is accomplished by conventional dimensioning. Remember, all dimensions are located either from surface to surface or from center line to center line.

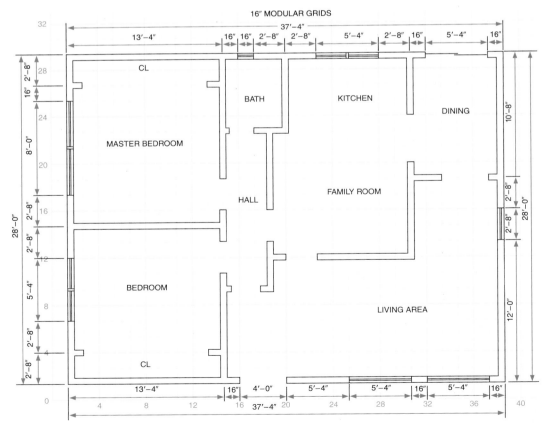

FIGURE 22.34 > Floor plan aligned with a 16″ modular grid. *Hepler/Wallach/Hepler © Cengage 2013*

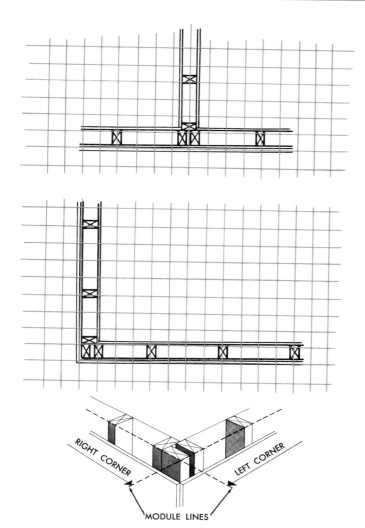

Modular Elevations

As with modular-component floor plans, the modular-component elevation drawings are prepared using a 16″ grid. Elevation drawings closely resemble conventional elevations, except the components are identified and shown in their proper relationship with modular alignments. See Figure 22.36. Likewise, elevation framing drawings must be aligned with established modular grid lines, as shown in Figure 22.37.

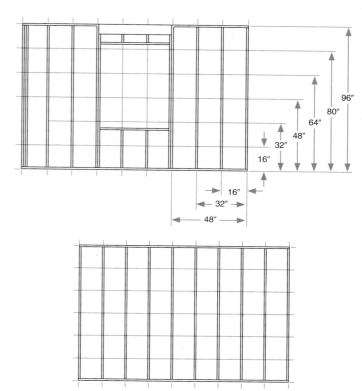

FIGURE 22.35 ⟩ Stud layout aligned on a 4″ modular grid with studs at 16″ on center. *Hepler/Wallach/Hepler © Cengage 2013*

FIGURE 22.37 ⟩ Vertical modular size increments for elevation framing drawings. *Hepler/Wallach/Hepler © Cengage 2013*

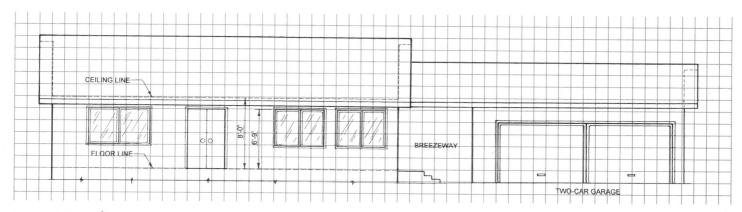

FIGURE 22.36 ⟩ Elevation drawing on a vertical modular grid. *Hepler/Wallach/Hepler © Cengage 2013*

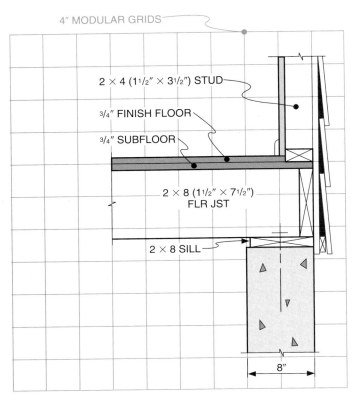

FIGURE 22.38 > Sill detail on a modular grid. *Hepler/Wallach/ Hepler © Cengage 2013*

Modular Details

Many modular drawings are needed to describe every detail of a modular plan. As in a conventional design, more detailed drawings improve the chances of achieving the desired outcomes. Details may be required for standard floor and roof components, as well as for many nonstandard components.

To ensure the correct alignment of key horizontal and vertical surfaces, such as walls and floors, detail drawings are often prepared using a modular grid. See Figure 22.38. An elevation detail section is used primarily to provide exact data for the establishment of critical height dimensions, such as for foundations and floors. Plan detail drawings are used to provide exact alignment data for width and length dimensions.

Modular Dimensioning

Dimensions that align with a module are known as *grid dimensions*. Dimensions that do not align with a module are known as *nongrid dimensions*. Figure 22.39 shows the two methods of indicating grid dimensions and nongrid dimensions. Grid dimensions may be shown by conventional arrowheads, and nongrid dimensions are shown by dots (small circles) instead of arrowheads.

Overall dimensions are first established. Next, nonmodular dimensions are incorporated into the plan. Then panels for exterior doors, windows, and exterior walls are located. The floor plan in Figure 22.34 has been properly fitted on the

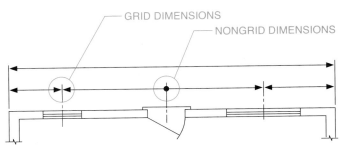

FIGURE 22.39 > Modular and nonmodular dimension methods. *Hepler/Wallach/Hepler © Cengage 2013*

modular grid. It is important for the foundation, floors, walls, windows, doors, partitions, and roof to be properly aligned. Establishing all of the dimensions and components precisely on the 16″, 24″, and 48″ spaces of the modular grid ensures the accurate fitting of the components.

In many construction detail drawings, it is possible to eliminate the placement of some dimensions by placing the grid lines directly over the drawing. When the grid lines coincide exactly with the material lines, no dimensions are needed. For example, using a 4″ module, each line represents 4″. Any building material that is an increment of 4″ will align exactly with a line on this grid.

When using a CAD system, grid dimensions are created on the monitor using the *Grid* command. All lines can be snapped to align with grid lines. Nongrid dimensions are drawn in the free pick mode.

CAD Grids and Modules

Grid spacing and points can be selected to align with any modular scale and spacing. Grids should be selected to represent the smallest grid unit to be used on the drawing. Typical grid sizes for floor plans and elevations are 12″ and 4″ or 8″ for details. Most modular grids are available with major and minor module lines in different weights or colors. To make lines conform to a vertical or horizontal axis with or without the use of grids, the *Ortho* command can be used.

CAD grid lines can also be placed on a different layer than object lines. Later these grid lines can be plotted either in a second color or with a fine line to show the modular alignment. Nonmodular lines must be drawn in a nongrid mode so that they do not automatically snap to the grid lines.

Manufactured Buildings

To some degree, all contemporary buildings are manufactured. Not every component is totally built on site, even for a completely custom home. The amount of manufactured components used in buildings varies. Most companies combine mass-production methods with other construction design to minimize custom-job work without sacrificing quality.

Precut Structures

Precut structures are built from materials that are cut to specification at a factory and then assembled on site by conventional ("stick-built") methods. Designing precut buildings involves using standard sizes of materials and components to eliminate waste and on-site labor time.

Prefabricated Homes

In the most common type of **prefabricated homes,** the major components, such as the walls, decks, and partitions, are assembled at a factory. The utility work, such as installation of electrical, plumbing, and heating systems, is completed on site. The final finishing work, such as installation of prehung doors, prefinished roof coverings, and prefinished walls, is also done on site.

Prebuilt Homes

Prebuilt homes are manufactured houses that are totally built in a factory. The first completely factory-built homes were mobile homes. The mobile home buyer, like the buyer of a conventional factory-built home, has little option to adjust or customize the basic design. Nonetheless, there are opportunities to select different sizes, models, and interiors. All electrical, plumbing, and HVAC units are built into the structure.

Prebuilt homes are made in widths up to 28'. Most states allow only 14' widths on roadways. A manufactured house can be produced in two 14' halves for shipping. Because most trucks can only haul a 48' length, a 28' × 48' building represents two 14' × 28' prefab units. To overcome these size restrictions, multiple prefab units are often combined on the job site to create large structures.

Prebuilt Modules

When larger structures or different configurations are needed, manufactured **prebuilt modules** can be combined to form a variety of floor plan shapes. Construction modules can be manufactured as complete rooms or as independent functional areas. In either case, they contain complete built-in components such as kitchen and bath cabinets, major appliances, and fixtures.

Area modules contain room clusters that can be combined with other room or room cluster modules.

Modules can be used as elements of an expandable plan that may take years to complete. Combinations of modules can be designed to attach horizontally side to side or end to end or vertically bottom to top. They may also be located separately in a campus plan and connected by walkways or lanais.

Designing prebuilt houses or modules involves conforming to factory production standards and sizes. Nevertheless, a designer can develop optional housing models and a wide variety of modular configurations for customer consideration.

In summary, the standardization of materials and building components into modular sizes has created many opportunities for architectural innovation while controlling the costs normally associated with creative designs.

Natural Building Types

Construction types that use minimal amounts of wood, steel, and composite members include straw-bale, rammed earth, adobe, cob construction, and wattle and daub.

Straw-bale construction consists of straw bales stacked between vertical posts to form a wall, as shown in Figure 22.40. Studs, sills, and headers (lintels) are only used to create door and window openings. Because tubular straw is strong and dense, a trussed roof can be supported by the straw walls. Additional stability is provided by steel rods placed vertically through the stack as shown in Figure 22.41.

Straw-bale construction is relatively inexpensive, fire resistant, tremor proof, soundproof, nonallergenic, and nontoxic and provides an insulation rating of R48. Disadvantages include potential rodent nesting and the development of fungi and mites. A stucco surface helps keep this wall dry and eliminates these potential problems. Two-foot-thick walls and restrictive building codes are also a problem in some locations.

Rammed earth construction uses a mixture of damp soil that is rammed (compacted) into 18" wide forms to create a sturdy masonry wall as seen in Figure 22.42. The walls are compacted between a post-and-beam framework that provides the major load-bearing support. Walls can be stuccoed, plastered, or unsurfaced. Maintained costs are low, and walls provide high levels of insulation. The R value range of an 18" wall is 4.5 to 81. The R value range of a 24" wall is 6 to 108.

Rammed earth buildings provide passive solar qualities, look similar to conventional buildings, are conducive to the

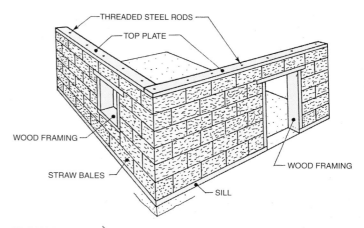

FIGURE 22.40 > Straw-bale wall construction. *Hepler/ Wallach/Hepler © Cengage 2013*

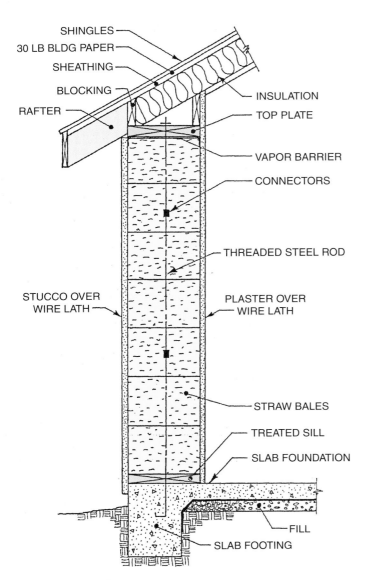

FIGURE 22.41 › Straw-bale wall section. *Hepler/Wallach/
Hepler © Cengage 2013*

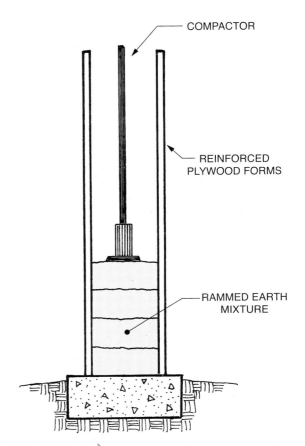

FIGURE 22.42 › Rammed earth compaction. *Hepler/Wallach/
Hepler © Cengage 2013*

use of radiant heat, and are free of paint, adhesives, and preservative toxins. However, costs are slightly higher than wood
frame construction and some local codes restrict use.

Other earth-based construction types and methods include adobe, cob, and waffle and daub. *Adobe block* construction is one of the world's oldest building types and uses blocks
of clay created in molds and baked in the sun. These are used
to create walls similar to the way concrete block walls are laid.
Cob construction uses a mixture of sand, clay, and straw applied in layers after each previous layer has dried. *Waffle and
daub* construction is similar but consists of interweaving rods
and twigs that are plastered with mud, clay, and straw.

Principles of Construction Exercises

1. Describe four structural forces and give an example of how each can be counteracted.
2. Explain what makes up a building load. How does it relate to construction?
3. Sketch an elevation that minimizes live loads from snow and wind. Explain why you chose that design.
4. Explain how size, shape, placement, and spacing of structural members affect the strength of a building.
5. After selecting the size of a roof and materials you would use, calculate the weight of the roof.
6. Use a grid snap function to design a modular floor plan and elevation.
7. Draw a floor plan of your own design based on the grid system in this chapter. Include modular dimensions.
8. Sketch an elevation that combines modular and nonmodular components. Use grid and nongrid dimensions.
9. Draw a construction detail using a 4″ grid system.
10. Design a small commercial building using a modular grid of 8′.
11. Draw Figure 22.43 at a scale of 1/4″ = 1′-0″ and align all exterior walls, doors, and windows to a 16″ modular grid.

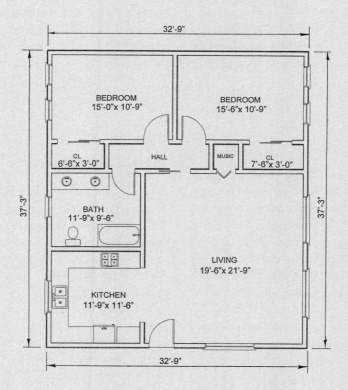

FIGURE 22.43 > Draw this plan on a modular grid. *Hepler/ Wallach/Hepler © Cengage 2013*

Foundations and Fireplace Structures

OBJECTIVES

In this chapter you will learn to:

> identify the components and materials used in foundations.

> describe the types of foundations.

> design a fireplace with sufficient structural support and appropriate safety components.

> relate the layout and excavations for a building to the type of foundation it will have.

> draw a T-foundation plan.

> draw slab foundation plans.

TERMS

bearing surface
damper
firebox (fire chamber)
flue
footings (footers)
foundation sills

pier-and-column foundations
pilasters
piles
pole/column foundation
Pythagorean theorem
reinforcing bars (rebars)

slab foundations
spark arrestors
stepped footings
T-foundation
wire mesh
wood foundations

INTRODUCTION

Every structure needs a foundation. The function of a foundation is to provide a level and uniformly distributed support for the structure. Foundations must be strong enough to support and distribute the load of the structure. Foundations must remain level to prevent the walls from cracking and the doors and windows from sticking. They also fulfill other functions, such as to help prevent cold air and dampness from entering buildings, to waterproof basements, and to form the supporting walls for basements.

The methods and materials used in constructing foundations vary greatly in different parts of the country and are continually changing. The basic principles of foundation construction are the same, though, regardless of the application.

FOUNDATION MATERIALS AND COMPONENTS

The components and materials in foundations vary, depending on the foundation type, size, and design. Materials used include concrete, concrete block, steel reinforcement bars, welded wire mesh, and a variety of composite form materials.

Bearing Surface

The transfer of all building loads ends at a stable ground level. There the support must be rigid and static enough to prevent the foundation base from excessive vertical or lateral movement. Without a stationary base the foundation would shift and the structural stability of the entire building would fail.

TYPE OF SOIL	BEARING CAPACITY (POUNDS PER SQ. FT.)
Soft clay, loose dirt, loam	2,000
Dry sand and hard clay	4,000
Hard sand or gravel	6,000
Partially cemented sand or gravel	20,000

FIGURE 23.1 > Bearing capacity of typical bearing soils.
Hepler/Wallach/Hepler © Cengage 2013

An important part of a foundation's base is the area underneath—the soil. The soil must be capable of bearing the load, or weight, of the foundation and the structure. This **bearing surface,** or bearing soil, must be compactable, contain no clays or organic matter, drain easily, and be freeze resistant in cold climates. See Figure 23.1.

Minimum amounts of settlement will occur in bearing soil regardless of the amount of compaction. Therefore, compaction must be uniformly distributed to eliminate voids. Compaction must also be level to ensure that the supported structure will not shift out of plumb. Poor soil conditions may require a larger footing to distribute the building load.

Concrete

The basic material used to pour foundation bases, walls, and floors is concrete. Concrete is a combination of cement (clay and limestone), water, stone aggregate, and chemicals that improve strength or workability. Solid concrete foundation walls, as shown in Figure 23.2, are more stable and less prone to leaks and cracks than block walls. However steel reinforcement rods are needed if the walls support heavy loads. See Chapter 26 for more information on concrete characteristics and related building systems.

Concrete Block

Economy and ease of construction have made concrete block a popular material for foundation walls. Foundation blocks are manufactured in lengths of 16″, heights of 8″, and in modular widths of 4″, 6″, 8″, 10″, and 12″. Each dimension is actually 3/8″ smaller than the listed size to allow for mortar. This keeps the finished dimensions in modular units. Figure 23.3 illustrates the difference between actual and nominal sizes of the concrete block used in foundations.

Building codes for concrete block foundation walls usually require the bottom course to rest on a reinforced concrete footing and the top course to be solid or filled with concrete. Sill plate anchor bolts must extend above the top course at least 2″ and penetrate down at least 10″ or two courses. Pre-

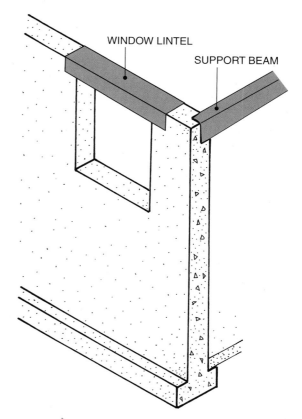

FIGURE 23.2 > Solid concrete linteled opening. *Hepler/Wallach/Hepler © Cengage 2013*

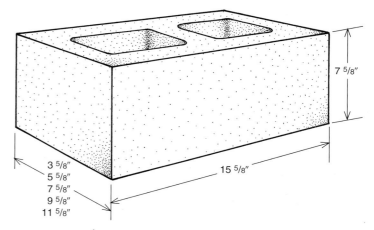

FIGURE 23.3 > Concrete block sizes. *Hepler/Wallach/Hepler © Cengage 2013*

cast solid masonry or wood lintels must be provided for all door and window openings, as seen in Figure 23.4.

Reinforcing Bars

Concrete resists compression very well. To resist tension forces, steel bars are added to concrete slabs, beams, and columns. These bars are known as **reinforcing bars** or **rebars.** Steel rebars are either smooth or deformed (grooved or embossed). Deformed bars create a stronger bond between bar and concrete because the concrete is held in place by

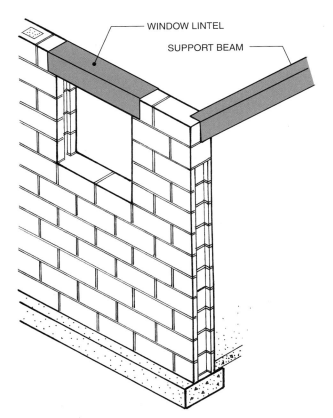

BAR SIZE	AREA (SQ. IN.)	WEIGHT (LBS. PER FT.)	DIAMETER (INCHES)
3	.11	.376	3/8"
4	.20	.668	1/2"
5	.31	1.043	5/8"
6	.44	1.502	3/4"
7	.60	2.044	7/8"
8	.79	2.670	1"
9	1.00	3.400	1 1/8"
10	1.27	4.303	1 1/4"
11	1.56	5.313	1 3/8"
14	2.25	7.650	1 3/4"
18	4.00	13.600	2 1/4"

FIGURE 23.5 > Rebar sizes. *Hepler/Wallach/Hepler © Cengage 2013*

the grooves or depressions. Rebars are sized by numbers (1 through 18) representing 1/8″ increments, up to 2 1/4″. See Figure 23.5. The bar size number, mill number, symbol of the steel type, and the grade are marked on each rebar.

Beam rebars are located horizontally near the bottom of beams to provide maximum tension resistance. Similarly, slab rebars are placed horizontally and in parallel rows close to the bottom of the slab. To prevent cracking due to temperature and moisture changes, some rebars are also placed perpendicular to the load-supporting bars. These rebars are known as temperature bars.

Because a minimum thickness (1″ to 3″) must be maintained between rebars and the concrete surface, fixtures known as *bolsters* (*saddles*) and *chairs* are used to hold the bars in place during slab pouring. U-shaped rods, known as *stirrups,* are used for this purpose in beam pouring. See Figure 23.6.

The exact positions of rebars in a slab are shown on a foundation sectional drawing. The locations of rebars in a wall are shown on a plan detail and/or elevation section.

Wire Mesh

Steel-welded **wire mesh** or wire fabric is often used in slabs in place of rebars. Square, rectangular, and triangular patterns are available. Square patterns are the most common. Wire mesh is specified on construction drawings by the spacing (in inches)

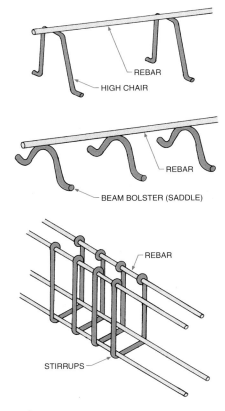

FIGURE 23.6 > Rebar support fixtures. *Hepler/Wallach/Hepler © Cengage 2013*

SQUARE PATTERN AND GAUGE	RECTANGULAR PATTERN AND GAUGE
6 × 6–10/10	6 × 12–4/4
6 × 6–8/8	6 × 12–2/2
6 × 6–6/6	6 × 12–1/1
6 × 6–4/4	
	4 × 12–8/12
4 × 4–10/10	4 × 12–6/10
4 × 4–8/8	
4 × 4–6/6	4 × 16–8/12
4 × 4–4/4	4 × 16–6/10

FIGURE 23.7 > Wire mesh spacing and gauges. *Hepler/ Wallach/Hepler © Cengage 2013*

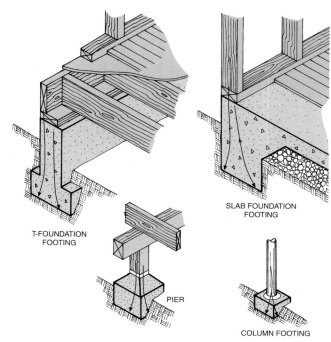

FIGURE 23.8 > Types of footings. *Hepler/Wallach/Hepler © Cengage 2013*

between wire strands and by the gauge of the wires. The term *gauge* refers to the diameter of the wire. Be aware, however, that the larger the gauge number, the smaller the wire diameter. The two intersecting wires in a pattern are known as longitudinal (long direction) and transverse (short direction) wires. A rectangular pattern with transverse wires spaced 4″ apart and longitudinal wires spaced 12″ apart is labeled 4 × 12. This is followed by the gauge of each of the wires. For example, a wire mesh may consist of a #6 gauge for the transverse wire and a #10 gauge for the longitudinal wire. All of this information is noted on drawings as follows: 4 × 12—6/10 to indicate spacing and wire gauges. See Figure 23.7.

Footings

Footings, or **footers,** are the bases of foundations and foundation walls. There are two types: continuous and individual. Continuous footings extend under walls and around the perimeter of a foundation. The footings are often wider than the foundation walls in order to distribute the weight of the building over a larger area. See Figure 23.8. Individual footings (called *piers*) support vertical structural members.

Concrete is commonly used for footings because it can be poured to maintain a firm contact with the supporting soil. Concrete is also effective because it can withstand heavy weights and is a decay-proof material. Steel reinforcements (rebars) are added to concrete footings to keep the concrete from cracking and to provide additional support. In some circumstances other masonry types and even treated wood can be used for temporary footings.

Footings must be laid on solid ground to support the weight of the building effectively and evenly. In cold climates,

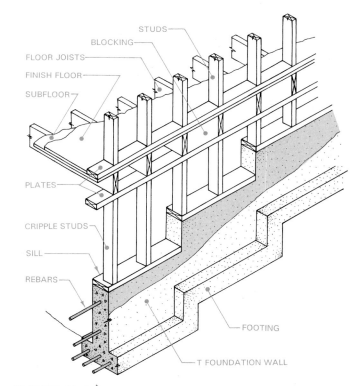

FIGURE 23.9 > Continuous stepped footing. *Hepler/Wallach/ Hepler © Cengage 2013*

footings must be placed below the frost line (the depth to which the soil freezes). Always consult the local building code for frost line foundation depth requirements before establishing footing depths. **Stepped footings** at different levels are used on sloping sites. See Figure 23.9. Figure 23.10 illustrates

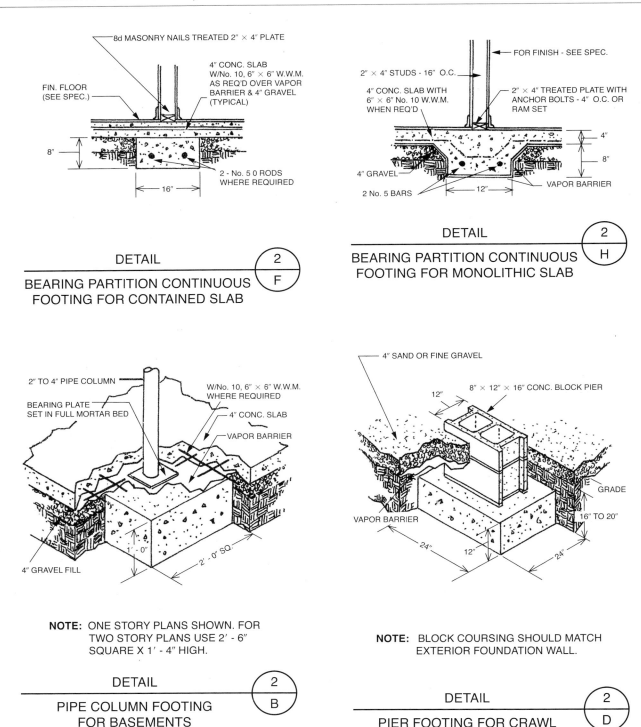

8d MASONRY NAILS TREATED 2″ × 4″ PLATE

4″ CONC. SLAB W/No. 10, 6″ × 6″ W.W.M. AS REQ'D OVER VAPOR BARRIER & 4″ GRAVEL (TYPICAL)

FIN. FLOOR (SEE SPEC.)

8″

16″

2 - No. 5 0 RODS WHERE REQUIRED

DETAIL ② F
BEARING PARTITION CONTINUOUS FOOTING FOR CONTAINED SLAB

FOR FINISH - SEE SPEC.

2″ × 4″ STUDS - 16″ O.C.

4″ CONC. SLAB WITH 6″ × 6″ No. 10 W.W.M. WHEN REQ'D

2″ × 4″ TREATED PLATE WITH ANCHOR BOLTS - 4″ O.C. OR RAM SET

4″ GRAVEL

2 No. 5 BARS

4″

8″

VAPOR BARRIER

12″

DETAIL ② H
BEARING PARTITION CONTINUOUS FOOTING FOR MONOLITHIC SLAB

2″ TO 4″ PIPE COLUMN

BEARING PLATE SET IN FULL MORTAR BED

W/No. 10, 6″ × 6″ W.W.M. WHERE REQUIRED

4″ CONC. SLAB

VAPOR BARRIER

4″ GRAVEL FILL

1′ - 0″

2′ - 0″ SQ.

NOTE: ONE STORY PLANS SHOWN. FOR TWO STORY PLANS USE 2′ - 6″ SQUARE X 1′ - 4″ HIGH.

DETAIL ② B
PIPE COLUMN FOOTING FOR BASEMENTS

4″ SAND OR FINE GRAVEL

8″ × 12″ × 16″ CONC. BLOCK PIER

12″

GRADE

VAPOR BARRIER

16″ TO 20″

24″

12″

24″

NOTE: BLOCK COURSING SHOULD MATCH EXTERIOR FOUNDATION WALL.

DETAIL ② D
PIER FOOTING FOR CRAWL

FIGURE 23.10 > Footing details. *Used by permission Hanley Wood, LLC*

several additional footing applications. These are typical footing designs and show the placement to support interior walls, columns, and block piers. Figure 23.11 shows a chart used with the *International Building Code* for determining footing dimensions as applied to one-, two-, and three-story light construction buildings. Regardless of the footing type, perforated PVC drain tiles should be laid outside and parallel to the perimeter footing, as emphasized in Figure 23.12. This type of drainage must be designed to drain into a street, catch basin, or storm system.

Composite Foundation Materials

In addition to concrete, foundation materials are also manufactured from ceramic, carbon fibers, and reinforced polymers.

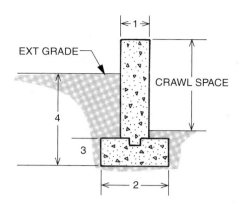

UNIFORM BUILDING CODE FOR T-FOUNDATIONS

BUILDING HEIGHTS	1. WIDTH OF WALL	2. WIDTH OF FOOTING	3. HEIGHT OF FOOTING	4. DEPTH OF FOOTING
ONE STORY	6"	12"	6"	12"
TWO STORY	8"	15"	7"	18"
THREE STORY	10"	18"	8"	24"

Girder spans (pier spacing) for non-bearing walls is 3'-0"
Girder spans for bearing walls is 5'-6"
All calculations are for Douglas fir #2 grade (DF #2)

FIGURE 23.11 > T-foundation code dimensions. *Uniform Building Code*

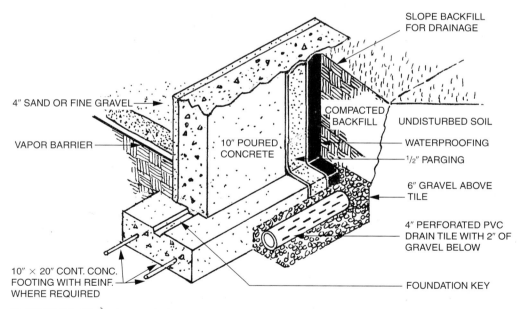

FIGURE 23.12 > Drain tile location. *Used by permission Hanley Wood, LLC*

TYPES OF FOUNDATIONS

The type of foundation depends on the nature of the soil, slope of the terrain, the size and weight of the structure, the climate, building laws, and the relationship of the floor line to the grade line (ground). See Figure 23.13. Foundations covered in this chapter are divided into four basic types: T-foundations, slab foundations, pier-and-column foundations, and temporary wood foundations. Figure 23.14 shows the relationship of these foundation types to the ground line, and Figure 23.15 shows how these foundation types are drawn as elevation sections.

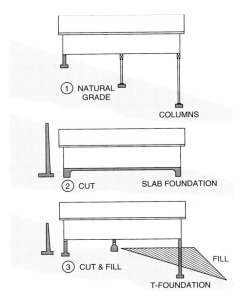

FIGURE 23.13A > Effect of grade contour on foundations.
Used by permission Hanley Wood, LLC

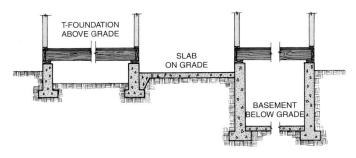

FIGURE 23.13B > Foundation position related to grade line.
Hepler/Wallach/Hepler © Cengage 2013

T-Foundations

A **T-foundation** consists of a footing and a poured concrete or concrete block wall that forms an inverted T. T-foundations are necessary in structures with basements or when the underside of the first floor must be accessible. See Figure 23.16. The details of construction related to several variations of a T-foundation plan are shown in Figure 23.17.

T-foundations are prepared by pouring the footing into an excavated trench, leveling the top of the footing, and erecting a concrete block or masonry wall on top of the footing. If poured concrete foundation walls are to be used, building forms are erected on top of the footing. Concrete is poured into these forms simultaneously for the walls and footing.

Foundation Walls

The function of foundation walls is to support the load of the structure and to transmit its weight to the footing. Foundation walls are normally made of poured concrete, concrete block, stone, or brick. See Figure 23.18. When a complete excavation is made for a basement, foundation walls also provide the walls of the basement, as shown in Figure 23.19. Figure 23.20 shows how foundation walls are drawn as plan and elevation sectional views.

Foundation walls can support heavy compression loads. However, lateral earth loads are often a problem. Imagine a sheet of paper standing on end. It will not stand unsupported. Fold it at right angles and it may stand, but wobble. Fold it again in the shape of a V and it gains great stability.

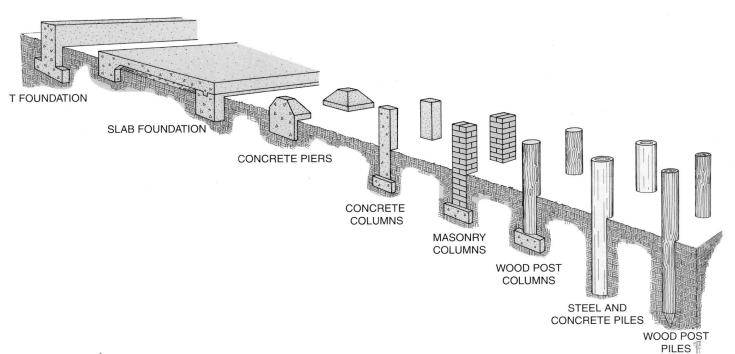

FIGURE 23.14 > Types of foundations. *Hepler/Wallach/Hepler © Cengage 2013*

TYPICAL RESIDENTIAL FOUNDATIONS

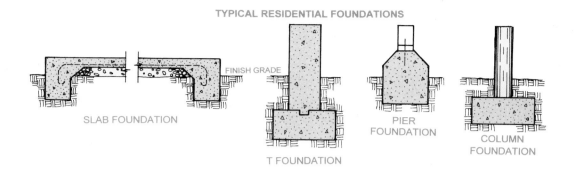

SLAB FOUNDATION

FINISH GRADE

T FOUNDATION

PIER FOUNDATION

COLUMN FOUNDATION

COMMON BUILDING MATERIALS FOR FOUNDATION WALLS

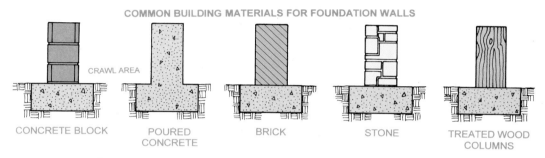

CRAWL AREA

CONCRETE BLOCK

POURED CONCRETE

BRICK

STONE

TREATED WOOD COLUMNS

FIGURE 23.15 > Sections of foundation types. *Hepler/Wallach/Hepler © Cengage 2013*

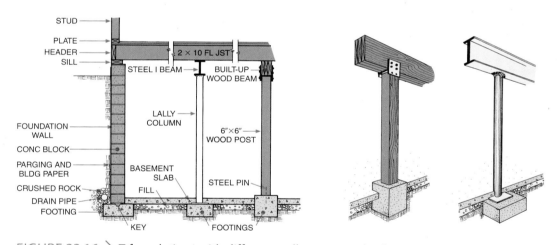

STUD
PLATE
HEADER
SILL
2 × 10 FL JST
STEEL I BEAM
BUILT-UP WOOD BEAM
LALLY COLUMN
6"×6" WOOD POST
FOUNDATION WALL
CONC BLOCK
PARGING AND BLDG PAPER
CRUSHED ROCK
DRAIN PIPE
FOOTING
BASEMENT SLAB
FILL
STEEL PIN
KEY
FOOTINGS

FIGURE 23.16 > T-foundations with different walls, posts, and columns. *Hepler/Wallach/Hepler © Cengage 2013*

See Figure 22.22 in Chapter 22. This same principle applies to foundation walls. Each corner, offset, or pilaster adds strength, as shown in Figure 23.21.

Pilasters are reinforcements in a wall designed to provide more rigidity without increasing the width of the entire wall. See Figure 23.22. Pilasters are also used to support girders (main horizontal members) instead of making girder pockets (inset spaces) in the foundation walls. Building codes normally specify the size and spacing of pilasters, depending on the wall width, height, and material.

🌐 **Preinsulated Concrete Walls** As an alternative to using removable forms, foundation walls can be poured into or around insulating material to create foundation walls that are structurally sound and totally insulated. Figure 23.23 on page 482, shows two polystyrene panels with concrete poured between them, making the panels a permanent part of the wall. In another version concrete is poured into polystyrene blocks to form a rigid wall. A third version involves suspending foam panels into a poured wall to create a core sandwich wall.

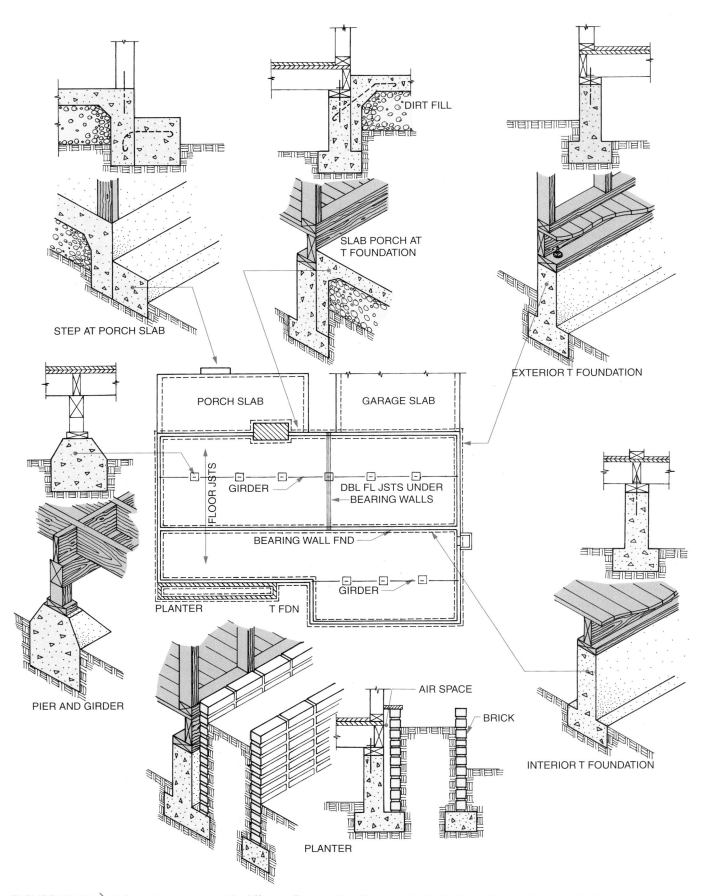

DIRT FILL

SLAB PORCH AT
T FOUNDATION

EXTERIOR T FOUNDATION

STEP AT PORCH SLAB

PORCH SLAB

GARAGE SLAB

FLOOR JSTS

GIRDER

DBL FL JSTS UNDER
BEARING WALLS

BEARING WALL FND

GIRDER

PLANTER

T FDN

PIER AND GIRDER

AIR SPACE

BRICK

INTERIOR T FOUNDATION

PLANTER

FIGURE 23.17 > T-foundation plan with different floor and wall types. *Hepler/Wallach/Hepler © Cengage 2013*

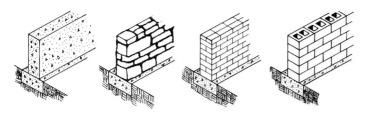

FIGURE 23.18 › Common foundation wall materials. *Hepler/Wallach/Hepler © Cengage 2013*

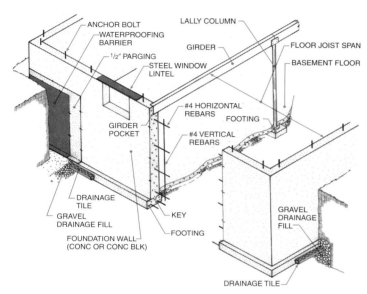

FIGURE 23.19 › Foundation walls may also serve as basement walls. *Hepler/Wallach/Hepler © Cengage 2013*

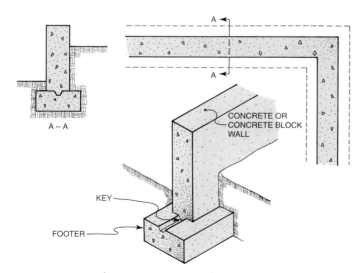

FIGURE 23.20 › T-foundation shown in plan and elevation section. *Hepler/Wallach/Hepler © Cengage 2013*

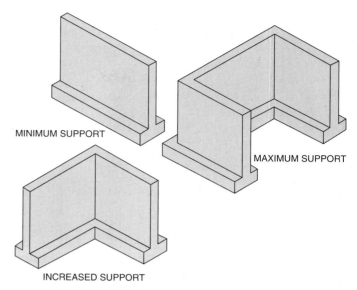

FIGURE 23.21 › Foundation shape affects lateral load resistance. *Hepler/Wallach/Hepler © Cengage 2013*

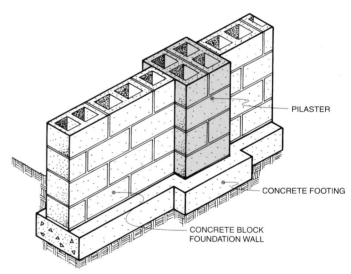

FIGURE 23.22 › Concrete wall pilaster. *Hepler/Wallach/Hepler © Cengage 2013*

Insulated concrete forms are energy efficient, soundproof, tornado resistant, and LEED approved. Figure 23.24 illustrates the construction and related drawing of an insulated concrete form (ICF) corner.

Precast foundation panels with integral foamboard can also be used as a substitute for poured concrete.

Moisture Control In addition to the previously covered construction methods using waterproofing barrier materials and gravel fill and drain tile, moisture can be controlled inside foundation walls. Figure 23.25 illustrates the use of embedded vertical cores and flashing pans to direct water to the exterior through weep spouts. Another moisture blocking method, shown in Figure 23.26, lets water migrate from the interior of block cells to the outdoors. This allows moisture to drain past excess mortar (droppings) and onto drainage strips.

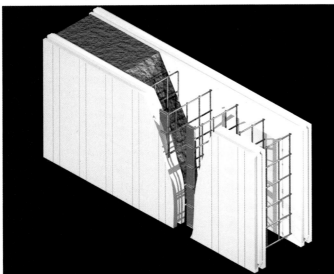

FIGURE 23.23 > Insulated concrete forms. *ARXX Steel*

Sills

Foundation sills are wood or steel members that are fastened to the top of foundation walls. Sills provide the base for attaching floor systems to foundations. See Figure 23.27. Wood sills must be pressure treated, or a sheet-metal termite shield must be placed between the wood and concrete (Figure 23.28). This applies to all wood in contact with masonry, concrete, or soil. Building laws specify the distance required from the bottom of the sill to the grade line inside and outside the foundation.

Anchor Devices

Anchor devices are embedded in the top of the foundation walls or piers. See Figure 23.29. The exposed part of the device is attached to the first wood member (the sill). Anchor bolts are used most often. Sizes of bolts typically used in residences are 1/2″ or 5/8″ in diameter and 10″ long. They are usually spaced 4′ apart, starting 1′ from each corner. Bolts

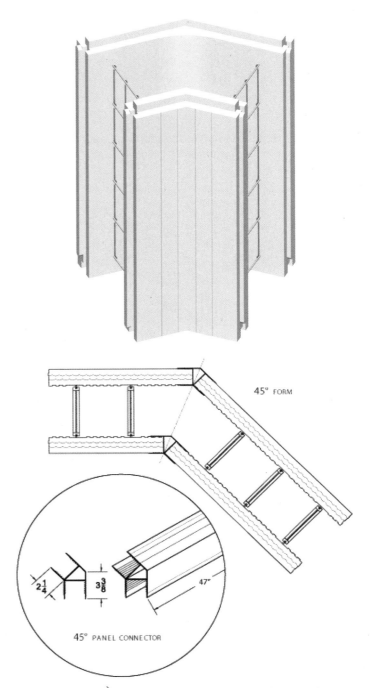

FIGURE 23.24 > Corner construction for an insulated concrete form. *ARXX Steel*

may be embedded into drilled holes or shot into the concrete with low-caliber, power-activated fastening guns while the concrete is still pliable.

Cripples

Cripples are used to raise floor levels without building a higher concrete foundation wall. See Figure 23.30. Because the load of the structure must be transmitted through the cripples, these are usually heavy members, often four-by-

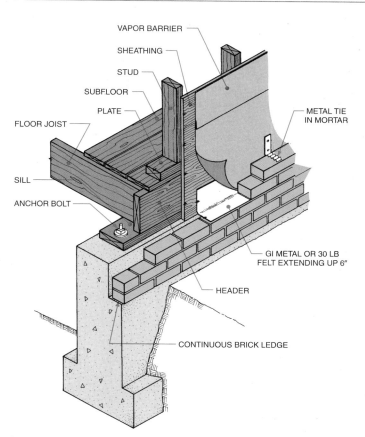

FIGURE 23.27 > Sills connect floors and walls to foundations. *Hepler/Wallach/Hepler © Cengage 2013*

FIGURE 23.25 > Blockflash moisture control system. *Mortar Net® USA, Ltd.*

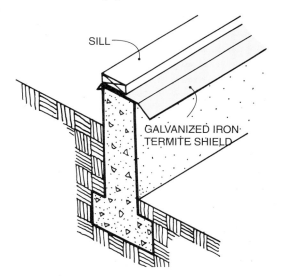

FIGURE 23.28 > Termite shield placed between the wood sill and concrete wall. *Hepler/Wallach/Hepler © Cengage 2013*

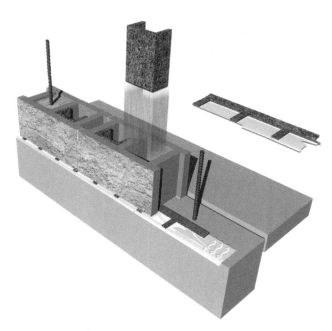

FIGURE 23.26 > Blocknet moisture control system. *Mortar Net® USA, Ltd.*

fours (4 × 4's) spaced closer than the normal 16″ on center. "On center" (OC) refers to the distance (spacing) measured from the center of one member to the center of the next. Shear stress plywood must be used over cripples to overcome lateral stresses.

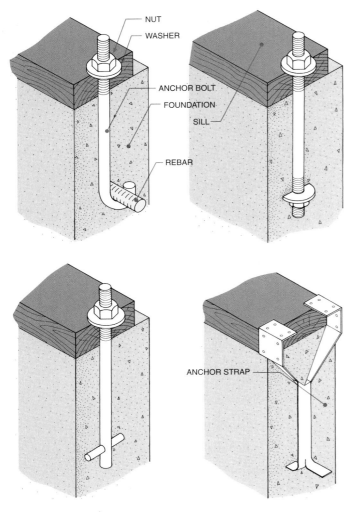

FIGURE 23.29 > Types of foundation-to-sill anchor devices. *Hepler/Wallach/Hepler © Cengage 2013*

Slab Foundations

Slab foundations, or *slabs,* are made of reinforced concrete. They are either monolithic (one piece) or separate footing and slab. In a monolithic slab, the slab floor and footing are poured as one piece. See Figure 23.31. Rebars, wire mesh, plumbing line risers, waterproof membranes, electrical con-

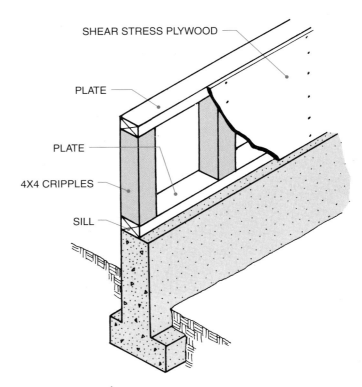

FIGURE 23.30 > Cripples extend foundation height. *Hepler/Wallach/Hepler © Cengage 2013*

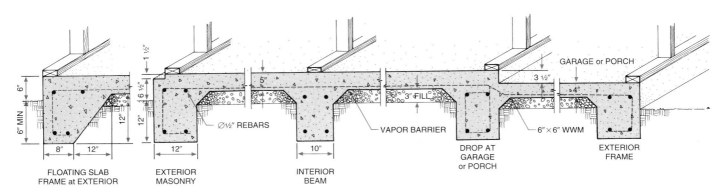

FIGURE 23.31 > Types of monolithic slab foundations. *Hepler/Wallach/Hepler © Cengage 2013*

duits (tubes for wires), and HVAC (heating, ventilating, and air conditioning) ducts (if in the floor) must all be securely in place on a compacted soil base before pouring. Figure 23.32 shows a sectional drawing of an HVAC duct embedded in a concrete slab. Figure 23.33 shows a chart of the number of cubic feet of concrete needed for different slab thicknesses.

In slab foundations that are not monolithic, the footings are poured separately from the floor slabs, as shown in Figure 23.34. Figure 23.35 shows a variety of slab floor, wall, and footing details used in different foundation situations.

Figure 23.36 shows how a concrete slab floor relates to the fill, vapor barrier, and rigid insulation below and the flooring above. Figure 23.37A shows the use of rebars to connect a slab with a foundation wall. Figure 23.37B shows a slab detail for a monolithic (solid, one-piece) pour with rebars and

welded wire mesh. Where concrete or concrete block is used separately for foundation walls, the slab and footing must be secured to the foundation wall with rebars.

Building codes specify the minimum distance from the top of a slab to the grade line, usually 6″ to 8″. Because slabs lose heat around the perimeter, adequate insulation is important.

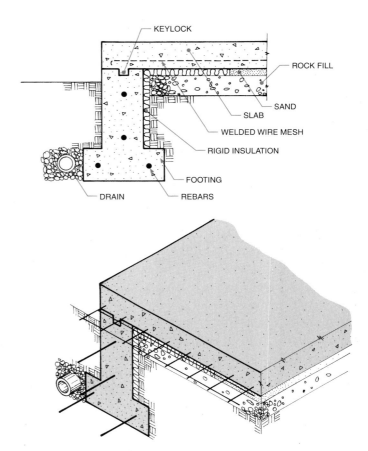

FIGURE 23.32 > HVAC duct embedded in a slab. *Hepler/ Wallach/Hepler © Cengage 2013*

FIGURE 23.34 > Two-piece slab. *Hepler/Wallach/Hepler © Cengage 2013*

CUBIC YARDS OF CONCRETE FOR SLABS						
SLAB THICKNESS	10 SQ. FT.	25 SQ. FT.	50 SQ. FT.	100 SQ. FT.	200 SQ. FT.	300 SQ. FT.
4 in.	.12 cu. yd.	.31 cu. yd.	.62 cu. yd.	1.23 cu. yd.	2.47 cu. yd.	3.7 cu. yd.
5 in.	.15 cu. yd.	.39 cu. yd.	.77 cu. yd.	1.54 cu. yd.	3.09 cu. yd.	4.63 cu. yd.
6 in.	.19 cu. yd.	.46 cu. yd.	.93 cu. yd.	1.85 cu. yd.	3.7 cu. yd.	5.56 cu. yd.

FIGURE 23.33 > Data for calculating amounts of slab concrete required for different slab thicknesses. *Hepler/Wallach/Hepler © Cengage 2013*

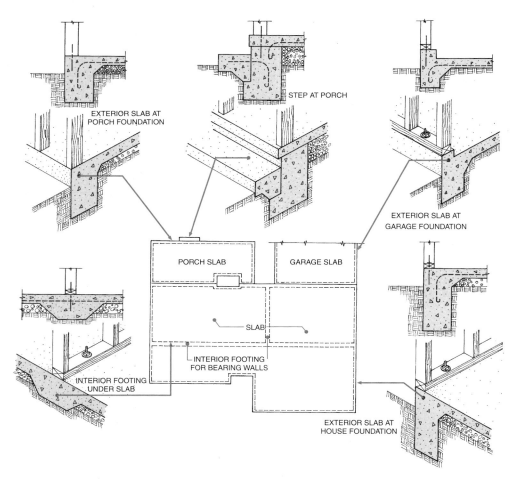

FIGURE 23.35 > Slab foundation details. *Hepler/Wallach/Hepler © Cengage 2013*

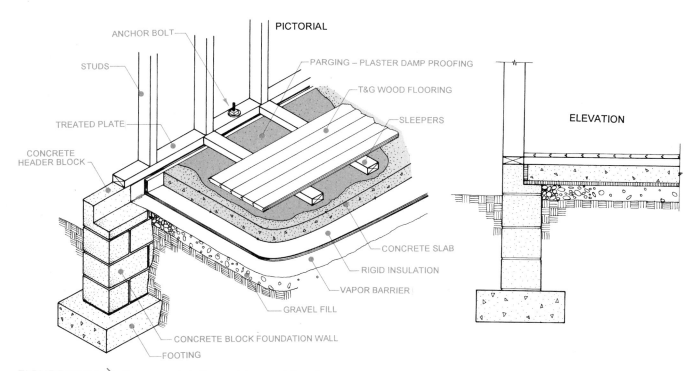

FIGURE 23.36 > Concrete slab floor with wood flooring system. *Hepler/Wallach/Hepler © Cengage 2013*

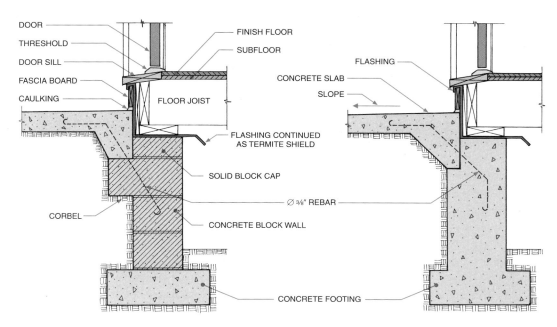

FIGURE 23.37A > Rebars connect the slab with the foundation. *Hepler/Wallach/Hepler © Cengage 2013*

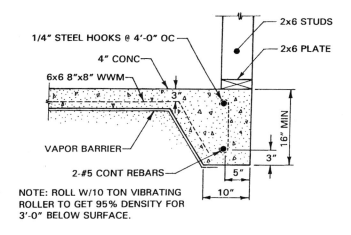

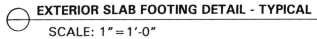

FIGURE 23.37B > Slab detail showing rebars and welded wire mesh placement. *Hepler/Wallach/ Hepler © Cengage 2013*

Pier-and-Column Foundations

Pier-and-column foundations consist of individual footings (piers) on which posts and columns are placed. Posts and columns are vertical members used to support floor systems. See Figure 23.38.

Individual footings are known as *piers*. Piers may be sloped or stepped in order to spread the load of the structure on a wider base. See Figure 23.39.

Posts and columns are vertical members that support girders and beams and transmit their weight and the weight of the entire building to the footings. See Figure 23.40. The terms *post* and *column* are often used interchangeably. Generally, short vertical supports are called *posts* and longer vertical members are known as *columns*. Posts are usually made of wood, and columns are usually steel or masonry. Figure 23.41 illustrates how concrete column footings act as piers in supporting foundation walls.

Pole foundations are designed with enough vertical stability to function as extended piers. Figure 23.42 shows a pictorial plan and elevation drawing of a typical **pole/column foundation.** These drawings show the pole intersections with the floor and roof structure. The size, type, and spacing of members depends on the soil slope, soil conditions, and the weight of the structure.

Piers and posts or columns may be used as the sole support of the structure. They also may be used in conjunction with foundation walls to provide intermediate support for horizontal members. See Figure 23.43 on page 490. Fewer materials and less labor are needed for pier-and-column foundations than other types of foundations, but they are seldom used in basements because they occupy needed open space.

Beams and Girders

Beams are horizontal structural members that support a load. Girders are large beams that are the major horizontal support members for a floor system. In common practice, the terms *beam* and *girder* are often used interchangeably. However, technically, girders are members that are supported by piers and columns and secured to foundation walls. Beams and girders may also be supported on foundation walls

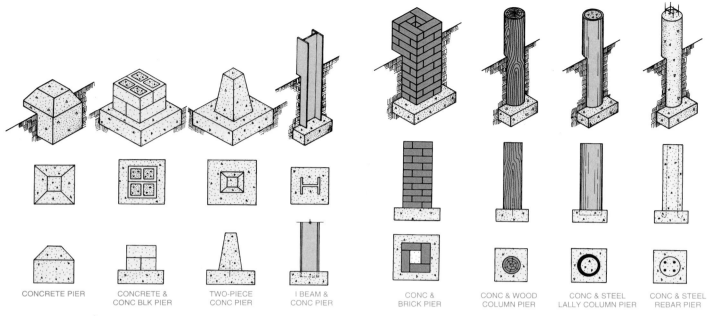

CONCRETE PIER   CONCRETE &   TWO-PIECE   I BEAM &   CONC &   CONC & WOOD   CONC & STEEL   CONC & STEEL
                 CONC BLK PIER CONC PIER   CONC PIER   BRICK PIER   COLUMN PIER   LALLY COLUMN PIER REBAR PIER

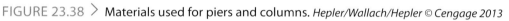

FIGURE 23.38 > Materials used for piers and columns. *Hepler/Wallach/Hepler © Cengage 2013*

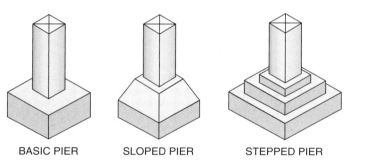

BASIC PIER   SLOPED PIER   STEPPED PIER

FIGURE 23.39 > Basic types of piers. *Hepler/Wallach/Hepler © Cengage 2013*

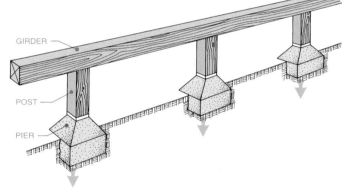

GIRDER

POST

PIER

FIGURE 23.40 > Posts transmit weight to footings. *Hepler/Wallach/Hepler © Cengage 2013*

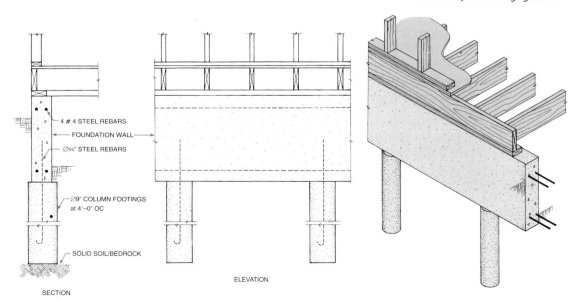

4 # 4 STEEL REBARS

FOUNDATION WALL

Ø⅝" STEEL REBARS

Ø9" COLUMN FOOTINGS
at 4'-0" OC

SOLID SOIL/BEDROCK

SECTION

ELEVATION

FIGURE 23.41 > Column footings. *Hepler/Wallach/Hepler © Cengage 2013*

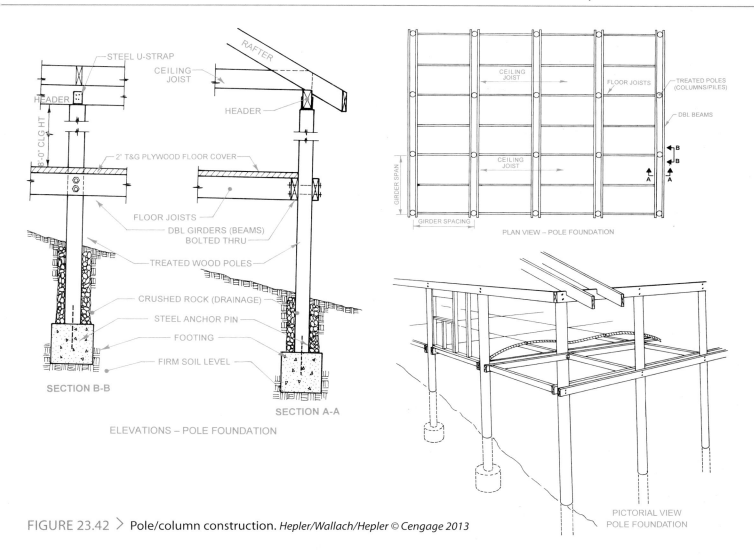

FIGURE 23.42 > Pole/column construction. *Hepler/Wallach/Hepler © Cengage 2013*

or indentations (pockets) in foundation walls as shown in Figure 23.44. Figure 23.45 illustrates another method of beam support on a concrete block wall pilaster.

Girder sizes are closely regulated by building codes. The allowable span of the girder depends on the size of the girder. A decrease in the size of a girder means that the span must be decreased. This is done by adding additional pier-and-column supports underneath the girder.

Most wood girders for residential construction are 4″ × 6″, 4″ × 8″, or 4″ × 10″. Steel beams or girders can perform the same function as wood beams or girders, but steel members can span larger distances than wood members of the same size.

Joists

Joists are the parts of the floor system that are placed perpendicular to the girders. See Figure 23.46. Joists span either from girder to girder or from girder to the foundation wall. The ends of the joists butt against a header rim joist or extend

to the end of the sill. Bridging is placed between joists. Bridging consists of smaller structural members fastened between the joists to add stability and keep spacing consistent.

Piles

When supports are driven into supporting soil or bedrock, without a separate footing, they are known as **piles.** Masonry, wood, steel, and concrete are used for piles, as shown in Figure 23.47. Piles are used to support large structures. They are driven deep into the soil to support structures on sites where stable soil conditions do not exist near the surface.

Piles support building loads in several ways: by friction with the soil, with self-contained footings as shown in Figure 23.48, or through contact with bedrock. Although bedrock may support the compression load of a building, sufficient soil-bearing capacity and/or horizontal ties must be used to prevent piles from drifting out of position. Figure 23.49 shows the use of horizontal ties to prevent the lateral drifting of piles.

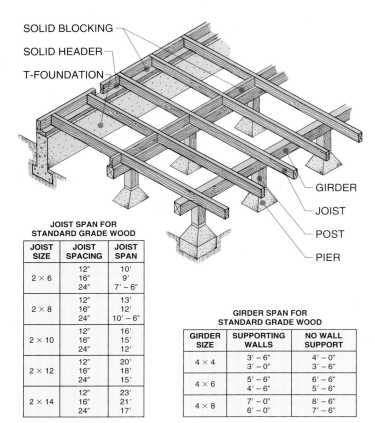

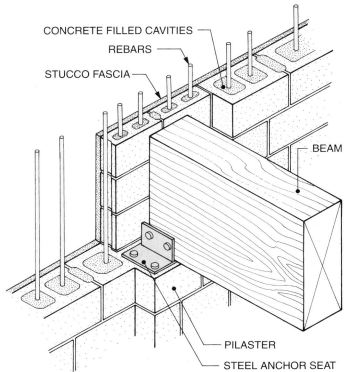

SOLID BLOCKING
SOLID HEADER
T-FOUNDATION
GIRDER
JOIST
POST
PIER

JOIST SPAN FOR STANDARD GRADE WOOD

JOIST SIZE	JOIST SPACING	JOIST SPAN
2 × 6	12″	10′
	16″	9′
	24″	7′ – 6″
2 × 8	12″	13′
	16″	12′
	24″	10′ – 6″
2 × 10	12″	16′
	16″	15′
	24″	12′
2 × 12	12″	20′
	16″	18′
	24″	15′
2 × 14	12″	23′
	16″	21′
	24″	17′

GIRDER SPAN FOR STANDARD GRADE WOOD

GIRDER SIZE	SUPPORTING WALLS	NO WALL SUPPORT
4 × 4	3′ – 6″	4′ – 0″
	3′ – 0″	3′ – 6″
4 × 6	5′ – 6″	6′ – 6″
	4′ – 6″	5′ – 6″
4 × 8	7′ – 0″	8′ – 6″
	6′ – 0″	7′ – 6″

FIGURE 23.43 > Piers and posts used for intermediate support. *Hepler/Wallach/Hepler © Cengage 2013*

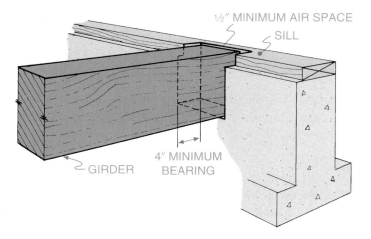

½″ MINIMUM AIR SPACE
SILL
GIRDER
4″ MINIMUM BEARING

FIGURE 23.44 > Girder pocket intersection. *Hepler/Wallach/ Hepler © Cengage 2013*

Wood Foundations

Wood foundations may be constructed similarly to wood frame walls in other parts of a building. See Figure 23.50. There is one important difference: The plywood and lumber components of permanent wood foundation walls are pressure treated with wood preservatives that become chemically bonded in the wood. This process protects the foundation

CONCRETE FILLED CAVITIES
REBARS
STUCCO FASCIA
BEAM
PILASTER
STEEL ANCHOR SEAT

FIGURE 23.45 > Beam support on a concrete block pilaster. *Hepler/Wallach/Hepler © Cengage 2013*

from fungi, termites, and other causes of decay. The lumber species used must be highly stress resistant.

Wood foundation walls are engineered to absorb and distribute loads and stresses that frequently crack and split other types of foundations. Another advantage of wood foundation design is that it prevents the types of moisture problems that typically plague conventional basements. The design incorporates moisture deflection and diversion features, such as vapor barriers, horizontally along the ground below the floor and vertically along the outside of the foundation. See Figure 23.51. Figure 23.52 shows a pictorial drawing of a wood foundation system. Note, however, that "permanent" wood foundations are not totally permanent. They may last for decades in dry climates, but not as long as masonry foundations. Many building codes prohibit wood foundations, which is why they are rarely used.

FIREPLACE CONSTRUCTION

The construction of masonry fireplaces is directly related to foundation design because structural support is essential for safe fireplace and chimney construction.

Today, fireplaces are rarely constructed completely on site. Most fireplaces consist of manufactured components around which framing or masonry walls are placed. Fireplace components are designed to produce fire and to provide for safety, convenience, and efficiency. Figure 23.53 reveals the components of a typical fireplace foundation and chimney structure.

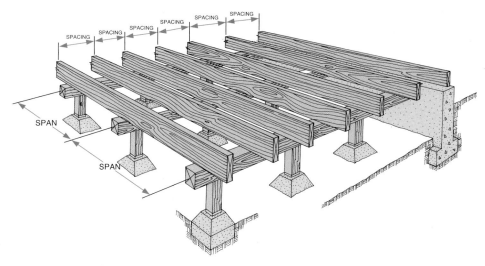

FIGURE 23.46 〉 Joists resting on girders. *Hepler/Wallach/Hepler © Cengage 2013*

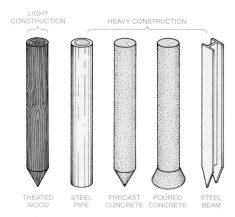

FIGURE 23.47 〉 Pile shapes and materials. *Hepler/Wallach/ Hepler © Cengage 2013*

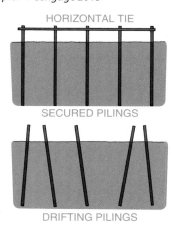

FIGURE 23.49 〉 Horizontal ties keep piles plumb. *Hepler/ Wallach/Hepler © Cengage 2013*

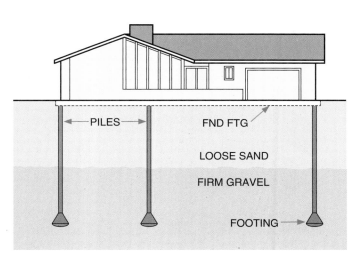

FIGURE 23.48 〉 Deep piles used in loose upper level soil. *Hepler/Wallach/Hepler © Cengage 2013*

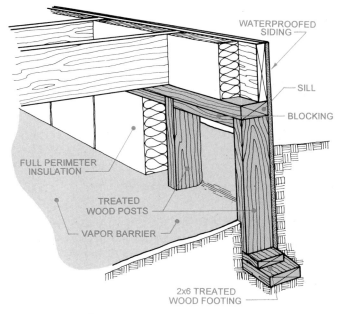

FIGURE 23.50 〉 Permanent wood foundation members. *Hepler/Wallach/Hepler © Cengage 2013*

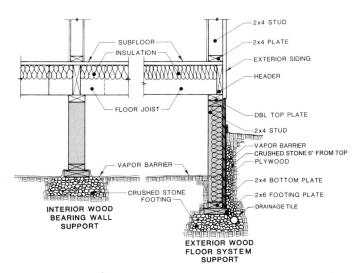

FIGURE 23.51 > Section of a wood foundation wall and posts. *Hepler/Wallach/Hepler © Cengage 2013*

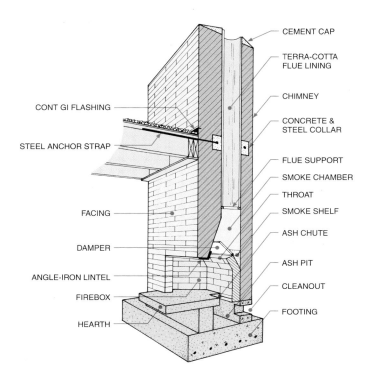

FIGURE 23.53 > Fireplace and chimney structure. *Hepler/Wallach/Hepler © Cengage 2013*

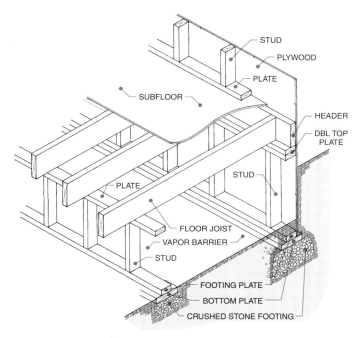

FIGURE 23.52 > Wood foundation details. *Hepler/Wallach/Hepler © Cengage 2013*

🌐 Fire-Producing Components

A firebox to support the fire, a damper to control air flow, and a flue system to exhaust fumes are necessary to create and maintain combustion in most fireplaces. A **firebox** (or **fire chamber**) contains and supports a fire while reflecting heat and exhausting smoke through a flue system. Fire chambers may be built on site, as shown in Figure 23.54, but most fireboxes are factory built with the adjoining structure, foundation, hearth, and chimney built on site.

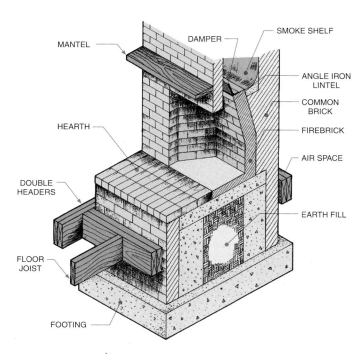

FIGURE 23.54 > Site-built firebox. *Hepler/Wallach/Hepler © Cengage 2013*

Manufactured fireboxes and fireplaces that adhere to environmental and LEEDs standards, produce maximum heat with minimum generation of environmentally unsafe particles. These open combustion units trap, ignite, and recycle polluted flue gases (carbon monoxide, sulfur dioxide, and benzopyrene) into additional heat.

The floor of the firebox or fire chamber (also called the internal hearth) and the wall surfaces that are in direct contact with the fire are covered with firebrick, which is laid in fire-repellent mortar known as fireclay.

If the depth of the firebox is excessive, only a small percentage of heat will be reflected into the room. Smoke may escape into the room if the depth is too shallow. The size and proportion of the firebox are critical to the effective operation of a fireplace. Therefore, care must be taken to use the most efficient ratio (usually 1:10) of flue size to firebox opening. These ratios are identified in Figure 23.55 as W, H, D, and T. The chart in Figure 23.56 lists the fireplace dimensions recommended for configuration. In this chart the symbol ⌗″ is used to represent square inches (sq. in.). The symbol ⌗′ is used to represent square feet (sq. ft.).

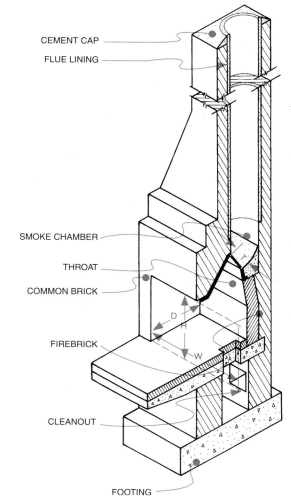

FIGURE 23.55 > Locations of ratios used to calculate fireplace dimensions. *Hepler/Wallach/Hepler © Cengage 2013*

FIREPLACE DIMENSIONS (INCHES)		FIREPLACE WIDTH W	RECTANGULAR FLUES (INCHES)			ROUND FLUES (INCHES)	
			Nominal or Outside Dimension	Inside Dimension	Effective Area	Inside Diameter	Effective Area
W	24 to 84						
H	2/3 to 3/4 W	24	8 1/2 × 8 1/2	7 1/4 × 7 1/4	41°″	8	50.3⌗″
D	1/2 to 2/3 H } 16 to 24 (Rec) for Coal / 18 to 24 (Rec) for Wood	30 to 34	8 1/2 × 13	7 × 11 1/2	70°″	10	78.54⌗″
FLUE (*effective area*)	1/8 WH for unlined flue / 1/10 WH for rectangular lining / 1/12 WH for circular lining	36 to 44 / 46 to 56	13 × 13 / 13 × 18	11 1/4 × 11 1/4 / 11 1/4 × 6 1/4	99°″ / 156°″	12 / 15	113.0⌗″ / 176.7⌗″
T (*area*)	5/4 to 3/2 flue area	58 to 68	18 × 18	15 3/4 × 5 3/4	195°″	18	254.4⌗″
T (*width*)	3″ minimum to 4 1/2″ minimum	70 to 84	20 × 24	17 × 21	278°″	22	380.13⌗″

FIGURE 23.56 > Key dimensions of efficient fireplaces. *Hepler/Wallach/Hepler © Cengage 2013*

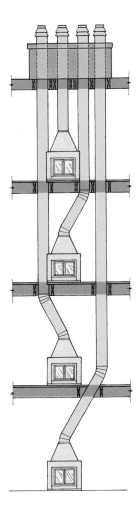

FIGURE 23.57 > Positioning of multiple flues in one chimney. *Hepler/Wallach/Hepler © Cengage 2013*

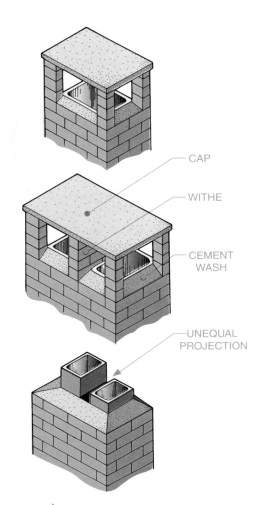

FIGURE 23.58 > Chimney tops designed for downdraft control. *Hepler/Wallach/Hepler © Cengage 2013*

The **flue** is the opening in a chimney through which smoke passes. A **damper** is a door that separates the firebox from the flue area. When the fireplace is in use, this door is opened to allow the upward flow of hot air to create a draft that expels smoke and gas from the firebox to the flue. When the fireplace is not in use, the damper is closed to prevent downdrafts from the flue to the firebox. Part of a damper system is a smoke shelf. When the damper is open, it deflects cold downdrafts into the rising warm air currents.

Designing for adequate warm-air rise (called *draw*) is critical for proper fireplace functioning. Inadequate draw, either from an undersized flue or from improper chimney placement, can result in smoke leaking into the room. The cross section area of a flue should be a minimum of one-tenth the area of the fireplace opening. One flue is necessary for each fireplace or furnace, but multiple flues can extend vertically through one chimney if properly offset, as shown in Figure 23.57. The recommended angle for flue offset ranges from 60% to a maximum of 45%. Often masonry caps or uneven flue projections prevent downdrafts. See Figure 23.58.

Chimneys

Masonry chimneys extend from the footing through the roof of a house. The chimney extends above the roof line to provide a better draft for drawing the smoke and to eliminate the possibility of sparks igniting the roof. **Spark arrestors** are required by many building codes. See Figure 23.59.

The minimum required height of a chimney above the roof line varies somewhat among local building codes. In most areas the minimum distance is 2′ (610 mm) if the chimney is closer than 15′ to the nearest ridge. Tie rods or saddles, as seen in Figure 23.60, may be required to stabilize an excessively high and narrow chimney and to divert water. The fireplace footing must also be designed to support the weight of the entire fireplace and chimney.

FIGURE 23.59 > Spark arrestor and enclosure prevents downdrafts. *Hepler/Wallach/Hepler © Cengage 2013*

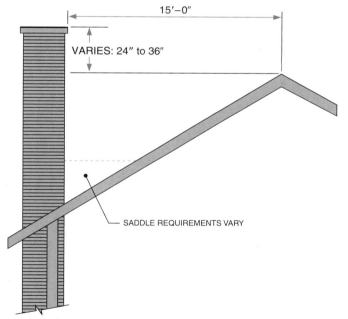

FIGURE 23.60 > Common chimney code requirements. *Hepler/Wallach/Hepler © Cengage 2013*

Safety Components

Some components are not necessary for fire maintenance, but are necessary to prevent flames from spreading outside the firebox. These are covered specifically in building codes.

Wood-burning fireplaces generate environmentally unsafe combustion by-products. Fireplaces must therefore be designed to control and separate these toxic fumes and sub-

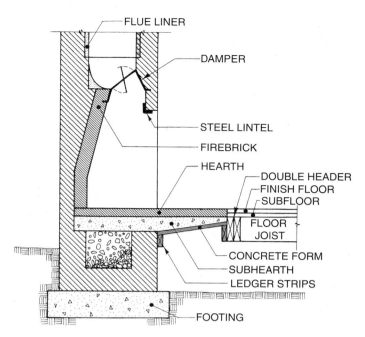

FIGURE 23.61 > Fireplace section. *Hepler/Wallach/Hepler © Cengage 2013*

stances from human exposure and inhalation. This is accomplished by adhering to the following design measures.

- *Dimensioning.* Fireplace detail dimensions must be determined according to fire codes or UL standards. Dimensions should be labeled "Minimum—must be held." This is because undersized or oversized openings may allow smoke and invisible fumes to escape into living areas.
- *Controls.* A high-temperature limit switch and a safety pilot valve should be specified for gas units.
- *Hearth.* Most codes require an extended hearth to cover a floor area 16″ beyond the firebox face and 6″ to each side of the firebox. Hearths may be elevated or flush with the floor level (Figure 23.61).
- *Materials.* Noncombustible materials must be used in fireplace components that will either be in contact with flames or be excessively heated during operation. Firebrick and fire mortar must be used in the firebox. Generally all masonry materials such as concrete, brick, stone, tile, or marble are safe for use outside the firebox. Wood products are not acceptable. Flues must be made of heat-reflecting materials such as terra-cotta, or a 3″ minimum air space must be provided between metal flues and wood framework. An airspace or fire-resistant wallboard or insulating material must be placed between any wood members and a firebox. Loose-fill insulation should not be specified for areas adjacent to the firebox. Manufactured fireboxes must be UL, ASA, and CGA certified.
- *Structural support.* During the structural design phase, provisions must be made so that the heavy masonry loads of the fireplace and chimney assemblies rest on a solid foot-

ing. A solid reinforced concrete footing is most often used for residential construction. Extra footing depth must be provided under the fireplace area. Wood-framed fireplaces with manufactured fireboxes and sheet-metal flues may require a shallow footing of 6″. A masonry fireplace and chimney may require a footing depth of 12″ to 24″, depending on the size of the chimney structure. Footings should extend at least 6″ beyond the fireplace outline on all sides. Wood-burning freestanding wood stoves do not require foundations. If the bottom of a chimney is located on a ridge line, the top must be at least 2′-0″ above the ridge. If the chimney intersection with the roof is located below the ridge line the top of the chimney must be at least 2′-6″ above the ridge. The top of chimneys located on a flat roof must be at least 3′-0″ above the roof.

- *Fire prevention.* Smoke detectors must be hard wired and located on a ceiling not separated by a wall from the fireplace. Each bedroom must include a smoke detector. To prevent sparks from projecting outside the firebox, a glass or wire safety screen should cover the fireplace opening.

Convenience Components

Some items are not necessary for fire production or safety but do add to the convenience and efficiency of using a fireplace:

- *Glass enclosures.* Glass enclosures allow the flames to be seen while preventing smoke or sparks from escaping. When glass enclosures are closed, most heat from the fire is transmitted to the room through vents in the firebox. Built-in glass screens block heat from entering the room, but also stop warm air from escaping as the fire diminishes. Figures 7.51, 7.52, and 7.53 show glass enclosures applied to several fireplace types.

- *Ash pit.* Except for fireplaces on a slab construction, a metal trap door can be placed on the inner hearth. Cold ashes are then dumped through this door to a metal container below. This container has a cleanout door for the removal of ashes. Ash pits are possible on a slab foundation if removal is possible from the top. Figure 23.56A shows a typical ash pit design drawing.

- *Blowers and remote outlets.* To project heat into a room beyond normal air movement, blowers can be used. They may direct air into the room or to remote outlets. See Figure 23.62.

- *Air intake ducts.* To add more oxygen to a firebox, a cold air return or an air duct connected directly from the outside can be added. See Figure 23.63. Some systems use blowers to accelerate air flow through the duct. The use of an outside duct is especially helpful in tightly insulated houses.

- *Freestanding fireplaces.* Freestanding metal fireplaces constructed of heavy-gauge steel are available in a variety of

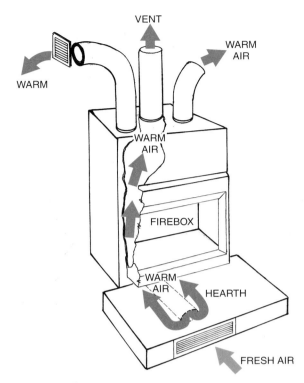

FIGURE 23.62 › Remote outlets used to distribute heat. *Hepler/Wallach/Hepler © Cengage 2013*

FIGURE 23.63 › Air duct installed to bring oxygen to the firebox. *Hepler/Wallach/Hepler © Cengage 2013*

shapes, as shown in Chapter 7. They are relatively light wood-burning stoves and therefore need no concrete foundation for support. A stovepipe leading into the chimney provides the exhaust flue. Because metal units reflect more heat than masonry, metal fireplaces are much more heat efficient, especially if centrally located. For safety, fire-resistant materials such as concrete, brick, stone, or tile must be used beneath and around these fireplaces.

Wood-burning and *pellet stoves* are classified as freestanding stoves. Therefore, codes do not require a foundation system. They are defined as space heaters that are intended to only heat a specific space, unlike a central heating system. They are available in many shapes (Figure 7.55), but the concept for heating is the same. The metal parts become heated and radiate heat into an area. The bottoms and backs are shielded to prevent overheating of floors and walls. Wood-burning stoves burn wood, coal, and almost any combustible material. Pellet stoves burn only small pellets of sawdust, wood chips, or other biomass such as shelled corn and wheat nuts.

FOUNDATION DRAWINGS

Before learning to draw foundation plans, an understanding of foundation layout and excavation is necessary.

Layout

Establishing the exact position of a building on a lot requires locating each building corner. This is done by measuring the distance from property lines to building corners on the plot plan. After one corner of a building has been located, the remaining corners can be located by turning angles with a transit.

Right angles (90°) can be plotted by using the *3.4.5 unit method*. See Figure 23.64. The **Pythagorean theorem** can be used to determine if the measurements are correct.

Pythagorean Theorem:

Square of the hypotenuse of a right triangle = the sum of the square of the two sides:

$$c^2 = a^2 + b^2$$

Example:

$$c^2 = a^2 + b^2$$
$$\sqrt{c^2} = \sqrt{a^2 + b^2}$$
$$c = \sqrt{32^2 + 24^2}$$
$$c = \sqrt{1,024 + 576}$$
$$c = \sqrt{1,600}$$
$$c = 40''$$

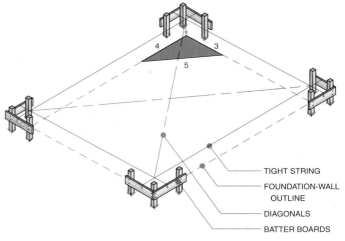

FIGURE 23.64 > Building layout using the 3.4.5 unit method. *Hepler/Wallach/Hepler © Cengage 2013*

TIGHT STRING
FOUNDATION-WALL OUTLINE
DIAGONALS
BATTER BOARDS

If only one point is dimensioned, the azimuth (compass heading) of one side needs to be known to accurately lay out the entire building parameters. For rectangular areas the accuracy of the layout can be checked by measuring across the diagonals. The diagonal distances will be equal if the measurements are true. Once each corner is located, string attached to batter boards is used to identify the corners during excavation, as shown in Figure 23.64.

Excavations and Forming

Foundation plans should clearly show whether areas of a foundation are to be completely excavated for a basement, partly excavated for a crawl space, or unexcavated. The depth of the excavation should also be indicated on the elevation drawings. If a basement is planned, the entire excavation for the basement is dug before the footings are poured. If there is to be no basement, a trench excavation is made. To calculate the volume of an excavation or the amount of concrete needed, use the following formula:

$$\text{Formula: cu. yds.} = \frac{W' \times L' \times D'}{27}$$

Example (see Figure 23.65):
(wall)

$$\text{cu. yds.} = \frac{.75' \times 40' \times 1.25'}{27} = \frac{37.5}{27} = 1.39 \text{ cu. yd.}$$

(footing)

$$\text{cu. yds.} = \frac{2' \times 40' \times .5'}{27} = \frac{40}{27} = 1.49 \text{ cu. yd.}$$

Total   = 2.88 cu. yd.

Although slab foundations may only require trench footings, all organic soil material must be removed and replaced with nonorganic, compactable soil. Excavation must always be made at least 6″ below the frost line. All HVAC, electrical, or plumbing lines which are to be embedded in or pass through a poured area must be installed when the form for the foundation is constructed.

Some foundation plans include the position of plumbing lines if lines are to pass under or through a poured area. Usually, though, the position of plumbing lines is read from wall and fixture positions on either a floor plan or plumbing plan.

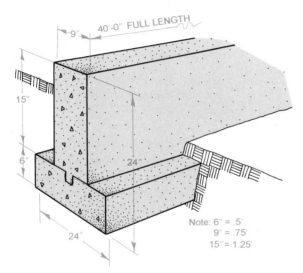

FIGURE 23.65 > Dimensions used to compute foundation and footing volume. *Hepler/Wallach/Hepler © Cengage 2013*

Foundation Plans

Foundation plan drawings are floor plans drawn at the foundation level. The same plan symbols, such as those for brick and concrete, show construction materials. Positions of footings and piers under the slab and soil line are shown with hidden lines on a foundation plan.

Foundation Drawing on CAD

Because foundation drawings align with the floor above, floor plans are used to establish the basic lines of the foundation plan. This is done by drawing the foundation plan on a separate layer over the floor plan layer, which can be shown with lighter lines or color. The floor plan layer can be periodically removed to check progress. The use of layers in this way also enables different drafters to accurately draw different layers for a complex design and then coordinate their work with ease.

T-Foundation Drawings

Figure 23.66 shows a T-foundation drawing. The sequence for drawing T-foundation or basement plans is similar to that for drawing floor plans. The outside perimeter line should be drawn first, then the interior partitions. Next, draw the parallel wall thickness lines, footing lines, and material symbols. Add the position of any piers, columns, footings, and beams. Include a floor joist directional label with dimensions. Figure 23.67 shows how different types of T-foundation walls and footings are drawn as plans and elevation sections. More detailed drawings, as shown in Figure 23.68, may include added dimensions and related components.

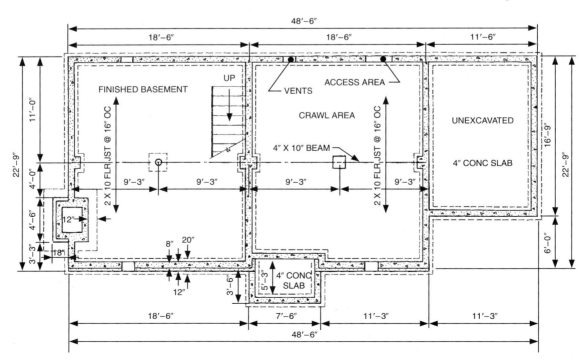

FIGURE 23.66 > T-foundation plan. *Hepler/Wallach/Hepler © Cengage 2013*

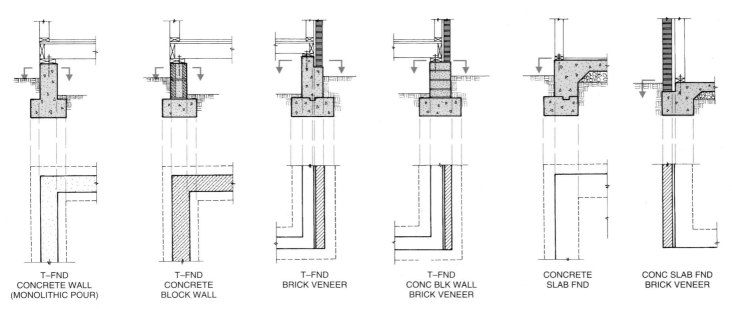

T–FND
CONCRETE WALL
(MONOLITHIC POUR)

T–FND
CONCRETE
BLOCK WALL

T–FND
BRICK VENEER

T–FND
CONC BLK WALL
BRICK VENEER

CONCRETE
SLAB FND

CONC SLAB FND
BRICK VENEER

FIGURE 23.67 > Methods of drawing foundation walls and footings details. *Hepler/Wallach/Hepler © Cengage 2013*

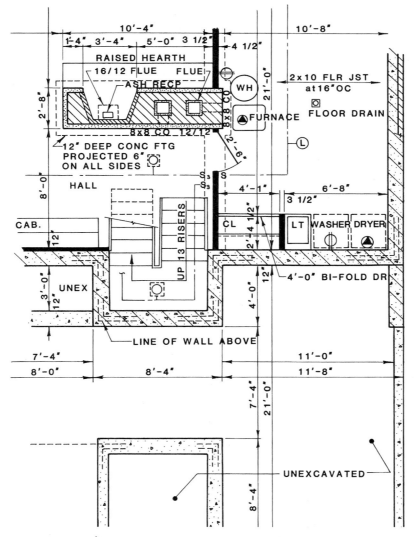

FIGURE 23.68 > Part of a detailed foundation plan. *Hepler/Wallach/ Hepler © Cengage 2013*

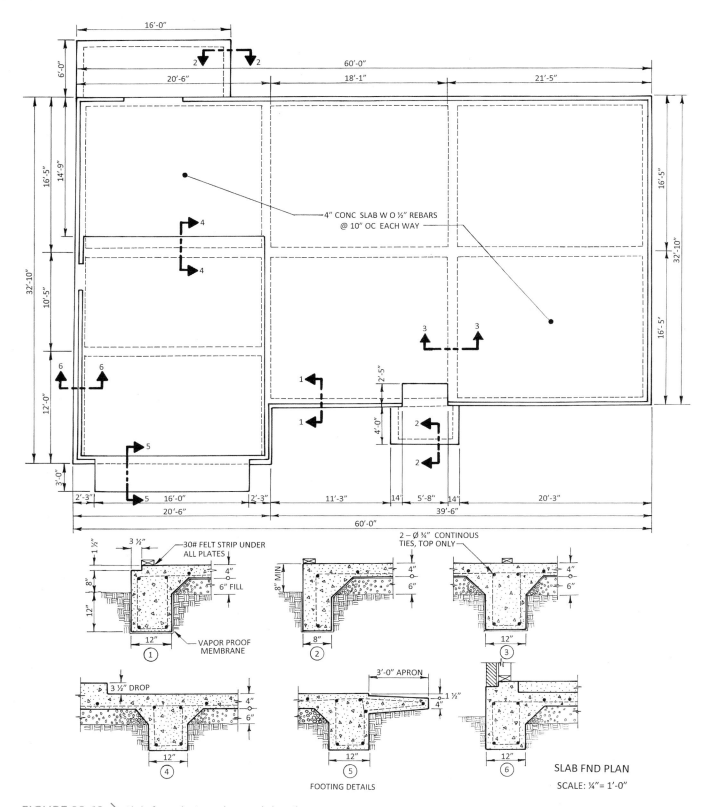

Slab Foundation Drawings

Slab drawings show only the outer perimeter line with hidden lines to represent footings. Slab drawings must note the thickness of the slab, size and spacing of reinforcing bars and wire mesh, concrete PSI (pounds per square inch), and thickness of waterproof membrane.

Footings for slab foundations are usually poured as part of the slab. The design of a monolithic (one-piece) pour and the location of rebars is indexed from the plan to detail drawings as shown in Figure 23.69.

Detail Drawings

Foundation plan views only show the location of all features on a horizontal plane. Most foundation details require elevation sections to convey vertical positions and dimensions. The most common types of foundation details are footing and sill details.

Footing detail drawings show a vertical section through the footing, foundation wall, or slab. These drawings show the size, shape, and material used in footings. See Figure 23.70. If many different footing types are used under partitions, an entire profile of the foundation may be drawn with the location of all footings, as shown in Figure 23.71. Figure 23.72 shows a detailed plan of the lanais and pool identified in the profile section shown in Figure 23.71. Where two slabs meet at different elevations, a step-down detail is needed as in Figure 23.73. Each surface height must be calculated to allow for the thickness of the flooring.

Another feature that needs to be detailed is the foundation support system for fireplaces. Figure 23.74 is a footer detail section designed to be located under a fireplace and chimney.

Sill detail drawings show how the foundation, exterior wall, and floor system intersect. Sill details are viewed in elevation, pictorial, or plan sections. See Figure 23.75.

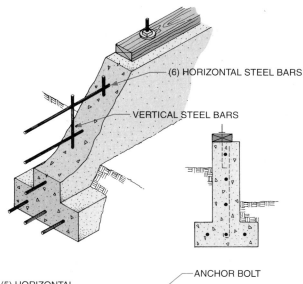
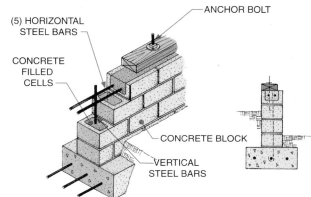

FIGURE 23.70 > Footing detail drawing. *Hepler/Wallach/Hepler © Cengage 2013*

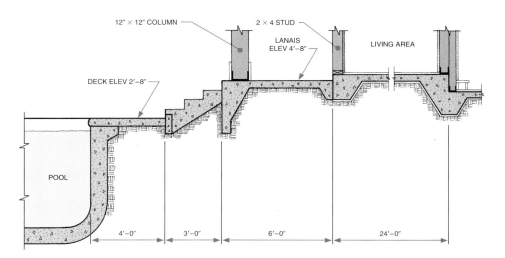

FIGURE 23.71 > Foundation profile section. *Hepler/Wallach/Hepler © Cengage 2013*

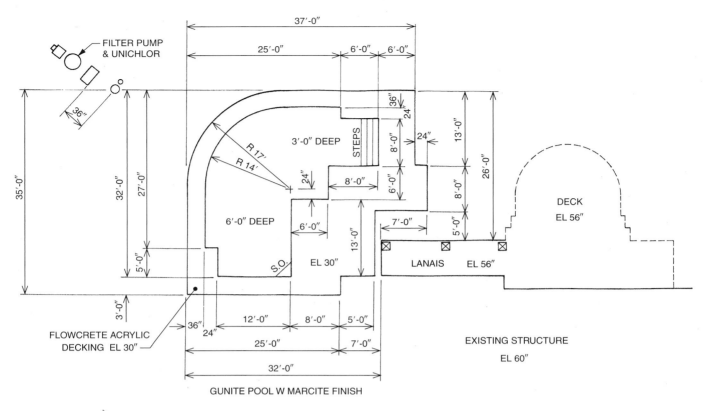

FIGURE 23.72 ❯ Pool and lanais plan to match Figure 23.71 and 13.51. *Hepler/Wallach/Hepler © Cengage 2013*

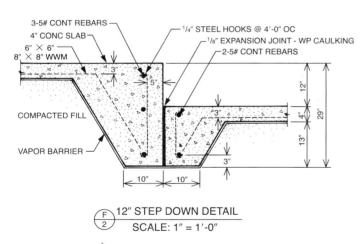

FIGURE 23.73 ❯ Step down concrete foundation section. *Hepler/Wallach/Hepler © Cengage 2013*

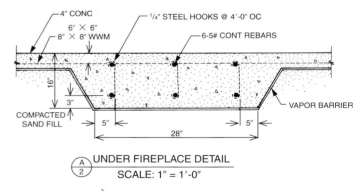

FIGURE 23.74 ❯ Fireplace support footing section. *Hepler/ Wallach/Hepler © Cengage 2013*

Fireplace Drawings

A horizontal section through a fireplace firebox is drawn on floor plan drawings. The vertical face outline of the fireplace opening and the chimney design are shown on interior elevation drawings. However, floor plan and elevation scales are not large enough to show the amount of detail needed to build a fireplace system, so enlarged sections are used to show construction details for masonry fireplaces. For fireplaces that include manufactured components in framed walls, enlarged orthographic views or sectional framing drawings are necessary to show construction details. See Figure 23.76.

If the entire fireplace and chimney structure is to be built on site, a complete set of plans, elevations, and sectional drawings should be prepared, as in Figure 23.76. Figure 23.77 shows working drawings for a fireplace design.

Floor and roof framing drawings related to fireplaces and chimney construction are covered in more detail in Chapters 28 and 30.

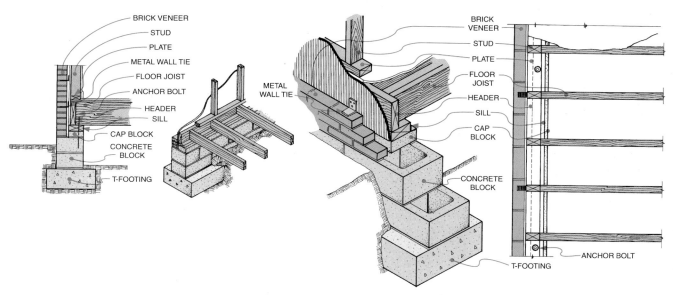

FIGURE 23.75 > Brick veneer foundation and sill details. *Hepler/Wallach/Hepler © Cengage 2013*

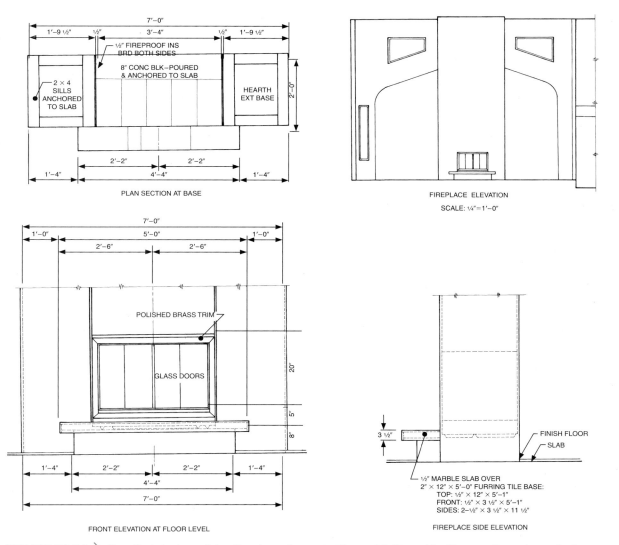

FIGURE 23.76 > Detailed design of the fireplace shown in Figure 23.63 and in Chapter 7. *Hepler/Wallach/ Hepler © Cengage 2013*

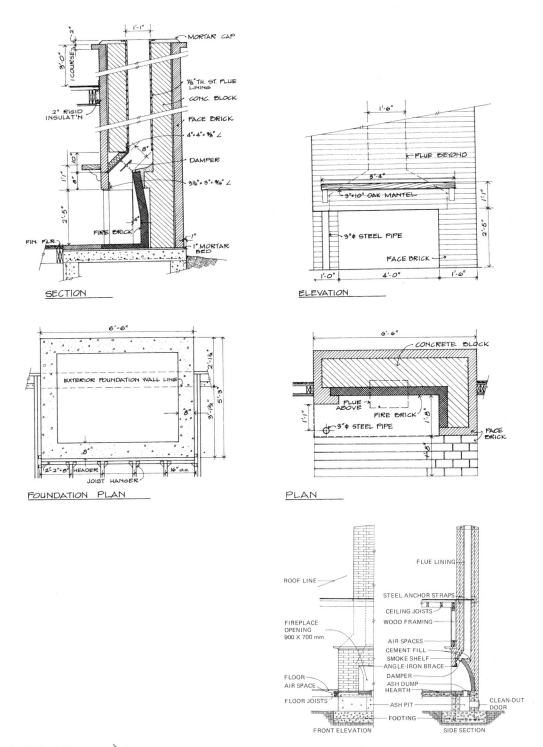

FIGURE 23.77 > Fireplace plan, elevation, and section details. *Hepler/Wallach/Hepler*
© Cengage 2013

Foundations and Fireplace Structures Exercises

1. Draw the foundation plan for the house you are designing.

2. Draw or sketch a foundation plan for the design shown in Figure 15.25 using the scale 1/4″ = 1′-0″ for a T-foundation.

3. Draw or sketch a foundation plan for the design shown in Figure 15.27, using the scale 1/4′ = 1′-0″ for a slab foundation.

4. Draw or sketch a foundation plan for the design shown in Figure 15.50 using the scale 1/4′ = 1′-0″ for a pier foundation.

5. Draw a plan and elevation view of the fireplace in Figure 23.54, using the scale 1/2′ = 1′-0″. See Figure 23.56B for typical dimensions.

6. Design a fireplace for the house you are designing.

7. Draw a sill and footing detail for the house you are designing.

8. Design a T-foundation for a rectangular 25′ × 30′ cabin with a 48″ crawl space. Use a 24″ cripple wall for the top of the T-foundation's wall.

9. Draw the construction details for Exercise 8.

10. Design a foundation for a building site with a consistent 30° slope.

11. Refer to Figure 23.78; draw and dimension sections A-A, B-B, and C-C.

    Perimeter footing to be 18″ deep and 12″ wide. Interior footing to be 18″ deep and 10″ wide. Include symbols and locations of rebars, welded wire mesh, base fill, insulation, and drain.

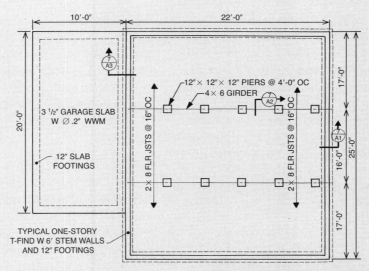

FIGURE 23.78 > Design and draw sections A-A, B-B, and C-C for the locations described by the cutting plane lines. *Hepler/Wallach/Hepler © Cengage 2013*

CROSS B
HEADER

OBJECTIVES

In this chapter you will learn to:

> differentiate between skeleton-frame and post-and-beam construction.

> identify major characteristics of lumber, plywood, and structural timber.

> calculate the number of board feet in a piece of lumber.

> identify steel connectors.

TERMS

balloon framing
board foot
finish size
grade marks
hardwoods

laminated timber
lumber
nominal size
plank-and-beam construction
platform framing

plywood
post-and-beam construction
skeleton-frame construction
softwoods
species

INTRODUCTION

In wood-frame construction, wood structural members are joined to make an open framework for the structure. This open framework is then covered with layers of other construction materials to form the solid surfaces of floors, walls, and roofs.

This chapter introduces the products and most common methods used in designing wood-framing systems. Specific applications and wood-framing details are found in Chapter 28, *Floor Framing Drawings,* Chapter 29, *Wall Framing Drawings,* and Chapter 30, *Roof Framing Drawings.*

Wood-frame systems are comprised of many components, as shown in the exploded view in Figure 24.1. These systems are also connected in various ways to other materials and components.

There are several varieties of wood-frame construction. This chapter covers skeleton-frame construction and post-and-beam construction.

SKELETON-FRAME CONSTRUCTION

In **skeleton-frame construction** (Figure 24.2), small structural members are joined in such a way that they share the loads of the structure. When the structural members are covered, they form complete walls, floors, and roofs. Because of the limited size of wood materials, the skeleton-frame method is considered light construction. See Figure 24.3.

Materials

Light construction materials include lumber, plywood, reconstituted wood, plastics, fasteners, and multimaterial components such as doors, windows, cabinets, plumbing, ductwork, and electrical fixtures.

Lumber

Lumber is wood that has been sawed, surfaced, and planed into specific sizes suitable for construction purposes. Construction lumber is classified according to grade, species, size, and whether it has been treated. **Species** is the designation that identifies the original tree source. Lumber grades are determined by the number and location of defects, such as knots, checks, and splits, and by the degree of warp (deviation from a flat, even surface). For grading purposes, lumber is divided into two broad categories: hardwoods and softwoods. Small, young trees up to 12″ in diameter are cut into 2 × 4

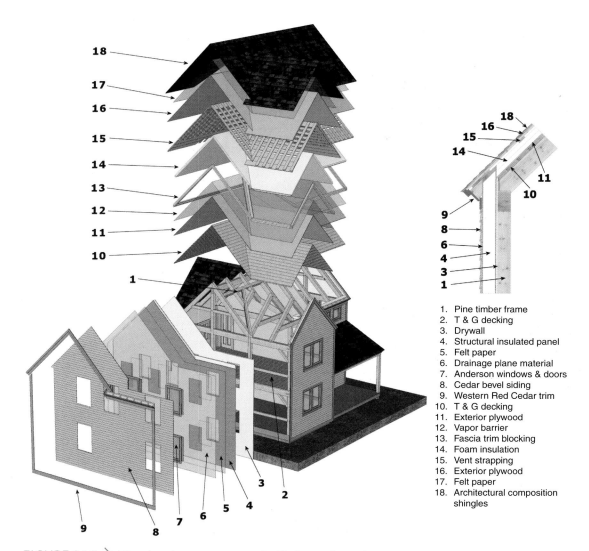

1. Pine timber frame
2. T & G decking
3. Drywall
4. Structural insulated panel
5. Felt paper
6. Drainage plane material
7. Anderson windows & doors
8. Cedar bevel siding
9. Western Red Cedar trim
10. T & G decking
11. Exterior plywood
12. Vapor barrier
13. Fascia trim blocking
14. Foam insulation
15. Vent strapping
16. Exterior plywood
17. Felt paper
18. Architectural composition shingles

FIGURE 24.1 > Wood systems components. *Timberpeg Inc.*

FIGURE 24.2 > Skeleton-frame building under construction.
Hepler/Wallach/Hepler © Cengage 2013

and 2 × 6 framing lumber. Large, older growth logs are cut to produce a wide range of lumber, from large beams to exterior siding boards.

Hardwoods are used for surfaces that must withstand much wear, such as flooring and railings. Hardwoods are also used to make items that require a fine natural finish, such as cabinets and furniture. Hardwoods come from broad-leaved trees. The species most commonly used in construction include oak, walnut, birch, cherry, mahogany, and maple. Hardwood is graded from highest quality to lowest quality depending on the amount of usable material in each piece. See Figure 24.4. Hardwood lumber is available in lengths of 4′ to 16′, widths up to 12″, and thicknesses up to 2″.

Softwoods come from coniferous (needle-bearing) trees such as Douglas fir, pine, or cedar. They are used for structural members such as joists, rafters, studs, sheathing, and formwork. Most skeleton-frame lumber is softwood. Softwood lumber is divided into three grading classes:

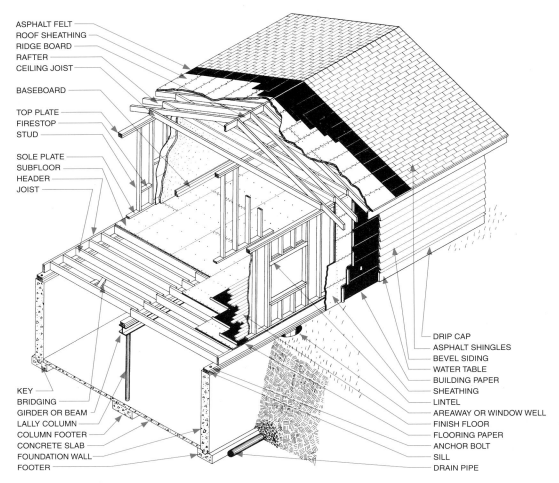

ASPHALT FELT
ROOF SHEATHING
RIDGE BOARD
RAFTER
CEILING JOIST

BASEBOARD

TOP PLATE
FIRESTOP
STUD

SOLE PLATE
SUBFLOOR
HEADER
JOIST

KEY
BRIDGING
GIRDER OR BEAM
LALLY COLUMN
COLUMN FOOTER
CONCRETE SLAB
FOUNDATION WALL
FOOTER

DRIP CAP
ASPHALT SHINGLES
BEVEL SIDING
WATER TABLE
BUILDING PAPER
SHEATHING
LINTEL
AREAWAY OR WINDOW WELL
FINISH FLOOR
FLOORING PAPER
ANCHOR BOLT
SILL
DRAIN PIPE

FIGURE 24.3 > Skeleton-frame building major components. *Hepler/Wallach/Hepler © Cengage 2013*

WOOD GRADE	QUALITY LEVEL
FAS firsts	High Quality
FAS seconds	
Select	
Common #1	
Common #2	
Common #3A	Poor Quality
Common #3B	

FIGURE 24.4 > Hardwood lumber grades. *Hepler/Wallach/ Hepler © Cengage 2013*

yard, structural, and factory (shop) lumber. *Yard lumber* is used for most light framing members, such as sheathing, bracing, subfloors, and casings. See Figure 24.5. *Structural lumber,* as the name implies, is used for load-bearing members and is classified according to use and grades. Grades are based on a lumber's stress resistance. See Figure 24.6.

Factory, or *shop, lumber* consists of light members that are finished at a mill and used for trim, molding, and door and window sashes.

Lumber is graded to provide a consistent standard for identifying vital characteristics among lumber-producing mills. All lumber is graded by an authority at a lumber mill according to the American Lumber Standards. These standards ensure that lumber is grade marked with a variety of appropriate, accurate information. Lumber is labeled with a mill identification number and a certification association logo. Other identifying information of the **grade mark** includes the grade number, moisture content, and species classification, as shown in Figure 24.7.

Structural lumber is defined as either rough or finished. Rough sizes represent the width and thickness of a piece of lumber as cut from a log. Rough lumber is also called **nominal size** lumber. Finished lumber sizes represent the actual dimensions of a member after final surfacing. See Figure 24.8. **Finish size** lumber is also known as surfaced, *dressed,* dimensional, or actual size lumber. Nominal lumber

GRADE	USE	GRADE	USE
Selects and finish	Graded from the best side. Used for interior and exterior trim, molding, and woodwork where appearance is important.	No. 2 common (WWPA) Construction (WCLIB)	All sound tight knots with some defects, such as stains, streaks, and patches of pitch, checks, and splits. Used as paneling and shelving, subfloors, and sheathing.
B & BTR	Used where appearance is the major factor. Many pieces clear, but minor appearance defects allowed which do not detract from appearance.		
		No. 3 common (WWPA) Standard (WCLIB)	Some unsound knots and other defects. Used for rough sheathing, shelving, fences, boxes, and crating.
C Select	Used for all types of interior woodwork. Appearance and usability slightly less than B & BTR.		
		No. 4 common (WWPA) Utility (WCLIB)	Loose knots and knotholes, up to 4" wide. Used for general construction purposes, such as sheathing, bracing, low-cost fencing, and crating.
D Select	Used where finishing requirements are less demanding. Many pieces have finish appearance on one side with larger defects on back.		
Boards	Lumber with defects that detract from appearance but suitable for general construction.	No. 5 common (WWPA) Economy (WCLIB)	Large knots or holes, unsound wood, massed pitch, splits, and other defects. Used for low-grade sheathing, bracing, and temporary construction. Pieces of higher grade wood may be obtained by crosscutting or ripping boards without defects.
No. 1 common (WWPA) Select merchantable (WCLIB)	All sound tight knots, with use determined by size and placement of knots. Used for exposed interior and exterior locations where knots are not objectionable.		

FIGURE 24.5 > Yard lumber grades. *Hepler/Wallach/Hepler © Cengage 2013*

GRADE	USE
LF (Light Framing)	Used in thicknesses from 2" to 4" and widths from 3" and 4", for studs, joists, and rafters in light framing.
JP (Joints and Planks)	Used in thicknesses from 2" to 4" and widths over 2", for joists and rafters to be loaded on either side, or for planking when laid flat.
B&S (Beams and Stringers)	Used in thicknesses from 2" to 4". Widths over 2" must be loaded on narrow edge.
P&T (Posts and Timbers)	Used for posts or columns 5" × 5" and larger or where bending resistance is not critical.

FIGURE 24.6 > Structural lumber grades. *Hepler/Wallach/Hepler © Cengage 2013*

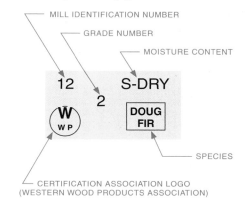

FIGURE 24.7 > Lumber grade marks. *Hepler/Wallach/Hepler © Cengage 2013*

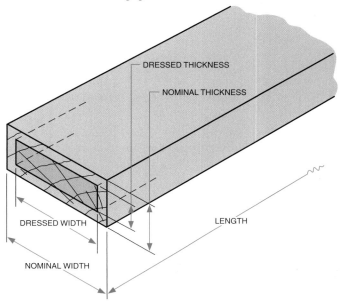

FIGURE 24.8 > Nominal and finished dressed lumber sizes. *Hepler/Wallach/Hepler © Cengage 2013*

LUMBER SIZES IN INCHES

Nominal Size	2 × 4	2 × 6	2 × 8	2 × 10	2 × 12	4 × 6	4 × 8	4 × 10	6 × 6	6 × 8	6 × 10	8 × 8	8 × 10
Dressed Size	1½ × 3½	1½ × 5½	1½ × 7¼	1½ × 9¼	1½ × 11¼	3⁹⁄₁₆ × 5½	3⁹⁄₁₆ × 7½	3⁹⁄₁₆ × 9½	5½ × 5½	5½ × 7½	5½ × 9½	7½ × 7½	7½ × 9½

BOARD SIZES IN INCHES

Nominal Size	1 × 4	1 × 6	1 × 8	1 × 10	1 × 12
Actual Size—Common	¾ × 3⁹⁄₁₆	¾ × 5⁹⁄₁₆	¾ × 7¼	¾ × 9¼	¾ × 11¼
Actual Size—Shiplap	¾ × 3	¾ × 4¹⁵⁄₁₆	¾ × 6⅞	¾ × 8⅞	¾ × 10⅞
Actual Size—T&G	¾ × 3¼	¾ × 5³⁄₁₆	¾ × 7⅛	¾ × 9⅛	¾ × 11⅛

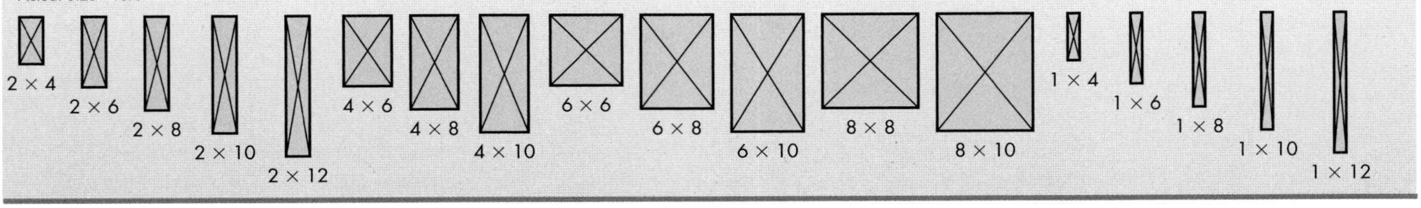

FIGURE 24.9 > U.S. customary lumber sizes. *Hepler/Wallach/Hepler © Cengage 2013*

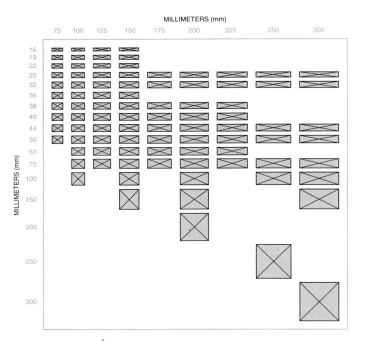

FIGURE 24.10 > Metric lumber sizes. *Hepler/Wallach/Hepler © Cengage 2013*

DESIGNATION	DESCRIPTION
S1S	Surfaced on one side
S2S	Surfaced on two sides
S1E	Surfaced on one edge
S2E	Surfaced on two edges
S1S1E	Surfaced on one side, one edge
S2S1E	Surfaced on two sides, one edge
S4S	Surfaced on four sides

FIGURE 24.11 > Lumber surfacing codes. *Hepler/Wallach/ Hepler © Cengage 2013*

surfaced, or dimensional sizes by some designers or builders. Figure 24.10 shows the range of metric lumber sizes. Because lumber is not always surfaced on all sides, symbols (or codes) designate the number of sides that are surfaced. See Figure 24.11. Figure 24.12 shows the nominal thickness of lumber sizes from 1″ to 16″ and the resulting (dressed) thickness of each.

Board-Foot Measure Lumber is purchased in bulk by the board foot. One **board foot** is 1″ × 12″ × 12″. To determine the number of board feet (BF) in a given piece of lumber, multiply the thickness (in inches) × width (in inches) × length (in feet) and divide the result by 12, as shown in Figure 24.13.

Plywood

Solid lumber is limited in width and has a tendency to warp, split, and check. **Plywood** can be made in wide sheets and is structurally stable. Plywood is manufactured from thin sheets (0.10″ to 1.25″) of wood laminated together with an adhesive under high pressure. The number of layers (plies) varies from

sizes, milled for light construction, are used in the design of skeleton-frame structures. These include:

Thicknesses: 1″, 2″, and 4″

Widths: 2″, 4″, 6″, 8″, 10″, and 12″

Lengths: 8′, 10′, 12′, 14′, 16′, 18′, and 20′

When a drawing callout reads 2 × 4, builders know that the rough size is 2″ × 4″ but the actual size of the member is 1 1/2″ × 3 1/2″. Figure 24.9 shows the U.S. customary range of rough (nominal) and finished (actual) standard lumber sizes. The nominal sizes in this chart may also be labeled as rough sizes. The dressed sizes may also be labeled as finished,

NOMINAL THICKNESS	DRESSED THICKNESS
1″	3/4″
2″	1 1/2″
3″	2 1/2″
4″	3 1/2″
5″	4 1/2″
6″	5 1/2″
8″	7 1/4″
10″	9 1/4″
12″	11 1/4″
14″	13 1/4″
16″	15 1/4″

FIGURE 24.12 > Standard lumber thicknesses. *Hepler/Wallach/Hepler © Cengage 2013*

$$\text{Formula: BF} = \frac{T'' \times W'' \times L'}{12}$$

$$\text{Example: BF} = \frac{2'' \times 10'' \times 3'}{12} = 5 \text{ BF}$$

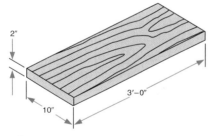

FIGURE 24.13 > Board feet measurement calculation example. *Hepler/Wallach/Hepler © Cengage 2013*

three to seven. The grain of each ply is laid perpendicular to the grain of each adjacent layer. The grain of both outside sheets always faces the same direction. This layering process greatly reduces the tendency of plywood to warp, check, split, splinter, and shrink. Plywood is available in individual 4′ × 8′ sheets or in continuous panels up to 50′ in length. It is made in thicknesses of 1/8″, 3/16″, 1/4″, 5/16″, 1/2″, 5/8″, 3/4″, 1″, 1 1/8″, and 1 1/4″.

All plywood is divided into two broad categories: exterior (waterproof) or interior. Plywood for structural use is made with surfaces of softwood. Plywood for cabinets and furniture has hardwood surfaces. Because of the different uses, construction-grade plywood (softwood) and veneer-grade plywood (hardwood) are identified by two different quality-rating systems.

GRADE	DESCRIPTION
Standard	For use as subflooring, roof sheathing, wall sheathing, and structural interior applications.
Structural Class I and II	For uses requiring resistance to tension, compression, and shear stress including box beams, stressed skin panels, and engineered diaphragms. High nail-holding quality and controlled grade and glue bonds.
CC Exterior	Meets all exterior plywood requirements.
BB Concrete-Form Panels, Class I and II	Edges sealed and oiled at the mill and used for concrete form panels.

FIGURE 24.14 > Plywood construction grades. *Hepler/Wallach/Hepler © Cengage 2013*

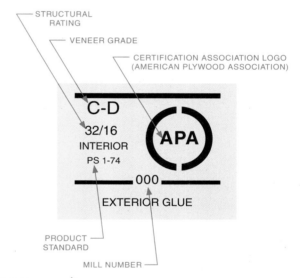

FIGURE 24.15 > Plywood construction grade marks. *Hepler/Wallach/Hepler © Cengage 2013*

Construction-grade plywood is unsanded. It is identified by grade levels based on structural strength, as shown in Figure 24.14. Because of live-load differences, plywood panels used for structural purposes are marked with two numbers indicating the structural rating. See Figure 24.15. The first number represents the maximum span (in inches) possible between supporting roof members. The second number represents the maximum allowable span when used for flooring.

Because hardwood plywood is used for making cabinets and furniture, veneer plywood grades (groups) are classified by a letter indicating the number of knots, checks, stains, and open sections in each panel, as shown in Figure 24.16. These letters are used to show only the group of the front and back plies (layers). When front and back plies are of different grades, letters are combined. For example, a grade of B–D means the front ply is B grade and the back ply is D grade. On Figure 24.17, notice the various kinds of information

GRADE	DESCRIPTION
Grade A	Paintable and smooth with no more than 18 neat boat, sled, or router type repairs made parallel with grain. Will accept natural finish.
Grade B	Solid surface with shims, circular repair plugs, or tight knots less than 1" wide permitted.
Grade C	Tight knots of less than 1 1/2", knotholes less than 1" wide, synthetic or wood repairs, limited splits, slices (gouges), discoloration, and sanding defects that do not impair strength permitted.
Grade C Plugged	Some broken grain, synthetic repairs, splits up to 1" wide, knotholes and bareholds (other holes) up to 1 1/4" × 1/2" permitted.
Grade D	Knots and knotholes up to 2 1/2" wide across grain or 3" wide if within limits permitted but restricted to interior use.

FIGURE 24.16 ❯ Hardwood veneer plywood grades. *Hepler/ Wallach/Hepler © Cengage 2013*

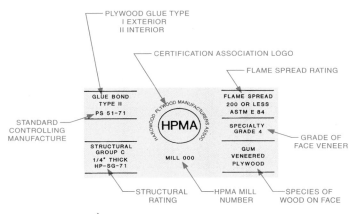

FIGURE 24.17 ❯ Typical hardwood plywood grade mark. *Hepler/Wallach/Hepler © Cengage 2013*

GROUP 1	GROUP 2	GROUP 3	GROUP 4	GROUP 5
Beech	Cedar, port	Alder, red	Aspen	Basswood
Birch	Cypress	Birch, paper	Cedar	Poplar, balsam
Sweet	Douglas fir 2	Cedar, alaska	Incense	
Yellow	Fir	Fir, subalpine	Western	
Douglas fir 1	Balsam	Hemlock	red	
Maple, sugar	California red	Maple, bigleaf	Cottonwood	
Pine	White	Pine	Pine, Eastern	
Caribbean	Hemlock	Jack	white	
Ocote	Lauan	Ponderosa	Sugar	
Pine, south	Maple, black	Spruce		
Loblolly	Pine	Redwood		
Longleaf	Red	Spruce		
Shortleaf	Western			
Slash	white			
	Spruce			
	Yellow poplar			

FIGURE 24.18 ❯ Wood species groups. *Hepler/Wallach/Hepler © Cengage 2013*

included in the hardwood plywood standard grade marks stamped on each sheet. Wood species are classified in groups. See Figure 24.18.

When selecting hardwood plywood, several characteristics must be considered. Grain patterns, color, and texture consistency, as well as specific species, smoothness, and finishability, should be matched for cabinets, paneling, and furniture.

🌐 Wood and the Environment

Efforts to balance architecture and building with nature are not new. Many decades ago Frank Lloyd Wright pioneered this cause with his principles of natural building. This primarily involved preserving natural sites and using local and natural building materials.

The term *green lumber* should not be confused with the term *green* as applied to environmentally safe products. In the language of construction, *green lumber* is unseasoned lumber that has a high moisture content, which must be reduced before the lumber can be used for construction purposes. This is done through air-drying or kiln drying (KD). This drying process converts green (unseasoned) lumber to seasoned lumber, which has a low moisture content. Lumber seasoned in this way is not the same as *treated lumber*.

Timber or lumber may be treated with a preservative that protects it from destruction by insects, fungus, or exposure to moisture. The preservative is usually applied through combined vacuum and pressure treatment, which removes air from wood fibers. The preservatives used to pressure treat lumber are classified as pesticides. Due to potential hazards to humans and the environment, many have been discontinued. Most newer preservatives are free of metallic compounds and are based on biodegradable organic chemical ingredients. Treating lumber provides long-term resistance to organisms that cause deterioration. If applied correctly, treatment significantly extends the productive life of lumber.

Treated wood products and untreated products share the same standard sizes, grades, and species classifications. However, some treated lumber and plywoods contain harmful chemicals that emit toxic vapors. These may include processed woods such as particleboard, pressboard, pressure-treated lumber, and some plywoods. Some wood preservatives, foam insulation, and plaster and cement products also contain formaldehyde, radon, benzene, arsenic, asbestos, and chlorofluorocarbons. These are especially harmful in enclosed spaces. For these reasons untreated products are usually preferred, but these are prone to develop dry rot, fungi, and insect infestation. Therefore, treated lumber should be restricted to applications near soil and well-ventilated areas.

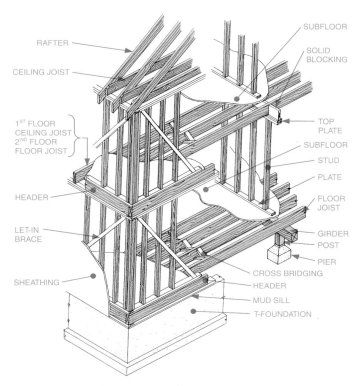

FIGURE 24.19 ❯ Platform framing components. *Hepler/Wallach/Hepler © Cengage 2013*

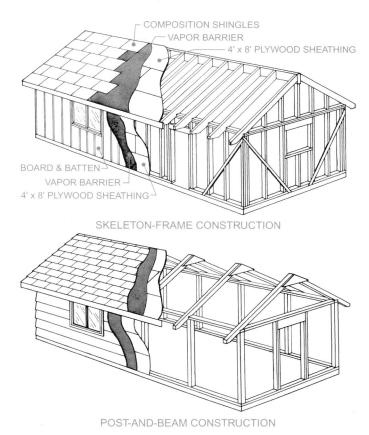

FIGURE 24.20 ❯ Skeleton-frame and post-and-beam comparison. *Hepler/Wallach/Hepler © Cengage 2013*

Wood-Framing Methods

When multiple-level buildings are constructed using a wood skeleton frame, either the **platform framing** method or the **balloon framing** method is used. In platform framing, the second floor rests directly on first-floor exterior walls. See Figure 24.19. Because of the difficulty involved in maintaining the plumb stability of continuous studs, balloon framing is seldom used.

POST-AND-BEAM CONSTRUCTION

Large timbers have been used in construction for centuries, usually for floor and roof systems in buildings of bearing wall design. The development of large glass sheets and sheathing materials as well as improvements in the manufacture and transport of large wood members have made new uses possible. Today, heavy timbers may be used for walls as well as for floors and roofs. When heavy timbers are used in this manner, the construction method is called **post-and-beam construction.**

Post-and-beam construction uses larger structural members than skeleton-frame construction, and they are spaced farther apart. See Figure 24.20. The larger spacing means that fewer members are needed. Labor savings are also considerable because fewer members are handled and fewer intersections connected.

Many post-and-beam members remain exposed in the finished building. To create a more pleasing visual effect, a better grade of lumber is specified. Rigid insulation must be used above, instead of under, the roof planks to expose the natural plank ceiling. Plumbing and electrical lines must be passed through cavities in columns or beams. Bearing partitions and other heavy dead loads, such as bathtubs, must be located over beams, or additional support framing must be used.

Post-and-beam construction relies on the relationship of three basic components: *posts* or *columns, beams,* and *planks.* Vertical columns support horizontal beams. These beams support planks placed perpendicular to the beams, as in Figure 24.21. Floor and roof systems are supported by the beams, which, as stated, are supported by posts or columns, which transfer loads to footings. Member sizes and spacing vary depending on load requirements.

Where loads are not a factor, beams can be hung on girders as shown in Figure 24.22. Wherever built-up members are supported by a beam, the joints must be staggered as shown in Figure 24.23. Figure 24.24 illustrates the use of a gusset plate and joist hanger to support a joist and control lateral movement of a column, joist, and beam.

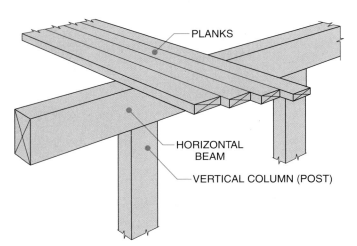

FIGURE 24.21 > Components of heavy timber, post-and-beam construction. *Hepler/Wallach/Hepler © Cengage 2013*

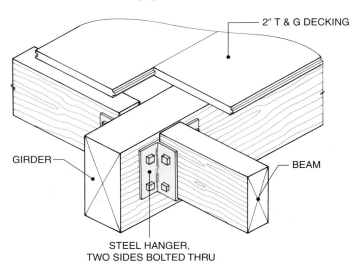

FIGURE 24.22 > Girder hung beams. *Hepler/Wallach/Hepler © Cengage 2013*

Floor Construction

Timber floor systems use heavy wood planks or tongue-and-groove plywood placed over widely spaced beams, as shown in Figure 24.25. This type of floor system is called **plank-and-beam construction.** Each evenly spaced beam replaces several of the intermediate joists that would be used in a skeleton-frame floor system. For example, a 24′ distance may require 19 conventional floor joists spaced at 16″ OC (on center). See Figure 24.26. However, only 7 joists placed 4′ OC may be needed to support the same loads in plank-and-beam construction. In this system, floor planks must be strong enough to avoid deflection at the middle or at the midspan. In this design, joists support heavy lumber planks and beams support the joists. The beams, in turn, are supported by piers and posts in the center and by the foundation wall on the perimeter.

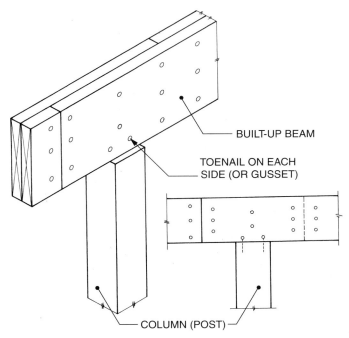

FIGURE 24.23 > Built-up beam with staggered joists. *Hepler/Wallach/Hepler © Cengage 2013*

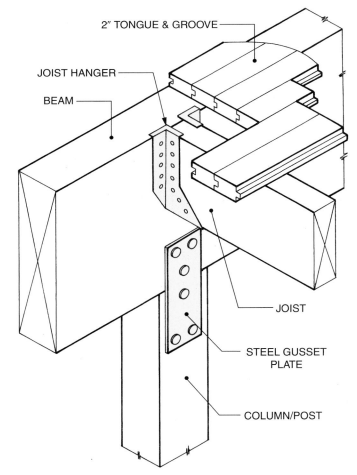

FIGURE 24.24 > Use of joist hanger and gusset plate. *Simpson Strong-Tie Company Inc.*

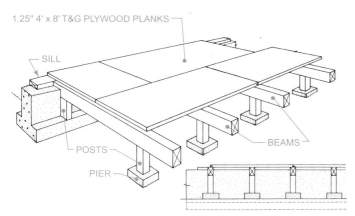

FIGURE 24.25 > Plank-and-beam floor construction using heavy plywood. *Hepler/Wallach/Hepler © Cengage 2013*

Wall Construction

Just as beams replace conventional joists in plank-and-beam floor systems, posts replace conventional studs in post-and-beam wall construction. The large open spans between the wall posts can be occupied by nonbearing material or components such as windows, doors, or insulating material. See Figure 24.27. For this reason nonstructural elements in a post-and-beam outside wall are known as *curtain walls*. An example of the construction of wall column framing intersections is detailed in Figure 24.28.

Roof Construction

In plank-and-beam roofs, beams replace conventional roof rafters and planks replace conventional roof sheathing. The two types of plank-and-beam roof systems are longitudinal and transverse.

In longitudinal systems, roof beams are aligned parallel with the long axis of the building. See Figure 24.29. In transverse systems, beams are aligned across the short width of the building, as illustrated in Figure 24.30. One end of the beam is supported by a post. The other end may rest on top of a ridge beam, or it may be butted and fastened against the side of the ridge beam. Transverse beams either intersect a ridge beam on pitched roofs or lie flat across the span on flat roofs.

Structural Timber Members

Three types of structural members are used in contemporary post-and-beam construction: *solid*, *laminated*, and *fabricated* components.

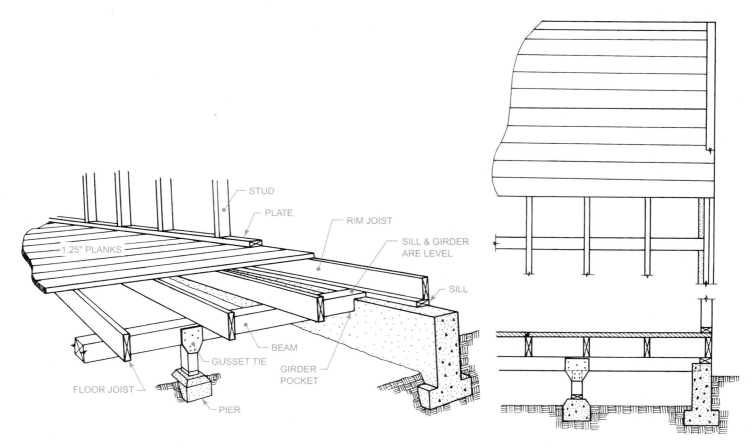

FIGURE 24.26 > Beam, girder, and plank floor system. *Hepler/Wallach/Hepler © Cengage 2013*

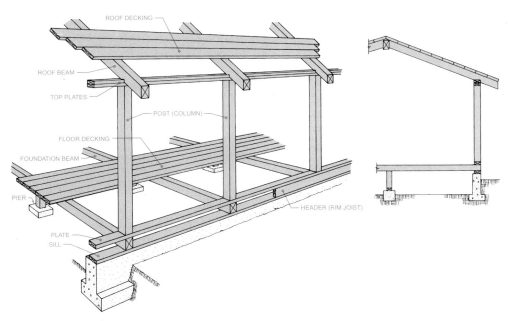

FIGURE 24.27 ⟩ Post-and-beam wall construction. *Hepler/Wallach/Hepler © Cengage 2013*

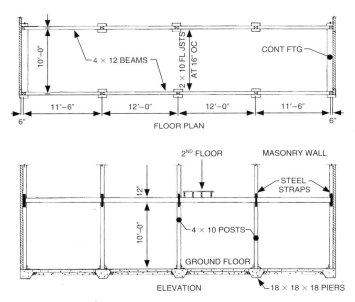

FIGURE 24.28 ⟩ Post-and-beam wall framing plan and elevation. *Hepler/Wallach/Hepler © Cengage 2013*

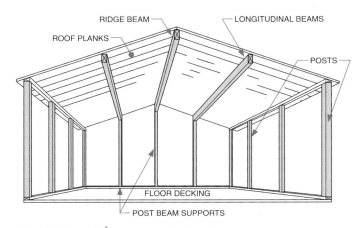

FIGURE 24.29 ⟩ Longitudinal roof framing. *Hepler/Wallach/ Hepler © Cengage 2013*

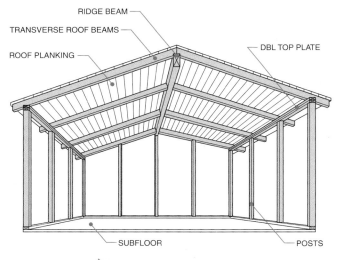

FIGURE 24.30 ⟩ Transverse roof framing. *Hepler/Wallach/ Hepler © Cengage 2013*

Solid Members

Solid wood timbers are available in thicknesses that range from 3″ to 12″. However, the use of sizes over 8″ is hampered by the tendency of large solid wood timbers to warp. One method of stabilizing larger solid wood members is to add steel plates to create *flitch beams,* which are stronger and remain straighter. See Figure 24.31.

Solid planking in small widths (2″ and 4″) is more commonly used than plywood panel flooring. Because of the impact of live-load thrusts, solid planking is usually speci-

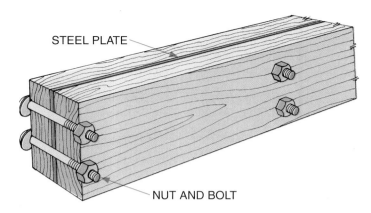

FIGURE 24.31 › Flitch beam. *Hepler/Wallach/Hepler © Cengage 2013*

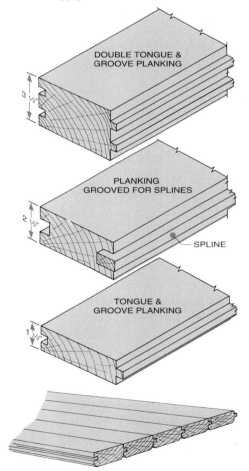

FIGURE 24.32 › Tongue-and-groove planking. *Hepler/ Wallach/Hepler © Cengage 2013*

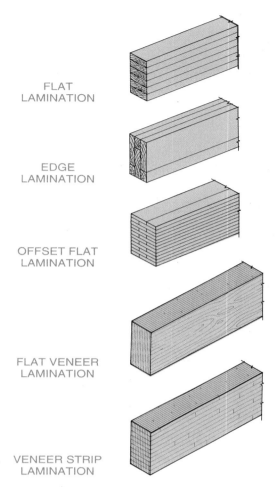

FIGURE 24.33 › Laminated beam types. *Hepler/Wallach/ Hepler © Cengage 2013*

fied as tongue and groove (T&G). The T&G joint reduces deflection by tying the flooring planks together into one monolithic unit, as shown in Figure 24.32.

Laminated Members

When larger timbers are needed to support heavier weights or greater spans, glue-laminated timbers are often used. **Lami-**

nated timbers are made from thin layers (less than 2″) of wood, glued together either vertically or in patterns. See Figure 24.33. Laminated timbers are stronger than solid timbers because the grain direction is reversed (180°) in alternate layers. Because of its more consistent moisture content, there is less expansion or contraction in a laminated wood member than in solid wood. Glue-laminated members (or *glulam* for short) are manufactured in forms for columns, beams, and arches.

Laminated construction members used for plywood include *parallel stranded lumber* (PSL) and *laminated veneer lumber* (LVL). PSL is made from parallel strands of wood and are manufactured in large sizes suitable for heavy load-bearing posts and beams. LVL is made from thin veneer layers glued together with phenolic glue. The joints are staggered over the length with the grain facing the same direction.

Oriented strand board (OSB) can be substituted for laminated plywood mainly for roof and wall sheathing. OSB is composed of shredded strands of wood that are glued and compressed into layers.

In addition to the laminated members for posts and beams, laminated decking is also available in 2″ thicknesses and in 6″

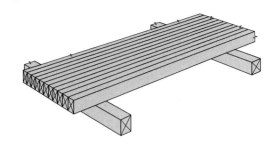

FIGURE 24.34 > Side-laminated decking. *Hepler/Wallach/ Hepler © Cengage 2013*

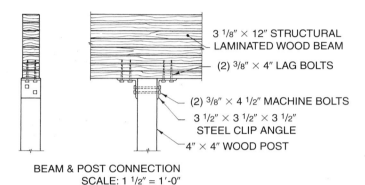

3 1/8″ × 12″ STRUCTURAL LAMINATED WOOD BEAM

(2) 3/8″ × 4″ LAG BOLTS

(2) 3/8″ × 4 1/2″ MACHINE BOLTS

3 1/2″ × 3 1/2″ × 3 1/2″ STEEL CLIP ANGLE

4″ × 4″ WOOD POST

BEAM & POST CONNECTION
SCALE: 1 1/2″ = 1′-0″

FIGURE 24.35 > Lag screws connecting post and beam. *Hepler/Wallach/Hepler © Cengage 2013*

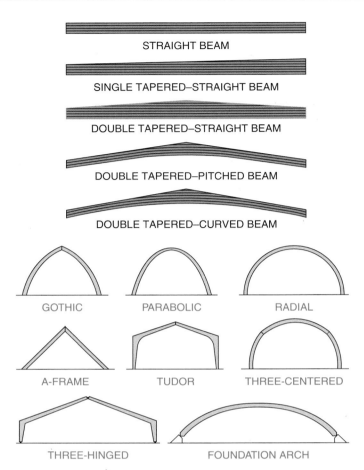

STRAIGHT BEAM

SINGLE TAPERED–STRAIGHT BEAM

DOUBLE TAPERED–STRAIGHT BEAM

DOUBLE TAPERED–PITCHED BEAM

DOUBLE TAPERED–CURVED BEAM

GOTHIC   PARABOLIC   RADIAL

A-FRAME   TUDOR   THREE-CENTERED

THREE-HINGED   FOUNDATION ARCH

FIGURE 24.36 > Laminated beam forms. *Hepler/Wallach/ Hepler © Cengage 2013*

to 12″ widths. Laminated decking is specified in nominal sizes on construction drawings. When decking material is laminated, the layers may be offset to fit together. Another method of lamination is to align the wood grain vertically and laminate the sides for greater resistance to loads. See Figure 24.34. Where flat laminated beams are supported by a wood post, lag bolts are required through at least three layers, as shown in Figure 24.35.

Although lamination can create stronger, larger, and more structurally stable members, its most popular feature is its ability to be bent into a wide variety of structurally sound and aesthetically pleasing shapes, as shown in Figure 24.36. Various beam forms, including arches, are created by first bending thin, parallel layers of wood to a desired shape. Then the layers are glued and clamped together under pressure. When the glue dries, the member retains the new, bent form.

To indicate arches on architectural drawings, the base location of each arch is shown in the ground floor plan. The profile shape, including height and width dimensions, is drawn on elevation drawings and elevation sectional drawings. See Figure 24.37.

🌐 Fabricated Members

Many fabricated products and materials are used in place of solid or laminated wood members. The term *engineered wood* is used to describe a wide array of wood-based products that are not 100% wood. This includes the blending and molding

of wood fibers with plastic resins to form construction members. The term is also used to describe the processing and use of solid wood shapes combined in unique ways as described later in this chapter. Engineered wood is straighter and stiffer than natural wood and can be produced in thicker, wider, and longer sizes. These include a variety of wood I-beams, truss joists, panels, box beams, strand lumber, and recycled plastic material.

Wood I-beams and also I-joists used for floor, ceiling joists, and rafters are constructed with a plywood or oriented strand board (OSB) that is inserted into a groove in a laminated or machine-stressed wood flange. This construction produces a very straight, lightweight, strong, and stable member. Knockout areas can be used to install HVAC, electrical, or plumbing lines without sacrificing strength. See Figure 24.38. Wood I-beams can span much longer distances than comparably sized solid members. Figure 24.39 shows examples of the various types of wood I-beams.

Truss joists are constructed like a truss, but with parallel flanges usually made of 2 × 4's. See Figure 24.40. These members can support greater loads than the same size solid lumber. Figure 24.41 shows truss joists used as a substitute for solid roof rafters or conventional trusses.

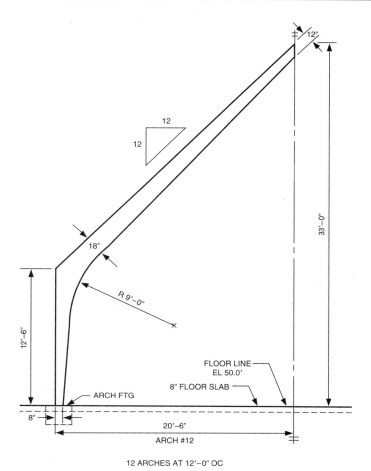

FIGURE 24.37 > Laminated arch dimensioning methods.
Hepler/Wallach/Hepler © Cengage 2013

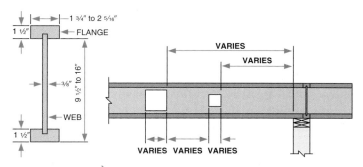

FIGURE 24.38 > Wood I-beam or truss details. *Weyerhaeuser*

o.c. spacing	JOIST DEPTH			
	A 9 1/2"	B 11 7/8"	C 14"	D 16"
12"	16'-10"	20'-0"	24'-6"	27'-1"
16"	15'-4"	18'-2"	22'-3"	24'-8"
19.2"	14'-5"	17'-1"	20'-11"	23'-2"
24"	13'-4"	15'-10"	19'-4"	21'-5"

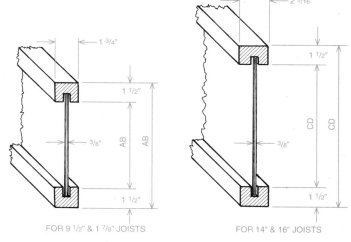

FOR 9 1/2" & 1 7/8" JOISTS   FOR 14" & 16" JOISTS

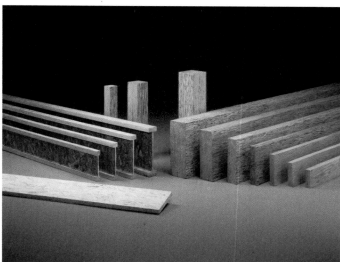

FIGURE 24.39 > Truss joists and stranded wood beams.
Weyerhaeuser

Stressed-skin or *sandwich panels* are often used in place of solid or laminated members. These lightweight prebuilt or site-built panels are made by gluing, screwing, or nailing plywood sheets to structural member frames. Because panels can easily be constructed using standard plywood sizes, they are often used for floor, wall, and roof panels. Stressed-skin panels can also be used to make box beams for spans up to 120' depending on load factors. Folded plate roofs and curved panels can also be either fabricated on the site or factory built.

The use of stressed-skin panels in roof construction is covered in Chapter 30.

Foam sandwich panels are made of two plywood or strand board sheets adhered to a core of polyurethane or polystyrene. These panels function as framing, insulation, and sheathing and may sometimes be used as finished interior wallboard.

Strand lumber is made of cellulose fiber strips. The fibers from poor-quality or small trees, rice, rye, wheat, or straw are crushed into long strands. After being combined with formaldehyde and adhesives, the strands are woven, compressed,

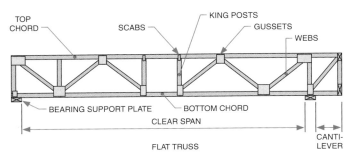

FIGURE 24.40 〉 Truss joist components.

FIGURE 24.41 〉 Truss joists used as roof rafters.

and heat treated. The finished shapes are very strong and can span long distances. They are used primarily for columns, beams, and large headers. In addition to the PSL, LVL, and OSB types of lumber described earlier in the plywood section, *laminated strand lumber* (LSL) is frequently used in place of solid wood as core material for doors and short span headers.

Recycled thermoplastics are plastics that are ground, glued, and pressed into structural members. These thermoplastic members are capable of being sawed and glued, and they can accept and hold screws and nails. Contrary to wood, thermoplastic members will not rot, absorb moisture, expand, contract, or be infested by insects.

Timber Connectors

Because of heavy timber sizes, concentrated loads, and lateral thrust from winds and earthquakes, special joints and fasteners are required to attach post-and-beam members to the foundation and to other members. Nails and screws are useful only as temporary holding devices, and lag bolts can

FIGURE 24.42 〉 Examples of heavy timber joint construction. *DAL-Tile*

only be used in areas of limited stress. Figure 24.42 shows the application of heavy timber joint construction at the base, intermediate, and roof levels. Figure 24.43 shows several typical post-and-beam intersections that do not make use of brackets or straps. Figures 24.44A and 24.44B show common joints and brackets used in post-and-beam construction.

Base anchors and plates are used to attach the base of heavy timber posts to the foundation and to prevent wood deterioration. See Figure 24.45. Timber brackets can be embedded into concrete as shown in Figure 24.46, or timber can be placed into an impact-fastener base anchor. See Figure 24.47.

Metal strap ties and gusset plates are used to attach posts to beams and to keep transverse roof beams aligned with ridge beams. Post-and-beam construction has excellent resistance to dead loads, which exert pressure directly downward. However, because of the large unsupported wall areas, lateral live loads can be a problem. Although diagonal ties or sheathing can help control the lateral thrust, angle brackets should be used to fasten perpendicular intersections to prevent lateral movement between members and help provide rigidity to joints.

Figure 24.48 shows the use of angle brackets to hold overlapping beams at right angles. Figure 24.49 shows the use of

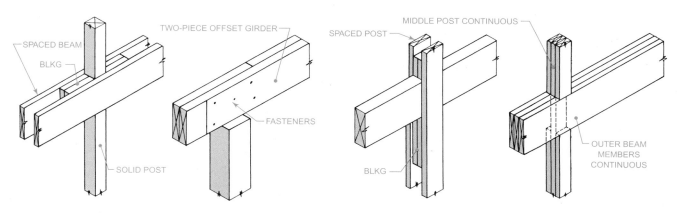

FIGURE 24.43 > Post-and-beam intersections without brackets. *Hepler/Wallach/Hepler © Cengage 2013*

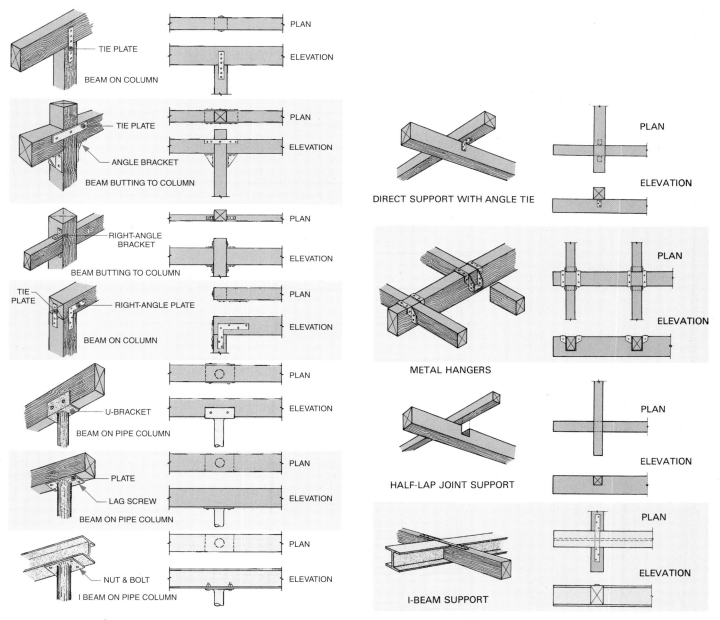

FIGURE 24.44A > Common post-and-beam joints. *Hepler/ Wallach/Hepler © Cengage 2013*

FIGURE 24.44B > Common post-and-beam brackets. *Hepler/Wallach/Hepler © Cengage 2013*

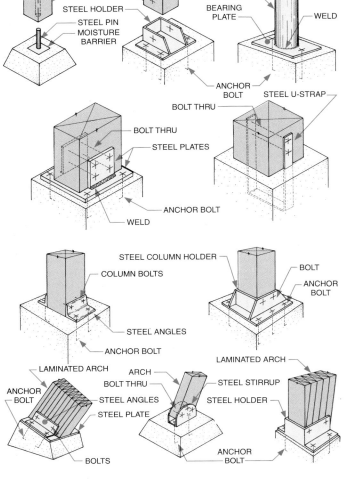

FIGURE 24.45 > Base anchors and plates. *Hepler/Wallach/ Hepler © Cengage 2013*

Before installation.

After installation.

FIGURE 24.47 > Timber base anchors. *Hepler/Wallach/Hepler © Cengage 2013*

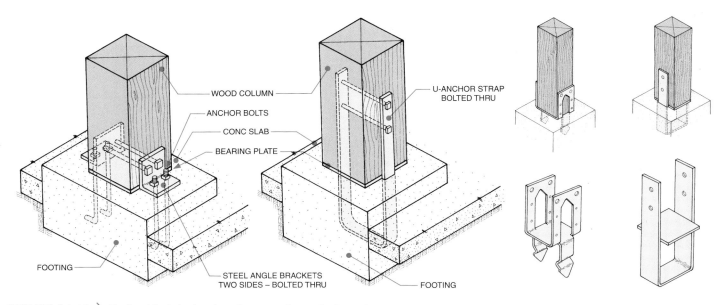

FIGURE 24.46 > Embedded timber brackets. *Hepler/Wallach/Hepler © Cengage 2013*

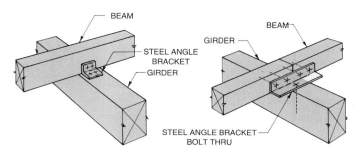

FIGURE 24.48 > Timber angle brackets. *Hepler/Wallach/Hepler © Cengage 2013*

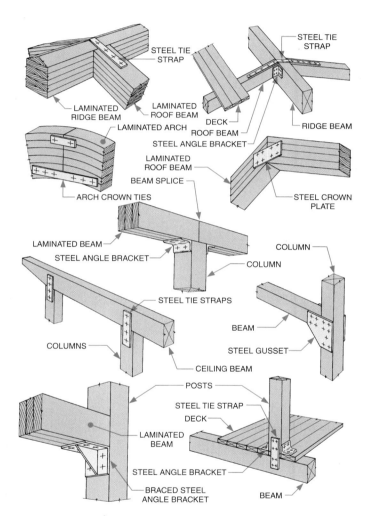

FIGURE 24.49 > Use of straps, gussets, and angle brackets. *Hepler/Wallach/Hepler © Cengage 2013*

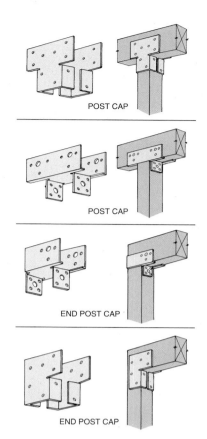

FIGURE 24.50 > Post caps. *Hepler/Wallach/Hepler © Cengage 2013*

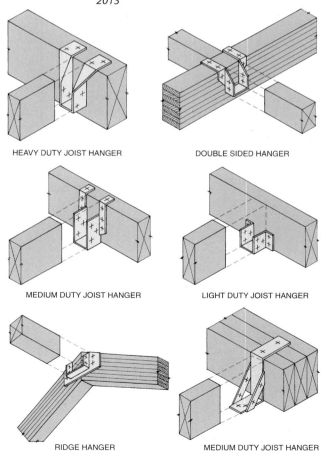

FIGURE 24.51 > Joist and beam hangers. *Hepler/Wallach/ Hepler © Cengage 2013*

straps, gussets, and angle brackets to hold beams and columns together.

Post caps are used extensively when posts intersect beams at a beam joint, as shown in Figure 24.50. When the end of a member intersects the side of another member, without resting on it, metal hangers are usually specified. See Figure 24.51. In some cases special truss clips may be used to prevent movement where two members intersect at

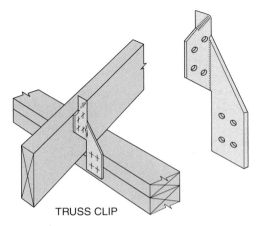

TRUSS CLIP

FIGURE 24.52 › Use of a truss clip. *Hepler/Wallach/Hepler*
© *Cengage 2013*

angles other than 90°. See Figure 24.52. Figure 24.53 shows the types of mortise and tenon joints available for use with exposed joints.

Because interior timber connections are sometimes exposed, hidden joints and fasteners using dowels, rods, and half-lap joints, as shown in Figure 24.54, are often used. Split-ring connectors (Figure 24.55), which are extremely strong and easy to assemble, are also used for this purpose. The inclusion of dado, rabbit, and mortise and tenon joints in a framing design greatly reduces the need for many connectors as shown in Figure 24.56.

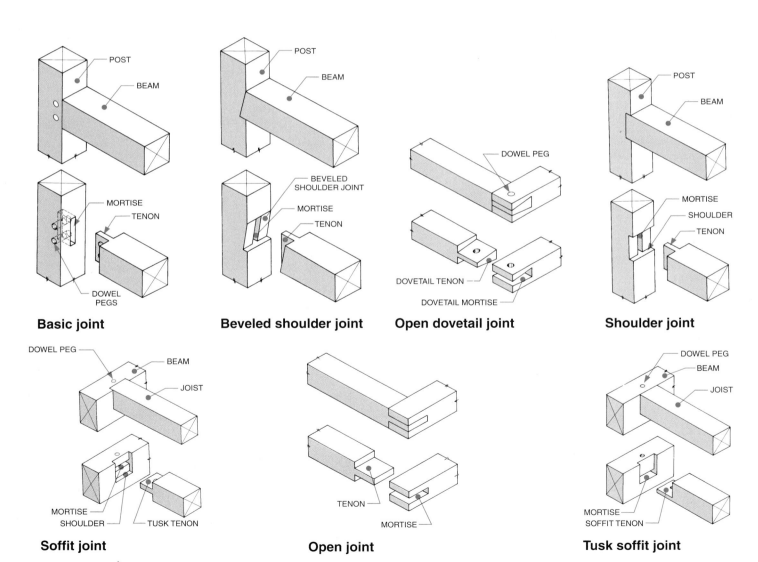

Basic joint

Beveled shoulder joint

Open dovetail joint

Shoulder joint

Soffit joint

Open joint

Tusk soffit joint

FIGURE 24.53 › Types of mortise and tenon joints. *Hepler/Wallach/Hepler* © *Cengage 2013*

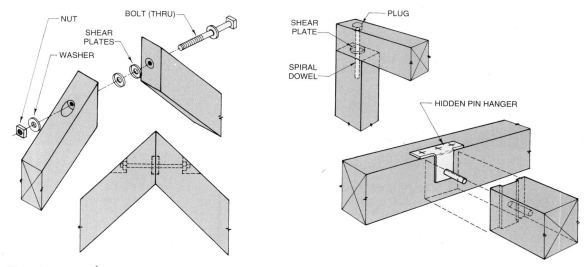

FIGURE 24.54 > Use of hidden fasteners. *Hepler/Wallach/Hepler © Cengage 2013*

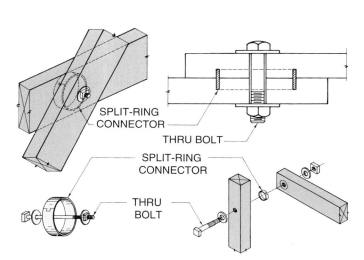

FIGURE 24.55 > Split-ring connector assembly. *Hepler/ Wallach/Hepler © Cengage 2013*

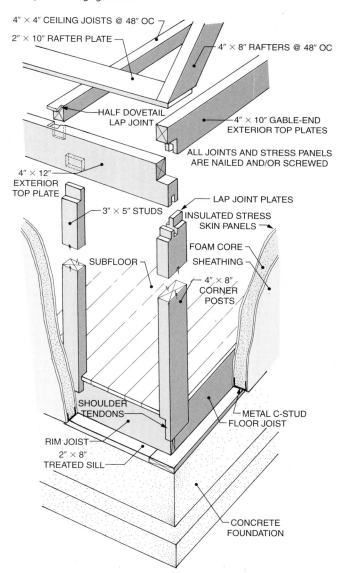

FIGURE 24.56 > Joints used to reduce connectors. *Hepler/ Wallach/Hepler © Cengage 2013*

Wood-Frame Systems Exercises CHAPTER 24

1. Describe the major differences between skeleton-frame and post-and-beam construction.
2. Describe the differences between platform and balloon framing.
3. List six types of fabricated members used in light construction.
4. Select the lumber grade you will specify for the studs, rafters, sheathing, and joists for the house you are designing.
5. Describe the difference between yard, structural, and factory lumber.
6. How many board feet are in 400 wood members 2″ × 6″ × 12′?
7. What is the dressed size of a 2 × 4 and a 4 × 6 wood member?
8. What is the range of plywood thicknesses?
9. Name three uses for hardwoods and softwoods in light construction.
10. Check the types of plywood in your local lumber store and write a short description of each.
11. Check the types of construction grades of lumber in your local lumber store and write a short description of each.

25

Masonry and Concrete Systems

OBJECTIVES

In this chapter you will learn to:

> identify the types of masonry materials used in construction.

> describe four types of masonry walls.

> describe ways to strengthen concrete and prevent deflection.

> explain how concrete is used for slabs and other structural components.

TERMS

aggregate
bonds
brick
cavity wall
flat slab
lally column

masonry bond
one-way slab system
pan slab
post-tensioning
prestressing

pretensioning
steel ties
two-way slab system
veneer wall
waffle slab

INTRODUCTION

Masonry is the term used to describe the shaping, arranging, and uniting of materials such as brick, stone, concrete block, or clay tile products. Masonry units are arranged, usually row upon row (courses), to form structures such as walls. *Concrete* construction systems use structural members made of poured or precast concrete.

MASONRY CONSTRUCTION SYSTEMS

Construction systems that use masonry are usually combined with other systems, such as structural steel or skeleton-frame construction. Buildings are not usually constructed only with masonry materials because wood, steel, or reinforced concrete is needed to strengthen large span floor and roof systems.

Masonry Materials

Masonry materials used for today's construction include a broad range of manufactured products made with a mixture of clay, cement, and gypsum. Masonry materials that are environmentally safe include adobe, brick, clay, slate, and stone.

The many different types, sizes, shapes, and grades of masonry materials serve a variety of purposes.

Brick

Bricks are divided into two general categories: *common brick* and *face brick*. Color, texture, and dimensional tolerance are less consistent and critical for common brick than for face brick. Common brick is therefore less expensive and is generally used in unexposed construction areas. Common brick is graded according to structural characteristics. See Figure 25.1.

Face brick is used in exposed areas that require dimensional accuracy and absorption control, as well as consistent color and texture. Face brick is therefore graded according to the characteristics shown in Figure 25.2. Many special types of face brick are available for specific construction needs; for example, glazed brick, fire brick, cored brick, and paving brick.

Bricks are also classified by their positioning in construction. See Figure 25.3.

Most bricks are rectangular. Special shapes for sills, corners, and thresholds are also available or can be made to order. Bricks usually have holes, as shown in Figure 25.4. Holes

GRADE	USE
SW	Used for maximum exposure to heavy snow, rain, and/or continuous freezing conditions.
MW	Used for average exposure to rain, snow, and moderate freezing conditions.
NW	Used for minimum exposure to rain, snow, and freezing conditions.

FIGURE 25.1 〉 Grades of common brick. *Hepler/Wallach/ Hepler © Cengage 2013*

TYPE	USE
FBX	Used where minimum size and color variations, and high mechanical standards are required.
FBS	Used where wide color variations and size variations are permissible or desired.
FBA	Used where wide variations in color, size, and texture are required or permissible.

FIGURE 25.2 〉 Grades of face brick. *Hepler/Wallach/Hepler © Cengage 2013*

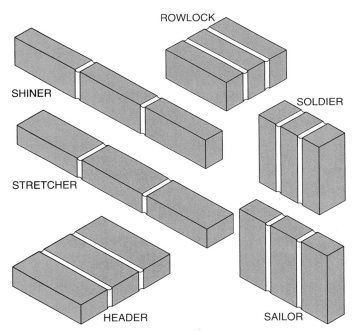

FIGURE 25.3 〉 Laid brick positions. *Hepler/Wallach/Hepler © Cengage 2013*

reduce the weight of bricks and increase bonding ability. Masonry bonds are discussed later in this chapter.

Sizes differ among brick types. Common bricks come in standard, oversized, and modular sizes. See Figure 25.5. Face bricks come in standard, Norman, and Roman sizes. See Figure 25.6. Modular bricks are standardized to align

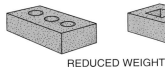
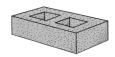

SOLID   REDUCED WEIGHT

FIGURE 25.4 〉 Holes reduce brick weight. *Hepler/Wallach/ Hepler © Cengage 2013*

TYPE	SIZE
Standard	2 1/2″ × 3 7/8″ × 8 1/4″ 2 1/4″ × 3 3/4″ × 8″
Oversized	3 1/4″ × 3 1/4″ × 10″
Modular (1/4″ Joints)	2 1/2″ × 3 3/4″ × 7 3/4″ 2 5/16″ × 3 3/4″ × 7 3/4″ 2 1/2″ × 3 3/4″ × 11 3/4″
Modular (1/2″ Joints)	2 1/4″ × 3 1/2″ × 7 1/2″ 2 1/4″ × 3 1/2″ × 11 1/2″ 2 1/16″ × 3 1/2″ × 7 1/2″
Modular (3/8″ Joints)	2 1/4″ × 3 5/8″ × 7 5/8″

FIGURE 25.5 〉 Common brick sizes. *Hepler/Wallach/Hepler © Cengage 2013*

TYPE	SIZE
Standard	2 1/2″ × 3 1/2″ × 11 1/2″
Norman	2 3/16″ × 3 1/2″ × 11 1/2″ 2 1/4″ × 3″ × 11 11/16″
Roman	1 1/2″ × 3 1/2″ × 11 1/2″

FIGURE 25.6 〉 Face brick sizes. *Hepler/Wallach/Hepler © Cengage 2013*

on 4″ grids after mortar joint dimensions have been added. Increments of 4″, 8″, 12″, and so forth, fit into established measured spaces.

Concrete Block

Concrete blocks are made in many different shapes for a wide variety of construction purposes, as shown in Figure 25.7. Concrete blocks are either solid, hollow-core, or split-face for exposed surfaces. The weight, texture, and color of each block are determined by the types of aggregate used. **Aggregate** is a combination of sand and crushed rocks, slate, slag, or shale. When concrete masonry does not need to support heavy loads, lightweight masonry blocks are ideal. Lightweight blocks are molded by adding fly ash cinders to concrete.

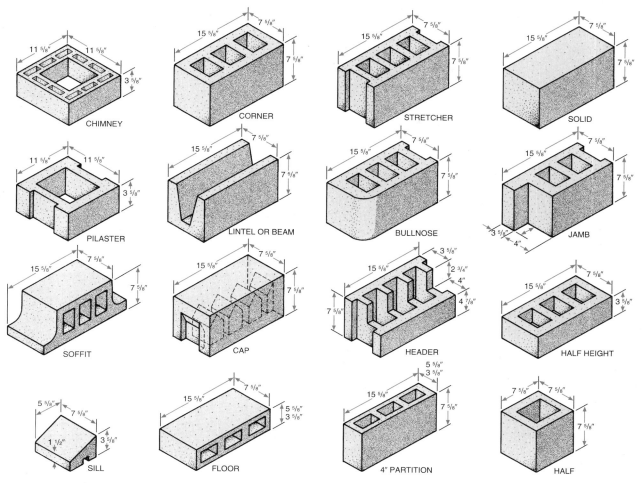

Similar to modular bricks, concrete block is manufactured in modular sizes. That means the actual size of each block is 3/8″ smaller than the space to be filled. The difference allows for the thickness of the mortar joint. For example, the dimensions of an 8″ × 8″ × 16″ concrete block are actually 7 5/8″ × 7 5/8″ × 15 5/8″. Because their sizes are standardized, modular concrete blocks can be used in conjunction with modular bricks. See Figure 25.8.

Some lightweight masonry blocks are made from a combination of sawdust and cement.

Stone

Stone is composed of natural inorganic minerals. In construction, field stone is used for aggregate, cut stone (dimensional, ashlar) is used for walls and wall veneers. Flagstone slabs are used for flooring and decks. Figure 25.9 shows the effective use of flagstone as a paving surface. Rubble, or irregularly shaped stone, can also be used for walls if laid horizontally. This is known as coursed rubble (Figure 25.10).

For centuries, natural stones were used as a major structural material. Today stone is primarily used decoratively except for landscape construction. The stone wall shown in

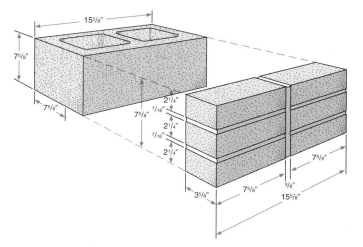

Figure 25.10 shows the use of walkway stone and garden border stone. The stone chimney in the background is structural. Stone masonry is classified by the type of material, shape of cut, finish, and laying pattern. The most common types of stone used in construction are sandstone, limestone, granite,

FIGURE 25.9 〉 Effective use of flagstone pavers. *Hepler/ Wallach/Hepler © Cengage 2013*

FIGURE 25.10 〉 Stone walkway surfaces and garden borders. *Hepler/ Wallach/Hepler © Cengage 2013*

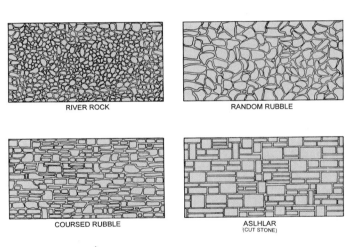

RIVER ROCK   RANDOM RUBBLE

COURSED RUBBLE   ASLHLAR (CUT STONE)

FIGURE 25.11 〉 Stonework patterns. *Hepler/Wallach/Hepler © Cengage 2013*

slate, and marble. These can be cut and arranged into a variety of patterns, as shown in Figure 25.11.

Masonry Paving Stones

Interlocking paving stones are made from a masonry composite that is molded into various shapes. These permeable pavers hold their position and resist high levels of compression. A wide variety of shapes and colors is available. Installation requires no mortar, only a locked-in base of compressed soil, aggregate, and sharp sand, as shown in Figures 25.12A and 25.12B.

Structural Clay Tile

Hollow-core structural tile units are larger than bricks and can be either load-bearing or non-load-bearing structures.

Structural tiles are used for partitions, fireproofing, surfacing, or furring. See Figure 25.13. Load-bearing tile is graded according to structural characteristics: LBX for tile exposed to weathering and LB for tile not exposed to weathering or frost. Non-load-bearing facing tile is graded by surface texture, its stain resistance, color consistency, and dimensional accuracy: FTX for high quality, FTS for low quality.

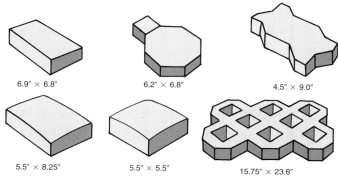

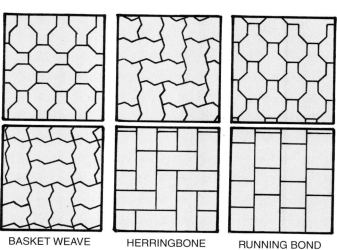

BASKET WEAVE   HERRINGBONE   RUNNING BOND

FIGURE 25.12A > Interlocking paver shapes. *Hepler/Wallach/ Hepler © Cengage 2013*

Masonry Walls

Four basic types of masonry wall construction are solid, cavity, facing, and veneer.

Solid Masonry Walls

Most masonry bearing-wall construction is solid. Solid masonry construction can utilize almost any masonry material if it is laid flat to support loads. However, the material used must be able to withstand the loads involved.

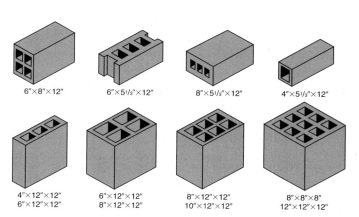

FIGURE 25.13 > Structural tile sizes and shapes. *Hepler/ Wallach/Hepler © Cengage 2013*

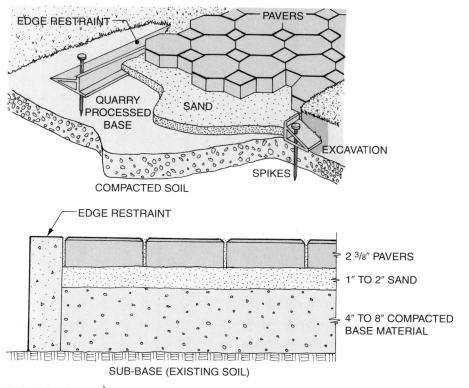

FIGURE 25.12B > Paver base ingredients. *Hepler/Wallach/Hepler © Cengage 2013*

Concrete block is commonly used for solid load-bearing walls. For heavy loads or high walls, steel reinforcing rods (rebars) are added and the block cells are filled with concrete for structural stability. See Figure 25.14. When solid masonry walls are constructed with combinations of materials (such as concrete block and brick), steel reinforcement is mandatory, as shown in Figure 25.15A. Figure 25.15B shows several alternative **steel ties** and flashing used to join two parallel solid brick walls. Reinforcement between courses or between materials is particularly necessary for walls subject to earthquakes, heavy storms, wind, and lateral earth loads. See Figure 25.16.

Figure 25.17 illustrates the use of wire **bonds** to support and control the placement of brick veneer walls. Figure 25.18 shows an anchoring system designed to stabilize triple veneer and concrete wall.

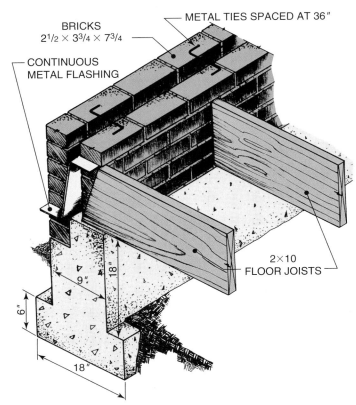

FIGURE 25.15B ❭ Steel ties and flashing on parallel brick walls. *Hepler/Wallach/Hepler © Cengage 2013*

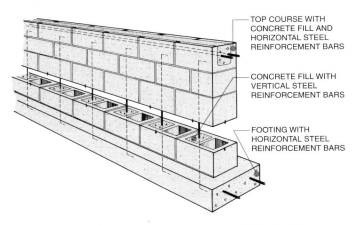

FIGURE 25.14 ❭ Rebars used in concrete block walls. *Hepler/ Wallach/Hepler © Cengage 2013*

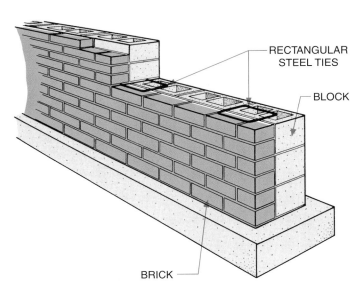

FIGURE 25.15A ❭ Steel tie reinforcement between masonry walls. *Hepler/Wallach/Hepler © Cengage 2013*

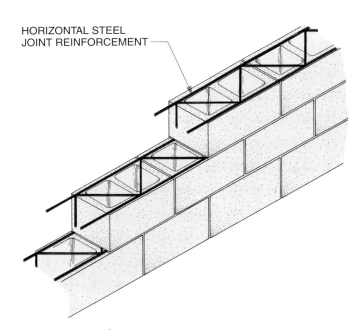

FIGURE 25.16 ❭ Steel reinforcement between masonry courses. *Hepler/Wallach/Hepler © Cengage 2013*

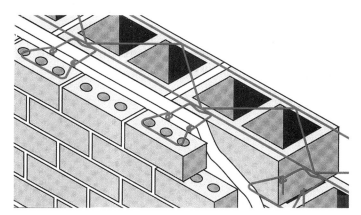

FIGURE 25.17 > Wire bonds used for veneer wall control.
Wire Bond

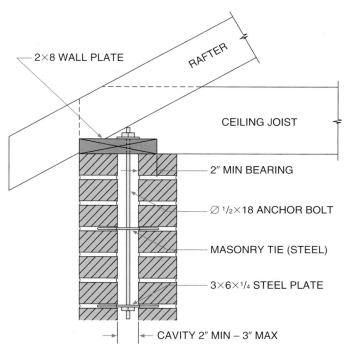

FIGURE 25.19 > Elevation section of a masonry-cavity wall.
Hepler/Wallach/Hepler © Cengage 2013

Masonry-Cavity Walls

To reduce dead loads and to improve temperature and humidity insulation, **cavity walls** are preferred to solid masonry walls. In cavity wall construction, two separate and parallel walls are built several inches apart. A structural tie, usually metal, bonds the walls together. See Figure 25.19.

Masonry-Faced Walls

Walls are often faced with different masonry materials. Any type of masonry wall can be faced with another facing material. For example, a faced wall may consist of common bricks faced with structural tile or concrete block faced with brick. Regardless of the material, the two walls are always bonded so that they become one wall structurally. The bonding material can be metal ties as shown in Figure 25.20, steel reinforcing rods, or long masonry units laid on end to intersect the opposite wall. Always remember that walls with different coefficients of expansion are never faced together. Their differing rates of expansion and contraction under extreme temperature-change conditions can cause cracks and damage the structural integrity of the wall.

Masonry-Veneer Walls

Veneer walls, like masonry-faced walls, include two separate walls constructed side by side. Unlike faced walls, the veneer wall is not tied to the other wall to form a single structural unit. The veneer wall is simply a non-load-bearing decorative

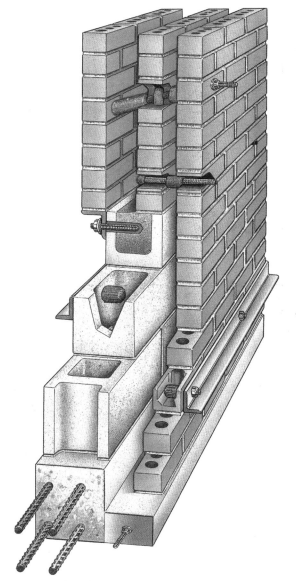

FIGURE 25.18 > Masonry anchoring systems. *Simpson Strong-Tie Company Inc.*

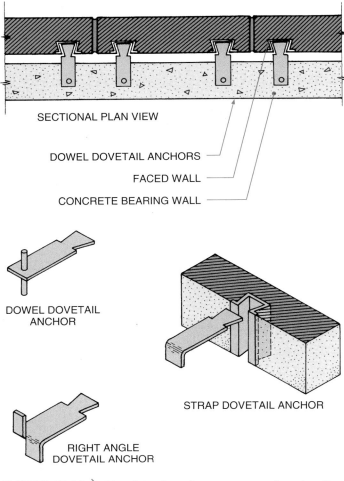

SECTIONAL PLAN VIEW

DOWEL DOVETAIL ANCHORS

FACED WALL

CONCRETE BEARING WALL

DOWEL DOVETAIL
ANCHOR

STRAP DOVETAIL ANCHOR

RIGHT ANGLE
DOVETAIL ANCHOR

FIGURE 25.20 > Metal ties bonding a masonry-faced wall.
Hepler/Wallach/Hepler © Cengage 2013

facade, although the two walls may be connected with masonry ties or adhesives. See Figure 25.21.

A veneer wall may include two different masonry materials or include a skeleton-frame wall veneered with a masonry material. In the latter case, the space between the wood and masonry walls (usually 1″) may remain empty or may be filled with insulation, depending on local climate conditions. A wall detail or sectional drawing is usually prepared to provide this information.

Masonry Bonds

A **masonry bond** for walls is the pattern formed by arranging and attaching masonry units in courses (rows). Masonry can be placed in a variety of bond patterns, as shown in Figure 25.22. Different patterns can make the same size, shape, and material appear completely different. Various types of mortar joints are specified on construction drawings. See Figure 25.23.

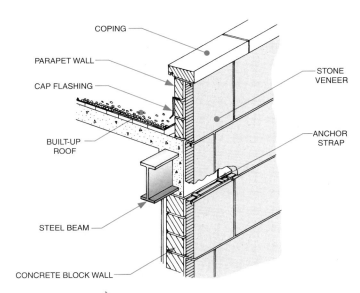

COPING

PARAPET WALL

CAP FLASHING

BUILT-UP
ROOF

STONE
VENEER

ANCHOR
STRAP

STEEL BEAM

CONCRETE BLOCK WALL

FIGURE 25.21 > Masonry-veneer wall. *Hepler/Wallach/Hepler © Cengage 2013*

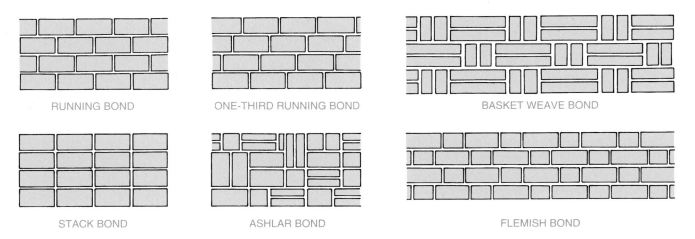

RUNNING BOND

ONE-THIRD RUNNING BOND

BASKET WEAVE BOND

STACK BOND

ASHLAR BOND

FLEMISH BOND

FIGURE 25.22 > Common types of masonry bonds. *Hepler/Wallach/Hepler © Cengage 2013*

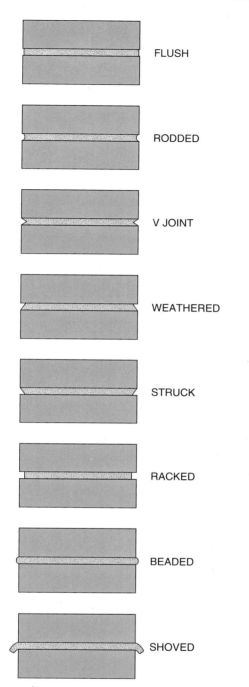

FLUSH

RODDED

V JOINT

WEATHERED

STRUCK

RACKED

BEADED

SHOVED

FIGURE 25.23 > Types of mortar joints. *Hepler/Wallach/Hepler*
© Cengage 2013

⊕ CONCRETE CONSTRUCTION SYSTEMS

Early Romans crushed and processed rocks to create cement for bonding their structures. Today, cement continues to be an important material in construction. Cement is manufactured primarily from clay and limestone. *Concrete,* so widely used in today's construction, is basically made from cement combined with water and aggregate (such as sand and gravel). Concrete is exceedingly strong against com-

pression and tensile forces and can be molded into a wide array of shapes. Concrete finished surfaces can be smooth or heavily textured.

A typical concrete mix consists of 41% crushed rock, 26% sand, 16% water, 11% portland cement, and 6% air. Engineering researchers are currently developing inexpensive and lightweight concrete made from recycled glass and dredge materials. The new material combines the advantages of thermal and sound insulation properties with adequate strength, durability, and reduced weight. Fiber-reinforced concrete provides greater fracture-resistant properties and energy absorption capacity than conventional concrete mixes.

Another development in concrete is autoclaved aerated concrete (AAC), which is produced from lime, sand, gypsum, cement, aluminum powder, and water. AAC has low heat conductivity (U) values and high thermal (R) values. It is also noncombustible, lightweight, insect proof, fire resistant, and can be sawed or drilled with common hand tools. See Chapter 32 for a detailed description of R and U values.

Different types of concrete are identified by their compressive strength, measured in pounds per square inch (PSI). Concrete strengths range from 2500 PSI to 4000 PSI for most residences and up to 8000 PSI for large industrial buildings. Concrete volume is measured in cubic yards.

Concrete can either be cast in place on site or precast off site and shipped to the site as finished girders, beams, slabs, columns, or other components. Either way, concrete may be reinforced, prestressed, or poured plain depending on the construction application or site conditions.

Reinforced Concrete

Because concrete is weak in tensile strength but has a strong resistance to compression stress, it was previously used only for nontension applications such as ground-level slabs, walks, or roadways. However, when materials with high tensile strength are added to the concrete, the tensile strength of the concrete is greatly increased, and its compression strength is doubled. Improvements in the reinforcing of concrete are mainly responsible for the increased use of concrete in all types of building construction.

Reinforcement Materials

Reinforcement bars (rebars) and welded wire mesh provide the steel that converts concrete into reinforced concrete. The use of rebars and welded wire mesh was described in relation to foundations and slabs in Chapter 23.

Welded wire mesh is designed to evenly distribute stress forces and prevent cracking. The spacing of the longitudinal and transverse wires and the size of each are noted in Figures 25.24 and 25.25.

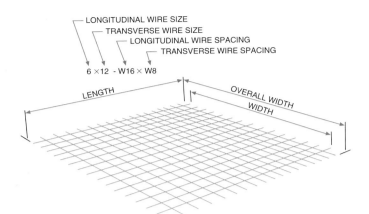

FIGURE 25.24 〉 Welded wire mesh specifications. *Hepler/Wallach/Hepler © Cengage 2013*

∅ IN INCHES		GAUGE NUMBER	ACTUAL SIZE OF WIRE	WOVEN WIRE MESH SPACING
DECIMAL	FRACTION			
.2437	1/4″	3	●	–
.2253	7/32″	4	●	–
.2070	13/64″	5	●	–
.1920	3/16″	6	●	2 1/2″
.1770	11/64″	7	●	2 1/4″
.1620	5/32″+	8	●	2″
.1483	5/32″–	9	●	1 3/4″
.1350	9/64″	10	●	1 1/2″
.1205	1/8″	11	●	1 1/4″
.1055	7/64″	12	●	1″
.0915	3/32″	13	●	–
.0800	5/64″	14	●	3/4″
.0625	1/16″	16	●	3/8″ or 1/2″

FIGURE 25.25 〉 Welded wire mesh sizes and spacing. *Hepler/Wallach/Hepler © Cengage 2013*

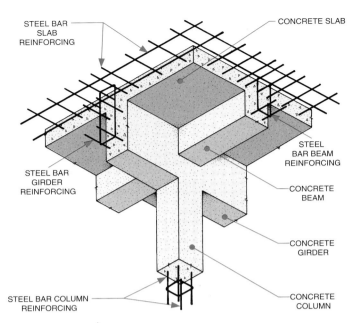

FIGURE 25.26 〉 Use of rebars in structural concrete. *Hepler/Wallach/Hepler © Cengage 2013*

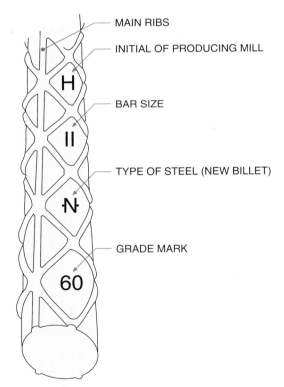

FIGURE 25.27 〉 Rebar marking system. *Hepler/Wallach/Hepler © Cengage 2013*

Prestressed Concrete

When loads are added to concrete members, some deflection (sag) occurs in the center of the member. This happens to all materials under load. However, because concrete has very low tensile strength, excessive deflection can result in

The use of rebars for reinforcing slabs was covered in Chapter 23. Rebars are also used in the formation of concrete beams, columns, walls, and suspended decks. See Figure 25.26. A marking system identifies rebar manufacturer, size, steel type, and PSI grade, as shown in Figure 25.27.

tension cracking or complete member failure. This is caused by compression of the upper side and tensioning (stretching) of the lower side. See Figure 25.28. To counteract these unstable compression and tension stresses, concrete is often prestressed.

Prestressing is a method of compressing concrete so that both the upper and lower sides of a member remain in compression during loading. Prestressing can be accomplished either by pretensioning or post-tensioning.

Pretensioning

In **pretensioning,** deformed steel bars called *tendons* are stretched (tensioned) between anchors and the concrete is poured around the bars. Once the concrete has cured, the tension is released and the bars attempt to return to their

original, shorter length. However, the concrete that has hardened around the bar grooves holds the deformed bars at nearly their stretched length. This creates a continual state of compressive stress that can be compared to pressing a row of blocks together. See Figure 25.29.

As a further aid to prevent bending, concrete members are prestressed by draping tendons near the bottom of the member, as shown in Figure 25.30. This bottom tension buckles the member upward so that when the anticipated loads are added, the beam straightens to a level position.

Post-Tensioning

Post-tensioning is done after concrete has cured. In post-tensioning, tendons are either placed inside tubes embedded in the concrete or the tendons are greased to allow slippage. The tendons are then stretched with hydraulic jacks and the ends anchored. This creates compressive stress because the ends of the tendons pull toward the center. Post-tensioning

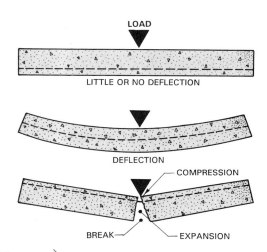

FIGURE 25.28 > Loading effect on concrete. *Hepler/Wallach/Hepler © Cengage 2013*

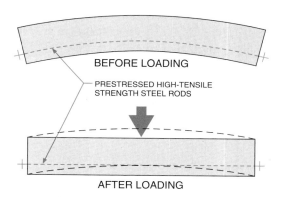

FIGURE 25.30 > Prestressing with draped tendons. *Hepler/Wallach/Hepler © Cengage 2013*

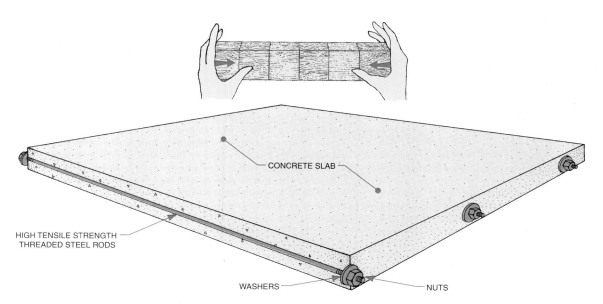

FIGURE 25.29 > Principle of pretensioning. *Hepler/Wallach/Hepler © Cengage 2013*

can be done at a factory or on site to reduce shipping weight, especially for large members.

Concrete Structural Members

Concrete has been used for centuries for foundations, walls, and ground-supported slabs. Not until low-tensile-strength concrete was reinforced with high-tensile-strength steel, however, could concrete be used for structural components such as columns, beams, girders, and suspended-slab floor and roof systems. Figure 25.31 shows a concrete floor system that combines concrete block, concrete reinforced ribs, and a reinforced-concrete slab.

Concrete Slabs

Once slabs could only be poured in place at grade level, as shown in Figure 25.32. However, with advances in steel reinforcement methods, structural slab members can now be manufactured off site. Lightweight slabs can also be poured on a wood floor system if proper underlayments are specified as shown in Figure 25.33. Elevated slabs can also be poured using parallel rebars aligned with the beam direction. To prevent cracking due to temperature and moisture changes, rebars are also placed perpendicular to the load-supporting rebars. These rebars are known as temperature rebars (or bars). See Figure 25.34.

Columns

Concrete columns are vertical members that support weights transferred from horizontal beams and girders. Concrete columns are made structurally sound by the addition of rebars. Another method is to fill a hollow steel column with concrete. Such a member is called a **lally column.** A large concrete column like that shown in Figure 25.35 is called a *caisson.*

Sectional drawings convey the exact relationship of column, beam, and reinforcement material, as shown in Figure 25.36. If the exact position of each column is not dimensioned on a floor plan, column schedules are prepared. Column schedules include coding that is indexed to a column plan.

Beams and Girders

Concrete girders are major horizontal members that rest on columns. Beams are horizontal members supported by girders or columns. Concrete beams and girders are reinforced

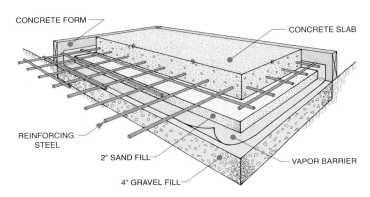

FIGURE 25.32 > Grade-level slab construction. *Hepler/ Wallach/Hepler © Cengage 2013*

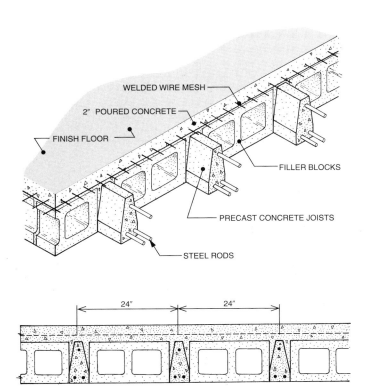

FIGURE 25.31 > Concrete block and rib floor system. *Hepler/ Wallach/Hepler © Cengage 2013*

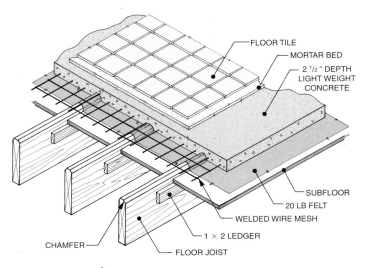

FIGURE 25.33 > Concrete slab on a wood floor system. *Hepler/Wallach/Hepler © Cengage 2013*

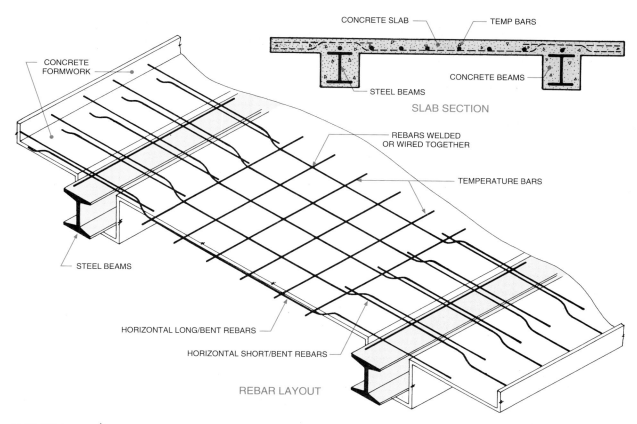

CONCRETE SLAB

TEMP BARS

CONCRETE BEAMS

STEEL BEAMS

SLAB SECTION

CONCRETE FORMWORK

REBARS WELDED OR WIRED TOGETHER

TEMPERATURE BARS

STEEL BEAMS

HORIZONTAL LONG/BENT REBARS

HORIZONTAL SHORT/BENT REBARS

REBAR LAYOUT

FIGURE 25.34 > Temperature and load-supported bar placement. *Hepler/Wallach/Hepler © Cengage 2013*

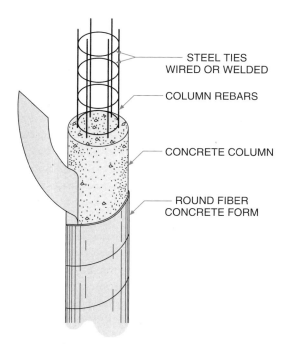

STEEL TIES WIRED OR WELDED

COLUMN REBARS

CONCRETE COLUMN

ROUND FIBER CONCRETE FORM

FIGURE 25.35 > Steel-reinforced concrete column. *Hepler/ Wallach/Hepler © Cengage 2013*

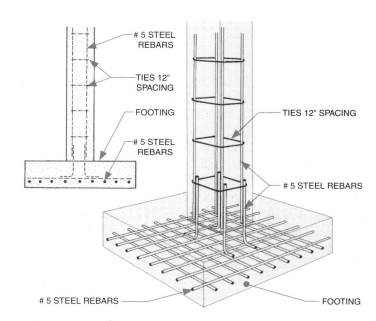

5 STEEL REBARS

TIES 12" SPACING

FOOTING

5 STEEL REBARS

TIES 12" SPACING

5 STEEL REBARS

5 STEEL REBARS

FOOTING

FIGURE 25.36 > Column and footing rebar placement. *Hepler/Wallach/Hepler © Cengage 2013*

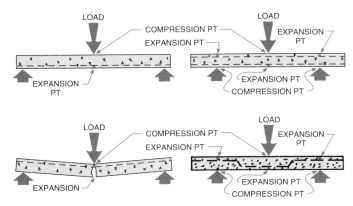

FIGURE 25.37 > Placement of rebars (WWM) in beams or suspended slabs to prevent cracking. *Hepler/Wallach/Hepler © Cengage 2013*

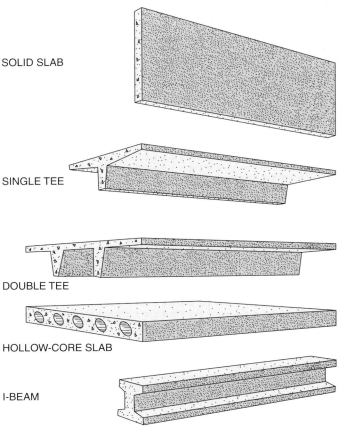

FIGURE 25.38 > Precast concrete members. *Hepler/Wallach/ Hepler © Cengage 2013*

with steel rebars to increase tensile strength. Some reinforced-concrete beams are rectangular, but most are wider at the top.

Rebars or welded wire mesh (WWM) placed low in a beam will prevent cracking. Rebars or WWM placed too high in a beam may bend and result in cracking. For very heavy loads, a top and bottom row may be needed. This practice is called *draped reinforcement*. See Figure 25.37.

The position of girders and beams is shown on floor plans with dotted lines or indexed to a beam schedule, similar to a column schedule.

Lintels

Short horizontal members that span the top of openings in a wall are known as lintels (headers). Concrete lintels are either poured in a form, into lintel blocks, or precast as a small concrete beam.

Cast-in-Place Concrete

Forms for pouring concrete for footings, foundations, slabs, and walls have been used for a long time. However, new developments in reinforced and prestressed concrete enable builders to erect structures with extremely complex contours. Concrete shells are a type of concrete system that uses poured reinforced concrete. A light steel structure is erected. Then concrete is poured or sprayed over the steel frame. The concrete holds the steel in place after hardening.

Drawings for cast-in-place concrete systems need to include the outline and dimensions of the finished job, including the position of rebars and joints.

Precast Concrete

Precast concrete is the opposite of cast-in-place concrete. Precasting of concrete involves pouring the concrete into wood, metal, or plastic molds. Precast concrete is usually reinforced.

Once set, the molded concrete is placed in position in its hardened form.

Although concrete block is the most commonly used precast concrete material, it is considered a masonry material, like brick. Precast concrete structural members include wall panels, girders, beams, and a variety of slabs. See Figure 25.38. Available in solid, hollow-core, and single and double tee shapes, precast slabs are used for walls, floors, and roof decks. Wall panels are solid precast units used either for bearing or nonbearing walls, depending on the amount of reinforcement. Because the exterior sides of concrete wall panels are usually exposed, special textured finishes are often applied during the casting process. These panels may be combined with layers of insulation to form a complete monolithic wall unit.

Concrete floor system drawings show only the dimensions of slabs that are poured in place. For precast systems, the locations of ribs or support beams are shown with dotted lines. See Figure 25.39. All other information is found on detail or sectional drawings keyed to the general floor framing plan. Precast concrete units with interlocking shapes can be used without mortar or reinforcement to build high retaining walls as shown in Figure 25.40.

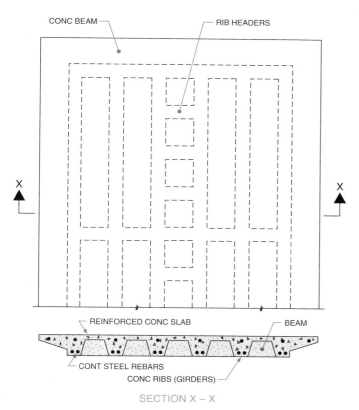

CONC BEAM — RIB HEADERS

X — X

REINFORCED CONC SLAB — BEAM

CONT STEEL REBARS

CONC RIBS (GIRDERS) —

SECTION X – X

FIGURE 25.39 > Method of drawing ribs and beams in a precast slab. *Hepler/Wallach/Hepler © Cengage 2013*

Slab Component Systems

Precast slab components or cast-in-place slabs are divided into two types: one-way systems and two-way systems.

One-Way Systems

In **one-way slab systems** the rebars are all parallel. One-way system girders, which rest on columns, are parallel to the rebar alignment. One-way solid slabs are extremely heavy and are therefore impractical for most spans over 12 feet. To lighten the dead load, ribbed one-way slabs are often used. See Figure 25.41. The ribbed slab is a thin slab (2″ to 3″ thick), supported by cast ribs. These units are constructed of precast slab tees or are cast in place. When ribbed slabs are to be poured in place, a ribbed slab plan shows the horizontal rib positions with dotted lines. Dotted lines are also used on detail drawings to show the position of rebars in the slab and in the ribs.

Two-Way Systems

In **two-way slab systems** the rebars, girders, and beams are placed in perpendicular directions. See Figure 25.42. When ribs extend in both directions, the system is known as a waffle slab.

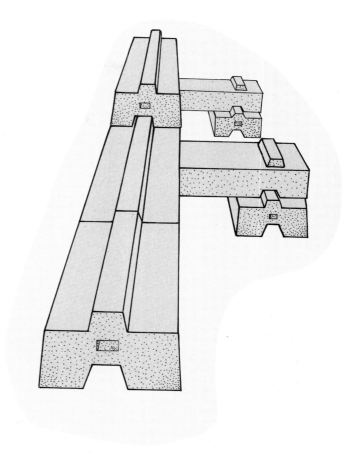

FIGURE 25.40 > Interfacing precast concrete units. *Hepler/Wallach/Hepler © Cengage 2013*

Pan and waffle slabs are used to cast suspended floor and roof systems for spans up to 60 feet. A **pan slab** is created by pouring concrete into molded fiberglass forms on site. **Waffle slabs** may be poured on site or prefabricated off site. Temporary fiberglass or metal pans (domes) are placed, open side down, 4″ to 7″ apart on a temporary floor. No pans are placed around columns. Rebars are added and concrete is then poured to a depth of several inches over the pans. After the concrete has cured, the pans and temporary flooring are removed, and a suspended waffled floor (or roof) results. See Figure 25.43. This type of cast-in-place system is lightweight (for concrete), sound resistant, fireproof, and economical.

A **flat slab** is a two-way slab unit that rests directly on columns without a girder or beam support. A flat slab floor (or roof) system is actually a series of individual slabs, with the slab resting independently on a column. All of the slab's weight is directed through the columns to footings. When the slabs are joined, a unified floor is created. In flat slab systems the supporting columns are strengthened by the addition of a thicker slab area (*drop panel*) around columns. See Figure 25.44. A column capitol, or flared head, also

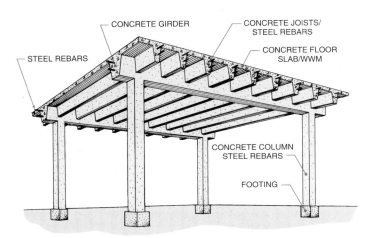

FIGURE 25.41 > Ribbed one-way concrete slab system. *Hepler/Wallach/Hepler © Cengage 2013*

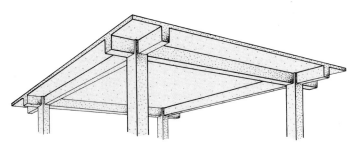

FIGURE 25.42 > Two-way concrete slab system. *Hepler/Wallach/Hepler © Cengage 2013*

helps spread the slab loads onto a column in this type of construction.

Concrete Joints

Because concrete expands and contracts with changes in moisture levels and temperature, relief joints (expansion joints) are required to allow for these fluctuations. Some construction

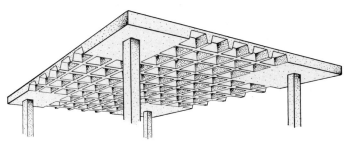

FIGURE 25.43 > Waffle concrete slab system. *Hepler/Wallach/Hepler © Cengage 2013*

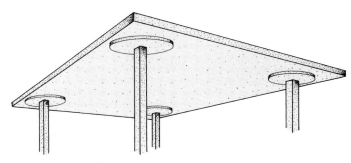

FIGURE 25.44 > Flat concrete slab system. *Hepler/Wallach/Hepler © Cengage 2013*

drawings and specifications indicate minimum dimensions for placement of expansion joints. Some drawings show specifically where joints are required or must be avoided.

Concrete Wall Systems

Conventional Concrete Walls

Most concrete walls are poured between wood or metal forms. Most walls of this type are used with wood-frame construction. Conventionally poured concrete blocks and walls, are shown in Figure 25.45. These include separate layers of in-

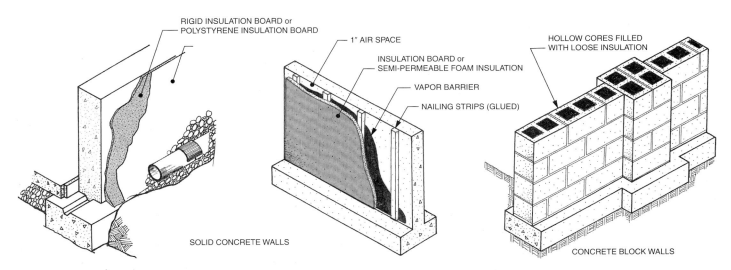

FIGURE 25.45 > Poured concrete and concrete wall with insulation. *Hepler/Wallach/Hepler © Cengage 2013*

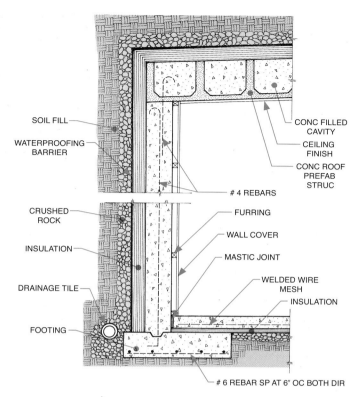

FIGURE 25.46 > Reinforced-concrete walls in an earth-sheltered building. *Hepler/Wallach/Hepler © Cengage 2013*

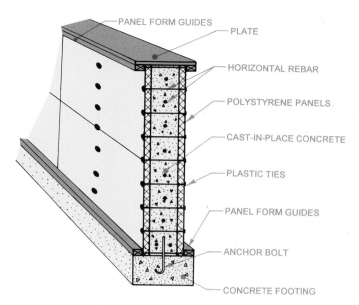

FIGURE 25.47 > Section of a preinsulated wall form system. *Hepler/Wallach/Hepler © Cengage 2013*

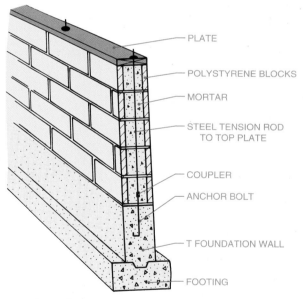

FIGURE 25.48 > Preinsulated block wall system. *Hepler/Wallach/Hepler © Cengage 2013*

sulation, a vapor barrier, and nailing strips. See Chapters 23 and 29 for more detailed information on poured concrete walls. Reinforced-concrete walls are also the main structural material for both walls and roofs in an earth-sheltered construction, as shown in Figure 25.46.

Insulated Concrete Walls

There are three types of preinsulated concrete walls. The first is an insulated wall form system that is composed of two polystyrene panels separated by plastic ties that provide the wall thickness needed, as shown in Figure 25.47. Once concrete is poured into the cavity, the forms remain as part of the finished wall and act as insulation. The second type is a polystyrene block wall system in which interlocking rigid foam blocks are stacked to form a wall, as shown in Figure 25.48. Concrete is poured into the hollow core of the blocks creating a rigid wall. The third type is a conventionally formed (with wood or metal) wall. A plastic foam panel is suspended in the center of the wall while concrete is poured, creating a foam core sandwich with concrete on the outside and insulation on the inside.

All preinsulated systems are strong, durable, fire resistant, and termite proof. They also provide excellent insulation and create an effective vapor barrier. In preparing floor plan drawings for preinsulated walls, the thickness of the insulation must be added to the thickness of the concrete, interior wall covering, and external siding to determine the total wall thickness. Figure 25.49A shows several types of preinsulated wall molds, and Figure 25.49B shows a sectional drawing of a finished preinsulated wall.

FIGURE 25.49A > Preinsulated concrete wall forms. *Korfil*

FIGURE 25.49B > Section through insulated concrete wall. *Korfil*

Masonry and Concrete Systems Exercises

CHAPTER 25

1. Name the types of bricks and their uses.

2. Describe the types of masonry walls.

3. Describe the types of concrete construction systems.

4. Draw a wall section for the house you are designing using a masonry system described in this chapter.

5. Draw a plan detail of the brick veneer wall shown in Figure 25.50.

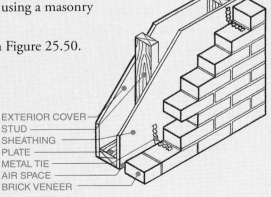

EXTERIOR COVER
STUD
SHEATHING
PLATE
METAL TIE
AIR SPACE
BRICK VENEER

FIGURE 25.50 > Brick veneer on stud wall. *Hepler/Wallach/Hepler © Cengage 2013*

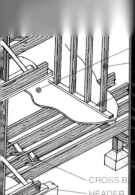

CHAPTER 26

Steel and Reinforced-Concrete Systems

OBJECTIVES

In this chapter you will learn to:

> describe three types of steel construction and explain the basic purpose of each.

> identify manufactured steel forms and their function as structural members.

> read and interpret steel symbols, weld notations, identification, and measurements for working drawings.

> relate the types of fasteners and intersections of steel members to construction methods.

TERMS

bars
bolts
brackets
built-up girders
channels
large-span construction

L-shapes (angles)
plates
rivets
rolled steel
spandrel
S-shapes

steel-cage construction
steel space frame
tees
tubing
welds
W-shapes

🌐 INTRODUCTION

Steel was first produced from iron and carbon in 1855. Since then steel has been the primary metal used for construction. Steel is environmentally safe because no chemical additives are needed, no offgases are produced, and it is easily recycled. Steel is also earthquake and hurricane resistant, noncombustible, and impervious to fungus growth and insect infestation. As a building material steel has the highest strength-to-weight ratio and will not warp, split, or rot. It also will not rust if it has an oxide or zinc (galvanized) coating. Steel framing can span large distances and support heavy loads. In contrast to rumors, steel structures do not attract lightning or interfere with electronic devices. Figure 26.1 is an example of steel frame construction applied to a residence.

Steel does have some disadvantages, including high levels of heat conduction and sound transmission, which require special insulation measures. Also, skilled steel erectors may be in short supply and materials labor and engineering costs tend to be high. Steel is extremely heavy and thus creates heavy dead loads. This chapter describes structural steel members and explains how they are fastened together to form building components.

🌐 STEEL BUILDING CONSTRUCTION

In steel construction, plates, bars, tubing, and rolled shapes are used for columns, girders, beams, and bases.

When steel members are used in a manner similar to skeleton-frame wood members, the system is known as **steel cage construction.** The terms *steel skeleton-frame* and *steel cage* construction are often used interchangeably. The major structural members used in steel cage construction are

FIGURE 26.1 > Residential steel framing. *Hepler/Wallach/Hepler © Cengage 2013*

545

columns, girders, and beams. The definitions of structural members are the same for steel construction as for other types of construction. Columns are vertical members that rest on footings or piers. Girders are horizontal members that extend between columns. They are sometimes called *spandrel beams,* or merely **spandrels,** if they connect to columns erected on the perimeter. See Figure 26.2. *Beams* are horizontal members placed on or between girders. *Purlins* are beams that connect roof trusses or rafters. Figure 26.3 shows an example of the many structural components used in steel cage construction.

Beams are supported by girders, which in turn are rigidly attached to columns, through which all loads are transmitted through bearing plates to footings. Because all live and dead loads are transmitted through the columns, there is no need for additional exterior or interior bearing walls in steel cage construction. This enables buildings to be built extremely high with a minimum of interior obstruction. It also allows for the use of large exterior wall panels that have no structural value. These walls, called curtain walls, are often constructed of glass.

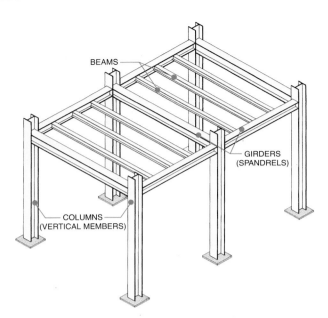

FIGURE 26.2 > Spandrels (girders) are connected to columns and beams are connected to spandrels. *Hepler/Wallach/Hepler © Cengage 2013*

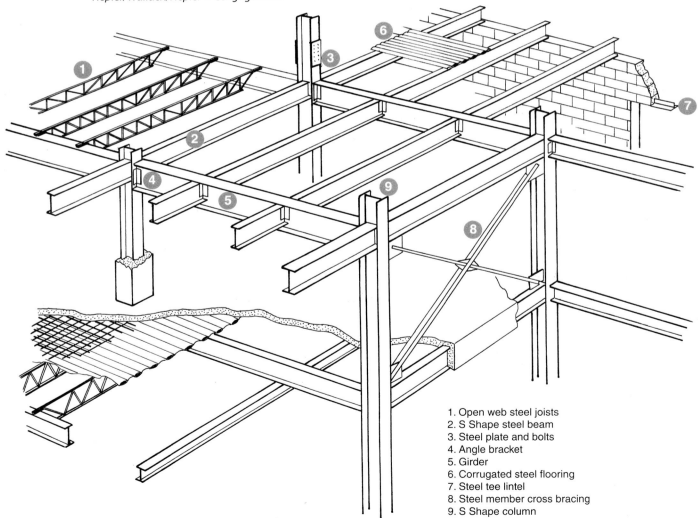

1. Open web steel joists
2. S Shape steel beam
3. Steel plate and bolts
4. Angle bracket
5. Girder
6. Corrugated steel flooring
7. Steel tee lintel
8. Steel member cross bracing
9. S Shape column

FIGURE 26.3 > Steel cage construction. *Hepler/Wallach/Hepler © Cengage 2013*

ASTM TYPE	MIN YIELD* STRESS POINT	MANUFACTURED FORMS	DESCRIPTION
A36	36,000 PSI	Sheets Plates Bars Shapes Rivets Nuts Bolts	A medium carbon steel that is the most commonly used structural steel. Suitable for buildings and general structures, and capable of welding and bolting.
A440	42,000 PSI	Plates Bars Shapes	A high-strength, low-alloy steel suitable for bolting and riveting, but not welding. Used for lightweight structures—high resistance to corrosion.
A441	40,000 PSI	Plates Bars Shapes	A high-strength, low-alloy steel modified to improve welding capabilities in lightweight buildings and bridges.
A572	41,000 PSI	Limited types of shapes Bars & plates	A high-strength, low-alloy economical steel suitable for boltings, riveting, and welding with lightweight high toughness for buildings and bridges.
A242	42,000 PSI	Plates Bars Shapes	A durable, corrosion-resistant, high-strength, low-alloy steel that is lightweight and used for buildings and bridges exposed to weather. Can be welded with special electrodes.
A588	42,000 PSI	Plates Bars Shapes	A lightweight, corrosion-resistant, high-strength, low-alloy steel with high durability in high thicknesses used for exposed steel.
A514	90,000 PSI	Limited shapes & plates	A quenched and tempered alloy steel with varying strength, width, thickness, and type.
A570	25,000 PSI	Plates Light shapes	A light gauge steel used primarily for decking, siding, and light structural members.
A606	45,000 PSI	Plates	A high-strength, low-alloy sheet and strip steel with high atmospheric corrosion resistance.

*American Society for Testing Materials

FIGURE 26.4 > Structural steel types. *Hepler/Wallach/Hepler © Cengage 2013*

Even steel cage construction cannot provide the enormous amount of unobstructed space needed in structures such as aircraft hangars, sports stadiums, and convention centers. For these structures, large trusses or arches are necessary to span long distances. Such construction is called **large-span construction.**

Structural steel types are specified by their metallurgical characteristics and minimum stress yields, as designated by the American Society for Testing and Materials (ASTM). See Figure 26.4.

Preparing structural steel drawings requires a working knowledge and understanding of steel symbols, weld notations, identifications, drawing conventions, and measurement, fastening, and intersection methods.

STEEL STRUCTURAL MEMBERS

Steel used for structural purposes is manufactured in plates, bars, pipes, tubing, and a variety of other shapes. Many of these items are formed by passing steel between a series of rollers. Steel that has been shaped by this method is called **rolled steel.** Rolled steel is formed into a variety of shapes and sizes, which are then dimensioned in multiples of inches. These are used for structural framing members. Lighter steel sheets are used for nonstructural purposes and are measured by gauge thicknesses. These lightweight members are pressed into sheet steel shapes and used for pipes, tubes, angles, and channels.

Plates

Structural **plates** are flat sheets of rolled steel. These sheets range in thickness from 1/8" to 3" and in width from 8" to 60". Plates are specified by thickness, width, and length in that order. See Figure 26.5.

Plates are used as webs in **built-up girders** and columns, as shown in Figure 26.6, and to reinforce other webs or flanges of structural steel shapes. Bearing plates provide bearing surfaces between columns and concrete footings. See Figure 26.7.

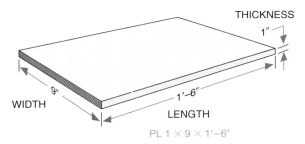

FIGURE 26.5 > Steel plate specifications. *Hepler/Wallach/ Hepler © Cengage 2013*

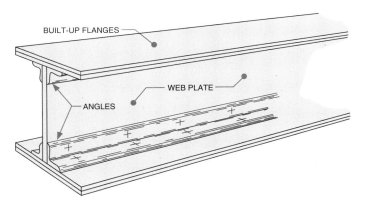

FIGURE 26.6 > Built-up steel plate girder. *Hepler/Wallach/ Hepler © Cengage 2013*

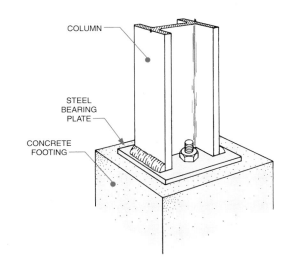

FIGURE 26.7 > Steel bearing plate. *Hepler/Wallach/Hepler © Cengage 2013*

Bars

Steel **bars** used for structural purposes are available in round, square, hexagonal, and flat (rectangular) cross-section shapes. See Figure 26.8. Square, hexagonal, and round bars are manufactured in 1/16" increments from 1/16" to 12". Flat bars are manufactured in 1/4" increments up to 8". Round bars are specified by diameter; square bars by width or gauge number; flat bars by width and thickness; and hexagonal bars by the distance across the flats (AF). Steel bars are used primarily for bracing other structural components and for concrete reinforcement.

Steel Pipe and Structural Tubing

Steel pipe and **tubing** are used extensively in exposed areas because of their clean, pleasing lines. Structural pipe and tubing are available in round, square, and rectangular cross-section shapes, as shown in Figure 26.9.

Hollow steel pipe is manufactured in sizes from 1/2" to 12" (inside diameter) and in three strength classes. Strength classes relate to wall thicknesses and are either standard weight (STO), extra strong (x-strong), or double extra strong (xx-strong). On structural drawings, pipe is specified by diameter and strength. Thus, a 4" double extra-strong steel pipe is labeled "Pipe 4 xx-strong." See Figure 26.10. When hollow steel pipe is used as a vertical structural support, it is called a lally column, as covered in Chapter 25.

Square structural tubing is specified by cross-section width and thickness and is available in sizes from 2" × 2" to 10" × 10"

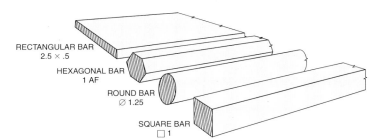

FIGURE 26.8 > Steel bar shapes. *Hepler/Wallach/Hepler © Cengage 2013*

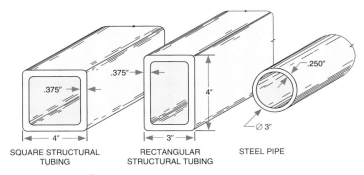

FIGURE 26.9 > Structural steel pipe and tube shapes. *Hepler/Wallach/Hepler © Cengage 2013*

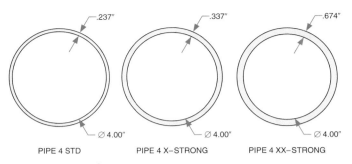

FIGURE 26.10 › Steel pipe thicknesses. *Hepler/Wallach/Hepler © Cengage 2013*

outside dimension (OD). Rectangular tubing sizes range from 3″ × 2″ to 12″ × 8″ OD. Structural tubing is specified by the symbol TS followed by the width, thickness, and wall thickness. A rectangular structural tube 4″ wide and 3″ thick with a wall thickness of 1/4″ is therefore labeled TS 4 × 3 × .25. All inch marks are omitted on shape notations used on structural drawings since all sizes are assumed to be in inches unless otherwise specified. Round structural tubing is specified by outside diameter and wall thickness. For example, a 4″ diameter tube with a 1/4″ wall thickness is labeled 4 OD × .25. The length dimension of tubing is placed at the end of the note or on the working drawing.

Other Structural Steel Shapes

In addition to plates and bars, steel is rolled into channels, tee sections, and a number of other shapes. Their designations are shown in Figure 26.11. Steel shapes are designated on construction drawings by shape symbol, depth in inches, and weight in pounds per foot. Figure 26.12 shows the most common structural steel shapes and the related drawing symbols.

L-shapes (angles) are structural steel members rolled in the (cross-section) shape of the letter L with legs of equal or unequal length. (Equal leg lengths are available in sizes of 1″ to 8″. Unequal leg lengths range from 1 3/4″ to 9″.) Whether equal or unequal, the thickness of each leg (called wall thickness) is always the same. L-shapes (angles) are specified on drawings by the symbol L followed by the length of each leg, followed by the wall thickness and length.

For example, an L-shape member with one 2″ and one 3″ leg and a wall thickness of 1/2″ is specified as L2 × 3 × .5. L-shape members are used as components in built-up beams, columns, and trusses. They are also used for connectors and as lintels in light- or short-span construction.

Channels are rolled into a cross-section shape resembling the letter U, with the inner faces of flanges shaped with a 2/12 pitch. Channels are classified by depth, from 3″ to 15″. Two types of channels are specified for structural use: American Standard channels (C) and Miscellaneous channels (MC). Channels are specified by symbol (C or MC), followed by the depth times the weight per foot. An 8″-deep Standard

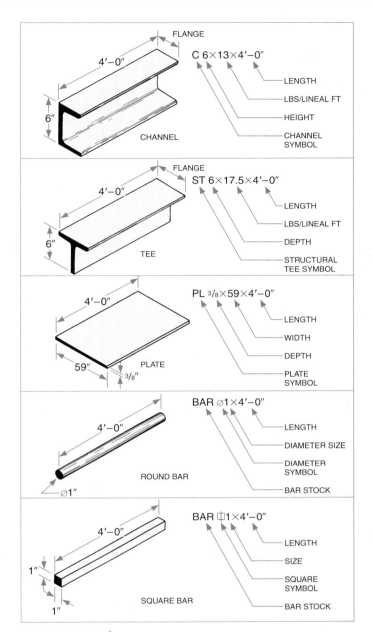

FIGURE 26.11 › Structural steel designations. *Hepler/ Wallach/Hepler © Cengage 2013*

channel that weighs 11.2 lb/ft is labeled C 8 × 11.2. Channels are used for roof purlins, lintels, and truss chords and to frame-in floor and roof openings. Steel components such as studs, joists, and rafters, use C-members and tracks with open slots designed for cable and piping insertion. For example, C-members are labeled as steel joists in Figure 26.13A because the channels function as joists. The floor system in Figure 26.13B illustrates the many C-member functions in a floor framing system.

S-shapes (formerly I-beams) are rolled in the shape of a capital letter I. American Standard shapes have narrow flanges with a 2/12 inside pitch. S-shapes are classified by the depth of the web and the weight per foot. The web is the portion between the flanges. Web depths range from 3″ to 24″. S-shapes

NAME	SECTIONAL FORM	SYMBOL	PICTORIAL
WIDE FLANGE		W	
AMERICAN STANDARD BEAM		S	
TEE		T	
ANGLE		L	
ZEE		Z	
AMERICAN STANDARD CHANNEL		C	
BULB ANGLE		BL	
LALLY COLUMN		◎	
SQUARE BAR		⌗	
ROUND BAR		φ	
PLATE		℔	

FIGURE 26.12 > Structural steel shapes. *Hepler/Wallach/ Hepler © Cengage 2013*

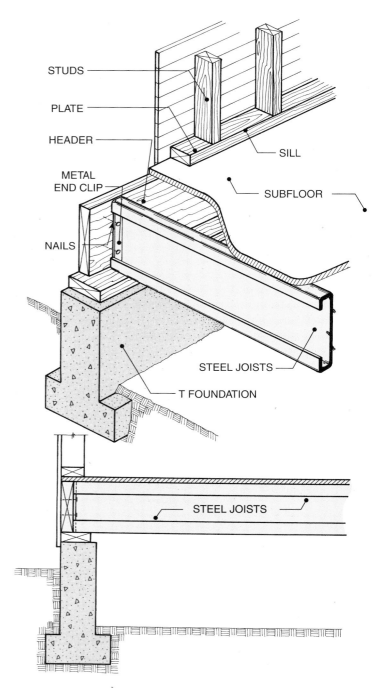

FIGURE 26.13A > C-members used as joist. *Hepler/Wallach/ Hepler © Cengage 2013*

are designated by their symbol (S) followed by the web depth and the weight per lineal foot. See Figure 26.14. For example, an S-shape member with a 14″-deep web that weighs 56 lb/ft is labeled S 14 × 56. On some drawings, the length may be added to the designation rather than as a dimension on the drawing. S-shapes are used extensively as columns because of their symmetry. Their narrow flanges are applicable to many designs where size restrictions are a problem.

W-shapes (formerly wide-flange or H-beams) are similar to S-shapes but with wider flanges and comparatively thinner webs. Their capacity to resist bending is greater than that of S-shapes. W-shapes are designated in the same manner as S-shapes. See Figure 26.15. For example, W 18 × 62 describes a W-shape member with an 18″-deep web weighing 62 lb/ft. W-shapes are available in depths from 4″ to 36″. Lighter weight versions of W-shape members are known as M-shapes.

Structural **tees** are made by cutting through the web of an S- or W-shape, although some tees are rolled to order. If the web is cut exactly through the center, two identical tees result. The symbol for a tee is the capital letter T. On structural drawings the tee symbol includes the shape from which the tee was cut (S, W, or M) followed by the letter T, the depth of cut (from web to flange), and the weight per foot. Therefore, a tee-shape member cut in half from a W 12 × 50 would be specified WT 6 × 25 (6 is half the depth and 25 is half

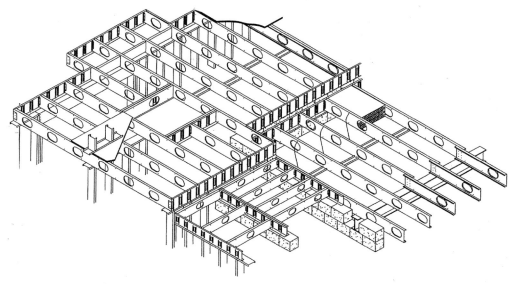

FIGURE 26.13B > Steel floor framing system. *Dietrich Industries, Inc.*

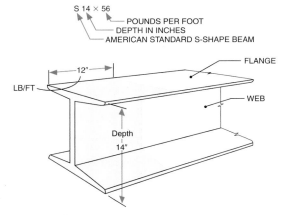

FIGURE 26.14 > S-shape designations. *Hepler/Wallach/Hepler © Cengage 2013*

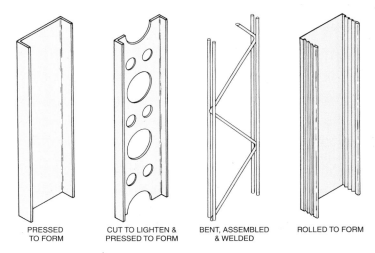

FIGURE 26.16 > Types of metal joists and studs. *Hepler/ Wallach/Hepler © Cengage 2013*

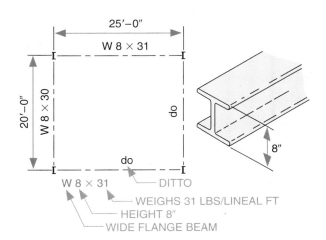

FIGURE 26.15 > W-shape designations. *Hepler/Wallach/Hepler © Cengage 2013*

the weight per foot). Tees are most commonly used for truss chords and to support concrete reinforcement rods.

Metal studs, as shown in Figure 26.16, are manufactured in different forms. Stud thicknesses vary depending on the loads to be supported. Widths are manufactured to be identical with conventional wood stud sizes. Fireproof steel framing can be substituted for conventional wood framing. It is insect proof and will not decay, expand, contract, warp, or split. Some steel studs can be attached to wood members by stapling or nailing flanges.

Steel space frame designs are created by adding a third dimension to open web trusses. These types of trusses weigh less than a comparable wood truss. Figure 28.19 shows a common open web truss, which can be used as a floor joist or roof

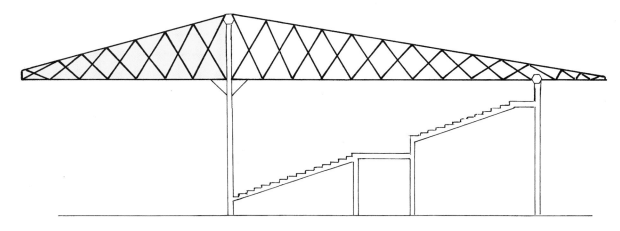

CANTILEVERED TRUSS OVER AN AMPHITHEATER

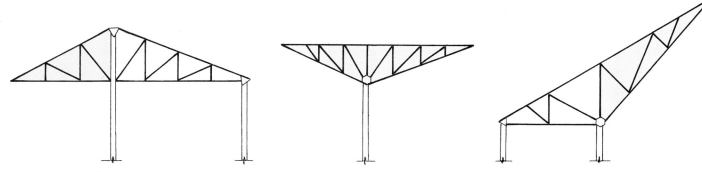

FIGURE 26.17 ⟩ Cantilevered open web trusses. *Hepler/Wallach/Hepler © Cengage 2013*

truss when cables and pipes are applied large cantilevering is possible, as shown in Figure 26.17. Steel space frame design is based on the premise that a triangle represents the strongest and most stable geometric form with the fewest members.

CAD Steel Member Symbol Library

Sections of structural steel members are included in most steel member symbol libraries including drawings of fasteners used in steel construction. These are stored in both plan and sectional views. The *Copy-Multiple* command is used on structural steel drawings to locate and draw evenly spaced rows of bolts and rivets. This command allows an evenly spaced member to be repeated by inputting the spacing, location, and number of copies needed.

STEEL FASTENERS AND INTERSECTIONS

Major structural steel members depend on a wide variety of joining methods and devices to function as a structurally stable frame. This includes the use of brackets, rivets, bolts, and welds to attach members to each other and to foundation piers and footings. See Figure 26.18. Some steel components can be assembled before shipment to a site. These are usually assembled

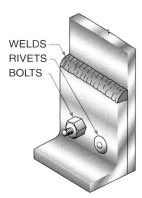

WELDS
RIVETS
BOLTS

FIGURE 26.18 ⟩ Methods of joining steel members. *Hepler/ Wallach/Hepler © Cengage 2013*

and welded at a fabrication shop. All other members are assembled and permanently fastened at the building site. For example, some brackets may be welded to a girder at a shop, then bolted or welded to a column at the site during construction.

Brackets

Most structural steel members intersect at right angles. Many different types of **brackets** are used to provide a perpendicular surface for bolting, riveting, or welding. Figure 26.19 shows

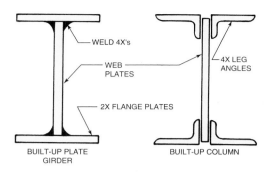

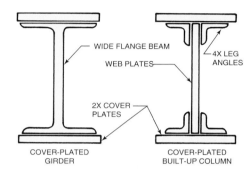

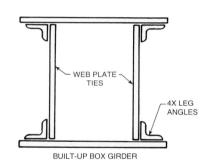

the use of brackets and welds to assemble built-up girders. Angles, L-shapes, and bent or welded plates are used for this purpose. Angles, welds, rivets, nuts, and bolts help join a girder to the top of a column. Figure 26.20 shows a typical connection between an S-shape steel beam and an S-shape steel column.

Bracket information on a structural drawing shows the size of the bracket legs followed by the thickness, width, shape symbol, and fastening device information. See Figure 26.21. If brackets are to be welded to a member at a fabrication shop, a detail drawing is not provided in the field. Only the assembled intersection of the joint is drawn for field reference. Only shop fabricators are provided with a complete set of details.

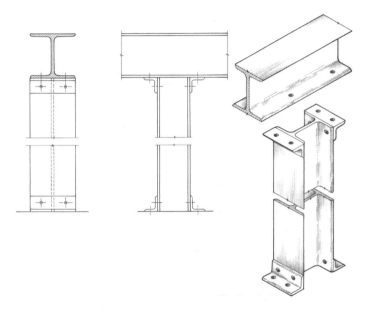

FIGURE 26.19 > Methods of joining built-up girders. *Hepler/Wallach/Hepler © Cengage 2013*

FIGURE 26.20 > S-shape beam and column connection. *Hepler/Wallach/Hepler © Cengage 2013*

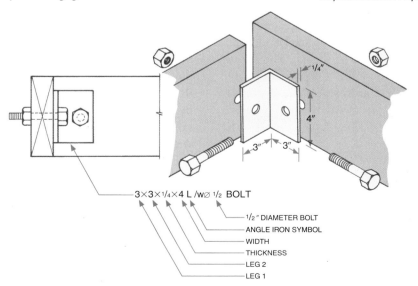

FIGURE 26.21 > Angle bracket and bolt notations. *Hepler/Wallach/Hepler © Cengage 2013*

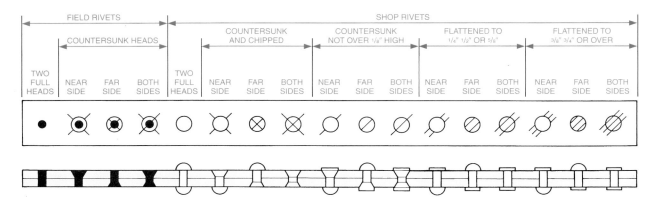

FIGURE 26.22 > Types of field and shop rivets. *Hepler/Wallach/Hepler © Cengage 2013*

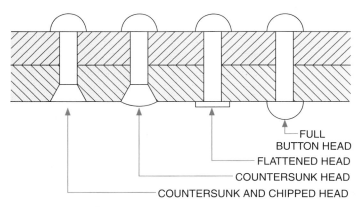

FIGURE 26.23 > Common types of rivet heads. *Hepler/ Wallach/Hepler © Cengage 2013*

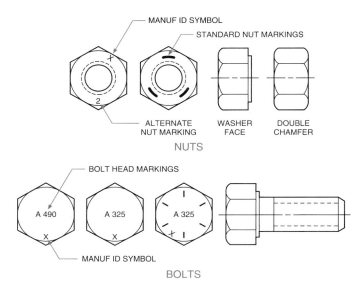

FIGURE 26.24 > High-strength nut and bolt notations. *Hepler/Wallach/Hepler © Cengage 2013*

Rivets

Rivets are used to connect steel members. Figure 26.22 shows the most common types of rivets use for shop fabrication and job-site use. Figure 26.23 shows the four basic types of rivet heads. The type of rivet specified is shown at the end of the drawing notation. Rivets are made of soft steel and, when heated and then cooled, rivets tend to shrink. The shrinking decreases the tightness of the joint. Consequently, bolts are now used more extensively than rivets in the erection of structural steel.

Bolts

High-strength **bolts** and nuts can carry loads equal to rivets of the same size, but they can be turned tighter because of their high tensile strength. Bolts used in steel construction are either high-strength or unfinished bolts. High-strength bolts are used to connect extremely heavy load-bearing members such as girders, beams, and columns. See Figure 26.24. They are also used to attach members where shear loads are transmitted through the bolts.

Unfinished bolts are used for lighter connections, where loads are transmitted directly from member to member. For example, unfinished bolts could be used when a beam rests directly on a girder because there is no vertical shear load on the bolts holding these two members together. Unfinished bolts are also used to anchor column base plates to footings, as shown in Figure 26.25, since there is also no shear stress at this location.

Welds

Welding is a popular method of connecting structural steel members. It has some advantage over bolting. For example, fabrication is simplified by reducing the number of individual parts to be cut, punched with holes, handled, and installed. The major types of **welds** and their symbols are illustrated in Figure 26.26.

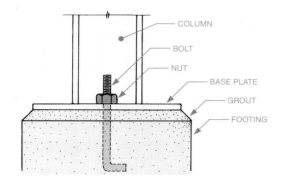

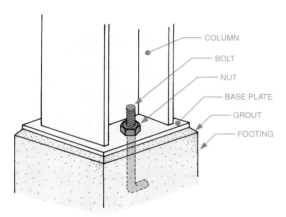

FIGURE 26.25 > Unfinished bolts used on a column base.
Hepler/Wallach/Hepler © Cengage 2013

WELD SYM	WELD NAME
◹	FILLET
▭	PLUG/SLOT
○	SPOT/PROJECTION
⊖	SEAM
▽	BACK/BACKING
⌣	SURFACING
//	SCARF (BRAZING)
⊔	FLANGE – EDGE
⊓	FLANGE – CORNER
||	GROOVE – SQUARE
V	GROOVE – V
/	GROOVE – BEVEL
Y	GROOVE – U
⊬	GROOVE – J
⫲	GROOVE – FLARE V
⫧	GROOVE – FLARE BEVEL

FIGURE 26.26 > Types of welds. *Hepler/Wallach/Hepler © Cengage 2013*

The convention used to locate welding information on drawings is a horizontal reference line with a sloping arrow directed to the joint. See Figure 26.27. The arrow may be directed right or left, upward or downward, but always at an angle to the reference line. If no extra marking is shown, a shop weld is assumed. Other drawing symbols give additional instructions. A triangular flag indicates a field weld. An open circle means weld all around the member.

The basic weld symbols or supplementary weld symbols are located midway on the horizontal reference line. See Figure 26.28. The symbol is located below the line if the weld is to be placed on the near side of the workpiece where the arrow points. The symbol is placed above the line if the weld is to be placed on the far side. The symbol is placed above and below if both sides are to be welded. The side of the weld (or its depth) is indicated to the left of the basic symbol. The length of the weld is shown to the right of the symbol. When long joints are used, intermittent welds are often specified. These are indicated by the length of weld followed by the center-to-center spacing (pitch). Such welds are usually staggered on either side of a joint. Figure 26.29 shows the position of supplementary weld symbols on the horizontal line.

WELD NOTATION:

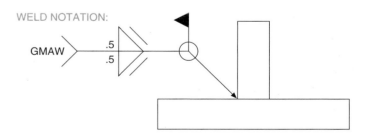

MEANS:
FILLET WELD BOTH SIDES
GAS METAL ARC WELDING
FLUSH CONTOUR
WELD ALL AROUND
WELD IN FIELD

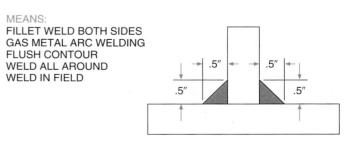

FIGURE 26.27 > Welding symbol used on a detail drawing.
Hepler/Wallach/Hepler © Cengage 2013

F – Finish symbol
⌒ – Contour symbol
A – Groove angle: included angle of countersink for plug welds
R – Root opening: depth of filling for plug and slot welds
S – Depth of preparation
 – Size or strength for specific welds
 – Height of weld reinforcement
 – Radii of flare-bevel grooves
 – Radii of flare-V grooves
 – Angle of joint (brazed welds)
(E) – Effective throat
T – Specific process or reference
L – Length of weld
 – Length of overlap (brazed joints)
P – Pitch of welds (center-to-center spacing)
1 – Weld located on opposite side of arrow
2 – Weld located on same side of arrow
(N) – Number of spot or projection welds
⌐ – Weld made in field
o – Weld all around

FIGURE 26.28 > **Common weld symbols.** *Hepler/Wallach/ Hepler © Cengage 2013*

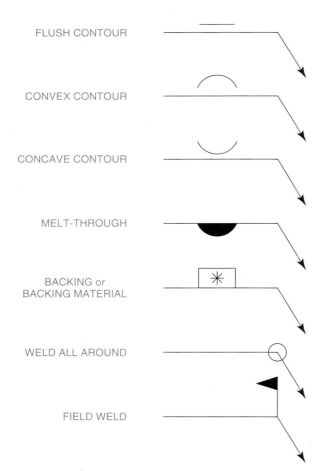

FIGURE 26.29 > Supplementary weld symbols. *Hepler/ Wallach/Hepler © Cengage 2013*

CAD Welding Symbol Library

Welding symbol libraries contain symbols for each type of weld. They also include the symbol line on which the welding symbols can be located with the cursor. These are stored as base symbol *Blocks.* The numerals related to each weld are then added using the *Text* command.

The tail of the reference line may contain information about the kind of material or process required. This feature is not often used on structural steel details. When no information is required, the tail is omitted. Figure 26.30 shows the position of information on a welding symbol.

Although steel construction is made extremely rigid through the use of regular fasteners, special cross bracing is often required to counteract lateral wind loads. Figure 26.31

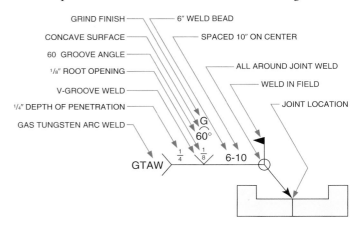

FIGURE 26.30 > Position of information on a weld symbol. *Hepler/Wallach/Hepler © Cengage 2013*

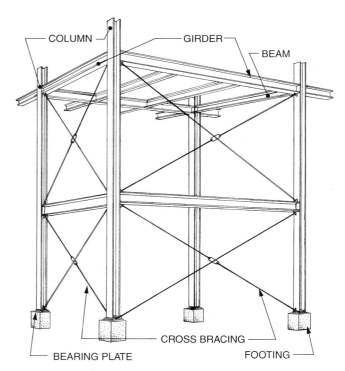

FIGURE 26.31 > Structural cross bracing. *Hepler/Wallach/ Hepler © Cengage 2013*

shows an example of cross bracing, which is designed to stabilize the position of columns, girders, and beams.

STRUCTURAL STEEL DRAWING CONVENTIONS

Structural steel drawings are of several types: design (schematic) drawings, working (shop) drawings, and erection drawings. *Design (schematic) drawings* are very symbolic and show only the position of each structural member with a single line. Notations describing each member's shape, size, and weight are included on each line. When several members with identical characteristics are aligned, the successive lines are labeled with a ditto symbol (do) indicating that the shape,

size, and weight of the member are identical to the previous member's. See Figure 26.32.

Working (shop) drawings are complete orthographic engineering drawings showing the exact size and shape of each member, including every cut, hole, and method of fastening. Figure 26.33 shows a structural steel working detail drawing that is part of a series of detail drawings indexed to an erection set of plans. Figure 26.34 is an example of an erection plan. *Erection drawings* show the method and order of assembling each member, which is coded for easy field identification. The specific methods used to prepare structural steel floor,

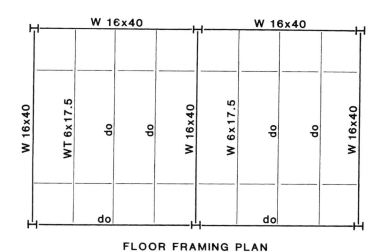

FIGURE 26.32 > Structural steel schematic drawing. *Hepler/Wallach/Hepler © Cengage 2013*

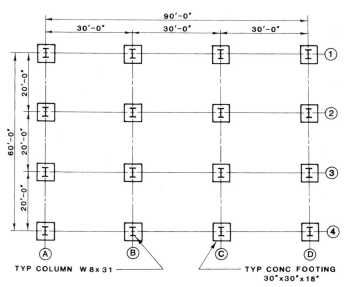

FIGURE 26.34 > Erection drawing showing column, footing, and girder locations. *Hepler/Wallach/Hepler © Cengage 2013*

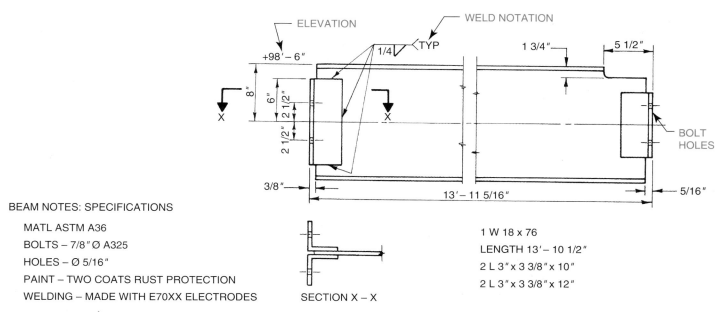

BEAM NOTES: SPECIFICATIONS

    MATL ASTM A36

    BOLTS – 7/8" Ø A325

    HOLES – Ø 5/16"

    PAINT – TWO COATS RUST PROTECTION

    WELDING – MADE WITH E70XX ELECTRODES

SECTION X – X

1 W 18 x 76

LENGTH 13' – 10 1/2"

2 L 3" x 3 3/8" x 10"

2 L 3" x 3 3/8" x 12"

FIGURE 26.33 > Steel detail working drawing. *Hepler/Wallach/Hepler © Cengage 2013*

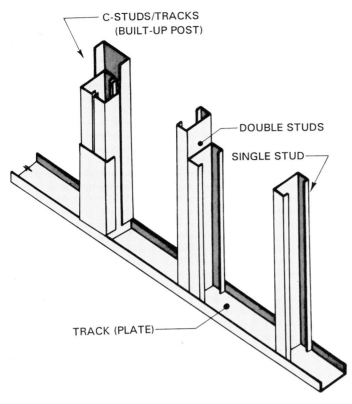

FIGURE 26.35 › Lightweight steel stud wall. *Hepler/Wallach/ Hepler © Cengage 2013*

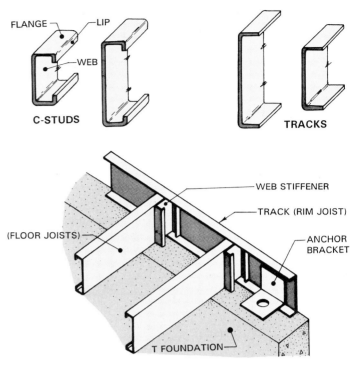

FIGURE 26.36 › Lightweight steel floor framing. *Hepler/ Wallach/Hepler © Cengage 2013*

wall, and roof framing drawings are covered in Chapters 28, 29, and 30.

Lightweight steel structures resemble wood skeleton-frame buildings. Light-gauge zinc-coated steel members (C-studs, U-shaped tracks, and angles) are manufactured in structural wood sizes such as 2 × 4, 2 × 6, and 2 × 8. Using screws, not nails, contractors can build structures in a manner similar to that used for building wood structures. Steel members can also be joined with bolts, welds, or power-driven fasteners. Figure 26.35 shows how C-studs intersect with tracks to form a stud wall. Lightweight steel floor systems also resemble conventional wood floor framing, as shown in Figure 26.36. Lightweight steel framing provides structurally stable door lintels, jambs, and wall intersections (Figure 26.37).

Lightweight steel members can be used with wood or concrete materials (Figure 26.38); however, special intersections must be designed in these cases. The most effective use of lightweight steel construction involves the use of steel for the entire structure. Figure 26.39 shows an example of an all-steel wall. This elevation section through an outside steel wall also shows the intersecting steel roof and floor members.

Nonferrous Metals

Copper, brass, aluminum, and lead are used in construction; however, their use is restricted to nonstructural applications. Copper, lead, and aluminum are used for flashing and roofing. Extruded aluminum is also used for window and door trim and also for reflective insulation. The use of brass is limited to plumbing and door and window trim applications.

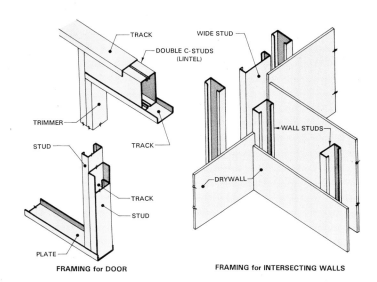

FIGURE 26.37 > Lightweight steel door and wall intersection framing. *Hepler/Wallach/Hepler © Cengage 2013*

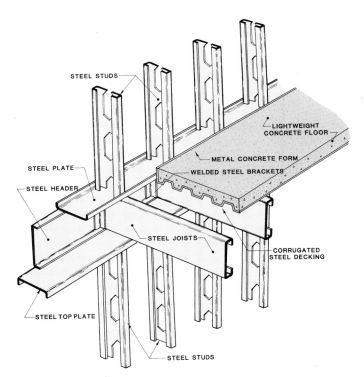

FIGURE 26.38 > Lightweight steel wall intersecting concrete floor. *Hepler/Wallach/Hepler © Cengage 2013*

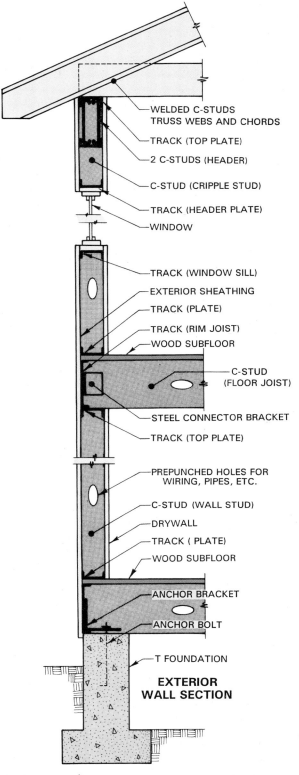

FIGURE 26.39 > Lightweight steel exterior wall section. *Hepler/Wallach/Hepler © Cengage 2013*

Steel and Reinforced-Concrete Systems Exercises

CHAPTER **26**

1. Describe the differences between a design drawing, shop drawing, and erection drawing.

2. List the advantages and disadvantages of steel construction.

3. Describe the difference between beams, girders, columns, and purlins.

4. Name the types of structural steel shapes.

5. Draw the symbols for these steel shapes: wide flange beam, tee, angle, channel, and square bar.

6. Draw a structural steel framing plan and elevation drawing for the house you are designing.

7. Name the types of fastening methods used in steel construction.

8. Draw the symbols for these welds: fillet, spot, seam, edge flange, V groove, and backing.

9. Design, draw, and dimension a steel erection plan for a 28′ × 56′ concrete block storage building. Show columns, footings, and girder locations. Allow 4′-0″ RO (rough opening) for window and 16′-0″ RO for an overhead door. Minimize interior columns and hold spans to a maximum of 18′-0″.

10. Draw a foundation plan to match the pier, column, and beam plan shown in Figure 26.40.

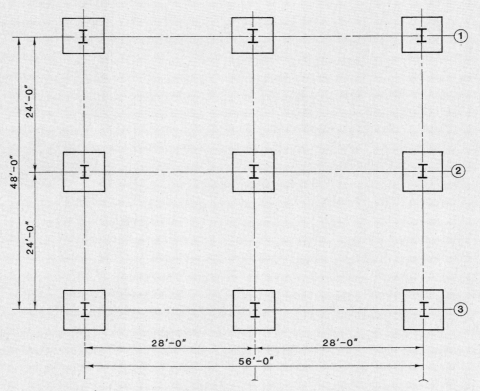

FIGURE 26.40 ⟩ Pier, column, and beam plan. *Hepler/Wallach/Hepler © Cengage 2013*

CROSS B
HEADER

CHAPTER 27
Disaster Prevention Design

OBJECTIVES

In this chapter you will learn to:

> describe the measures that can be taken during construction to minimize potential damage from natural disasters.

> describe how to prevent gas leaks.

> name ways to provide fire protection for a structure and its residents.

> discuss methods for ensuring clean air and water in a building.

TERMS

cannon test
carbon monoxide
Environmental Protection Agency (EPA)

propane
radon
safe room

straps
toxic materials
Underwriters Laboratories (UL)

INTRODUCTION

Hurricanes, tornadoes, high winds, floods, blizzards, gas leaks, wildfires—some disasters are not preventable. However, precautions can be designed into a structure to prevent or minimize the damage they cause. Architectural features are a key factor in disaster prevention. Site and building plans can either help or hinder efforts to maintain a safe environment.

PREVENTING WIND DAMAGE

The causes of wind-related structural damage range from routine thunderstorms to severe tornadoes and hurricanes. Winds do not need to reach hurricane or tornado velocity to seriously damage or demolish a structure. To design a structure with maximum wind resistance, a continuous and strong structural link must be made from the foundation to the roof. Reinforced concrete construction resists wind loads better than does concrete block; and concrete block resists wind loads better than wood construction. However, these differences can be minimized through the use of structural ties and reinforcements (Figure 27.1). Many methods are employed to achieve a strong structure—depending on the type of construction.

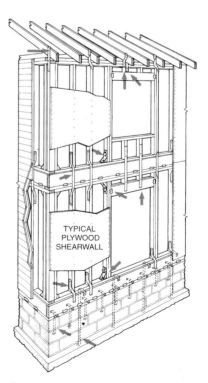

TYPICAL PLYWOOD SHEARWALL

FIGURE 27.1 > Continuous load transfer design using structural link connectors. *Simpson Strong-Tie Company, Inc.*

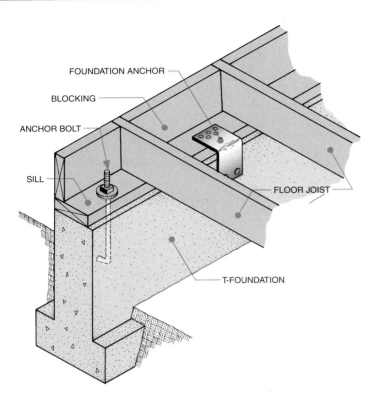

FIGURE 27.2 > Fastening sill to foundation with anchors. *Hepler/Wallach/Hepler © Cengage 2013*

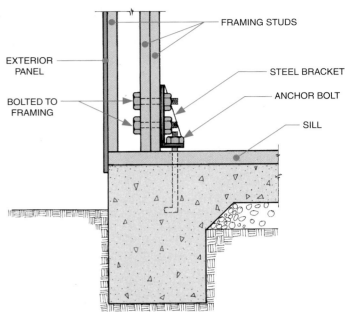

FIGURE 27.3 > Fastening sill to a slab foundation with a steel bracket. *Hepler/Wallach/Hepler © Cengage 2013*

To maximize structural strength, vertical rebars and continuous pours are required during concrete construction. In wood construction, the effective placement of timber connectors is vital.

Sills must be fastened firmly to the foundation using foundation anchors. Figure 27.2 shows how this is done using anchor bolts and L-brackets. Figure 27.3 shows the use of a steel bracket bolted to studs and a slab foundation anchor. Long anchors provide additional control by connecting the foundation and sill to the wall framing.

Tie straps help hold the wall framing to the sill, as shown in Figure 27.4. Additional blocking on the sill and between vertical members also helps with anchoring and provides added rigidity to the structural frame. Steel ties between joists and the foundation provide protection against uplift. See Figure 27.5.

Nonmasonry wall surfaces should be covered with shear stress plywood or a material with equal wind velocity rating. Figure 27.6 illustrates the use of anchors, clips, and shear stress sheathing panels to protect against wind damage. Shear stress panels are usually placed on the outside of an exterior stud wall, but can also be placed on the inside. For most residential applications shear stress panels must be at least 5/8″ thick, applied in a vertical orientation, and lapped over sills and plates. Figure 27.7A shows the load forces acting on a shear paneled wall, and Figure 27.7B shows two alternative

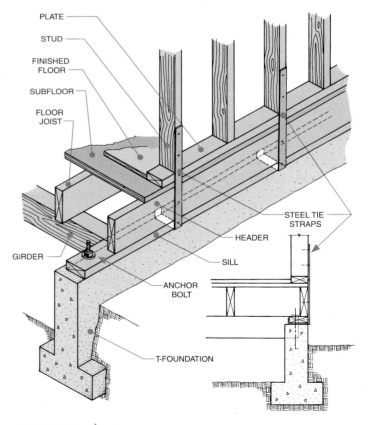

FIGURE 27.4 > Sill tie-down straps. *Hepler/Wallach/Hepler © Cengage 2013*

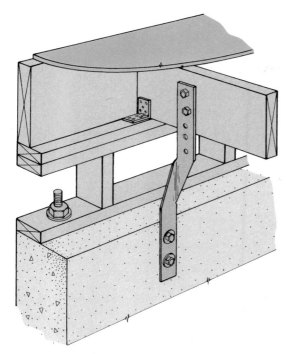

FIGURE 27.5 〉 Ties used to prevent uplift. *Hepler/Wallach/ Hepler © Cengage 2013*

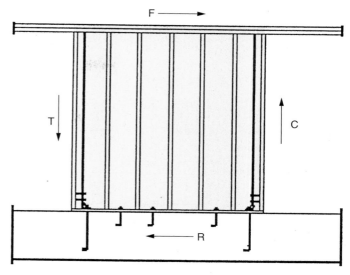

C = COMPRESSION WILL RESIST ROTATION
T = TENSION WILL RESIST ROTATION
F = APPLIED LOAD
R = RESISTANCE AT FOUNDATION

FIGURE 27.7A 〉 Shear wall stress framing. *Hepler/Wallach/ Hepler © Cengage 2013*

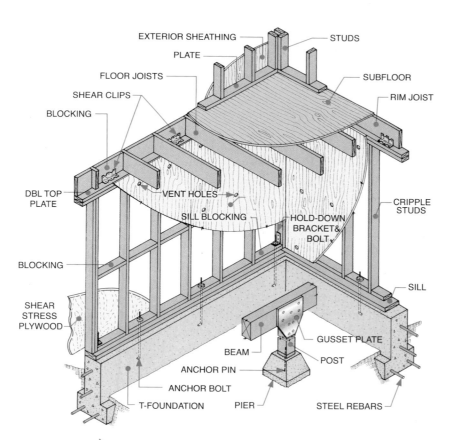

FIGURE 27.6 〉 Methods used to protect against wind and seismic (earthquake) damage. *Hepler/Wallach/Hepler © Cengage 2013*

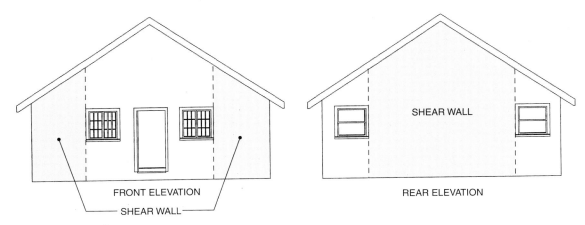

FIGURE 27.7B ⟩ Shear wall panel placement. *Hepler/Wallach/Hepler © Cengage 2013*

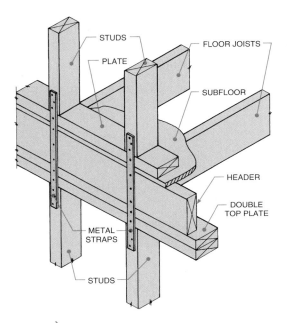

FIGURE 27.8 ⟩ Platform framing connectors that use metal straps. *Hepler/Wallach/Hepler © Cengage 2013*

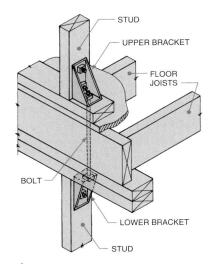

FIGURE 27.9 ⟩ Platform framing connectors that use bolts and brackets. *Hepler/Wallach/Hepler © Cengage 2013*

placements for shear panels. Windows must also be rated to withstand the same wind velocity as the structural walls.

On bilevel platform framed structures, the two levels must be connected in order to transfer the uplift forces from the upper studs to the lower studs. Figure 27.8 shows how metal **straps** are used, and Figure 27.9 shows bolts and brackets connecting two levels. Bolt holddowns between joists and studs on both levels will also provide this connection. When bolts are used, they must be firmly fastened with nuts.

Connections between wall framing and roof framing are also critical. Winds trapped under overhangs can damage or break away soffits and fascia boards. Any area where wind can be trapped should be ventilated to allow the air to escape. Roof sheathing should be connected with hurricane

clips. Roofing screws or nails, not staples, must be used to attach sheathing, soffit, and fascia materials. In high-wind areas, wind shear shingles should be specified and connectors used to attach top plates to studs, floor joist to girders, and top plates to rafters as shown in Figure 27.10. Structural straps as shown in Figure 27.11 hold studs and ceiling joists together.

In addition to structural reinforcements, other design features must be considered to minimize wind damage. Hip roofs withstand more wind force than gable roofs, so gable roof trusses should be cross braced to add more resistance. Large roof overhangs should be vented to prevent uplift. Specifying #1 grade lumber for roof trusses also helps stabilize roof structures during high wind gusts.

Doors and windows must be designed to withstand wind gusts that could expose the inside of the structure to severe damage. Window glass, in hurricane- or tornado-prone areas,

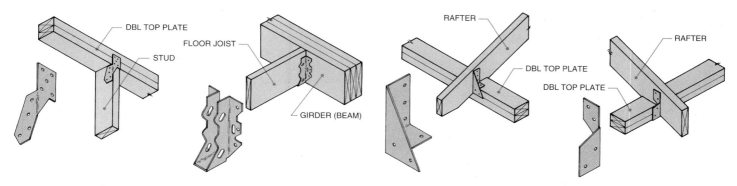

FIGURE 27.10 ❯ Rafter-plate and hanger connectors. *Hepler/Wallach/Hepler © Cengage 2013*

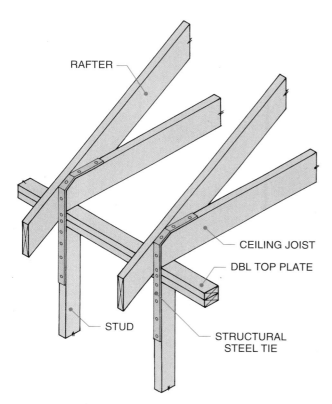

FIGURE 27.11 ❯ Ceiling joist-stud structural tie. *Hepler/Wallach/Hepler © Cengage 2013*

should withstand the **cannon test**—a 2 × 4 stud hurled at least 34 feet before impacting the window. Windows in these areas should also be equipped with roll-down hurricane shutters or fitted with supports to which plywood covers can be attached. Doors in hurricane areas should open out and have top and bottom reinforcement bolts with welded hinge pins.

PREVENTING EARTHQUAKE DAMAGE

The structural features designed to reduce wind damage also apply to earthquake damage reduction. Figure 27.12 illus-

trates many of these features. However, special attention is needed for earthquake protection as follows:

1. Design a continuous structural link from foundation to roof according to the same principles as those described earlier for preventing wind damage.
2. Specify that all gas appliances are to be connected with flexible lines.
3. Fasten built-in appliances to structural members, not to wallboard or trim.
4. Do not use masonry chimney materials above the eave line.
5. Specify push-type cabinet latches (touch latch).
6. Specify hook-and-eye latches on workshop cabinets.
7. Specify security film for windows—to prevent shattering.
8. Use ball-bearing supports to prevent columns from flexing or bending. See Figure 27.13.
9. Footings should cover the largest horizontal area possible.
10. Avoid using suspended floor systems, which could collapse as one unit on the floors below.
11. Ensure good site drainage.
12. Strap all gas, water, and vent pipes to framing members.
13. Bolt all tall, heavy wall furniture into wall studs.
14. Ensure that footings are poured on highly compacted, nonorganic soil or preferably on bedrock.

🌐 GAS CONTAINMENT AND VENTING

The nature of gas is to expand and fill all available space. When gas escapes from containers, pipes, or the soil, it may be trapped in a sealed building. These trapped gases can seriously injure or kill people who inhale them. If ignited with a spark or flame, some gases can explode and cause great damage.

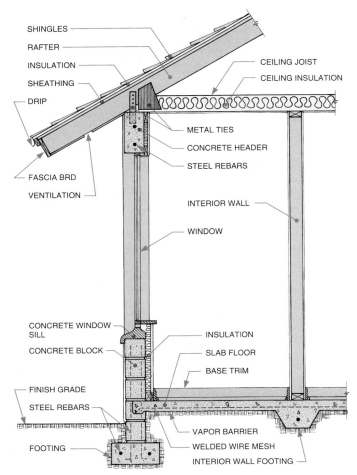

FIGURE 27.12 > Hurricane- and earthquake-resistant design components. *Hepler/Wallach/Hepler © Cengage 2013*

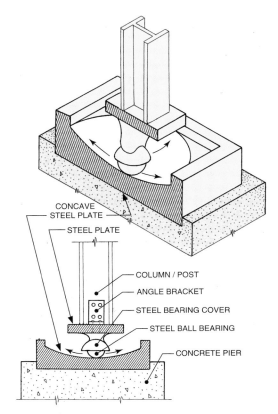

FIGURE 27.13 > Ball-bearing supports help absorb ground movement during an earthquake. *Hepler/Wallach/Hepler © Cengage 2013*

Natural gas, propane, carbon monoxide, and radon are potentially the most dangerous gases. Gas fumes created by many synthetic building materials and interior furnishings are suspected of causing long-range health problems.

Natural and Propane Gas

Flammable liquids such as gasoline will burn when ignited. Gases, however, will explode when ignited. Because flammable gases are used in buildings for the functioning of a wide variety of fixtures, these gases must be tightly contained to be safe. Providing a high level of safety is the role of effective plumbing, HVAC, and supporting structural systems. To minimize the potential for gas leakage, specify that pipes should be strapped to structural members and use flexible lines to all appliances. Locate **propane** containers as far from structures as possible. Specify electronic pilots for all gas appliances.

Carbon Monoxide (CO)

Carbon monoxide is a colorless, odorless gas produced by combustion. Poisoning from CO gas causes brain damage at low levels and death at high levels. Carbon monoxide is produced by all combustion appliances including wood fireplaces, stoves, grills, gas or oil furnaces, clothes dryers, and water heaters. Designers must ensure that gas emissions from these devices are adequately vented to the exterior and cannot escape into the interior. An inspection should be specified and CO detectors installed during construction. Detectors should be mounted on ceilings on each story.

Radon

Radon is a colorless gas produced by the natural decay of the element radium. This radioactive gas enters a structure from the soil. To minimize risk, specify polyethylene membranes under floors and slabs. Specify total parging (protective plasterwork) on all walls below the grade line. Specify a separate outside air source for fireplaces to avoid drawing radon in through small cracks in the floor system. Ventilate crawl spaces to the outside. Provide a fan-driven air escape vent to

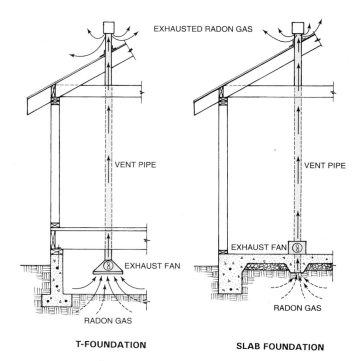

FIGURE 27.14 > Radon prevention methods. *Hepler/Wallach/ Hepler © Cengage 2013*

avoid gas buildup. Test for radon before construction begins. In areas with high levels of radon, a central vent from under the foundation through the roof can divert gas from entering the house. See Figure 27.14.

FIRE PREVENTION AND CONTROL

Fire is a valuable asset when contained in a furnace, heater, fireplace, or incinerator. However, fire can be a most destructive force when not contained. Other than personal destructive accidents, unwanted fires are caused by nature (lightning), by overheated devices, and by loss of fuel containment. Good architectural plans cannot control bad personal habits, but good designs can prevent or mitigate the result of natural causes. Good design and construction can also ensure that heat-producing devices and support structures can control and direct fires where needed—and away from areas where fires are not wanted.

The best fire protection is to observe building codes and to use electrical and gas devices approved by **Underwriters Laboratories (UL),** a nonprofit agency that tests products for safety. All electrical wiring and devices must also be grounded according to UL standards. Fire control measures include specifying smoke detectors on each level of the house. Fire extinguishers and water hoses should be specified to reach the kitchen, laundry, garage, and workshops. Designing buildings that are "firesafe" also involves following fire code structural design such as specifying firewalls in hazardous areas,

and roof venting construction. Other architectural features used to prevent fires or manage fire damage include use of fireplace screens, fire-retardant roofing materials, chimney spark arrestors, and adequate circuit capacities.

AIR PURIFICATION

Pure air is an inert gas and is vital for life, but polluted air has the reverse effect. As covered in Chapter 6, architectural designs should exclude materials and conditions that lead to the development of mold, bacteria, and volatile chemicals that can become airborne. Plans must also include air filtering systems and ultraviolet lights in ducts to kill mold. For more information on air quality control measures, refer to Chapter 32, *Comfort Control Systems (HVAC)*.

To provide the maximum amount of pure air, specify electronic air filters for all furnaces, air conditioners, and the air exchangers inside heat pumps. Air cleaners may be installed to quickly purify contaminated air.

WATER CONTROL

Structural and site damage from the presence of unwanted water and moisture can originate from external or internal sources. Internal problems are primarily caused by leaking pipes or malfunctioning appliances and equipment. External problems are caused by soil saturation and water penetrating unsealed surfaces. Both problems can be avoided by effective architectural planning.

Water quality varies greatly from one community to another. To maximize water purity, specify a water system that may include a water softener, carbon filter, ultraviolet filter, and/or reverse osmosis unit. You may specify a dedicated faucet for pure drinking water and for the refrigerator ice-maker.

In addition to purifying consumable water, unwanted water in the form of moisture (humidity) must be controlled to eliminate the formation of fungi, mold, dry rot, and termite attacks. Diverting water from foundation walls through the use of swales, trenches, berms, or drains is the first line of defense. See Figure 27.15. The use of vents, fans, cross ventilation, and dehumidifiers is most effective inside a structure.

CONTROLLING TOXIC POLLUTANTS

High levels of toxic pollutants in buildings (sometimes termed *sick building syndrome*) result from external airborne sources or internal offgasing of building products, finishing materials, and furnishings. Both problem sources can be eliminated or minimized by careful HVAC planning (see Chapter 32) and material specifications (see Chapter 35).

EXTERIOR WATER CONTROL METHOD

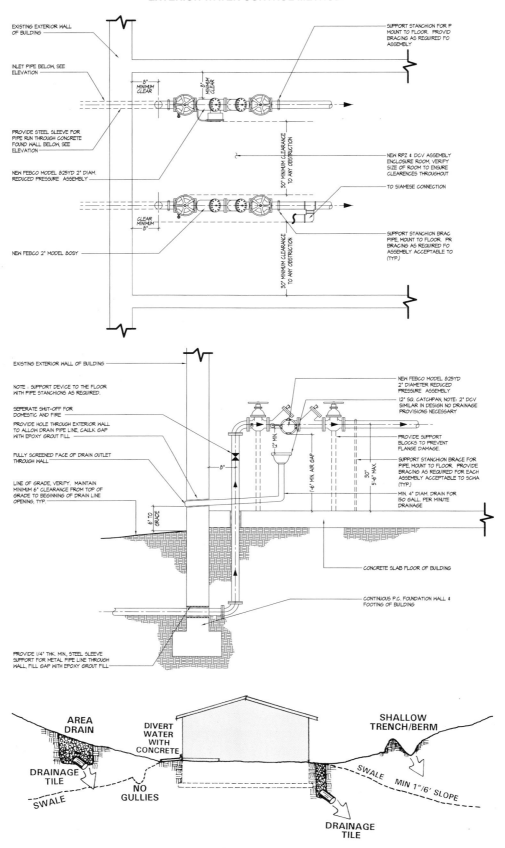

FIGURE 27.15 > Exterior water control methods. *Hepler Associates LA, PC*

Although the effects of synthetic material vapors may not be immediate, the long-range health effects may be serious. Avoid the interior use of exposed and unsealed building materials that contain toxic materials as defined by the **Environmental Protection Agency (EPA),** a federal regulatory agency. Use solid wood products where possible.

Toxic materials such as arsenic, asbestos, and formaldehyde are found in many building materials including structural members, coatings, insulation, and wall and floor coverings. Where possible designers should specify nontoxic materials recommended by the EPA.

PEST CONTROL

Wood-boring insects and wall-dwelling rodents can be controlled by specifying treatment barriers under the slab and footing. Ensuring that the foundation connections to the upper structure have no openings for insect penetration must be part of the structural design.

Household pests can be controlled in the design process by specifying the installation of wall cavity tubes. These tubes safely carry pesticides into the walls without detectable levels entering living spaces. Designing facilities for the short- and long-range secure storage of trash and garbage is one of the most effective methods of pest control. See Chapter 11and 27 for more detailed information.

ELECTRICAL HAZARDS

Electricity, like water and gas, is an asset that must be controlled and contained to be safely usable. The main methods for preventing electrical accidents consist of isolating "live" wires and avoiding wattage overloads on circuits and electrical devices. For more details about eliminating electrical shock and fire hazards, follow the electrical design practices outlined in Chapter 31. This means designing circuits that will not overload such as GFCI. Providing outlets that minimize the use of long extension cords, and switches and wires with the capacity to carry the maximum load as designed is also mandatory.

PREVENTING PERSONAL INJURY

Many accidents within buildings, or on sites, result from personal misconduct. However, many are caused by architectural design features that violate established safety codes.

Designers can prevent accidents by following the national code and Occupational Safety and Health Administration requirements that are appropriate to the building type. Special consideration should be given to the design of the areas where most accidents occur—stairs, kitchens, baths, children's bedrooms, and swimming pools.

SECURITY

Two types of security measures involve architectural design. First, disaster prevention includes the topics previously covered. The second is security from intruders, which includes site, building, and personal security.

Site security includes visible or invisible fence and gate alarms, closed-circuit television outposts, and perimeter communication facilities.

General building security includes an alarm system, closed-circuit monitors, a valuables safe, dead-bolted doors, and break-proof windows, as detailed in Chapter 31.

Personal security is largely dependent on the effectiveness of site and building security measures. If these two measures fail, a third personal safety option is available: the **safe room.** Safe rooms are either deigned for disaster safety or security from intruders. These functions can be combined, but in either case their design assumes a short stay.

Safe rooms designed for disaster protection must be structurally stable and flame retardant with a fireproof door that opens inward. The room should be located near the center of the building, have no or unprotected windows, and be designed to withstand 260-mph winds as shown in Figure 27.16.

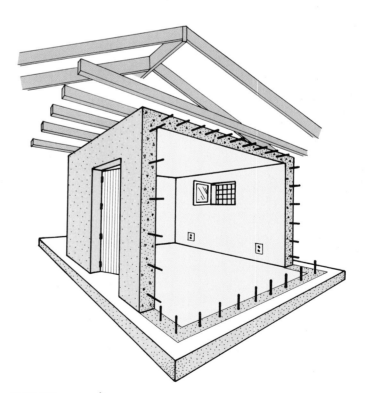

FIGURE 27.16 > Disaster protection safe room. *Hepler/ Wallach/Hepler © Cengage 2013*

Safe rooms designed for intruder protection must be bulletproof, flame retardant, hidden from visitors' view, and with a crash-proof door as shown in Figure 27.17. These two types of protection may be incorporated into one room.

Equipment for a safe room should include a cell phone, air purifier (scrubber), closed-circuit TV for the outside, battery radio, and generator with connecting lights. Furniture can include chairs, a table, and bed. Supplies should include water, nonperishable food, flashlights, and sleeping bags.

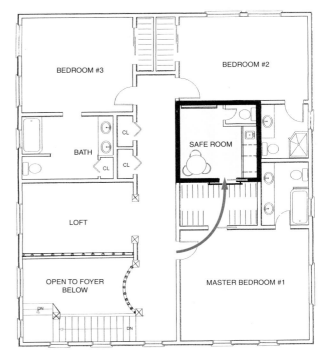

FIGURE 27.17 > Intruder protection safe room. *Hepler/ Wallach/Hepler © Cengage 2013*

Disaster Prevention Design Exercises

CHAPTER 27

1. List the design features you will use in the house you are designing to prevent wind, earthquake, gas, and fire damage.

2. Sketch structural methods of preventing wind and earthquake damage.

3. Explain how you will provide for water and air purification in the house you are designing.

4. Sketch the ventilating system you will use to provide adequate ventilation in the house you are designing.

5. Locate and design a safe room for your house.

6. List the disaster prevention measures you would include in the plan shown in Figure 27.18.

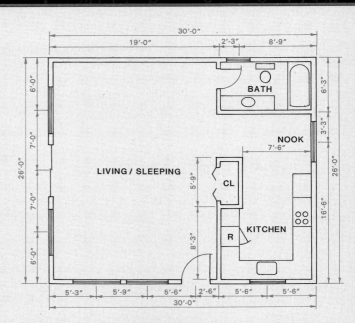

FIGURE 27.18 > Add disaster prevention methods to this plan. *Hepler Associates LA, PC*

FRAMING SYSTEMS

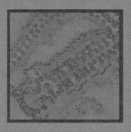

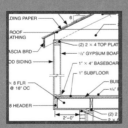

(2) 2 × 4 TOP PLATE
½" GYPSUM BOARD
1" × 4" BASEBOARD
1" SUBFLOOR
BUIL
¾" F
2'-0"
(2) 2 ×
2 × 6
ATING
HING
2 × 4

CHAPTER 28

Floor Framing Drawings

OBJECTIVES

In this chapter you will learn to:

> identify the components of floor systems.

> draw a floor framing plan that shows all structural parts.

> draw details of sills, supports, and stairwells.

> draw a floor framing plan with steel joists.

TERMS

bays
blocking
bridging
decking
firecuts
grid column identification system

headers
I-joists
lookout joists
panelized floor systems
plank-and-beam floor system
platform floor systems

sequential column identification system
splices
trimmers
truss joists

INTRODUCTION

The design of floor framing systems demands careful calculation of the live and dead loads acting on a floor. The exact size and spacing of floor framing members, plus the most appropriate materials, must be selected. Through floor systems, live and dead loads are transferred to foundations through multilevel bearing walls while providing lateral support to columns and beams. Floor systems must be rigid with space to accommodate plumbing, HVAC ducts, and electrical lines.

Some floor framing drawings are prepared to show only the structural support for the floor platform. Others may illustrate details of construction, such as the attachment of the floor frame to the foundation.

TYPES OF PLATFORM FLOOR SYSTEMS

Systems that are supported by foundation walls and beams or girders are called **platform floor systems.** These differ from ground-level slab floors that are structurally part of the foundation. (Refer back to Chapter 23.) Platform floor systems are divided into three types: conventional, heavy timber (plank-and-beam), and panelized floor systems.

• *Conventional systems.* Conventionally framed platform systems provide a flexible method of floor framing for a wide variety of designs. Floor joists are usually spaced 16″ (406 mm) apart and are supported by the side walls of the foundation and/or by girders. See Figure 28.1.

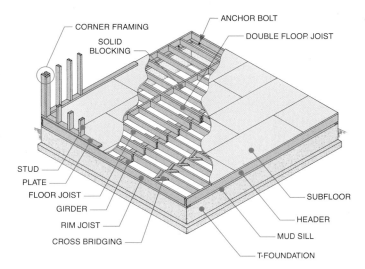

FIGURE 28.1 > Conventional platform floor system. *Hepler/Wallach/Hepler © Cengage 2013*

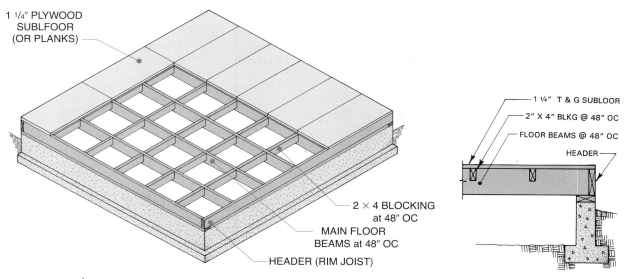

1 ¼" PLYWOOD
SUBLFOOR
(OR PLANKS)

1 ¼" T & G SUBLOOR

2" X 4" BLKG @ 48" OC

FLOOR BEAMS @ 48" OC

HEADER

2 × 4 BLOCKING
at 48" OC

MAIN FLOOR
BEAMS at 48" OC

HEADER (RIM JOIST)

FIGURE 28.2 〉 Panelized platform floor system. *Hepler/Wallach/Hepler © Cengage 2013*

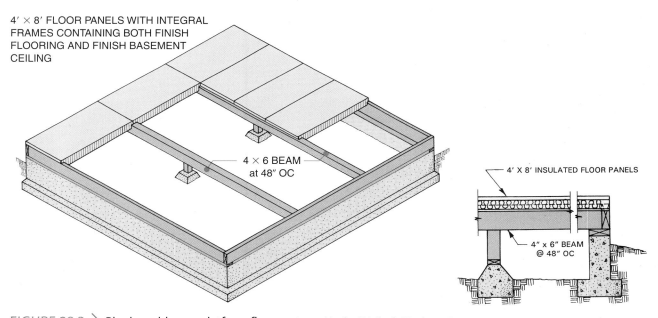

4′ × 8′ FLOOR PANELS WITH INTEGRAL
FRAMES CONTAINING BOTH FINISH
FLOORING AND FINISH BASEMENT
CEILING

4 × 6 BEAM
at 48" OC

4′ X 8′ INSULATED FLOOR PANELS

4" x 6" BEAM
@ 48" OC

FIGURE 28.3 〉 Plank-and-beam platform floor system. *Hepler/Wallach/Hepler © Cengage 2013*

- *Panelized systems.* **Panelized floor systems** are composed of preassembled sandwich panels. The panels are made from a variety of skin and core materials. Panelized systems are used for long, clear spans. The main advantage of panelized systems is the reduction of on-site construction costs. See Figure 28.2.

- *Plank-and-beam systems.* The **plank-and-beam floor system** method of framing uses fewer and larger members than conventional framing. Because of the increased size and rigidity of the larger members, longer distances can be spanned. Unlike conventional framing, no cross bridging is needed between joists. See Figure 28.3.

FLOOR FRAMING MEMBERS

Floor framing members for platform floors consist of decking, subfloor, finish floor, sills, headers, joists, and girders or beams. Supporting walls, posts, and columns also are part of the floor framing system.

Decking

Decking is the surface of a floor system. Decking usually consists of a subfloor and a finish floor, although in some plank systems these are combined. Subfloor decking materials range

FIGURE 28.4 > Aerial view of decking with fire pit. *Photo courtesy of Trex Company*

from wood boards and plywood sheets to concrete slabs or corrugated steel sheets. The finish floor may be wood, ceramic, vinyl, concrete, or carpeting.

Deck surfaces that are applied to an external structure, such as a porch or elevated deck, will usually not include a subfloor. Larger members (1 1/4″ to 1 5/8″ × 4″ to 8″) can be used to provide the extra strength and stability, as shown in Figure 28.4. Spacing between these members can be 1/8″ × 1/4″ to allow for drainage. If members are butted together a slope of 1/32″ per 1′-0″ is needed for adequate drainage where tongue-and-groove joints are necessary to avoid buckling. Figure 28.5 shows several optional interlocking designs. Engineered wood planks are also available in a variety of tongue-and-groove designs.

Wood Decking

Boards or plywood sheets used as subflooring are placed directly over the joists. See Figure 28.6. Unless the edges of plywood sheets are tongue and grooved,, the edges (joints) must be placed over a joist or blocking. **Blocking** consists of short pieces of lumber nailed between the joists. The blocking provides additional support for the subfloor joints. Sole plates (sills) for exterior and interior walls are laid directly on the subflooring.

The functions of the subfloor are to:

- Increase the strength of the floor and provide a surface for laying a finished floor.
- Help to stiffen the position of floor joists.
- Serve as a working surface during construction.
- Help deaden sound.
- Prevent dust from rising through the floor.
- Help insulate.
- Act as a buffer to soften and reduce the hard impact of slab floor construction. See Figure 28.7.

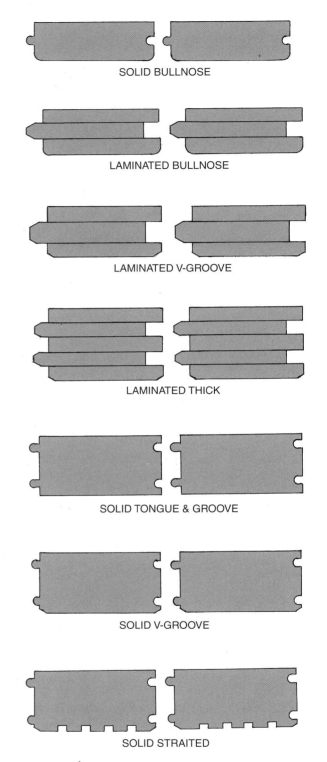

FIGURE 28.5 > Interlocking plank design. *Hepler/Wallach/Hepler © Cengage 2013*

Finish flooring is installed over the subfloor. The finish floor provides a wearing surface over the subfloor. If there is no subfloor, the finished floor must be tongue-and-groove boards 1 1/2″ to 2″ thick. Hardwoods such as oak, maple, beech, and birch are used for finish floors over wood subfloors. Vinyl, ceramic tile, marble, and carpeting are also used.

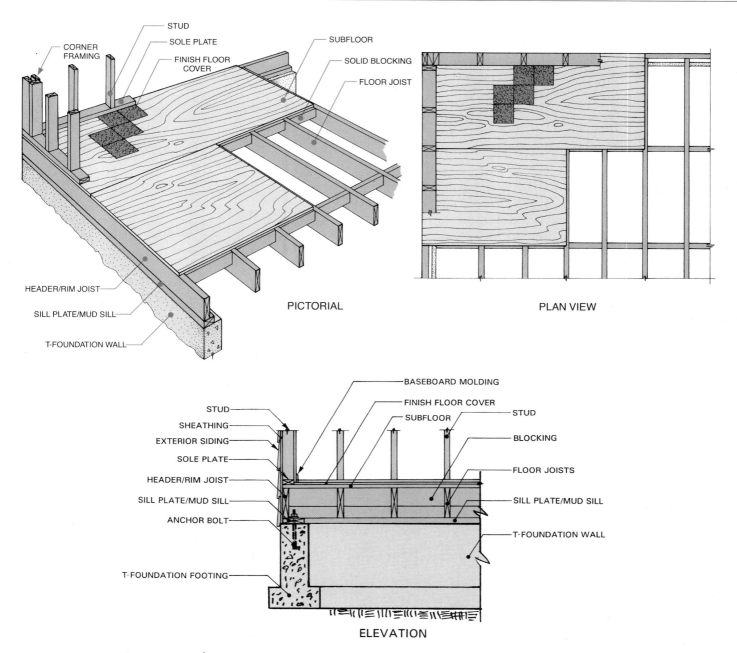

FIGURE 28.6 > Plywood subfloor system. *Hepler/Wallach/Hepler © Cengage 2013*

Some wood-framing members are treated with preservatives that release toxic vapors, as described in Chapters 6 and 24. Treated members should only be used on exterior structures. Although most toxic structural materials are now removed during processing, some floor finishing products do contain toxic substances. These should only be specified for use in highly ventilated areas or substituted with natural finishing materials.

Steel and Concrete Decking

Steel decks for floors (and roofs) use corrugated sheets, interlocking galvanized steel panels, or cellular units over steel beams. As shown in Figure 28.8, steel deck details or sectional drawings are usually prepared to show the relationship of the decking to the structural support members. When steel subfloors are used, they are usually constructed of corrugated sheet steel. These subfloors act as platform surfaces during construction and also provide the necessary subfloor surface for a concrete slab floor. Steel subfloors are not always needed with concrete slabs. The concrete slabs can function as subfloors if the concrete floor is precast with reinforcement bars.

Joists

Floor joists are horizontal members that rest on a foundation wall and/or girder (beam) and support the floor decking.

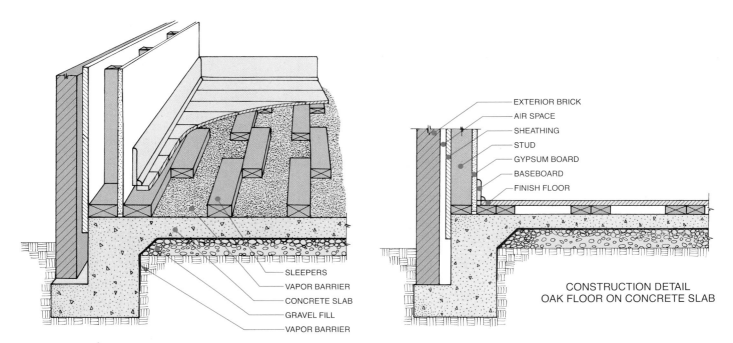

EXTERIOR BRICK
AIR SPACE
SHEATHING
STUD
GYPSUM BOARD
BASEBOARD
FINISH FLOOR

CONSTRUCTION DETAIL
OAK FLOOR ON CONCRETE SLAB

SLEEPERS
VAPOR BARRIER
CONCRETE SLAB
GRAVEL FILL
VAPOR BARRIER

FIGURE 28.7 > Subfloor sleepers on a concrete slab. *Hepler/Wallach/Hepler © Cengage 2013*

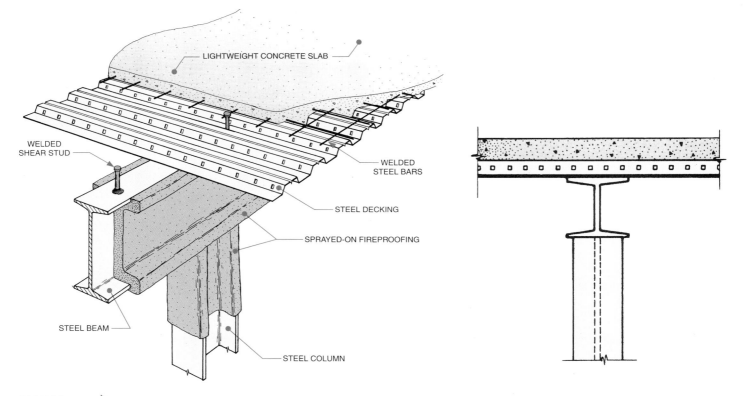

LIGHTWEIGHT CONCRETE SLAB

WELDED
SHEAR STUD

WELDED
STEEL BARS

STEEL DECKING

SPRAYED-ON FIREPROOFING

STEEL BEAM

STEEL COLUMN

FIGURE 28.8 > Steel deck details. *Hepler/Wallach/Hepler © Cengage 2013*

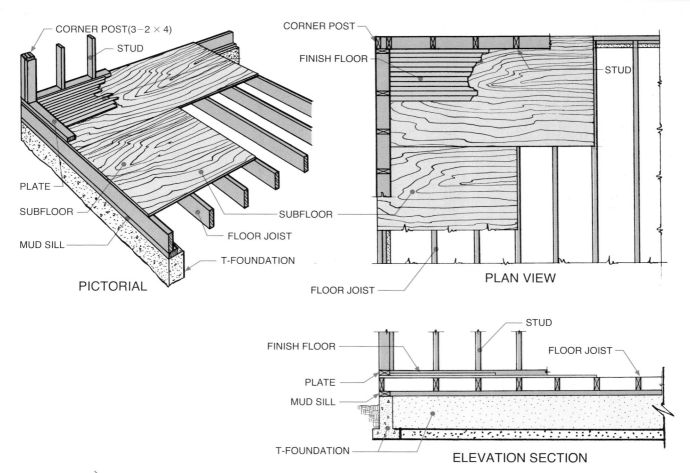

FIGURE 28.9 〉 Solid lumber joists in a conventionally framed floor system. *Hepler/Wallach/Hepler © Cengage 2013*

Floor joists must support the maximum live and dead load of the floor. Many variations of wood and steel joists are manufactured.

Solid Lumber Joists

Conventional residential framing normally uses solid lumber joists. Solid lumber joists are most commonly made from Douglas fir, pine, spruce, or hemlock. All joists must rest directly on a girder, beam, wall, or foundation wall. See Figure 28.9.

Double joists, known as **trimmers,** are used under partitions (interior walls) that are parallel to the floor joist. This provides additional support. Usually a small space is allowed between these joists to provide a channel for electrical or piping access. See Figure 28.10. Where joists are to be level with the top of a girder, as shown in Figure 28.11, they should rest on a ledger board attached to the girder or hung on joist hangers. Trimmers are also used for framing openings for stairwells, chimneys, and skylights.

Joists are often intersected at right angles to reduce spans. At these intersections, double joists and joist hangers are used to attach joists to trimmers or beams. See Figure 28.12. Wood cross bracing is seldom used; however, solid bridging is required by some codes as a firestop. To prevent drift and warp and to distribute loads more evenly, **bridging** is used between joists. See Figure 28.13.

Wherever joists need to be cut for an opening, such as a stairwell, chimney, or hearth, it is necessary to provide trimmers and auxiliary joists called **headers.** Headers are placed at right angles to trimmers to support the ends of joists that are cut. A header cannot be of a greater depth than a joist. Therefore, headers are usually the same width as floor joists and are doubled (placed side by side) to compensate for additional loads. See Figure 28.14. Double headers and trimmers are also used for floor, ceiling, and roof openings around fireplaces, chimneys, stairwells, and skylights. Figure 28.15A, shows a typical chimney opening through a floor structure. Note that all openings are double framed on all sides.

The size, spacing, and strength needed for joists depend on the loads acting on a floor. (Review Chapter 22 and see Appendix B for the physical effects of loads, material strength, member size, and spacing.) Standard joist sizes for most residential construction range from 2″ × 6″ to 2″ × 14″. Normal

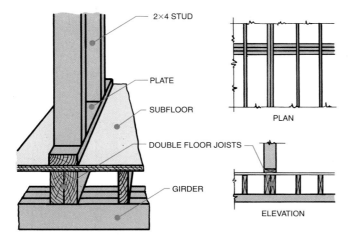

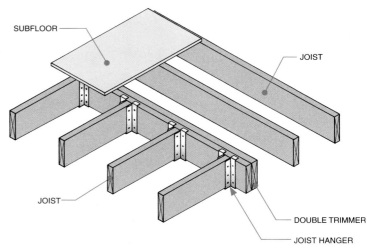

FIGURE 28.12 > Joist hangers used to change joist direction. *Hepler/Wallach/Hepler © Cengage 2013*

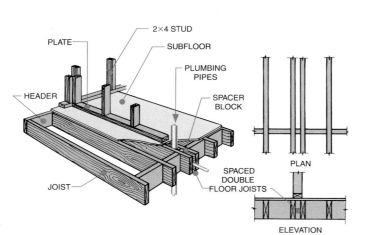

FIGURE 28.10 > Trimmer detail drawings. *Hepler/Wallach/ Hepler © Cengage 2013*

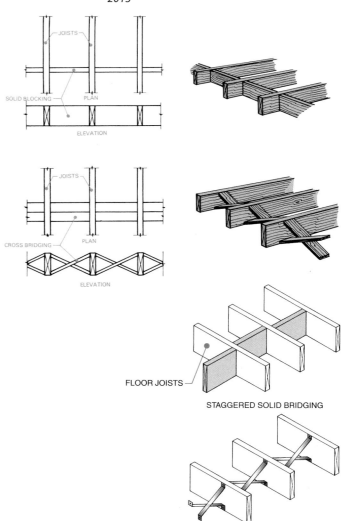

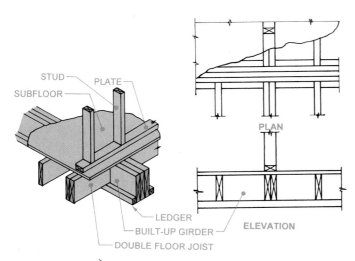

FIGURE 28.11 > Double joists aligned with girder top. *Hepler/Wallach/Hepler © Cengage 2013*

FIGURE 28.13 > Types of joist bridging. *Hepler/Wallach/Hepler © Cengage 2013*

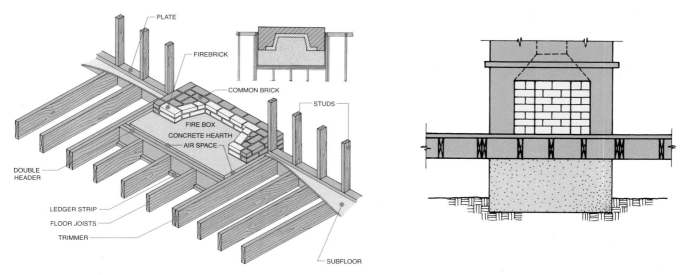

FIGURE 28.14 ❯ Trimmers and headers around a fireplace structure. *Hepler/Wallach/Hepler © Cengage 2013*

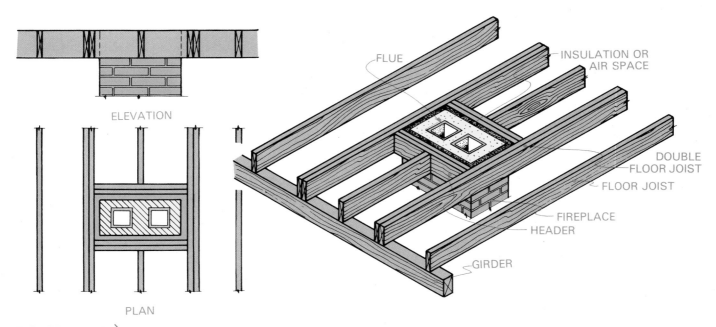

FIGURE 28.15A ❯ Trimmers and headers used around chimney opening. *Hepler/Wallach/Hepler © Cengage 2013*

residential spacing of wood joists is from 12″ to 24″ OC (on center). The most common spacing is 16″ OC.

Floor framing systems are horizontal planes. Floor loads must be transferred horizontally to vertical supports such as columns or bearing walls. To support both fixed (dead) loads and moving (live) loads, a floor system must be rigid yet elastic enough to absorb bending stresses.

Live loads bear directly on the decking and joists. Therefore, the total live load for the room containing the heaviest furniture and the heaviest traffic should be used to compute the total load for the entire floor.

To find the live load in pounds per square foot, divide the total room load in pounds by the number of square feet supporting the load. The average live load for most residences is between 40 and 50 pounds per square foot. See Appendix B, *Mathematical Calculations,* for calculating loads.

Engineering tables are used to select sizes and spacing for structural members of standard grade wood. Figure 28.15B is a table used to select joist sizes and spacing. To use this table, first select the wood group. Numbers indicate the quality of lumber. Number 1 is top quality, and number 4 is lowest quality. Next select the shortest span the joists must cross.

ALLOWABLE SPANS FOR FLOOR JOISTS USING NONSTRESS-GRADED LUMBER

SIZE OF FLOOR JOISTS (INCHES)	SPACING OF FLOOR JOISTS (INCHES)	MAXIMUM ALLOWABLE SPAN (FEET AND INCHES)							
		GROUP I		GROUP II		GROUP III		GROUP IV	
		PLASTERED CEILING BELOW	WITHOUT PLASTERED CEILING BELOW	PLASTERED CEILING BELOW	WITHOUT PLASTERED CEILING BELOW	PLASTERED CEILING BELOW	WITHOUT PLASTERED CEILING BELOW	PLASTERED CEILING BELOW	WITHOUT PLASTERED CEILING BELOW
2 × 6	12	10-6	11-6	9-0	10-0	7-6	8-0	5-6	6-0
	16	9-6	10-0	8-0	8-6	6-6	7-0	5-0	5-0
	24	7-6	8-0	6-6	7-0	5-6	6-0	4-0	4-0
2 × 8	12	14-0	15-0	12-6	13-6	10-6	11-6	8-0	8-6
	16	12-6	13-6	11-0	11-6	9-0	10-0	7-0	7-6
	24	10-0	11-0	9-0	9-6	7-6	8-0	6-0	6-6
2 × 10	12	17-6	19-0	16-6	17-6	13-6	14-6	10-6	11-6
	16	15-6	16-6	14-6	15-6	12-0	13-0	9-6	10-0
	24	13-0	14-0	12-0	13-0	10-0	10-6	7-6	8-6
2 × 12	12	21-0	23-0	21-0	21-6	17-6	19-0	13-6	14-6
	16	18-0	20-0	18-0	19-6	15-6	16-6	12-0	13-0
	24	15-0	16-6	15-0	16-6	12-6	13-6	10-0	16-6

FIGURE 28.15B ❯ Engineering tables used to select size and spacing of floor members. *Hepler/Wallach/Hepler © Cengage 2013*

GIRDER SIZE	SUPPORTING WALLS	NO WALL SUPPORT
4 × 4	3'-6"	4'-0"
	3'-0"	3'-6"
4 × 6	5'-6"	6'-6"
	4'-6"	5'-6"
4 × 8	7'-0"	8'-6"
	6'-0"	7'-6"

FIGURE 28.16 ❯ Allowable girder spans in light construction. *Hepler/Wallach/Hepler © Cengage 2013*

The joist size and spacing (distance between members) is shown on the left. For example, for group 1, if the shortest span is 12′-0″, then the smallest joist that can be used is a 2″ × 8″ at 16″ OC, which has a maximum span of 12′-6″. Figure 28.16 shows the normal girder spans for standard grade wood, with and without supporting walls. Always remember: As the size, spacing, and load vary, the spans must vary accordingly. If the span is changed, the joist and girder spacing must also change. You should not overdesign but you MUST NOT underdesign.

Floor structures can be cantilevered perpendicular to joists (Figure 28.17) or parallel to joists as shown in Figure 28.18. For parallel cantilevered joists, overhangs are usually one-third the length of the floor stringers.

Floor Truss Joists

Truss joists have long, usually parallel, top and bottom chords connected by shorter pieces called *webs*. Triangular web patterns give truss joists the ability to span long distances, and they weigh less than solid lumber. Truss joists have open spaces through which plumbing and electrical lines can be placed. They can also be designed to accommodate heating and air-conditioning ducts. Truss joists are manufactured using stress grade lumber or a combination of wood, metal, and/or composite materials. See Figures 28.19 and 28.20. Figure 28.21 shows 12 configurations of truss joist designs and intersections with different wall supports.

Floor I-Joists

Structurally, **I-joists** are similar to steel I-beams (S-beams). The weight of I-joists weight is reduced through the use of thin webs without sacrificing the strength and stability provided by

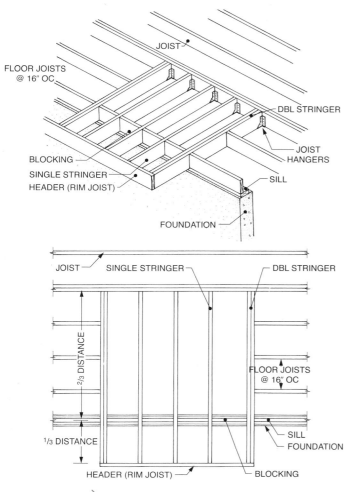

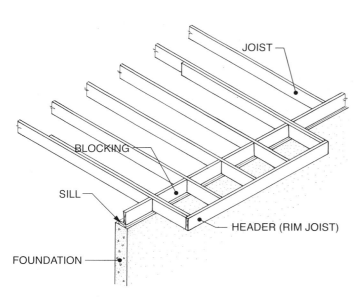

FIGURE 28.17 > Floor system cantilevered perpendicular to joists. *Hepler/Wallach/Hepler © Cengage 2013*

FIGURE 28.18 > Floor system cantilevered parallel to joists. *Hepler/Wallach/Hepler © Cengage 2013*

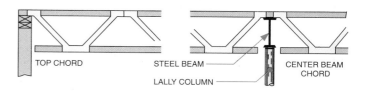

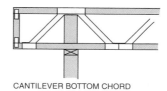

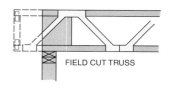

FIGURE 28.19 > Fabricated truss joists with lumber flanges (chords) and steel webs. *Hepler/Wallach/Hepler © Cengage 2013*

FIGURE 28.20 > Use of truss joists. *Hepler/Wallach/Hepler © Cengage 2013*

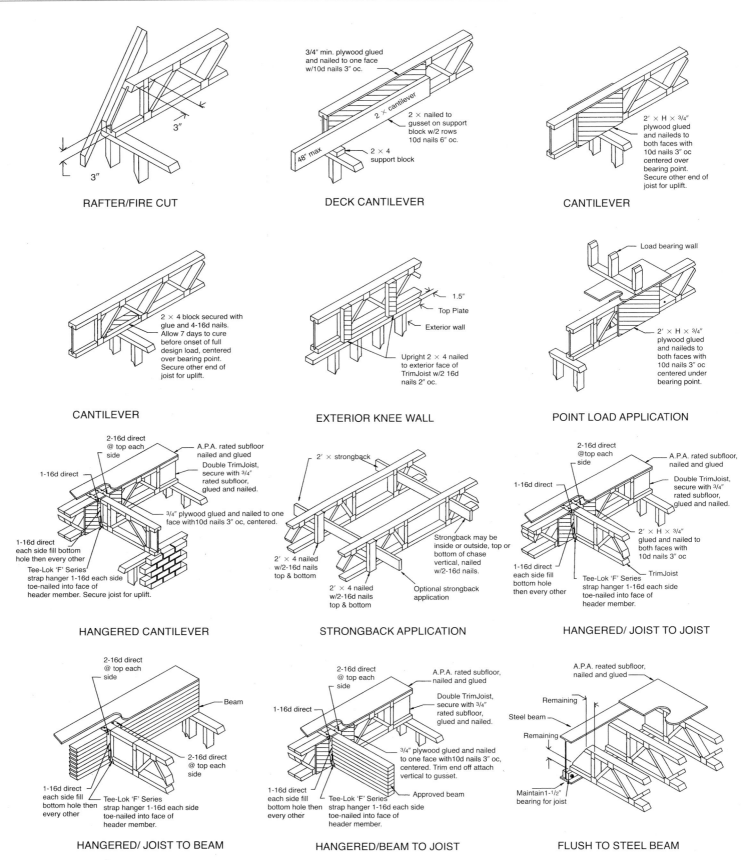

RAFTER/FIRE CUT

DECK CANTILEVER

CANTILEVER

CANTILEVER

EXTERIOR KNEE WALL

POINT LOAD APPLICATION

HANGERED CANTILEVER

STRONGBACK APPLICATION

HANGERED/ JOIST TO JOIST

HANGERED/ JOIST TO BEAM

HANGERED/BEAM TO JOIST

FLUSH TO STEEL BEAM

FIGURE 28.21 ⟩ **Truss joist configurations.** *Trimjoist*

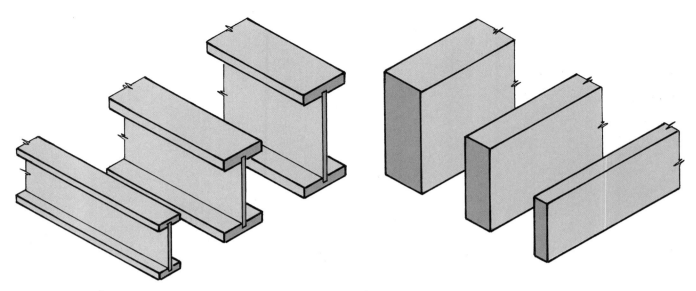

FIGURE 28.22 > I-joists compared to solid joists. *Hepler/Wallach/Hepler © Cengage 2013*

the flanges. See Figure 28.22. I-joist webs are usually made of stranded wood. The flanges are made of laminated or solid lumber. The web area can be cut, within limits, to receive HVAC, plumbing, or electrical lines without sacrificing strength. Figure 28.23 shows the use of I-joists with laminated headers and beams in a platform floor system.

Laminated Joists

Laminated members are made by bonding layers of material together. Solid and parallel strand lumber are used to manufacture laminated joists, headers, and beams that can span up to 60'. Laminated joists are extremely straight, dimensionally stable, and without checks, cracks, or twists.

Laminated joists can be positioned and fastened in a manner identical to that used for solid lumber, but using different bearing length requirements. Some design restrictions may occur because some laminated members cannot be cut, notched, or drilled. Nail and screw placement is also limited.

Steel Joists

Joists made of steel are either *bent sheet steel* or *open web* joists. See Figure 28.24. Open web steel joists are more common than the bent sheet type and consist of angles and bars welded into truss shapes. Open web steel joists or bar joists are actually trusses fabricated with angles and steel bars welded together. In addition to floor framing, open web steel trusses are used extensively for roof framing. See Figure 26.13B shows a typical application of steel columns, beams, joists, and fabric components in a floor system.

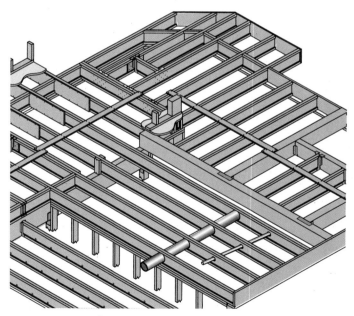

FIGURE 28.23 > I-joists in a platform system. *Hepler/Wallach/ Hepler © Cengage 2013*

The three types of steel open web joists are short-span, long-span, and deep long-span joists. Because joists are closely spaced, one note is used on drawings to give the number, classification, spacing, and length of all joists in a series. Only the first few joists in a series are usually noted. For example, if there are eight short-span joists spaced at 3' intervals over a 24' distance, the note should read 8SP @ 3'-0" = 24'-0". In addition, a notation is placed on a line representing the joists' direction and includes the length, class of the joist, and load range.

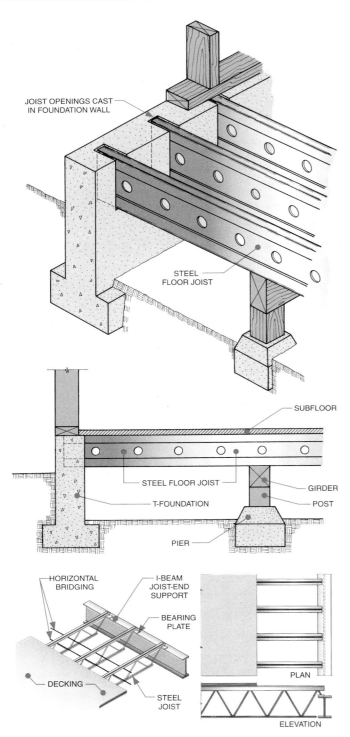

FIGURE 28.24 > Types of steel joists. *Hepler/Wallach/Hepler © Cengage 2013*

Girders and Beams

As explained previously, girders are the largest horizontal support members and rest on columns, posts, or exterior walls. In heavy construction, beams are the members that span the distances between girders. In residential timber construction, the terms *beam* and *girder* are often used interchangeably.

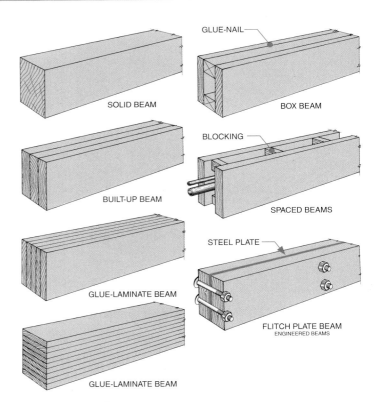

FIGURE 28.25 > Types of beams or girders and connectors. *Hepler/Wallach/Hepler © Cengage 2013*

Wood Girders and Beams

Girders (or beams) used in wood construction are either built up from solid lumber, laminated, or fabricated as shown in Figure 28.25. Girders are connected to joists with a variety of connectors. Figure 28.26 shows how wood girders (beams) are used to support joists. Note how joist hangers, scabs, and notches are used to maintain a constant height (flush) of intersecting joists. Figure 28.27 shows the use of a lally (pipe) column, built-up wood girder, and plate to support floor joists. Floor joists may be positioned on top of a girder or hung from the side of a girder with joist hangers.

When two or more girders are placed end to end to span the distance between outside supports, the joints between the girders must be placed directly over supporting columns or posts. Built-up girder members must be overlapped. Heavy timber girders should be half-lapped over columns. Members can be spliced to reduce compression, tension, bending, and torque forces. Figure 28.28 shows various types of **splices.** Second-floor and higher level girders are supported by bearing partitions or by columns aligned with lower level columns. See Figure 28.29.

Post-and-beam floor systems have no joists. In this system, the girders and blocking perform the function of joists. Girders rest directly on posts, and the subflooring rests directly on girders. In this type of construction, girders need to be spaced more closely together than girders that support joists. See Figure 28.30.

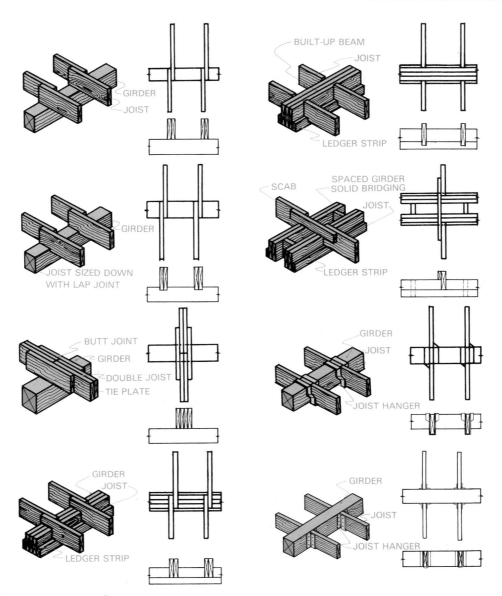

FIGURE 28.26 〉 Use of wood girders and connectors to support joists. *Hepler/ Wallach/Hepler © Cengage 2013*

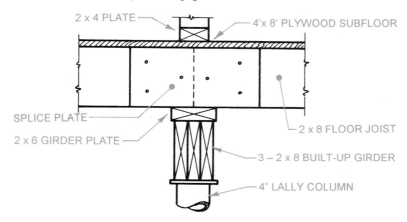

SECTION THROUGH GIRDER

FIGURE 28.27 〉 Built-up wood girder intersection with joist and column. *Hepler/Wallach/Hepler © Cengage 2013*

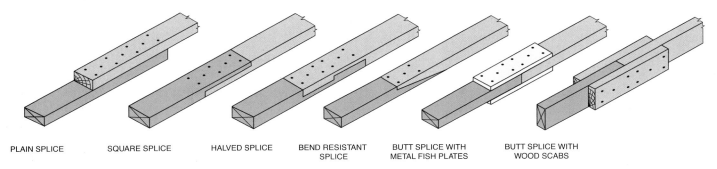

PLAIN SPLICE   SQUARE SPLICE   HALVED SPLICE   BEND RESISTANT SPLICE   BUTT SPLICE WITH METAL FISH PLATES   BUTT SPLICE WITH WOOD SCABS

FIGURE 28.28 > Splices used to reduce structural forces. *Hepler/Wallach/Hepler © Cengage 2013*

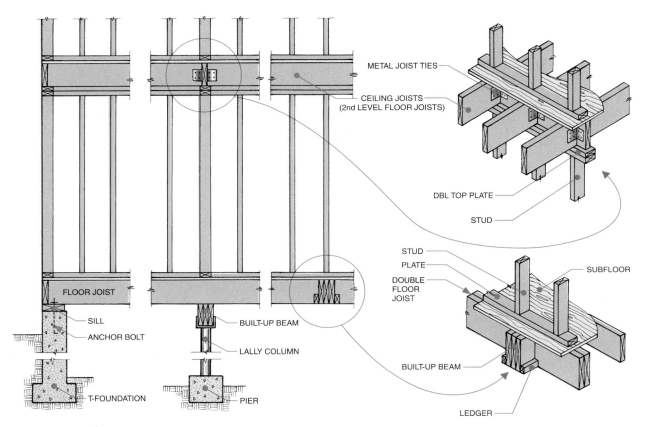

FIGURE 28.29 > Beam and bearing wall support of first- and second-floor framing systems. *Hepler/Wallach/Hepler © Cengage 2013*

Steel Girders and Beams

Steel girders and beams may be solid steel (S-shape or W-shape) or built-up steel assemblies. They are bolted or welded together to form the major structural element of steel cage construction. See Figure 28.31. In wood construction, steel girders may be used in addition to or in place of wood girders. Figure 28.32 shows the use of a steel I-beam and lally column to support floor joists. Review Chapter 26 for more details on steel girders and beams.

FLOOR FRAMING PLANS

If a set of architectural drawings does not include a separate floor framing plan, the builder, not the designer, determines the framing design. Floor framing plans for wood framing and steel framing use different conventions and symbols.

Floor Framing Plans on CAD

Floor framing plans are prepared on a separate layer using the corresponding floor plan layer as the base drawing. The joist framing can be represented as either a single-line or a double-line drawing. Use the *Line* command to create both, but note that the *Offset* command is used to create the second line of the joist. The *Copy-Multiple* command can be used to space the joists evenly. Interruptions for such features as stairwells and chimney openings can be deleted by identifying the segment to be removed and using the *Trim* command.

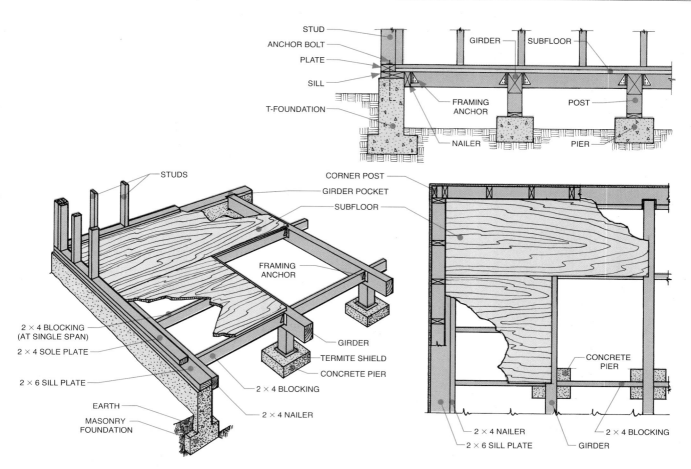

FIGURE 28.30 > Post-and-beam floor support system. *Hepler/Wallach/Hepler © Cengage 2013*

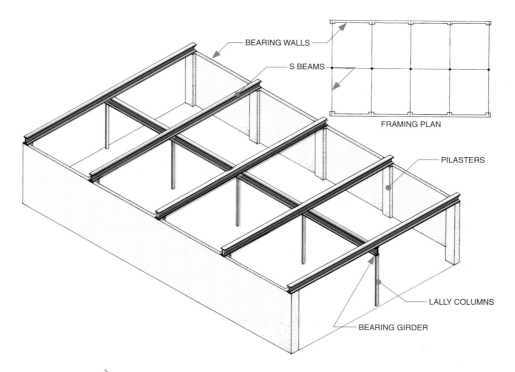

FIGURE 28.31 > Girders, beams, and columns used in steel cage construction. *Hepler/ Wallach/Hepler © Cengage 2013*

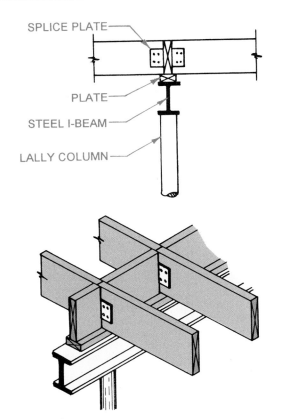

SPLICE PLATE

PLATE

STEEL I-BEAM

LALLY COLUMN

FIGURE 28.32 ⟩ Steel I-beam support detail. *Hepler/Wallach/Hepler © Cengage 2013*

Floor Framing Plans for Wood

Floor framing plans for wood structures range from simple to very detailed, as shown in Figure 28.33. In some cases, the direction of joists and girders may simply be shown on the floor plan (Figure 28.33A). In the most detailed floor framing plans, each structural member is represented by a double line to show exact thicknesses (Figure 28.33B).

The more simplified plan (Figure 28.33C) is a short-cut method of drawing floor framing plans. A single line is used to designate each member. Chimney and stair openings are shown by diagonals. Only the outline of the foundation and post locations is shown.

The abbreviated floor framing plan (Figure 28.33D) simply shows the entire area where the uniformly distributed joists are placed. The direction of joists is shown with arrows, and notes indicate the size and spacing of the joists. This type of framing plan is usually accompanied by numerous detail drawings.

While floor framing may be shown on drawings, the method of cutting and fitting the subfloor and finished floor panels is usually left up to the builder. When off-site or mass-produced floor systems are to be installed, a floor panel layout may be prepared. Its purpose is to ensure maximum use of materials and minimum waste.

Second-Floor Framing Plans

Floor framing details for second floors or above are usually shown with a full section through an exterior wall. For platform framing construction, second-floor joists are placed on a top plate and the subfloor is located on top of the second floor joist, as shown in Figure 28.34.

When a combination of exterior covering materials is used, the relationship between the floor system and the exterior wall is shown on an elevation section. See Figure 28.35.

If an upper-level floor is cantilevered over the first floor, the second-floor joists are either parallel or perpendicular to the first-floor top plate that supports the second floor. See Figure 28.36. When the joists are parallel to the joists, the construction is simple. When the joists are perpendicular to the joists, **lookout joists** must be used to support the cantilevered second-floor extension.

Details

Most floor framing plans are easily interpreted by experienced builders. Some plans may require additional details to explain a construction method. Details are drawn to eliminate the possibility of error in interpretation or to explain a unique condition. Details may be enlargements of what is already on the floor framing plan. They may be prepared for dimensioning purposes, or they may show a view from a different angle for better interpretation. On the floor framing plan in Figure 28.37, the circles indicate areas for which detail drawings have been made.

Sill Support   The sill is the transition between the foundation and the exterior walls of a structure. Drawings that show sill construction details reveal not only the construction of the sill, but also the method of attaching the sill to the foundation. A sill detail is included in most sets of architectural plans. Some sill details are drawn in pictorial form. Pictorial drawings are easy to interpret but are difficult and time consuming to draw and dimension, and they are not orthographically accurate. Therefore, most sill details are prepared in two-dimensional sectional form. Figure 28.38 shows a detail drawing in plan, elevation, and pictorial form.

The floor area in a sill detail sectional drawing usually shows at least one joist. This is done to show the direction of the joist and its size and placement in relation to the placement of the subfloor and finished floor. For example, the floor framing plan in Figure 28.38 on page 592, is needed to indicate the spacing of joists and studs. The elevation section shows the intersections between the foundation sill and exterior wall. For this reason, both a plan and an elevation are needed to more fully describe this type of construction.

Finished detail drawings include all dimensions and notations necessary for construction.

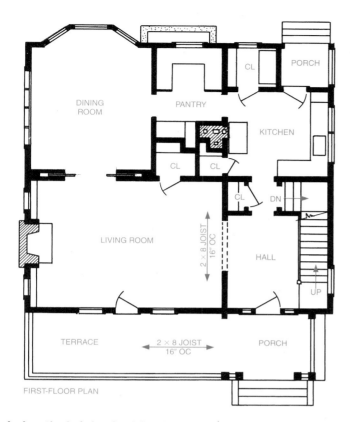

A. A method of showing joist direction on a floor plan.

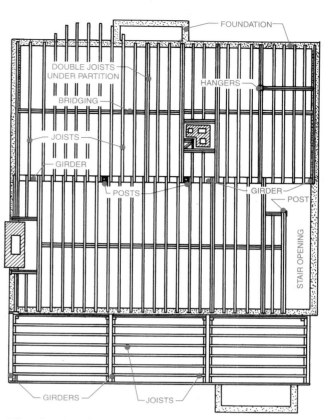

B. A floor framing plan showing material thickness with double lines.

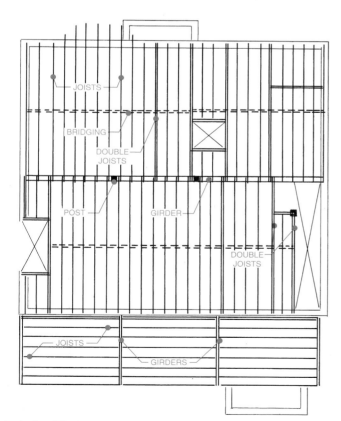

C. A simplified method of drawing floor framing plans with single lines.

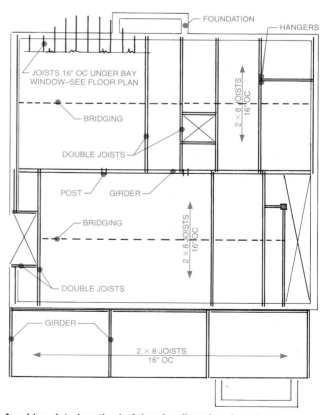

D. An abbreviated method of drawing floor framing plans.

FIGURE 28.33 ❭ Types of floor framing plans. *Hepler/Wallach/Hepler © Cengage 2013*

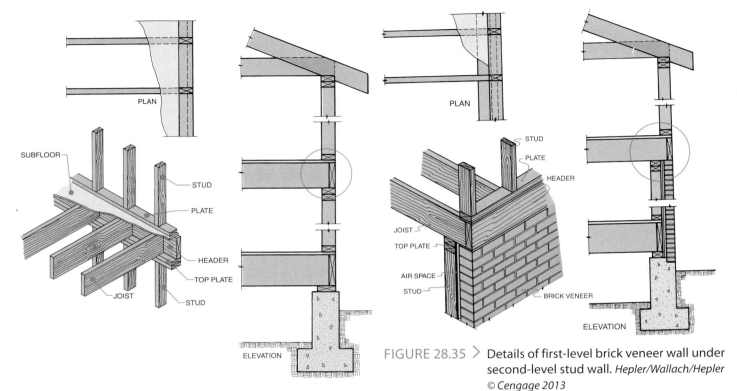

SUBFLOOR
STUD
PLATE
HEADER
TOP PLATE
JOIST
STUD

PLAN

ELEVATION

FIGURE 28.34 > Platform framing second-floor details.
Hepler/Wallach/Hepler © Cengage 2013

PLAN

STUD
PLATE
HEADER

JOIST
TOP PLATE

AIR SPACE
STUD
BRICK VENEER

ELEVATION

FIGURE 28.35 > Details of first-level brick veneer wall under second-level stud wall. *Hepler/Wallach/Hepler © Cengage 2013*

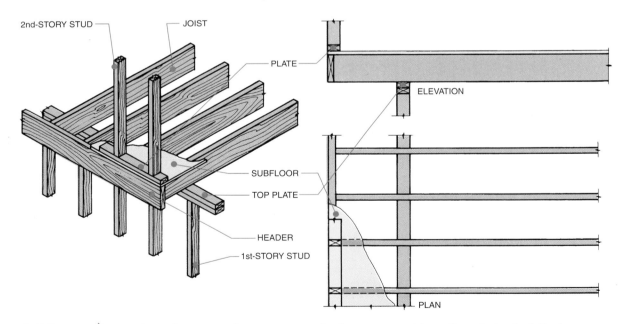

2nd-STORY STUD
JOIST
PLATE
ELEVATION
SUBFLOOR
TOP PLATE
HEADER
1st-STORY STUD
PLAN

FIGURE 28.36 > Cantilever framing with joists perpendicular to an outside wall. *Hepler/Wallach/Hepler © Cengage 2013*

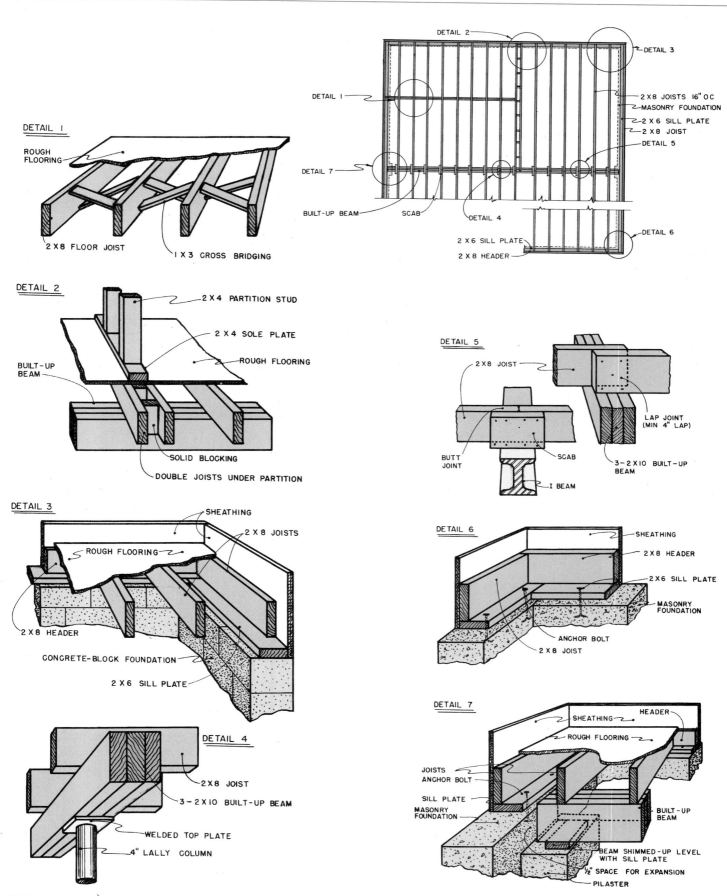

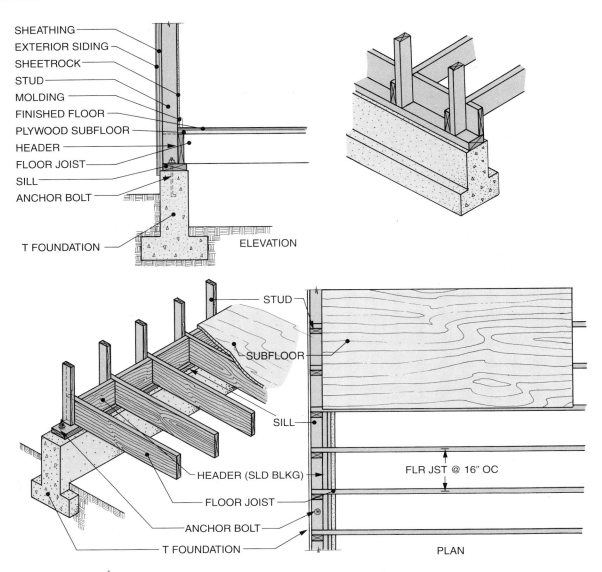

SHEATHING
EXTERIOR SIDING
SHEETROCK
STUD
MOLDING
FINISHED FLOOR
PLYWOOD SUBFLOOR
HEADER
FLOOR JOIST
SILL
ANCHOR BOLT

T FOUNDATION

ELEVATION

STUD

SUBFLOOR

SILL

HEADER (SLD BLKG)

FLOOR JOIST

ANCHOR BOLT

T FOUNDATION

FLR JST @ 16" OC

PLAN

FIGURE 28.38 > Sill framing details. *Hepler/Wallach/Hepler © Cengage 2013*

Sill details are also required to show how materials are joined, such as masonry, wood, precast concrete, and structural steel. Special design features are shown on details as well. For example, **firecuts** are necessary in masonry walls. See Figures 28.39 and 28.40. Sill details are also needed to show the intersection between girders and foundation walls. Figure 28.41 shows a box sill used to support a girder. Some details may be expanded to show additional details. See Figure 28.42.

Intermediate Support   If girders cannot safely span the distance between exterior supports, intermediate vertical supports (such as wood or steel columns, piers, or bearing walls) must be used to reduce the span. Detail drawings for intermediate supports consist of sections, elevations, or plan views to show the intersections between footings, vertical members, and horizontal girders. Intermediate framing details are

needed, for example, where level changes in floor level require special support. See Figure 28.43.

Headers and Trimmers   Building codes require headers and trimmers around stairwells, chimneys, and hearths and under heavy dead loads such as bathtubs, waterbeds, and masonry furniture. To ensure headers and trimmers are used correctly, detail drawings are prepared. For example, stairwell floor framing details show the position of joists, headers, and trimmers. Figure 28.44 shows these components; note, however, that ledger strips have mostly been replaced with joist hangers.

Stairwell Framing

Because stairwell openings must be precisely shown on the floor framing plan, the stair system must be designed before the floor framing plan is drawn. Figure 28.45 shows the

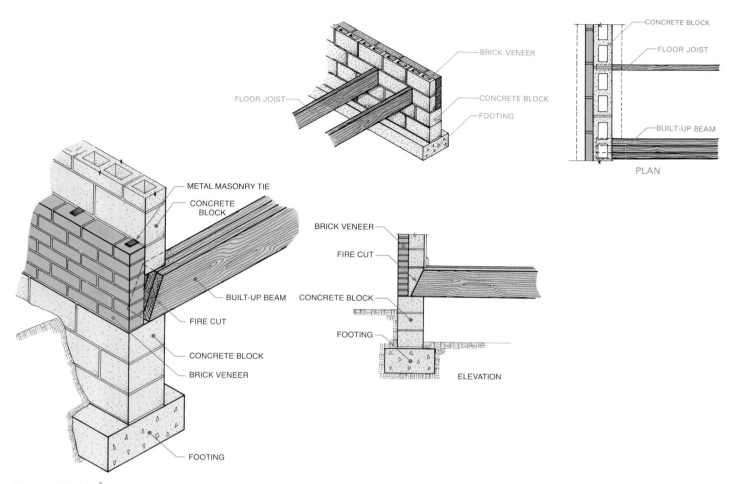

FIGURE 28.39 〉 Brick veneer sill details showing firecut beams. *Hepler/Wallach/Hepler © Cengage 2013*

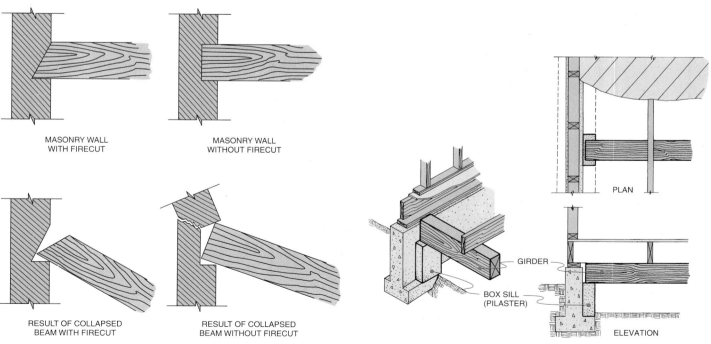

FIGURE 28.40 〉 Firecut used to prevent masonry collapse. *Hepler/Wallach/Hepler © Cengage 2013*

FIGURE 28.41 〉 Use of box sill to support girder. *Hepler/ Wallach/Hepler © Cengage 2013*

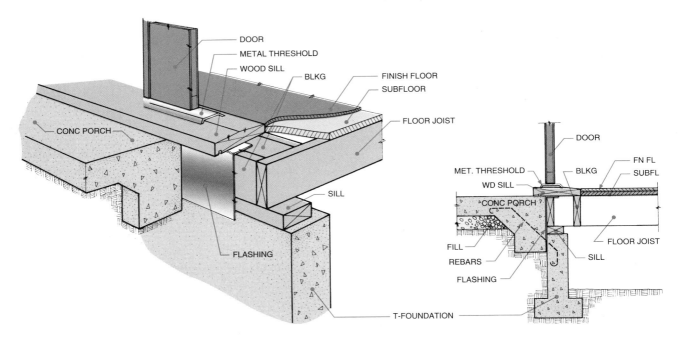

FIGURE 28.42 ⟩ Relationship of door sill to interior wood floor and exterior concrete porch slab. *Hepler/Wallach/ Hepler © Cengage 2013*

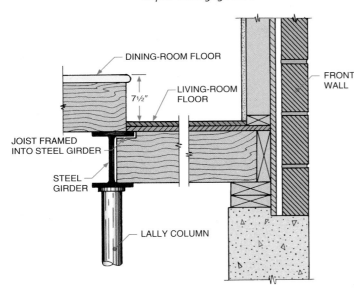

FIGURE 28.43 ⟩ Method of creating a one-step floor-level change. *Hepler/Wallach/Hepler © Cengage 2013*

floor framing plan and a matching pictorial view of a stair system. Elevation views of stairwell systems are covered in Chapter 29, *Wall Framing Drawings*.

FLOOR FRAMING PLANS FOR STEEL

Steel floor framing plans are similar to other plans, except the exact position of every column, girder, and beam is classified and dimensioned. Grid systems are used to identify the position of each member.

The **sequential column identification system** is shown in the top drawing in Figure 28.46. In this system steel columns are identified by numbering them in sequence from left to right and from the rear to the front.

The **grid column identification system** is shown in the bottom drawing in Figure 28.46. In this system column rows are numbered on the horizontal perimeter from left to right. Letters are used to identify column rows on the vertical perimeter from front to rear.

Floor framing systems for steel construction include girders, beams, joists, and decking materials. A separate floor framing plan is prepared for each floor of a multilevel building. Although the framing for many floors may be nearly identical, this cannot be assumed unless specified. Usually there are slight differences on each floor plan. For this reason CAD layering is especially ideal for preparing high-rise structural steel floor framing plans.

Layering (or pin graphics) as used in conventional drafting allows the drafter to draw a base floor plan and make specific floor changes without redrawing each floor separately. This is done by drawing each floor plan on clear acetate. Drawings are aligned in layers with registration pins. The same spacing must be used between grid lines to ensure alignment of columns and other vertically oriented features, such as stairwells, plumbing lines, HVAC ducts, and electrical conduits. Each floor framing plan shows the position of each column that passes through the floor.

In drawing steel floor framing plans, major members, such as girders and beams, are shown with a solid heavy line with the identifying notation placed directly on or under the line. The length of each line represents the length of each member.

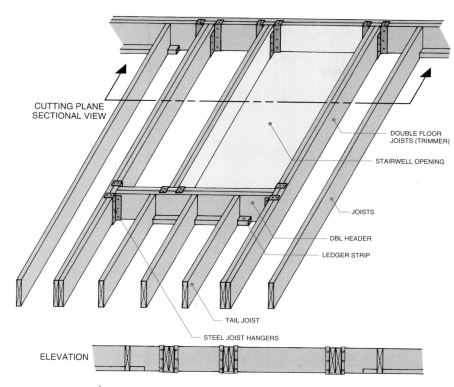

FIGURE 28.44 > Stairwell opening detail. *Hepler/Wallach/Hepler © Cengage 2013*

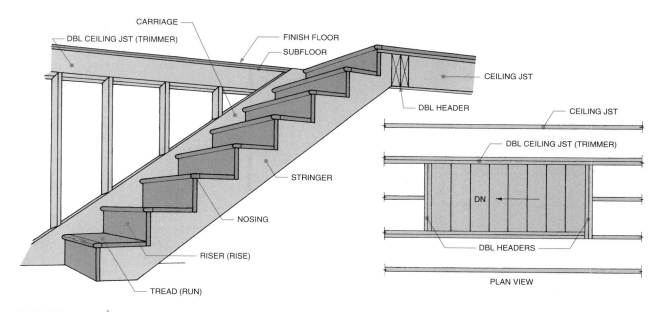

FIGURE 28.45 > Stairwell pictorial and plan view. *Hepler/Wallach/Hepler © Cengage 2013*

If a continuous beam passes over a girder, a solid unbroken line is drawn through the girder line. However, if the beam stops and is connected to the girder, the beam line is broken. See Figure 28.47. Remember that solid lines represent continuous members. Broken lines indicate that the member intersects or is under a continuous member.

Spaces created between rows of members in two directions are known as **bays.** See Figure 28.48.

Three methods of dimensioning are used on structural steel drawings. In the first method, a description of each member includes the length placed directly on each schematic line. The second method uses notations to show only the shop size (width) and weight. Dimension lines are used to show the position and length of each member. The third method uses a coding system that relates each member to a schedule containing all pertinent information. See Figure 28.49.

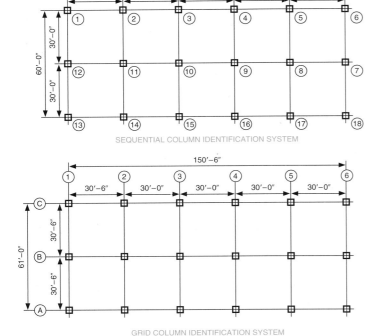

SEQUENTIAL COLUMN IDENTIFICATION SYSTEM

GRID COLUMN IDENTIFICATION SYSTEM

FIGURE 28.46 > The sequential column identification system (top) and the grid column identification system (bottom). *Hepler/Wallach/Hepler © Cengage 2013*

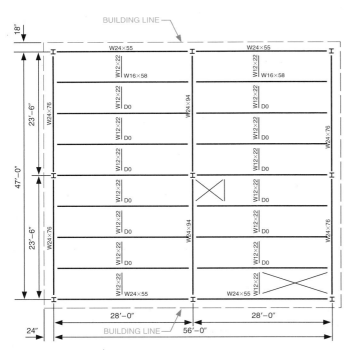

FIGURE 28.47 > Use of solid and broken lines to show locations and intersections of steel members. *Hepler/Wallach/Hepler © Cengage 2013*

Bays with framing plan.

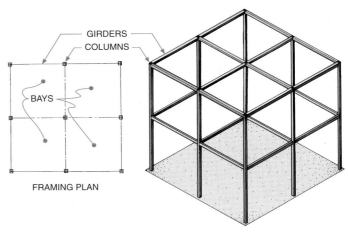

GIRDERS
COLUMNS
BAYS

FRAMING PLAN

Dimensioning of bay modules.

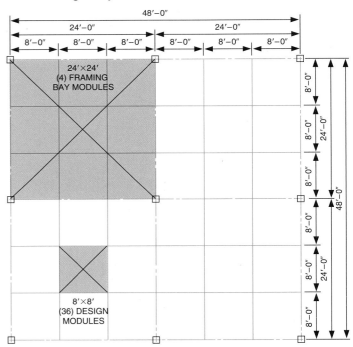

24'×24'
(4) FRAMING BAY MODULES

8'×8'
(36) DESIGN MODULES

FIGURE 28.48 > Use of construction bays in steel floor framing plans. *Hepler/Wallach/Hepler © Cengage 2013*

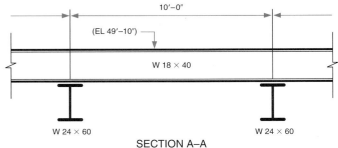

(EL 49'–10")

W 18 × 40

W 24 × 60   W 24 × 60

SECTION A–A
4th FLOOR ELEVATION
TOP OF BEAM ELEVATION SHOWN WITH (EL)

FIGURE 28.49 > Method of identifying steel beam size, type, and weight. *Hepler/Wallach/Hepler © Cengage 2013*

Floor Framing Drawings Exercises

1. Draw a simplified plan view of a floor system shown in this chapter.

2. Redesign a floor framing plan shown in this chapter with the joists aligned in the opposite direction.

3. Add a 12′ × 14′ loft to the upper left corner of a 30′ × 35′ room. Draw the outlines and show the position of joists and girders.

4. Develop a complete floor framing plan for one floor or as many floors as you are designing in your house plans.

5. Complete an abbreviated floor framing plan.

6. Draw the detail of a sill support in this chapter. Include callouts for all of the parts shown.

7. Draw a slab foundation for a 30′ × 40′ house. Install sleepers for a wood finish floor.

8. Draw the construction details for a concrete porch adjacent to a T-foundation.

9. Draw two different construction details showing the end support of a girder at a T-foundation wall.

10. Draw a floor framing plan for the plan shown in Figure 28.50.

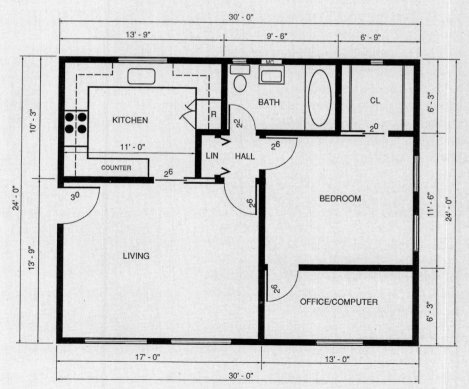

FIGURE 28.50 > Draw a floor framing plan for this design. *Hepler Associates LA, PC*

(2) 2 × 4 TOP PLATE
½" GYPSUM BOARD
1" × 4" BASEBOARD
1" SUBFLOOR
BUIL
¾" F
2'-0"
(2) 2
2 × 6
ATING
HING
2 × 4

CHAPTER 29
Wall Framing Drawings

OBJECTIVES

In this chapter you will learn to:

> draw an exterior wall framing elevation and plan.

> draw an interior wall framing elevation and plan.

> draw details and sections of walls.

> draw wall intersections.

TERMS

column schedules
cripple
curtain walls
diagonal bracing

drywall construction
framing elevation drawings
headers
masonry veneer walls

medal siding
rough openings
siding
stud layout

INTRODUCTION

Wall framing provides the base to which coverings, such as siding and drywall, are attached. Typically, exterior wall framing supports the roof and ceiling loads. In most designs, the interior walls also help support these loads. Wall framing drawings may consist of exterior and interior elevations, column and stud layouts, and details. Framing methods covered in this chapter adhere to most municipal building codes. Nevertheless, codes do vary somewhat throughout the country. Therefore, the framing examples shown in the following drawings also vary.

EXTERIOR WALLS

Exterior wall systems provide support, rigidity, and lateral stability to a structure through the use of posts, columns, studs, and shear panels. Exterior walls also function as a protective barrier to resist damage from wind, precipitation, and the sun. In addition exterior walls, with adequate insulating, mitigate excessive outside temperature and sound transmissions.

Exterior walls for most wood-frame residential buildings use either skeleton-frame or post-and-beam construction. The typical method of erecting walls for most buildings follows a conventional braced-frame system. Figure 29.1 shows the skeleton framing of a residence and the finished building. Prefabricated components have led to variations in the meth-

A. Skeleton frame of a house.

B. The completed house.

FIGURE 29.1 > A skeleton-frame structure (A) under construction and (B) completed. *Hepler/ Wallach/Hepler © Cengage 2013*

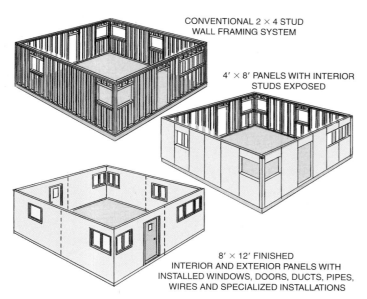

CONVENTIONAL 2 × 4 STUD WALL FRAMING SYSTEM

4′ × 8′ PANELS WITH INTERIOR STUDS EXPOSED

8′ × 12′ FINISHED INTERIOR AND EXTERIOR PANELS WITH INSTALLED WINDOWS, DOORS, DUCTS, PIPES, WIRES AND SPECIALIZED INSTALLATIONS

FIGURE 29.2 > Basic types of wall construction. *Hepler/ Wallach/Hepler © Cengage 2013*

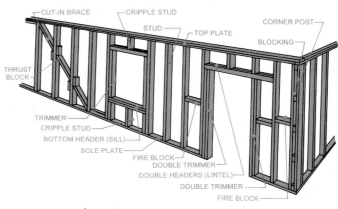

FIGURE 29.3 > Wall framing members. *Hepler/Wallach/Hepler © Cengage 2013*

ods of erecting walls. Manufactured panels range from a basic wall-frame panel to a completed wall that includes plumbing, electrical work, doors, and windows. See Figure 29.2. Whether a structure is prefabricated or field constructed, the preparation of exterior framing drawings is the same.

Framing Elevations

Exterior walls are best constructed when a framing elevation drawing is used as a guide. A wall framing elevation drawing is the same as an elevation of a building with the siding materials removed. Figure 29.3 shows the basic wood wall framing members included in framing elevations.

To draw framing elevations, project lines from floor plans and elevations, as shown in Figure 29.4. The elevation supplies all of the projection points for the horizontal framing members. The floor plan provides all of the points of projection for locating the vertical members. Because floor plan wall thicknesses normally include the thickness of siding materials, care should be taken to project the outside of the *framing* line to the framing elevation drawing and not the outside of the siding line.

When drawing door and window **rough openings** (framing openings), first check the sizes listed in manufacturing specifications or door and window schedules. Rough openings are

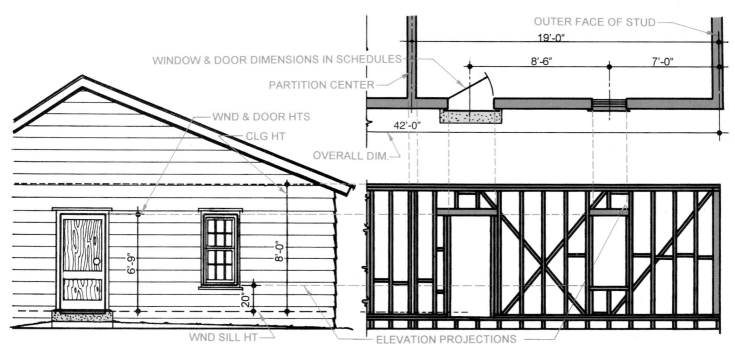

FIGURE 29.4 > Projections of a wall framing elevation. *Hepler/Wallach/Hepler © Cengage 2013*

slightly larger than the size of doors or windows. Then project the position of the top, bottom, and sides of the door and window openings from the floor plan and elevation. Figure 29.5 shows a typical example of exterior wall framing elevations.

In drawing steel-framed elevations, only the locations of structural members are included. See Figure 29.6. Information about coverings for curtain walls is shown on wall elevations and wall details.

In Figure 29.7A, the framing elevation is incorporated into a sectional drawing of the structure. This drawing shows an elevation section of the framing from the foundation-floor system to the roof construction. In a sectional drawing, any members intersected by a cutting-plane line, such as the joists, are indicated by crossed diagonals.

Wall Framing Drawings on CAD

Using an exterior or interior wall elevation layer as a guide, repetitive vertical wall framing members, such as studs, can be drawn on a superimposed layer using the *Copy-Multiple*

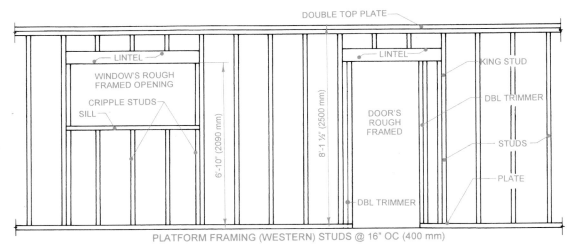

FIGURE 29.5 > Platform wall framing elevations. *Hepler/Wallach/Hepler © Cengage 2013*

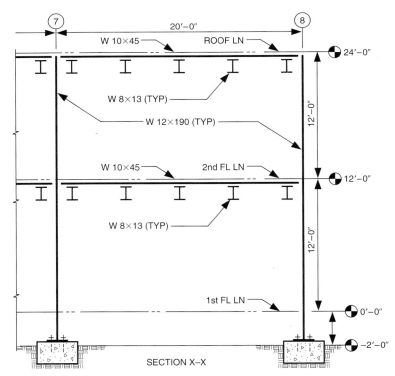

FIGURE 29.6 > Structural steel elevation. *Hepler/Wallach/ Hepler © Cengage 2013*

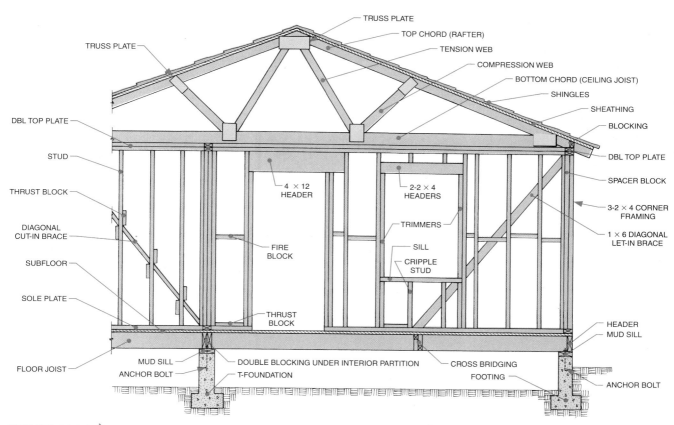

TRUSS PLATE
TOP CHORD (RAFTER)
TENSION WEB
COMPRESSION WEB
BOTTOM CHORD (CEILING JOIST)
SHINGLES
SHEATHING
BLOCKING
DBL TOP PLATE
SPACER BLOCK
3-2 × 4 CORNER FRAMING
1 × 6 DIAGONAL LET-IN BRACE
HEADER
MUD SILL
ANCHOR BOLT

TRUSS PLATE
DBL TOP PLATE
STUD
THRUST BLOCK
DIAGONAL CUT-IN BRACE
SUBFLOOR
SOLE PLATE
FLOOR JOIST
MUD SILL
ANCHOR BOLT
T-FOUNDATION
DOUBLE BLOCKING UNDER INTERIOR PARTITION
CROSS BRIDGING
FOOTING

4 × 12 HEADER
2-2 × 4 HEADERS
TRIMMERS
SILL
CRIPPLE STUD
FIRE BLOCK
THRUST BLOCK

FIGURE 29.7A > Complete wall framing elevation. *Hepler/Wallach/Hepler © Cengage 2013*

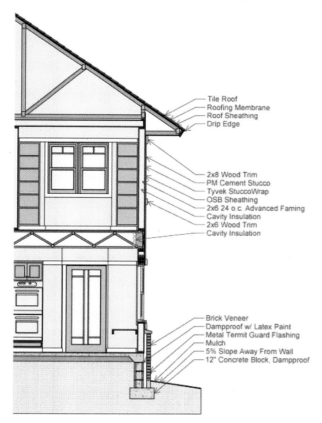

Tile Roof
Roofing Membrane
Roof Sheathing
Drip Edge

2x8 Wood Trim
PM Cement Stucco
Tyvek StuccoWrap
OSB Sheathing
2x6 24 o.c. Advanced Faming
Cavity Insulation
2x6 Wood Trim
Cavity Insulation

Brick Veneer
Dampproof w/ Latex Paint
Metal Termit Guard Flashing
Mulch
5% Slope Away From Wall
12" Concrete Block, Dampproof

FIGURE 29.7B > Combination exterior section and interior elevation. *Chieft Architect Software*

command. This is done using the *Single Line* or *Polyline* task as with floor framing plans. The *Trim* command can be used to eliminate framing members from openings for doors, windows, or chimneys. Lintels, jambs, and plates can then be added individually. Figure 29.7B combines an exterior wall section with a partial interior elevation drawing.

Bracing

To make wall frames rigid, structural lumber members are attached at an inclined angle. Members used for this purpose are labeled **diagonal bracing.** The bracing may be placed on the inside or outside of the wall, or between the studs. See Figure 29.8. Steel straps are often used for corner bracing, as shown in Figure 29.9.

Difficulties may occur in interpreting the true position of **headers, cripple** studs, plates, and trimmers. Figure 29.10 shows how to illustrate the position of these members on framing elevation drawings to ensure proper interpretation.

Panels

Panel elevations show the attachment of the sheathing panels to the framing. To show the relationship between the panel layout and the framing, panel drawings and framing

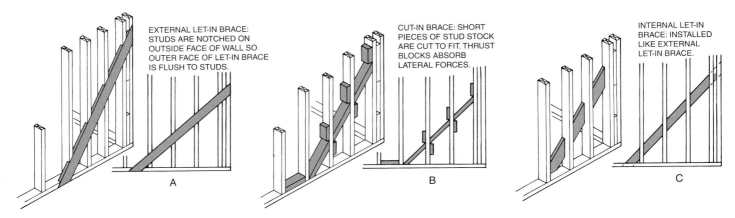

FIGURE 29.8 > Diagonal bracing on wall framing elevation. *Hepler/Wallach/Hepler © Cengage 2013*

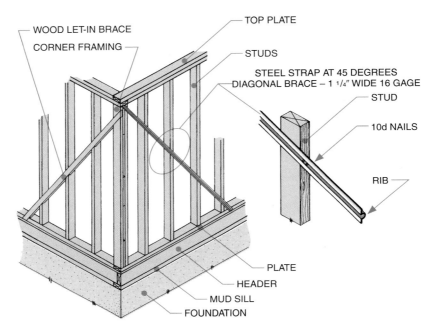

FIGURE 29.9 > Use of steel strap corner brace. *Hepler/Wallach/Hepler*
© Cengage 2013

drawings are usually combined in one drawing. Diagonal dotted lines indicate the position of the panels. When only the panel layouts are shown, the outline of the panels and the diagonals are drawn solid. See Figure 29.11.

Dimensions

Dimensions on framing elevations include overall widths, heights, and spacing of all studs. See Figures 29.12, 29.13A. Control dimensions for the sizes of all rough openings (RO) for windows are also indicated. If the spacing of studs does not automatically provide the rough opening necessary for windows, the RO width of windows must

also be dimensioned. Framing dimensions for a standard 8' ceiling height using standard 93" studs are shown in Figure 29.14.

The finished ceiling height will vary depending on the thickness of the finished floor and ceiling covering materials. A typical calculation of ceiling height is shown here:

Typical standard stud height		92 1/2"
Double top plate		3"
Bottom plate		1"
	Total	97"
Less the finish floor (1/2") and ceiling cover (1/2")		−1"
Typical finished ceiling height		96" = 8'-0"

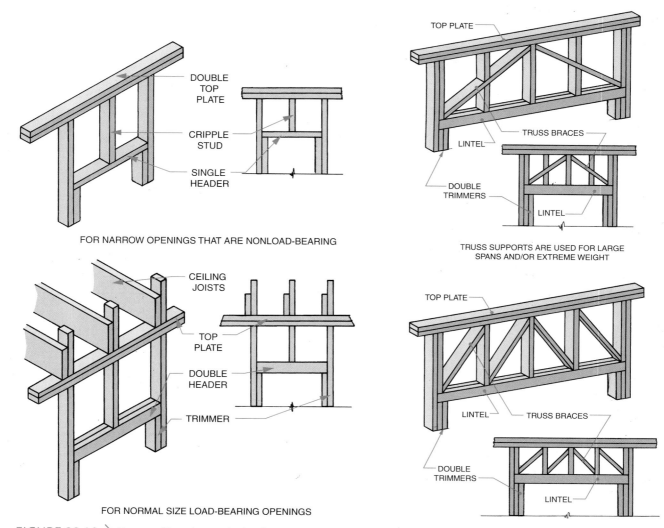

DOUBLE TOP PLATE

CRIPPLE STUD

SINGLE HEADER

FOR NARROW OPENINGS THAT ARE NONLOAD-BEARING

TOP PLATE

TRUSS BRACES

LINTEL

DOUBLE TRIMMERS

LINTEL

TRUSS SUPPORTS ARE USED FOR LARGE SPANS AND/OR EXTREME WEIGHT

CEILING JOISTS

TOP PLATE

DOUBLE HEADER

TRIMMER

FOR NORMAL SIZE LOAD-BEARING OPENINGS

TOP PLATE

LINTEL   TRUSS BRACES

DOUBLE TRIMMERS

LINTEL

FIGURE 29.10 > Types of header and cripple construction. *Hepler/Wallach/Hepler © Cengage 2013*

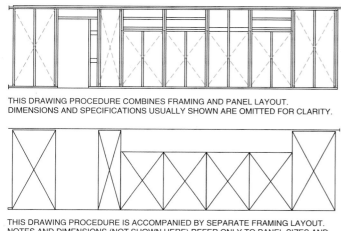

THIS DRAWING PROCEDURE COMBINES FRAMING AND PANEL LAYOUT. DIMENSIONS AND SPECIFICATIONS USUALLY SHOWN ARE OMITTED FOR CLARITY.

THIS DRAWING PROCEDURE IS ACCOMPANIED BY SEPARATE FRAMING LAYOUT. NOTES AND DIMENSIONS (NOT SHOWN HERE) REFER ONLY TO PANEL SIZES AND SPECIFICATIONS.

FIGURE 29.11 > Use of diagonal lines to show panel positions. *Hepler/Wallach/Hepler © Cengage 2013*

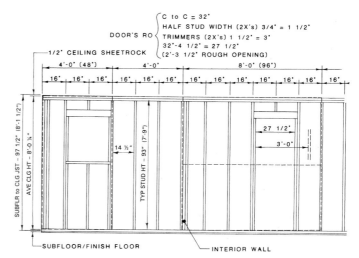

FIGURE 29.12 > Fully dimensioned wall framing elevation. *Hepler/Wallach/Hepler © Cengage 2013*

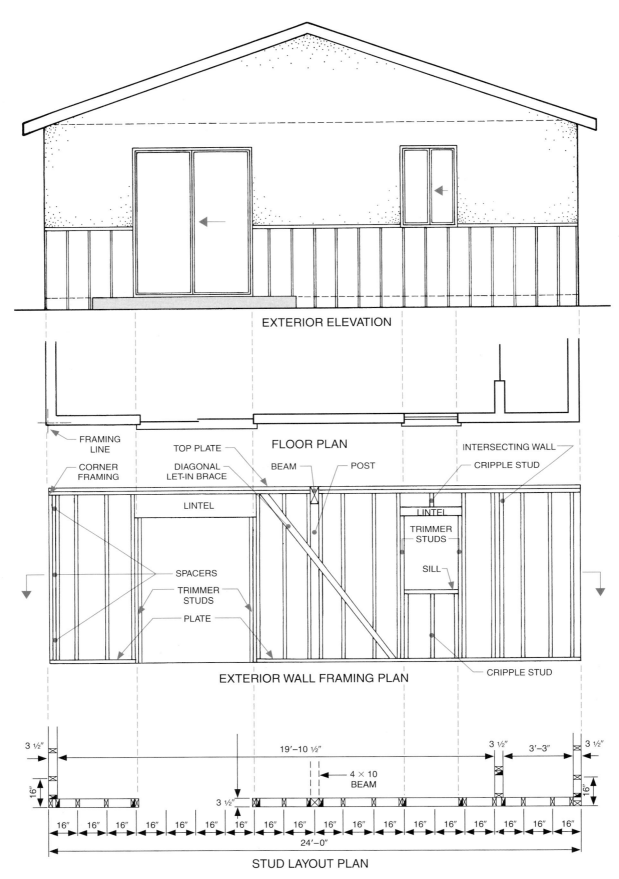

EXTERIOR ELEVATION

FLOOR PLAN

FRAMING
LINE

CORNER
FRAMING

TOP PLATE

DIAGONAL
LET-IN BRACE

LINTEL

BEAM

POST

INTERSECTING WALL

CRIPPLE STUD

LINTEL

TRIMMER
STUDS

SILL

SPACERS

TRIMMER
STUDS

PLATE

CRIPPLE STUD

EXTERIOR WALL FRAMING PLAN

3 ½"

19'-10 ½"

3 ½"   3'-3"   3 ½"

4 × 10
BEAM

16"

3 ½"

16"

16" 16" 16" 16" 16" 16" 16" 16" 16" 16" 16" 16" 16" 16" 16" 16" 16"

24'-0"

STUD LAYOUT PLAN

FIGURE 29.13A > Related exterior elevation, floor plan, framing elevation, and stud layout. *Hepler/Wallach/ Hepler © Cengage 2013*

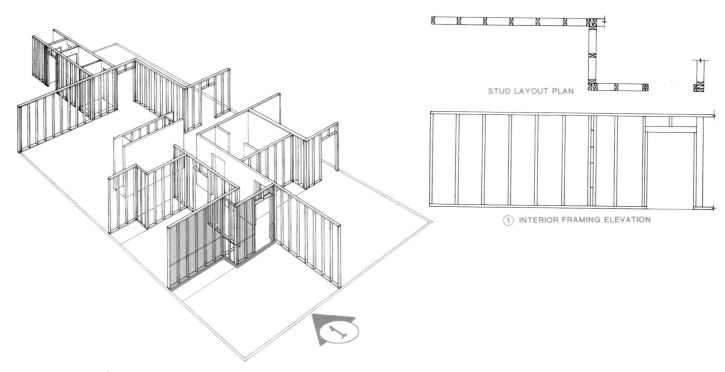

FIGURE 29.13B ⟩ Related interior framing elevation and stud layout. *Hepler/Wallach/Hepler © Cengage 2013*

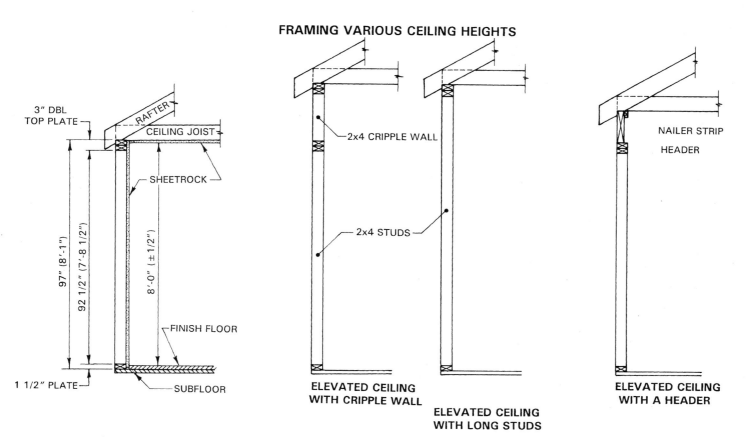

FIGURE 29.14 ⟩ Framing elevation for 8'-0" ceiling heights. *Hepler/Wallach/Hepler © Cengage 2013*

Detail Drawings

Not all of the information needed to frame an exterior wall can be shown on a framing elevation. Many details must be shown through the use of sectional drawings, exploded views, or pictorials.

Sections

Sectional wall framing drawings are either complete sections, partial sections, or removed sections indexed from a plan or elevation drawing.

Figure 29.15 shows a section through an outside wall from the foundation to the roof. This is the most used wall section type because it often represents a typical section through most outside walls of a structure. When only a portion of a wall is unique, a partial section, as shown in Figure 29.16, is often used. Framing drawings do not include assembly devices such as nails, screws, glue, or sealers unless unique to the building process. These are included in the set of specifications.

With sectional breaks, such as on a full wall section, a larger scale can be used to allow more detail. Elevation drawings may also be partially sectioned to reveal construction details, as shown in Figure 29.17.

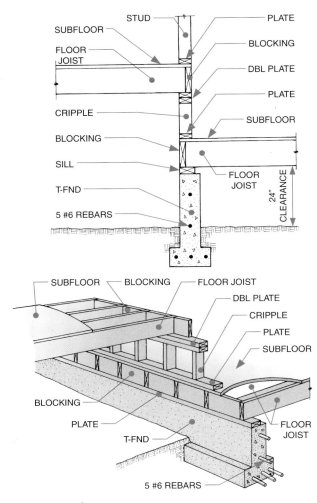

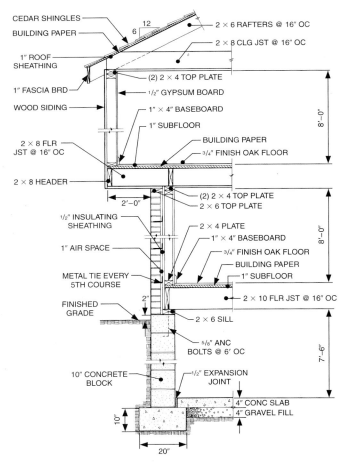

FIGURE 29.15 > Complete exterior wall section. *Hepler/ Wallach/Hepler © Cengage 2013*

FIGURE 29.16 > Method of drawing a floor-level change. *Hepler/Wallach/Hepler © Cengage 2013*

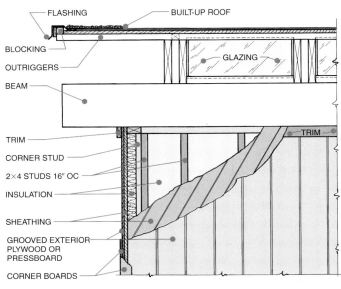

FIGURE 29.17 > Use of an exposed section to show framing details. *Hepler/Wallach/Hepler © Cengage 2013*

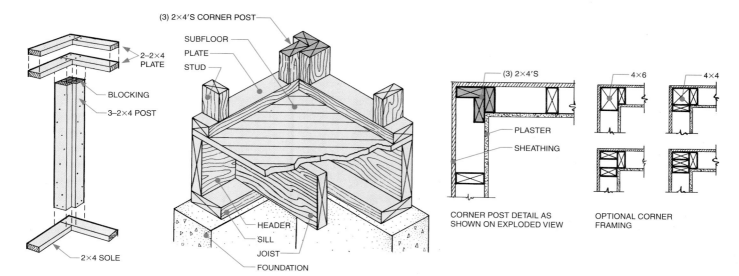

FIGURE 29.18 ＞ Use of exploded views. *Hepler/Wallach/Hepler © Cengage 2013*

Exploded Views

Occasionally, exploded views are drawn to show internal wall framing construction if it is hidden when the total assembly is drawn. Figure 29.18 shows an exploded view of a corner-post construction, and Figure 29.19 shows an exploded view of the wall intersection of two stud partitions. Some drawings simply remove individual members of an assembly to reveal more detail, as was done in Figure 29.20. Exploded views are generally used in cabinet work.

Pictorials

Pictorial framing drawings help eliminate construction errors due to misreading of plans, elevations, and sectional drawings. Full wall framing pictorial drawings may be used for this purpose. However, most pictorial details are limited to a single intersection detail or a complex construction feature.

🌐 Finish Wall and Siding Details

The covering for an exterior wall is called **siding.** Siding is the face of a structure and occupies more square footage than any other area. Siding is classified by material type, shape, and pattern. Materials used for siding include solid wood, plywood, plastic, masonry, and metal. Siding forms include lap, board and batten, panels, shingles, stucco, solid masonry, masonry veneer, metal, curtain walls, and log walls. Figure 29.21 shows a comparison of the types of materials used for siding.

Siding material details are most often shown on elevation views. Plan and pictorial sections help builders to interpret drawings.

- *Lap siding* is horizontal siding applied over sheathing. Each piece covers (overlaps) part of the piece below it. Lap siding is available in solid redwood, cedar, and pine. Other

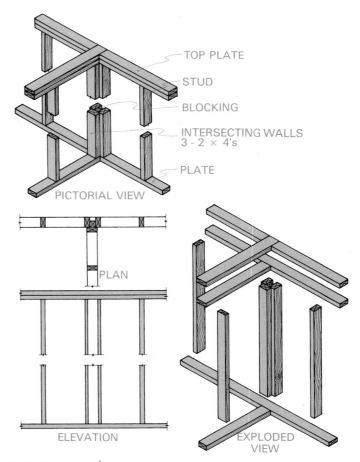

FIGURE 29.19 ＞ Exploded views of wall intersection. *Hepler/ Wallach/Hepler © Cengage 2013*

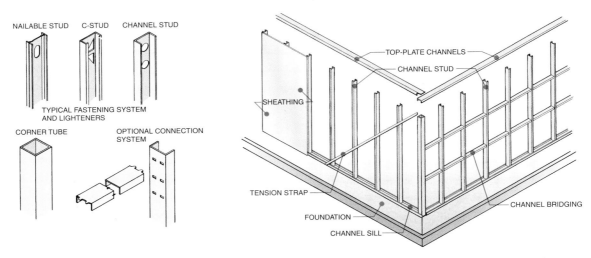

FIGURE 29.20 > Steel wall framing drawing with removed members. *Hepler/Wallach/Hepler © Cengage 2013*

SIDING	CHARACTERISTICS
Three-coat stucco	A durable siding material, but may develop hairline cracks. Installation cost largely dependent on local labor market.
EIFS stucco	Durable and more resistant to hairline cracking than traditional stucco. Has excellent insulating qualities and color retention.
Brick	The most durable siding material. May develop efflorescence (white powder), which must be cleaned.
Cedar shingles	Durable shingles if they are stained, sealed, or painted about every five years.
Redwood lap siding	Durable shingles if they are stained, sealed, or painted about every five years.
White pine siding	Less durable than cedar or redwood, but provides good performance if painted, or stained with a heavy oil about every five years.
Southern yellow pine tap siding	Good durability if painted or stained about every five years. Only the rough sawn variety holds paint well.
Exterior plywood panels	The durability is comparable to pine when painted, sealed, or stained about every five years. Installation is easy and fast.
Oriented strand board	A durable siding if properly installed and sealed or painted for moisture control.
Aluminum	Durable, but may be easily dented. Requires no maintenance other than cleaning once a year.
Steel	Very durable and dent resistant. Scratches through its finish to bare steel must be painted to stop rusting. Requires cleaning once a year.
Vinyl	Very durable and dent resistant. Requires cleaning once a year. It is the most widely used siding.
Polypropylene	Very durable and dent resistant. Various molded textures as shingles or stone are available. Requires cleaning about once a year. The coarse texture makes cleaning more difficult than with the smooth siding surfaces.

FIGURE 29.21 > Siding characteristics. *Hepler/Wallach/Hepler © Cengage 2013*

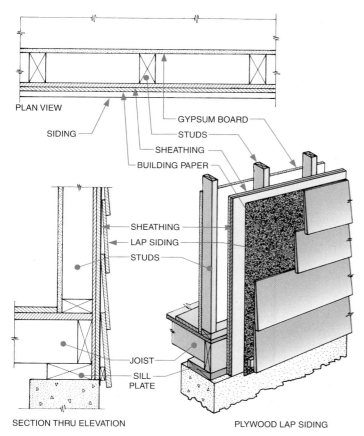

PLAN VIEW

SIDING
GYPSUM BOARD
STUDS
SHEATHING
BUILDING PAPER

SHEATHING
LAP SIDING
STUDS

JOIST
SILL PLATE

SECTION THRU ELEVATION

PLYWOOD LAP SIDING

FIGURE 29.22 › Lap siding framing plan and elevation. *Hepler/Wallach/Hepler © Cengage 2013*

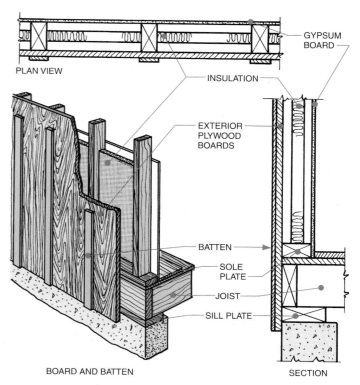

PLAN VIEW

GYPSUM BOARD
INSULATION
EXTERIOR PLYWOOD BOARDS
BATTEN
SOLE PLATE
JOIST
SILL PLATE

BOARD AND BATTEN

SECTION

FIGURE 29.23 › Board-and-batten wall framing plan and elevation. *Hepler/Wallach/Hepler © Cengage 2013*

materials are also used extensively, such as aluminum, steel, and fabricated boards. See Figure 29.22.

- *Board-and-batten siding* is vertical siding that originally consisted of a vertical board placed over studs, with the joints covered with vertical batten boards. Today, many board and batten sidings are made of plywood or strand board sheets with battens over the 48″ joints. Additional battens at 16″ intervals add consistency and help stabilize the vertical sheets. See Figure 29.23.

- *Panel boards* are a variation of vertical sheet siding. These are available with a wide variety of surface textures, grades, and groove sizes and shapes. Panels are also available that simulate the appearance of bevel or lap siding or shingles. *Structural insulated panels* are comprised of oriented strand board (OSB) laminated to a core of polyurethane foam. Panelized stress skin panels can carry the load of a structure equal to or greater than a wood stud wall.

- *Shingle siding* is applied as individual overlapped shingle boards, like roofing shingles, or as prefabricated shingle siding sheets. In either case they are applied over insulated sheathing.

- *Stucco siding* (Figure 29.24) is applied with a trowel or sprayed. In wood construction, steel mesh must be applied

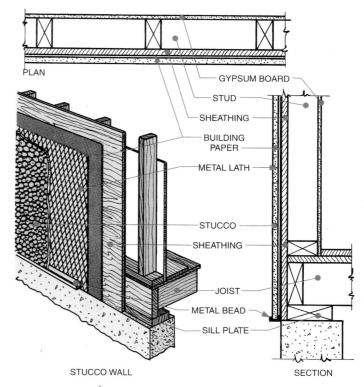

PLAN

GYPSUM BOARD
STUD
SHEATHING
BUILDING PAPER
METAL LATH
STUCCO
SHEATHING
JOIST
METAL BEAD
SILL PLATE

STUCCO WALL

SECTION

FIGURE 29.24 › Stucco wall framing plan and elevation. *Hepler/Wallach/Hepler © Cengage 2013*

over insulated sheathing to provide a base for the stucco. Stucco can be applied directly to concrete block without the use of mesh. Stucco is available in traditional three-coat portland cement formula or in synthetic or insulated-finish system (EIFS) form. The formula for synthetic stucco replaces portland cement with acrylic resins, and color can be blended into the final finished coat. Figure 29.25 shows the sequence of stucco coat applications.

- *Solid masonry wall facing materials* include stucco, brick, stone, aluminum, steel, vinyl, or polyurethane siding. These are applied over solid concrete or concrete block.

Brick is a material that can be used both structurally and as a finished wall face. Figure 29.26 shows a solid brick wall that performs both functions. Some brick walls include cavities (spaces between) to provide insulation, reduce dead loads, and lower material costs. Figure 29.27 illustrates the use of insulation seals around wall ties to create a monolithic layer that also serves as an air barrier. Figure 29.28 shows typical wall sections of different types of brick veneer, solid brick, and brick-cavity walls. Concrete block can be used in place of brick in these types of construction.

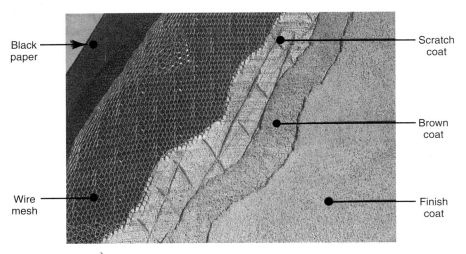

FIGURE 29.25 > Stucco coats. *Dietrich Industries, Inc*

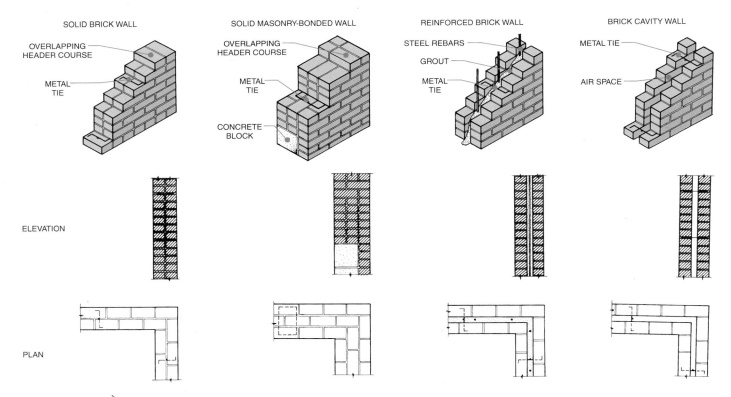

FIGURE 29.26 > Types of brick or concrete block construction. *Hepler/Wallach/Hepler © Cengage 2013*

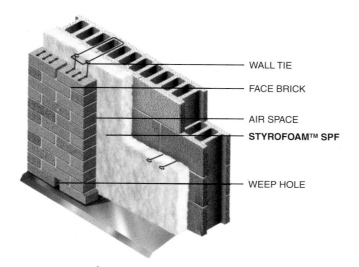

- **Masonry veneer walls** add brick or stone facing to wood-framed walls. See Figure 29.29. An entire wall may be a brick veneer wall. Where added support is required, a concrete block veneer wall may be used.

- **Metal siding** is manufactured from steel or aluminum sheets and given a factory-applied coating. Both may be formed to appear as wood siding. Both types of metal are durable; however, steel is more dent and warp resistant than aluminum.

- **Curtain walls** are used where structural steel or post-and-beam construction provides large wall spaces that are not part of the load-bearing structure. See Figure 29.30. One of the greatest advantages of steel cage construction is the unobstructed space provided by curtain walls. Building loads are transmitted through columns, so the remaining open wall space can be filled with any type of nonbearing

FIGURE 29.27 > Insulated masonry wall. *Hepler/Wallach/Hepler © Cengage 2013*

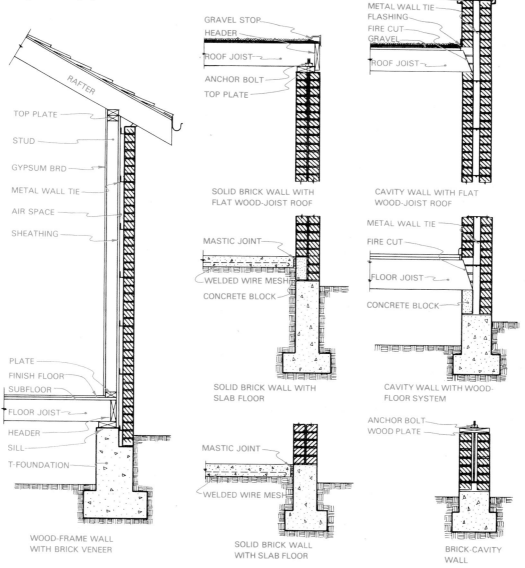

FIGURE 29.28 > Types of masonry construction with related roof and foundation intersections. *Hepler/Wallach/Hepler © Cengage 2013*

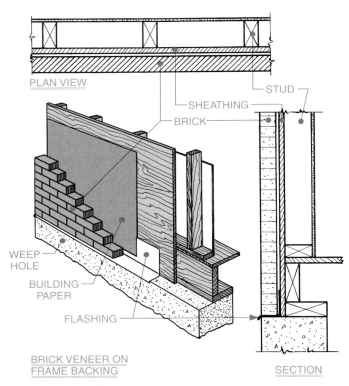

FIGURE 29.29 > Brick veneer wall plan and elevation.
Hepler/Wallach/Hepler © Cengage 2013

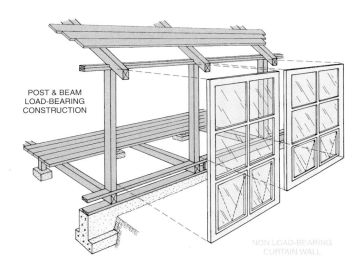

FIGURE 29.30 > Curtain wall construction. *Hepler/Wallach/ Hepler © Cengage 2013*

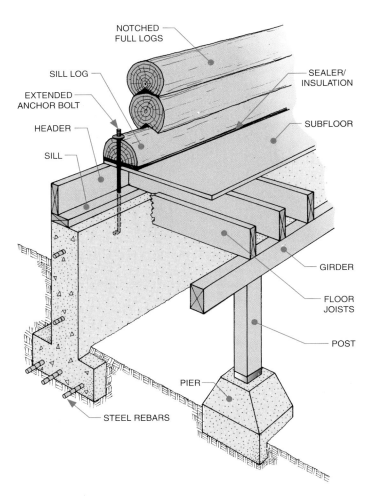

FIGURE 29.31 > Log wall construction. *Hepler/Wallach/Hepler © Cengage 2013*

Window Framing Drawings

Windows primarily provide light views and may allow for ventilation. Windows must be insulated, weather tight, and resistant to condensation. Framing members around a window must not transfer structural movement or thermal stress to the glass. To design the framing needed to support each window, an understanding of the major components of a window is necessary.

Figure 29.32 shows the relationship between the rough opening for a window and the window frame, sash, and trim. Typical rough opening framing is shown in Figure 29.33. Rough opening dimensions are usually 1″ to 3″ larger than the window dimensions. Window manufacturers often use exploded views to show these relationships, as seen in Figure 29.34.

Window construction details are usually shown on a head, jamb, rail, or sill section. See Figure 29.35. The positions of trimmers, headers, and sill-support members may also be shown on these details. The method of weatherproofing between window components and panel framing is illustrated in Figure 29.36.

(curtain) panels. These panels are usually prefabricated in modular units. Therefore, wall framing plans show only the position of modular units and not the construction details.

- *Log wall* construction is shown in Figure 29.31. To further insulate and finish interior stud walls, sheathing and drywall panels are normally attached to plumbed furring strips. Is is unnecessary to add vinyl sheet wraps to log walls. The thickness of the logs density provide natural insulation which protects against heat, cold, and moisture.

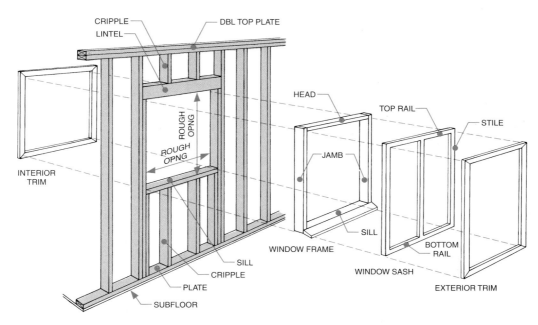

FIGURE 29.32 > Basic elements of a window assembly. *Hepler/Wallach/Hepler © Cengage 2013*

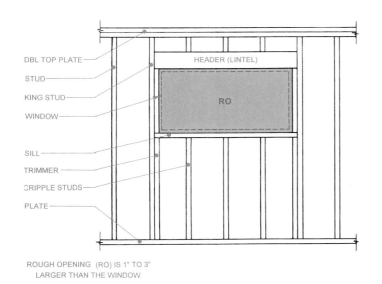

FIGURE 29.33 > Typical rough opening framing. *Hepler/ Wallach/Hepler © Cengage 2013*

There are hundreds of window manufacturers, and each has hundreds of window styles and sizes. It is therefore necessary to refer to the manufacturer to determine the exact dimensions and rough opening for each window. This information is usually included in schedules and specifications. (See Chapter 35.)

Rough opening dimensions vary among window manufacturers for each window style. Figures 29.37 and 29.38 show common charts used by manufacturers to summarize rough opening requirements.

When fixed windows or unusual window shapes or sizes are to be constructed in the field—or even at a factory— complete framing details must be drawn. Unusual and

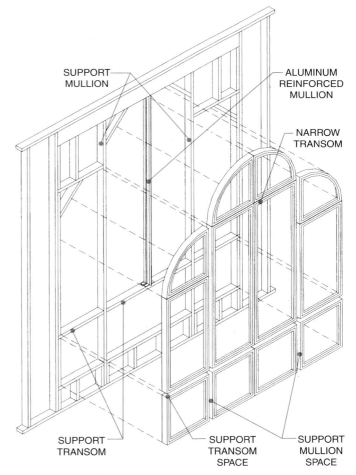

FIGURE 29.34 > Exploded view of manufactured window components. *Image courtesy of Andersen Windows*

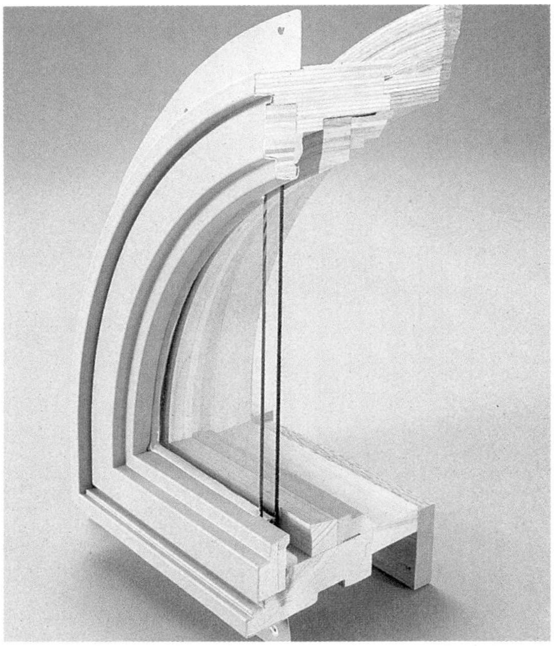

FIGURE 29.35 > Sill and head section of a manufactured window. *Image courtesy of Andersen Windows*

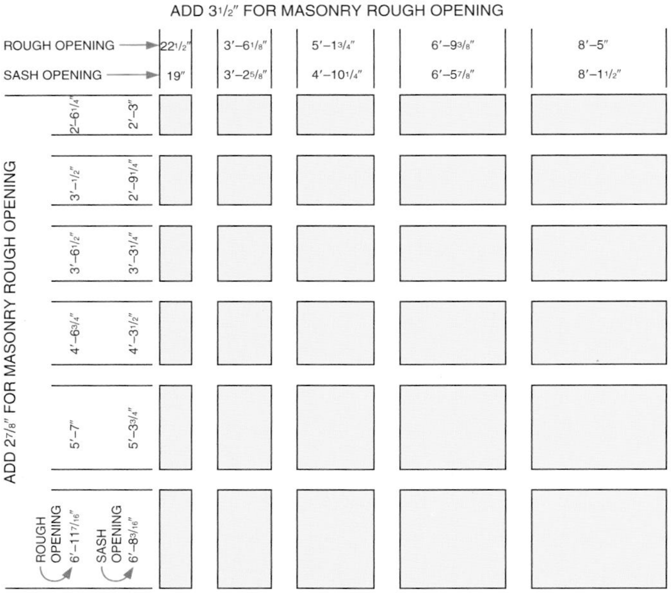
ADD 3 1/2" FOR MASONRY ROUGH OPENING
ROUGH OPENING
SASH OPENING
22 1/2"
19"
3'–6 1/8"
3'–2 5/8"
5'–1 3/4"
4'–10 1/4"
6'–9 3/8"
6'–5 7/8"
8'–5"
8'–1 1/2"
ADD 2 7/8" FOR MASONRY ROUGH OPENING
2'–6 1/4"
2'–3"
3'–1/2"
2'–9 1/4"
3'–6 1/2"
3'–3 1/4"
4'–6 3/4"
4'–3 1/2"
5'–7"
5'–3 3/4"
ROUGH OPENING 6'–11 7/16"
SASH OPENING 6'–8 3/16"

FIGURE 29.37 > Common rough opening dimensions for windows. *Hepler/Wallach/Hepler © Cengage 2013*

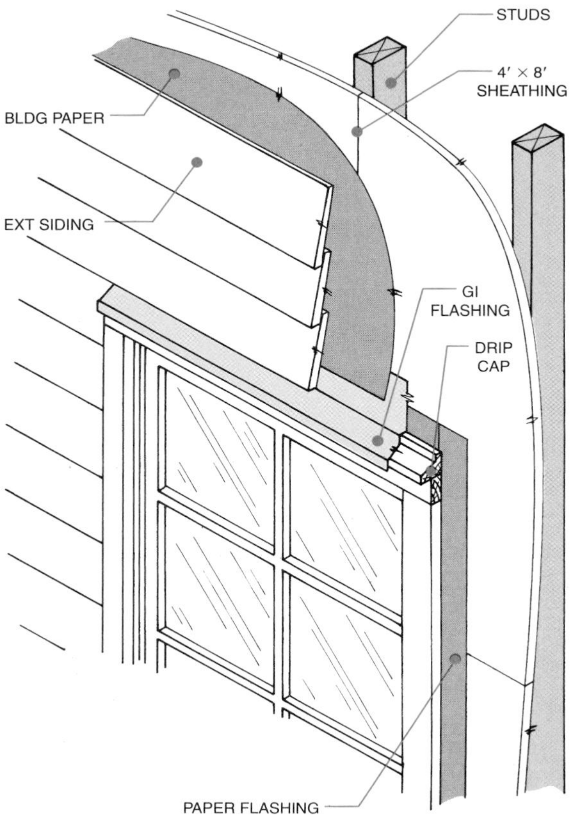
STUDS
4' × 8' SHEATHING
BLDG PAPER
EXT SIDING
GI FLASHING
DRIP CAP
PAPER FLASHING

FIGURE 29.36 > Typical window waterproofing installation. *Hepler/Wallach/Hepler © Cengage 2013*

nonstandard sizes and components require more complete detail drawings. *Fixed sheet glass* thicknesses range from 3/32" to 7/16". Widths range from 40" to 60" and lengths range from 50" to 120". *Plate glass,* 1/8" to 1/2" thick, ranges from 80" × 130" to 125" × 280". *Glass block* windows also require a head, jamb, and sill detail. See Figure 29.39. Glass block rough openings must allow space for the block size, plus mortar or channel space as prescribed by the manufacturer. Figure 29.40 shows a sectional drawing of the glass block installation shown in Figure 17.40A.

Head and sill sections are vertical sections. Figure 29.41 shows the relationship between the cutting-plane line and the head and sill sections. The circled areas in Figure 29.42 show the areas that are removed when a separate head and sill section is prepared. When a cutting-plane line is extended horizontally across the entire window, the resulting sections are knows as jamb sections. Jamb details (horizontal sections) are projected from the window elevation drawing. See Figure 29.43.

Door Framing Drawings

Component drawings of a door assembly include the wall framing with the rough opening, the door frame (head and side jambs), the door, and sometimes the sill (threshold). Doors are usually prehung (attached with hinges) to the jamb, and the complete assembly is fit into a rough opening in a framed wall. See Figure 29.44. A lintel must be placed above the head jamb to prevent the jamb from sagging.

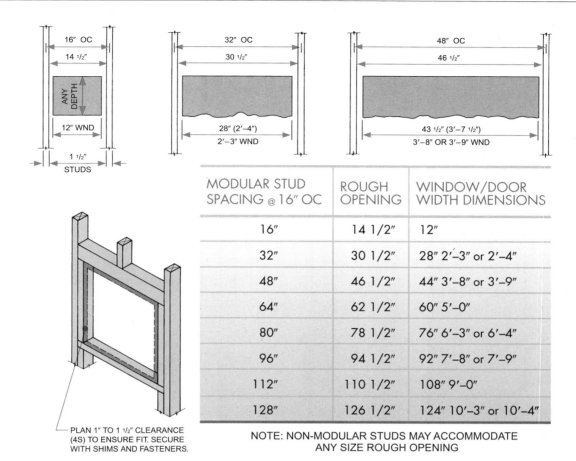

MODULAR STUD SPACING @ 16" OC	ROUGH OPENING	WINDOW/DOOR WIDTH DIMENSIONS
16"	14 1/2"	12"
32"	30 1/2"	28" 2'–3" or 2'–4"
48"	46 1/2"	44" 3'–8" or 3'–9"
64"	62 1/2"	60" 5'–0"
80"	78 1/2"	76" 6'–3" or 6'–4"
96"	94 1/2"	92" 7'–8" or 7'–9"
112"	110 1/2"	108" 9'–0"
128"	126 1/2"	124" 10'–3" or 10'–4"

PLAN 1" TO 1 1/2" CLEARANCE (4S) TO ENSURE FIT. SECURE WITH SHIMS AND FASTENERS.

NOTE: NON-MODULAR STUDS MAY ACCOMMODATE ANY SIZE ROUGH OPENING

FIGURE 29.38 > Rough openings and modular sizes for windows and doors. *Hepler/Wallach/Hepler © Cengage 2013*

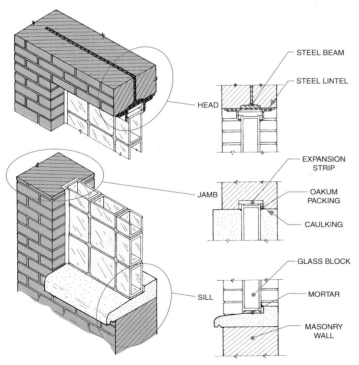

FIGURE 29.39 > Glass block wall section. *Hepler/Wallach/Hepler © Cengage 2013*

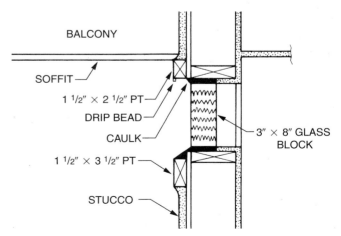

FIGURE 29.40 > Glass block application detail related to the elevation in Figure 17.40A. *Hepler/Wallach/Hepler © Cengage 2013*

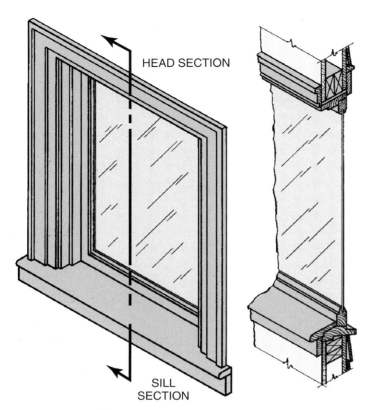

FIGURE 29.41 > Window head and sill sections in the vertical plane. *Hepler/Wallach/Hepler © Cengage 2013*

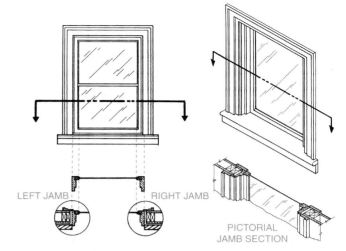

FIGURE 29.43 > Projection of window jamb sections. *Hepler/Wallach/Hepler © Cengage 2013*

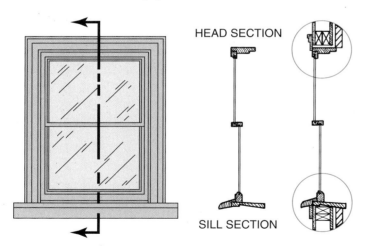

FIGURE 29.42 > Projection of window head and sill sections. *Hepler/Wallach/Hepler © Cengage 2013*

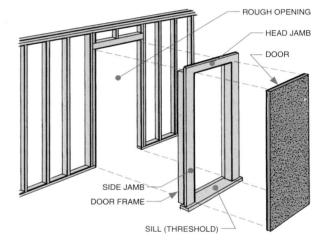

FIGURE 29.44 > Relationship of door jamb and rough opening. *Hepler/Wallach/Hepler © Cengage 2013*

Lintels distribute building loads to vertical support members such as studs, trimmers, or posts.

Figure 29.45 shows a large solid lintel with future trimmer positions noted by dashed lines. Solid wood lintel sizes and makeup vary greatly. Figure 29.46 illustrates the typical lintel sizes used in light wood construction. Lighter weight headers that use strand board and insulation cores, as shown in Figure 29.47, are energy efficient (R20) and dimensionally stable.

Modular door units may include multiple doors and windows that must also fit into a rough opening. See Figure 29.48. Specifying, dimensioning, and constructing accurate rough openings for these units is critical to successful door functioning. See Figure 29.49. Like windows, rough openings for doors are found in manufacturer's specifications, which are included in a door schedule. (See Chapter 35.) Add 3 1/2″ to the width and 1 1/2″ to the height of a door if the rough opening is not specified by a manufacturer.

Head, sill (threshold), and jamb sections are just as effective in describing door framing construction as they are in showing window framing details (Figure 29.50). Because doors extend to the floor, the relationship of the floor framing system to the position of the door is critical.

FIGURE 29.45 > Solid door lintel. *Hepler/Wallach/Hepler © Cengage 2013*

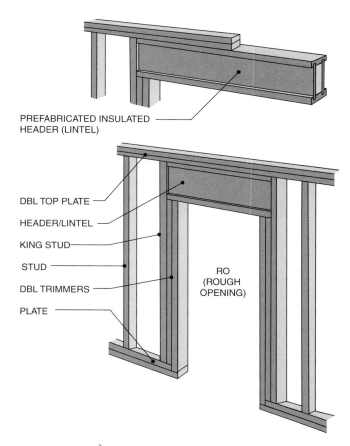

FIGURE 29.47 > Preinsulated lightweight header. *Hepler/ Wallach/Hepler © Cengage 2013*

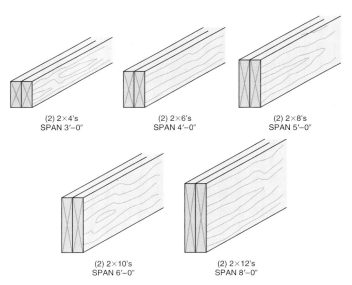

(2) 2×4's
SPAN 3'-0"

(2) 2×6's
SPAN 4'-0"

(2) 2×8's
SPAN 5'-0"

(2) 2×10's
SPAN 6'-0"

(2) 2×12's
SPAN 8'-0"

FIGURE 29.46 > Typical wood lintel sizes and types. *Hepler/ Wallach/Hepler © Cengage 2013*

Framing for sliding doors (Figure 29.51) also requires careful detailing of intersections. See Figure 29.52. If thresholds are not part of the exterior door assembly, a separate detail is necessary to show the exact type and alignment of the door and the threshold. See Figure 29.53.

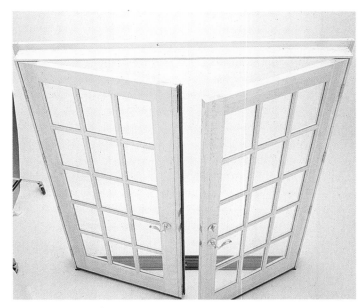

FIGURE 29.48 > Prehung double French doors. *Marvin Windows & Doors*

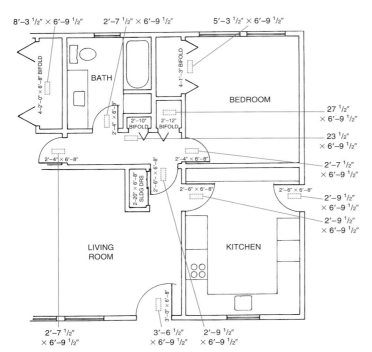

8'–3 ½" × 6'–9 ½"   2'–7 ½" × 6'–9 ½"   5'–3 ½" × 6'–9 ½"

4–2'–0" × 6'–8" BIFOLD

BATH

4–1'–3" BIFOLD

BEDROOM

27 ½" × 6'–9 ½"

2'–4" × 6'–8"

2'–10" BIFOLD   2'–12" BIFOLD

23 ½" × 6'–9 ½"

2'–4" × 6'–8"

2'–4" × 6'–8"

2'–7 ½" × 6'–9 ½"

2–20" × 6'–8" SLDG DRS

2'–6" × 6'–8"

2'–6" × 6'–8"

2'–6" × 6'–8"

2'–9 ½" × 6'–9 ½"

2'–9 ½" × 6'–9 ½"

LIVING ROOM

KITCHEN

3'–0" × 6'–8"

2'–7 ½" × 6'–9 ½"

3'–6 ½" × 6'–9 ½"   2'–9 ½" × 6'–9 ½"

FIGURE 29.49 > Rough opening dimensions for standard door sizes. *Hepler/Wallach/Hepler © Cengage 2013*

FIGURE 29.51 > Typical sliding glass door components. *Marvin Windows & Doors*

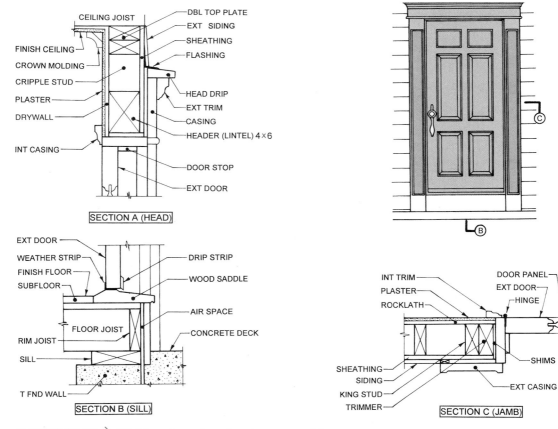

CEILING JOIST

DBL TOP PLATE

EXT SIDING

SHEATHING

FINISH CEILING

FLASHING

CROWN MOLDING

CRIPPLE STUD

PLASTER

HEAD DRIP

EXT TRIM

DRYWALL

CASING

HEADER (LINTEL) 4×6

INT CASING

DOOR STOP

EXT DOOR

SECTION A (HEAD)

EXT DOOR

WEATHER STRIP

DRIP STRIP

FINISH FLOOR

SUBFLOOR

WOOD SADDLE

AIR SPACE

FLOOR JOIST

RIM JOIST

CONCRETE DECK

SILL

T FND WALL

SECTION B (SILL)

INT TRIM

DOOR PANEL

PLASTER

EXT DOOR

ROCKLATH

HINGE

SHEATHING

SIDING

KING STUD

TRIMMER

SHIMS

EXT CASING

SECTION C (JAMB)

FIGURE 29.50 > Wall framing related to door assembly. *Hepler/Wallach/Hepler © Cengage 2013*

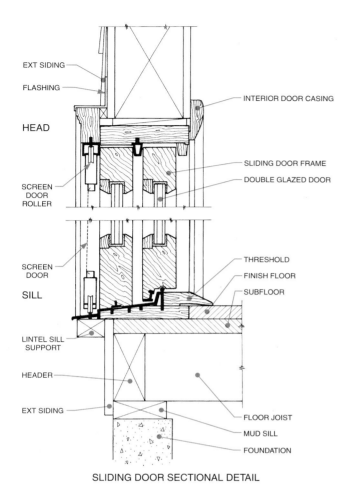

FIGURE 29.52 > Typical sliding door framing. *Hepler/Wallach/ Hepler © Cengage 2013*

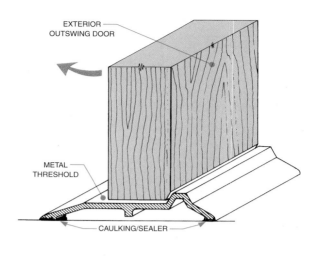

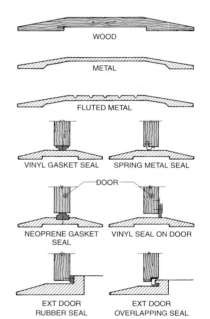

FIGURE 29.53 > Common types of thresholds and weather stripping. *Hepler/Wallach/Hepler © Cengage 2013*

A horizontal section of all doors is indicated on detail drawings. When a cutting-plane line is extended vertically through the sill and head, a section is revealed, as shown in Figure 29.54. Because doors are normally not as wide as they are high, an adequate jamb detail can be projected without the use of break lines or removed sections. See Figure 29.55.

INTERIOR WALLS

Interior walls may be load bearing or non-load bearing. Openings in load-bearing walls, such as door cavities and windows, must be linteled (Figure 29.47) to support the loads above. Interior framing drawings include plan, elevation, and pictorial drawings of partitions and wall coverings. Detail drawings of interior partitions may also show intersections between walls and ceilings, floors, windows, and doors. See Figure 29.56A.

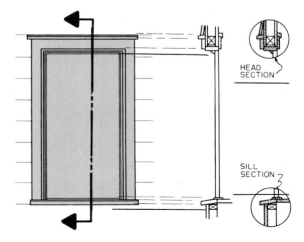

FIGURE 29.54 > Projection of door head and sill sections. *Hepler/Wallach/Hepler © Cengage 2013*

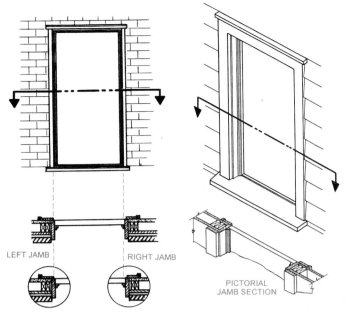

FIGURE 29.55 ⟩ Projection of left and right jamb sections.
Hepler/Wallach/Hepler © Cengage 2013

Framing Elevations

Framing elevation drawings show direct two-dimensional views of the framing. They are most effective for showing the construction of interior partitions. Interior partitions are projected from a floor plan, as viewed from the center of the room. To ensure the correct interpretation of the partition, each interior elevation drawing should include a label that indicates the room, a reference number, and the compass orientation of the wall. If either the room name, compass direction, or reference is omitted, the elevation may easily be misinterpreted and confused with a similar wall in another

room. Interior framing elevations are also shown in full building sections as shown in Figure 29.56B.

A complete study of the floor plan, elevation, plumbing diagrams, and electrical plans should be made prior to the preparation of interior wall framing drawings. Provisions must be made in framing drawings for special needs and to allow openings for electrical, plumbing, and HVAC installations. Figure 29.57 shows a framing drawing used to accommodate plumbing pipes. Figure 29.58 shows how wall framing is adjusted to allow space for heating registers.

When a stud must be broken to accommodate various items, the framing drawing must show the recommended construction. A structural stud should never have more than half its thickness removed.

Built-in wall items may only require a partial framing elevation if location dimensions are included in an interior wall elevation or the heights are noted on a floor plan. See Figure 29.59.

Columns

Steel, concrete, masonry, and wood columns or posts perform the function of load-bearing partitions. They support the girders and beams on which floor decking and ceiling systems rest. Horizontal members can also be supported at the same height as the post through the use of hangers or blocking. See Figure 29.60.

Steel column positions may be shown on floor plans or on elevation drawings. On a floor plan, the style, size, and weight may be noted on each outline of the column, such as A1, B2. This information may also be shown on a column schedule. **Column schedules** are schematic elevation drawings showing the entire height of a building and the elevation of each floor, base plate, and column splice. The type, depth, weight, and length of columns with common characteristics are shown

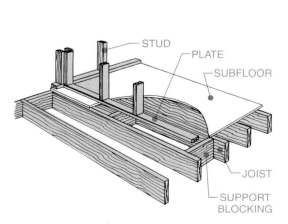

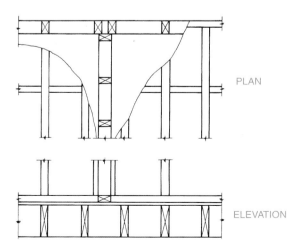

FIGURE 29.56A ⟩ Wall framing intersection detail. *Hepler/Wallach/Hepler © Cengage 2013*

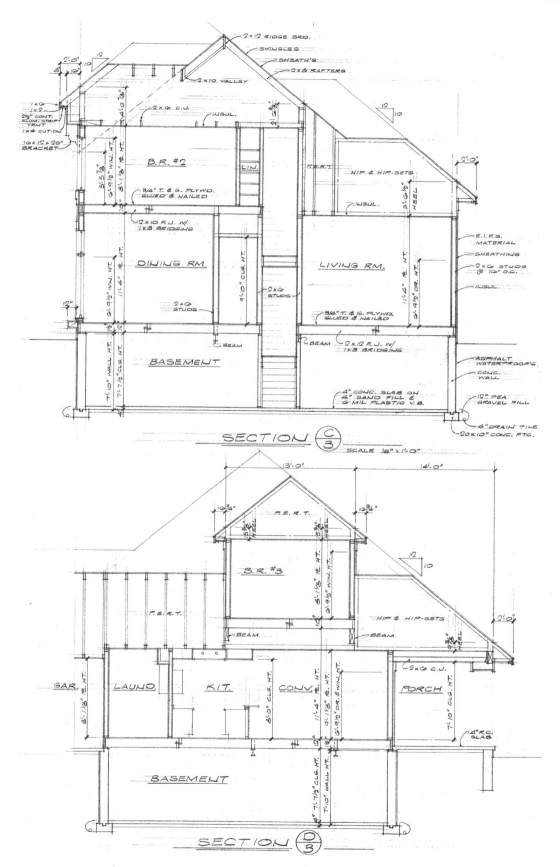

FIGURE 29.56B ❭ A full section showing interior framing. *Used by permission Hanley Wood, LLC*

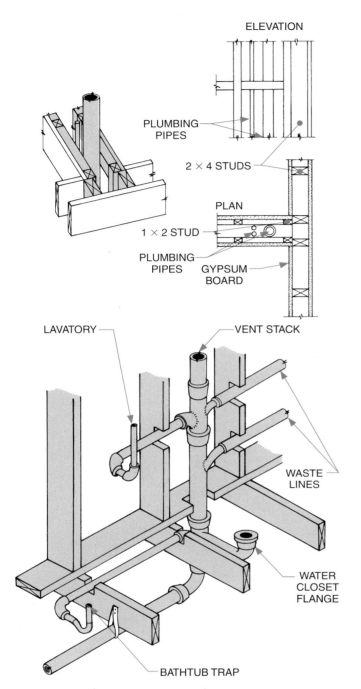

ELEVATION

PLUMBING PIPES

2 × 4 STUDS

PLAN

1 × 2 STUD

PLUMBING PIPES

GYPSUM BOARD

LAVATORY

VENT STACK

WASTE LINES

WATER CLOSET FLANGE

BATHTUB TRAP

FIGURE 29.57 > Framing adjustments for plumbing lines. *Hepler/Wallach/Hepler © Cengage 2013*

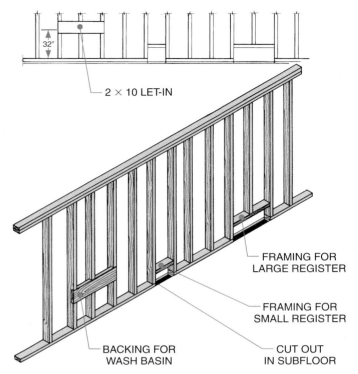

2 × 10 LET-IN

FRAMING FOR LARGE REGISTER

FRAMING FOR SMALL REGISTER

BACKING FOR WASH BASIN

CUT OUT IN SUBFLOOR

FIGURE 29.58 > Framing adjustments for heating outlets. *Hepler/Wallach/Hepler © Cengage 2013*

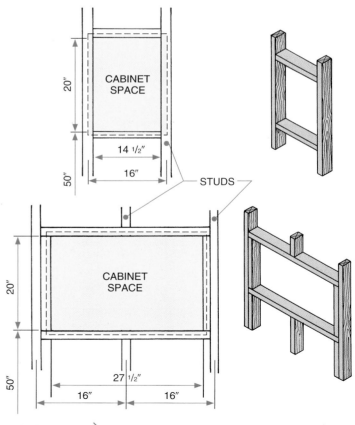

CABINET SPACE

20″

50″

14 1/2″

16″

STUDS

CABINET SPACE

20″

50″

27 1/2″

16″

16″

FIGURE 29.59 > Framing for built-in items. *Hepler/Wallach/Hepler © Cengage 2013*

under the column mark for each column. Figure 29.61 shows a pictorial plan with column locations related to the column schedule shown in Figure 29.62. There are 13 columns in this plan.

Large structures are designed with curtain walls (see Figure 29.30) on the exterior. In this type of construction, nonstructural wall panels cover the structural steel frame. Figure 29.63 shows the common types of steel-frame structures and their relationship to building height.

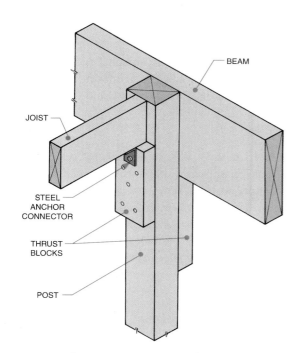

FIGURE 29.60 > Post support for joists and beams. *Hepler/ Wallach/Hepler © Cengage 2013*

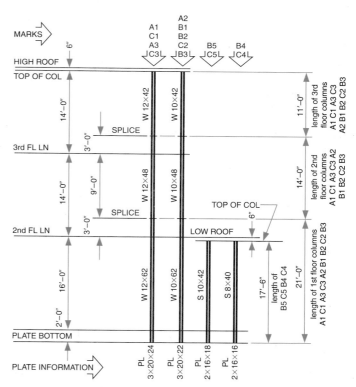

FIGURE 29.62 > Column schedule showing height, size, and column type. *Hepler/Wallach/Hepler © Cengage 2013*

Columns with common specifications are grouped together at the top of the schedule. Under each grouping a heavy vertical line represents the height of each column, with the type, size, and weight noted on each. For example, the columns with marks A1, C1, A3, and C3 in Figure 29.62 are all 12″-wide flange shapes and extend vertically 46′-0″ from

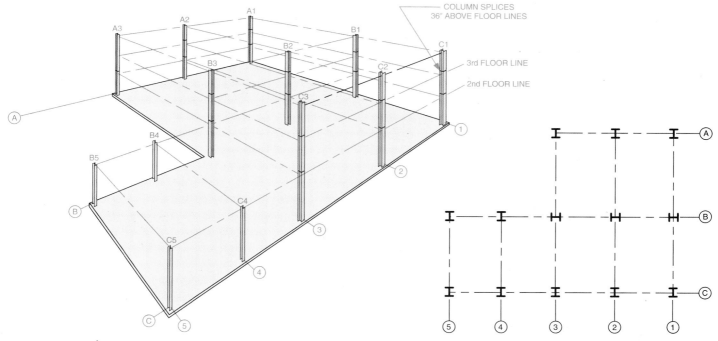

FIGURE 29.61 > Method of marking column locations. *Hepler/Wallach/Hepler © Cengage 2013*

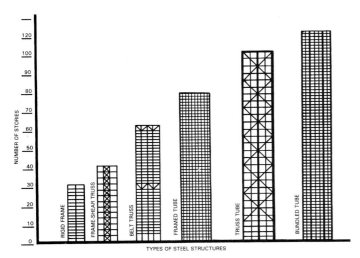

FIGURE 29.63 > Types of high-rise steel framing. *Hepler/ Wallach/Hepler © Cengage 2013*

base to top (2′ + 16′ + 14′ + 14′). Three individual columns comprise each of these. The bottom length is 21′-0″ from plate to first splice (3′ above the floor line), 14′-0″ from the first splice to the second splice, and 11′-0″ from the second splice to the top. The second row of column marks (A2, B1, B2, C2, and B3) have the same lengths as the first row but are 10″-wide flange shapes. The third-row and fourth-row columns are continuous 17′-6″, without splices; the row 3 columns (B5, C5) are 10″ S-shapes and the row 4 columns (B4, C4) are 8″ S-shapes.

Details

Additional drawings of wall construction, besides the basic structural framing, are often needed. Such drawings include molding and trim, interior doors, stair elevations, and wall coverings.

Molding and Trim

For finished interior walls, moldings and trim are used at intersections. Small-scaled drawings cannot accurately show the size, shape, position, and material used for molding and trim. Thus, detail sections are prepared, such as for crown (ceiling) and base (floor) moldings. Where more intricate designs are used, a detail drawing that shows the different molding segments is prepared. See Chapter 18.

Interior Doors

Pictorial or orthographic section drawings of the jamb, sill, and head should be prepared to illustrate the framing around interior doors. See Figure 29.64. A detailed drawing need not be prepared for each door, but one should be prepared for each *type* of door used. The position of headers, particularly to support closet sliding doors and pocket doors, is critical to the vertical fit and horizontal matching of doors. Drawings should also be keyed to the door schedule for identification. A pictorial section of an interior-door jamb may be needed to show variations because of different wall coverings, such as plaster, gypsum board, or paneling. See Figure 29.65.

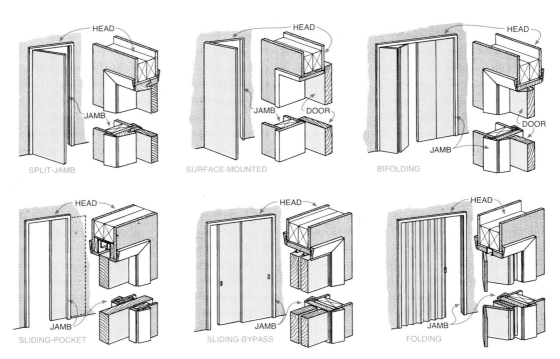

FIGURE 29.64 > Framing and trim details for interior doors. *Hepler/Wallach/Hepler © Cengage 2013*

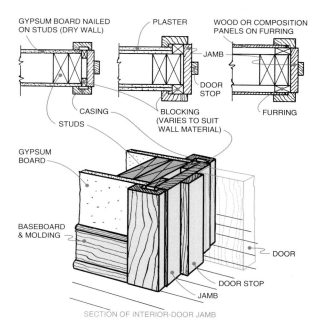

GYPSUM BOARD NAILED ON STUDS (DRY WALL)

PLASTER

WOOD OR COMPOSITION PANELS ON FURRING

JAMB

DOOR STOP

FURRING

CASING

BLOCKING (VARIES TO SUIT WALL MATERIAL)

STUDS

GYPSUM BOARD

BASEBOARD & MOLDING

DOOR

DOOR STOP

JAMB

SECTION OF INTERIOR-DOOR JAMB

Stairs

Just as rough openings must be allowed for windows and doors, floor openings must be planned for stairwell framing. A sectional wall elevation drawing is often used to describe the vertical distances that are not found on floor framing plans. Figure 29.66 shows an elevation, plan, detail, and perspective drawing of a typical stair system designed for an 8-foot ceiling height.

Abbreviated elevation sections are adequate to show only the rise, run, stairwell opening, and headroom clearance. Figure 29.67 shows an abbreviated elevation for a straight-run stair system. Figure 29.68 shows the same type of drawing for an L-shaped system. Where more detailed information is needed, a completely dimensioned elevation drawing is used, as shown in Figure 29.69.

FIGURE 29.65 $>$ Types of door trim for different wall coverings. *Hepler/Wallach/Hepler*
© *Cengage 2013*

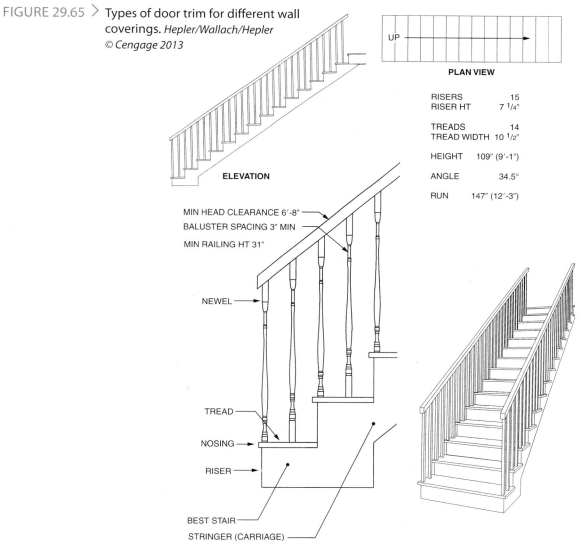

UP

PLAN VIEW

RISERS	15
RISER HT	7 1/4"
TREADS	14
TREAD WIDTH	10 1/2"
HEIGHT	109" (9'-1")
ANGLE	34.5°
RUN	147" (12'-3")

ELEVATION

MIN HEAD CLEARANCE 6'-8"

BALUSTER SPACING 3" MIN

MIN RAILING HT 31"

NEWEL

TREAD

NOSING

RISER

BEST STAIR

STRINGER (CARRIAGE)

FIGURE 29.66 $>$ Stair plan and elevation drawings. *Hepler/Wallach/Hepler*
© *Cengage 2013*

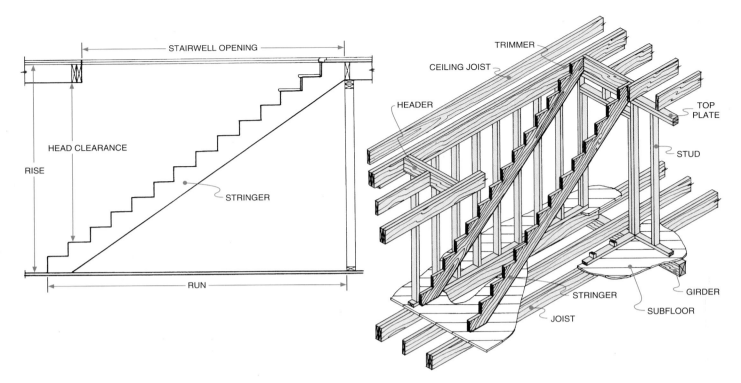

FIGURE 29.67 > Straight-run stair system. *Hepler/Wallach/Hepler © Cengage 2013*

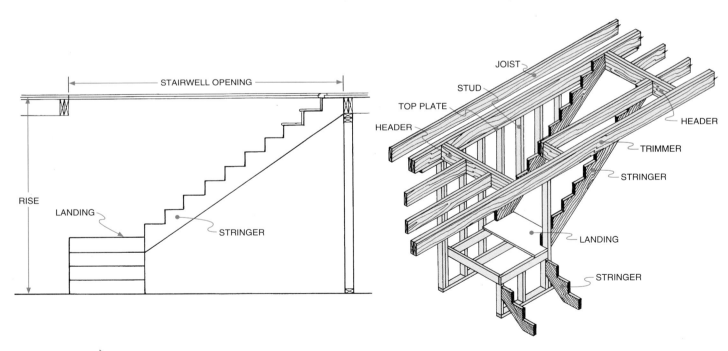

FIGURE 29.68 > L-shaped stair system. *Hepler/Wallach/Hepler © Cengage 2013*

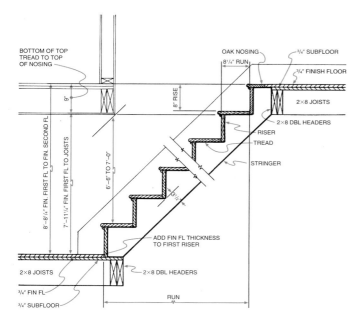

FIGURE 29.69 > Fully dimensioned stair assembly drawing.
Hepler/Wallach/Hepler © Cengage 2013

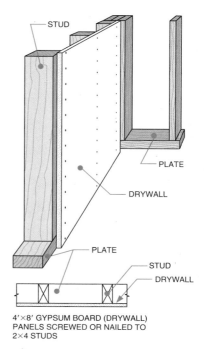

FIGURE 29.70 > Drywall construction. *Hepler/Wallach/Hepler © Cengage 2013*

Tread width is shown on floor plans and floor framing plans. Riser height may be shown on interior elevations and on stair framing drawings. To calculate the number of risers use the following formula:

Formula:

$$\frac{\text{height of stairs}}{\text{height of each riser}} = \text{riser number}$$

Example:

$$108'' + 4'' = 112''$$

$$\frac{112''}{7''} = 16 \text{ risers}$$

(adjust 16 risers at 7.15" each)

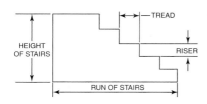

See Chapter 15 for more information about calculating stair dimensions.

Wall Coverings

Basic types of wall-covering materials used for finished interior walls include plaster, drywall, paneling, tile, and masonry as covered in Chapter 25. Each type requires a different method of attachment to the wall.

- *Plaster* is applied to interior walls over wire lath or gypsum sheet lath. Plaster walls are very strong and sound absorbing. Plaster is also decay proof and termite proof. However, plaster walls crack easily, they take months to dry, and the installation costs are high.

- **Drywall construction** is a system of interior wall finishing using prefabricated sheets of materials. A variety of manufactured materials may be used, such as fiberboard, gypsum wallboard, stranded lumber sheets, Sheetrock, and plywood. Drywall is nailed or screwed directly to studs. Then the drywall joints are finished. See Figure 29.70. Drywall thicknesses range from 1/4" to 1". Width and length range from 2' × 8' to 4' × 14'. Moisture-resistant drywall, called *greenboard,* is specified where walls are to be exposed to excess moisture such as on baths and kitchens. *Blueboards* are specified where plaster is to be applied.

- When *paneling* is used as an interior finish, horizontal furring strips should be placed on the studs to provide a nailing or gluing surface for the paneling. See Figure 29.71. The type of joint used between panels should be determined by developing a separate detail. The method of intersecting the outside corners of paneling should also be detailed. Outside corners can be intersected by mitering, overlapping, or exposing the paneling. Corner boards, metal strips, or molding may be used on the intersections. Inside corner intersections can be constructed by butting the wall coverings or by using corner moldings.

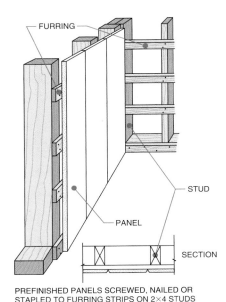

PREFINISHED PANELS SCREWED, NAILED OR
STAPLED TO FURRING STRIPS ON 2×4 STUDS

FIGURE 29.71 ❯ Furring strips provide a horizontal surface for attaching paneling. *Hepler/Wallach/Hepler © Cengage 2013*

STUD LAYOUTS

A horizontal framing section called a **stud layout** is a plan similar to a floor plan, except it shows the position of each wall framing member, exterior and interior. Stud layouts are used to show how studs are spaced on the plan and how interior partitions fit together.

A stud layout is a horizontal section through walls and partitions. The cutting-plane line is placed approximately at the midpoint of the panel elevation. See Figure 29.72. The exact position of each stud that falls on an established modular center (16″, 32″, or 48″) is usually shown by diagonal crossed lines. Studs other than those on 16″ centers are shown with different symbols. Different symbols also identify studs that are different in size, such as blocking and short pieces of stud stock. See Figure 29.73. A coding system of this type eliminates the need for dimensioning the position of each stud. The practice of coding studs and other members in the stud plan also eliminates the need for repeating the dimensions of each stud.

To conserve space, nonbearing studs are sometimes turned so they are flat. This rotation should be reflected in the stud layout. See Figure 29.74.

Dimensions

Detailed dimensions are normally shown on a stud symbol key or on a separate enlarged detail. Distances that are typically dimensioned on a stud layout include the following:

- Inside framing dimensions of each room (stud to stud)
- Framing width of the halls

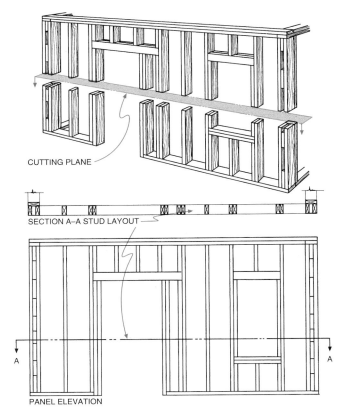

FIGURE 29.72 ❯ A stud layout is a section through wall framing. *Hepler/Wallach/Hepler © Cengage 2013*

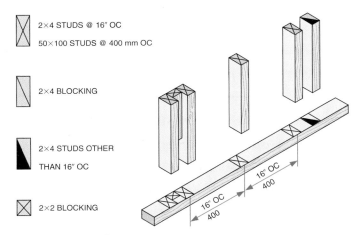

2×4 STUDS @ 16″ OC
50×100 STUDS @ 400 mm OC

2×4 BLOCKING

2×4 STUDS OTHER THAN 16″ OC

2×2 BLOCKING

FIGURE 29.73 ❯ Stud layout symbols. *Hepler/Wallach/Hepler © Cengage 2013*

- Rough opening for doors and arches
- Length of each partition
- Width of partition where dimension lines pass through from room to room. This provides a double check to ensure that the room dimensions plus the partitioned dimensions add up to the overall dimension. When a stud layout is available, it may be used on the job to establish partition positions.

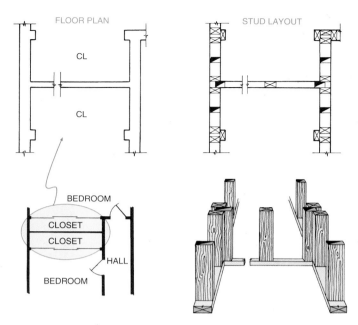

FIGURE 29.74 > Stud position shown on a stud layout.
Hepler/Wallach/Hepler © Cengage 2013

Stud Details

Stud layouts are of two types: *complete plans* (Figure 29.75), which show the position of all framing members on the floor plan, and *stud details,* which show only the position and relationship of some studs or framing intersections. For example, the position of each stud in a corner-post layout is frequently detailed in a plan view. See Figure 29.76. Builders often prefer different corner-post layouts. If corners do not intersect at right angles, different positioning of studs is also necessary as shown in Figure 29.77.

Occasionally, siding and inside-wall covering materials are included on this plan. Preparing this type of drawing without showing the covering materials, however, is the quickest way to show corner-post construction. If wall coverings are included on the detail, the complete wall thickness can be drawn. In a plan section, care should be taken to show the exact position of blocking because it may not pass through the cutting-plane line. Blocking should be labeled or a symbol used to prevent the possibility of mistaking it for a full-length stud. See Figure 29.78.

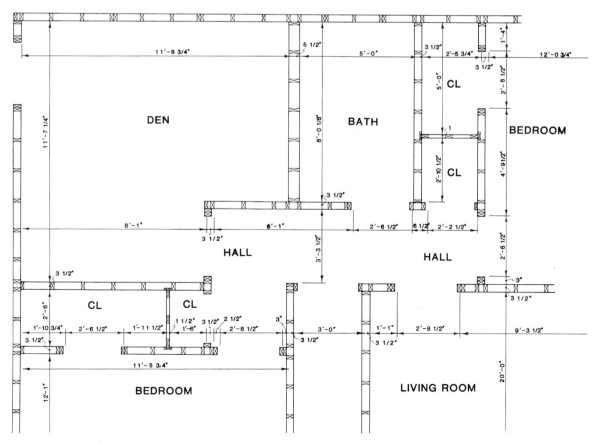

FIGURE 29.75 > Dimensioned stud layout. *Hepler/Wallach/Hepler © Cengage 2013*

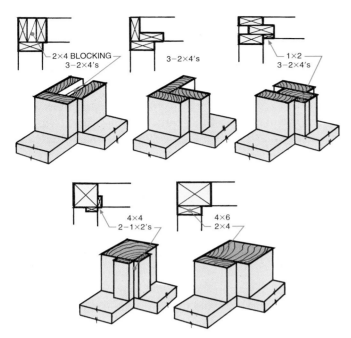

FIGURE 29.76 > Corner-post layout shown with stud layouts. *Hepler/Wallach/Hepler © Cengage 2013*

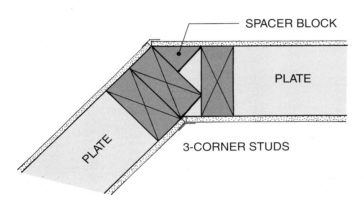

FIGURE 29.77 > Angular corner framing. *Hepler/Wallach/ Hepler © Cengage 2013*

When laying out the position of all studs, remember that the finished dimensions of a 2 × 4 stud are actually 1 1/2″ × 3 1/2″. For the exact dressed sizes of other rough stock, refer back to Chapter 24. Stud lengths precut to 92 1/2″ and combined with a double top plate (3″) and a bottom plate (1 1/2″) yield a floor-to-ceiling length of 8′-1″.

Modular Plans

In drawing modular framing plans, partitions should fit in relation to modular grid lines. Allowances for exterior wall thicknesses need to be indicated. See Figure 29.79. Space must also be provided for such items as door and window placement, medicine cabinets, closets, fireplaces, and plumbing runs.

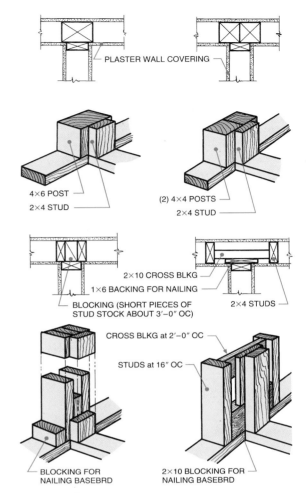

FIGURE 29.78 > Wall covering relationship with stud positions. *Hepler/Wallach/Hepler © Cengage 2013*

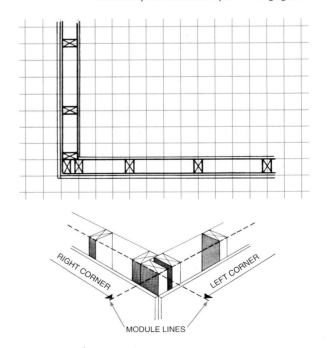

FIGURE 29.79 > Stud layout applied to a 4″ modular grid. *Hepler/Wallach/Hepler © Cengage 2013*

Steel Studs

Steel studs are one-half the weight of wood studs. They are stronger, won't warp or split, and are moisture and insect resistant. Some nonbearing steel studs are prepunched for electrical or plumbing lines or for attachment to wood sills, plates, or other studs. See Figure 29.80. A steel symbol should be added to stud layouts where steel studs are specified.

FIGURE 29.80 > Nonbearing steel stud wall. *Hepler/Wallach/ Hepler © Cengage 2013*

Wall Framing Drawings Exercises

CHAPTER 29

1. Prepare an exterior wall framing plan for a home of your own design.
2. Draw a stud layout (16″ OC) for a floor plan. Indicate the rough openings for doors and windows. Show a corner post and intersection detail.
3. Draw a floor plan and a wall framing elevation using post-and-beam construction for a 15′ × 10′ storage shed.
4. Draw a window detail of a window component: a head, jamb, or sill.
5. Draw a framing elevation for a double swinging door. Include the dimensions.
6. Project a framing elevation drawing of a kitchen wall in Figure 17.41.
7. Draw an interior wall framing plan for a kitchen, bath, and living area wall of the house you are designing.
8. Prepare a stud layout (1/4″ = 1′−0″). for a home of your own design.

9. Complete a stud detail of a corner post and another intersection.

10. Draw the construction details for a standard T-foundation with a brick veneer wall.

11. Draw a wall framing elevation of the four exterior walls shown in Figure 29.81.

12. Draw the plan in Figure 29.82 on a 4" modular grid.

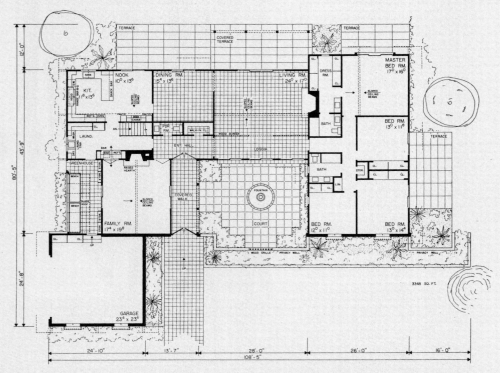

FIGURE 29.81 > Draw exterior wall framing elevations of all exterior walls shown on this plan. *Hepler/Wallach/Hepler © Cengage 2013*

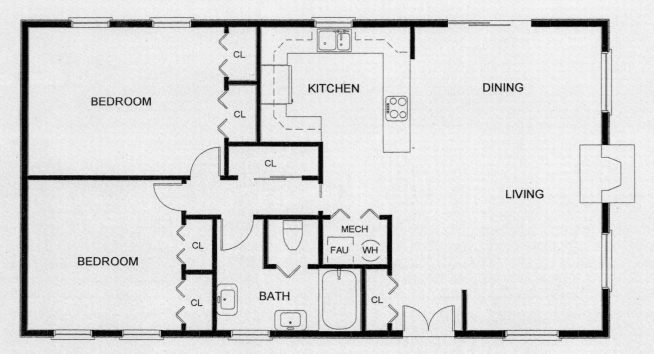

FIGURE 29.82 > Draw this plan on a 4" modular grid. *Hepler Associates LA, PC*

OBJECTIVES

In this chapter you will learn to:

> describe roof framing members, components, and methods.

> understand how roof appendages are used.

> calculate roof pitch.

> know roof covering material options.

> draw a roof framing plan showing structural members, sizes, pitch, and spacing.

> draw roof framing details and elevations.

TERMS

cant strip	knee wall	shingles
collar beam	lookout rafters	slope
common rafter	pitch	slope diagram
cornice	purlin	soffit
fascia	rafter	span
flashing	ridge board	stressed skin
folded plate	rise	truss
gusset plate	run	valley rafter
hip roof	shakes	

INTRODUCTION

Roof styles and construction methods developed through the centuries. Pitches (slopes) of roofs were changed, gutters and downspouts were added for better drainage, and overhangs were extended to provide more protection from the sun. The size of roofs and the types of materials changed accordingly. The structure and architectural style of a roof affects the choice of roof framing and roof covering materials, as well as the interior ceiling systems.

ROOF FUNCTION

The main function of a roof is to provide protection from rain, snow, sun, and hot or cold temperatures. In a cold climate, a roof is designed to withstand heavy snow loads. In a tropical climate, a roof provides protection mainly from sun and rain.

The walls of a structure are given stability by their attachment to the ground and to the roof. Buildings are not structurally sound without roofs. As explained in Chapter 22, live loads that act on the roof include wind loads and snow loads, which vary greatly from one geographical area to another. Dead loads that bear on roofs include the weight of the structural members and coverings. All loads are computed by pounds per square foot, or kilograms per square meter if metric measurements are used. (For computation of loads for an entire structure, see Appendix B, *Mathematical Calculations*.)

ROOF FRAMING MEMBERS

The structural members of a roof must be strong enough to withstand many types of loads. Members used for structural roof support include rafters, girders, beams, joists, purlins, trusses, slabs, and shells. These members can be made from wood, steel, or concrete.

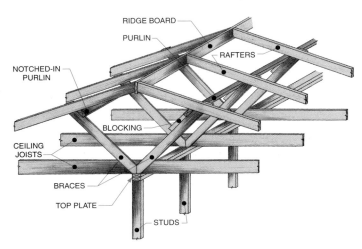

FIGURE 30.1 > Ridge board related to roof framing members. *Hepler/Wallach/Hepler © Cengage 2013*

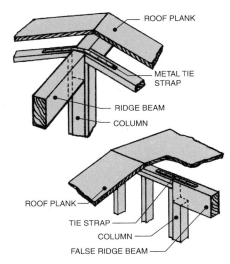

FIGURE 30.2 > Roof sheathing shown over ridge beam. *Hepler/Wallach/Hepler © Cengage 2013*

Wood Roof Members

Structural members used for roof construction may be made from the same materials used for floor framing. A **ridge board,** is the top member in the roof assembly. See Figure 30.1. In post-and-beam construction, the ridge board may be exposed on the inside of a building. Figures 30.2 and 30.3 show various intersections of ridge boards and rafters with roof sheathing.

Roof **rafters** intersect ridge boards and rest on the tops of outside walls. Rafters may be selected from the same materials as floor joists. They may be solid lumber, truss joists, I-joists, laminated, or stranded lumber. Figure 30.4 shows how rafters are joined to the ridge board and top wall plate. Refer to Chapter 24 to review the fundamentals of wood framing.

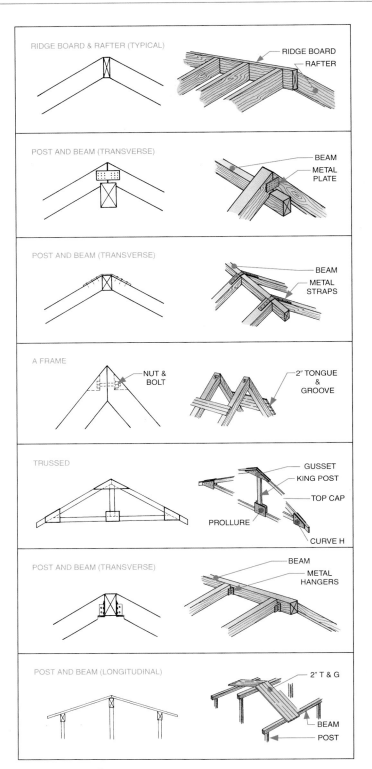

FIGURE 30.3 > Typical ridge board intersections. *Hepler/Wallach/Hepler © Cengage 2013*

(SOME DETAILS LEFT OUT FOR CLARITY.)

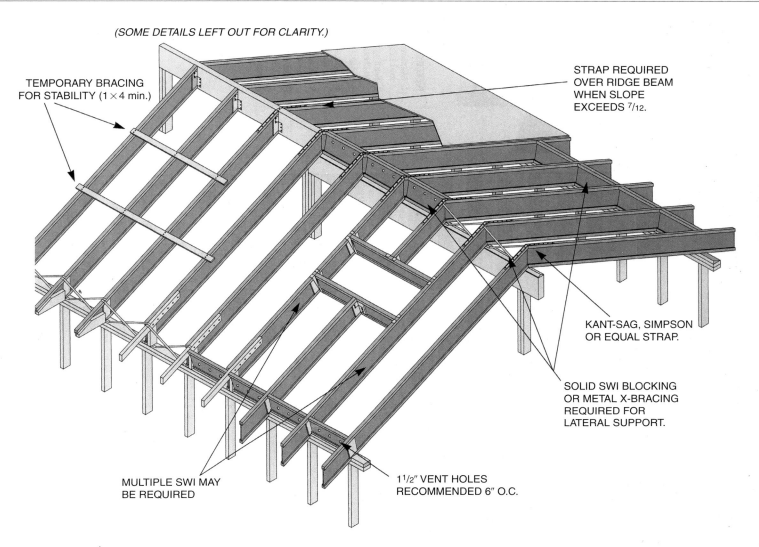

TEMPORARY BRACING
FOR STABILITY (1 × 4 min.)

STRAP REQUIRED
OVER RIDGE BEAM
WHEN SLOPE
EXCEEDS 7/12.

KANT-SAG, SIMPSON
OR EQUAL STRAP.

SOLID SWI BLOCKING
OR METAL X-BRACING
REQUIRED FOR
LATERAL SUPPORT.

MULTIPLE SWI MAY
BE REQUIRED

1 1/2" VENT HOLES
RECOMMENDED 6" O.C.

FIGURE 30.4 › Rafter connections—Simpson strong tie. *Hepler/Wallach/Hepler © Cengage 2013*

Steel Roof Members

Roofs framed with structural steel use steel girders, beams, and joists in the same manner as steel-framed floors. Most steel-framed roofs are flat, but steel ridge boards and rafters are also used on pitched roofs with large spans. Steel construction as applied to roofs is covered later in this chapter. Refer to Chapter 26 to review the fundamentals of steel building methods.

Concrete Roof Members

To be used for roof members, concrete must be reinforced with rebars and may be prestressed, precast, or sprayed into shells. To better understand the use of concrete for roofs, refer to *concrete construction systems* in Chapter 25 for a review of the fundamentals of concrete construction.

Reinforced concrete is either precast or poured on site with steel reinforcing rods inserted for stability and rigidity. Reinforced concrete slabs are used extensively for roof systems with short spans. A wood-framed flat roof and a reinforced-concrete roof may have the same covering as demonstrated in Figure 30.5.

A *prestressed concrete* roof beam or slab is made by stretching steel rods and pouring concrete around the rods. When the stretched rods are released, the rods try to return to their original shape; this creates tension on the concrete. Prestressing strengthens the concrete and allows beams to have a lower ratio of concrete depth to span.

Precast concrete roof members may be poured off site into molds and then reinforced with steel. When high stresses will not be incurred, precasting without prestressing is acceptable for most short span construction. Precast concrete joists are used where heavy loads are supported by short

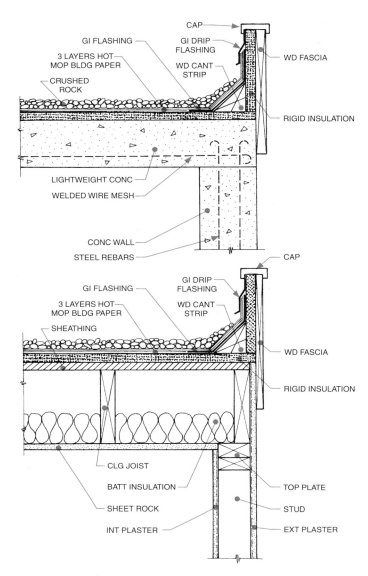

FIGURE 30.5 > Wood-framed roof compared to reinforced-concrete roof construction. *Hepler/Wallach/ Hepler © Cengage 2013*

spans. Filler blocks and a poured slab then create a mono-lithic roof or floor.

Concrete shells are created from lightweight concrete poured or sprayed onto steel mesh and bars that provide a temporary form until the concrete hardens. Once the concrete hardens, the steel is locked in place and the structure becomes rigid and stable. Concrete shells are used to create continuous roofs that can be contoured to whatever extent the mesh can be formed.

ROOF FRAMING COMPONENTS

The materials required for framing roofs are extensive and include trusses, roof panels, cornices, collar beams and knee walls, and dormers.

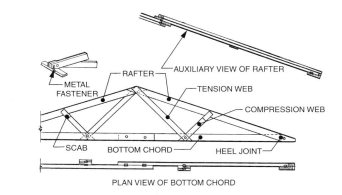

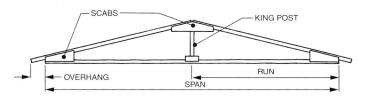

FIGURE 30.6 > Common truss members. *Hepler/Wallach/ Hepler © Cengage 2013*

Trusses

Trusses for roofs are prefabricated components that perform all the functions of rafters, collar beams, and ceiling joists. Trusses consist of a top chord and bottom chord, joined with diagonal and vertical members called *webs*. Figure 30.6 shows the members of a truss, and Figure 30.7 shows examples of common lightweight wood trusses. These truss members are held rigidly in place with bolts, ring connectors, and gussets. Plywood gussets or sheet-metal gussets are used in light construction. Figure 30.8 shows the common locations of **gusset plate** connectors on a typical wood truss. Figure 30.9 illustrates the application of gussets on truss joists. Heavy-timber trusses require heavy-duty steel gusset plates that are welded together and bolted through the top and bottom chords. These connections are shown in Figure 30.10.

Widely spaced trusses are tied together horizontally to make the roof system structurally stable. Horizontal members known as **purlins** are used for this purpose. See Figure 30.11. On a trussed roof, purlins perform the same function as joist bridging in conventional floor construction.

Trusses prevent sags and cracks in the roof system because they are structurally independent and resist both compression and tension forces. See Figure 30.12. Nontrussed members tend to flex. Because trusses can span larger distances than conventionally framed roofs, the spacing of interior partitions is more flexible. Fewer or no bearing partitions may be needed. Trusses may rest on steel beam columns, masonry walls, or on exterior wood-framed walls. Figure 30.13 shows trusses located

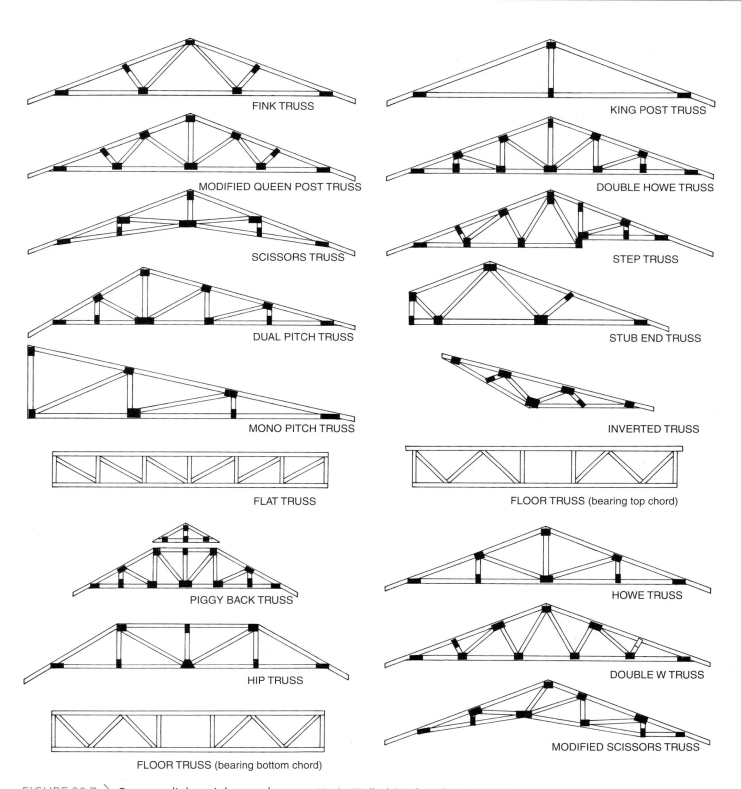

FIGURE 30.7 › Common lightweight wood trusses. *Hepler/Wallach/Hepler © Cengage 2013*

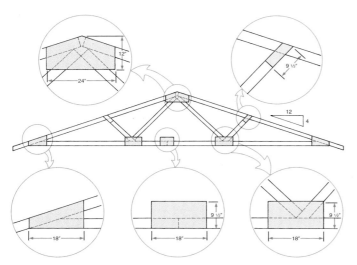

FIGURE 30.8 > Location of gusset plate connectors on a truss. *Hepler/Wallach/Hepler © Cengage 2013*

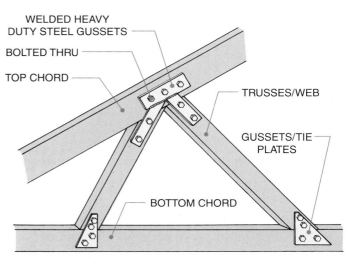

FIGURE 30.10 > Heavy-duty bolted steel gusset plates. *Hepler/Wallach/Hepler © Cengage 2013*

FIGURE 30.9 > Gusset plate applications on trusses. *Hepler/Wallach/Hepler © Cengage 2013*

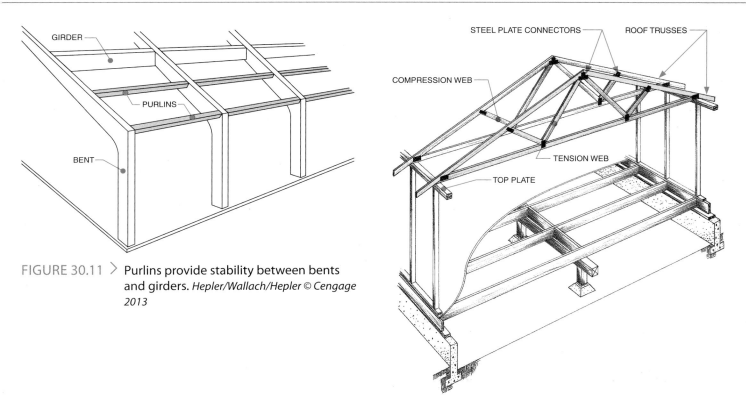

FIGURE 30.11 > Purlins provide stability between bents and girders. *Hepler/Wallach/Hepler © Cengage 2013*

FIGURE 30.13 > Trusses bearing on exterior skeleton-frame walls. *Hepler/Wallach/Hepler © Cengage 2013*

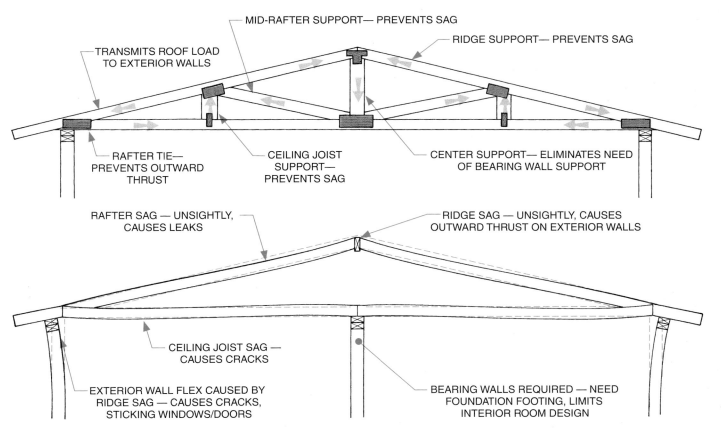

FIGURE 30.12 > Trusses resist tension and compression forces. *Hepler/Wallach/Hepler © Cengage 2013*

on an exterior wood-framed wall, and Figure 30.14 shows trusses supported by masonry walls with pilasters located at the support points.

Truss type and design depend on the length of span, room height requirements, spacing, roof pitch, live and dead loads, and cost factors. Several types of trusses are manufactured for structural steel construction. Figure 30.15 shows the most common types of steel trusses. Figure 30.16 shows the outline of a vast variety of steel trusses. One of the most used steel truss types is the open web steel truss as shown later in Chapter 26 and in Figure 28.19. Open web trusses can be used for floors, flat roofs, and as a bottom chord for gable-type roofs. More complete truss specifications are found in Appendix B, *Mathematical Calculations.*

Roof Panels

Figure 30.17 shows some of the many forms of lightweight, prefabricated roof panels. These units can be designed to resist loads, span great distances, or eliminate the need for trusses.

Stressed Skin Roof Panels

Stressed skin roof panels are constructed of plywood or stranded panels and seasoned lumber. The framing plywood skin acts as a unit to resist loads. Glued joints transmit the shear stresses, making it possible for the structure to act as one piece.

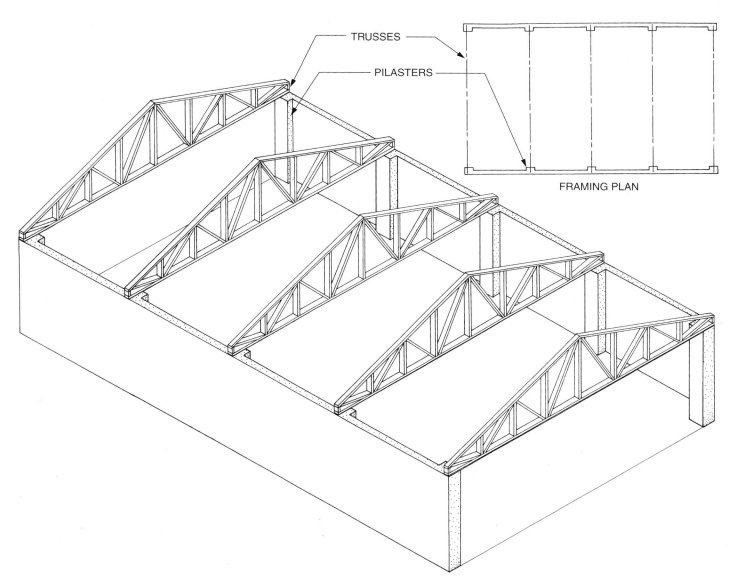

FIGURE 30.14 > Trusses supported by masonry pilasters. *Hepler/Wallach/Hepler © Cengage 2013*

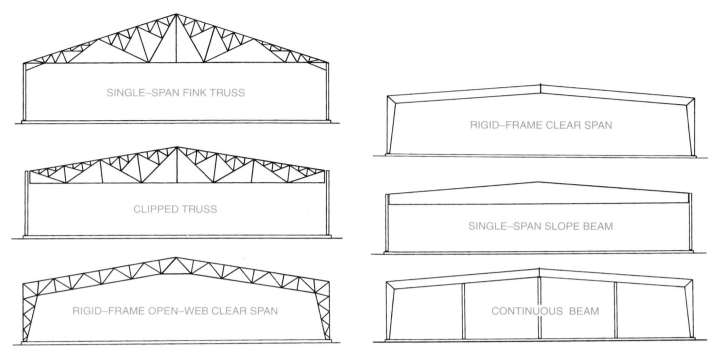

FIGURE 30.15 > Common types of steel trusses. *Hepler/Wallach/Hepler © Cengage 2013*

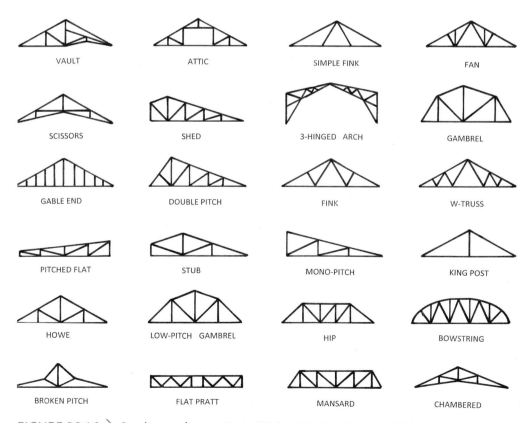

FIGURE 30.16 > Steel truss shapes. *Hepler/Wallach/Hepler © Cengage 2013*

FOLDED PLATES

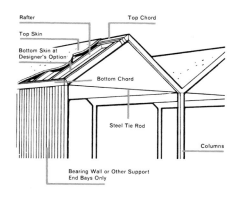

RADIAL FOLDED PLATES

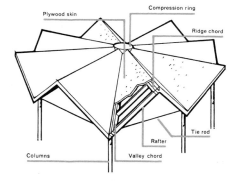

CURVED PANELS

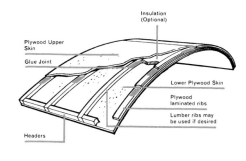

STRESSED SKIN PANELS

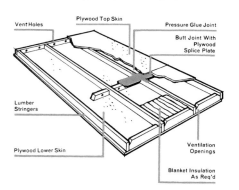

BOX BEAMS

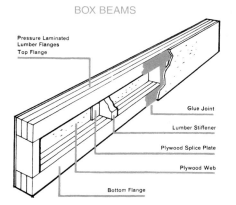

SPACE PLANES

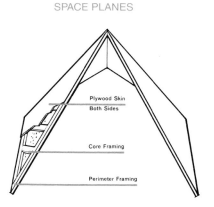

FIGURE 30.17 > Prefabricated roof units. *Hepler/Wallach/Hepler © Cengage 2013*

Curved Panels

The three types of curved panels are the sandwich (or honey-comb paper-core) panel, the hollow-stressed end panel, and the solid-core panel. The arching action of these panels permits spanning across great distances with a relatively thin cross section, as illustrated by the curved ceiling in Figure 30.18.

Folded Plate Roofs

Folded plate roofs are thin skins of plywood reinforced by rafters to form shell structures that can utilize the strength of plywood. The use of folded plate roofs eliminates trusses and other roof members. The tilted plates lean against one another, acting as giant V-shaped beams that are supported by walls or columns.

Cornices

The area of the roof that intersects with the outside walls and extends to the end of the roof overhang is the **cornice.** Detail drawings are necessary to show cornice areas. These detail drawings include part of the wall framing, roof framing, and methods of attaching the roof structure to the wall. Figure 30.19 shows several methods used to draw cornice details.

FIGURE 30.18 > Curved ceiling. *Hepler/Wallach/Hepler © Cengage 2013*

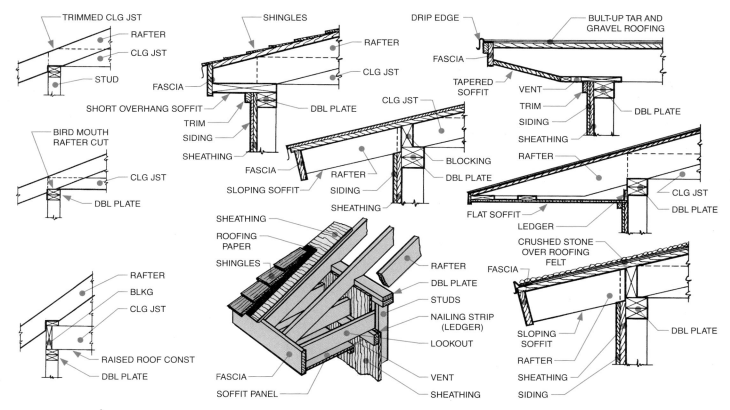

FIGURE 30.19 › Typical cornice framing details. *Hepler/Wallach/Hepler © Cengage 2013*

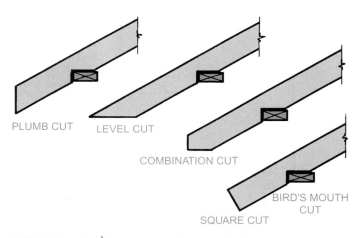

FIGURE 30.20 › Bird's-mouth cut and rafter tail cuts. *Hepler/ Wallach/Hepler © Cengage 2013*

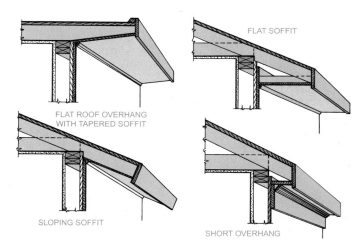

FIGURE 30.21 › Types of soffit design. *Hepler/Wallach/Hepler © Cengage 2013*

Several types of rafter cuts are used at intersections to help hold rafters onto top wall plates. For example, the *bird's mouth cut* shown in Figure 30.20 provides a level surface for the intersection of the rafters and top plates. The area bearing on the plate should not be less than 3″ (76 mm).

The outer vertical surface of an overhang is the **fascia,** and the horizontal bottom of an overhang structure is the **soffit.** There are two types of soffit design: open and closed. In open soffit construction the rafters are exposed and no soffit panel is used. In closed soffit construction a soffit panel is added using one of

the four types shown in Figure 30.21. Figure 30.22 shows the details of a closed soffit design. Figure 30.23 shows one method of designing overhangs with optional rafter lengths.

Soffits are made from plywood or sheet-metal panels. Soffit panels should contain screened openings to allow air to pass between rafters or to circulate through attics or crawl spaces if rafters are exposed. Sheet-metal soffits are available in solid or ventilated designs to allow air flow. Regardless of the material used, some air flow must be designed into the cornice, as shown in Figure 30.24, to provide ventilation and

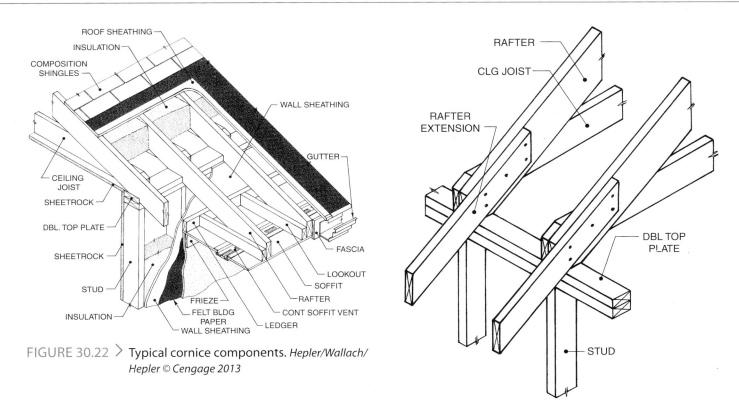

ROOF SHEATHING
INSULATION
COMPOSITION SHINGLES
WALL SHEATHING
GUTTER
CEILING JOIST
SHEETROCK
DBL. TOP PLATE
SHEETROCK
STUD
INSULATION
FASCIA
LOOKOUT
SOFFIT
RAFTER
CONT SOFFIT VENT
FRIEZE
FELT BLDG PAPER
WALL SHEATHING
LEDGER

FIGURE 30.22 > Typical cornice components. *Hepler/Wallach/ Hepler © Cengage 2013*

RAFTER
CLG JOIST
RAFTER EXTENSION
DBL TOP PLATE
STUD

FIGURE 30.23 > Adjustable rafter length design. *Hepler/ Wallach/Hepler © Cengage 2013*

2x10 RAFTERS at 16" OC NOTCHED TO 6 1/2" AT OVERHANG
METAL DRIP EDGE
1x8 FASCIA
1 1/2"x3"x5" BLOCKING
SOFFIT
1x4 TRIM
ALUM SCREEN
air flow
1x12 FRIEZE
2-2x8 HEADERS

SECTION D/2
SCALE: 1"=1'-0"

1x3
DRIP FLASHING
SCREEN FLASHING
air flow
2x10 RAFTERS at 16" OC NOTCHED TO 5 1/2" AT OVERHANG
3/4" QUARTER ROUND
3 1/8"x12" STRUC LAM WOOD BEAM
6" THICK INSULATION
5"
METAL DRIP EDGE
1x8 FASCIA

SECTION E/4
SCALE: 1"=1'-0"

3/8" EXTERIOR GRADE SOFFIT WITH CONTINUOUS 2" WIDE ALUM SCREENED VENT

CONT ALUM SCREEN OPENING
DRIP FLASHING
FLASHING
ASPHALT SHINGLES
15# FELT
1/2" PLYWOOD SHEATHING
2x10 RAFTERS at 16" OC
CONT AIR SPACE
6" THICK INSULATION
3 1/8"x12" STRUC LAM WOOD BEAM
2x4x2 7/16"-4 3/16" BLOCKING EACH RAFTER
VENT
8'-1 1/2" PLATE HT
1x3 TRIM
2x6 RAFTERS at 12" OC

SECTION F/1
SCALE: 1"=1'-0"

2x10 RAFTERS at 16" OC NOTCHED TO 6 1/2" AT OVERHANG
CONT AIR FLOW SPACE OVER 6" THK INSULATION
3/4" QUARTER ROUND
1x8
SOFFIT WITH CONTINOUS 2" WIDE SCREENED VENT
2-2x8 HEADER
FIXED GLASS
2'-0"

SCALE: 1"=1'-0"   **SECTION** C/2

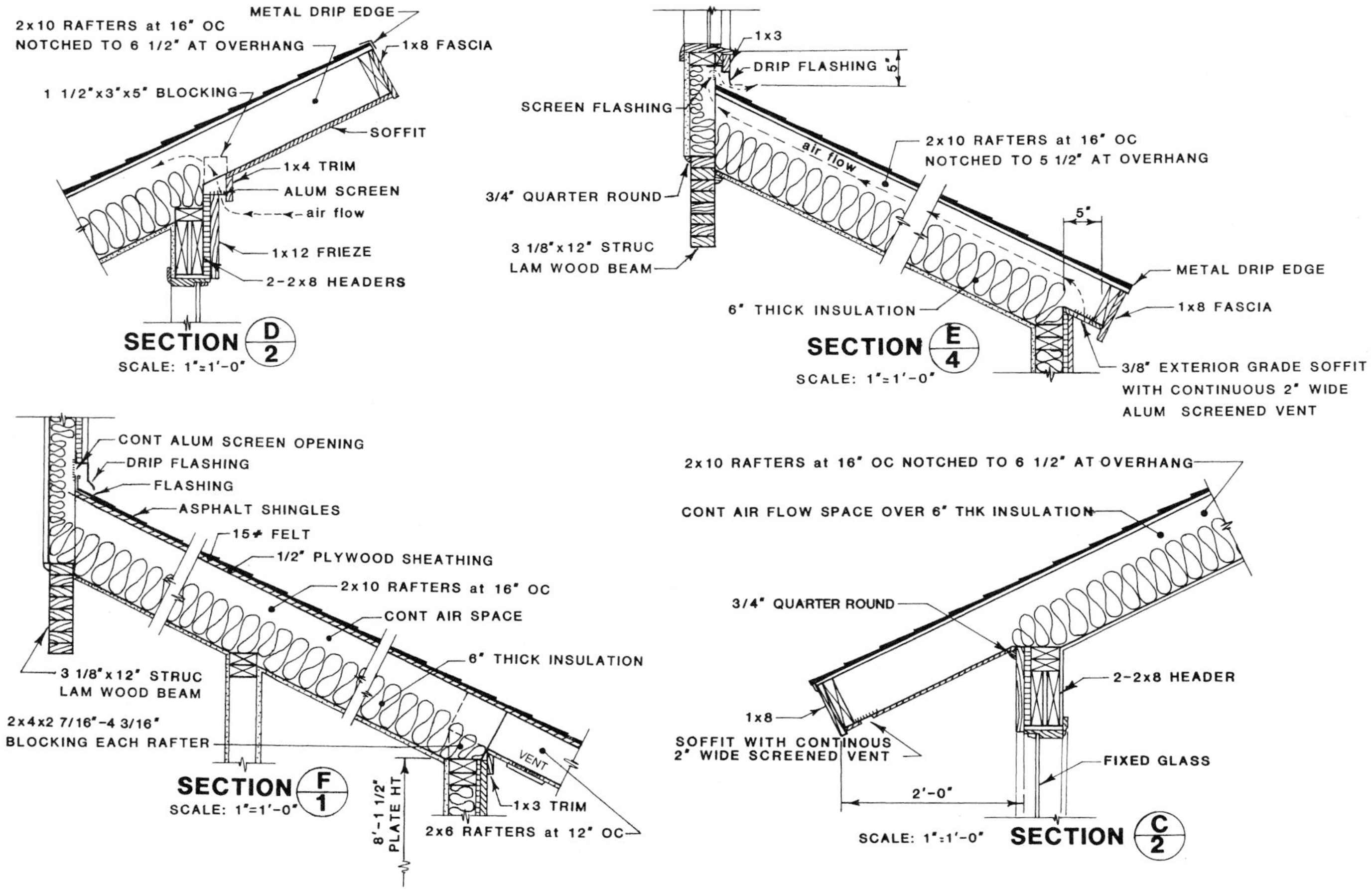

FIGURE 30.24 > Cornice design incorporating insulation and air flow. *Hepler/Wallach/Hepler © Cengage 2013*

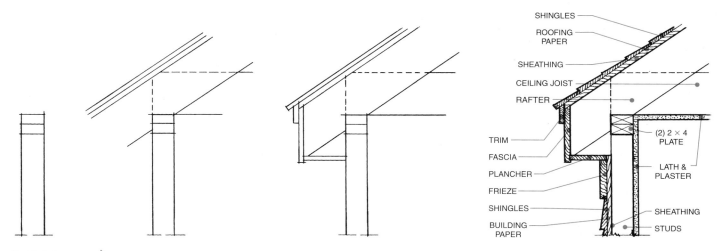

FIGURE 30.25 > Steps in drawing cornice details. *Hepler/Wallach/Hepler © Cengage 2013*

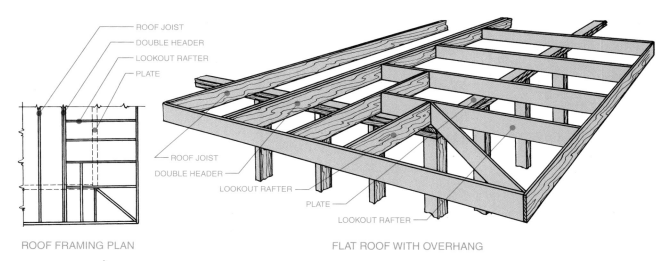

ROOF FRAMING PLAN

FLAT ROOF WITH OVERHANG

FIGURE 30.26 > Lookout rafters used to extend overhang. *Hepler/Wallach/Hepler © Cengage 2013*

prevent dry rot. Figure 30.25 outlines the steps in laying out and drawing a typical cornice elevation detail.

Cantilevered **lookout rafters** are used where **common rafter** extensions cannot create an overhang. See Figure 30.26. Lookout rafters are placed perpendicular to the common rafter direction.

Collar Beams and Knee Walls

Collar beams provide a tie between opposing rafters. They are usually placed on every rafter or every second or third rafter. Collar beams are used to reduce the rafter stress that occurs between the top plate and the top of the rafter. They lock rafters into position. They may also act as ceiling joists for finished attics. On low-pitched roofs, collar beams may be required to counteract the lateral (outward) thrust of joists. See Figure 30.27.

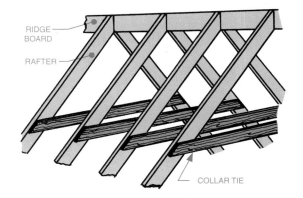

FIGURE 30.27 > Collar beams (ties) are used to reduce stress on rafters. *Hepler/Wallach/Hepler © Cengage 2013*

Knee walls are vertical studs that project from ceiling joists or attic floors to roof rafters. See Figure 30.28. Knee walls add rigidity to the rafters and may also provide half-wall framing for finished attics.

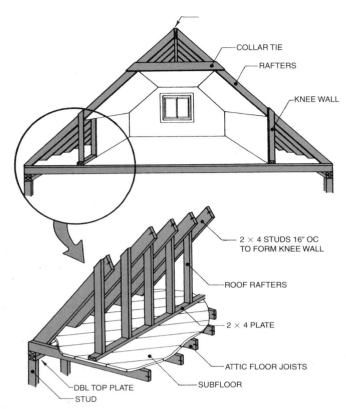

FIGURE 30.28 > Knee walls add rigidity to rafters. *Hepler/ Wallach/Hepler © Cengage 2013*

Dormers

As covered in Chapter 16, there are two basic types of dormers. These have different framing requirements. Figure 30.29 illustrates the methods of drawing a related plan view and front elevation of a gable dormer. Figure 30.30 shows a side eleva-

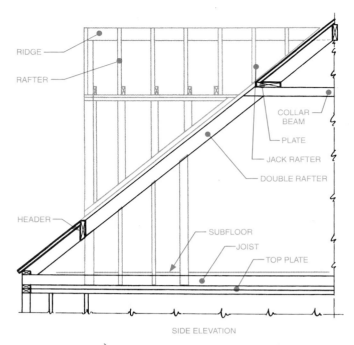

FIGURE 30.30 > Gable dormer side elevation. *Hepler/ Wallach/Hepler © Cengage 2013*

tion of a different gable end dormer, and Figure 30.31 shows a framing elevation for a typical individual shed dormer. See Chapter 16 for a review of dormer types. All of these require the same type of detail drawing.

Dormer rafters and walls do not lie in the same plane as the remainder of the roof rafters. A framing elevation drawing is therefore needed to show the exact position of the dormer members and their tie-in with the common roof rafters and with other roof framing members.

Chimney Details

If a detailed roof framing plan is not prepared, chimney framing roof details are often provided on separate detail drawings. Chimney details include the position of ceiling joists, roof rafters, and type of construction. See Figure 30.32.

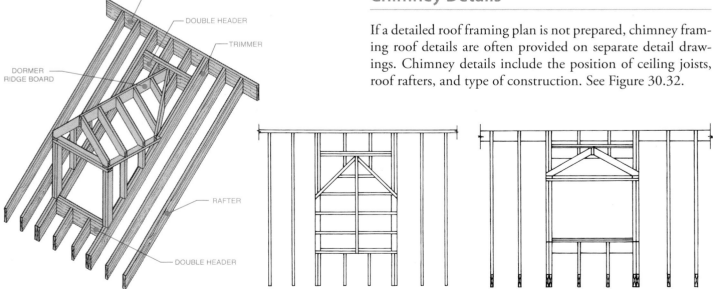

FIGURE 30.29 > Gable dormer framing plan and elevation. *Hepler/Wallach/Hepler © Cengage 2013*

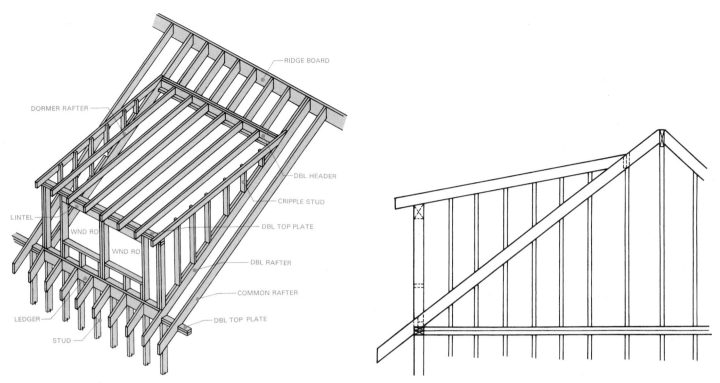

FIGURE 30.31 > Shed dormer framing elevation. *Hepler/Wallach/Hepler © Cengage 2013*

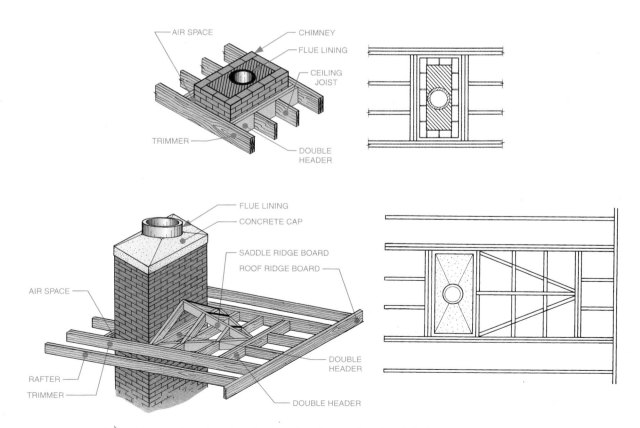

FIGURE 30.32 > Chimney framing details showing flat roof and saddle intersections. *Hepler/Wallach/Hepler © Cengage 2013*

ROOF FRAMING DRAWINGS

Plans

To convey the structural design of a roof to a builder, plans must be accurate and complete. A *roof plan* shows only the outline and the major object lines of the roof. A roof plan is not a framing plan but a plan view of the roof. A roof plan can be used, however, as the basic outline to develop a roof framing plan. A *roof framing plan* exposes the exact position and spacing of each structural member. See Figure 30.33A.

Roof Framing Plans on CAD

The top-level floor plan layer is used as a guide in preparing the roof framing plan. The *Line* command or the *Polyline* command is then used to draw rafters. The *Trim* or *Break* command is then used to remove portions of rafters for openings such as chimneys, skylights, and dormers. Double rafters and special framing for intersections are then added using the *Offset* command. Figure 30.33B shows a roof framing plan.

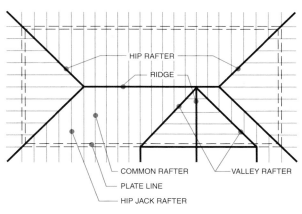

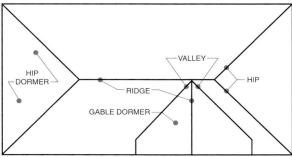

FIGURE 30.33A ⟩ Single-line roof framing plan compared to a roof plan. *Hepler/Wallach/Hepler* © Cengage 2013

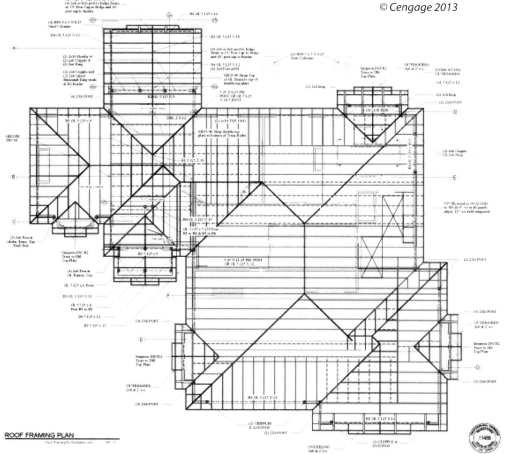

FIGURE 30.33B ⟩ Roof framing plan with pictorials. *Chief Architect Software*

Roof framing plans may be either simplified, single-line plans or detailed, double-line plans. A single-line plan (Figure 30.33A) is acceptable only to show the general relationship and spacing of the structural members. When more details concerning the exact construction of intersections and joints are needed, a double-line plan showing the thickness of each member should be prepared. See Figure 30.34. This type of plan is necessary to indicate the relative placement of one member compared with another; that is, to indicate whether one member passes over or under another. In a double-line plan, the width of ridge boards, rafters, headers, and plates should be drawn to the exact scale. If a complete roof framing plan of this type cannot fully describe construction framing details, then additional removed pictorial or elevation drawings should be prepared. See Figure 30.35. These separate drawings can also be enlarged, for example, to better show detail dimensions.

In contrast to a true orthographic projection, only the outline of the top of the rafters is shown on roof framing plans. Areas underneath the rafters are shown by dashed lines. Roof framing plans show only horizontal relationships of members, such as thickness, length, and horizontal spacing. See Figure 30.36. In a top view (plan view), you cannot show vertical dimensions such as structural heights and pitches. Some plan sets show the position of ridge, hip, valley, jack rafters, and plate lines directly on the floor plan with dashed lines as shown in Figure 30.37.

Roof plans are also used as a basis for preparing *roof drainage plans,* which show the direction of water runoff in all directions. Arrows show the runoff direction and are located within each flat segment of a roof, as shown in Figure 30.38.

Elevations

The angle or vertical position (roof pitch) of any roof framing member should be shown on a roof framing elevation. If a comparison of different heights and pitches is desired, a *composite framing-elevation* drawing should be prepared. This elevation is one that can be projected from the roof framing plan or from corresponding lines on the elevation drawings. Figure 30.39 on page 652, shows the projection of an elevation framing drawing from a framing elevation plan. In Figure 30.40 on page 652, the roof framing details are abbreviated on a sectional elevation drawing of the entire structure.

Dimensions

Dimensions on roof framing plans usually include the size and spacing of framing members and the major distances (spans) between framing components. See Figure 30.41 on page 653. Regular spacing of structural members, such as roof rafters, floor joists, and wall studs, is not dimensioned if these fall on modular increments. Notes may be used on framing drawings to show the size and spacing of members.

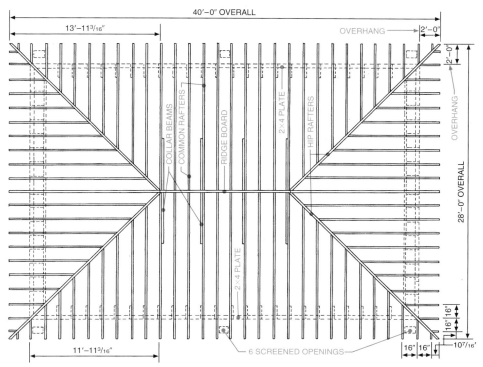

FIGURE 30.34 > Double-line roof framing plan. *Hepler/Wallach/Hepler © Cengage 2013*

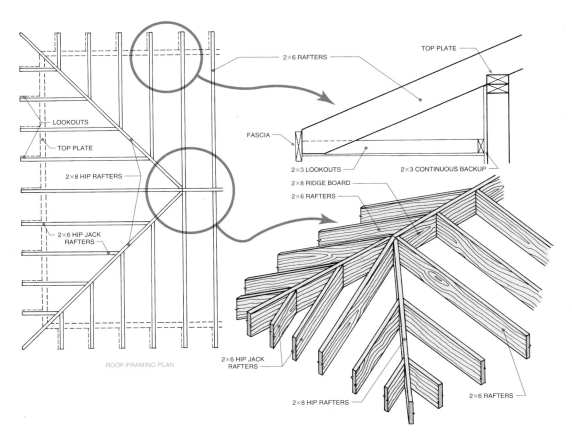

FIGURE 30.35 > Roof framing detail plan and elevation. *Hepler/Wallach/Hepler © Cengage 2013*

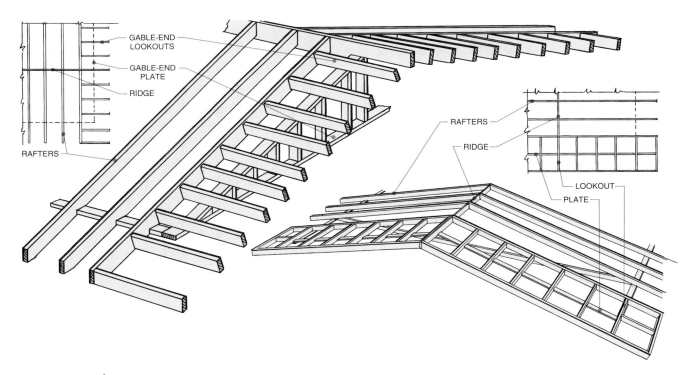

FIGURE 30.36 > Roof framing plans show only tops of members. *Hepler/Wallach/Hepler © Cengage 2013*

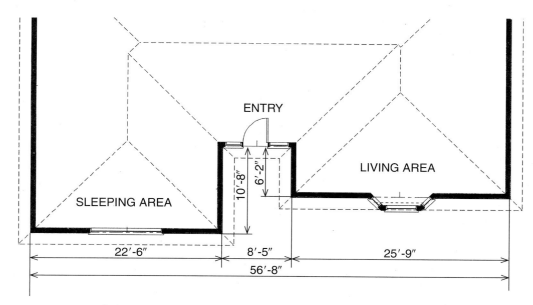

FIGURE 30.37 > Roof plan lines on a floor plan. *Hepler/Wallach/Hepler © Cengage 2013*

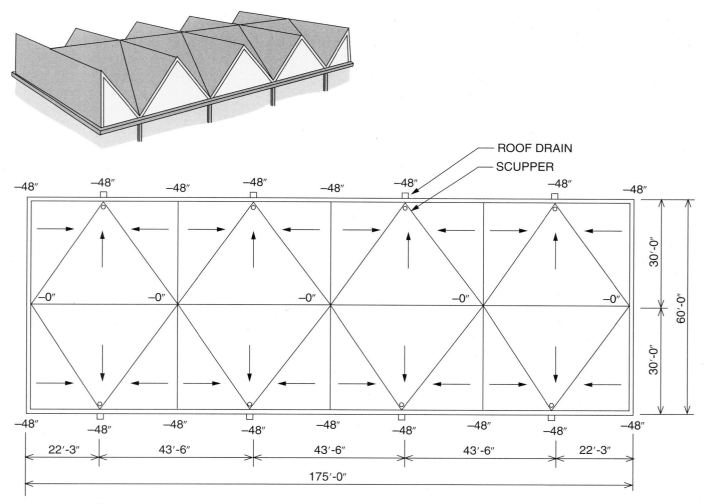

FIGURE 30.38 > Roof drainage plan. *Hepler/Wallach/Hepler © Cengage 2013*

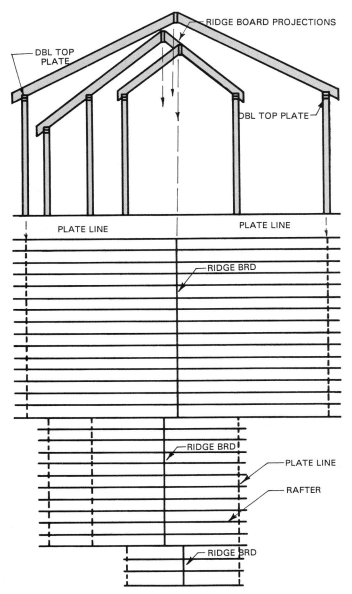

FIGURE 30.39 > Projection of roof framing elevation from framing plan. *Hepler/Wallach/Hepler*
© Cengage 2013

Figure 30.42 shows examples of three ceiling joist framing plans using different size and spacing notations. On detail drawings, overall dimensions are not given. Instead, the key distances between structural levels and horizontal span distances are shown.

If material sizes are not given on the framing drawing, they should be included in the specifications. Figure 30.43 shows typical light construction roof rafter sizes, spans, and spacing for a slope of 4/12 or more. Similar information for ceiling joists is listed in Figure 30.44. All roof member sizes and spacing are based on spans and loads. Refer to Appendix B, *Mathematical Calculations*.

ROOF SLOPE AND PITCH

Slope is the angle between the roof's surface (top plate to ridge board) and the horizontal plane. The **rise** is the vertical distance from the top plate to the roof's ridge. The **run** is the horizontal distance from the top plate to the ridge. It is expressed in units of 12. The **span** is the full horizontal distance between outside supports. It is double the run.

Figure 30.45 shows how units of rise (vertical distance) per units of run (horizontal distance) determine roof pitch. On drawings, this figure is shown in a **slope diagram** (triangle) near the line of the roof along with the run unit number. The roof pitch on this drawing is 8/24 or 1/3.

Pitch is the ratio of the *actual* rise to the *actual* span. It is also the ratio of the *units* of rise to *units* of span (double the units of run). Refer again to Figure 30.45. In the drawing on the left, the pitch is 10/30 (ratio of actual rise to actual span), which reduces to 1/3. It is also 8/24 (unit ratio of rise to span), which also reduces to 1/3.

Slope is the incline of a roof. Slope is expressed as a ratio of the vertical (rise) to the horizontal (run). This may be a fraction or may be noted as (*x* in 12) as shown in the triangle dimension in Figure 30.45. Slope data is converted to a roof

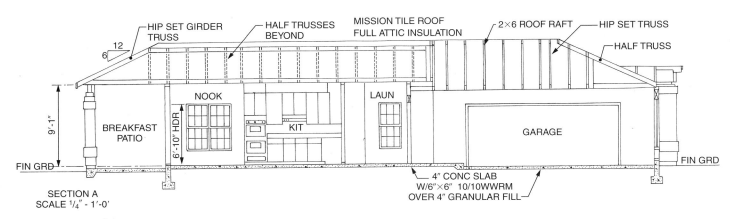

FIGURE 30.40 > Roof framing shown on a simplified full elevation section. *Hepler/Wallach/Hepler* © Cengage 2013

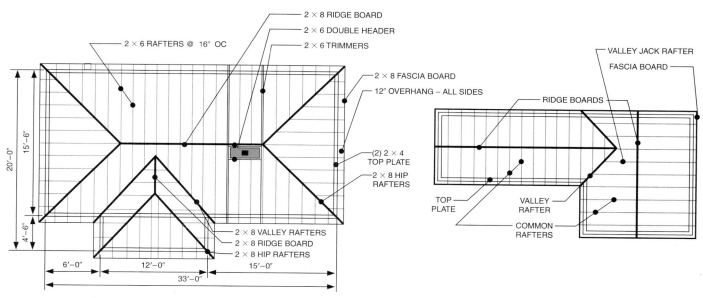

FIGURE 30.41 ❯ Dimensioned roof framing plan. *Hepler/Wallach/Hepler © Cengage 2013*

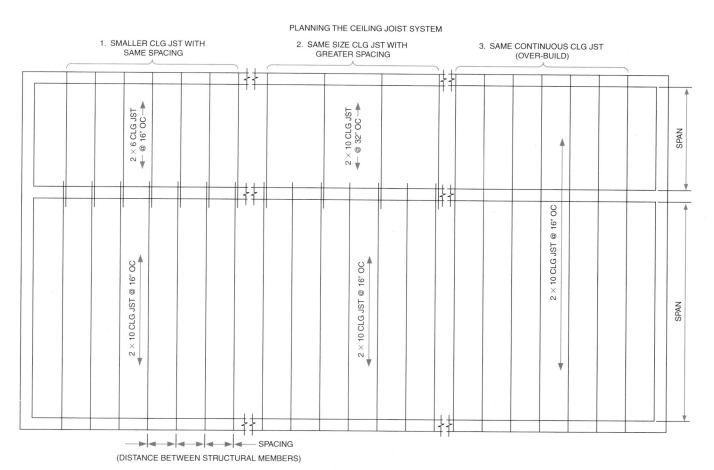

FIGURE 30.42 ❯ Notation methods used to define roof and ceiling framing. *Hepler/Wallach/Hepler © Cengage 2013*

SIZE RAFTER	SPACING RAFTER	MAXIMUM SPAN			
		GROUP 1	GROUP 2	GROUP 3	GROUP 4
2 × 4	12"	10'–0"	9'–0"	7'–0"	4'–0"
	16"	9'–0"	7'–6"	6'–0"	3'–6"
	24"	7'–6"	6'–6"	5'–0"	3'–0"
	32"	6'–6"	5'–6"	4'–6"	2'–6"
2 × 6	12"	17'–6"	15'–0"	12'–6"	9'–0"
	16"	15'–6"	13'–0"	11'–0"	8'–0"
	24"	12'–6"	11'–0"	9'–0"	6'–6"
	32"	11'–0"	9'–6"	8'–0"	5'–6"
2 × 8	12"	23'–0"	20'–0"	17'–0"	13'–0"
	16"	20'–0"	18'–0"	15'–0"	11'–6"
	24"	17'–0"	15'–0"	12'–6"	9'–6"
	32"	14'–6"	13'–0"	11'–0"	8'–6"
2 × 10	12"	28'–6"	26'–6"	22'–0"	17'–6"
	16"	25'–6"	23'–6"	19'–6"	15'–6"
	24"	21'–0"	19'–6"	16'–0"	12'–6"
	32"	18'–6"	17'–0"	14'–0"	11'–0"

FIGURE 30.43 > Common wood rafter spans for pitch greater than 4/12. *Hepler/Wallach/Hepler © Cengage 2013*

SIZE OF CEILING JOISTS	SPACING OF CEILING JOISTS	MAXIMUM SPAN			
		WOOD GROUP 1	WOOD GROUP 2	WOOD GROUP 3	WOOD GROUP 4
2 × 4	12"	11'–6"	11'–0"	9'–6"	5'–6"
	16"	10'–6"	10'–0"	8'–6"	5'–0"
2 × 6	12"	18'–0"	16'–6"	15'–6"	12'–6"
	16"	16'–0"	15'–0"	14'–6"	11'–0"
2 × 8	12"	24'–0"	22'–6"	21'–0"	19'–0"
	16"	21'–6"	20'–6"	19'–0"	16'–6"

FIGURE 30.44 > Common wood ceiling joist spans. *Hepler/ Wallach/Hepler © Cengage 2013*

ROOF FRAMING METHODS WITH WOOD

Wood framing can be divided into conventional and heavy-timber methods, regardless of roof shape. Both methods are used extensively in all types of wood-framed roofs.

The *conventional* method of roof framing consists of spacing roof rafters or trusses at small intervals, such as 16" on center. See Figure 30.47. These roof rafters are perpendicular to the ridge board and align with the exterior studs placed on the same centers.

An adaptation of this conventional method of constructing roofs is to substitute roof trusses for conventional rafters and ceiling joists. Trusses create a much more rigid roof, but they make it impossible to use space between ceiling joists and rafters for an attic or crawl-space storage. See Figure 30.48.

Heavy-timber construction, another method of roof framing, consists of posts that support beams. Longitudinal beam

pitch number by doubling the run (24) and positioning the rise as a numerator (x/24).

Roof slopes vary greatly. For example, a roof with a slope of 12/12 is steep. The pitch would be 12/24 or 1/2 (45°). A roof with a slope of 8/12 is moderately sloped. The pitch would be 8/24 or 1/3. A roof with a slope of 2/12 is nearly flat. The pitch would be 2/24 or 1/12. See Figure 30.46.

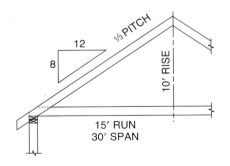

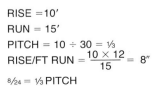

RISE = 10'
RUN = 15'
PITCH = 10 ÷ 30 = 1/3
RISE/FT RUN = $\frac{10 \times 12}{15}$ = 8"

8/24 = 1/3 PITCH

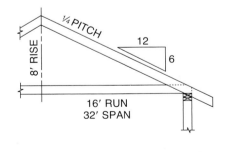

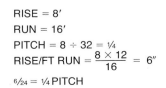

RISE = 8'
RUN = 16'
PITCH = 8 ÷ 32 = 1/4
RISE/FT RUN = $\frac{8 \times 12}{16}$ = 6"

6/24 = 1/4 PITCH

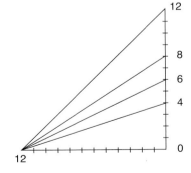

TYPICAL PITCHES	1/2	1/3	1/4	1/6
RISE/FT RUN	12	8	6	4

FIGURE 30.45 > Methods of determining roof pitch and slope. *Hepler/Wallach/Hepler © Cengage 2013*

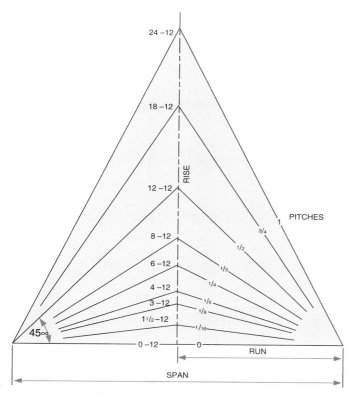

FIGURE 30.46 > Commonly used roof pitches. *Hepler/ Wallach/Hepler © Cengage 2013*

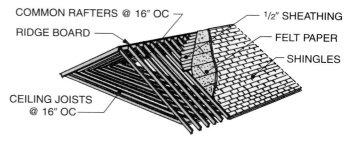

FIGURE 30.47 > Conventionally framed wood roof. *Hepler/ Wallach/Hepler © Cengage 2013*

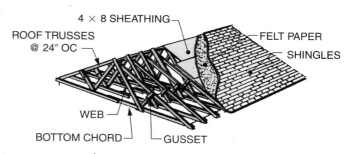

FIGURE 30.48 > Wood-trussed roof. *Hepler/Wallach/Hepler © Cengage 2013*

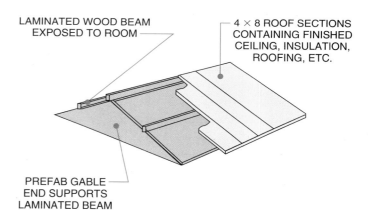

FIGURE 30.49 > Longitudinal post-and-beam roof construction. *Hepler/Wallach/Hepler © Cengage 2013*

sizes vary with the span and spacing of beams. The beams are installed parallel to the ridge board, or ridge beam. This is called longitudinal roof beam construction. See Figure 30.49. Beams may also be placed perpendicular (transverse) to the ridge beam. Planks are then placed across the beams. The planks can serve as a ceiling, as well as a base for roofing that will shed water. When planks are selected for appearance in an open-beam ceiling, the only ceiling treatment needed may be a protective finish on the planks and beams.

ROOF FRAMING TYPES

The most common roof framing types are *gable, hip, shed,* and *flat*. Other types, such as mansard, gambrel, and A-frame, are variations of the four common types and are framed in a similar manner.

Gable Roof Framing

Gable roof framing consists of rafters that form two inclined planes extending from the outer walls to a ridge. Figure 30.50 shows gable roof framing on the right end and hip roof framing on the left end. When trusses or truss joists are substituted for common rafters, a ridge board is not used and purlins are used to stabilize the trusses.

A-frame roof framing is similar to conventional gable framing, except the roof rafters rest on a foundation rather than on top plates.

Gable slopes range from 2/12 to 12/12. The angle of a gable roof does not show on a roof framing plan. A framing elevation is necessary to show slope and gable framing. See Figure 30.51. A *gable end* is the side of a building that rises to meet the ridge. See Figure 30.52. In some cases, especially on low-pitch roofs, the entire gable-end wall from the floor to the ridge can be framed with varying lengths of studs. However, it is more common to prepare a rectangular wall panel and erect separate cripple studs on gable ends.

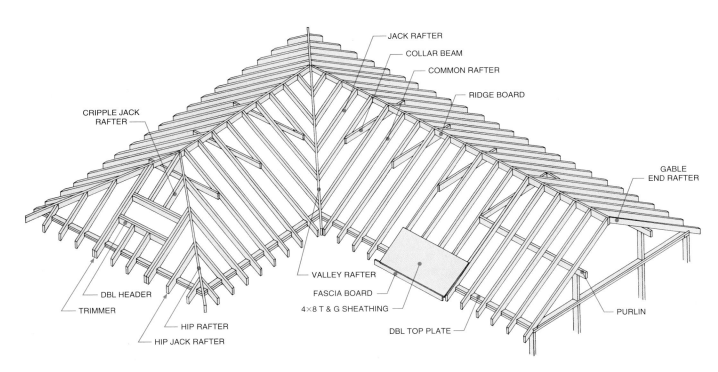

JACK RAFTER

COLLAR BEAM

COMMON RAFTER

RIDGE BOARD

GABLE END RAFTER

CRIPPLE JACK RAFTER

VALLEY RAFTER

FASCIA BOARD

4×8 T & G SHEATHING

DBL TOP PLATE

PURLIN

DBL HEADER

TRIMMER

HIP RAFTER

HIP JACK RAFTER

FIGURE 30.50 › Gable roof framing on the right and hip roof framing on the left. *Hepler/Wallach/Hepler © Cengage 2013*

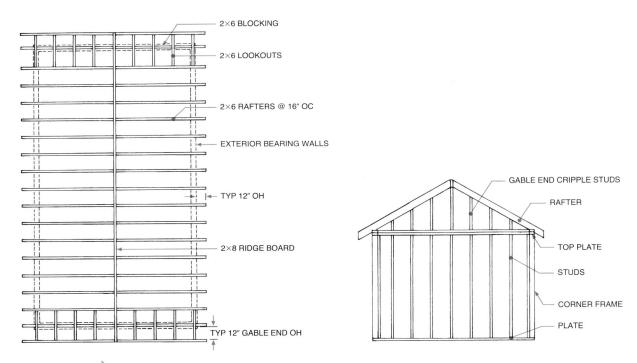

2×6 BLOCKING

2×6 LOOKOUTS

2×6 RAFTERS @ 16″ OC

EXTERIOR BEARING WALLS

TYP 12″ OH

2×8 RIDGE BOARD

TYP 12″ GABLE END OH

GABLE END CRIPPLE STUDS

RAFTER

TOP PLATE

STUDS

CORNER FRAME

PLATE

FIGURE 30.51 › Relationship of a gable framing plan and a gable framing elevation. *Hepler/Wallach/Hepler © Cengage 2013*

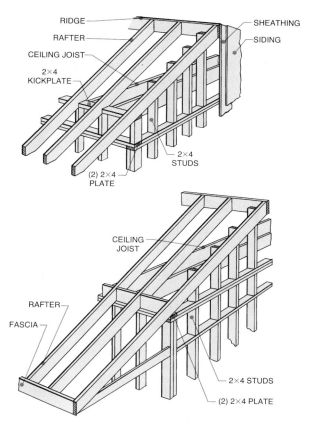

FIGURE 30.52 > Gable-end overhang construction. *Hepler/ Wallach/Hepler © Cengage 2013*

To design a large overhang on a gable end, *lookouts* are used. Lookouts are short rafters placed perpendicular to the first or second common rafters. *Winged gables,* as shown in Figure 30.53, use lookouts of varying lengths to form a triangular overhang. Large winged gable ends may require additional columns and beam support. See Figure 30.54.

Gambrel Roof Framing

Gambrel (barn) roofs are a variation of gable roofs but use double-pitched rafters. See Figure 30.55. The slope of the lower part is always steeper than the slope of the top part. Purlins may be used to stabilize pitch intersections. Prefabricated truss joists may also be used for gambrel roofs.

Hip Roof Framing

Hip roof framing is similar to gable roof framing except that the roof slopes in four directions instead of intersecting gable-end walls. See Figure 30.56. Where two adjacent slopes meet, a hip is formed on the external angle. A *hip rafter* extends from the ridge board, over the top plate, and to the edge of the overhang. A hip rafter supports the ends of the shorter hip-jack rafters.

The internal angle formed by the intersection of two slopes of the roof is known as the *valley.* A **valley rafter** is used to form this angle. See Figure 30.57.

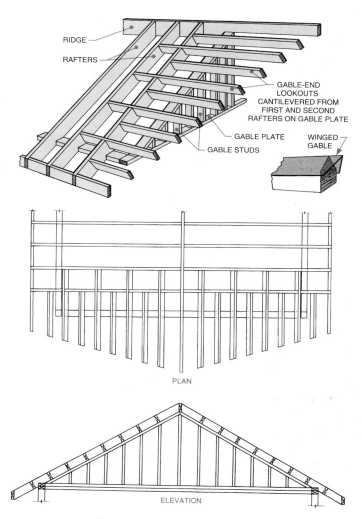

FIGURE 30.53 > Winged-gable lookout construction. *Hepler/ Wallach/Hepler © Cengage 2013*

Hip rafters and valley rafters are normally 2″ (50 mm) deeper or 1″ (25 mm) wider than the common rafters, for spans up to 12′ (3,658 mm). This is needed to allow for the angular cut of the jack rafters. For spans over 12′, the hip and valley rafters should be double the width of common rafters.

Jack rafters are rafters that extend from the wall plate to the hip or valley rafter. They are always shorter than common rafters, which extend from the top plate to the ridge.

Dutch Hip Roof Framing

A Dutch hip roof is framed in the same way as a gable roof in the center and as a partial hip roof on the ends. See Figure 30.58. One of the many variations of this basic design is a mansard roof, a double-pitched hip roof. Refer back to the many roof styles shown in Chapter 16.

Shed Roof Framing

A shed roof is a roof that slants in only one direction. A gable roof is actually two shed roofs, sloping in opposite directions.

FIGURE 30.54 > Double winged gable. *Eagle Window & Door Inc.*

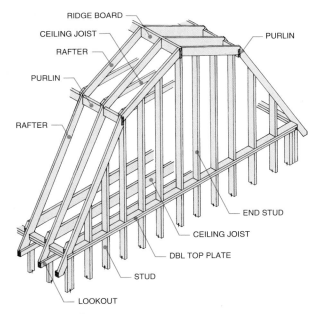

FIGURE 30.55 > Gambrel roof framing members. *Hepler/ Wallach/Hepler © Cengage 2013*

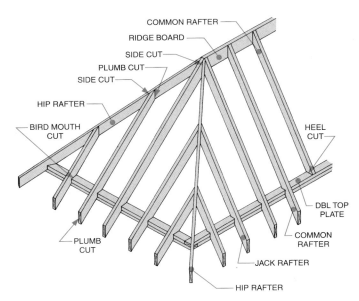

FIGURE 30.56 > Hip roof framing members. *Hepler/Wallach/ Hepler © Cengage 2013*

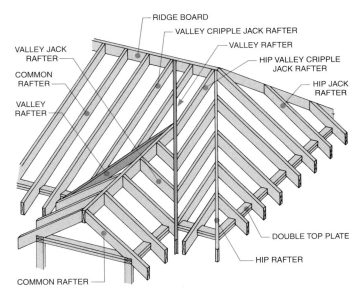

FIGURE 30.57 > Comparison of valley rafters and hip rafters. *Hepler/Wallach/Hepler © Cengage 2013*

Shed roof rafter design is the same as rafter design for gable roofs except that the run of the rafter is also the span. Some shed roofs are nearly flat. Shed roof slopes range from 2/12 to 12/12. Figure 30.59 shows the basic elements of shed roof framing.

Flat Roof Framing

In conventional flat roof framing, rafters are similar to ceiling joists. Rafters span from wall to wall or from exterior wall to bearing partitions, columns, or beams. In heavy-timber construction, beams may be used to span these distances. Roof decking is laid directly on the beams. The members of a

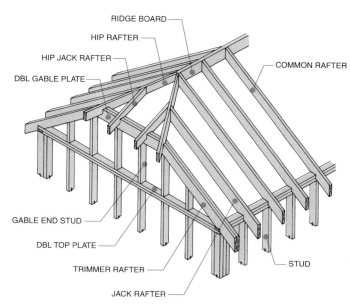

FIGURE 30.58 > Dutch hip roof framing. *Hepler/Wallach/ Hepler © Cengage 2013*

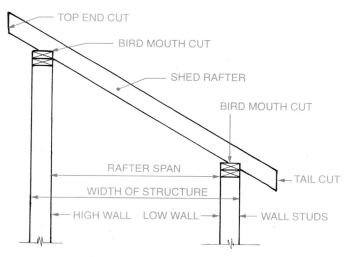

FIGURE 30.59 > Elements of shed roof framing. *Hepler/ Wallach/Hepler © Cengage 2013*

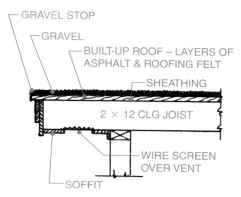

FIGURE 30.60 > Lightweight flat roof framing elevation. *Hepler/Wallach/Hepler © Cengage 2013*

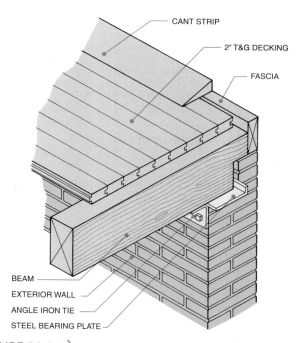

FIGURE 30.61 > Heavy-timber flat roof construction. *Hepler/ Wallach/Hepler © Cengage 2013*

conventional wood-frame flat roof are shown in the elevation section in Figure 30.60. Figure 30.61 shows one method of intersecting heavy-timber roof framing with a solid masonry wall. Flat roofs must be designed to withstand maximum snow loads since the snow will not slide off. Instead, it must melt and drain away.

Flat roofs are usually not absolutely flat. Most flat roofs have a slight slope of 1/8″ to 1/2″ per foot to allow for drainage. Some roofs are flat to allow a specific level of water to remain for insulation. Flat roof drainage must be provided through internal or external downspouts. A **cant strip** should be located on flat roof perimeters to stop water from flowing over the sides rather than through downspouts.

A flat roof may stop at a wall intersection or continue to form an overhang. Overhangs may include lookouts on the cantilevered ends, perpendicular to the rafter direction. A flat roof may also intersect a *parapet* (short wall). Figure 30.62 shows a cornice section with a wood-framed parapet wall and intersecting steel joists. Parapets are used to hide the roofing surface vents and any roof-mounted mechanical equipment. Parapets are also used to simulate mansard roofs, although true mansard roofs are double-pitched hips.

ROOF FRAMING METHODS WITH STEEL

Structural steel framing methods are especially appropriate for flat roofs. Steel roof framing systems are very similar to wood floor framing systems. Only the covering and cornice intersecting details are different.

Steel joists, in light construction, rest directly on concrete or on masonry bearing walls. See Figure 30.63. In heavy

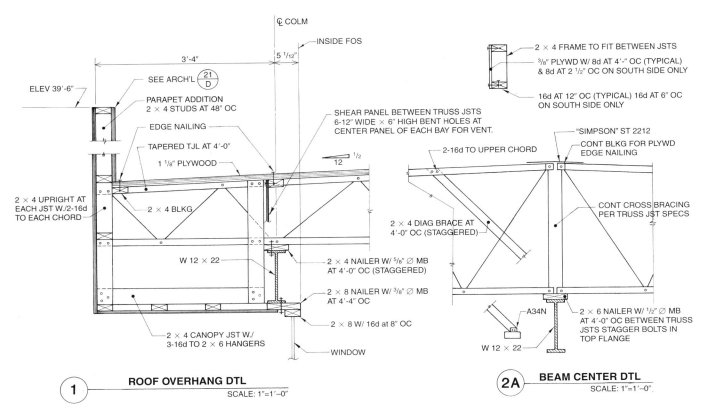

1 **ROOF OVERHANG DTL**
SCALE: 1"=1'-0"

2A **BEAM CENTER DTL**
SCALE: 1"=1'-0"

FIGURE 30.62 ⟩ Wood and steel cornice and parapet construction. *Hepler/Wallach/Hepler © Cengage 2013*

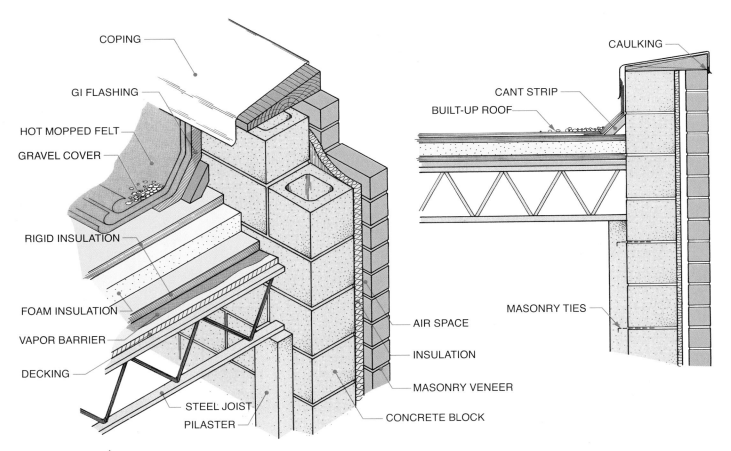

FIGURE 30.63 ⟩ Steel roof and masonry wall construction. *Hepler/Wallach/Hepler © Cengage 2013*

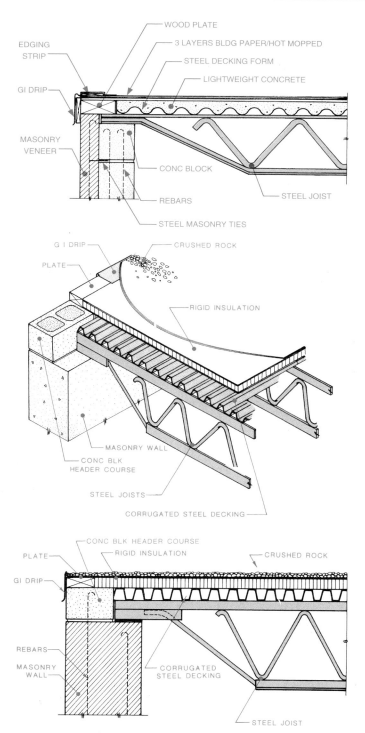

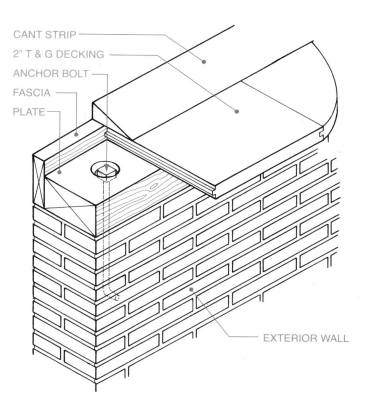

FIGURE 30.65 ⟩ Masonry wall and roof intersection. *Hepler/ Wallach/Hepler © Cengage 2013*

FIGURE 30.64 ⟩ Open web steel joists intersecting masonry exterior walls. *Hepler/Wallach/Hepler © Cengage 2013*

foundation to the roof are prepared, they are indexed to the basic plan and elevation drawings as shown in Figure 30.66.

Steel joists can span up to 40 feet or more depending on load factors. For larger spans, steel bents, arches, or space frames are used. Rigid steel bent frames are either straight single-span, shaped, or multiple-span frames as shown in Figure 30.67.

Arches are bent (curved) trusses. See Figure 30.68. Arch details are shown on structural drawings the same way truss details are shown. Space frames are three-dimensional trusses formed by connecting series of triangular polyhedrons. See Figure 30.69. Space frames, because of their light weight and ability to resist bending, can span extremely large distances. Because their load-bearing capacity is limited, they are used primarily for roof and not floor systems.

Steel roof framing plans are prepared similar to single-line wood framing plans. The lines, symbols, and notations are the same as those used on steel floor framing plans.

ROOF COVERING MATERIALS

Roof coverings protect buildings from rain, snow, wind, heat, and cold. Materials used to cover pitched roofs include wood, fiber, cement, asphalt, and composition shingles. On heavier roofs, ceramic tile, fiberglass, or slate may also be used, but these require stronger framing systems to compensate for the

construction, steel joists rest on steel girders or columns. Similar to other flat roofs, steel flat roof construction may meet an outside wall to form a right angle, intersect an outside wall to form a parapet, or extend to form an overhang. See Figure 30.64. Figure 30.65 shows a common roof intersection with a masonry wall. When complete wall sections from the

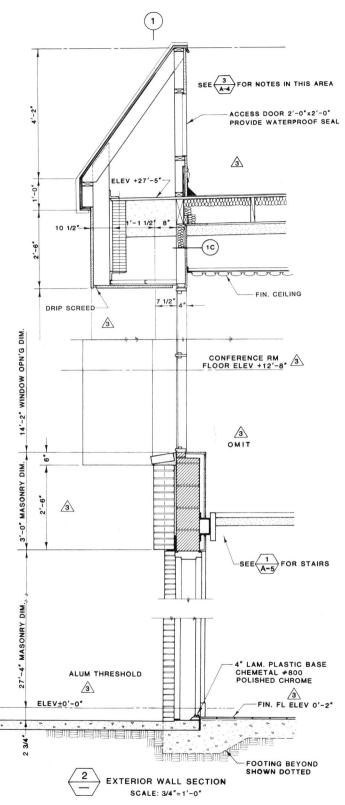

SEE ③/A-4 FOR NOTES IN THIS AREA

ACCESS DOOR 2'-0"x2'-0" PROVIDE WATERPROOF SEAL

③

4'-2"

1'-0"

ELEV +27'-5"

2'-6"

10 1/2"   1'-1 1/2"   8"

1C

FIN. CEILING

DRIP SCREED

7 1/2"   4"

③

CONFERENCE RM FLOOR ELEV +12'-8"   ③

14'-2" WINDOW OPN'G DIM.

③
OMIT

6"

3'-0" MASONRY DIM.

2'-6"

③

SEE ①/A-5 FOR STAIRS

27'-4" MASONRY DIM.

ALUM THRESHOLD

③

4" LAM. PLASTIC BASE CHEMETAL #800 POLISHED CHROME

③
FIN. FL ELEV 0'-2"

ELEV±0'-0"

2 3/4"

FOOTING BEYOND SHOWN DOTTED

②/— EXTERIOR WALL SECTION
SCALE: 3/4"=1'-0"

FIGURE 30.66 > Complete wall section including roof intersections and foundation. *Hepler/Wallach/Hepler © Cengage 2013*

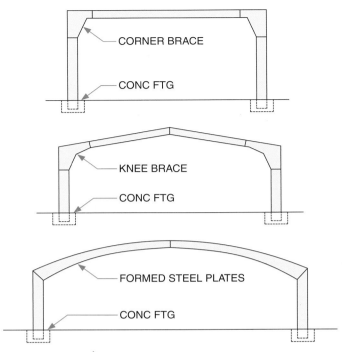

CORNER BRACE

CONC FTG

KNEE BRACE

CONC FTG

FORMED STEEL PLATES

CONC FTG

FIGURE 30.67 > Rigid steel bents. *Hepler/Wallach/Hepler © Cengage 2013*

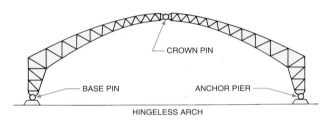

CROWN PIN

BASE PIN   ANCHOR PIER

HINGELESS ARCH

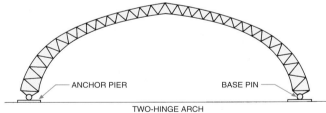

ANCHOR PIER   BASE PIN

TWO-HINGE ARCH

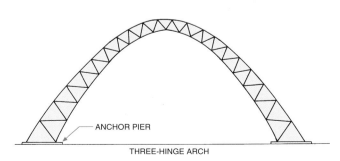

ANCHOR PIER

THREE-HINGE ARCH

FIGURE 30.68 > Types of steel arches. *Hepler/Wallach/Hepler © Cengage 2013*

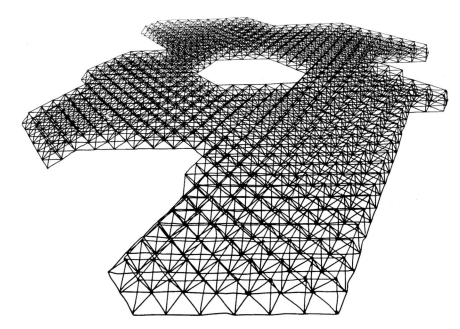

FIGURE 30.69 > Space frame construction. *Hepler/Wallach/Hepler © Cengage 2013*

increased weight. Roll roofing or other sheet material, such as galvanized iron, aluminum, copper, and tin, may also be used for all roofs. Built-up roofing of felt and gravel is used extensively on flat or low-pitched roofs. Most light weight shingles cannot be used on low pitch roofs due to wind lifing. Figure 30.70 lists the characteristics of common roof covering materials. Figure 30.71 shows an example of the many components used for roof coverings.

The weight of roofing materials is important in computing dead loads. A heavier roofing surface makes the roof more permanent than does a lighter surface. Generally, heavier roofing materials last longer than lighter materials but require heavier framing. Roof covering materials are classified by their weight per 100 square feet (100 sq. ft. equals 1 *square*). Thus 30-lb. roofing felt weighs 30 lbs. per 100 sq. ft.

Sheathing

Roof sheathing may consist of lumber boards, gypsum, fiberboard, or plywood sheets nailed directly to roof rafters. Sheathing adds rigidity to the roof and provides a surface for the attachment of waterproofing materials. In humid areas, sheathing boards or panels are sometimes spaced slightly apart to provide ventilation and to prevent shingle rot. The thickness of roof (and wall) sheathing depends on rafter (or stud) spacing and the amount of insulation required. Tongue-and-groove plywood provides the greatest amount of rigidity.

The joints of sheathing panels are always staggered, and only exterior grade material is used. Most codes in high-wind areas require sheathing to be applied with nails or screws not staples. Hurricane clips that lock the panels together may also be required. In large housing developments, a *roof sheathing plan* is often prepared to plan the best possible arrangement with the minimum amount of waste. See Figure 30.72.

Roll Roofing

Roll (continuous membrane) roofing may be used as an underlayment for shingles or as a finished roofing material for slopes less than 3/12. Roll roofing material used as an underlayment includes asphalt and saturated felt. The underlayment serves as a barrier against moisture and wind. As a finished roofing material, copper, aluminum, or galvanized steel is used in rolls or sheets. Seams are sealed to provide a watertight surface especially for valleys and ridges.

Roof Shingles, Shakes, and Tiles

Roof **shingles** and **shakes** are made from asphalt, cement, fiberglass, cedar, or bonded wood fibers. Shingles are laid over building felt that covers the sheathing. See Figure 30.73. Shingles are available in a variety of patterns. See Figure 30.74 on page 666. Most shingles are not recommended for slopes less than 4/12 because of the danger of wind lift.

TYPE	TYPICAL FORM DESCRIPTION	TYPICAL WIDTHS	AVERAGE SPANS	REMARKS
TONGUE & GROOVE (T & G) WOOD-LAMINATE	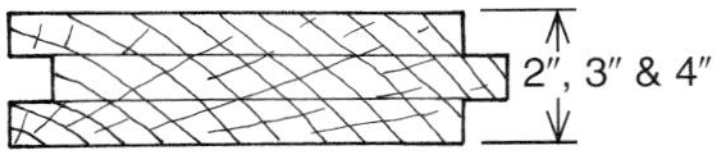 2″, 3″ & 4″	6″	8′ 14′ 20′	MAXIMUM 20′ LENGTHS
T & G PRECAST GYPSUM	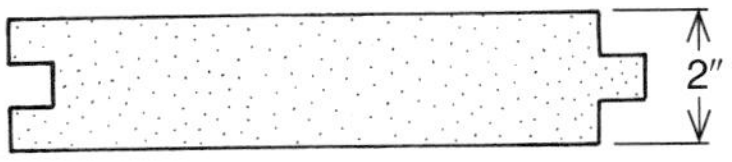 2″	15″	7′	MAXIMUM 10′ LENGTHS AVAILABLE WITH METAL EDGES
CORRUGATED STEEL	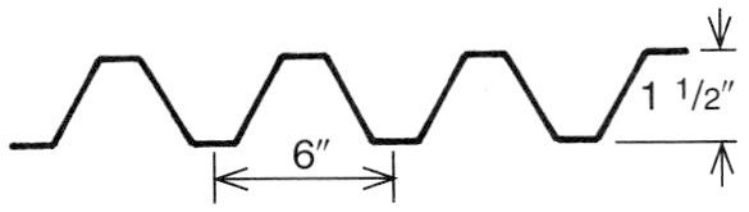 1 1/2″ 6″	24″	8′	MAXIMUM 12′ LENGTHS
CORRUGATED STEEL	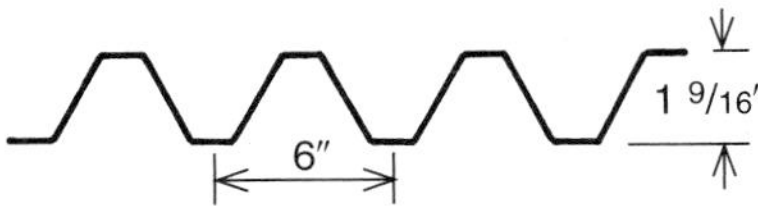 1 9/16″ 6″	24″	7′	LENGTH ONLY LIMITED BY SHIPPING. MAY BE USED INVERTED.
CORRUGATE STEEL	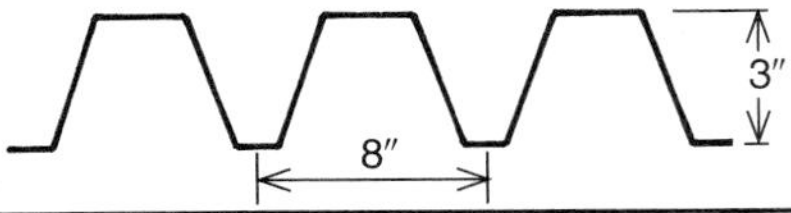 3″ 8″	24″	15′	MAXIMUM 10′ LENGTHS ECONOMICAL FOR MEDIUM SPANS
CORRUGATE STEEL	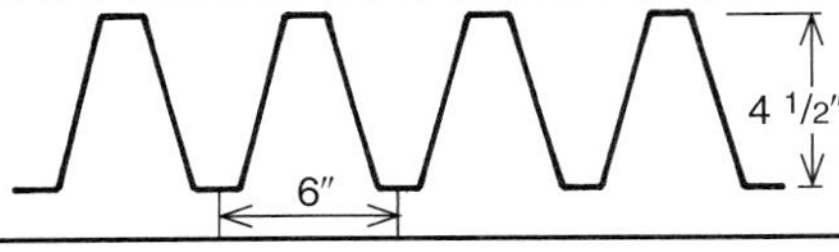 4 1/2″ 6″	12″	25′	MAXIMUM 30′ LENGTHS ECONOMICAL FOR LONG SPANS
CORRUGATE STEEL FORM AND CONCRETE	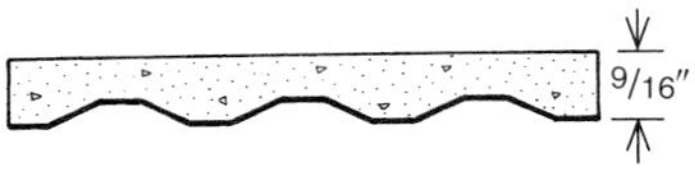 9/16″	24″	4′	MAXIMUM 20′ LENGTHS
PRECAST CONCRETE	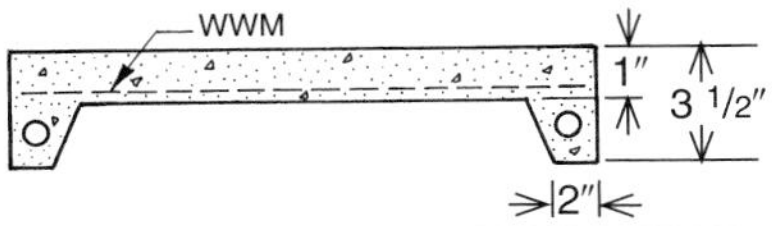 WWM 1″ 3 1/2″ 2″	24″	8′	MAY BE CAST TO ANY LENGTH AND THICKNESS
TONGUE & GROOVE PRECAST CONCRETE	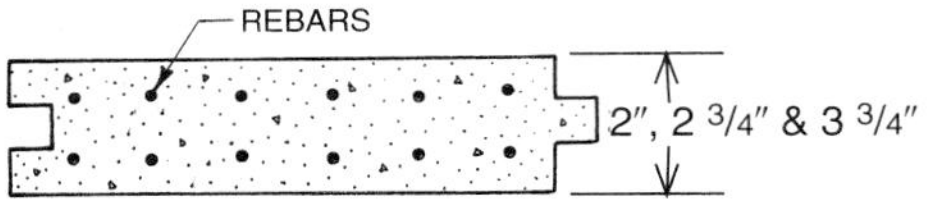 REBARS 2″, 2 3/4″ & 3 3/4″	24″	7′ 8′ 10′	MAY BE CAST TO ANY LENGTH AND THICKNESS
PRECAST CONCRETE	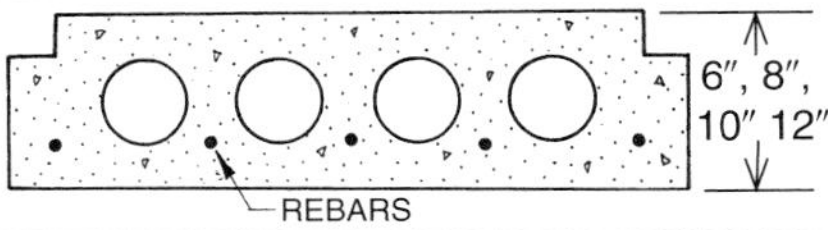 6″, 8″, 10″ 12″ REBARS	24″	24′ 32′ 40′ 48′	MAY BE CAST TO ANY LENGTH AND THICKNESS
TONGUE & GROOVE PRECAST WOOD FIBER & CEMENT	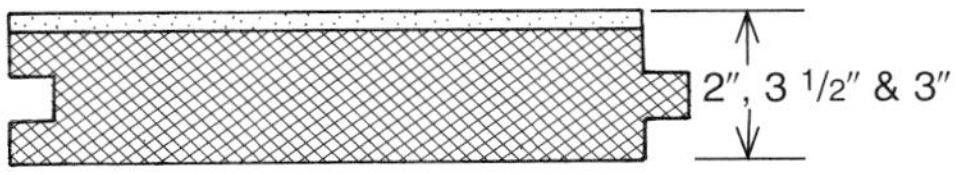 2″, 3 1/2″ & 3″	30″	3′ 4′ 5′	MATERIAL IS NAILABLE
TONGUE & GROOVE WOOD FIBER BOARD	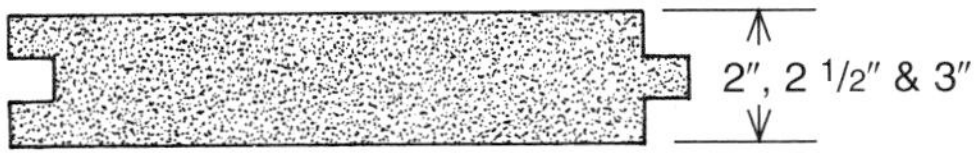 2″, 2 1/2″ & 3″	32″ 48″	4′ 8′ 12′	MATERIAL IS NAILABLE
CANE FIBER WITH CEMENT ASBESTOS FACING	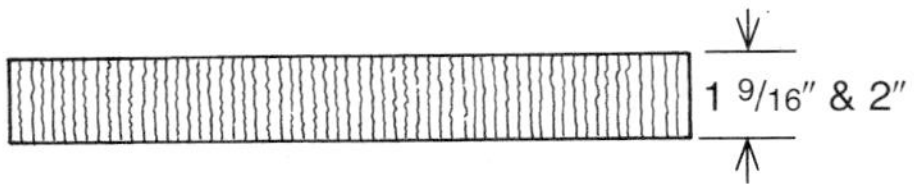 1 9/16″ & 2″	48″	6′ 8′	MATERIAL IS NAILABLE

FIGURE 30.70 > Characteristics of common roof covering materials. *Hepler/Wallach/Hepler © Cengage 2013*

FIGURE 30.71 ❯ Examples of roof covering materials. © *used under license from Shutterstock, 2011/Karamysh*

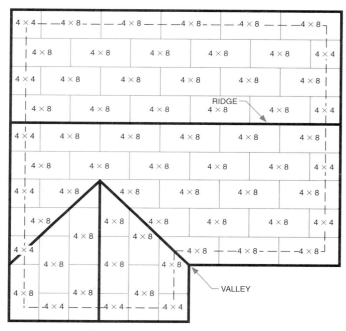

FIGURE 30.72 ❯ Roof sheathing plan. *Hepler/Wallach/Hepler © Cengage 2013*

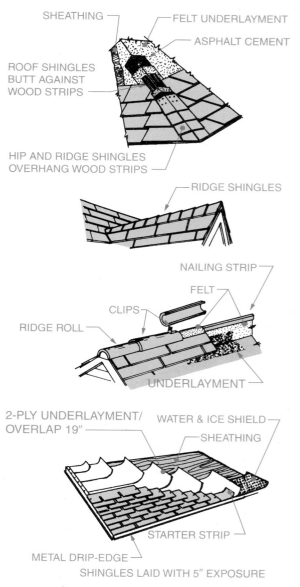

FIGURE 30.73 ❯ Methods of shingle application. *Hepler/ Wallach/Hepler © Cengage 2013*

Shingles are classified by weight per 100 sq. ft. Shingles range in weight from 180 to 390 lbs. per square (i.e., per 100 sq. ft.) for residential roofs. The average residential asphalt shingle is 245 lbs. per square. Shingles are also classified by special features such as their resistance to mildew, wind, hail, and water and by their life expectancy, usually 20 or 25 years.

Tiles are manufactured from clay-ceramic, cement, and polystyrene. Copper, aluminum, and galvanized steel panels are also available in patterns that simulate shingles and tiles.

Built-Up Roofs

Built-up roof coverings are used on flat or extremely low-pitched roofs. Built-up roofing cannot be used on high-pitched roofs because the gravel will wear off during rain and high winds. Because rain may be driven into gravel crevices and snow may not quickly melt from these roofs, complete waterproofing is essential.

Built-up roofs may have three, four, or five layers of roofing felt, sealed with hot-mopped tar or asphalt between coatings.

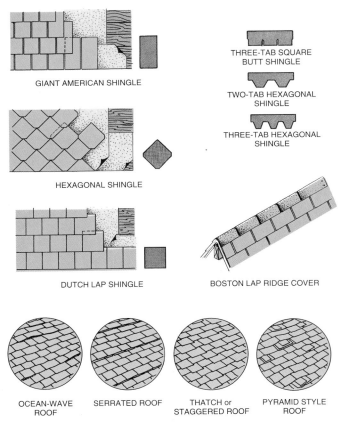

GIANT AMERICAN SHINGLE

HEXAGONAL SHINGLE

DUTCH LAP SHINGLE

THREE-TAB SQUARE
BUTT SHINGLE

TWO-TAB HEXAGONAL
SHINGLE

THREE-TAB HEXAGONAL
SHINGLE

BOSTON LAP RIDGE COVER

OCEAN-WAVE
ROOF

SERRATED ROOF

THATCH or
STAGGERED ROOF

PYRAMID STYLE
ROOF

FIGURE 30.74 > Common shingle patterns. *Hepler/Wallach/
Hepler © Cengage 2013*

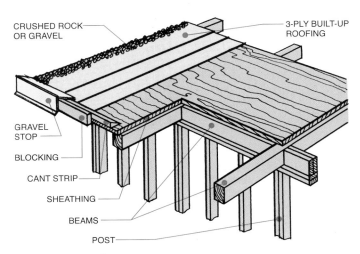

FIGURE 30.75 > Built-up roof covering. *Hepler/Wallach/Hepler
© Cengage 2013*

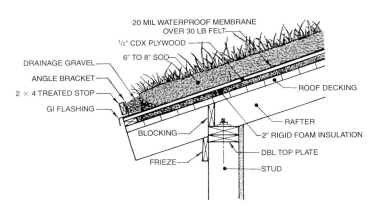

FIGURE 30.76 > Sod roof covering. *Hepler/Wallach/Hepler
© Cengage 2013*

The final layer of tar or asphalt is then covered with roofing gravel or a top sheet of roll roofing. See Figure 30.75. Even sod can be used as a roof covering as shown in Figure 30.76. This "green" roof provides protection from the elements and is also compatible with efforts to use environmentally safe materials.

ROOF APPENDAGES

Gutters, downspouts, flashing, vents, and skylights are additions to many roofs. Appendages perform several functions, such as control of water flow, ventilation of building fumes, admitting of light, and covering of rough lumber.

Gutters and Downspouts

Gutters are troughs designed to carry water to downspouts, where it can be emptied into a sewer system or away from the building. Materials used most commonly for gutters are sheet metal, cedar, redwood, and plastic. Gutters may be built into

the roof structure, as shown in Figure 30.77. Sheet-metal or wood gutters may also be attached or hung from the fascia board. All gutters should be pitched at least 1:20 to provide for drainage to downspouts. See Figure 30.78. Gutters must be kept below the roof eave line to prevent snow and ice from accumulating. In selecting gutters and downspouts, care must be taken to ensure that their size is adequate for the local rainfall.

Flashing

Joints where roof covering materials intersect a ridge, hip, valley, chimney, wall, vent, skylight, or parapet must be flashed. **Flashing** is additional covering used under a joint to provide complete waterproofing. Roll roofing, galvanized sheet steel, copper sheeting, or bituthene (polyethylene film or rubber-

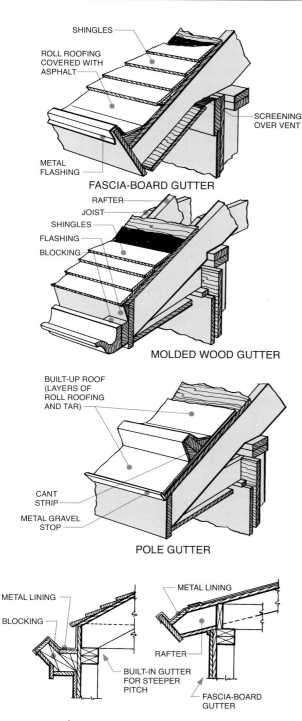

SHINGLES
ROLL ROOFING COVERED WITH ASPHALT
SCREENING OVER VENT
METAL FLASHING
FASCIA-BOARD GUTTER

RAFTER
JOIST
SHINGLES
FLASHING
BLOCKING
MOLDED WOOD GUTTER

BUILT-UP ROOF (LAYERS OF ROLL ROOFING AND TAR)
CANT STRIP
METAL GRAVEL STOP
POLE GUTTER

METAL LINING
BLOCKING
METAL LINING
RAFTER
BUILT-IN GUTTER FOR STEEPER PITCH
FASCIA-BOARD GUTTER

FIGURE 30.77 > Built-in gutters. *Hepler/Wallach/Hepler © Cengage 2013*

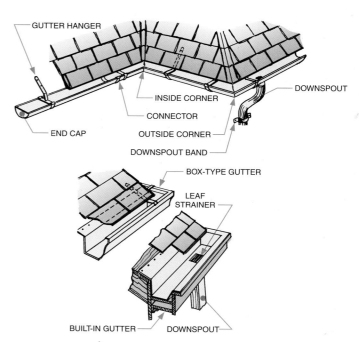

GUTTER HANGER
DOWNSPOUT
INSIDE CORNER
CONNECTOR
END CAP
OUTSIDE CORNER
DOWNSPOUT BAND
BOX-TYPE GUTTER
LEAF STRAINER
BUILT-IN GUTTER
DOWNSPOUT

FIGURE 30.78 > **Hanging gutter and downspout assembly.** *Hepler/Wallach/Hepler © Cengage 2013*

metal joints should be made. Chimney flashing is frequently bonded into the mortar joint and caulked under shingles to provide a waterproof joint.

Roof Ventilation

Areas where excessive heat and moisture are trapped must be ventilated. Attics and crawl spaces can be ventilated by installing fans, roof vents, or a cupola connecting the inside area with the outside. Vent areas should be 1/150th of the enclosed area or 1/300th if the area has a vapor barrier. The space between rafters also needs to be ventilated. This is best done by creating an air flow between ventilated soffits and a continuous ridge vent. See Figure 30.80. Designing roof louvers (vents) into the roof plan also helps exhaust excess heat from attics or roof crawl spaces. These may be gravity or fan powered. Gable end or wall louvers are also advised for enclosed nonliving spaces.

Skylights

Roof framing for skylights, as shown in Figure 30.81, is the same as for chimneys or any mechanical equipment requiring a break in the rafter and roof covering pattern. All openings must be flashed. See Chapter 16 for a review of skylight locations and design.

ized asphalt) can be used as flashing material. On sloped roofs step flashing is used, as shown in Figure 30.79. Unless seams are sealed watertight, a second layer of counter flashing may be needed. For sheet-metal flashing, watertight sheet-

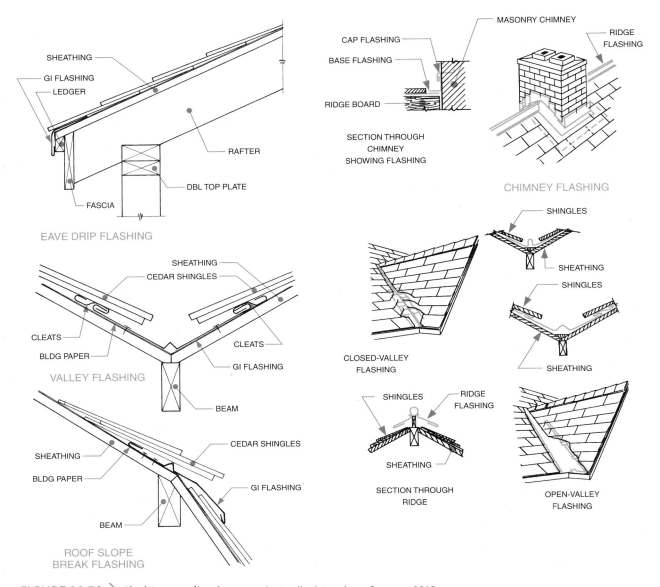

SHEATHING

GI FLASHING

LEDGER

RAFTER

DBL TOP PLATE

FASCIA

EAVE DRIP FLASHING

CAP FLASHING

MASONRY CHIMNEY

BASE FLASHING

RIDGE FLASHING

RIDGE BOARD

SECTION THROUGH CHIMNEY SHOWING FLASHING

CHIMNEY FLASHING

SHEATHING

CEDAR SHINGLES

CLEATS

CLEATS

BLDG PAPER

GI FLASHING

VALLEY FLASHING

BEAM

SHINGLES

SHEATHING

SHINGLES

CLOSED-VALLEY FLASHING

SHEATHING

SHEATHING

BLDG PAPER

CEDAR SHINGLES

GI FLASHING

BEAM

ROOF SLOPE BREAK FLASHING

SHINGLES

RIDGE FLASHING

SHEATHING

SECTION THROUGH RIDGE

OPEN-VALLEY FLASHING

FIGURE 30.79 > Flashing applications. *Hepler/Wallach/Hepler © Cengage 2013*

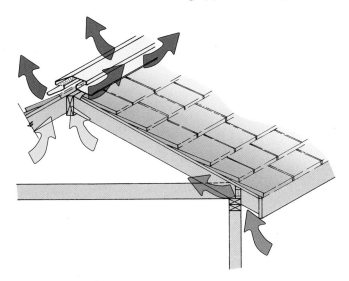

FIGURE 30.80 > Continuous ridge vents exhaust trapped air. *Hepler/Wallach/Hepler © Cengage 2013*

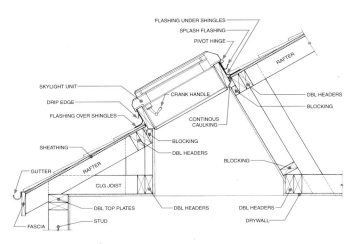

FLASHING UNDER SHINGLES

SPLASH FLASHING

PIVOT HINGE

RAFTER

SKYLIGHT UNIT

CRANK HANDLE

DBL HEADERS

DRIP EDGE

BLOCKING

FLASHING OVER SHINGLES

CONTINOUS CAULKING

SHEATHING

BLOCKING

DBL HEADERS

BLOCKING

GUTTER

RAFTER

CLG JOIST

DBL TOP PLATES

DBL HEADERS

DBL HEADERS

DRYWALL

FASCIA

STUD

FIGURE 30.81 > Skylight roof framing. *Hepler/Wallach/Hepler © Cengage 2013*

Roof Framing Drawings Exercises

CHAPTER **30**

1. What are the two types of roof framing plans? What is the function of each?

2. Prepare a roof framing plan for a house of your own design. Label the members.

3. What is the roof pitch of a roof with a rise of 8′ and a run of 24′?

4. Name the main parts of a roof truss. List the advantages and disadvantages of using trusses for roofs.

5. Draw a cornice section for the house of your design. Include the roof covering.

6. Sketch a roof plan for a gable roof. Include the dimensions of the run.

7. Specify the type of roofing to be used in the house of your design. Explain why you chose a particular type of material: wood, steel, and/or concrete.

8. How many square feet of shingles are in four squares?

9. Prepare a roof framing plan for a house of your own design, a cornice section, and/or a roof detail in this chapter.

10. Draw a gable roof framing plan for the cabin shown in Figure 30.82.

11. Draw a hip roof framing plan for the cottage shown in Figure 30.83.

12. Draw a flat roof framing plan for the cottage shown in Figure 30.84.

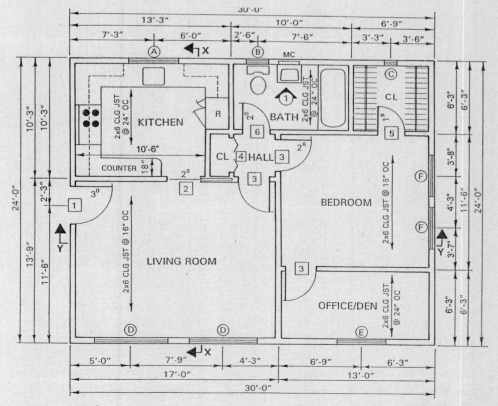

FIGURE 30.82 > Draw a gable roof framing plan for this cabin. *Hepler Associates LA, PC*

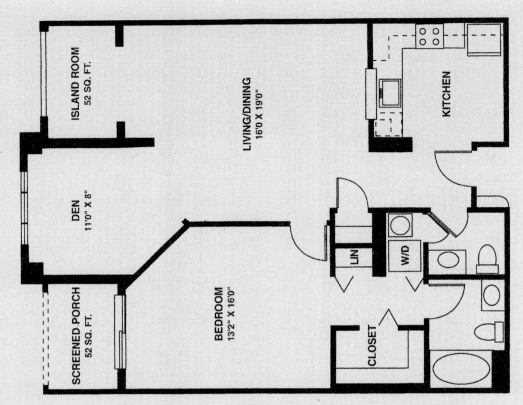

FIGURE 30.83 > Draw a hip roof framing plan for this collage. *Hepler Associates LA, PC*

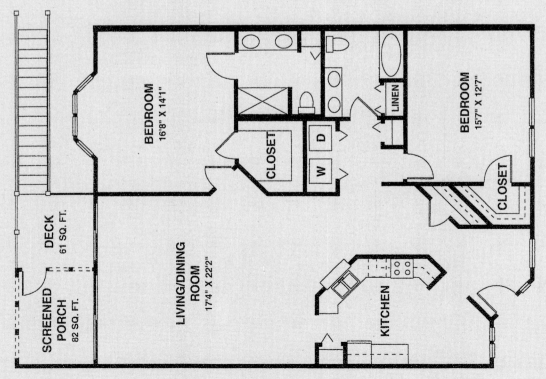

FIGURE 30.84 > Draw a flat roof framing plan for this collage. *Hepler Associates LA, PC*

PART 8

8

ELECTRICAL AND MECHANICAL DESIGN DRAWINGS

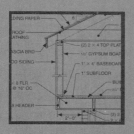

CHAPTER 31

Electrical Design and Drawing

OBJECTIVES

In this chapter you will learn to:

> plan and draw electrical circuits for a house on a floor plan.

> calculate electrical measurements for each circuit.

> plan and draw lighting for each room in a house.

> use electrical symbols.

> design and draw an electronic building control system.

> understand how energy conservation, efficiency, and sustainability is achieved through electronic and electrical design.

> plan the use of lighting fixtures.

TERMS

alternating current (AC)	ground fault circuit interrupter (GFCI)	National Electrical Code (NEC)
ampere	hard-wired	ohms
circuit	incandescent	outlet
circuit breaker	insulators	receptacle
conductors	integrated system	resistance
current	kilowatt-hour	service entrance
direct current (DC)	lightning rod	surge
distribution panel	load	switch
fixture	lumens	volt
fluorescent	lux	watt
footcandle		

INTRODUCTION

Buildings cannot function as they should without electricity. The design and drawing of electrical systems requires a knowledge of electrical power distribution, wiring circuits, lighting methods, and electrical symbols and conventions. A knowledge of electronic systems, such as security, safety, communications, and integrated controls systems, is vital. In addition, energy efficiency and conservation are becoming increasingly important in our efforts to maintain a healthy environment.

⊕ ENERGY CONSERVATION, EFFICIENCY, AND SUSTAINABILITY

Architectural designers have little or no effect on the original sources of the electricity delivered to a site or building except to endorse the development of new technologies. These include solar, land gas, nuclear fusion, artificial photosynthesis, and wind- and water-generated energy. The electricity generated is **direct current (DC)** which is changed to **alternating current (AC)** by an inverter for architectural use. A designer can, however, specify certain electrical energy sources for a site or building plan as appropriate. Such sources include photovoltaic solar cells, windmill/wind helix devices, and geothermal sources, passive solar construction, and active solar panels.

Figure 31.1 is a drawing of a double wind helix that can generate electricity using wind originating from any direction. Because of its small size, this device can be located on any structural high point and not negatively affect the elevation's appearance.

Unlike the active solar systems covered in Chapter 32, solar electrical systems use semiconductor devices to convert

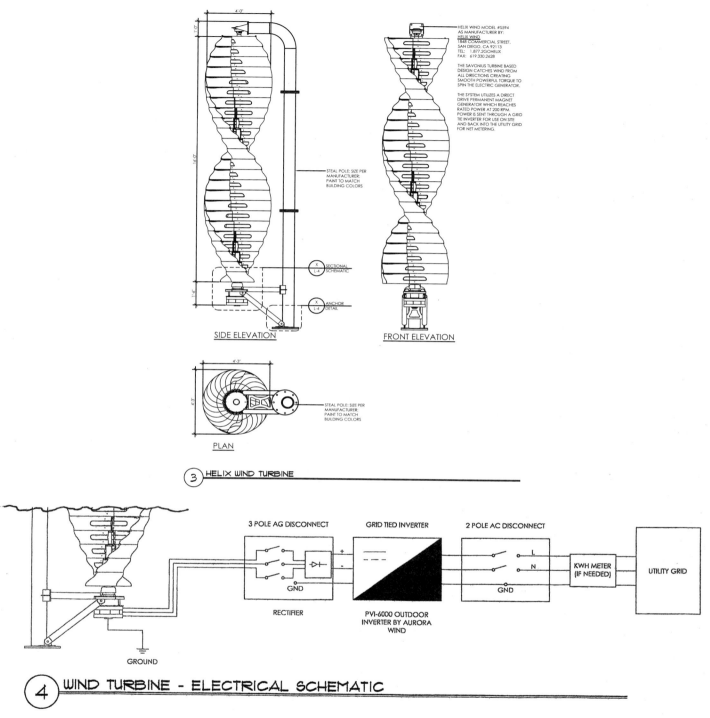

FIGURE 31.1 > Wind helix. *Hepler/Wallach/Hepler © Cengage 2013*

sunlight directly into electricity. Figure 31.2 shows how panels of photovoltaic cells create direct current (DC). The cells are connected to an inverter that changes the direct current into usable alternating current (AC). Some areas allow net metering, which permits power to flow back to the power grid if more energy is generated than is used.

Other electrical energy conservation and efficiency measures are an integral part of the electricity and electronics

design process. These are covered in the descriptions and guidelines that follow.

Energy Star is an international standard that identifies energy-efficient consumer products and construction methods. Under this program the Environmental Protection Agency and Department of Energy have developed an energy performance rating system for buildings. This involves energy guidelines for insulation, windows, HVAC systems, and tight

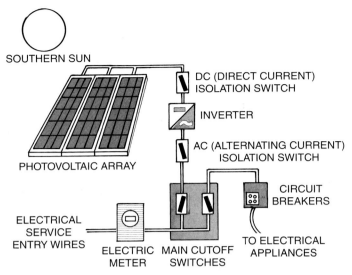

FIGURE 31.2 ＞ Solar electrical system. *Hepler/Wallach/Hepler*
© *Cengage 2013*

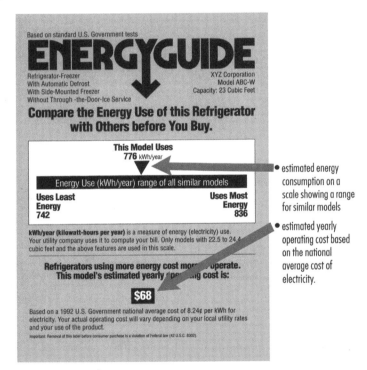

FIGURE 31.3 ＞ Energy Star rating guide. *Hepler/Wallach/
Hepler © Cengage 2013*

construction standards. In the electrical area, the Energy Star program provides ratings for appliances and lighting fixtures. Figure 31.3 shows a typical energy guide to be used by designers and consumers alike.

ELECTRICAL PRINCIPLES

Understanding electrical principles is vital to designing safe and efficient architectural electrical systems.

Power Distribution

As mentioned, electric power is generated from several sources of energy: wind, water, nuclear, fossil fuel, solar (photovoltaic), and geothermal. Photovoltaic cells convert solar energy directly into an electric current. All other energy sources are harnessed to produce a rotary mechanical motion that drives electric generators. The generators convert movement into electricity. Transformers are used to "step up" (increase) the electrical power to very high voltages (hundreds of thousands of volts) for transmission by wires over long distances. Wherever the transmission lines enter an industrial or residential community for local power distribution, large transformers are used to "step down" the voltage to a few thousand volts. Smaller transformers set on poles or in underground vaults are used for final distribution to small groups of houses or individual factories. Voltages of 120 and 240 are delivered to residences and small buildings. Because a voltage drop occurs in a building's circuit during delivery, 110 or 220 volts (V) are actually delivered to electrical outlets.

Electrical Measurements

Electrical properties can be measured with instruments. The terms used to describe units of electricity—*volt, ampere,* and *watt*—are used in both metric and customary systems.

A **volt** is a unit of electrical *pressure* or potential. This pressure makes electricity flow through a wire. For a particular electrical **load,** the higher the voltage, the greater the amount of electricity that will flow.

The term for flow of electricity is **current.** An **ampere,** or amp, is the unit used to measure the magnitude of an electric current. An ampere is defined as the specific quantity of electrons passing a point in 1 second. The amount of current, in amperes, that will flow through a circuit must be known in order to determine proper wire sizes and the current rating of circuit breakers and fuses.

The amount of *power* (energy) required to light lamps, heat water, turn motors, and do all types of work is measured in **watts.** Wattage depends on both voltage and amperage. Current (in amperes) multiplied by potential (in volts) equals power (in watts):

$$\text{amperes} \times \text{volts} = \text{watts}$$

The actual energy used (the watts utilized) for work performed is the basis for figuring the cost of electricity. The unit used to measure the consumption of electrical energy is the **kilowatt-hour.** A kilowatt is 1,000 watts. An hour, of course, is a unit of time. A 1,000-watt hand iron operating for 1 hour consumes one kilowatt-hour (1 kWh). The device used to measure the kilowatt-hours consumed is the watt-hour meter.

Electricity flowing through a material always meets with some **resistance.** Materials such as wood, glass, and plastic have a high resistance. They are good **insulators.** Copper, aluminum, and silver have low resistance and are therefore good **conductors** of electricity. Most electrical wiring consists of copper or aluminum surrounded by plastic insulation. The plastic keeps the electricity from flowing where it is not wanted.

The amount of electrical resistance is measured in **ohms.** The electron flow (or current in amperes) through a circuit is equal to the voltage (number of volts) divided by the resistance (ohms). This can be expressed by the following formula:

$$I = \frac{E}{R} \text{ or } E = IR \text{ or } R = \frac{E}{I}$$

where

I = current (amperes)
E = electromotive force (volts)
R = resistance (ohms).

Example:

If the current is 10 amperes and the electromotive force is 120 volts, what is the resistance?

$$R = \frac{E}{I}$$
$$R = \frac{120}{10}$$
$$R = 12 \text{ ohms}$$

Electrical Service Entrance

Power is supplied to a building through a **service entrance.** Three heavy wires, together called the *drop,* extend from a utility pole or an underground source to the structure. These wires are twisted into a cable. At the building, overhead wires are fastened to the structure and spliced to service entrance wires that enter a conduit through a service head, as shown in Figure 31.4.

In planning overhead service drop paths, minimum height requirements for connector lines must be carefully followed, as shown in Figure 31.5. Designers must locate the service heads on buildings to ensure compliance with all local codes. If the minimum height distances cannot be maintained, underground rigid conduit, electrical metallic tubing, or busways (channels, ducts) must be used.

If the service is supplied underground, three wires are placed in a rigid conduit. An underground service conduit is brought to the meter socket. All service entrances includes a watt-hour meter, main breaker, and lightning protection. Automatic brownout equipment is also required by many codes for new construction. All electrical systems must be grounded through the service entrance. The location and path of underground service conduits is shown on plot or site plans. Figure 31.6 shows a typical underground installation. The use of underground connections greatly beautifies an area and also eliminates power outages due to downed power lines. Figure 31.7

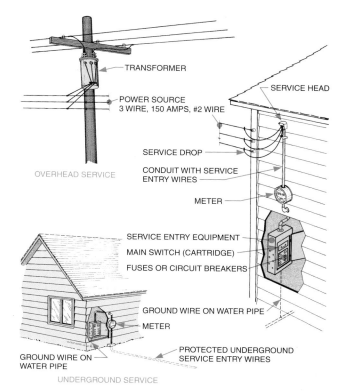

FIGURE 31.4 > Electrical distribution to buildings. *Hepler/Wallach/Hepler © Cengage 2013*

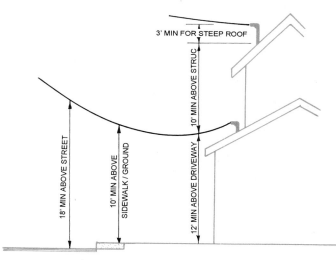

FIGURE 31.5 > Minimum overhead service line clearances. *Hepler/Wallach/Hepler © Cengage 2013*

displays the details of the main surface connections between the power source and the building electrical system.

Safety

In designing electrical systems it is important to protect wiring and equipment from damage. It is even more important to protect people from electrical shock and possible electrocution. The major design flaws that result in electrical damage

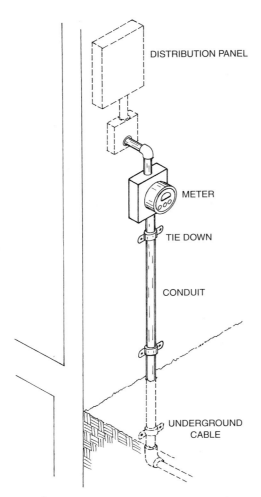

FIGURE 31.6 › Underground service connection. *Hepler/ Wallach/Hepler © Cengage 2013*

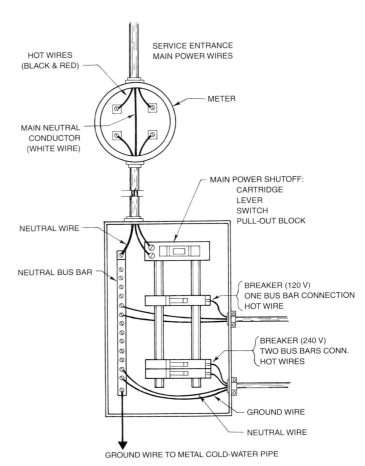

FIGURE 31.7 › Power source connection details. *Hepler/ Wallach/Hepler © Cengage 2013*

and injury are current overloads, improper grounding, exposed wiring, wrong wire size, and insufficient outlets, which can result in an excessive use of extension cords. Adequate circuit design and the use of the correct circuit breakers and wiring sizes, sufficient outlets, and code-approved grounding methods can eliminate these potential problems.

Surge Protection

Surges are microsecond increases in voltage that are significantly above the capacity of a wiring system or device to withstand. This increase can result in burned wires and damaged circuits and equipment. Very high surges can ignite wires and adjacent materials. Surge protection can be provided through the service entrance, distribution panel, and lightning rod grounding. Added surge protection should also be provided at the power supply connection to sensitive electronic devices such as computers, televisions, and phone service systems. Where water may contact outlets, **ground fault circuit interrupters (GFCIs)** must also be provided.

Grounding is a method of ensuring that excess voltage surges are dissipated harmlessly into the earth. The **National**

Electrical Code (NEC) requires ground rods to penetrate at least 8 feet deep. In dry sandy soil penetration to at least 30 feet may be necessary. This is because the code requires a resistance of 25 ohms or less.

Lightning rods were invented by Benjamin Franklin more than 200 years ago. The basic theory has not changed since then: Lightning rods direct lightning strike voltage to the ground and away from the building. This is done by providing a path of least electrical resistance from the strike point to the ground.

A lightning rod system consists of a series of short lightning rods (air terminals) located above the highest point of a building. These are connected through cables to at least two grounding rods, located a minimum of 2 feet from the foundation. The location of these rods and grounding points is shown on the site plan.

Buildings not surrounded by other buildings or tall trees (nature's lightning rods) should include a lightning rod system.

Electrical Service Distribution

Electrical current is delivered throughout a building through a **distribution panel,** or service panel. See Figure 31.8. The size of

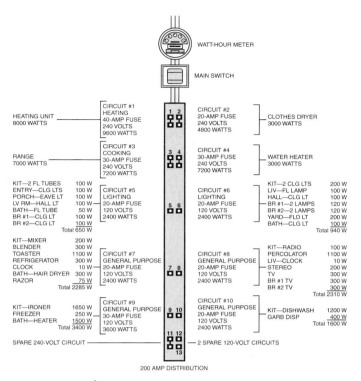

FIGURE 31.8 > Distribution panel circuits. *Hepler/Wallach/ Hepler © Cengage 2013*

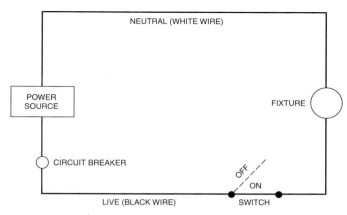

FIGURE 31.9 > Simple electrical circuit. *Hepler/Wallach/Hepler © Cengage 2013*

a distribution panel (in amperes) is determined by the total load requirements (watts) of the entire building. Watts can be converted to amperes by dividing the total (and future) watts needed by the amount of voltage delivered to the distribution box:

$$\text{Formula: } \frac{\text{watts}}{\text{volts}} = \text{amperes}$$

where

$$W = \text{symbol for watts}$$
$$V = \text{symbol for volts}$$
$$A = \text{symbol for amperes.}$$

$$\text{Example: } \frac{35{,}000\text{W}}{240\text{V}} = 145\text{A}$$

Most residences require a distribution panel with a capacity of 100 to 200 amps. The NEC minimum for new residential construction is 60 amps. To compute the total load requirements, the watts needed for each circuit must first be determined.

Branch Circuits

From the distribution panel, electricity is routed to the rest of the building through *branch circuits.* A **circuit** is a circular path that electricity follows from the power supply source to a light, appliance, or other electrical device and back again to the power supply source. See Figure 31.9. If the electrical load for an entire building were placed on one circuit, overloading would leave the entire building without power. Thus branch

circuits are used. Each circuit delivers electricity to a limited number of outlets or devices.

Each circuit is protected with a **circuit breaker.** A circuit breaker is a device that opens (disconnects) a circuit when the current exceeds a certain amount. Without a circuit breaker, excessive electrical loads could cause the wiring to overheat and start a fire. When a breaker opens, or "trips," the power to the branch circuit is disconnected. Similarly, if the sum of the current drawn by the branch circuits exceeds the rating of the main circuit breaker, the main breaker will trip. This protects the service-entrance wires and equipment from overheating and damage. Older homes often have fuses instead of circuit breakers. They serve the same purpose, but overloaded fuses must be replaced. In contrast, circuit breakers that trip can easily be reset.

Branch circuits are divided into three types by the NEC: lighting circuits, small-appliance circuits, and individual circuits (dedicated circuits).

Lighting Circuits

Lighting circuits are connected to lighting outlets for the entire building. Different lights in each room are usually on different circuits so that if one circuit breaker trips, the room will not be in total darkness.

In all dwellings other than hotels, the NEC requires a minimum general lighting load of 3 watts per square foot of floor space. However, the amount of wattage demanded at one time (demand factor) is calculated at 100% only for the first 3,000 watts; 35% is used for the second 17,000 watts; and 25% is used for commercial demands over 120,000 watts. Thus, the general lighting load planned for a 1,500-sq.-ft. house would be 3,525 watts, not the full 4,500 watts. It is calculated as follows:

$$1{,}500 \text{ sq. ft.} \times 3\text{ W} = 4{,}500\text{W}$$

First 3,000	× 100% =	3,000W
Next 1,500W ×	35% =	525W
Total		3,525W

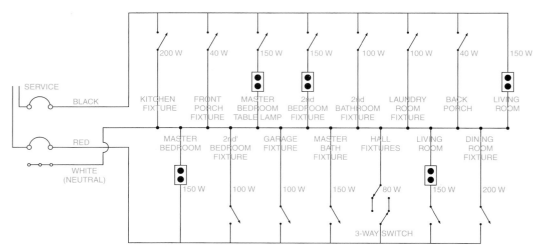

FIGURE 31.10 > Typical breakdown of lighting loads. *Hepler/Wallach/Hepler © Cengage 2013*

If each branch circuit supplies 2,400 watts (120V × 20A = 2,400W), a 1,500-sq.-ft. house should have two 2,400-watt general lighting circuits. See Figure 31.10. Lighting circuits may also be used for small devices such as clocks and radios. However, because all lights and other items on the circuit are probably not going to be used at the same time, it is not necessary to provide a service capable of supplying the full load.

Small-Appliance Circuits

These circuits provide power to outlets wherever small appliances are likely to be connected. Small appliances include items such as toasters, electric skillets, irons, electric shavers, portable tools, and computers. Appliance circuits are not designed to also support lighting needs. See Figure 31.11. The NEC requires a minimum of two small-appliance circuits in a residence. Each circuit is usually computed as a 3,600-watt load (30A × 120V = 3,600W).

Individual Circuits

Individual dedicated circuits are designed to serve a single large electrical appliance or device, such as electric ranges, automatic heating units, built-in electric heaters, and workshop outlets. Large motor-driven appliances, such as washers, garbage disposals, and dishwashers, also use individual circuits. These circuits are designed to provide sufficient power for starting loads. When a motor starts, it needs an extra surge of power to bring it to full speed. This is called a *starting load.*

Separate circuits (20 amps) are required in a laundry area to provide power for the washing machine and the dryer. Because of the danger of water leakage, a GFCI receptacle is recommended.

Ground-Fault Circuit Interrupter (GFCI)

A GFCI receptacle must be located wherever there is a possibility for people to ground themselves and be shocked by the

electrical current flowing through their body to the ground. The purpose of a GFCI receptacle is to cut off the current at the outlet. When the GFCI receptacle senses any change of current, it immediately trips a switch to interrupt the current. It operates faster and is safer than the circuit breaker switch or fuse at the power entry panel. A GFCI switch will trip in 1/40th second when an extremely small current variation (ground fault) of 0.005 amp is reached.

In new and remodeling construction, GFCI receptacles must be located with each convenience outlet near water sources or pipes in the bathroom, kitchen, garage, laundry, and outdoors. Any receptacle located within 10′ (15′ according to some codes) of the inside of a permanently installed swimming pool must also be wired through a GFCI. GFCIs are also required if outlets are placed in unfinished crawl spaces below grade level. Some codes require *all* GFCI outlets in any room containing a water supply.

Electrical Conductors

Wires used to conduct electricity are classified by the type of wire material, the insulation material, and the wire size. Wire size is classified by number in reverse order of size, as shown in Figure 31.12. Figure 31.13 shows the ampere rating of different wire sizes from 0000 to 14. The size of the wire used in a circuit depends on the current to be carried by the circuit. Although the meter voltage is 120V and 240V, wiring resistance reduces the voltage at the receptacles to approximately 110V and 220V. Wire sizes 6 through 2/0 are used for 240V service entrance and circuits. The exact size depends on the capacity of the service panel. Sizes 10 through 14 are used for 120V and 240V lighting and small-appliance circuits. Sizes 16 and 18 are used for low-voltage items such as thermostats and doorbells.

A low-voltage switching system may be used to turn on or off any **fixture,** appliance, or light. Because of the low

	TYPICAL CONNECTED WATTS	VOLTS	WIRES	CIRCUIT BREAKER OR FUSE	OUTLETS ON CIRCUIT	OUTLET TYPE	NOTES
KITCHEN							
Range	12,500	120/240	3 #6 + GND	50A	1	14-50R	
Oven (built-in)	4,500	120/240	3 #10 + GND	30A	1	14-30R	#1
Range top	6,000	120/240	3 #10 + GND	30A	1	14-30R	#1
Dishwasher	1,500	120	2 #12 + GND	20A	1	5-15R	#2
Waste disposer	800	120	2 #12 + GND	20A	1	5-15R	#2
Trash compactor	1,200	120	2 #12 + GND	20A	1	5-15R	#2
Microwave oven	1,450	120	2 #12 + GND	20A	1 or more	5-15R	
Broiler	1,500	120	2 #12 + GND	20A	1 or more	5-15R	#3
Fryer	1,300	120	2 #12 + GND	20A	1 or more	5-15R	#3
Coffeemaker	1,000	120	2 #12 + GND	20A	1 or more	5-15R	#3
Refrigerator/freezer 16–25 cubic feet	800	120	2 #12 + GND	20A	1 or more	5-15R	#4
Freezer chest or upright 14–25 cubic feet	600	120	2 #12 + GND	20A	1 or more	5-15R	#4
Roaster-broiler	1,500	120	2 #12 + GND	20A	1 or more	5-15R	#3
Waffle iron	1,000	120	2 #12 + GND	20A	1 or more	5-15R	#3
FIXED UTILITIES							
Fixed lighting	1,200	120	2 #12	20A	1 or more		#10
Window air conditioner 14,000 Btu	1,400	120	2 #12 + GND	20A	1	5-15R	#11
25,000 Btu	3,600	240	2 #12 + GND	20A	1	6-20R	#11
29,000 Btu	4,300	240	2 #10 + GND	30A	1	6-30R	#11
Central air conditioner 23,000 Btu	2,200	240					#6
57,000 Btu	5,800	240					#6
Heat pump	14,000	240					#6
Sump pump	300	120	2 #12	20A	1 or more	5-15R	#1
Heating plant oil or gas	600	120	2 #12	20A	—	—	#6
Fixed bathroom heater	1,500	120	2 #12	20A	—	—	#6
Attic fan	300	120	2 #12	20A	1	5-15R	
Dehumidifier	350	120	2 #12	20A	1 or more	5-15R	#1

continued

FIGURE 31.11 > Load requirements of common electrical appliances. *Hepler/Wallach/Hepler © Cengage 2013*

	TYPICAL CONNECTED WATTS	VOLTS	WIRES	CIRCUIT BREAKER OR FUSE	OUTLETS ON CIRCUIT	OUTLET TYPE	NOTES
LAUNDRY							
Washing machine	1,200	120	2 #12 + GND	20A	1 or more	5-15R	#5
Dryer all-electric	5,200	120/240	3 #10 + GND	30A	1	14-30R	#1
Dryer gas/electric	500	120	2 #12 + GND	20A	1 or more	5-15R	#5
Ironer	1,650	120	2 #12 + GND	20A	1 or more	5-15R	
Hand iron	1,000	120	2 #12 + GND	20A	1 or more	5-15R	
Water heater	3,000–6,000					DIRECT	#6
LIVING AREAS							
Workshop	1,500	120	2 #12 + GND	20A	1 or more	5-15R	#7
Portable heater	1,300	120	2 #12 + GND	20A	1	5-15R	#3
Television	300	120	2 #12 + GND	20A	1 or more	5-15R	#8
Portable lighting	1,200	120	2 #12 + GND	20A	1 or more	5-15R	#9
Band saw	300	120	2 #12 + GND	20A	1 or more	5-15R	#6
Table saw	1,000	120/240	2 #12 + GND	20A	1	5-15R	#6

NOTES
#1 May be direct-connected.
#2 May be direct-connected on a single circuit; otherwise, grounded receptacles required.
#3 Heavy-duty appliances regularly used at one location should have a separated circuit. Only one such unit should be attached to a single circuit at a time.
#4 Separate circuit serving only refrigerator and freezer is recommended.
#5 Grounding-type receptacle required. Separate circuit is recommended.
#6 Consult manufacturer for recommended connections.

#7 Separate circuit recommended.
#8 Should not be connected to appliance circuits.
#9 Provide one circuit for each 500 sq. ft (46 m^2). Divided receptacle may be switched.
#10 Provide at least one circuit for each 1,200 watts of fixed lighting.
#11 Consider 20-amp, 3-wire circuits to all window-type air conditioners. Outlets may then be adapted to individual 120- or 240-volt units. This scheme will work for all but the very largest units.

FIGURE 31.11 ⟩ Continued.

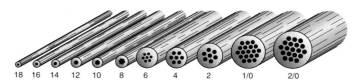

FIGURE 31.12 ⟩ Sizes of copper wire conductors. *Hepler/ Wallach/Hepler © Cengage 2013*

voltage, extremely small wires are used to attach the switch to the fixture. A step-down transformer at the switch reduces the voltage. A step-up transformer at the fixture raises the voltage level back.

Wire size is critical. If a wire is too small for the current applied, excessive resistance (overload) can result. This may cause the insulation to overheat and break down, causing a potential fire hazard. When selecting or preparing to use appliances, it is important to check the UL (Underwriters Laboratories) ratings to learn the proper wiring requirements.

Aluminum wire is lighter and less expensive than copper, but many codes apply stricter rules to the use of aluminum for residential work. Insulation is available in flexible metal armored or nonmetal sheathed form. For underground or exterior exposed wiring, wires must be encased in rigid or flexible metal or PVC (plastic) conduits.

BRANCH CIRCUITS

SIZE	DIAMETER (INCHES)	CURRENT RATING (AMPERES)	
		COPPER	ALUMINUM
14	0.064	15	—
12	0.081	20	15
10	0.102	30	25
8	0.129	40	30
6	0.162	55	40
4	0.204	70	55
3	0.229	80	65
2	0.258	95	75
1	0.289	110	85
0	0.325	125	100
00 (2/0)	0.365	145	115
000 (3/0)	0.410	165	130
0000 (4/0)	0.460	195	155

SERVICE ENTRANCE

3-WIRE SERVICE SIZE (EACH WIRE)	SERVICE RATING CURRENT (AMPERES)	
	COPPER	ALUMINUM
4	100	—
3	110	—
2	125	100
1	150	110
0	175	125
00 (2/0)	200	150
000 (3/0)	—	175
0000 (4/0)	—	200

FIGURE 31.13 > Relationship of amperes to wire size.
Hepler/Wallach/Hepler © Cengage 2013

Calculating Total System Requirements

The installation of the proper size of service entrance equipment and branch circuits depends on the square footage of the residence, number of appliances, lighting, and future expansion allowances. To find the total amp service needed for an entire building, first determine the total number of watts needed for each circuit. Add these to find the total watts needed for the building. For example, to calculate the size of the service entrance for a 2,000-sq.-ft. residence, list the amount of wattage to be used as follows:

- Lighting circuits (typical):
  2,000 sq. ft. uses 3 watts per square foot = 4,050 watts.

- Convenience outlets:
  Two circuits in service area (120V × 20A) = 4,800 watts
  Two circuits in sleeping area (120V × 20A) = 4,800 watts
  Two circuits in living area (120V × 20A) = 4,800 watts.

- Dedicated circuits:

Central AC	=	4,000 watts
Microwave oven	=	1,200 watts
Electric range	=	10,000 watts
Electric dryer	=	5,000 watts
Washing machine	=	1,000 watts
Dishwasher	=	1,000 watts
Forced air unit	=	1,000 watts
Electric water heater	=	2,000 watts
	Total	43,650 watts

To find the required service panel amps needed, divide the total watts by the available voltage (240V):

$$43,650 \text{ watts} \div 240 = 182 \text{ amps}$$

Service panels are available with capacities of 30, 40, 50, 60, 70, 100, 125, 150, 175, and 200 amps. The next highest panel above the required amps should be chosen to allow for future expansion. In this case, the next highest is 200-amp service.

LIGHTING DESIGN

Functional lighting design involves the interaction among eyesight, objects, and light sources. Good lighting design provides sufficient but not excessive light. Glare from unshielded bulbs or improperly placed lighting should be avoided. Excessive contrast between light and shadows within the same room should also be avoided, especially in work areas.

For centuries, candles and oil lamps were the major source of artificial light. Although candles continue to function for special effects, the major sources of light today are incandescent and fluorescent lamps. **Incandescent** lamps have a filament (a very thin wire) that gives off light when heated. **Fluorescent** lamps have an inner coating that gives off visible light when exposed to ultraviolet light. The ultraviolet light is released by a gas inside the fluorescent tube when an electronic circuit is passed through the tube. Incandescent lamps concentrate the light source, whereas fluorescent lamps provide linear patterns of light. Fluorescent lamps give a uniform glareless light that is ideal for large working areas. Fluorescent lamps give more light per watt, last seven times longer, and generate less heat than incandescent lamps. Light emitting diodes (LED) lamps use less energy, last up to 50,000 hours.

Light Measurements

Human eyes adapt to varying intensities of light. However, they must be given enough time to adjust slowly to different light

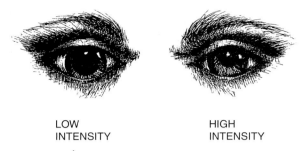

LOW
INTENSITY

HIGH
INTENSITY

FIGURE 31.14 > Pupil dilation under different light conditions. *Hepler/Wallach/Hepler © Cengage 2013*

levels. Sudden extreme changes of light may cause discomfort. Figure 31.14 shows the effect of light on pupil dilation under high and low intensities of light.

Light intensity is measured in units called footcandles. A **footcandle** (candela) is equal to the amount of light a candle casts on an object 1 foot away. The intensity of light is inversely proportional to an object's distance from the light source. See Figure 31.15. Ten footcandles (10 fc) equals the amount of light that 10 candles throw on a surface 1 foot away. In the metric system, the standard unit of illumination is the **lux** (lx). One lux is equal to 0.093 fc. To convert footcandles to lux, multiply by 10.764. Figure 31.16 shows the accepted artificial light levels for common tasks compared to natural sunlight levels. Light bulbs are rated by watts (energy used) and also by **lumens** (brightness).

Types of Lighting

The three basic types of lighting are general lighting, specific lighting, and decorative lighting. Examples of all three types of lighting can be found throughout the illustrations in Part 3. Figure 31.17 shows an example of an excellent combination of all three types of lighting combined with an effective use of color. Note how this design eliminates glare and unwanted shadows.

General Lighting

General lighting provides overall illumination and radiates a comfortable level of brightness for an entire room. General lighting replaces sunlight and is provided primarily with chandeliers, ceiling or wall-mounted fixtures, and track lights. To avoid contrast and glare, general lighting should be diffused through the use of fixtures that totally hide the light source or that spread light through panels. Close spacing of hanging fixtures also creates diffuse lighting. Another solution is to use adjustable fixtures so that the light can be directed away from eye contact.

Where possible, daylight should be included as a part of the general lighting plan during daylight hours. If ad-

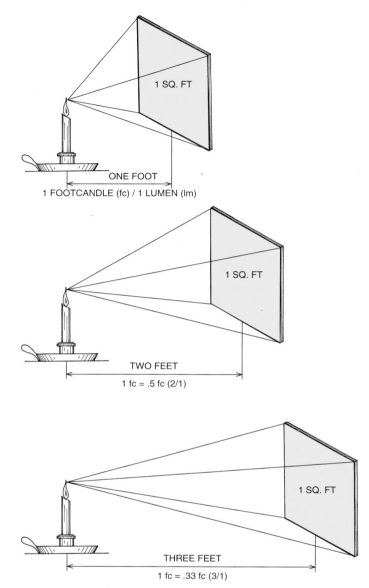

1 SQ. FT

ONE FOOT
1 FOOTCANDLE (fc) / 1 LUMEN (lm)

1 SQ. FT

TWO FEET
1 fc = .5 fc (2/1)

1 SQ. FT

THREE FEET
1 fc = .33 fc (3/1)

FIGURE 31.15 > Footcandle definition. *Hepler/Wallach/Hepler © Cengage 2013*

equate window light is not available, the use of skylights should be considered. The general and decorative lighting in Figure 31.18 blends with natural lighting to create a well-diffused lighting environment.

The intensity of general lighting should be between 5 and 10 fc (54 and 108 lx). A higher level of general lighting should be used in the service area and bathrooms. Many general lighting fixtures can also be used for decorative lighting by a connection to dimmer switches.

Specific Lighting

Light directed to a specific area or located to support a particular task is known as specific, local, or task lighting. Specific lighting helps in performing such tasks as reading, sewing, shaving,

SUNLIGHT

Beaches, open fields	10,000 fc (107,640 LX)
Tree shade	1,000 fc (10,764 LX)
Open park	500 fc (5,382 LX)
Inside 3' from window	200 fc (2,153 LX)
Inside center of room	10 fc (108 LX)

ACCEPTED ARTIFICIAL LIGHT LEVELS

Casual visual tasks, conversation, watching TV, listening to music	10–20 fc (108–215 LX)
Easy reading, sewing, knitting, house cleaning	20–30 fc (215–323 LX)
Reading newspapers, kitchen & laundry work, keyboarding	30–50 fc (323–538 LX)
Prolonged reading, machine sewing, hobbies, homework	50–70 fc (538–753 LX)
Prolonged detailed tasks such as fine sewing, reading fine print, drafting	70–200 fc (753–2,153 LX)

FIGURE 31.16 > Comparison of sunlight and artificial light levels. *Hepler/Wallach/Hepler © Cengage 2013*

FIGURE 31.18 > Well-diffused indoor general lighting combined with natural lighting. *Carl's Furniture Showrooms Inc.—Teri Kennedy, ASID, and Linda Dragin, ASID, Designers; Lee Gordon, Photographer*

FIGURE 31.17 > Combined effect of lighting types and color. *Marc-Michaels Interior Design, Inc.*

computer work, and home theater viewing. It also adds to the general lighting level. Track lighting and portable lamps provide sources of specific indoor lighting. The placement of light sources is critical in preventing shadows in task areas.

Decorative Lighting

Bright lights are stimulating, while low levels of light are quieting. Decorative lighting is used to create atmosphere and interest. Indoor decorative lights are often directed on plants, bookshelves, pictures, wall textures, fireplaces, or any architectural feature worthy of emphasis. Some decorative lighting can be used as general lighting through the use of dimmer switches.

Outdoor decorative lighting can be most dramatic. Exterior structural and landscape features can be accented by well-placed lights. Outdoor lighting is used to light and accent wall textures, trees, shrubs, architectural features, pools, fountains, and sculptures. Outdoor lighting is especially needed to provide a safe view of stairs, walks, and driveways. The exterior lighting design in Figure 31.19 provides both decorative and specific lighting for outdoor activities.

Remember to conceal light sources and do not overlight. Use waterproof devices and an automatic timing device to turn lights on and off.

Light Dispersement

Light from any artificial source can be distributed (dispersed or directed) in five different ways: direct, indirect, semidirect, semi-indirect, and diffused. See Figure 31.20. *Direct* light shines directly on an object from a light source. *Indirect* light

FIGURE 31.19 ⟩ Dramatic example of outdoor decorative and specific lighting. *Isleworth—Tom Price, Architect; Phil Eschbach, Photographer*

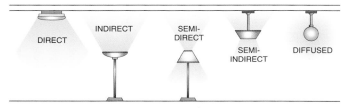

FIGURE 31.20 ⟩ Methods of light dispersement. *Hepler/Wallach/Hepler © Cengage 2013*

is reflected from surfaces. *Semidirect* light shines mainly down as direct light, but a small portion of it is directed upward as indirect light. *Semi-indirect* light is mostly reflected, but some light shines directly. *Diffused* light is spread evenly in all directions with the light source (bulb) not visible.

Reflection

All objects absorb and reflect light. Some white surfaces reflect 94% of the light that strikes them. Some black surfaces reflect only 2%. The remainder of the light is absorbed. All surfaces in a room act as a secondary source of light when light is reflected. Excessive reflection causes glare. Glare can be eliminated from this secondary source by using matte (dull) finish surfaces and by avoiding exposed light bulbs. Eliminating excessive glare is essential in designing adequate lighting.

Structural Light Fixtures

Light fixtures are either portable plug-in lamps or structural fixtures. Structural fixtures are wired and built into a build-

ing's **hard-wired** system. These must therefore be shown on electrical plans and specifications. Structural fixtures may be located on ceilings, on interior and exterior walls, and on the grounds around the building.

Different light patterns are produced, depending on the type of light fixture. Figure 31.21 illustrates the types of structural light fixtures:

- *Soffit lighting* is used to direct a light source downward to "wash" over wall surfaces as general and decorative lighting. Soffit lighting can also be designed to provide light for horizontal surfaces, such as kitchen and bath countertops, wall desks, music centers, and computer centers.

- *Cove lighting* directs light (usually fluorescent) onto ceiling surfaces and indirectly reflects light into the center of a room. Soffits should hide the fixtures from view from any position in the room.

- *Valance lighting* directs light upward to the ceiling and down over the wall or window treatment. Valance faceboards can be flat, scalloped, notched, perforated, papered, upholstered, painted, or trimmed with molding.

- *Cornice lighting* directs all light downward. It is similar to soffit lighting, except cornice lights are totally exposed at the bottom.

Wall Fixtures

Wall fixtures are used as a source of general lighting, as well as decorative lighting when attached to a dimmer switch. Wall spotlights or fluorescent fixtures may also be used as

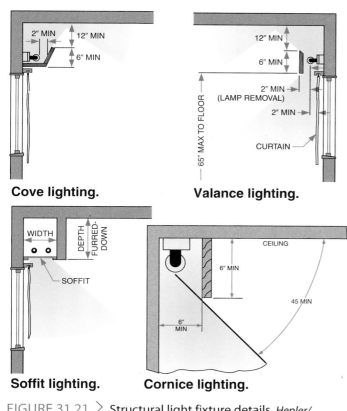

Cove lighting.

Valance lighting.

Soffit lighting.

Cornice lighting.

FIGURE 31.21 > Structural light fixture details. *Hepler/Wallach/Hepler © Cengage 2013*

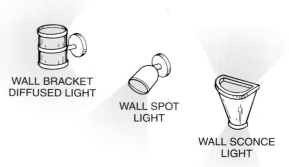

WALL BRACKET DIFFUSED LIGHT

WALL SPOT LIGHT

WALL SCONCE LIGHT

FIGURE 31.22 > Spot and sconce wall fixtures. *Hepler/Wallach/Hepler © Cengage 2013*

task lighting. Wall spotlights for accents, diffusing fixtures for general lighting, and sconces are used extensively on walls. See Figure 31.22. Vanity lights and concealed fluorescent tube lights are also used on walls as task lighting.

Ceiling Fixtures

A wide variety of lighting fixtures are designed for ceiling installation. Many optional designs are possible within each type. See Figure 31.23. Likewise, track-lighting units are available in a variety of shapes, materials, and colors. Because track light units can be moved and rotated, the track should be placed to take full advantage of these features. The path of irregular curved tracks (wave rails) can be designed to

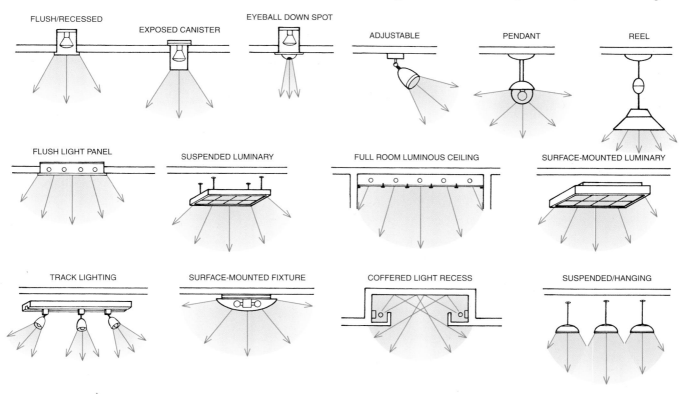

FIGURE 31.23 > Types of ceiling fixtures. *Hepler/Wallach/Hepler © Cengage 2013*

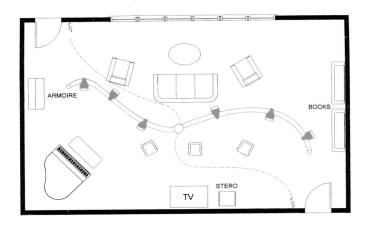

FIGURE 31.24 > Ceiling wire rail fixture. *Hepler/Wallach/ Hepler © Cengage 2013*

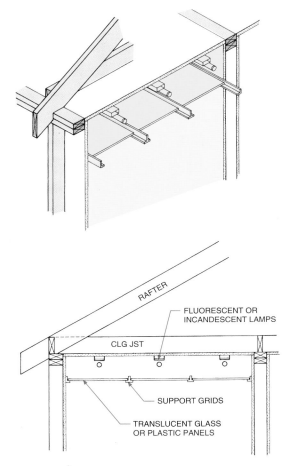

FIGURE 31.25 > Luminous ceiling design. *Hepler/Wallach/ Hepler © Cengage 2013*

provide lighting for any area in a room. The path outline of these tracks is drawn on a reflected ceiling plan as shown in Figure 31.24.

When entire ceilings are to be illuminated, fluorescent fixtures are ceiling mounted. Translucent or open mesh panels are suspended below the fixtures. The position of the fixtures should be shown on a reflected ceiling plan and detailed as shown in Figure 31.25. To provide adequate lighting, downlights are often used for specific task lighting over reading, dining, and work areas. To provide the correct location for these fixtures, junction boxes in the ceiling must be dimensioned. Task lighting is ineffectual if in the wrong position.

Exterior Lighting Fixtures

Waterproof spotlights, floodlights, and wall bracket lights are used on exterior walls for both general lighting and decorative lighting. See Figure 31.26. Exterior wall lights are often connected to motion detectors for security purposes. Lighting fixtures are used for landscaping, driveways, and walkways. These fixtures are designed to direct light at any angle to illuminate design features. Some fixtures, such as post lamps (lanterns), are designed to emit light in all directions. Other

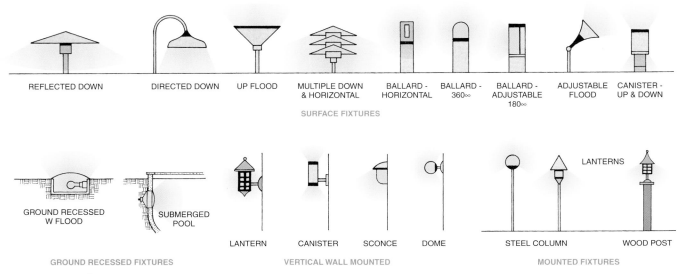

FIGURE 31.26 > Types of exterior lighting fixtures. *Hepler/Wallach/Hepler © Cengage 2013*

post (ballard) lights are designed with shields that can be adjusted to direct light 360 degrees or to any smaller segment. Swimming pool lights can also be used effectively for landscape lighting since the entire pool becomes a large light source when illuminated.

DEVELOPING AND DRAWING ELECTRICAL PLANS

Wiring methods are regulated by building codes, and wiring is approved and installed by licensed electricians. However, wiring plans are prepared by designers. For large structures, a consulting electrical contractor may prepare the final detailed electrical plans. Electrical plans include data on the type and location of all fixtures, devices, switches, and outlets.

Fixture and Device Selection

Before placing fixture locations on a floor plan, the number and type of fixtures needed for each room should be determined and listed. See Figure 31.27. In addition to lighting fixtures, all electrical or electronic devices should also be listed. This list becomes the basis for developing an electrical fixture and device schedule. (Schedules are discussed in Chapter 35.)

Switches

The number, type, and location of switches depend on the fixtures and devices. **Switches** control the flow of electricity to outlets and to individual devices.

Types of Switches

Small-appliance circuits and individual circuits are usually "hot," meaning that electricity is available in the outlet at all times. Lighting circuits, however, may be either hot (active) or controlled by switches. See Figure 31.28. *Single-pole switches* control fixtures, devices, or outlets. To control lights from two different switches, a *three-way* switching circuit (three wires and two switches) is used. A three-way switching circuit is often installed for the top and bottom of stairways and at the end of long halls. Figure 31.29 lists the major types of switch mechanisms and the circuits or devices they control.

A four-way switch and two three-way switches are used to control fixtures from three different locations. Additional four-way switches may be added to the circuit in Figure 31.30, allowing the lights to be controlled from any switch. The low-voltage method of switching offers convenience and flexibility. A relay (step down transformer) isolates all switches from the 120V system. The voltage from the switch to the appliance is only 24V, or less. At the appliance, a *magnetic-controlled switch* (step up transformer) opens the full 120V to the appliance.

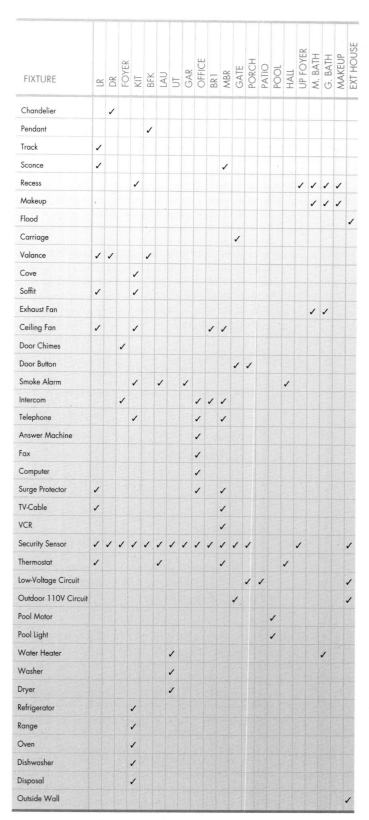

FIGURE 31.27 > Electrical fixture and device listing. *Hepler/ Wallach/Hepler © Cengage 2013*

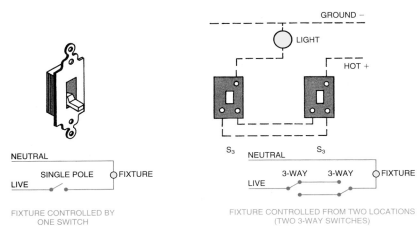

FIXTURE CONTROLLED BY
ONE SWITCH

FIXTURE CONTROLLED FROM TWO LOCATIONS
(TWO 3-WAY SWITCHES)

FIXTURE CONTROLLED FROM THREE LOCATIONS
(TWO 3-WAY AND ONE 4-WAY SWITCH)

FIGURE 31.28 ❯ Types of switching controls. *Hepler/Wallach/Hepler © Cengage 2013*

TYPE	DESCRIPTION
Toggle Switches	Used to control circuits; available in single-pole, three-way, or four-way types.
Mercury Switches	Silent, shockproof; available in single-pole, three-way, and four-way types.
Automatic Cycle Controls	Installed on appliances to control their functions on a time cycle.
Photoelectric Cells	Control switching by blocking a beam of light.
Automatic Controls	Adjust heating and cooling systems to desired temperatures.
Clock Thermostats	Adjust heating and cooling by adjusting both temperature and time.
Aquastats	Keep water heated or cooled to selected temperatures.
Dimmer Switches	Control the intensity of light; can be controlled by touch, slider, or rotary controls.
Time Switches	Cause lights or devices to switch on and off at specified time intervals.
Safety Alarm	Systems and switches that activate bells or lights when a circuit, usually on a door or window, is broken.
Master Switches	Control circuits throughout an entire building or area from one location.
Low-Voltage Switching	Systems that provide economical long runs for low-voltage lighting.
Computer Systems	Control all mechanical-electrical devices from a master control unit.
Delayed Action Switches	Allow current to flow for a limited time (usually one minute) after the switch has been turned off.

FIGURE 31.29 ❯ Switch mechanisms. *Hepler/Wallach/Hepler © Cengage 2013*

Dimmer switches allow the light intensity of light fixtures to be controlled. This can be done with touch dimmers, slide dimmers, or rotary dimmers. Integrated dimming systems can also be used to make light adjustments from a wireless remote control. Many circuits can also be controlled with wireless switches.

Switch Locations

Switch symbols are located on floor plans. Connections to the outlet, fixture, or device controlled by each switch are shown with a dashed line. See Figure 31.30. Use the following guidelines in planning switch locations:

1. Include a switch for all structural fixtures and devices that need to be turned on or off.

2. Indicate the height of all switches, which is usually 4' to 5' above floor level (3'-6" for wheelchair access). Figure 31.31 illustrates the recommended code heights of switches and receptacles.

3. Locate switches on the latch side of doors, no closer than 2 1/2″ from the casing. See Figure 31.32.

4. Exceptions to any standard should be dimensioned on the plan or elevation drawing.

5. Select the type of switch, switch mechanism, switch plate cover, and type of finish for each switch.

6. Plan a switch to control at least one light in each room.

7. Use three-way switches to control lights at both ends of stairwells, halls, and garages.

8. Locate garage door-closer switches at the house entry and within easy reach inside the garage door.

9. Control bedroom lights with a three-way switch at the entry and at the bed.

10. Use timer switches for garage general lighting, bathroom exhaust fans, and heat lights.

11. Use three-way switches for all large rooms that have two exits. Use additional four-way switches for rooms with more than two exits.

12. Use timer switches on closet and storage areas.

13. Specify timer switches for pool motors.

14. Locate safety alarm switches for a security system in the master control unit and in the master bedroom.

15. Switches for outdoor security lighting (motion detector lights) should be installed on all levels.

16. Locate lighted switches in all rooms to ensure that a person need not enter or leave a room in the dark.

17. Specify timed occupancy (motion controlled) sensors for light fixtures for non-essential fixtures.

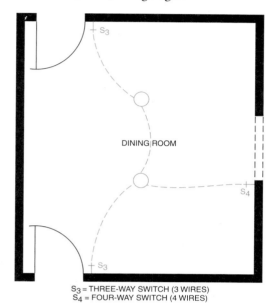

S_3 = THREE-WAY SWITCH (3 WIRES)
S_4 = FOUR-WAY SWITCH (4 WIRES)

FIGURE 31.30 > Three- and four-way switches used to control fixtures from three locations. *Hepler/Wallach/Hepler © Cengage 2013*

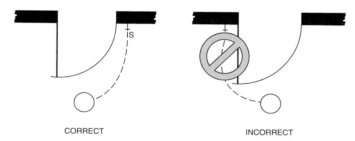

CORRECT   INCORRECT

FIGURE 31.32 > Light switches must be located on the latch side of doors. *Hepler/Wallach/Hepler © Cengage 2013*

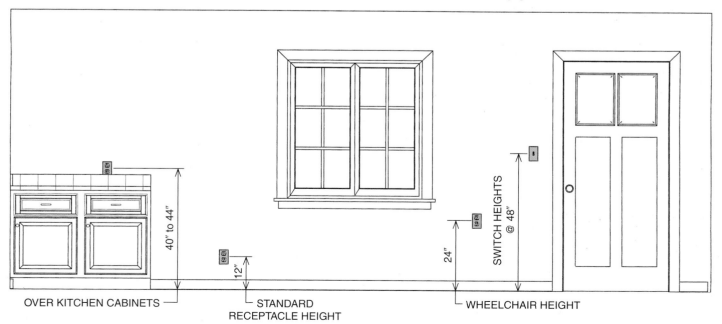

OVER KITCHEN CABINETS   STANDARD RECEPTACLE HEIGHT   WHEELCHAIR HEIGHT

FIGURE 31.31 > Recommended code heights for switches and receptacles. *Hepler/Wallach/Hepler © Cengage 2013*

Electrical Outlets and Receptacles

The terms **outlet** and **receptacle** are often used interchangeably. The NEC defines an outlet as a point in a circuit where other devices can be connected. A receptacle is a device (at an outlet box) to which any plug-in extension line, appliance, or device can be connected.

Types of Outlets and Receptacles

Different types of electrical outlets and receptacles serve different functions.

- Convenience receptacles are used for small appliances and lamps. These are available in single, double, or multiple units. See Figure 31.33.

- Lighting outlets are for the connection of lamp-holders, surface-mounted fixtures, flush or recessed fixtures, and all other types of lighting fixtures.

- Special-purpose (dedicated circuit) receptacles are the connection point of a circuit for only one electrical device.

- Special-purpose outlets and convenience outlets are connected to hot circuits, while lighting outlets are controlled with a switching device. Remember, GFCIs are installed in all outlets and switches located near water sources.

Outlet Locations

The positioning of outlets must be consistent with local codes. In addition to code requirements, the following guidelines should be used for locating outlets:

1. Outlets (except in the kitchen) should average one every 6′ (1,829 mm) of wall space.

2. Kitchen appliance outlets (with GFCIs) should average one every 4′ of wall space, be located over countertops, and include at least one countertop outlet between major appliances.

3. Hall outlets should be placed every 15′.

4. An outlet should be placed no further than 6′ from each room corner, unless a door or built-in feature occupies this space.

5. GFCI outlets should be placed as described earlier in this chapter.

6. One switch-controlled (split) outlet should be provided in each room to control plug-in lamps.

7. Consider furniture placement and positioning of portable lamps when placing lighting outlets. Room-centered furniture may need floor outlets.

8. An outlet should be placed on any wall between doors regardless of space.

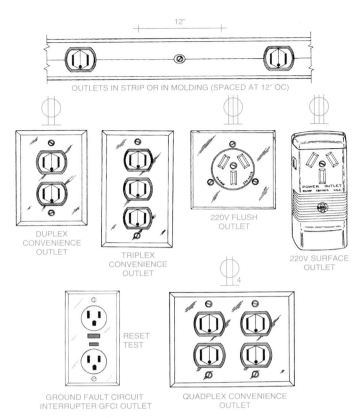

FIGURE 31.33 > Types of receptacles. *Hepler Associates LA, PC*

9. The height of all outlets should be noted on the electrical plan. Exceptions to standard dimensions should be noted at each outlet or referenced on an interior wall elevation. Normal code height for wall outlets is 10″ to 12″ from the floor. Countertop switch heights are normally 40″ to 44″ above the floor line. See Figure 31.31.

10. All individual outlets and dedicated circuits should be labeled with the appliance or device served.

11. At least one GFCI outlet should be placed above each bathroom countertop or vanity table. A minimum of two GFCI outlets should be in each bathroom.

12. Provide an outlet for each fixture, device, or appliance in the plan.

13. An outdoor weatherproof outlet (GFCI) should be provided on each side of a house. Position a waterproof outlet for a patio, pool, and a grill. Position outside outlets for decorative lighting and for low-voltage circuits, such as for entry doors, garage door, and security lights.

14. The location of all common and special-purpose outlets, such as cable and telephone jacks, is shown on floor plans using one of the symbols from Figure 31.34. See Chapter 14 for standards for people with disabilities.

NAME	ABBREV	SYMBOL	ELEVATION	PICTORIAL
SWITCH SINGLE-POLE	S	OR / S $		
SWITCH DOUBLE-POLE	S₂	S₂		
SWITCH THREE-WAY	S₃	S₃		
SWITCH FOUR-WAY	S₄	S₄		
SWITCH WEATHERPROOF	Swp	Swp		
SWITCH AUTOMATIC DOOR	SD	SD		
SWITCH PILOT LIGHT	Sp	Sp		
SWITCH LOW-VOLTAGE SYSTEM	S	S		
SWITCH LOW-VOLTAGE MASTER	MS	MS		
TWO SWITCHES	SS	S S		
THREE SWITCHES	SSS	S S S		
TELEVISION AERIAL OUTLET	TV AER	TV		
DUPLEX OUTLET	DUP OUT			
SINGLE OUTLET	S OUT	1		

NAME	ABBREV	SYMBOL	ELEVATION	PICTORIAL
TRIPLE OUTLET	TR OUT	3		
WEATHERPROOF OUTLET	WP OUT	WP		
SPLIT WIRE OUTLET	SPT WR OUT			
FLOOR OUTLET	FL OUT			
OUTLET WITH SWITCH	OUT/S	S		
HEAVY-DUTY OUTLET 220 VOLTAGE	HVY DTY OUT			SURFACE FLUSH
SPECIAL-PURPOSE OUTLET 110 VOLTAGE	SP PUR OUT	x x		
RANGE OUTLET	R OUT	R		
REFRIGERATOR OUTLET	REF OUT	R		
WATER HEATER OUTLET	WH OUT	WH		
GARBAGE-DISPOSAL OUTLET	GD OUT	GD		
DISHWASHER OUTLET	DW OUT	DW		
WASHER OUTLET	W OUT	W		
DRYER OUTLET	D OUT	D		

FIGURE 31.34 > Commonly used electrical symbols. *Hepler/Wallach/Hepler © Cengage 2013*

continued

NAME	ABBREV	SYMBOL	ELEVATION	PICTORIAL
MOTOR OUTLET	M OUT	M		
STRIP OUTLET	STP OUT	6"		
GROUNDED OUTLET	GRD OUT	GR		
LIGHTING OUTLET–CEILING	LT OUT CLG			
LIGHTING OUTLET–RECESSED	LT OUT REC			
LIGHTING OUTLET–WALL	LT OUT WALL			
LIGHTING OUTLET–CEILING PULL SWITCH	PS	PS		
FLOOD LIGHT	FL			
SPOT LIGHT	SL			
LIGHTING OUTLET–VAPOR PROOF	VP	VP		
WALL BRACKET LIGHT WITH SWITCH	WL BRK LT/S	S		
LIGHTING FLUORESCENT	LT FLUOR			
EXIT LIGHT	EXT LT	X	EXIT	EXIT
ILLUMINATED HOUSE NUMBER	ILL HSE NO	N	1234	1234

NAME	ABBREV	SYMBOL	ELEVATION	PICTORIAL
CLOCK OUTLET	CLK OUT	C		
BUZZER	BZR			
CHIME	CH	CH		
FIRE ALARM	FA	F		
FAN	F			
SERVICE PANEL WITH SWITCHES	SERV PN/SW			
ELECTRIC HEATER	ELEC HTR			
JUNCTION BOX	JUNC BX	J		
PUSH BUTTON	PB			
ELECTRIC DOOR OPENER	ELEC DR OP	D		
INTERCOMMUNI-CATION	INTER-COM			
TELEPHONE OUTLET	TEL OUT			
TELEPHONE JACK	TEL JK			
DIMMER SWITCH	DM SW	S_DM		

FIGURE 31.34 > Continued.

ELECTRICAL WORKING DRAWINGS

The use of complete electrical plans ensures that electrical equipment and wiring are installed exactly as planned. If electrical plans are incomplete and sketchy, the installation depends on the judgment of the electricians. Designers should not rely on electricians to design the electrical system, only to install it. Conversely, designers do not plan the position of every wire, only the position and relationship of all fixtures, devices, switches, and controls. This is done with the use of electrical symbols. Hundreds of electrical symbols are used on floor plans to describe what and where electrical elements will be installed. About 60 symbols apply to residential or light construction. Some of the more commonly used symbols are shown in Figure 31.34.

Completely true wiring diagrams are not used on electrical floor plans. The abbreviated architectural method, which shows only the position of fixtures, switches, and connecting lines, is drawn as shown in Figure 31.35. Figure 31.36 shows a comparison of a circuit diagram with the same circuit as shown on an electrical floor plan. Locations and connections of electrical devices are also shown on the floor plan. The control button or switch is shown on the plan in the same way switches are shown connected to lights. Lines also connect master and remote devices such as intercoms, speakers, television, DVRs, and DVD players.

Preparing Electrical Plans

The following sequence is recommended for drawing electrical floor plans:

1. Prepare the base floor plan by one of the following methods:
   a. Print a reproducible floor plan without dimensions or notes. Print a CAD layer if available; or trace a floor plan including only walls and major features.
   b. Print a reproducible floor plan with very thin lines so electrical symbols will have high contrast.
   c. Add electrical symbols and features on a print in red. This is the least desirable method, but is often practiced by electrical contractors in the field.
2. Select fixtures using Figure 31.27 as a guide.
3. Locate and draw the service entrance and distribution panel.
4. Locate and draw the position of each fixture and device on the plan using the symbols shown in Figure 31.34. Ceiling fixtures may be shown on a reflected ceiling plan.
5. Select the switch type (see Figure 31.29) needed to control each fixture or device.

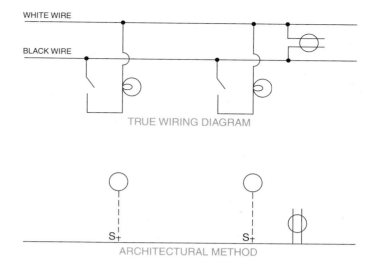

FIGURE 31.35 > True wiring diagram compared to the architectural method. *Hepler/Wallach/Hepler* © Cengage 2013

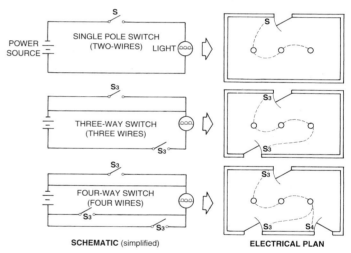

FIGURE 31.36 > Circuit diagram compared to same circuit shown on a floor plan. *Hepler/Wallach/Hepler* © Cengage 2013

6. Locate and draw the switch symbol for each switch on the plan.
7. Draw a curved solid or dashed line connecting each switch with the fixture the switch controls. The connecting lines may be curved to avoid confusion with straight object lines.
8. Locate the position and draw the symbol for each outlet on the plan.
9. Where switches or lines are connected to a different level or another drawing, draw the line outside the plan and label the line destination as shown in Figure 31.38A and Figure 31.38B.

10. Show the location of devices and control centers that require special wiring such as TV or computer network cables. Connect these with curved lines.

11. Add notes to clarify any wiring needs not obvious on the drawing.

12. List the number and type of switches, outlets and fixtures found in Figure 31.38A on page 697 and B on page 698.

Figures 31.37A through 31.37M show the common application of electrical symbols to various residential room plans. Figures 31.37L and 31.37M show an example of a two-level electrical plan that includes appliance and individual lighting circuits. Figures 31.38A and 31.38B show a more complete plan, which also includes cable wiring and electrical connections to the site, gate, and pool devices. In this plan the connections to be second level are shown with bold lines to separate them from other first-floor lines.

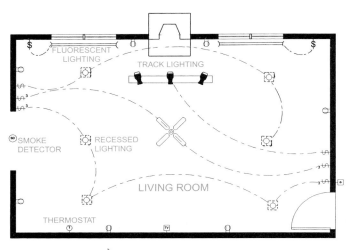

FIGURE 31.37A > Living room wiring plan. *Hepler/Wallach/ Hepler © Cengage 2013*

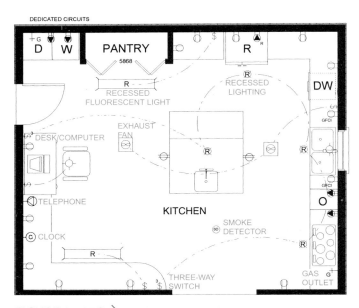

FIGURE 31.37B > Kitchen wiring plan. *Hepler/Wallach/Hepler © Cengage 2013*

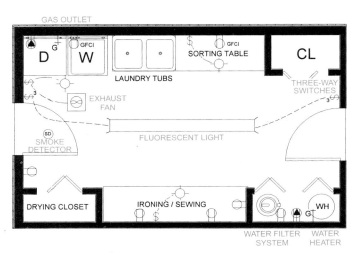

FIGURE 31.37C > Utility room wiring plan. *Hepler/Wallach/ Hepler © Cengage 2013*

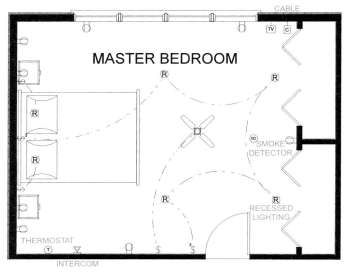

FIGURE 31.37D > Bedroom wiring plan. *Hepler/Wallach/ Hepler © Cengage 2013*

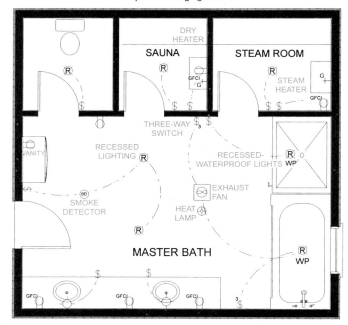

FIGURE 31.37E > Bathroom wiring plan. *Hepler/Wallach/ Hepler © Cengage 2013*

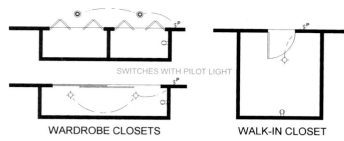

FIGURE 31.37F > Closet wiring plan. *Hepler/Wallach/Hepler © Cengage 2013*

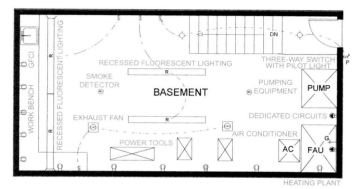

FIGURE 31.37G > Basement wiring plan. *Hepler/Wallach/ Hepler © Cengage 2013*

FIGURE 31.37H > Hall and stairs wiring plan. *Hepler/Wallach/ Hepler © Cengage 2013*

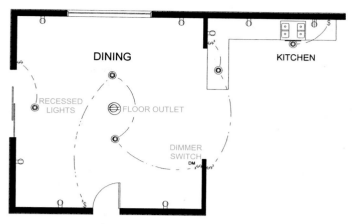

FIGURE 31.37I > Dining room wiring plan. *Hepler/Wallach/ Hepler © Cengage 2013*

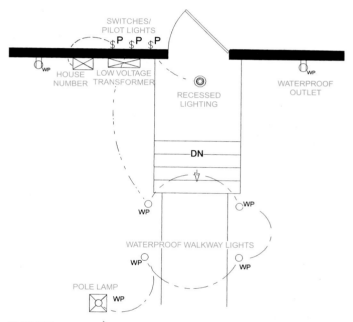

FIGURE 31.37J > Entry wiring plan. *Hepler/Wallach/Hepler © Cengage 2013*

Refer ahead to Figures 34.6 through Figure 34.13 and study the relationship between the drawing set and the electrical plans shown in Figures 31.38C and 31.38D. The complete floor plan used as a base for this electrical plan is found in Figure 15.28. The electrical plan in Figures 31.38A and 31.38B was drawn on a floor plan printed before most dimensions, notes, or reference symbols were added.

The height of switches, outlets, or fixtures is usually noted on electrical plans. A note such as "all convenience outlets to be 12″ above floor line unless otherwise noted" is common. This means only outlets that are not 12″ above the floor line need to be dimensioned. Also dimension the height of outlets above countertops. If interior wall elevations are prepared, the position of electrical components should be included and dimensioned as described in Chapter 17.

Because many kitchen appliances produce heat and therefore require high wattages, kitchen outlets are divided among several circuits. Otherwise, two or more appliances could

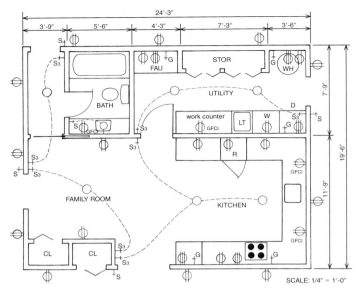

FIGURE 31.37K > Service area wiring plan. *Hepler/Wallach/ Hepler © Cengage 2013*

overload a circuit when used at maximum load. Most codes require dedicated circuits for all major appliances. Utility rooms also require heavy-duty outlets for motor-driven and heat-producing appliances. On the other hand, bedrooms, bathrooms, and closets require comparatively low wattage levels. Stairs and halls present special problems in electrical planning. Three-way and four-way switches must be carefully located to provide control at many locations and thus eliminate unnecessary backtracking.

CAD Electrical Symbols Library

On simple plans, electrical symbols are often included as a separate layer on the floor plan. For larger or more complex structures, a separate plan is prepared. In either case electrical features or plans should be prepared on a separate level. Fixtures, switches, and devices are moved from the electrical symbol library to their position on a drawing, and then connected with a curved line using the *Spline* task.

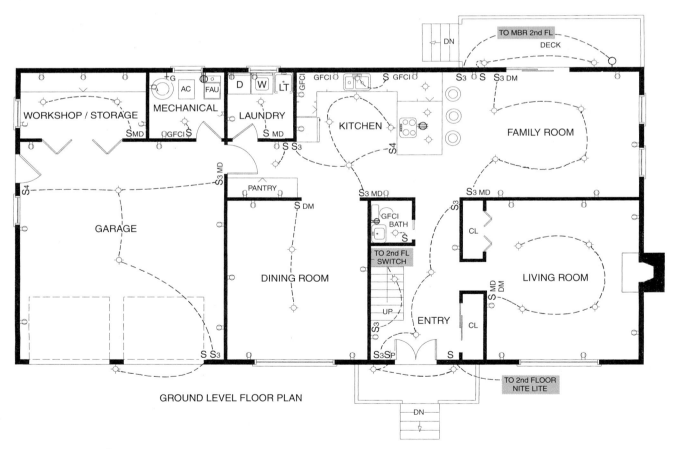

FIGURE 31.37L > First-level electrical plan showing lighting, appliance, and individual circuits. *Hepler/Wallach/Hepler © Cengage 2013*

SECOND LEVEL FLOOR PLAN

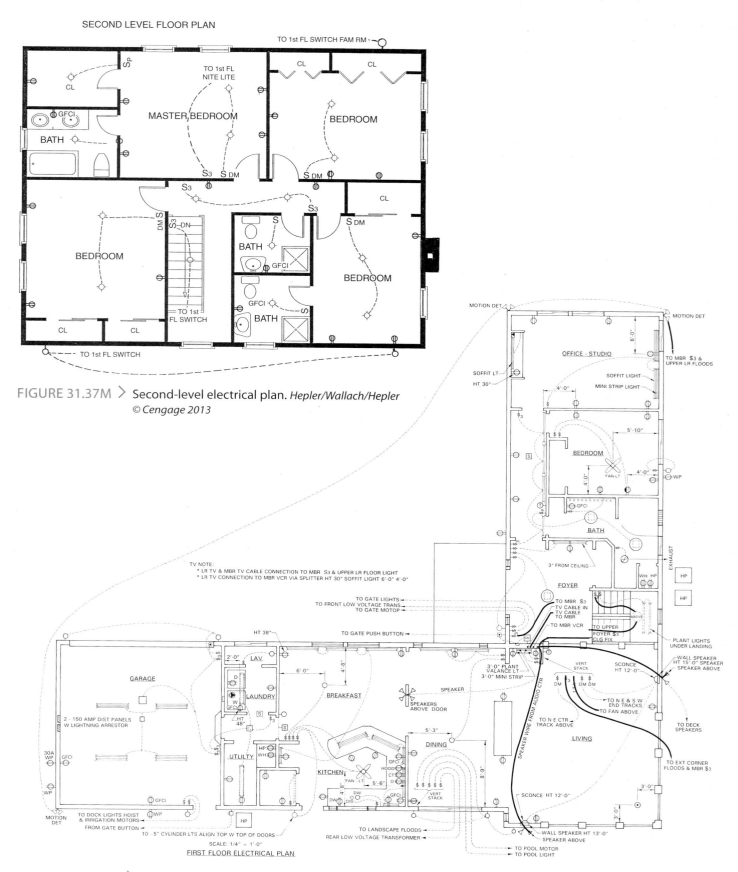

FIGURE 31.37M > Second-level electrical plan. *Hepler/Wallach/Hepler*
© Cengage 2013

FIGURE 31.38A > First-level electrical plan showing wiring to the site and the second level. *Hepler/Wallach/Hepler*
© Cengage 2013

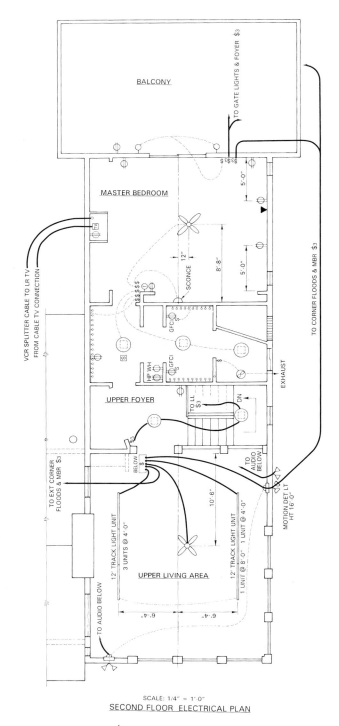

SCALE: 1/4" = 1'-0"
SECOND FLOOR ELECTRICAL PLAN

FIGURE 31.38B ⟩ Second-level electrical plan
showing wiring from first level.
Hepler/Wallach/Hepler © Cengage 2013

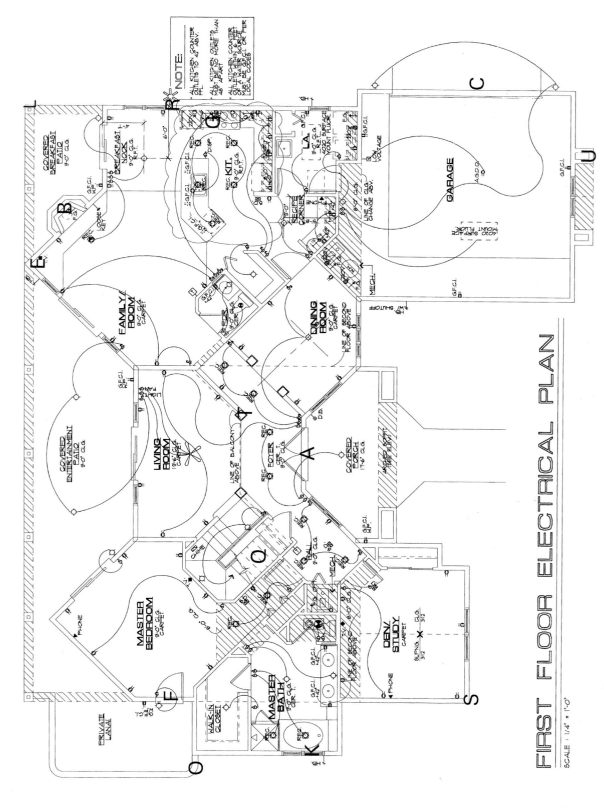

FIRST FLOOR ELECTRICAL PLAN

SCALE : 1/4" = 1'-0"

FIGURE 31.38C 〉 First-floor electrical plan for Chapter 34 set. *Hepler/Wallach/Hepler © Cengage 2013*

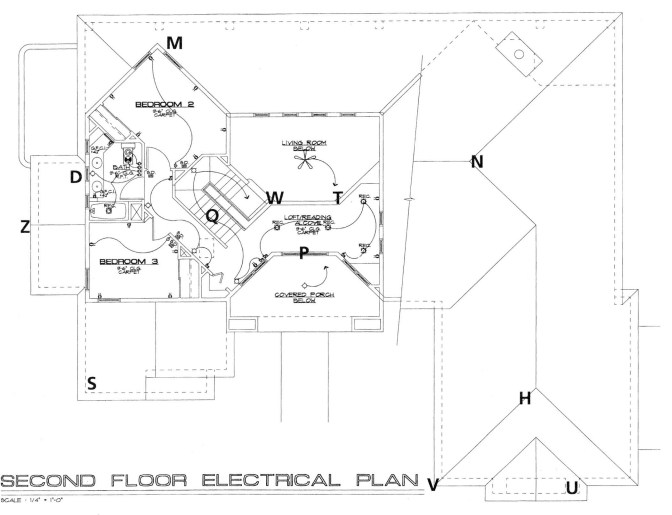

FIGURE 31.38D > Second-floor electrical plan for Chapter 34 set. *Hepler/Wallach/Hepler © Cengage 2013*

ELECTRONIC BUILDING SYSTEMS

Operating systems for buildings can be controlled electronically through the use of centralized computer control units. The major areas of control include safety, security, communication, and convenience systems. Any or all of these separate systems can be combined and connected into one integrated system. Segments can be automatically turned on or off, up or down, and/or used to activate an alarm. These systems use critical point sensors connected to control panels. When sensors are activated, preprogrammed responses create actions or send messages to a wide range of locations. Houses with these features are often called *smart houses.*

Safety and Security Systems

Some components of electronic systems are designed for protection from natural elements; some are designed as protection from intruders. These include:

1. Door and window sensors activate alarms or a call to the police when contact points are separated and break a circuit.

2. Sonic detectors trigger an alarm when the noise of breaking glass, voices, or forced entry is sensed.

3. Motion detectors record movement when the path of an intruder interrupts the lines of the sensor, as shown in Figure 31.39.

4. Fire detectors react to smoke or heat to activate an alarm and/or sprinkler system and to notify the police and fire departments.

5. Carbon monoxide and radon detectors are programmed to sound an alarm (some may also open windows) when levels exceed established limits.

6. Natural gas sensors detect leaks from appliances or furnace burners and shut off the gas line, open windows, and sound an alarm.

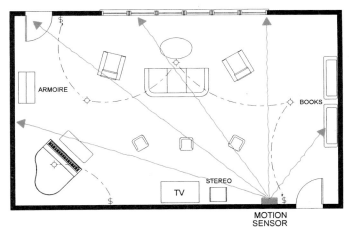

FIGURE 31.39 > Motion detector positioning. *Hepler/ Wallach/Hepler © Cengage 2013*

7. Water sensors sense water leakage and sound an alarm and notify a public utility or service company.

8. Cut-line monitors alert occupants with an alarm when an outside electrical service, telephone, or cable line has been cut or disconnected.

9. Low-temperature sensors alert residents in a remote location of the potential danger of pipes freezing.

10. Visual monitors link remote cameras (still and/or video) to a video-monitoring station. Images can be recorded for future reference using a closed-circuit video interfaced with a computer program. An alarm and call to the police can also be programmed.

11. Intruder action can be monitored through body heat sensors (infrared) and alarms activated when set levels are exceeded.

12. Light interruption sensors can activate alarms when light beams are blocked between two points.

13. A variety of audible alarms can be activated when any electronic sensor is activated. These include high-level decibel sound (+100 dB), strobe lights, voice commands, and indoor and outdoor sirens.

14. Systems can be programmed to automatically call the police, fire department, emergency medical unit, or technical support personnel for any of the functions.

15. A visitor code, which is changed frequently, allows access to selected visitors who possess the proper code.

Electronic Convenience Systems

In addition to the need for security and communications, electronic systems can remotely control otherwise manual functions for convenience. These electronic functions are designed solely for the convenience of the occupant, although some do overlap with other systems. These include:

1. Monitor and adjust heating and cooling temperatures. Also monitor humidity and air quality levels.

2. Open or close, lock and unlock doors, windows, vents, gates, and drapes.

3. Turn on or off appliances and audio or visual systems. Control station selection and volume on/off timing.

4. Monitor and/or set timing for the opening and closing of solar devices.

5. Control lighting circuits and on/off timing sequences.

6. Activate all functions on a timed basis, including night, morning, midday, evening, workday, workweek, weekend, or vacation modes.

7. Turn on/off water irrigation system.

8. Announce arrival of mail or deliveries through a sensor in the mail receptacle.

9. Warn occupants of malfunctioning of appliances, fixtures, or devices by calling a repair service company.

10. Start water at shower or bath at a preprogrammed water temperature.

11. Remotely block the use of selected devices.

12. With alarms off, activate a chime each time a door or window opens.

13. Assist people with physical impairments by adjusting the height of working surfaces, fixtures, and appliances.

14. Activate preprogrammed functions via phone signals.

Communication Systems

Communication systems include the networking of audio and visual functions. Some are one directional; others include two-way or multiple station communications. The functions available for electronic communication systems, include the following:

1. Videophone intercom communications between entrances and other locations including recorded and synthesized voice responses to visitors.

2. Activation of internal and external fixture and appliances by remote phone commands.

3. Ability to remotely activate all audio and visual functions.

4. Video viewing of key building areas.

5. Audio system distribution of prerecorded or radio reception throughout the system or to selected areas.

6. Video distribution that permits multiple television sets to receive cable or satellite TV and also view Blu-ray discs and DVDs simultaneously on multiple sets.

7. Integrated multiline phone systems that support personal, business, fax, and Internet services.

8. Internet sharing device (router) that allows web surfing on multiple computers simultaneously.

9. Networking that allows communication between multiple computers, printers, and scanners.

10. Monitoring of children's TV watching through a master TV unit.

11. Intercom with multiple-station talk and listen units.

12. Video conferencing among multiple computers.

13. Voice activation and command systems that can be applied to all audio functions.

In addition to these systems, a true environmentally safe *smart house* design includes the maximum use of natural daylight, surface insulations, nontoxic sustainable materials, solar heating, cooling, and orientation.

Integrated Systems

Electronic systems can be controlled remotely as an independent unit, or can be combined into one network that is controlled through a central computer station. This networked system is known as an **integrated system,** sometimes called a *smart building system* or *home automation system.*

Integrated systems operate through the placement of digital chips (sensors) embedded into each component to be controlled. These chips are linked together into a single network and electronically connected to one or more preprogrammed sensing devices, such as keypads, touch screens, remote controllers, coded telephones, or other telecommunication units. These sensing devices can be centrally located or separated by function or location. They are coded to identify a designated person and can be hard wired or wireless.

All of the specific functions previously described for the safety/security, convenience, and communication system units can be controlled through an integrated system. Figure 31.40 shows a drawing of an electronic system floor plan combined with wiring circuits and related data.

Electronic Systems Drawings

Electronic control units and device locations are either added to the electrical plan or included on a separate drawing as shown in Figure 31.40. If hard wired, the wiring connections between the control unit and all devices should be included

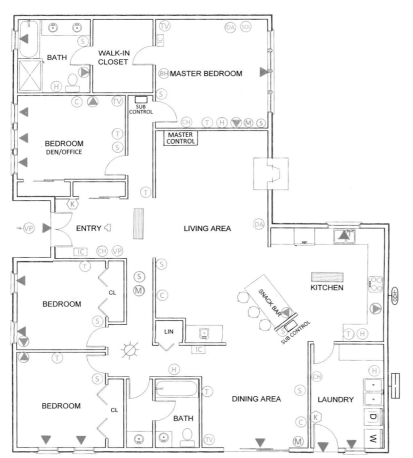

SAFETY & SECURITY SYSTEMS	
▲	DOOR AND WINDOW CONTACTS
Ⓢ	SONIC DETECTORS
Ⓜ	MOTION DETECTORS
▭	SMOKE, HEAT, CARBON MONOXIDE, AND NATURAL GAS DETECTORS
TV▷	CLOSED CIRCUIT TV
☼	LIGHT INTERRUPTION SENSOR
△	ALARM
-T	LOW TEMPERATURE SENSOR
BH	BODY HEAT SENSOR
▭	CUT LINE SENSOR
Ⓚ	KEYPAD CONTROL
H2O	WATER LEAK SENSOR
CONVENIENCE & COMFORT SYSTEMS	
Ⓣ	TEMPERATURE CONTROL
Ⓗ	HUMIDIITY CONTROL
CH	CHIMES
Ⓢ	LIGHTING, CIRCUITS AND AUTO CONTROL POINTS
SOL	SOLAR DEVICE CONTROLS
DA	DEVICE AND APPLICANTE REMOTE CONTROL
COMMUNICATION SYSTEMS	
TV	MULTIPLE NETWORKED TV UNITS
VP	VIDEO PHONE CIRCUIT
IC	INTERCOM STATION CENTER (VOICE, VIDEO, MUSIC)
▼	MULTIPLE PHONE LOCATIONS
Ⓒ	NETWORKED COMPUTER STATIONS

FIGURE 31.40 ⟩ Electronic system floor plan and data. *Hepler/Wallach/Hepler © Cengage 2013*

on the plan. If wireless, only the position of the control unit and the device need to be drawn.

Wiring for Electronics

Many electronic devices can operate wirelessly and most will be capable of wireless operation in the near future. Nevertheless, some electronics systems and networks will continue to require high-speed cable connections into the near future.

In placing electronic devices and control symbols on a drawing, the same guidelines used for electrical fixtures and switches apply here. Designers determine the present and possible future locations of all electronic devices and develop wiring plans to service these locations.

In specifying cable options, speed (bandwidth) is the most important factor. Bandwidth is measured in megabits per second (Mbps), which indicates how much information can be transmitted through a cable line in a given time frame.

Cable-type options include fiber optics and coaxial cable (CAT). Fiber-optic cables use light, not electronic, impulses to transmit information. These cables carry more information than copper and are less subject to magnetic interference. CAT 3 and CAT 5 cables are shielded with four twisted wires. CAT 5e (enhanced) cables have the greatest bandwidth. Shielded coaxial cables are the best type of cable for preventing leakage of signals. All cables should be encased in metallic conduits.

Electrical Design and Drawing Exercises

CHAPTER 31

1. What size distribution panel is required for a total of 40,000 watts with a service supply of 240 volts?

2. How many amperes flow through a 120-volt circuit with 14 ohms resistance?

3. What are the three types of branch circuits?

4. Name five locations where GFCI receptacles are required.

5. What unit is used to measure light intensity?

6. Name the three types of lighting functions.

7. Name five methods of light distribution.

8. Plan the lighting needs and fixtures of the house you are designing. Specify fixtures by room.

9. Determine the location and size of the distribution box for the house you are designing.

10. Locate the position of all lights, small appliances, and individual circuits on the floor plan you are designing.

11. Calculate the number of lighting circuits needed for a three-bedroom house.

12. Describe three types of switching mechanisms that you would select. Why?

13. Describe three types of outlets and receptacles that you would select. Why? What guidelines would you follow?

14. What is the normal spacing and height of outlets?

15. Draw an automated electronic system and a lighting system with lines between the switch symbol(s) and the fixture symbol(s) represented.

16. Draw 10 commonly used electrical symbols for a kitchen, bedroom, and bathroom.

17. Draw the complete electrical plan for the house you are designing. Show all circuits and label the capacity of each. Identify the circuits protected by a GFCI device.

Comfort Control Systems (HVAC)

OBJECTIVES

In this chapter you will learn to:

> plan and draw a mechanical heating and cooling system on a floor plan.

> use appropriate symbols to draw devices, ductwork, or piping for heating and cooling systems.

> calculate heat loss to design HVAC systems needed for specific situations.

> plan and draw passive and active solar heating and cooling systems.

TERMS

active solar systems
BTU
conduction
convection
forced-air systems
geothermal
glazing
greenhouse effect
heat gain

heat loss
heat pumps
HVAC
hydronic systems
infiltration
passive solar systems
photovoltaic film (PV)
plenum

radiation
resistivity
R-value
thermal conductivity
thermal mass
thermostats
trombe wall
U-value

⊕ INTRODUCTION

Comfort control requires more than just providing warmth or coolness. True comfort control includes providing the correct temperature, correct humidity, and a constant supply of clean, fresh, and odorless air in motion. This is accomplished through the use of a heating system, a cooling system, air filters, ventilation, and humidifiers. Climate control plans show the systems and devices used to maintain temperature, moisture, and the exchange and purification of the air supply. Effective solar design may also increase the efficiency of all climate control systems.

Air temperature, movement, pollutants, humidity, and odors can all be controlled through the use of mechanical systems. Heating, ventilating, and air conditioning (**HVAC**) systems include a wide variety of devices and delivery systems, which are shown on HVAC plans.

The types of systems designed to bring comfort control to a building are forced-air, hydronics, radiant, steam, and passive and active solar systems.

⊕ HVAC PLANS AND CONVENTIONS

Before HVAC plans can be drawn, heat loss must be calculated and the type and size of the HVAC system must be determined. The location of devices, pipes, and ducts must also be determined prior to drawing an HVAC plan.

It is necessary to know the standard HVAC symbols and abbreviations. Symbols are used to show the location and type of equipment. See Figure 32.1. Arrows are used to show the movement of hot and cold air and water.

The location of horizontal ducts is shown on an HVAC duct plan by outlining the position of the ducts. Because vertical ducts pass through the plane of projection, diagonal lines are used to indicate the position of vertical ducts. HVAC plans also show the position of all control devices, outlets, pipes, and heating and cooling units. See Figure 32.2. Figure 32.3 shows a typical combined electrical and heating system plan in which dashed lines are used to identify the position of ducts.

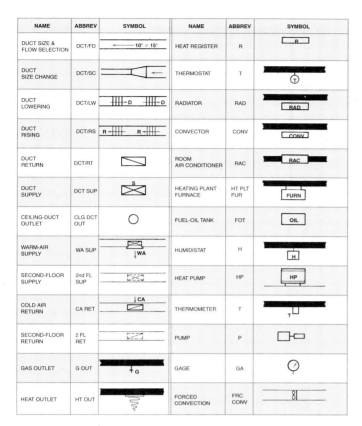

NAME	ABBREV	SYMBOL	NAME	ABBREV	SYMBOL
DUCT SIZE & FLOW SELECTION	DCT/FD	⟵ 10" × 15"	HEAT REGISTER	R	R
DUCT SIZE CHANGE	DCT/SC	⟵	THERMOSTAT	T	T
DUCT LOWERING	DCT/LW	D ⟶ D	RADIATOR	RAD	RAD
DUCT RISING	DCT/RS	R ⟶ R	CONVECTOR	CONV	CONV
DUCT RETURN	DCT/RT		ROOM AIR CONDITIONER	RAC	RAC
DUCT SUPPLY	DCT SUP	S	HEATING PLANT FURNACE	HT PLT FUR	FURN
CEILING-DUCT OUTLET	CLG DCT OUT	○	FUEL-OIL TANK	FOT	OIL
WARM-AIR SUPPLY	WA SUP	WA	HUMIDISTAT	H	H
SECOND-FLOOR SUPPLY	2nd FL SUP		HEAT PUMP	HP	HP
COLD AIR RETURN	CA RET	CA	THERMOMETER	T	T
SECOND-FLOOR RETURN	2 FL RET		PUMP	P	
GAS OUTLET	G OUT	G	GAGE	GA	○
HEAT OUTLET	HT OUT		FORCED CONVECTION	FRC CONV	

FIGURE 32.1 ⟩ Climate control symbols. *Hepler/Wallach/Hepler © Cengage 2013*

CAD HVAC Symbol Library

Because most HVAC symbols are located on or near duct-work, all duct lines should be drawn first. This is done using the *Line* task with the *Offset* command because parallel duct lines are usually too far apart to use the *Polyline* command. Once the ducts are drawn, the symbols can be moved from the HVAC library and placed on the drawing using the *Insert* command.

PRINCIPLES OF HEAT TRANSFER

As preparation for drawing HVAC plans, this chapter describes the principles of heat transfer and design considerations for an efficient HVAC system. Heat inside a building is created not only by solar heat through roofs, windows, and walls but also by heat-producing equipment and by human activity. Whether heat is inside or outside a building, it always travels from a warm area to a cool area.

Methods of Heat Transfer

Heat travels in three ways: by radiation, convection, and conduction. See Figure 32.4. In **radiation,** heat travels as waves through space in the same manner that light travels.

WARM AIR FLOOR SUPPLY
COLD AIR RETURN
HORIZONTAL DUCT
FLOW DIRECTION
8" × 18" WIDTH × DEPTH
Ø 6" DUCT DIAMETER

FIGURE 32.2 ⟩ Typical HVAC floor plan. *Hepler/Wallach/Hepler © Cengage 2013*

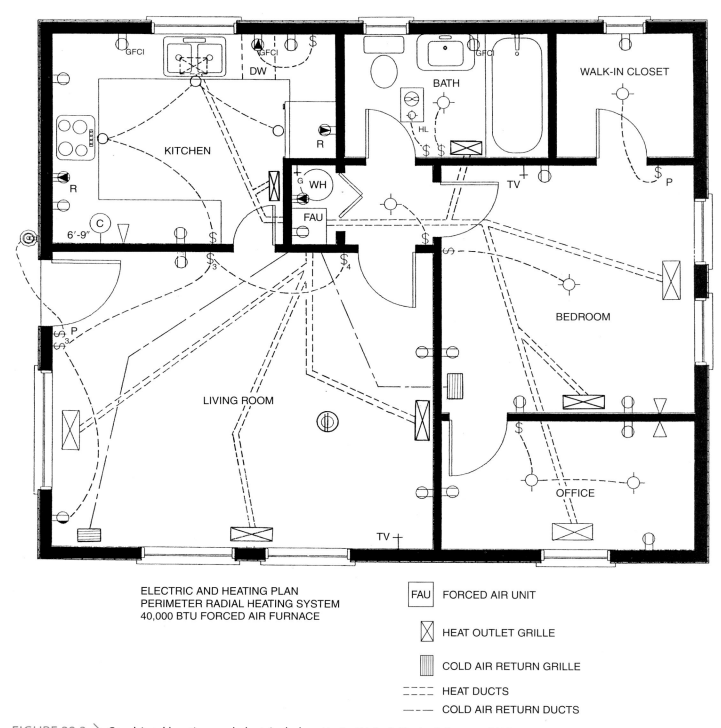

ELECTRIC AND HEATING PLAN
PERIMETER RADIAL HEATING SYSTEM
40,000 BTU FORCED AIR FURNACE

FAU	FORCED AIR UNIT
⊠	HEAT OUTLET GRILLE
▥	COLD AIR RETURN GRILLE
====	HEAT DUCTS
—·—	COLD AIR RETURN DUCTS

FIGURE 32.3 > Combined heating and electrical plan. *Hepler/Wallach/Hepler © Cengage 2013*

In **convection,** heat travels through liquids or gases. For example, a warm surface heats the air around it. The warmed air rises, and cool air moves in to take its place, causing a convection current. In **conduction,** heat moves through a solid material. The denser the material, the better it will conduct heat. The difference between the amount of heat on the inside of a building compared to the outside is known as **heat loss** or **heat gain. Infiltration** is the heat lost through openings in a building and **resistivity** is the ability of a material or component to resist heat transfer.

Heat Measurement

A measurement standard for heat levels is needed to determine the effectiveness of building materials and components in transmitting or blocking the transmission of heat.

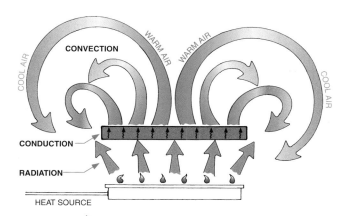

FIGURE 32.4 > Methods of heat transfer. *Hepler/Wallach/ Hepler © Cengage 2013*

British Thermal Units

The standard unit of measurement for heat is the British thermal unit (BTU). A **BTU** is the amount of heat needed to raise the temperature of 1 pound of water 1 degree Fahrenheit, at a constant pressure of 1 atmosphere (air pressure at sea level). The metric unit for heat is *joules* (J). To convert BTUs to joules, multiply the BTU value by 1055. This method of warm air heat transfer is shown in Figure 32.4.

A BTU is a measure of heat *generated*. The measure of heat *flow* is thermal conductivity. **Thermal conductivity** is the amount of heat that flows from one face of a material, through the material, to the opposite face. Thermal conductivity is defined mathematically as the amount of heat transferred through a 1-square-foot area, 1 inch thick, for each 1 degree Fahrenheit temperature difference between the faces of the material. See Figure 32.5.

Materials that transfer heat readily are known as *conductors*. Some materials resist the transfer of heat. Materials with high resistance are known as *insulators*. The effectiveness of a material in resisting heat transfer is indicated by its R-value.

⊕ R-Values and U-Values

The **R-value** is a uniform method of rating the resistance of heat flow through building materials. The higher the R number, the greater the resistance to heat flow. All building materials have been tested and assigned a thermal resistance number, or R-value. For example, the R-value of 2.5″ thick fiberglass insulation is R-7. The R-value of 6.5″ thick fiberglass insulation is R-22.

When building materials are combined in layers, the sum of their R-values is the total R-value for the component. See Figure 32.6.

To more accurately indicate the combined thermal conductivity of all materials in a structure, including air spaces, the U-value is used. **U-value** is the amount of heat conducted in 1 hour, through a 1-square-foot area, for each degree Fahrenheit difference in temperature between inside and outside

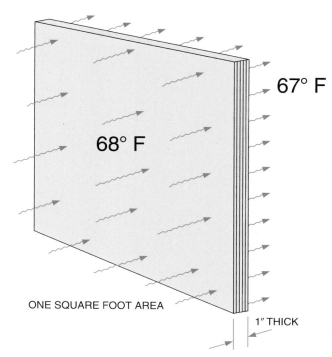

ONE SQUARE FOOT AREA

68° F

67° F

1″ THICK

FIGURE 32.5 > Measurement of thermal conductivity. *Hepler/Wallach/Hepler © Cengage 2013*

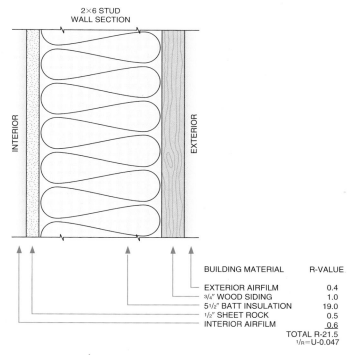

2×6 STUD WALL SECTION

INTERIOR

EXTERIOR

BUILDING MATERIAL	R-VALUE
EXTERIOR AIRFILM	0.4
³/₄″ WOOD SIDING	1.0
5¹/₂″ BATT INSULATION	19.0
¹/₂″ SHEET ROCK	0.5
INTERIOR AIRFILM	0.6
TOTAL R-21.5	
1/R=U-0.047	

FIGURE 32.6 > Calculating R- and U-values. *Hepler/Wallach/ Hepler © Cengage 2013*

air. The U-value is the reciprocal (1/R) of the R-value. High R-values and low U-values indicate greater efficiency.

Different climates and seasons require different R-value (or heat resistance) levels to maintain desired indoor temperatures. R-values must be chosen for the average low temperature of a geographic area. See Figure 32.7.

OUTDOOR TEMP	INDOOR SURFACE TEMPERATURE				
	COOL 60°F	FAIR 64°F	MEDIUM 66°F	WARM 68°F	MIN FOR FLOOR
+30°F	R-2.3	R-3.4	R-5.1	R-10.0	R-1.7
+20°F	R-2.8	R-4.2	R-6.4	R-12.5	R-2.2
+10°F	R-3.4	R-5.1	R-7.8	R-14.5	R-2.6
0°F	R-3.9	R-6.0	R-9.2	R-17.0	R-3.0
−10°F	R-4.4	R-6.8	R-10.1	R-20.0	R-3.4
−20°F	R-5.1	R-7.8	R-11.3	R-23.0	R-3.9
−30°F	R-5.7	R-8.4	R-12.8	R-25.0	R-4.4
−40°F	R-6.4	R-10.2	R-14.5	R-28.0	R-4.8

FIGURE 32.7 > R-values required to maintain indoor temperatures. *Hepler/Wallach/Hepler © Cengage 2013*

The R-value of building materials varies greatly. For example, the R-value of a 1″ face brick is only 0.11, whereas the R-value of 1″ pine is 1.25. See Figure 32.8. As the thickness of any material is increased, so is its R-value.

Windows and doors account for most heat loss in cold climates. Windows alone can allow 25% of the heat within a house to transfer to the outside. Heat flows through windows, in both directions, through radiation, convection, conduction, and infiltration (air leakage). Double window panes (double **glazing**) with argon or krypton gas between the window panes slow heat transfer. See Figure 32.9. Low-emissivity (low-E) coatings reflect heat energy (invisible solar radiation) yet transmit visible light. These measures increase R-values and decrease U-values. See Figure 32.10. R- and U-values also vary greatly among exterior door types. The door surface and core material greatly affect R- and U-values. See Figure 32.11.

Figure 32.12 shows a window section that identifies four segments that affect the window's energy performance ratings. These segments plus condensation resistance, solar heat gain, and U-values are described in Figure 32.13. A sample rating report issued by the National Fenestration Rating Council (NFRC) is shown in Figure 32.14.

In a poorly constructed building, cracks around doors, windows, and fireplaces can allow all internal heat to escape in less than an hour. Effective orientation and design can help prevent much of this heat loss. The use of insulation is also a significant deterrent to heat loss or heat gain.

⊕ INSULATION

Insulation is any material that is used to slow the transfer of heat. When effectively placed, insulation not only retards the transfer of heat but also stops moisture, sound, fire, and

TYPE OF BUILDING MATERIAL	THICKNESS	CONDUCTANCE R-VALUE (HIGH VALUE IS MORE EFFICIENT)	RESISTANCE U-FACTOR (LOW VALUE IS MORE EFFICIENT)
Roof decking insulation	2″	5.56	0.18
Mineral wool fibrous insul	1″	3.12	0.32
Loose fill insulation	1″	3.00	0.33
Acoustical tile	1″	2.86	0.35
Carpet and pad, fibrous	1″	2.08	0.48
Wood fiber sheathing	25/32″	2.06	0.49
Wood door	1 3/4″	1.96	0.51
Fiber board sheathing	1/2″	1.45	0.69
Softwoods (pine, fir, etc.)	1″	1.25	0.80
Wood subfloor	25/32″	0.98	1.02
Hardwoods	1″	0.91	1.10
Wood shingles, 16″ 7 1/2″ exp	standard	0.87	1.15
Hardboard, wood fiber	1″	0.72	1.39
Plywood	1/2″	0.65	1.54
Asphalt shingles	standard	0.44	2.27
Sheet rock/ plasterboard	1/2″	0.44	2.27
Built-up roofing	3/8″	0.33	3.03
Concrete/stone	4″	0.32	3.13
Gypsum plaster (light weight)	1/2″	0.32	3.13
Common brick	1″	0.20	5.00
Stucco	1″	0.20	5.00
Cement plaster	1″	0.20	5.00
Face brick	1″	0.11	9.10
Felt building paper (15 lb)	standard	0.06	16.67
Steel	1″	0.0032	312.50
Aluminum	1″	0.0007	1428.57

FIGURE 32.8 > R- and U-values for typical building materials. *Hepler/Wallach/Hepler © Cengage 2013*

insect penetration. To be effective, insulation must be placed everywhere that heat loss (or gain) will occur.

Without insulation an HVAC system must work harder to overcome the loss of warm air or cool air through walls, floors, and ceilings. The use of the proper insulation in walls and floors can reduce 25% of the heat transfer. Because most roofs cannot be sheltered from the sun, 40% of most heat

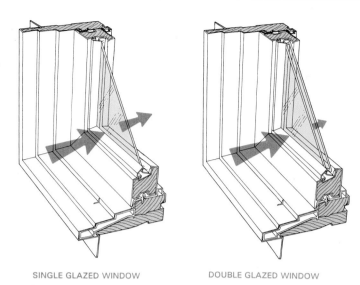

SINGLE GLAZED WINDOW   DOUBLE GLAZED WINDOW

FIGURE 32.9 > Effect of double glazing on heat transfer.
Hepler/Wallach/Hepler © Cengage 2013

TYPE OF DOOR	U-FACTOR	R-VALUE
Hollow core wood	1.00	R-1.0
Hollow core wood and storm door	0.67	R-1.5
Solid core wood	0.43	R-2.3
Solid core wood and storm door	0.28	R-3.5
Metal with urethane core	0.07	R-13.5

FIGURE 32.11 > R- and U-values for exterior doors. *Hepler/ Wallach/Hepler © Cengage 2013*

MATERIAL	U-FACTOR		R-VALUE	
	COLD CLIMATE (WINTER)	WARM CLIMATE (SUMMER)	COLD CLIMATE (WINTER)	WARM CLIMATE (SUMMER)
SINGLE GLASS	1.13	1.06	0.88	0.94
INSULATED GLASS 1/4" Air space 1/2" Air space	0.65 0.58	0.61 0.56	1.54 1.72	1.64 1.79
STORM WINDOWS 1"–4" Air space	0.56	0.54	1.79	1.85
LOW EMITTANCE 1/2" Air space $\varepsilon = .20$ $\varepsilon = .60$	0.32 0.43	0.38 0.51	3.13 2.33	2.63 1.96
GLASS BLOCK 6" × 6" × 4" 12" × 12" × 4"	0.60 0.52	0.57 0.50	1.67 1.92	1.76 2.00

FIGURE 32.10 > R- and U-values for window types. *Hepler/ Wallach/Hepler © Cengage 2013*

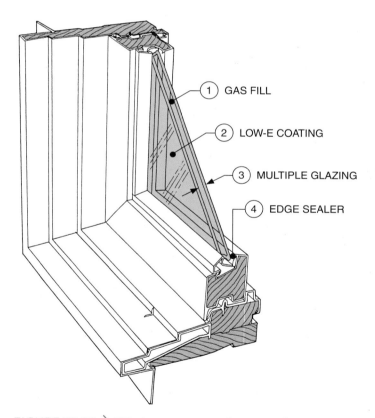

1 GAS FILL

2 LOW-E COATING

3 MULTIPLE GLAZING

4 EDGE SEALER

FIGURE 32.12 > Window energy performance factors.
Hepler/Wallach/Hepler © Cengage 2013

WINDOW SEGMENT	CHARACTERISTICS
Gas fill	The gas fill in double and triple paned windows is usually argon. It is nontoxic, clear, and odorless. It also is an excellent insulator.
Multiple glazing	Double or triple glazed windows greatly improves heat and sound insulation. The optimal space between the window panes is 1/2 inch.
Low-emittance coating (low-e)	These coatings are metal films that are transparent. The low-e coatings reflects radiant heat in the summer back outside. In the winter, it reflects the home's heat back inside the home.
Edge sealers	These sealers are made from rubber, foam, or plastic. They will improve the insulation quality of the window and stop condensation.
Condensation resistance	A window may be rated that will predict the amount of condensation. The index rating is 0 to 100. The larger the number, the less is the condensation.
Solar heat gain coefficient (SHGC)	This measure's the amount of the sun's radiant heat that will enter the home. A low number as 0.4 is excellent. A higher number of 0.6 plus will mean a large heating and air-conditioning bill.
U-factor	The U-factor measures the energy efficiency of the insulation. The lower the U-factor number the better the insulation efficiency. The U-factor is the inverse ratio of the R-value. The higher the R number, the better is the insulation.
Special performance ratings	There are additional special ratings for air leakage, water leakage, and insulating features. These ratings are offer required by the local buildings codes.

FIGURE 32.13 > Energy performance descriptions. *Hepler/Wallach/Hepler © Cengage 2013*

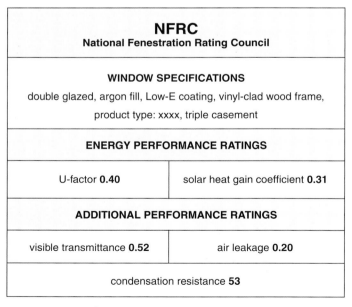

FIGURE 32.14 > Sample NFRC ratings report. *Hepler/Wallach/Hepler © Cengage 2013*

transfer is through the roof. Insulation, with an area for ventilation above the insulation, is most effective.

To prevent excessive heat transfer, a layer of insulation should be placed between the foundation and the earth below. This insulation should be outside and surround the structure, thus placing the building in an insulation envelope. Such an arrangement not only conserves heat but prevents rapid changes in inside temperature in all seasons. Figure 32.15 shows a house totally enveloped in insulation.

Insulation is most critical in cornice areas and exposed foundation walls and sills. Figure 32.16 shows methods of insulating cornice areas, and Figure 32.17 shows areas of greatest perimeter heat loss on a T-foundation.

Insulation is produced in a wide variety of vegetable, mineral, plastic, paper, and metal materials and in several different forms. These are described and illustrated in Figure 32.18. Insulation R-values vary greatly depending on the type of insulating material and thickness. See Figure 32.19.

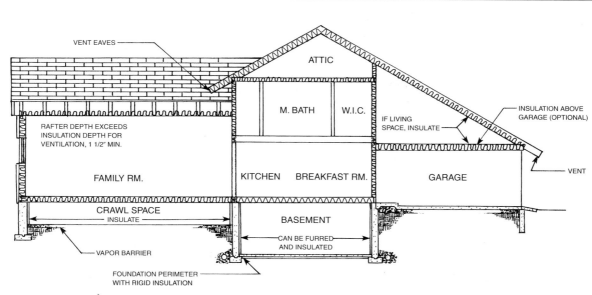

FIGURE 32.15 ⟩ Insulation locations. *Hepler/Wallach/Hepler © Cengage 2013*

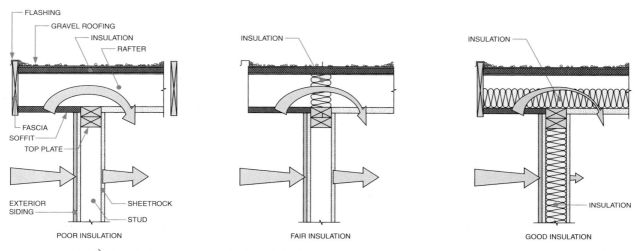

FIGURE 32.16 ⟩ Insulation at wall and ceiling joist junction. *Hepler/Wallach/Hepler © Cengage 2013*

Surface air film is a thin covering of air that clings to surfaces. The amount of air clinging varies with the amount of air movement and the type of surface. Figure 32.20 shows R- and U-values for 3/4″ air spaces with reflective and nonreflective surfaces. Vapor barrier films are nonreflective, plastic fabric sheets used to totally wrap the exterior and prevent air vapors and water from penetrating exterior walls. If a nonreflective foil sheet is specified as a vapor barrier, the foil shiny side should face the inside of the building.

Foam core insulated roof, wall, and floor structural panels provide adequate insulation without adding a separate layer of insulation to framing. See Figure 32.21. Structural foam core insulated components, including columns and beams,

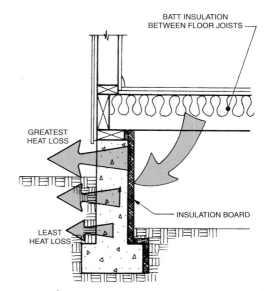

FIGURE 32.17 ⟩ Heat loss on a T-foundation. *Hepler/Wallach/Hepler © Cengage 2013*

Flexible batt	Paper-covered insulating materials that are attached between structural members. The batts are 2″ to 6″ (50 mm to 150 mm) thick.	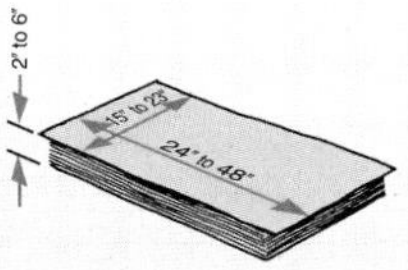
Flexible blanket	Paper-covered insulating materials that are attached between structural members. The blankets are long sheets, 1″ to 3″ (25 mm to 75 mm) thick.	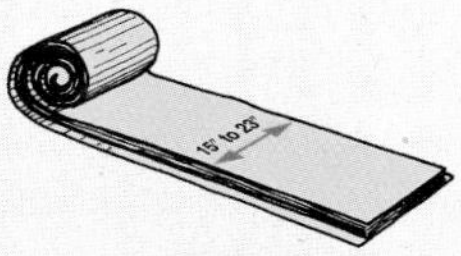
Reflective	Reflecting metal foil attached between construction members. Reflective material is often mounted on other types of insulation. It is excellent for reflecting heat, for retarding fire and decay, and for keeping out insects.	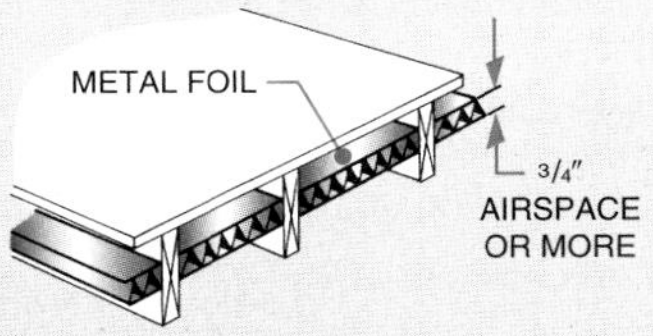
Additives	Lightweight aggregates that are mixed with construction materials to increase their insulating properties.	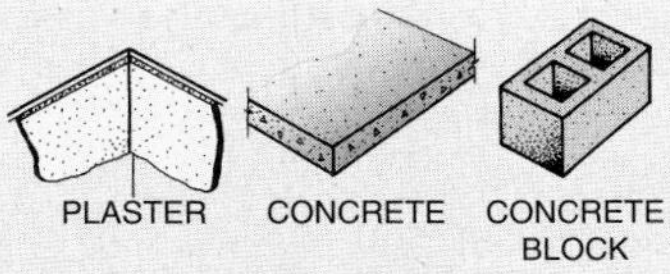
Corrugated paper	Multiple layers of corrugated paper that are easy to cut and install.	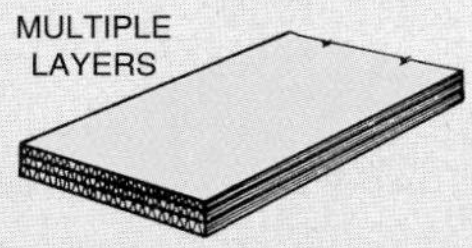
Loose fill	Materials poured or blown into walls or attic floors.	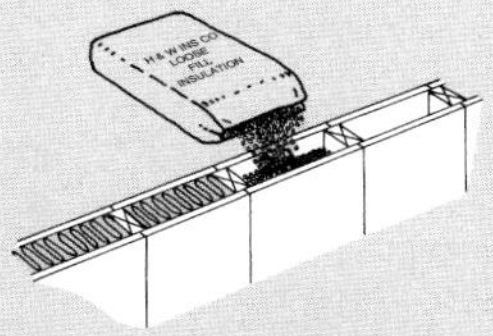
Rigid board	Thin insulating sheathing cover that is manufactured in varying sizes.	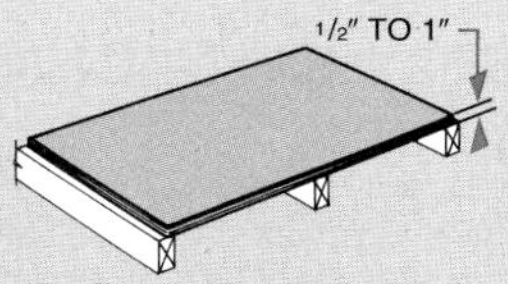
Spray-on	Insulating materials mixed with an adhesive.	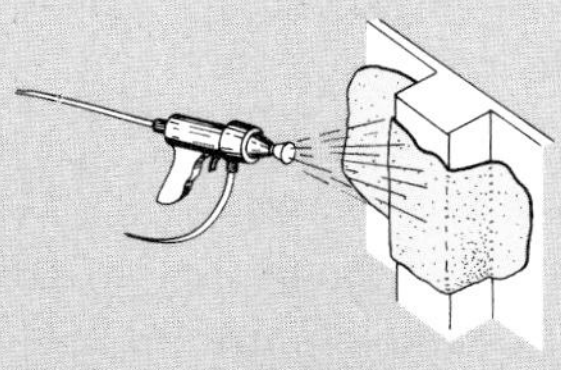

FIGURE 32.18 > Types of insulation. *Hepler/Wallach/Hepler © Cengage 2013*

INSULATING MATERIALS	MATERIAL'S THICKNESS	R-VALUE
Glass fiber	2" Batt	R-7
	4" Batt	R-11
	6" Batt	R-19
	6" Blown	R-13
	8 1/2" Blown	R-19
	12" Blown	R-26
	18" Blown	R-38
	2-6" Batts	R-38
Rock wool	4" Blown	R-11
	6 1/2" Blown	R-19
	13" Blown	R-38
Cellulose fiber	4" Blown	R-11
	6 1/2" Blown	R-19
	13" Blown	R-38
Polystyrene	1" Board	R-5
	1 1/2" Board	R-7.5
Polyurethane foam	1 1/2" Board	R-9.3
	4" Injected	R-25
Expanded MICA	Loose	R-2.5

FIGURE 32.19 > R-values of insulating materials. *Hepler/ Wallach/Hepler © Cengage 2013*

3/4" AIR SPACES	R-VALUE	U-FACTOR
Heat flow up		
Nonreflective	R-0.87	1.15
Reflective, one surface	R-2.23	0.45
Heat flow down		
Nonreflective	R-1.02	0.98
Reflective, one surface	R-3.55	0.28
Heat flow horizontal Nonreflective (also same for 4" thickness)	R-1.01	0.99
Reflective, one surface	R-3.48	0.29
AIR SURFACE FILMS (INSIDE STILL AIR)	R-VALUE	U-FACTOR
Heat flow up (through horizontal surface)		
Nonreflective	R-0.61	1.64
Reflective	R-1.32	0.76
Heat flow down (through horizontal surface)		
Nonreflective	R-0.92	1.09
Reflective	R-4.55	0.22
Heat flow horizontal (through vertical surface)		
Nonreflective	R-0.68	1.47

FIGURE 32.20 > R- and U-values for air spaces. *Hepler/ Wallach/Hepler © Cengage 2013*

can be specified in place of separate insulation as shown in Figure 32.22.

Heat Loss Calculations

Heat loss or heat gain is the amount of heat that passes through an exterior surface of a building. Regardless of the material, some heat gain or heat loss always occurs. When this happens, the temperature transfer is always from hot to cold. Factors influencing the amount of heat transfer in a building include the difference between indoor and outdoor temperatures, the type and thickness of building materials, the amount and type of insulation, and the amount of air leakage (infiltration) into or out of a structure.

To compute heat loss in BTUs, multiply the total interior surface areas (floors, walls, ceilings) by the U-value and by the temperature difference through materials (HL = A × U × T). See Figure 32.23. Special software is available to aid in computing heat loss.

HEATING SYSTEMS

Conventional heating methods use a variety of devices and distribution (delivery) systems. Devices that produce heat include warm-air, hot-water, steam, electricity, solar, fireplace, and heat pump units. Heat pumps may also be used for cooling.

FIGURE 32.21 > Foam core structural insulated panel. *Dietrich Industries, Inc.*

FIGURE 32.22 > Foam core structural insulated column. *MURUS*

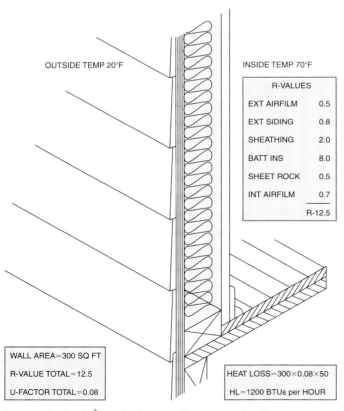

FIGURE 32.23 > Calculating a heat loss of 50°. *Hepler/ Wallach/Hepler © Cengage 2013*

Heat is usually distributed throughout buildings by ducts, pipes, or wires. Ducts are used to move both heated and cooled air. Ducts are either round, square, or rectangular. They are made of sheet metal, wood, or flexible foil-covered fiberglass. Ducts can harbor many allergens and toxins, so high-efficiency particulate air (HEPA) filters should be specified to ensure indoor air quality. Pipes are used to carry hot water or steam to radiators or base units within each room. Electric resistance wires are embedded in ceilings or floors for radiant heating.

Forced Warm-Air Units

In warm-air systems, air is heated in a furnace. Air ducts distribute the heated air to outlets throughout the building. Air filters and humidity control devices can be combined with warm-air units. Cooling systems can use the same ducts as the heating system if the ducts are rustproof.

Warm-air units operate either by gravity or by forced air. *Gravity systems* rely on allowing warm air to rise naturally without the use of a fan. Therefore, furnaces in a gravity system must be located on a level lower than the area to be heated. Gravity systems are rarely used today.

In **forced-air systems,** air is blown through the ducts by a fan located in the heating or cooling device. See Figure 32.24. In a downflow furnace, cool air enters the top and warmed

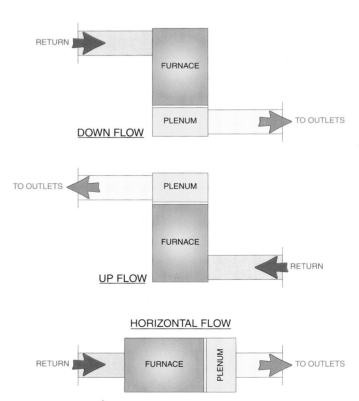

FIGURE 32.24 > Types of forced-air flow systems. *Hepler/ Wallach/Hepler © Cengage 2013*

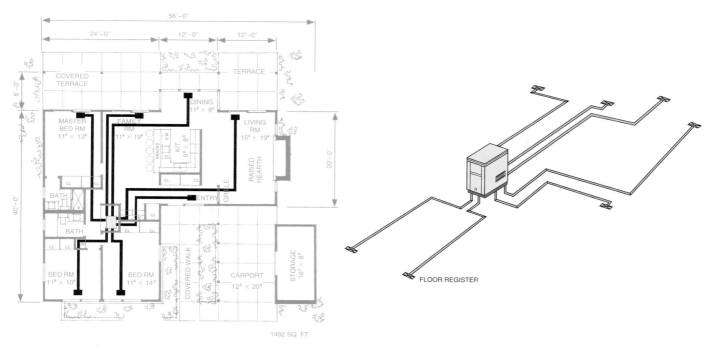

FIGURE 32.25 > Individual duct system. *Hepler/Wallach/Hepler © Cengage 2013*

air exits the bottom. Conversely, cool air enters the bottom of an upflow furnace. Cool and warm air moves through a horizontal flow furnace at the same level.

Forced-air ducts are either distribution ducts or return ducts. The return ducts bring cool air back to the furnace to be warmed. Most open-plan small buildings or buildings with many zones do not require return ducts. Instead, return air moves to the return side of a furnace.

Forced-air distribution ducts are connected to a plenum chamber, an enclosed space located between the furnace and distribution ducts. The plenum is larger than any duct and slows the flow of air through the ducts. For heating, ducts lead to floor outlets usually on the floor. At least one outlet is needed for every 15 feet of exterior wall space. For cooling, ceiling locations are preferred.

Duct systems vary in patterns as they connect a furnace to outlets throughout a building. Figure 32.25 shows an *individual duct system,* in which separate ducts directly link the furnace, or AC unit, with each outlet. These systems provide well-balanced heat but require more duct length. Figure 32.26 illustrates an *individual plenum system.* In some systems of this type, the plenum size is the same through its length. In some systems the plenum size is reduced because fewer outlets are served.

In a *perimeter loop system,* as shown in Figure 32.27, the perimeter duct is connected to a furnace with a feeder duct on each side. In a variation of the perimeter loop, the ducts of a *perimeter radial system* (Figure 32.28) radiate directly to each outlet like the spokes of a wheel.

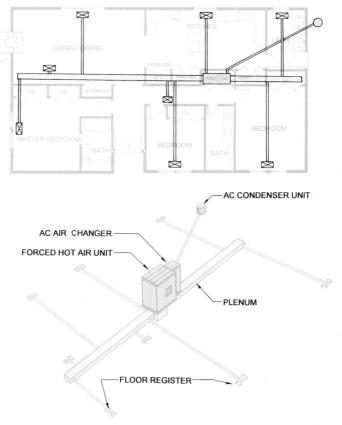

FIGURE 32.26 > Individual plenum system. *Hepler/Wallach/ Hepler © Cengage 2013*

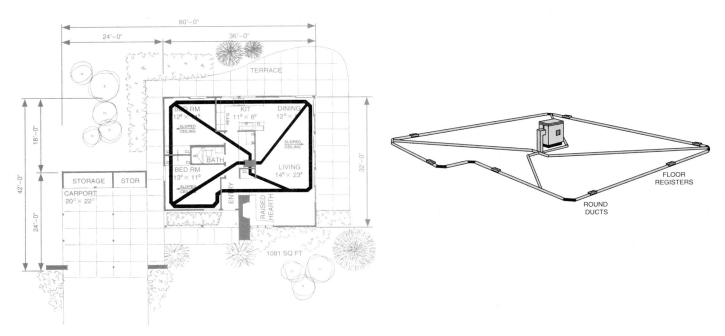

FIGURE 32.27 > Perimeter loop system. *Hepler/Wallach/Hepler © Cengage 2013*

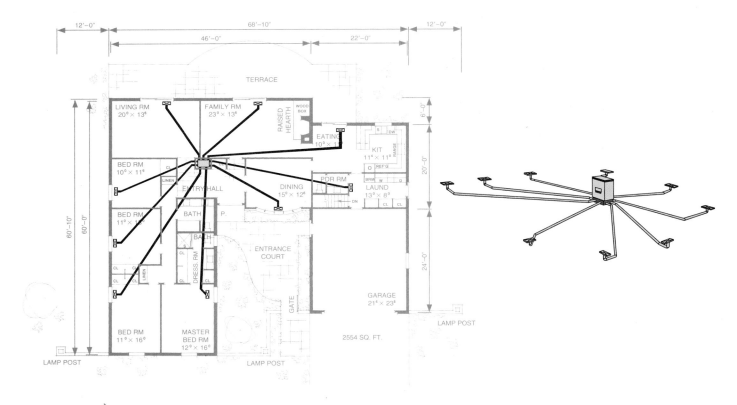

FIGURE 32.28 > Perimeter radial system. *Hepler/Wallach/Hepler © Cengage 2013*

Where heating ducts also serve an air-conditioning system, the locations of the inside air handlers, exhaust hoses, and outside compressor units need to be added. All warm-air systems include room outlets (registers) on either the floor, ceiling, or wall as shown in Figure 32.29. Figure 32.30 shows methods of connecting ducts to floor, wall, or ceiling outlets.

Wood Plenum System

In light-frame construction, a wood plenum system may be specified. A **plenum** is an enclosed space in which the air pressure is greater than it is outside. Plenum systems constructed of wood are based on a simple concept. Instead

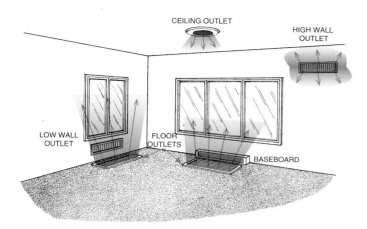

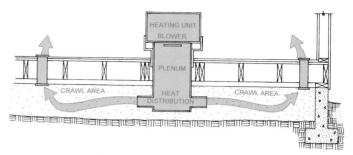

FIGURE 32.31 › Wood plenum system. *Hepler/Wallach/Hepler © Cengage 2013*

FIGURE 32.29 › Outlet types and locations. *Hepler/Wallach/ Hepler © Cengage 2013*

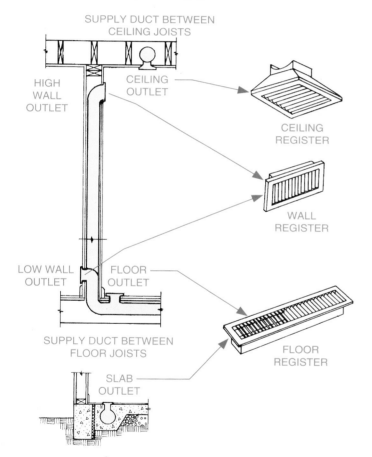

FIGURE 32.30 › Outlet connections. *Hepler/Wallach/Hepler © Cengage 2013*

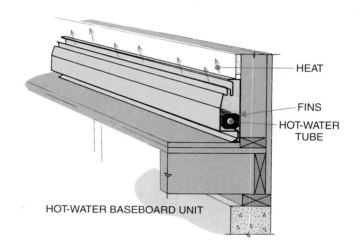

FIGURE 32.32 › Hydronic baseboard outlet. *Hepler/Wallach/ Hepler © Cengage 2013*

few or no added supply ducts. This type of system was used by the early Romans.

Forced-air system drawings include the location of each furnace, outlet, and duct. Buildings requiring more than one furnace are drawn in zones with a separate furnace and duct-work system for each. Abbreviated forced-air system drawings show only the position of outlets. This means the builder must determine the type, size, and location of all devices, controls, and ductwork.

Hydronic (Hot-Water) Units

Hot-water heating systems use an oil or gas boiler to heat water and a water pump to send the water to radiators, finned tubes, or convectors. **Hydronic systems** provide even heat and are quiet and clean. These systems do not provide air filtration or circulation and are not compatible with cooling systems that require air ducts.

The most common and most effective hydronic outlet type is the baseboard outlet shown in Figure 32.32. Some hydronic system units, such as domestic oil burners, also provide hot water for sinks and showers, thus eliminating the need for a separate water heater unit.

of using heating and cooling ducts, the entire underfloor space (crawl space) is used as a sealed plenum chamber to distribute warm or cool air to floor registers in the rooms above. See Figure 32.31. The plenum consists of wood floor construction with sealed and insulated foundation walls. A forced-air mechanical heating and/or cooling unit maintains slight air pressure in the plenum. This ensures a uniform distribution of conditioned air throughout the building with

There are several types of hot-water systems: the series-loop system, the one-pipe system, the two-pipe system, and the radiant system.

The series-loop system as shown in Figure 32.33, is a continual loop of pipes containing hot water. Hot water flows continually from the boiler through outlet units and back again to the boiler for reheating. The heat in a series-loop system cannot be controlled except at the source of the loop.

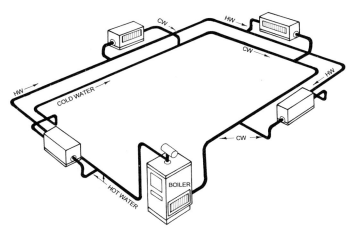

FIGURE 32.33 > Hydronic series-loop system. *Hepler/Wallach/Hepler © Cengage 2013*

In one-pipe systems (Figure 32.34), heated water is circulated through pipes that are connected to radiators or convectors by means of bypass pipes. This allows each radiator to be individually controlled by valves. Water flows from one side of each radiator to the main line and returns to the boiler for reheating.

A *two-pipe system,* as shown in Figure 32.35, has two parallel pipes: one for the supply of hot water from the boiler to each radiator, and the other for the return of cooled water from each radiator to the boiler. The heated water is directed from the boiler to each radiator but returns from each radiator through the second pipe to the boiler for reheating.

A *radiant system* distributes hot water through a series of continual pipes in floors and sometimes ceilings. Ceiling systems are not often used because of the weight of the filled pipes. A radiant floor system consists of pipes laid on a concrete base then covered with a finished concrete slab. The hot pipes conduct heat to the surface where convection currents take over. Figure 32.36 shows a radiant hot-water system and a related floor plan.

Steam Units

A steam-heating unit operates from a boiler that makes steam. The steam is transported by pipes to radiators or convectors

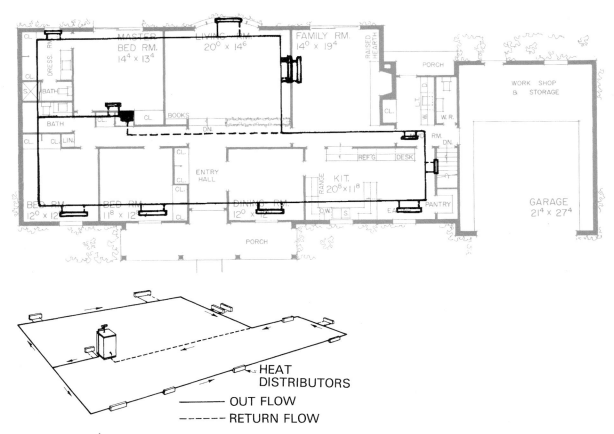

FIGURE 32.34 > Hydronic one-pipe system. *Hepler/Wallach/Hepler © Cengage 2013*

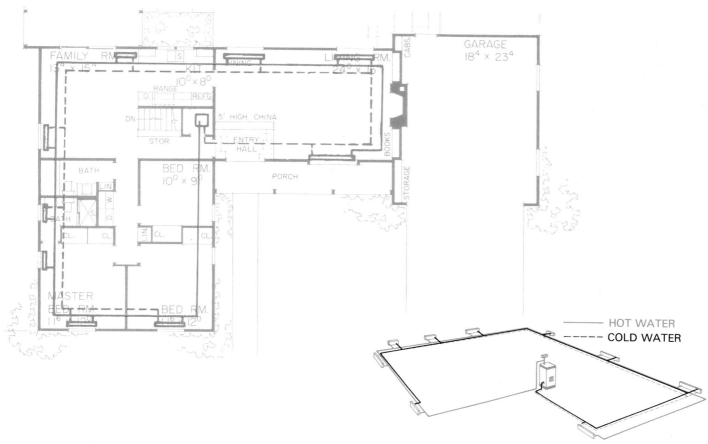

FIGURE 32.35 > Hydronic two-pipe system. *Hepler/Wallach/Hepler © Cengage 2013*

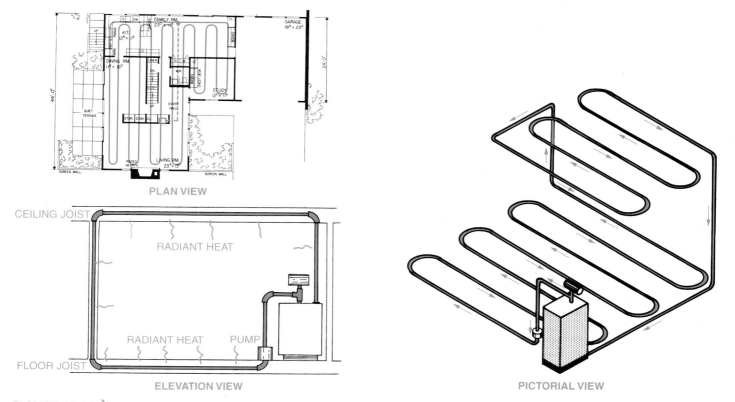

FIGURE 32.36 > Hydronic radiant system. *Hepler/Wallach/Hepler © Cengage 2013*

and baseboards that give off the heat. The steam condenses to water, which returns to the boiler to be reheated to steam.

Although steam-heating systems function on water vapor rather than hot water, drawings for steam systems are identical with those prepared for hot-water systems. Steam systems are easy to install and maintain, but they are not suitable for use with most convector radiators. They are most popular for large apartments, commercial buildings, and industrial complexes where separate steam generation facilities are provided. Steam heat is delivered through either perimeter or radial systems.

Electric Heat

Electric heat is produced when electricity passes through resistance wires. This heat is usually radiated, although it could be fan-blown (convection). Resistance wires can be placed in panel heaters installed in the wall or ceiling, placed in baseboards, or set in plaster to heat the walls, ceilings, or floors. See Figure 32.37. Electric heaters use very little space. Because there is no flame, they require no air for combustion. Electric heat is very clean. It requires no storage or fuel and no ductwork. Complete ventilation and humidity control must accompany electric heat, however, because it provides no air circulation and tends to be very dry.

Separate plans are seldom drawn for electric heat, but notations are made on floor plans, or reflective ceiling plans, concerning the location of either resistance wires or electric panels. On electrical plans, the location of facilities for the power supply and the thermostat is shown.

COOLING SYSTEMS

Buildings are cooled by removing heat. Heat can be transferred in one direction only, from warmer objects to cooler objects. Therefore, to cool a building, warm air is carried away from rooms to an air-conditioning unit, where a filter removes dust and other impurities. A cooling coil containing refrigerant absorbs heat from the air passing around it. Then the same blower that pulled the warm air from the rooms pushes cooled air back to the rooms. The size of air-conditioning equipment is rated in BTUs. A 2,000-sq.-ft. building can be comfortably cooled with a central air conditioning unit of 24,000 to 36,000 BTUs. Larger homes may require 60,000 or more BTUs, depending on the components specified. Air cooling systems are designed to control temperature and humidity and also supply fresh filtered air to an interior. As with heating systems, design begins with the identification of zones to be serviced, followed by the sizing and location of distribution and return air ducts. Next the type, size, and location of the central control unit (or units), including piping and wiring support, are planned. This includes provisions for air movement, filtering, and outside power, fuel, and air supply, as shown in Figure 32.38.

Combined Heating and Cooling Systems

Cooling systems may be separate systems or may share blowers, ducts, and outlets with heating systems. See Figure 32.39A. Like heating units, cooling units are positioned as upflow, downflow, or horizontal flow. Drawings of

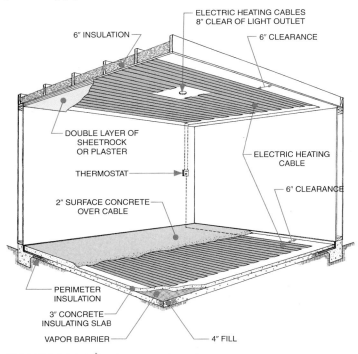

FIGURE 32.37 > Electric radiant heating system. *Hepler/ Wallach/Hepler © Cengage 2013*

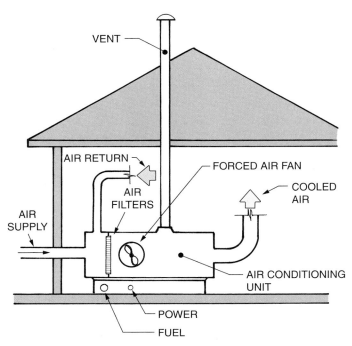

FIGURE 32.38 > Cooling system components. *Hepler/ Wallach/Hepler © Cengage 2013*

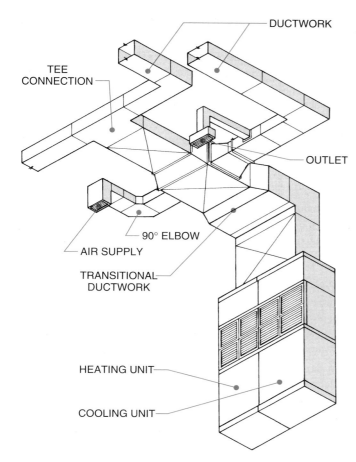

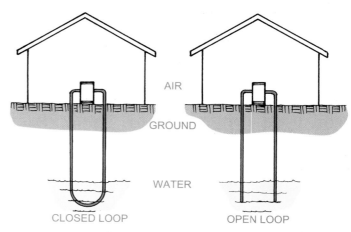

FIGURE 32.39B > Geothermal heating and cooling. *Hepler/Wallach/Hepler © Cengage 2013*

FIGURE 32.39A > Combined heating and cooling units. *Hepler/Wallach/Hepler © Cengage 2013*

combined systems use the same duct patterns and outlets as for heating plans.

Heat Pumps

Heat pumps work like a reversible refrigeration system. They can both cool and heat buildings. In warm weather a heat pump removes heat from indoor air through a closed refrigerant cycle. In cold weather, heat is drawn from the outside air, water, or ground and transferred inside. Air-source heat pumps are effective in warm or mild climates but do not heat efficiently when temperatures are below 30°F. A supplementary heat source is necessary below that temperature.

Ground-source **(geothermal)** or water-source heat pumps extract heat from the earth and are therefore more effective for heating in cooler climates. In a geothermal system, heat exchangers extract heat from the subterranean ground or water. The heat is then delivered from the geothermal source to the heating system of a building. Geothermal systems are either open-loop or closed-loop systems as shown in Figure 32.39B.

Heat pump systems consist of an inside air handler, outside heat pump, ducts, and outlets. When a building is designed, space must be allowed for the heat pump and air handler. Ducts and outlets are drawn the same as warm air ducts and outlets. Heat pumps are classified by BTU capacity or by tons. Typical units range from 1 to 5 tons.

ENVIRONMENTAL COMFORT FACTORS

Many natural factors that affect comfort levels can be controlled through effective environmental design, as covered in Chapter 6. Depending on the climate additional mechanical methods may be needed to ensure optimum levels of comfort. These include ventilation, air filtration, and temperature and humidity control.

Ventilation

Comfort control involves more than just temperature control. All buildings must be well ventilated with clean air that contains acceptable levels of humidity. Ventilation is necessary to keep fresh air circulating. Effective ventilation also controls moisture and keeps air relatively dry. All air in a building must be circulated gently and constantly, 24 hours a day, or physical discomfort may result. Air movement is provided by cross ventilation and by blowers in heating and cooling units. Ceiling fans and exhaust fans circulate air in attics, crawl spaces, and bathrooms.

Although structural tightness is a desirable method of controlling heat loss, some amount of fresh air must be allowed to enter the structure. See Figure 32.40. Attics and upper-level room ventilation helps circulate out hot air. Crawl space ventilation helps remove excessive moist air. Structural methods that create good ventilation patterns are shown in the structural framing chapters. High ceilings, as shown in Figure 32.41, help provide added space for adequate air circulation.

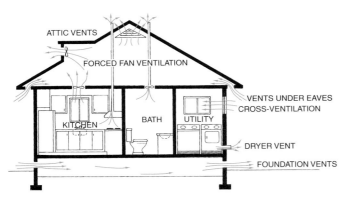

FIGURE 32.40 > Effective ventilation locations. *Hepler/ Wallach/Hepler © Cengage 2013*

FIGURE 32.41 > High ceilings help air circulation. *Oak Leaf Conservatories, www.oakleafconservatories.com*

Air Filtration

Simply moving air does not create a supply of fresh air. To create clean air, airborne pollutants such as dust mites, chemical fumes, pollen, bacteria, mold spores, and mildew must be constantly removed from the air inside a building. Some airborne particles are just a nuisance, whereas others—such as benzene, ammonia, radon, chloroform, formaldehyde, and carbon monoxide gases—are very dangerous. Many of these gases come from building materials, tobacco smoke, paint, carpet, and upholstery.

Air filters are placed in heating or cooling devices on the air return side of a furnace or air exchanger. Several levels of filters are available to trap pollutants. Mechanical filters filled with fiberglass or charcoal may only remove 15% of all pollutants. Electrostatic filters with ionizing wires can trap over 99%. Combinations of these filter types are most effective. To further purify the air, devices called ozonators may be used to introduce low levels of ozone (electronically charged oxygen) into the air supply.

Temperature Control

Thermostatic controls keep buildings at a constant temperature by turning climate control systems on or off when a set temperature is reached. Thermostatic controls may be used with any heating or cooling system. **Thermostats** are sensing devices and should be located on interior walls away from sources of heat or cold such as fireplaces or windows. Larger buildings may need two or more separate heating or cooling zones that work on separate thermostats. One advantage of electrical heating is that each room may be thermostatically controlled. The location of thermostats is shown on electrical floor plans with a thermostat symbol.

Humidity Control

Humidity is moisture in the air. The proper amount of moisture in the air is important for good climate control. Adequate ventilation, especially in attics and crawl spaces, and the use of vapor barriers in construction will prevent excessive moisture. Excess humidity can result in the growth of fungi, mold, amoebas, and bacteria.

Forced-air systems reduce humidity levels. Cooling units remove humidity from the air through condensation. If the air is too dry, a humidifier may be necessary to add more humidity to the air. Hot-water systems generally add humidity to the air.

To add more moisture to the air, a *humidifier* can be added to the plenum of a forced-air system. To remove excessive moisture from the air, a *dehumidifier* passes damp air over cold coils. The moisture-laden air then deposits excess moisture on the coils by condensation. *Humidistats* are sensing devices that monitor humidity levels and signal for more or less moisture in the air. Planning buildings in which the correct amount of humidity can be controlled is essential for creating healthy building environments.

PASSIVE SOLAR SYSTEMS

Solar heating and/or cooling involves using the sun's energy to the fullest extent possible. **Passive solar systems** are integrated with the basic design of a structure. Passive systems operate without the use of special mechanical or electronic devices to heat or cool a structure.

Passive solar heating or cooling of a building includes four steps: collecting, storing, distributing, and controlling. These steps occur in active solar systems as well. However, the equipment, materials, and devices used differ greatly.

The earth's annual revolution around the sun and daily rotation on its axis determine how much solar energy is available at any time in any location on earth. Both the daily and seasonal paths of the sun over a building site are the first consideration in passive solar planning.

The amount of solar radiation reaching a site also depends on the atmosphere. When the sun is directly overhead, its rays travel the shortest distance through the atmosphere. As the sun moves closer to the horizon, the amount of atmosphere through which the rays travel increases. See Figure 32.42. This greater amount of atmosphere decreases the amount of solar radiation reaching the site. This means winter, early morning, and evening rays travel greater distances than summer and midday rays.

The slope of a site also affects the amount of usable solar radiation. This is because the sun's rays when striking a surface perpendicularly are more concentrated than when they intersect the surface at an angle. Therefore, solar radiation striking a south-facing slope, as shown in Figure 32.43, will be more concentrated than rays striking a nearly flat terrain.

In addition to the atmospheric relationship of the sun to the earth's surface, two other principles of solar physics are used in passive solar planning. One is the **greenhouse effect.** The other is the natural law of rising warm air.

Greenhouse Effect

A car parked in direct sunlight with the windows closed illustrates the greenhouse effect. As sunlight enters through the windows, heat is absorbed by the interior surfaces of the car and trapped inside the car as stored heat. This heat cannot leave the car as easily as sunlight can enter. The interior temperature of the car may reach 200°F (93°C) or more. This is why it is against the law in many states to leave children and pets in a parked car in the summer. They could die from the intense heat.

The greenhouse effect can be dangerous in a car, but it can be useful in a building. Heat from the sun that enters a building through windows can be stored to be used later when the sun's heat is not available.

A **thermal mass** is any material that will absorb heat from the sun and later radiate the heat back into the air. The seats

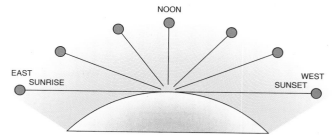

FIGURE 32.42 > Sun ray distances at different times of the day. *Hepler/Wallach/Hepler © Cengage 2013*

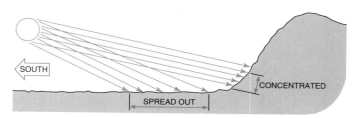

FIGURE 32.43 > South-facing slopes receive concentrated solar energy. *Hepler/Wallach/Hepler © Cengage 2013*

in a car act as a thermal mass. Walls, floors, and masonry can also function as a thermal mass in a building designed for maximum solar effectiveness. In passive solar systems, the thermal mass is both the storage and the distribution system. Storing and dissipating the trapped heat, to either lower or raise the temperature of a building as needed, is one of the most important features of passive solar design. See Figure 32.44. A greenhouse should be attached to the south side of a structure to gain heat in winter and yet repel most of the solar heat during the summer.

Rising Warm Air

Figure 32.45 shows how heat is transferred in a passive solar system. Heat from the sun (radiation) enters the attic or upper level by conduction through the roof. The heat is radiated into the attic and generates convection currents, which spread the heat throughout the confined space. This heat is conducted through the ceiling to the lower levels, where it again radiates into the rooms, causes convection currents, and natural circulation begins.

Heated air will always rise by convection until trapped. Therefore, recirculating heated air from high places to cooler lower areas helps heat living space. Likewise, expelling high-level warm air that would otherwise move downward helps reduce living-level temperatures. Roof-level vents may be used. Also, the level and placement of windows can provide natural convection and ventilation to both circulate and exhaust the warm air. See Figure 32.46.

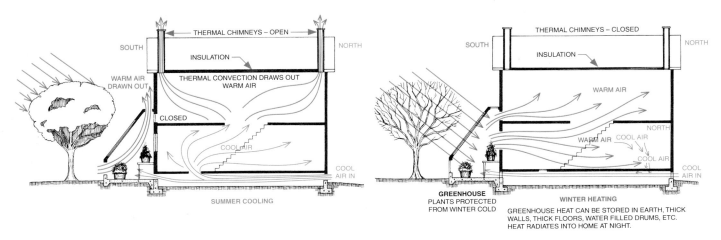

FIGURE 32.44 > Greenhouse effect on temperatures. *Hepler/Wallach/Hepler © Cengage 2013*

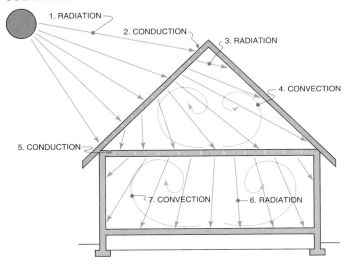

FIGURE 32.45 > Heat transfer in a passive solar system. *Hepler/Wallach/Hepler © Cengage 2013*

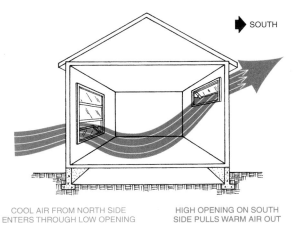

FIGURE 32.46 > Exhausting warm air by convection. *Hepler/ Wallach/Hepler © Cengage 2013*

Ceiling heights affect room temperature. Low ceilings tend to trap warm air in the living space. High ceilings allow it to rise.

Passive Solar Methods

Passive solar methods depend on environmental elements without additional assistance from electromechanical devices. Both the *direct-gain* method and the *indirect-gain* method are designed to take full advantage of the sun to provide heat when and where it is needed and to block the sun's heat when and where it is not wanted.

Direct-Gain Method

In the direct-gain method, the inside of a building is heated by the sun's rays directly as they pass through large glass areas or structural materials. See Figure 32.47. To maximize the amount of winter heat directly entering a building, large south-facing glass areas are used. Once the winter sun's rays enter a building through windows, they are absorbed. The heat is stored in thermal-mass objects to be used later. These objects include floors, walls, and furniture. At night, or when clouds block the sun's rays, the stored heat in the thermal masses is slowly released, keeping the inside temperature higher. Large thermal masses such as masonry floors, walls, and fireplaces store more heat than wood and other fibrous materials do. Therefore, they hold heat longer for later use.

Water has the highest capacity for retaining heat, followed by steel, then aluminum. Masonry and rock are next. Wood and similar porous materials are the poorest retainers of heat.

Some direct-gain walls are designed with reflective insulating units that provide insulation and retain inside heat in cold weather, yet reflect most of the sun's heat in hot seasons. The effective design of south-facing roof overhangs, as illustrated

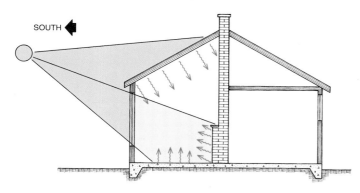

FIGURE 32.47 > Direct-gain heating. *Hepler/Wallach/Hepler © Cengage 2013*

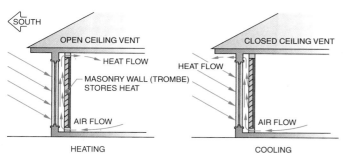

FIGURE 32.48 > Trombe wall indirect-gain method. *Hepler/ Wallach/Hepler © Cengage 2013*

in Chapter 6, provides one of the best methods for passive solar planning.

Indirect-Gain Method

The indirect-gain method uses a thermal mass placed between the sun and the inside of a building. This thermal mass is heated directly by the sun. When heat is needed, the thermal mass is exposed to the inside of the building and heats the inside air. A **trombe wall** is a type of indirect-gain system. See Figure 32.48. In this system the temperature is controlled by directing varying amounts of the rising warm air to the inside or to the exterior of the building, depending on the comfort level needed. Warm air is directed outside during warm weather.

Not all passively planned buildings use all of the passive solar features available. Designers should, however, include as many features as the design situation allows.

🌐 ACTIVE SOLAR SYSTEMS

Planning for **active solar systems** requires knowledge of both mechanical systems and thermal principles. Active solar systems use mechanical devices to drive the components needed for solar heating or cooling. This includes devices and facilities for collection, storage, distribution, and control of heat.

Active solar systems operate more effectively when combined with passive solar features.

Active systems can be designed for comfort control for an entire structure, but this usually requires some support from conventional HVAC systems depending on the climate. The most frequent use of active systems are for heating hot water for consumption or for heating pool and spa water.

Collection

Active solar systems use south-facing solar collectors, which contain circulating water, oil, or air. Heating collectors should be set at an angle perpendicular to the rays of the sun for the maximum number of hours each day. Unless rotating collectors are used, this position is ideal for only a short time each day. Therefore, fixed collectors are usually positioned to face the midday sun (between 10 a.m. and 2 p.m.), since air temperatures are usually higher during these hours. In most North American areas, this means the collectors will face south-southwest.

In addition to the orientation of the collectors, the vertical tilt angle should be the same as the local latitude for maximum year-round effect. A tilt angle of 15° greater than the local latitude is best for winter heating because of the lower path of the sun.

Storage

Because heat is needed when the sun is not producing heat, storage of the absorbed heat (usually in rocks or water) is necessary. Stored heat is limited to the capacity of the storage unit. The larger the storage unit, the longer the solar system will operate without the use of auxiliary heating devices. Maximum insulation is needed in storage containers to minimize the loss of stored heat.

Distribution and Control

Thermostatic controls activate solar distribution systems. Signals cause hot air to be blown or hot water to be pumped from storage containers to parts of a building where heat is needed.

Types of Active Solar Systems

Several types of systems convert sunlight into usable energy. These systems may rely on a solar furnace, photovoltaic (solar cell) technology, or liquid-based or air-based collectors.

Solar Furnace

A solar furnace is a collection of mirrors that focuses the sun's heat on a concentrated area. Temperatures as high as

3500°F (1927°C) can be attained, and the energy can heat or cool large clusters of buildings. Solar furnace collectors must move with the sun and are not currently practical for single-building use.

Photovoltaics

Photovoltaic film (PV), known as a solar cells, converts sunlight directly into direct current (DC) electricity with the use of silicon wafers. The direct current (DC) is converted by inverters to alternating current (AC). In some locations solar cells can power all household electrical appliances including swimming pool operations. See Chapter 31 for coverage of the electrical functioning of solar cells. See Figure 32.49.

Liquid-Based Systems

Liquid-based systems are most popular for light construction. They use liquid to trap and distribute the heat from the sun. In designing or choosing an active solar-heating system, the type of fluid used to transport the heat must be determined. Water, antifreeze solutions, or oils may be used. See Figure 32.50. If the building is in a cold climate and antifreeze is not used, a draindown system must be installed that empties the pipes when the system is not needed.

The liquid is heated in collectors. The collectors may be attached to a roof or installed near the structure. Collectors consist of *absorbers* placed over a layer of insulation to help prevent heat loss. Liquid system absorbers may be plastic or metal sheets (plates) over which a liquid flows. Absorber plates are constructed of steel, copper, aluminum, or plastic because of their heat conductivity. Other absorbers consist of a network of pipes containing a circulating liquid. The absorber pipes (or tubes) are arranged in panels, usually 4′ × 10′ or 4′ × 12′. Even on a very cold day, with bright or filtered sunlight, absorbers can be heated to 200°F (93°C).

Absorbers are designed to retain a maximum amount of heat. The amount of heat retained is called *absorptance,* whereas *emittance* is the amount of heat reflected. An efficient absorber has high absorptance and low emittance.

The heated liquid from the collectors is pumped to insulated storage tanks. In large multiple-building complexes, heated liquid is stored seasonally in underground units. When needed, the heated liquid is pumped to liquid-to-air heat exchangers, water heaters, or swimming pools.

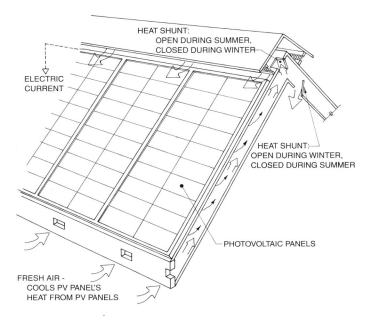

FIGURE 32.49 > Photovoltaic panels. *Hepler/Wallach/Hepler*
© *Cengage 2013*

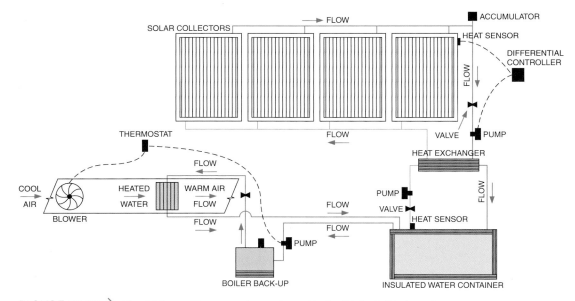

FIGURE 32.50 > Liquid-based house solar system. *Hepler/Wallach/Hepler © Cengage 2013*

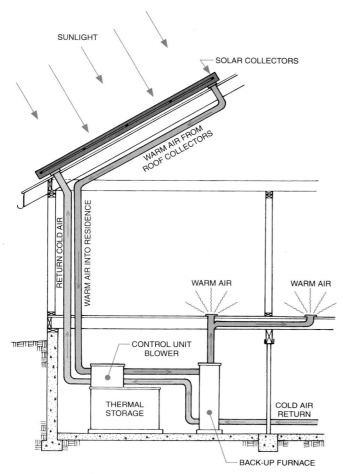

Air-Based Systems

Active solar systems that use air as the heating element are more simple to construct and operate than liquid systems. See Figure 32.51. They are effective for space heating. However, their inability to provide very hot water is a serious drawback. Air system collectors use sheet-metal (copper, aluminum, or steel) plates to heat the air trapped between a cover plate and the absorber. This heated air is then blown to the storage facility where heat is stored in a thermal mass (rocks, water, masonry). From the storage area, the heated air can be blown directly through a duct system to appropriate rooms, as in a conventional warm-air system. Because air systems have low heat capacity, large ducts (up to 6″) must be used.

With air systems, heat can also be stored in rocks or gravel. The heat can be distributed through convection or radiant panels. For each square foot of collector for heat storage, 80 to 400 lbs. (36 to 180 kg) of rock are required. Rock storage requires 2 1/2 times as much volume as water to store the same amount of heat over the same temperature rise.

⊕ Auxiliary Heating Systems

Except in very mild climates, most solar heating storage facilities cannot keep constant pace with peak demand. For this reason, an auxiliary heating system is usually recommended, especially for hot-water production. An active solar heating unit is often integrated with an auxiliary heating unit. When public utilities are not available, a self-sustaining energy system may be used. See Figure 32.52.

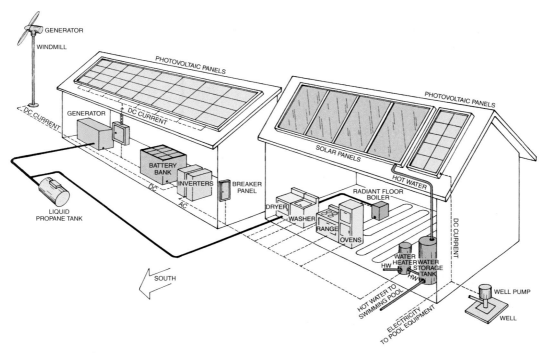

FIGURE 32.52 ❯ Self-sustaining energy system. *Hepler/Wallach/Hepler © Cengage 2013*

🌐 Solar Cooling

Solar cooling is possible through the same absorption-cooling method that is used in gas refrigerators. However, at present, the equipment for an active solar heating and cooling system is very expensive. See Figure 32.53.

Collectors can be used minimally to help cool buildings by exposing them to cooler night air and closing them to daytime exposure. The cooler night air cools the liquid or air that returns to the storage area. The cooled liquid is then released the next day to augment passive or conventional cooling systems.

Backup Power

Because all active solar systems rely on the availability of direct sunlight, backup systems are needed for cloudy periods. This requires alternate connections to the building's electrical power supply but that can be minimized through the use of wind-driven generators. Stored liquid propane can also power backup generators for appliances that operate on natural gas.

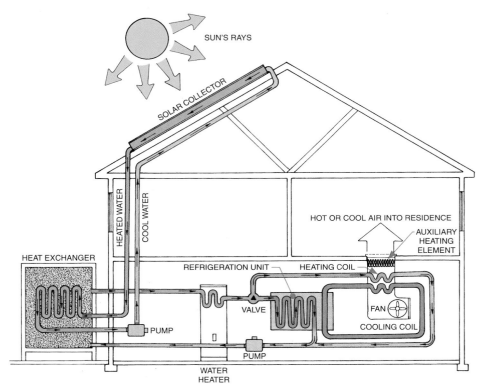

FIGURE 32.53 > Solar heating and cooling system. *Hepler/Wallach/Hepler © Cengage 2013*

Comfort Control Systems (HVAC) Exercises

CHAPTER 32

1. What factors affect heat measurement for a building's HVAC system?
2. Plan and draw an HVAC system for the plan shown in Figure 15.28.
3. Name the four steps in the solar heating process.
4. Sketch a plan or elevation showing use of passive solar principles.
5. Evaluate the active solar systems in this chapter in terms of advantages and disadvantages.
6. Sketch a floor plan showing a mechanical heating system with a heating unit in the most appropriate location. Locate the ducts and outlets for this system.
7. Draw a plan of a climate control system appropriate for the house of your design.
8. Sketch a floor plan and indicate the best location for solar collectors and storage.
9. List the passive and/or active solar features you would include in a residence of your own design.
10. Draw the following HVAC plans on a Figure 32.54 overlay.
    - Gas heating system
    - One pipe hot water system
    - Radiant heating system
    - Active solar system

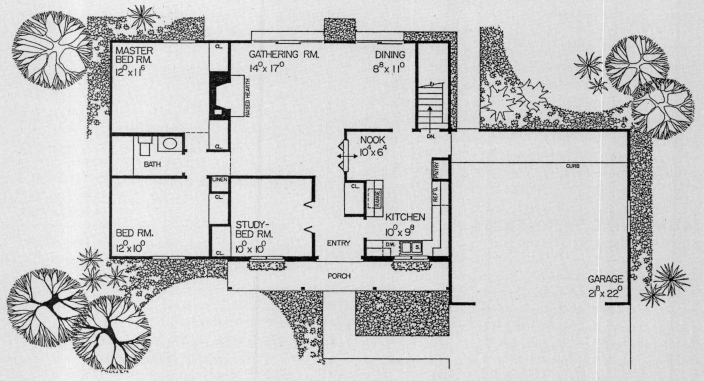

FIGURE 32.54 > Draw different HVAC plans for this design. *Hepler Associates LA, PC*

CHAPTER 33

Plumbing Drawings

OBJECTIVES

In this chapter you will learn to:

> draw plumbing fixtures on a floor plan.

> draw the water supply lines and waste discharge system on a floor plan.

> draw the water supply lines and waste discharge system on an elevation.

TERMS

branches	gray-water waste	shut-off valve
building main	hose bibs	soil lines
cleanouts	percolation	stacks
distribution field	riser diagrams	valve
drainage fields	schematic	vent stack
fixture trap	septic tank	waste discharge

INTRODUCTION

The history of plumbing extends back to 500 B.C.E. when the ancient Greeks and Egyptians built tile-lined bathtubs. However, the Romans are credited with creating the first plumbing systems through their development of aqueducts. The term *plumbing* is derived from the Latin *plumbum*. Today plumbing refers to the supply, distribution, control, and discharge of all liquid and gases. Plumbing drawings are needed to design and locate the type and position of all plumbing fixtures, devices, and pipes within a plumbing system.

PLUMBING CONVENTIONS AND SYMBOLS

Like most architectural drawings, plumbing drawings must be prepared to a very small scale. Therefore, schematic symbols are used as a substitute for drawing plumbing lines, joints, and intersections. See Figure 33.1. These schematic symbols show the type and location of fixtures, joints, valves, and other plumbing devices that control the flow of liquids. Figure 33.2 shows the plan and elevation symbols for pipe intersections,

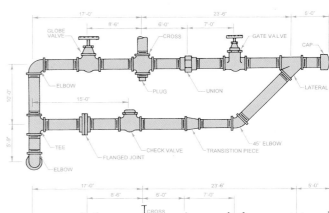
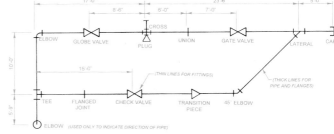

FIGURE 33.1 > Orthographic drawing compared to a schematic plumbing drawing. *Hepler/Wallach/Hepler © Cengage 2013*

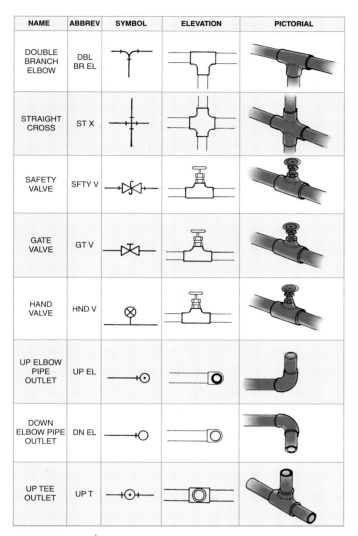

FIGURE 33.2 > Pipe intersection symbols. *Hepler/Wallach/Hepler © Cengage 2013*

Figure 33.3 shows pipe joint symbols, Figure 33.4 shows sanitary facility symbols, and Figure 33.5 shows the symbols for the types of piping lines used on plumbing drawings. Symbols used to represent bath and kitchen plumbing fixtures are part of a complete floor plan and are shown in Chapter 15.

Plumbing drawings include the following symbol and line conventions:

- Cold-water supply lines
- Lines to each outlet
- Drains
- Plumbing fixtures
- Gas supply line
- Soil lines
- Cleanouts
- Vents
- Hot-water lines

- Water heater
- Furnace
- Hose bibs
- Pipe sizes, fittings, and valves
- Stack lines—on elevation
- Vent lines—on elevation
- Shut-off valves
- Oil supply lines
- Purification devices.

CAD Plumbing Library

Plumbing drawings contain series of lines representing pipes. Special line types for gas, water, oil, etc., can be selected using the *Line* task. The *Break* command can also be used to interrupt a solid line at regular intervals by inputting the gap

NAME	ABBREV	SYMBOL	ELEVATION	PICTORIAL
FLANGED FITTING	FL FT			
SCREWED FITTING	SC FT			
BELL & SPIGOT FITTING	BL/SP FT			
WELDED FITTING	WLD FT			
SOLDERED FITTING	SLD FT			
EXPANSION JOINT	EXP JT			

FIGURE 33.3 > Pipe joint symbols. *Hepler/Wallach/Hepler © Cengage 2013*

NAME	ABBREV	SYMBOL	ELEVATION	PICTORIAL
METER	M			
FLOOR DRAIN	FD			
CESS POOL	CP			
DRY WELL	DW			
SEPTIC TANK	SEP TNK			
SEPTIC-TANK DISTRIBUTION BOX	SEP TANK DIS BX			
SUMP PIT	SP			

FIGURE 33.4 > Sanitary facility symbols. *Hepler/Wallach/ Hepler © Cengage 2013*

NAME	ABBREV	SYMBOL	NAME	ABBRV	SYMBOL
COLD-WATER LINE	CW		AIR-PRESSURE RETURN LINE	APR	
HOT-WATER LINE	HW		ICE-WATER LINE	IW	
GAS LINE	G		DRAIN LINE	D	
VENT	V		FUEL-OIL RETURN LINE	FOF	
SOIL STACK PLAN VIEW	SS		FUEL-OIL FLOW LINE	FOR	
SOIL LINE ABOVE GRADE	SL		REFRIGERANT LINE	R	
SOIL LINE BELOW GRADE	SL		STEAM LINE MEDIUM PRESSURE	ST	
CAST-IRON SEWER	S-CI		STEAM RETURN LINE—MEDIUM PRESSURE	ST	
CLAY-TILE SEWER	S-CT		PNEUMATIC TABE	PT	
LEACH LINE	LEA		INDUSTRIAL SEWAGE	IS	
SPRINKLER LINE	SPR		CHEMICAL WASTE LINE	CW	
VACUUM LINE	VAC		FIRE LINE	F	
COMPRESSED AIR LINE	COMA		ACID WASTE LINE	AC WST	
AIR-PRESSURE LINE FLOW	APF		HUMIDIFICATION LINE	HUM	

FIGURE 33.5 > Piping line symbols. *Hepler/Wallach/Hepler © Cengage 2013*

size, spacing, and number. Once lines are drawn, symbols for valves, unions, elbows, tees, and joints are added. This is done by first breaking the line using the *Trim* task. Then symbols from the plumbing symbol library are inserted in the gap with the cursor.

PLUMBING SYSTEMS

Plumbing drawings are prepared to describe types of plumbing systems, such as water supply, waste discharge, and hydronic heating systems. Plumbing drawings are also used to show gas appliance lines, built-in vacuum systems, and pest-control piping.

Water Supply System

A water supply system consists of a network of pipes, under pressure or gravity feed, that carry fresh water to appliances, sinks, water closets (toilets), tubs, filters, showers, and water heaters. See Figure 33.6.

Water systems are designed for consumption such as drinking and washing or for circulation such as heating and humidity control. The storage of water for fire recharge control and swimming pools must also be considered. Figure 33.7 shows a pictorial, plan, and elevation drawing

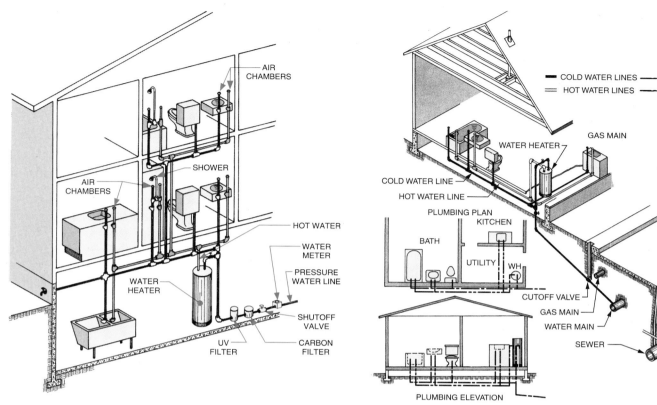

FIGURE 33.6 > Water supply system. *Hepler/Wallach/Hepler © Cengage 2013*

FIGURE 33.7 > Pressurized water supply system. *Hepler/Wallach/Hepler © Cengage 2013*

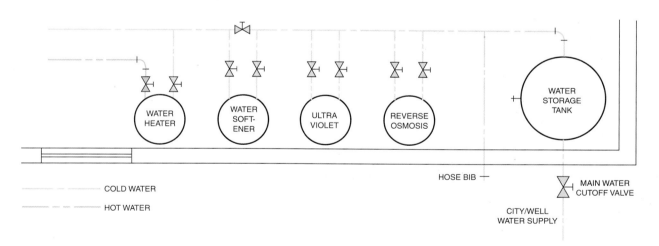

FIGURE 33.8 > Filter system branching. *Hepler/Wallach/Hepler © Cengage 2013*

of a pressurized water supply system. This illustration shows the circulation of cold water, then hot water, through and to all plumbing fixtures.

Water is brought from a well or public water supply main through a **building main.** If public water is used, this main contains a utility company **valve,** a meter, and a building main valve. If water filters are used, the water may be directed through carbon filters, a reverse osmosis system, ultraviolet (UV) filters, and/or softeners. Figure 33.8 shows a typical pipe branching arrangement for a water filtering system.

Figure 33.9 shows a water filtration, softening, and purifying system combined with the electrical delivery system for a residence. A separate line may be used to connect a special drinking water supply through a purification system. **Hose bibs** or sprinkler system lines not requiring purification may be connected before filtering.

After filtering, water is branched into a cold-water main and a hot-water main, which passes through a water heater or heaters. Because this water is under pressure, pipes can be located in any direction. Lines that pass vertically through

FIGURE 33.9 > Filtering, softening, and purifying equipment. *Hepler/Wallach/Hepler © Cengage 2013*

FIGURE 33.10 > Riser pipes extending above a slab. *Hepler/Wallach/Hepler © Cengage 2013*

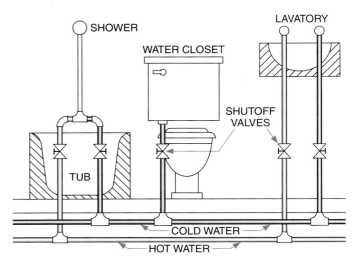

FIGURE 33.11 > Shut-off valve locations. *Hepler/Wallach/Hepler © Cengage 2013*

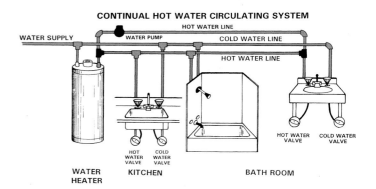

FIGURE 33.12A > A continuous hot-water circulating system. *Hepler/Wallach/Hepler © Cengage 2013*

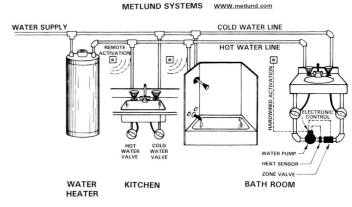

FIGURE 33.12B > Metlund hot-water system. *Hepler/Wallach/Hepler © Cengage 2013*

slabs must be precisely located to align with partition and fixture locations. See Figure 33.10.

Hot-water branch lines are normally located 6″ from cold-water branch lines and parallel with them. Where a fixture includes both hot- and cold-water outlets, the hot-water line and valve are located to the left of the cold-water line and valve. Every fixture must include a **shut-off valve.** See Figure 33.11.

The continuous hot-water circulating system shown in Figure 33.12A delivers hot water immediately because hot water remains in the pipes continuously. In the Metlund system shown in Figure 33.12B, cold water is stored in the hot-water system until the hot-water valve is turned on and the water pump is activated. This pumps the stored cold water back into the water heater where the thermostat controls the water temperature. Tankless water heaters and domestic oil burners provide immediate hot water only when a hot-water valve is opened. Consequently, these types of heaters require less energy than other types.

Pipes used for light construction plumbing are available in copper, plastic, brass, iron, and steel. In addition to pipe

FIXTURE	COLD WATER	HOT WATER	SOIL, WASTE	VENT
Sinks (Lav)	1/2″	1/2″	1 1/2″–2″	1 1/4″–1 1/2″
Lavatory	1/2″–3/8″	1/2″–3/8″	1 1/4″–2″	1 1/4″–1 1/2″
Water closet	1/2″–3/8″	—	3″–4″	2″
Tub (Bath)	1/2″	1/2″	1 1/2″–2″	1 1/4″–1 1/2″
Shower	1/2″	1/2″	2″	1 1/4″
Water heater	3/4″	3/4″	—	4″
Washer	1/2″	1/2″	2″	1 1/2″
Lau Sink	1/2″	1/2″	1 1/2″	1 1/4″

FIGURE 33.13 > Minimum pipe sizes for common fixtures. *Hepler/Wallach/Hepler © Cengage 2013*

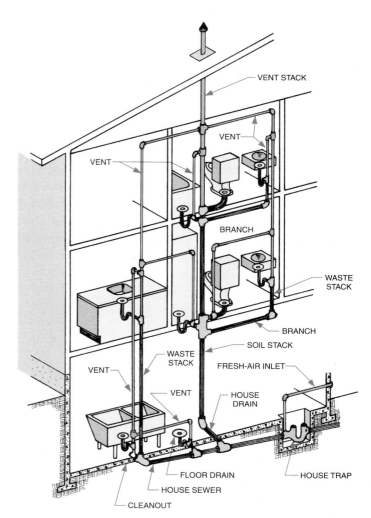

FIGURE 33.14 > Gravity waste system. *Hepler/Wallach/Hepler © Cengage 2013*

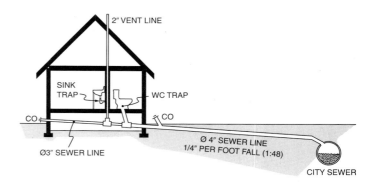

FIGURE 33.15 > Waste discharge system. *Hepler/Wallach/ Hepler © Cengage 2013*

material, pipe size is important. The size of all residential water supply lines ranges from 1/2″ to 3/4″ for water lines, from 1 1/4″ to 4″ for soil lines and vent lines; as shown in Figure 33.13. All fixtures must have a free-flowing supply of water at all times. Lines that are too small cause a whistling sound as water flows through at high speeds. *Air-cushion chambers* stop hammering noises.

The location of pipes must be carefully planned. Excessive changes of pipe direction cause friction and reduce water pressure. Designers must also consider energy conservation. Placing insulation around hot-water lines conserves hot water and reduces the total cost of fuel for heating the water. A note on the plumbing diagram should specify the type and thickness of the desired insulation material.

🌐 Waste Discharge System

Although water is supplied to a building under pressure, wastewater is discharged through a disposal gravity system. Figure 33.14 summarizes the functioning of a gravity waste system which shows a vertical view of the stacks, branches, vents, and drains that comprise a waste discharge system. **Waste discharge** lines are not under pressure. They are empty except when waste is flushed through them. Therefore, all pipes in a discharge system must slope in a downward direction, usually 1/4″ per foot as shown in Figure 33.15. Because of this gravity flow, waste (soil) lines are larger than water supply lines that are under pressure.

There are two types of **soil lines:** branches and stacks. **Branches** are nearly horizontal lines that carry waste from each fixture to the **stacks.** Stacks are vertical lines. Soil stacks (3″ to 4″ in diameter) carry waste to the house sewer. The portion of the soil stack above the highest branch intersection

is known as a **vent stack.** Some vent stacks are separate and parallel to soil stacks. Vent stacks are dry pipes that extend through the roof (a minimum of 6″) and provide ventilation for the discharge system. Vent stacks permit sewer gases to escape to the outside and equalize the air pressure in the system. Stacks that intersect house sewer lines under a slab must be dimensioned accurately to ensure alignment of the stack pipe with the partition location.

The flow of all wastewater begins at a fixture. Each fixture contains a **fixture trap** to prevent the backflow of sewer gas from the branch lines. Fixture traps are exposed for easy maintenance, except for water-closet traps, which are built into the fixture. A total system house trap is provided in the house sewer line outside the perimeter of the building. **Cleanouts** must be provided for the main house sewer drain. Some municipalities allow sewer waste to be separated from **gray-water waste** as shown in Figure 33.16. Filtered gray water (nonsewage wastewater) is usually acceptable for landscape irrigation.

When municipal sewage-disposal facilities are available, connection of the house sewer to the public system is shown on the plot plan. When public systems are not available, a septic system is used. Either a separate drawing is provided or the location of the system is included on the plot plan. This location drawing is usually required by local building codes and the local board of health before a building permit is issued.

Septic Tanks

In a septic system, building wastes flow from the house sewer into a **septic tank** that is buried a prescribed distance from the building. A septic tank is a tank in which solid waste is processed by bacteria. Lighter liquids flow out of the septic tank into drainage fields through porous pipes. These pipes spread over an area to allow wide distribution of liquids. The solid wastes, which settle to the bottom of the septic tank, are converted to liquids by bacterial action.

The size and type of septic tank and drainage field are specified by code. Many factors such as the number of occupants, baths, bedrooms, and kitchen disposals, as well as soil type and topography, are considered to determine the size and type specified. Septic tank size in gallons must be at least 50% larger than the daily sewage output. Because detergents interfere with bacterial action, some codes allow smaller septic tanks if laundry waste is diverted to a separate discharge system. Septic tanks must be completely watertight to prevent contamination of surrounding soils. See Figure 33.17. Flood

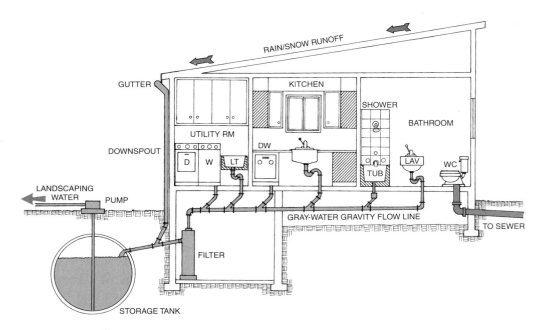

FIGURE 33.16 〉 Gray-water collection system. *Hepler/Wallach/Hepler © Cengage 2013*

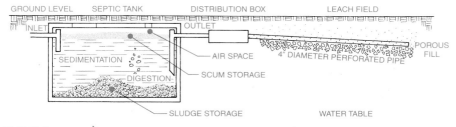

FIGURE 33.17 〉 Septic system components. *Hepler/Wallach/Hepler © Cengage 2013*

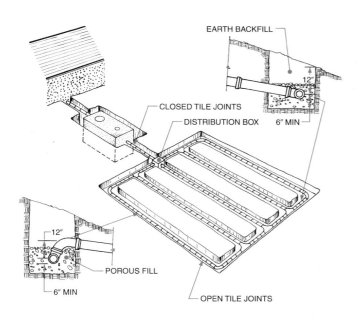

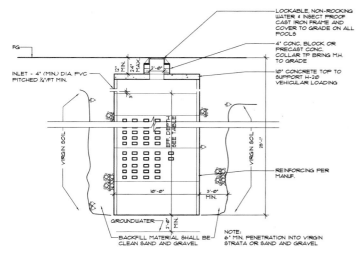

FIGURE 33.18 > Trench-type distribution field. *Hepler Associates LA, PC*

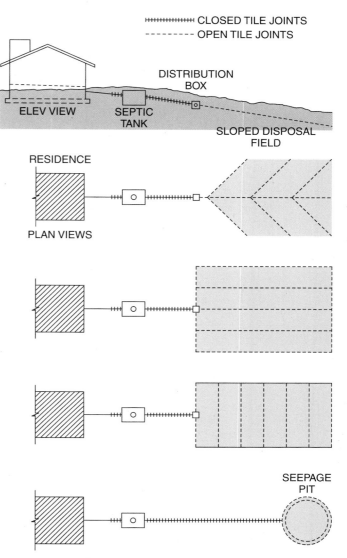

FIGURE 33.19 > Common distribution field shapes. *Hepler/ Wallach/Hepler © Cengage 2013*

drains or rainwater runoff lines should not be connected to the septic system.

🌐 Drainage Fields

Liquid waste (effluent) flows from the top of the septic tank to a drainage field. **Drainage fields,** also called leaching or absorption fields, consist of plastic (PVC) perforated pipe or agricultural open-joint pipe laid in coarse gravel. These seepage pipes are either laid in trenches or in continuous beds 12″ to 16″ below grade level. See Figure 33.18. Fields are arranged in a variety of shapes and patterns depending on the site contour and restrictions, such as building or tree locations. See Figure 33.19.

Local codes specify the minimum distance allowed between the end of a drainage field and bodies of water, wells, roadways, right-of-ways, buildings, and property lines. See Figure 33.20. Fields cannot be located under paved areas or uphill from a well or water supply. Codes also require that fields be placed no closer than 5′ from a water table level. Regardless of location, soil under the field must be porous enough to absorb the effluent. The process of liquid absorption is known as **percolation.** A percolation test consists of digging a hole in the drainage field area and filling the hole with water. Then the rate at which the water is absorbed into the soil is measured. A percolation rate of 40 to 60 minutes per inch is considered acceptable, depending on local code requirements. Figure 33.21 shows the required rates for determining **distribution field** size for a two-bedroom residence with good percolation. Figure 33.22 shows a section through a typical cylindrical septic tank.

738 Electrical and Mechanical Design Drawings

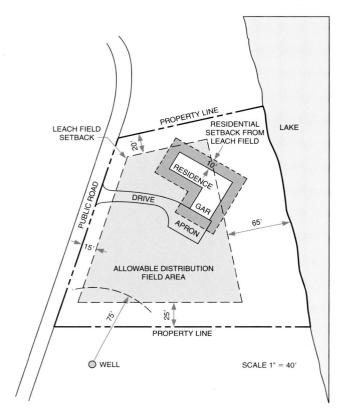

FIGURE 33.20 > Typical minimum code requirements for distribution field location. *Hepler/Wallach/Hepler © Cengage 2013*

PERCOLATION RATE (MIN. PER INCH)	RATING	REQUIRED SQ. FT. (PER BEDROOM)
Over 45 minutes	Will not pass code	Unacceptable
31 to 45 minutes	Poor	600 square feet
16 to 30 minutes	Acceptable	400 square feet
15 or less minutes	Very good	300 square feet

FIGURE 33.21 > Percolation test standards. *Hepler/Wallach/Hepler © Cengage 2013*

PLUMBING DRAWINGS

Drawings used to describe plumbing systems include plumbing plans, elevations, and in some cases pictorial drawings. Figure 33.23 shows an example of these three types of plumbing drawings.

Plumbing Floor Plans

Plumbing lines and symbols are usually added to a reproducible print of a complete floor plan. If this is not done, a separate abbreviated floor plan is prepared. See Figure 33.24.

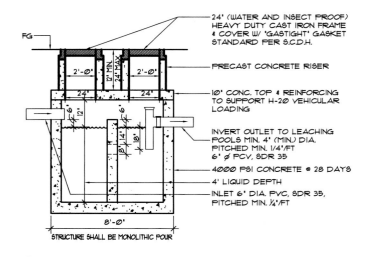

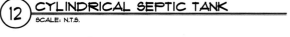

FIGURE 33.22 > Cylindrical Septic Tank Details. *Hepler Associates LA, PC*

Many floor plan details and dimensions can be eliminated, except for partition positions that require the placement of stacks before a slab or foundation is poured. Some plumbing plans are combined with HVAC or electrical floor plans. Some plumbing lines and facilities are located on a plot plan or site plan. These include septic system, including lines, septic tank, and field; irrigation system lines and devices; swimming pool or spa lines, heaters, filters, and pump; and public water and discharge lines and connections.

Water Supply Lines

Plumbing lines are drawn to represent connections in the water supply system. Begin with the house main and draw cold-water branch lines to each fixture. Next, draw the hot-water lines from the house main to water heaters and then to each fixture requiring hot water. Draw the hot-water lines parallel to the cold-water lines where possible. Add symbols for shut-off valves on each fixture line.

Wastewater Lines

Draw heavy wastewater branch lines connecting fixtures with soil stacks. Add a trap symbol at each fixture. The positions of stack and vertical pipes are shown with a circle inside walls and partitions. The main house drain and sewer lines are next drawn connected to the stack symbols and cleanouts, and the house trap is added. The house drain is then connected to either a municipal sewer line or to a septic tank. The outline of septic tanks and fields is normally drawn on the plot plan or site plan. Because soil lines are larger than supply lines, they should be drawn heavier to show the difference.

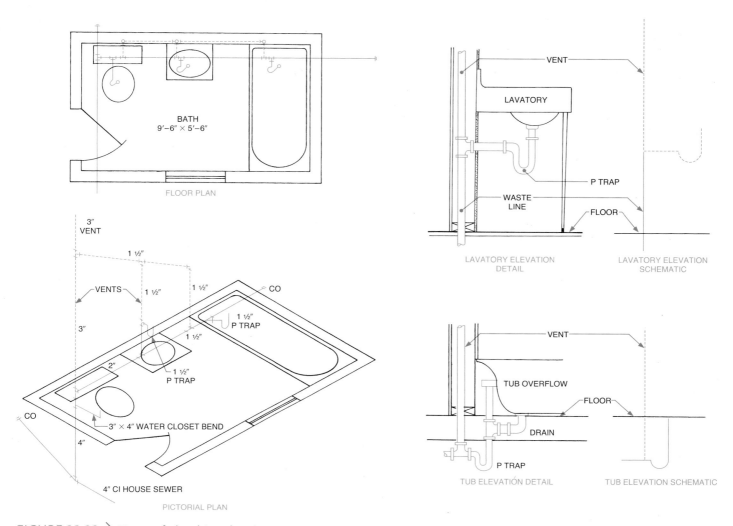

FIGURE 33.23 > Types of plumbing drawings. *Hepler/Wallach/Hepler © Cengage 2013*

The waste line diameter from the farthest fixture in a building to a public sewer or septic tank is 4″. Waste lines behind the farthest fixture has a diameter solid of 3″ because it does not carry waste products. Horizontal waste lines slope at 1/4″ per foot to ensure an adequate flow. A cleanout pipe opening must be located at the end of each waste line and at each pipe turn of more than 45°.

Gas Lines

If gas appliances or hydronic heating systems are used, lines with cut-off valves are drawn from the main gas line or gas tanks to each device.

Plan Details

Because most plumbing fixtures are concentrated in the bathroom, laundry, and kitchen, detailed **schematic** plans are sometimes prepared for those rooms. See Figure 33.25. This enables areas where piping is dense to be drawn at a larger

scale for easier reading. Where special framing adjustments must be made for piping, framing details should be prepared. See Figure 33.26.

The maximum outside pipe diameter that will fit inside a standard 4″ stud wall or an 8″ concrete block is 2″. For a 2″ × 6″ stud wall, the maximum diameter is 3″; for a 12″ concrete block it is 6″. Also note that pipe notches cut into any structural member must be replaced with blocking to avoid structurally weakening the member, as shown in Figure 33.27.

Notes are added to plumbing plans to label the type and size of pipes. Dimensions are added to critical positions, such as locations of underground valves and cleanouts.

Plumbing Elevations

Plumbing plans show only the horizontal positioning of pipes and fixtures. Therefore, the amount of rise above floor level and the flow of fresh water and wastes between levels are difficult to read. Elevation drawings show vertical distances and angles that cannot be shown in floor plans.

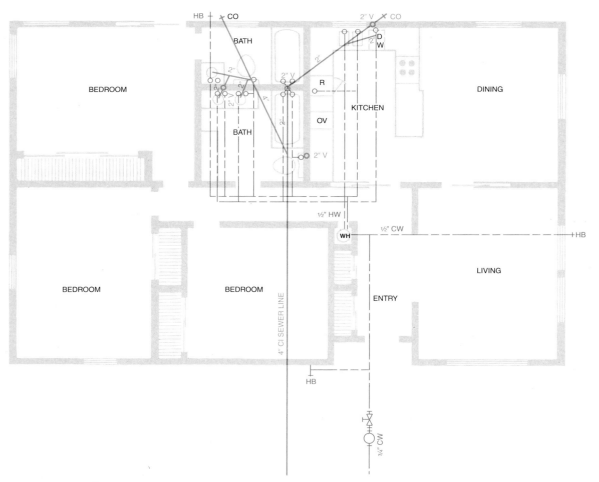

FIGURE 33.24 > Plumbing system floor plan. *Hepler/Wallach/Hepler © Cengage 2013*

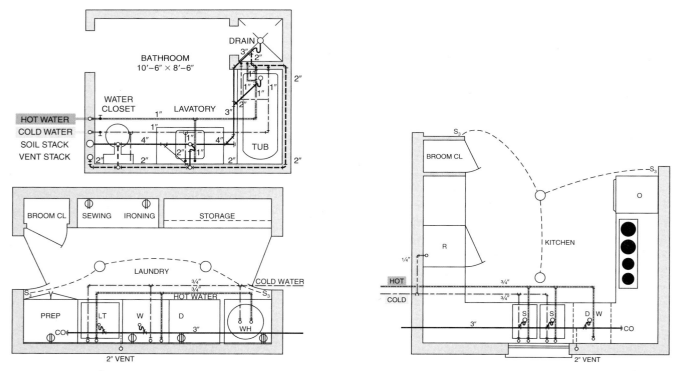

FIGURE 33.25 > Room plumbing plan details. *Hepler/Wallach/Hepler © Cengage 2013*

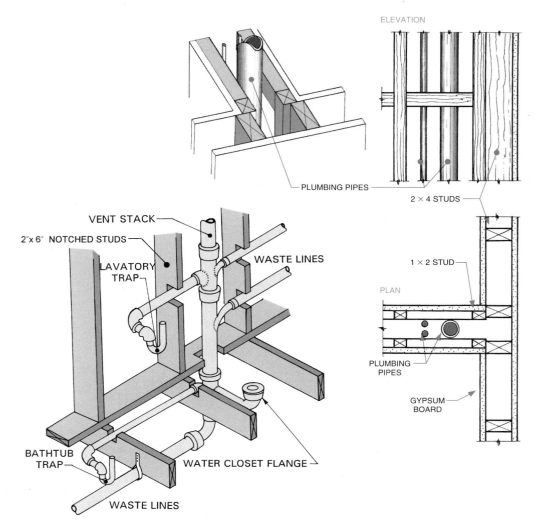

FIGURE 33.26 > Framing allowances for piping. *Hepler/Wallach/Hepler © Cengage 2013*

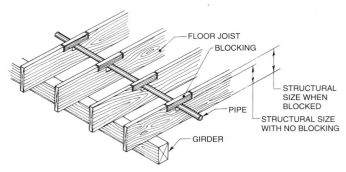

FIGURE 33.27 > Blocking of pipe notching. *Hepler/Wallach/ Hepler © Cengage 2013*

Plumbing elevations are sometimes called **riser diagrams** because they show the amount of vertical rise of each pipe. Figure 33.28 includes a plumbing elevation drawing with the matching plumbing plan. Note that the supply system in this drawing is shown with red lines and the waste discharge system in blue.

Because much of a waste discharge system contains vertical stacks, elevations are very effective for describing these systems. Figure 33.29 shows a riser diagram on a full building elevation. In this illustration of a waste discharge system, the waste lines are shown in red and the vent pipes in blue. A full building riser diagram of a water supply system is shown in Figure 33.30. In this drawing hot-water lines are shown in red and cold-water lines in blue.

Because plumbing fixtures are located in different parts and on different planes of a building, detail riser diagrams are often required. A detailed drawing of this type is shown in Figure 33.31. Fixture valves are usually located on vertical pipes, so valve positions are also shown on elevations. Multiple plumbing vents often create an unattractive appendage to a roof line. To overcome this feature, vents should be combined wherever possible. See Figure 33.32 on page 744.

As a summary, study the drawing in Figure 33.33 on page 744, to observe the functioning of all plumbing systems. This drawing illustrates pictorially the different residential

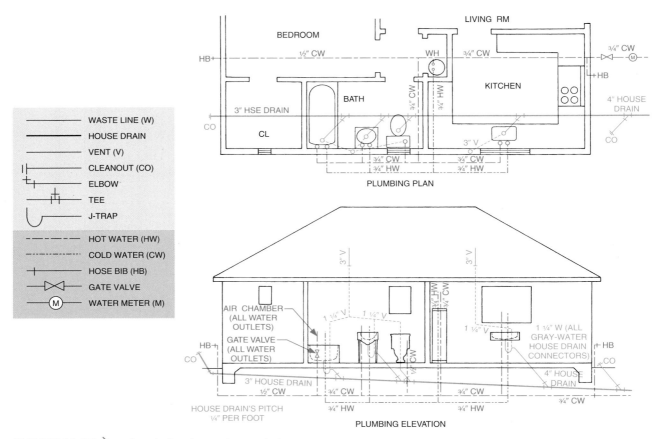

FIGURE 33.28 > Related plumbing plan and elevation. *Hepler/Wallach/Hepler © Cengage 2013*

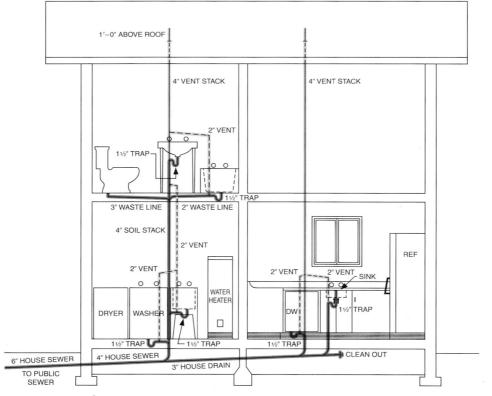

FIGURE 33.29 > Waste discharge system elevation. *Hepler/Wallach/Hepler © Cengage 2013*

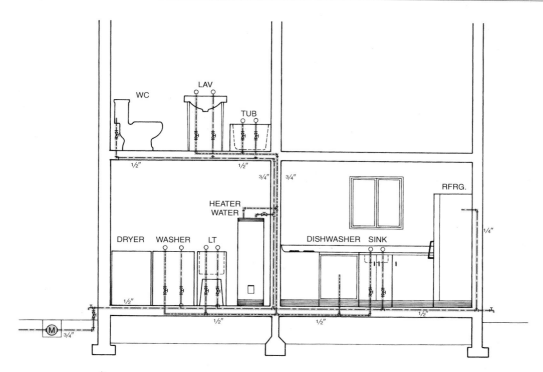

FIGURE 33.30 > Water supply system elevation. *Hepler/Wallach/Hepler © Cengage 2013*

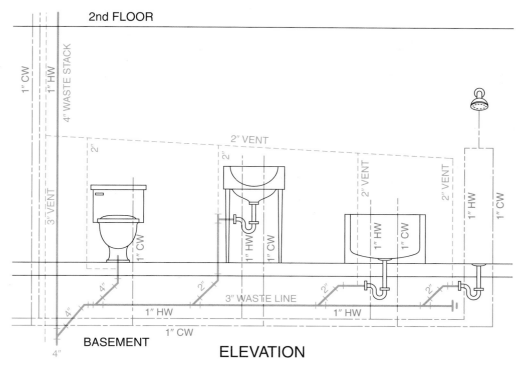

FIGURE 33.31 > Bath piping elevation. *Hepler/Wallach/Hepler © Cengage 2013*

plumbing systems described in plan and elevation drawings for construction purposes.

Pool and Spa Plumbing

Pool design is covered in Chapter 8; however, a pool is just a box of water without a plumbing system. This includes piping and devices for filtering, purifying, circulating, and heating, as shown in Figure 33.34. This illustration shows the route through which water is pumped to pool outlets and recycled through an intake skimmer. Filters may be sand,

cartridge, or diatomaceous earth types. Water may be fully or partially purified by chlorine feeders, electrolytic chlorine generators, ionization systems, or ozone generators.

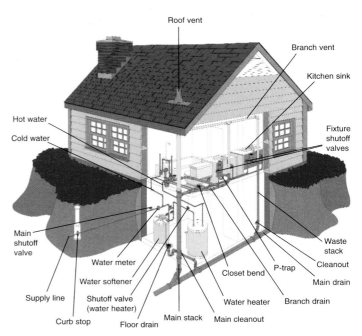

FIGURE 33.33 > Total residential plumbing system. *Roto-Rooter Services Company*

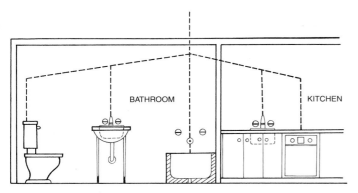

FIGURE 33.32 > Multiple plumbing vents. *Hepler/Wallach/ Hepler © Cengage 2013*

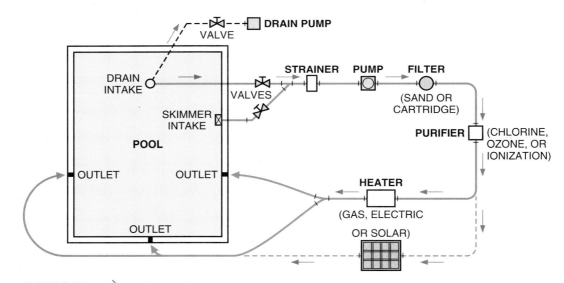

FIGURE 33.34 > Pool plumbing system components. *Hepler/Wallach/Hepler © Cengage 2013*

Plumbing Drawings Exercises

1. Briefly describe a water supply system.

2. Name the types of lines in a waste discharge system.

3. Make a sketch of a plumbing plan in this chapter. Draw the position of all plumbing fixtures. Sketch water supply lines on an overlay. Sketch a waste discharge system on another overlay.

4. Add plumbing fixture symbols to your library.

5. Prepare a plumbing plan for a house of your own design.

6. Draw an elevation of a waste discharge system.

7. Add plumbing lines to the plan in Figure 33.35, using an overlay.

8. Using an overlay, add plumbing lines to the plan in Figure 33.36.

9. Add plumbing lines in color to an overlay of the plan shown in Figure 33.37.

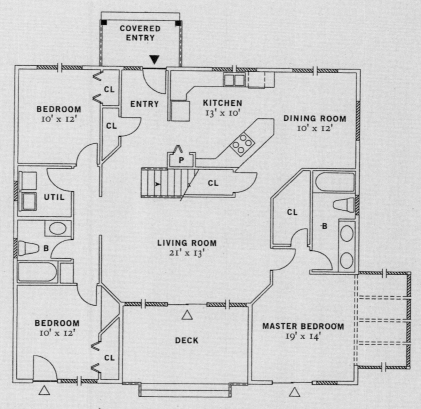

FIGURE 33.35 > Add colored plumbing lines to an overlay of this plan.
Hepler Associates, LA PC

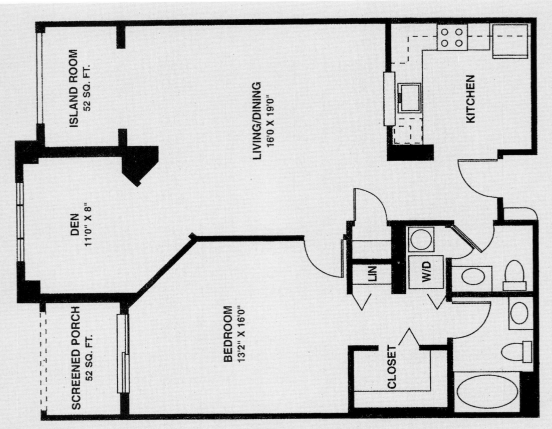

FIGURE 33.36 > Add plumbing lines to an overlay of this plan. *Hepler Associates, LA PC*

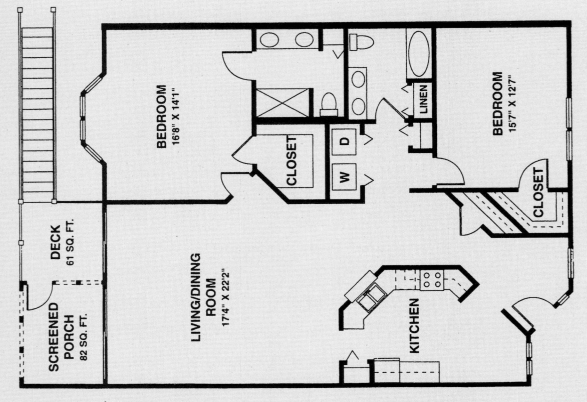

FIGURE 33.37 > Add plumbing lines to an overlay of this plan. *Hepler Associates, LA PC*

PART 9

DRAWING MANAGEMENT AND SUPPORT SERVICES

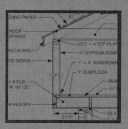

OBJECTIVES

In this chapter you will learn:

> to organize and check a complete set of architectural drawings.

> how drawings in a set are related.

> to identify identical locations on all drawings in a set.

> to select the drawings needed to complete a set of architectural drawings.

> methods of drawing and recording changes on drawings according to change orders.

> understand the functions of drawing management.

TERMS

building information modeling (BIM)
change orders
checker

combination plans
cross-referencing symbols
discipline

drawing code
drawing sequence
layering

INTRODUCTION

Architectural drawings must be complete and accurate. Each drawing must also be easily retrievable and consistent with all other drawings in a set. To accomplish this, drawings are organized and arranged according to established standards. The standards for coordinating, cross-referencing, coding, checking, and making changes are presented in this chapter.

DRAWING SEQUENCE

The **drawing sequence,** that is, the order in which drawings are prepared in a set, is important because the preparation of some drawings depends on the completion of others. The site plan should be developed first because the location and orientation of buildings depends on the design of the site. Next, the first-level floor plan is prepared because other level plans, specialized plans, and elevations are projected from or overlaid on this plan. Once these are complete, framing plans and construction detail drawings can be drawn. Bearing partitions, plumbing walls, stairwells, chimney openings, and other components should align vertically with the first-floor plan. Therefore, this base plan must be carefully checked for accuracy before other drawings are completed.

MANAGEMENT METHODS

Drawing management involves the control of drawing sequences and the coordination of drawings through the use of codes, cross-referencing symbols, and cross checking of drawings in a set. Managing changes and corrections and relating drawings to other construction documents such as schedules, specifications, cost estimates, codes, and contracts is also a vital function. These functions can be performed through the use of checkers, take-off specialists, and accountants. Most of the tasks that link drawings to supporting construction documents can be performed electronically by coding the materials and components located on a drawing.

Building information modeling (BIM) is a process of creating and managing construction data for a specific project. The American Institute of Architects (AIA) defines BIM as a "model-based technology linked with a database of project information." BIM designs are created in 3D using building modeling software. This enables everybody involved in a building project to view a three-dimensional building and simultaneously observe volume and relationships among potentially conflicting elements such as plumbing, electrical, and HVAC systems. In addition BIM drawings can be directly linked to the automatic electronic generation of construction documents such as specifications, purchasing orders, cost estimates, and scheduling details. Construction management data, financial analyses, and environmental code monitoring are also made possible by the use of BIM.

DRAWING COORDINATION

Large construction projects may require thousands of drawings and documents, and even a set of residential plans may include dozens of drawings. A system of codes and symbols is necessary to organize and relate each drawing to other drawings and documents in a set.

Drawing Codes

The AIA recommends the use of an alphanumerical **drawing code** system to identify and classify drawings in a set. An alphabetical prefix is used to denote a specific discipline as introduced in Chapter 4 and as shown in Figure 34.1. Each **discipline** is divided into groups. For example, the architectural discipline (A) is further divided into 10 specific groups, A0 through A9, as shown in Figure 34.2. Drawings are also identified by the work phase of each project in order to separate preliminary drawings from final working drawings. These include:

Cross-Referencing Symbols

To further relate drawings to other drawings or details in a set, **cross-referencing symbols** are used. Although different versions of these symbols are used, the symbols shown in Figure 34.3 are the most common. Review Chapter 4, *Conventions and Procedures,* as a basis for studying the next segment of this chapter. Figure 34.4 shows a specific application of the cross-reference symbols shown in Figure 34.3.

CAD Layering

"Two objects cannot occupy the same space at the same time." The practice of **layering** drawings is designed to ensure that two objects will not be in the same place in a set of plans. For example, a plumbing plan may include a furnace located in the exact position as a distribution panel on an electrical plan. Checking for this type of conflict can be done by manually looking through drawing tracings but this can only be done one drawing at a time. In CAD layering different colors are used for each layer. The colored layers can then be viewed or printed to show the separate layers on one sheet or in combination with other layers.

Layering of Plan Sets

Each drawing in a set should be prepared as a separate layer. Each plan is also the sum total of many sublayers. Therefore, plans need to be movable with or without all sublayers included. This layering process applies to floor plan levels and also to the specialized floor plans such as the electrical, framing, plumbing, HVAC, and interior design plans to ensure that their features are compatible. Elevation drawings in a set should also be layered to allow for cross-checking with floor plans, framing plans, and other elevation drawings. Remember, when changes are made on one drawing, many other drawings in the set are affected and must be changed.

CHECKING DRAWINGS

Each completed drawing should be checked carefully by the designer and drafter. In large firms, this function is performed by a **checker.** In smaller firms, designers and drafters exchange drawings for checking. At the completion of each set or subset of drawings, another complete check should be made.

Checking Methods

One of the most effective tools to use when checking architectural drawings is a colored pencil. See Figure 34.5. A checker draws a line through each dimension, label, and symbol as it is checked. This eliminates rechecking the same dimensions and missing many others. Many architectural offices use a color-coding system to indicate the checker's reaction to the drawing. In one system, a yellow pencil is used for checking items that are correct, a red pencil for checking errors, and a blue pencil for marking recommended changes. There are many variations of this practice. It is critical that

A	Architectural
C	Civil
D	Interior design (color schemes, furniture, furnishings)
E	Electrical
F	Fire protection (sprinkler, standpipes, CO_2, and so forth)
G	Graphics
K	Dietary (food service)
L	Site
L	Landscape
M	Mechanical (heating, ventilating, air-conditioning)
P	Plumbing
S	Structural
T	Transportation/conveying systems

FIGURE 34.1 > Letters used to identify drawing disciplines.
Hepler/Wallach/Hepler © Cengage 2013

SYSTEM CODE:

A2.1
—DRAWING NUMBER
—GROUP NUMBER
—DISCIPLINE PREFIX

Architectural Drawings
A0.1,2,3—General (Index, Symbols, Abbrev. notes, references)
A1.1,2,3—Demolition, Site Plan, Temporary Work
A2.1,2,3—Plans, Room Material Schedule, Door Schedule, Key Drawings
A3.1,2,3—Sections, Exterior Elevations
A4.1,2,3—Detailed Floor Plans
A5.1,2,3—Interior Elevations
A6.1,2,3—Reflected Ceiling Plans
A7.1,2,3—Vertical Circulation, Stairs (Elevators, Escalators)
A8.1,2,3—Exterior Details
A9.1,2,3—Interior Details

Structural Drawings
S0.1,2,3—General Notes
S1.1,2,3—Site Work
S2.1,2,3—Framing Plans
S3.1,2 —Elevations
S4.1,2 —Schedules
S5.1,2 —Concrete
S6.1,2 —Masonry
S7.1,2 —Structural Steel
S8.1,2 —Timber
S9.1,2 —Special Design

Mechanical Drawings
M0.1,2 —General Notes
M1.1,2 —Site/Roof Plans
M2.1,2 —Floor Plans
M3.1,2 —Riser Diagrams
M4.1,2 —Piping Flow Diagram
M5.1,2 —Control Diagrams
M6.1,2 —Details

Plumbing Drawings
P0.1,2 —General Notes
P1.1,2 —Site Plan
P2.1,2 —Floor Plans
P3.1,2 —Riser Diagram
P4.1,2 —Piping Flow Diagram
P5.1,2 —Details

Electrical Drawings
E0.1,2 —General Notes
E1.1,2 —Site Plan
E2.1,2 —Floor Plans, Lighting
E3.1,2 —Floor Plans, Power
E4.1,2 —Electrical Rooms
E5.1,2 —Riser Diagrams
E6.1,2 —Fixture/Panel Schedules
E7.1,2 —Details

FIGURE 34.2 ⟩ Code system used to identify drawing groups. *Hepler/Wallach/Hepler © Cengage 2013*

the sum of each subdimension row be identical to the related overall dimension as shown in Figure 34.5.

During the design stage, drawings are first checked for their agreement with established goals and objectives. Working drawings are checked for correctness, accuracy, and completeness with special attention paid to:

- Structural design
- Materials
- Projections of views
- Indexing between views (cross-referencing)
- Symbols
- Changes
- Legibility
- Title blocks
- Dimensions and notes
- Agreement among drawings
- Code compliance
- Agreements and contract
- Specifications
- Schedules
- Contracts.

Agreement among Drawings

The elements of each drawing must agree. For example, the dimensions on one side of a rectangular area must agree with the total shown on the opposite side. It is also important for related features on different drawings to agree.

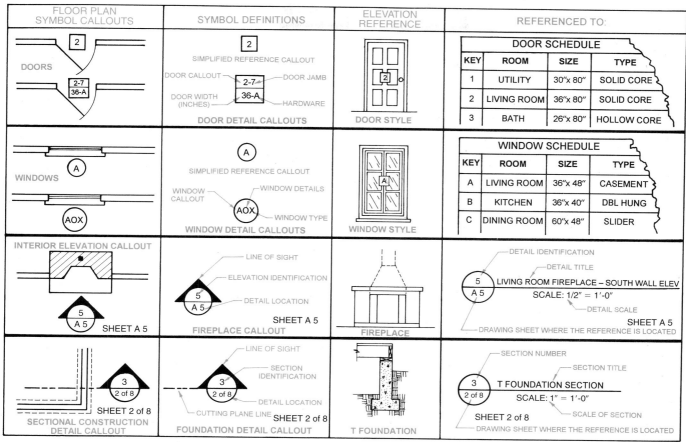

FIGURE 34.3 > Cross-reference symbols used to relate drawings.
Hepler/Wallach/Hepler © Cengage 2013

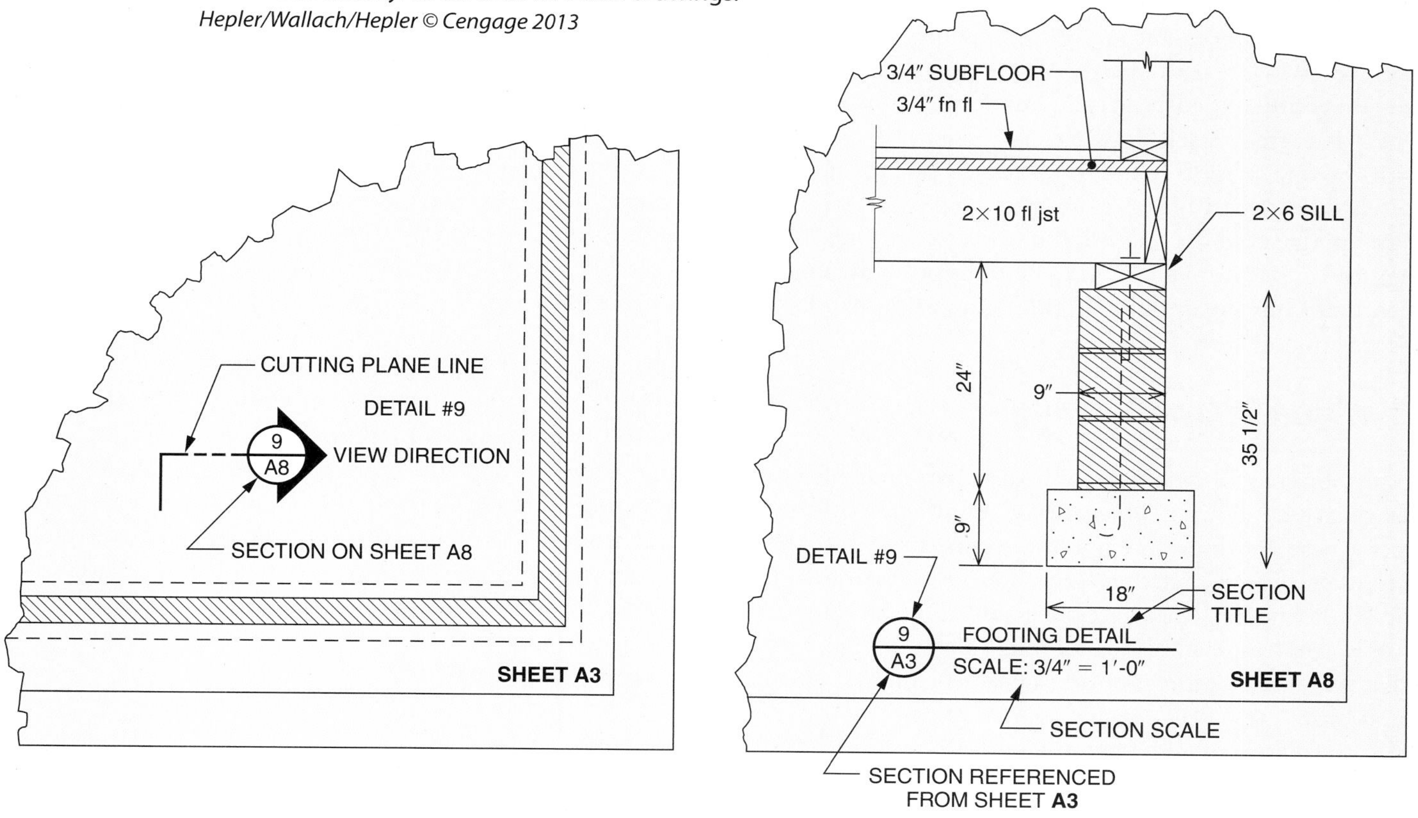

FIGURE 34.4 > Application of cross-reference symbols. *Hepler/Wallach/Hepler © Cengage 2013*

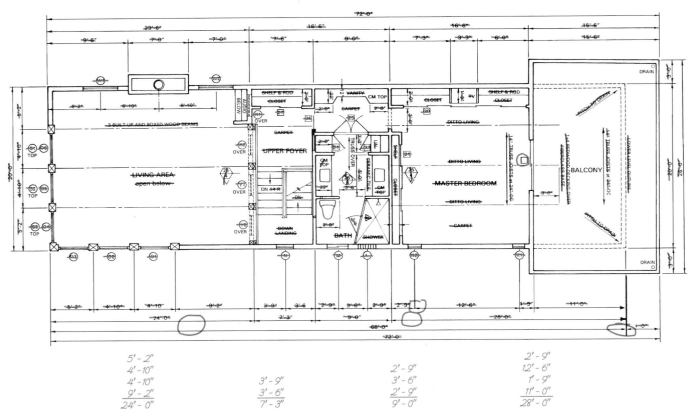

FIGURE 34.5 > Example of dimensional and symbol checking. *Hepler/Wallach/Hepler © Cengage 2013*

Checking the agreement of the same components on different drawings is extremely critical. If drawings in a set contain conflicting information, builders cannot determine which is correct. Each drawing in the set must also agree with the information found in all related documents, such as schedules, specifications, budgets, and legal forms. To better understand the relationship between drawings, study Figures 13.51A through 13.51E. Find the other drawings or pictures of this site or building throughout this text. Locate the position of each.

Checking Sets of Drawings

In previous chapters, samples of many kinds of architectural drawings were shown to illustrate various principles and practices. These drawings were not of the same structure. Therefore, there was no relationship among the electrical plan, the plumbing plan, the survey plan, and so forth. Drawings in this chapter are all of the building shown in Figure 34.6. Therefore, the interpretation and agreement among plans can be studied.

Selected areas of each plan in Figures 34.7 through 34.16 have been marked with large red letters. Identical letters identify the same area as shown on any other drawing. By studying the position of these symbols on each drawing, you can observe how a specific area appears on other drawings in the set. For example, the position of the red letter "A" on the foundation plan Figure 34.9 marks the same area covered with the red letter "A" on the first-floor plan Figure 34.7 and the front elevation Figure 34.10. The relationship between sectional views and basic drawings can also be followed by finding these letters on the drawings and by locating the position of the cutting-plane line on the basic plans. For example, in Figure 34.15 locate the section C/10 cutting-plane line on the floor plan and compare the position of the red letter "A" on that plan and on the section drawing (Figure 34.15).

Study the residential plans in the sequence shown on the following pages and follow the location of each of the large letters throughout the set of plans. Commercial and industrial buildings normally have more floors, cover wider areas, and use heavier construction members. Nevertheless, the sequence of studying and understanding the relationship between drawings is identical.

Combination Plans

Sometimes, several plans in a complete set of plans are combined.

Combination plans are more difficult to read than separate specialized plans. A combination floor plan may include

FIGURE 34.6 › Rendering of the structures described in the drawings in Figures 34.7 through 34.16. *Used by permission Hanley & Wood, LLC*

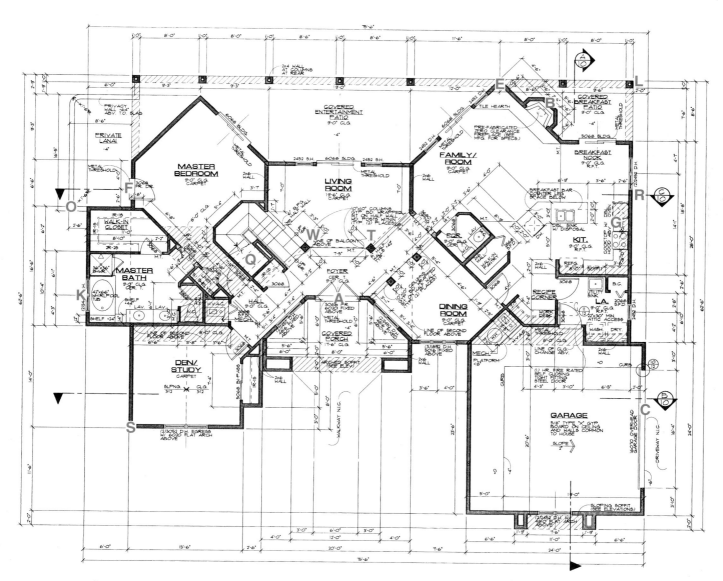

FIGURE 34.7 › First-floor plan. *Used by permission Hanley & Wood, LLC*

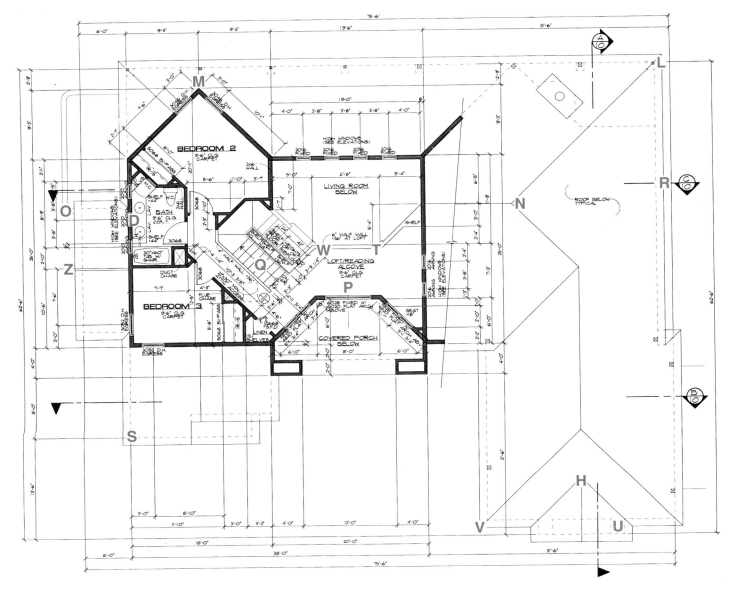

FIGURE 34.8 > Second-floor plan. *Used by permission Hanley & Wood, LLC*

not only information normally found on a floor plan, but also electrical, air-conditioning, plumbing systems, and some landscape, survey, and plot plan symbols. CAD layering of these systems and symbols, in addition to interior design facilities and furniture, can eliminate much of the confusion of interpreting combination plans.

In studying this kind of plan, the specialized elements must be separated out of the plan through spatial perception to eliminate confusion. Study only one element at a time. If you are studying the electrical part, refer only to as much of the remainder of the plan as necessary to orient the position of switches, outlets, and so forth. Imagine that the plan is only an electrical plan and that all other features do not exist.

CORRECTIONS AND CHANGES

Changes are made to a finished drawing either to correct errors, alter designs, substitute materials, or add design features. Once a set of plans is completed and part of a contract, any changes to a drawing or document must be recorded. Changes made on site during construction must also be drawn and recorded on the master set of drawings.

Change Orders

Requests for changes can be initiated by the client, architect, contractor, or subcontractor. Most changes require an increase or decrease in the original cost estimate. **Change orders** are

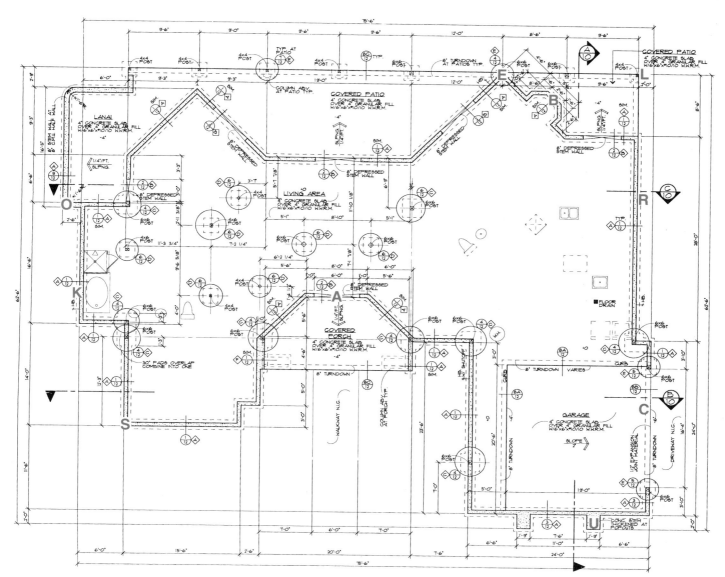

FIGURE 34.9 ＞ Foundation plan. *Used by permission Hanley & Wood, LLC*

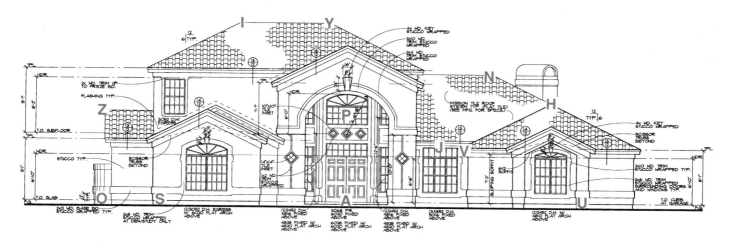

FIGURE 34.10 ＞ Front elevation. *Used by permission Hanley & Wood, LLC*

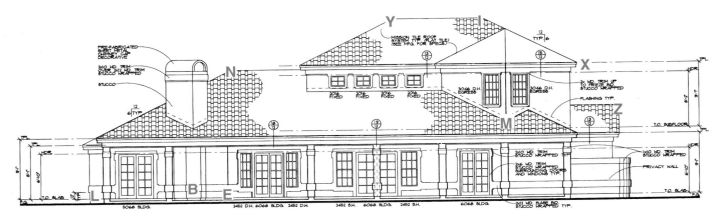

FIGURE 34.11 ⟩ Rear elevation. *Used by permission Hanley & Wood, LLC*

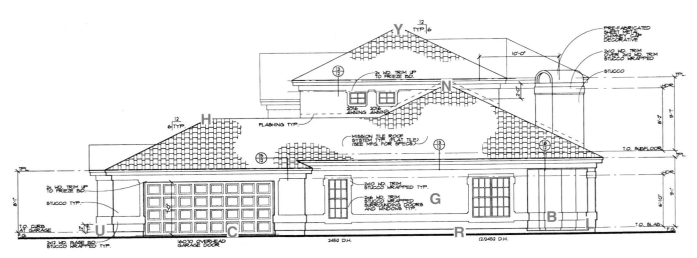

FIGURE 34.12 ⟩ Right elevation. *Used by permission Hanley & Wood, LLC*

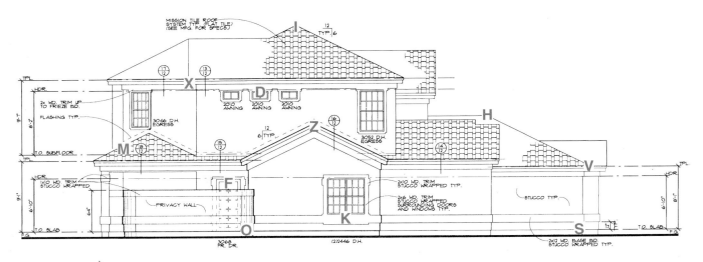

FIGURE 34.13 ⟩ Left elevation. *Used by permission Hanley & Wood, LLC*

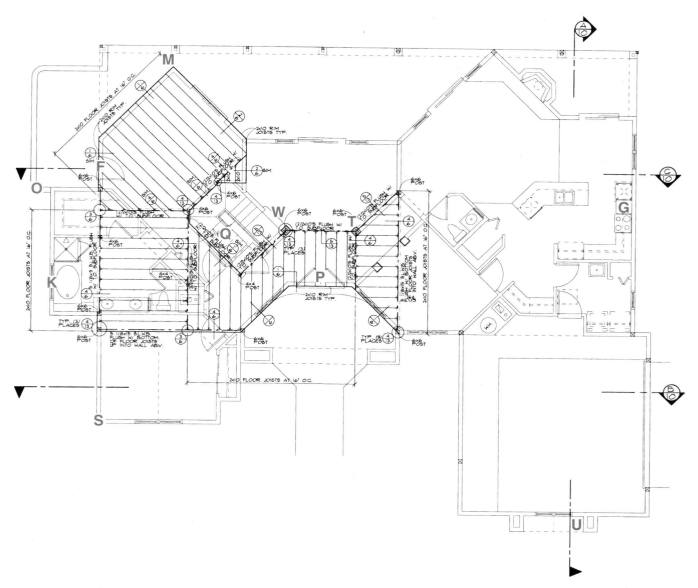

FIGURE 34.14 > Floor framing plan. *Home Planners Inc. Used by permission Hanley & Wood, LLC*

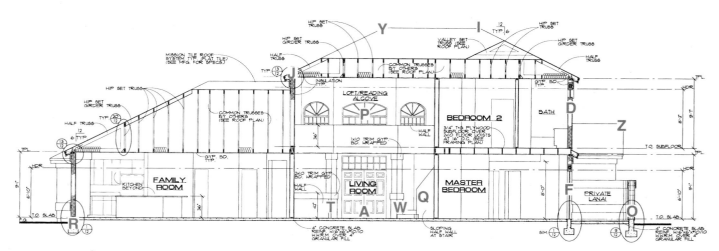

FIGURE 34.15 > Longitudinal section C. *Used by permission Hanley & Wood, LLC*

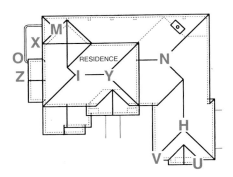

FIGURE 34.16 > Roof plan. *Used by permission Hanley & Wood, LLC*

used to record change requests. See Figure 34.17. A change order includes a description of the change, the adjustment in cost, and usually a print of the altered design or correction.

Changes on Drawings

Each change made on a drawing after the entire set is complete sets up a chain reaction throughout the entire set. For example, if a stairwell opening is enlarged after the electrical, plumbing, and HVAC drawings have been completed, wiring, pipe, and duct locations may also need to be changed.

CONSTRUCTION CHANGE AUTHORIZATION

Owner	☐
Architect	☒
Consultant	☐
Contractor	☐
Field	☐
Other	☐

AIA DOCUMENT G713

PROJECT: SMITH RESIDENCE
(name, address)   87 OAK DRIVE

CONSTRUCTION CHANGE
AUTHORIZATION NO: #7

OWNER: RA SMITH

DATE OF ISSUANCE: 4-16-11

TO:
(Contractor) WILSON CONST CO

ARCHITECT: WOMACK

CONTRACT FOR: 1-10-11

ARCHITECT'S PROJECT NO: P-324

In order to expedite the Work and avoid or minimize delays in the Work which may affect Contract Sum or Contract Time, the Contract Documents are hereby amended as described below. Proceed with this Work promptly. Submit final costs for Work involved and change in Contract Time (if any), for inclusion in a subsequent Change Order.

Description: GUEST BATH CHANGES:
        REDUCE WINDOW J WIDTH TO 2'-0"
        MOVE WINDOW K ℄ TO 5'-3" FROM NORTH WALL ℄
        CHANGE WINDOW M TO GLASS BLOCK
        REPLACE CI TUB W CM TUB
        REPLACE TUB FITTINGS AS SHOWN ON PLUMB SCHEDULE

Attachments: *(Here insert listing of documents that support description.)*
⚠ REVISIONS DRAWN & NOTED ON SHEET A-11

The following is based on information provided by the Contractor:

Method of Determining Change in the Contract Sum: ___COSTS + 10%___
        (lump sum, unit prices, cost plus fee or other)

☒ Fixed	Change in Contract Sum of $ + 1813.66	☒ Fixed	Change in Contract Time of + 4
☐ Estimated		☐ Estimated	
☐ Maximum		☐ Maximum	days

ISSUED:	AUTHORIZED:	CONFIRMED:
BY S. WOMACK	BY R. SMITH	BY R. WILSON
Architect *S Womack*	Owner *(signature)* 4/20 Date	Contractor *(signature)* 4/22 Date

FIGURE 34.17 > AIA change order. *Hepler/Wallach/Hepler © Cengage 2013*

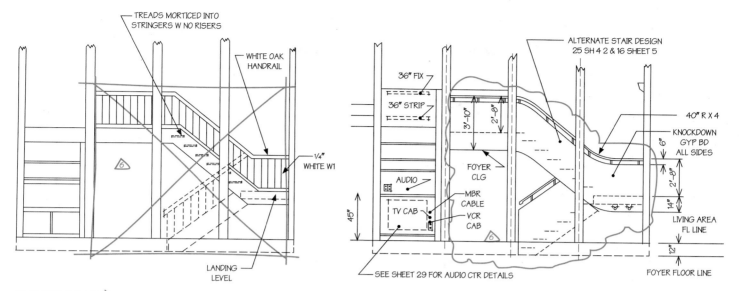

FIGURE 34.18 > Method of marking drawing deletions and changes. *Hepler/Wallach/Hepler © Cengage 2013*

In addition, changes may be required on the floor plan, floor framing plan, stud layout, framing elevations, and specifications. Every drawing must be reviewed after each change to determine what other drawings need to be changed and what redesign may be needed. Each change must also be evaluated to determine if the improvement created by the change is worth the additional cost or secondary problems created.

No portion of a final drawing should be erased to make a change. Changes are drawn as a separate detail and identified with a number inside a triangle to indicate the change order number. A hand-drawn, scalloped line is used to enclose the new or revised portion of the drawing. See Figure 34.18. The numbered triangle is drawn within this enclosure. The number, date, and title of the change is also added to the revision record at the bottom of the drawing. See Figure 34.19. Revisions (changes) are listed on the drawing sheet from the bottom upward. The list also includes general revisions that may predate the final drawing and contract date.

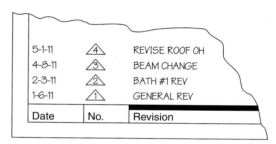

FIGURE 34.19 > Method of recording changes and revisions on a drawing sheet. *Hepler/Wallach/Hepler © Cengage 2013*

Many changes are made by photocopying details onto a transparent appliqué, which is then attached (applied) to the original drawing. This enables future prints to include the change without drawing the changes on the original. This is especially helpful when working with old or worn originals.

Drawing Management Exercises

CHAPTER 34

1. List the types of information that must be checked on architectural drawings.

2. Check the set of house plans you have developed in earlier exercises for the types of information you listed in Exercise 1. Prepare at least one change order and make a drawing of it as shown in Figure 34.18.

3. Design a drawing format with a border and title block. Call up all the drawings of your design and place them on this format. Index all drawings and details.

4. List the drawings in the set of plans in Figure 34.7 through 34.16 that show information about the main entrance, the kitchen island, the family room fireplace, and the overhead garage door.

5. Name the persons who may initiate a change order.

6. What information is included in a change order?

7. What symbol is used on a drawing to label a change or revision?

8. Check the agreement of dimensions in Figure 34.20.

9. Complete a change order for extending the living room wall 2'-0" to the right.

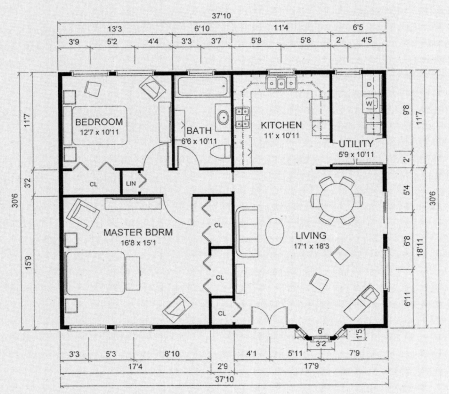

FIGURE 34.20 > Check the dimensions and complete a change order for moving the living room wall 2'-0" to the right. *Hepler Associates, LA PC*

CHAPTER 35

Schedules and Specifications

OBJECTIVES

In this chapter you will learn to:

> create schedules for a set of architectural drawings.

> develop a material list for on-site construction.

> create a set of specifications in a standard format.

TERMS

collage
color coding
CSI format

door hand
material list
schedules

specifications
Sweets Catalog Files
takeoff

INTRODUCTION

Architectural drawings describe the design features of a building, but they do not describe the hundreds of products, legal documents, and structural and financial calculations needed for construction. Furthermore, architectural drawings cannot be drawn large enough to include the hundreds of details needed for construction. This information is supplied through the use of schedules, specifications, and financial and legal documents. **Schedules** are charts that contain product or component information. **Specifications** are written materials that describe what materials are to be used and how. Information is often repeated among schedules, specifications, and drawings. These data *must* be consistent throughout all documents.

Construction drawings include sizes, locations, and general material notations. Many drawings may not specify the exact finishing materials to be used or their composition. Therefore, to ensure that environmentally safe materials and components are used, schedules, specifications, and material lists must be prepared that allow the specific requirements of the designer to be met.

Using 5D building information modeling (BIM) modeling software, many construction documents can be created directly from an original BIM design model. These include schedules, specifications, time projections, and lists of the quantity and properties of materials and components to be used.

SCHEDULES

A wide range of component or material information is recorded on schedules. Schedules provide a concentration of information about specific items. Schedules are used for estimating and ordering purposes. They are also used to conserve drafting time and drawing sheet space.

Schedules are developed mainly for items to be installed in a building, such as windows, doors, and fixtures. Finishing materials are also listed on a schedule. However, building materials for components constructed on site are shown on a **material list,** rather than on a schedule. Material lists are discussed later in this chapter. Some common types of schedules are described next.

Spreadsheet Schedules

The *Bill of Materials CAD* command is used to prepare spreadsheet output spreadsheet computer software. This enables items to be entered and their cost, square footage, volume, model numbers, and quantity to be tracked by the *Attributes* command feature within the inserted blocks on a drawing. Once in a spreadsheet format, the values can be mathematically tabulated.

Window Schedules

Each line on a window schedule represents a specific type of window on a plan. To relate each window on a plan to a line on a window schedule, a key symbol is used in both places. The key symbol, or mark, for windows is usually a letter within a circle. See Figure 35.1. Other geometric symbols may be used. Some drafters use a split circle with a W for window or a D for door. A number (or sometimes a letter) is assigned to each window and door. A window schedule includes a wide range of data headings. See Figure 35.2. Each window's reference letter is located in the left column of the schedule. Each line on a window schedule represents a window with the exact same list of specifications. Even if only one detail is different, another key symbol must be assigned.

If a manufacturer's data are very detailed, catalog or code numbers can be substituted for many of the schedule headings. Care must be taken to ensure that the information in the schedule matches each window shown on a drawing.

Window key symbols may also be added to elevation drawings. In some cases, the outline of a window, with a key symbol in the opening, substitutes for drawing a detail on the elevation. In other cases, a separate elevation detail is used to show the style and form of the window. (Refer back to Chapter 17.) Many details of this type are drawn using a larger scale than can be used on the elevation drawings.

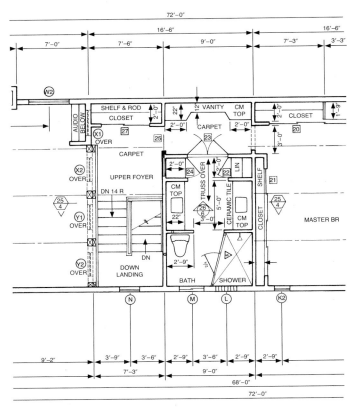

FIGURE 35.1 > Floor plan showing door and window schedule symbols. *Hepler/Wallach/Hepler © Cengage 2013*

KEY	QUANTITY	ROOM	TYPE	MATERIAL	COLOR	GLAZING	WIDTH HEIGHT	SCREEN	ROUGH OPENING	MANUFACTURER CAT#
A	1	Kitchen	Slider	Aluminum	White	Double insulated	6'–0" × 4'–6"	16 × 16 Fiberglass (Fbg.)	6'–0" × 4'–6⅜"	Peachtree Co. 1824 A
B	6	Living	Fixed	Aluminum	White	Double insulated	3'–6" × 7'–0"	—	3'–6¾" × 7'–1⅜"	Kinco 6013 R
C	6	Living	Hopper	Aluminum	White	Double insulated	3'–6" × 1'–0"	16 × 16 Fbg.	3'–6¾" × 1'–0⅜"	Kinco 16-32 A
D	3	Living	Fixed round top	Aluminum	White	Double insulated	3'–6" × 3'–4"	—	3'–6¾" × 3'–4⅜"	Kinco RT 105
E	1	M. bath	Fixed	Glass block	Clear	High diffusion	2'–0" × 4'–6"	—	2'–1" × 4'–7"	H & W Windows GB 870
F	2	Dining	Fixed	Aluminum	White	Double insulated	1'–0" × 6'–8"	—	1'–1" × 6'–8⅜"	T & D Glass Co. F 304
G	1	Breakfast nook	Awning	Aluminum	White	Double insulated	4'–0" × 1'–0"	—	4'–0¾" × 1'–0⅜"	T & D Glass Co. A103-6
H	1	Lavatory	Slider	Aluminum	White	Double insulated	4'–0" × 1'–0"	16 × 16 Fbg.	4'–0¾" × 1'–0⅜"	T & D Glass Co. S 713 A

FIGURE 35.2 > Partial window schedule. *Hepler/Wallach/Hepler © Cengage 2013*

KEY	ROOM	TYPE	HAND	MATERIAL	COLOR	GLASS	THICK WIDTH HEIGHT	SCREEN	THRESHOLD	ROUGH OPENING	MANUFACTURER CAT #
1	Garage	Ext. flush	Right hand	Insulated steel	White	—	1¾" × 3'–0" × 6'–8"	16 × 16 Fiberglass (Fbg.)	Aluminum	3'–1¼" × 6'–8⅜"	Peachtree Co. B 3285
2	Kitchen	Ext. French	Right hand	Steel	White	Double insulated	1¾" × 2'–8" × 6'–8"	—	Aluminum	2'–8⅜" × 6'–8⅜"	Peachtree Co. F 3604
3	Dining	Ext. slider	—	Aluminum	White	Double insulated	8'–0" × 6'–8"	16 × 16 Fbg.	Aluminum	8'–1⅝" × 6'–8¾"	A & N Door Co. S 106-A
4	Foyer	Ext. panel	Double	Insulated steel	White	—	1¾" × 3'–0" × 6'–8"	—	Oak	6'–2" × 6'–8¾"	A & N Door Co. P 113 S
5	Foyer	French storm	Double	Aluminum	White	Single	1¼" × 3'–0" × 6'–8"	16 × 16 Fbg.	Oak	6'–2" × 6'–8¾"	A & N Door Co. F 204
6	Garage	4-section rollup	—	Ribbed steel	White	—	1¾" × 18'–0" × 6'–6"	—	Vinyl	18'–3" × 6'–7½"	Curtis Doors R 10 W
7	Kitchen–Utility	Flush	Left hand	Lauan	White	—	1⅜" × 2'–8" × 6'–8"	—	—	2'–8¾" × 6'–8¾"	Kinco FR 34
8	Dining–Breakfast nook	Low slide	Quadruple	Pine	White	—	1⅛" × 2'–10" × 6'–8"	—	Floor guides	2'–11" × 6'–8¾"	Kinco LS 132
9	Bedroom 1	Flush slide	Double	Lauan	White	—	1¼" × 2'–8" × 6'–8"	—	Floor guides	5'–4¾" × 6'–8¾"	Curtis Doors F 2842

FIGURE 35.3 > Partial door schedule. *Hepler/Wallach/Hepler © Cengage 2013*

Door Schedules

A variety of information about doors can be included on a door schedule. See Figure 35.3. Door schedules are similar to window schedules except door schedules must include the **door hand** of each swinging door. To determine hand, an observer views the door from the hinge side. If the right hand can open the door inward the door is right handed. The opposite is true for left-hand doors as shown in Figure 35.4. On a door schedule, as on a window schedule, the entries are often manufacturers' codes.

Numbers placed inside a square or rectangle are most often used for door key symbols. Other geometric shapes may be used, but they should be different from those used to identify windows.

Exterior door designs are shown on exterior elevations. Unless interior door designs are shown on interior elevations, separate drawings may be needed to describe interior door design details. See Figure 35.5. Some detailed door schedules also include designations for exterior/interior, swing, core type, jamb size and type, bore and setback, strike plate, trim, and finish.

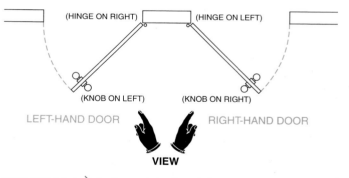

FIGURE 35.4 > Left- or right-hand door determination. *Hepler/Wallach/Hepler © Cengage 2013*

Hardware and Fittings

Hardware and fittings schedules include door hardware, cabinet hardware, plumbing controls and fittings, and hardware accessories. These are prepared as separate schedules or may be combined in one schedule.

- *Door hardware schedules* include information on locks, levers, hinges, escutcheons (keyhole plates), and kickplates. This information is sometimes included on door schedules.

However, a separate schedule is usually prepared, using the same key symbol as used on the door schedule.

- *Cabinet hardware schedules* include information on knobs, hinges, catches, and drawer glides. This information may be included in a cabinet schedule by adding headings for knobs, hinges, catches, and glides.
- *Plumbing hardware schedules* include all controls and fittings attached to fixtures. These include faucets, sprays, dispensers, drain covers, strainers, shower heads, and controls. See Figure 35.6.
- *Hardware accessories schedules* include items such as shelves, robe hooks, towel racks, shower rods, and soap, toothbrush, and bath tissue holders. These items are sometimes included in the plumbing hardware schedule.

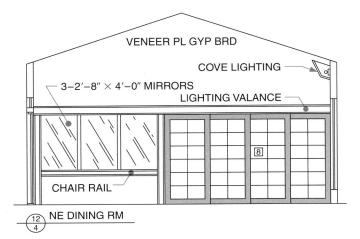

FIGURE 35.5 ❯ Door design and symbol shown on an interior elevation. *Hepler/Wallach/Hepler © Cengage 2013*

Plumbing Fixtures

Schedules for plumbing fixtures include all items to which water lines are connected, such as sinks, water closets, and bathtubs. Manufacturers often use different terms for describing styles and colors. Each manufacturer may offer dozens of styles and colors for each fixture. For this reason, schedule entries must be consistent with the terms used in manufacturers' specifications. See Figure 35.7.

Electrical Fixture Schedules

An average residential plan may contain over a hundred electrical fixtures. Electrical plans show only the location and symbol for each electrical fixture. Schedules are needed to fully describe the exact specifications of each fixture. See Figure 35.8. Electrical fixture schedules include convenience outlets, switches, lighting fixtures, fans, heaters, chimes, bells, and smoke alarms. They may also include audio, video, and security system fixtures, if separate schedules are not prepared for these.

Appliance Schedules

Appliance schedules provide information and dimensions for cabinet spacing, and appliance model numbers for purchasing, electrical requirements for wiring, and often prices for budgeting. See Figure 35.9. If there is only one of each appliance type in a room, key symbols are not needed. The identification of the room is the only information required.

HARDWARE/FITTING	ROOM	QUANTITY	TYPE	MANUFACTURER	CODE	SIZE	MATERIAL	FINISH
Sink faucet	Kitchen	1	W/spray	Kohler	15176	1 hole	Brass	Polished
Showerhead	M. bath	1	Compound arm	Universal	NE01	12″	Brass	Polished
Shower control	M. bath	1	Knob	Delta	2882	—	Brass	Polished
Tub faucet	M. bath	1	Roman	Delta	2780	8″ spout	Brass	Polished
Lav. faucet	Laundry	1	Bar	Sterling	083-99	4″ space	Chrome	Bright
Strainer	Kitchen	1	Adjustable	Kohler	8803	—	Brass	Polished
Door & drawer knobs	Kit. cabinets	36	Button	Belwith	722-UB	1¼″ D	Brass	Polished
Door & drawer knobs	Baths	34	Dome	Belwith	770-LP	1⅛″ D	Brass	Polished
Ext. door knobs w/key	Foyer/Kitchen	3	Lever set	Kwikset	400 DL	1⅜″ to 1¾″	Brass	Polished
Dead bolt locks	Kit./Lau.	2	Dead bolt	Kwikset	600 CB	1⅜″ to 1¾″	Brass	Polished

FIGURE 35.6 ❯ Partial hardware and fittings schedule. *Hepler/Wallach/Hepler © Cengage 2013*

FIXTURE	ROOM	QUANTITY	MANUFACTURER	SIZE	STYLE	MODEL	COLOR	MATERIAL
Sink	Kitchen	1	Eljer	22″ × 33″	HiLo ½ bowl	212-3086	White	Cast iron–porcelain
Lavatory	Laundry	1	American Standard	31″ × 34″	Drop-in	0179-015	White	Vitreous china
Lavatory	M. bath	2	Milesco	17¾″ × 27⅜″	Diamond	9402	Bone	Cultured marble
Lavatory	Bath 2	2	Milesco	18″ × 22″	Waikiki	9401	Peach	Cul. marble
Water closet	M. bath	1	American Standard	22″ × 20″	Roma	2012.14	Bone	Vit. china
Water closet	Bath 2	1	American Standard	22″ × 20″	Round front	Cadet	Peach	Vit. china
Shower	M. bath	1	Custom	60″ × 64″	Clear glass	Custom	Gold	Anodized aluminum
Tub	Bath 2	1	Milesco	60″ × 60″	Mark 26	4115	Peach	Cul. marble

FIGURE 35.7 > Partial plumbing fixture schedule. *Hepler/Wallach/Hepler © Cengage 2013*

FIXTURE	ROOM	QUANTITY	TYPE	SIZE	COLOR	MANUFACTURER	MODEL	MATERIAL	LIGHT BULBS
Hanging	Breakfast nook	1	Chandelier	14″ D	Clear	Besco	M2061	Brass-Glass	6-40 Ctc
Track	Living	12	Flush	2–8′ units	White	Moe	L730	Steel	12-75 Par
Wall	Lavatory	1	Makeup	16″	Chrome	Besco	SL7413	Steel	3-60
Flood	Outside	4	Motion	150 W	White	Lowes	BC800	Vinyl	8-150 Parfl
Recessed	M. bath	2	Recessed	7″ trim	White	Peachtree Co.	Pic-7	Steel	2-60A
Post	Gate posts	2	Carriage	12″	White-Brown	Besco	93-1500	Brass	2-60AC
Exhaust fans	Baths	2	Ducted	10″	White	Nutone	671	Steel	—
Ceiling fans	Liv./Kit./Brs.	5	5-blade	52″	White	Gulf	Riveria	Fabric blades	20-60A
Chimes	Foyer	1	Wall	16″	White	Nutone	LA107	Aluminum	—
Smoke alarms	Gar./Bfk./Halls	4	Ceiling	6″	White	1st Alert	SA670	Vinyl	—
Fluorescents	Hall/Foyer/ Kit./Din./Stairs/ Office/Gar.	25	Single exposed	4′–0″	White	Besco	SA140	Steel	25-3212 CW

FIGURE 35.8 > Partial electrical fixture schedule. *Hepler/Wallach/Hepler © Cengage 2013*

Surface Coverings Schedules

Floor, wall, and ceiling coverings are included in coverings schedules. There are two types of coverings schedules: *general design schedules* and *material specification schedules*. General design schedules are developed during the design process. See Figure 35.10. Material specification schedules are developed once working drawings have been completed and covering materials can be finalized. See Figure 35.11. Different materials or colors are often used on the same floor, wall, or ceiling.

APPLIANCE	ROOM	QUANTITY	MANUFACTURER	MODEL	SIZE	TYPE	COLOR	MATERIAL	ELECTRICAL REQUIREMENTS
Dishwasher	Kitchen	1	Maytag	DWU 7400	24″	4 cycle	White on white (w/w)	Steel porcelain	110 V
Garbage disposal	Kitchen	1	GE	GFC 1000	¾ HP	Automatic	—	Stainless steel	110 V
Refrigerator	Kitchen	1	GE	TBH 25 PAS	25 CF	Top mount	w/w	Steel porcelain	110 V
Cooktop	Kitchen	1	Amana	AKE 30	30″	Euro	White	Cast iron heat units	220 V
Oven	Kitchen	1	GE	JPP11 WP	30″	Built-in	w/w	Steel porcelain	220 V
Range hood	Kitchen	1	Kenmore	52 391	30″	Ducted	w/w	Steel procelain	110 V
Washer	Laundry	1	Amana	LW 4303	7 CF	Top load	White	Steel porcelain	110 V
Dryer	Laundry	1	Amana	LE 4407	Max cap	Front load	White	Steel porcelain	220 V

FIGURE 35.9 > Appliance schedule. *Hepler/Wallach/Hepler © Cengage 2013*

	FLOOR									CEILING					WALL			WAINSCOT			BASE							
ROOMS	Asphalt tile	Ceramic tile	Cork tile	Vinyl tile	Wood strip—Oak	Woods sqs.—Oak	Plywood panel	Carpeting	Slate	Diazo	Plaster	Wood panel	Acoustical tile	Exposed beam	Plaster	Wood panel	Wall paper	Wood	Ceramic tile	Paper	Asphalt tile	Stone veneer	Sheet vinyl	Wood	Rubber	Tile—Ceramic	Asphalt	REMARKS
Entry									✓	✓		✓				✓								✓				Diazo step covering
Hall			✓								✓				✓					✓				✓				
Bedroom 1					✓								✓		✓			✓									✓	Mahogany wainscot
Bedroom 2					✓								✓		✓		✓	✓									✓	Mahogany wainscot
Bedroom 3							✓	✓					✓			✓								✓				See owner for grade carpet
Bath 1		✓									✓				✓				✓							✓		Water-seal-tile edges
Bath 2	✓										✓				✓				✓							✓		Water-seal-tile edges
Kitchen				✓							✓				✓						✓		✓					
Dining				✓								✓	✓	✓	✓						✓		✓					
Living					✓	✓					✓												✓	✓				See owner for grade carpet

FIGURE 35.10 > General floor and wall covering schedule. *Hepler/Wallach/Hepler © Cengage 2013*

COVERING	ROOMS	MANUFACTURER	MATERIAL	UNIT SIZE	COLOR	STYLE	BORDER	GROUT	AREA (SQ. FT.)
Carpet	Liv./Din./Bfk.	Aladdin	Nylon 100%	—	White 717	Virtuous	—	—	80
Carpet	Brs./Halls/Stairs	Aladdin	Nylon 100%	—	Beige 618	Virtuous	—	—	84
Floor tile	Foyer	Fl Tile	Shellstone	16″ × 16″	Coral	Honed	—	White	154
Floor tile	Bath 2	Fl Tile	Ceramic	12″ × 12″	Coral	1212	—	Peach 610	70
Floor tile	M. bath	Watson	Ceramic	12″ × 12″	Cameo	8402	—	White	70
Floor tile	Lau.–Util.	Natura	Vinyl	12″ × 12″	Sand	8425	—	Sand	130
Floor tile	Kitchen	Saloni	Ceramic	12″ × 12″	Beige	03016	—	White	110
Wall tile	Bath 2	Watson	Ceramic	4″ × 4″	Peach	1428	LD 474	White	40
Wall tile	M. bath	Watson	Ceramic	8″ × 10″	Uropa	180RD	½″ × 8″ gold	White	42
Wall tile	Kitchen	Watson	Ceramic	6″ × 8″	3202	190RD	½″ × 8″ gold	White	34
Int. wall paint	All	S & W Co.	Latex	—	1108	Flat	—	—	All
Door & wood paint	All	S & W Co.	Oil	—	1641	Semigloss	—	—	All
Ceilings	All	S & W Co.	Latex	—	1004	Flat	—	—	All
Ext. Stucco	All	S & W Co.	Latex	—	SW 2060	Semigloss	—	—	All
Decks	All	Thompson's	Water Seal®	—	Clear	2-coat	—	—	All

FIGURE 35.11 > Floor and wall covering schedule with material specifications. *Hepler/Wallach/Hepler © Cengage 2013*

In these cases, **color coding** can be used on a floor plan or wall elevation to indicate the exact position and pattern of each material. Figure 35.12 shows a floor plan with different floor covering materials shown in colors that can be keyed to a floor covering schedule. Figure 35.13 shows an interior wall elevation with different materials labeled and also shown in color.

Often fabric, texture, color, or other features cannot be accurately described in words on a schedule. In these cases, swatches of material or paint color chips are often attached to a sample **collage.** These are either keyed to the floor plan by number or related directly with leaders. This identification method eliminates much misunderstanding in the final execution of the design and selection of materials.

Paint and Finishing Schedules

Many types and grades of finishes are used on the interior and exterior of a building. A finishing schedule condenses all of this information into one chart. A preliminary interior finish schedule shows the color classification for each wall in each room for design purposes. See Figure 35.14. In a final and more complete schedule, each manufacturer's exact color code, numbers of coats, and surface preparation are added to the schedule.

In specifying finishing materials, avoid environmentally hazardous products that contain formaldehyde, acrylonitrile, polyethylene, butadiene styrene, polyamide, or toluene. This is necessary to ensure personal safety and provide maximum environmental protection.

Furniture and Accessory Schedules

Furniture and accessories are usually not included as part of a basic architectural design. When a totally integrated plan involves all aspects of the design, furniture schedules are developed and the information is keyed to a floor plan that includes furniture outlines.

Furniture and accessory schedules include all pertinent information relating to each piece of furniture. Numbers

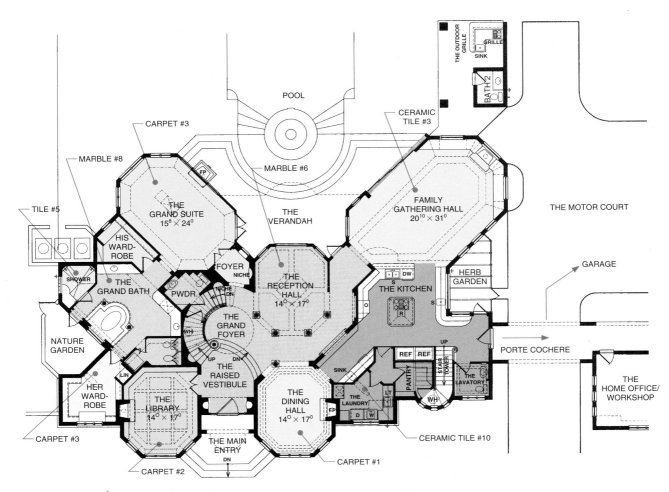

FIGURE 35.12 〉 Color coding of floor coverings. *Home Design Services, Inc.*

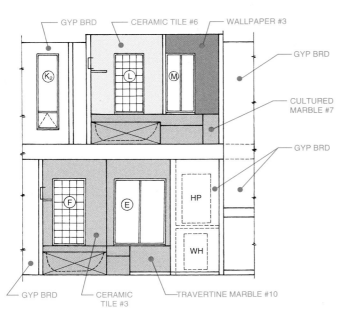

FIGURE 35.13 〉 Color coding of wall coverings. *Hepler/ Wallach/Hepler © Cengage 2013*

corresponding to the schedule numbers are placed on or near the appropriate item on the floor plan. Figure 35.15A shows an abbreviated floor plan with numbered furniture outlines keyed to the furniture schedule shown in Figure 35.15B. This system allows the interior designer and the contractor to find both the locations and specifications for each item. Photographs, catalog illustrations, or collage boards are sometimes keyed or attached to furniture schedules. See Figure 35.16.

In specifying furniture and accessory materials, be aware of the personal and environmentally toxic materials that may be used in some fabrics. Refer to Chapters 6 and 27 for more information on the use of toxic materials in building.

Construction Component Schedules

Separate schedules can be developed for a wide variety of construction components. For example, schedules may be prepared for cabinetry, moldings and trims, beams, lintels, footings, and any other category that includes a number of different, but related items.

ROOMS	FLOOR			CEILING					WALL					BASE					TRIM					REMARKS
	Floor varnish	Unfinished	Waxed	Enamel gloss	Enamel semigloss	Enamel flat	Flat latex	Stain	Enamel gloss	Enamel semigloss	Enamel flat	Flat latex	Stain	Enamel gloss	Enamel semigloss	Enamel flat	Flat latex	Stain	Enamel gloss	Enamel semigloss	Flat latex	Enamel flat	Stain	
Entry			✓				Off Wht					Off Wht		Off Wht						Off Wht				Oil stain
Hall			✓				Lt Brn		Tan						Drk Brn					Drk Brn				Oil stain
Bedroom 1	✓						Off Wht					Off Wht			Grey					Grey				One coat primer & sealer—painted surface
Bedroom 2	✓						Off Wht				Lt Yel				Yel							Yel		One coat primer & sealer—painted surface
Bath				Wht					Wht							Lt Blue				Lt Blue				Water-resistant finishes
Closets	✓					Brn					Brn				Brn						Brn			
Kitchen			✓	Wht							Yel					Yel					Yel			
Dining			✓				Tan		Yel							Yel				Yel				Oil stain
Living		✓					Tan						Lt Brn					Lt Brn					Lt Brn	Oil stain

FIGURE 35.14 › Paint and finishing schedule. Hepler/Wallach/Hepler © Cengage 2013

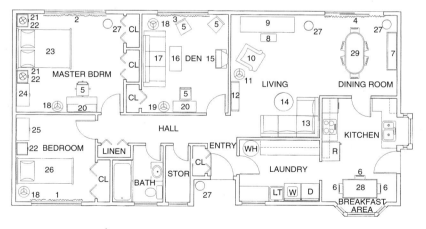

FIGURE 35.15A › Floor plan with furniture keyed to the schedule in Figure 35.15B. Hepler/Wallach/Hepler © Cengage 2013

MATERIAL LISTS

Schedules are prepared for building components that are available in their final form or that describe a final outcome. However, raw building materials merely represent the ingredients for construction. Building *materials* used for on-site construction, such as lumber, concrete, masonry, and steel, are shown on material lists. These lists include descriptions, quantities, and costs of materials. The creation of a material list from working drawings is known as a **takeoff.** Figure 35.17A shows a portion of a 48-page material list for the house plan featured in Chapter 35. Planting schedules

SYMBOL	ITEM	ROOM	LENGTH	WIDTH	HEIGHT	MATERIAL	COLOR	QUANTITY	MANUFACTURER	CAT. #	COST	REMARKS
1	Drapes	Bedroom	11'	—	7'	Cotton blend	Brown	1 set	Sears	CD101	$75	Lined
2	Drapes	M. bedroom	12'	—	7'	Cotton blend	Yellow	1 set	Sears	CD107	$85	Lined
3	Drapes	Den	7'	—	7'	Cotton blend	Yellow	1 set	Sears	CD106	$65	Lined
4	Drapes	Living/Dining	24'	—	7'	Acrylic	Brown pat.	1 set	Sears	CD203	$150	Lined
5	Chair	Mbr./Den	18"	18"	18"	Plastic	Brown	4	ID Furn. Co.	X117	$45 ea.	
6	Chair	Din./Kit.	18"	18"	18"	Oak	Natural	9	ID Furn. Co.	L217	$65 ea.	Oil finish
7	China cab.	Dining	6'	18"	5'–6"	Oak	Natural	1	Danish Furn. Co.	13712	$650	Oil finish
8	Piano bench	Living	33"	18"	18"	Mahogany	Brn. stain	1	Music Co. Inc.	23L19	$50	Piano finish
9	Piano	Living	5'–6"	2'–0"	5'–6"	Mahogany	Brn. stain	1	Music Co. Inc.	P17731	$1750	Piano finish
10	Up. wing ch.	Living	33"	30"	20"	Leather	Natural	1	Danish Furn. Co.	18979	$575	
11	Fl. lamp	Living	14" dia.	—	4'–6"	Metal/Cloth	Tan	1	Danish Furn. Co.	37111	$85	
12	Stereo	Living	30"	11"	5'–0"	Teak	Brown	1	Danish Furn. Co.	60701	$450	
13	Sofa/sec.	Living	14'	30"	18"	Velveteen	Red	3 pcs.	Danish Furn. Co.	42107	$1200	
14	Coffee tbl.	Living	30" dia.	—	15"	Teak	Natural	1	Danish Furn. Co.	77310	$110	Natural oil finish
15	Television	Den	21"	18"	30"	21" color	Brown	1	Sony	XL19	$675	
16	Coffee tbl.	Den	48"	15"	15"	Oak	Natural	1	Danish Furn. Co.	78325	$80	Natural oil finish
17	Sofa	Den	6'–6"	30"	18"	Cotton blend	Tan	1	Danish Furn. Co.	59781	$800	
18	Fl. lamp	Mbr./Br./Den	15" dia.	—	5'–0"	Wood/Cloth	Brown	3	Danish Furn. Co.	66362	$75 ea.	
19	Fl. lamp	Den	12" dia.	—	4'–6"	Wood/Plastic	Yellow	1	Danish Furn. Co.	65731	$50	
20	Desk	Mbr./Den	39"	18"	29"	Oak	Natural	2	Danish Furn. Co.	47772	$225 ea.	Natural oil finish
21	Nightstand	Mbr./Br.	18"	15"	24"	Oak	Natural	3	Danish Furn. Co.	64991	$45 ea.	Natural oil finish
22	Tbl. lamp	Mbr.	9" dia.	—	30"	Wood/Plastic	Brown	2	Danish Furn. Co.	65820	$35 ea.	
23	Full bed	Mbr.	6'–9"	46"	20"	Standard	—	1	Acme Bed Co.	AC12	$235	Box spring/Mat./Frame
24	Dresser	Mbr.	39"	20"	48"	Oak	Natural	1	Danish Furn. Co.	37452	$125	Dbl. dresser
25	Dresser	Bedroom 1	48"	20"	52"	Oak	Natural	1	Danish Furn. Co.	37471	$200	Triple dresser
26	Twin bed	Bedroom 2	6'–9"	42"	20"	Standard	—	1	Acme Bed Co.	AC08	$190	Box spring/Mat./Frame
27	Planter	Mbr./Liv./Den/Porch	10" dia.	—	12"	Terra cotta	Brown	4	Flowers Inc.	23FP	$10 ea.	
28	Table	Kitchen	36"	22"	30"	Teak	Natural	1	Danish Furn. Co.	17832	$110	Natural oil finish
29	Table	Dining	5'–0"	3'–3"	30"	Teak	Natural	1	Danish Furn. Co.	17876	$235	Natural oil finish

FIGURE 35.15B > Furniture and accessory schedule keyed from the floor plan in Figure 35.15A. *Hepler/Wallach/Hepler © Cengage 2013*

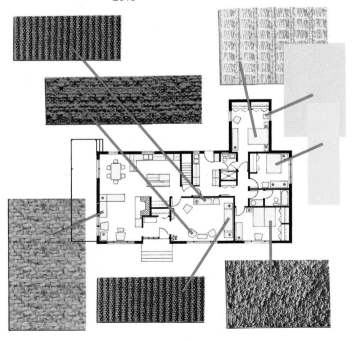

FIGURE 35.16 > Collage board related to floor plan furniture. *Hepler/Wallach/Hepler © Cengage 2013*

(Figure 35.17B) are often called the *plant material lists* and show the botanical name, common name, quantity, and size plus the abbreviations used on the drawing. These are sometimes keyed to a landscape plan with numerals.

SPECIFICATIONS

Specifications (specs) are detailed written instructions describing the requirements for construction. Information not included in a drawing, schedule, material list, or legal document is included in the specifications.

Purpose and Use

Specifications are written to ensure that a building will be constructed as designed and to prevent misunderstandings between the client, the architect, and the builder. Specifications are also used by lending institutions to evaluate the quality of construction. Along with drawings and schedules, specifications are used by contractors to make construction estimates and bids. Material lists are sometimes included in specifications.

MATERIAL LOCATION	SIZE		MATERIAL DESCRIPTION		UNIT	QUANTITY	MATERIAL COST	INSTALLATION COST	TOTAL COST	GRADE OPTIONS
GARAGE										
Bottom plate	2	4	Standard & better, pressure treated Douglas fir	R/L	Lin. feet	66	33.33	20.85	54.18	
Studs & cripples	2	4	#2 & btr., D. fir	10	Lin. ft.	160	58.62	40.44	99.06	
Studs & cripples	2	4	#2 & btr., D. fir	12	Lin. ft.	132	48.36	33.35	81.71	
Studs & cripples	2	4	#2 & btr., D. fir	14	Lin. ft.	126	46.16	31.84	78.00	
Fire blocking	2	4	#2 & btr., D. fir	R/L	Lin. ft.	24	8.79	7.75	16.54	
Trimmers	2	4	Stud grade D. fir (81 ⅛″)	8	Lin. ft.	16	5.86	5.05	10.91	
Top plate	2	4	#2 & btr., D. fir	R/L	Lin. ft.	138	50.56	43.59	94.15	
Ext. wall sheathing	4	8	½″ APA RS ³²⁄₁₆ CDX ply	—	Sq. ft.	320	142.58	74.12	216.70	
Bottom plate	2	6	Std. & btr. PT D. fir	R/L	Lin. ft.	24	18.30	8.09	26.39	
Studs & cripples	2	6	#2 & btr., D. fir	10	Lin. ft.	110	56.64	31.65	88.29	
Studs & cripples	2	4	#2 & btr., D. fir	12	Lin. ft.	60	21.98	15.16	37.14	
Fire blocking	2	4	#2 & btr., D. fir	R/L	Lin. ft.	12	4.40	3.88	8.28	
Top plate	2	6	#2 & btr., D. fir	R/L	Lin. ft.	52	26.77	17.52	44.29	
Ext. wall sheathing	4	8	½″ APA RS ³²⁄₁₆ CDX ply	—	Sq. ft.	160	71.29	37.07	108.36	
Bottom plate/Int. wall	2	4	Std. & btr. PT D. fir	R/L	Sq. ft.	14	7.07	4.43	11.50	

FIGURE 35.17A > Partial material list for the plan shown in Chapter 35. *Reed Construction Data, a division of Reed Elsevier, Inc.*

KEY	BOTANICAL NAME	COMMON NAME	QTY.	SIZE
As	Alnus serrulata	Smooth Alder	1	12–18″ #1 cont.
Aa	Aronia arbutifolia	Red Chokeberry	4	12–18″ b/b
Bh	Baccharus halimifolia	Groundsel Tree	4	2 qt.
Co	Cephalanthus occidentalis	Buttonbush	14	12–18″ b/b
Cl	Clethra alnifolia	Sweet Pepper Bush	3	18–24″ b/b
Cr	Cornus amomum	Silky Dogwood	6	18–24″ b/b
Cor	Cornus sericca 'Ruby'	R.R. Osier Dogwood	1	18–24″ b/b
Hib	Hibiscus moschentos	Marsh Hibiscus	3	1 qt. pot
If	Iva frutenscens	High-Tide Bush	39	18–24″ stab.
Jun	Juneus rocmerianus	Black Needle Rush	35	1 qt. pot
Kos	Kosteletzkya virginica	Seashore Mallow	5	1 qt. pot
Le	Leersia oryzoides	Rice Cutgrass	3	1–3/4″ peat pot
Lc	Limonium Carolinianum	Sea Lavender	65	1 qt. pots

FIGURE 35.17B > Plant material list. *Hepler/Wallach/Hepler © Cengage 2013*

Specifications must agree with the information shown in the set of plans, because the specs and the drawings become a legal part of the construction contract. If a discrepancy does exist, then the information on the specs usually takes precedence over a drawing. The specs generally contain more product and material details than do drawings.

Specifications are intended to simplify and clarify, not to make the construction process longer and more difficult. Accurate specifications are critical to the contractor. If the contractor's materials or methods are inferior to those specified, the project will not be approved. If the contractor uses materials and methods that exceed the limits described in the specifications, material expense and labor costs may increase.

Specific product decisions are often not finalized at contract time. This is common for categories such as appliances, electrical fixtures, floor and wall coverings, plumbing fixtures, hardware, and fittings. If this is the case, a cost allowance is applied to each category. After the items are purchased, the contract amount is adjusted up or down.

Guidelines for Writing Specifications

Specification entries are written to describe minimum acceptable standards or to describe specific end results. When writing specifications, these guidelines should be followed:

1. Use accepted terms and abbreviations consistently.
2. Use a consistent format.
3. Explain any unique or obscure terms.
4. Specify standard or manufactured components wherever possible.
5. Indicate "or equal" alternatives that do not sacrifice quality.
6. Use each manufacturer's latest catalog code numbers and titles.
7. Write brief simple sentences.
8. Use the accepted standards in each field for specifying materials.
9. Be very specific when writing criteria for work quality standards.
10. Number all pages and indicate the end page.
11. Include a table of contents with headings and page numbers.
12. Capitalize proper names, rooms, legal documents, and material grades.
13. To define an obligation, use the term "shall," not "will."
14. Avoid specifying nonstandard or obscure sizes or materials.
15. Avoid repetition or overlapping.
16. Ensure specifications are reasonable, practical, and possible.
17. Be sure specs include only areas that are part of the contract.
18. Ensure that all specifications adhere to current code regulations. If they do not, an "exception" must be requested from the local municipal building department.

Organization

The types of information contained in the specifications include the scope of the work, product descriptions, and methods of execution. The scope of the work describes the amount of work to be completed in each category of construction. Product descriptions, including schedules and material lists, detail the characteristics of every component and material that will become part of the finished structure. The methods of execution describe the approved methods of construction to be used. Specifications may also include restrictions on the use of some equipment or devices, delivery access routes, or processes that may potentially interfere with public safety or convenience during the construction process.

Of the many types of specifications forms, the most popular for small residences is the FHA-VA form. This form does not include scope-of-the-work or methods of execution. It is primarily a material list that includes several abbreviated schedules. It is, therefore, not comprehensive or expandable enough to use for larger residences or commercial buildings.

To ensure consistency among specifications, the American Institute of Architects (AIA) and the Construction Specification Institute (CSI) have approved and recommended the use of the **CSI format** for construction specifications. This flexible and expandable format organizes information under 16 major divisions with subdivisions included under each, as shown in Figure 35.18.

Computer-Generated Specifications

Specifications are based on the content of the drawings and standards of construction. Master specifications programs are available from the Construction Specifications Institute in the CSI format. Specific data can then be inserted and a total set of specifications printed.

A five-digit numbering system is used to identify each specification heading. The first two digits represent the major division. The last three digits represent sections within each division. Sections are further divided into detail sections using a progression of the last three digits. In using this system only the headings that apply to each project are listed. If an entire division is not used, it is still included, but is listed with a note: "no listing in this division."

DIVISION	SECTION		DETAIL	
1. General requirements				
2. Site work				
3. Concrete	06100 Rough carpentry		06131	Timber trusses
4. Masonry	06130 Heavy timber		06132	Mill-framed structures
5. Metals	06150 Trestles		06133	Pole construction
6. Wood and plastics	06170 Prefab. structural wood			
7. Thermal protection	06200 Finish carpentry			
8. Doors and windows	06300 Wood treatment			
9. Finishes	06400 Architectural woodworks			
10. Specialities	06500 Prefab. structural plastics			
11. Equipment	06600 Plastics fabrications			
12. Furnishings				
13. Special construction				
14. Conveying systems				
15. Mechanical				
16. Electrical				

FIGURE 35.18 > CSI specification divisions with a sample section and detail. *Hepler/ Wallach/Hepler © Cengage 2013*

A complete set of residential specifications may cover 20 or more pages. Very brief specifications must be offset by including a great deal of detail in drawings and schedules. Specifications for large commercial or industrial designs may require hundreds of pages.

Reference Sources

Most specifications are developed using computer libraries of descriptions, material data, and cost information. This database must be updated with each project. Information to be included in sets of specifications and schedules can be found in manufacturers' literature, in trade association publications, or through sources such as **Sweets Catalog Files.** "Sweets files" are a compilation of information on a wide variety of architectural products from most major manufacturers. These files are organized under categories used by designers in all areas of construction. Example areas include general building, engineering, interiors, home building, industrial building, electrical, kitchen, and bath design.

Accessing Resource Information on CAD

Manufacturers' websites contain valuable design information that can shorten and improve the design and drawing process. This information takes the form of drawings and data.

DWG file drawings, when downloaded, can be altered to fit specific design needs. PDF or JPEG files cannot be modified and usually must be used exactly as received. However, JPEG files can be imported as stand-alone images such as details or pictures and used as guides similar to underlays. Using AutoCAD® a drafter can then "draw over" the JPEG to create a new drawing. This is typically done when an architect has hand-drawn a design that is scanned and imported into AutoCAD® as a JPEG. This version is then "drawn over" to create a computer-generated drawing. The JPEG image can then be deleted.

Manufacturers websites are also good sources for detailed product specification options such as size, color, configuration, and materials.

A CD-ROM product known as *Sweet-Source®* is an electronic supplement to Sweets files. This product compresses 250,000 pages of product information onto a single compact disc. The disc contains high-resolution color images and CAD interfacing capabilities. The Sweets System also provides an electronic information link between design professionals and manufacturers of architectural products. Through this service designers and contractors can obtain specific product specifications, pricing information, product availability status, CAD details, and environmental information.

Schedules and Specifications Exercises

1. What kinds of information are included on a schedule? How is a schedule referenced on a drawing?

2. Prepare a door and window schedule for the plan shown in Figure 13.51.

3. Name four uses for specifications.

4. Explain the differences between material lists, schedules, and specifications.

5. What are the three categories used in the CSI format? How would details be added to a list?

6. What division and section are represented by the CSI code 06133?

7. Name three reference sources for information that is included on specifications and schedules.

8. Complete schedules for the home of your design. Include as much manufacturing information as you can obtain.

9. Make a specifications list for the home of your own design.

10. Prepare a schedule for the furniture in Figure 35.19.

11. Prepare a collage board for the finishes and materials in Figure 35.19.

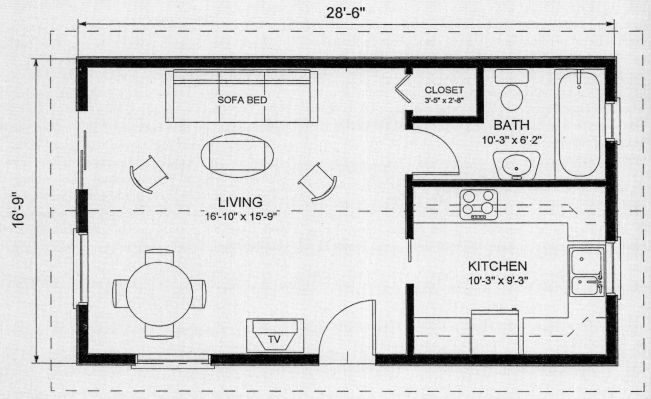

FIGURE 35.19 > Prepare a collage board and furniture schedule for this plan. *Hepler Associates, LA PC*

Building Costs and Financial Planning

OBJECTIVES

In this chapter you will learn to:

> estimate building costs by the square-foot method.

> estimate building costs by cubic volume.

> make up a home budget.

> calculate monthly payments.

TERMS

closing costs
cubic-foot method
down payment
escrow accounts
foreclose

interest
labor-and-materials method
labor costs
materials costs
mortgage

principal
rough estimates
soft costs
square-foot method

INTRODUCTION

The total cost of building and financing a new structure must be estimated before construction contracts are signed. Budgets are the framework within which designers must plan most buildings. A designer must work to create architectural plans that will provide the best facilities for the finances available.

BUILDING COSTS

Many factors affect building costs. The costs of real estate, labor, materials, and financing influence the total cost of any building.

The location of the site is extremely important. An identical house built on an identical lot can vary thousands of dollars in cost, depending on whether it is located in a city, in a suburb, or in the country. Labor costs also vary greatly from one part of the country to another. Material costs vary depending on the quality, unit cost, and quantity of materials used. These costs also depend on whether the materials are available locally or must be shipped in.

Rough estimates are usually developed using either the square-foot or cubic-foot method. When these methods are used, the costs of unique materials or components are added. Adjustments in cost may be required, such as inflationary

costs and labor and material cost differences because of the geographical location.

Cost Estimating Methods

Different types of cost estimates are developed during various design phases. Very rough estimates are needed during the conceptual design phase. Very accurate and precise estimates are required on completion of working drawings. Accurate estimates are also needed during the bidding process.

Three basic methods of estimating building costs are the square-foot method, the cubic-foot method, and the labor-and-materials method. The square-foot and cubic-foot methods are quick rule-of-thumb methods. These two methods are not as accurate as itemizing the costs of labor and materials. The cost of the property, landscaping, drive, walkways, special features such as pools, and fees must be added to any of these estimates to arrive at a total cost.

Square-Foot Method

The **square-foot method** simply involves multiplying the square footage of a structure (sq. ft. = W × L) by a cost per square foot figure. The averages of square-foot costs according to building type are published for each geographical area.

SQUARE-FOOT METHOD

Building Dimensions: W = 30′   L = 40′   H = 12′
Construction costs: $100 per square foot
Square footage: 30′ × 40′ = 1,200 square feet
Cost: 1,200 × $100 = $120,000

CUBIC-FOOT METHOD

Building Dimensions: W = 30′   L = 40′   H = 12′
Construction cost: $8.33 per cubic foot
Cubic volume: floor area (square feet) × height
Cubic volume: 1,200 × 12 = 14,400 cubic feet
Total cost: cubic volume × cost per cubic foot
Total costs: 14,400 × $8.33 = $119,952 = $120,000
(round off for estimate)

FIGURE 36.1 > Square-foot and cubic-foot methods of estimating building costs. *Hepler/Wallach/ Hepler © Cengage 2013*

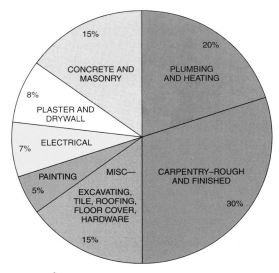

FIGURE 36.2 > Breakdown of construction labor and material costs. *Hepler/Wallach/Hepler © Cengage 2013*

If the cost of a building type is $200 per square foot, then the cost of a 3,000-sq.-ft. building would be $600,000 ($200 × 3,000). Figure 36.1 shows the calculation of building costs for a 30′ ×40′ cabin with 12′ ceilings using both the square-foot and cubic-foot method.

If the building shape is not a square or rectangle, divide the space into rectangles, compute the square feet in each, and then add them to arrive at the total square feet.

Cubic-Foot Method

The **cubic-foot method** adds a height dimension factor to the square-foot method (sq. ft. = W × L × H). The height of ceilings is included in the computations. In buildings with average ceiling heights, cost estimates made by the two methods will vary little. See Figure 36.1. Much variation will occur where ceiling heights are high or rooms are double level. In such cases, the estimate prepared using the cubic-foot method will be much more precise. The cubic-foot method is especially helpful if different ceiling heights are used within a structure. The cubic footage of each area with the same ceiling height must first be calculated separately. These figures are then added to arrive at the total cubic feet. This method is particularly useful when estimating the costs of multilevel structures with various ceiling heights.

Labor-and-Materials Method

The **labor-and-materials method** adds the total cost of all materials to the estimated cost of labor to arrive at a total cost estimate. Approximately 40% of the cost of a residence is for materials. Labor costs account for another 40%. The remainder is used for real estate purchase, site development, and

fees. These building costs are broken down into construction divisions. See Figure 36.2.

Labor costs are determined by multiplying the estimated number of hours needed to build a structure by the hourly rate. Hourly rates by trade are published for each geographical region. The number of hours needed by each trade for different building types and sizes is also available. Labor costs also vary greatly from one part of the country to another and from urban to rural areas. The amount of customized construction, such as nonmodular or unique component construction, affects the labor cost of a building most dramatically. Not only does customizing take additional time, but it requires the services of more highly paid technicians.

Building **material costs** vary greatly among regions. In some areas, brick is a relatively inexpensive building material. In other parts of the country, a brick home may be one of the most expensive. Climate also has an effect on the cost of building. In moderate climates, many costs are eliminated because large heating plants and frost-deep foundations are not needed. In other climates, maximum installation of heating and/or cooling equipment is needed.

One of the greatest variables affecting the cost of building materials is the use of standard sizes versus custom construction of components. For example, if built-in components are constructed on the site, the cost could easily double the price of preconstructed factory units. Designing the residence in modular units also enables the builder to use standard sizes of framing materials with a minimum of waste. A direct relationship exists between the amount of on-site construction and the cost of factory-built modular components.

Specifying unique or exotic materials, such as rare stone or paneling, adds considerably to the cost. Material lists (take-offs) are used to estimate the cost of material. The totals for each category are combined with the labor costs to arrive at

COST BY CATEGORY			PER SQ. FT. OF LIVING AREA		
			MAT.	LABOR	TOTAL
1	Site Work	Excavation for the slab and footings.		.49	.49
2	Foundation	Main house – 8″ and 12″ wide reinforced concrete stem wall on 16″ × 10″ reinforced concrete perimeter footings. Trench footings – 8″ wide reinforced concrete. Slabs – 4″ thick steel trowel finished reinforced concrete over compacted gravel.	4.78	5.58	10.36
3	Framing	Exterior walls – 2 × 6 studs, 16″ on center with ½″ plywood sheathing. Floor – 2 × 10 joists, 16″ on center with ¾″ plywood sheathing. Garage – 2 × 4 studs, 16″ on center with ½″ plywood sheathing. Roof – pre-engineered trusses and site cut rafters with ⅝″ plywood sheathing.	9.27	10.21	19.48
4	Ext. Walls	Stucco veneer siding over 15# felt building paper with R-19 and R-11 insulation. Wood awning, double-hung and fixed windows, and sliding glass patio doors.	6.72	2.46	9.18
5	Roofing	Field tile shingles over 30# felt roofing paper. Aluminum gutters, downspouts, drip edge, and flashings.	3.29	3.05	6.34
6	Interiors	Walls and ceilings – ½″ and ⅝″ taped and finished gypsum wallboard, primed and painted with one coat latex. Finger-jointed interior trim with one coat paint. Flooring – 68% carpet, 16% vinyl, and 16% ceramic tile.	9.94	10.36	20.30
7	Specialties	Plastic laminated particle board case kitchen cabinets and bathroom vanities with plastic laminate countertops. Washer, dryer, cooktop with hood, dishwasher, and refrigerator. One masonry fireplace.	3.43	.70	4.13
8	Mechanical	Oil-fired forced hot-air heat with central air conditioning. One full bath, one ½ bath, and a master suite with whirlpool, shower, and an outdoor spa. Stainless steel kitchen sink with disposal.	4.43	2.47	6.90
9	Electrical	200-amp service, branch circuit wiring with romex cable. Exterior and interior lighting fixtures, receptacles, and switches.	.88	1.53	2.41
10	Overhead Architect's Fees and Profit	Contractor's overhead and profit.	10.26	8.85	19.11
		Total Cost/sq.ft.	53.00	45.70	**98.70**

FIGURE 36.3 ⟩ Breakdown of labor and material costs by category. *Reed Construction Data, a division of Reed Elsevier, Inc.*

the cost per square foot. See Figure 36.3. The cost of some materials and components includes labor costs. Therefore, care must be taken to ensure labor costs are not duplicated in the final sum. The Federal Housing Administration and other funding agencies provide an itemized list of construction categories to be used for estimating. See Figure 36.4. These lists contain hundreds of line items that describe all of the materials, finishes, fixtures, and components needed for the projects, including the cost of each. A typical material estimating list is organized under the categories shown later in Figure 36.4.

Database Estimating

Estimating costs with computers is possible using estimating software that stores prices and descriptions of thousands of building materials and components. These figures can be combined with wage rates for each trade within each zip code area. Other factors such as building size, shape, and type can then be used to compute an estimate for the entire project. On CAD-generated drawings, precise takeoff cost estimates can easily be completed by scanning directly from the drawing into the computer database. Using 4D Building Information Modeling (BIM) software, cost management data and

TYPICAL ESTIMATING CATEGORIES

- General Information
  - Building Areas
  - Site Related Data
- Masonry
- Lumber
- Millwork
- Cabinetry & Millwork
- Drywall—Plaster
- Finish Flooring by Room
- Painting and Finishing
- Miscellaneous Hardware

- Finish Hardware
- Sheet Metal Work
- Floor Finishing Material
- Wall Finishing Materials
- Plumbing
- Electrical Wiring
- Phone Wiring
- Computer and Media Wiring
- HVAC

FIGURE 36.4 > Typical material estimating list categories.
Hepler/Wallach/Hepler © Cengage 2013

accurate cost estimates can be directly generated from an original BIM design model.

Minimizing Costs

The following construction methods and material utilization can greatly reduce the ultimate cost of a home:

1. Square or rectangular homes are less expensive to build than irregular-shaped homes.

2. Building on a flat lot is less expensive than building on a sloping or hillside lot.

3. Using locally manufactured or produced materials cuts costs greatly.

4. Using standard stock materials and standard sizes of components takes advantage of mass-production cost reductions.

5. Using materials that can be installed quickly cuts labor costs. Prefabricating large sections or panels saves time on the site.

6. Using prefinished materials saves labor costs.

7. Using prehung doors cuts considerable time from on-site construction.

8. Designing the home with a minimum amount of hall space increases the usable square footage and provides more living space for the cost.

9. Using prefabricated fireboxes for fireplaces cuts installation costs.

10. Investigating existing building codes before beginning construction eliminates unnecessary changes as construction proceeds.

11. Refraining from changing the design or any aspect of the plans after construction begins will keep costs from increasing.

12. Minimizing special jobs or custom-built items keeps costs from increasing.

13. Designing the house for short plumbing lines saves on labor and materials.

14. Using proper insulation will save heating and cooling costs.

15. Using passive solar features such as correct orientation reduces future utility costs.

FINANCIAL PLANNING

Few people can accumulate enough money to pay for a home in one installment. Therefore, most home buyers pay a percentage of the cost of a home at the time of purchase. This initial or first payment is referred to as a **down payment** and is typically 5% to 20% of the cost of a home. The balance of the cost is usually acquired through a loan. After analyzing the construction costs for a home, the availability and costs of financing must be established. This includes information concerning mortgages, interest, taxes, insurance, and fees. Then a builder or owner can determine the financial feasibility of a building project.

Mortgage

A **mortgage** is an agreement (contract) for a loan. The purpose of a *mortgage loan* is to finance the purchase of a parcel of real estate, including any structures located on the property. Home mortgages usually require the loan amount, or **principal,** to be repaid over a period ranging from 15 to 30 years. A mortgage can be obtained from different sources: a mortgage company, a bank, a savings and loan association, or the seller of the property. Mortgage agreements are complex contracts that require the assistance of an attorney. Most mortgages allow the buyer (*mortgagor*) to pledge the property to the institution or individual providing the loan (*mortgagee* or *mortgage holder*). This means that if the loan payments are not made as agreed, the mortgage holder has the right to **foreclose,** or take possession of the property.

Interest

In addition to paying back the exact amount of the loan, the buyer must also pay additional amounts over the life of the loan for the use of the borrowed money. These charges are known as **interest.** The amount of interest paid is a percentage of the loan outstanding at a given point in time. Interest rates (percentages) depend on many factors, such as the availability of money and risk factors associated with the mortgagee and/or property. Interest rates can be established to remain unchanged over the life of the mortgage. These are referred to as *fixed rates.*

Interest rates can change, based on various fluctuations in lending institution rates. These rates, which change with time, are referred to as *adjustable, variable,* or *floating rates.* During the past several decades, the typical fixed interest rates for home mortgages have ranged from 5% to 15%. Interest rates can be a key factor in the ability to purchase a home.

The monthly payments and total amounts paid over the life of a loan differ depending on interest rates. Figure 36.5 shows how much a buyer would pay for a $200,000 home, assuming a 10% down payment with a $180,000 mortgage, based on various interest rates. As this table demonstrates, interest rates can have a significant impact on both the monthly mortgage payments and the total amounts paid over the life of the mortgage. Looking at the examples for a 30-year mortgage, monthly payments jump from $1079 at 6% interest rate to $2133 at 14%. This represents a 98% increase. The higher payments can easily disqualify many potential buyers who might be able to afford monthly payments at lower interest rates. In the same example, total payments over the life of the loan climb from $388,509 at 6% to $767,797 at 14%. A higher down payment will reduce monthly payments and total payments over the life of the loan. For typical down payments of 20% or less, interest rates have the greatest influence on the ability to afford a home.

Taxes

Owners of property must also pay property taxes imposed by the various local governments. Property tax rates are usually expressed at a rate per $100 of assessed value of the property (*mil rate*). Rates vary greatly from location to location and depend on numerous factors. Property taxes can add a significant annual cost to property ownership. Residential property taxes are often several thousand dollars or more a year. For this reason, many mortgage holders collect property taxes on a regular monthly basis and hold these funds in special accounts called **escrow accounts.** This ensures that tax payments are made on a timely basis.

Insurance

The purchase of a home is a large investment and it should be insured for the protection of the home buyer and the mortgagee. Most mortgage agreements require that homeowner's insurance be maintained. Like property taxes, many mortgage holders collect insurance on a monthly basis, hold it in escrow accounts, and make payments directly to the insurer.

LOAN DURATION	MONTHLY PAYMENT	TOTAL PAYMENTS
Interest rate of 6%		
15 years	1,519	273,410
30 years	1,079	388,509
Interest rate of 8%		
15 years	1,720	309,631
30 years	1,321	475,479
Interest rate of 10%		
15 years	1,934	348,172
30 years	1,580	568,666
Interest rate of 12%		
15 years	2,160	388,854
30 years	1,852	666,541
Interest rate of 14%		
15 years	2,397	431,484
30 years	2,133	767,797

FIGURE 36.5 > Mortgage costs at different interest rates for a $180,000 mortgage. *Hepler/Wallach/Hepler* © Cengage 2013

Insurance rates vary greatly depending on the cost of a home, location, type of construction, proximity to potential natural disasters, and the availability of firefighting equipment. Homes should be insured against fire, public liability, property damage, vandalism, natural destruction, and accidents to trespassers and workers.

Institutional lenders may also require insurance on their behalf, in the event a borrower is unable to pay a loan under the terms of the mortgage. This insurance is called *private mortgage insurance* (PMI). PMI is usually required when a home buyer makes less than a 20% down payment. Rates vary depending on the amount of the down payment, location of the property, and credit standing of the borrower.

Soft Costs and Closing Costs

The purchase of a home requires the buyer to incur many costs other than the actual cost of the property. **Soft costs** is a term used to describe the costs incurred prior to the actual construction of a home or other structure. Soft costs may include permits, surveyor's fees, soil test fees, engineering studies, legal fees, and architectural fees. Architects usually work on an 8% commission for their design work and may charge fees as high as 10% of the project cost if they also supervise construction.

The term **closing costs** refers to costs paid by the purchaser before taking possession of a property. Closing costs do not include the actual cost of the property or buildings. Closing costs typically include legal fees, transfer taxes, title insurance, settlement of taxes and insurance with the seller, and financing fees.

Acquiring a mortgage may require the buyer to pay a fee up front to the mortgagee referred to as *points*. Points are a percentage of the mortgage amount, usually between 1% and 2%.

Budgets and Financing Qualifications

The ability to purchase a residence is dependent on the ability to acquire the down payment and also pay the soft costs and closing costs. The ability to make the required ongoing payments once a home is purchased is also vital.

Consider the purchase of a $200,000 home by a potential buyer who will make a 10% down payment as shown in Figure 36.6. The down payment in this case is 10% of $200,000, or $20,000. Assume there will be no soft costs. The closing costs are estimated at 2% of the purchase price ($4,000). Points may be 1% to 2% of the mortgage amount ($1,800 to $3,600). Finally, the lending institution will usually require the buyer to have money saved in an amount equal to two or three months of mortgage payments, taxes, and interest. Assuming a 10% interest rate on the mortgage, the buyer would need to have roughly $4,000 to $6,000 of savings after the home was purchased. Approximately $33,600 would be needed by the buyer to complete the purchase.

After gathering the necessary initial funds, potential home buyers must also determine their ability to meet the ongoing payments associated with owning the property. The monthly expenses associated with the home purchase include principal and interest of the monthly mortgage payment, plus property taxes and insurance (homeowner's insurance and, if applicable, PMI). Together, these items are referred to as "PITI" for principal, interest, taxes, and insurance. Most institutional lenders require that PITI not exceed 28% of

the monthly household income. They will also require that PITI plus other monthly loan payments not exceed 36% of the household income. Figure 36.7 shows the breakdown of monthly PITI expenses based on a $180,000 mortgage over 30 years at an interest rate of 10%. All lending institutions require documentation to ensure that the proposed building is structurally viable, legally sound, and financially secure. A typical list of institutional requirements needed for construction approval is shown in Figure 36.8.

Principal and interest (1)	$1,580
Taxes (2)	250
Insurance (3)	50
Total monthly PITI	$1,880
Minimum required monthly income (4)	$6,714

1) Monthly principal and interest on a $180,000 mortgage over 30 years at an interest rate of 10%.
2) Monthly tax escrow assuming a mil rate of $1.50 and an assessed value of $200,000.
3) Monthly homeowners insurance escrow assuming an annual premium of $600.
4) Monthly PITI cannot exceed 28% of monthly income.

FIGURE 36.7 > Monthly PITI requirements. *Hepler/Wallach/Hepler © Cengage 2013*

LENDING INSTITUTION APPROVAL REQUIREMENTS

- Application Form
- Complete Set of Architectural Drawings
  - Types of Drawings Required
  - Number of Sets Required
- Total Project Cost
- Cost Breakdown of Labor and Materials by Sub—Category
- Signed Contract Between Contractor and Client
- Signed Schedule and Specification Forms
- Legal Property Description from Deed
- Initial Fees Paid
- Contractor Firm Information
  - Resume of Principle and Firm
  - Contractors License
  - History of Building Projects
  - Financial Statement
  - Overhead Expense for this Project
  - Profit Projected for this Project

Down Payment
10% or $200,000 of the purchase price	$20,000

Closing Costs
| 2% of purchase price | 4,000 |

Points
| 2% of mortgage amount (mortgage amount = $180,000) | 3,600 |

Required Savings after Purchase
| Estimate of 3 months mortgage payments, plus taxes and insurance | 6,000 |

Total Accumulated Funds Needed | $33,600 |

FIGURE 36.6 > Funds needed to purchase a $200,000 house. *Hepler/Wallach/Hepler © Cengage 2013*

FIGURE 36.8 > Lending institution approval requirements. *Hepler/Wallach/Hepler © Cengage 2013*

Building Costs and Financial Planning Exercises

CHAPTER 36

1. At $275 a square foot, how much will a 30′ × 40′ home cost?

2. At $25 per cubic foot, how much will a one-level, flat-roof 30′ × 40′ × 12′ home cost?

3. Find the cost per square foot or cubic foot of a building in your area. Then compute the cost of the house you designed, based on this cost.

4. Refer to the numbers shown in Figure 36.7 and assume the potential buyer earns the minimum monthly income stated. How much could this potential buyer have in other monthly loan payments and still qualify for a mortgage?

5. What will be your total monthly PITI on a $300,000 home if you make a 10% down payment and pay interest of 10% on a 30-year mortgage for the balance? Your property taxes are $2,000 and your homeowner's insurance is $450.

6. What are property tax rates in your area? How did you find this information?

7. What is the single most significant factor affecting the ability to afford a home? Why?

8. What would your monthly mortgage payment be if you bought a 30′ × 40′ house for $400 per square foot? Your interest is 7% for 25 years. You make a down payment of 9% of the total price. Your closing costs total $2,000.

9. What is the cost of a home you could afford in your area if you were earning $50,000 a year? $90,000 a year? $150,000 a year?

10. At $450 per square foot, what is the cost of the house in Figure 36.9?

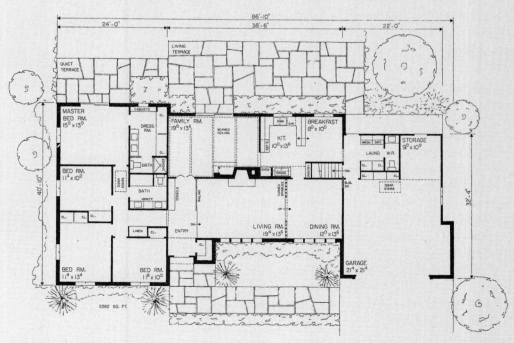

FIGURE 36.9 > Calculate the cost of this house at $450 per square foot. *Hepler Associates, LA PC*

Codes and Legal Documents

OBJECTIVES

In this chapter you will learn to:

> consider building codes in architectural design.

> determine legal documents needed for building construction.

TERMS

agreements
bid forms
bids
bond
certificate of occupancy (CO)

codes
contracts
deed
environmental codes
lien

model codes
restrictive codes
structural codes
treatment of materials
zoning codes

INTRODUCTION

Different types of building codes are used to control the design and construction of buildings. Governmental laws are imposed on the design and construction of a building through the use of codes. Legal documents are created to protect the architect, builder, client, and the general public.

BUILDING CODES

Governmental laws that regulate building construction have existed for hundreds of years. Building codes are collections of laws (codified) that ensure that minimum building standards are met. These laws are enacted to safeguard life, health, property, and the public welfare. The enforcement of codes has reduced the loss of life and property from earthquakes, storms, fires, and floods.

Many aspects of building construction are regulated and controlled by building codes. See Figure 37.1. Building codes include information and ordinances related to building permits, fees, inspections, zoning, drawings, and legal documents required for approval. Building codes are presented through printed materials, charts, detailed drawings, and specification lists. These laws must be conformed to before beginning to design any structure.

Types of Codes

Several types of **codes** are used to control building construction. These include zoning regulations, structural codes, site-related codes, restrictive codes, model codes, health and safety codes, and environmental codes.

Zoning Ordinances

Zoning codes define and restrict the occupancy and use of buildings. Zoning laws may also prescribe the type, style, and location of structures on a site. (Refer back to Chapter 13 for a more detailed coverage of zoning.)

Typical zoning categories controlled by codes include:

- District classification: residential (single family, duplex, apartment), commercial, industrial
- Minimum lot size
- Minimum lot frontage
- Minimum setbacks
- Percentage of lot coverage
- Maximum height of building
- Accessory structures
- Driveways

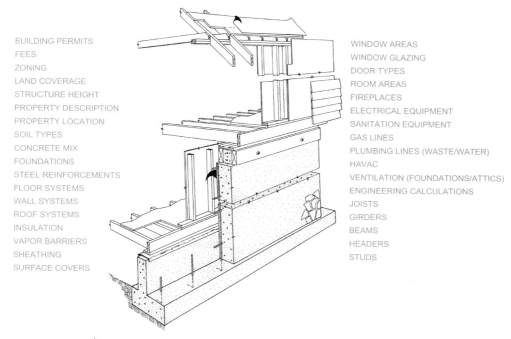

BUILDING PERMITS
FEES
ZONING
LAND COVERAGE
STRUCTURE HEIGHT
PROPERTY DESCRIPTION
PROPERTY LOCATION
SOIL TYPES
CONCRETE MIX
FOUNDATIONS
STEEL REINFORCEMENTS
FLOOR SYSTEMS
WALL SYSTEMS
ROOF SYSTEMS
INSULATION
VAPOR BARRIERS
SHEATHING
SURFACE COVERS

WINDOW AREAS
WINDOW GLAZING
DOOR TYPES
ROOM AREAS
FIREPLACES
ELECTRICAL EQUIPMENT
SANITATION EQUIPMENT
GAS LINES
PLUMBING LINES (WASTE/WATER)
HAVAC
VENTILATION (FOUNDATIONS/ATTICS)
ENGINEERING CALCULATIONS
JOISTS
GIRDERS
BEAMS
HEADERS
STUDS

FIGURE 37.1 > Common construction components controlled by codes. *Hepler/Wallach/ Hepler © Cengage 2013*

- Remodeling conditions
- Architectural style
- Daylight planes.

Structural Codes

Structural codes deal with the loading capacity of materials and the structural integrity of the construction. These codes involve detailed regulations related to excavations, foundations, floors, roofs, stairs, and bearing-wall construction. There are two general types of structural codes: performance criteria codes and specification criteria codes.

Performance-oriented codes do not limit or specify the use of most construction materials or methods. They establish only safety and performance requirements for the finished building.

Specification-type codes include very specific requirements for the use and location of materials and methods of construction. Options are included in many codes for some materials, but any substitutions must be equal to or better than what is specified by the code. Most building codes are specification-type codes. They are easier to enforce, and they provide a way to measure compliance and make evaluations.

The maximum permissible loads for each type of structure are listed in structural building codes. The required sizes of structural members to support various loads are also included. Calculations included in a set of permitted drawings must be made by a licensed architect or engineer.

When material and size regulations are compiled for building codes, they are computed on the basis of maximum allowable loads. Engineers determine the correct size of construction members for supporting maximum loads. A safety factor is then added to the size of each material to eliminate the possibility of building failure.

Structural sizes required by building codes not only provide for the support of all weight in a vertical direction, but also allow for all possible horizontal loads, such as high winds and earthquakes. Codes include precalculated sizes and types of materials used for studs, sheathing, roofing, foundations, and footings. Codes also include spans and spacing required for normal loads on rafters, joist beams, girders, and lintels.

For unique building sizes or designs, there are no predetermined code entries. In these cases, materials, sizes, and spans must be calculated by a licensed architect or engineer. See Appendix B, *Mathematical Calculations*.

Site-Related Codes

Many code items relate to the building site. These include specifications on soil percolation (drainage), soil support capabilities, test boring, and water runoff. Others cover environmental topics such as endangered species habitats, wetlands protection, and zoning density. (See Chapter 6 on environmental codes.)

The location and size of buildings, setbacks, driveways, and road right-of-ways are also covered in these codes. When these items are not covered, a licensed landscape architect or civil engineer must approve the design.

Restrictive Codes

Some code items specify materials, processes, sizes, and locations of building materials within the structure that are prohibited. These types of **restrictive codes** are imposed because of potential structural or environmental safety problems.

Model Codes

To provide local authorities with a consistent and current source of code information, several organizations prepare **model codes.** These codes are not intended to be adopted intact. They are designed to be used as a base or guide from which local codes can be developed. The most widely used national codes are the National Building Code and the National Electrical Code. Specialized codes are also available for plumbing, HVAC, and related trades. None of these codes is legally binding until passed into law by a municipality.

🌐 Health and Safety Codes

In addition to the safety prevention implications of good structural, mechanical, and electrical design, codes also contain sections dealing specifically with personal and public safety. These cover such areas as electrical hazards, swimming pool enclosures, scaffolds, elevators, number and sizes of exits, air and water pollution, health and disease prevention, and fire prevention and control. Fire codes define **treatment of materials,** building material size, sprinkler systems, escape routes, site security, and the functioning of alarm systems.

Building codes are extremely strict in stipulating the location, traffic patterns, and structural quality of buildings designed for public occupancy. Rigid code controls are placed on facilities such as factories and garages that may contain flammable substances or that may emit pollutants into the atmosphere. Structures in this risk category must also adhere to Environmental Protection Agency standards in addition to local codes.

🌐 Environmental Codes

Environmental codes are an expansion of health and safety codes. These "green" codes specify the materials that can or cannot be used on a project. They also require high levels of energy efficiency, reduction of water usage, and control of contaminants. In addition to protecting the individual, these codes are written to protect and sustain all aspects of the total environment.

Conforming to environmental codes is not always a yes or no matter—there are many levels of compliance. Determining the degree to which a design, and ultimately a structure, adheres to environmental codes requires a description of the levels of conformity. The *Leadership in Energy and Environmental Design* (LEED) organization provides a rating service

to evaluate these levels. LEED is the only nationally recognized green building rating system for evaluating the performance of buildings from a "whole building" perspective. It provides designers and project developers with a checklist of criteria by which a building and site can be judged in six categories: sustainable sites, water efficiency, energy and atmosphere, materials and resources, indoor environmental quality, and innovation and design. The largest point-generating category is the energy and atmosphere category. There are four possible LEED ratings: Certified, Silver, Gold, and Platinum. A LEED Gold-rated building is estimated to have reduced its environmental impact by 50% compared with an equivalent conventional building.

The measures required by these codes are covered in Chapters 6 and 27 and are included in the applications shown in the framing chapters.

Code Compliance

Codes are enforced through a series of legal controls. *Building permits* are issued after working drawings and specifications have been approved by a municipal building department. Then a *notice of commencement* must be completed when construction begins.

During the building process inspections are made of the setbacks, the foundation, electrical system, HVAC, plumbing, framing work, and insulation materials. After a final inspection, including a review of all change orders, a **certificate of occupancy (CO)** is issued. This certificate allows the building to be occupied.

Most sets of plans must contain a licensed architect's or engineer's seal. On larger projects, the seal of a licensed landscape architect may also be required on all site design drawings. All plot plans or surveys registered with the local municipality must also contain the seal of a licensed surveyor. Licensed subcontractors are specified and are required by code on most construction jobs. Licensing requirements are usually specified in construction contracts.

Downloading Codes

Specific portions of most municipal codes can be downloaded online by architects, builders, or owner-clients. Detail drawings showing minimum structural requirements can often be downloaded and incorporated into a CAD-prepared set of plans. Many municipalities store their codes, including building, zoning, planning, and land-use regulations, on a website. Downloading access to these codes is normally restricted to licensed professionals who work in these areas. Master contract documents relating to the bidding of projects can also be downloaded by approved licensed professionals.

LEGAL DOCUMENTS

A number of legal documents are used to protect the property and rights of architects, builders, and clients. Unlike codes, which apply to all projects within a given area, these legal documents are created individually for each building project. Documents include contracts, bonds, deeds, liens, and bids. Working drawings and specifications also become legal documents when tied to a contract.

Contracts

Legal agreements between two or more parties are known as **contracts** or **agreements.** Agreements are made between architects, builders, and owners. Architects may contract directly with an owner and/or with a builder. When an architect contracts only with an owner, the role of the architect and the builder must then be separately defined before the building process begins. Many architects supervise or monitor construction for a percentage of the construction cost.

Architects further contract with specialized consultants such as landscape architects, structural and mechanical engineers, or interior designers. General contractors and builders sign separate agreements with subcontractors. All of these documents must be consistent with the contents of the architect, owner, or builder contract. Agreements among architects, owners, and builders must also be consistent with the codes in force for the project at the time the contract is signed.

Contracts include many conditions besides costs, such as schedules, bonds, and the responsibilities of each party. Restrictions on each party, criteria for handling changes, warranties, acts of God (unforeseeable events such as tornadoes), and cancellation conditions are also described in detail. In addition, contracts include very specific references to working drawings and specifications by citing the title, sheet numbers, and latest revision dates of each.

Bonds

A **bond** is a binding agreement that ensures that obligations are met. Two types of bonds used in construction are performance bonds and labor-and-materials bonds.

A *performance bond* is offered by the contractor and guarantees that the performance of responsibilities as builder will be in accordance with the conditions of the contract.

A *labor-and-materials bond* posted by the contractor guarantees that invoices for materials, supplies, and services of subcontractors will be paid for by the prime contractor, according to the terms of the contract.

Deeds

A **deed** is a legal certificate of property ownership. Building code requirements are sometimes repeated in the contents of the property deed. These restrictions often describe the minimum setbacks, utility easements, building areas, and building types that are permitted. When building codes are updated, deeds are not necessarily changed. For this reason, many deeds may contain requirements that are no longer in existence. Designers must therefore check deeds for any obsolete restrictions and have the deed updated before proceeding with the design process.

Liens

A legal document used to take or hold the property of a debtor is known as a **lien.** Any architect, builder, subcontractor, consultant, or material supplier can petition the court to place a lien on real property for nonpayment of fees or material costs. A certificate of occupancy or bill of sale cannot be issued while a lien is in force.

Construction Bids

Bids are legal proposals to construct a project as defined in a contract, in specifications, and in a set of drawings. Contractors receive invitations to bid by mail, through newspaper advertisements, or through private resources.

Construction **bid forms** announce the availability of documents and tell when these documents can be examined. They also provide for the resolution of questions, approval for submission of materials, specific dates for bid submission, and the form for preparing bids. The bid form includes specific instructions to the bidders, the price of the bid, allowable substitutions, restrictions, and the involvement of subcontractors.

The bid form is a letter sent from the owner or architect to bidders. The letter covers the following points: verification of receipt of all drawings and documents, specific length of time the bid will be held open, price quotation for the bid, and a listing of substitute materials or components if any item varies from specified requirements. When bidders sign a bid form, they are also agreeing to abide by all conditions of the bid, including the price, time, quality of work, and materials as specified in the contract documents and drawings.

Codes and Legal Documents Exercises

CHAPTER 37

1. What are building codes and what is their purpose?

2. Describe the two types of structural codes.

3. Determine any code restrictions in building the residence of your design in your community. Make and record needed changes.

4. What is the difference between codes and agreements or contracts?

5. Explain the purpose of bid forms.

6. List the local codes that apply to the house in Figure 37.2.

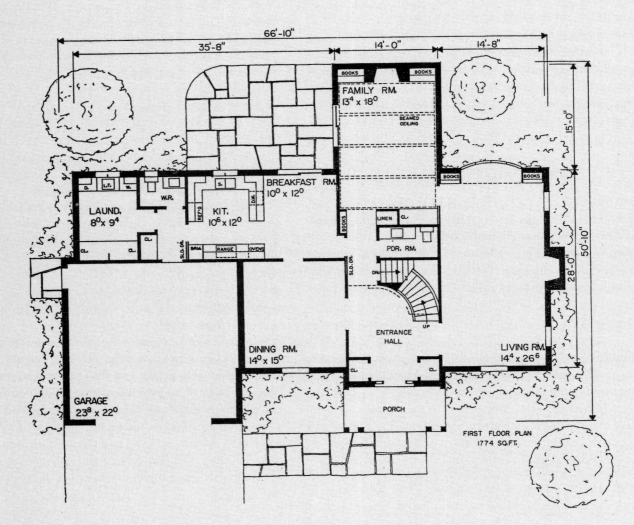

FIGURE 37.2 > List the codes in your area that apply to this plan. *Hepler Associates, LA PC*

APPENDIXES

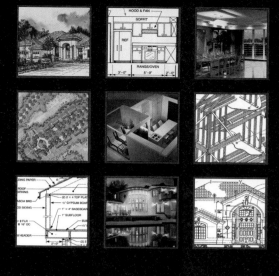

The following appendixes contain information that does not align specifically with any single chapter. The scope of the following appendix material applies, in different degrees, to many subjects covered in this text. This includes information on architectural careers, mathematical calculations, architectural abbreviations, synonyms, and a glossary.

Career opportunities are available in a wide variety of fields related to architecture. These include careers in architectural design, engineering, and construction. Regardless of the specific field you choose, there is great reward in seeing your creations and efforts take form in structures that become part of the physical environment. A brief introduction to the major careers in architectural design, engineering, and construction is presented here.

Additional information concerning educational requirements, licensing, and professional organizations is included on the CD-ROM.

CAREERS IN ARCHITECTURAL DESIGN

Architectural design is a very broad field with many areas of specialization. Specialists contribute in different ways to the creation and completion of an architectural design. In a small firm, many tasks may overlap and be performed by one person—usually the architect. In larger firms a specialist can concentrate on one aspect of architectural design.

Before becoming licensed, professionals must first work under the direction of a licensed professional for several years and pass a state board examination. Once licensed, the professional can design and certify the safety of each design by affixing a license stamp on each approved drawing.

The following information applies to the major careers involved in the architectural design process.

Architects

Architects create original designs and are responsible for preparing and/or supervising the preparation of all construction drawings and documents. An architect is part artist, part engineer, and part executive. A successful architect must be able to design highly functional, aesthetically pleasing structures and also possess skills in problem solving, planning, and management. A talent for creative design and skills in math and science are also vital.

Architects must be state licensed to design buildings and sites that meet established standards of structural, environmental, and health safety. Licensing involves completion of a five-year program in an accredited college of architecture, passing a rigorous licensing examination (AIA), and working under the direction of a licensed architect for several years.

Landscape Architects

Landscape architects design all aspects of a building site, from individual lots to massive building developments. Landscape architecture is a very diverse profession, involving the analyzing, designing, managing, and preserving of land. Landscape architects are involved with environmental hazards, civil engineering, building locations, street layout, earthwork, wildlife management, drainage, and horticulture. They must therefore be talented designers with skills in math and environmental science. Site plans, plot plans, plats, landscape plans, and details are prepared by landscape architects as covered in Chapters 13 and 14.

To become a landscape architect it is necessary to complete a bachelor's program at an accredited college of landscape architecture (ASLA), pass a state licensing examination, and work under a licensed landscape architect for several years.

Naval Architect

A naval architect is a professional engineer and architect who is responsible for the design and construction of ships and boats, including tankers, cargo ships, passenger liners, ferries, warships, submarines, fishing boats, rescue boats, and pleasure craft. A naval architect may also be responsible for the design and construction of marine shore installations.

City Planners

City planners plan for the overall growth and development of cities and communities. This includes planning for transportation, utilities, schools, housing, and shopping centers. City planners complete general master plans. Specific plans are then developed by architects, landscape architects, and engineers. City planners are usually architects or landscape architects who specialize in urban planning and development.

Architectural Designers

Architectural designers design buildings under the directions of an architect but are not licensed. They usually work on residential projects but may design small areas of a large project. Architectural designers must be versed in all aspects of basic architectural design including the use of CAD programs. Architectural designers are trained in colleges or technical institutes offering curricula in architectural and/or construction technology.

Architectural Drafters

Architectural drafters prepare all types of architectural drawings and documents. They draw with instruments and/or with a CAD system. Drafters work from sketches, partial or preliminary drawings, notes, and reference materials supplied by a supervising architect or architectural designer. Drafters are trained in vocational schools, technical schools, or colleges.

Architectural Detailers

Architectural detailers are drafters who specialize in preparing detail drawings such as sectional and framing drawings. They must be familiar with building codes, materials specifications, and construction methods. In small architectural offices the architectural designer, drafter, and detailer may be the same person, often the architect.

CAD Specialists

CAD specialists convert design sketches or manual drawings into CAD-generated drawings. Although architects, engineers, or drafters may prepare drawings directly on a CAD system, a CAD specialist inputs these drawings into a networked system and may further detail their work. A CAD specialist establishes and manages the information system of a firm and provides consistent standards for all drafters. More detailed drawings including 3D models and virtual reality programs are frequently developed by a CAD specialist. CAD specialists come from the ranks of architects, designers, or drafters who choose to specialize in CAD work.

Interior Designers

Interior designers deal with all aspects of a building's interior including wall construction, built-in components, and lighting systems. They also function as *interior decorators* in designing the surface treatments for walls, floors, and ceilings. This includes the design and selection of colors, fabrics, trim and moldings, window treatments, and furniture. Some specialize in designing baths and kitchens, which involves the selection and placement of fixtures, appliances, and cabinets. Interior designers receive their training as part of a college art program or in a professional school of interior design, vocational or technical school, or community college.

Interior Decorators

Interior decorators design the interior surface treatments of walls, floors, and ceilings, including counters and interior window treatments. This involves the selection of color schemes, fabric type and color, and hard surface materials and finishes. Interior decorators may also select shelving, cabinetry, furniture, fixtures, and appliances and lay out the position of each.

Lighting Designers

Lighting designers specialize in planning the illumination of building interiors, exteriors, and total sites. They must be familiar with lighting fixtures and devices and also understand electrical systems. On small projects, architects and/or interior designers design the lighting system; lighting designers are most often employed on large projects. Lighting designers are often interior designers or architectural designers who choose to specialize in lighting.

Architectural Illustrators

Architectural illustrators prepare renderings used for presentations. They need a knowledge of architectural drawings, commercial art techniques, 3D computer modeling, and often photography and photo retouching. Rendering is done manually, with a CAD system or a combination of both. The task of an illustrator is to present an architectural design as realistically as possible. Illustrators are trained as artists who specialize in architectural rendering. They hold two- or four-year art degrees from art institutes.

Architectural Model Makers

Architectural model makers construct scale models of buildings and sites to show the actual appearance of the finished project in three dimensions. Model makers must work from architectural drawings and be familiar with many model materials including wood, plastics, paper, and sheet metal. Model makers are drafters, finish carpenters, or cabinetmakers who specialize in model making.

Architectural Photographers

Architectural photographers specialize in producing photographs of structures or sites that reveal the best features of a design. Architectural photographers usually photograph finished buildings but models are also photographed and the photographs retouched to add realism to the environment. Photographers work closely with illustrators and work for publishers, newspapers, magazines, and real estate agencies. Many of the photographs in this text were prepared by professional photographers.

Architectural Educators

Architectural educators teach a variety of subjects relating to architecture in secondary school technology programs, vocational and technical schools, and community colleges. They also teach in colleges of architecture and engineering. A bachelor's degree and a state teaching certificate are required for teaching technology education. Vocational school teachers must hold a teaching certificate and a license in the subject taught. College professors usually hold a master's degree or doctorate.

RELATED ENGINEERING CAREERS

A wide array of engineering and engineering technology careers are directly related to architectural and construction practices. The number of engineering disciplines involved in an architectural project increases in proportion to the size and complexity of a project. Engineering careers that most directly apply to architecture and construction are introduced here. Education and licensing requirements are included with professional organizations in the accompanying CD-ROM.

Civil Engineers

Civil engineers design and calculate the structural requirements for large-scale projects such as roads, utilities, bridges, airfields, tunnels, harbors, sewage plants, and high-rise structures. Civil engineers also do site engineering and surveying. They must hold a degree in engineering and be licensed as a professional engineer to "stamp" public project drawings.

Structural Engineers

Structural engineers are civil engineers who specialize in designing the structural framework of a building. They calculate and specify the types and sizes of building materials required to insure that structures function safely as designed. Structural engineers must possess a high level of competency in math and science. Education and licensing requirements are the same as those for civil engineers.

Mechanical Engineers

Mechanical engineers design the mechanical components of a building. This includes the use of power and machines such as elevators, conveyer systems, and heating and cooling systems. To design components for public buildings, a degree in engineering and a state license are required.

Heating, Ventilating, and Air-Conditioning (HVAC) Engineers

HVAC engineers are mechanical engineers who specialize in designing HVAC systems. Such systems include oil or gas furnaces, cooling and refrigeration equipment, and air-movement systems. Education and licensing requirements are the same as those for mechanical engineers.

Surveyors

Surveyors prepare drawings that describe the size, position, and topography of a specific land area. These drawings (surveys) show the legal subdivision (plats) of properties. Surveyors also write legal descriptions of properties. This means surveyors must be proficient in geometry and site engineering and be dedicated to a high degree of accuracy in preparing documents. Surveyors are licensed civil engineers or landscape architects, who specialize in surveying.

Electrical Engineers

Electrical engineers design and draw the electrical components of a building and site. This involves preparing electrical floor plans, details, and elevations that conform to electrical building codes. Details include lighting, power distribution, and special needs such as sound systems and computer networking. A licensed electrical engineer must hold a degree in electrical engineering and pass a state licensing board examination.

Acoustical Engineers

Acoustical engineers design shapes, materials, and devices used to control sound in a structure. This is critical in the design of concert halls and auditoriums. Noise suppression in shopping malls, medical facilities, and factories is also a vital part of this discipline. Acoustical engineers need a degree in electrical engineering with emphasis on physics, math, and architecture.

Structural Drafters

Structural drafters are architectural drafters who specialize in steel-cage construction. They create the structural drawings of a building under the direction of an architect or civil engineer. Structural drafters usually prepare drawings of large structural steel buildings. A knowledge of steel structural members, fastening devices, and building methods is essential.

Estimators

Estimators compute the cost of a construction project by determining the amount and cost of all materials and labor based on a set of drawings and specifications. Estimators prepare *take-offs*, which are compiled from the estimating information "taken off" the drawings and specifications. Estimators must have good math skills and be dedicated to accuracy in their work. Estimators should have an associate degree in construction technology and several years of experience in construction.

Specification Writers

Specification writers analyze construction drawings and documents and prepare written descriptions (specifications) of the materials, construction methods, finishes, and performances required in the building of a structure. Standards outlined by the Construction Specifications Institute (CSI) must be followed. A knowledge of construction materials and processes, good analytical skills, and experience in construction are necessary. A degree in construction technology is recommended.

Military Engineers

Military engineers perform the same functions as their counterparts in the civilian construction world when not in a combat zone. Military schools teach the basics in all areas of construction in addition to combat engineering operations. During combat operations military engineers design, build, and maintain fortifications, roads, and structures. They also provide engineering intelligence. Military engineers serve in the Army Corps of Engineers, the marine engineers, or navy construction battalions.

RELATED CONSTRUCTION CAREERS

Personnel on a construction site range from supervisors to highly skilled technicians to unskilled laborers. All but the unskilled laborers work with architectural drawings in the conduct of their jobs. The careers that are served by vocational, technical, or civil technology programs are described here.

Contractors

The two types of contractors are *general contractors* and *subcontractors*. General contractors are responsible for all phases of a construction project. General contractors must ensure that all schedules are met within the budget and according to the architectural plans and specifications. They do this by supervising subcontractors who are responsible for just one phase of construction such as carpentry, electrical, plumbing, site work, masonry, or HVAC. General contractors need years of experience in construction and come from the ranks of subcontractors, engineers, or architects. A general contractor must hold either a light construction or a commercial building license. Subcontractors need construction experience plus expertise and often licensing in their area of specialization. All must be familiar with all types of architectural plans and documents.

Carpenters

Carpenters are divided into two groups: *rough carpenters* and *finish carpenters*. Rough carpenters build the wood framing of buildings, which includes flooring, walls, and roofs. Finish carpenters complete the trim, molding, paneling, door and window installations, and on-site cabinetry. Skill in measuring and using hand and power tools is essential as is the ability to accurately read architectural plans. Completion of a vocational, construction technology, or apprenticeship program is recommended.

Cabinetmakers

Cabinetmakers, unlike carpenters, work mainly off-site. Cabinets and other built-in wood components are built off-site and installed during the final construction phase. This is because the precision of stationary woodworking power equipment is needed to ensure accuracy and save time in the construction and finishing of high-quality cabinetry. Training in all aspects of fine woodworking and the ability to read architectural detail drawings are essential.

Masons

Masons are divided into three categories: *stone masons, cement masons,* and *bricklayers*. Stone masons build stone walls, piers, patios, and walkways. They also build solid or stone veneer siding. Cement masons pour, shape, and smooth concrete surfaces to form slab foundations, walls, patios, walkways, and driveways. Bricklayers build chimneys, fireplaces, patio surfaces, and solid brick and brick veneer walls. Masons must be familiar with the workability and characteristics of each material. Completion of vo-tech and apprenticeship programs is recommended.

Plumbers and Pipefitters

Plumbers and pipefitters install piping systems. They must understand hydraulics and the types of piping systems needed for water, gas, steam, air, and waste disposal systems. Completion of vo-tech and apprenticeship programs is recommended.

Electricians

Electricians install the electrical system in a building including distribution panels and all wiring. They must read electrical plans to determine the type and location of all electrical fixtures and outlets. They also need to install specialized systems such as communication, security, entertainment, and computer networks. Electricians must be licensed.

HVAC Technicians

HVAC technicians install refrigeration, air flow, and heating equipment and systems. This includes creating and installing ductwork, furnaces, heat pumps, air exchangers, motors, filters, and purification systems. Vo-tech and apprenticeship programs are recommended.

Schedulers

Schedulers study the completion date goals and the set of specifications for a project. From this information they prepare timetables for the delivery of building materials and equipment to the building site. Schedulers plan for the delivery of materials prior to the date needed. If materials are delivered too soon, the site can become cluttered and work progress can be slowed. Also early delivery of materials increases the early outlay of money. If delivered too late, schedules and expensive labor time will be wasted. Schedulers must have experience in construction. Many come from the ranks of subcontractors or drafters.

Expediters

Expediters are troubleshooters who identify potential scheduling problems and communicate with materials suppliers to ensure that delivery schedules are maintained. On small projects the job of an expediter is performed by the scheduler.

Marketing Specialists

Marketing specialists sell real estate in different forms. Some sell parcels in housing developments using models, plats, abbreviated floor plans, and pictorial drawings to communicate with customers. Some sell manufactured buildings to be shipped to any location. Real estate agents resell existing properties. All need a working knowledge of architectural styles, floor plan alternatives, and basic construction types.

Building Inspectors

Building inspectors check construction progress at regular intervals to determine adherence to codes and the approved set of plans and specifications. They perform a final inspection, which must be passed before the inspector approves issuing a certificate of occupancy (CO) as described in Chapter 37. Inspectors are usually builders or engineers with experience as general contractors, drafters, or specification writers.

Environmental Inspectors

Many types of inspectors fall into this category. They most often work for a government agency: municipal, regional, state, or national. The requirements for this position vary among agencies and areas of scientific, technical, and management abilities. The higher levels in each category usually require academic credentials and or certification in the specialized field.

Mathematical Calculations

Throughout the text mathematical formulas are presented where needed. Appendix B contains a summary of these for easy reference as well as other formulas not presented in the text. These are divided into arithmetic, structural, and geometric calculations.

ARITHMETIC CALCULATIONS

The majority of errors in architectural drawings are mathematical errors. Yet most calculations performed in the process of preparing architectural working drawings involve the arithmetic functions of adding, subtracting, multiplying, and dividing. Basic arithmetic follows that can be applied to the preparation of architectural drawings.

Conversions

Part of the construction industry uses customary foot, inch, and fractional dimensions. Other parts use decimal or metric dimensions. It is, therefore, important to be able to convert dimensions from one system to another.

To convert inch fractions to decimals, divide the numerator of the fraction by the denominator. For example, $7/8'' = 7 \div 8 = .875$.

To convert decimals to fractions, use the decimal number as the numerator and place a number 1 in the denominator, followed by as many zeros as there are places in the nominator. For example, $.3 = 3/10$, $.45 = 45/100$, and $.675 = 675/1000$. See Figure B.1.

To convert inches to millimeters (mm), multiply inches by 25.4 ($1'' = 25.4$ mm). For example, $1/2'' = .5''$; therefore, $1/2'' = 12.7$ mm ($.5 \times 25.4$). Likewise $6'-6'' = 6.5'$ ($78''$); therefore, $6'-6'' = 1981.2$ mm. Conversions commonly applied in architecture are listed in Figure B.2.

Adding Dimensions

Most rows of dimensions include both feet and inches and may include fractional inches. In adding rows of mixed numbers such as these, add the feet and inches separately, convert the inch total to feet and inches and then re-add the foot total. For example, in the plan shown in Figure B.3, three

1/64	0.015625		33/64	0.515625
1/32	0.03125		17/32	0.53125
3/64	0.046875		35/64	0.546875
1/16	0.0625		9/16	0.5625
5/64	0.078125		37/64	0.578125
3/32	0.09375		19/32	0.59375
7/64	0.109375		39/64	0.609375
1/8	0.1250		5/8	0.6250
9/64	0.140625		41/64	0.640625
5/32	0.15625		21/32	0.65625
11/64	0.171875		43/64	0.671875
3/16	0.1875		11/16	0.6875
13/64	0.203125		45/64	0.703125
7/32	0.21875		23/32	0.71875
15/64	0.234375		47/64	0.734375
1/4	0.2500		3/4	0.7500
17/64	0.265625		49/64	0.765625
9/32	0.28125		25/32	0.78125
19/64	0.296875		51/64	0.796875
5/16	0.3125		13/16	0.8125
21/64	0.328125		53/64	0.828125
11/32	0.34375		27/32	0.84375
23/64	0.359375		55/64	0.859375
3/8	0.3750		7/8	0.8750
25/64	0.390625		57/64	0.890625
13/32	0.40625		29/32	0.90625
27/64	0.421875		59/64	0.921875
7/16	0.4375		15/16	0.9375
29/64	0.453125		61/64	0.953125
15/32	0.46875		31/32	0.96875
31/64	0.484375		63/64	0.984375
1/2	0.5000		1	1.0000

FIGURE B.1 > Decimal equivalents. *Hepler/Wallach/Hepler © Cengage 2013*

dimensions combine to equal $13'-3''$: these are $5'-3''$, $5'-3''$, and $2'-9''$. To add these:

$$5'-3''$$
$$5'-3''$$
$$\underline{2'-9''}$$

STEP 1 $12'-15''$ (Total)
STEP 2 $15'' = 1'-3''$
STEP 3 $12' + 1'-3'' = 13'-3''$ (Total)

MULTIPLY	BY	TO OBTAIN
Angles:		
Degrees	60	Minutes
Degrees	3600	Seconds
Area:		
Square feet	2.296×10^5	Acres
Square feet	144	Square inches
Square feet	3.587×10^8	Square miles
Square inches	6.452	Square centimeters
Square inches	6.944×10^3	Square feet
Square inches	645.2	Square millimeters
Square meters	2.471×10^4	Acres
Square meters	10.76	Square feet
Square miles	27.88×10^6	Square feet
Energy:		
British thermal units	2.928×10^4	Kilowatt-hours
Kilowatts	56.92	BTU/minute
Kilowatt-hours	3415	BTU
Kilowatt-hours	2.655×10^6	Foot-pounds
Watts	0.05692	BTU/minute
Watt-hours	3.415	BTU
Length:		
Centimeters	0.3937	Inches
Feet	30.48	Centimeters
Feet	0.3048	Meters
Inches	2.540	Centimeters
Kilometers	3281	Feet
Kilometers	0.6214	Miles
Meters	3.281	Feet
Meters	39.37	Inches
Miles	5280	Feet
Miles	1.609	Kilometers
Millimeters	0.03937	Inches
Volume:		
Board feet	144×1	Cubic inches
Cubic centimeters	3.531×10^5	Cubic feet
Cubic centimeters	6.102×10^2	Cubic inches
Cubic feet	1728	Cubic inches
Cubic feet	0.02832	Cubic meters
Cubic inches	5.787×10^4	Cubic feet
Cubic inches	1.639×10^5	Cubic meters
Cubic meters	35.31	Cubic feet
Cubic meters	61.023	Cubic inches
Weight/Mass:		
Pounds	16	Ounces
Tons (long)	2240	Pounds
Tons (metric)	2205	Pounds
Tons (short)	2000	Pounds

FIGURE B.2 > Conversion table. *Hepler/Wallach/Hepler*
© Cengage 2013

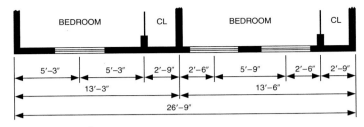

FIGURE B.3 > Adding dimensions. *Hepler/Wallach/Hepler*
© Cengage 2013

If fractions are involved, find the lowest common denominator in the adding the fractions column. Then add the feet, inches, and fractions separately, convert the results to inches and feet and inches, and re-add. For example:

STEP 1 $1'\text{-}7\frac{7}{8}'' = 1'\text{-}7\frac{14}{16}''$
$2'\text{-}8\frac{1}{4}'' = 2'\text{-}8\frac{4}{16}''$
$\underline{6'\text{-}10\frac{9}{16}'' = 6'\text{-}10\frac{9}{16}''}$
STEP 2 $9'\text{-}25\frac{27}{16}''$ (Total)
STEP 3 $\frac{27}{16}'' = 27 \div 16 = 1\frac{11}{16}''$
STEP 4 $25'' = 25 \div 12 = 2'\text{-}1''$
STEP 5 $9' + 1\frac{11}{16}'' + 2'\text{-}1'' = 12'\text{-}2\frac{11}{16}''$ (Total)

Modular Measurements

In designing a structure, it is often necessary to establish modular units. To determine the number of modular units needed for a given area, divide the planned overall size by the modular unit size. For example, if a window opening dimension is planned to be approximately $36'' \times 66''$ and the modular unit is $16''$, proceed as follows:

$$36'' \text{ (height)} \div 16'' = 2.25 \text{ units}$$

Use two $16''$ units to cover $32''$ height.

$$66'' \text{ (width)} \div 16'' = 4.125 \text{ units}$$

Use four $16''$ units to cover $64''$, as shown in Figure B.4. A $32'' \times 64''$ modular window is the nearest modular size window that will fit into the opening. The rough opening should be closer to $34'' \times 66''$ for the modular unit.

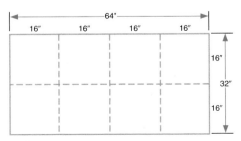

FIGURE B.4 > Modular fitting. *Hepler/Wallach/Hepler*
© Cengage 2013

Construction Material Calculations

Construction materials are packaged or sold in a wide variety of set quantities and standard sizes. It is, therefore, important to convert the total amount of material needed into the standard marketing measures. Doing this often requires converting small volume units into larger units. See Figure B.5.

Board-Foot Measure

Lumber is purchased in bulk by the board foot (BF). One board foot is $1'' \times 12'' \times 12''$. To determine the number of

MATERIAL	MEASUREMENT	PACKAGED OR SOLD BY
Cement	Bag (1 cubic foot)	Bag
Concrete	Cubic foot	Cubic yard
Sand	Ton/pounds	Ton/cubic yards
Blacktop	Square yards (after installation)	Cubic yard
Gravel	Size of stone ¼" to 3"	Ton/cubic yard
Concrete block	Standard height 7⅝" Standard length 15⅝" Widths 2" to 14"	Pallet, 100-block Piece
Mortar	Bag (1 cubic foot)	Bag (1 cubic foot)
Reinforcing rods	Width of bar ¼" to 1⅝"	Pounds
Welded wire mesh	Size of wire 1⁄16" to ¼" Size of grid	Roll 5 × 100 feet Sheet 5 × 20 feet
Asphalt, static	Pounds	Bucket or barrel
Plywood	Thickness of sheet ¼" to 1"	Sheet 4' × 8'
Paneling	Thickness of sheet ¼" to ⅜"	Sheet 4' × 8'
Lumber: 2 × 4	Board foot/piece	Board foot/piece 6' to 20' lengths
2 × 6	Board foot/piece	6' to 20' lengths
2 × 8	Board foot/piece	6' to 20' lengths
2 × 10	Board foot/piece	6' to 20' lengths
2 × 12	Board foot/piece	6' to 20' lengths
Particleboard panels	Thickness of sheet 3⁄16" to 1"	Sheet 4' × 8'
Hardboard panels	Thickness of sheet ⅛" to ⅜"	Sheet 4' × 8'
Roof shingles	Square (100 square feet)	⅓ square foot per bundle or 33.3 square feet per bundle
Tar paper	Square feet	Rolls 3' × 100'
Rain gutters	Depth 4" to 5"	Lineal foot
Aluminum siding	Square foot	Square (100 square feet = 1 square)
Cedar siding	Square foot	Square foot
Drywall wallboard	Square foot	Sheet 4' × 8'
Sheathing panels	Square foot	Sheet 4' × 8'
Nails	Pennyweight	Box or keg 1 pound to 25 pounds
Pipe, copper	¼" to 2½" diameter	Lineal feet, 20' length
Pipe, plastic	¼" to 4" diameter	Lineal feet
Pipe, iron	¼" to 4" diameter	Lineal feet
Pipe, galvanized	¼" to 4" domestic ¼" to 36" industrial	Lineal feet Lineal feet
Pipe, cast iron	4" diameter	Lineal feet
Wire	Wire diameter/gauge	Roll 50' to 100'
Conduit	½" to 4" diameter	Lineal feet, 10' lengths

FIGURE B.5 > Standard marketing measures for construction materials. *Hepler/Wallach/Hepler © Cengage 2013*

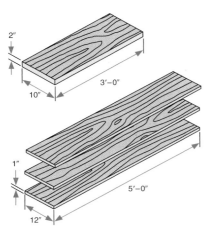

FIGURE B.6 > Board-foot measure. *Hepler/Wallach/Hepler*
© *Cengage 2013*

board feet in a given piece of lumber, multiply the thickness (in inches) × width (in inches) × length (in feet) ÷ 12:

$$BF = \frac{T'' \times W'' \times L'}{12}$$

The board feet in the top piece of lumber shown in Figure B.6 is calculated as follows:

$$\frac{2'' \times 10'' \times 3'}{12} = 5\,BF$$

When dealing with multiple pieces of the same size, either compute the BF for one piece and multiply it by the number of pieces or treat the entire package as one piece as follows:

$$\frac{1'' \times 12'' \times 5'}{12} = 5\,BF\ each \times 3\ pieces = 15\,BF$$

Or

$$\frac{3 \times 1'' \times 12'' \times 5'}{12} = 15\,BF$$

STRUCTURAL CALCULATIONS

A well-designed building *should* be aesthetically pleasing but it *must* be structurally sound. A functional working knowledge of engineering mechanics and mathematics is a necessity in architectural design and drafting.

The field of engineering mechanics is divided into two main parts: *statics* and *dynamics*. Statics deals with objects at rest. Dynamics deals with objects in motion or potential motion. The principles of statics and dynamics are combined with information on the strength of construction materials in the design of structures that will withstand the forces of nature and human use.

An effective structural design is neither underdesigned or overdesigned. When structures are underdesigned, systems fail and buildings sag or collapse. However, if a building is overdesigned, materials are wasted and costs increase greatly. Thus the primary task of the structural designer is to design all components to meet and/or exceed the safety factor without excessively increasing the cost of the building.

Designers and drafters must determine the most appropriate materials, sizes, spacing, and construction methods for an architectural design. Determining building loads is the first step in this process.

Building Loads

The weight of all movable items, such as the occupants, wind, snow, and furniture, make up the *live load*. The weight of all materials used in the construction of a building, including all permanent structures and fixtures, make up the *dead load* of a building. The total weight of the live load plus the weight of the dead load is called the *building load*. (Building loads were introduced in Chapter 22.) Loads are measured in pounds per square foot (lb/ft² or PSF). See Figure B.7.

Live loads act on floors, walls, ceilings, and roofs. Floor live loads include persons and furniture. Live loads acting on walls are wind loads. Materials differ greatly in their ability to withstand loads. See Figure B.8. Lateral loads from earthquakes must also be considered. Live loads acting on roofs include wind and snow loads. Roof loads are comparatively light, but vary according to the pitch of the roof. Flat roofs offer more resistance to live loads than do pitched roofs. To meet building codes, low-pitched roofs (those below 3/12 pitch) must often be designed to support wind and snow loads of 30 lb/ft². See Figure B.9. The wind load increases as the pitch increases. The snow load increases as the pitch decreases. High-pitched roofs (those over 3/12 pitch) need be designed to support live loads of 15 lb/ft².

CONSTRUCTION AREA	LIVE LOAD PSF	DEAD LOAD PSF
Roof	30	20
Ceiling joists attic/heavy storage	30	20
Ceiling joists attic no floor	0	10
Ceiling joists attic habitable rooms	30	10
Floors of rooms	40	20
Floors of bedroom	30	20
Partitions	0	20

FIGURE B.7 > Typical loads for a two-level frame construction building. *Hepler/Wallach/Hepler*
© *Cengage 2013*

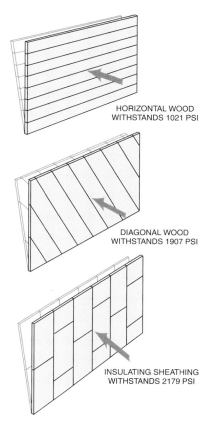

FIGURE B.8 > Ability of sheathing to resist horizontal (wind) loads in pressure (pounds per square inch). *Hepler/Wallach/Hepler © Cengage 2013*

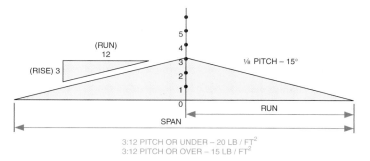

FIGURE B.9 > Typical roof design requirements set in building codes. *Hepler/Wallach/Hepler © Cengage 2013*

Structural calculations are used to determine the size and type of materials to be used in foundations to support all *dead loads*. The size, spacing, and type of materials used in walls that support the roof load must also be determined. Dead loads become greater from the top of the structure to the bottom footing. For example, the load on an attic may be only 25 lb/ft², whereas the load on the first floor of the same building may be 45 lb/ft². The typical floor load for an average residential room usually varies from 30 to 40 lb/ft². Loads vary depending on the materials used in construction. See Figure B.10.

Tributary Areas

The *tributary area* of a structural member is the area of weight transmitted to a vertical support. The total load of the building is transmitted into the ground from footings and piers through a series of tributary areas. Because footings are continuous, load distribution is calculated as one lineal foot, as shown in Figure B.11.

The tributary area over each pier extends one-half the distance to the next structural support members. For example, pier B in Figure B.12 is 9'-0" from the nearest pier on both sides. The tributary area for pier B, therefore, extends one-half the distance to piers A and C (4'-6") and one-half the distance to each foundation wall (5'-0"). The tributary area is, therefore, 4'-6" + 4'-6" by 5'-0" + 5'-0", or 9'-0" × 10'-0", or 90 ft². (See p. **807** for calculating the tributary area of girders and beams.)

Soil Conditions

The safety of a structure also depends on the type of soil and the load it can support. It is important that the weight of the structure not exceed the safe load capacity of the soil. See Figure B.13. Footing sizes are critical because they spread the weight of the structure on the soil. Footing area (FA) is calculated as follows (Figure B.14):

Pier footing calculation:
Structure load = 5,000 lb/ft²
Soil-bearing capacity = 6,000 lb/ft² (given)

$$\text{Footing area} = \frac{\text{structure load}}{\text{soil-bearing capacity}}$$

$$FA = \frac{5000}{6000} = 0.83$$

(Convert sq. ft. to sq. in.)*
$$FA = 0.83 \text{ ft}^2 \times 144^* = 119.52 \text{ in}^2$$
(Find sq. root: $\sqrt{119.52} = 10.9$)

$$FA = 11'' \times 11'' \times 121 \text{ in}^2$$
(closest square to 119.52 in²)

* 144 sq. in. = 1 sq. ft.

Perimeter foundation wall footing calculation:
Structure load   = 4,500 lb/ft²
Soil-bearing capacity = 6,000 lb/ft² (given)

$$\text{Footing area} = \frac{\text{structure load}}{\text{soil-bearing capacity}}$$

$$FA = \frac{4,500}{6,000}$$

$$FA = 0.75 \text{ ft}^2$$

MATERIALS	WEIGHT	MATERIALS	WEIGHT
Roofs		**Woods** (12% moisture content, lb/ft^3)	
built-up roofing, 3-ply and gravel	6 PSF	cedar	22 PCF
rafters, 2 × 4 at 16″ oc	2 PSF	douglas fir	34 PCF
rafters, 2 × 6 at 16″ oc	2.5 PSF	maple	42 PCF
rafters, 2 × 8 at 16″ oc	3.5 PSF	oak	47 PCF
sheathing, ½″ fiberboard	0.75 PSF	pine	27 PCF
sheathing, ½″ gypsum	2 PSF	poplar	28 PCF
sheathing, 1″ wood	3 PSF	redwood	28 PCF
shingles, asbestos	4 PSF	**Glass**	
shingles, asphalt	2 PSF	double strength, 1–8″	1.5 PSF
shingles, wood	2.5 PSF	insulating plate, ⅛″ with air space	3.25 PSF
skylights, glass 7 frame	11 PSF	glass block, 4″	20 PSF
tile, cement	15 PSF	plate glass, ¼″	3.25 PSF
tile, mission	13 PSF	plastic, ¼″ acrylic	1.5 PSF
Ceilings		**Masonry**	
acoustical tile, ½″	0.8 PSF	brickwork, 4″	35 PSF
plaster on wood lath	8 PSF	concrete wall, 6″	75 PSF
suspended metal lath and cement plaster	15 PSF	concrete wall, 8″	100 PSF
suspended metal lath and gypsum plaster	10 PSF	poured concrete	150 lbs./cu. ft.
Walls		concrete block, lightweight 4″	22 PSF
brick wall, 4″	35 PSF	concrete block, lightweight 6″	31 PSF
brick wall, 8″	74 PSF	concrete block, stone 6″	50 PSF
brick (4″) on 6″ concrete block	80 PSF	concrete block, stone 8″	58 PSF
brick (4″) on wood frame with sheathing & plaster	45 PSF	facing tile, 2″	16 PSF
building board, ½″	0.8 PSF	facing tile, 4″	30 PSF
concrete block wall, 6″	40 PSF	facing tile, 6″	41 PSF
concrete block wall, 8″	55 PSF	marble, 1″	13 PSF
concrete wall, 8″	100 PSF	slate, 1″	14 PSF
concrete wall, 10″	125 PSF	stone, 1″	12 PSF
gypsum block, 2″	9.5 PSF	tile, structural clay, 4″ hollow	23 PSF
gypsum block, 4″	12.5 PSF	tile, structural clay, 6″ hollow	33 PSF
gypsum wallboard	2.5 PSF	tile, structural clay, 8″ hollow	42 PSF
plaster, ½″	4.5 PSF	**Metals** (lb/ft^3)	
plywood, ½″	1.5 PSF	aluminum, cast	165 PCF
tile, facing, 2″	15 PSF	brass, yellow	528 PCF
tile, glazed ⅜″	3 PSF	bronze, commercial	552 PCF
wood siding, 1″	3 PSF	copper, cast or rolled	556 PCF
wood stud wall	5 PSF	iron, cast	450 PCF
wood stud wall, plastered one side	12 PSF	iron, wrought	485 PCF
wood stud wall, plastered two sides	20 PSF	lead	710 PCF
Floors		steel, rolled	490 PCF
cement finish 1″	12 PSF	tin, cast or hammered	459 PCF
clay tile on 1″ mortar base	23 PSF	steel beam S7 × 15.3	15.3 lbs./lineal foot
concrete slab, 4″	48 PSF	**Soil, Sand, & Gravel** (lb/ft^3)	
hardwood flooring, 25/32″	4 PSF	clay, damp	110 PCF
floor joist, 2 × 8, 16″ oc/subflooring	6 PSF	clay, dry	63 PCF
floor joist, 2 × 10, 16″ oc/subflooring	6.5 PSF	clay and gravel, dry	100 PCF
floor joist, 2 × 12, 16″ oc/subflooring	7.0 PSF	earth, loose, dry	76 PCF
marble on 1″ mortar base	28 PSF	earth, packed, dry	95 PCF
plywood subflooring, ½″	1.5 PSF	earth, moist, loose	78 PCF
quarry tile, ½″	6 PSF	earth, moist, packed	96 PCF
terrazzo, 2″	25 PSF	earth, mud, packed	115 PCF
vinyl asbestos tile, ⅛″	1.3 PSF	sand/gravel, dry, loose	110 PCF
4 × 6 girder/post	4 PSF	sand/gravel, dry, packed	120 PCF
Insulation		sand/gravel, wet	120 PCF
bats, blankets, 3″	0.5 PSF		
boards, vegetable fiber	1.7 PSF		
cork board, 1″	0.6 PSF		
foam board, 1″	0.1 PSF		

FIGURE B.10 > Building material loads. *Hepler/Wallach/Hepler © Cengage 2013*

$$FA = 0.75 \text{ ft}^2 \times 144 = 108 \text{ in}^2$$
(Width is number of in^2 per lineal foot.)

$$FA = \frac{108 \text{ in}^2}{12}$$

FA = 9″ wide × 1 lineal foot

Use next standard size—12″.

Roof Members

The size of all structural members used for roof framing depends on the combined loads bearing on the member and the spacing and span of each member. If the load is increased, either the spacing or span must be decreased or the size of the member increased to compensate for the increased load. If the size of a member is decreased, the members must also be spaced more closely or the span must be decreased. If the length of the span is increased, the size of the members must be increased or the spac-

FIGURE B.11 > Loads are calculated on one lineal foot of footing. *Hepler/Wallach/Hepler © Cengage 2013*

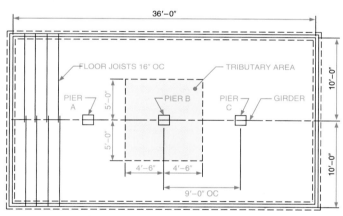

FIGURE B.12 > Foundation tributary area. *Hepler/Wallach/ Hepler © Cengage 2013*

ing made closer. To compute the most appropriate size of roof rafter for a given load, spacing, and span, follow these steps:

STEP 1 To determine the total load per square foot of the roof space, add the live load and the dead load. See Figure B.15. Figure B.16 shows the components of a roof system that are used in the calculation of roof loads as follows:

Component	Dead Load (lb/ft^2)
2 × 6 roof rafter	3
3 layers felt, tar, and gravel (or shingles)	5
1″ insulation	1
1/2″ plywood sheathing	2
5/8″ gypsum board ceiling	2
2 × 6 ceiling joist	3
Dead load	16
Live load	20
Total load	36 lb/ft^2

SOIL TYPE	SAFE LOAD PSF
Soft clay; sandy loam	1,000
Firm clay; sand and clay mix; fine sand, loose	2,000
Hard dry clay; fine sand, compact; sand and gravel mixtures; coarse sand, loose	3,000
Coarse sand, compact; stiff clay; gravel, loose	4,000
Gravel; sand and gravel mixtures, compact	6,000
Soft rock	8,000
Exceptionally compacted gravels and sands	10,000
Hardpan or hard shale; sillstones; sandstones	15,000
Medium hard rock	25,000
Hard, sound rock	40,000
Bedrocks, such as granite, gneiss, traprock	100,000

FIGURE B.13 > Safe soil loads. *Hepler/Wallach/Hepler © Cengage 2013*

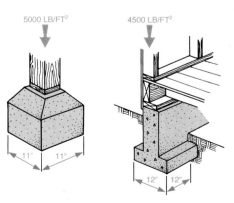

FIGURE B.14 > Footing sizes required by loads. *Hepler/ Wallach/Hepler © Cengage 2013*

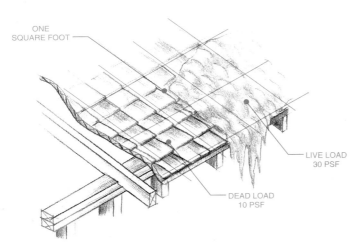

DEAD LOAD PLUS LIVE LOAD EQUALS THE TOTAL LOAD
10 PSF + 30 PSF = 40 PSF

FIGURE B.15 ⟩ Determining the total roof load per square foot. *Hepler/Wallach/Hepler © Cengage 2013*

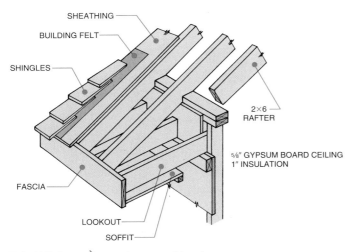

FIGURE B.16 ⟩ Roof material loads. *Hepler/Wallach/Hepler © Cengage 2013*

In roof design, the live load is the safety factor allowing for weather conditions and movable objects the structure must support.

STEP 2 Loads are expressed in lineal feet. See Figure B.17. To determine the load per lineal foot on each rafter, multiply the load per square foot by the spacing of the rafters. If rafters are spaced at 12″ intervals, then the load per square foot and the load per lineal foot will be the same. However, if the rafters are spaced at 16″ intervals, then each lineal foot of rafter must support 1 1/3 of the load per square foot. See Figure B.18.

STEP 3 To find the total load each rafter must support, multiply the load per lineal foot by the length of the span in feet. See Figure B.19.

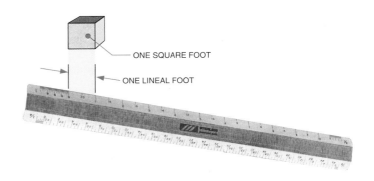

FIGURE B.17 ⟩ Comparison of a square foot and a lineal foot. *Hepler/Wallach/Hepler © Cengage 2013*

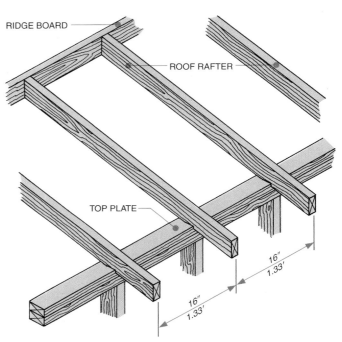

TOTAL LOAD × RAFTER SPACING (FT) = LOAD/LINEAL FT
40 LB/FT² × 1.33′ = 53 LB/LINEAL FT

FIGURE B.18 ⟩ Determining the load per lineal foot on each rafter. *Hepler/Wallach/Hepler © Cengage 2013*

STEP 4 To compute the bending moment in inch-pounds, multiply total load × rafter span × 12. Divide this figure by 8 (a constant). The *bending moment* (BM) is the force needed to bend or break the rafter. When the length of the span in pounds is multiplied by the length of the span in feet, the result is expressed in foot-pounds. The span must be multiplied by 12 to convert the bending moment into inch-pounds. See Figure B.20. The bending moment (in inch-pounds) of a structural member (in pounds)

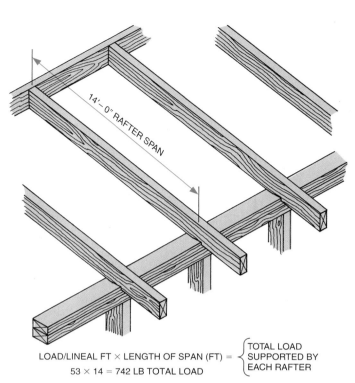

LOAD/LINEAL FT × LENGTH OF SPAN (FT) = $\left\{\begin{array}{l}\text{TOTAL LOAD}\\ \text{SUPPORTED BY}\\ \text{EACH RAFTER}\end{array}\right.$

53 × 14 = 742 LB TOTAL LOAD

FIGURE B.19 > Determining the total load each rafter must support. *Hepler/Wallach/Hepler © Cengage 2013*

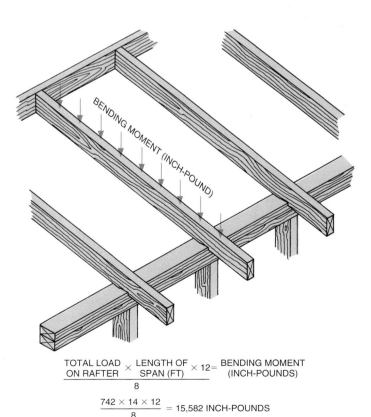

$$\frac{\dfrac{\text{TOTAL LOAD}}{\text{ON RAFTER}} \times \dfrac{\text{LENGTH OF}}{\text{SPAN (FT)}} \times 12}{8} = \begin{array}{l}\text{BENDING MOMENT}\\ \text{(INCH-POUNDS)}\end{array}$$

$$\frac{742 \times 14 \times 12}{8} = 15,582 \text{ INCH-POUNDS}$$

FIGURE B.20 > Computing the bending moment. *Hepler/Wallach/Hepler © Cengage 2013*

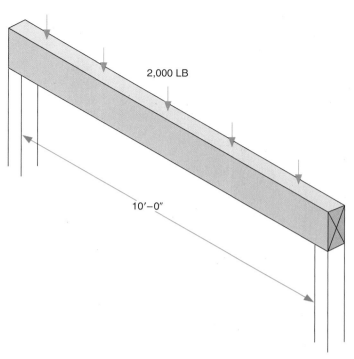

FIGURE B.21 > Bending forces on a horizontal member. *Hepler/Wallach/Hepler © Cengage 2013*

equals the load (in pounds) times the length (in feet) times 12 divided by 8:

$$\text{Bending Moment} = \frac{L \times D \times 12}{8}$$

The bending moment of a beam is calculated in the same way. See Figure B.21.

$$\text{Bending Moment} = \frac{2{,}000 \text{ lb} \times 10' \times 12}{8}$$

$$\text{BM} = \frac{240{,}000}{8}$$

$$\text{BM} = 30{,}000 \text{ in-lb}$$

STEP 5   Set up the equation to determine the resisting moment. The *resisting moment* is the strength or rafter resistance the rafter must possess to withstand the force of the bending moment of the rafter. See Figure B.22. If the resisting moment is less than the bending moment, the member will bend or break. The resisting moment is determined by multiplying the fiber stress by the rafter width by the rafter depth squared. This figure is divided by 6 (a constant). Rafter widths should be expressed in the exact dimensions of the finished lumber. For example, if a rafter is a standard size 2″ × 4″, its actual width is 1 1/2″ and its depth is 3 1/2″. See Figure B.23.

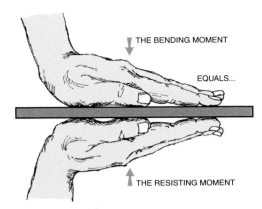

FIGURE B.22 > The resisting moment must equal or be greater than the bending moment. *Hepler/Wallach/Hepler © Cengage 2013*

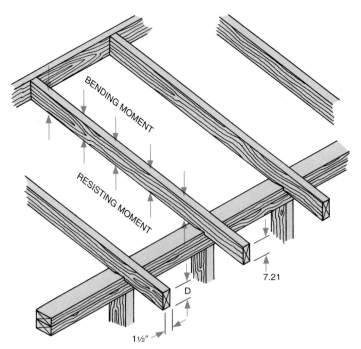

BENDING MOMENT = RESISTING MOMENT

$$\frac{742 \times 14 \times 12}{8} = \frac{1200 \times 1\frac{1}{2}'' \times D^2}{6}$$

$$15{,}582 = 300\,D^2$$

$$\frac{15{,}582}{300} = \frac{300\,D^2}{300}$$

$$51.94 = D^2$$

$$\sqrt{51.94} = \sqrt{D^2}$$

$$7.21 = D$$

FIGURE B.24 > To compare the bending moment with the resisting moment, use the formulas together. *Hepler/Wallach/Hepler © Cengage 2013*

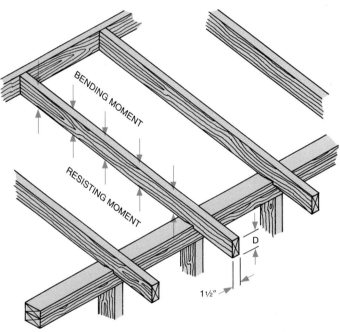

$$\text{RESISTING MOMENT} = \frac{\underset{\text{STRESS}}{\text{FIBER}} \times \underset{\text{WIDTH}}{\text{RAFTER}} \times \left(\underset{\text{DEPTH}}{\text{RAFTER}}\right)^2}{6}$$

$$\text{Example: Resisting Moment} = \frac{1200 \times 1\frac{1}{2}'' \times D^2}{6} \begin{pmatrix} \text{D Unknown} \\ \text{Solve for} \\ \text{rafter depth} \end{pmatrix}$$

$$\text{RM} = \frac{1200 \times 1\frac{1}{2}'' \times (7\frac{1}{2})^2}{6} \ (\text{D is known})$$

$$\text{RM} = \frac{1200 \times 1.5 \times 56.25}{6}$$

$$\text{RM} = 16{,}875$$

FIGURE B.23 > Determining the resisting moment. *Hepler/Wallach/Hepler © Cengage 2013*

The *fiber stress* is the tendency of the fibers of the wood to bend and become stressed as the member is loaded. Fiber stresses range from 1,750 lb/in² for southern dense pine select to 600 lb/in² for red spruce. Dense Douglas fir and southern pine have average fiber stresses of 1,200 lb/in². The rafter depth is squared, since the strength of the member increases by squares. For example, a raf-

ter 12″ deep is not three times as strong as a rafter 4″ deep. It is nine times as strong.

STEP 6   Because the bending moment is compared with the resisting moment, the formulas can be combined, to find rafter depth, as shown in Figure B.24. The formula can then be followed for any of the variables, preferably for the depth of the rafter, since varying the depth will alter the resisting moment more than any other single factor. An example of figuring the bending moment and the resisting moment together for a 2″ × 6″ beam is shown in Figure B.25. Computations are summarized as follows:

$$\text{Bending moment} = \text{Resisting moment}$$

$$\frac{3{,}000 \times 10 \times 12}{8} = \frac{1{,}500 \times 1.5 \times 5.5^2}{6}$$

$$\frac{360{,}000}{8} = \frac{68{,}062.5}{6}$$

$$45{,}000 = 11{,}343.75$$

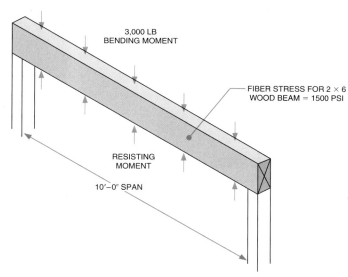

FIGURE B.25 > Factors used in bending and resisting moment calculations. *Hepler/Wallach/Hepler*
© *Cengage 2013*

The bending moment is greater than the resisting moment. Increase the beam size to 4 × 8 (standard size).

$$45,000 = \frac{1,500 \times 3.5 \times 7.5^2}{6}$$
$$45,000 = 49,218.75$$

The beam is now in equilibrium. The resisting moment is greater than the bending moment. The member will not bend or break.

Care should be taken in establishing all sizes to ensure that the sizes of materials conform to manufacturers' standards and building-code allowances. Figure B.26 shows a typical rafter space-span chart based on common lumber sizes and spacing. Figure B.27 shows common truss specifications based on normal loading for residential work.

Ceiling Joists

Information needed when selecting standard ceiling joists is provided in Figure B.28. (See Chapter 24 for information about wood groups.) If there is to be an attic above the ceiling for storage or living area, then the ceiling joists must be treated as floor joists using a floor joist table to calculate the joists for a heavier load. For a flat or very low-pitched roof, where the roof rafters are also the ceiling joists, then the table in Figure B.29 must be used.

Floors

The major structural parts of the floor system are the floor joists. See Figure B.30. The other parts of the floor system are the subfloor, finish floor, and supporting structural members. When the live load is determined, the size and spacing of joists can be established by referring to Figure B.31.

PITCH: 5/12		LOAD: 40 PSF
FIBER STRESS, 1,200 POUNDS, FOR DOUGLAS FIR AND SOUTHERN YELLOW PINE		
LUMBER SIZE	SPACING, IN INCHES	MAXIMUM SPAN
2 × 4	24	6'–6"
	20	7'–3"
	16	8'–1"
	12	9'–4"
2 × 6	24	10'–4"
	20	11'–4"
	16	12'–6"
	12	14'–2"
2 × 8	24	13'–8"
	20	15'–2"
	16	16'–6"
	12	18'–4"

FIGURE B.26 > Rafter spans. *Hepler/Wallach/Hepler*
© *Cengage 2013*

This table is based on No. 1 southern white pine with a fiber stress of 1,200 lb/in² and a modulus of elasticity (ratio of stress and strain) of 1,600,000 lb/in². For other materials, such as spruce or Douglas fir lumber, with a different fiber stress and different modulus of elasticity, a different table must be used.

An example of the use of Figure B.31 is as follows: If the live loads are approximately 40 lb/ft² and 16" spaces are desired between joists, a 2 × 8 is good for a span of only 12'-1". For a span larger than 12'-1", a joist with a larger cross section is necessary.

Headers and Lintels

The header (structural wood member) and the lintel (structural steel member) are horizontal structural supports. They support the openings of windows and doors. Engineering tables for their selection are shown in Figure B.32. This table is for celling and roof support only.

Girders and Beams

All the weight of the floor system, including live loads and dead loads, is transmitted to bearing partitions. These loads are then transmitted either to the foundation wall, to intermediate supports, or to horizontal supports known as joists, girders, or beams.

To determine the exact spacing, size, and type of girder to support the structure, follow these steps:

STEP 1   Determine the total load acting on the entire floor system in pounds per square foot. Divide the total live and dead loads by the number of square

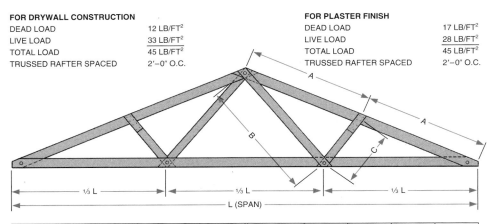

FOR DRYWALL CONSTRUCTION
DEAD LOAD   12 LB/FT²
LIVE LOAD   33 LB/FT²
TOTAL LOAD   45 LB/FT²
TRUSSED RAFTER SPACED   2'–0" O.C.

FOR PLASTER FINISH
DEAD LOAD   17 LB/FT²
LIVE LOAD   28 LB/FT²
TOTAL LOAD   45 LB/FT²
TRUSSED RAFTER SPACED   2'–0" O.C.

SLOPE	SPAN "L"	A	B	C	SLOPE	SPAN "L"	A	B	C
4/12 1/6 PITCH	20'–0"	5'–3 1/4"	4'–8 3/16"	2'–3 15/16"	6/12 1/4 PITCH	20'–0"	5'–7 1/16"	5'–11 11/16"	2'–11 5/8"
	22'–0"	5'–9 9/16"	5'–1 7/8"	2'–6 3/4"		22'–0"	6'–1 13/16"	6'–6 7/8"	3'–3 1/4"
	24'–0"	6'–3 7/8"	5'–7 1/2"	2'–9 9/16"		24'–0"	6'–8 1/2"	7'–2 1/8"	3'–6 7/8"
	26'–0"	6'–10 3/16"	6'–1 3/16"	3'–0 7/16"		26'–0"	7'–3 3/16"	7'–9 5/16"	3'–10 7/16"
	28'–0"	7'–4 9/16"	6'–6 13/16"	3'–3 1/4"		28'–0"	7'–9 15/16"	8'–4 9/16"	4'–2 1/16"
	30'–0"	7'–10 7/8"	7'–0 1/2"	3'–6 1/16"		30'–0"	8'–4 5/8"	8'–11 3/4"	4'–5 11/16"
	32'–0"	8'–5 3/16"	7'–6 3/16"	3'–8 7/8"		32'–0"	8'–11 15/16"	9'–6 15/16"	4'–9 1/4"
5/12 5/24 PITCH	20'–0"	5'–5"	5'–3 3/8"	2'–7 5/8"	7/12 7/24 PITCH	20'–0"	5'–9 7/16"	6'–8 3/16"	3'–3 7/8"
	22'–0"	5'–11 1/2"	5'–10 1/16"	2'–10 13/16"		22'–0"	6'–4 7/16"	7'–4 1/4"	3'–7 15/16"
	24'–0"	6'–6"	6'–4 7/16"	3'–2"		24'–0"	6'–11 3/8"	8'–0 3/8"	3'–11 15/16"
	26'–0"	7'–0 1/2"	6'–10 7/8"	3'–5 1/4"		26'–0"	7'–6 5/16"	9'–8 3/8"	4'–4"
	28'–0"	7'–7"	7'–5 1/4"	3'–8 7/16"		28'–0"	8'–1 1/4"	9'–4 7/16"	4'–8"
	30'–0"	8'–1 1/2"	7'–11 11/16"	3'–11 5/8"		30'–0"	8'–8 3/16"	10'–0 1/2"	5'–0 1/16"
	32'–0"	8'–8"	8'–6 1/16"	4'–2 13/16"		32'–0"	9'–3 1/8"	10'–8 9/16"	5'–4 1/16"

RISE	MAXIMUM SPAN (APPROX)	
	2 × 4 SPAN	2 × 6 SPAN
3	25'–0"	30'–0"
4	27'–0"	32'–0"
5	30'–0"	35'–0"
6	32'–0"	38'–0"

Specifications for a Fink truss.

FIGURE B.27 > Truss specifications. *Hepler/Wallach/Hepler © Cengage 2013*

SIZE	SPACING (OC)	GROUP I	GROUP II	GROUP III	GROUP IV
	MAXIMUM ALLOWABLE SPAN (FEET AND INCHES)				
2 × 4	12"	11'–6"	11'–0"	9'–6"	5'–6"
	16"	10'–6"	10'–0"	8'–6"	5'–0"
2 × 6	12"	18'–0"	16'–6"	15'–6"	12'–6"
	16"	16'–0"	15'–0"	14'–6"	11'–0"
2 × 8	12"	24'–0"	22'–6"	21'–0"	19'–0"
	16"	21'–6"	20'–6"	19'–0"	16'–6"

FIGURE B.28 > Ceiling joists. *Hepler/Wallach/Hepler © Cengage 2013*

SIZE	SPACING (OC)	GROUP I	GROUP II	GROUP III	GROUP IV
	MAXIMUM SPAN				
2 × 6	24"	9'–4"	7'–10"	7'–2"	6'–6"
	16"	11'–4"	9'–8"	8'–8"	8'–0"
2 × 8	24"	13'–0"	11'–2"	10'–0"	9'–4"
	16"	15'–10"	13'–8"	12'–4"	11'–6"
2 × 10	24"	17'–4"	14'–10"	13'–8"	12'–4"
	16"	19'–2"	18'–2"	16'–8"	15'–0"

FIGURE B.29 > Low-slope roof joists. *Hepler/Wallach/Hepler © Cengage 2013*

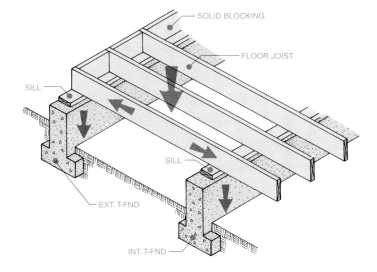

FIGURE B.30 > Floor joist loads are transmitted to intermediate supports. *Hepler/Wallach/Hepler © Cengage 2013*

feet of floor space. For example, if the combined load for the floor system shown in Figure B.33 is 48,000 lb, then there are 50 lb/ft² of load acting on the floor (48,000 lb divided by 960 ft² equals 50 lb/ft²).

STEP 2   Lay out the proposed position of all columns and beams. It will also help to sketch the position of the joists to be sure that the joist spans are correct.

LIVE LOAD—POUNDS PER SQUARE FOOT	SPACING	2 × 6	2 × 8	2 × 10	2 × 12	2 × 14
10	12" 16" 24"	12'–9" 11'–8" 10'–3"	16'–9" 15'–4" 14'–6"	21'–1" 19'–4" 17'–3"	24'–0" 23'–4" 20'–7"	— 24'–0" 24'–0"
20	12" 16" 24"	11'–6" 10'–5" 9'–2"	15'–3" 13'–11" 12'–3"	19'–2" 17'–6" 15'–6"	23'–0" 21'–1" 18'–7"	24'–0" 24'–0" 21'–9"
30	12" 16" 24"	10'–8" 9'–9" 8'–6"	14'–0" 12'–11" 11'–4"	17'–9" 16'–3" 14'–4"	21'–4" 19'–6" 17'–3"	24'–9" 22'–9" 20'–2"
40	12" 16" 24"	10'–0" 9'–1" 7'–10"	13'–3" 12'–1" 10'–4"	16'–8" 15'–3" 13'–1"	20'–1" 18'–5" 15'–9"	23'–5" 21'–5" 18'–5"
50	12" 16" 24"	9'–6" 8'–7" 7'–3"	12'–7" 11'–6" 9'–6"	15'–10" 14'–7" 12'–1"	19'–1" 17'–6" 14'–7"	22'–4" 20'–5" 17'–0"
60	12" 16" 24"	9'–0" 8'–1" 6'–8"	12'–0" 10'–10" 8'–11"	15'–2" 13'–8" 11'–3"	18'–3" 16'–6" 13'–7"	21'–4" 19'–3" 15'–11"
70	12" 16" 24"	8'–7" 7'–8" 6'–5"	11'–6" 10'–2" 8'–5"	14'–6" 12'–10" 10'–7"	17'–6" 15'–6" 12'–9"	20'–6" 18'–3" 15'–0"

FIGURE B.31 > Maximum floor joist spans. *Hepler/Wallach/Hepler © Cengage 2013*

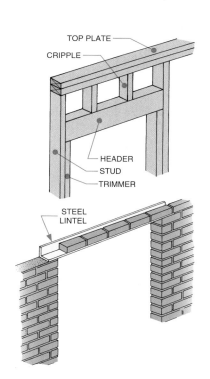

SAFE SPANS FOR WOOD HEADERS

SIZE	SPAN
4 × 4	3'–6"
4 × 6	4'–6"
4 × 8	6'–0"
4 × 10	7'–6"
4 × 12	9'–0"

SAFE SPANS FOR STEEL LINTELS 4" MASONRY WALL

SIZE	SPAN
L-3½" × 3½" × ¼"	3'–0"
L-3½" × 3½" × 5/16"	4'–0"
L-4" × 3½" × 5/16"	6'–0"
L-5" × 3½" × 5/16"	8'–0"
L-6" × 3½" × 3/8"	10'–0"

FIGURE B.32 > Lintel and header spans. *Hepler/Wallach/Hepler © Cengage 2013*

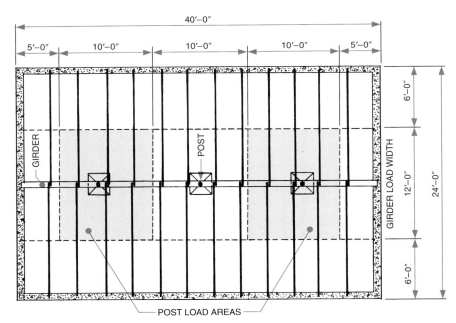

FIGURE B.33 > Tributary area of a girder. *Hepler/Wallach/Hepler © Cengage 2013*

STEP 3 Determine the number of square feet supported by the girder (girder load area). The girder load area is determined by multiplying the length of a girder, from column to column, by the girder load width. The *girder load width* is the distance extending on both sides of the center line of the girder, halfway to the nearest support, as shown in Figure B.33. The remaining distance from a girder load area to the outside wall is supported by the outside wall.

STEP 4 To find the load supported by the girder load area, multiply the girder load area by the load per square foot. For example, the girder load in Figure B.33 is 6,000 lb (120 ft^2 × 50 lb/ft^2).

STEP 5 Select the most suitable material to carry the load at the span desired. Built-up wood girders will span a greater length than a solid wood girder. However, I-beams will span a greater length without intervening support.

To compute the minimum cross-section area of a girder or beam, divide the load (in pounds) by the material's coefficient of elasticity:

$$CS = \frac{I}{E}$$

For example, if the combined live and dead loads on a wood member is 50,000 lb and the coefficient of elasticity is 1,600, the cross-section area is:

$$CS = \frac{50,000}{1,600}$$
$$= 31.25 \text{ in}^2$$

The cross section of a 4 × 10 member surfaced to 3.5″ × 9.5″ is 33.25 in^2. A 4 × 10 beam is therefore the smallest member possible above the minimum 31.25 in^2.

STEP 6 Select the exact size and classification of the beam or girder. Use Figure B.34 to select the most appropriate wood girder. For example, to support 6,000 lb over a 10′ span, either an 8 × 8 built-up girder or a 6 × 10 solid girder would suffice. The girder should be strong enough to support the load, but any size larger is a waste of materials. The only alternative to increasing the size of the girder is to decrease the size of span.

Calculating the Tributary Area Calculating the tributary area that is supported by a girder or beam is shown in Figure B.35. The factors used to calculate a safe load that is evenly distributed on a beam are shown in Figure B.36.

The allowable fiber stress of the beam must be known to complete these calculations. For example, if a 6 × 12 beam (actual size 5 1/2″ × 11 1/4″) spans 18′-0″ and the beam has a fiber stress rated at 1,800 lb/ft^2, the safe evenly distributed weight is calculated as follows:

$$W = \frac{f \times b \times d^2}{9 \times L}$$

where W = weight evenly distributed, lb
f = allowable fiber stress, lb/in^2
b = width of beam, in
d = depth of beam, in
L = span, ft

$$W = \frac{1,800 \times 5.5 \times 11.25^2}{9 \times 18} = 8,082 \text{ lb}$$

GIRDER SIZE	SAFE LOAD IN POUNDS FOR SPANS FROM 6 TO 10 FEET				
	6 FT	7 FT	8 FT	9 FT	10 FT
6 × 8 built-up	8,306	7,118	6,220	5,539	4,583
6 × 8 solid	7,359	6,306	5,511	4,908	4,062
6 × 10 built-up	11,357	10,804	9,980	8,887	7,997
6 × 10 solid	10,068	9,576	8,844	7,878	7,086
8 × 8 built-up	11,326	9,706	8,482	7,553	6,250
8 × 8 solid	9,812	8,408	7,348	6,544	5,416
8 × 10 built-up	15,487	14,732	13,608	12,116	10,902
8 × 10 solid	13,424	12,768	11,792	10,504	9,448

FIGURE B.34 > Wood girder safe loads. *Hepler/Wallach/Hepler © Cengage 2013*

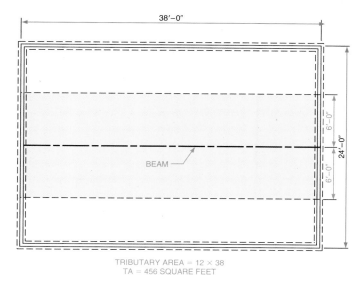

TRIBUTARY AREA = 12 × 38
TA = 456 SQUARE FEET

FIGURE B.35 > Calculating the tributary area of a beam. *Hepler/Wallach/Hepler © Cengage 2013*

Calculating Deflection The stiffness of a beam, or any structural member, is the ability of the member to resist bending. The bending force that changes a straight member to a curved member is *deflection*. The calculation of the deflection of a structural member uses Hooke's law. Data for a supported beam are shown in Figure B.37. For example, the deflection of a 2 × 6 (1 1/2″ × 5 1/2″) joist that has a 20′ span and supports a central load of 4,000 lb is calculated as follows:

$$D = \frac{PL^3}{48EI}$$

where   D = deflection
        P = force in center of span, lb
        L = length of beam span, ft
        E = modulus of elasticity (given in table as 1,000,000)
        I = moment of inertia (for rectangular members)
        I = bd/12; b = width, d = height

$$D = \frac{PL^3}{48EI}$$

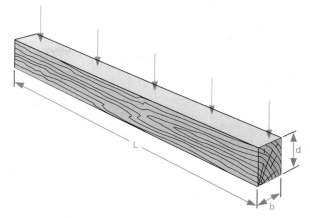

FIGURE B.36 > Evenly distributed load on a girder or beam. *Hepler/Wallach/Hepler © Cengage 2013*

FORCE: 4,000 LB
SPAN: 20′
BEAM: 2 × 6

FIGURE B.37 > Beam deflection. *Hepler/Wallach/Hepler © Cengage 2013*

D = the amount of deflection
P = the concentrated load, lb
L = the span, ft
E = the modulus of elasticity (1,000,000) and the moment of inertia
I = the cross section of the beam

$$D = \frac{4,000 \times 20^3}{48 \times 1,000,000 \times 1.5 \times 5.5}$$

$$D = \frac{32,000,000}{396,000,000}$$

$$D = 0.08″$$

Steel Beams The method of determining the size of steel beams is the same as for determining the size of wood beams. As wood beams vary in width for a given depth, steel beams vary in weight, depth, and thickness of webs and flanges. Classifications vary accordingly. Figure B.38 shows the relationship of the span, load, depth, and weight of standard steel W-shape beams and channels. A steel beam may be selected by referring to the desirable span and load and then choosing the most appropriate size (depth and weight) for the W-shape beam. For example, a 4″ × 8″, 18.4-pound W-shape beam will support 8.5 kips at a span of 20 ft. A *kip* is equal to 1,000 lb.

Columns and Posts

When girders or beams do not completely span the distance between foundation walls, then wood posts, steel-pipe columns, masonry columns, or steel columns must be used for

SIZE OF BEAM	2⅝″ × 4″	2¾″ × 4″	3″ × 5″	3¼″ × 5″	3⅜″ × 6″	3⅝″ × 6″	3⅝″ × 7″	3⅞″ × 7″	4″ × 8″	4⅛″ × 8″	4⅝″ × 10″	5″ × 10″
WEIGHT PER FOOT IN POUNDS	7.7	9.5	10.0	14.75	12.5	17.25	15.3	20.0	18.4	23.0	25.4	35.0
Span in feet												
4	9.0	10.1	14.5	18.0	21.8	26.0	31.0	36.0	42.7	48.2	73.3	87.5
5	7.2	8.0	11.6	14.4	17.4	20.8	24.8	28.7	34.1	38.5	58.6	70.0
6	6.0	6.7	9.7	12.0	14.5	17.3	20.7	24.0	28.5	32.1	48.8	58.3
7	5.1	5.7	8.3	10.3	12.5	14.9	17.7	20.5	24.4	27.5	41.9	50.0
8	4.5	5.0	7.3	9.0	10.9	13.0	15.5	18.0	21.3	24.1	36.6	43.7
9	4.0	4.5	6.5	8.0	9.7	11.6	13.8	16.0	19.0	21.4	32.6	38.9
10	3.6	4.0	5.8	7.2	8.7	10.4	12.4	14.4	17.1	19.3	29.3	35.0
11	—	—	5.3	6.5	7.9	9.5	11.3	13.1	15.5	17.5	26.6	31.8
12	—	—	—	—	7.3	8.7	10.3	12.0	14.2	16.1	24.4	29.2
13	—	—	—	—	6.7	8.0	9.5	11.1	13.1	14.8	22.5	26.9
14	—	—	—	—	6.2	7.4	8.9	10.3	12.2	13.8	20.9	25.0
15	—	—	—	—	—	—	8.3	9.6	11.4	12.8	19.5	23.3
16	—	—	—	—	—	—	7.7	9.0	10.7	12.0	18.3	21.9
17	—	—	—	—	—	—	—	—	10.0	11.3	17.2	20.6
18	—	—	—	—	—	—	—	—	9.5	10.7	16.3	19.4
19	—	—	—	—	—	—	—	—	9.0	10.1	15.4	18.4
20	—	—	—	—	—	—	—	—	8.5	9.6	14.7	17.5

FIGURE B.38 ❯ Safe loads for steel W-shape beams in kips (1,000 lb). *Hepler/Wallach/Hepler © Cengage 2013*

intervening support to reduce the span. To determine the most appropriate size and classification of posts or columns to support girders or beams, follow these steps:

STEP 1   Determine the total load in pounds per square foot for the entire floor area. Calculate this amount by multiplying the total load by the number of square feet of floor space.

STEP 2   Determine the spacing of posts necessary to support the ends of each girder. Great distances between posts should be avoided because excessive weight would concentrate on one footing. Long spans also require extremely large girders. For example, it is possible to span a distance of 20′, but to do so a 5″ × 10″ W-shape beam would be needed. The extreme weight and cost of this beam would be prohibitive. On the other hand, if only a 6′ span is used, the close spacing might greatly restrict the flexibility of the internal design. As a rule, use the shortest span that will not interfere with the design function of the area.

STEP 3   Find the number of square feet supported by each post. A post will carry the load on a girder to the midpoint of the span on both sides. The post also carries half the load to the nearest support wall on either side of the post tributary area.

STEP 4   Find the load supported by the post support area. Multiply the number of square feet by the load per square foot for the total tributary load.

STEP 5   Determine the height of a post. The height of the post is related to the length of the column. The

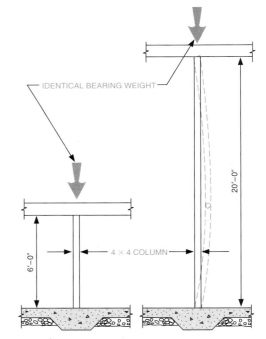

FIGURE B.39 ❯ Relationship of height to stability. *Hepler/Wallach/Hepler © Cengage 2013*

4 × 4 post shown in Figure B.39 may be more than adequate to support a given weight if the height of the post is 6′. However, the same 4 × 4 post may be totally inadequate to support the same weight when the length is increased to 20′.

STEP 6   Determine the thickness and width of the post needed to support the load at the given height by referring to Figure B.40 for wood posts, Figure B.41 for W-shape columns, or Figure B.42

STANDARD SIZE, INCHES	3×4	4×4
ACTUAL SIZE, INCHES	$2\frac{1}{2} \times 3\frac{1}{2}$	$3\frac{1}{2} \times 3\frac{1}{2}$
AREA IN SQUARE INCHES	9.51	13.14
Height of column:		
4 feet	8 720	12 920
5 feet	7 430	12 400
6 feet	5 630	11 600
6 feet 6 inches	4 750	10 880
7 feet	4 130	10 040
7 feet 6 inches	—	9 300
8 feet	—	8 350
9 feet	—	6 500
10 feet	—	—
11 feet	—	—
12 feet	—	—

FIGURE B.40 > Maximum loads for wood posts. *Hepler/ Wallach/Hepler © Cengage 2013*

DEPTH IN INCHES	10	9
WEIGHT PER POUND PER FOOT	25.4	21.8
Effective length:		
3 feet	110.7	94.8
4 feet	110.7	94.8
5 feet	109.5	91.2
6 feet	101.7	83.9
7 feet	93.8	76.7
8 feet	86.0	69.7
9 feet	78.7	63.2
10 feet	71.8	57.2
11 feet	65.5	51.8
12 feet	59.7	47.0
Area in square inches	7.38	6.32

FIGURE B.41 > Safe loads for I-columns. *Hepler/Wallach/ Hepler © Cengage 2013*

for the diameter of steel-pipe columns. To check the compressive stress on posts, apply the following formula:

$$F = \frac{P}{A}$$

where   F = compressive stress
P = compressive force
A = cross-section area

For example, in Figure B.43 the compressive force on the 4×4 post (16 in²) is 32,000 lb; therefore,

$$F = \frac{32,000}{16}$$

F = 2,000 lb/in² or 2 kips (compressive stress)

Calculating Deformation *Deformation* is the change of form, shape, or dimensions of the structural members of a building caused by load stress. The amount of deformation on a member is calculated as follows:

$$\delta = \frac{PL}{AE}$$

where δ = deformation, in.
P = force, lb
L = length, in.
A = cross-section area
E = modulus of elasticity, lb/in² (for structural steel, E = 29,000,000 lb/in²; for wood, E = 1,000,000 to 2,000,000 lb/in²)

For example, a tensile force P of 50,000 lb is acting on the structural column shown in Figure B.44. The deformation is, therefore,

$$\delta = \frac{50,000 \text{ lb} \times 100''}{16'' \times 29,000,000} = \frac{5,000,000}{464,000,000}$$

$$\delta = .0108''$$

NOMINAL SIZE, INCHES	6	5	4½	4	3½	3	2½	2	1½
EXTERNAL DIAMETER, INCHES	6.625	5.563	5.000	4.500	4.000	3.500	2.875	2.375	1.900
THICKNESS, INCHES	.280	.258	.247	.237	.226	.216	.203	.154	.145
Effective length:									
5 feet	72.5	55.9	48.0	41.2	34.8	29.0	21.6	12.2	7.5
6 feet	72.5	55.9	48.0	41.2	34.8	28.6	19.4	10.6	6.0
7 feet	72.5	55.9	48.0	41.2	34.1	26.3	17.3	9.0	5.0
8 feet	72.5	55.9	48.0	40.1	31.7	24.0	15.1	7.4	4.2
9 feet	72.5	55.9	46.4	37.6	29.3	21.7	12.9	6.6	3.5
10 feet	72.5	54.2	43.8	35.1	26.9	19.4	11.4	5.8	2.7
11 feet	72.5	51.5	41.2	32.6	24.5	17.1	10.3	5.0	—
12 feet	70.2	48.7	38.5	30.0	22.1	15.2	9.2	4.1	—
Area in square inches	5.58	4.30	3.69	3.17	2.68	2.23	1.70	1.08	0.80
Weight per pound per foot	18.9	14.6	12.5	10.7	9.11	7.58	5.79	3.65	2.72

FIGURE B.42 > Safe loads for steel-pipe columns. *Hepler/Wallach/Hepler © Cengage 2013*

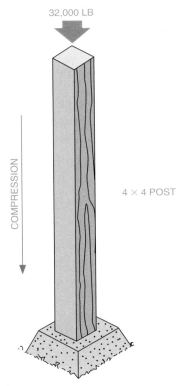

32,000 LB

COMPRESSION

4 × 4 POST

FIGURE B.43 > Compressive-force stress. *Hepler/Wallach/ Hepler © Cengage 2013*

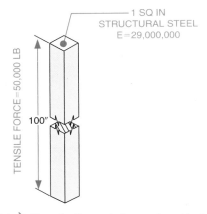

1 SQ IN
STRUCTURAL STEEL
E=29,000,000

TENSILE FORCE=50,000 LB

100″

FIGURE B.44 > Tensile-force deformation. *Hepler/Wallach/ Hepler © Cengage 2013*

Foundation Footings and Piers

The basic T-foundation has an exterior concrete footing and a wall forming an inverted T. Generally, builders follow the sizes for foundation walls and footings as recommended in Figure B.45. Piers support the interior floor system. In residential construction, piers are usually 12″ high and 12″ to 18″ square at the base. To compute the bearing area of the pier or the T-foundation's footing, divide the total load on one square foot by the soil-bearing capacity per square foot.

Footing Size Calculations To calculate the minimum bearing size of a lineal foot of footing refer to A through F in Figure B.46 and Figure B.47. First find the total load on the footing as follows:

A. Roof live load . . . . .30 lb/ft²
   Roof dead load
       Rafters 2 × 8, 16″ OC . . . . 3.5 lb/ft²
       Wood sheathing 1″ . . . . .3 lb/ft²
       Wood shingles . . . . .3 lb/ft²
       Batt insulation 3″ . . . . 0.5 lb/ft²
                                                                    40 lb/ft²
       Rafter length = 20′ (1 side) × 40 lb/ft² = 800 lb

B. Attic floor/ceiling live load . . . . .20 lb/ft²
   Attic floor/ceiling dead load
       Joists 2 × 10, 16″ OC . . . . 6.5 lb/ft²
       Lath and plaster (one side) . . . . .10 lb/ft²
       Batt insulation 3″ . . . . 0.5 lb/ft²
                                                                    37 lb/ft²
              Tributary length 9′ × 37 = 333 lb
   (See Figure B.48.)

C. Exterior wall live load 0
   Exterior wall dead load
       2 × 4 wood stud/plaster one side . . . . .12 lb/ft²
       Exterior wood siding 1″ . . . . .3 lb/ft²
       Batt insulation 3″ . . . . 0.5 lb/ft²
                                                                    15.5 lb/ft²
              Wall height 8′ × 15.5 = 124 lb

HEIGHT DESCRIPTION	WOOD FRAME HOUSE		MASONRY HOUSE	
	MINIMUM FOUNDATION WALL THICKNESS	FOOTING PROJECTION EACH SIDE OF FOUNDATION WALL	MINIMUM FOUNDATION WALL THICKNESS	FOOTING PROJECTION EACH SIDE OF FOUNDATION WALL
One story–no basement	6″	2″	6″	3″
One story–with basement	6″	3″	8″	4″
Two story–no basement	6″	3″	6″	4″
Two story–with basement	8″	4″	8″	5″

FIGURE B.45 > Footing and foundation wall sizes. *Hepler/Wallach/Hepler © Cengage 2013*

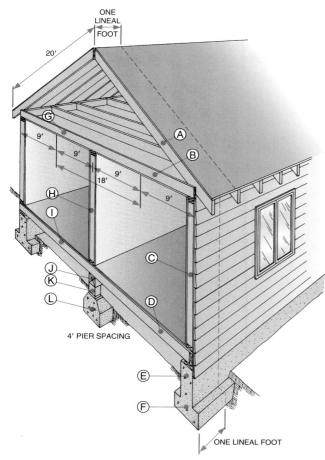

FIGURE B.46 > Structural members used in calculating dead loads. *Hepler/Wallach/Hepler © Cengage 2013*

D. Floor live load. . . . . .40 lb/ft²
Floor dead load 2 × 10 floor joists
16″ OC/subfloor. . . . . 6.5 lb/ft²
Hardwood floor 25/32″ . . . . .4 lb/ft²
                                                      50.5 lb/ft²
          Tributary length 9′ × 50.5 = 454.5 lb
(See Figure B.48.)

E. Foundation wall live load 0
Foundation wall dead load
6″ concrete . . . . .75 lb/ft²
          Approximately 3′ height × 75 = 225 lb

F. Footing live load. 0
Footing dead load 6″ concrete. . . . . .75 lb/ft²
          Approximately 1′-0″ wide × 1 lineal foot × 75 = 75 lb

Total load per lineal foot of footing is 2,011.5 lb

To calculate the minimum footing size apply the following formula:

Total load on footing:. . . . . 2,011.5 lb
Soil-bearing capacity (given): . . . . 2,500 lb/ft²
                              = .80 ft² (115.2 in²)

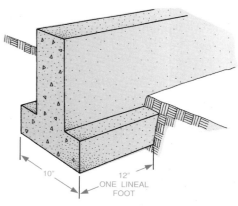

FIGURE B.47 > Loads calculated on one lineal foot of footing. *Hepler/Wallach/Hepler © Cengage 2013*

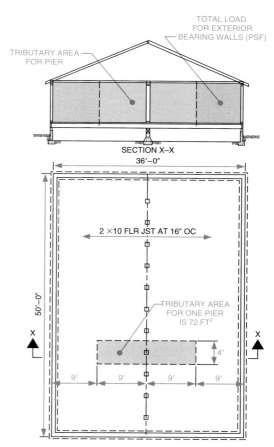

FIGURE B.48 > Tributary areas for exterior walls and piers. *Hepler/Wallach/Hepler © Cengage 2013*

A lineal foot is 12″, therefore the width of the footing should be 10″ (10″ × 12″ = 120 in²) minimum. A 12″ standard footing width can therefore be used. (See Figure B.47.)

Pier Size Calculation To calculate the size of a pier needed to support the structure refer to items G through L in Figure B.46 and calculate the load as follows:

G. Attic floor/ceiling live load . . . . 20.0 lb/ft²
   Attic floor/ceiling dead load
      Floor joists 2 × 10, 16″ OC. . . . . 6.5 lb/ft²
      Lath and plaster, one side . . . . 10.0 lb/ft²
      3″ batt insulation . . . . <u>0.5 lb/ft²</u>
                                                        37.0 lb/ft²
   (Again, see Figure B.46.)
             Tributary area (72 ft²); 72 × 37 = 2,664 lb

H. Wall live load 0
   Wall dead load
      2 × 4 wood stud/plaster
      Two sides . . . . .20 lb/ft²
              Wall height 8′; 8 × 20 = 160 lb

I. Floor live load. . . . . 40.0 lb/ft²
   Floor dead load
      2 × 10 floor joist, 16″
      OC/subfloor. . . . . 6.5 lb/ft²
      Hardwood floor 25/32″ . . . . <u>4.0 lb/ft²</u>
                                             50.5 lb/ft²
      Tributary area 4′ × 18′ = 72 ft²; 72 × 50.5
        = 3,636 lb

J. Girder 4 × 6, dead load . . . . .4 lb/ft²
                    4′ length × 4 = 16 lb

K. Post 4 × 6, dead load .4 lb/ft²
            Approximately 1.5′ × 4 = 6 lb

L. Concrete pier . . . . .150 lb/ft²
        Approx. 24″ × 24″ × 12″ = 4 cubic feet
            4 ft³ × 150 = 600 lb
   Total load per pier = 7,082 lb (Total load on one pier)

Now use the total load and apply the following formula to compute the minimum size of pier:

$$\text{Total load on pier: } \frac{7,082}{2,500} = 2.8 \text{ ft}^2$$

Soil-bearing capacity (given): 2.8 ft² × 144 = 403.2 in²

21″ × 21″ (441 in²) is the next largest standard square size.

Stress and Dynamics Formulas

Some structural designs include construction methods, materials, and sizes that are not standard. These cannot be determined by using precalculated charts. The following formulas relate to these aspects of structural design.

Direct Stress Calculations To calculate the unit of direct stress on a structural member, divide the load by the cross-section area of the member:

$$F = \frac{P}{A}$$

where F = unit of stress
       P = load/force
       A = cross-section area

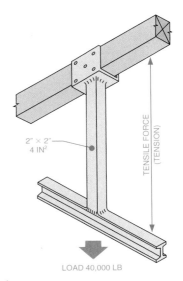

FIGURE B.49 > Direct stress calculation. *Hepler/Wallach/ Hepler © Cengage 2013*

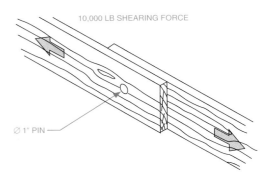

FIGURE B.50 > Shearing stress. *Hepler/Wallach/Hepler © Cengage 2013*

For example, in Figure B.49, a 40,000 lb I-beam is supported by a 2″ × 2″ (4 in²) support rod. The stress on the rod is

$$F = \frac{40,000}{4} = 10,000 \text{ lb/in}^2 \text{ or } 10 \text{ kips}$$

Shearing Stress A unit of shearing stress equals the shearing force (in pounds) divided by the cross-section area (in square inches) of the structural member:

$$S = \frac{P}{A}$$

where S = unit of shearing stress
      P = shearing force
      A = cross-section area of stress object

For example, in Figure B.50, a shearing force of 10,000 lb is acting on a 1″ pin. Therefore,

$$S = \frac{10,000}{\pi r^2} = \frac{10,000}{3.14 \times .5^2} = \frac{10,000}{.785}$$
$$= 12,739 \text{ lb/in}^2 \text{ or } 12.7 \text{ kips}$$

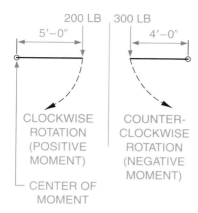

FIGURE B.51 〉 Moment of force. *Hepler/Wallach/Hepler*
© *Cengage 2013*

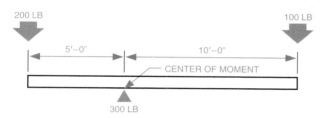

FIGURE B.52 〉 Forces in equilibrium. *Hepler/Wallach/Hepler*
© *Cengage 2013*

Moment of Force To calculate the moment of force (in foot-pounds) on a structural member, multiply the force (in pounds) by the length of the moment arm (in feet):

Moment of force = force × arm length

For example, the left side of Figure B.51 shows a force of 200 lb acting on the end of a 5′ arm. The moment of force is, therefore, 1,000 lb (200 lb × 5′). The moment of force on the 4′ arm on the right is 1,200 ft-lb.

Forces in Equilibrium The positive and negative moment values are always equal when there is no movement—that is, when the system is in equilibrium. The beam in Figure B.52 is in balance (equilibrium). The moment of force on the right of the center of moment is 200 × 5 = 1,000 ft-lb. The moment of force to the left of center is 100 lb × 10 = 1,000 ft-lb.

GEOMETRIC CALCULATIONS

Architectural drawings contain a wide variety of geometric shapes. Some are two-dimensional flat surfaces and others are three-dimensional volume-containing areas. In estimating amounts of construction materials, labor, and costs, the ability to compute distances, areas, and volumes is important. The majority of geometric calculations in architectural work falls into these categories.

Perimeter of a Polygon

Formula: Perimeter = sum of all sides
Examples: See Figure B.53.

Top drawing: 10 + 20 + 10 + 20 = 60′

Bottom drawing: 25′-0″
60′-6″
40′-0″
20′-0″
15′-0″
40′-6″
200′-12″ = 201′

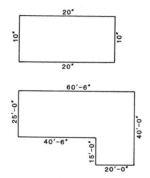

FIGURE B.53 〉 Perimeter of a polygon. *Hepler/Wallach/Hepler*
© *Cengage 2013*

Circumference of a Circle

Formula: Circumference = π × diameter

$$C = \pi D$$

Example: See Figure B.54. π = 22/7 or 3.14.

Left drawing: $C = 3.14 \times 7 = 22'$

Right drawing: Inside-diameter

$C = 3.14 \times 14' = 43.96'$

Outside-diameter

$C = 3.14 \times 22' = 69.08'$

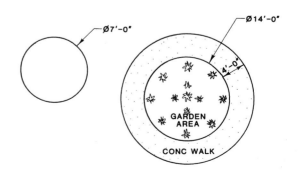

FIGURE B.54 〉 Circumference of a circle. *Hepler/Wallach/Hepler* © *Cengage 2013*

Area of a Square or Rectangle

Formula: Area = side × side

$$A = SS$$

Example: See Figure B.55.

Top drawing:	50′ × 25′	=	1,250 ft²
Center drawing:	175′ × 60′	=	10,500 ft²
Bottom drawing:	60′ × 60′	=	3,600 ft²
	20′ × 20′	=	400 ft²
			4,000 ft²

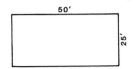

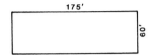

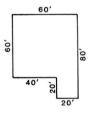

FIGURE B.55 > Area of a square or rectangle. *Hepler/Wallach/ Hepler © Cengage 2013*

Area of a Triangle

Formula: Area = 1/2 base × altitude

$$A = 1/2 \ BA$$

Example: See Figure B.56.

Top drawing: $\dfrac{20'' \times 10''}{2} = 100 \ in^2$

Center drawing: $\dfrac{15' \times 30'}{2} = 225 \ ft^2$

Bottom drawing: $\dfrac{100 \times 70}{2} = 3,500 \ ft^2$

$\dfrac{100 \times 40}{2} = 2,000 \ ft^2$

$3,500 + 2,000 = 5,500 \ ft^2$

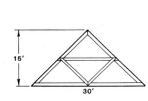 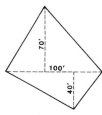

FIGURE B.56 > Area of a triangle. *Hepler/Wallach/Hepler © Cengage 2013*

Surface Area of a Cylinder

Formula: Surface area = π × diameter × height

$$A = \pi DH$$

Example: See Figure B.57.

Top: 3.14 × 10″ × 20″ = 628 in²

Center: 3.14 × 2″ × 100″ = 628 in²

Bottom: Well is 10′-0″ deep and 5′-0″ wide
3.14 × 10′ × 5′ = 157 ft²

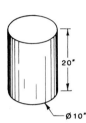 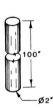 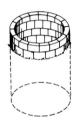

FIGURE B.57 > Surface area of a cylinder. *Hepler/Wallach/ Hepler © Cengage 2013*

Area of a Circle

Formula: Area = π × radius squared

$$A = \pi R^2$$

Example: See Figure B.58.

Top drawing: 3.14 × 7²
3.14 × 49 = 153.86 in²

Center drawing: 10 ÷ 2 = 5
3.14 × 5²
3.14 × 25 = 78.5 in²

Bottom drawing: 20 ÷ 2 = 10
3.14 × 10² = 314
314 ÷ 2 = 157 in²

FIGURE B.58 > Area of a circle. *Hepler/Wallach/ Hepler © Cengage 2013*

Volume of a Cube

Formula: Volume = length × height × width

$$V = LHW$$

Example: See Figure B.59.

Top: $V = 10'' \times 5'' \times 12''$
$= 600 \text{ in}^3$

Center: $V = 6 \times 50 \times 60$
$= 18{,}000 \text{ ft}^3$
$= 18{,}000 \div 27 = 666 \text{ yd}^3$

Bottom: $40' \times 24'' \times 6''$
$40' \times 2' \times 0.5' = 40 \text{ ft}^3$
$40' \times 6'' \times 12$
$40' \times 0.5' \times 1' = \underline{20 \text{ ft}^3}$
60 ft^3

$60 \div 27 = 2.2 \text{ yd}^2$

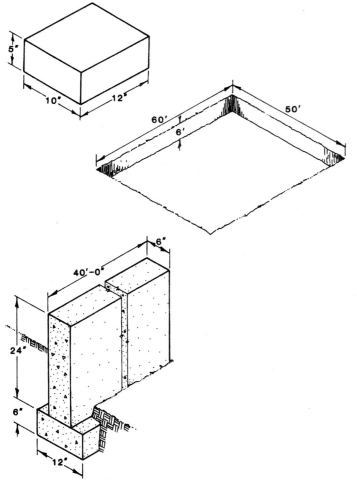

FIGURE B.59 ⟩ Volume of a cube. *Hepler/Wallach/Hepler*
© *Cengage 2013*

Volume of a Cylinder

Formula: Volume = π × radius squared × height

$$V = \pi R^2 H$$

Example: See Figure B.60.

Top: $V = 3.14 \times 5^2 \times 7$
$= 3.14 \times 25 \times 7$
$= 550 \text{ in}^3$

Center: $V = 3.14 \times 10^2 \times 14$
$= 3.14 \times 100 \times 14$
$= 4396 \text{ ft}^3$

Bottom: $V = 3.14 \times 3^2 \times 7$
$= 3.14 \times 9 \times 7$
$= 197.8 \text{ yd}^3$

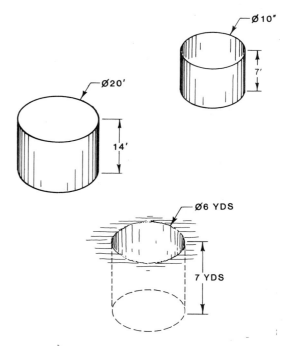

FIGURE B.60 ⟩ Volume of a cylinder. *Hepler/Wallach/Hepler*
© *Cengage 2013*

Volume of a Square Pyramid

Formula: Volume = 1/3 × width of base × depth of base × height

$$V = 1/3 WDH$$

Example: See Figure B.61.

Top: $V = 0.33 \times 25'' \times 30'' \times 20''$
$= 4{,}950 \text{ in}^3$

Bottom: $V = 0.33 \times 10' \times 12' \times 10'$
$= 396 \text{ ft}^3$

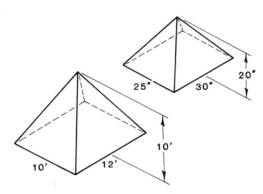

FIGURE B.61 > Volume of a square pyramid. *Hepler/Wallach/ Hepler © Cengage 2013*

Volume of a Cone

Formula: Volume = 1/3 × π × radius squared × height

$$V = 1/3 \pi R^2 H$$

Example: See Figure B.62.

Top: $V = .33 \times 3.14 \times 10^2 \times 30$
$= .33 \times 3.14 \times 100 \times 30 = 3{,}108.6 \text{ in}^3$

Bottom: $V = .33 \times 3.14 \times 20^2 \times 20 \text{ ft}^2$
$= .33 \times 3.14 \times 400 \times 20 \text{ ft}^2$
$= 8{,}289.6 \text{ ft}^3$

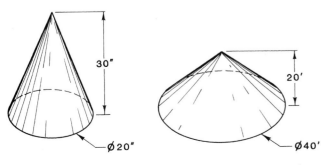

FIGURE B.62 > Volume of a cone. *Hepler/Wallach/Hepler © Cengage 2013*

Volume of a Sphere

Formula: Volume = 1/6 × π 3 diameter cubed

$$V = 1/6 \pi D^3$$

Example: See Figure B.63. 1/6 = .166.

$$V = .166 \times 3.14 \times 3^3$$
$$= .166 \times 3.14 \times 27$$
$$= 14.07 \text{ in}^3$$

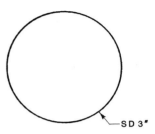

FIGURE B.63 > Volume of a sphere. *Hepler/Wallach/Hepler © Cengage 2013*

Right Triangle Law (Pythagorean Theorem)

Formula: Square of the hypotenuse = the sum of the square of the two sides

$$C^2 = A^2 + B^2$$

Example: See Figure B.64.

$$C^2 = A + B^2$$
$$C = \sqrt{A^2 + B^2}$$
$$C = \sqrt{32^2 + 24^2}$$
$$C = \sqrt{1{,}024 + 576}$$
$$C = \sqrt{1{,}600}$$
$$C = 40''$$

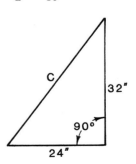

FIGURE B.64 > Hypotenuse of a right triangle. *Hepler/ Wallach/Hepler © Cengage 2013*

APPENDIX C

Architectural Abbreviations and Professional Organizations

ARCHITECTURAL ABBREVIATIONS

Hundreds of words may be lettered on an architectural drawing. Abbreviations are often used to conserve space. By using standard abbreviations, drawings can be interpreted consistently.

Capital letters are used for abbreviations on drawings. Periods are used only when the abbreviation may be confused with a whole word. The same abbreviation is used for both singular and plural terms, and two or more terms may use the same abbreviation.

Access panel	**AP**
Acoustic	**ACST**
Actual	**ACT.**
Addition	**ADD.**
Adhesive	**ADH**
Aggregate	**AGGR**
Air condition	**AC**
Alternate	**ALT**
Alternating current	**AC**
Aluminum	**AL**
Amount	**AMT**
Ampere	**AMP; AP**
Anchor bolt	**AB**
Apartment	**APT.**
Approved	**APPD**
Approximate	**APPROX**
Architectural	**ARCH**
Area	**A**
Asbestos	**Asb**
Asphalt	**ASPH**
At	**@**
Automatic	**AUTO**
Auxiliary	**AUX**
Avenue	**AVE**
Average	**AVG**
Azimuth	**AZ**
Balcony	**BALC**

Base	**B**
Basement	**BSMT**
Bathroom	**B**
Bathtub	**BT**
Batten	**BATT**
Beam	**BM**
Bearing	**BRG**
Bedroom	**BR**
Benchmark	**BM**
Bending moment	**BM**
Between	**BET**
Bill of material	**B/M**
Block	**BLK**
Blocking	**BLKG**
Blower	**BLO**
Blueprint	**BP**
Board	**BD**
Board feet	**BF**
Boiler	**BLR**
Both sides	**BS**
Bottom	**BOT**
Brick	**BRK**
British thermal units	**BTU**
Bronze	**BRZ**
Broom closet	**BC**
Building	**BLDG**
Building line	**BL**
Cabinet	**CAB.**
Carbon monoxide	**CO**
Casing	**CSG**
Cast concrete	**C CONC**
Cast iron	**CI**
Catalog	**CAT.**
Caulking	**CLKG**
Ceiling	**CLG**
Cement	**CEM**
Center	**CTR**
Center to center	**C to C**
Centerline	**CL**
Centimeter	**CM**
Ceramic	**CER**
Certificate of occupancy	**CO**

Chamfer	**CHAM**
Channel	**CHAN**
Check	**CHK**
Chimney	**CHIM**
Chord	**CHD**
Circle	**CIR**
Circuit	**CIR; CKT**
Circuit breaker	**CIR BKR**
Circumference	**CIRC**
Cleanout	**CO**
Clear	**CLR**
Closet	**CL**
Coated	**CTD**
Cold water	**CW**
Column	**COL**
Combination	**COMB.**
Common	**COM**
Composition	**COMP**
Computer-aided drafting	**CAD**
Computer Aided Drafting and Design	**CADD**
Concrete	**CONC**
Construction	**CONST**
Continue	**CONT**
Continuous	**CONT**
Contractor	**CONTR**
Copper	**Cu**
Corrugate	**CORR**
Counter	**CTR**
Course	**C**
Cross section	**X-SECT; CS**
Cubic foot	**CU FT**
Cubic inch	**CU IN.**
Cubic yard	**CU YD**
Damper	**DMPR**
Dampproofing	**DP**
Datum	**DAT**
Dead load	**DL**
Decibel	**DB**
Decking	**DK**
Deflection	**D**

Degree	(°) DEG	Finish	FIN.	House	HSE
Department	DEPT	Finished grade	FG	Hundred	C
Design	DSGN	Fire hydrant	FH	Hypotenuse	H
Design development	DD	Fireproof	FPRF	I-beam	I
Detail	DET	Fixture	FIX.	Inch	(″) IN.
Diagonal	DIAG	Flashing	FL	Inch-pounds	IN-LB
Diagram	DIAG	Floor	FL	Incinerator	INCIN
Diameter	D; Ø	Floor drain	FD	Include	INCL
Dimension	DIM	Flooring	FLG	Information	INFO
Dining room	DR	Fluorescent	FLUOR	Inside diameter	ID
Direct current	DC	Foot	(′) FT	Install	INST
Dishwasher	DW	Footcandle	FC	Insulate	INS
Disk operating system	DOS	Footing	FTG	Intercommunication	INTER-COM
Disposal	DISP	Footing area	FA	Interior	INT
Ditto	DO.	Foot-pounds	FT-LB	Iron	I
Division	DIV	Forced-air unit	FAU	Jamb	JMB
Door	DR	Foundation	FDN	Joint	JT
Double	DBL	Frame	FR	Joist	JST
Double-hung	DH	Freshwater wetlands	FWW	Joule	J
Dowel	DWL	Full size	FS	Junction	JCT
Down	DN	Furnace	FURN	Kilowatt	KW
Downspout	DS	Furnished by others	FBO	Kilowatt-hour	KWH
Drafting	DFT	Furred (ing)	FUR	Kip (1,000 lb)	K
Drain	DR	Furred ceiling	FC	Kitchen	KIT
Drawing	DWG	Gage	GA	Knockout	KO
Dryer	D	Gallon	GAL	Laminate	LAM
Drywall	DW	Galvanize	GALV	Latitude	LAT
Duplicate	DUP	Galvanized iron	GI	Laundry	LAU
Each	EA	Garage	GAR	Lavatory	LAV
East	E	Gas	G	Leader	LDR
Elbow	EL	Girder	G	Left	L
Electric	ELEC	Glass	GL	Left hand	LH
Elevation	EL	Glue laminated	GLULAM	Length	L; LG
Enamel	ENAM	Grade	GR	Length overall	LOA
Engineer	ENGR	Grade line	GL	Level	LEV
Entrance	ENT	Ground	GND	Light	LT
Equal	EQ	Ground-fault circuit interrupter	GFCI	Linear	LIN
Equipment	EQUIP.	Grout	GR	Linen closet	L CL
Estimate	EST	Gypsum	GYP	Lintel	LTL
Excavate	EXC	Hall	H	Live load	LL
Existing	EXIST.	Hardware	HDW	Living room	LR
Expansion joint	EXP JT	Head	HD	Long	LG
Exposed	EXPO	Header	HDR	Louver	LV
Extension	EXT	Heater	HTR	Lumber	LBR
Exterior	EXT	Heating/ventilating/air conditioning	HVAC	Lumen	LW
Fabricate	FAB	Height	HT	Main	MN
Face brick	FB	Hollow core	HC	Manual	MAN.
Face of studs	FOS	Horizontal	HOR	Manufacturing	MFG
Face to face	F to F	Hose bib	HB	Master plan	MP
Fahrenheit	F	Hot water	HW	Material	MATL
Feet	(′) FT	Hour	HR	Maximum	MAX
Feet board measure	FBM			Medicine cabinet	MC
Figure	FIG				

Membrane	**MEMB**	Prefabricated	**PREFAB**	South	**S**
Metal	**MET.**	Preferred	**PFD**	Specification	**SPEC**
Meter	**M**	Preliminary	**PRELIM**	Square	**SQ**
Mile	**MI**	Pressure treated	**PT**	Square foot	**FT²;**
Millimeter	**MM**	Programming	**PR**		**SQ FT; ′**
Minimum	**MIN**	Property	**PROP**	Square inch	**IN²;**
Minute	**MIN**	Property line	**PL**		**SQ IN; ″**
Miscellaneous	**MISC**	Quality	**QUAL**	Stainless steel	**SST**
Mixture	**MIX.**	Quantity	**QTY**	Stairs	**ST**
Model	**MOD**	Radiator	**RAD**	Standard	**STD**
Modular	**MOD**	Radius	**R**	Steel	**STL**
Moisture resistant	**MR**	Random-access memory	**RAM**	Stock	**STK**
Molding	**MLDG**	Range	**R**	Storage	**STG**
Motor	**MOT**	Read-only memory	**ROM**	Storm drain	**SD**
Mullion	**MULL**	Receptacle	**RECP**	Street	**ST**
National Electrical Code	**NEC**	Reference	**REF**	Structural	**STR**
Natural	**NAT**	Refrigerator	**REF**	Substitute	**SUB**
Nominal	**NOM**	Register	**REG**	Supply	**SUP**
North	**N**	Reinforce	**REINF**	Surface	**SUR**
Not applicable	**NA**	Reinforcing bar	**REBAR**	Switch	**S**
Not to scale	**NTS**	Reproduce	**REPRO**	Symbol	**SYM**
Number	**NO.**	Required	**REQD**	Symmetrical	**SYM**
Obscure	**OB**	Resistance moment	**RM**	System	**SYS**
On center	**OC**	Return	**RET**	Tangent	**TAN.**
Opening	**OPNG**	Revision	**REV**	Tar and gravel	**T & G**
Opposite	**OPP**	Ridge	**RDG**	Tarpaulin	**TARP**
Ounce	**OZ**	Right hand	**RH**	Tee	**T**
Out to out	**O/O**	Riser	**R**	Telephone	**TEL**
Outside diameter	**OD**	Roof	**RF**	Television	**TV**
Overall	**OA**	Roofing	**RFG**	Temperature	**TEMP**
Overhead	**OVHD**	Room	**RM**	Terra-cotta	**TC**
Panel	**PNL**	Rough opening	**RO**	Terrazzo	**TER**
Parallel	**PAR.**	Round	**RD**	Thermostat	**THERMO**
Part	**PT**	Safety	**SAF**	Thick	**THK**
Particle board	**PBD**	Sanitary	**SAN**	Thickness	**THK**
Partition	**PTN**	Scale	**SC**	Thousand	**M**
Penny (nails)	**d**	Schedule	**SCH**	Threshold	**THR**
Per	**/**	Schematics	**SC**	Through	**THRU**
Permanent	**PERM**	Second	**(″) SEC**	Toilet	**T**
Perpendicular	**PERP**	Section	**SECT**	Tongue and groove	**T & G**
Pi (3.1416)	π	Select	**SEL**	Total	**TOT.**
Piece	**PC**	Service	**SERV**	Tread	**TR**
Plaster	**PL**	Sewer	**SEW.**	Tubing	**TUB.**
Plate	**PL**	Sheathing	**SHTHG**	Typical	**TYP**
Plumbing	**PLMG**	Sheet	**SH**	Unfinished	**UNFIN**
Plywood	**PLY**	Shower	**SH**	Urinal	**UR**
Pound	**LB**	Side	**S**	Valve	**V**
Pounds per square		Siding	**SDG**	Vanishing point	**VP**
foot	**LB/FT²;**	Similar	**SIM**	Vapor barrier	**VB**
	PSF	Sink	**S**	Vaporproof	**VAP PRF**
Pounds per square inch	**LB/IN²;**	Sketch	**SK**	Vent pipe	**VP**
	PSI	Soil pipe	**SP**	Vent stack	**VS**
Precast	**PRCST**	Solid core	**SC**	Ventilate	**VENT.**

Vertical . . . . **VERT**
Vinyl . . . . **VIN**
Volt . . . . **V**
Volume . . . . **V; VOL**
Washing machine . . . . **WM**
Waste stack . . . . **WS**
Water . . . . **W**
Water closet . . . . **WC**

Water heater . . . . **WH**
Watt . . . . **W**
Weather stripping . . . . . . **WS**
Weatherproof . . . . **WP**
Weep hole . . . . **WH**
Weight . . . . **WT**
West . . . . **W**

Width . . . . **W**
Window . . . . **WDW**
With . . . . **W/**
Without . . . . **W/O**
Wood . . . . **WD**
Wrought iron . . . . **WI**
Yard . . . . **YD**

PROFESSIONAL ORGANIZATIONS RELATED TO ARCHITECTURE

American Congress on Surveying and
  Mapping . . . . **ACSM**
American Institute of Architects . . . . **AIA**
American Institute of Building Designers . . . . . . **AIBD**
American Institute of Steel Construction . . . . . . **AISC**
American Institute of Timber Construction . . . . **AITC**
American National Standards Institute . . . . **ANSI**
American Planning Association . . . . **APA**
American Plywood Association . . . . **APA**
American Society for Testing and Materials . . . . **ASTM**
American Society of Civil Engineers . . . . **ASCE**
American Society of Heating, Refrigeration,
  and Air Conditioning Engineers . . . . **ASHRAE**
American Society of Interior Designers . . . . **ASID**
American Society of Landscape Architects . . . . . **ASLA**
American Society of Mechanical Engineers . . . . . **ASME**
American Society of Professional Estimators . . . . **ASPE**
Americans with Disabilities Act . . . . **ADA**
Associated Builders and Contractors . . . . **ABC**
Associated General Contractors of America . . . . **AGCA**
Associated Landscape Contractors of America . **ALCA**
Association of Women in Architecture . . . . **AWA**
Congress for New Urbanism . . . . **CNU**
Construction Management of America . . . . **CMA**
Construction Specification Institute . . . . **CSI**
Department of Environmental Conservation . . . **DEC**
Department of Environmental Protection . . . . . **DEP**

Environmental Protection Agency . . . . **EPA**
Federal Housing Authority . . . . **FHA**
Institute of Electrical and Electronics
  Engineers . . . . **IEEE**
International Brotherhood of Electrical
  Workers . . . . **IBEW**
International Code Council . . . . **ICC**
International Masonry Institute . . . . **IMI**
International Standards Organization . . . . **ISO**
National Association of Home Builders . . . . **NAHB**
National Association of Plumbing, Heating,
  and Cooling Contractors . . . . **NAPHCC**
National Association of Trade and Technical
  Schools . . . . **NATTS**
National Homebuilders Association . . . . **NHA**
National Lumber Manufacturers Association . . . **NLMA**
National Society of Professional Engineers . . . . . **NSPE**
Occupational Safety and Health
  Administration . . . . **OSHA**
Society of Women Engineers . . . . **SWE**
United Brotherhood of Carpenters and
  Joiners of America . . . . **UBCJA**
U.S. Department of Energy . . . . **DOE**
U.S. Department of Housing and Urban
  Development . . . . **HUD**
U.S. Geological Survey . . . . **USGS**
U.S. Green Building Council . . . . **USGBC**
Urban Land Institute . . . . **ULI**
Women Contractors Association . . . . **WCA**

Architectural terms are standard. Nevertheless, architects, drafters, and builders often use different terms for the same object. Geographic location can influence a person's word choice. For instance, what is referred to as a *faucet* in one area of the country is called a *tap* in another area. What one person calls an *attic*, another calls a *garret*, and still another a *loft*. Each boldface term below is followed by a word or words that may be used in some areas to refer to that term.

Anchor bolt   securing bolt, sill bolt

Apartment   tenement, multiple dwelling, condominium

Arcade   corridor, passage

Arrowhead   slash, dot

Attic   garret, loft, half story

Awning   overhang, canopy, blind

Back plaster   parget, parging

Baffle   screen, barrier, shield

Baked clay   terra-cotta

Base mold   shoe mold, base cap

Baseboard   mopboard, finish board, skirting, scrubboard

Basement   cellar

Batten   cleat, wood strip

Bead   thin molding

Bearing partition   support partition, bearing wall

Bearing plate   sill, load plate

Bearing soil   compact soil

Bibs   faucets, taps, bibcock

Bird's-mouth   seat of a rafter

Blanket insulation   sheet insulation

Bridging   bracing, joining, cross supports, strutting

Buck   door frame

Building area   setback, building lines

Building code   building regulations, building ordinances

Building lines   setback, building area

Building paper   felt, tar paper, sheathing paper, construction paper, roll roofing, underlayment

BX   conduit, tubing, metal casing, armored cable

Caps   coping, capital

Carriage   stringer, rough string

Casement window   hinged window

Casing   window frame, shell, trim

Catch basin   cistern, dry well, reservoir

Caulking compound   grout, cogging

Cavity wall   hollow wall

Ceiling clearance   headroom

Cesspool   sewage basin, seepage pit

Circuit box   fuse box, power panel, distribution panel

Cleat   batten

Colonnade   portico, porch, poticus

Column   post, pillar, cylinder, pile

Common wall   party wall

Coping   caps

Cursor   pointer, crosshairs

Door frame   buck

Doorsill   saddle, threshold, cricket

Double hung   double sashed

Drainage hole   weep hole

Drywall   gypsum board, sheetrock, wallboard, building board, rocklath, plasterboard, composition board, insulating board

Duct   pipeline, vent, raceway, plenum

Dwelling   home, house, residence

Easement   right of way

Egress   exit, outlet

Escalator   motor stairs

Exit   egress, outlet

Eyebrow dormer

Facade exterior facing

Filler stud trimmer

Fillers shims, extender

Finish work trim, millwork

Firebrick adobe brick, fireclay brick

Flow line drain line

Flue chimney opening, flueway

Flue cap chimney pot

Flush plate switch plate

Footer anchorage, footing

Foundation sill mudsill, sill plate

Gallery ledge, platform, corridor, arcade

Gazebo pavilion, belvedere

Glazing bar muntin, pane frames, sash bars

Grade ground level, ground line, grade line, material classification, slope

Ground line grade, grade line, ground level

Hatchway opening, trapdoor, scuttle

Header lintel, bandstone, trimmer joist

Hoist lift, elevator, dumbwaiter

Hung ceiling drop ceiling, clipped ceiling, suspended ceiling

Jalousies louvers

Lacing lattice bars

Lavatory sink, basin

Lintel header, door head

Live load moving load

Load weight, force

Load plate bearing plate, sill

Lobby vestibule, stoop, porch, portal, entry

Lot plot, property, site

Louvers jalousies, slats

Mantel shelf

Millwork trim, finish work

Modern contemporary

Module standard unit

Moving load live load

Muntin glazing bar, pane frame, sash bar

Particleboard composition board, fiberboard, chipboard

Partition wall

Pilaster wall column

Plank and beam post and beam, post and lintel

Plate cut bird's-mouth, seat cut, seat of a rafter

Platform framing western framing

Plenum pipeline, vent, raceway, duct, chamber

Pressure stress, force

Profile elevation section

Rough floor subfloor, base floor

Rough lumber undressed lumber

S-Straps I beam

Sill bearing plate, load plate

Sill bolt securing bolt, anchor bolt

Slope slant, grade, incline, pitch

Soffit underside

Spar lumber, beam, wood, timber, common rafter

Spiral stairs screw stairs, winding stairs

Standard unit module

Step tread

Strutting bridging, bracing, stiffener

Terra-cotta baked clay

Threshold saddle, doorsill, cricket

Tower turret

Tread step

Undressed lumber rough lumber, unsurfaced timber

Veranda passageway, balcony, porch

Water closet toilet, W.C.

Water table water level

Weight load

Western framing platform framing

Window panes window lites, glazing

A

Abbreviated floor plans Two-dimensional drawings of a floor plan with minimum information.

Acoustical ceilings Ceilings designed to help keep noise from spreading from one room to others.

Acrylics Water-based permanent paints.

Active solar design Design method using mechanical or electrical devices to make use of the sun's energy in heating and cooling.

Active solar systems Cooling and heating systems that use mechanical devices to collect the sun's energy.

AEC The acronym for Architecture, Engineering, and Construction.

Aesthetic value Value placed on an object because of its form, beauty, or uniqueness.

Aggregate A combination of sand and crushed rocks, slate, slag, or shale added to concrete to improve its strength or workability.

Aging in place The design concept based on maximizing the length of time people can live safely and independently in their own residence.

Agreements A signed legal agreement for a transaction.

Air-dried lumber Lumber that is left in the open to dry rather than being dried by a kiln.

Alternating current An electric current that changes value and direction periodically.

Ambient light Light that illuminates all surfaces of an object with equal intensity.

Ampere Unit used to measure the rate of flow of an electric current; the specific quantity of electrons passing a point in one second.

Apron Part of the driveway leading to a garage.

Aquifer A water-bearing geologic formation, sometimes confined between clay layers as in permeable rock or sand.

Arc A portion of a circle that forms a curve.

Arch Structure in the shape of an inverted U around an opening.

Architect Person who plans and designs buildings and oversees their construction.

Architect's scale Tool used in preparing scale drawings and in checking existing architectural plans and details.

Array Multiple copies of selected objects in a rectangular or polar pattern.

Arrowhead A terminator, such as an arrowhead, slash, or dot, at the end of a dimension line showing where a dimension begins and ends.

Atrium Open court within a building.

Attic The space between the roof and the ceiling of a building.

Auxiliary elevation An additional view drawn to the true scale of the base elevation.

Awnings A roof like covering over windows or doors for protection from the elements. Usually canvas or other materials may be used.

Azimuth The measurement of an angle measured clockwise from the north.

B

Baffle A flat deflector to shield light and/or noise.

Balance Any design that is symmetrical.

Balcony Porch having no access from the outside that is suspended from an upper level of a structure.

Balloon framing Framing in which the first-floor joists rest directly on a sill plate, and the second-floor joists bear on ribbon (ledger) strips set into continuous studs. Also called *eastern framing*.

BANANA Acronym for "build absolutely nothing anywhere near anything."

Bars Thin strips of wood, metal, or plastic that creates divisions in windows and doors.

Base map Map that shows all fixed factors related to the site that must be accommodated in the site plan.

Baseboard The finish board covering the interior wall where the wall and floor meet.

Basic layout model Model of a structure built to the same scale as the floor plans and elevations.

Bay window A window projecting out from the wall to form a recess in the room.

Bays Spaces between rows of structural-steel framing members.

Beam A horizontal structural member that carries a load.

Bearing surface Any structural member that rests on supports.

Bearing-wall structures Structures with solid walls that support the weight of the walls, floors, and roof.

Bearing walls Solid walls that provide support for each other and for the roof of a structure.

Berm A continuous mound of earth.

Bidet A low, basin-like plumbing fixture used to wash the feet and posterior when seated.

Bid forms A form that offers to perform the work described in a contract at a specified cost.

Bids Legal proposals from contractors to construct a project according to conditions defined in a contract, in specifications, and in a set of drawings.

Biomass Energy produced from organic matter.

Bioretention System A storm water management practice to treat water runoff. Often referred to as a "rain garden."

Bird's-eye cut A notch cut in a rafter to fit onto a top plate.

Bird's-eye view A 2D or 3D view directly from the top of an object.

Blight Physical and economic conditions within an area that cause a reduction of or lack of proper utilization of that area.

Block A generic term for one or more objects that are combined to create a single object.

Blocking Short pieces of lumber nailed between joists to add support to joists.

Board-and-batten siding Vertical siding consisting of a vertical board placed over studs, with the joints covered with vertical batten boards: plywood or strandboard sheets with battens over the 48″ joints.

Board foot Unit of measure for lumber that equals 1″ × 12″ × 12″.

Bolts A metal rod with a head on one end and an external thread on the other end for screwing on a nut.

Bond Legally binding agreement that ensures that obligations are met; bonds used in construction are performance, labor, and materials bonds.

Bracing Framing members attached at an angle to make walls rigid.

Brackets Any overhanging member attached to a vertical support used to support a shelf or object.

Branches In a sewerage system, horizontal lines that carry waste from fixtures to the stacks.

Break line A line used to show a deleted part of a drawing.

Breakout sectional drawings Separate drawings that show the internal construction of a component.

Break-out sections A segment from sectional drawing that is a separate drawing.

Brick A hollow or solid rectangular masonry unit of clay that was hardened in a kiln.

Bridging A system of bracing placed between structural members to stiffen and reinforce.

British Thermal Unit (BTU) Standard unit of measurement for heat; the amount of heat needed to raise the temperature of one pound of water one degree Fahrenheit at sea level.

Buffer zone A strip of land created to separate and protect one type of land use from another.

Building codes A series of building regulations to enforce all areas of safety and health for construction.

Building envelope Three dimensional space surrounding a building.

Building information modeling (BIM) A software program that permits a seamless integration for the complete design and manufacturing of building projects.

Building load Total weight or mass of all live and dead loads on a building.

Building main Line through which water is brought from a well or public water supply to a building.

Building permit Document issued by the local government that allows construction of a structure or dwelling.

Built-in components Any components that are structurally built into the structure such as furniture or cabinets.

Built-up girders Several layers of wood are laminated or fastened together to complete a structural member.

Buttress Protruding structure added at the base of an arch or wall to give support.

C

Cabinet coding system Shortcut method of dimensioning cabinets using code numbers for standard modular units.

CAD models Computer generated three-dimensional drawings of objects.

CADD An acronym for Computer-Aided Drafting and Design.

Cannon test A test where a heavy object is hurled against an object, testing it for its resiliency.

Cantilever Protruding horizontal structural member supported in the center or at one end.

Cant strip A long beveled piece of wood used at gable ends to establish the correct slant for the shingles.

Cape Cod Style An early American architectural style of a small, rectangular home with a large central chimney.

Carbon Footprint/CO₂ Footprint A building projects impact which includes the amount of gaseous emissions (CO₂/methane etc) produced by the project. This includes all the renewable and non-renewable resources used as well as the ongoing operation of the project.

Carbon monoxide Colorless, odorless, poisonous gas produced by combustion.

Carport Garage minus one or more of the exterior walls.

Cartesian coordinate system Means by which CAD computer software locates points on a two-dimensional surface using an imaginary grid and two axes at right angles to each other.

Casements A metal framed window that swings opens with the use of hinges.

Catch basin A conventional structure for the capture of storm water utilized in street and parking areas. It includes an inlet, sump, and outlet and provides minimal removal of suspended solids. In most cases a hood also is included to separate oil and grease from storm water. Catch basins are differentiated from drainage "inlets" which do not contain sumps or hoods.

Cavity wall A double exterior masonry wall with a dead air space between.

CD-ROM Abbreviation for "compact disc, read-only memory."

Ceiling lines A line depicting the overhead surface of a room on an elevation drawing.

Central bath Bathroom for general use that is accessible from all bedrooms in a house.

Central heating Single source of heat that is distributed by pipes or ducts.

Central processing unit (CPU) Basic control center hardware for a computer system.

Certificate of occupancy Legal document that allows a building to be occupied and used.

Change orders Requests for alteration or correction in a building under construction.

Channel Structural steel shape resembling the letter U that is used for roof purlins, lintels, truss chords, and to frame-in floor and roof openings.

Charrette A planning session in which participants brainstorm and visualize solutions to a design issue.

Checker In large architectural firms, the person who carefully checks completed drawings for errors and omissions.

Chord Principal member of a roof or bridge truss; a straight line that connects adjacent points on curved members connecting opposite rafters.

Circuit Continuous path that electricity follows from the power source to a light, appliance, or other device and back again.

Circuit breaker Device that opens (disconnects) an electrical circuit when current exceeds a certain amount.

Civil engineer's scale Instrument used in preparing drawings to a decimal scale.

Cleanouts A removable fitting at the end of a pipe which provides access for inspection or cleaning. Also an opening at the base of a chimney for the removal of ash.

Clerestory windows A high strip of windows.

Climate change Any long-term significant change in the weather patterns of an area, which can occur naturally or by changes people have made to the land or atmosphere.

Closed plan Building plan in which rooms are completely separated by partitions and doors.

Closing costs Legal, tax, insurance, and lender's costs paid by the owner before taking possession of a property.

Cluster housing Dwellings that are closely grouped together, leaving large areas for recreation and pedestrian use.

Codes Laws that ensure minimum building standards are met that safeguard life, health, property, and the public welfare.

Coding system System identifying every drawing and detail for a project; method of keeping similar drawings together and organized.

Collage A composition of various materials attached to a surface.

Collar beam A horizontal tie beam connecting roof rafters or trusses.

Color coding Using various colors to identify segments on a drawing.

Column Vertical supporting member.

Column schedule Schematic elevation drawings and specifications showing the entire height of a building and the elevation of each floor, base plate, and column splice, as well as the type, depth, weight, and length of columns.

Combination plans Plans in a set that include specialize drawings such as electrical HVAC and plumbing on one plan.

Commands Instructions given to a CAD system by means of a keyboard or mouse that tell the software what function to perform.

Common rafter A full length roof rafter that runs from the ridge board to the top plate of a wall.

Common wall A single wall that serves two dwelling units.

Compact building design The act of designing vertically oriented buildings to a neighborhood scale, in order to efficiency use land and resources.

Compact discs (CDs) An optical disc that uses laser technology to read and write data from the long and short pits on the surface of the disc.

Compaction The act of compressing the volume of soils or aggregates.

Compartment plan Plan that uses partitions, such as sliding doors, glass dividers, louvers, or plants, to divide areas by function.

Compass Tool used to draw circles, arcs, radii. Instrument used to show geographic angles and directions.

Compression force Type of force that exerts a crushing pressure on a structure.

Computer-aided design and drafting (CADD) CAD program which include design components.

Computer-aided drafting (CAD) Computer drafting software that automates many tasks.

Conceptual design Final best design response to the information from a site analysis and on a user analysis chart.

Condensation The formation of frost or drops of water on inside walls when warm vapor inside a room meets a cold wall or window.

Condominium A structure of two or more units in which the interior space is owned by the inhabitants of individual units.

Conduction Process by which heat moves through a solid material. The denser the material, the better it will conduct heat.

Conductors Materials that have a low resistance to electricity flowing through them, such as copper, aluminum, and silver.

Conservation areas Environmentally sensitive and valuable lands protected from any activity that would significantly alter their ecological integrity, balance, or character.

Construction document Document that contains facts, figures, and legal and financial information related to the building process.

Contour interval Vertical distance represented by the space between contour lines.

Contour-interval model Model that represents the shape and slope of a site.

Contour lines On maps or drawings showing terrain, lines which connect points at the same elevation above or below a datum line.

Contour numbering The elevation height from a datum that is shown on a contour line.

Contractor Manager of a construction project.

Contracts Legally binding agreements between two or more parties.

Convection Process by which heat travels through liquids or gases.

Cornice Area of the roof that intersects with the outside walls and extends to the end of the roof overhang.

Corridor kitchen Kitchen having two walls that include appliances and work areas that are directly across from one another.

Court An open, uncovered area that is surrounded by a structure.

Creativity Having the ability and power to be creative with original ideas.

Cripple A short stud in a building frame usually over a door or window or in a T foundation to raise the floor level.

Crosshatching Lines drawn closely together at an angle to show a section cut or shading.

Cross-referencing symbols Callout symbols placed on a working drawing informing the location of the detail on other pages of the drawing set.

Cross ventilation The circulation of fresh air through open doors and/or windows.

CSI format Format approved by the American Institute of Architects (AIA) and the Construction Specification Institute (CSI) to ensure consistency among specifications; information is organized under 16 major divisions with subdivisions under each.

Cubic-foot method A mathematical method of calculating the volume of an area by multiplying the length, width, and height.

Cul-de-sac End of a street with a circular turnaround.

Cupola Small domed structure built on top of a roof to provide ventilation.

Current The flow of electrical energy through a circuit.

Cursor Device, controlled with a mouse, used to identify locations on a monitor screen, also may be cross hairs or pointer.

Curtain walls Exterior walls used to cover a structure that provides no structural support.

Cutting plane In section drawings, an imaginary plane that passes through the drawing of a building; used to show interior construction.

Cutting-plane line Long, heavy line broken by two dashes and used to indicate the cutting plane in a drawing.

D

Damper In a fireplace, a door that control the amount of draft.

Datum line Reference line on a drawing that remains constant; sea level is commonly used.

Daylight plane A building code ordinance controlling the size and height of multiple stories.

Dead load Weight of building materials and permanently installed components on a structure.

Deciduous tree A tree that loses its leaves in the winter.

Deck Open, elevated platform attached to a building.

Decking Surface of a floor system that usually consists of a subfloor and a finished floor.

Decor General style of decoration in a room or building.

Deed Legal certificate of property ownership.

Default A predefined image or detail for a program.

Deflection Stress that results from both compression and tension forces acting on a structural member at the same time. Also called *bending*.

Den Room that may function as a reading room, writing room, hobby room, or study.

Density The average number of people, families, or housing units on one unit of land.

Depreciation Loss of value.

Depressions A lower depressed area of land such as a hollow.

Design standards Design standards or guidelines can serve as a community's desire to control its appearance from within and without, through a series of standards that govern site planning policies, densities, building heights, traffic, and lighting.

Design study model Model used to check the form of a structure, verify the basic layout, clarify construction methods, finalize orientation, or show interior design options.

Detached garage Garage that is not connected to a main structure.

Detail drawings Drawings that reveal more precise information about a portion of a structure.

Detail section Section drawing that shows specific parts of a building in greater detail than the base drawing.

Detailed model Model of a building that shows features in detail.

Detention ponds An area surrounded by an embankment, or an excavated pit, designed to temporarily hold storm water. Also called *extended detention basins*.

Diagonal bracing An angular structural brace on a stud framed wall that helps to strengthen the walls against horizontal forces such as winds and earthquakes.

Dimension lines In drawings, unbroken lines with arrowheads on both end used to show the distance between two points.

Direct current (DC) An electrical circuit that flows in one constant direction.

Display To show or exhibit.

Distribution field The area where a septic system discharges its flow.

Dividers (1) Planters, half walls, louvered walls, or furniture used to separate areas without the use of solid walls. (2) Instrument used to divide a drawn object into an equal number of parts and transfer measurements.

Dome Hemispherical roof form.

Dome huts A small structure covered by a curved spherical roof.

Door handle The handle that is used to open, close, or lock a door.

Door swing The path of a hinged door.

Dormer A structure projecting from a sloping roof.

Double header Two or more timbers joined for strength.

Double hung windows A window that has two vertical sliding sashes.

Downdraft exhaust A system of venting an exhaust down through the floor and exiting outside.

Down payment Percentage of the cost of a property paid by the buyer at the time of purchase.

Downzone To change the zoning of a tract or parcel of land from a greater to a lesser category of use or development.

Drafting brush Brush used to remove eraser and graphite particles and to keep them from being redistributed on a drawing.

Drafting machine Mechanical tool that can serve as an architect's scale, triangle, protractor, T-square, or parallel slide all in one.

Drainage field Plastic (PVC) perforated pipes or agricultural open-joint pipes laid in coarse gravel, which disperse liquid waste from a septic tank into the soil. Also called *leaching field* or *absorption field*.

Drawing sequence The orderly steps to sketch a design, make changes, and then to complete the set of working drawings.

Drawing Web Format (DWF) Autocad files used to display drawings on the Internet.

Drop-leaf workbench Workbench with drop leaves that can be extended for increased work space.

Drywall construction Method of interior wall finishing using prefabricated sheets of material, such as fiberboard, gypsum wallboard, stranded lumber sheets, sheetrock, and plywood.

Dutch Colonial Style An early American architectural style whose main characteristic is a double pitched gable roof. The design was brought here by German immigrants (Deutsch).

Dutch hip This roof style is similar to a gable roof with a small hip construction at the top end of each gable.

E

Early American style Term for all styles of architecture that developed in the various regions of the American colonies.

Earth-sheltered homes Homes designed to be partially covered with earth or partially submerged.

Easements Right-of-way across private land, such as for utility lines or roads.

Eave Part of a roof that projects over the outer wall. Also called a *cornice*.

Eave line Design term for the horizontal line created by the eaves of a structure; can be used to create emphasis on the horizontal in a design.

Eclectic The mixing of architectural styles and/or periods in one design.

Eclectic design An architectural style that mixes several characteristics of other styles into its design.

Ecological footprint The impact of humans on ecosystems created by their use of land, water, and other natural resources.

Ecological planning Planning done during the design stage of a structure to protect or improve the environment.

Ecosystem The species and natural communities of a specific location interacting with one another within the physical environment.

Egress An exit.

Elements of design Elements including line, form, space, color, light (value), texture, and materials, that are used to create designs.

Elevation drawings A two-dimensional drawing of the exterior of a structure.

Elevations Drawings that show the front, back, or sides of a structure or interior walls.

Eminent domain The legal right of government to take private property for public use, provided the owner is offered just compensation for the taking of property.

Emphasis To point out a specific area by repetition, positioning of objects and/or color.

Energy efficiency Using less energy to achieve the same outcome.

Energy Star A rating system that measures the energy usage of electric appliances.

Engineering scale A measuring scale that divides the inch into ten parts.

English style Architectural style developed in England that features such things as high-pitched roofs, massive chimneys, half-timber siding, small windows, and exterior stone walls; Elizabethan and Tudor are examples.

Enlarged scale A large scaled drawing, usually of a small item drawn larger for easier comprehension.

Entertainment room A separate room for games, music, TV, and/or movies.

Entourage People or objects drawn as part of a building's surroundings in order to enhance its size or realism.

Environmental codes The building code ordinances that control the use of toxic materials and the release of pollution in to air, water, and land.

Environmental impact statement (EIS) A comprehensive study of likely environmental impacts that can result from major federally assisted projects; as required by the National Environmental Policy Act (NEPA).

Environmental Protection Agency (EPA) The federal body charged with responsibility for natural resource protection and oversight of the release of toxins and other threats to the environment.

Erasing shield Thin piece of metal or plastic having small, different-shaped openings through which pencil lines on a drawing can be erased without disturbing the rest of the drawing.

Ergonomics Practices that deals with designing and arranging components, devices, tools, etc., for personal convenience and comfort.

Escrow Money collected and held by mortgage holder to ensure that tax payments on property are made on a timely basis.

Estuary A water body where saltwater and freshwater meet, resulting in brackish water.

Explode To disassemble an object, into simpler objects.

Extension lines Thin lines locating two specific points on a drawing to which a dimension line with the dimension is drawn.

Exterior elevation drawings Orthographic representations of the exterior of a structure to show the design, materials, dimensions, and final appearance of doors, windows, outer surfaces, and roof.

F

Facade Exterior face of a building.

Fair Market Value The price at which an owner is willing to sell and a buyer is willing to pay without the compulsion of either party.

Family kitchen Open kitchen that provides a meeting place for the entire family in addition to the usual kitchen services.

Family room Room in a home designed for family-centered activities.

Fascia Vertical board nailed on the ends of the rafters at the eave line.

Federal Housing Administration (FHA) A government agency that insures loans made by regular lending institutions.

Felt markers Pens with various sizes of felt tips that come in a large selection of colors. They may be water base or permanent ink base.

Fenestration Arrangement of windows or openings in a wall.

Feng shui An ancient Asian method of interior design based on concepts of energy flow, balance, and harmony with nature.

Fill A solid color covering an area bounded by lines or curves.

Finished dimensions Actual distances between finished features in a structure, such as from the finished floor to finished ceiling levels.

Finish size The final size of structural wood after it has been surfaced smooth.

Firebox Portion of the fireplace that contains and supports the fire. Also called *fire chamber.*

Firecuts If a floor system collapses the angular cuts of the floor joists ends will not cause damage to the foundation wall.

Fixture trap Device linked to a plumbing fixture that prevents the backflow of sewer gas from the branch lines.

Fixtures Permanent items connected to plumbing, such as lavatory, water closet, bathtub, and shower.

Flashing Additional covering used over roof, door, or window joints to provide complete waterproofing, such as roll-roofing, shingles, or sheet metal.

Flat slab A concrete foundation that is laid directly on the ground.

Flexible curve Drawing tool used to draw irregular curves that have no true radius or series of radii and cannot be drawn with a compass.

Floodplain The land adjacent to a water body, stream, river, lake, or ocean that experiences occasional flooding.

Floor lines A dotted line on a 2D elevation representing the level of the floor.

Floor plans Plans showing locations, sizes, materials, and components contained in the interior design of a building as viewed from the top.

Floor plan sketches The freehand drawings for the development of the architectural design.

Flue Controllable opening in a chimney through which the smoke passes.

Fluorescent Type of lighting created when an electric arc passes through a tube filled with mercury vapor and coated with phosphor which converts ultraviolet light into useable light.

Flush The surfaces of two or more objects in the same plane that have a smooth level.

Flying buttress Protruding structure that helps support the sides of a wall without adding additional weight.

Folded-plate construction Construction materials, usually concrete, which are firmly connected at an angle forming a strong cross section (similar to an accordion fold) that is capable of carrying roof loads over a long span.

Font A unique design set for letters, numbers, punctuation marks, and symbols of a distinctive proportion and design.

Footcandle Unit of measurement for light intensity; the amount of light a candle casts on an object one foot away.

Footings Bases of foundations and foundation walls that are of two types: continuous and individual. Also called *footers.*

Footing section A sectional drawing of a concrete support in the ground that supports foundation walls and columns.

Forced-air systems A warm-air furnace with blower to circulate the warm air through structures ductwork.

Foreclose To take possession of the property if loan payments are not made as agreed.

Foreshortened A line or area that does not represent the true dimensions of a drawing.

Form A removable form of wood or fiberglass used to hold the desired shape of the concrete pour.

Foundation Lowest structural component of a building upon which all other members rest.

Foundation sills Wood or steel members that are fastened to the top of foundation walls as a base for attaching floor systems.

Foyer Inside area in a house adjacent to the main entrance. Also called an *entrance hall.*

Framing Structural skeleton of a building.

Framing dimensions In drawings, numbers used to show the actual distances between framing members.

Framing elevation drawings Drawings that show vertical views of the wall framing.

French curve Tool used to make curved lines that are not part of an arc, such as an irregular curve.

French door A double door sharing the same top and bottom rail. French doors are usually glass paneled.

French style Type of architecture originating in France. French provincial houses feature steeply pitched hip roofs, long projecting windows, corner quoins, curved lintels, and towers.

Frontage The continuous linear distance along any approved way, measured on the street line, between the side lot lines.

Full scale A drawing drawn to the full size of the object.

Full section Sectional drawing prepared for the entire length of a structure.

Fully divided scale A measuring scale that shows all of the divisional marking of the ruler.

Functionalism An architectural style asserting that the design, form, and construction are functional and express the nature of the building materials.

Function keys The programmable keys located typically across the top of the keyboard.

Functional The quality of being useful, of serving a purpose other than that of beauty or aesthetic value.

Furniture templates Furniture cut-outs of stiff paper, plastic, or metal, of the same scale as the floor plan drawing. The templates are used to confirm room size and traffic patterns in a room.

G

Gable The triangular end of an exterior wall just above the eaves.

Gable roof Pitched roof that takes its shape from the triangular ends of exterior walls.

Gambrel roofs A gable roof that has a double pitch on each side.

Gas pilot A continuing small gas flame in gas-operated appliances that will instantly turn on the main gas burners.

General-purpose drawings Architectural drawings used for sales promotion or preliminary planning purposes.

Geodesic dome A dome created by the connection of multiple triangles that combine to form hexagons and pentagons.

Geothermal Refers to the heat in the earth.

GFCI outlets An acronym for a Ground Fault Circuit Interrupter. It is an electrical outlet that turns off instantly when there is a problem with the circuit, insuring that no injuries will occur to an individual.

Glazing Placing of glass in windows or doors.

Global warming An ongoing increase in the average temperature of the earth's surface.

Gothic arch A pointed arch very popular in cathedrals.

Grade marks An identification marking on building materials specifying the materials qualities.

Graphic Information System (GIS) A system used to develop maps that depict and link multiple resources and features.

Graphic accelerator Circuit board added to a computer to increase the amount of Random Access Memory (RAM) and processing power dedicated to displaying graphic images.

Graphics card An electronic component that fits into the motherboard of a computer that will produce quality graphic images.

Gray water waste All the water used, except the water with solid wastes, for watering the landscape.

Great room A large open gathering room consisting of the kitchen, dining, living, and family areas.

Green design Building design that yields environmental benefits, such as saving energy and building materials and reducing water consumption. Also referred to as *green building*.

Greenhouse effect Buildup of heat created when sunlight falls on trapped air; used in passive solar planning.

Greenhouse gas Gases such as carbon dioxide, methane, nitrous oxide, and ozone that contribute to the greenhouse effect.

Grid An area covered with regularly spaced dots to aid drawing.

Grid column identification system A numbering system used to identify and locate the structural columns in a structure.

Ground line On an elevation drawing, a horizontal line that represents ground level.

Groundwater All water below the surface of the land.

Groups A command used to manipulate a group of objects.

Gusset plate A wood or metal plate used to fasten members of a roof truss together.

Gutter A trough for carrying off water.

H

Habitat Living environment of a species that provides whatever that species needs for its survival, such as nutrients, water and living space.

Half-bath Bathroom that contains only a lavatory and water closet.

Half-timber construction Frame construction of heavy timbers in which the spaces are filled in with masonry or wattle.

Hand tools Basic tools necessary for any type of hobby or home maintenance work, such as a claw hammer and screwdrivers.

Hardware (1) Mechanical devices used in a computer system. (2) Metal attachments to doors, windows, and cabinetry.

Hard-wired When an electrical appliance is permanently fastened to the electrical circuit.

Hardwoods Woods that come from deciduous trees, such as oak, walnut, birch, cherry, and mahogany.

Hatch A pattern of lines or symbols used to fill a closed boundary.

Header (1) Horizontal supporting member above openings that serves as a lintel. (2) One or more pieces of lumber supporting ends of joists.

Headroom The ceiling heights of rooms and stairs. The minimum for rooms is 8 feet, and 6′-9′ for stairs.

Head sections Sectional drawings of the tops of structural members of windows and doors.

Heat gain The increased temperature within a structure due to the transmission of heat from the outside through the walls, windows, doors, and infiltration.

Heat loss The decreased temperature within a structure due to the transmission of heat from inside through the walls, windows, and doors to the outside.

Heat pumps A heating unit that draws the heat from the ground and/or air and transfers it into a structure.

Hip roof Roof with four sloping sides.

Hilltop summits The top height of a hill or mountain.

Home office A separate room exclusively used for office work and/or study.

Home theater A separate room used exclusively for the watching of movies or TV on a large screen.

Hopper windows A swinging window that is at the bottom.

Horizon line On perspective drawings, a line where earth and sky meet; the observer's eye level.

Horizontal wall sections Sections of exterior and interior floor plan walls drawn to clarify how the walls are constructed.

Hose bib An outside water faucet that is threaded so a hose may be attached.

Hue The reference to a color.

Humidistat A device for measuring and controlling humidity levels.

HVAC Mechanical systems that control heating, ventilation, and air conditioning.

Hydronic systems A heating system that uses hot water to transfer heat into a structure.

I

I-beam Steel beam with an I-shaped cross section.

Icon A small picture or symbol used to identify commands.

Idealized drawings Sketches that designate ideal spatial relationships of the user elements as determined by the user analysis.

I-joists Joists that are structurally similar to steel I-beams (S-beams); joist weight is reduced through the use of thin webs made of stranded wood.

Incandescent Lighting produced by a filament of very thin wire that glows when heated.

Infiltration The seepage of air into or out of a structure through spaces around windows, doors, or wall cracks.

Infrastructure The water and sewer lines, roads, urban transit lines, schools, and other public facilities needed to support developed areas.

Ink-jet printers An electronic printer that is connected to a computer that prints a document or drawing with a small jet spray system.

Insulation Material used to stop the transfer of heat and sound in a structure.

Insulators Materials that have good resistance to the flow of electricity, such as wood, glass, and plastic.

Integral garage Garage that is connected to the house.

Intensity The intensiveness or the denseness of colors.

Interest Charges that a borrower must pay for the use of the borrowed money; usually a percentage of the loan.

Interior decoration The finish of interior surfaces and the placement of furniture and accessories.

Interior design model Model used to show individual room designs.

Interior elevation drawings Drawings that show the vertical design of interior walls.

Island kitchen Kitchen having a freestanding structure usually located in the center of the room, that offers additional work space, a range top, or a sink.

Isometric drawings Pictorial drawings in which the receding lines are projected at an angle of 30° from the horizon so that they are parallel.

Istea/Tea-21 (Transportation Efficiency Act for the 21st Century) Federal legislation that encompasses all transportation regulations and funding and to which federally funded projects must conform.

Italian style Architectural style originating in Italy and featuring columns and arches that are generally part of an entrance and windows or balconies that open onto a loggia.

J

Jalousie window A window consisting of a number of long, thin, hinged panels with several horizontal glass planes that open and close.

Jamb Sides of a doorway or window opening.

Jamb sections A sectional drawing of the vertical side construction of windows and doors.

Joist Horizontal structural member that supports a floor or ceiling.

K

Keystone Wedge-shaped center stone that locks the other stones in an arch in place.

Kilowatt (KW) A measure of electric power consumption or production, equal to 1,000 watts.

Kinetic energy The energy of motion, or the amount of work needed to accelerate a body of a given mass from a resting situation to its current velocity.

Knee wall A short wall, usually in an attic, between the floor and rafters.

L

Labor costs The financial expense for the payments of the labor force for a building project.

Labor and materials cost The financial expenses for the payment of the labor force and all the building materials for a building project.

Lally column Hollow steel column filled with concrete and used to support weight transferred from horizontal beams and girders.

Laminated timber Timber made from thin layers of wood glued together; used to support heavy weights or over long spans.

Lanai Covered exterior passageway; Hawaiian word for "porch."

Land integrity Keeping the natural land form during construction.

Land use The manner in which a parcel of land is used or occupied.

Landfill gas Methane gas that forms in landfills from the decay of organic materials. The gas can be collected and used for power generation.

Landform model A scaled model of the building site.

Landings Points at which stairs change direction.

Landscape plans Drawings that show the types and locations of vegetation on a building site. They may also show contour changes and building position.

Lap siding Horizontal siding applied so that each piece overlaps part of the piece below it.

Large-span construction Type of construction that uses large trusses or arches to span long distances, such as in aircraft hangars, sports stadiums, and convention centers.

Laser printer A high quality printer that uses a laser beam to copy text and drawings.

Lateral load Horizontal forces on a building.

Lavatory Bathroom washbasin; sink.

Layering The superimposition of drawings separating levels or components.

Layers A system of separate drawings (layers) of an object that may be viewed separately or combined together.

LED Light-emitting diode.

LEED Acronym for Leadership in Energy and Environmental Design, a green building rating system that is a nationally accepted benchmark for the design, construction, and operation of high-performance green buildings.

Libraries The areas of a CAD program in which frequently used symbols or details are stored.

Lien Legal document used to take or hold the property of a debtor in default of payment.

Lightning rod An electrical conductor placed on the highest part of a structure that will conduct lightning strikes harmlessly into the ground.

Line conventions Series of lines having special meanings that are used in architectural drawings; sometimes called the *alphabet of lines.*

Line of sight The direction a viewer has when observing an object.

Lintel Horizontal piece of wood, stone, or steel across the top of door and window openings that bears the weight of the wall above the opening.

Live load Force on a building that includes the weight of all nonpermanent, movable objects, such as people and furniture, snow, and the force of wind.

Living area Area in a house where the family entertains, relaxes, dines, and participates in recreational activities.

Load The weight of all the building materials and moveable objects in a structure that is transmitted into the ground.

Load-bearing wall See *Bearing walls.*

Loggia Open passage covered by a roof.

Longitudinal section Section drawing in which a cutting plane extends the full length or major axis of a building.

Lookout joists Joists that are used to support a cantilevered second-floor extension.

Lookouts Short members used to support the overhang of a roof and to act as a fastener base for the soffit cover.

Lot A real estate property defined by surveyed property lines, ownership, acreage, or square footage.

Low-E windows Windows that have very thin coatings on them that reduce heat flow through the windows. "Low-E" stands for "low emittance."

L-shape Structural steel member rolled in the shape of the letter L with legs of equal thickness but varying length. Also called an *angle beam.*

L-shaped kitchen Kitchen design that features continuous counters, appliances, and equipment on two adjoining, perpendicular walls.

Lumber Refers to all the wood used in construction.

Lumens A unit of a luminary's intensity. Light sources are now measured by lumens instead of the amount of energy using watts.

Lux In the metric system, the standard unit of illumination; one lux (lx) is equal to 0.093 footcandle.

L-winder stairs Stairs constructed in an L shape with a landing at the 90 degree turn.

M

Major module Large (48″) base unit in a construction system in which parts are standardized.

Mansard roof A formal style of roof design that has a double pitch on all the side of the structure.

Mantel A shelf over a fireplace.

Marquee Covering over an entrance that is not supported by posts or columns and that connects the building with the street.

Masonry Stone, brick, tiles, or concrete.

Masonry bond Pattern of masonry units in courses (rows).

Masonry veneer walls A thin masonry cover over a wood frame wall or permanent structural masonry wall.

Master bath Bathroom that is accessible only from the master bedroom.

Master bedroom Bedroom that may also have an adjacent bath and dressing room.

Master plan A statement made using text, map(s), illustrations, or other forms of communication that is designed to provide a basis for decision making regarding the long-term physical development of a municipality.

Material list List of all materials, used for on-site construction; includes descriptions, quantities, and costs.

Media room A separate room devoted to all sorts of entertainment. Similar to the home theater room.

Mediterranean style Italian and Spanish architectural styles. Also called *Southern European* style.

Megawatt (MW) A measure of instantaneous electric power consumption or production. Equal to 1,000 KW (or 1,000,000 watts).

Menu Listings of all mode and command options within a software program.

Metric scale Scale used with metric drawings; units are in millimeters (mm).

Mid-Atlantic style Architectural style common in the Mid-Atlantic states and resulting from the availability of brick, a seasonal climate, the influence of Thomas Jefferson and early Greek and Roman architecture; also known as *classical revival.*

Minor module Small (24″) base unit in a construction system in which parts are standardized.

Mirror Command used to create a new version of an existing object by reflecting it symmetrically.

Mitigation Process or projects designed to replace lost or degraded resources.

Mixed-use development Developed area that is intended to satisfy the demands of different users, including residential, commercial, and business users.

Mode A software setting or operating state.

Models Three-dimensional replicas of a structure.

Modem Telecommunications device that allows computer operators to send and receive digital information. (*mo*dulator/*dem*odulator).

Modular A construction system using a basic modular unit that will ensure that all mating parts will fit together.

Modular cabinetry A building system for cabinet makers using the basic modular unit that will ensure that the components from different manufactures will fit together.

Modular components Standardized building parts or sections constructed away from the building site; typical components include preassembled wall sections, windows, and molded bathrooms.

Module Base unit in a construction system in which parts are standardized.

Moisture barrier Material that retards the passage of vapor or moisture and prevents condensation.

Molding A decorative trim that outlines windows, doors, and the junctions of the walls at the ceiling and floor.

Monitor A peripheral part of a computer system that is a flat liquid crystal display (LCD) that resembles a television screen.

Mortar Mixture of cement, sand, and water, used to bond bricks and stone.

Mortgage Contract for a loan for the purchase of real estate property.

Mortgagee The party who lends money for property purchase.

Mortgagor The party who borrows money to build or purchase property.

M-shape Lightweight steel beam.

Mouse A device used to control a cursor, activate commands, and select options.

Mud room A small entry room in the front or back of a home where muddy footwear and dirty clothes may be changed.

Mullion A vertical member that separates a series of windows or panels.

Multiple-level floor plans Floor plans for structures having more than one floor.

Muntin Thin strip separating panes (lights) in a window.

N

Nation Electric Code (NEC) A national building code ordinance that controls the proper installation of all electric work and safety preparations.

Networking A system where computers stations are connected together so that they share data.

New England Colonial style Architectural style developed by the colonists who settled the New England coastal areas.

New urbanism Neighborhood design trend used to promote community and livability; includes narrow streets, wide sidewalks, porches, and homes located closer together than typical suburban designs.

NIMBY Acronym for the "not in my backyard" sentiment that exists among some people who do not want any type of change in their neighborhood.

Noise abatement Any method used to stop the movement of unwanted noise.

Nominal size The dimensions of cut lumber into standard sizes before it is dried and surfaced. An example: a 2" × 4" nominal stud when finished is 1½" × 3½".

Non–load-bearing wall Interior wall that does not support a load.

O

Object One or more graphical elements such as text, dimensions, lines, circles, or polylines, treated as a single element.

Object lines Heavy lines on a drawing, such as those describing the main exterior walls and interior partitions.

Oblique drawings Drawings in which all receding lines are commonly drawn at a very low angle of 10° to 15° from a front view.

Offset Recessed portion of a wall.

Ohm The unit of electrical resistance within a conductor.

On-center (OC) Measurement from the center of one member to the center of another.

One-point perspective Drawing in which the front view is drawn to its true scale and all receding sides are projected to a single vanishing point on the horizon line.

One-wall kitchen Kitchen plan in which work centers are located in a row along one wall.

One-way slab system Foundation in which the rebars are all parallel.

Open-divided scales Measuring scales where only the first unit of measurement is divided into parts.

Open plan Floor plan design in which partitions do not completely separate the rooms from one another.

Open Space Undeveloped land used for recreation, farmland and natural habitats.

Organic design Architectural concept in which all materials, forms, and surroundings are coordinated and in harmony with nature.

Oriental architecture The styles of architecture developed throughout Asia.

Orientation A building's position in relation to sun, wind, view, and noise.

Ortho A drawing command used to create objects at right angles.

Orthographic projection Multiview drawing-related views of an object so as to show it from all sides.

Outlet A connection for an electrical appliance that is plugged into.

Overall dimensions Total length and width of a building.

Overhang Portion of a roof that projects beyond the outer walls.

Overlay A transparent piece of paper that is placed over an existing drawing so design changes may be quickly made.

Overlay districts Zoning districts in which additional regulatory standards are superimposed on existing zoning.

P

Pan To move the view of a drawing without changing magnification.

Pan slab Also called a waffle slab; it is a suspended floor or roof system that is created by pouring concrete into molded fiberglass forms on site or prefabricated off site.

Panel door A door with stiles and rails which form one or more frames around recessed door panels.

Panelized floor systems Preassembled sandwich panels that are placed over the floor framing system forming a finished floor. They are made from a variety of materials for the skin and core. A typical size is 4' × 8'.

Parallel angle projection A pictorial drawing, as in isometric, where the receding lines of the object are parallel.

Parallel slide Drawing tool used as a guide for drawing horizontal lines and a base for aligning triangles. Also called a *parallel rule.*

Parlor An old term designating a room primarily for entertaining and conversing with guests.

Partition Wall that subdivides an area into rooms.

Passive solar systems Heating and cooling systems that use the power of the sun without the aid of mechanical devices.

Pastels Light-colored, water-based drawing medium similar to chalk; pale and subdued colors.

Patio Open area adjacent or directly accessible to the house; courtyard.

Peninsula kitchen Kitchen design in which a work center projects into the room like a piece of land into a body of water.

Peninsula workbench Workbench that projects into the room so that it has three working surfaces with storage compartments on each of its sides.

Percolation test A test to determine the rate at which soil will absorb water from a hole in the ground.

Perimeter The outside surfaces of a geometric space.

Permanent wood foundations Foundations constructed similarly to wood frame walls except that the lumber components are pressure treated with wood preservatives.

Perspective drawings Drawing in which receding lines appear to meet at a point on the horizon; the object appears as it would to the eye.

Phasing Planning the completion of a project in several steps spread over a specified period of time.

Phi The golden ratio of design 1:1.618.

Photovoltaic (PV) Literally, "light" (photo) and "electricity" (voltaic). The class of equipment used to generate electricity directly from sunlight.

Picture plane Imaginary plane between the station point and the object in a perspective view drawing.

Pier-and-column foundations Foundations having individual footings (piers) on which posts and columns are placed.

Pilasters A column or a shaft that is attached to a vertical surface for strengthening and/or decoration with a capital and a base.

Pile A large vertical shaft driven into the ground to support a structure. They may be made of wood, steel, concrete, or a steel pipe filled with concrete.

Pilotless gas appliance A gas appliance that has an immediate start with an electric sparking switch instead of a gas pilot light.

Pitch Angle between the roof's surface (top plate to ridge board) and the horizontal plane; ratio of the rise to the span.

Pit house Also called an American Indian earth lodge. An excavated area is covered with branches and foliage for the roof.

Plank-and-beam construction Floor construction system in which heavy wood planks are placed over widely spaced beams.

Plank-and-beam floor system A floor framing system made up of large girders and a floor covering of thick wood planks.

Plans Drawings that show views of an object from the top down.

Planting schedules In a landscape plan, guides for the purchase and placement of each size and species of plant material.

Plat Map or chart of an area showing boundaries of lots and other parcels of property.

Plates The first wood member at the bottom of stud framed wall.

Platform floor systems A raised wood framing system over a T foundation. Also called Western floor framing.

Platform framing Framing of a multiple-level building in which the second floor rests directly on first-floor exterior walls. Also called *western framing.*

Plenum Enclosed space inside of which the air pressure is greater than it is outside; used in a type of heating and cooling system.

Plot plans Plans used to show the size and shape of a building site and the location and size of all buildings, walks, drives, pools, streams, patios, and courts on it.

Plywood Wood product manufactured from thin sheets of wood laminated together with an adhesive, under high pressure; the grain of each ply is laid perpendicular to the grain of each adjacent layer, which reduces the tendency to warp, check, split, splinter, and shrink.

Polar coordinate system Input method based on an established angle and a distance.

Pole foundation Also called a column foundation, are the major support for a structure. See pile.

Polyline An object composed of one or more connected line segments or circular arcs treated as a single object.

Porch Covered platform leading into the entrance of a building and which may be enclosed by glass, screens, or posts and railings.

Post-and-beam construction Type of construction that uses heavy timbers for wall framing, as well as for floors and roofs.

Post-and-lintel construction Type of construction in which a horizontal beam, called a *lintel,* is placed across two vertical posts, such as for a door or window.

Post-tensioning Method used to prestress, or compress, concrete so that both the upper and lower sides of a member remain in compression during loading; tendons are placed inside tubes embedded in the concrete and then stretched with hydraulic jacks while the ends are anchored.

Power tools Electrically powered tools, such as electric drills, belt sanders, and drill presses.

Prebuilt home House that is totally built in a factory.

Prebuilt module Construction module in the form of a complete room or independent functional area and having built-in components, such as cabinets, major appliances, and fixtures.

Prefabricated home Home in which the major components, such as the walls, decks, and partitions, are assembled at a factory; installation of electrical wiring, plumbing, and heating systems is completed on site, as is finishing work.

Presentation drawing Drawings having realistic features added for the purpose of showing the building to clients.

Presentation model Building model used for sales purposes.

Prestressing Compressing concrete so that a member remains in compression during loading.

Pretensioning Method of prestressing concrete in which tendons are stretched between anchors and the concrete is poured around them.

Principal Amount of a loan, minus any interest.

Principles of design Guidelines for how to combine the elements of design.

Prints Reproductions of architectural drawings used by contractors and workers to guide the building process. Formerly called *blueprints.*

Profile A vertical section through a site.

Profile drawings Elevation drawings of a site showing a section cut through the terrain.

Prompt A message that asks for information or requests action.

Propane A gaseous hydrocarbon of methane found in petroleum that is stored in tanks and used like natural gas.

Proportion The relationship of size, numbers, and colors to each other.

Protractor Instrument used to measure angles.

Purlin Horizontal structural members that tie together and strengthen roof rafters and/or trusses.

Q

Quoin Support or decorative masonry applied to building corners.

R

Radiant heat Method of heating surfaces to create area heat through radiation.

Radiation Process by which heat travels as waves through space.

Radon Colorless, toxic gas produced by the natural decay of radium; radon enters a structure from the soil.

Rafter Structural member used to frame a roof.

RAM Abbreviation for "random-access memory," a computer's temporary memory that determines the amount of software data the CPU can process at one time.

Ranch style Architectural style adapted to the needs of settlers as they moved west, featuring a single-level, rambling plan.

Receptacle A device that will accept a plug from an electrical appliance. It is similar to the outlet box.

Recharge Water that infiltrates into the ground, usually from above, and replenishes groundwater reserves.

Recreation room Room designed specifically for active play, exercise, and recreation. Also called a *game room* or *playroom.*

Redraw To clean up blip marks in the current viewport without updating the drawing's database.

Reflected ceiling plans Drawings which show ceiling designs and/or multiple-lighting fixtures or levels using the floor plan as a base; the ceiling appears as though reflected in a mirror.

Reinforcing bars Steel bars added to concrete slabs, beams, and columns to help them resist tension forces. Also called *rebars.*

Removed section Section drawing done at a larger scale to clarify small details.

Render To add shading and texture to a drawing to add realism.

Rendering Process of adding lights and shadows to a drawing so that it looks more realistic; usually done to one-, two-, or three-point perspective drawings to show how the finished building is expected to look.

Renewable energy Energy that is generated from naturally replenished resources such as sunlight, wind, and tides.

Reversed plans Plans that are a mirror image of the originals; created to provide more plan choices.

Rhythm All segments of the interior design of a room flows smoothly together.

Ridge board Top member in the roof assembly. Also called a *ridge beam.*

Ridge line Line formed by the roof ridge on an elevation; accenting the ridge line places emphasis on the horizontal.

Rise Vertical height of a roof from the top of the wall plate to the roof's ridge.

Riser The vertical height of a stair.

Rolled steel Steel that has been shaped by passing it between a series of rollers.

ROM Abbreviation for "read-only memory," the memory in a computer that contains the fixed data that the computer uses while it is operating.

Room template A paper cut-out of a scaled outline of a room.

Rough opening Opening in the framing that is slightly larger than the door or window that will fit into it.

Run Horizontal distance covered by a roof.

Runoff The water that flows off the surface of the land.

R-value Rating of resistance of heat flow through building materials, such as insulation; the higher the number, the greater the resistance to heat flow.

S

Safe room A structurally reinforced room where occupants could be safe from intruders, fire, and earthquakes.

Sauna A steam bath, of Finnish origin, in which steam is produced by spraying water on very hot stones. Commercial saunas insert the steam directly from a boiler's steam unit.

Scale A mathematical ratio between the design's full size or real size and the size needed to fit on a drawing sheet.

Schedules (1) Detailed lists that contain descriptive information, such as size and type of windows, appliances, fixtures, or doors. (2) Charts that provide information for estimating purposes or to conserve drafting time and drawing space.

Schematic A simplified diagram of an architectural or electronics design.

Sectional drawings Drawings that reveal interior construction by showing a "slice" of a planned structure. Also called *sections.*

Septic tank Tank in which sewage is processed by bacteria and the liquids are allowed to seep into the ground into a leech field.

Service entrance The back entrance to a home, usually in the outdoor service area.

Serving walls Walls that have openings with countertops for passing items back and forth between a kitchen and dining room.

Setbacks Minimum distances structures must be located from property lines as stated in zoning laws.

Shear force Force that tends to make one part of an object slide on or past another part; excessive shear may cause fractures.

Sheathing Structural covering of boards or wallboards, placed over exterior studding or rafters of a structure.

Shed roof A flat roof with a pitch of more than 3 to 12 inches.

Shingles Thin pieces of wood or other materials that overlap each other in covering walls and/or roofs.

Siding Covering for an exterior wall.

Sill sections Sectional drawings of the bottom framing of a window and the bottom of a stud wall.

Single-line drawing Simple, scaled drawing that does not show wall thickness.

Site analysis Study of a site that helps the designer take advantage of its positive features and minimize its negative features; site analysis also helps ensure appropriate land use.

Site plan A scaled plan showing proposed uses and structures for a parcel of land. A site plan also describes the location of lot lines, the layout of buildings, open spaces, parking areas, landscape feature, and utility lines.

Site-related drawing Drawing that matches the idealized drawing of a structure to its site and introduces size requirements.

Situation statement Formal statement of the agreement between a client and a designer regarding the purpose, theme, scope, budget, and schedules of the project.

Skeleton-frame construction Construction in which wall coverings are attached to an open frame in which small structural members share the loads.

Skylight Glazed opening in a roof.

Slab foundations Foundations made of reinforced concrete, which may be either in one piece or in several pieces and rest directly on the ground.

Slope Ratio of the rise of a roof to the run.

Slope diagram Diagram drawn on the elevation after the roof pitch is established to aid the carpenter in determining the angle of the rafters.

Smart energy Energy generated from renewable resources that is used to create electricity and to heat and cool buildings.

Smart growth Refers to a well-planned development that protects open space and farmland, revitalizes communities, keeps housing affordable, and provides various transportation choices.

Snap The command which enables the selection of a specific point on a grid.

Soffit Underside of an overhang.

Software Instructions through which a computer performs tasks.

Softwoods Woods from coniferous (needle-bearing) trees such as Douglas fir, pine, or cedar.

Soil line A horizontal pipe which conveys the discharge of toilets and fixtures to a city sewer line or a septic tank.

Solar Power (or energy) Use of sunlight, and solar energy to heat and light buildings, generate electricity, heat hot water, and for a variety of commercial and industrial uses.

Solid form model Model of a structure created without final dimensions.

Solid model Type of three-dimensional drawing done on a computer; created when properties such as mass must be considered.

Southern Colonial An early American style of architecture. Its characteristics are a large two-story rectangular house with front columns and a large balcony.

Span The horizontal distance between supporting vertical members.

Spandrel The triangular form between the meeting of two arches.

Specifications Detailed written instructions describing materials and requirements for construction.

Spiral stairs Circular stairs supported by a central column.

Splices The joint attaching two structural members.

Sprawl Development pattern where rural land is converted to urban/suburban uses more quickly than needed to house new residents and support new businesses.

Square-foot method Estimating construction costs by calculating the square footage (width × length).

S-shape Steel rolled into the shape of a capital letter I. Formerly called an *I-beam.*

Stacks In a sewerage system, vertical lines 3″ to 4″ in diameter that carry waste.

Stairs Series of steps that provide access from one level of a structure to another.

Stamps Images in rubber that can be coated with ink and transferred to drawings; used for architectural features that are often repeated, such as landscape features, people, and cars.

Station point Location of the observer in a perspective drawing.

Steel-cage construction Type of construction in which steel members are used in a manner similar to skeleton-frame wood members. Also called *steel skeleton-frame construction.*

Steel space frame A ridged framed three-dimensional structural framework.

Steel ties A metal plate that ties two components together.

Stepped footing Foundation footings built on a sloped site. They appear as large concrete steps.

Stoop Projection from a building that provides shelter and access to a landing at the entrance.

Straps Long metal straps used to attach two structural members together.

Streetscape The space between the buildings on either side of a street that defines its character.

Stress Any force acting on a part or member of a structure.

Stressed skin construction A strong thin panel material is attached to the exterior framing of a structure to strengthen the wall to carry heavier loads.

Structural model Model of a building in which only the structural members are shown in order to check unique structural methods or to study framing options.

Stud Vertical structural member in the framework of a building.

Stud layout Horizontal framing section showing the position of all wall members, both exterior and interior.

Studio Room in which an engineer, architect, drafter, or artist works.

Study Room used for reading, writing, hobbies, or office duties.

Subdimensions Architectural dimensions that are subdivisions of the overall dimensions.

Subdivision A subdivision occurs as the result of dividing land into lots for sale or development.

Subterranean garage A parking place for an automobile that is partially or fully underground.

Surface model Three-dimensional computer drawing that consists of solid plane surfaces.

Survey Drawing showing the exact size, shape, and levels of a property; when prepared by a licensed surveyor, a survey is a legal document that establishes property rights.

Sustainable development Development that has the goal of preserving environmental quality, natural resources, and livability for present and future generations.

Swale A continuous depression in a flat site to control rain runoff.

Swim-out In a swimming pool, an elevated platform below the water level that allows the swimmer to get out without using a ladder.

Symbol library Computer software that contains ready-to-use symbols for doors, windows, fixtures, and other items that can be inserted easily into any computer drawing.

Symbols Standardized elements on a drawing used to identify fixtures, doors, windows, stairs, partitions, and other common items.

T

Take-off Creation of a material list from working drawings.

Tangent A location on a circle where a line and a circle touch at one point and only one point.

Technical pens Ink pens used for drafting.

Tee Structural steel shape usually made by cutting through the web of an S-, W-, or M-shape and used for truss chords and to support concrete reinforcement rods.

Templates Cutouts in plastic, paper, cardboard, or metal that represent various symbols, furniture, and fixtures, which can then be traced on a drawing by following the outline with a pencil or pen.

Tension force Force that pulls on objects causing them to stretch.

T-foundations Combinations of footings and a poured concrete or concrete block wall that form an inverted T; necessary in structures with basements or when the underside of the first floor crawl space must be accessible.

Thermal conductivity Ability of a material to allow heat to flow through it.

Thermal mass Any material that will absorb heat from the sun and later radiate that heat back into the air.

Thermostats An instrument that responds to the changes of temperature and turns the heating unit on or off to maintain a required temperature.

Three-point perspective Drawing in which horizontal lines recede to vanishing points in the distance and vertical lines also recede slightly so that the tops of extremely tall buildings appear smaller.

Timber Piece of lumber with a cross section larger than $4'' \times 6''$.

Title block Written information on an architectural drawing that identifies it.

Topography Drawing showing land elevation details; surface configuration of an area.

Torsion force Force that twists an object, causing it to fracture or become misshapen.

Toxic materials Any material that is harmful to human's health.

Traffic areas Areas of a building or room through which people pass to get from one place to another.

Transformer A device that converts electricity from one voltage to another.

Transverse section Section drawing, the cutting plane of which lies across the shorter, or minor axis, of the building.

Tread Horizontal part of a step; the part on which one walks.

Treatment of materials The addition of any foreign materials to structural members that make them stronger, last longer, or make attractive.

Triangle Drafting tool used with either a T-square or other horizontal guide to draw vertical and diagonal or inclined lines.

Trimmers A vertical structural member that supports lintels for windows and doors.

Trombe wall Device used in passive solar heating, consisting of an outer glass surface, a masonry wall, and vents that channel heated air through the building or to the outside.

Truss Prefabricated, triangular-shaped unit used to support roof loads over long spans.

Truss joists Joists having long, usually parallel, top and bottom chords connected by shorter pieces called webs that give the ability to span long distances.

T-square Drafting tool used primarily as a guide for drawing horizontal lines and as a base for the triangle used to draw vertical and inclined lines.

Two-point perspective Drawing in which receding lines are projected to two vanishing points on opposite ends of the horizon line.

Two-way slab system Foundation in which the rebars, girders, and beams are placed perpendicular to one another; when ribs extend in both directions, the system is known as a "waffle slab."

U

Underlays Drawings or parts of drawings that are placed under an original drawing and traced onto it.

Underwriters Laboratory (UL) A nonprofit agency that tests products for safety.

United States Geological Survey (USGS) Federal agency that provides mapping of topography of the entire country.

Upzone To change the zoning of a tract or parcel of land from a lesser to a greater category of use or development.

User analysis Study of a structure in which each goal is further refined in terms of space elements, usage, size, and the relationships between areas.

U-shaped kitchen Kitchen design in which the sink is located at the bottom of a U-shape, and the range and the refrigerator are on opposite sides.

Utility room Room that may include facilities for washing, drying, ironing, sewing, storage, and heating and air-conditioning equipment. Also called *service room, all-purpose room,* and *laundry room.*

U-value Proportional heat transfer difference between inside and outside air temperatures; reciprocal (1/R) of the R-value. A low U-value is most efficient.

V

Valley rafter A structural member that helps form a valley. A valley is formed when two roofing segments meet at an angle to form the valley.

Valve A device for controlling the flow of gas or a liquid in a pipe line.

Vanishing point On a perspective drawing, the point in the distance at which horizontal lines appear to meet and disappear.

Variance The temporary changing of the requirements of a zoning district for a specific parcel or tract of land.

Vault Passageway or room formed by a series of arches.

Vaulted roofs A curved roof built on the principle of the arch.

Vellum Transparent medium on which architectural drawings are prepared.

Vent stack In plumbing, the portion of the soil stack above the highest branch intersection that permits sewer gases to escape to the outside and equalize the air pressure in a waste system.

Ventilated shelving Shelving made by welding steel rods together at intervals and then coating them with vinyl.

Ventilation Circulation of air through a structure.

Veranda Large porch extending around several sides of the home.

Vertical wall sections Section drawings that show exposed construction members in a wall.

Victorian Era Highly decorative style of the nineteenth-century era named for Queen Victoria.

Viewpoint The location in 3D model space from which the observer views.

Volt Unit of electrical pressure or potential; the higher the voltage, the greater the flow of electricity.

W

Walk-in closet Recessed storage facility large enough to walk around in.

Wall closet A shallow closet in the wall for cupboards, shelves, and drawers, providing access to all stored items without using an excessive amount of floor area.

Wall-storage cabinet A built-in storage unit, usually reaching the ceiling.

Wardrobe closet Storage facility for clothing that may be either built-in or offset.

Wash drawing Rendering in which only black and gray tones are used.

Waste discharge The solid wastes from the water closet that is dispersed through the sewer lines.

Water closet Toilet fixture.

Watershed The geographic area that drains into a specific body of water.

Watt Unit of measurement for electrical power; current (in amperes) multiplied by potential (in volts), equals watts (W = A × V).

Welds To unite metals together by heating them until they melt together or by applying a filler metal.

Wetlands Areas having specific hydric soil and water table characteristics that support or are capable of supporting wetlands vegetation.

Wind power Harnessing the wind to generate electricity.

Wind turbine A machine that converts kinetic energy in wind into mechanical energy that produces electricity.

Wireframe drawing Three-dimensional computer drawing of an object that looks as though the object were shaped out of wire mesh.

Wireframe A term used with CAD drawings that only show the outlines of an item. No solid surfaces are shown.

Wire mesh An oversized "chicken wire" that is in place for the pouring of concrete. It adds strength and helps to stop the concrete from cracking.

Wood foundations Heavily treated wood to stop decay and termites are sometimes used for the foundation of a structure.

Work triangle Area connecting the three main work centers of a kitchen.

Working drawings Drawings that contain all of the information needed to build a structure.

W-shape Type of I-beam, similar to an S-shape but with wider flanges and comparatively thinner webs, which gives a greater capacity to resist bending.

Z

Zero-lot-line development A development option in which side yard restrictions are reduced and the building abuts a side lot line.

Zoning Classification of land in a community into different areas and districts.

Zoning codes Building ordinances that control the size, height, and location of a structure on the building site.

Zoning ordinance Law or regulation defining the type of structure that can be built in a certain area in order to provide safety and convenience for the public and to preserve or improve the environment.

Zoom A command which enables increasing or decreasing the size of the image on the monitor screen.

INDEX

Entries in **bold** represent a figure (diagrams or tables).

Abbreviated floor plans, 304
Absorptance, 726
AC; *See* alternating current (AC)
Accessories
    in bathrooms, 221
    schedules for, 767–768, **769**, **770**
Accident prevention, 569
Acid rain, 113
Acoustical engineering; *See* noise abatement/control
Acrylics, 431, 433
Active solar (design) systems, 99, 725–728; *See also* solar energy
Activities room; *See* family rooms
Adobe construction, 98, 105, 470
Aesthetic value, 21
A-frame roof, 339, **340**, 457
Aggregate, 528
Agreements; *See* contracts, as legal document
AIA; *See* American Institute of Architects (AIA)
Air conditioning; *See* cooling systems; HVAC
Air ducts, for fireplaces, 496, **496**
Air filtration, 567, 722
Air purification, 567
Air quality, design considerations and, 111–112, **112**, 567
Air-based solar system, 727, **727**; *See also* solar energy
Air-cushion chambers, 735
Align command, in CAD, 77, **77**, 86, 369
All-night lighting, 168
All-purpose room; *See* utility rooms
Alphabet of lines, 51
Alternating current (AC), 672
American architecture, 10–16
    early styles, 10–12, **11**, **12**
    influences on, 7–10, **8–10**, 13–16
    later styles, 12–16, **13–16**
American Institute of Architects (AIA)
    BIM practices, 92, 748
    change orders, **758**
    coding system of, 45–46
    dimensioning style, 328
    specification consistency, 772
American National Standard Alphabet, 53, **53**
American Society for Testing and Materials (ASTM), 547, **547**
American system, 238, **238**
Amperes (amps), 674, 675, 677, 681, **681**
Anchor devices
    for foundations, 482, **484**, **562**

for masonry walls, 532, **533**
    for post-and-beam construction, 520, **522**
Angle brackets; *See* brackets
Angular Dimension command, in CAD, 328
Appliances
    circuits for, 678, **679–680**
    dimensioning for, **188**, 407, **409**
    door opening for, **193**
    electrical plan for, 695–696
    Energy Star rating, 192, 673–674, **674**
    pilotless gas appliances, 192
    schedules for, 764, **766**
    symbols for, 310, 312, **312**
    utility room equipment and appliances, 197, **197**
Apron; *See* parking apron
Arc command, in CAD, 70
Arches; *See also* Gothic arches
    ceiling usage, **22**
    evolution of, 5, **5**, **6**
    laminated beams as, 518, **518**, **519**
    roof trusses as, 661, **662**
Architect's scale, 34–38, **35–38**
Architectural CAD, 63; *See also* computer-aided drafting (CAD)
Architectural conventions and procedures, 43–62
    abbreviations for, 817–820
    architectural disciplines, 749, **749**
    conventions, 51–54
    correction equipment, 59
    curved lines, instruments for, 58–59
    drafting aids, 59–61
    drafting pencils and pens, 55–56
    drawing techniques, 54–55
    drawings, 43–51
    glossary of terms, 823–832
    guides for straight lines, 57–58
    for HVAC, 704–705, **705**
    for plumbing, 730–732, **731**, **732**
    plumbing conventions, 730–732
    professional organizations, 820
    steel drawing, **557**, 557–558
    synonyms for, 821–822
Architectural design
    careers in, 788–790
    drawings for; *See* sectional drawings
    excellence in, 17
    fundamentals of; *See* design, fundamentals of
    future of, 16–18
    process of, 275, **276**
Architectural forms, 2–7, **3–6**; *See also specific forms*

Architectural history and styles, 2–19
    decor and, 121–122
    forms, development of, 2–7
    the future, 16–18
    influences on; *See* American architecture styles, development of, 7
Architectural lettering; *See* lettering
Architectural models, 443–454, **444**; *See also* models
Architectural rendering, 431–442; *See also* renderings
Area
    CAD command, 86
    of a circle, 156, 814, **814**
    of a polygon, 813, **813**
    of a square or rectangle, 155, 814, **814**
    of a triangle, 814, **814**
Areaways, 330
Arithmetic calculations, 793–796
Array Polar command, in CAD, 77, **77**
Arrowhead styles, in dimensioning, **330**, **331**
Asbestos, renovation and, 112
Ash pit, 496
ASTM; *See* American Society for Testing and Materials (ASTM)
Attics
    guest bedroom in, **216**
    loading calculations, 810, 812
    ventilation in, 112
Attributes command, in CAD, 761
Autoclaved aerated concrete, 535
Auxiliary elevations, 362, 365, **366**
Average density, 233, **233**
Awning windows, 349, **350**, **351**, 372
Azimuth system, 238, **238**, **239**, 497

Backups
    for electrical power, 728
    for security, 67
Baffles
    in dining area, 127–128
    at entrance, 167
    on patios, 150
    privacy, 529, **530**
    sun baffling, 102, **102**, 146
    vegetation as, 150, 269
Balance, 26, **26**, **27**, 46, 336
Balconies, 144, **145**, **146**, 347, 463
Ball-bearing supports, 565, **566**
Balloon framing, 513
Barrel vault, 5, 6
Bars, steel, 548, **548**

Base maps, 279, **279**
Base molding, 400, **401**, 624
Base plans, 748
Base sections, 399
Basements
  electrical plan for, **695**
  excavations for, **497**
  foundation walls as, 478–482, **481**
  as part of living area, **118**
  for recreation room, 135
Basic eraser, 59
Basic floor plans, 275, **276**, **287**
Basic layout models, 445, **445**
Bathrooms
  accessibility, 223, **298**, **299**, **300**
  accessories, 221
  cabinets, 225
  decor, 223–225
  electrical plan for, **694**
  fixtures, 219–223, **220–222**, 225, 297, **297**, **298**
  function of, **219**, 219–222
  heating, 222
  layout, 221–222
  lighting, 222
  location, 223
  plumbing lines, 223, **743**
  size and shape, **224**, 225, **226**
  special needs planning, 300, **300**
  storage areas, 205
  symbols for, 312, **313**
  ventilation, 222, 721
Bay windows, **350**
Bays, 595, **596**
Beam compass, 58
Beam details, 399
Beams; *See also* girders; rafters
  in concrete construction, 538, 540
  moment calculations, 800–803, **801–803**
  in pier-and-column construction, 487, 489, **490**
  post support for, 620, **623**
  in post-and-beam construction, 513, **514**, **515**
  for steel construction, 546, **546**, 586, **587**, **588**
  tributary area of, **807**
  wooden, 584, **585**, **586**
Bearing, 238
Bearing plates, 548, **548**
Bearing surface, 472–473
Bearing walls (structures)
  buttress support for, 6, **6**
  construction of, 456–457, **457**, **586**
  evolution of, 3, **4**
Bedrooms
  children's rooms, 214, **215**
  decor, 215, **216**, **218**
  doors, 211, **212**
  dressing areas, 214
  electrical plan for, **694**
  function, 209
  location, 209–215
  master suite, **215**
  noise control, 210–211, **211**
  size and shape, 215–219, **216**
  special needs planning, 300
  storage areas, 205, 212–213
  ventilation, 212
  wall space, 211
  wheelchair accessibility, **301**
  windows, 211–212
Bending moment, 800–803, **801–803**
Berms, 105, 113
Bermuda roofs, 340, **340**, **342**
Bid forms, for construction, 785
Bidets, 220
Bill of Materials command, in CAD, 92, 761
BIM; *See* building information modeling (BIM)

Bird's mouth cut, on rafters, 643, **643**
Bird's-eye view, 44, 426, **428**, **429**
Block command, in CAD, 81, 84, 246, 285–286, 312
Blockflash moisture control system, 481, **483**
Blocking, 574, 629
Blocknet moisture control system, 481, **483**
Blueboard, 627
Board foot, 510, **511**, 795–796, **796**
Board-and-batten siding, 609, **609**
Bolsters, 474
Bolts
  lag screws, 518, **518**, 520
  for platform framing, **564**
  for steel construction, 554, **554**, **555**
Bonds
  as legal document, 785
  for masonry, 534, **534**, **535**
Bookshelves, built-in, 169, **408**
Border lines, 52, **52**
Bow compass, 58
Bow windows, **350**
Box sill; *See* pilasters
Bracing
  cross bracing/bridging, **556**, 556–557, 564, 573
  diagonal, 601, **602**
Brackets
  for steel construction, **523**, 552–553
  for wood construction, 520, **521–523**, 523
Branch circuits, 677, **677**
Branches, and waste discharge system, 735
Break command, in CAD, 75, 78, **78**, 648, 731–732
Break lines
  drawing stair systems, 320
  function of, 52, **52**, 393–394, **394**, 397, **398**
Breakfast room, **377**
Break-out sections, 403
Breezeways, 147, 198; *See also* patios
Brick veneer
  detail drawings for, **503**, **590**, **593**
  elevation plan for, **612**
  support for walls, 532, **533**
  use of, 611, **612**
Bricks, 527–528, **528**, 532
British thermal unit (Btu), 707
Budgeting, 780
Building codes, 782–784
  chimneys, 494, **495**
  compliance to, 784
  concrete block foundations, 473
  design requirements and, 295
  fire safety, 567
  frost line depth for footings, 475
  girder sizes, 489
  plumbing drainage fields, 737
  roof design, **797**
  roof sheathing, 663
  slab foundations, 485
  switch height, **689**
  T-foundations, **477**
  types of, 783–784
  zoning ordinances and, 229, 782–783
Building envelope, 231–232
Building information modeling (BIM)
  construction documents and, 761
  elevation plans, **94**
  estimating with, 777–778
  floor plans, 93, **93**, **94**, 382
  perspective drawings, **95**
  pictorial drawings, **95**, **383**
  purpose of, 92–95, 332, 382, 748–749
Building main, 733
Building user analysis; *See* user analysis
Buildings; *See also* residences
  cost of, 775–778

  fire protection in, 115
  loads supported by, 458; *See also* loads (loading)
  materials for; *See* materials
  orientation of, **361**
  permits for, 235–236, **236**, **237**
  position of, wetlands and, **109**
  venturi effect, 109, **109**
Built-in components
  bookshelves as, **169**, **407**
  dimensioning, 407, **408**
  framing for, 620, **622**
Built-up roofs, 665–666, **666**
Burnishing plates, 60
Butt joints, 406, **406**
Butterfly roofs, 339, **340**
Buttress, **6**, 6–7

Cabinetry
  architectural scale and, **39**
  in bathrooms, 225, **225**
  careers in, 791
  coding system for, 411, **413**
  component drawings, **404**, 404–413
  construction of, 404–406, **405**
  dimensioning, 404–405, **405**, 407, **409**, 409–412
  door openings, **192**
  in entertainment rooms, 137
  floor plan for, **305**
  joints for, 405–406, **406**
  for kitchen storage, 174
  modeling, 452
  as room separator, **129**
  schedules for, 764
  sizing, 186, **187**
Cable systems, 703
CAD; *See* computer-aided drafting (CAD)
CADD; *See* computer-aided drafting and design (CADD)
Caissons, 538, **539**
Calculations
  arithmetic, 793–796
  geometric, 813–816
  heat loss, 713, **714**
  overhang size, **101**
  roof slope and pitch, 367–368, **368**, 652, **654**
  stair dimensions, 320, 322, 324, **325**
  structural, 796–813
  swimming pool size, **155**, 155–156, **156**
Callouts, in drawings, 49, **51**
Cane sweep width, 295, **296**
Cannon test, 565
Cant strip, 659
Cantilever structure
  floor joists and, 580, **581**, 588, **590**
  support for, 463, **463**, **464**
  trusses for, 552, **552**
Capacity, 156; *See also* volume
Cape Cod style, 11, **11**
Carbon monoxide, leak prevention, 566
Carports, 198; *See also* garages
Cartesian coordinates, in CAD, 69, **69**, **82**, 82–83, 87, **88**, 327
Casement windows, 349, **350**, **351**, **372**
Cathedral windows, **350**
Cavity walls, 533, **533**
CDs; *See* compact disks (CDs)
Ceiling lines, 355
Ceilings
  acoustical, 132, 211
  curved, **642**
  framing for, **653**
  height of, 602, **605**, 724
  insulation for, **711**
  joists, loading for, 803, **804**

Ceilings (*continued*)
light fixtures, **685**, 685–686, **686**
for living rooms, 124
luminous design of, 686, **686**
reflected ceiling plans, 324, **326**
skylights and, 346
surface coverings schedules for, 765
types of, **348**
ventilation for, 721, **722**
windows and, **124**
Cement, 535; *See also* concrete
Centerlines, 51, **52**, **53**
Center-supported cantilever structure, 463, **463**
Central bath, 223, **223**
Central processing unit (CPU), 65, 66, 67
Certificate of occupancy, 784
Chairs, for concrete, 474
Chamfer command, in CAD, 72, **72**
Change orders, 754, 758, **758**
Channels, 549, **550**
Charcoal rendering, **432**
Checking drawings, 749–750, 752, **752**
Children's rooms, 214, **215**
Chimneys
building codes, 494, **495**
elevation details, 345, **345**, **360**
flashing for, 667, **668**
footings for, 494, 495–496, **502**
framing details, 646, **647**
masonry construction, 124, 139
structure for, **346**
Chroma, color and, 24
Circle command, in CAD, 70, **70**
Circles
area of, 156, 814, **814**
circumference of, 813, **813**
template for, 58
Circuit breaker, 677
Circuits, 677, **677**, 681, **693**
City planning models, **444**
Classical Colonial, 7
Classical revival, 11
Clay tile, structural, 530, **531**
Cleanouts, and waste discharge system, 736
Cleanup center, in kitchen, 175
Clerestory windows, 339, **350**
Climate control symbols, **705**
Closed planning, 118, 120, **120**, 167
Closets; *See also* storage areas
configuration of, **205**
doors for, **170**, **212**
electrical plan for, **695**
at entrance, 165, 169
types of, 203–205; *See also specific closet types*
Closing costs, 779–780
Cluster housing, 233, **233**
Cob construction, 470
Coding system, for architectural drawings, 45–46, **49**
Collage, 767, **770**
Collar beams, 645, **645**
Colonial architecture, 7, 11, **12**, 336, 338–339
Colonial-styled kitchen, 184, **185**
Color
coding of, 767, **768**
color wheel, **23**
decor and, 122
as design element, 22–24, **23–25**, 29, **154**
elevation and, 336–337, **338**
harmony and, 24, 197, 203
hatching, 73, 93, **94**
layering, 67, **80**, **84**
lighting and, 153, **154**, 682, **683**
Column orders, 4, **4**
Columns; *See also* posts
column schedules, 620, **623**, 623–624

footings, **475**, **476**, **478**, **479**, **488**
function of, 620
insulation for, **714**
lally columns, 538, **539**
loading calculations, 807–809
pier-and-column foundations, 487–490
pole/column foundations, 487, **489**
rebars for support, 538, **539**
for slabs, 541–542, **542**
for steel construction, 546, **546**, **553**
Combination plans, 752, 754
Combined heating and cooling system, 720–721, **721**, **728**
Combined living plan, 120–121
Common brick, 527, **528**
Common rafters, 645
Communication systems, 701–702
Communities, smart growth in, 16–18, **18**
Community development models/plans, **236**, **448**
Compact disks (CDs), 67, 406
Compartment plan, **221**, 221–222, **222**
Compass, 58
Compass orientation, 361–362, **361–364**
Complementary (color harmony), 24
Component drawings for cabinetry, **404**, 404–413
Composite analysis, 281–282, **284**
Composite framing-elevation drawing, 649
Compression forces, 458, **458**, 473–474, 489, **639**, **810**
Computer-aided drafting (CAD), 63–96
advantages of, 64
applications of, 91–96, **93–96**
Bill of Materials command, 761
BIM practices, 92–95, 332
building codes and, 784
building materials library, 392
for cabinetry, 404–407
careers in, 789
characteristics of, 63–65
coding for, 46
commands for, 68–81; *See also specific CAD commands*
consistency in, 64
continuing education, 95
correcting, 59
cost estimates from, 777–778
curved lines, 58–59, 70
detail (section) drawings in, 87, **88**
document management, 93, 95, **95**
electrical symbol library, 74–75, 696
elevation rendering, 369, 375, **385**
elevation symbol library, 369, 375
engineering studies, 92
entourage libraries, 441, **441**
examples of, **64**
exterior elevation projection, 369, 383
file types and transmission, 64, 92, 406–407, 773
floor framing, 586
floor grids, 327–328
floor plan design with; *See* floor plans
for foundations, 498
full section drawing, 392
functionality of, 65
HVAC symbol library, 705
interior elevation projection, 378
layering in; *See* layering
lettering in, 54
manual versus, 63, 64, **65**
modeling, 44, **48**, 443
modular design, 468
pictorial contours, 246, **246**
pictorial drawings, 426, 429
plumbing drawings, 731–732
reflecting plans, 324
renderings, 431, **432**, 435, 441, **442**

resource information for, 773
reversing plans, 324
ribbon menu, **69**
roof framing, 648
scales for, 41–42
sequencing in, **316–317**
software for, 68, 95, **96**
in steel construction, 552
symbol blocks, 312, **313**
symbol libraries, 74–75, **81**, **313**, 369, 375, 392, 696, 705
system components of, 65–68, **66**, **67**
three-dimensional drawings, 87–91
topographic symbols, 246, **247**
two-dimensional drawings, 82–87, **82–88**
types of, 82–91
wall framing, 600–601
welding libraries, 556
for windows, **85**
workstation, **66**
Computer-aided drafting and design (CADD), 63
Concept study, 394, **395**
Conceptual design, 282–283, **285**
Concrete
autoclaved aerated concrete, 535
block size, **473**, 529, **529**
block types, 528, **529**
building codes for, 473
cast-in-place, 540
composition of, 535
concrete decking, 575, **576**
concrete shells, 636
construction systems, 535–544, **610**
for foundations, 473, **485**
joints for, 542
precast, 540, **540**, **541**
preinsulated walls, 479, 481, **482**
prestressed, 536–538
rebars for; *See* rebars
reinforced, 535–536, **536**, 561
for roofs, 635–636
slab components, 541–542
structural members, 538–540
walls, 116, 540, **542**, 542–544
wind resistance of, 561
Conduction, 706
Conductors
electrical, 675, 678, 680
heat, 707
Connectors
joints and, 524, **525**
joists and, 584, **585**
split-ring connectors, 524, **525**
timber connectors, 520–525, **521–525**, 562
Conservation, of electricity, 673–674
Conservatories, 149–150, **150**
Construction, 456–471; *See also* curtain-wall construction; frame construction; modular construction; post-and-beam construction; skeleton-frame construction
adobe construction, 98, 105, 470
of bearing walls (structures), 456–457, **457**, **586**
bid documents, 785
building codes and, 782, **783**
of cabinetry, 404–406, **405**
careers in, 791–792
of chimneys, 124, 139
component schedules for, 768
concrete systems, 535–544, **610**
documents for, 44–45, 65, 382
drywall construction, 627, **628**
heavy-timber construction, 654–655, **659**
labor costs, 776, **776**, **777**
large-span construction, 547
masonry systems, 527–535, **611**
materials for, 13, 16; *See also specific kinds*

methods of, 16
plank-and-beam construction, 514
rammed earth construction, 105, 469–470, **470**
scale and details, **39**
section drawings, **397**
steel-cage construction, 457, 545–547, **546**, 558, 586, **587**
straw-bale construction, 469, **469**, **470**
structural design, 456–463
stud layout, 467, **604**, **605**, 628–631, **628–631**
systems of, 456
Construction Line command, in CAD, 86, 369
Construction (layout) lines, 52, **52**
Construction Specifications Institute (CSI)
drawing format, 772, 773
master programs, 91, 772
specification consistency, 772, **773**
Contemporary Colonial, 7
Contemporary influences on architecture, 13–16, **15**, **16**
Contemporary-styled kitchen, **179**, 179–180, **180**, 184, **185**
Continuous stepped footing; *See* stepped footings
Continuous-arch construction, 457
Contours, 239–247
anatomy of, 244–246, **245**
contour altering lines, 243
contour intervals, 244, **244**
contour lines, 240–243, **240–243**, 251–252, 280
grade contours, **478**
numbering, 244, 246
pictorial, in CAD, 246, **246**
symbols, **246**, 246–247, **247**
Contracts, as legal document, 785
Convection, 706, 720, **724**
Conventional framing, for roofs, 654, **655**
Conventions; *See* architectural conventions and procedures
Cooking center, in kitchen, 174–175
Cooling systems
energy efficient designs, 116
mechanics of, 720–721, **721**, 728
return system, 112, **112**
Coordinate pair, 82
Copy, multiple, or repeat command, in CAD, 76, **76**
Copy-Multiple command, in CAD, 75, 84, 91, 441, 552, 586, 600–601
Corner wall sections, **400**
Corner windows, **350**
Cornice lighting, 684, **685**
Cornices
detail drawings, 368, **371**, 642, **643**, 645, **645**, **660**
section drawings, 397, **398**
Corridor kitchen, 177, **178**
Costs
estimating, 775–778, **776**, **777**
minimizing, 778
Coursed rubble; *See* rubble
Court, 148; *See also* patios
Courtyard; *See* patios
Cove lighting, 684, **685**
CPU; *See* central processing unit (CPU)
Crawl spaces, ventilation in, 112
Creativity in design, 22
Crests, **240**; *See also* hilltop summits
Cripples, 482–483, **484**, 601
Cross bracing/bridging, **556**, 556–557, 564, 573
Cross vault, 6, **6**
Cross ventilation, 212, **212**
Cross-referencing, 46, **50**
Crown molding, 125, 400, **401**, 624
Crown sections, 399
CSI; *See* Construction Specifications Institute (CSI)
Cubic-foot method, for estimating costs, 776, **776**

Current, 674
Curtain-wall construction
coverings for, 600
glass walls, 546
post-and-beam outside walls, 515, 611–612
steel skeleton construction, 457, 611–612
wood construction, **612**
Curved lines, 58–59, 70
Curved panels, 642, **642**
Curves, in Cartesian coordinate systems, 83, **83**
Cut-off valves, 739
Cutting planes, 389, **389**, **390**, 399
Cutting-plane lines
for beams, 399
for doors, 403
in floor plans, 389, **389**, 399, 402, 403, 614
function of, 51, **52**, **53**
transverse versus longitudinal sectioning, 389, **389**
for windows, 402, 614
Cylinders
surface area of, 814, **814**
volume of, 815, **815**

Dampers, 494
Data storage devices, for CAD workstation, 67
Databases, for estimating costs, 777–778
Datum, 239, 240
Datum line, 380
Daylight plane, 230, **230**
DC; *See* direct current (DC)
Dead loads, 458, **458**
calculating, 799, **811**
in frame construction, **796**, 796–797
headers and trimmers and, 592
on roofs, 460, 633
Decibels, 114, **114**
Deciduous trees, 105, **107**, 269; *See also* trees
Decking, 573–575, **574**
Deck/patio lines, 355
Decks; *See also* porches
cantilever structure, 463
elevation and, 347
as extension of living area, **145**
flooring, 149
laminated timbers for, 517–518, **518**
multilevel, **145**
pool deck, 154
sun baffling, **146**
Decor
architectural styles and, 121–122
bathroom, 223–225
bedrooms, 215, **216**, **218**
dining rooms, 128–129
entrances, 166–168
family rooms, 131–132
fireplaces, 124
garages, 198–199
halls, **169**
kitchen, 182–185
lanais, 152
living rooms, 121–125, **123**
patios, 149–151
porches, 146
recreation rooms, **134**, 135
utility rooms, 196–197
for walls, 122, 131
for windows, 122–124, **124**
workshops, 203
Decorative lighting, 125, 682, 683, **683**, **684**
Deeds, as legal document, 238, **239**, 785
Deflection, 461, **462**, 463, **463**
calculating, 807, **807**
prestressed concrete and, 536–537, **537**
tongue-and-groove planking and, 517
Deformation, 461, 809, **810**

Dehumidifier, 722
Delete command, in CAD, 75, **76**
Dens, 205
Density, 233–234, **233–235**, 444, **444**
Depressions, **242**, 244
Design, fundamentals of, 20–31
applications of, 30–31
careers in, 788–789
creativity in, 22
elements of, 22–26, **22–26**
feng shui, 301
form follows function, 20–21, **21**
goals and objectives, 277, **277**
horizontal emphasis, **335**
interior design, **21**, 21–22
needs versus wants, 277, **277**
principles of, 26–30, **26–30**
structural calculations, 796–813
vertical emphasis, **335**
visual perception and, 113–114
Design (schematic) drawings, 557, **557**, **730**, 739, **739**
Design study models, 443–447, **444**
Detached garage, 198
Detail (section) drawings, 396–404
for brick veneer, **503**, **590**, **593**
in CAD, 87, **88**
for cornices, 368, **371**, 397, **398**, 642, **643**, 645, **645**, **660**
door sections, 402–403
examples of, **46**, **399**
exploded views, 607, **607**, **613**
for exterior walls, 396–399, 400, **400**, **559**, **606**, 606–612
for fireplaces, **495**, **503**, **504**
for framing, 588, **590**
for headers and trimmers, 592
horizontal wall sections, **399**, 399–402
for interior walls, 399, 400, **401**, **402**, 619, **620**, **621**, 624–627
for intermediate support, 592
in modular construction, 466, 468
purpose of, 44
riser diagrams, 741
for septic tanks, **738**
for sills, 397, **398**, 402, **468**, 501, **503**, 588, 592, **592**
for slab foundations, 484–485, **485–487**, **500**, 501
stud layout, 629–630
symbols for, **396**
vertical wall sections, 396–399
for walls, **662**
welding presentation, 555, **556**
window sections, 402
for wood foundations, **492**
Detailed models, 447–448, **449**
Development marketing plat, **254**
Development plan, **264**
Diffused light, 684
Dim command, in CAD, 315
Dimension lines, 51, **52**, **53**, 328
Dimensioning
for appliances, **188**, 407, **409**
arrowhead styles in, **330**, **331**
built-in components, 407, **408**
for cabinetry, 404–405, **405**, 407, **409**, 409–412
CAD command, 72–74, **74**, 84, 86, 328
for construction studs, 628, **629**
for doors, 328
in electrical plans, 695
for elevations, **365**, **366**, 380–382, **381**, **383**, 602, **603**
of fences, **366**
of fireplaces, 400, **403**, 493, **493**, 495
in floor plans, 324, **327**, 327–332, **329–333**

Dimensioning (*continued*)
  for foundations, **498**
  framing versus finished, 382
  in full section drawings, 394, **395**
  guidelines for, 380, 382
  human factors, **110**
  in kitchens, **186**, 186–187, **187**
  in modular construction, 332, **465**, 466, 468
  of railings, **296**
  for roofs, 649, 652, **653**
  of rooms, 328
  rough openings, 602, 612, 613, **613–615**
  for stairs, **172**, 322, **324**, 627, **627**
  for steel structural drawings, 595, **596**
  for swimming pools, 332
  of walls, 328, 330, **331**
  of windows, 328, 373–375, **374**
Dimmer switch, 688, **688**; *See also* rheostat
Dining patio, 147, **147**, **148**, 166
Dining rooms/areas, 127–130
  decor, 128–129
  electrical plan for, **695**
  elevations, 376, **377**
  function and location, 127–128, **127–128**, 184
  kitchen and, 174, **184**
  size and shape, 129–130, **130**
  storage areas, 205
Direct current (DC), 672
Direct light, 683
Direct stress, 812, **812**
Direct-gain method, 724–725, **725**
Disaster prevention, 561–570
Dispersement, of light, 683–684, **684**
Display; *See* liquid-crystal display (LCD)
Display Grid command, in CAD, 80, 328
Distance command, in CAD, 86
Distribution fields, 737, **737**, **738**
Distribution panel, 676, **677**
Dividers
  in architectural drafting, 58
  in bathrooms, 221
  in bedrooms, 214
  in dining area, 129
  in halls, 169
  for rooms, 205
Dome (dome-shaped) roofs, **340**, 340–341
Domed hut, 2, **3**
Domes, 6, **6**, **30**
Door hand, 763, **763**
Door swing, 307, **315**
Doors
  bedroom doors, 211, **212**
  codes for, 375, **375**
  dimensioning, 328
  door sill, **594**
  elevation and, **337**, 351, **764**
  at entrances, 160
  exterior doors, **352**, 375, **709**
  fireproof, 198
  floor plan guides, **315**
  framing drawings, 614, 616–619, **616–619**, 624, **624**, **625**
  framing for, 599–600
  garage doors; *See under* garages
  integrated design of, 349
  modular construction, 465
  openings to hall/closets, **170**, 204–205, **307**
  orientation of, **763**
  R-values/U-values for, **709**
  schedules for, 763, **763**
  section drawings for, 402–403
  sizes of, 308–309, **310**, **376**
  special needs planning, 295–296, **296**, 300
  steel framing for, 558, **559**
  symbols for, 306–309, **307–310**

  trim for, **625**
  types/styles of, 307–308, **308**, **309**, **352**
  wind gusts and, 564, 565
Dormers, 343, **344**, 646, **646**, **647**
Double bevel, **36**
Double joists, 577, **578**, **579**
Double-glazed windows, 211, 708, **709**
Double-hung windows, 349, **350**, **351**, **372**
Double-line drawings, 649, **649**
Dove joints, 406, **406**
Down payment, 778
Downdrafts
  fireplaces and, 112, 494, **494**, **495**
  kitchen exhausting, 192
Downspouts; *See* gutters and downspouts
Drafting; *See also* computer-aided drafting (CAD)
  aids for, 59–61
  careers in, 789, 790
  correction equipment, 59
  media for, 60–61, **61**
  pencils and pens for, **55**, 55–56, **56**
  scales and measurements, 34–42
  tape for, 60
Drafting brush, 59
Drafting machine, 57
Drag command, in CAD, 81
Drainage fields, 736, 737–738
Drainage tile, for foundations, 113, 476, **477**
Draped reinforcement, 540
Draw (warm-air rise), 494
Drawing code, 749, **750**
Drawing commands in CAD
  Arc command, 70
  Chamfer command, 72, **72**
  Circle command, 70, **70**
  dimensioning commands, 72–74, **74**, 84, 86, 328
  Draw command, 369
  Ellipse command, 70, **71**
  Fillet command, 72, **72**
  Hatch command, 72, **72**, **73**, 75, 86, 88, **89**, 90, 246, 375, 392, 431, 440
  Line command, 57–58, **69**, 69–70, 75, 84, 315, 586, 648, 731
  Polygon command, **71**, 71–72
  Rectangle command, 70–71, **71**
  Spline command, 70, **71**, 92, 246, 696
  Text command, 70, **71**, 75, 84, 86, 315, 556
Drawing sequence, 748
Drawing versus drafting, 54
Drawings, architectural, 59–61
  agreement among, 750, 752
  changing, 754, **758**, 758–759, **759**
  checking, 749–750, 752, **752**
  coding system for, 45–46, **49**
  coordination of, 749
  cross-referencing symbols, 749, **751**
  documents, **44**, 44–45
  line weights, 51
  management of, 748–760
  reading, 45–50, **49–51**
  schedules, 761–769
  sequencing, 748, 752
  specifications, 761, 770, 772–774
  types of, **44**, **44–48**
Drawings, component; *See* component drawings
Drawings, detail; *See* detail (section) drawings
Drawings, orthographic; *See* orthographic drawings
Drawings, section; *See* sectional drawings
Dressing areas, 214
Driveways, 162, **162–164**, 201, **201**, **202**
Drop-box compass, 58
Dropleaf workbench, 202
Dry cleaner bags, 59
Drywall construction, 627, **628**

Ductwork
  for dryer, 194–195
  for fireplaces, 496, **496**
  in HVAC system, 704, **705**, **715**, 716, **717**
Dutch Colonial, 11, 339
Dutch doors, 351
Dutch hip roof, 339, **340**, 657, **659**
Dynamics, 796

Early American architecture, 10–12, **11**, **12**
Early Colonial, 7
Earthquakes, design prevention, **563**, 565, **566**
Earth-sheltered homes, 104–105, **105–107**, **543**
Easements, 279
Eave line, 335, **343**, 666
Eccentric cantilever structure, 463
Eclectic design, 22
Ecology, design considerations and, 110–117
  air quality, 111–112, **112**, 567
  fire protection, 115; *See also* fire protection/ prevention, design considerations and
  land integrity, 111, **111**
  noise, **114**, 114–115, **115**
  visual perception, 113–114
  water quality, 112–113, **113**, 567
Editing commands in CAD, 75–78
  Align command, 77, **77**, 86, 369
  Array Polar command, 77, **77**
  Break command, 75, 78, **78**, 648, 731–732
  Copy, multiple, or repeat command, 76, **76**, 84
  Delete command, 75, **76**
  Erase command, 75, 76, 86
  Join command, 77, **78**, 84, 315
  Lengthen command, 78, **78**
  Mirror command, 76–77, **77**, 324
  Move command, 76, **76**, 91, 441
  Offset command, 77, **77**, 84, 85, 315, 586, 648
  Redo command, 75
  Redraw command, 78
  Rotate command, 76, 77
  Set Variable command, 76, 324
  Stretch command, 77, **78**, 91, 441
  Trim command, 78, **78**, 84, 586, 601, 648
  Undo command, 75
Egress, 135, **137**
Electric eraser, 59
Electric heating system, 720, **720**
Electrical plans, 687–692
  design considerations, 116, 135
  floor plans for, **696–700**
  outlets and receptacles, 690, **690**
  schedules for, 764, **765**
  special needs planning, 300–301
  switches, 687–689, **688**, **689**
  symbols for, 74–75
  wiring plans, 693–700, **694–696**
Electricity
  careers in, 790, 792
  design considerations, 116
  design drawings for, 672–703
  distribution of, 674, **675**, 676–677, **677**
  dryer requirements for, 194–195
  generation of, 672, **673**
  hazards from, 116, 569
  plans for; *See* electrical plans
  post-and-beam construction and, 513
  power tool requirements for, 202
  principles of, 674–681
  symbols for, **691–692**
  system requirements for, 681
  UL standards for, 115
  working drawings for, 693–700
Electromagnetic force (EMF), 116
Electronics
  building systems, 700–703

in entertainment rooms, 137–138
hazards, design considerations and, 116
in home office, 135
wiring for, 703
Elevations; *See also* exterior elevations; interior elevations
architectural character and, 113
auxiliary elevations, 362, 365, **366**
BIM, 93, **94**, 382, **383**
in CAD, **86**, 86–87, **87**
color coding for, 767
concept study for, 394, **395**
construction and, 365, **381**
design elements in, 335–337
designing, 334–352
dimensioning, **365**, **366**, 380–382, **381**, **383**, 602, **603**
doors and, **337**, 351, **764**
drawing, 353–387, 396–397, **467**, 646, **646**, **759**
for fireplaces, **504**
framing elevations, **599–604**, 599–605
for garages, 198
height factors and, **336**
height lines, **368**
interior elevations, **601**
for kitchens, **180**, 187, **187**, **190–192**, 376, **377–380**, 410, **411**, **413**
LEEDs features in, 116
for masonry-cavity walls, **533**
modeling, **450**, **451**
modular construction and, 467, **467**
orientation for, 360–362, **360–364**
panel elevations, 601–602, **603**
plot plans and, 251
for plumbing, 739, **739**, 741–744, **742**, **743**
post-and-beam construction, **516**
presentation drawings, 382–386
profiles, 239, **240**, **242**, 269, **270**, **755**, **756**
projections, 353–370
purpose of, 44, **45**
relationship with floor plans, 334, 355, **358**, 362, **365**, 376
rendering, 269–271, **269–271**, 339, 385
roof design and, 338–347, 368, **369**, **370**, 649, **650**, **656**
room spacing and, 293
section drawings of, **396**, **652**
on site development plans, 239
for stairways, **171**, 625, **625**, **626**
in survey plans, 239
symbols for, 75, 369, 371–375, **382**
true size of, 362, **366**
for waste discharge system, 741, **742**
wind helix and, 672
window style and, 347–351
Elevators, 171–172
Elizabethan style, **8**
Ellipse command, in CAD, 70, **71**
Embellished floor plan, **287**
EMF; *See* electromagnetic force (EMF)
Emittance, 726
Emphasis, 27, **28**, **29**, **123**
Energy, design considerations and, 98–99, **99**, 116, **710**
Energy Star rating, 192, 673–674, **674**
Engineered wood, 518
Engineering
CAD in, 63
careers in, 790–791
engineering studies, 92
Engineer's scale, 38, **38**, **39**
English style, 8
Englund, Mark, 434
Enlarged scale, 34

Entertainment center, **138**, **205**
Entertainment rooms, 137–139
Entities, in CAD, 69
Entourage, rendering of, 266, **267**, 269, **270**, **440**, 440–441, **441**
Entrance hall, 160
Entrances
closed planning, 167
decor, 166–168
electrical plan for, **695**
function and types, 160, **161**
lighting, 168, **168**
location, 165–166
main, 162, 165, **165**
open planning, 167, **168**
service, 165, 166, **167**
site, 162
size and shape, 168–169
special needs planning, 295, 298–299
special purpose, 165, 166, **167**
storage areas, 205
surface materials, 167–168
symbols for, 256, **257**
and traffic areas, **161**
Environmental analysis, 228–229, **229**, 792
Environmental design, 98–117
carbon monoxide, 566
drainage fields, 737
ecology and, 110–117
energy guide, 674
energy orientation, 98
ergonomic planning, 109–110
fabricated members, 518
fire protection, 115
gas containment and venting, 565
greenhouse effect, 723
HVAC plans, 704
insulation, 708
land and structure, 102
LEED design considerations, **116**, 116–117
LEEDs, 117
National Fenestration Rating Council, 710
natural and propane gas, 566
orientation, 98–109
overhang and baffle protection, 100
passive solar methods, 724
protective measures, 108
radon, 566
R-values and U-values, 707
self-sustaining energy system, 727
site and environmental analysis, 228
solar heat and cooling system, 736
steel building construction, 545
structural types, 229
suitability levels, 229
toxic by-products, 192
water quality, 112
water supply systems, 732
windows and ventilation, 211
wind helix, 673
wood and the environment, 512
zoning laws, 231
Environmental Protection Agency (EPA)
air quality and, 112, 569
building codes and, 784
energy ratings by, 673
Erase command, in CAD, 75, 76, 86
Erasing shield, 59
Erection drawings, 557, **557**
Ergonomics, design considerations, 109–110
Escrow accounts, 779
Estate master plan, **47**, 258
Excavations, 497–498
Exercise room, 133, 207, **207**
Exhaust system/duct, **112**, 195

Extension lines, 51, **52**, **53**, 328
Exterior elevations
common features of, **356**, **357**, **359**, **360**, **364**, **386**
presentation drawings, 383, **384**, 385
projecting, in CAD, 369
purpose of, 353
scale and, **39**
Exterior perspective drawings, 420–423, **421–423**
Exterior walls, 598–619; *See also* walls
detail section drawings; *See under* detail (section) drawings
door framing, 614, 616–617, 619; *See also* doors
elevations for, **599–604**, 599–605
framing members, 599, **599**
loading calculations, 810
methods of erecting, 598–599, **599**
tributary area of, **811**
window framing, 612–614; *See also* windows

Fabricated timbers, 518–525
Face brick, 527, **528**
Factory lumber, 508
Fads versus trends, 22
Family kitchen, 180, **181**, 182, **182**
Family rooms, 130–132
decor, 131–132
function, 131
kitchen adjacent to, **185**
location, **127**, 131, **132**
size and shape, 132, **132**
storage areas, 132, 205
Fascia, 340, 343, 643
Fasteners; *See also* timber connectors
for post-and-beam construction, 524, **525**
for steel construction, **552–556**, 552–557
Felt (felt-tip) markers, 434
Fences
dimensioning of, **366**
landscape features, **271**
for pools, 156
wind deflection by, **109**
Fenestration, 349, **349**
Feng shui, 301
Filtration
air purification, 567, 722
plumbing design, 113, 733, **733**, **734**, 744, **744**
in swimming pools, 157
Financial planning, 778–781, **779**, **780**
Finish flooring, 574
Finish size, 508, **509**
Finished dimensions, 382
Finishing schedules, 767, **769**
Fire chamber; *See* firebox
Fire pit, **574**
Fire protection/prevention, design considerations and, 115, 198, 496, 567, 700, 784
Firebox, 492, **492**, **493**, **496**
Firecuts, 592, **593**
Fireplaces, 490–497, 502–505
construction of, 490, **492**
convenience components, 496–497
decor and, 124
design of, 139–141
dimensioning of, 400, **403**, 493, **493**, 495
downdrafts and, 112
drawings for, 502, **503**, **504**
fire-producing components, 492–494
as focal point in living room, **125**
footings for, 494, 495–496, **502**
freestanding, 496–497
fuel types for, 139–140
LEEDs standards, 492
openings for, 140, **140**, **141**
as room separator, **129**

Fireplaces (*continued*)
  safety components, 495–496
  symbols for, 312, **313**
  two-sided, 120, **120**, 128, **141**
  types of, **141, 346**
  ventilation for, 115
Fitness facilities; *See* exercise room
Fixed dormer windows, **350**
Fixed double awning windows, **372**
Fixed hopper windows, **372**
Fixed sheet glass, 614
Fixture lines, 52, **53**
Fixtures
  in bathrooms, 219–223, **220–222, 225**, 297,
    297, 298
  device listing, 687, **687**
  light fixtures, 684–687, **685, 686**
  modeling, 452, 453
  schedules for, 764, **765**
  switching system for, 678, 680
  symbols for, 310, 312, **312**
  traps for, in waste discharge system, 736, 738
Flagstone, 529, **530**
Flashing, 666–667, **668**
Flat roofs, 339, **340**, 658–659, **659**
Flat slabs, 541, **542**
Flexible curve, 58
Flitch beams, 516, **517**
Floor lines, 355
Floor plans, 275–302, 303–333
  abbreviated, 304
  for apartments, 288, **290**
  basic plans, 275, **276, 287**
  for bathrooms, **219–221**
  for bedrooms, **210, 217**
  BIM, 93, **93, 94**, 382
  bird's-eye view, **44**
  for cabinetry, **305**
  in CAD, 83–85, **85**, 285–286, 303, 315–319,
    316–318
  calculations in, 86
  for closets, **213**
  color coding in, 767, **768**
  conceptual, 282–283, **285**
  designing for later expansion, 288, **291**
  designing versus drawing, 275, 303
  dimensioning in, 324, **327**, 327–332, **329–333**
  for dining rooms, **127** ·
  door guides, **315**
  for earth-sheltered home, **106**
  for electricity, **696–700**
  electronics, **702**, 702–703
  elevations, relationship with, 334, 355, **358**, 362,
    365, 376
  for elevators, **172**
  embellished, **287**
  for entrances, **161**
  evaluating, 283–284, 292–295
  for family rooms, **127, 132**
  for foundations, **498**
  for foyer, **165, 166**
  framing model, 446
  functional space planning, 286–295
  for garage, **198**
  for great rooms, **121, 133, 134**
  for home offices, **137**
  for HVAC, **705**, 738
  idealized, 282–283, **284**
  isometric, **320**
  joist direction on, 588, **589**
  for kitchens, **177, 180, 182–184, 189, 191,
    193, 377, 410, 410, 412**
  for laundry room, **200**
  layering, 85–86, 315, 498
  for living rooms, **121**
  for mobile homes, 288, **288, 289**

modeling, 449
  for modular construction, 466, **466**, 468
  for mud room, **200**
  for multiple levels, 319–324, **321–323, 753, 754**
  outdoor living areas, **143, 147, 158**
  panning, **80**
  pictorial, 84, **84, 88, 91**, 93, **95**
  for plumbing, 738–739, **739, 740**
  presentation floor plans, 304, **304**
  process development, 275, **276**, 284–286,
    303, 314
  project analysis, 278–282, **278–284**
  project definition, 277
  purpose of, 275, 304
  reflected ceiling plans, 324, **326**
  reversed plans, 324, **326**
  roof plan on, **651**
  scale and, 38, **38, 39**, 304–305, **305, 318**, 319
  schedules for, **762**
  sequencing, 314
  single-line, **48**
  site-related drawings, 283, **285**
  sketching, 287–288, 292, 304
  space planning, 288–295
  special needs planning, 295–301
  steps in drawing, 315–317
  symbols for, 306–314, **306–314**
  templates for, **294**
  three-dimensional design, 319, **319**
  two-dimensional design, 83–85, 86
  types of, 304–305
  for utility room, **177, 195, 198**
  without textures, 304
Floors
  block and rib system, 538, **538**
  concrete floors, 540
  for entrances, 167–168
  for family room, 131
  framing drawings for, 572–597
  framing members, 573–586, 594–595, **596**
  framing plans, 586–594
  for garages, 198–199, **199**
  level changes in, 592, **594, 606**
  for living room, 124
  loading on, 579, 797, 803, **804, 805**, 811,
    811, 812
  for patios, 149
  plank-and-beam construction, 514, **515**
  platform systems, **572**, 572–573
  post-and-beam construction, 514, **515**,
    584, **587**
  special needs planning, 296, 299
  steel framing system, **551, 558**
  surface coverings schedules for, 765, **766, 767**
Flues, 494, **494**
Fluorescent lamps, 681
Flush doors, 351
Flying buttress, 6–7, **7**
Foam sandwich panels, 519
Focal point
  fireplace as, **125**
  purpose of, 27, **28**
  water as, 269
Folded plate roofs, 642
Foliage; *See* landscape plans
Footcandle, 682, **682**
Footing lines, 355
Footings (footers)
  area calculations for, 797–798
  drawings for, 501, **501**
  excavations and, 497
  for fireplaces/chimneys, 494, 495–496, **502**
  function of, 475–476, **476**
  load calculations, 797–798, **799**, 810–811, **811**
  sections for, 398, **399**
  sizing, 810, **810**

for steel construction, 546
  types of, **475, 476**
Force
  bending force, **801**
  equilibrium of, 813, **813**
  loads and, **458**, 458–460, **460**, 800–813
  major lines of, 457
  resistance, **460**, 460–461, **461**
  types of, 458, **458**
Forced-air system, **714**, 714–716
Foreclosure, 778
Foreshortening, 353, 355, **355**
Form, 22, **23, 122**, 336; *See also* architectural forms
Form follows function, 20–21, **21**
Formal balance, 26, **26, 27**, 336, **336**
Formal living room, 125
Formaldehyde, 112
Foundation lines, 355
Foundations, 472–490, 497–505
  anchor devices, 482, **484**
  brick veneer, **503**
  components of, 472–477
  composite materials for, 476, **481**
  cripples, 482–483, **484**
  dimensioning for, **498**
  drainage tile for, 113, 476, **477**
  drawings for, 497–504
  excavations and, 497–498
  foundation sills, 482, **483, 484**
  foundation walls, 478–482, **479**
  grade contour and, **478**
  loading calculations, 810–812
  moisture control, 481, **483**
  plans for, **39, 755**
  purpose of, 472
  sizing, **810**
  tributary area of, **799**
  types of, 477–490, **478**
Four-way switches, 687, **689**
Foyer, 160, **165**, 167, **168, 169, 170**
Frame construction; *See also specific types of
    construction*
  dimensioning, 328, **330**
  floor planning, **757**
  plumbing allowances in, 739, **741**
  post-and-beam construction, 513–525
  skeleton-frame construction, 506–513
  wood-frame systems, 506–526
Framing dimensions, 382
Framing drawings
  for doors, 614, 616–619, **616–619**, 624,
    624, 625
  for elevations, **599–604**, 599–605
  for floors, 572–597
  for roofs, 633–669
  for walls, 598–631
  for windows, 612–614, **615, 616**
Framing members
  exterior walls, 599, **599**
  floors, 573–586, 594–595, **596**
  roofs, 633–636
French curve, 58
French doors, 351, **617**
French provincial style, **9**, 339
French style, 8
French windows, **350**
Front elevation, **356, 359, 360**
Full scale, 37
Full section drawings, 388–396
  cutting plane, 389, **389, 390, 399**
  dimensioning in, 394, **395**
  scale, 393–394, **394**
  steps in, 395–396, **396**
  symbols for, 389–390, **391–393**
Fuller, R. Buckminster, 340
Fully-divided scale, 34, **35**

Functionalism, 21
Furnace, ventilation for, 115
Furniture
    architectural style and, 125
    bathroom, 221
    bedroom, 212, 215, 217–218, **218**
    collage board for, **770**
    dining room, 129, 130, **130**, **131**
    home office, **136**
    living room, 125, **125**, **126**
    modeling, 452, **453**
    patio/porches, 146, 148, **151**
    planning space for, 288–290
    recreation room, **135**
    schedules for, 767–768, **769**, **770**
    special needs planning, 299, 300
    as storage, 205, **205**
    symbols for, 312, **323**
    templates for, 126, **126**, **130**, **131**, **218**,
        289–291, **291**, **292**
Furring strips, **628**

Gable roofs
    dormers and, **646**
    structure, 338–339, **340**, **341**, 655, **656–658**
    wind resistance and, 564
Gallery, 169, **169**
Gambrel roofs, 339, **340**, **341**, 657, **658**
Gap command; *See* Trim command, in CAD
Garages, 197–200
    decor, 198–199
    design of, 198, **199**
    doors, 198, 199, **199**, 351, 403
    entrance symbol for, **257**
    floor, 198–199, **199**
    function and location, 197–198, **198**
    living quarters above, **206**
    size, 199–200, **200**
    storage design, 199, 205, **206**
Garbage facilities, design of, 114, 206–207
Gas
    containing and venting, 565–566
    fireplaces, 140, 495
    leak detection, 700
    leak prevention, 566
    lines for, 739
    pilot for, 192, 566
    shutoff valve for, 115
Gated entrance, 162, **164**, **366**
Gathering room; *See* great rooms
Gauge, 475
GCFI; *See* ground fault circuit interrupter (GFCI)
    outlet
General building security, 569
General contractor, 791
General design schedules, 765
General lighting, 125, 129, 151, 168, 203,
    682, **683**
Geodesic dome, 340–341
Geographical survey maps, 249, **250**, **251**
Georgian style, 11
Geothermal heating/cooling, 721, **721**
Girders; *See also* beams
    in concrete construction, 538, 540
    in floor construction, **515**
    intermediate support for, 592, **593**
    load calculations, 803–807, **807**, 812
    for pier-and-column foundations, 487, 489,
        490, **491**
    in post-and-beam constructions, 513, **514**
    sizing and loads, 803–804, 806–807
    steel, 546, **546**, 548, **548**, **553**, 586, **587**
    tributary area of, 806, **806**
    wooden, 584, **585**, **586**
Glass enclosures, for fireplaces, 496
Glass roofs, 341, **342**

Glass walls
    in bathrooms, 222, **222**
    curtain walls as, 546
    detailing, 614, **615**
    green design and, **125**
    open plans and, **119**
    in solariums, 150, **150**
Glazing, 708, **709**
Glossary of architectural terms, 823–832
Golden rule, 28
Gothic arches, **6**, 6–7, **7**
Grade marks, 508, **509**, **511**, 511–512
Grading, 236
Graphics card, for CAD workstation, 66–67
Gravity system, for heating, 714
Gravity waste system, **735**
Gray water
    irrigation with, 113
    waste discharge system for, 113, 736, **736**
Great rooms, **121**, 133, **133**, **134**, **183**
Greek and Roman column orders, 4
Green Building Council, 116
Green lumber, 512
Green spaces, 111, 230
Greenboard, 627
Greenhouse effect, 723, **724**
Grid column identification system, 594, **596**
Grid command, in CAD, **80**, 80–81, 367, 468
Grid paper, 61, **61**
Grids
    for modular construction, **464**, 464–465
    patterns, **81**
    for steel flooring systems, **594**
    window grids, **373**
Grooved panels; *See* tongue-and-groove planking
Ground fault circuit interrupter (GFCI) outlet, 116,
    192, 301, 569, 676, 678, 690
Ground lines, 335, 355, 360, **361**, 367
Grounding, 676
Groundwater, earth-sheltered homes and, 104
Growth; *See* smart growth
Guest closet, 202
Guidelines (line type), 52, **52**, 53
Gum eraser, 59
Gusset plates, 520, **523**, 636, **638**
Gutters and downspouts, 113, 666, **667**

Half-baths, 223
Halls, **169**, 169–170, **170**, 205, **695**
Hand tools, 202
Handicapped parking areas, 295, **295**
Hangers, 523, **523**, **565**, 577, **578**, 592
Hard disk drive, 67
Hardness, of pencils, 55–56, **56**
Hardware, for CAD system, 66, **66**
Hard-wired system, 684, 702–703
Hardwoods, 506–508, 511, **512**
Harmony
    of color, 24
    of design, 28–29
Hatch command, in CAD, 72, **72**, **73**, 75, 86, 88,
    89, 90, 246, 375, 392, 431, 440
Head sections, 402
Headers
    for chimneys and fireplaces, **579**
    construction types, 601, **603**
    for joists, 577
    for stairwells, 592, **595**
    for windows and doors, 614, 616, **616**, **617**,
        619, 803, **805**
Headroom, 171
Health and safety codes, 784
Hearth, internal, 492–493, 495
Heat loss, calculating, 713, **714**
Heat loss/gain, 706
Heat pumps, 721

Heat transfer
    common deterrents, **100**
    insulation and, 708–713, **711–713**
    in passive solar systems, 723, **724**
    principles of, 705–708, **707**
    remote, for fireplaces, 496, **496**
Heating, ventilating, and air conditioning systems
    (HVAC); *See* HVAC
Heating systems; *See also* HVAC
    in bathrooms, 222
    energy efficient designs, 116
    outlets for, framing and, **622**
    plan for, **706**
    return system, 112, **112**
    types of, 713–721
Heavy-timber construction, 654–655, **659**
Height
    of ceilings, 602, **605**, 724
    elevations and, **336**
    height lines, 368
    stability and, 808, **808**
    switch height, building codes and, **689**
HEPA filters; *See* high-efficiency particulate air
    (HEPA) filters
Hidden lines, 51, **52**, **53**
High-efficiency particulate air (HEPA) filters, 714
Hilltop summits, 244
Hip rafter, 657, **657**
Hip roofs, 339, **340**, **341**, 564, 657, **658**
Hollow core doors, 351
Home automation system, 702
Home offices, 135–136, **135–137**
Home owner's insurance, 779, 780
Home theaters, 137, **139**; *See also* Entertainment
    rooms
Hopper windows, 349, **351**, **372**
Horizon lines, 418, **418–420**, **422**
Horizontal loads; *See* lateral loads
Horizontal sections, **399**, 399–402
Hose bibs, 733
Hot water; *See also* hydronic systems
    continuous circulating system, 734, **734**
    Metlund system, 734, **734**
    supply lines for, 738
Housing development, **233**, 233–234
Hue, 24
Human factors
    building design considerations, 110, 186, **186**
    human dimensions, 110, **110**
    kitchen planning, 110, 186, **186**
    room sizes and, 291, **293**
    site design and, 229
Humidifier, 722
Humidistats, 722
Humidity control, for heating, 720, 722
Hurricane clips, 564
HVAC, 704–729
    for bedrooms, 212
    careers in, 790, 792
    cooling systems; *See* cooling systems
    environmental comfort factors, 721–722
    equipment housing, 195
    heating systems, 713–721, **714–720**
    insulating, 115, 708
    ozone depletion and, 116, 117
    plans and conventions, 704–705, **705**
    principles of heat transfer, 705–708
    sick building syndrome and, 112, 567
    slab foundation and, 485, **485**
    solar systems, 723–728
Hydronic systems, 717–718, **717–719**

I-beams, **514**, 518, **519**, 588
ICF; *See* insulated concrete forms (ICFs)
I-columns, **809**
Idealized designs, 282–283, **284**

I-joists, 580, 583, **583**, 634
Imhoptep, 3
Incandescent lamps, 681
Indirect light, 683–684
Indirect-gain method, 725, **725**
Individual footings; *See* piers
Indoor living areas, 118–142; *See also specific rooms*
   closed plan, 118, 120, **120**
   combined plan, 120–121
   dining rooms, 127–130
   entertainment rooms, 137–139
   family rooms, 130–132
   fireplaces; *See* fireplaces
   great rooms, 133
   home offices, 135–136
   kitchens, 174–193
   living rooms, 121–126
   open plan, 118–120, **119**, **121**
   plans for, 118–121
   special purpose rooms, 133–139
Infiltration, 706
Informal balance, 26, **27**
Infrastructure, 17
Ink renderings; *See* pens
Ink-jet printer, 68
Input devices, for CAD workstation, 67–68
Insert command, in CAD, 375
Insulated concrete forms (ICFs), 479, 481, **482**
Insulation, 708–713
   asbestos, 112
   for ceilings, **711**
   for columns, **714**
   for concrete walls, 542, **542–544**, 543
   in garages, 112
   heat transfer and, 708–713, **711–713**
   for HVAC systems, 115, 708
   insulated concrete forms, 479, 481, **482**
   for joists, **711**
   noise reduction, 114, **115**, 210
   for plumbing systems, 735
   in post-and-beam construction, 513
   for roofs, 116
   structural insulated panels, 609
   types of, **712**, **713**
Insulators, 675, 707
Insurance, 779, 780
Integral garage, 198
Integrated electronic system, 702
Intensity, 24
Interest rates, of mortgage, 778–779, **779**
Interior design
   careers in, 789
   decoration versus design, 21–22
   exterior design, relationship to, **21**
   modeling, 446–447, **447**
   symmetry in, **27**
Interior elevations
   drawings of, 375–380, **377–381**
   presentation drawings, 385, **386**
   scale and, **39**
Interior perspective drawings, 423, **425**, 425–426, **426**
Interior walls; *See also* walls
   detail drawings for; *See under* detail (section)
     drawings
   elevation drawings, 620–624
   elevations for, **601**, **605**
   loading and, 619
International Building Code, 476, **477**
Interpolation, in topography, 244, **245**
Irregular curve; *See* French curve
Irregular lines; *See* curved lines
Irrigation plan, 273
Island kitchen, 178–180, **178–181**, **183**
ISO standards, for architectural drawings, 41

Isometric drawings, 416, **417**, 418, **418**
Isometric floor plans, **320**
Italian style, 9, 10, 144

Jack rafters, 657
Jalousie windows, 212, 349, **350**, **351**, 372
Jamb sections, 402, 403
Join command, in CAD, 77, **78**, 84, 315
Joints
   butt joints, 406, **406**
   concrete, 542
   connectors and, 524, **525**
   dove joints, 406, **406**
   mortar joints, 534, **535**
   mortise and tenon joints, 406, **406**, 524, **524**
   in walls, 400, **401**
Joists, 575, 577–584; *See also* rafters; truss joists
   blocking and, 574
   bridging for, 489, 577, **578**
   cantilever structure and, 580, **581**, 588, **590**
   for ceilings, **565**, **654**, 803, **804**
   connectors and, 584, **585**
   double joists, 577, **578**, **579**
   on floor plans, direction of, 588, **589**
   for floors, 803, **804**, **805**
   function of, 489, 575, 577
   hangers for, 523, **523**, 577, **578**, 592
   headers for, 577
   I-joists, 580, 583, **583**, 634
   insulation for, **711**
   laminated, 583
   loading and, 513, **514**, 803, **804**
   lookout joists, 588
   metal, **551**
   post support for, **623**
   sizing of, 577, 579, 580
   solid lumber as, **577**, 577–580
   spacing for, **491**
   steel joists, 583, **584**, 640, 659, 661, **661**, **662**

Keystone, 5, **5**
Kilowatt-hour, 674
Kips, 807
Kitchens, 174–193
   appliances, 184, **188**, **192**
   decor, 182–185
   design considerations, 174–187
   dimensions of, 186, 186–187, **187**
   dining area in, **131**
   drawings of, 186–187, **410**, **411**, **412**
   electrical plan for, **694**
   elevations; *See* elevations
   floor plans for; *See under* floor plans
   functionality, 174–175, **175**, 187
   location, **131**, 189
   planning guidelines, 187–193
   size, shape, and space, 186, **186**, **187**, 190
   special needs planning, 299–300
   storage areas, 205; *See also* Cabinetry
   types of, 175–182
   utilities, 182, 192
Kneaded eraser, 59
Knee walls, 645, **646**

Labor costs; *See under* construction
Labor-and-materials bond, 785
Labor-and-materials method, for estimating costs, 776, **776**
Lag screws/bolts, 518, **518**, 520
Lally columns, 538, **539**, 584, **585**, 586
Laminated joists, 583
Laminated strand lumber, 520
Laminated timbers, 517–518, **518**, 634
Laminated veneer lumber, 517
Lanais, **148**, 151–153, **152**, **153**, 160, **502**

Land
   integrity of, design considerations and, 111, **111**
   preservation of, 111
   topographical design considerations, 102–104, **103**, **104**
Landfills, 111, 117
Landform models, 241, **241**, **243**, 447, **448**
Landings, and stairs, 170, 171, **172**
Landscape plans, 257–264
   careers in landscape design, 788
   elevation rendering, 269–271, **269–271**, 382–383, **384**, **385**
   examples of, **263**, **268**
   guidelines for drawing, 260–261, 263–264
   modeling, 453
   phasing, 264
   plan rendering, **264–268**, 264–269, 441, **441**, **442**
   rendering media, 264
   site rendering, 264–271
   specifications, **771**
   symbols, 258, **259**, **262**
   visual perception and, 114
Lap siding, 607, 609, **609**
Large-span construction, 547
Laser printer, 68
Later American architecture, 12–16, **13–16**
Lateral loads, 458, **458**, 478–479, **481**
Laundry rooms, 194–195
Lavatories (sinks), 219, 297, **298**, 300
Layer command, in CAD, 47, 49, 79–80, **80**, 85–86, 315
Layering
   in conventional drafting, 594
   in floor plans, 85–86, 315, 498
   method of, 46, **51**, 749
   for plywood, 510–511
Layout, for foundations, 497, **497**
Layout lines; *See* construction (layout) lines
LCD; *See* liquid-crystal display (LCD)
Leach field, 113, **736**; *See also* drainage fields
Leader lines, 52, **52**
Leadership in Energy and Environmental Design (LEED), **116**, 116–117, 784
LEDs; *See* light emitting diodes (LEDs)
LEED; *See* Leadership in Energy and Environmental Design (LEED)
Left-side elevation, **357**, **359**, **360**
Legal documents, 782–786
   bonds, 785
   construction bids, 785
   contracts, 785
   deeds, 238, **239**, 785
   liens, 785
   plats, 250–251, **251–255**
   property description, 238, **239**
Lengthen command, in CAD, 78, **78**
Lettering, **52**, 53–54, **54**
Library, 169, **169**
Liens, as legal document, 785
Light and shadow
   in CAD, 89, **91**
   design fundamentals, **24**, 24–25, **25**
   effects on renderings, 437–438, **437–439**
   elevation design, 336–337, **337**
   local versus general, **129**
   at night, **153**
   opposition with, **30**
Light emitting diodes (LEDs), 681
Lighting
   in bathrooms, 222
   careers in, 789
   circuits for, 677–678, **678**
   design considerations, 681–687
   in dining room, **129**

dispersing, 683–684, **684**
energy efficiency, 117
in entrances, 167, 168, **168**
in halls, 169, **169**, **170**
in living room, 125
measuring, 681–682
natural versus artificial, 682, **683**
on patio, 151
reflected ceiling plans, 324, **326**
special needs planning, 297
on stairs, 170
for swimming pools, **153**, 687
types of, 681, 682–683
in utility room, 197
for workshops, 203
Lightning rod, 115, 676
Line drawings, 435, **436**, **439**
Line of sight, 418, **419**
Line weights, 51
Lineal foot, 800, **800**
Linear dimensions, 328
Lines; *See also specific types of lines*
    CAD commands, 57–58, **69**, 69–70, 75, 84,
       315, 586, 648, 731
    characteristics of, 69–70
    conventions for, 51–52, **52**, **53**
    as a design element, **21**, 22, **22**, **30**, **124**, **150**
    drafting pens and pencils, 55–56, **56**
    elevations and, **335**, 335–336, **336**
    guides for, **57**, 57–59, **58**
    topographic (contour lines); *See* contours
    types of, **70**
    utility, 675, **675**, **732**, 738, 739
Lintels
    concrete lintels, **473**, **474**, 540
    door framing, 614, 616, **617**
    wall framing, 601, **603**, 803, **805**
Liquid-based solar system, 726, **726**; *See also* solar
    energy
Liquid-crystal display (LCD), 67
Live loads, 458, **458**; *See also* loads (loading)
Living areas
    electrical plans for, **696–700**
    entrance symbol for, **257**
    indoor, 118–142
    outdoor, 100, 107
    special needs planning, 299
Living patio, 147, **147**, 148, 166, **257**
Living rooms, 121–126
    decor, 121–125, **123**
    electrical plan for, **694**
    fireplace in, 124, **125**
    function of, 121, **126**
    location of, 121, **121**
    orientation, 121, **122**
    size and shape, 125–126, **126**
    storage areas, 205
Loads (loading)
    calculations, **796**, 796–797, **797**, 800–813
    for concrete, **537**, **540**
    electrical, 674, **679–680**
    for floor systems, 579, 797
    lighting, 677, **678**
    for masonry, 530
    material loads, **798**
    on roofs, **457–461**, 459–461, 633, 796, 799–
       800, **800**
    transfer design, **561**
Local lighting, 125, 129, 135, 151
Location zones, 281–282, **284**
Log walls, 612, **612**
Loggia; *See* patios
Long break lines, **52**, **53**
Longitudinal roof system, 515, **516**
Longitudinal section, 389, **389**, **391**, 757

Lookout joists, 588
Lookout rafters, 645
Lot areas
    lot shape, **231**, **232**, **253**
    setback requirements, 231, **231**, **232**
    sign design and, 104, **256**
Louvered windows, 349, **350**, **351**
Louvers, **102**, **103**, 221
L-shaped kitchen, 176, **177**
L-shaped (angle) steel, 549, **550**
Lumber, 506–510; *See also* plywood
    connectors for, 520–525, **521–525**, 562
    grades of, **508**, **509**, 579, **580**
    sizing, **510**, **511**
    solid, for joists, **577**, 577–580
    solid, for rafters, 634
    types of, 508, **509**, 512, 517, 519–520
Lumens, 682
Lux, 682
L-winder stairs, 170

Magnetic-controlled switch, 687
Main entrance, 162, 165, **165**
Mansard, François, 8
Mansard roofs, 9, 339, **340**
Manufactured buildings, 468–469
Manufactured homes, 288, **289**
Marquee, 160
Masonry
    adobe construction, 105
    bonding for, 534, **534**, **535**
    careers in, 791
    construction systems, 527–535, **611**
    elevation rendering, 271
    firecuts in walls, 592, **593**
    fireplace/chimney construction, 124, 139
    longevity of, 490
    materials for, 527–531
    paver blocks, **202**
    as siding, 610–612, **611**
    texture of, **26**
    for walls, 531–534, **532–534**, **660**
Master bath, 223
Master bedroom, 209, **210**
Master plan, **445**; *See also* estate master plan;
    multiuse master plan
Master suite, **215**
Material command, in CAD, 90
Materials
    for bathrooms, 223–224
    calculations for, 794–795, **795**
    categories of, 778
    composite, for foundations, 476
    cost of, **776**, 776–777, **777**
    development of, 13, **15**, 16
    for driveways, 201
    elevation and, 337
    for entrances, 167–168
    for exterior wall siding, 607, **608**, 609, 610–612
    for finish flooring, 574
    for fireplaces, 140, 495
    for garage doors, 199
    for kitchens, 184
    loads for, **798**
    masonry, 527–531
    for models, 448, **449**, **450**
    for patios, 149
    patterning on exterior perspective drawings, **73**
    placement of, 462, **463**
    for plumbing, 734–735
    R-values for, **708**
    schedules for, 761, 765, 769–770
    shape of, 462, **462**
    for shelving, 205
    for skeleton-frame construction, 506–512

specifications for, 765, 767, **767**, 771
strength of, 461–463, **462**
for swimming pools, 154
symbols for, 369, 371, **371**, 390, **391–392**
texture, 25–26, **26**, 119
treatment of, 784
for walls, 122, 131
waterproof, 199
Maximum plans, 43
Measurement; *See* scales and measurements
Mechanical pencils, 55, **55**
Mechanical room; *See* utility rooms
Media rooms; *See* entertainment rooms
Mediterranean style, 9, **10**
Memory, for CAD workstation, 67
Meridian, 238
Metal doors, 351
Metal siding, 611
Metal strap ties
    corner bracing with, 601, **602**
    for masonry walls, 532, 533, **534**
    for platform framing, 564, **564**, 601, **602**
    for sills, 562, **562**, 564, **564**, 601, **602**
    for timber, 520, **523**
Metal studs, 551, **551**, 558, **558**, 631, **631**
Metlund system for hot water, 734, **734**
Metric scale, 38–41, **40–42**, 332, **333**, 465
Mid-Atlantic Colonial, 11, **12**
Minimum plans, 43
Mirror command, in CAD, 76–77, **77**, 324
Mirrored doors, 211, **212**
Mixed-use development, density zoning and,
    233, **233**
Model codes, 784
Models
    careers in modeling, 789
    constructing, 448–453
    design study models, 443–447, **444**
    three-dimensional, 44, **48**, 87
    types of, 443–447
Modem, 68
Modify command, in CAD, 324
Modular construction, 463–470
    as contemporary method, 16, **16**
    dimensioning, 332, **465**, 466, 468
    drawings for, 466, 468
    elevations and, 467, **467**
    floor plans for, 466, **466**, 468
    manufactured buildings, 468–469
    measurements for, 794, **794**
    natural building types, 469–470
    size standardization, **464**, 464–465, **465**
Modular prefabricated homes; *See* prefabricated
    homes
Module, 464, **465**
Molding, **401**, **402**, 624; *See also* crown molding
Moment of force, 813, **813**
Monitor, for CAD workstation, 66–67
Monochromatic (color harmony), 24
Monolithic slab foundation, 484, **484**, 501
Mortgages, 778–779, **779**, **780**
Mortise and tenon joints, 406, **406**, 524, **524**
Motion detectors, 700, **701**
Mouse, 68
Move command, in CAD, 76, **76**, 91, 441
Mud rooms, 200, **200**
Mullions, **124**, 309, 347, **349**
Multiactivity rooms; *See* family rooms
Multiline command, in CAD, 84, 315
Multiline text, in CAD, 70, **71**
Multi-lite doors, 351
Multimedia computers, 68
Multimedia rendering, 434–436, **436**
Multiple level floor plans, 319–324, **321–323**
Multiple workstation kitchens, 182, **184**

Multiuse master plan, 258, **261**
Multiview projection; *See* orthographic projection
Muntins, 309, 347, **349**, **373**
Music center, 137, **138**

National Building Code, 784
National Electrical Code (NEC), 676, 677, 784
National Fenestration Rating Council (NFRC),
    708, **710**
Natural building types, 469–470
Natural gas; *See* gas
Natural light
    in home office, 135
    in stairwells, 170
    and windows, **122**
Natural resources, preserving, 17
NEC; *See* National Electrical Code (NEC)
Needs versus wants, 277, **277**
Networking, 68
Neutral, 23
New England Colonial, 11
NFRC; *See* National Fenestration Rating Council
    (NFRC)
Noise abatement/control
    in bedrooms, 210–211, **211**
    design considerations and, **114**, 114–115, **115**
    in utility room, 197
    in workshops, 203, **204**
Nominal size, 508, **509**

Object lines, 51, **52**, 315
Objects, in CAD, 69
Oblique drawings, 416, **417**
Occupational Safety and Health Administration, 569
Offset command, in CAD, 77, **77**, 84, 85, 315,
    586, 648
Ohms, 675
Oils, for rendering, 431, 433
One-point perspective, 420, **422**, 423, 425, **425**,
    428, **429**
One-wall kitchen, 177–178, **178**
One-way slab systems, 541, **542**
Open planning
    in bathrooms, 221–222
    for entrances, 167, **168**
    for kitchens, **183**
    purpose of, 118–120, **119**, **121**
Open space, preserving, 17
Open webs; *See* webs
Open-divided scale, 34, **35**, 35–36
Open-riser stairs, 170
Opposition, 29, **30**, **124**
Organic design, 21
Oriental architecture, 5, **5**
Orientation, 98–109
    for doors, **763**
    earth-sheltered homes, 104–105, **105–107**
    for elevation, 360–362, **360–364**
    energy, 98–99, **99**
    land area and, 102–104, **103**, **104**
    of living room, 121
    noise and, 114
    overhangs and baffles, 100–102, **101**, **102**
    pollution and, 112
    protective measures, **108**, 108–109, **109**
    rooms and outdoor areas, 100, **101**
    solar, **99**, 99–100, **100**
    vegetation and, 105, **107**
    wind control, 107–109, **108**, **109**
Oriented strand board, 517, 518, 609
Ortho command, in CAD, 81, **81**, 328
Orthographic drawings, for plumbing, **730**
Orthographic projection, 353, **354**, **355**, **386**, 649
Outdoor living areas
    design considerations and, 100, 107
    lanais, 151–153

lighting for, 683, **684**, **686**, 686–687
modeling, 452
patios, 147–151
porches, 143–147
recreational facilities, 158
storage areas, 205
swimming pools, 153–157
Outdoor special needs planning, 295
Outlets and receptacles, for electricity, 690, **690**
Output devices, for CAD workstation, 68
Overall dimensions, 328
Overhangs
    calculating, **101**
    cantilevering of, 463, 645, **645**
    function of, 100–102, **101**, **102**, **342**,
        342–343, **343**
    on hip roofs, 339
    lookout rafters and, 645, **645**, 657
    openings in, 343, **343**
    steel roof construction and, 661
    support for, **344**
    for walkways, **164**
    wind resistance and, 564
Overhead service lines, for electricity, 675, **675**
Overlay drafting, 46, 59
Overlays; *See* pressure-sensitive overlays

Paint
    air quality and, 112
    schedules for, 767, **769**
    storage of, 115
Pan command, in CAD, 79, **80**, 238
Pan slabs, 541, **542**
Panel boards (siding), 609, **713**
Panel doors, 351
Panel elevations, 601–602, **603**
Paneling, 627, **628**
Panelized platform floor system, 573, **573**
Parallel angle projection, 416
Parallel slide, 57
Parallel stranded lumber, 517
Parapets, 659, **660**, 661
Parking apron, 162, **201**
Parking facilities, 295, **295**
Parlor; *See* formal living room
Partitions, **118**, 120, 124
Passive solar (design) systems, 99, 100, 723–725; *See*
    also solar energy
Pastels, 24, **24**, 432, **435**
Patios, **146**, 147–151; *See also* porches
    coverings for, 149, **150**
    decor, 149–151
    as extension of recreation room, 135, 151
    function and types, 147–148, **149**, **151**
    location, 148–149, **149**
    protecting, **150**, 150–151
    size and shape, 151
    storage areas, 205
Paving
    interlocking paving stones, 530, **531**
    plan rendering of, 265, **266**
    site details for, **272**
PDD; *See* planned development district (PDD)
Pellet stoves, 497
Pencils
    drawing checking with colored, 749
    hardness of, 55–56, **56**
    rendering techniques, 432, **432**, **439**
    types of, **55**, 55–56, **56**
Peninsula kitchen, 175–176, **176**, **177**
Peninsula workbench, 202
Pens
    ink renderings, 433, **433**
    technical pens, 56, **56**
Percolation, 737, **738**
Performance bond, 785

Performance-oriented codes, 783
Perimeter loop system, 715, **716**
Perimeter radial system, 715, **716**
Personal security, 569–570
Perspective drawings, 418–426; *See also* one-point
        perspective; three-point perspective; two-
        point perspective
    BIM imaging, **95**
    exterior, 420–423, **421–423**
    grid paper for, 61, **61**
    interior, 423, **425**, 425–426, **426**
    material patterns of, **73**
    renderings as, 44, **46**, **47**
Pest control, 569
Pet support facilities, 207
Phantom lines, 52, **52**
Phasing, 264
Phi; *See* golden rule
Photovoltaic film, 726
Pictorial drawings, 416–430
    BIM imaging, **95**, **383**
    for columns, 622, **623**
    detail drawings, **397**
    in elevation rendering, **385**
    for floor framing plan, **591**
    floor plans, 84, **84**, **88**, **91**, 93, **95**
    for kitchens, **191**
    perspective, 418–426
    for plumbing, 732–733, **733**, **739**
    projection methods for, 426–429
    purpose of, 607
    for roofs, **648**
    sections, 388, 402, **404**
    for stairwells, **595**
    types of, 416–418, **417**
Picture plane, 418
Picture windows, **350**
Pier-and-column foundations, 487–490
Piers
    floor construction, 514
    footing type, 475, **475**, 476
    foundation type, 478, **479**, 487
    for intermediate support, **490**
    materials for, **488**
    sizing, 811–812
    tributary area of, 797, **811**
    types of, **488**
Pilasters, 479, **481**, **490**, **593**, **640**
Piles, 489, **491**
Pin drafting, 46
Pin graphics; *See* layering
Piping; *See* plumbing
Pit house, 2
Pitch, of roofs
    determining, 343, 367, **368**, 652, **654**
    high versus low pitch, 459, **460**
    types of, **655**
Pivoting windows, 349
Plank-and-beam construction, 514
Plank-and-beam floor system, 573, **573**
Planks, 513, **514**, **515**
Planned development district (PDD), 234, **235**
Plans; *See also specific plans*
    bird's-eye view, 44
    CAD library items for, **81**
    closed planning, 118, 120, **120**, 167
    combined, 120–121
    compartment, **221**, 221–222, **222**
    electrical, 687–692
    for foundations, **498–500**, 498–501
    open; *See* open planning
Planting schedules, 271
Plaster, 627
Plate glass, 614
Platform floor systems, **572**, 572–573, 588, **590**
Platform framing, 513, **513**, 564, **564**

Plats, 250–251, **251–255**
Play patio, 147, 148
Pleated roofs, 339, **340**
Plenum system, **715**, 715–717, **717**
Plot plans, 251–257
    guidelines for drawing, 252, **255–257**, 256
    plumbing on, 738
    symbols, **255**, **257**
    variations, 257, **258**, **259**
Plot-Print command, in CAD, 79
Plotter, 68
Plumbing
    in bathrooms, 220, 221, 223, **223**
    CAD drawings, 731–732
    careers in, 791
    conventions and symbols, 730–732, **731**, **732**
    drainage fields, 737–738
    drawings for, 75, 730–745
    elevations for, 739, **739**, 741–744, **742**, **743**
    excavations for, 498
    filtration design, 113, 733, **733**, **734**, 744, **744**
    floor plans for, 738–739, **739**, **740**
    framing and, **622**
    history of, 730
    insulation for, 735
    materials for, 734–735
    notching for, 739, **741**
    orthographic drawings for, **730**
    pictorial drawings for, 732–733, **733**, **739**
    pipe sizes for, 735, **735**, 739
    on plot plans, 738
    post-and-beam construction and, 513
    residential, total system for, **744**
    schedules for, 764, **764**, **765**
    schematic details for, **730**, 739, **739**
    shutout valves for, 734, **734**, 738
    site plans and, 738
    for slab foundations, 733–734, **734**
    for swimming pools, 157, 744, **744**
    symbols for, 730–732, **731**, **732**
    valves for, 733
    waste discharge system, 735–737
    water filtration in, 113, 733, **733**, **734**, 744, **744**
    water flow in, 117
    water supply system, 732–735, **733**, **734**, 738, **743**
Plywood
    composition of, 510–511
    grades of, **511**, 511–512, **512**
    laminated timbers, 517
    plank-and-beam construction with, **515**
    subflooring with, 574, **575**
    wind resistance with, 562
PMI; See private mortgage insurance (PMI)
Points, in a mortgage, 780
Polar coordinate system, 238; See also Cartesian coordinates
Pole/column foundations, 487, **489**
Pollution
    air quality and, 111–112
    construction and, 117
    land integrity and, 111
    noise, **114**, 114–115
    toxic by-products, 192, 567, 569, 767, 768
    water quality and, 112–113, **113**
Polygon command, in CAD, **71**, 71–72, **72**
Polygrams, 82, **82**
Polyline command, in CAD, 84, 92, 246, 601, 648
Polylines, in CAD, 69, 75, 84, 246, 601, 648
Pools; See swimming pools
Porches, 143–147
    decor, 146
    as dining area, 128
    door sill for, **594**
    elevation and, 347
    entrances and, 160, **167**

function and types, 144
    location, 144, **144**
    patio versus, 143, **144**
    size and shape, 147
    storage areas, 205
Positioning drawings, 409, **409**
Post-and-beam construction, 513–525
    components of, 513, **514**
    floors, 514, **515**, 584, **587**
    history of, 4, **4**
    lag screws, 518, **518**
    ridge board and, 634, **634**
    roofs, 515, **516**, 654–655, **655**
    skeleton versus, 513, **513**
    structural components, 515–525
    timber connectors, 520, **521–525**, 524, 562
    walls, 515, **516**
Post-and-lintel construction, 4, 4–5
Posts; See also columns, footings; post-and-beam construction
    caps for, **523**
    function of, 487, 513
    as intermediate support, **490**
    loading calculations, 808–809, **809**, 812
    weight transmission, **488**
Post-tensioning, 537–538
Powder, use of, 59
Power tools, 202–203
Prebuilt homes, 469
Prebuilt modules, 469
Precast concrete, 540, **540**, **541**, 635–636
Precast foundation walls, 481
Precut structures, 469
Prefabricated homes, 288, **289**, **290**, 469
Prefabricated roof units, **642**
Prefabrication, 16, **16**
Presentation elevation drawings, 382–383, **383–386**, 385
Presentation floor plan, 304, **304**
Presentation model, 443, 447–448, **448**
Presentation section, **389**, **394**
Preservatives, for wood, 512, 575
Pressure-sensitive overlays, 59, 434
Prestressed concrete, 536–538, 635
Pretensioning, 537, **537**
Primary color, 23
Principal, of mortgage, 778
Principles of design; See design, fundamentals of
Printed pressure-sensitive tapes, 59–60
Private mortgage insurance (PMI), 779, 780
Profiles (profile drawings)
    contour lines and, **240**, **242**
    for foundations, 501, **501**
    function of, 239
    site development plans and, 269, **270**, 355, **358**
Progressive realization, 264
Projections (elevations), 353–370; See also elevations
    auxiliary, 362, 365
    construction and, 365
    exterior, in CAD, 369
    orientation, 360–362
    orthographic, 353, **354**, 355
    process of, 355, **356–360**, **367**, 367–369
Propane gas, 565
Property description, 238, **239**
Property taxes, 779, 780
Proportion
    in design, 28, **29**
    selecting a scale, 37
Protective measures, design considerations and, **108**, 108–109, **109**
Protractor, 58
Public buildings
    mechanical engineering, 790; See also specific utilities
    special needs planning for, 295–298

Pull-down bed, 214, **216**
Pupils, dilation of, 682, **682**
Purlins, 546, 636, **639**
Pythagorean theorem, 497, 816, **816**

Quadruple bevel, **36**
Queen Anne style, **13**
Quiet patio, 147, **147**, 148, 166, **257**

Radiant system, 718, 719
Radiation
    electrical, 116
    heat, 705–706
    solar, 100–102, **103**
Radon
    exhausting, 112
    ventilation of, 566–567, **567**
Rafters; See also roofs
    adjustable design, **644**
    connections for, **565**, 634, **635**
    cuts of, 643, **643**
    loading calculations, 800–803, **800–803**
    spans for, **654**, 803, **803**, **804**
    truss joists as, 518, **520**, 634
Railings
    dimensions of, **296**
    porch, 146, **147**
    stairway, 170, 171
Rammed earth construction, 105, 469–470, **470**
Ranch style, 13, **14**, 16
Random-access memory (RAM), 67
Ratios, 34, 38, 39, 41, **41**
Read-only memory (ROM), 67
Rear elevation, **356**, **360**
Rebars
    column supports, 538, **539**
    function of, 473–474, **474**, 535, 536, **536**, 540, 562
    for masonry walls, 532, **532**
    sizing, 474
    in slab foundations, 487, 538, **538**, **539**
Recreation rooms, 133–135
    decor, **134**, 135
    function of, 133
    location of, 135
    size and shape, 135, **135**
    storage areas, 205
Recreational facilities, as outdoor living areas, 158, **158**
Rectangle command, in CAD, 70–71, **71**
Recycled thermoplastics, 520
Redo command, in CAD, 75
Redraw command, in CAD, 78
Reduced scale, 34, **35**, **38**
Reflected ceiling plans, 324, **326**
Reflection, 684
Reinforced concrete, 635, **636**; See also precast concrete
Reinforcement, 459, **459**
Reinforcing bars; See rebars
Removed sections, 394, 397, **398**, 403
Render command, in CAD, 90
Renderings
    in CAD, 89, **90**, 90–91, **91**, 431
    development plan, **264**
    elevations, 269–271, **269–271**, 382–383, **84**, **385**
    entourage, **440**, 440–441, **441**
    landscape, 441, **442**; See also landscape plans
    light effects and, 437–438, **437–439**
    perspective drawings, 44, **46**, **47**
    plan rendering, 264–269, **265–268**, **753**
    plat plans, 250, **253**, **254**
    rendering media, 264, 431–436
    techniques for, 54
    texture effects and, 438–440, **439**, **440**

Repetition, 26–27, **27**, **28**, **122**
Residences
  basic structural types, **336**
  financial planning for, 778–781
  special needs planning, 298–301
  specifications for, 772, 773
  styles of; *See* American architecture
Resistance
  electrical, 674–675
  force and, **460**, 460–461, **461**
Resisting moment, 801, **802**, **803**
Resistivity, 706
Restrictive codes, 784
Retaining walls, 269
Reversed plans, 324, **326**
Rheostat, 129
Rhythm, 26, **27**, **28**, **122**
Ribbon windows, 211–212, **350**
Ridge board, for roof, 634, **634**
Ridge lines, **240**, 244, 335
Right triangle law; *See* Pythagorean theorem
Right-side elevation, **357**, **359**
Rigidity, 458, **458**; *See also* stability
Rise, for roofs, 343, 367, **368**, 652
Riser, stairs and, 170, 171, **172**
Riser diagrams, 741
Rivets, 554, **554**
Roll roofing, 663
Rolled steel, 547
ROM; *See* read-only memory (ROM)
Roman columns; *See* Greek and Roman
      column orders
Roof lines, 355
Roofs; *See also* rafters
  appendages for, 666–668
  chimneys; *See* chimneys
  covering materials, 661, 663–666, **664–666**
  dimensioning, 649, 652, **653**
  dormers, 343, **344**, 646, **646**, **647**
  drainage plan for, 649, **651**
  elevation and, 338–347, 368, **369**, **370**, 649,
      650, **652**
  forces and loading, **457–461**, 459–461, 633,
      796, 798–803, 810
  framing, 446, 515, **516**
  framing components, 636–647
  framing drawings, 633–669
  framing members, 633–636
  framing types, 655–659
  function of, 633
  insulation value of, 116
  lightning rod for, 115, 676
  modeling, 452, **453**
  new technology in, 341–342
  outline of, **343**
  overhangs; *See* overhangs
  panels for, 640, **642**, 642–645
  pitch of; *See* pitch
  plan for, 648–652, **648–652**, **758**
  post-and-beam construction, 515, **516**
  resisting wind damage, 564, **565**
  rise of, 343, 367, **368**, 652
  roll roofing, 663
  run of, 343, 367, **368**, 652
  sheathing, 663, **665**
  skylights, 99, **100**, 346, **346**, **347**
  slop diagram for, 367, **368**
  slope of, 343, 652, **654**
  span of; *See* span
  steel framing methods, 659, **660**, 661
  symbols for, **371**
  types of, **9**, **107**, 338–342, **340–342**
  wind protection and, 108
  window mullions and, **124**
  wood framing methods, 654–655

Rooms; *See also specific rooms*
  design considerations and, 100, **101**, 120
  dimensioning, 328
  earth-sheltered homes and, 105, **106**
  size and shape, 291, **293**, **305**
Rotate command, in CAD, 76, **77**
Rotation modeling, **90**
Rough openings
  dimensioning, 602, 612, 613, **613**, **614**, **615**
  door framing, 614, **615**, 616, **616**, **618**
  framing details, 599–600
  window framing, 612, 613, **613**, **614**, **615**
Rounds, 72
Rubble, 529, **530**
Run, 343, 367, **368**, 652
R-value, 707, **707**, **708**, **709**, 711, **713**

Saddles; *See* bolsters
Safe rooms, **569**, 569–570, **570**
Safety
  alarms, 112, 115, 700–701
  bathroom safety, 220
  building systems, 700–701
  design considerations, 110, 170
  electrical, 675–676
  fire prevention and control, 567
  fireplace issues, 495–496
  porch railing, 146, **147**
  preventing personal injury, 569
  special needs planning, 296–297, 301
  on stairs, 170
  swimming pool, 153, 156–157, **157**
Sandwich panels, 519
Sanitary land filling, 111
Sauna, 219, 225, **226**
Save command, in CAD, 81
Scale
  floor plans and, 38, **38**, **39**, 304–305, **305**,
      318, 319
  modeling and, **446**
  site-related drawings and, 283, **285**, **286**
Scale command, in CAD, 81, **81**, 91
Scales and measurements, 34–42
  architect's scale, 34–38, **35–38**
  CAD scale, 42–43
  engineer's scale, 38, **38**, **39**
  metric scale, 38–41, **40–42**, 332, **333**, **510**
  selecting, 37, 38
  unit conversions, 793, **794**
Schedules, 761–769
  appliances, 764, **766**
  cabinetry, 764
  careers in construction scheduling, 792
  column schedules, 620, **623**, 623–624
  construction component, 768
  door, 763, **763**
  electrical fixtures, 764, **765**
  finishing, 767, **769**
  for floor plans, **762**
  furniture and accessories, 767–768, **769**, **770**
  for materials, 761, 765, 769–770
  paint and finishing, schedules, 767, **769**
  plumbing, 764, **764**, **765**
  site details and, 271–273, **272**, **273**, 308, 312
  site development plans and, 271–273
  surface covering, 765, **766**, **767**
  vegetation planting, 271
  windows, 762, **762**
Screened doors, 351
Seawalls, 149, **149**, 153
Secondary color, 23
Section lines, 52, **52**, **53**
Sectional drawings, 388; *See also* detail (section)
      drawings; full section drawings
Section-lining symbols, 389

Sections, 44, **45**; *See also* detail (section) drawings;
      full section drawings
Security
  design and, **569**, 569–570, **570**
  at main entrance, 162; *See also* gated entrance
Self-sustaining energy system, 727, **727**
Semidirect light, 684
Semi-indirect light, 684
Septic distribution field area, 249, **250**
Septic system/tanks, 113, **256**, **736**, 736–737, **738**
Sequential column identification system, 594, **596**
Service areas, 194–208
  driveways, 201
  electrical plan for, **696**
  entrance symbol for, **257**
  garages and carports, 197–200
  mud rooms, 200
  specialized areas, 206–207
  storage areas, 203–206
  utility rooms, 194–197
  workshops, 201–203
Service entrance
  for electricity, 675, **676**, 681
  residential, 165, 166, **167**, 182
Service patio, **257**
Service room; *See* utility rooms
Serving walls, 128
Set Variable command, in CAD, 76, 324
Setbacks, 230–231, **230–232**, 279
Settling, 473
Sewer system; *See* waste discharge system
Shade
  color and, 24
  patterns of, **100**
  for pool deck, 153
  rendering methods, 437, **437**
Shading, **55**
Shadowing; *See also* light and shadow
  building shape and, 438, **438**
  in CAD, 72, **72**, 90–91
  daylight plane, 230, **230**
  elevations and, 337, **338**
  entourage and, **267**
  rendering effects, 438, **439**
  vanishing point and, **438**
Shakes, 663
Shear forces (stress), 458, **458**, 812, **812**
Shear stress panels, 562, **563**, **564**
Sheathing, 663, **665**, **797**
Shed roofs, 339, **340**, 657–658, **659**
Sheet overlays, 59
Shingle siding, 609
Shingles, 663, 665, **665**, **666**
Shipping container, conversion to house, 288, **288**
Shop drawings; *See* working drawings
Shop lumber; *See* factory lumber
Short break lines, **52**, **53**
Shrubs; *See* trees and shrubs
Shutout valves, for plumbing, 734, **734**, 738
Sick building syndrome, 112, 567
Siding, 607, **608**, 609
Sills
  foundation sills, 482, **483**, **484**, 562, 588
  section drawings, 397, **398**, 402, **468**, 501, **503**,
      588, 592, **592**
Simpson strong tie, for rafters, **635**
Single awning windows, **372**
Single bevel, **36**
Single lines, in CAD, 69, 84, 601
Single-Line command, in CAD, 84, 601
Single-line drawings, 285, **286**, **648**, 649
Single-line text, in CAD, 70, **71**
Single-pole switches, 687
Sinks; *See* lavatories (sinks)
Site analysis, 229, 278–282, **284**

Site and environmental analysis, 228–229, **229**
Site details and schedules, 271–273, **272, 273,** 308, 312
Site development plans, 228–274
　landscape plans, 257–264
　plot plans, 251–257
　site and environmental analysis, 228–229
　site details and schedules, 271–273
　site drawings, rendering, 264–271
　survey plans, 238–251
　topographic drawings, 236–238
　zoning ordinances, 229–236
Site drawings
　conceptual floor plans, 283, **285**
　rendering, 264–271; *See also* landscape plans
Site plans; *See also* site development plans
　CAD symbols for, 75
　dimensioned site plan, **259**
　drainage plans, 111
　geological surveys, 92
　hatching in, **73**
　lighting and, 117
　modeling and, 444, **444**
　plumbing and, 738
　profile drawings and, 355
　scale and, **39**
　site characteristics and, 102–104, **104**
　suitability analysis, 229
　Zoom and Pan on, 238
Site profile, 355, **358**
Site security, 569
Site-related codes, 783
Situation statement, 277, **277**
Skeleton-frame construction, 506–513, 598, **598**
　components of, **508**
　design of, 457, **457,** 506, **507**
　lumber, 506–510
　plywood, 510–512
　post-and-beam versus, 513, **513**
　steel for, 545–546, 558
　trusses for, **639**
Sketching, 54–55, **55**
Skylights, 99, **100,** 346, **346, 347,** 667, **668**
Slab foundations, 484–487
　door sill for, **594**
　drawings for, **500,** 501
　excavations for, 497–498
　footings for, **475, 478, 479**
　grade-level, 538, **538**
　plumbing lines for, 733–734, **734**
　radon prevention, **567**
　types of, **484,** 484–485, **485,** 538, **538**
　vent stacks for, 735–736
　wood flooring for, **486, 538**
Sleeping areas, 209–226; *See also* bedrooms
Slide-out rooms, 288, **288**
Sliding doors, 617, **618, 619**
Sliding (slider) windows, **350,** 351, **372**
Slope, 280, 343, 652, **654,** 723, **723**
Slope analysis, 280–281, **281, 282**
Slope diagram, 367, **368,** 652
Smart building system, 110, 702
Smart growth, 16–18, **18,** 111
Snap Grid command, in CAD, 80, **80,** 328
Snap-Off command, in CAD, 80, **80,** 328
Sod roof, **107,** 666, **666**
Soffit lighting, 684, **685**
Soffits, 643, **643**
Soft costs, 779
Softwoods, 506–508, 511
Soil; *See also* topsoil
　bearing capacity of, **473,** 489, 797, 812
　compaction of, 111
　earth-sheltered homes and, 104
　erosion of, 111

footing area calculations, 797–798
　safe loads, **799**
Soil analysis, 279–280, **280**
Soil lines, and waste discharge system, 735
Solar cooling system, 728
Solar energy
　active solar (design) systems, 99, 725–728
　auxiliary heating and, 727
　design considerations and, **99,** 99–100, **100**
　electricity generation with, 672–673, **674**
　passive solar (design) systems, 99, 100, 723–725
　sun position and, 723, **723**
　swimming pool temperature and, 157, **157**
Solar furnace, 725–726
Solariums, 150, **150**
Solid form models, 443–444, **444**
Solid models, 89, **90,** 426, 429
Solid timbers, 516–517, **517**
Solid wood doors, 351
Southern Colonial, 11, **12**
Southwestern ranch, 13, **14**
Spa, 157, **157**
Space, 22, **23,** 28, **119,** 336
Space frames, 661, **663**
Space planning, functionality of, 286–295
Span
　calculating, 367, 368, 652
　loading and, 462–463, **463, 464**
　rafter sizing and, **654,** 803, **803, 804**
Spandrels, 546, **546**
Spanish architecture, 9, **10,** 144, 147, 148
Spark arrestors, 494, **495**
Special needs planning, 295–301
　public buildings, 295–298
　residences, 298–301
Special purpose entrances, 165, 166, **167**
Special purpose rooms, 133–139
Species, 506
Specific lighting, 682–683, **684**
Specifications, 761, 770, 772–774
　careers in specification writing, 791
　CSI format, 772, **773**
　guidelines for writing, 772
　organization, 772
　purpose and use, 770, 772
　reference sources, 773
Specification-type codes, 783
Spiral stairs, 170; *See also* stairs/stairwells
Splices, for beams, 584, **586**
Spline command, in CAD, 70, **71,** 92, 246, 696
Split-ring connectors, 524, **525**
Spot lighting, 168
Sprawl, 16
Square footing slab foundation, 484; *See also* footings (footers)
Squared paper, 60
Square-foot method, for estimating costs, 775–776, **776**
S-shaped steel, 549–550, **551, 553**
Stability, 457, **457, 458,** 459
Stacks, and waste discharge system, 735, 738, 741
Stair lifts, 172
Stairs/stairwells, 170–171
　calculating dimensions for, 320, 322, 324, **325**
　dimensioning, **172,** 322, **324,** 627, **627**
　electrical plan for, **695**
　floor framing plans, 592, 594, **595**
　framing for, **626**
　lighting for, **170**
　template for, **325**
　types of, **171**
Stampat, 59
Stamps, 60, **60**
Stand-alone features, 440, **440**
Statics, 796

Station point, 418, **418**
Steam system, 718, 720
Steel; *See also* beams; girders
　advantages/disadvantages of, 545
　bars, 548, **548**
　dimensioning for, structural drawings, 595, **596**
　drawing conventions for, 557, 557–558
　floor framing plans for, 594–596
　high-rise framing with, 622, **624**
　nonferrous, 558
　reinforcement for masonry walls, 532, **532,** 533
　for roofs, 635, 659, **660,** 661
　safe loads for, **808, 809**
　shapes of, 547–552, **548, 550, 551**
　sizing calculations, 807
　space frames, 551–552, 661, **663**
　steel decking, 575, **576**
　steel joists; *See* joists
　steel straps; *See* metal strap ties
　steel studs, 551, **551,** 558, **558,** 631, **631**
　steel-cage construction, 457, 545–547, **546,** 558, 586, **587**
　structural elevations for, 600, **600**
　structural members, 547–552
　trusses, 641
　types of, 547, **547**
Stepped footings, 475, **475, 476**
Stirrups, 474
Stitch lines, **52**
Stone, 529–530, **530**
Stoop, 143; *See also* porches
Storage areas
　for active solar systems, 132
　in bedrooms, 212–213
　in family room, 132
　function and types, 203–205
　in garages, 199
　in home offices, **135**
　in kitchen, 174
　location, 205
　for paint, 115
　service areas as, 203–206
　special needs planning, 300, **301**
Storage center, in kitchen, 174
Storm water flow, 117
Straight lines, guides for, 57–58
Strand board; *See* oriented strand board
Strand lumber, 519–520, 634
Strap ties; *See* metal strap ties
Straw bale walls, 105
Straw-bale construction, 469, **469, 470**
Stressed-skin panels, 519, 640
Stretch command, in CAD, 77, **78,** 91, 441
Structural calculations, 796–813
Structural clay tile, 530, **531**
Structural codes, 783; *See also* building codes
Structural design, 456–463
　forces and, **457, 458,** 458–461, **461**
　loads and, **458,** 458–460, **460, 461**
　material strength, 461–463, **462**
　resistance and, **460,** 460–461, **461**
Structural insulated panels, 609
Structural lumber, 508, **509**
Structural members
　concrete, 538–540
　steel, 547–552
Structural models, 445–446, **446, 447**
Structural plates, 548, **548**
Structural tees, 550–551
Stucco siding, **609,** 609–610, **610**
Stud layout, 467, **604, 605,** 628–631, **628–631**
Studio/study; *See* home offices
Subcontractors, 791
Subdimensions, 330–331
Subfloor, 574, 575

Subordination, 27, **123**
Suitability levels, 229
Sullivan, Louis, 20
Sun baffling, **102**
Support beam, for foundations, **473**, **474**
Surface area of a cylinder, 814, **814**
Surface models, 89, **89**, **90**, 426, 429, **429**
Surfacing code, for lumber, 510, **510**
Surge protection, 676
Surveys/survey plans, 238–251
    careers in surveying, 790
    contours, 239–247
    dimensions, establishing, 238, **238**, **239**
    elevations, 239
    geographical survey maps, 249, **250**, **251**
    guidelines for drawing, 247, 249, **249**
    plats, 250–251, **251**–**255**
    symbols, 246, **247**, **248**
Swales, 113, 117; *See also* valleys
Sweets Catalog Files, 773
Swimming pools, 153–157
    construction of, 154–156
    dimensioning, 332
    drawing for, **502**
    equipment for, 157
    function, 153
    lighting for, **153**, 687
    location and orientation, 153, **154**
    plumbing for, 157, 744, **744**
    safety devices, 153, 156–157, **157**
    shapes of, 155, **155**
    size calculations, **155**, 155–156, **156**
Swim-out, 156
Switches, 687–689, **688**, **689**
Symbols
    for appliances, 310, 312, **312**
    for bathrooms, 312, **313**
    CAD libraries; *See under* computer-aided
        drafting (CAD)
    for climate control, **705**
    for contours, **246**, 246–247, **247**
    cross-referencing, in architectural drawings,
        749, **751**
    for detail (section) drawings, **396**
    for doors, 306–309, **307**–**310**
    for electrical plans, 74–75
    for electricity, **691**–**692**
    for elevations, 75, 369, 371–375, **382**
    for entrances, 256, **257**
    for fireplaces, 312, **313**
    for fixtures, 310, 312, **312**
    for floor plans, 306–314, **306**–**314**
    for full section drawings, 389–390, **391**–**393**
    for furniture, 312, **323**
    for garage entrances, **257**
    for landscape plans, 258, **259**, **262**
    for materials, 369, 371, **371**, 390, **391**–**392**
    for plot plans, **255**, **257**
    for plumbing, 730–732, **731**, **732**
    for roofs, **371**
    section-lining symbols, 389
    for site plans, 75
    for surveys/survey plans, 246, **247**, **248**
    for walls, 306, **307**, **371**
    for windows, 309–310, **310**, **311**, 371, **372**,
        373, **373**, 375
Symmetry of design, **27**
Synonyms in architecture, 821–822
Synthetic materials, for fireplaces, 140

T square, 57, **57**
Tail cuts, on rafters, **643**
Takeoffs, 769, 776, 791
Tapes, 59–60
Tax maps, 229
Taxes, property, 779, 780

Technical pens, 56, **56**
Tees; *See* structural tees
Temperature control, 157, **157**, 722; *See also* climate
    control symbols
Temperature rebars, 538, **539**
Templates
    architectural, 59
    circle, 58
    floor plan, **294**
    furniture; *See under* furniture
    human, **293**
    room, 291, **293**
    space planning, 287
    stair design, **325**
Tendons, in prestressed concrete, 537
Tension forces, 458, **458**, 473–474, **639**, **810**
Termite shield, **483**
Terrace, as extension of recreation room, 135; *See also*
    patios
Terrain texture, 243, **243**
Tertiary color, 23
Text command, in CAD, 70, **71**, 75, 84, 86, 315,
    556; *See also* lettering
Texture
    design fundamentals, 25–26, **26**
    elevation design, 337, **338**
    rendering effects, 438–440, **439**, **440**
T-foundations, 478–484
    drawings for, 498, **498**
    floor types, **480**
    footings for, **475**, **477**–**479**, 810
    heat loss with, **711**
    radon prevention, **567**
    walls, **479**–**481**
Thermal conductivity, 707, **707**
Thermal mass, 723
Thermostatic controls, 722, 725
Thermostats, 722
3.4.5 unit method, 497, **497**
3D Orbit command, in CAD, 246
Three-dimensional design, 87–91
    BIM models, 382, **383**, 748–749
    CAD models, 287
    Cartesian coordinates, 87, **88**
    floor plans, 319, **319**
    models, 44, **48**, 87
    modular components and, **465**
    pictorial rendering, **91**
    solid models, 89, **90**
    surface models, 89, **89**, **90**
    virtual reality (VR) systems, 89–90
    wireframe drawings, 88, **88**, **89**
Three-point perspective, 343, 367, **368**, 423, **425**
Three-season rooms; *See* conservatories
Three-way switches, 170, 687, **689**
Thresholds, 617, **619**; *See also* sills
Tie straps; *See* metal strap ties
Timber; *See* lumber
Timber connectors, 520–525, **521**–**525**, 562
Tint, 24
Title block, 46, **49**
Toilets, 113, 219–220, 297, **297**, 300
Tone, 24
Tongue-and-groove planking, 516–517, **517**
Tools menu, in CAD, 312
Topography/topographic drawings, 236–238
    base map for, 279, **279**
    contours; *See* contours
    design considerations, 102–104, **103**, **104**
    recontouring and, 111
    water flow and, 113, **113**
Topsoil; *See also* soil
    recycling of, 111
    removal of, 105
Torsion forces, 458, **458**
Toxic by-products, 192, 567, 569, 575, 767, 768

Trace, 59, **60**; *See also* sheet overlays
Traditional Colonial, 7
Traffic areas/patterns, 160–173
    elevators, 171–172
    entrances, 160–169
    halls, 169–170
    in kitchens, 189
    special needs planning, 296, 299
    stairs, 170–171
Transformers, 674, 680, 687
Transition, 29–30, **31**, **164**
Transportation
    LEED practices, 117
    smart growth and, 17
Transverse roof system, 515, **516**
Transverse section, 389, **389**, 391
Trash; *See* garbage facilities, design of
Tread, stairs and, 170, 171, **172**, 322
Treated lumber, 512, 575
Trees and shrubs; *See also* deciduous trees; vegetation;
        wooded areas
    hardwoods versus softwoods, 506–508, **512**
    plan rendering, 265, **266**, 269, **269**
    preservation details, **272**
    shading with, 116
Trellis, **271**
Trenches, 113
Trends versus fads, 22
Triadic (color harmony), 24
Triangles, 57, **58**, 814, **814**
Triangular bevel, **36**
Tributary area, 797, **799**, 806, **806**, **807**
Trim command, in CAD, 75, 78, **78**, 84, 586,
        601, 648
Trimmers, 592, **595**, 601, **603**; *See also* double joists
Trombe wall, 725, **725**
True size, 362, **366**
Truss clips, 523–524, **524**
Truss joists
    cantilevered, **552**
    components of, **520**
    configurations of, **581**, **582**
    detail drawings, **519**
    function of, 518, 580
    roof rafters and, 518, **520**, 634
    webs in, 580, **581**, 636
Trusses, 636–640, **636**–**641**, 661, **662**, 804
Tubing, 548, 548–549, **549**
Tubs and/or showers, 220, 220–221, **221**, 300
Tudor style, **8**
Two-dimensional design, 82–87
    Cartesian coordinate systems, 82–83; *See also*
        Cartesian coordinates
    elevations in, 44, **86**, 86–87, **87**
    floor plans, 83–85, 86
    proportion in, **29**
Two-point perspective
    comparing perspective types, **425**
    drawing, **424**, **428**
    exterior, 418, 420, 423
    grid for, **429**
    horizon line, 418, **418**, **419**, **420**
    interior, 426
    positioning options, **422**
    visualization, **418**
Two-way slab systems, 541, **542**

Underlays, 60
Underwriters Laboratories (UL), 115, 567, 680
Undo command, in CAD, 75
Unity of design, 28–29, **30**, **129**
Uplift, preventing, 563, **563**, 564
U.S. Geological Survey
    geographical survey maps, 249, **250**, **251**
    pictorial contours and, 246
    site plans from, 92, 280

User analysis, 278, **278**
U-shaped kitchen, 175, **175**, **176**
Utilities; *See also specific utilities*
   kitchen design considerations, 182, 192
   underground versus above ground, 114
Utility commands in CAD, 78–81
   Block command, 81, 84, 246, 285–286, 312
   Display Grid command, 80, 328
   Drag command, 81
   Grid command, **80**, 80–81, 367, 468
   Layer command, 79–80, **80**, 85–86, 315
   Ortho command, 81, **81**, 328
   Pan command, 79, **80**, 238
   Plot-Print command, 79
   Save command, 81
   Scale command, 81, **81**, 91
   Snap Grid command, 80, **80**, 328
   Snap-Off command, 80, **80**, 328
   View command, 79, 238
   Wblock command, 81, 285–286, 404
   Zoom command, 78–79, **79**, 238
Utility rooms
   appliances and equipment, 197
   electrical plan for, **694**
   equipment sizes, **197**
   floor plan for, **177**, **195**, **198**
   function, 194–195
   location of, 182, 195–196
   size and shape, 197
   storage areas, 205
   style and decor, 196–197
U-value, 707, **707**, **708**, **709**, 711, **713**

Valance lighting, 684, **685**
Valley rafter, 657, **657**
Valleys, 244, 338; *See also* swales
Value (of a color), 24
Valves, for plumbing, 733
Vanishing point
   exterior perspectives, 418, 420, **421**, 423,
     423, **424**
   interior perspectives, 423, 425–426,
     425–427, **429**
   shadow pattern and, 438, **438**
Vapor barrier films, 711
Variety in design, 29, **124**
Vaulted roofs, 339–340, **340**
Vaults, **5**, 5–6, **6**
Vegetation; *See also* landscape plans
   as baffling, 150, 269
   design considerations and, 105, **107**, 111,
     114, 117
   noise reduction with, 211
   planting schedules, 271
   on plat plans, **254**
   removal of, 111
Vellum, 60, 61
Veneer walls, 533–534, **534**; *See also* brick
   veneer
Vent stacks, and waste discharge system,
   735–736, **744**
Ventilated shelving, 205
Ventilation
   attic and crawl space, 112, 721
   in bathrooms, 222
   in bedrooms, 212
   in cornices, 643, **644**
   for electric heat, 720
   gases, 565–567
   and HVAC, 112, 721, **722**
   in roofs, 667, **668**
   for waste discharge system, 735–736, 741
Venturi effect, 109, **109**
Verandas, 144, **144**; *See also* porches
Vertical sections, **395**, 396–399
Vertical shear, 461, **461**

Victorian era, 12–13, **13**
View command, in CAD, 79, 238
Virtual reality (VR) systems, 89–90
Visible lines, 51, **53**, 315
Visual analysis, 281, **283**
Visual perception
   design considerations and, 113–114
   in open plan, 118, **119**
Voice recognition, 68
Volts, 674, 675, 677
Volume
   of a cone, 816, **816**
   of a cube, 815, **815**
   of a cylinder, 815, **815**
   of a rectangle, 156
   of a sphere, 816, **816**
   of a square pyramid, 816, **816**
VPOINT command, in CAD, 92, 246
VR systems; *See* virtual reality (VR) systems

Waffle and daub construction, 470
Waffle slabs, 541, **542**
Walk-in closets, 203, 204, 212–213, **213**, **214**
Walkways, 162, **164**, 198
Wall closet, 204–205
Wall storage cabinet, 212
Wall windows, **350**
Walls
   color coding for, **768**
   concrete, 116, 540, 542–544, **542–544**
   coverings for, 627
   decor for, 122, 131
   detail section drawings for, 396–401, **662**
   dimensioning, 328, 330, **331**
   dividers, 169
   in earth-sheltered homes, **105**, 107
   elevations and, 375, 376, **377**, 378, **378**
   entrance, 168
   exterior; *See* exterior walls
   fire protection and, 115
   foundation walls, 478–482, **479**, **499**
   framing, 446, **446**
   framing drawings for, 598–631
   furniture contouring with, **126**
   glass; *See* glass walls
   interior; *See* interior walls
   joints and corners, 400, **401**
   light fixtures for, 684–685, **685**
   loading calculations, 812
   log walls, 612, **612**
   masonry, 105, 531–534, **532–534**, 611,
     611, **660**
   modeling, 449–451, **452**
   modular construction, 465
   patio walls, 150
   post-and-beam construction, 515, **516**
   preinsulated concrete walls, for foundations, 479,
     481, **482**
   rammed earth, 105, 469–470, **470**
   rendering of, 269, **271**
   resisting wind damage, 564, **565**
   retaining walls, 269
   roof loading and, 457, **457**
   serving walls, 128
   space usability of, 211, **216**
   special needs planning, **296**
   stability of, 459
   steel framing for, 558, **559**
   supporting, **458**, 458–459, 584, **586**
   surface coverings schedules for, 765, **766**, **767**
   symbols for, 306, **307**, 371
   wind deflection by, **109**
Wardrobe closet, 203, **204**, 212, **213**, **214**
Wash drawings, 433
Waste discharge system, **735**, 735–737, **736**,
   738, 741

Water
   air movement and, **108**
   exterior control methods, **568**
   hot water recirculating system, 734, **734**
   interior control methods, 567
   leak detection, 701
   plan rendering of, 265, **267**, 269, **269–271**
   quality of, 112–113, **113**, 567
   supply system plumbing, 732–735, **733**, **734**,
     738, **743**
Water closet, 219–220; *See also* toilets
Water meter, 113
Water purification system
   design considerations, 113, 567
   filtration; *See* filtration
   softening, 733, **734**
   for swimming pools, 157, 744, **744**
   waste discharge system, **735**, 735–737, **736**, 738
Watercolors, 433, **434**, 435, **435**
Waterproofing, in earth-sheltered homes, 104
Watts, 674, 677, 681
Wblock command, in CAD, 81, 285–286, 404
Weatherstripping, **619**
Webs
   in steel joists, 583, 640, **661**
   in truss joists, 580, **581**, 636
Welding, in steel construction, 554–555, **555**, **556**
Wells, 113
Wetlands, structural positioning and, 113
Wheelchair accessibility, planning for, 297, 298–299,
   298–301, 309
Whirlpool tub, 219
Wind
   design considerations and, 107–109, **108**, **109**
   preventing damage from, 561–565, 569, **797**
Wind eddies, 108–109
Wind helix, 672, **673**
Windchill effect, 107
Window option, in CAD, 238
Window seats, 205
Windows
   assembly of, **613**
   decor for, 122–124, **124**
   details for, **373**
   dimensioning, 328, 373–375, **374**
   elevation and, **337**, 347–351, 371, **372**, **373**
   framing drawings, 612–614, **615**, **616**
   framing for, 599–600
   function of, 124, 612
   furniture contouring with, **126**
   glass block, 169, 337, 614, **615**
   guidelines for, 347–349
   integrated design of, **349**
   lintel for, **473**, **474**
   manufactured, **613**, **614**
   modeling, **451**
   modular construction, 465, **466**
   natural light and, **122**
   NFRC rating report for, **710**
   noise reduction and, 114
   overhangs and baffles, **101**, 101–102, **102**
   performance factors for, **709**, **710**
   renderings of, 438, **439**
   R-values/U-values for, **709**
   schedules for, 762, **762**
   section drawings, 402, **403**, 404
   shape of, 121, **122**, **124**
   sizes of, 310, **311**, 614
   symbols for, 309–310, **310**, **311**, 371, **372**, 373,
     373, 375
   types of, 349, **350**, **351**
   view and, **349**
   waterproofing installation, 614
   wind gusts and, 564–565
Windscreens, 151
Winged gable roof, 657, **657**

Wire bonds, 532, **533**
Wire mesh, 474–475, 535, **536**, 540, **540**
Wireframe drawings, 88, **88**, **89**, 93, 426, **429**
Wiring, 678, 680, **680**, **681**, 693–700, **694–696**, 703
Wiring method, **693**
Wood
    carpentry careers, 791
    environmental issues, 512
    for fireplace, 139, 497
    floor framing plans for, 588–592
    frame construction, 506–526
    hardwoods versus softwoods, 506–508, 511
    lightwood trusses, **637**
    lumber, 506–510
    post-and-beam construction, 513–525
    for roofs, 634, **634**, **635**, **636**, 654–655

    sheathing, 663, **797**
    skeleton-frame construction, 506–513
    trusses, **637**, **639**, **640**, **655**
    wind resistance of, 561
    wood decking, 574–575
Wood foundations, 490, **491**, **492**
Wood plenum system, 716–717, **717**
Wood rail and stile doors, 351
Wooded areas, **108**, **111**, **154**
Wooden pencils, 55, **55**
Work triangle, 175, **175**
Working drawings
    in CAD, 83
    electrical plans, 693–700
    perspective drawings and, 420
    purpose of, 44, 304
    for structural steel, 557, **557**

Workshops, 201–203, **202**, **203**, 205
World's tallest structures, **17**
Wright, Frank Lloyd, 20, **21**, 30, 118, 512
W-shaped steel, 550, **551**

Yard lumber, 508, **509**

Zero-lot-line properties, 233–234, **234**
Zoning ordinances, 229–236, 782–783
    building permits, 235–236, **236**, **237**
    commercial, 229
    density zoning, 233–234, **233–235**
    land coverage and setbacks, 230–232, **230–232**
    structural types, 229–230, **230**
    types of, 229
    variances to, 236
Zoom command, in CAD, 78–79, **79**, 238